W9-AHN-451

Organic Chemistry

Principles and Mechanisms

Organic Chemistry

Principles and Mechanisms

Joel M. Karty

Elon University

W · W · NORTON

NEW YORK · LONDON

W. W. Norton & Company has been independent since its founding in 1923, when William Warder Norton and Mary D. Herter Norton first published lectures delivered at the People's Institute, the adult education division of New York City's Cooper Union. The firm soon expanded its program beyond the Institute, publishing books by celebrated academics from America and abroad. By mid-century, the two major pillars of Norton's publishing program—trade books and college texts—were firmly established. In the 1950s, the Norton family transferred control of the company to its employees, and today—with a staff of four hundred and a comparable number of trade, college, and professional titles published each year—W. W. Norton & Company stands as the largest and oldest publishing house owned wholly by its employees.

Editor: Erik Fahlgren
Assistant Editor: Renee Cotton
Developmental Editor: John Murdzek
Manuscript Editor: Julie Henderson
Project Editor: Christine D'Antonio
Marketing Manager, Chemistry: Stacy Loyal
Science Media Editor: Rob Bellinger
Associate Media Editor: Jennifer Barnhardt
Assistant Media Editor: Paula Iborra
Production Manager: Eric Pier-Hocking
Photo Editor: Michael Fodera
Photo Researcher: Jane Miller
Permissions Manager: Megan Jackson
Text Design: Lisa Buckley
Art Director: Hope Miller Goodell

Composition: codeMantra
Illustrations: Imagineering
Manufacturing: Courier, Kendallville

Library of Congress Cataloging-in-Publication Data has been applied for.

ISBN 978-0-393-91904-2

W. W. Norton & Company, Inc., 500 Fifth Avenue, New York, NY 10110
www.wwnorton.com

W. W. Norton & Company Ltd., Castle House, 75/76 Wells Street, London W1T 3QT

1 2 3 4 5 6 7 8 9 0

To Pnut, Fafa, and Jakers

About the Author

JOEL KARTY earned his B.S. in chemistry at the University of Puget Sound and his Ph.D. at Stanford University. He joined the faculty of Elon University in 2001, where he currently holds the rank of associate professor. He teaches primarily the organic chemistry sequence and also teaches general chemistry. In the summer, Joel teaches at the Summer Medical and Dental Education Program through the Duke University medical center. His research interests include investigating the roles of resonance and inductive effects in fundamental chemical systems and studying the mechanism of pattern formation in Liesegang reactions. He has written a very successful student supplement, *Get Ready for Organic Chemistry*, Second Edition (formerly called *The Nuts and Bolts of Organic Chemistry*).

Brief Contents

Contents

1 Atomic and Molecular Structure 1

Nomenclature

1 Introduction: The Basic System for Naming Simple Organic Compounds

Alkanes, Cycloalkanes, Haloalkanes, Nitroalkanes, and Ethers 54

Nomenclature

3 Considerations of Stereochemistry
R and S Configurations about Tetrahedral Stereocenters and Z and E Configurations about Double Bonds   276

6 The Proton Transfer Reaction
An Introduction to Mechanisms, Thermodynamics, and Charge Stability   295

7 An Overview of the Most Common Elementary Steps   351

Interchapter

1 Molecular Orbital Theory and Chemical Reactions   388

11 Electrophilic Addition to Nonpolar π Bonds 1
Addition of a Brønsted Acid   570

12 Electrophilic Addition to Nonpolar π Bonds 2
Reactions Involving Cyclic Transition States   610

16 Structure Determination 2
Nuclear Magnetic Resonance Spectroscopy and Mass Spectrometry 757

17 Nucleophilic Addition to Polar π Bonds 1
Addition of Strong Nucleophiles 815

18   Nucleophilic Addition to Polar π Bonds 2
Weak Nucleophiles and Acid and Base Catalysis   861

22 Electrophilic Aromatic Substitution 1
Substitution on Benzene; Useful Accompanying Reactions   1065

23 Electrophilic Aromatic Substitution 2
Substitution on Mono- and Disubstituted Benzene and Other Aromatic Rings   1104

Interchapter

2 Fragmentation Pathways in Mass Spectrometry   1245

26 Polymers   1255

List of Interest Boxes

List of Mechanisms

Preface

More Than an *Emphasis* on Mechanisms: *Organized* by Mechanism

During my first year of teaching organic chemistry, I taught what I have come to learn is a traditional approach. I organized the course the way that most textbooks are organized, with reactions pulled together according to the functional groups involved. Moreover, because I wanted my students to understand and not just memorize the material, I emphasized mechanisms very heavily. Despite my best efforts, the majority of my students struggled with even the basics of mechanisms and, consequently, turned to flashcards as their primary study tool. They tried to memorize their way through the course, which made matters worse.

My goal in writing this book was to solve the problem of memorization by grouping reactions according to similarities in their *mechanisms*. Thus, while the content of this book is the same as in other mainstream textbooks, the different *organization* establishes a coherent story of chemical reactivity. The story begins with molecular structure and energetics, and then guides students into reaction mechanisms with a few transitional chapters. Thereafter, students study how and why reactions take place as they do, focusing on one type of mechanism at a time. Ultimately, students learn how to intuitively use reactions in synthesis.

As mechanisms are central to the story of the book, students are naturally deterred from overlooking them. Students are made to feel more comfortable with mechanisms, and are clearly shown how the material builds from one chapter to the next, providing the foundation for understanding mechanisms in later chapters. Consequently, early in the course, students naturally embrace mechanisms as a learning tool, which I believe is vital to their success throughout the entire course and later—including on admission exams such as the MCAT.

Advantages of a Mechanistic Organization

In terms of student success, an organization by mechanism type offers two main advantages over the traditional organization by functional group. First, it allows students to focus more on reaction mechanisms within each chapter. This is because, once students are introduced to a particular reaction type, they get to apply those mechanisms across various functional groups. For example, after learning nucleophilic substitution reactions, students see that the mechanism applies to alkyl halides, alcohols, ethers, ketones, aldehydes, amines, and carboxylic acids. Second, as students begin to see the mechanistic patterns that unfold in one chapter, they will develop a better toolbox of mechanisms to draw upon in subsequent chapters. Students will therefore be better able to predict what will happen and why.

An organization by functional group, on the other hand, makes it very difficult for students to recognize patterns because each functional group chapter presents disjointed pieces of information related to that functional group. A functional group chapter discusses aspects of nomenclature, physical properties, synthesis, and spectroscopy in addition to new reactions and mechanisms. As a result, students find themselves overwhelmed and most will see no option but to memorize. Specifically, they will memorize what they perceive to be most important—predicting products of reactions—and will typically ignore, or give short shrift to, fundamental concepts and mechanisms.

I have now taught organic chemistry using a mechanistic organization for nearly a decade, during which time I have seen student performance and outlook improve dramatically.[1] I believe it all begins with students having a better handle on concepts and reactions early on. In my experience, the greatest motivator for students to put forth effort is the feeling of *understanding* the material—the feeling of being in *control* over the material. Students who feel that they "get" it are vastly more motivated to put in an even greater effort. The better understanding that a mechanistic organization affords students at the outset, therefore, paves the way for their success throughout the entire course.

Details about the Organization

The book is divided into three major parts:

Part I: Atomic and molecular structure

- Chapter 1: Atomic structure, Lewis structures and the covalent bond, and resonance theory, culminating in an introduction to functional groups
- Chapter 2: Aspects of three-dimensional geometry and its impacts on intermolecular forces
- Chapter 3: Structure in terms of hybridization and molecular orbital (MO) theory
- Chapters 4 and 5: Isomerism in its entirety, including constitutional isomerism, conformational isomerism, and stereoisomerism

Part II: Developing a toolbox for working with mechanisms

- Chapters 6 and 7: Ten elementary steps of mechanisms are examined.
- Chapter 8: Beginnings of multistep mechanisms using S_N1 and E1 reactions as examples

Part II provides a transition into Part III, which deals more intently with reactions.

Part III: Major reaction types

- Chapters 9 and 10: Nucleophilic substitution and elimination
- Chapters 11 and 12: Electrophilic addition
- Chapters 17 and 18: Nucleophilic addition
- Chapters 20 and 21: Nucleophilic addition–elimination
- Chapters 22 and 23: Electrophilic aromatic substitution
- Chapter 24: Diels–Alder reactions and other pericyclic reactions
- Chapter 25: Radical reactions
- Chapter 26: Polymerization

Notice that several of these chapters come in pairs. The first chapter in each pair is used to introduce key ideas about the reaction/mechanism and the second chapter explores the reaction/mechanism to greater depth and breadth.

Interspersed in Part III are chapters dealing with synthesis (Chapters 13 and 19), conjugation and aromaticity (Chapter 14), and spectroscopy (Chapters 15 and 16). The spectroscopy chapters are self contained and can be taught earlier, at the instructor's discretion.

Another major structural component of the book pertains to nomenclature. Nomenclature is separated out from the main chapters, in four relatively short units. Each unit focuses on specific rules of nomenclature, as opposed to specific functional groups. With each new nomenclature unit, new rules are introduced, which increases the complexity of the material discussed. These units can be covered in lecture or easily assigned for self study.

---

[1]Bowman, B. G.; Karty, J. M.; Gooch, G. "Teaching a Modified Hendrickson, Cram and Hammond Curriculum in Organic Chemistry." *J. Chem. Ed.* **2007**, *84*, 1209.

Finally, the application of MOs toward chemical reactions is separated from the main reaction chapters, and is presented, instead, as an optional, self-contained interchapter. This interchapter appears just after Chapter 7, the overview of the 10 most common elementary steps. Each elementary step from Chapter 7 is revisited from the perspective of MO theory—more specifically, frontier MO theory. Because this interchapter is optional, chapters later in the book do not rely on coverage of this material.

A Better Tool for Students

While the organization provides a coherent story, other aspects of the book make it an excellent learning tool for students.

Extended coverage of general chemistry topics. The early chapters provide extended coverage of a variety of general chemistry topics. This is deliberate because I believe most students need a review of several of these topics upon entering organic chemistry. For example, I have found that most students do not have a firm grasp of Lewis structures, intermolecular forces, and equilibria and thermodynamics. Rather than assume that students will dive into their general chemistry textbook to review these topics, I have provided this additional material, with an organic focus, as a convenient student resource. Instructors can tailor their in-class coverage of this material as they deem necessary.

Strategies for Success. In addition to reviewing important general chemistry topics, I have provided Strategies for Success sections to help students build specific skills they need in this course. For example, Chapter 1 provides strategies for drawing all resonance structures of a given species, and sections in Chapters 2 and 3 are devoted to the importance of molecular modeling kits in working with the three-dimensional aspects of molecules and also with the different rotational characteristics of σ and π bonds. In Chapter 4, students are shown how to draw chair conformations and how to draw all constitutional isomers of a given formula. Chapter 5 provides help with drawing mirror images of molecules. One Strategies for Success section in Chapter 6 helps students estimate pK_a values and another helps students rank acid and base strengths based only on their Lewis structures. In Chapter 14, I include a section that shows students how to use the Lewis structure to assess conjugation and aromaticity, and Chapter 16 has a section that teaches students the chemical distinction test for nuclear magnetic resonance.

Your Turn exercises. Getting students to read *actively* can be challenging, so I wrote the Your Turns in each chapter to motivate this type of behavior. Your Turns are basic exercises that ask students to either answer a question, look something up in a table from a previous chapter, construct a molecule using a model kit, use a table in the chapter, or interact with art in a figure or data in a plot. In addition to getting students active when they read, these exercises are intended to be "reality checks" for students as they read. Your Turns should be used as indicators to students as to whether they understand what they have just read. If they cannot solve/answer a Your Turn exercise easily, students should interpret this as a signal that they need to either reread the previous section(s) or seek help. Short answers to all Your Turns are provided in the back of the book and complete solutions to these exercises are provided in the Study Guide and Solutions Manual.

Consistent and effective problem-solving approach. Helping students become expert problem-solvers, in this course and beyond, is one of my major goals. I have developed the Solved Problems in the book to train students how to think as they approach a problem. On average, there are seven Solved Problems per chapter and each one is broken down into two parts: *Think* and *Solve*.

7.2 YOUR TURN

Use the box provided to draw the product suggested by the faulty curved arrow notation in the following chemical equation. What is unacceptable about the product you drew?

SOLVED problem 7.10 Draw the S_N2 step that would occur between $C_6H_5CH_2I$ and CH_3SNa.

Think Which species is the nucleophile? Which is the substrate? What do we do with the metal atom? Which species is electron rich? Electron poor?

Solve $C_6H_5CH_2I$ will behave as the substrate because it possesses as I, a good leaving group that departs as I^-. The conjugate acid of I^-, HI, is a very strong acid. CH_3SNa has a metal atom that can be treated as a spectator ion and thus ignored. The nucleophile is therefore CH_3S^-. In an S_N2 step, a curved arrow is drawn from the lone pair of electrons on the electron-rich S atom to the electron-poor C atom bonded to I. A second curved arrow must be drawn to indicate that the C—I bond is broken (otherwise that C would have five bonds).

In the Think part, students are provided a handful of *questions* that I want them to be asking as they approach the problem. In the Solve part, those questions are answered and the problem is solved. This mirrors the strategy I use to help students during office hours, and we have used these same steps for *every* problem in the Solutions Manual that accompanies the book.

Another excellent training tool is SmartWork, Norton's online tutorial and homework system. SmartWork allows students to practice their problem-solving skills and receive hints and answer-specific feedback that reinforce what students see in the book.

Developing a toolbox of mechanisms. Understanding the common elementary steps that make up mechanisms is a crucial part of solving organic chemistry problems. The elementary steps introduced in Chapters 6 and 7 effectively provide students a toolbox for working comfortably with mechanisms later on. Moreover, students will find that many reactions they encounter throughout the course have mechanisms that comprise just these steps. This makes it more transparent to students how seemingly different reactions can, in fact, be very closely related—through the mechanism.

Separating nomenclature. As I discussed earlier, nomenclature is presented in four separate units, interspersed between chapters in the first half of the book. These units are self-contained and they can be covered where they are located in the textbook or any point after. One of the main reasons for presenting nomenclature separately is that it helps minimize distractions. A second reason for separate coverage of nomenclature is that nomenclature is among the most straightforward topics students will encounter. Naming a molecule requires memorizing certain rules and then practicing applying those rules. This is something that students are quite comfortable with, so instructors have the option of holding students accountable for learning nomenclature on their own or covering it in class.

Biochemistry and MCAT 2015. Most organic chemistry students are biology majors and/or are seeking a career in a health profession. They appreciate seeing how organic chemistry relates to their interests and look for ways in which this course will prepare them for the admissions exams (such as the MCAT) that may determine their future.

Rather than relegating biochemistry to the end of the book, I have placed the Organic Chemistry of Biomolecules in self-contained sections at the ends of several chapters, beginning with Chapter 1. The topics chosen for these sections cover many of the topics outlined in the MCAT 2015 Preview Guide, which means that the Organic Chemistry of Biomolecules sections are not *in addition to* what students are expected to know for the MCAT; they are topics that students *should know* for the test. In even the earliest of chapters, students have the tools to start learning aspects of this traditional biochemistry coverage. More importantly, these sections provide reinforcement of topics. In each biomolecules section, the material is linked directly back to concepts encountered earlier in the chapter.

These Organic Chemistry of Biomolecules sections are both optional and flexible. Instructors can decide to cover only a few of these topics or none at all, and can do so either as they appear in the book or as special topics at the end of the second semester.

A range of interesting applications. In addition to the Organic Chemistry of Biomolecules sections, most chapters have two special interest boxes. These boxes apply a concept in the chapter to a discovery or process that students can relate to. In addition to reinforcing concepts from the chapter, these boxes are intended to provide *meaning* to what students are learning, and to motivate students to dig deeper.

A focus on synthesis. Synthesis problems represent one of the greatest challenges undergraduates face in this course. Not only must students have a command of the reactions they have learned, but they must also be able to think critically to find the right combination of those reactions that will transform the starting mate-

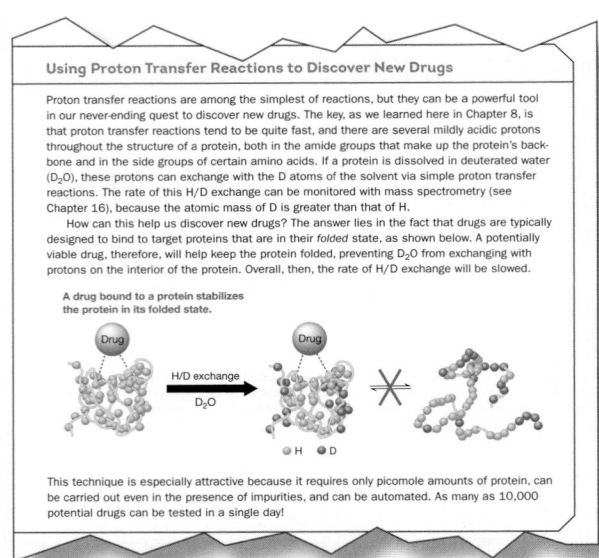

Using Proton Transfer Reactions to Discover New Drugs

Proton transfer reactions are among the simplest of reactions, but they can be a powerful tool in our never-ending quest to discover new drugs. The key, as we learned here in Chapter 8, is that proton transfer reactions tend to be quite fast, and there are several mildly acidic protons throughout the structure of a protein, both in the amide groups that make up the protein's backbone and in the side groups of certain amino acids. If a protein is dissolved in deuterated water (D_2O), these protons can exchange with the D atoms of the solvent via simple proton transfer reactions. The rate of this H/D exchange can be monitored with mass spectrometry (see Chapter 16), because the atomic mass of D is greater than that of H.

How can this help us discover new drugs? The answer lies in the fact that drugs are typically designed to bind to target proteins that are in their *folded* state, as shown below. A potentially viable drug, therefore, will help keep the protein folded, preventing D_2O from exchanging with protons on the interior of the protein. Overall, then, the rate of H/D exchange will be slowed.

A drug bound to a protein stabilizes the protein in its folded state.

H/D exchange
D_2O

○ H   ● D

This technique is especially attractive because it requires only picomole amounts of protein, can be carried out even in the presence of impurities, and can be automated. As many as 10,000 potential drugs can be tested in a single day!

rial into the desired compound. I provide a thorough introduction to organic synthesis in two chapters—Chapters 13 and 19. Chapter 13 discusses introductory topics in synthesis, including the basics of retrosynthetic analysis and the idea of cataloging reactions according to what they accomplish. Chapter 19 presents more challenging topics in synthesis, such as the use of protecting groups and how to place functional groups strategically within a carbon backbone. Therefore, whereas Chapter 13 ought to be covered by most mainstream classes, instructors can choose to cover only certain sections of Chapter 19 or skip it entirely.

I have found that treating synthesis in dedicated chapters makes it more meaningful to students. When I taught synthesis under a traditional functional group organization, it became a distraction to the reactions that students are simultaneously learning. I also found that students often associated a synthetic strategy only with the functional group for which it was introduced. For example, when the idea of protecting groups is introduced in the ketones/aldehydes chapter of a textbook traditionally organized by functional group, students tend to associate protecting groups with ketones/aldehydes *only*. My dedicated synthesis chapters help students focus on synthesis without compromising their focus on reactions. Furthermore, synthesis strategies are discussed more holistically, so students can appreciate them in a much broader context rather than being applicable to a single functional group.

Optional interchapter on the application of MO theory toward reactions. Under an organization according to functional group, the roles of MOs in chemical reactions typically appear integrated into several different functional group chapters. For example, the role of orbitals in an S_N2 reaction is typically integrated in an alkyl halides chapter, and the role of orbitals in a nucleophilic addition reaction is typically integrated into the ketones/aldehydes chapter. For instructors who do not teach this aspect of MOs in their course, these discussions can represent distractions and are potentially counterproductive to student learning.

Presenting this material together in an optional interchapter, as I have done in this book, offers two main advantages to students. One is that it removes a potential distraction from the main reaction chapters and, being optional, instructors have the choice of not covering it at all. Another advantage comes from the fact that the MO pictures of all 10 common elementary steps appear together in the interchapter. Therefore, instructors who wish to cover this interchapter can expect their students to come away with a better understanding of the bigger picture of MO theory as it pertains to chemical reactions.

Acknowledgments

There are many people who have been a part of or impacted by my work on this textbook, and my gratitude for all of them is immense. First and foremost, however, I must acknowledge my family—Valerie, Joshua, and Jacob. When I began work on this book, Joshua wasn't even walking; now he's 11 and Jacob is 9. It has been a long and arduous endeavor. But it is also one that has been worthwhile and has helped shape who I am today. Thank you for your love and support and for standing by me the entire way.

I must also acknowledge my parents, Alec and Maraline, who instilled in me the importance of academics and the love of learning. My thanks goes out to them especially for the numerous sacrifices they made so that I could have the best in education and the best in life.

I owe a lot to my brothers and sister, too—Ben, Kevin, and Sarah. Those sibling rivalries we had growing up certainly brought out the best in me. More recently, thank you for tolerating me throughout this process.

Thanks must also go to my colleagues in the chemistry department here at Elon, for your support and understanding of how important this book has become. I especially must acknowledge Gene and Marcia Gooch. Gene was initially part of this project but died tragically in a bicycle accident.

To my teachers through the years, thank you for your wisdom and encouragement, as well as your friendship. Jack Howard, you got me started in chemistry in high school, and I still remember when you said: "I'm not a smart man, but I do know how to convert units." John Hanson and Bill Dasher, you "showed me the way" in organic chemistry, and Ken Rousslang, you taught me to appreciate the finer details in physical chemistry and turned me on to the world of research. Tim Hoyt, thank you for being the "Wiz" that you are. And to John Brauman, it is because of you that I can truly call myself a scientist.

My students deserve a lot of credit, too. Even though I am the teacher (technically), I continue to learn from my students year in and year out. Thank you, especially, for being guinea pigs at times, giving me the chance to learn how to become a better teacher.

To Maureen Cullins, thank you for letting me be creative. It was in my first year teaching at the Summer Medical and Dental Education Program that I "discovered" a better way to teach organic chemistry, and you have always been one of my biggest fans.

A tremendous amount of thanks goes to the many members of the Norton team. Erik Fahlgren, thank you for believing in me, and for taking on this project with as much passion as I have given it. Your balance of optimism and critique has truly brought the book to a whole new level of quality. John Murdzek, your help through the developmental editing process has been priceless, and your humor has helped me keep things in perspective. Renee Cotton and Christine D'Antonio, I continue to be amazed with your attention to detail and ability to stay on top of things. Jane Miller, I appreciate the hours you have spent researching photos and bearing with me when things are not precisely to my liking. And to Stacy Loyal, I admire the work you've done in marketing this book. Changing a paradigm that's over half a century old is no small task. Kudos to you.

A special thanks to Steve Pruett and Marie Melzer. Steve, your patience with me and with this book has been incredible, and your insights have been tremendously appreciated. Thank you, in particular, for your commitment through it all. Marie, I appreciate your help creating the Your Turn answers at the end of the book and greatly value the energy and the insight that you have brought to the Study Guide and Solutions Manual.

Finally, I am indebted to the many class-testers and reviewers, whose feedback has been invaluable in the evolution of this book. I am especially grateful to Larry French, Laurie Witucki, and Steve Miller, who accuracy-checked the *entire* book. A tremendous—and tremendously important—undertaking, indeed!

Adopters of the Preliminary Edition

California State University–Los Angeles

Catawba College

Catawba Valley Community College

DePauw University

Des Moines Area Community College

Eastern Mennonite University

Elon University

Jefferson Technical and Community College

Lakeland College

Luther College

Malcolm X College

Pasadena City College

St. Catherine University

SUNY—Potsdam

Texas Lutheran University

University of Minnesota–Morris

University of Tennessee–Knoxville

Upper Iowa University

Wake Forest University

Reviewers

Robert Allen, Arkansas Tech University

Herman Ammon, University of Maryland

Carolyn Anderson, Calvin College

Aaron Aponick, University of Florida

Phyllis Arthasery, Ohio University

Jared Ashcroft, Pasadena City College

Athar Ata, University of Winnipeg

Jovica Badjic, Ohio State University

John Belizzi, University of Toledo

Daniel Berger, Bluffton University

Anthony Bishop, Amherst College

Rebecca Broyer, University of Southern California

Larry Calhoun, University of New Brunswick

Shawn Campagna, University of Tennessee–Knoxville

Nancy Carpenter, University of Minnesota–Morris

Brad Chamberlain, Luther College

Robert Coleman, Ohio State University

Tammy Davidson, University of Florida

Lorraine Deck, University of New Mexico

Jeff Elbert, University of Northern Iowa

Seth Elsheimer, University of Central Florida

Eric Finney, University of Washington

Andrew Frazer, University of Central Florida

Larry French, St. Lawrence University

Gregory Friestad, University of Iowa

Brian Frink, Lakeland College

Anne Gorden, Auburn University

Christopher Gorman, North Carolina State University

Oliver Graudejus, Arizona State University

Robert Grossman, University of Kentucky

Daniel Gurnon, DePauw University

Jeffrey Hansen, DePauw University

Bryan Hanson, DePauw University

Andrew Harned, University of Minnesota

Stewart Hart, Arkansas Tech University

John Hershberger, Hamilton College

Gail Horowitz, Brooklyn College

Roger House, Auburn University

Philip Hultin, University of Manitoba

Kevin Jantzi, Valparaiso University

Amanda Jones, Wake Forest University

Jeff Jones, Washington State University

Paul Jones, Wake Forest University

Robert Kane, Baylor University

Arif Karim, Austin Community College

Steven Kass, University of Minnesota

Stephen Kawai, Concordia University

Valerie Keller, University of Chicago

Mark Keranen, University of Tennessee–Knoxville

Kristopher Keuseman, Mount Mercy College

Angela King, Wake Forest University

Jesudoss Kingston, Iowa State University

Francis Klein, Creighton University

Jeremy Klosterman, Bowling Green State University

Dalila Kovacs, Grand Valley State University

Jason Locklin, University of Georgia

Brian Long, University of Tennessee–Knoxville

Claudia Lucero, California State University–Sacramento

David Madar, Arizona State University–Polytechnic

Kirk Manfredi, University of Northern Iowa

Eric Masson, Ohio University

Anita Mattson, Ohio State University

Gerald Mattson, University of Central Florida

Jimmy Mays, University of Tennessee–Knoxville

Alison McCurdy, California State University–Los Angeles

Dominic McGrath, University of Arizona

Mark McMills, Ohio University

Marie Melzer, Old Dominion University

Ognjen Miljanic, University of Houston

Justin Miller, Hobart and William Smith Colleges

Stephen Miller, University of Florida

Barbora Morra, University of Toronto

Joseph O'Connor, University of California–San Diego

James Parise, University of Notre Dame

Noel Paul, Ohio State University

James Poole, Ball State University

Christine Pruis, Arizona State University

Harold Rogers, California State
University–Fullerton

Sheryl Rummel, Pennsylvania State
University

Nicholas Salzameda, California State
University, Fullerton

Adrian Schwan, University of Guelph

Colleen Scott, Southern Illinois
University–Carbondale

Sergei Dzyuba, Texas Christian University

Alan Shusterman, Reed College

Joseph Simard, University of New England

Chad Snyder, Western Kentucky University

John Sorensen, University of Manitoba

Levi Stanley, Iowa State University

Laurie Starkey, California State
University–Pomona

Tracy Thompson, Alverno College

Nathan Tice, Butler University

John Tomlinson, Wake Forest University

Melissa VanAlstine-Parris, Adelphi
University

Nanine Van Draanen, California
Polytechnic State University–San Luis
Obispo

Quin Wang, University of South Carolina

Don Warner, Boise State University

Haim Weizman, University of California–
San Diego

Lisa Whalen, University of New Mexico

James Wilson, University of Miami

Laurie Witucki, Grand Valley State
University

James Wollack, St. Catherine University

Andrei Yudin, University of Toronto

Michael Zagorski, Case Western Reserve
University

Rui Zhang, Western Kentucky University

Regina Zibuck, Wayne State University

Eugene Zubarev, Rice University

James Zubricky, University of Toledo

Additional Resources

For Students

Study Guide and Solutions Manual

by Joel Karty, Elon University, and Marie Melzer, Old Dominion University

Written by two dedicated teachers, this guide provides students with fully worked solutions to all unworked problems in the text. Every solution follows the Think/Solve format used in the textbook, so the approach to problem-solving is modeled consistently.

SmartWork

Created by chemistry educators, SmartWork is the most intuitive online tutorial and homework system available for organic chemistry. A powerful engine supports and grades an unparalleled range of problems written for Karty's text, including numerous arrow-pushing problems. *Every* problem in SmartWork has hints and answer-specific feedback to coach students and provide the help they need, when they need it. Problems in SmartWork link directly to the appropriate page in the electronic version of Karty's text so students have an instant reference and are prompted to read.

Instructors can draw from Norton's bank of more than 2000 high-quality, class-tested problems, or use our innovative authoring tools to easily modify existing problems or write new ones. Instructors can sort problems by learning goal and create assignments to assess any learning goals, concepts, or skills that they choose.

The Karty SmartWork course also features:

- **An expert author team.** The organic SmartWork course was authored by instructors who teach at a diverse group of schools: Arizona State University, Florida State University, Brigham Young University, and Mesa Community College. The authors have translated their experience in teaching such a diverse student population by creating a library of problems that will appeal to instructors at all schools.

- **Pooled drawing and nomenclature problems.** SmartWork features sets of pooled problems for drawing and nomenclature to promote independent work. Groups of similar problems are "pooled" into one problem so different students receive different problems from the pool. Instructors can choose our preset pools or create their own.

For Instructors

Instructor's Guide

by Stephen R. Pruett, Jefferson Community and Technical College

Written by one of the first users of Joel's material, each chapter in the Instructor's Guide begins with a brief overview of the chapter, followed by a more detailed section-by-section discussion that includes information on differences between this book and textbooks with a functional group organization.

Pruett includes suggestions on how to present difficult concepts to students, and, incorporating his experiences teaching with the preliminary edition of the textbook, documents information about his and his students' experiences with the material. Suggested clicker questions from *Clickers in Action: Active Learning in Organic Chemistry* and a section on additional resources (such as specific websites, articles, or books) that instructors can incorporate into their course round out each chapter. The Instructor's Guide includes a chapter for each of the 26 chapters in the textbook, plus a chapter for the MO theory interchapter and a chapter for the nomenclature units.

Clickers in Action: Active Learning in Organic Chemistry

by Suzanne M. Ruder, Virginia Commonwealth University

This instructor-oriented resource provides information on implementing clickers in organic chemistry courses. Part I gives instructors information on how to choose and manage a classroom response system, develop effective questions, and integrate the questions into their courses. Part II contains 140 class-tested, lecture-ready questions. Most questions include histograms that show actual student response, generated in large classes with 200–300 students over multiple semesters. Each question also includes insights and suggestions for implementation. The 140 questions from the book and an additional 100 lecture-ready questions are available in PowerPoint, sorted to correspond to the chapters in the textbook, at wwnorton.com/instructors.

Test Bank

by Amy M. Deveau, University of New England; James Wollack, St. Catherine University; and Alexandra Jones, St. Catherine University

The Test Bank contains more than 1300 multiple-choice and short-answer questions. Questions are organized by chapter section, and each question is ranked by difficulty and type. Questions are further classified by learning objectives. The list of learning objectives provided at the beginning of each chapter makes it easy to find questions that test each objective.

The Test Bank is available in print, *ExamView* Assessment Suite, Word RTF, and PDF formats.

ExamView Test Generator Software

All Norton test banks are available with ExamView Test Generator software, allowing instructors to effortlessly create, administer, and manage assessments. The convenient and intuitive test-making wizard makes it easy to create customized exams with no software learning curve. Other key features include the ability to create paper exams with algorithmically generated variables and export files directly to Blackboard, WebCT, and Angel.

Instructor's Resource Disc

This helpful classroom presentation tool features:

- Select photographs and every piece of line art in JPEG format
- Select photographs and every piece of line art in PowerPoint
- Lecture PowerPoint slides with integrated figures from the book
- Clicker questions from *Clickers in Action: Active Learning in Organic Chemistry*

Downloadable Instructor's Resources (wwnorton.com/instructors)

This instructor-only, password-protected site features instructional content for use in lecture and distance education, including test-item files, PowerPoint lecture slides, images, figures, and more. The instructor's website includes:

- Select photographs and every piece of line art in JPEG format
- Select photographs and every piece of line art in PowerPoint
- Lecture PowerPoint slides with integrated figures from the book
- Clicker questions from *Clickers in Action: Active Learning in Organic Chemistry*
- Instructor's Guide in PDF format
- Test bank in PDF, Word RTF, and ExamView formats

Author Blog: www.teachthemechanism.com

Starting in July 2012, Joel Karty and Steve Pruett started blogging about Joel's approach and their experience teaching a course organized by mechanism. In addition to written posts, the blog includes videos of Joel discussing his approach and answering some of the questions he is frequently asked about teaching a mechanistically organized course. Guest bloggers, most of whom have adopted Karty's text, have also contributed their stories and insights about teaching a course organized by mechanism. You are encouraged to visit the blog and submit comments of your own.

Preface for the Student

Organic Chemistry and You

You are taking organic chemistry for a reason—you might be pursuing a career in which an understanding of organic chemistry is crucial, or the course might be required for your particular field of study, or both. You might even be taking the course simply out of interest. Regardless of the reason, organic chemistry impacts your life in significant ways.

Consider, for example, the growing concern about the increasing resistance of bacteria to antibiotics. Perhaps no germ has caused more alarm than methicillin-resistant *Staphylococcus aureus* (MRSA), a type of bacteria responsible for "staph" infections. Methicillin is a member of the penicillin family of antibiotics, and resistance to methicillin in these bacteria was first observed in 1961. Today MRSA, which has been called a "superbug," is resistant to most antibiotics, including *all* penicillin-derived antibiotics.

A breakthrough in the fight against MRSA occurred in 2006 with the discovery of a compound called platensimycin, isolated from *Streptomyces* spores.

Streptomyces spores   **Platensimycin**

The way that platensimycin targets bacteria is different from that of any other antibiotic in use and, therefore, it is not currently susceptible to bacterial resistance.

Platensimycin is found in a type of South African mushroom, *Streptomyces platensis*, and was discovered by screening 250,000 natural product extracts for antibacterial activity. Sheo B. Singh (Merck Research Laboratories) and coworkers determined the structure of platensimycin using a technique called nuclear magnetic resonance (NMR) spectroscopy, which we discuss in Chapter 16. Not long after, K.C. Nicolaou and coworkers from The Scripps Research Institute (La Jolla, California) and the University of California, San Diego, were the first to devise a synthesis of platensimycin from other readily available chemicals.

The story of platensimycin, from discovery to synthesis, involves several of the subdisciplines that make up the field of organic chemistry.

- **Biological chemistry (biochemistry):** The study of the behavior of biomolecules and the nature of chemical reactions that occur in living systems.
- **Structure determination:** The use of established experimental techniques to determine the structure of newly discovered compounds.
- **Organic synthesis:** The design of pathways for making new compounds from existing, readily available compounds by means of known organic reactions.

Because each of these areas typically focuses on solving existing and practical problems, they are considered to be *applied* areas of organic chemistry. However, other areas of organic chemistry, considered to be *theoretical* in nature, provide the foundations on which such applications rest. They focus on answering questions about the "how" and "why" of chemical processes. For example, an understanding of the basic principles of NMR spectroscopy (an analytical technique discussed in Chapter 16) underlies our ability to determine molecular structure. Understanding the principles that govern organic reactions (such as those involved in the synthesis of platensimycin) may allow us to enhance yields, not only by altering reaction conditions, but also perhaps by devising entirely new synthesis schemes. And understanding platensimycin's specific mode of attack on bacteria will likely guide us in modifying its chemical structure to make it even more effective.

The story of platensimycin showcases the importance of organic chemistry in the pharmaceutical industry, but organic chemistry is at the center of other high-profile areas as well, including the fabrication of new materials such as plastics (the topic of Chapter 26). The durability and chemical stability of plastics have made them excellent choices for use in food packaging (Fig. P.1a) and the fabrication of the artificial heart (Fig. P.1b). Plastics are the source of synthetic fibers such as nylon and polyester, which are often used in the clothing industry, as well as Kevlar®, which is used to make body armor (Fig. P.1c). Composite materials made from plastic and carbon fibers are so strong that some commercial jets are now constructed with a body made largely from plastics (Fig. P.1d).

FIGURE P.1 Some uses of plastics Plastics, which are designed and created in the laboratories of organic chemists, are found in a wide range of products, such as (a) food packaging, (b) an artificial heart, (c) body armor made from Kevlar®, and (d) a Boeing 787, a commercial jet whose body consists largely of composite materials made from plastics and carbon fibers.

(a)

(b)

(c)

(d)

FIGURE P.2 Organic chemistry in the electronics industry (a) A full-color digital display made from organic light-emitting diodes. (b) A molecular switch in which an organic molecule joins together two carbon nanotubes—sheets of carbon in the form of cylinders with a diameter on the order of 10^{-9} meters.

(a)

(b)

Organic chemistry has also been at the forefront of generating new materials for electronic devices. Organic light-emitting diodes (OLEDs) are the main components of full-color electronic displays (Fig. P.2a), and single organic molecules can be used to make electronic switches tens of thousands of times smaller than those used in today's integrated circuits (Fig. P.2b).

Perhaps even more important to our lives is the impact that organic chemistry can have on our ability to understand, and solve, environmental problems, such as overflowing landfills (Fig. P.3a), the destruction of the stratospheric ozone layer (Fig. P.3b), and global warming (Fig. P.3c). Organic chemistry, for example, is helping

(a)

Sept. 22, 2012

(b)

FIGURE P.3 Organic chemistry and the environment Organic chemistry continues to play a significant role in solving environmental problems, such as (a) overflowing landfills, (b) ozone depletion (the area in blue represents the "ozone hole" over Antarctica), and (c) global warming (the ice sheets in Montana's Glacier National Park have been melting at a dramatically accelerating rate over the past 90 years).

(c)

provide new ways to recycle waste materials. Additionally, organic chemistry has been used to engineer new coolants that are safer for the environment than the chlorofluorocarbons (CFCs) used in the late 20th century in refrigerators and air conditioners. Finally, organic chemistry may lead us to economically feasible processes by which we can synthesize hydrogen gas, a fuel whose combustion product is only water. This could be a welcome alternative to coal and oil, whose combustion products not only cause air and water pollution, but also generate carbon dioxide, one of several greenhouse gases responsible for global warming.

Because organic chemistry is important in so many ways, you will find interest boxes in each chapter, which show how the material in the chapter directly connects to issues that you might find more relevant or more interesting. Take the time to read those boxes, and consider researching them even further.

Some Suggestions for Studying

Perhaps you have heard that organic chemistry is difficult. Perhaps you have heard that it requires an enormous amount of *memorization*. Are these statements true? It depends on how you approach the course. What is true is that this book contains a lot of information—much more than you can memorize. There is a better way.

Organic chemistry can be *understood* through models and theories that are built upon *fundamental concepts*. Consider, for example, that when two compounds react under a given set of conditions, the outcome of that reaction is precisely the same each and every time. Is this because the reactant molecules have *memorized* what products they are supposed to make? No—they are obeying certain chemical laws, and those laws can be learned.

You will spend considerable effort throughout this course developing those models and theories. Reaction mechanisms—detailed steps that show how reactions take place—are among the most important ideas to develop. If you devote your time and energy to understanding them and learning how they are applied toward solving problems, you will find that much of organic chemistry can be conquered without rote memorization, and you will find the course to be quite rewarding and enjoyable. The skills you develop in organic chemistry will apply to complex situations you will face beyond this course.

If you are planning on a career in a health profession, it is particularly important for you to focus on understanding and applying concepts as opposed to memorizing. On standardized exams like the MCAT, you will often need to choose between answers that are designed to look equally good to students who have memorized the material. To a student who is well versed in applying concepts and mechanisms toward solving problems, on the other hand, those choices are more easily discernible.

In light of how important it is to understand concepts and mechanisms, your success in this course will demand a lot of time and devotion. The following are suggestions for using that time, and this book, most efficiently:

- **Read actively and diligently.** You should try to read the assigned sections before class if possible. Reading prior to lecture means that lecture is the second time you'll be presented the material. This will allow you to better process information and give you ample opportunity to ask pertinent questions. When you read, you should have a pen or pencil in hand so you can underline or highlight what you feel is important, and take notes about what you find enlightening or confusing. When the text refers to a figure or reaction mechanism, take that as a cue to study that figure now. Be sure that what the text is describing makes sense to you before you move on. If you are referred to a previous chapter, flip to the appropriate page to refresh your memory.
- **Your Turns.** The "Your Turn" exercises are relatively short activities that ask you to complete a task based on what you have just read. These exercises were developed to help you remain *actively engaged* while you read. They should also help you quickly

evaluate whether or not you understand the topic at hand. I encourage you to *work through all Your Turn exercises in each chapter* and quickly check the answers in the back of the book. Feedback from students who have used this book supports this advice.

- **Problems.** As with anything new you attempt, mastery requires practice. Most of your practice should come from solving problems. I have included nearly 2000 problems throughout this book. Many are integrated into the chapters, but most are gathered at the end of each chapter. Take the time to work through as many problems as possible, and use them to assess areas of strength and weakness.

That said, it's time to get started. Keep your focus on concepts and mechanisms, and work hard!

Organic Chemistry

Principles and Mechanisms

CHAPTER 1

Atomic and Molecular Structure

Very likely you have heard that organic chemistry is "the chemistry of life." Inherent to this description is the idea that certain types of compounds, and the reactions they undergo, are suitable to sustain life, while others are not. If so, what are the characteristics of such compounds and what advantages do those compounds afford living organisms? Here in Chapter 1 we will begin to answer these questions.

We will review several aspects of atomic and molecular structure typically covered in a general chemistry course, including ionic and covalent bonding, the basics of Lewis structures, and resonance theory. With such a general foundation, we will then begin to tighten our focus on organic molecules. We will present various types of shorthand notation that organic chemists often use and we will introduce you to functional groups commonly encountered in organic chemistry.

Toward the end of this chapter, we will shift our focus to examining specific classes of biomolecules: amino acids, monosaccharides, and nucleotides. Not only will such a discussion provide insight into the relevance of organic chemistry to biological systems, but it will also reinforce specific topics discussed in the chapter, such as functional groups.

> Organic chemistry is often referred to as the chemistry of life, because biological compounds such as DNA, proteins, and carbohydrates are themselves organic molecules. In this chapter, we will examine some of the bonding characteristics of these and other organic molecules, which are constructed primarily from carbon, hydrogen, nitrogen, and oxygen.

Upon completing Chapter 1 you should be able to:

- Distinguish organic compounds from inorganic ones.
- Explain the advantages brought about by the fact that carbon is the basis of organic molecules.
- Describe the basic structure of an atom and understand that the vast majority of its volume is taken up by electrons.
- Determine the ground state electron configuration of any atom in the first three rows of the periodic table and distinguish valence electrons from core electrons.
- Define bond length and bond energy and understand how these two quantities change with increasing numbers of bonds between a given pair of atoms.
- Draw the Lewis structure of a species, given only its connectivity and total charge.
- Differentiate between a nonpolar covalent bond, a polar covalent bond, and an ionic bond, and distinguish a covalent compound from an ionic compound.
- Assign a formal charge and an oxidation state to any atom in a molecular species, given only its Lewis structure.
- Describe what a resonance structure is and explain the effect that resonance has on a species' stability.
- Draw all resonance structures of a given species, as well as its resonance hybrid, and determine the relative stabilities of resonance structures.
- Draw and interpret Lewis structures, condensed formulas, and line structures.
- Explain why functional groups are important and identify functional groups that are common in organic chemistry.

1.1 What Is Organic Chemistry?

Before beginning our study of organic chemistry, we ought to have an idea of what organic chemistry is. Very crudely, **organic chemistry** is the branch of chemistry involving *organic compounds*. What, then, is an organic compound?

In the late 1700s, scientists defined an **organic compound** as one that could be obtained from a *living* organism, whereas **inorganic compounds** encompassed everything else. It was believed that organic compounds could *not* be made in the laboratory; instead, only living systems could summon up a mysterious "vital force" needed to synthesize them. This belief was called **vitalism**. Using this definition, many familiar compounds, such as glucose (a sugar), testosterone (a hormone), and deoxyribonucleic acid (DNA) are *organic* (Fig. 1-1).

This definition of organic compounds broke down in 1828, when Friedrich Wöhler (1800–1882), a German physician and chemist, synthesized urea (an organic compound known to be a major component of mammalian urine) by heating a solution of ammonium cyanate (an inorganic compound; Equation 1-1). Within a couple of decades after Wöhler's discovery, vitalism was dead.

An inorganic compound

An organic compound

$$(NH_4)^+(NCO)^- \xrightarrow{\text{Heat}} \underset{\text{Urea}}{H_2N-\overset{\displaystyle O}{\overset{\|}{C}}-NH_2} \qquad (1\text{-}1)$$

Ammonium cyanate

With the end of vitalism, another definition was needed to describe the many compounds that were already labeled as organic. Gradually, chemists arrived at our modern definition:

An **organic compound** is composed primarily of carbon and hydrogen.

This definition, however, is still imperfect, because it leaves considerable room for interpretation. For example, many chemists would classify carbon dioxide (CO_2) as *inorganic* because it

FIGURE 1-1 Some familiar organic compounds Glucose, testosterone, and DNA are organic compounds produced by living organisms.

Glucose

Testosterone

DNA

Bolinus brandaris

"Royal purple"

FIGURE 1-2 Royal purple Ancient Phoenicians processed about 10,000 aquatic snails, *Bolinus brandaris* (left), to yield 1 g of "royal purple" dye. The structure of the molecule responsible for the dye's color is shown at right.

does not contain any hydrogen atoms, whereas others would argue that it is *organic* because it contains carbon and is critical in living systems. In plants, it is a starting material in photosynthesis, and in animals, it is a byproduct of respiration. Similarly, tetrachloromethane (carbon tetrachloride, CCl_4) contains no hydrogen, but many would classify it as an organic compound. Butyllithium (C_4H_9Li), on the other hand, is considered by many to be inorganic, despite the fact that 13 of its 14 atoms are carbon or hydrogen. (In Chapter 17, we will learn that the behavior of such compounds containing metal atoms, called *organometallic* compounds, is substantially different from that of typical organic compounds.) Rest assured that although this definition of an organic compound has its inadequacies, it does allow chemists to agree on the classification of most molecules.

Historians place the birth of organic chemistry as a distinct field around the time that vitalism was dismissed, thus making the discipline less than 200 years old. However, humans have taken advantage of organic reactions and the properties of organic compounds for thousands of years! Since about 6000 b.c., for example, civilizations have fermented grapes to make wine. Some evidence suggests that Babylonians, as early as 2800 b.c., could convert oils into soaps.

Many clothing dyes are organic compounds. Among the most notable of these dyes is **royal purple**, which was obtained by ancient Phoenicians from a type of aquatic snail called *Bolinus brandaris* (Fig. 1-2). These organisms produced the compound in such small amounts, however, that an estimated 10,000 of them had to be processed to obtain a single gram of dye. This effectively limited the availability of the dye only to those who had substantial wealth and resources—royalty.

Organic chemistry has matured tremendously since its inception. Today, we can not only use organic reactions to reproduce complex molecules found in nature, but also engineer new molecules never before seen.

1.2 Why Carbon?

There is no question that the chemistry of life is centered primarily around the carbon atom. The backbones of familiar biomolecules like DNA, proteins, and carbohydrates are all composed primarily of carbon. Why does the carbon atom play this central role and what is so special about it?

One of the main reasons must be the *diversity* of compounds possible when carbon is their chief structural component. As we will see in Section 1.6, the carbon atom is capable of forming four covalent bonds to other atoms—especially other carbon atoms. Consequently, carbon atoms can link together in chains of almost any length, allowing for an enormous range in molecular size. Moreover, the ability to form four bonds means there is potential for *branching* at each carbon in the chain. And each carbon atom is capable of forming not only single bonds, but double and triple bonds as well. These characteristics make possible a tremendous number of compounds, even with a relatively small number of carbon atoms. Indeed, to date, tens of millions of organic compounds are known, and the list is growing rapidly as we continue to discover or synthesize new compounds.

A chain of carbon atoms with single bonds only	A chain of carbon atoms with a double and triple bond	A *branched* chain of carbon atoms

With a chain of oxygen atoms, no double bonds, triple bonds, or *branching* is possible.

This same kind of diversity would not be possible in compounds based on another element, such as oxygen. Oxygen atoms are most stable when they form two covalent bonds, so they could form a linear chain only (as shown in the hypothetical example above). No branching could occur, nor could other groups or atoms be attached to the chain except at the ends. Furthermore, the atoms along the chain could not participate in either double or triple bonds.

If carbon works so well, then why *not* silicon, which appears just below carbon in the periodic table? Elements in the same group (column) of the periodic table tend to exhibit similar chemical properties, so silicon, too, can form four covalent bonds, giving it the same potential for diversity as carbon.

The answer is *stability*. As we will see in Section 1.4, the carbon atom tends to form rather strong bonds with a variety of atoms, including other carbon atoms. For example, it takes 339 kJ/mol (81 kcal/mol) to break an average C—C single bond, and 418 kJ/mol (100 kcal/mol) to break an average C—H bond. By contrast, it takes only 223 kJ/mol (53 kcal/mol) to break a typical Si—Si bond. The strength of typical bonds involving carbon atoms goes a long way toward keeping biomolecules intact—an essential characteristic for molecules whose job it is to store information or provide cellular structure.

Even though organic molecules are based on the carbon atom, what would life be like if silicon atoms were to replace carbon atoms in biomolecules such as glucose $(C_6H_{12}O_6)$? Glucose is broken down by our bodies through respiration to extract energy, according to the overall reaction in Equation 1-2. One of the byproducts is carbon dioxide, a gas, which is exhaled from the lungs. In a world in which life is based on silicon, glucose would be $Si_6H_{12}O_6$, and its byproduct would be silicon dioxide (SiO_2), as shown in Equation 1-3. Silicon dioxide, a solid, is the main component of sand; in its crystalline form, it is known as quartz (Fig. 1-3).

FIGURE 1-3 Quartz crystal Quartz (silicon dioxide) is the silicon analog of carbon dioxide. Whereas carbon dioxide is gaseous, silicon dioxide is a solid.

$$C_6H_{12}O_6 + 6\ O_2 \longrightarrow 6\ CO_2 + 6\ H_2O \qquad (1\text{-}2)$$

$$Si_6H_{12}O_6 + 6\ O_2 \longrightarrow 6\ SiO_2 + 6\ H_2O \qquad (1\text{-}3)$$

1.3 Atomic Structure and Ground State Electron Configurations

In Section 1.2, we saw that carbon's bonding characteristics are what give rise to the large variety of organic molecules. Those bonding characteristics, and the bonding characteristics of all atoms, are governed by the electrons that the atom has.

With this in mind, Section 1.3 is devoted to the nature of electrons in atoms. We first review the basic structure of an atom, followed by a discussion of orbitals and shells. Finally, we review electron configurations, distinguishing between *valence electrons*—electrons that can be used for bonding—and *core electrons*.

Even though carbon takes center stage in organic chemistry, organic molecules invariably include other atoms as well, such as hydrogen, nitrogen, oxygen, and halogen atoms. Some of the most exciting chemistry today, however, involves extended frameworks of *only* carbon. A single flat sheet of such a framework is called **graphene**, and resembles molecular chicken wire. Wrapped around to form a cylinder, a graphene sheet forms what is called a **carbon nanotube**. Pure carbon can even take the form of a soccer ball—the so-called **buckminsterfullerene**.

A sheet of graphene   A carbon nanotube   Buckminsterfullerene

These structures themselves have quite interesting electronic properties, giving them a bright future in nanoelectronics. Carbon nanotubes and buckminsterfullerenes have high tensile strength, moreover, giving them potential use for structural reinforcement in concrete, sports equipment, and body armor. Chemical modification gives these structures an even wider variety of potential uses. Graphene oxide, for example, has promising antimicrobial activity, and attaching certain molecular groups to the surface of a carbon nanotube or buckminsterfullerene has potential for use as drug carriers for cancer therapeutics.

1.3a The Structure of the Atom

At the center of an atom (Fig. 1-4) is a positively charged nucleus, composed of *protons* and *neutrons*. Surrounding the nucleus is a cloud of negatively charged *electrons*, attracted to the nucleus by simple **electrostatic forces** (the forces by which opposite charges attract one another and like charges repel one another). It is important to realize that individual electrons are incredibly small, even much smaller than the nucleus.

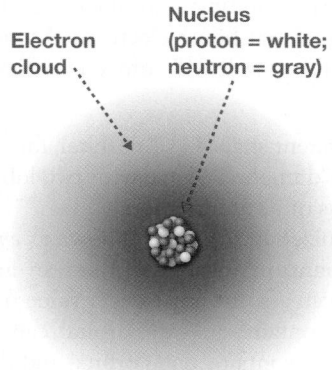

Electron cloud

Nucleus (proton = white; neutron = gray)

FIGURE 1-4 **Basic structure of the atom** Atoms are composed of a nucleus surrounded by a cloud of electrons. Protons (white) and neutrons (gray) make up the nucleus. (This figure is not to scale. If it were, the size of the electron cloud is so much larger than the size of the nucleus that its radius would be on the order of 500 meters!)

TABLE 1-1 Charges and Masses of Subatomic Particles

Particle	Charge (a.u.)	Mass (a.u.)
Proton	+1	~1
Neutron	0	~1
Electron	−1	~0.0005

a.u. = atomic units

However, the space that electrons occupy (i.e., the *electron cloud*) is much larger than the nucleus. In other words,

- The size of an atom is essentially defined by the size of its electron cloud.
- The vast majority of an electron cloud (and thus the vast majority of an atom) is empty space.

Table 1-1 lists the mass and charge of each of these elementary particles. Notice that the masses of the proton and neutron are significantly greater than that of the electron, so the mass of an atom is essentially the mass of just the nucleus.

An atom, by definition, has no net charge. Consequently, the number of electrons in an atom must equal the number of protons. The number of protons in the nucleus, called the **atomic number** (Z), defines the element. For example, a nucleus that has six protons has an atomic number of 6, and can only be a carbon nucleus.

If the number of protons and the number of electrons are not equal, then the entire **species** (that particular combination of protons, neutrons, and electrons) bears a net charge, and is called an **ion**. A negatively charged ion, an **anion** (pronounced AN-ion), results from an excess of electrons. A positively charged ion, a **cation** (pronounced CAT-ion), results from a deficiency of electrons.

SOLVED problem 1.1 How many protons and electrons does a cation of the carbon atom have if its net charge is +1?

Think How many protons are there in the nucleus of a carbon atom? Does a cation have more protons than electrons, or vice versa? How many more, given the net charge of the species?

Solve A carbon atom's nucleus has six protons. A cation with a +1 charge should have one more proton than it has electrons, so this species must have five electrons.

---

problem 1.2 **(a)** How many protons and electrons does an anion of the carbon atom have if its net charge is −1? **(b)** How many protons and electrons does a cation of the oxygen atom have if its net charge is +1? **(c)** How many protons and electrons does an anion of the oxygen atom have if its net charge is −1?

1.3b Atomic Orbitals and Shells

Electrons in an isolated atom reside in **atomic orbitals**. As we shall see, the exact location of an electron can never be pinpointed. An orbital, however, specifies the region of space where the probability of finding a given electron is high. More simplistically, we can view orbitals as "rooms" that house electrons. Atomic orbitals are examined in greater detail in Chapter 3; for now, it will suffice to review some of their more basic concepts.

1. Atomic orbitals have different shapes. An *s* orbital, for example, is a sphere, whereas a *p* orbital has a "dumbbell" shape with two lobes (Fig. 1-5). Each orbital is centered on the nucleus of its atom or ion.
2. Atomic orbitals are organized in *shells* (also known as *energy levels*). A **shell** is defined by the **principal quantum number, *n*.** There are an infinite number of shells in an atom, given that *n* can assume any integer value from 1 to infinity.
   a. The first shell ($n = 1$) contains only an *s* orbital, called "1*s*."
   b. The second shell ($n = 2$) contains one *s* orbital and three *p* orbitals, called "2*s*," "2p_x," "2p_y," and "2p_z."
   c. The third shell ($n = 3$) contains one *s* orbital, three *p* orbitals, and five *d* orbitals.

s Orbital   *p* Orbital

FIGURE 1-5 Orbitals Orbitals represent regions in space where an electron is likely to be. An *s* orbital is spherical, and a *p* orbital is a dumbbell.

3. Up to two electrons are allowed in any orbital.
   a. Therefore, the first shell can contain up to two electrons (a **duet**).
   b. The second shell can contain up to eight electrons (an **octet**).
   c. The third shell can contain up to 18 electrons.
4. With increasing shell number, the *size* and *energy* of the atomic orbital increases. For example, comparing *s* orbitals in the first three shells, the size and energy increase in the order 1*s* < 2*s* < 3*s*, as shown in Figure 1-6. Similarly, a 2*p* orbital is smaller in size and lower in energy than a 3*p* orbital.
5. Within a given shell, an atomic orbital's energy increases in the following order: *s* < *p* < *d*, etc. In the second shell, for example, the 2*s* orbital is lower in energy than the 2*p*.

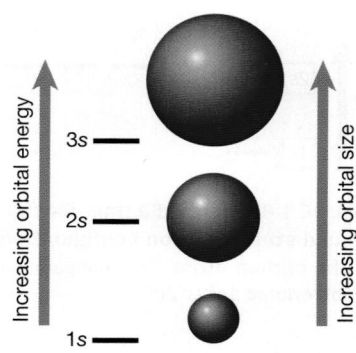

FIGURE 1-6 Relationship between principal quantum number, orbital size, and orbital energy As the shell number of an orbital increases, its size and energy increase, too. The horizontal black lines indicate each orbital's energy.

1.3c Ground State Electron Configurations: Valence Electrons and Core Electrons

The way in which electrons are arranged in atomic orbitals is called the atom's **electron configuration**. The *most stable* (i.e., the lowest energy) electron configuration is called the **ground state** configuration. Knowing an atom's ground state configuration provides insight into the atom's chemical behavior, as we will see.

With the relative energies of atomic orbitals established, an atom's ground state electron configuration can be obtained by applying the following three rules:

1. **Pauli's exclusion principle:** No more than two electrons (i.e., zero, one, or two electrons) can occupy a single orbital; two electrons in the same orbital must have opposite spins.
2. **Aufbau principle:** Each successive electron must fill the lowest energy orbital available.
3. **Hund's rule:** All orbitals *at the same energy* must contain a single electron before a second electron can be paired in the same orbital.

According to these three rules, the first 18 electrons fill orbitals as indicated in Figure 1-7.

1.1 YOUR TURN

In Figure 1-7, place a box around all of the orbitals in the second shell and label them.

Answers to Your Turns are in the back of the book.

In the ground state, the six electrons found in a carbon atom would fill the orbitals as shown in Figure 1-8, with two electrons in the 1*s* orbital, two electrons in the 2*s* orbital, and one electron in each of two different 2*p* orbitals (it doesn't matter which two). The shorthand notation for this electron configuration is $1s^2 2s^2 2p^2$.

Knowing the ground state electron configuration of an atom, we can distinguish *valence* electrons from *core* electrons.

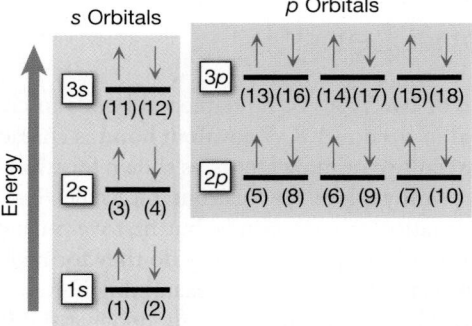

FIGURE 1-7 Energy diagram of atomic orbitals for the first 18 electrons The order of electron filling is indicated in parentheses. Each horizontal black line represents a single orbital. Each successive electron fills the lowest energy orbital available. Notice in the 2*p* and 3*p* sets of orbitals that no electrons are paired up until after the addition of the fourth electron.

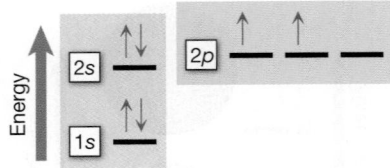

FIGURE 1-8 Energy diagram for the ground state electron configuration of the carbon atom This configuration is abbreviated $1s^2 2s^2 2p^2$.

- **Valence electrons** are those occupying the highest energy (i.e., valence) shell. For the carbon atom, the valence shell is the $n = 2$ shell.
- **Core electrons** occupy the remaining lower energy shells of the atom. For the carbon atom, the core electrons occupy the $n = 1$ shell.

The notion of valence electrons is important because, as we discuss in Section 1.5, *bonding is governed primarily by the valence electrons.* As we can see in Figure 1-8, for example, carbon has four valence electrons and two core electrons, so bonding involving carbon is governed by those four valence electrons.

YOUR TURN 1.2

In Figure 1-8, place a circle around the valence electrons and label them. Place a box around all of the core electrons and label them.

The periodic table is organized in such a way that *the number of valence shell electrons of an atom can be read from the element's group number* (a copy of the periodic table appears inside the book's front cover). Carbon is located in group 4A, consistent with its four valence electrons. Similarly, chlorine is found in group 7A, so it has seven valence electrons. Its ground state electron configuration is $1s^2 2s^2 2p^6 3s^2 3p^5$; that is, seven electrons occupy the third shell (the valence shell).

Atoms are especially stable when they have completely filled valence shells. This is exemplified by the **noble gases** (group 8A), such as helium and neon, because they have completely filled valence shells and they do *not* form bonds to make compounds. Although the specific origin of this "extra" stability is beyond the scope of this book, the consequences are the basis for the "octet rule" and the "duet" rule we routinely use when drawing Lewis structures (Section 1.5).

SOLVED problem 1.3 Write the ground state electron configuration of the nitrogen atom. How many valence electrons does it have? How many core electrons does it have?

Think How many total electrons are there in a nitrogen atom? What is the order in which the atomic orbitals should be filled (see Fig. 1-7)? What is the valence shell and where do the core electrons reside?

Solve There are seven total electrons ($Z = 7$ for N). The first two are placed in the 1s orbital and the next two in the 2s orbital, leaving one electron for each of the three 2p orbitals. The electron configuration is $1s^2 2s^2 2p^3$. The valence shell is the second shell, so there are five valence electrons and two core electrons.

---

problem 1.4 Write the ground state electron configuration of the oxygen atom. How many valence electrons are there? How many core electrons are there?

1.4 The Covalent Bond: Bond Energy and Bond Length

A *covalent bond* is one of two types of fundamental bonds in chemistry; the other, an ionic bond, is discussed in Section 1.8. A **covalent bond** is characterized by the *sharing of valence electrons* between two or more atoms, as shown for two H atoms in Figure 1-9.

In Section 1.5, we will explore how various molecules can be constructed from atoms through the formation of such bonds, but first we must examine more closely the nature of covalent bonds. In particular, why do they form at all?

We can begin to answer these questions by examining Figure 1-10a, which illustrates how the energy of two H atoms changes as a function of the distance between their

Each electron belongs to an isolated H atom.   A covalent bond

$$H\cdot \quad \cdot H \longrightarrow H\!:\!H$$

FIGURE 1-9 A covalent bond A covalent bond is the sharing of two electrons between nuclei.

FIGURE 1-10 Formation of a chemical bond (a) Plot of energy as a function of internuclear distance for two H atoms. The H atoms are most stable at the distance at which energy is a minimum. (b) A ball at the top of a hill becomes more stable at the bottom of the hill, and therefore tends to roll downhill.

nuclei. Namely, when two H atoms separated by a large distance are brought together, their total energy begins to decrease. At one particular internuclear distance, the energy of the molecule is at a minimum, while at shorter distances the energy rises dramatically.

The two H atoms are most stable at the internuclear distance that corresponds to that energy minimum, a distance called the **bond length** of the H—H bond. The energy that would be required to remove the H atoms from that internuclear distance to infinity is the **bond strength**, or **bond energy**, of the H—H bond.

We can relate this process to a more familiar one of a ball rolling down a hill (Fig. 1-10b). A ball at the top of a hill has more potential energy than the ball at the bottom of the hill. This is why the ball at the top of the hill will tend to roll downhill, coming to rest at the bottom. By the same token, it requires energy to roll the ball from the bottom of the hill back to the top.

1.3 YOUR TURN

Estimate the bond energy of the bond represented by Figure 1-10a.

The shape of the energy curve in Figure 1-10a is similar to that which describes the stretching and compressing of a spring connecting two masses (Fig. 1-11). The minimum in energy corresponds to the spring's rest position—that is, when it is neither stretched nor compressed. Both stretching and compressing the spring from its rest position require energy. Thus, it is often convenient to *think of a covalent bond as a spring that connects two atoms.* (This view of the chemical bond is extended further in the discussion of infrared spectroscopy in Chapter 15.)

SOLVED problem 1.5 In the diagram below, which curve represents a stronger covalent bond?

Think How can bond breaking be represented for each curve? Which of those processes requires more energy?

FIGURE 1-11 The spring model of a covalent bond The energy curve of a spring connecting two masses resembles that of the covalent bond shown in Figure 1-10a. Both stretching and compressing the spring from its rest position cause a rise in energy.

Solve Bond breaking is represented by climbing from the bottom of the curve toward the right (i.e., the internuclear bond distance increases toward the right). For this process, more energy is required for the red curve, so the red curve represents a stronger bond.

problem **1.6** Which of these two curves represents a longer bond?

FIGURE 1-12 Stabilization of electrons in a covalent bond In an isolated H atom (left), the electron is attracted to a single nucleus. In a covalent bond (right), electrons are attracted simultaneously to two H nuclei, thus lowering the energy of each electron.

Less stable/higher energy

Each electron is attracted to one nucleus.

More stable/lower energy

Each electron is attracted to two nuclei.

We have yet to explain what gives rise to the stability of a covalent bond. Why are two hydrogen atoms in close proximity lower in energy than two isolated hydrogen atoms? It is due largely to the additional electrostatic attraction experienced by electrons when they are *shared* between nuclei. In each *isolated* hydrogen atom, the lone electron "sees" only a single proton (at the nucleus); that is, a single negative charge is attracted to a single positive charge (Fig. 1-12, left). On the other hand, when the two hydrogen atoms are close together, each of the two electrons "sees" two protons; that is, each negative charge is attracted *simultaneously* to two positive charges (Fig. 1-12, right). This additional attraction lowers each electron's energy.

Although single bonds are the most common type of bond found in organic molecules, we frequently encounter double bonds and triple bonds as well. The main difference among single, double, and triple bonds is the number of electrons involved. In a single bond, two electrons are shared between different nuclei; in a double bond, four electrons are shared; and in a triple bond, six electrons are shared.

Tables 1-2 and 1-3 list average bond energies for a variety of common bonding partners found in organic species. Table 1-2 contains those of single bonds. Table 1-3 lists bond energies and lengths for some common single, double, and triple bonds, as well.

Notice in Table 1-3 that *as the number of bonds increases between a pair of atoms, bond energy increases and bond length decreases.* That is, multiple bonds can be viewed as shorter, stronger springs than single bonds (Fig. 1-13).

Bond strength increases; bond length decreases

X—Y	X=Y	X≡Y
Single bond	**Double bond**	**Triple bond**

FIGURE 1-13 Bond strength and bond length For a particular pair of atoms that are covalently bonded (X and Y), a triple bond is shorter and stronger than a double bond, which is shorter and stronger than a single bond.

YOUR TURN 1.4

Refer to Tables 1-2 and 1-3 to answer the following questions, which are designed to acquaint you with the range of strengths of common bonds.

(a) What is the value of the strongest *single* bond listed? _____

(b) What bond does that correspond to? _____

(c) What is the value of the weakest *single* bond? _____

(d) What bond does that correspond to? _____

(e) What is the value of the strongest bond of any type? _____

(f) What bond does that correspond to? _____

Turning an Inorganic Surface into an Organic Surface

Gold (Au) is a relatively unreactive metal, which is one reason it is widely used in jewelry and high-end electronic components. Gold, however, has a relatively high affinity for sulfur (S); the Au—S bond energy is roughly 120–150 kJ/mol (~30–40 kcal/mol), or nearly half the strength of a typical C—C bond. Sulfur also forms a relatively strong bond to carbon, about 290 kJ/mol (70 kcal/mol). This dual affinity of sulfur has enabled chemists to use sulfur to anchor organic groups to the surface of gold. The reaction is extremely easy to carry out: A sample of gold metal is simply immersed in a solution of an alkanethiol [$CH_3(CH_2)_n$—SH] and the Au—S bond forms spontaneously, yielding a **self-assembled monolayer** in which the alkyl chains extend away from the gold surface. Effectively, then, the inorganic gold surface is converted into an organic one.

These self-assembled monolayers have a wide variety of applications, such as protecting the gold surface from substances that would otherwise cause corrosion. More interestingly, by changing the organic portion to which the thiol group is attached, the gold surface can be programmed to have specific affinity for other molecules. Gold nanoparticles (GNP), for example, have been capped with thiol-terminated molecules that enable the GNP to form bonds to epidermal growth-factor receptor antibodies. The resulting antibody–GNP conjugates have been used to image cancer cells.

TABLE 1-2 Average Bond Energies of Common Single Bonds

	H		C		N		O		F		Cl		Br		I		Si	
	kJ/mol	kcal/mol	kJ/mol	kcal/mol	kJ/mol	kcal/mol	kJ/mol	kcal/mol	kJ/mol	kcal/mol	kJ/mol	kcal/mol	kJ/mol	kcal/mol	kJ/mol	kcal/mol	kJ/mol	kcal/mol
H	436	104	418	100	389	93	460	110	569	136	431	103	368	88	297	71	301	72
C	418	100	339	81	289	69	351	84	439	105	331	79	280	67	238	57	289	69
N	389	93	289	69	159	38	180	43	272	65	201	48	243	58	169	40	355	85
O	460	110	351	84	180	43	138	33	209	50	209	50	222	53	238	57	430	103
F	569	136	439	105	272	65	209	50	159	38	251	60	251	60	280	67	586	140
Cl	431	103	331	79	201	48	209	50	251	60	243	58	222	53	209	50	402	96
Br	368	88	280	67	243	58	222	53	251	60	222	53	192	46	180	43	289	69
I	297	71	238	57	169	40	238	57	280	67	209	50	180	43	151	36	209	50
Si	301	72	289	69	355	85	430	103	586	140	402	96	289	69	209	50	188	45

TABLE 1-3 Average Bond Energies and Bond Lengths of Single and Multiple Bonds

Atoms Bonded Together	Single Bond			Double Bond			Triple Bond		
	Energy		Length	Energy		Length	Energy		Length
	$\frac{kJ}{mol}$	$\frac{kcal}{mol}$	picometers (pm)	$\frac{kJ}{mol}$	$\frac{kcal}{mol}$	pm	$\frac{kJ}{mol}$	$\frac{kcal}{mol}$	pm
C—C	339	81	154	619	148	134	812	194	120
C—N	289	69	147	619	148	129	891	213	116
C—O	351	84	143	720	172	120	1072	256	113
N—N	159	38	145	418	100	125	946	226	110
O—O	138	33	148	498	119	121			

1.5 Lewis Dot Structures and the Octet Rule

To understand a molecule's chemical behavior, it is necessary to know its **connectivity**—that is, which atoms are bonded together, and by what types of bonds (single, double, or triple). It is also useful to know which valence electrons participate in bonding and which do not. **Lewis dot structures** (or, **Lewis structures**) are a convenient way to convey this information. Let's review some basic conventions of Lewis structures.

- **Lewis structures take into account only valence electrons.** In a complete Lewis structure, such as the following for isolated C and Cl atoms, *all* valence electrons are shown:

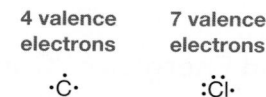

4 valence electrons   7 valence electrons

- **Bonding and nonbonding electrons are clearly shown.**
  - Single, double, and triple bonds are indicated by one, two, or three lines (i.e., —, =, or ≡), respectively, which represent the sharing of two, four, or six electrons, respectively. Thus, each line represents a shared pair of electrons.
  - Nonbonding electrons are indicated by dots, and are usually paired (:). These are called **lone pairs** of electrons. In some species, nonbonding electrons are **unpaired**, and are represented by single dots. These species are called **free radicals** and are discussed in greater detail in Chapter 25.

- **Atoms in Lewis structures obey the duet rule and the octet rule.** Atoms are especially stable when they have complete valence shells: two electrons (a duet) for hydrogen and helium, and eight electrons (an octet) for atoms in the second row of the periodic table. These duets and octets can be achieved through the formation of covalent bonds. Examples are shown in Figure 1-14.

Each H has a share of 2 electrons.

Each H has a share of 2 electrons.

Carbon's octet = 8 shared e⁻

Nitrogen's octet = 6 shared e⁻ + 2 unshared e⁻

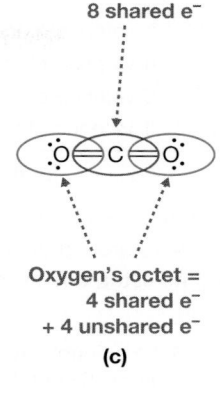

Carbon's octet = 8 shared e⁻

Oxygen's octet = 4 shared e⁻ + 4 unshared e⁻

Each H has a share of 2 electrons.

(a)   (b)   (c)   (d)

FIGURE 1-14 Covalent bonding: Sharing electrons to produce full valence shells In each of these molecules, all atoms have completely filled valence shells: H has a share of two electrons [(a), (b), and (d)], whereas C [(b) and (c)], O [(c)], and N [(d)] each have an octet of electrons made up of a total of 8 shared and unshared valence electrons.

1.5 YOUR TURN

In the following Lewis structure, circle the electrons that compose carbon's octet, oxygen's octet, and each hydrogen's duet.

$$H-\overset{\displaystyle H}{\underset{\displaystyle H}{C}}-\ddot{O}-H$$

   Because of their widespread use in organic chemistry, you must be able to draw Lewis structures quickly and accurately. The following steps allow you to do so in a systematic way.

Steps for Drawing Lewis Structures

1. Count the total number of *valence* electrons in the molecule.
   a. The number of electrons contributed by each atom is the same as its group number (H = 1, C = 4, N = 5, O = 6, F = 7).
   b. Each negative charge increases the number of electrons by one; each positive charge decreases the number of electrons by one.
2. Write the skeleton of the molecule, showing only the atoms and the single bonds required to hold them together.
   a. If molecular connectivity is not given to you, the central atom (the one with the greatest number of bonds) is usually the one with the smallest electronegativity. (Electronegativity is reviewed in Section 1.7.)
3. Subtract two electrons from the total in Step 1 for each single covalent bond drawn in Step 2.
4. Distribute the remaining electrons as lone pairs.
   a. Start with the outer atoms and work inward.
   b. Try to achieve a filled valence shell on each atom—namely, an octet on each atom other than hydrogen, and a duet on each hydrogen atom.
5. If there is an atom with less than a filled valence shell, convert lone pairs from *neighboring* atoms into **bonding pairs** of electrons, thereby creating double and/or triple bonds.

SOLVED problem 1.7 Draw a Lewis structure of HCO_2^-, where carbon is the central atom.

Think Consider the steps for drawing Lewis structures. Which atoms must be bonded together?

Solve First, count total valence electrons.

$$\begin{aligned}
\text{1 H atom: } & 1 \times 1 \text{ valence electorn } = 1 \text{ valence electron} \\
\text{1 C atom: } & 1 \times 4 \text{ valence electrons} = 4 \text{ valence electrons} \\
\text{2 O atoms: } & 2 \times 6 \text{ valence electrons} = 12 \text{ valence electrons} \\
-1 \text{ charge: } & 1 \times 1 \text{ valence electron } = 1 \text{ valence electron}
\end{aligned}$$

Total: = 18 valence electrons

Six electrons must be used to connect C to the other three atoms, leaving 12 more that can be placed as lone pairs around the O atoms to achieve octets, as shown below on the left. To give C its octet, one of those lone pairs is converted to a C=O double bond.

Make a double bond
to give C its octet.

problem **1.8** Draw a Lewis structure for C_2H_3N. A carbon that is bonded to three hydrogen atoms is bonded to the second carbon, which is bonded, in turn, to the nitrogen.

Deficient of octet

(a)

Expanded octet

(b)

FIGURE 1-15 The octet rule
(a) Elements in the second row must not exceed an octet. (b) Elements in the third row and below may exceed an octet.

In some molecular species, not all atoms have a complete valence shell. In borane (BH_3, Fig. 1-15a), for example, the B atom has a share of only six valence electrons, two electrons short of an octet. This is because there are not enough valence electrons available to achieve a complete valence shell for all atoms in the molecule. Similarly, in thionyl chloride ($SOCl_2$, Fig. 1-15b), the S atom has a share of 10 electrons but, being in the third row of the periodic table, its valence shell can contain up to 18 electrons. There are even some examples of molecular species in which hydrogen has a share of fewer than two electrons, but these are somewhat rare. (Such examples will be discussed as necessary.)

Because of the emphasis that organic chemistry has on atoms from the second row of the periodic table, we often talk about atoms from the third row and below in terms of an octet. In $SOCl_2$, for example, both Cl atoms have a share of eight valence electrons, so we say that each Cl atom has an octet. The S atom, on the other hand, has what is called an **expanded octet**, given its share of 10 electrons. Be careful, however, when using this terminology, because only atoms in the third row and below can have an extended octet.

Atoms in the second row are forbidden to exceed the octet!

problem **1.9** For each structure below, determine whether it is a legitimate Lewis structure. If not, explain why not.

1.6 Strategies for Success: Drawing Lewis Dot Structures Quickly

After you've used the systematic steps (p. 13) to construct Lewis structures of several species, you may begin to notice that each type of atom tends to form a specific number of bonds and to have a specific number of lone pairs of electrons. Table 1-4 summarizes these patterns. What you will learn later is that *those are the number of bonds and lone pairs for atoms that bear no formal charge.* Atoms that are charged have combinations of bonds and lone pairs that are different from those in Table 1-4.

We can now use the patterns shown in Table 1-4 to complete the Lewis structure in Solved Problem 1.10.

TABLE 1-4 Common Number of Covalent Bonds and Lone Pairs for Selected Uncharged Atoms

Atom	Number of Bonds	Number of Lone Pairs	Examples
H	1	0	—H
C	4	0	
N	3	1	
O	2	2	
X (X = F, Cl, Br, I)	1	3	
Ne	0	4	:Ne:

SOLVED problem 1.10 Complete the Lewis structure for the compound whose skeleton is shown at the right. Assume that all atoms are uncharged.

Think Which atoms (other than hydrogen) have an octet and which atoms don't? How many bonds and lone pairs are typical for each element?

Solve The atoms not shown with an octet are highlighted in red below on the left. Based on Table 1-4, we need to add two lone pairs to O, convert two C—C single bonds of the ring into double bonds, convert the C—N single bond into a triple bond, and add a lone pair to N.

Red atoms = no octet

All atoms have octet.

problem **1.11** Complete the Lewis structure for the molecule with the following connectivity. You may assume that all atoms have the number of bonds and lone pairs listed in Table 1-4.

1.7 Electronegativity, Polar Covalent Bonds, and Bond Dipoles

We've seen that covalent bonds are characterized by the sharing of electrons between two atomic nuclei. If the atoms are identical, the electrons are shared equally. Otherwise, one nucleus will attract electrons more strongly than the other. That ability to attract electrons in a covalent bond is defined as the element's **electronegativity (EN)**.

There are a variety of different electronegativity scales that assign values to each element, but the one devised by Linus Pauling, which ranges from 0 to about 4, is perhaps the most well known. Figure 1-16 shows that for main group elements, electronegativity follows a clear trend in the periodic table. Namely,

- Within the same row, electronegativity values tend to increase from left to right across the periodic table.
- Within the same column, they tend to increase from bottom to top.

As a result, the elements with the largest electronegativities (not counting the noble gases) tend to be in the upper right corner of the periodic table (e.g., N, O, F, Cl, and Br), whereas the elements with the smallest electronegativities tend to be in the lower left corner (e.g., K, Rb, Cs, Sr, Ba).

Electronegativity (EN) using the Pauling scale

Group (vertical)	1	2	3	4	5	6	7	8	9	10	11	12	13	14	15	16	17	18
Period (horizontal)	IA																	VIIIA
1	H 2.20	IIA											IIIA	IVA	VA	VIA	VIIA	He
2	Li 0.98	Be 1.57											B 2.04	C 2.55	N 3.04	O 3.44	F 3.98	Ne
3	Na 0.93	Mg 1.31											Al 1.61	Si 1.90	P 2.19	S 2.58	Cl 3.16	Ar
4	K 0.82	Ca 1.00	Sc 1.36	Ti 1.54	V 1.63	Cr 1.66	Mn 1.55	Fe 1.83	Co 1.88	Ni 1.91	Cu 1.90	Zn 1.65	Ga 1.81	Ge 2.01	As 2.18	Se 2.55	Br 2.96	Kr 3.00
5	Rb 0.82	Sr 0.95	Y 1.22	Zr 1.33	Nb 1.6	Mo 2.16	Tc 1.9	Ru 2.2	Rh 2.28	Pd 2.20	Ag 1.93	Cd 1.69	In 1.78	Sn 1.96	Sb 2.05	Te 2.1	I 2.66	Xe 2.60
6	Cs 0.79	Ba 0.89	La 1.1	Hf 1.3	Ta 1.5	W 2.36	Re 1.9	Os 2.2	Ir 2.20	Pt 2.28	Au 2.54	Hg 2.00	Tl 1.62	Pb 2.33	Bi 2.02	Po 2.0	At 2.2	Rn 2.2

EN = 1.0   EN = 4.0

FIGURE 1-16 Pauling's electronegativity scale for the elements Electronegativity generally increases from left to right across a row of the periodic table, and it increases up a column.

In a covalent bond, electrons are much more likely to be found near the nucleus of the more electronegative atom and less likely near the nucleus of the less electronegative atom. This creates a separation of partial positive and negative charges along the bond, called a **bond dipole**. More specifically,

- The more electronegative atom of a covalent bond bears a partial negative charge (δ^-, "delta minus").
- The less electronegative atom bears a partial positive charge (δ^+, "delta plus").
- A **dipole arrow** ($\longleftrightarrow$) can be drawn from the less electronegative atom (δ^+) toward the more electronegative atom (δ^-).

These ideas are shown for HF, CH_4, and CO_2 in Figure 1-17.

FIGURE 1-17 **Bond dipoles in various molecules** The dipoles are represented by the red arrows. Each arrow points from the less electronegative atom (δ^+) to the more electronegative atom (δ^-). The length of the arrow indicates the relative magnitude of the bond dipole. EN = electronegativity.

The magnitude of a bond dipole depends on the *difference* in electronegativity between the atoms involved in the bond. A larger difference in electronegativity yields a larger bond dipole. Relative magnitudes of a bond dipole are often depicted by the lengths of dipole arrows. For example, the difference in electronegativity between hydrogen and fluorine is larger than that between carbon and hydrogen. In Figure 1-17, therefore, the dipole arrow along the H—F bond is longer than the bond dipole arrows along the H—C bonds.

The Lewis structure of BH_3 is shown below. Write the electronegativity next to each atom. Along one of the B—H bonds, draw the corresponding dipole arrow, and add the δ^+ and δ^- symbols.

problem **1.12** For each *uncharged* molecule below,

(a) complete the Lewis structure by adding multiple bonds and lone pairs, and
(b) draw dipole arrows along each **polar covalent bond**. Pay attention to the lengths of the arrows.

(i)   (ii)   (iii)

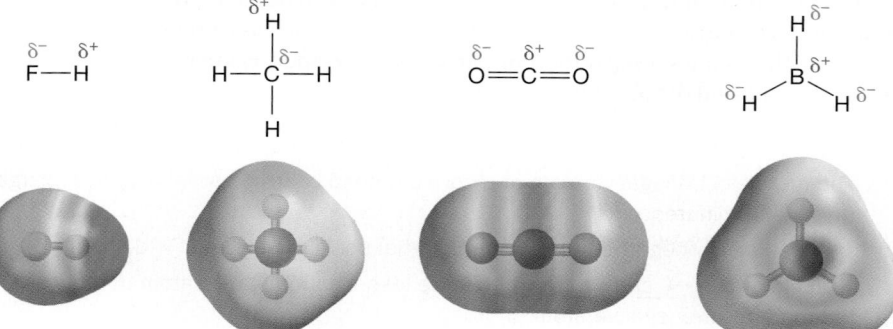

FIGURE 1-18 Electrostatic potential maps of four molecules Red indicates a buildup of negative charge; blue indicates a buildup of positive charge.

Another useful way to illustrate the distribution of charge along a covalent bond is with an **electrostatic potential map**, examples of which are shown in Figure 1-18. An electrostatic potential map depicts a molecule's electron cloud in colors that indicate its *relative* charge. Red corresponds to a buildup of negative charge, whereas blue represents a buildup of positive charge. Colors in between red and blue in the spectrum, such as green, represent a more neutral charge.

More positive charge   More negative charge

Some of the structures in Figures 1-17 and 1-18 demonstrate that a single molecule can possess more than one bond dipole. When this occurs, the bond dipoles' orientations and relative magnitudes dictate the overall distribution of charge within the molecule. We explore this concept further in Chapter 2.

---

problem **1.13** Which of the compounds below is consistent with the electrostatic potential map shown? Explain.

1.8 Ionic Bonds

When elements in a compound have large enough differences in electronegativity, **ionic bonding** can occur. Rather than sharing electrons, as in a covalent bond, the more electronegative atom acquires electrons given up by the less electronegative atom, forming oppositely charged ions. The electrostatic attraction between the positively charged cations and the negatively charged anions constitutes the ionic bond. Sodium chloride (NaCl), for example, consists of sodium cations (Na$^+$) and chloride anions (Cl$^-$). Sodium is a metal from the left side of the periodic table, so it has a small electronegativity, whereas chlorine is a nonmetal from the right side of the periodic table, so it has a large electronegativity. Most ionic compounds (e.g., MgBr$_2$)

consist of a metal and a nonmetal, whereas most covalent compounds (e.g., CH_4) consist of nonmetals.

While covalent compounds are generally found as discrete uncharged molecules, the ions in an ionic solid are arranged in a regular array, called a **crystal lattice**, as shown for NaCl in Figure 1-19.

Polyatomic ions, such as the hydroxide (OH^-), methoxide (CH_3O^-), and methylammonium ($CH_3NH_3^+$) ions shown in Figure 1-20, contain more than one atom. Polyatomic ions usually consist only of nonmetals, and their atoms are held together by covalent bonds.

Figure 1-20 also shows that polyatomic ions can be either *anions* (HO^- and CH_3O^-) or *cations* ($CH_3NH_3^+$). Most polyatomic cations possess a nitrogen atom participating in four covalent bonds (another common example is NH_4^+). The reasons why are discussed in Section 1.9. Interestingly, the existence of these kinds of polyatomic cations makes possible the formation of ionic compounds composed entirely of nonmetals (e.g., CH_3NH_3Cl or NH_4Br).

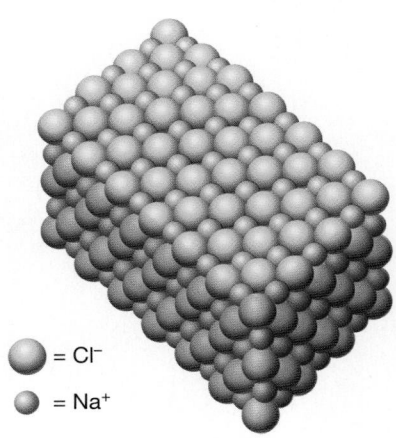

= Cl⁻

= Na⁺

FIGURE 1-19 Representation of the solid crystal structure of NaCl(s) The Na⁺ and Cl⁻ ions are held together by electrostatic forces, called ionic bonds, whereby opposite charges attract. As a solid, the ions form a regular array called a crystal lattice.

SOLVED problem 1.14 Identify each of the following as either an ionic compound (i.e., one containing ionic bonds) or a covalent compound (i.e., one containing only covalent bonds).

(a) NH_4CHO_2 **(b)** $LiOCH_2CH_3$ **(c)** $CH_3CH_2CH_2OH$

Think Does the compound contain elements from both the left and right sides of the periodic table? Does the compound contain any recognizable polyatomic cations?

Solve

(a) Ionic compound: NH_4^+ (ammonium ion) is a common polyatomic cation, leaving CHO_2^- as the remainder.

(b) Ionic compound: Li is a metal from the left side of the periodic table, and the remaining elements, which compose the ethoxide anion ($CH_3CH_2O^-$), are nonmetals from the right side.

(c) Covalent compound: All elements are nonmetals from the right side of the periodic table, and no recognizable polyatomic ions are present.

Atoms found on the left side of the periodic table

Atoms found on the right side of the periodic table (except for H)

FIGURE 1-20 Polyatomic ions in ionic compounds Polyatomic ions are usually composed of nonmetal elements from the right-hand side of the periodic table. Those elements have similar electronegativities, allowing covalent bonds to form between them.

problem **1.15** Which of the following are ionic compounds (i.e., ones containing ionic bonds) and which are covalent compounds (i.e., ones containing only covalent bonds)?

| (a) | (b) | (c) |

1.9 Assigning Electrons to Atoms in Molecules: Formal Charge and Oxidation State

In an *isolated* atom or atomic ion, charge is determined by the difference between the atom's group number and the actual number of valence electrons it possesses. A carbon atom, for example, has zero charge if it possesses four valence electrons—its group number is 4. It carries a charge of -1 if it has five valence electrons and it carries a charge of $+1$ if it has only three valence electrons.

YOUR TURN 1.8

Fill in the table below for a carbon atom.

Number of Valence Electrons	Total Number of Electrons	Number of Protons	Charge
3			
4			
5			

In a molecule or polyatomic ion, we can also assign a charge to an individual atom by computing the difference between the atom's group number and the number of valence electrons it possesses. But how do we assign electrons to atoms involved in covalent bonds, where electrons are being *shared*?

Two methods are used: **formal charge** and **oxidation state**. In both methods, lone pairs are assigned to the atom on which they appear in the Lewis structure. The methods differ, however, in how they treat covalent bonds.

Formal Charge

In a given covalent bond, half the electrons are assigned to each atom involved in the bond.

Oxidation State

In a given covalent bond, all electrons are assigned to the more electronegative atom. If the two atoms are identical, the electrons are split evenly.

Because the methods of assigning formal charge and oxidation state differ simply by the way in which they assign valence electrons to atoms within a molecule or ion, the methods are essentially different ways of distributing the net charge of the species. Therefore,

- The formal charges of all atoms must sum to the total charge of the species.
- The oxidation states of all atoms must sum to the total charge of the species.

SOLVED problem **1.16** Determine the formal charge and the oxidation state on every atom in the methanoate anion (formate anion, HCO_2^-).

Think How are lone pairs assigned? In determining how to assign bonding pairs of electrons, when is electronegativity relevant and when is it not?

Solve The figure below on the left shows how valence electrons are assigned according to the formal charge method. Notice that each pair of electrons in a covalent bond is split evenly. The H atom, the C atom, and the top O atom are assigned formal charges of 0 because they have the same number of valence electrons as their corresponding group numbers. The O on the right is assigned seven valence electrons, which is one more than its group number of 6. It is therefore assigned a formal charge of −1.

| Formal charges | Oxidation states |

The oxidation state method for assigning valence electrons is shown on the right. In this case, the electrons in each covalent bond are assigned to the more electronegative atom. H (EN = 2.20) is assigned zero electrons, one less valence electron than its group number, for an oxidation state of +1. C (EN = 2.55) is assigned two electrons, two fewer electrons than its group number, giving it an oxidation state of +2. Each O atom (EN = 3.44) is assigned eight electrons, two more electrons than its group number, giving each an oxidation state of −2.

1.9 **YOUR TURN**

Sum the formal charges and oxidation states assigned in Solved Problem 1.16. What do you notice?

Formal charges _____   Oxidation states _____

Determine the formal charge and oxidation state of each atom in the molecule below. (*Note:* You may assume each atom has a filled valence shell.)

The two methods for assigning covalently bonded electrons represent extremes. The formal charge method, for example, assumes perfect covalency, where the electrons are shared equally. The oxidation state method, on the other hand, assumes perfect ionic character, where the electrons belong entirely to the more electronegative atom. Neither is exactly correct; the true behavior of the electrons is somewhere in between.

Despite the fact that neither method is exactly correct, *formal charges are used much more often than oxidation states* because, in most cases, formal charge provides more insight into a species' reactivity. Therefore,

Unless otherwise stated, you may assume that any charges that appear in a Lewis structure are formal charges.

1.10 Resonance Theory

Some species are not described well by Lewis structures. In HCO_2^-, for example, both of the C—O bonds are identical experimentally; they have the same bond length and bond strength (Fig. 1-21a), intermediate between those of a single and a double bond. However, the Lewis structure of HCO_2^- (Fig. 1-21b) shows one C—O single bond and one C=O double bond, suggesting that the two bonds are different. Furthermore, it has been shown experimentally that both oxygen atoms carry an identical partial negative charge (δ^-), whereas the Lewis structure suggests that one oxygen atom bears a negative charge and the other is uncharged.

How, then, do we reconcile the differences between the Lewis structures of species like HCO_2^- and their observed characteristics? The answer is through **resonance theory**, the key points of which can be summarized as follows:

Rule 1

Resonance occurs in species for which there are two or more valid Lewis structures.

For such species, each valid Lewis structure is called a **resonance structure** or a **resonance contributor**. In HCO_2^-, for example, there are two possible resonance

FIGURE 1-21 Limitations of Lewis Structures The actual features of HCO_2^- (left) disagree with those suggested by the Lewis structure (right). In the actual structure, both C—O bonds are identical and the charge on each O atom is the same. The Lewis structure indicates a C—O single bond and a C=O double bond, as well as charges on each O atom that are different.

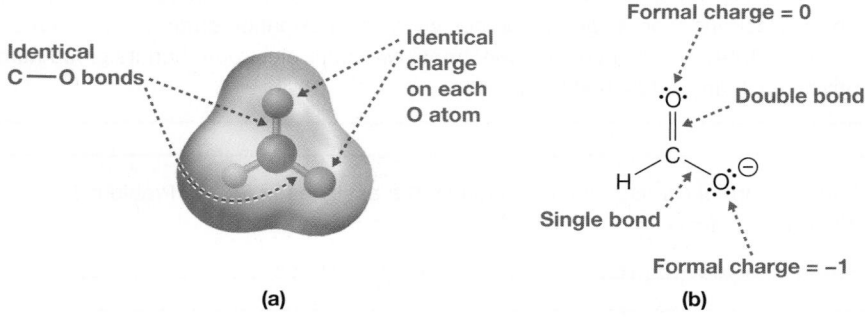

(a)

(b)

contributors. As we can see below, they differ in how we apply Step 5 for drawing Lewis structures—that is, which neighboring lone pair is used to construct the C═O double bond. As a result,

> Resonance structures differ only in the placement of their valence electrons, not their atoms.

Steps 1–4   Steps 1–4

Step 5   Step 5

Two resonance structures

Rule 2
Resonance structures are imaginary; the one, true species is represented by the resonance hybrid.

A **resonance hybrid** is a *weighted average of all resonance structures*. In the case of HCO_2^-, the hybrid is an average of two resonance structures:

Two resonance structures

Take the average

Single bond plus a partial bond

Partial charges

Resonance hybrid

Averaging the two C—O bonds (a single and a double bond) makes each one an identical 1.5 bond—that is, more than a single bond, but less than a double bond. (A partial bond is represented in a resonance hybrid by a dashed line connecting the two atoms.) Averaging the charge on each oxygen atom gives each an identical -0.5 charge. This resonance hybrid is now consistent with experimental results.

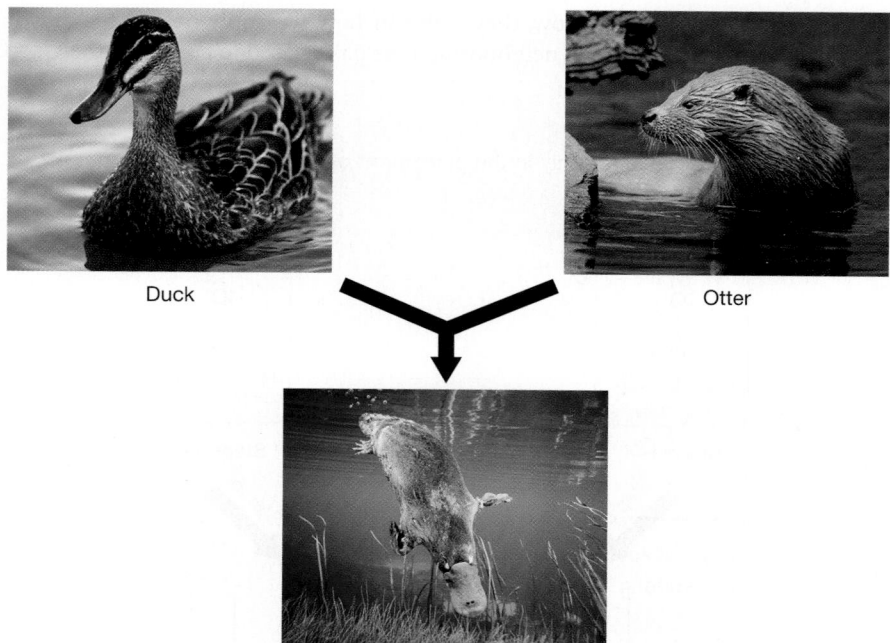

Duck

Otter

Duck-billed platypus

FIGURE 1-22 An analogy of resonance A duck-billed platypus has characteristics of a duck and an otter, just as a resonance hybrid has characteristics from its resonance contributors.

To help us remember that resonance structures are imaginary, we draw square brackets [ ] around the group of resonance structures and we place double-headed single arrows ($\leftrightarrow$) between them. These are *not* equilibrium arrows ($\rightleftharpoons$), which indicate a chemical reaction, because a compound does not rapidly interconvert between its resonance structures. Rather, think of a resonance hybrid in the same way as you might think of a duck-billed platypus (Fig. 1-22). That is, a platypus has characteristics of both a duck and an otter, but it is a unique species that does *not* rapidly interconvert between those two animals!

Rule 3

The resonance hybrid looks most like the lowest energy (most stable) resonance structure.

The two resonance contributors of HCO_2^- are *equivalent*. Each is composed of one H—C bond, one C=O double bond, and one C—O single bond, and each has a -1 formal charge on the singly bonded O. As a result, each structure contributes equally to the resonance hybrid.

In cases in which resonance structures are inequivalent, we must be able to determine their relative energies to determine their relative contributions to the hybrid. In general,

---

A resonance structure is lower in energy (i.e., more stable) with:

- a greater number of atoms with filled valence shells
- more covalent bonds
- fewer atoms with formal charges other than zero

The issues surrounding formal charge and resonance structures are covered in greater detail in Chapter 6.

SOLVED problem **1.18** Which of the following resonance structures makes a greater contribution to the resonance hybrid?

$$\left[\begin{array}{c} H_3C \underset{:O:}{\overset{H}{\underset{|}{C^{\oplus}}}} CH_3 \end{array} \longleftrightarrow \begin{array}{c} H_3C \overset{H}{\underset{O^{\oplus}}{\overset{|}{C}}} CH_3 \end{array} \right]$$

Think Are both resonance structures equivalent? If not, which has more atoms with filled valence shells? Which structure has more bonding pairs of electrons? Which structure has fewer atoms with formal charges other than zero?

Solve The resonance structures are inequivalent. In the structure on the right, all atoms have filled valence shells, whereas in the structure on the left, the central carbon atom has less than an octet. Additionally, there is one additional bond in the structure on the right, between the C and O. For both of these reasons, the structure on the right makes a much greater contribution to the overall resonance hybrid.

problem **1.19** Which of the following resonance structures makes a greater contribution to the resonance hybrid? *Note:* Formal charges are not shown.

$$\left[\begin{array}{c} H_3C \overset{:O:}{\underset{.Cl:}{\overset{||}{C}}} \end{array} \longleftrightarrow \begin{array}{c} H_3C \overset{:O:}{\underset{Cl:}{\overset{|}{C}}} \end{array} \right]$$

Rule 4

Resonance provides stabilization.

The stabilization of resonance structures results from the **delocalization** of electrons; that is, electrons have lower energy when they are less confined. In HCO_2^-, for example, four electrons are delocalized over three atoms (Fig. 1-23). In a single resonance structure, on the other hand, those electrons are **localized**. The extent by which a species is stabilized in this fashion is called its **resonance energy** or its **delocalization energy**.

FIGURE 1-23 Resonance stabilization In each resonance structure of HCO_2^- (top), the four electrons in red are localized. In the resonance hybrid (bottom), those four electrons are delocalized over three atoms. That delocalization results in lower energy and greater stability.

Rule 5

Resonance stabilization is usually large when resonance structures are equivalent.

FIGURE 1-24 Equivalent resonance structures The two resonance structures of benzene (left) are equivalent, so they contribute equally to the resonance hybrid (right).

Six electrons are *delocalized* over six carbon atoms.

This is an outcome of Rules 3 and 4. If resonance structures are equivalent, then they will contribute equally to the hybrid, allowing electrons the greatest possible delocalization. The two resonance structures of HCO_2^- are equivalent, so its resonance energy is quite substantial. Another example is benzene, whose resonance structures and hybrid are shown in Figure 1-24. Benzene's resonance energy is estimated to be about 150 kJ/mol (36 kcal/mol), which is nearly half the energy of a C—C bond![1] We discuss benzene in much greater depth in Chapter 14.

Acetic acid (CH_3CO_2H) is *not* stabilized greatly by resonance. Although it has two resonance structures, only the one on the left in Figure 1-25 contributes significantly. And, in that resonance structure, electrons are localized.

FIGURE 1-25 Nonequivalent resonance structures The two resonance structures of acetic acid are nonequivalent, so they contribute unequally to the resonance hybrid. The one on the left is lower in energy, so it has the greater contribution.

Small contribution to hybrid

> **Rule 6**
>
> All else being equal, the greater the number of resonance structures, the greater the resonance stabilization.

Rule 6 is an outcome of Rule 4, because additional resonance structures mean electrons are more widely delocalized, which leads to additional lowering of energy.

1.11 Strategies for Success: Drawing All Resonance Structures

It is important to be able to draw *all* of a species' resonance structures for two reasons: all resonance structures contribute to the features of the resonance hybrid, and the total number of resonance structures is related to the species' stability. We therefore devote Section 1.11 to providing practice in drawing resonance structures.

To begin, remember that any two resonance structures differ only in where the *electrons* are located. *Atoms must remain frozen in place!* We emphasize this explicitly by using **curved arrows**.

> A curved arrow ⌒ illustrates the *imaginary* electron movement necessary to change one resonance structure into another.

---

[1] As discussed in Chapter 14, much of benzene's resonance energy is due to a phenomenon called *aromaticity*.

This is shown for HCO_2^- in Figure 1-26.

Supply the necessary curved arrows to go from the resonance structure on the left to the resonance structure on the right.

With this brief introduction to curved arrows, we will now examine four basic situations that are commonly encountered.

1.11a Lone Pair Adjacent to a Multiple Bond

The two curved arrows used to interconvert the resonance structures of HCO_2^- (Fig. 1-26) can be used whenever a Lewis structure exhibits a lone pair of electrons on an atom connected to a double bond or triple bond. The first curved arrow is used to convert the lone pair into a covalent bond, and the second curved arrow is used to convert a pair of electrons from the original multiple bond into a lone pair. The general scenarios are as follows:

Each curved arrow illustrates the movement of an electron pair.

FIGURE 1-26 Curved arrow notation in resonance Shifting the four electrons indicated by the two curved arrows converts the resonance structure on the left into the one on the right.

Notice how the formal charges on the atoms change.

- When a lone pair of electrons on an atom is converted into a bonding pair (as on atom W above), the formal charge of that atom becomes more positive by 1.
- When a bonding pair involving an atom becomes a lone pair on that atom (as on atom Y above), the formal charge of that atom becomes more negative by 1.

SOLVED problem 1.20 Draw all resonance structures of the following species.

Think Which atom corresponds to W above? To X? To Y? What happens to the respective formal charges?

Solve The C with the lone pair corresponds to atom W and the right-most C corresponds to atom Y. The curved arrows are applied, yielding the additional resonance structure shown at right.

---

problem **1.21** The following compound belongs to a class of compounds called *enamine*s. Draw all additional resonance structures, including minor contributors.

The same strategy can be used when there is a chain of alternating multiple bonds and single bonds, as shown in Figure 1-27. The set of curved arrows is simply repeated time after time to get from one resonance structure to the next.

FIGURE 1-27 Multiple resonance structures involving an atom with a lone pair Each resonance structure has an atom with a lone pair adjacent to a double bond, so two curved arrows convert one resonance structure into another.

YOUR TURN 1.11

Draw in the curved arrows needed to convert the third resonance structure in Figure 1-27 to the fourth.

1.11b An Incomplete Octet Adjacent to a Multiple Bond

Resonance does not always involve lone pairs. If an atom with an unfilled valence shell is adjacent to a double bond or triple bond, then we can convert from one resonance structure to another using a single curved arrow:

In each case, a bond between X and Y is converted into a bond between W and X. Although W initially lacks an octet, it achieves an octet when it gains a share of two additional electrons. Meanwhile, Y initially has an octet, but then loses it. Moreover, notice once again that the formal charges change from one resonance structure to the other.

- When an atom lacking an octet gains a bond to achieve an octet (as with atom W), the atom's formal charge becomes more negative by 1.
- When a bond is removed from an atom that initially has an octet (as with atom Y), the formal charge of the atom becomes more positive by 1.

This concept is illustrated below with the allyl cation ($CH_2\!=\!CH\!-\!CH_2^+$).

$$\left[\ H_2\overset{\oplus}{C}\!-\!\underset{H}{C}\!=\!CH_2 \quad\longleftrightarrow\quad H_2C\!=\!\underset{H}{C}\!-\!\overset{\oplus}{C}H_2\ \right]$$

1.12 **YOUR TURN**

The resonance structures of the propargyl cation are shown below. Draw the curved arrows necessary to convert the first resonance structure into the second.

$$\left[\ HC\!\equiv\!C\!-\!\overset{\oplus}{C}H_2 \quad\longleftrightarrow\quad \overset{\oplus}{H}C\!=\!C\!=\!CH_2\ \right]$$

problem **1.22** Draw all resonance structures of the following species:

$$HC\!\equiv\!C\!-\!BH_2$$

Just as in the previous case involving lone pairs, several resonance structures can be obtained when a set of alternating multiple bonds and single bonds is attached to the atom lacking an octet. We simply move electrons multiple times, as shown in the following example:

1.13 **YOUR TURN**

Draw the necessary curved arrows on the preceding structures to transform the fourth resonance structure into the fifth.

problem **1.23** Draw the resonance hybrid of the following species (if necessary, consult the individual resonance structures shown just before Your Turn 1.13):

1.11c A Lone Pair Adjacent to an Atom with an Incomplete Octet

In Sections 1.11a and 1.11b, we worked with two structural features to convert from one resonance structure to another by rearranging valence electrons over three atoms at a time. When an atom with a lone pair (W) is attached to an atom lacking an octet (X), however, we can convert from one resonance structure to another by involving just two atoms:

In each case, the lone pair on W is converted into an additional bond between W and X. By doing so, X gains a share of two electrons, which gives it a complete octet and decreases its formal charge by 1. Even though W loses its lone pair, it retains a share of those electrons and thus keeps its octet. The formal charge on W, however, increases by 1.

Specific examples of resonance involving this structural feature are as follows:

In the example above on the left, the atom with the lone pair is initially attached by a single bond to the atom lacking an octet. In the example on the right, the two atoms are initially connected by a double bond.

---

problem **1.24** Draw all resonance structures of the following species. Remember to keep formal charges to a minimum when possible. Which do you think makes the greatest contribution to the resonance hybrid?

1.11d A Ring of Alternating Single and Multiple Bonds

With a ring of alternating single and multiple bonds, a pair of electrons from each multiple bond can shift around the ring—either clockwise or counterclockwise—to arrive at a new resonance structure. This is exemplified with benzene:

In the case of benzene, all six electrons must be shifted at the same time. If only two or four electrons are shifted, we end up with a Lewis structure that is not valid.

Octet exceeded!

Octet exceeded!

1.14 **YOUR TURN**

Draw the resonance structure indicated by the curved arrows in the structure above on the right. Label the atom with more than an octet of electrons.

problem **1.25** Draw all resonance structures of the following compound, making certain to keep formal charges to a minimum. Be sure to include curved arrows to indicate which pairs of electrons are being shifted around.

problem **1.26** Draw all resonance structures of each of the following species. Be sure to include curved arrows to indicate which pairs of electrons are being shifted.

(a)   (b)   (c)   (d)

1.12 Shorthand Notations

Learning organic chemistry requires drawing numerous molecules, but drawing the complete, detailed Lewis structure each and every time becomes tedious and quite cumbersome. For this reason, organic chemists have devised various shorthand notations that save time but do not entail any loss of structural information.

We will use these shorthand notations throughout the rest of the book. You should therefore take the time now to become comfortable with them.

1.12a Lone Pairs and Charges

Lone pairs of electrons are frequently omitted. They have vital roles in many organic reactions, however, so you must be able to put them back in as necessary. Doing so requires knowledge of how formal charge relates to the numbers of bonds and lone pairs on various atoms, as illustrated in Table 1-5. Pay particular attention to the fact that the atoms have octets in all of the scenarios shown, except in the case of a carbocation (C^+).

---

problem **1.27** Draw the complete Lewis structure of each of the following species, including lone pairs.

(a)   (b)   (c)   (d)   (e)

TABLE 1-5 Formal Charges on Atoms with Various Bonding Scenarios

Atom	Formal Charge		
	−1	0	+1
Carbon	$-\overset{..}{\underset{|}{C}}{}^{\ominus}-$	$-\overset{|}{\underset{|}{C}}-$	No octet! $-\overset{|}{C}{}^{\oplus}-$
Nitrogen	$-\overset{..}{\underset{..}{N}}{}^{\ominus}-$	$-\overset{..}{\underset{|}{N}}-$	$-\overset{|}{\underset{|}{N}}{}^{\oplus}-$
Oxygen	$-\overset{..}{\underset{..}{O}}{}^{\ominus}:$	$-\overset{..}{\underset{..}{O}}-$	$-\overset{..}{\underset{|}{O}}{}^{\oplus}-$
Halogen (X = F, Cl, Br, and I)	$:\overset{..}{\underset{..}{X}}{}^{\ominus}:$	$-\overset{..}{\underset{..}{X}}:$	$-\overset{..}{\underset{..}{X}}{}^{\oplus}-$

1.12b Condensed Formulas

Condensed formulas allow us to include molecules and molecular ions as part of regular text. Each nonhydrogen atom is written explicitly, followed immediately by the number of hydrogen atoms that are bonded to it. Adjacent nonhydrogen atoms in the condensed formula are interpreted as being covalently bonded to each other. For example, the condensed formula CH_3CHN^- indicates that there is a central C atom bonded to another C atom and to a N atom, giving rise to a skeleton that appears on the left in Figure 1-28. The structure is completed by adding the electrons shown in red—namely, a double bond between C and N and two lone pairs on N.

FIGURE 1-28 Lewis structure of CH_3CHN^- **To convert from the skeleton on the left to the Lewis structure on the right, the C—N bond and two lone pairs in red must be added.**

SOLVED problem 1.28 Draw the Lewis structure for $CH_3CHCHCHO$.

Think Which nonhydrogen atoms are bonded together? How can we add bonds and lone pairs to maximize the number of octets and also conform to the total charge of zero? How many total valence electrons must be accounted for in this compound?

Solve The condensed formula indicates the atoms should be connected as shown below on the left. To arrive at a total charge of zero, we give each C atom a total of four bonds and we give the O atom two bonds and two lone pairs. Double checking the structure, notice that it has 28 valence electrons, as it should.

Often a molecule will have multiple CH_2 groups bonded together. In these cases, we can simplify a condensed formula using the notation $(CH_2)_n$. Thus, $CH_3CH_2CH_2CH_2CH_3$ would simplify to $CH_3(CH_2)_3CH_3$.

Condensed formulas become somewhat cumbersome when three or four groups are attached to a single atom. In 2-methylbutane (Fig. 1-29a), for example, the second C atom from the left is bonded to two CH_3 groups and one CH_2CH_3 group. Parentheses can be used to account for these additional groups, in which case 2-methylbutane can be written as $CH_3CH(CH_3)CH_2CH_3$.

Many common organic compounds contain structures in which a carbon atom is bonded to one oxygen atom by a double bond and to a second oxygen atom by a single bond. We can write this group in the form $—CO_2—$, as shown in Figure 1-29b for acetic acid (CH_3CO_2H).

FIGURE 1-29 Condensed formulas and branching Condensed formulas are shown on the top and Lewis structures are shown on the bottom. (a) Parentheses in a condensed formula denote that the group is attached to the previous C. (b) The CO_2 notation represents a C atom that is doubly bonded to one O atom and singly bonded to another. (c) Rings are generally not shown in their condensed formulas, but are commonly shown in their partially condensed form.

$CH_3CH(CH_3)CH_2CH_3$

2-Methylbutane

(a)

CH_3CO_2H

Ethanoic acid (Acetic acid)

(b)

$[—CH_2CH_2CH_2CH_2CH_2CH_2—]$

Cyclohexane

(c)

Cyclohexane (partially condensed form)

Condensed formulas for cyclic structures, like cyclohexane (Fig. 1-29c), are problematic. If the formula is written on a single line of text, the left-most C atom must be bonded to the right-most one to complete the ring. This would appear as follows: $\boxed{-CH_2CH_2CH_2CH_2CH_2CH_2-}$. Because this is so cumbersome, we generally do not represent rings in their fully condensed forms. Instead, rings are often depicted in their *partially* condensed form, as shown at the right in Figure 1-29c.

1.12c Line Structures

Line structures, like condensed formulas, are compact and can be drawn quickly and easily. Unlike condensed formulas, however, they are not intended to be written as part of text. The rules for drawing line structures are as follows:

Rules for Drawing Line Structures

- Carbon atoms are not drawn explicitly, but they are implied at the intersection of every two or more lines and at the end of every bond that is drawn, unless another atom is written there already.
- Hydrogen atoms bonded to carbon are *not* drawn, but hydrogen atoms bonded to all other atoms are.
- All noncarbon and nonhydrogen atoms, called **heteroatoms**, are drawn.
- Bonds to hydrogen are not drawn, but all other bonds are drawn explicitly.
- Several carbon atoms bonded in a single chain are represented by a zig-zag structure.
- Enough hydrogen atoms are assumed to be bonded to each carbon atom to fulfill the carbon atom's octet, with close attention paid to the formal charge on carbon (see Table 1-5).
- Lone pairs of electrons are generally not shown unless they are necessary to emphasize an important aspect of an atom.

The line structure of $CH_3CH_2CH_2CH_2CH_2CH_2CH_2NH_2$ is shown in Figure 1-30. Notice that the intersection of each bond line represents a C atom, as does the end of the bond on the left side of the molecule. The NH_2 group is written in explicitly.

FIGURE 1-30 Line structure of $CH_3(CH_2)_6NH_2$ In the line structure on the right, neither the C atoms nor the H atoms attached to C are shown. However, the N atom is shown, and so are the H atoms attached to N.

A line structure omits all carbons and all hydrogens bonded to carbon.

---

problem **1.29** Redraw each of the following Lewis structures as the corresponding line structure.

(a)   (b)   (c)

problem **1.30** For each of the following line structures, draw in all carbon atoms, hydrogen atoms, and lone pairs.

(a)   (b)   (c)

SOLVED problem **1.31** Why is it incorrect to draw resonance structures for propene as shown below?

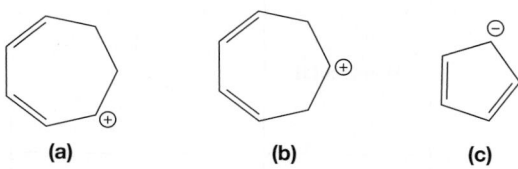

Think What would the structures look like as complete Lewis structures (i.e., with all hydrogen atoms drawn in)? What rule is broken in the transformation from the first structure to the second?

Solve If all atoms and bonds are included, then the transformation would appear as follows.

Bond broken   Bond formed

This process entails the breaking of a C—H bond on the left and the formation of a C—H bond on the right. When drawing resonance structures, however, all atoms must remain frozen in place.

---

problem **1.32** Draw all resonance structures for each of the following ions using only line structures.

(a)   (b)   (c)

1.13 An Overview of Organic Compounds: Functional Groups

Certain types of bonds and arrangements of atoms tend to appear quite frequently in organic chemistry. The most common types of bonds we encounter are C—C and C—H single bonds. Compounds consisting of nothing but C—C and C—H single bonds are called **alkanes**. Examples include methane, 2-methyloctane, and ethylcyclohexane (Fig. 1-31).

Alkanes tend to be among the most *unreactive* organic compounds. For this reason, liquid alkanes (e.g., pentane and hexane) are often used as solvents in which to carry out organic reactions.

Alkanes are relatively inert in part because C—C and C—H bonds are so strong; and, for a reaction to occur, these bonds must break. In fact, the data in Table 1-2 show that these are among the strongest single bonds we will encounter.

CH_4
Methane

2-Methyloctane

Ethylcyclohexane

FIGURE 1-31 **Some alkanes** Alkanes consist entirely of carbon and hydrogen and have only single bonds.

Alkanes also tend to be unreactive because C—C and C—H bonds are either non-polar or are only very slightly polar. In Chapter 7, we explain that a large driving force behind many chemical reactions depends upon the existence of substantial bond dipoles.

When other types of bonds and other types of atoms are introduced, then organic molecules tend to be more reactive. In some cases, this may be due to the introduction of weaker bonds, and in other cases, to the presence of large bond dipoles.

As we will learn throughout the remainder of this book, specific arrangements of atoms connected by specific types of bonds tend to react in characteristic ways. In fact, even though such groups may consist of only two or three atoms, they dictate the *function* (i.e., the behavior) of the entire molecule. Hence these structural components are called **functional groups**. Consequently,

A molecule's reactivity is governed by the functional groups it has.

The functional groups that appear most often are listed in Table 1-6. Take the time to commit these functional groups to memory and review them frequently.

TABLE 1-6 Common Functional Groups[a]

Functional Group	Name	Functional Group	Name	Functional Group	Name
	Alkene		Ether		Ketone
	Alkyne		Acetal		Aldehyde
	Aryl		Hemiacetal		Carboxylic acid
(X = F, Cl, Br, I)	Alkyl halide		Epoxide		Ester
	Alcohol		Amine		Amide
	Thiol		Nitrile		

[a] "R" indicates an alkyl group and the absence of an atom at the end of a bond indicates that either "R" or H may be attached.

There are two important features in Table 1-6 that you must understand. The first is "R," which is used to represent an **alkyl group**—that is, a portion of the molecule that consists only of carbon and hydrogen. As with alkanes, alkyl groups tend to be unreactive. The benefit of replacing large, unreactive alkyl groups with "R" is that it allows us to focus on just the reactive portion of the molecule (i.e., the functional group).

We will learn more about alkyl groups in the first **nomenclature** unit, following this chapter, which pertains to the naming of molecules. For the purpose of Table 1-6, *a bond to R specifically indicates a bond to a carbon atom.*

The second important feature in Table 1-6 is the absence of atoms at the ends of specific bonds. This indicates that the bond may be either to an alkyl group (R) or to a hydrogen atom.

With these in mind, the following conclusions can be drawn from Table 1-6:

1. **Functional groups can differ in type of atom.** R—S—H and R—O—H, for example, are different functional groups.
2. **Functional groups can differ in type of bond.** C=C and C≡C are different functional groups.
3. **Some functional groups contain the bonding arrangements of smaller functional groups.** A carboxylic acid functional group (—CO$_2$H), for example, contains both the C—O—H and C=O arrangements, but is not an alcohol, ketone, or aldehyde.
4. **Alkanes are considered to have no functional groups.** They are relatively un-reactive, and when they do react, the reaction tends to be very unselective (see Chapter 25).
5. **Rings generally do not constitute new functional groups.** In most cases, the reactivity of a functional group in a ring is very similar to its reactivity in an open chain.
   a. One exception is the *aryl group*, which is discussed in depth in Chapter 14.
   b. Another exception is an *epoxide*, which is a cyclic *ether*. As we discuss in Chapters 2 and 4, this is because three- and four-membered rings are highly strained, which tends to make them more reactive than an analogous open-chain ether group.

problem **1.33** In Table 1-6, there are six functional groups that contain the bonding arrangements of simpler functional groups. A carboxylic acid is one of them, as noted previously. What are the other five and what functional groups do they appear to contain?

SOLVED problem **1.34** Will the chemical reactivity of the following compounds be similar or significantly different? Explain.

Cyclohexanone   Hexan-3-one

Think Are their functional groups the same or different? What impact will the ring have on the reactivity of cyclohexanone?

Solve Both compounds are ketones. Although cyclohexanone's ketone group is part of a ring, this should not significantly alter its reactivity relative to that of the open-chain hexan-3-one (see Rule 5). As a result, both compounds should behave similarly.

problem **1.35** Will the chemical reactivity of the following compounds be similar or significantly different? Explain.

YOUR TURN 1.15

Circle and identify the functional groups present in cyclohexanone and hexan-3-one in Solved Problem 1.34.

SOLVED problem **1.36** Circle each of the functional groups present in Ebalzotan, which was developed as an antidepressant and an antianxiety agent.

Ebalzotan

Think Are there *bonds* present other than C—C and C—H single bonds? Are there *atoms* present other than carbon and hydrogen? Are there any special rings present?

Solve There are three C═C double bonds present, suggesting the alkene functional group. They comprise three alternating single bonds and double bonds in a ring, however, so the functional group is in fact an aryl group. There are two O atoms and two N atoms present, each providing the potential for another functional group. The N atom at the right is part of an amine (in this case, R_3N). The O atom at the top is part of an ether (R—O—R). The O atom and the N atom at the bottom left are part of an amide.

Aryl   Ether

Amine

Amide

problem **1.37** Circle each of the functional groups present in ciprofloxacin, a powerful antibiotic more commonly known as "Cipro".

Ciprofloxacin
"Cipro"

1.14 Wrapping Up and Looking Ahead

We began Chapter 1 by examining what organic chemistry is and why it is important. In short, organic chemistry is the branch of chemistry that involves compounds made principally of carbon and hydrogen, and is primarily responsible for the chemical processes that take place in living organisms. Moreover, organic chemistry is instrumental in the engineering of a variety of new compounds and materials, including pharmaceutical drugs, polymers, and electronics. Such applications are afforded by the relative stability and structural diversity of organic compounds.

With this in mind, much of Chapter 1 focused on aspects of molecular structure and stability. We reviewed the general structure of atoms, including ground state electron configurations. We discussed the stability that arises from an octet, which forms the basis of covalent bonding and Lewis dot structures. We saw that moderate differences in the electronegativity of atoms bonded together give rise to polar covalent bonds, whereas large differences result in ionic bonding. Resonance theory was introduced as a way to compensate for a shortcoming of Lewis dot structures. And we ended by introducing some shorthand notations, as well as by examining common functional groups—bonding scenarios that govern the reactivity of an entire molecule.

The topics discussed in this chapter will have important applications in the chapters throughout the rest of the book. Lewis structures provide insight into whether valence electrons are part of single bonds, double bonds, triple bonds, or lone pairs. As we will see in Chapter 2, that information will provide us with insight into three-dimensional geometry. And in Chapter 7, we will see that it helps us understand relative reactivities of molecules. Furthermore, electronegativity and resonance are key concepts that govern the distribution of charge within a molecular species, which, as we will discuss in Chapter 7, is instrumental in the outcome of many chemical reactions.

THE ORGANIC CHEMISTRY OF BIOMOLECULES

1.15 An Introduction to Proteins, Carbohydrates, and Nucleic Acids: Fundamental Building Blocks and Functional Groups

Organic chemistry plays a pivotal role in biology at the molecular level. As we saw in Section 1.1, the original definition of an "organic" compound was a substance of plant or animal origin, and even today many people still think of "organic" as synonymous with "natural," and distinct from a "synthetic" substance made in a laboratory or an industrial plant. To the chemist, however, organic molecules are those composed chiefly of carbon and hydrogen, and usually other nonmetals such as oxygen, nitrogen, phosphorus, sulfur, and the halogens (fluorine, chlorine, bromine, and iodine). The particular organic molecules found almost exclusively in living organisms are more properly called **biomolecules**.

Because they are generally the products of millions of years of evolution, the structures of biomolecules are highly adapted to serve specific biological functions. Some of these functions are described here in Section 1.15. It is important to realize, however, that these molecules behave basically the same way in living cells as they do in a flask. The chemistry of living cells is thus ultimately the chemistry of organic compounds.

There are four major classes of biomolecules that we will discuss: *proteins, carbohydrates, nucleic acids,* and *lipids.* In this unit on biomolecules, we discuss aspects of proteins, carbohydrates, and nucleic acids, leaving lipids to Section 2.10. We choose this organization because proteins, carbohydrates, and nucleic acids have a common structural attribute that lipids do not.

> Proteins, carbohydrates, and nucleic acids are typically very large structures, with molar masses that can reach the millions, but are constructed from relatively few types of small organic molecules.

By contrast, lipids are relatively small- to medium-sized molecules, characterized by their insolubility in water—a topic that is covered in Chapter 2.

Moreover, the small organic molecules that compose large proteins, carbohydrates, and nucleic acids, are recognizable because they have distinct structural features containing a variety of the characteristic functional groups listed in Table 1-6. We will examine some of those features in the remainder of this unit.

1.15a Proteins and Amino Acids

Proteins are quite versatile biomolecules. Some, such as actin and myosin, are responsible for the mechanical processes involving muscle tissue. Others, such as cortactin, are responsible for regulating cell shape. There are also many different proteins that are classified as **enzymes**, which act as catalysts for biological reactions—that is, enzymes facilitate biological reactions, but are not consumed while doing so. Acetylcholinesterase, for example, catalyzes the breakdown of acetylcholine, a neurotransmitter.

Proteins are constructed from relatively few types of small organic molecules called **α-amino acids**. As a result, we can view proteins on different size scales, as shown in Figure 1-32. The general structure of an amino acid is shown in Figure 1-33.

FIGURE 1-32 The size hierarchy of proteins Amino acids (left) are covalently bonded together to form proteins (second from left). Proteins, in turn, are integrated into more familiar structures, such as the muscle tissue shown on the right.

α-Amino acids are the small organic molecules used to build proteins.

Proteins have a variety of functions, including those involving the mechanical processes of muscle tissue.

| A typical α-amino acid | A segment of a protein | A typical fibrous protein | Muscle tissue |

As we can see, all α-amino acids have in common certain functional groups with specific relative locations:

- An **α-amino acid** contains both an amine and a carboxylic acid functional group.
- The amine group is attached to the C atom that is *adjacent to* the C of the carboxylic acid—that is, both the amine and the carboxylic acid are attached to the **alpha carbon**.
- The alpha carbon is also covalently bonded to a **side chain**, which is sometimes called an **R group**.

The side chain of an amino acid is what distinguishes one amino acid from another. In particular, there are 20 naturally occurring side chains, so there are 20 different naturally occurring amino acids. Their structures are shown in Table 1-7.

Notice in Table 1-7 that amino acid side chains contain a variety of functional groups. The side chain of serine, for example, contains an alcohol (R—OH) group, whereas aspartic acid has a carboxylic acid group (R—CO$_2$H) and lysine has an amine (R—NH$_2$). Some amino acids, such as phenylalanine, contain an aryl ring.

1.16 YOUR TURN

Circle and label the following groups:

(a) The alcohol group in the side chain of serine.
(b) The carboxylic acid group in the side chain of aspartic acid.
(c) The amine group in the side chain of lysine.
(d) The aryl ring in the side chain of phenylalanine.

The identity of the side chain is what distinguishes the properties of one amino acid from another, such as whether the amino acid is polar or nonpolar, or acidic or basic. These properties, in turn, are largely what govern an amino acid's function when it forms part of a protein. We explore these kinds of ideas in greater depth as we continue our discussion of biomolecules throughout this book.

problem **1.38** Identify all of the amino acids that contain the following functional groups in their side chains:

(a) Alcohol   **(b)** Amide   **(c)** Carboxylic acid
(d) Amine   **(e)** Aryl ring

1.15b Carbohydrates and Monosaccharides

Carbohydrates, also called **saccharides**, serve a variety of biological functions. Sugar, starch, and glycogen, for example, are fuels for primary metabolic pathways such as glycolysis, and cellulose is the structural component of the cell walls in plants. Derivatives of carbohydrates are involved in many important processes, too, such as those related to blood clotting and the immune system.

TABLE 1-7 The 20 Naturally Occurring Amino Acids

Name (abbreviation)[a]	Side Chain	Name (abbreviation)[a]	Side Chain
Nonpolar Amino Acids		**Polar Amino Acids**	
Alanine (Ala or A)	—CH₃	Asparagine (Asn or N)	
Glycine (Gly or G)	—H	Cysteine (Cys or C)	—CH₂SH
Isoleucine (Ile or I)		Glutamine (Glu or Q)	
Leucine (Leu or L)		Serine (Ser or S)	—CH₂OH
Methionine (Met or M)	—(CH₂)₂SCH₃	Threonine (Thr or T)	
Phenylalanine (Phe or F)		Tryptophan (Trp or W)	
Proline (Pro or P)		Tyrosine (Tyr or Y)	
Valine (Val or V)			
Acidic Amino Acids		**Basic Amino Acids**	
Aspartic acid (Asp or D)		Arginine (Arg or R)	
Glutamic acid (Glu or E)		Lysine (Lys or K)	
		Histidine (His or H)	

[a] Each amino acid has a one- and three-letter abbreviation.

Carbohydrates can form larger organized structures such as those in cellulose.

A polysaccharide

A monosaccharide (a simple sugar)

Monosaccharides are relatively small molecules that are used to construct larger carbohydrates.

Microfibrils of cellulose in a plant cell wall

FIGURE 1-34 The size hierarchy of carbohydrates Monosaccharides (far left) are used to construct polysaccharides (middle). Cellulose (far right) is a polysaccharide that makes up the cell walls in plants.

Carbohydrates are characterized by their chemical composition:

- Carbohydrates are composed of only carbon, oxygen, and hydrogen.
- There are two hydrogen atoms for every oxygen atom, regardless of how many carbon atoms are present, giving each carbohydrate the molecular formula $C_xH_{2y}O_y$.

Thus, hydrogen and oxygen are present in the same ratio as in water, so the general formula of a carbohydrate can be written $C_x(H_2O)_y$. This is why "hydrate" appears in the name. (The bonding scheme in a carbohydrate, however, is quite different from that in water.)

Carbohydrates are frequently very large molecules, called **polysaccharides**. Polysaccharides are constructed from just a few types of smaller molecules, called **monosaccharides** or **simple sugars**. Thus, carbohydrates can be viewed on the different size scales shown in Figure 1-34.

Monosaccharides are themselves carbohydrates, with one additional restriction to their chemical composition:

In a monosaccharide, the number of oxygen atoms is the same as the number of carbon atoms, giving it the general formula $C_xH_{2x}O_x$.

As shown in Figure 1-35, for example, ribose has the molecular formula $C_5H_{10}O_5$, whereas glucose and fructose have the formula $C_6H_{12}O_6$.

FIGURE 1-35 Some monosaccharides Functional groups such as alcohols, aldehydes, and ketones are common to monosaccharides.

An alcohol functional group

An aldehyde functional group

A ketone functional group

Ribose, $C_5H_{10}O_5$

Glucose, $C_6H_{12}O_6$

Fructose, $C_6H_{12}O_6$

Notice in these structures that each C atom is bonded to a single O atom. One O atom in each monosaccharide is part of either an aldehyde or a ketone functional group. The remaining O atoms are part of alcohol functional groups.

YOUR TURN 1.17

Circle and label all of the alcohol groups in the structure of glucose in Figure 1-35. Is the carbonyl (C=O) group in glucose part of an aldehyde or a ketone functional group?

Monosaccharides are cyclic when linked together in naturally occurring polysaccharides, as shown previously in Figure 1-34. Individual monosaccharides, on the other hand, continually interconvert between their cyclic and acyclic forms, as shown for ribose in Figure 1-36. This process is discussed in greater detail in Section 18.16.

An acyclic monosaccharide   **A cyclic monosaccharide**

FIGURE 1-36 Cyclization of a monosaccharide A monosaccharide such as ribose can equilibrate between its cyclic and acyclic forms.

Ribose, $C_5H_{10}O_5$

problem **1.39** Which of the following molecules are carbohydrates? Which are monosaccharides?

(a)   (b)

(c)   (d)

1.15c Nucleic Acids and Nucleotides

A **nucleic acid** is a large molecular chain that is primarily associated with the storage and transfer of genetic information. A pair of intertwined nucleic acids form the double-helical **deoxyribonucleic acid** (DNA), which stores genetic information, and **ribonucleic acid** (RNA), which participates in protein synthesis. Each strand of the double helix is a nucleic acid. Nucleic acids themselves are constructed from relatively

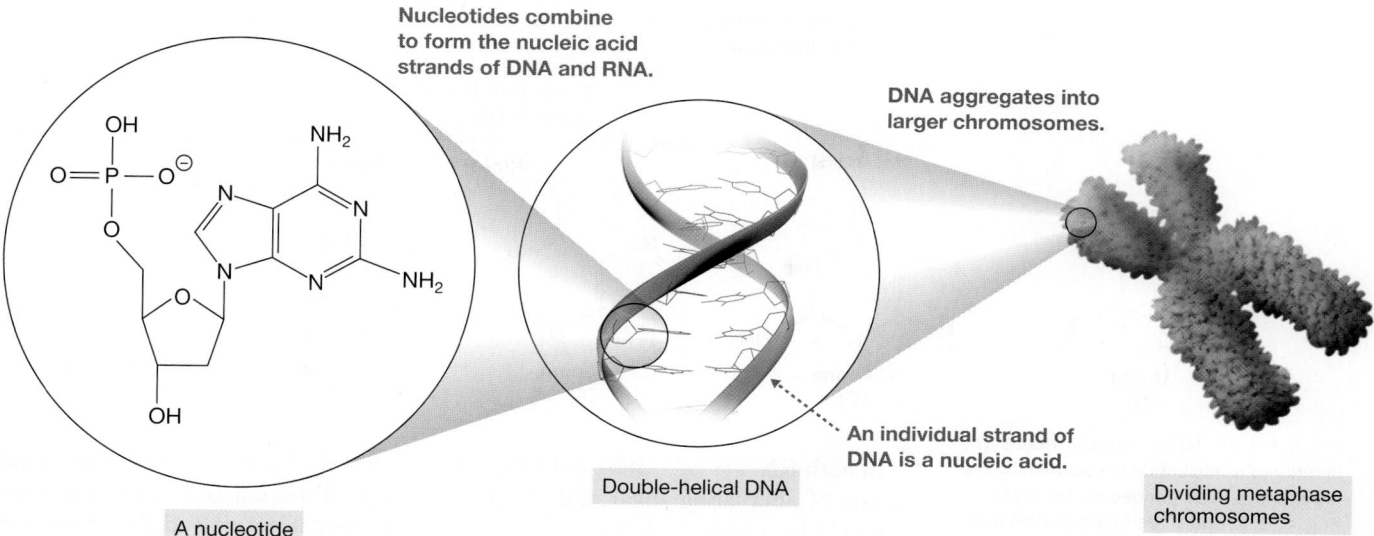

Nucleotides combine to form the nucleic acid strands of DNA and RNA.

A nucleotide

DNA aggregates into larger chromosomes.

An individual strand of DNA is a nucleic acid.

Double-helical DNA

Dividing metaphase chromosomes

FIGURE 1-37 The size hierarchy of nucleic acids Nucleotides (far left) are used to construct nucleic acids, which form the strands of RNA and DNA (middle). Chromosomes (far right) are aggregates of DNA.

small molecular units called **nucleotides**. Thus, just as we saw with proteins and carbohydrates, nucleic acids can be viewed on different size scales, as shown in Figure 1-37.

All nucleotides have three distinct components (Fig. 1-38a):

1. an inorganic phosphate (PO_4) group,
2. a cyclic monosaccharide (or sugar), and
3. a nitrogenous base.

FIGURE 1-38 Composition of a nucleotide (a) Every nucleotide has three components: a phosphate group (green), a cyclic sugar (red), and a nitrogenous base (blue). The backbone of a nucleic acid consists of alternating sugar and phosphate groups. (b) A nucleotide in a ribonucleic acid. The cyclic sugar component must be ribose. The nitrogenous base shown is guanine, but could also be adenine, cytosine, or uracil. (c) A nucleotide in a deoxyribonucleic acid. The cyclic sugar must be deoxyribose, in which the ribose oxygen indicated is not present. The nitrogenous base shown is cytosine, but could also be adenine, guanine, or thymine.

These four bases are found in RNA.

These four bases are found in DNA.

Uracil
(U)

Guanine
(G)

Adenine
(A)

Cytosine
(C)

Thymine
(T)

FIGURE 1-39 Nitrogenous bases The identity of a nucleotide is specified by the nitrogenous base bonded to the sugar ring. U, G, A, and C are found in RNA and G, A, C, and T are found in DNA.

In both RNA (Fig. 1-38b) and DNA (Fig. 1-38c), the backbone of a single strand consists of alternating sugar and phosphate groups. Thus, adjacent nucleotides are connected by a bond between the sugar group of one nucleotide and the phosphate group of another.

Nucleotides are distinguished from one another by the identity of the nitrogenous base. In both RNA and DNA, it is the specific sequence of the nitrogenous bases that determines the genetic information that is stored or carried. There are four types of nitrogenous bases that appear in RNA: uracil, guanine, adenine, and cytosine (abbreviated U, G, A, and C, respectively; Fig. 1-39). There are four types of nitrogenous bases that appear in DNA, too: G, A, C, and thymine (T). Thus, three of the bases in DNA are the same as the bases in RNA. Only T (in DNA) is different from U (in RNA).

Notice the functional groups in these nitrogenous bases. An alkene group appears, for example, in uracil and thymine, and an amide group can be found in guanine.

YOUR TURN 1.18

Circle and label the alkene and amide functional groups that appear in U, G, A, C, and T in Figure 1-39.

In subsequent units on biomolecules, we examine some of the details of the chemical processes involving DNA and RNA. In Section 14.12, for example, we examine the issue of complementarity among the nitrogenous bases.

problem **1.40** For each of the following nucleotides:

(a) Circle and label the phosphate group, the sugar group, and the nitrogenous base.
(b) Determine whether it can be part of RNA or DNA.
(c) Identify the nitrogenous base that it contains.

(i)

(ii)

Chapter Summary and Key Terms

- **Organic chemistry** is the subdiscipline of chemistry in which the focus is on compounds containing carbon atoms. **(Section 1.1)**

- Carbon's ability to form four strong covalent bonds is what gives rise to the great variety of organic compounds known. **(Section 1.2)**

- No more than two electrons can occupy an atom's first shell. As many as eight electrons can occupy an atom's second shell. A second or higher-level shell containing eight electrons (an **octet**) is especially stable (the octet rule). **(Section 1.3b)**

- An atom's ground state (i.e., lowest energy) electron configuration is derived using the following three rules **(Section 1.3c)**:
  - **Pauli's exclusion principle:** Up to two electrons, opposite in spin, may occupy an orbital.
  - The **aufbau principle:** Electrons occupy the lowest energy orbitals available.
  - **Hund's rule:** If two orbitals have the same energy, then each orbital is singly occupied before a second electron fills it.

- Sharing of a pair of electrons by two atoms lowers the energy of the system, creating a **covalent bond**. Some atoms can share two or three pairs of electrons, creating double and triple bonds, respectively. Double bonds are shorter and stronger than single bonds, and triple bonds are shorter and stronger than double bonds. **(Section 1.4)**

- **Lewis structures** illustrate a molecule's **connectivity**. They account for *all* valence electrons, differentiating between **bonding pairs** and **lone pairs** of electrons. They indicate, moreover, which atoms are bonded together and by what types of bonds. Lewis structures are constructed so that the maximum number of atoms have filled valence shells (i.e., duets or octets). **(Section 1.5)**

- Electrons in a bond are not always shared equally. **Polar covalent bonds** arise when atoms with moderate differences in electronegativity are bonded together **(Section 1.7)**. **Ionic bonds** arise when there are large differences in electronegativity. **(Section 1.8)**

- **Formal charges** and **oxidation states** reflect different ways in which valence electrons are assigned to individual atoms within a molecule. **(Section 1.9)**
  - To determine formal charges, each pair of electrons in a covalent bond is split evenly between the two atoms bonded together.
  - To determine oxidation states, each pair of electrons that make up a covalent bond is assigned to the more electronegative atom.

- **Resonance** exists when two or more valid Lewis structures can be drawn for a given molecular species. **(Section 1.10)**
  - Each Lewis structure is imaginary and is called a **resonance structure**.
  - The properties of the one, true species—the **resonance hybrid**—represent a weighted average of all the resonance structures.
  - The greater the stability of a given resonance structure, the greater its contribution to the resonance hybrid.

- Resonance structures are related by hypothetically shifting around lone pairs of electrons, and pairs of electrons from double and triple bonds. The atoms in the molecule must remain frozen in place. **(Section 1.11)**

- Shorthand notation is used throughout organic chemistry to draw molecules more quickly and efficiently. Lone pairs are often omitted. **(Section 1.12)**
  - **Condensed formulas** are used primarily to write a molecule in a line of text, with hydrogens written adjacent to the atom to which they are bonded.
  - **Line structures** show all bonds explicitly, except bonds to hydrogen. Hydrogen atoms are omitted if they are bonded to carbon. Carbon atoms are not drawn explicitly, but are assumed to reside at the intersection of two lines and at the end of a chain (unless otherwise indicated).

- **Functional groups** are common bonding arrangements of relatively few atoms. Functional groups dictate the behavior of entire molecules—that is, molecules with the same functional groups tend to behave similarly. **(Section 1.13)**

Problems

1.41 In which of the following orbitals does an electron possess the most potential energy?

2s   3s   4s   3p   2p

1.42 In which of the following orbitals does an electron possess the most potential energy?

4s   5s   5p   5d   4d

1.43 At which of the following separations would two cations have the most potential energy? Explain.

0.1 nm   1000 nm   100 nm   10 nm   1 nm

1.44 At what distance from a cation would an anion have the most potential energy? Explain.

0.1 nm   1000 nm   100 nm   10 nm   1 nm

1.45 Write the ground state electron configuration of each of the following atoms. For each atom, identify the valence electrons and the core electrons.
(a) Al   (b) S   (c) O   (d) N   (e) F

1.46 Which of the following contains an ionic bond?
(a) H_2 (b) NaCl   (c) NaOH   (d) CH_3ONa (e) CH_4
(f) $HOCH_2CH_3$ (g) $LiNHCH_3$ (h) $CH_3CH_2CO_2K$ (i) $C_6H_5NH_3Cl$

1.47 Rank the following in order of increasing negative charge on carbon, assuming that a bond exists between the atoms indicated.

$CH_3—CH_3$ $CH_3—MgBr$ $CH_3—Li$ $CH_3—F$ $CH_3—OH$ $CH_3—NH_2$

1.48 Which of the following electrostatic potential maps best represents nitromethane (CH_3NO_2)? Explain.

A   B   C

1.49 Draw Lewis structures for each of the following molecules:
(a) CH_5N (contains a bond between C and N)
(b) CH_3NO_2 (contains a bond between C and N but no bonds between C and O)
(c) CH_2O (d) CH_2Cl_2 (e) BrCN

1.50 In the methoxide anion (CH_3O^-), is it possible for a double bond to exist between C and O, given that the negative charge resides on O? Explain why or why not.

1.51 Draw Lewis structures for each of the following ions. One atom in each ion has a formal charge that is not zero. Determine which atom it is, and what the formal charge is.
(a) the C_2H_5 anion   (b) the CH_3O cation   (c) the CH_6N cation   (d) the CH_5O cation
(e) the C_3H_3 anion (all three H atoms are on the same carbon)

1.52 Complete the Lewis structure for each of the following molecules using the information provided in Table 1-4. You may assume that no formal charges exist on any atoms. (All H atoms are shown; add only bonding pairs and lone pairs of electrons.)

(a)

(b)

(c)

1.53 Determine the oxidation states for all atoms other than hydrogen in each of the following species.

(a)
(b)
(c)

(d)

1.54 Identify the formal charge and oxidation state on each atom in the following species. Assume that all valence electrons are shown.

1.55 The following is a skeleton of a molecular anion having the overall formula $C_6H_6NO^-$. The hydrogen atoms are not shown.

(a) Draw a complete Lewis structure of the species in which the -1 formal charge is on N. Include all H atoms and valence electrons.
(b) Do the same for the species with the -1 formal charge on O.
(c) Do the same for the species with the -1 formal charge on the C atom that is bonded to three other C atoms.

1.56 **(a)** Redraw the following structure of glucose as a line structure.
(b) Circle and identify each functional group in glucose.

Glucose

1.57 **(a)** Redraw the following line structure of sucrose as a complete Lewis structure. Include all hydrogen atoms and lone pairs.
(b) Circle and identify each functional group in sucrose.

Sucrose

1.58 Which of the following pairs are *not* resonance structures of one another? All lone pairs of electrons may or may not be shown. Identify where each lone pair that is not shown belongs.

A   **B**   **C**   **D**

E   **F**   **G**   **H**

1.59 Which of the following species is responsible for the electrostatic potential map provided? Explain.

A   **B**   **C**

D   **E**

1.60 Dimethyl sulfide contains a functional group that is not listed in Table 1-6. Which functional group in Table 1-6 do you think its reactivity might resemble?

$$H_3C-S-CH_3$$
Dimethyl sulfide

1.61 Which one of the following compounds do you think will behave most similarly to ethanol (CH_3CH_2OH)? Explain.

H_2O CH_3OCH_3 CH_3CO_2H $H_3C-\overset{O}{\overset{\|}{C}}-H$

A   **B**   **C**   **D**

1.62 Identify and name all functional groups that are present in strychnine, a highly toxic alkaloid used as a pesticide to kill rodents, whose connectivity is shown below.

Strychnine

1.63 Identify and name all functional groups that are present in doxorubicin, a drug used as an antibiotic and cancer therapeutic, whose line structure is shown below.

Doxorubicin

1.64 Draw the structure of a molecule with formula C_5H_9N that contains
(a) an alkyne functional group and an amine functional group.
(b) two alkene functional groups and an amine functional group.
(c) a nitrile functional group.

1.65 Draw all resonance contributors for each of the following molecules or ions. Be sure to include the curved arrows that indicate which pairs of electrons are shifted in going from one resonance structure to the next.
(a) CH_3NO_2
(b) $CH_3CO_2^-$
(c) $CH_3CHCHCH_2^+$ (the ion has two C—C single bonds)
(d) C_5H_5N (a ring is formed by the C and N atoms, and each H is bonded to C)
(e) C_4H_5N (a ring is formed by the C and N atoms, the N is bonded to one H, and each C is bonded to one H)

1.66 Draw the resonance hybrid of species (a) in Problem 1.65.

1.67 (a) Draw all resonance contributors of sulfuric acid, H_2SO_4 (the S atom is bonded to four O atoms).
(b) Which resonance structure contributes the most to the resonance hybrid?
(c) Which resonance structure contributes the least to the resonance hybrid?

1.68 Draw all of the resonance structures for each of the following species. Be sure to include the curved arrows that indicate which pairs of electrons are shifted in going from one resonance structure to the next. Draw the resonance hybrid of each species.

(a)

(b)

(c)

(d)

1.69 Draw the following Lewis structures using condensed formulas.

(a)

(b)

(c)

(d)

(e)

(f)

1.70 Draw the molecules in Problem 1.69 using line formulas.

1.71 Draw Lewis structures for the following molecules. Include all lone pairs and H atoms.

(a)

(b)

(c)

(d)

(e)

(f)

(g)

(h)

(i)

1.72 Draw each of the species in Problem 1.71 as a condensed formula.

1.73 Diazomethane has the formula H_2CN_2. Draw all valid resonance contributors for diazomethane and, using Table 1-3, propose which one contributes more to the resonance hybrid. (*Hint:* There are no structures that avoid charged atoms.)

1.74 **(a)** Draw all resonance contributors of the following ion. In drawing each additional resonance structure, used curved arrows to indicate which pairs of electrons are being shifted.
(b) Draw the resonance hybrid.
(c) Which C—C bond is the longest?

1.75 **(a)** Draw all valid resonance contributors for the following ion. Show how the electrons can be moved using curved-arrow notation.
(b) Draw the resonance hybrid.

1.76 Redraw the given line structure of cholesterol as a condensed formula. What advantages do line structures have?

Cholesterol

1.77 Experiments indicate that the C—C bonds in cyclobutadiene are of two different lengths. Argue whether or not cyclobutadiene has a resonance structure.

132 pm

160 pm

1.78 The two species shown are structurally very similar. Draw all resonance structures for each species and determine which is more stable. Explain.

A

B

1.79 The two species shown are structurally very similar. Draw all resonance structures for each species and determine which is more stable. Explain.

C

D

If your instructor assigns problems in smartwork, log in at **smartwork.wwnorton.com**.

Introduction: The Basic System for Naming Simple Organic Compounds

Alkanes, Cycloalkanes, Haloalkanes, Nitroalkanes, and Ethers

N1.1 The Need for Systematic Nomenclature: An Introduction to the IUPAC System

Learning how to name specific substances—that is, learning **nomenclature**—is a necessary task. Nomenclature, moreover, is not unique to chemistry. Every discipline of study, from accounting and art history to zoology, has its specialized vocabulary. Many terms in these fields can be as technical as anything in the physical sciences.

Biology, for example, includes taxonomy, the classification of organisms and the study of their relationships. The number of species living on our planet is conservatively estimated at 1.5 million (some estimates exceed 30 million). To deal with this great diversity, scientists have adopted, by international agreement, a single language to be used on a worldwide basis. All organisms are given a specific name in Latin, such as *Homo sapiens* (human beings), *Canis familiaris* (the domestic dog), or *Escherichia coli* (a common bacterium of the human digestive system). Like any language, the language of taxonomy evolves with use; some terms commonly used at one time are later declared obsolete and replaced with more precise or more relevant ones.

Organic chemistry is not very different. There are millions of organic compounds, a number comparable to the number of known biological species. In ancient and medieval times, chemists or alchemists sometimes assigned a name to a newly discovered substance that indicated its natural animal or plant source. Vanillin, for example, is the primary component extracted from the vanilla bean, and geraniol is derived from the geranium flower (Fig. N-1a and N-1b). In other cases the name for a substance was suggested by its physical characteristics or by its chemical or medicinal properties. For example, azulene, which is a dark blue crystalline solid, derives its name from *azul*, the Spanish word for "blue" (Fig. N-1c). And morphine, a powerful sedative isolated from opium, was named for Morpheus, the Greek god of dreams (Fig. N-1d).

FIGURE N1-1 Naturally occurring organic compounds (a) Vanillin can be isolated from vanilla extract. (b) Geraniol occurs in geranium oil. (c) Azulene gives this mushroom its blue color. (d) Morphine is derived from opium.

Almost 100 years ago, chemists realized the necessity for standardizing the names of substances, as well as other chemical terms. Consequently, the **International Union of Pure and Applied Chemistry** (**IUPAC**) was founded about 1920. Immediately work began on adopting a system of nomenclature for both inorganic and organic substances. Today, while many substances are known by more than one name, there is one *standard* chemical name.

The basis of the IUPAC system is that a molecule's name describes its structure. More specifically, an IUPAC name can have a root, one or more prefixes, and one or more suffixes, with each part of the name describing a particular part of the molecule. Therefore, if you know the IUPAC rules, then you can translate a name into a structure, and vice versa.

- Given a molecular structure, you can derive its IUPAC name using a set of straightforward rules.
- Given the IUPAC name, you can draw its structure, piece by piece.

This is especially convenient when communicating to others (e.g., in class or in a research paper), because you have the option of either drawing an explicit structure or simply providing the name of a substance.

To illustrate this point, consider the alkane whose IUPAC name is 2,2,4-trimethylpentane (we discuss how to name molecules like this in greater detail later in this nomenclature unit):

$$\begin{array}{c} CH_3 \quad\quad CH_3 \\ \mid 2 \quad\quad 4\mid \\ H_3C-\underset{1}{\overset{}{C}}-\underset{3}{CH_2}-\underset{}{CH}-\underset{5}{CH_3} \\ \mid \\ CH_3 \end{array}$$

2,2,4-Trimethylpentane

The "pentane" portion of the name indicates a chain of five carbons (which are numbered), and "trimethyl" indicates that there are three methyl (CH_3) groups attached to that chain. Finally, "2,2,4-" indicates where those methyl groups are located on the chain: Two methyl groups are bonded to C2 and one is bonded to C4.

Although the preceding example illustrates the general idea behind the IUPAC system of nomenclature, it does not provide all of the information you need to name such compounds systematically. There are several rules that must be learned first, some of which are discussed throughout the rest of this unit on nomenclature. In Section N1.2, for example, we explain how a molecule's root is derived. In that section and in the remaining sections of this unit on nomenclature, you will learn how to append basic prefixes to a root to describe specific structural features of a molecule. More complex aspects of the IUPAC system are discussed in future units on nomenclature.

N1.2 Alkanes and Cycloalkanes: Roots, Prefixes, Isomers, and Number Locators

Recall from Section 1.11 that *alkanes* comprise the simplest class of compounds, because they contain only C—C and C—H single bonds—that is, they contain *no functional groups*. As a result, alkanes tend to be the simplest molecules to name under the IUPAC system.

The most basic alkanes are the **straight-chain alkanes**, or **linear alkanes**, some of which are listed in Table N1-1.

All of the carbons in each of the compounds in Table N1-1 form *one continuous chain* from one end of the molecule to the other: no carbon atom is bonded to more than two others. The following two rules govern the naming of such compounds:

- All straight-chain alkanes have the *suffix* "ane."
- A *prefix* (e.g., meth-, eth-, prop-, but-) is used to identify the number of carbon atoms in the chain.

For alkanes that are not straight-chain alkanes, the name must be modified to account for the pertinent structural features. This is the case with **cycloalkanes**, in which

TABLE N1-1 Straight-Chain Alkanes

Name	Molecule	Number of Carbon Atoms	Name	Molecule	Number of Carbon Atoms
Methane	CH_4	1	Hexane		6
Ethane	H_3C—CH_3	2	Heptane		7
Propane		3	Octane		8
Butane		4	Nonane		9
Pentane		5	Decane		10

TABLE N1-2 Cycloalkanes

Name	Molecule	Number of Carbon Atoms	Name	Molecule	Number of Carbon Atoms
Cyclopropane	△	3	Cyclohexane	⬡	6
Cyclobutane	▢	4	Cycloheptane	⬡	7
Cyclopentane	⬠	5	Cyclooctane	⬢	8

the two ends of an alkane chain are bonded together to form a ring. Cycloalkanes with three to eight carbon atoms in the ring are listed in Table N1-2.

Because of the ring, a new rule is required to name cycloalkanes:

> A cycloalkane is named by placing the *prefix* "cyclo" in front of the name that corresponds to the straight-chain alkane containing the same number of carbon atoms.

Thus, cyclo*propane* contains three carbon atoms and cyclo*butane* contains four carbon atoms. How many carbon atoms does cyclo*pentane* have?

---

problem **N1.1** Draw the structure of cyclononane. (*Hint:* Consult Table N1-1.)

---

problem **N1.2** What is the name of the following cycloalkane? (*Hint:* Consult Table N1-1.)

Alkanes and cycloalkanes form the basis for the entire system of naming organic compounds, because these molecules provide the *roots* of the names for all molecules. For this reason, you should take the time to commit them to memory.

N1.2a Alkyl Substituents, Substituted Alkanes, and Substituted Cycloalkanes

If a hydrogen atom of an alkane is replaced by another atom or group of atoms, we say that the alkane is *substituted*, and we call the atom or group of atoms that replaces the hydrogen a **substituent**. In the molecule on the left in Figure N1-2, for example, a

A generic substituent G

A substituted butane

A substituted cyclohexane

FIGURE N1-2 Substituted compounds G represents a generic substituent on butane (left) and cyclohexane (right).

generic substituent G has replaced one H atom of butane, so the molecule is a *substituted* butane. Similarly, the molecule on the right is a *substituted* cyclohexane.

problem **N1.3** How would you describe each of the following molecules, where G is a generic substituent?

(a)   (b)

problem **N1.4** Using G as a generic substituent, draw **(a)** a substituted cyclobutane and **(b)** a substituted heptane.

Among the most common substituents are **alkyl substituents**, so called because they structurally resemble alkanes and cycloalkanes. Alkyl substituents are constructed by removing a single hydrogen atom from an alkane or cycloalkane, thus making available a bond by which the substituent can be attached to another atom. The names of alkyl substituents, therefore, are quite similar to those of alkanes, the primary difference being the suffix.

- Alkyl substituents have the *suffix* "yl."
- Straight-chain alkyl substituents are named with the same *prefix* as their corresponding alkanes (i.e., meth-, eth-, etc.).
- Cyclic alkyl substituents are named with the same *prefix* as their corresponding cycloalkanes (i.e., cycloprop-, cyclobut-, etc.).

Thus, a CH_3— group is called a **methyl group** (abbreviated **Me**), because it is constructed from *methane* (CH_4) by removing a single H. Similarly, CH_3CH_2— is called an **ethyl group** (**Et**) and $CH_3CH_2CH_2$— is called a **propyl group** (**Pr**). Other common alkyl substituents comprising up to six carbon atoms are as follows:

Alkyl substituents

Bond available to attach substituent

Methyl (Me)   Ethyl (Et)   Propyl (Pr)   Butyl (Bu)   Pentyl   Hexyl

Cyclopropyl   Cyclobutyl   Cyclopentyl   Cyclohexyl

problem **N1.5**

(a) What is the name of the following alkyl substituent? (*Hint:* See Table N1-2.)

(b) Draw the structure of a cycloheptyl substituent.

Often, we encounter situations in which we are not concerned about the specific structure of an alkyl substituent. For these situations, we can indicate the presence of a *generic* alkyl group using "R". We can think of R as standing for the "rest" of the molecule. In the following generic alkyl-substituted cycloheptane, for example, R can represent anything from a simple methyl group to the most complicated alkyl group you can imagine:

A generic alkyl group

An alkyl-substituted cycloheptane

problem **N1.6** How would you describe each of the following molecules?

(a)   (b)

problem **N1.7** Draw the structures of **(a)** a generic alkyl-substituted pentane and **(b)** a generic alkyl-substituted cyclopropane.

N1.2b Naming Substituted Alkanes

To name an alkane that contains a single alkyl substituent, we must distinguish the substituent from the *main chain*.

- In an alkane possessing an alkyl substituent, the **main chain** is the *longest continuous chain of carbon atoms*.
- The molecule is assigned a **root**, which is the name of the alkane containing the same number of carbon atoms as the main chain.
- The name of the substituent appears as a *prefix* before the root.

In methylbutane, for example, the longest continuous chain contains four C atoms (so the root is butane) and a CH_3 group (a methyl group) appears as a substituent on the chain as shown on the left in Figure N1-3.

Longest continuous chain contains 4 carbon atoms = *butane*

A *methyl* substituent

Methylbutane

Longest continuous chain contains 6 carbon atoms = *hexane*

An *ethyl* substituent

Ethylhexane

NOT the longest continuous chain of carbons

Not ethybutane

Longest continuous chain contains 5 carbon atoms = *pentane*

A methylpentane

FIGURE N1-3 Main chains In the first structure, the main chain has four C atoms. In the second structure, it has six. In the third structure, the longest chain is *not* the four-carbon chain highlighted, but rather is the five-carbon chain highlighted in the last structure.

Similarly, in ethylhexane, the root is hexane (because the longest continuous chain has six C atoms) and the substituent is an ethyl group (because it contains two C atoms).

Sometimes, the way in which a molecule is drawn can make it challenging to identify the main chain. In the third molecule in Figure N1-3, for example, it might initially appear that the root is butane, because the horizontal chain contains four C atoms. The substituent, then, would be an ethyl group, making the molecule ethylbutane. As shown on the right, however, the longest continuous chain contains five C atoms, so the root is *pentane* and the substituent is a *methyl* group. The proper name for the molecule, therefore, is methylpentane.

---

problem **N1.8** For each of the molecules below, how many carbons are in the longest continuous chain? What is the *root* for each molecule?

(a)   (b)   (c)

---

problem **N1.9** What is the IUPAC name for each of the following alkanes?

(a)   (b)

N1.2c Naming Substituted Cycloalkanes

The rules for naming a substituted cycloalkane are similar to those for naming a substituted alkane.

> ▪ A cycloalkane with an alkyl substituent is assigned a *root*, which is simply the name of the corresponding unsubstituted cycloalkane.
> ▪ The name of the substituent appears as a *prefix* before the root.

Two examples—methylcyclobutane and ethylcyclopentane—are shown in Figure N1-4. In methylcyclobutane, the ring contains four C atoms (so its root becomes cyclobutane) and the substituent is a CH_3 group (so the prefix is methyl). Similarly, in ethylcyclopentane, the ring contains five C atoms and the substituent is an ethyl group.

When a compound contains one or more substituted rings, two different names are possible because a ring can also be viewed as a substituent. In ethylcyclopentane, for example, the five-membered ring could be a cyclopentyl substituent, making the two-carbon piece the main chain (i.e., the root) of the molecule. The name, therefore,

FIGURE N1-4 Substituted cycloalkanes (left) A methyl substituent is attached to a cyclobutane ring. (right) An ethyl group is attached to a cyclopentane ring.

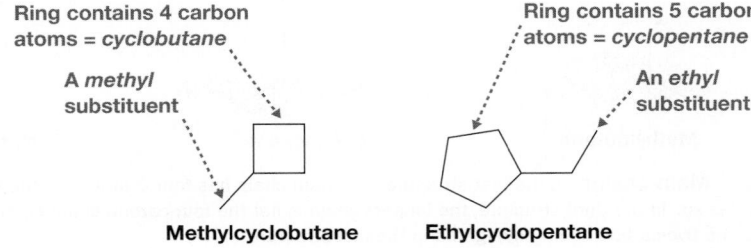

Ring contains 4 carbon atoms = *cyclobutane*

A *methyl* substituent

Ring contains 5 carbon atoms = *cyclopentane*

An *ethyl* substituent

Methylcyclobutane   **Ethylcyclopentane**

would be cyclopentylethane instead of ethylcyclopentane. To eliminate this ambiguity, the following rule is applied:

> In general, the chain or ring that corresponds to the root of a molecule should have more carbons than any of the substituents.

If a molecule consists of a ring and a chain containing the same number of carbons, the root is derived from the ring, and the chain is designated the substituent. (As we will see later, this can greatly simplify the name.) Thus, a ring of three C atoms attached to a chain of three C atoms is named propylcyclopropane (not cyclopropylpropane), and a ring of six C atoms attached to a chain of six C atoms is named hexylcyclohexane (not cyclohexylhexane):

Propylcyclopropane   **Hexylcyclohexane**

The preceding rule also applies to situations in which a cycloalkyl substituent is attached to a cycloalkane:

The cyclopropyl group is a substituent because it has fewer carbons than the cyclopentyl group.

Cyclopropylcyclopentane

This compound is called cyclopropylcyclopentane instead of cyclopentylcyclopropane, because the cyclopropyl group has fewer C atoms (three) than the cyclopentyl group (five). As a result, cyclopentane is the root and cyclopropyl is the substituent.

This rule is sometimes relaxed, however, if it is significantly easier to name a molecule by classifying the ring as a substituent. In the following molecule, for example, the root is methane (left), despite the fact that the cyclohexane ring has six C atoms:

Cyclohexyl substituent

Methane is the root

Cyclopropyl substituent

Cyclopentyl substituent

(Cyclohexylcyclopentylcyclopropylmethane)

Cyclohexane is the root

Difficult to name this as a substituent

The result is strictly not an IUPAC name, but rather is a *trivial name* (Section N1.3) and is indicated here by the parentheses. If cyclohexane were the root, instead, as indicated on the right, there would be a single substituent, but it would be more cumbersome to name due to its complex structure.

problem **N1.10** Name the following molecules.

(a)   (b)

problem **N1.11** Draw the structures for **(a)** cyclobutylcyclohexane and
(b) pentylcycloheptane.

N1.2d Naming Alkanes and Cycloalkanes with Multiple Alkyl Substituents

Only one substituent has been present in each of the examples we have encountered so far. Frequently, however, molecules may have two or more of a particular substituent. To indicate the number of substituents in a molecule, we must include a *prefix* in the name that corresponds to the appropriate number, as shown in Table N1-3.

TABLE N1-3 Prefixes Used to Identify Two or More of a Particular Substituent

Number of substituents	2	3	4	5	6	7	8	9	10
Prefix	di-	tri-	tetra-	penta-	hexa-	hepta-	octa-	nona-	deca-

The following examples illustrate how to use the prefixes listed in Table N1-3:

Two methyls attached to a chain of three carbons

Two ethyls attached to a chain of five carbons

Three methyls attached to a chain of four carbons

Four methyls attached to a chain of four carbons

Dimethylpropane   **Diethylpentane**   **Trimethylbutane**   **Tetramethylbutane**

In the first molecule, the longest continuous chain contains three C atoms, so the root is propane, and there are two CH_3 substituents, which require the prefix "di." This molecule is dimethylpropane. Similarly, in the second molecule, the two ethyl substituents on the five-carbon chain require the prefix "di," so the name is diethylpentane. In the third example, the three methyl groups on the four-carbon chain require the prefix "tri," and in the fourth molecule, the four methyl substituents require the prefix "tetra."

If two or more different *kinds* of substituents are present, additional rules are necessary.

- For molecules with two or more different kinds of substituents, each substituent (with its corresponding prefix di-, tri-, etc.) must appear in the name.
- The substituents should appear in alphabetical order according to the names of the substituents (i.e., ignoring any prefixes such as di-, tri-, etc.).

Pentaethylmethylcyclopropane and diethyltetramethylpentane illustrate both of these new rules:

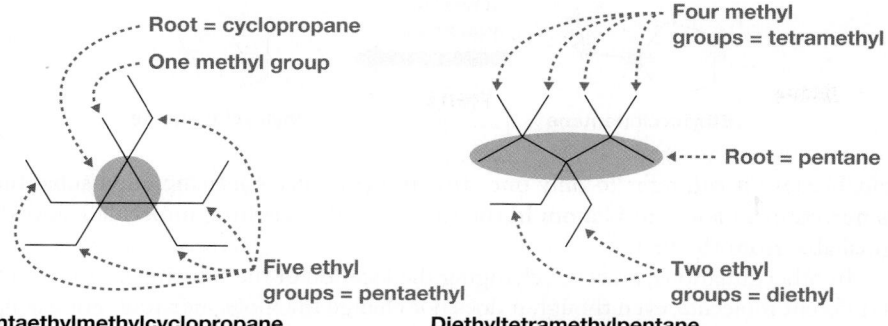

Pentaethylmethylcyclopropane   **Diethyltetramethylpentane**

Pentaethylmethylcyclopropane has five ethyl substituents and one methyl substituent, so both pentaethyl and methyl must appear as prefixes in the name. Pentaethyl comes before methyl in the name, moreover, because, alphabetically, ethyl comes before methyl. For the same reason, diethyl (indicating the presence of two ethyl groups) comes before tetramethyl (indicating four methyl groups) in diethyltetramethylpentane.

problem **N1.12** What is the IUPAC name for the following compound?

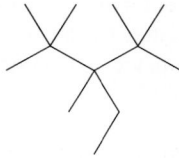

problem **N1.13** Draw the structures of **(a)** hexamethylcyclopropane and **(b)** cyclobutyldicyclopentylmethane.

N1.2e Isomers and Numbering Systems

In many of the examples so far, it has been unnecessary to indicate explicitly the locations of the substituents that appear along the main chain or ring because the names of the molecules lead to a single structure only. In methylbutane, for example, the methyl group must be attached to an internal C atom to maintain four C atoms in the longest continuous chain (Fig. N1-5). If the methyl group were attached to a terminal C, the compound would be pentane, because the longest continuous chain would have five C atoms, not four. Ethylcyclopentane is an unambiguous name,

FIGURE N1-5 Methylbutane is unambiguous Changing the location of CH_3 yields a different root.

Ethyl group moved to adjacent C

Ethylcyclopentane → **Ethylcyclopentane**

FIGURE N1-6 Ethylcyclopentane is unambiguous Changing the location of the ethyl group yields the same molecule.

CH₃ bonded to the central C   **CH₃ bonded to the noncentral C**

FIGURE N1-7 Isomers of methylpentane Different locations of CH₃ along the main chain yield different molecules.

too, because it can refer to only one structure (Fig. N1-6). If the ethyl substituent is moved to an adjacent C atom in the ring, then the resulting molecule is indistinguishable from the first.

In other situations, however, changing the location of the substituent *does* result in a different molecule, even though it does not change the molecule's root. For example, there are two different molecules that could be named methylpentane (Fig. N1-7). The two methylpentane molecules are **isomers** because they have the same formula, but the atoms are connected together differently. (We discuss isomers in greater depth in Chapter 4.) Each isomer of a particular compound is different, so each must have a unique name. As a result, methylpentane by itself is an unacceptable name.

To remedy this problem, we incorporate the following numbering system:

> ▪ If different locations of substituents on a main chain or ring correspond to different isomers, the carbon atoms of the main chain or ring must be numbered sequentially.
>
> ▪ The carbon number to which a substituent is bonded, called a **locator**, appears in the molecule's name immediately before the name of the substituent, and is separated from the rest of the molecule's name by hyphens.

For the two isomers of methylpentane, we can number the five-carbon chain from left to right as shown here:

Methyl group attached to C2

Methyl group attached to C3

3-Methylpentane   **2-Methylpentane**

The first isomer is named 3-methylpentane, because the methyl substituent appears on C3, whereas the second isomer is named 2-methylpentane, because the methyl group is attached to C2.

When more than one substituent is present, as in the following examples, each is given its own locator number.

3-Cyclopentyl-2-methylhexane   **1-Cyclopropyl-2-methylbutane**

Notice that if a substituent's locator appears in the middle of a name, then a hyphen is placed on *both* sides of the number. Notice, too, that according to our previous rule, the substituents are listed in alphabetical order.

An additional rule for incorporating substituent locators is necessary if a molecule contains more than one of a particular type of substituent.

> ■ For molecules that contain two or more of a particular substituent, a locator for *each* substituent appears immediately before the name of the substituent.
> ■ Numbers are separated from each other by commas.

For example, the "1,4" in "1,4-dicyclopropylbutane" indicates that two cyclopropyl groups are bonded to C1 and C4. Similarly, the "2,3,3" in "4-ethyl-2,3,3-trimethylhexane" indicates that there are three methyl groups, one of which is attached to C2 and two of which are attached to C3.

1,4-Dicyclopropylbutane   **4-Ethyl-2,3,3-trimethylhexane**

There is usually more than one way to devise a numbering system for a molecule. For any straight-chain alkane, C1 could be at either end of the chain, and numbering would increase toward the other end. For a cycloalkane, there are even more possibilities. Any carbon atom of the ring could be designated C1, and the numbering could increase either clockwise or counterclockwise around the ring. Under the IUPAC system, however, only one numbering system is acceptable.

> ■ The numbering system for a main chain or ring is chosen in such a way as to minimize the value of the *lowest* substituent locator.
> ■ If two such numbering systems result in a "tie" for the value of the lowest substituent locator, choose the system that minimizes the *next lowest* substituent locator.
> ■ If a tie persists, continue comparing sequentially higher substituent locators for each system until the tie is broken.

2-Methylpentane, for example, has two possible numbering systems:

CORRECT   NOT CORRECT

Substituent locator
not minimized

2-Methylpentane   **4-Methylpentane**

In the first case, C1 is the left-most carbon, and numbering increases from left to right along the carbon chain. This places the methyl substituent on C2, giving rise to the name 2-methylpentane. In the second case, C1 is the right-most carbon, and numbering increases from right to left, placing the methyl substituent on C4. The resulting name, 4-methylpentane, is incorrect, however, because the locator is higher than it has to be.

In a substituted cycloalkane, minimizing the substituent locators invariably means numbering the carbons so that at least one substituent is attached to C1. Minimizing the remaining substituent locators means that numbering should increase in such a way (i.e., clockwise or counterclockwise) that the greatest number of substituents get the lowest numbers possible.

In 1,1-dimethylcyclopentane, for example, C1 is chosen to be the one bonded to both methyl groups to minimize the values of the locators:

1,1-Dimethylcyclopentane **1,2-Dimethylcyclopentane** **1,5-Dimethylcyclopentane**

In 1,2-dimethylcyclopentane, C1 is bonded to one of the methyl groups, and numbering increases counterclockwise around the ring so that the second methyl group is bonded to C2. If numbering were instead to increase clockwise, the second methyl group would be attached to C5.

For the following trimethyl-substituted cyclohexane, numbering clockwise or counterclockwise results in a tie for two of the three substituent locators. The tie is broken with the third locator. As indicated, the correct numbering system is that on the right, which minimizes the third locator.

1,2,5-Trimethylcyclohexane **1,2,4-Trimethylcyclohexane**

---

problem **N1.14** What is the correct IUPAC name for the following trimethylcyclopentane molecule, shown with three different numbering systems?

1,3,3-Trimethylcyclopentane **1,1,3-Trimethylcyclopentane** **1,1,4-Trimethylcyclopentane**

---

problem **N1.15** What is the IUPAC name for each of the following molecules?

(a)   (b)

---

problem **N1.16** What is the IUPAC name of each of the following molecules?

(a)   (b)   (c)   (d)

The numbering system is also useful to emphasize the locations of generic alkyl substituents, R, as illustrated with the following molecules:

A 2,3-dialkylheptane   **A 1,2,3-trialkylcyclohexane**

The molecule on the left can be called a 2,3-dialkylheptane because the alkyl groups are attached to C2 and C3. Similarly, the molecule on the right can be called a 1,2,3-trialkylcyclohexane because the alkyl substituents are attached to C1, C2, and C3.

problem **N1.17** Draw each of the following generic alkyl-substituted molecules.

(a) A 3,3-dialkylpentane
(b) A 1,2,3,4-tetraalkylcyclopentane

Work through the following problems to become as comfortable as possible with what you have learned about nomenclature so far.

problem **N1.18** Given each of the IUPAC names provided, draw the corresponding structure.

(a) 2-Methylhexane   **(b)** 3-Methylhexane
(c) 2,3-Dimethylbutane   **(d)** 1,1-Dimethylcyclohexane
(e) 2,2,3-Trimethylbutane   **(f)** 2,2,4-Trimethylpentane
(g) 3-Ethyl-2,3-dimethylpentane   **(h)** 1,2-Dimethylcyclohexane
(i) 1,2,3-Trimethylcyclobutane   **(j)** 2,2,3,3-Tetramethylhexane

problem **N1.19** Given each of the structures provided, write the corresponding IUPAC name.

(a)   **(b)**   **(c)**   **(d)**

(e)   **(f)**   **(g)**   **(h)**

N1.3 Trivial Names and Common Alkyl Substituents

In Section N1.1, we mentioned that most organic compounds were originally given names based on their properties or origin. As the language of chemistry grew and evolved, a sizeable number of these **common names** or **trivial names** came into regular use, and eventually many of them became integrated into the IUPAC system. Nonscientists sometimes know these trivial names (e.g., "formaldehyde," "glucose," or "adrenalin") as well as any chemist or biologist does, even if they don't know the exact structures of these substances.

Although we will minimize the use of trivial names in this book, they are so frequently employed by a substantial number of professionals that you will need to know them to communicate effectively. Thus, where appropriate, we will provide both the IUPAC name and the trivial name together.

Trivial names will generally appear in parentheses, whereas IUPAC names will not.

One of the ways in which trivial names are used in connection with alkanes is to emphasize a particular isomer of a given formula as shown in Figure N1-8.

For example, the trivial names of both isomers of C_4H_{10} contain "butane" to expressly indicate that the molecule contains four C atoms. The prefix "*n*-" in *n*-butane stands for "normal" and is meant to indicate that the molecule is a straight-chain alkane. The prefix "iso" in isobutane, on the other hand, stands for "isomer" and is meant to distinguish it from the straight-chain alkane.

The pattern is similar with alkanes containing other numbers of carbon atoms. Two of the isomers with the formula C_5H_{12} are *n*-pentane (a straight-chain alkane) and isopentane (a branched isomer). The third common isomer, neopentane, is doubly branched. In addition, two of the C_6H_{14} isomers are *n*-hexane (a straight-chain alkane) and isohexane (a branched isomer).

A much more common use of trivial names involves alkyl substituents. The examples shown in Figure N1-9 should be committed to memory.

Note that we have already learned the IUPAC names of the straight-chain alkyl substituents—namely, propyl, butyl, and pentyl. The IUPAC system also has rules for naming the branched alkyl substituents (i.e., the names in parentheses), but they aren't discussed in this text.

To learn the trivial names for the above alkyl substituents, it may help to recognize some specific structural features of the groups. First, notice that all of the

FIGURE N1-8 Trivial names of alkanes Trivial names of these molecules appear in parentheses.

C_4H_{10}

IUPAC name ----→ **Butane**
Trivial name ---→ **(*n*-Butane)**

Methylpropane (Isobutane)

C_5H_{12}

Pentane (*n*-Pentane)

Methylbutane (Isopentane)

Dimethylpropane (Neopentane)

C_6H_{14}

Hexane (*n*-Hexane)

2-Methylpentane (Isohexane)

FIGURE N1-9 Trivial names of common alkyl substituents Trivial names of these substituents appear in parentheses.

alkyl substituents that begin with "iso," which can be called **isoalkyl groups**, possess a C atom bonded to two CH_3 groups. That C atom is at the end of the substituent opposite the point of attachment. In other words, all three isoalkyl groups have the following general form:

A generic isoalkyl substituent

$$\left(\begin{array}{c}\end{array}\right)_{n = 0,1,2, \ldots}$$

Second, the prefixes "*sec-*" and "*tert-*" stand for "secondary" and "tertiary," respectively, and reflect the type of carbon atom at the point of attachment. In general, we distinguish carbon atoms based on the number of *other* carbon atoms to which they are directly bonded (Fig. N1-10).

- A *primary carbon* (abbreviated 1°) is directly bonded to one other carbon atom.
- A *secondary carbon* (abbreviated 2°) is directly bonded to two other carbon atoms.
- A *tertiary carbon* (abbreviated 3°) is directly bonded to three other carbon atoms.
- A *quaternary carbon* (abbreviated 4°) is directly bonded to four other carbon atoms.

A 1° carbon

A 2° carbon

A 3° carbon

A 4° carbon

FIGURE N1-10 Types of carbon Carbons in a molecule are classified by the number of other carbon atoms that are directly attached.

Thus, examining the *sec*-butyl and *tert*-butyl substituents more closely, we can see that the C atom at the point of attachment in the former is a secondary (2°) carbon, whereas in the latter, it is a tertiary (3°) carbon (Fig. N1-11).

problem **N1.20** How would you classify the other three C atoms (not labeled) in the *sec*-butyl and *tert*-butyl groups in Figure N1-11?

The examples in Figure N1-12 illustrate how trivial names for alkyl substituents can be incorporated into a molecule's name. For the first molecule, cyclohexane is the root, and the two isopropyl groups are attached at C1 and C4. For the second

A secondary (2°) carbon

A tertiary (3°) carbon

(*sec*-Butyl)
(*s*-Butyl)

(*tert*-Butyl)
(*t*-Butyl)

FIGURE N1-11 Isomeric butyl groups A *sec*-butyl group is attached by a secondary C. A *tert*-butyl group is attached by a tertiary C.

Isopropyl groups

A *t*-butyl group

A neopentyl group

(1,4-Diisopropylcyclohexane)   **(4-*t*-Butylheptane)**   **(1-Cyclopropyl-2,2-dimethylpropane)**
                                                                (Neopentylcyclopropane)

FIGURE N1-12 Using trivial names of alkyl substituents (left) Isopropyl groups are attached to C1 and C4. (middle) A *t*-butyl group is attached to C4. (right) A neopentyl group is attached to a cyclopropane ring.

molecule, the longest continuous chain contains seven C atoms, so the root is heptane. A *tert*-butyl group is attached at C4 along that chain. For the third molecule, cyclopropane can be taken as the root, in which case the substituent is a neopentyl group. Strictly speaking, each of these names is a trivial name, indicated by the parentheses, due to the use of trivial names for the substituents.

---

problem **N1.21** Name the following molecules, using trivial names for the various alkyl substituents where possible.

(a)   (b)   (c)

Work through the practice problems for this section to become as comfortable as possible with what you have learned about nomenclature so far.

---

problem **N1.22** Draw the structures of the following compounds.

(a) Isodecane (the IUPAC name is 2-methylnonane)
(b) *tert*-Butylcyclopentane
(c) 3,3-Diisopropyloctane

---

problem **N1.23** What is the correct IUPAC name for the compound 3-*tert*-butylhexane?

---

problem **N1.24** In the compound 3-*tert*-butylhexane, how many primary carbons are there? How many secondary carbons are there?

---

problem **N1.25** Draw the structure that corresponds to each of the following names.

(a) 1-Methyl-4-neopentylcyclohexane
(b) Isobutylcyclobutane
(c) 5-*sec*-Butylnonane

---

problem **N1.26** Write the name of each molecule, using trivial names for substituents where appropriate.

(a) **(b)** **(c)**

(d) **(e)** **(f)**

N1.4 Substituents Other Than Alkyl Groups: Naming Haloalkanes, Nitroalkanes, and Ethers

The nomenclature rules we have examined so far have addressed alkanes exclusively. If *functional groups* are present, however, the rules can be modified and/or added to properly reflect the presence of those groups. As we discuss in future nomenclature units, the presence of some functional groups requires major modifications to the molecule's name, even including the root. Compounds containing other functional groups, such as haloalkanes (i.e., R—F, R—Cl, R—Br, and R—I), nitroalkanes $(R—NO_2)$, and ethers $(R—OR')$, can be named using just the rules we have learned so far, with only minor modifications (Fig. N1-13). For this reason, the nomenclature of these compounds is discussed here in Section N1.4.

We will begin in Section N1.4a with the IUPAC system for naming these compounds. Following that, we introduce some important trivial names for haloalkanes and ethers in Section N1.4b.

N1.4a IUPAC Names of Haloalkanes, Nitroalkanes, and Ethers

Under the IUPAC system, haloalkanes and nitroalkanes are treated in precisely the same way as alkanes, except in the prefix.

- A haloalkane requires the prefix "fluoro," "chloro," "bromo," or "iodo" to indicate the presence of F, Cl, Br, or I, respectively.
- A nitroalkane requires the prefix "nitro" to indicate the presence of an NO_2 group.

The numbering systems for the main chains or rings of these compounds are no different from those used with alkanes or cycloalkanes containing alkyl substituents.

X = F, Cl, Br, I

R—X   R—NO₂   R—OR′

Haloalkane   Nitroalkane   Ether

FIGURE N1-13 Functional groups discussed in Section N1.4 These functional groups are named using alkane suffixes.

Similarly, the substituents must appear in alphabetical order. The following examples illustrate these rules for haloalkanes and nitroalkanes:

Two iodo groups on C1

Nitro groups on C1 and C2

Chlorine on C4

Cyclopropyl group on C1

Bromine on C2

4-Chloro-3,3-dimethylhexane **1,2,-Dinitrocyclopentane** **2-Bromo-1-cyclopropyl-1,1-diiodobutane**

In 4-chloro-3,3-dimethylhexane, the main chain is numbered from right to left as drawn because this scheme places two substituents on C3 (the methyl groups) and one on C4 (the chloro substituent). If the main chain was numbered from left to right, instead, then only one substituent would be on C3 (the chloro substituent) and two would be on C4 (the methyl groups). In 1,2-dinitrocyclopentane, the "di" prefix is used to indicate the presence of two nitro groups, and the locators indicate that those groups are on C1 and C2. In 2-bromo-1-cyclopropyl-1,1-diiodobutane, the main chain is numbered from left to right as drawn to place three groups (the two iodo substituents and the cyclopropyl group) on C1. Additionally, the substituents—bromo, cyclopropyl, and iodo—are listed in alphabetical order.

---

problem **N1.27** What is the IUPAC name for each of the following molecules?

(a) (b) (c)

---

problem **N1.28** Draw the structure of each of the following compounds.

(a) 1,1-Dichloro-2-nitrocyclohexane
(b) 1,1,1,2-Tetrabromopropane

The same IUPAC rules apply to naming ethers (R—OR′), with additional considerations required to identify and name substituents. These additional rules are needed because an O atom separates two alkyl groups, R and R′, so either of these alkyl groups could conceivably be the main chain of the molecule and establish the root. According to our previous rule, however, the root is derived from the alkyl group that contains the longer chain of carbon atoms, which we shall assume is R. The other alkyl group, R′, is part of the substituent OR′, and is called an **alkoxy group** (Fig. N1-14).

The alkyl group with the longer chain of C atoms establishes the root.

The other alkyl group is viewed as part of an alkoxy substituent.

R—O—R′

An ether

FIGURE N1.14 Naming an ether An ether is composed of an alkyl group (R) and an alkoxy group (OR′)

- An ether requires a prefix of the form "alkoxy" to indicate the presence of the OR′ substituent.
- The specific name of an alkoxy substituent (OR′) is obtained by removing the suffix "yl" from the name of the corresponding alkyl group (R′) and adding the suffix "oxy."

Thus, —OCH$_3$ is a **methoxy** group and —OCH$_2$CH$_3$ is an **ethoxy** group. The following examples demonstrate this naming system:

Methoxymethane
(Dimethyl ether)

Ethoxyethane
(Diethyl ether)

Propoxycyclohexane
(Cyclohexylpropyl ether)

In methoxymethane and ethoxyethane, the O atom is attached to identical alkyl groups, so either alkyl group in each molecule can establish the root. In methoxymethane, each alkyl group has one C, whereas in ethoxyethane, each alkyl group has two C atoms. Therefore, the roots are methane and ethane, respectively, and they are attached to methoxy and ethoxy groups, respectively. In propoxycyclohexane, cyclohexane is the root because it has more C atoms (six) than the propyl group (three).

In the preceding compounds, numbers are unnecessary to locate the alkoxy substituents because attaching the substituent to a different carbon of the main chain or ring results in an identical molecule. In the following examples, however, substituent locators are needed to distinguish between isomers:

1-Methoxypropane
(Methyl propyl ether)

2-Methoxypropane
(Isopropyl methyl ether)

1,4-Diethoxycyclohexane

3,3-Dichloro-2-methoxypentane

2-Methoxy-1,1,3,3-tetramethylcyclopentane

1-Methoxypropane and 2-methoxypropane are isomers of each other: The methoxy group is attached to C1 in the first molecule and to C2 in the second. 1,4-Diethyoxycyclohexane requires the prefix "di" to indicate the presence of two ethoxy groups. The last two compounds illustrate how to incorporate alkoxy prefixes into the name of a molecule when other substituents, like halo and alkyl substituents, are present.

---

problem **N1.29** What is the IUPAC name for each of the following ethers?

(a)   (b)   (c)

---

problem **N1.30** Draw the structure for each of the following compounds.

(a) 3,3-Diethoxypentane   **(b)** 1-Chloro-5-methoxyhexane
(c) 1,2-Diethoxy-1-methylcyclopentane

N1.4b Trivial Names of Haloalkanes and Ethers

As mentioned in Section N1.3, many trivial names are in common use today. This is particularly true for haloalkanes (R—X) and ethers (R—OR′), because prior to the advent of the IUPAC system, compounds in each of these classes were named under a different system.

The system for haloalkanes mirrored the one for naming *ionic* compounds like NaCl (sodium chloride), where the metal is a cation (positively charged) and the halogen is a halide anion (negatively charged). This system, then, names haloalkanes as alkyl halides.

> ■ The name of the alkyl group (R) is followed by the name of the halide anion (F^- = fluoride, Cl^- = chloride, Br^- = bromide, I^- = iodide).
>
> ■ A space separates these two parts of the trivial name.

Notice in the following examples that, in addition to the trivial name derived from the above system, some haloalkanes have other trivial names as well. Important ones include methylene chloride (CH_2Cl_2), chloroform ($CHCl_3$), bromoform ($CHBr_3$), and carbon tetrachloride (CCl_4).

IUPAC and trivial names for some haloalkanes

Iodomethane
(Methyl iodide)

Chloroethane
(Ethyl chloride)

Dichloromethane
(Methyl dichloride)
(Methylene chloride)

Trichloromethane
(Methyl trichloride)
(Chloroform)

Tribromomethane
(Methyl tribromide)
(Bromoform)

Tetrachloromethane
(Methyl tetrachloride)
(Carbon tetrachloride)

2-Bromobutane
(*sec*-Butyl bromide)
(*s*-Butyl bromide)

2-Chloro-2-methylpropane
(*tert*-Butyl chloride)
(*t*-Butyl chloride)

Notice, too, that the trivial names for 2-bromobutane and 2-chloro-2-methylpropane use the trivial names "*sec*-butyl" and "*tert*-butyl," respectively, for the alkyl groups.

---

problem **N1.31** What are the IUPAC names for **(a)** cyclopentyl iodide and **(b)** *n*-hexyl bromide?

The trivial names for ethers, which take the form alkyl alkyl ether, also treat alkyl groups as substituents.

> The trivial name for an ether (R—O—R′) lists the names of each alkyl group attached to O first, followed by "ether." If the alkyl groups are identical, the prefix "di" is used.

The trivial name for ethoxyethane ($CH_3CH_2OCH_2CH_3$), for example, is diethyl ether. Often simply referred to as "ether," it is a very common organic solvent that was once used to anesthetize patients during medical procedures.

As with haloalkanes, trivial names for the alkyl groups are often used in the trivial names of ethers. The trivial name for 2-methoxypropane, for example, is isopropyl methyl ether. Other examples include di-*tert*-butyl ether and isobutyl methyl ether:

Trivial names for ethers

(Di-*tert*-butyl ether)　**1-Methoxy-2-methylpropane**
(Isobutyl methyl ether)

problem **N1.32** What is the IUPAC name for **(a)** *tert*-butyl ethyl ether; **(b)** cyclohexyl methyl ether; **(c)** *sec*-butyl propyl ether?

Work through the practice problems for this section to become as comfortable as possible with what you have learned about nomenclature so far.

problem **N1.33** Draw structures for the following haloalkanes.

(a) 1,2,3-Tribromohexane
(b) 2,2,3,3,4-Pentachlorohexane
(c) 1,2-Dichloro-4-nitrohexane
(d) 1,2-Dichloro-3-methoxycyclopentane
(e) 2,2-Dichloro-3-cyclopropylbutane
(f) 1-Bromo-1-chloro-1-iodobutane
(g) 1,1-Dibromo-2-methylcyclobutane
(h) 1,1,2,2-Tetrabromo-3-propoxypropane

problem **N1.34** Draw structures and provide IUPAC names for each of the following.

(a) *n*-Butyl chloride
(b) *s*-Butyl iodide
(c) *t*-Butyl fluoride
(d) Neopentyl bromide
(e) Diisopropyl ether

problem **N1.35** What are the common names for **(a)** 1-propoxybutane and **(b)** 2-ethoxybutane?

problem **N1.36** What is the IUPAC name for each of the following compounds?

(a)　　　(b)　　　(c)　　　(d)　　　(e)

(f)　　　(g)　　　(h)　　　(i)　　　(j)

Three-Dimensional Geometry, Intermolecular Interactions, and Physical Properties

Carbon dioxide (CO_2) and formic acid (HCO_2H) are similar in their chemical makeup, but the boiling point of CO_2 is -78 °C, whereas that of HCO_2H is 101 °C. Moreover, CO_2 is only slightly soluble in water, whereas HCO_2H is infinitely soluble in water. Why are their physical properties so vastly different?

$$O=C=O$$
Carbon dioxide

Molar mass = 44 g/mol
Boiling point = −78°C
Water solubility = slight

**Formic acid
(Methanoic acid)**

Molar mass = 46 g/mol
Boiling point = 101°C
Water solubility = infinite

As we discuss here in Chapter 2, these compounds behave differently because they experience different **intermolecular interactions**. Those intermolecular interactions are governed, in turn, by a variety of factors, including the three-dimensional shapes of the molecules and the functional groups they contain. We begin, therefore, with a review of the factors that determine molecular geometry and then discuss the different types of intermolecular interactions that are important in organic chemistry.

Geckos can climb effortlessly on almost every surface. Their ability to do so is attributed to ultrafine hairs on their feet, which give rise to a very large contact surface area. This allows for rather strong dispersion forces, one of the intermolecular interactions we will examine in this chapter.

These topics have a broad relevance to many aspects of organic chemistry. Toward the end of this chapter, we explain that intermolecular interactions determine how soaps and detergents function and contribute to the properties of cell membranes. In Chapter 5, we explain how molecular geometry is central to the important concept of *chirality*—that is, whether a molecule is different from its mirror image. And, in Chapter 9, we explain how intermolecular interactions can have a dramatic effect on the outcome of chemical reactions.

2.1 Valence Shell Electron Pair Repulsion (VSEPR) Theory: Three-Dimensional Geometry

To understand many aspects of molecular geometry, chemists routinely work with two models. One, which we discuss here, is **valence shell electron pair repulsion (VSEPR) theory**. The other, which we discuss in Chapter 3, uses the concepts of *hybridization* and *molecular orbital (MO) theory*. Although hybridization and MO theory constitute a more powerful model than VSEPR theory, VSEPR theory remains extremely useful because of its simplicity: Its concepts are easier to grasp and it allows us to arrive at answers much more quickly.

2.1a Basic Principles of VSEPR Theory

The basic ideas of VSEPR theory are as follows:

1. Electrons in a Lewis structure are viewed as groups.
   - A lone pair of electrons, a single bond, a double bond, and a triple bond each constitute one *group* of electrons (Table 2-1).

TABLE 2-1 Various Types of Electron Groups in VSEPR Theory

Type of Group	Total Number of e⁻	Number of Groups
1 Lone pair	2	1
1 Single bond	2	1
1 Double bond	4	1
1 Triple bond	6	1

Electron geometry = linear
Molecular geometry = linear

(a) Ethanenitrile (acetonitrile)

Electron geometry = trigonal planar
Molecular geometry = trigonal planar

(b) Propanone (acetone)

Electron geometry = tetrahedral
Molecular geometry = tetrahedral

(c) Ethane

FIGURE 2-1 Compounds in which the central atom lacks lone pairs Because there are no lone pairs about the central atom, the molecular geometries of acetonitrile, acetone, and ethane are identical to their electron geometries.

2. The negatively charged electron groups strongly repel one another, so they tend to arrange themselves as far away from each other as possible.
   - Two electron groups tend to be 180° apart (a linear configuration).
   - Three groups tend to be 120° apart (a triangular, planar configuration).
   - Four groups tend to be 109.5° apart (a tetrahedral configuration).
3. **Electron geometry** describes the orientation of the *electron groups* about a particular atom. These configurations are summarized in Table 2-2.
4. **Molecular geometry** describes the arrangement of *atoms* about a particular atom. Because atoms must be attached by bonding pairs of electrons, an atom's molecular geometry is governed by its electron geometry.

TABLE 2-2 Correlations between Electron Geometry and Bond Angle in VSEPR Theory

Number of Electron Groups	Electron Geometry		Approximate Bond Angle
2	Linear		180°
3	Trigonal planar		120°
4	Tetrahedral		109.5°

The common molecular geometries, which are summarized in Table 2-3, lead to the following conclusions:

- If all the electron groups are bonds (depicted in gray), then there is an atom attached to each electron group and the molecular geometry is the *same* as the electron geometry.
- If one or more of the electron groups is a lone pair (depicted in yellow and red), then the molecular geometry is *different* than the electron geometry.

Some examples of molecules containing central atoms without lone pairs are shown in Figure 2-1. In acetonitrile, CH_3—C≡N (Fig. 2-1a), the triply bonded C has two electron groups about it: a single bond and a triple bond. According to Table 2-2, its electron geometry is linear. Both electron groups are bonds, moreover, so the molecular geometry about that C atom is also linear, making the C—C—N bond angle 180°.

In acetone, $(CH_3)_2$C=O (Fig. 2-1b), the central C atom has three electron groups about it: two single bonds and a double bond. As a result, both the electron and molecular geometries are trigonal planar.

In ethane, CH_3—CH_3 (Fig. 2-1c), each C atom is surrounded by four electron groups—the four single bonds—so the electron and molecular geometries are tetrahedral at each carbon.

TABLE 2-3 Molecular Geometries in VSEPR Theory[a]

Number of Bonded Atoms/Groups	Electron Geometry		
	Linear (180°)	Trigonal Planar (120°)	Tetrahedral (109.5°)
2	Linear	Bent	Bent
3		Trigonal planar	Trigonal pyramidal
4			Tetrahedral

[a] Bonding electron groups are depicted with gray sticks; nonbonding electron groups are depicted as yellow sticks terminating in a red lone pair.

In 2-aminoethanol (ethanolamine), which is commonly used as feedstock for the production of a variety of industrial compounds (Fig. 2-2), the N and O atoms have molecular geometries that are different than their electron geometries. The electron geometry of the N atom is tetrahedral because it is surrounded by four electron groups: three single bonds and a lone pair. Its molecular geometry, however, which describes only the orientation of the three single bonds, is trigonal pyramidal. Likewise, the O atom of the OH group has a tetrahedral electron geometry (two single bonds and two lone pairs), but its molecular geometry is bent.

SOLVED problem 2.1 Imines, which are characterized by a C=N double bond, are commonly used as intermediates in organic synthesis. Use VSEPR theory to predict the electron and molecular geometries about the nitrogen atom in the acetone imine molecule below.

$$\overset{\cdot}{\underset{H_3C}{\overset{NH}{\underset{\|}{C}}} CH_3}$$

Think How many electron groups surround the N atom? Are any of them lone pairs?

2-Aminoethanol

Electron geometry = tetrahedral
Molecular geometry = bent

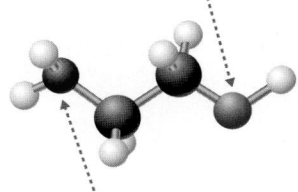

Electron geometry = tetrahedral
Molecular geometry = trigonal pyramidal

FIGURE 2-2 Lewis structure and VSEPR geometries about atoms in a molecule of 2-aminoethanol The electron geometries about the NH_2 nitrogen and the OH oxygen atoms in 2-aminoethanol are both tetrahedral, but they have different molecular geometries because N has one lone pair and O has two.

Electron geometry = trigonal planar
Molecular geometry = trigonal planar

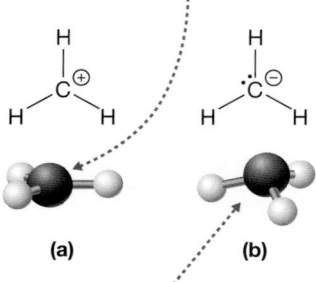

(a)   (b)

Electron geometry = tetrahedral
Molecular geometry = trigonal pyramidal

FIGURE 2-3 Lewis structures and three-dimensional geometries of the methyl cation and methyl anion The central C atom in CH_3^+ is surrounded by three single bonds only (i.e., no lone pairs), so its electron and molecular geometries are the same. The central C atom in CH_3^-, on the other hand, is surrounded by three single bonds and one lone pair, so its electron and molecular geometries are different.

Solve There are three groups of electrons around the nitrogen atom: one double bond, one single bond, and one lone pair. According to VSEPR theory, therefore, its electron geometry is trigonal planar, and its molecular geometry is bent.

~120°

Electron geometry = trigonal planar
Molecular geometry = bent

problem **2.2** Prop-2-yn-1-ol (propargyl alcohol) is used as an intermediate in organic synthesis and can be polymerized to make poly(propargyl alcohol). Use VSEPR theory to predict the electron and molecular geometries about each nonhydrogen atom in the molecule.

OH

Prop-2-yn-1-ol
(Propargyl alcohol)

The rules of VSEPR theory apply equally well to ions. Figure 2-3a shows, for example, that the methyl cation, CH_3^+, has a trigonal planar electron geometry, consistent with a carbon atom that is surrounded by three groups of electrons (i.e., three single bonds). The methyl anion (Fig. 2-3b), on the other hand, is surrounded by four groups of electrons (i.e., three single bonds and a lone pair). Its electron geometry therefore, is tetrahedral, and its molecular geometry is trigonal pyramidal.

YOUR TURN 2.1

Circle each electron group in the Lewis structures of CH_3^+ and CH_3^- in Figure 2-3.

Answers to Your Turns are in the back of the book.

2.1b Angle Strain

Geometric constraints can force an atom to deviate significantly from its **ideal bond angle**—that is, the bond angle predicted by VSEPR theory. Most commonly this happens in ring structures. For example, the carbon atoms in a molecule of cyclopropane (Fig. 2-4a) should have an ideal bond angle of 109.5°, given that each carbon is surrounded by four groups of electrons (four single bonds). To form the ring, however, the C—C—C bond angles must be 60°. Similarly, each carbon atom of cyclobutadiene (Fig. 2-4b) has an ideal bond angle of 120°, but geometric constraints force the angles in the ring to be 90°.

FIGURE 2-4 Examples of angle strain In cyclopropane (a), the ideal C—C—C bond angle is 109.5°, but the actual angle is 60°. In cyclobutadiene (b), the ideal C—C—C bond angle is 120°, but the actual angle is 90°.

Cyclopropane
Ideal bond angle = 109.5°
Real bond angle = 60°
(a)

Cyclobutadiene
Ideal bond angle = 120°
Real bond angle = 90°
(b)

The deviation of a bond angle from its ideal angle results in an increase in energy, called **angle strain**. Angle strain weakens bonds and makes a species more reactive. In some cases, excessive angle strain can preclude the existence of a molecule altogether.

2.2 Dash–Wedge Notation

Although molecules are three-dimensional, representing them on paper is confined to the two dimensions of the page. To work around this problem, we introduce **dash–wedge notation**, which provides a means to represent atoms both in front of and behind the plane of the paper. Dash–wedge notation has three components:

Rules for dash–wedge notation

1. A straight line (—) represents a bond that is in the plane of the paper.
   - Atoms at either end of the bond are also in the plane of the paper.
2. A wedge (◄) represents a bond that comes out of the plane of the paper and points toward you.
   - In general, the atom at the thinner end of the wedge is in the plane of the paper, whereas the atom bonded at the thicker end is in front of the page.
3. A dash (⑊) represents a bond that is pointed away from you.
   - In general, you may assume in this book that the atom bonded at the thicker end of this dash is behind the plane of the paper. (You may see different conventions in other books.)

Using the dash–wedge notation, there are two common ways of representing a tetrahedral carbon atom like that in CH_4. They are illustrated in Figure 2-5a and 2-5b. Both illustrations represent the same molecule with the same 109.5° bond angles. The only difference is the vantage point from which you view the molecule. The vantage point giving rise to the depiction of CH_4 in Figure 2-5b is the basis for a shorthand representation of tetrahedral carbon atoms, called the *Fischer projection*, which we introduce in Chapter 5.

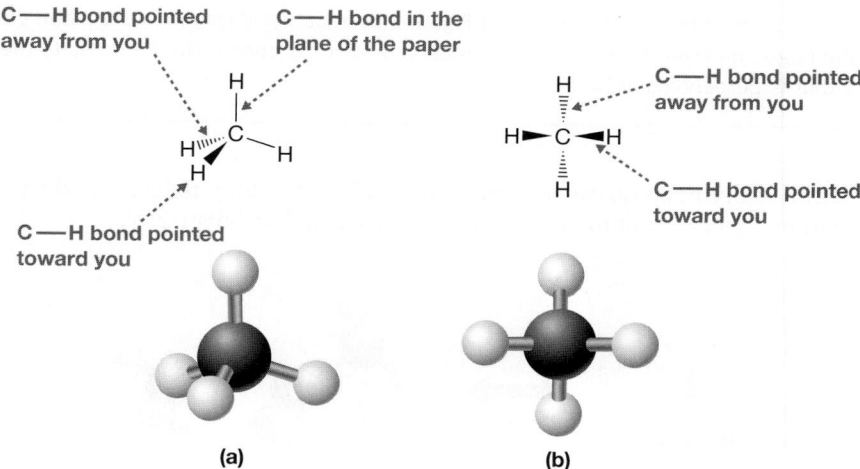

(a)   (b)

FIGURE 2-5 Representations of CH₄ using dash–wedge notation The two different depictions imply views of the molecule from different vantage points.

FIGURE 2-6 Tetrahedral geometry viewed as two perpendicular Vs
A tetrahedral atom can be viewed as the fusing together of two V shapes that are in perpendicular planes—one in the plane of the paper and one perpendicular to the plane of the paper. The two Vs must "open" in opposite directions.

This V is in the plane of the paper.

This V is perpendicular to the plane of the paper.

The two Vs open in opposite directions.

YOUR TURN 2.2

Ball-and-stick representations of NH_4^+ from two different vantage points are provided. Next to each one, draw the corresponding cation using dash–wedge notation.

Notice in Figure 2-5 that the four bonds of a tetrahedral atom define two *perpendicular* Vs. It is helpful to think of this whenever drawing the dash–wedge notation in Figure 2-5a.

- One V is in the plane of the paper, whereas the other is perpendicular to the plane of the paper.
- The Vs must open in *opposite directions*! This is explicitly shown in Figure 2-6.

YOUR TURN 2.3

In the Lewis structure at the right of Figure 2-6, trace the V that is in the plane of the page and draw an arrow in the direction in which it opens. Do the same to the V that is perpendicular to the page.

Dash–wedge notation can be combined with line structures to illustrate the three-dimensional geometry of more complex molecules, such as butan-2-ol:

OH
Butan-2-ol

In the box provided, draw the line structure of butan-2-ol using dash–wedge notation. Note that the C—O bond points away from you.

problem 2.3 Draw line structures of each of the following molecules using dash–wedge notation. Assume that no atoms have formal charges. (*Note:* Black = carbon, white = hydrogen, green-yellow = chlorine, and blue = nitrogen.)

(a) (b) (c)

problem 2.4 The following is a common mistake made with dash–wedge notation. Explain what is incorrect about it and then fix it.

1. Construct model

2. Rotate model
180°

3. Redraw

FIGURE 2-7 Model kits and 3D manipulations To draw the molecule in Equation 2-1 after it is flipped 180°, (1) construct the molecule with a model kit, (2) flip the molecule over 180°, and (3) use the model as a guide to redraw the molecule in dash–wedge notation.

2.3 Strategies for Success: The Molecular Modeling Kit

Much of organic chemistry requires us to manipulate molecules in three dimensions. Unfortunately, we are limited to two dimensions when we represent a molecule on paper, even when we use dash–wedge notation. **Molecular modeling kits** can help. Instead of having to rotate a three-dimensional image mentally, you can construct real models and rotate them in your hands. For example, let's use a modeling kit to determine what the following cyclopentane derivative looks like after it has been flipped over vertically.

Flip 180°   ?   (2-1)

You may develop your own process for these kinds of manipulations, but for now carry out the following steps, which are depicted in Figure 2-7:

1. Construct the molecular model exactly as indicated in the accurate dash–wedge notation.

2. Rotate the molecule as indicated.
3. Redraw the molecule in its dash–wedge notation using the rotated model as a guide.

Construct the following molecule using a molecular modeling kit and then draw its structure using dash–wedge notation after each successive 180° flip.

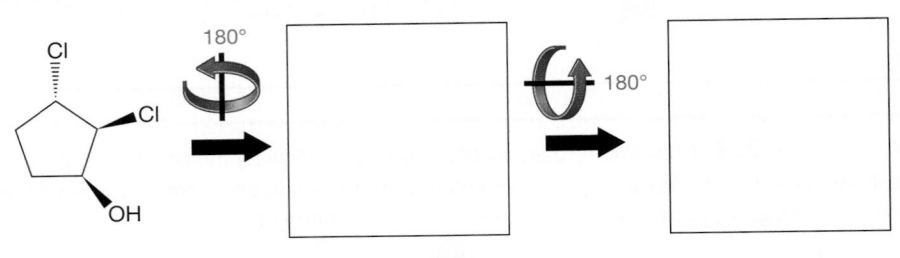

2.4 Net Molecular Dipoles and Dipole Moments

Recall from Section 1.7 that electrons involved in covalent bonds are not shared equally when the atoms they bind together have different electronegativities. Those electrons are more likely to be found nearer the atom with the higher electronegativity. The result is a *bond dipole*: a separation of charge along the bonding axis. An excess of negative charge resides on the more electronegative atom, while an excess of positive charge resides on the less electronegative atom. The resulting bond is called a *polar covalent bond*.

If there is only one polar covalent bond in the entire molecule, as in HF, then the molecule will have a **net molecular dipole**, or **permanent dipole**. That is, one end of the molecule will bear a partial positive charge and the other a partial negative charge of equal magnitude. In HF, the partial negative charge builds up on the side with the fluorine atom (electronegativity = 3.98; see Fig. 1-17), while the partial positive charge is left in the vicinity of the hydrogen atom (electronegativity = 2.20).

The **dipole moment** (μ) of a molecule is a measure of the *magnitude* of its dipole. For a pair of opposite charges, the dipole moment increases with both the magnitude of the charges (q) and the distance by which they are separated (r), according to Equation 2-2.

$$\mu = qr \tag{2-2}$$

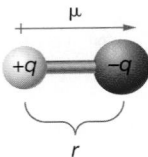

FIGURE 2-8 A dipole moment When equal but opposite charges ($+q$ and $-q$) are separated by a distance r, the resulting dipole moment (μ) points toward the negative charge.

By convention, the direction of a dipole moment is toward the negative charge (Fig. 2-8), and the magnitude is reported in units of **debye** (**D**). A dipole moment of 1 D is equal to 3.33564×10^{-30} coulomb meter (C · m). To gain a feel for units of debye, we can compute the dipole moment of a hypothetical species in which q is an atomic unit of charge, 1.602×10^{-19} C (i.e., the charge on a proton) and r is 154 pm, which is the length of an average carbon–carbon single bond. The dipole moment for such a species is 5.76 D:

$$\mu = (1.602 \times 10^{-19}\,\text{C})(154 \times 10^{-12}\,\text{m})\left(\frac{1\,\text{D}}{3.33564 \times 10^{-30}\,\text{C} \cdot \text{m}}\right) = 5.76\,\text{D}$$

Because we assumed that q was a full charge in the preceding example (not the partial charges, δ^+ or δ^-, that typically result in bond dipoles), a dipole moment of 5.76 D is rather large. Typical molecules have dipole moments that are considerably smaller, as shown in Figure 2-9. The dipole moment of HF, for example, is 2.2 D and points toward the F atom (Fig. 2-9a).

A dipole moment is a **vector**, which has both magnitude and direction, so we arrive at the net dipole moment of a molecule by adding the vectors of the bond dipoles

FIGURE 2-9 Bond dipole moments and net molecular dipole moments Bond dipole moments are shown as thin black arrows and net molecular dipole moments as thick red arrows. (a) HF is a polar molecule, with a net dipole moment of 2.2 D pointing toward the F atom. (b) CO_2 is nonpolar, because the two C=O bond dipoles point in exactly opposite directions and cancel each other out by vector addition. (c) Water is a polar molecule, with a net dipole moment of 1.8 D pointing from the point midway between the H atoms toward the O atom.

Hydrogen fluoride
Net dipole = 2.2 D
(a)

Carbon dioxide
Net dipole = 0 D
(b)

Water
Net dipole = 1.8 D
(c)

together. CO_2 (Fig. 2-9b), for example, is **nonpolar** (i.e., has no net dipole moment), despite having two polar covalent bonds. CO_2 is linear and symmetrical, so the dipole moments of the two C=O bonds point in exactly opposite directions. As a result, the dipole moment of one C=O bond exactly cancels the dipole moment of the other (i.e., the two vectors sum to zero).

2.6 YOUR TURN

Like CO_2, BeH_2 is a linear nonpolar molecule. Using the structure of BeH_2 below, draw dipole arrows above each Be—H bond. (For electronegativities, see Fig. 1-16.)

H—Be—H

A bent molecule, such as water (Fig. 2-9c), is polar. The dipole moments from the two O—H bonds do *not* point in opposite directions and therefore do *not* sum to zero. Rather, the net dipole moment points from the center of the H atoms toward the O atom.

Tetrahedral molecules, like linear molecules, can have polar bonds and still be nonpolar. All four C—Cl bonds in CCl_4 are polar, for example, but the vector addition of their bond dipoles results in no net molecular dipole moment (0 D); they cancel each other due to the symmetry of the tetrahedral geometry (Fig. 2-10a). Tetrahedral molecules that are not perfectly symmetrical, however, must be polar. Examples include CH_2Cl_2 (Fig. 2-10b) and CH_3Cl (Fig. 2-10c).

With CH_2Cl_2, vector addition is easiest if we mentally break the four bonds into two perpendicular Vs, as discussed in Section 2.2 and shown in Figure 2-10b. One V consists of H—C—H and the other of Cl—C—Cl. Because both Vs are symmetrical, each one's net dipole moment bisects the angle made by the three atoms, similar to what we observe in H_2O (Fig. 2-9c). The Vs' net dipole moments can then be added to yield the entire molecule's net dipole moment, which points from the C atom toward the Cl atoms.

2.7 YOUR TURN

A molecule of CCl_4 is decomposed into its two Vs below. Draw the bond dipoles along the two C—Cl bonds in each V and also draw the net dipole resulting from their vector addition.

(a) Tetrachloromethane (b) Dichloromethane (c) Chloromethane

Add together bond dipoles of separate fragments.

Add together net dipoles of each fragment.

FIGURE 2-10 Vector addition of bond dipoles in tetrahedral molecules (a) Tetrachloromethane. (b) Dichloromethane. (c) Chloromethane. Individual bond dipoles of the entire molecule are shown in the top row. The tetrahedral molecule is split into its constituent parts in the second row, and the net dipole of each part is indicated by the thick red arrow. In the third row, the net dipole of the entire molecule is indicated by the thick red arrow. The bottom row shows each species' electrostatic potential map. The details are described in the main part of the text.

Net dipole = 0 D   Net dipole = 2.1 D   Net dipole = 2.3 D

In CH_3Cl (Fig. 2-10c), treat the net molecular dipole moment as the sum of the net dipole moments from the CH_3 group and the C—Cl bond. The dipole moment resulting from the CH_3 group is relatively small and points from the center of the H atoms toward the C atom. The one from the C—Cl bond is larger and points from the C atom to the Cl atom. The result is a net molecular dipole moment that points from the center of the H atoms toward the Cl atom.

problem **2.5** Which of the following molecules are polar? For those that are, draw a dipole arrow indicating the direction of the net molecular dipole.

(a)   (b)   (c)   (d)

2.5 Physical Properties, Functional Groups, and Intermolecular Interactions

Now that we have reviewed some of the basic concepts of molecular geometry and polarity, it's time to explore how these factors affect the physical properties of compounds. We can begin to understand these influences by examining the boiling points, melting points, and water solubilities of some representative compounds, as shown in Table 2-4.

All of the compounds in Table 2-4 are covalent except sodium methanoate (sodium formate, $NaOCH{=}O$). The covalent compounds are all also similar in size, shape, and molar mass, with the exception of ethane, which is roughly 30% lighter. Therefore, differences in their physical properties largely reflect differences in their structures. More specifically, their structural differences are due to differences in the *functional groups* that they contain. Thus,

> Different functional groups in a compound can lead to significantly different physical properties.

> Identify the functional group that is present in each covalent compound in Table 2-4.

2.8 YOUR TURN

Why do functional groups have such a profound effect on the physical properties of organic compounds? The structural differences among the various functional groups give rise to differences in the *distribution of charge* (either partial or full) in the molecules of which they are a part. This should not be surprising because functional groups can differ in the identities and arrangements of the atoms they possess, and the electronegativities of those atoms can vary widely. Because of these different charge distributions, the various functional groups have different effects on how species interact. These types of interactions are known as **intermolecular interactions** (also called **intermolecular forces**). We will examine the following types of intermolecular interactions:

- Ion–ion interactions,
- Dipole–dipole interactions,
- Hydrogen bonding,
- Induced dipole–induced dipole interactions (or London dispersion forces), and
- Ion–dipole interactions.

These five kinds of intermolecular interactions are defined based on the mechanism involved in the attraction between molecules. Bear in mind, however, that all of these intermolecular interactions originate from the same fundamental law—namely, *opposite charges attract*. As a result, the strength of each intermolecular interaction depends on the concentrations of charge involved.

> All else being equal, the greater the concentrations of charge that are involved in an intermolecular interaction, the stronger the resulting attraction.

TABLE 2-4 Physical Properties of Representative Compounds

Compound	Molar Mass (g/mol)	Boiling Point (°C)	Melting Point (°C)	Solubility in Water (g/100 g H$_2$0)	Dipole Moment (D)	Dominant Intermolecular Interaction
Sodium methanoate (Sodium formate)	N/A	>253	253	77	N/A	Ion–ion
Methanoic acid (Formic acid)	46	101	8	Infinite	1.4	Hydrogen bonding
Ethanol	46	78	−114	Infinite	1.7	Hydrogen bonding
Ethanal (Acetaldehyde)	44	−19	−117	>100	3.0	Dipole–dipole
Dimethyl ether	46	−25	−139	6.9	1.3	Dipole–dipole
Propene	42	−48	−185	0.00061	0.3	Induced dipole–induced dipole
Propane	44	−45	−188	0.00039	0	Induced dipole–induced dipole
Ethane	30	−89	−183	0.006	0	Induced dipole–induced dipole

The first four of the intermolecular interactions on the previous page are discussed in Section 2.6 in the context of boiling points and melting points. We examine the fifth intermolecular interaction in Section 2.7 in the context of a compound's solubility in a given solvent.

2.6 Melting Points, Boiling Points, and Intermolecular Interactions

Melting points and boiling points can provide a wealth of information about intermolecular interactions. To help you see why, study Figure 2-11 to review what the different phases of a compound look like on the molecular level.

- Solids consist of atoms, ions, or molecules that are in contact with one another and are essentially immobile; they can only vibrate in place. If the solid is crystalline, then the components form a highly ordered structure.
- In a liquid, the components are also in close contact, but they can move around, rotating and sliding past one another.
- In a gas, the components are far enough apart that they are effectively isolated from one another and so can move freely.

These properties of the three phases of matter are an outcome the intermolecular interactions that are present. In solids, intermolecular interactions are maximized, and movement is limited only to vibration. In liquids, the intermolecular interactions are less substantial, so additional freedom of motion is available to the molecules. And in gases, where the molecules are widely separated, intermolecular interactions are essentially nonexistent.

Melting, therefore, decreases the intermolecular interactions that exist in the solid phase, and boiling effectively overcomes the remaining intermolecular interactions that exist in the liquid phase. Consequently, as the strength of the intermolecular interactions that exist in a particular substance increases, more energy (in the form of heat) is required for the substance to melt or boil.

- Melting points increase as the intermolecular interactions in a solid increase.
- Boiling points increase as the intermolecular interactions in a liquid increase.

FIGURE 2-11 Microscopic structure of the three phases of matter (a) In a crystalline solid, molecules or ions form a well-ordered structure called a crystal lattice, in which movement is limited to vibration and intermolecular forces are maximized. (b) In a liquid, molecules are free to move about, because intermolecular forces are somewhat less substantial. (c) In the gas phase, molecules are so far apart that intermolecular forces are effectively absent.

Solid: intermolecular interactions are maximized.

(a)

Melting

Liquid: intermolecular interactions are less substantial.

(b)

Boiling

Gas: intermolecular interactions are effectively absent.

(c)

In the sections that follow, we apply these ideas to interpret the relative strengths of (1) ion–ion interactions, (2) dipole–dipole interactions, (3) hydrogen bonding interactions, and (4) induced dipole–induced dipole interactions (London dispersion forces). Although we could use either melting points or boiling points for this purpose, we will focus primarily on trends in boiling points, because they reflect the relative strengths of intermolecular interactions more accurately than melting points do. The reason is that solids in their crystal form may have significantly different crystal structures, and the specific nature of the crystal structure can have a large impact on the solid's melting point.

2.6a Ion-Ion Interactions

Of all the compounds in Table 2-4, sodium methanoate (sodium formate, $NaOCH{=}O$) has the highest melting point and boiling point, suggesting that it has particularly strong intermolecular attractions in both its solid and liquid phases. Sodium methanoate is an ionic compound, composed of Na^+ and HCO_2^- ions held together (as we saw in Chapter 1) by the electrostatic attraction of oppositely charged ions, called *ionic bonds* or, more generally, **ion–ion interactions**.

> Ion–ion interactions are the strongest intermolecular interactions because ions have very high concentrations of positive and negative charge.

FIGURE 2-12 Dipole–dipole interaction The dominant intermolecular force between two molecules of dimethyl ether is a dipole–dipole interaction. The positive end of one ether molecule attracts the negative end of the other.

Partial positive

Attraction

Partial negative

2.6b Dipole-Dipole Interactions

Referring again to Table 2-4, notice that the compounds with significant dipole moments—methanoic acid (formic acid), ethanol, ethanal (acetaldehyde), and dimethyl ether—have boiling points and melting points that are significantly higher than those of the remaining compounds, which are essentially nonpolar.

> Polar molecules are attracted to each other more strongly than similar nonpolar molecules.

The basis for this behavior is dipole–dipole interactions. **Dipole–dipole interactions** arise because the positive end of one molecule's *net dipole* is attracted to the negative end of another's, as shown in Figure 2-12. Therefore,

> All other factors being equal, the strength of dipole–dipole interactions increases as the dipole moment increases.

And, in fact, Table 2-4 shows that ethanal ($CH_3CH{=}O$; $\mu = 3.0$ D) has a higher melting point and a higher boiling point than dimethyl ether (CH_3OCH_3; $\mu = 1.3$ D).

YOUR TURN 2.9

The boiling point of CH_3CH_2F is -37.1 °C. How does this compare to the boiling point of $CH_3CH{=}O$? Which compound has a greater dipole moment?

Although dipole–dipole interactions can be quite strong, they are not the strongest intermolecular interaction.

> Dipole–dipole interactions are generally much weaker than ion–ion interactions, because dipole–dipole interactions involve only *partial charges*, whereas ion–ion interactions involve *full charges*.

As a result, compounds such as ethanal (acetaldehyde) melt and boil at lower temperatures than ionic compounds.

SOLVED problem 2.6 Which compound has a higher melting point, $NaOCH_2CH_3$ or $CH_3CH{=}O$?

Think What is the strongest intermolecular interaction in $NaOCH_2CH_3$? In $CH_3CH{=}O$? Which type of interaction is stronger? How does that affect the melting point?

Solve $NaOCH_2CH_3$ is an ionic compound. It consists of $CH_3CH_2O^-$ and Na^+, which are held together by ion–ion interactions. $CH_3CH{=}O$, on the other hand, is a polar covalent compound, so it experiences dipole–dipole interactions. Because ion–ion interactions are stronger than dipole–dipole interactions, the melting point of $NaOCH_2CH_3$ is higher than the melting point of $CH_3CH{=}O$.

---

problem 2.7 Which compound has a higher boiling point, $NaOCH_2CH_3$ or $CH_3CH{=}O$?

---

problem 2.8 Which compound has a higher boiling point, CH_4 or CH_3F?

2.6c Hydrogen Bonding

Dipole–dipole interactions alone cannot account for some of the data in Table 2-4. Methanoic acid (formic acid, HCO_2H; $\mu = 1.4$ D) and ethanol (CH_3CH_2OH; $\mu = 1.7$ D), for example, have substantially higher melting points and boiling points than ethanal (acetaldehyde, $CH_3CH{=}O$; $\mu = 3.0$ D), despite having smaller net dipole moments. These apparent anomalies arise because methanoic acid and ethanol can form *hydrogen bonds*, whereas acetaldehyde cannot.

Each **hydrogen bond** (H bond) requires a *hydrogen-bond donor* and a *hydrogen-bond acceptor*, as shown in Figure 2-13.

> ■ A **hydrogen-bond donor** is *a hydrogen atom covalently bonded to either fluorine, oxygen, or nitrogen*. These three elements are so highly electronegative that they pull a large amount of the electron density from the bond toward themselves, in which case any hydrogen atom bonded to them acquires a large partial positive charge.
>
> ■ A **hydrogen-bond acceptor** can be any atom with a large concentration of negative charge and a lone pair of electrons. The atom may bear a full negative charge or a large partial negative charge. The only uncharged atoms that can function as H-bond acceptors are fluorine, oxygen, and nitrogen, due to their high electronegativities.

Hydrogen-bond donor (D = N, O, or F) **Hydrogen-bond acceptor** (A = N, O, or F)

$$\overset{\delta^-}{D}{-}\overset{\delta^+}{H}{--------}\overset{\delta^-}{:}A$$

A hydrogen bond

FIGURE 2-13 Hydrogen bonds A hydrogen bond consists of a hydrogen-bond donor (D—H) and a hydrogen-bond acceptor (:A). If the D and A atoms are uncharged, they must be nitrogen, oxygen, or fluorine.

FIGURE 2-14 Hydrogen bonding between ethanol molecules and between methanoic acid molecules (a) A H bond can form between two molecules of ethanol because the H atom is covalently bonded to an O atom in one molecule, and the second O atom to which the H atom is attracted has a lone pair of electrons and a significant partial negative charge. The H bond is indicated by a dashed line. (b) Two different H bonds between two molecules of methanoic acid (formic acid). In each H bond, one O atom is bonded to a H and the O atom of the second molecule has a significant partial negative charge and an available lone pair.

(a)

(b)

The actual H bond is created when a lone pair of electrons from the H-bond acceptor is partially shared with the hydrogen atom of the H-bond donor. This sharing of electrons is facilitated by the strong attraction between the large partial positive charge (δ^+) on the hydrogen atom of the H-bond donor and the negative charge (either − or δ^-) on the H-bond acceptor.

In ethanol, for example, a H bond (represented by a dotted line) can form between the H atom from the OH group on one molecule and the O atom from the OH group on a second molecule, as shown in Figure 2-14a. In methanoic acid (formic acid, Fig. 2-14b), the OH group can once again act as a H-bond donor, and either of the two O atoms can act as a H-bond acceptor, because both have available lone pairs.

YOUR TURN 2.10

Circle and label the H-bond donor and H-bond acceptor in *each* H bond shown in Figure 2-14b.

Notice that hydrogen bonding in uncharged species involves only *partial* charges. Therefore, like dipole–dipole interactions,

Hydrogen bonds are weaker than ion–ion interactions.

Hydrogen bonds are distinct from dipole–dipole interactions for two main reasons: (1) fluorine, oxygen, and nitrogen are highly electronegative, so the partial charges involved are large, and (2) the hydrogen atom is very small, which allows the partial positive charge on hydrogen to be very close to the partial negative charge on the H-bond acceptor. For these reasons,

Hydrogen bonds are often stronger than dipole–dipole interactions.

problem **2.9** Which pair of species will give rise to the strongest intermolecular interactions? The weakest?

A with **B** with **C**

It is important to be able to have a sense of the *extent* of hydrogen bonding among molecules of a particular compound—that is, the collective strength of all the hydrogen bonds present. As a general rule,

> The extent of hydrogen bonding increases as the total number of *potential* H-bond donors and H-bond acceptors increases.

With more potential donor–acceptor pairs involving two molecules, there are more ways in which hydrogen bonding can take place.

This explains why methanoic acid (formic acid) has a higher boiling point and melting point than ethanol (Fig. 2-15). A single molecule of ethanol (Fig. 2-15a) has one potential H-bond donor (the H atom of the OH bond) and one potential H-bond acceptor (the O atom). In a pair of ethanol molecules, therefore, there are two potential donors and two potential acceptors. In a single molecule of methanoic acid (formic acid, Fig. 2-15b), on the other hand, there is one potential donor (the H atom of the OH bond) and *two* potential acceptors (the two O atoms). In a pair of methanoic acid molecules, therefore, there are two potential donors and *four* potential acceptors. As a result, there are more ways in which hydrogen bonding can take place in methanoic acid than in ethanol.

(a) Ethanol

(b) Methanoic acid (formic acid)

FIGURE 2-15 **Potential hydrogen bond donors and acceptors** (a) In a molecule of ethanol, the OH bond is a potential H-bond donor and the O atom is a potential H-bond acceptor. (b) In a molecule of formic acid, the OH bond is a potential H-bond donor and each O atom is a potential H-bond acceptor.

2.11 **YOUR TURN**

High levels of cholesterol, a naturally occurring steroid, have been linked to cardiovascular disease. Octanoic acid (caprylic acid) is a fatty acid found in milk and is used commercially to manufacture perfumes and dyes. Identify the number of *potential* H-bond donors and acceptors in each of these compounds.

Cholesterol

Octanoic acid (Caprylic acid)

---

problem **2.10** How many potential H-bond donors and H-bond acceptors are there in each of the following molecules?

(a)   (b)   (c)   (d)   (e)

---

problem **2.11** Which functional groups in Table 1-6 possess at least one H-bond acceptor, but no H-bond donors? Which functional groups possess at least one H-bond donor and one H-bond acceptor? Which functional groups possess no H-bond donors and no H-bond acceptors?

SOLVED problem 2.12 1,2-Ethanediol (ethylene glycol, **A**) is historically used as an automotive antifreeze. Hydroxyacetaldehyde (**B**) is believed to be an intermediate in the metabolism of proteins and carbohydrates. Which of these compounds would you expect to have a higher boiling point? Why?

Think What is the most important intermolecular interaction that will occur between two molecules of **A**? Between two molecules of **B**? Which interaction is stronger, and how would it affect the boiling points of **A** and **B**?

Solve Both **A** and **B** are polar molecules, so dipole–dipole interactions should be present in a pair of each type of molecule. However, hydrogen bonding, which is often stronger than dipole–dipole interactions, is also possible: Each molecule contains at least one potential H-bond donor (an OH bond) and at least one potential H-bond acceptor (an O atom). To estimate which pair of molecules has greater hydrogen bonding, we count the total number of potential H-bond donors and acceptors. In two molecules of **A**, there are four potential donors and four potential acceptors, because there are two donors and two acceptors from each molecule. In two molecules of **B**, there are two potential donors and four potential acceptors, because there are one donor and two acceptors from each molecule. As a result, we would expect compound **A** to have more hydrogen bonding, and thus a higher boiling point.

problem 2.13 Which compound, **C** or **D**, would you expect to have a higher boiling point? Why?

The strength of hydrogen bonding also depends upon the concentrations of charge in the H-bond donors and H-bond acceptors. Ethanamine ($CH_3CH_2NH_2$), for example, has a lower boiling point than ethanol (CH_3CH_2OH), because N is less electronegative than O, resulting in smaller concentrations of charge in ethanamine than in ethanol (Fig. 2-16).

problem 2.14 Which H bond would you expect to be stronger? Why?

FIGURE 2-16 **Hydrogen bonding and electronegativity** Hydrogen bonding is stronger in ethanol (left) than in ethanamine (right), because O is more electronegative than N, thus giving rise to larger concentrations of positive and negative charges. As a result, the boiling point of ethanol is higher than the boiling point of ethanamine.

O is more electronegative than N, so this hydrogen bond is stronger.

Ethanol
Boiling point = 78 °C

Ethanamine
Boiling point = 17 °C

2.6d Induced Dipole–Induced Dipole Interactions (London Dispersion Forces)

Nonpolar molecules must also be able to attract each other through intermolecular interactions; otherwise, nonpolar compounds could never condense from gas to liquid. The dominant intermolecular interaction between nonpolar molecules is called **induced dipole–induced dipole interactions**, or **London dispersion forces**.

How do induced dipole–induced dipole interactions arise? Although the *average* electron distribution in a molecule such as propane ($CH_3CH_2CH_3$) does not give rise to a significant permanent dipole (Fig. 2-17a), electrons are constantly moving around. At some instant in time, there may be more electrons on one side of the molecule than there are on the other. The extra electrons on that one side give rise to an **instantaneous dipole** (Fig. 2-17b), which can alter the electron distribution on a second molecule by repelling or attracting nearby electrons. The second molecule then develops an **induced dipole** that is attracted to the first molecule (Fig. 2-17c).

Any electron cloud can be distorted—that is, any electron cloud is **polarizable**. Therefore,

> Although they are most pronounced in nonpolar molecules, induced dipole–induced dipole interactions are present when any two species interact.

To gain a sense of the relative strength of induced dipole–induced dipole interactions, notice in Table 2-4 that the nonpolar compounds tend to have the lowest boiling points and melting points. This suggests that

> Induced dipole–induced dipole interactions are generally the weakest of all intermolecular forces.

Their strength is highly variable, however, and can sometimes be quite significant, as shown in Table 2-5. A small molecule like CH_4 is a gas at room temperature, indicating that its induced dipole–induced dipole interactions are extremely weak. A much heavier molecule like I_2, on the other hand, is a solid at room temperature, indicating that its induced dipole–induced dipole interactions are quite strong—even stronger than the hydrogen bonding in water, which is a liquid at room temperature!

FIGURE 2-17 Induced dipole–induced dipole interaction (a) On average, two isolated molecules of propane are nonpolar. (b) Electrons are not static, however, so electron density can build up on one side of a molecule at some instant in time, resulting in a temporary dipole. (c) That temporary dipole, in turn, can alter the electron distribution of the second molecule, giving the second molecule an induced dipole. The oppositely charged ends of these induced dipoles attract one another.

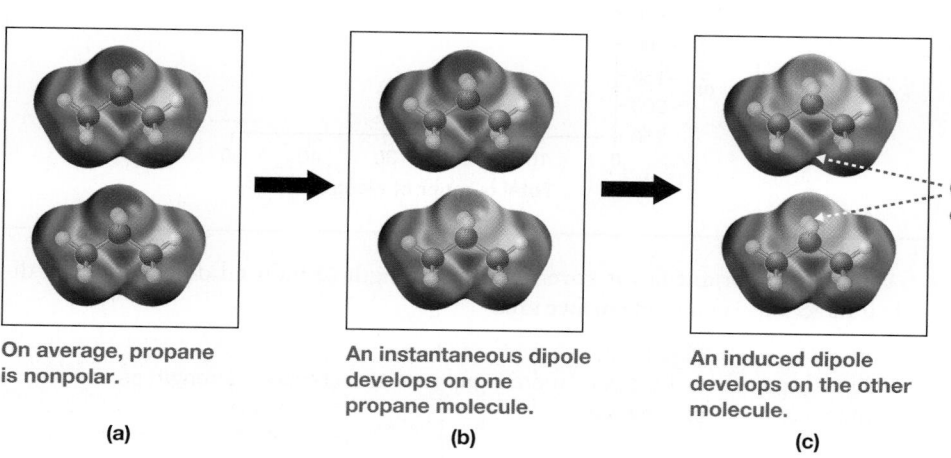

On average, propane is nonpolar.

(a)

An instantaneous dipole develops on one propane molecule.

(b)

An induced dipole develops on the other molecule.

(c)

Opposite partial charges attract.

TABLE 2-5 Melting Points and Boiling Points of Nonpolar Compounds

Molecule	Total Number of Electrons	Melting Point (°C)	Boiling Point (°C)	Molecule	Total Number of Electrons	Melting Point (°C)	Boiling Point (°C)
CH₄ Methane	10	−182	−161	Dimethylpropane	42	−17	10
CH₃CH₃ Ethane	18	−183	−89	Pentane	42	−130	36
Propane	26	−188	−42	Br₂ Bromine	70	−7	59
Cl₂ Chlorine	34	−101	−34	I₂ Iodine	106	114	184
Butane	34	−138	−1				

The precise strength of induced dipole–induced dipole interactions in a particular species depends on its **polarizability**, which can be defined as the ease with which its electron cloud is distorted. Polarizability tends to increase, in turn, as the total number of electrons increases. A species with relatively few electrons like CH_4 (melting point = $-182\,°C$, boiling point = $-161\,°C$), which has 10 total electrons, is not very polarizable. This is why its induced dipole–induced dipole interactions are weak. Pentane ($CH_3CH_2CH_2CH_2CH_3$), on the other hand, with 42 electrons, is significantly more polarizable, giving it a higher melting point ($-130\,°C$) and boiling point ($36\,°C$).

YOUR TURN 2.12

In the space provided, construct a rough plot of the boiling point as a function of total number of electrons for the straight-chain alkanes in Table 2-5 (i.e., methane, ethane, propane, butane, and pentane). What do you notice?

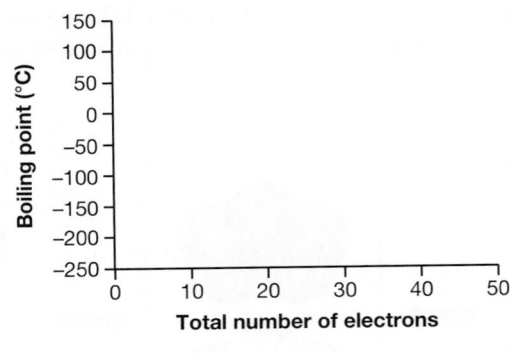

Another important factor governing the strength of induced dipole–induced dipole interactions is **contact surface area**.

Induced dipole–induced dipole interactions tend to increase in strength as the contact surface area increases.

Room-Temperature Ionic Liquids

The compound consisting of the 1-butyl-3-methylimidazolium (BMIM) cation and the bis(trifluoromethylsulfonyl)imide (NTf_2) anion, which is a *liquid* at room temperature, is one of a class of compounds called room-temperature ionic liquids (RTILs).

**1-Butyl-3-methylimidazolium cation
(BMIM)**

**bis(Trifluoromethylsulfonyl)imide anion
(NTf_2)**

Most ionic compounds, such as NaCl, are solids, given that ion–ion interactions are the strongest of the intermolecular interactions. How, then, can [BMIM][NTf_2], an ionic compound, be a liquid? It must be that the ion–ion interactions, in this particular case, are relatively weak. Examining the structures of the ions, we can see that there are two contributing factors. The first is the fact that the charges are not highly concentrated. In the BMIM cation, the positive charge is shared over both N atoms via resonance (prove this to yourself by drawing the other resonance structure). The negative charge in the NTf_2 anion is also delocalized via resonance. Moreover, the highly electronegative O and F atoms remove some negative charge from the N atom by polarizing the covalent bonds. The second reason is that the atoms that bear the formal charges are surrounded by bulky groups, which keep the opposite charges relatively far apart.

More than just being novel, RTILs have a variety of applications. RTILs typically have high electrical conductivity and are relatively stable, making them useful for electrochemical devices such as batteries, fuel cells, and electrochemical sensors. Additionally, chemists are increasingly finding RTILs to be useful in organic synthesis. Some organic reactions are catalyzed in RTILs relative to traditional molecular solvents, and the unique properties of RTILs can make it easier to recover products and catalysts, allowing reactions to be carried out in a "greener" fashion.

For instance, pentane [$CH_3(CH_2)_3CH_3$] and dimethylpropane [neopentane, $C(CH_3)_4$] have the same formula (C_5H_{12}), and thus the same number of electrons, but pentane has a more extended shape than the relatively compact dimethylpropane. As a result, two molecules of pentane have a greater surface area available for interaction than two molecules of dimethylpropane do, as shown in Figure 2-18 (p. 98). Greater contact surface area means that pentane has more effective induced dipole–induced dipole interactions and a higher boiling point than dimethylpropane.

In Figure 2-18, shade in the contact surface area for each pair of molecules. Indicate which pair of molecules has greater contact surface area.	**2.13 YOUR TURN**

problem **2.15** For each pair of molecules, which would you expect to have a higher boiling point? Why?

2,2-Dimethylbutane   2-Methylpentane
A

1,1-Dimethylcyclopropane
B

1,2-Dimethylcyclopropane

2.7 Solubility

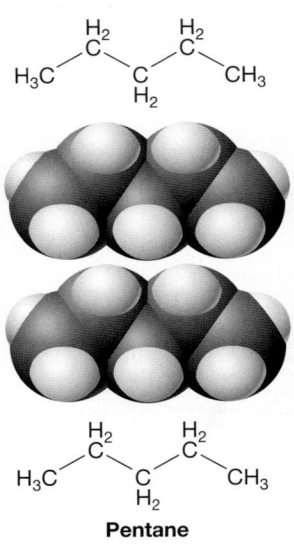

Pentane
Boiling point = 36 °C

(a)

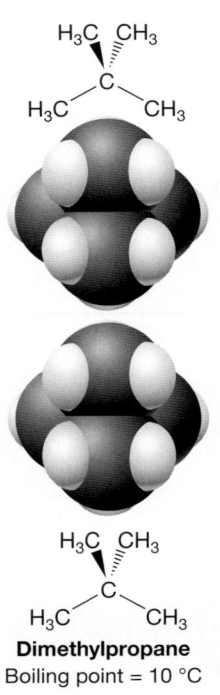

Dimethylpropane
Boiling point = 10 °C

(b)

FIGURE 2-18 Contact surface area
The contact surface area between two molecules of pentane (a) is greater than that between two molecules of dimethylpropane (b). The result is stronger induced dipole–induced dipole interactions in pentane and a correspondingly higher boiling point.

The general rule of solubility is that "like dissolves like." This means that polar compounds tend to be soluble in polar solvents but insoluble in nonpolar solvents, and nonpolar compounds tend to be soluble in nonpolar solvents but insoluble in polar solvents. As discussed in this section, the above rule is an outcome of the intermolecular interactions at play when two substances are mixed. Seeing how this is so will allow you to gain a deeper understanding of solubility and will also enable you to make predictions about solubility in situations where the rule of "like dissolves like" breaks down.

To understand how intermolecular interactions affect solubility, let's begin by examining the solubility of hexane ($CH_3CH_2CH_2CH_2CH_2CH_3$) in water. Hexane is nonpolar and essentially insoluble in water. We can understand why by comparing the intermolecular interactions that exist in the pure separated substances to those that exist between the solute and solvent (i.e., in the solution), as shown in Figure 2-19. In the pure substances, the strongest intermolecular interaction is the hydrogen bonding in water. Between a molecule of hexane and a molecule of water, however, the strongest intermolecular interaction involves an induced dipole on hexane, which is much weaker than hydrogen bonding. In other words, significantly stronger intermolecular interactions exist in the separated substances than exist in the solution, so the separated substances are more stable (i.e., lower in energy) than the solution.

The example of hexane and water (Fig. 2-19) can be generalized as follows:

> Two compounds are insoluble if the intermolecular interactions that exist in the pure separated substances are sufficiently stronger than those that exist in the solution.

Next, let's examine what happens when ethanol (CH_3CH_2OH) mixes with water. According to Table 2-4, ethanol is infinitely soluble in water. Once again, the strongest intermolecular interaction that exists in the pure separated substances is hydrogen bonding, both in water and in ethanol (Fig. 2-20). Unlike the hexane–water example, however, substantial hydrogen bonding also exists between the solute and solvent molecules in the ethanol–water solution. In this case, intermolecular interactions favor both the separated substances and the solution roughly equally.

If neither the separated state nor the solution is particularly favored by intermolecular interactions, why is ethanol soluble in water? The answer largely has to do with *entropy*. As you may recall from general chemistry, **entropy** is a thermodynamic quantity that increases with the number of equivalent ways the energy in a system can be arranged. Many people like to think of entropy as a *measure of disorder*, and a system with greater entropy (i.e., one that is more disordered) tends to be more likely to occur than one with less entropy.

A mixture of two compounds has more entropy than when the two compounds are separated. We say, therefore, that *entropy favors the mixed state.*

> In general, two compounds (e.g., ethanol and water) are soluble in each other if the intermolecular interactions that exist in the pure separated substances are roughly the same strength as those that exist in the solution (i.e., ethanol–water). In such cases, the major driving force for solubility is the increase in *entropy*.

Significantly stronger intermolecular interactions exist in the separated substances, so the compounds are not miscible.

Pure hexane

Induced dipole–induced dipole interactions

Pure water

Hydrogen bonding

Solution of hexane and water

Induced dipole–induced dipole interactions

FIGURE 2-19 Intermolecular interactions and the insolubility of hexane in water The dominant intermolecular interactions present in pure hexane are induced dipole–induced dipole interactions. The dominant intermolecular interaction in pure water is hydrogen bonding. When the compounds are mixed, induced dipole–induced dipole interactions remain but hydrogen bonding is diminished. This favors the pure substances on the left, and thus explains why hexane is insoluble in water.

By the same token, nonpolar compounds tend to be soluble in nonpolar solvents because the intermolecular interactions between species in the separated substances (mainly induced dipole–induced dipole interactions) are similar in strength to those in the solution. Once again, therefore, the major driving force for solubility is greater entropy in the solution compared to when the compounds are separated.

Which compound, **A** or **B**, do you expect to be more soluble in H_2O?

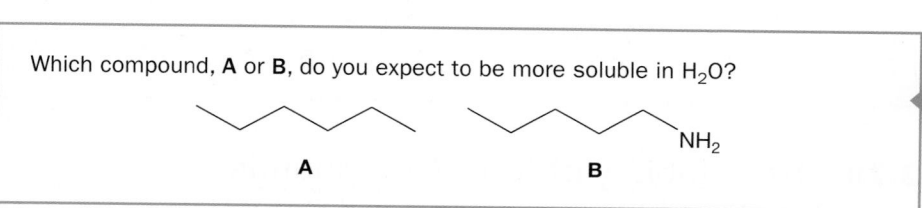

A　　　**B**

2.14 YOUR TURN

Similar intermolecular interactions exist in the mixed and unmixed states, so the mixed state is favored due to entropy.

Pure ethanol

Hydrogen bonding

Pure water

Hydrogen bonding

Solution of ethanol and water

Hydrogen bonding

FIGURE 2-20 Intermolecular interactions and the solubility of ethanol in water Hydrogen bonding exists in pure ethanol (blue) and in pure water (red). Substantial hydrogen bonding also exists between molecules of water and ethanol (right), allowing the two substances to dissolve readily and in any proportion.

SOLVED problem 2.16 Which of the following compounds would you expect to be more soluble in toluene ($C_6H_5CH_3$)? Explain.

Butan-1-ol or **Diethyl ether**

Think What are the relative strengths of the intermolecular interactions in the pure substances that would be disrupted upon mixing? How do these compare to the strengths of the intermolecular interactions that would be gained?

Solve Toluene is nonpolar, so the only interactions that can exist between toluene and each of the given compounds involves a relatively weak induced dipole on toluene. As such, the solute–solvent interactions are roughly the same in both solutions. In pure diethyl ether, on the other hand, dipole–dipole interactions are present, and in pure butan-1-ol, hydrogen bonding is present. Therefore, the mixing of diethyl ether with toluene would disrupt the dipole–dipole interactions, whereas the mixing of butan-1-ol with toluene would require the destruction of hydrogen bonding interactions. Because hydrogen bonding is often stronger than dipole–dipole interactions, mixing is less favorable for butan-1-ol, making diethyl ether more soluble in toluene.

---

problem 2.17 Which compound, **C** or **D**, would you expect to be more soluble in toluene ($C_6H_5CH_3$)? Explain.

C   **D**

2.7a The Solubility of Ionic Compounds: Ion–Dipole Interactions and Solvation

Based on what we've discussed so far, it might seem peculiar that an ionic compound such as sodium methanoate (sodium formate, $NaHCO_2$) dissolves in a polar solvent like water, because doing so eliminates ion–ion interactions, the strongest of the intermolecular forces. Yet sodium methanoate *does* dissolve in water, and its water solubility (like that of most ionic compounds) is quite high (Table 2-4). Why?

When an ionic compound like sodium methanoate dissolves in water, it does so as its individual ions, Na^+ and HCO_2^-, not as uncharged formula units, $NaHCO_2$. These free ions can interact with water molecules via **ion–dipole interactions**, a kind of intermolecular interaction different from the ones we have examined thus far. In these particular ion–dipole interactions, the positive end of water's dipole attracts a HCO_2^- anion and the negative end attracts a Na^+ cation (Fig. 2-21).

A variety of factors govern the strength of an individual ion–dipole interaction. One important factor is the solvent's polarity:

> All else being equal, the strength of the ion–dipole interaction increases as the dipole moment of the solvent molecule increases.

Other factors include the size and shape of the solvent molecule, as well as the nature of the ion, but we address these issues where appropriate in the context of specific examples.

FIGURE 2-21 Ion–dipole interactions (a) An ion–dipole interaction between HCO_2^- and H_2O. The anion interacts with the positive end of water's dipole. (b) An ion–dipole interaction between Na^+ and a molecule of water. The cation interacts with the negative end of water's dipole.

It is helpful to compare the strength of ion–dipole interactions with those of the other intermolecular interactions we have examined previously. To do so, consider the concentrations of charge involved in each type. An ion–ion interaction involves the attraction between two full charges, an ion–dipole interaction involves the attraction between a full charge and a partial charge, and a dipole–dipole interaction involves the attraction between two partial charges. Therefore,

> The strength of an ion–dipole interaction is intermediate between that of an ion–ion interaction and that of a dipole–dipole interaction.

How, then, are ion–dipole interactions capable of overcoming the stronger ion–ion interactions to dissolve an ionic compound? A major factor is **solvation**, depicted in Figure 2-22, in which an individual ion participates in *multiple* ion–dipole interactions with the solvent. When this occurs, the ions are said to be **solvated**. The collective stability from all of these ion–dipole interactions can be substantially greater than that of the ion–ion interactions in an ionic compound, thus making the mixture more favored (more stable).

In Figure 2-22, label every ion–dipole interaction.

2.15 YOUR TURN

Because the strength of an individual ion–dipole interaction depends upon the polarity of the solvent, the collective stability provided by solvation does, too

The positive end of water's dipole solvates the methanoate anion via multiple ion–dipole interactions.

The negative end of water's dipole solvates the sodium cation via multiple ion–dipole interactions.

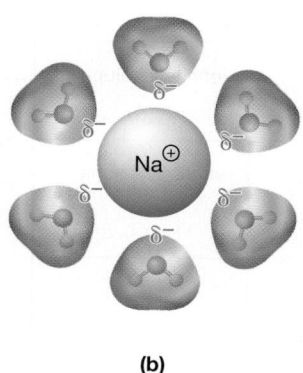

(a)

(b)

FIGURE 2-22 Solvation Ionic compounds can dissolve in water as a result of solvation of the respective ions. (a) The positive end of water's dipole solvates the methanoate anion. (b) The negative end of water's dipole solvates the sodium cation.

(see Solved Problem 2.18). This is important because, as we discuss in Chapter 9, a solvent's ability to solvate ions can have a dramatic effect on the outcome of reactions.

SOLVED problem 2.18 In which solvent would you expect NaCl to be more soluble, dimethyl ether (CH_3OCH_3) or ethanal ($CH_3CH{=}O$)?

Think What are the most important intermolecular interactions that would be disrupted upon dissolution? What intermolecular interactions would be gained in the solution? In which solvent are those intermolecular interactions stronger?

Solve Very strong ion–ion interactions are disrupted when NaCl dissolves. Both dimethyl ether and ethanal are polar molecules, so ion–dipole interactions would be present in solution with both solvents. Ethanal, however, has a much larger dipole moment (Table 2-4), so it will better solvate the Na^+ and Cl^- ions.

problem 2.19 NaCl is more soluble in methanol (CH_3OH) than in propan-1-ol ($CH_3CH_2CH_2OH$). Which solvent is better at solvating ions?

2.7b The Effect of Hydrocarbon Groups on Solubility

The types of functional groups present in a molecule have a direct effect on its solubility in various solvents, because the functional groups determine the kinds of intermolecular interactions that are available to the molecule. Alcohols, for example, can always hydrogen bond with water due to the presence of the **hydrophilic** ("water loving") OH group. Thus, small alcohols like methanol (CH_3OH), ethanol (CH_3CH_2OH), and propan-1-ol ($CH_3CH_2CH_2OH$) are infinitely soluble in water, as shown in Table 2-6.

Notice in Table 2-6, however, that the water solubility *decreases* as the size of the R group—the hydrocarbon portion of the molecule—*increases*. This is because R groups are highly nonpolar; they are **hydrophobic** ("water fearing"). We can generalize this trend as follows:

A species behaves more like a nonpolar alkane as the size of its alkyl group increases.

TABLE 2-6 Water Solubility of Some Simple Alcohols (R—OH)

Alcohol	Water Solubility (g/100 g H_2O)	Alcohol	Water Solubility (g/100 g H_2O)
CH_3OH Methanol	Infinitely soluble	Pentan-1-ol	2.7
CH_3CH_2OH Ethanol	Infinitely soluble	Hexan-1-ol	0.6
Propan-1-ol	Infinitely soluble	Heptan-1-ol	0.1
Butan-1-ol	7.7		

problem 2.20

Which functional groups in Table 1-6 would be considered hydrophilic? Which would be considered hydrophobic?

Mono-alcohols (i.e., alcohols with one OH group) are generally considered insoluble in water if they contain six or more carbons. The effect of the alkyl group, however, can be overcome by increasing the number of hydrophilic functional groups. Typically, molecules are found to be highly water soluble if their ratio of the total number of carbon atoms to the number of hydrophilic functional groups is less than about 3:1. Sucrose (table sugar), for example, is highly soluble in water despite the fact that it has 12 carbon atoms (Fig. 2-23). This is because it contains eight hydrophilic alcohol groups, so its ratio of carbon atoms to hydrophilic groups is 1.5:1, which is less than 3:1.

Sucrose

FIGURE 2-23 The water solubility of sucrose The large number of OH groups compared to C atoms makes sucrose soluble in water.

SOLVED problem 2.21

Would you expect each of the following compounds to be soluble in water? Why or why not?

Think What is the ratio of alcohol groups to the number of carbon atoms of the hydrocarbon group?

Solve Molecule **A** contains six carbons and two hydrophilic alcohol groups, giving it a ratio of 3:1. We therefore expect this molecule to be highly soluble in water. For molecules **B** and **C**, however, the ratios are 6:1 and 10:1, both of which exceed the 3:1 cutoff. We therefore expect these two molecules to be insoluble in water.

problem 2.22

Like alcohols, aldehydes (R—CH═O) become less soluble in water as the number of carbon atoms in the alkyl group R increases. Do you think the maximum number of carbon atoms for water-soluble aldehydes will be greater than or less than that for water-soluble alcohols? Explain.

2.8 Strategies for Success: Ranking Boiling Points and Solubilities of Structurally Similar Compounds

One of the more difficult tasks asked of students is to rank the boiling points, melting points, or solubilities of a set of compounds, given only their molecular structures. For example, suppose that we are asked to rank the boiling points of the following compounds, from lowest to highest:

To solve this problem you need to understand intermolecular interactions and how they relate to boiling points. This can be a daunting problem, however, because it requires us to consider several factors simultaneously. Therefore, we must find ways to break this large problem into smaller pieces. What follows is one way to do so.

1. Identify the most important intermolecular interaction between molecules of each compound and group the compounds accordingly. Which compounds are nonpolar? Which ones are polar? Which ones contain functional groups that can participate in hydrogen bonding?

   ■ Molecules **A, D, F,** and **G** are all nonpolar, so the primary intermolecular interactions governing their boiling points are induced dipole–induced dipole interactions.

   ■ Molecules **B** and **E** each contain alcohol groups and therefore can hydrogen bond with other molecules of the same compound.

   ■ Molecule **C** cannot hydrogen bond with another molecule of the same compound. It is polar, however, so it can participate in dipole–dipole interactions.

A   D   F   G

Group 1 = nonpolar
Induced dipole-induced dipole interactions

B   E

Group 2 = polar
Hydrogen bonding

C

Group 3 = polar
Dipole-dipole interactions

2. Rank each group by boiling point.

   ■ Of the intermolecular interactions we have identified, hydrogen bonding is the strongest. As a result, the compounds in Group 2 have the highest boiling points.

   ■ Dipole–dipole interactions are the next strongest, giving the compound in Group 3 the next highest boiling point.

   ■ The compounds in Group 1 have the lowest boiling points because induced dipole–induced dipole interactions are the weakest.

3. Rank the boiling points within each group.

   ■ In Group 1, boiling points are ranked by *polarizability*, which generally increases with the total number of electrons. Therefore, the boiling points increase in the order **D < A < F < G**.

   ■ In Group 2, boiling points are ranked by the extent of hydrogen bonding, which is approximated by the number of potential H-bond acceptors and donors. In a pair of molecules of **B**, there are two potential donors and two potential acceptors. In a pair of molecules of **E**, there are four potential donors and four potential acceptors. Consequently, **E** should have a higher boiling point than **B**.

4. Put it all together.

   ■ Keeping in mind that boiling points should increase in the order Group 1 < Group 3 < Group 2, we arrive at the following order (actual boiling points are included for comparison):

Boiling point:

| 80 °C | 111 °C | 139 °C | 173 °C | 181 °C | 202 °C | 281 °C |

D   **A**   **F**   **G**   **C**   **B**   **E**

   In solving the above problem, we assumed that induced dipole–induced dipole interactions were the weakest, even though they can be stronger than dipole–dipole interactions or hydrogen bonding if polarizability is sufficiently large. In the problem at hand, we would not expect such exceptions because the total number of electrons possessed by each compound in Groups 2 and 3 is not much different from the total number in the most polarizable compound in Group 1—namely, 1,4-dichlorobenzene (molecule **G**). As Solved Problem 2.23 shows, however, we must take into account significant differences in polarizability.

SOLVED problem **2.23** 1,4-Dibromobenzene, a nonpolar compound, has a boiling point of 219 °C. The boiling point of 1,2-dichlorobenzene, a polar compound, is 181 °C. Explain.

1,4-Dibromobenzene
Boiling point = 219 °C

1,2-Dichlorobenzene
Boiling point = 181 °C

Think What intermolecular interactions are present in each compound? Are they of similar strength in each compound?

Solve 1,4-Dibromobenzene is nonpolar, so it has only induced dipole–induced dipole interactions. 1,2-Dichlorobenzene, on the other hand, is polar, so it has both dipole–dipole interactions and induced dipole–induced dipole interactions. Dipole–dipole interactions are generally assumed to be the more important interaction when comparing compounds with similar polarizabilities. In this case, however, the two molecules should have different polarizabilities. More specifically, 1,4-dibromobenzene should be significantly more polarizable than 1,2-dichlorobenzene because it has 36 more electrons (146 vs. 110, the difference between two Br atoms and two Cl atoms). As a result, induced dipole–induced dipole interactions are more extensive in 1,4-dibromobenzene— enough to give 1,4-dibromobenzene a higher boiling point.

---

problem **2.24** Which of the following compounds do you think has a higher boiling point? Explain.

A   **B**

We can apply a similar strategy to predicting relative solubilities in a given solvent. However, we must consider the disruption of the intermolecular interactions in the pure separated substances as well as the formation of the intermolecular interactions in the mixture. This is illustrated in Solved Problem 2.25.

SOLVED problem 2.25 Rank the following compounds from lowest solubility in hexane, $CH_3(CH_2)_4CH_3$, to highest solubility in hexane.

Think What are the most important intermolecular interactions that exist in the isolated substances and what are their relative strengths? What are the most important intermolecular interactions that exist in the solution?

Solve In a solution of each compound **A–E** with hexane, the intermolecular interactions will be quite weak because hexane is nonpolar. Solubility, therefore, will decrease as the strength of the intermolecular interactions in the pure substances increase. As pure substances, the relative strengths of the intermolecular interactions are as follows:

Increasing strength of intermolecular interactions

Compound **D** is nonpolar and therefore possesses only induced dipole–induced dipole interactions. Compounds **A** and **E** are polar and therefore possess dipole–dipole interactions; those interactions are stronger in **E**, moreover, because it has a larger dipole moment. Compounds **B** and **C** both undergo hydrogen bonding, but it is more extensive in **C** because **C** has more H-bond donors and acceptors. Solubility in hexane will increase in the reverse order:

Increasing solubility in hexane

---

problem 2.26 Rank the compounds in Solved Problem 2.25 in order from least soluble in water to most soluble.

2.9 Protic and Aprotic Solvents

As we learned in Section 2.7, ionic compounds can have very high solubility in a polar solvent such as water, due to the strong solvation set up by ion–dipole interactions in the resulting solution. For example, about 35 g of sodium chloride (NaCl) dissolve in

Enzyme Active Sites: The Lock-and-Key Model

Enzymes are proteins that catalyze biological reactions. In many cases, enzymes enhance reaction rates by several orders of magnitude. What is particularly fascinating is how specific each enzyme's role is, so much so that the name of an enzyme derives from the reaction that it catalyzes. Consider the enzyme fructose 1,6-bisphosphate aldolase, which is responsible for cleaving 1,6-bisphosphate into dihydroxyacetone phosphate and glyceraldehyde-3-phosphate—a reaction that is integral in breaking down sugars.

The general model that accounts for the specificity of enzymes is called the "lock-and-key" model. An enzyme is typically a rather large molecule, but assumes a three-dimensional shape that defines an active site—a pocket in which the reaction takes place. The substrate (i.e., the reactant) docks with the enzyme in the active site to form an enzyme–substrate complex, the reaction takes place, and subsequently the products are released. Under the lock-and-key model, only certain, specific substrates "fit" into the active site.

Fitting into an active site means more than just spatial fitting; it means that the active site is specially designed to provide optimal intermolecular interactions for the substrate. We can see this by examining a portion of the enzyme–substrate complex in the fructose 1,6-bisphosphate aldolase reaction.

On the left, 1,6-fructose bisphosphate appears as a space-filling model, and the important side groups of some amino acids in the active site appear as ball-and-stick models. On the right, the substrate and side groups appear as line structures. Notice how the locations of the side groups are optimal for numerous specific hydrogen-bonding and ion–ion interactions.

100 mL of water. Quite interestingly, however, NaCl is rather insoluble in the solvent dimethyl sulfoxide (DMSO): only about 0.4 g dissolve in 100 mL of DMSO.

$$NaCl(s) \xrightarrow{H_2O} Na^{\oplus} + Cl^{\ominus}$$

NaCl is highly soluble in water, but not in DMSO.

$$NaCl(s) \xcancel{\xrightarrow{DMSO}} Na^{\oplus} + Cl^{\ominus}$$

What makes this result peculiar is that DMSO's dipole moment (4.0 D) is significantly greater than that of water (1.9 D). With a larger dipole moment, shouldn't DMSO give rise to stronger solvation, and thus dissolve more NaCl?

The negative end of water's dipole solvates the sodium cation strongly.

The positive end of water's dipole solvates the chloride anion strongly.

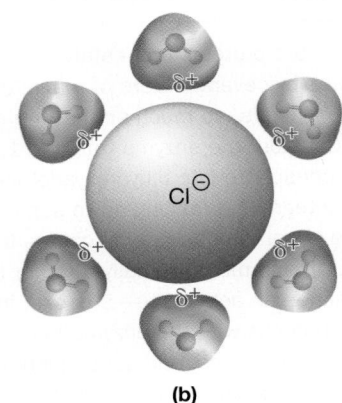

(a)

(b)

FIGURE 2-24 Solvation of NaCl in water (a) The Na^+ ion is strongly solvated because the partial negative charge of water is well exposed. (b) The Cl^- ion is strongly solvated because the partial positive charge of water is well exposed.

Clearly there must be more to the issue than just the magnitude of the solvent's dipole moment. What also comes into play is how close the ions can approach the partial charges of the polar solvent molecule. As shown in Figure 2-24, the partial negative and partial positive charges of water are well exposed to the Na^+ and Cl^- ions, respectively, so these ions are able to approach the partial charges rather closely. Thus, the solvation of each ion is relatively strong.

As shown in Figure 2-25a, the partial negative charge of DMSO is also well exposed, so Na^+ is strongly solvated by DMSO, just as it is strongly solvated by water. However, notice that the partial positive charge of DMSO is flanked by two bulky CH_3 groups. This makes it difficult for the Cl^- ion to approach DMSO's partial positive charge (Fig. 2-25b). We say that the methyl groups introduce **steric hindrance**. Thus, the Cl^- is not solvated very strongly.

These characteristics of water and DMSO, as they pertain to the solubility of ionic compounds, are not unique to just these two solvents. Water is an example of a **protic solvent**, because it possesses a H-bond donor—in this case, an O—H covalent bond.

(a)

(b)

FIGURE 2-25 Solvation in DMSO (a) Na^+ is solvated strongly in DMSO because the partial negative charge of DMSO is well exposed. (b) Cl^- is not solvated very strongly by DMSO, because DMSO's partial positive charge is buried inside the molecule.

TABLE 2-7 Common Polar Protic Solvents and Polar Aprotic Solvents

Polar Protic Solvents		Polar Aprotic Solvents	
Structure	Name	Structure	Name
H—O—H δ^+	Water	H_3C—S(=O)—CH_3 δ^+	Dimethyl sulfoxide (DMSO)
H_3C—CH$_2$—O—H δ^+	Ethanol	H_3C—C(=O)—CH_3 δ^+	Propanone (Acetone)
H_3C—C(=O)—O—H δ^+	Ethanoic acid (Acetic acid)	H—C(=O)—N(CH_3)CH_3 δ^+	N,N-Dimethyl formamide (DMF)

DMSO has no H-bond donors, so it is an **aprotic solvent**. Other examples of polar protic solvents and aprotic solvents are shown in Table 2-7.

2.16 **YOUR TURN**

In Table 2-7, circle and label each potential hydrogen-bond donor.

Like water, all polar protic solvents have easily accessible partial negative and positive charges. Polar aprotic solvents, on the other hand, have well-exposed partial negative charges, but their partial positive charges tend not to be easily accessible. Therefore,

- Polar protic solvents tend to solvate both cations and anions very strongly.
- Polar aprotic solvents tend to solvate cations very strongly, but not anions.

The importance of these attributes extends far beyond understanding the solubilities of ionic compounds. As we will see in Chapter 9, the solvation characteristics of protic and aprotic solvents play a major role in governing the outcomes of reactions.

problem **2.27** Will $KSCH_3$ be more soluble in ethanol or acetone? Explain.

2.10 Soaps and Detergents

Soaps and enzyme-free detergents remove dirt and oil from our hands, clothes, and dinnerware, all with no chemical reaction occurring in the process (i.e., no covalent bonds are broken or formed). Instead, the cleansing ability of soaps and detergents depends entirely on intermolecular interactions.

The molecules specifically responsible for a soap's cleansing properties are typically *salts of fatty acids*, which are ionic compounds of the form $RCO_2^- Na^+$ or $RCO_2^- K^+$.

FIGURE 2-26 Electrostatic potential map of a fatty acid carboxylate anion A significant negative charge is located on the portion of the ion containing the —CO_2^- group (the ionic head group), making it hydrophilic. At the same time, the hydrocarbon tail is very nonpolar, making it hydrophobic.

In these compounds, R is a long hydrocarbon chain, generally containing from 12 to 18 carbons. Examples include potassium oleate and sodium palmitate:

Potassium oleate
$C_{18}H_{33}O_2K$

Sodium palmitate
$C_{16}H_{31}O_2Na$

When a soap dissolves in water, it does so as its individual ions: the metal cation (Na^+ or K^+) and the carboxylate anion (RCO_2^-). Of these two species, the carboxylate anion is the one that is directly responsible for the soap's cleansing properties, because it has vastly different characteristics at its two ends (Fig. 2-26). Specifically,

> Soaps work because one end of the molecular species is very hydrophilic and the other end is very hydrophobic.

In the case of fatty acid carboxylates, the hydrophilic end is the one with —CO_2^-, called the **ionic head group**, and the hydrophobic end is the one with the nonpolar **hydrocarbon tail**.

YOUR TURN 2.17

On the electrostatic potential map in Figure 2-26, circle the hydrophilic region and label it. Circle the hydrophobic region and label it.

In water, the carboxylate anions from the fatty acid salts form spherical aggregates, called **micelles** (Fig. 2-27). In a micelle, the nonpolar tails are on the inside of the sphere, where they can interact with one another via extensive induced dipole–induced dipole interactions, whereas the charged head groups are on the outside, where they can form the greatest number of ion–dipole interactions with the surrounding water molecules. In other words, micelles are highly solvated (Section 2.7a).

problem **2.28** Ethyl ethanoate (ethyl acetate, $CH_3CO_2CH_2CH_3$), a relatively small ester, is insoluble in water. Knowing this, would you expect the following compound to form micelles in water? Explain.

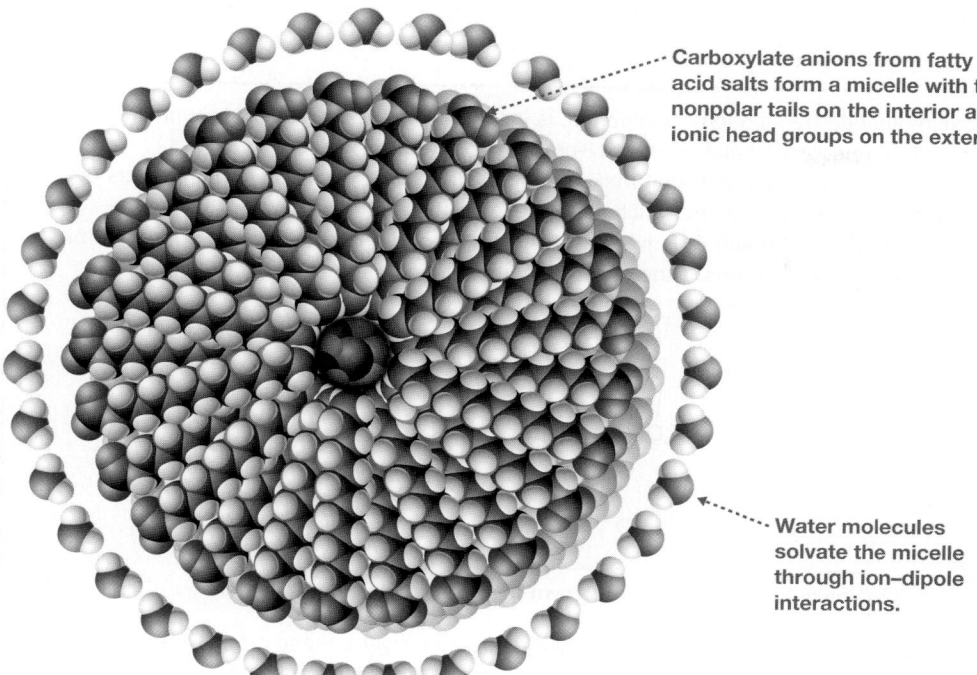

Carboxylate anions from fatty acid salts form a micelle with the nonpolar tails on the interior and ionic head groups on the exterior.

Water molecules solvate the micelle through ion–dipole interactions.

FIGURE 2-27 The structure of a micelle In water, carboxylate anions (RCO_2^-) from fatty acid salts form *micelles*, in which the nonpolar tails are on the inside and the ionic head groups are on the outside. Micelles are soluble in water because they are heavily solvated due to the extensive ion–dipole interactions that take place between the ionic head groups and the water molecules of the surrounding solution.

Dirt, grease, and oils are nonpolar substances, and so are insoluble in pure water. They are soluble in a solution of soap and water, however, because the hydrophobic tails of the soap molecules dissolve in droplets of oil or grease, leaving their hydrophilic head groups available for solvation by water (Fig. 2-28). Soap is said to **emulsify** such substances—that is, it disperses them in a solvent (water) in which they are normally insoluble. Such emulsified particles can then be washed away by water.

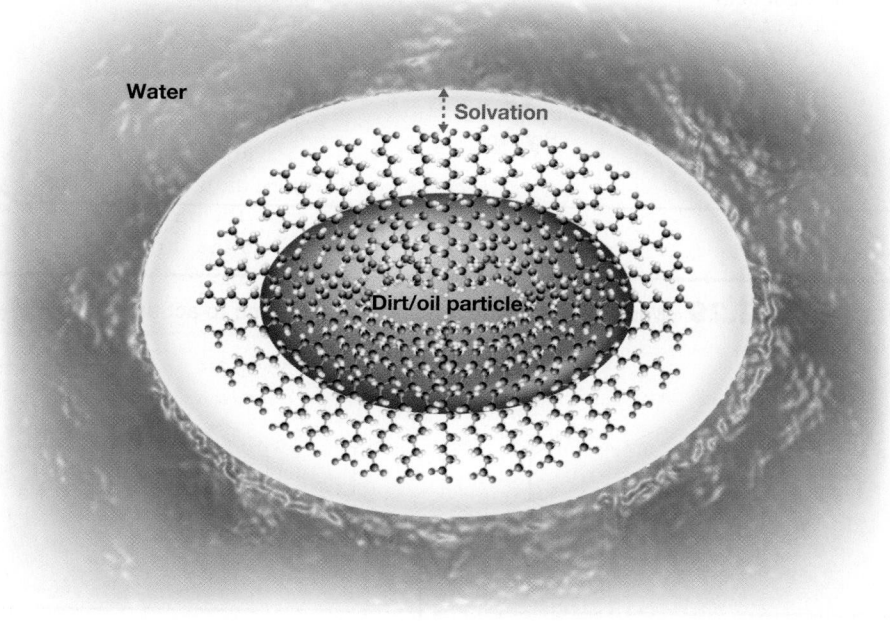

Water

Solvation

Dirt/oil particle

FIGURE 2-28 An oil droplet emulsified in water by soap The nonpolar tails of soap molecules dissolve in a droplet of oil, which can then be washed away by water molecules interacting with the exposed hydrophilic head groups.

Phase Transfer Catalysts

Potassium fluoride (KF), an ionic compound, is insoluble in benzene (C_6H_6), a nonpolar compound, but becomes soluble when 18-crown-6 is added.

What is it about 18-crown-6 that makes this possible? 18-Crown-6 is one of a class of compounds called crown ethers, which are cyclic polyethers of repeating —CH_2CH_2O— groups. The numbers 18 and 6 derive from the fact that the ring consists of 18 total atoms, six of which are ether oxygens. Other crown ethers include 12-crown-4 and 15-crown-5.

12-Crown-4 **15-Crown-5** **18-Crown-6**

With each O atom bearing a partial negative charge directed inward, these crown ethers have the ability to bind metal cations rather strongly. 18-Crown-6, in particular, strongly binds K^+ ions.

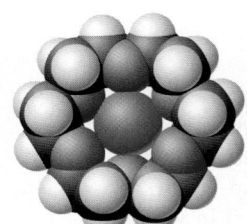

With the K^+ ion bound to the crown ether, the charge is localized on the inside of the ring. The outside of the ring, which is what is predominantly exposed to the solvent, is relatively nonpolar, making the complex soluble in nonpolar solvents such as benzene.

This property makes it possible to use crown ethers as phase-transfer catalysts. A phase-transfer catalyst helps move a reactant from one phase to another, where a reaction occurs. When 1-bromooctane is treated with KF in the presence of 18-crown-6, for example, fluoride replaces bromide to produce 1-fluorooctane in relatively high yield. In the absence of the crown ether, essentially no KF can dissolve in the benzene solvent, so no reaction occurs because the reactants never meet.

1-Bromooctane → KF / Benzene / 18-Crown-6 → **1-Fluorooctane** 92%

1-Bromooctane → KF / Benzene (crossed out)

---

problem **2.29** Would you expect the following compound to act as a soap? Why or why not?

Sodium dodecyl sulfate

Hard water poses a problem for soaps. Water is considered *hard* if it contains a significant concentration of Mg^{2+} or Ca^{2+} ions. These ions bind more strongly to the negative charge of the carboxylate than Na^+ or K^+ ions do (see Problem 2.52 at the end of the chapter), causing the long-chain fatty acids to precipitate from water (Equation 2-3).

$$Ca^{2+} \;+\; 2\left({}^{\ominus}O{-}C{=}O \cdots \right) \longrightarrow Ca\left(O{-}C{=}O \cdots \right)_2$$

Soap scum
(Insoluble in water)

(2-3)

Not only does this decrease the effectiveness of soap, but the precipitate is a nuisance known as **soap scum**. A common way to combat this problem is to run water through a *water softener*, which contains an ion-exchange resin that removes hard water ions and replaces them with Na^+ ions.

Unlike soaps, **detergents** are effective cleansers in hard water. Detergents such as sodium dodecyl sulfate are similar in structure to soaps, containing an ionic head group (sulfate, $—OSO_3^-$) and a nonpolar hydrocarbon tail (Fig. 2-29). As a result, they behave in much the same way as soaps, forming micelles in solution that are capable of emulsifying dirt, grease, and oils. The main difference is that metal ions in hard water do not bind to the sulfate portion of detergents as strongly as they do to the carboxylate portion of soaps. (See Problem 2.53 at the end of the chapter.) This allows detergents to remain active even in hard water.

2.11 Wrapping Up and Looking Ahead

This chapter began with VSEPR theory, a simple model that allows us to predict the electron and molecular geometries about atoms. With an understanding of molecular geometry, we saw how bond dipoles can be added together to determine a molecule's polarity, which is a key factor in governing physical properties such as boiling points, melting points, and solubility.

We next proceeded to discuss various intermolecular interactions in the context of physical properties. We saw that the attraction of opposite charges is the basis for all intermolecular interactions, and the greater the concentration of charge, the stronger the attraction. Thus, ion–ion interactions are generally regarded as the strongest intermolecular interaction, followed by ion–dipole interactions, hydrogen bonding, dipole–dipole interactions, and, finally, induced dipole–induced dipole interactions (a.k.a. London dispersion forces). Knowing how to determine the presence of intermolecular interactions from just Lewis structures, we then learned how to predict relative boiling points and solubilities of structurally similar compounds. These ideas also helped us understand the importance of the solvent type—protic or aprotic—in governing the solubility of ionic compounds, and they further helped us understand how soaps and detergents work.

Although physical properties provided the context of intermolecular interactions in this chapter, the importance of intermolecular interactions extends much further. We will see in Chapter 7, for example, that these intermolecular interactions provide a framework with which we can describe electron flow in chemical reactions, which will prove to be important throughout the rest of this book. In Chapter 9, we will see that the choice of solvent type—protic or aprotic—can have a dramatic effect on the products that are produced in a chemical reaction. And in Chapter 17, we will see that the insolubility of a particular product helps drive a reaction to completion.

THE ORGANIC CHEMISTRY OF BIOMOLECULES

2.12 An Introduction to Lipids

Section 1.15 introduced three of the four major classes of biomolecules: proteins, carbohydrates, and nucleic acids. Here in Section 2.12, we introduce the fourth: *lipids*.

Recall that proteins, carbohydrates, and nucleic acids can be very large molecules, but are constructed from just a few types of small organic molecules. Proteins, for instance, are constructed from α-amino acids, polysaccharides are constructed from monosaccharides (simple sugars), and nucleic acids are constructed from nucleotides. Moreover, these smaller molecules can be unambiguously identified by the specific functional groups they possess.

Lipids, on the other hand, are characterized by their solubility:

> **Lipids** are biomolecules that are relatively *insoluble in water*.

Thus, most lipids are highly nonpolar, consisting primarily of carbon and hydrogen, with very little oxygen or nitrogen content. Consequently, they tend to be soluble in relatively nonpolar organic solvents, such as ether.

Because lipids are characterized in such a broad way, they can have a variety of structures. Nevertheless, lipids can be divided into subclasses based on the structural features they have in common. We examine four of these subclasses in this unit: namely, *fats*, *steroids*, *phospholipids*, and *waxes*. In particular, we describe some of the biological functions that lipids have, along with the structural features shared by lipids within a specific subclass.

2.12a Fats, Oils, and Fatty Acids

Fats can be of animal or plant origin. Animal fats, such as lard, are generally solids at room temperature. Fats from plants, however, are generally liquids at room temperature, and are thus more properly called **oils**. These include corn oil, olive oil, peanut oil, safflower oil, coconut oil, and many more.

Perhaps the most common biological function of fats and oils is to store energy, but they have a variety of other functions as well. For example, fats are needed for the intake of so-called fat-soluble vitamins, such as vitamins A, D, E, and K. Fats can be used to insulate the body against heat loss and to insulate internal organs against physical impact. Fats can help protect organisms from foreign substances—either chemical or biological—by temporarily locking them away in new fat tissue. And, fats contain *fatty acids*, which help regulate blood pressure and blood lipid levels, and play major roles in the inflammatory response to injury.

Although there are several different kinds of fats and oils (having different plant or animal sources), they all share the same general structure:

Notice that a fat or oil contains three adjacent ester groups, each of which can be produced from a *fatty acid* and an alcohol group from *glycerol*. (The chemical reactions involved are discussed in Chapter 21.) Thus, a fat or oil is often described as a **triacyl glycerol** or a **triglyceride**.

2.18 YOUR TURN

Circle and label the other two ester functional groups in the fat molecule in Equation 2-4.

Because all fats and oils have the same general structure, the identities of the fatty acids are what distinguish one fat or oil from another. Some of these differences are discussed in Section 4.16.

Table 2-8 provides some examples of fatty acids. All natural fatty acids have an even number of carbons because, in nature, fatty acids are synthesized from *acetyl coenzyme A*, a two-carbon-atom source. Most fatty acids have between 14 and 18 carbon atoms, although they can have as few as 4 and as many as 22.

2.12b Phospholipids and Cell Membranes

Like a fat or an oil, a **phospholipid** consists of a glycerol backbone attached to fatty acids via ester linkages, as shown in Figure 2-30a.

Whereas a fat or an oil has three fatty acids, however, making it a *triglyceride*, a phospholipid has only two fatty acids, making it a **diglyceride**. In place of a third fatty acid, the glycerol backbone is bonded to a phosphate (PO_4) group, the same type of group that appears in nucleotides in DNA and RNA (Section 1.15). Typically, as indicated, the phosphate group is bonded to a second molecular fragment, such as choline, shown in the middle structure in Figure 2-30b.

Because of their structural similarities to fats and oils, phospholipids can efficiently store energy, too. More importantly, however, phospholipids are integral in the formation of cell membranes. Such a role is possible because the two ends of a phospholipid have vastly different properties—namely, the phosphate portion is ionic (hydrophilic), whereas the alkyl chains of the fatty acids are nonpolar (hydrophobic). These characteristics are highlighted by the simplified representation of a phospholipid in Figure 2-30b on the right, in which the red sphere represents the ionic head group and the two blue wavy lines represent the two nonpolar alkyl chains.

TABLE 2-8 Some Common Fatty Acids

Fatty Acid	Structure	Sources
Lauric acid $C_{12}H_{24}O_2$		Coconut oil, palm kernel oil
Palmitic acid $C_{16}H_{32}O_2$		Coconut oil, palm oil, palm kernel oil, meats, cheeses
Stearic acid $C_{18}H_{36}O_2$		Animal fats, cocoa butter
Oleic acid $C_{18}H_{34}O_2$		Olive oil, pecan oil, peanut oil
Linoleic acid $C_{18}H_{32}O_2$		Safflower oil, grape seed oil
Linolenic acid $C_{18}H_{30}O_2$		Kiwifruit seeds, flax

Under physiological conditions, the hydrophobic tails from different phospholipids associate into a **lipid bilayer** to escape the aqueous environment, as illustrated in Figure 2-31a. This lipid bilayer is the basis of a cell membrane, the basic features of which are shown in Figure 2-31b.

The cell membrane's hydrophobic interior and hydrophilic exterior are critical, because they restrict the free passage of most molecules from one side of the membrane to the other. Hydrophobic molecules, such as cholesterol, reside in the membrane's interior. Hydrophilic molecules, such as carbohydrates, are located on the outside. As we can see, however, some specialized proteins can exist in both regions simultaneously. Although we will not examine the details, these proteins tend to have separate hydrophilic and hydrophobic regions, and some are capable of shuttling molecules across the membrane.

FIGURE 2-30 Diglycerides in nature A phospholipid (a) and a phosphatidylcholine (b) are both diglycerides because two fatty acids (blue) are linked to the glycerol backbone (red).

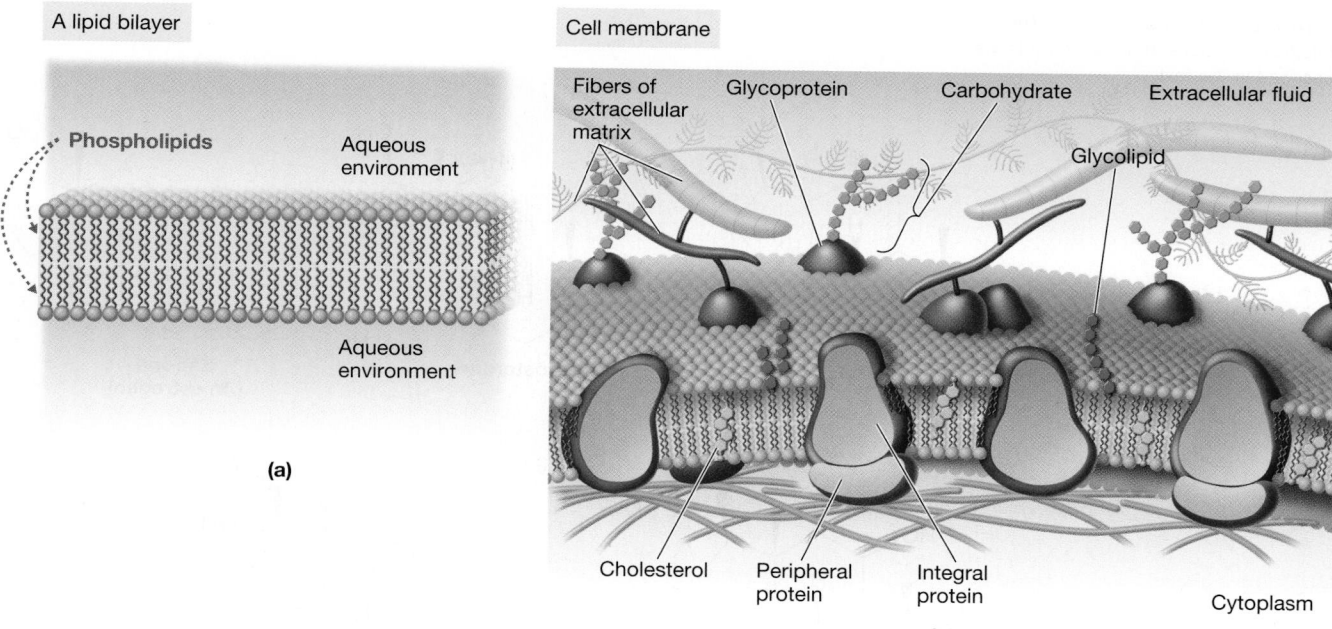

A lipid bilayer

Phospholipids

Aqueous environment

Aqueous environment

(a)

Cell membrane

Fibers of extracellular matrix

Glycoprotein

Carbohydrate

Extracellular fluid

Glycolipid

Cholesterol

Peripheral protein

Integral protein

Cytoplasm

(b)

FIGURE 2-31 The lipid bilayer and cell membrane (a) In an aqueous environment, phospholipids organize into a lipid bilayer. The ionic head groups remain on the exterior of the membrane and are thus stabilized by solvation. The hydrophobic tails aggregate together to escape the aqueous environment. (b) The lipid bilayer is the basis of cell membranes.

2.12c Steroids, Terpenes, and Terpenoids

Steroids (Fig. 2-32) have a wide variety of biological functions. *Cholesterol*, one of the most well-known steroids, is responsible for maintaining the permeability of cell membranes in mammals. *Testosterone* and *estrone* (an estrogen), on the other hand, are male and female sex hormones, respectively. *Aldosterone* helps regulate blood pressure, and *cortisone* suppresses the immune system and is used to treat inflammatory conditions.

All steroids have the same basic ring structure (Fig. 2-33), in which three six-membered rings and one five-membered ring are fused together. These are designated simply as the A, B, C, and D rings.

This ring structure is a common feature among steroids because every steroid is produced from chemical modification to *lanosterol*, as shown in Equation 2-5. Lanosterol, in turn, is produced from *squalene*.

The entire family of steroids is derived from squalene.

Squalene

Multiple steps

Other steroids (2-5)

Lanosterol

HO

FIGURE 2-32 Some steroids The biological functions of these steroids are described in the text.

Cholesterol

Testosterone

**Estrone
(An estrogen)**

Aldosterone

Cortisone

FIGURE 2-33 The steroid ring system All steroids are made of three six-membered rings and one five-membered ring, designated as the A, B, C, and D rings, fused in this arrangement.

Squalene itself belongs to a family of compounds called *terpenes*. A **terpene** is a naturally produced hydrocarbon whose carbon backbone can be divided into separate five-carbon units called **isoprene units**:

An isoprene unit   **Isoprene**

Isoprene units are highlighted in red.

Squalene

They are called isoprene units because they resemble the molecule whose common name is isoprene. Terpenes such as squalene, however, are not actually synthesized from isoprene.

Terpenes (Fig. 2-34) are abundant in nature and are frequently responsible for the aromas of natural products. α-Pinene, for example, is a constituent of pine resin; limonene, which can be isolated from the rinds of lemons, has a strong citrus odor; zingiberene is a component of ginger oil; and cembrene A is isolated from certain corals.

A monoterpene
α-Pinene

A monoterpene
Limonene

A sesquiterpene
Zingiberene

A diterpene
Cembrene A

FIGURE 2-34 Some terpenes The natural sources of these terpenes are described in the text. The separate isoprene units are highlighted in red.

Terpenes are classified according to the number of pairs of isoprene units they contain.

- A **monoterpene** contains one pair of isoprene units (i.e., two isoprene units), for a total of 10 carbons.
- A **sesquiterpene** contains one-and-a-half pairs of isoprene units (i.e., three isoprene units), for a total of 15 carbons. (*Note:* The prefix *sesqui* means $1\frac{1}{2}$.)
- A **diterpene** contains two pairs of isoprene units (i.e., four isoprene units), for a total of 20 carbons.
- A **triterpene** contains three pairs of isoprene units (i.e., six isoprene units), for a total of 30 carbons.

With these definitions, we can see that α-pinene is a monoterpene because it contains two isoprene units; zingiberene is a sesquiterpene because it contains three isoprene units; cembrene A is a diterpene because it contains four isoprene units; and squalene is a triterpene because it contains six isoprene units.

Many natural products are produced from chemical modifications to terpenes, in which the carbon backbone is altered and/or atoms other than just carbon or hydrogen are introduced. All of these compounds are called **terpenoids**. Thus, squalene is a terpene, but the steroids that are derived from it are terpenoids. Other terpenoids are shown in Figure 2-35, including menthol (obtained from mint oil), geraniol (a main constituent of rose oil), and retinal (involved in the chemistry of vision).

2.12d Waxes

Waxes are secretions from plants and animals that are solid at room temperature, but melt at relatively low temperatures. They are typically soft and malleable and are very hydrophobic. Bees use beeswax (Fig. 2-36a) to store honey and protect their eggs. More commonly in nature, however, wax is used as a protective coating. Carnuba wax, for example, which is found on the leaves of the carnuba palm tree native to Brazil

Menthol

Geraniol

Retinal

FIGURE 2-35 Some terpenoids Terpenoids are derived from chemical modifications of terpenes. The sources of these terpenoids are described in the text.

FIGURE 2-36 Sources of various kinds of natural wax (a) Beeswax is used as cells for storing honey and protecting eggs. (b) The carnuba palm has wax deposits on its leaves to protect from excessive water loss. (c) Wool has a coating of lanolin, which helps shed water. (d) Human earwax protects the ear canal from foreign substances.

(a)   (b)   (c)   (d)

(Fig. 2-36b), helps prevent excessive loss of water. Lanolin, found on wool, helps sheep shed water from their coats (Fig. 2-36c). Finally, earwax in the human ear canal (Fig. 2-36d) helps protect against bacteria and other foreign substances.

Waxes are generally mixtures of compounds, consisting principally of long-chain esters and alkanes. Tetracosyl hexadecanoate, for example, is a main constituent of beeswax:

Principal component of beeswax

Tetracosyl hexadecanoate

As we can see, any polarity that might originate from the ester functional group is overwhelmed by the large, highly nonpolar alkyl groups, making the molecule hydrophobic overall.

Chapter Summary and Key Terms

- **Valence shell electron pair repulsion (VSEPR)** theory can be used to predict the electron and molecular geometries about individual atoms. **(Section 2.1)**
  - Electron groups (i.e., lone pairs or bonds) repel each other to yield an atom's **electron geometry; molecular geometry**, which is derived from the electron geometry, is the molecule's geometry based solely on the orientation of its chemical bonds.
  - Two electron groups yield a **linear** electron geometry, three electron groups yield a **trigonal planar** electron geometry, and four electron groups yield a **tetrahedral** electron geometry.
- Bond angles that deviate from the ideal VSEPR geometry possess **angle strain**. **(Section 2.1)**
- **Dash–wedge notation** is used to depict three-dimensional molecular geometry; a wedge (◄) represents a bond pointing toward you, whereas a dash (⑊⑊⑊) represents a bond pointing away from you. **(Section 2.2)**
- A **dipole moment, μ,** is generated when opposite charges, $+q$ and $-q$, are separated by a distance, r. The magnitude of a dipole moment is the product $\mu = qr$ and is reported in units of **debye**, where $1 \text{ D} = 3.33564 \times 10^{-30} \text{ C} \cdot \text{m}$. **(Section 2.4)**
- Bond dipole moments are treated as vectors and added together to yield a molecule's **net molecular dipole moment**, or **permanent dipole moment**. If bond dipoles do not perfectly cancel, then the molecule is **polar**. If the bond

dipoles perfectly cancel, then the molecule is **nonpolar**. **(Section 2.4)**

- Differences in the physical properties of compounds similar in their makeup and mass are due to the presence of different functional groups. **(Section 2.5)**
- Boiling points and melting points increase as the **intermolecular interactions** increase in the liquid and solid phases, respectively. **(Section 2.6)**
- Two compounds are generally soluble in each other if the intermolecular forces in the solution are roughly the same strength as those that exist in the pure separated compounds. **(Section 2.5)**
- Intermolecular forces differ in strength. The greater the concentration of charges on the molecules interacting, the stronger their attraction. The strength of intermolecular forces generally decreases in the following order:
  - **Ion–ion interactions:** Attraction between two oppositely charged ions. **(Section 2.6a)**
  - **Ion–dipole interactions:** Attraction between a positive or negative ion and a net molecular dipole. **(Section 2.7a)**
  - **Hydrogen bonding:** Takes the form X–H⋯Y, where X and Y are each either nitrogen, oxygen, or fluorine (i.e., highly electronegative elements that have available lone pairs). The extent of hydrogen bonding increases as the number of potential **hydrogen-bond donors** and **hydrogen-bond acceptors** increases. It also increases as

the magnitudes of the bond dipoles involved increase. **(Section 2.6c)**

- **Dipole–dipole interactions:** Attractions between two net molecular dipoles. The strength of the attraction increases as the magnitudes of the molecular dipoles involved increase. **(Section 2.6b)**
- **Induced dipole–induced dipole interactions (London dispersion forces):** Attraction between two temporary dipoles. These interactions increase as the **polarizabilities** of the species involved increase and as the **contact surface area** increases. Polarizability generally increases with the number of total electrons. **(Section 2.6d)**
- With enough electrons, induced dipole–induced dipole interactions can become the dominant intermolecular force. **(Section 2.6d)**
- Even though ion–dipole interactions are weaker than ion–ion interactions, ionic compounds may dissolve in polar solvents as a result of **solvation** of the ions. **(Section 2.7a)**

- As the size of hydrocarbon groups increases, compounds become less polar and less soluble in polar solvents. Mono-alcohols with six or more carbon atoms are insoluble in water. **(Section 2.7b)**
- A **protic solvent** contains a H-bond donor, whereas an **aprotic solvent** does not. Ionic compounds tend to be much more soluble in protic solvents due to the strong solvation of anions. **(Section 2.9)**
- **Soaps** and **detergents** are long-chain molecules that are ionic on one end and highly nonpolar on the other. They dissolve in water as **micelles**, in which the **ionic head groups** are arranged on the outside and the **hydrocarbon tails** are grouped on the inside. **(Section 2.10)**
- When molecules of soaps and detergents encounter a particle of dirt, grease, or oil, the nonpolar tails dissolve in the particle, while the ionic head groups remain exposed to water. The **emulsified** particle is water soluble and can be washed away by water. **(Section 2.10)**

Problems

2.30 Brassinolide, a naturally occurring steroid derivative found in a wide variety of plants, is thought to promote plant growth. Identify the electron geometry for each atom indicated. Where applicable, describe the atom's molecular geometry and estimate the bond angle.

Brassinolide

2.31 Falcarinol, a naturally occurring pesticide found in carrots, is being studied as an anticancer agent. Identify the electron geometry for each atom indicated. Where applicable, describe the atom's molecular geometry and estimate the bond angle.

Falcarinol

2.32 Identify the electron geometry about each charged atom. Where appropriate, indicate the molecular geometry and approximate bond angle as well.

2.33 Which molecule, **A** or **B**, has greater angle strain? Explain.

A   **B**

2.34 Add dash–wedge notation to each line structure provided to accurately depict the ball-and-stick model appearing above it.

(a)   **(b)**   **(c)**   **(d)**

2.35 The structure of D-glucose using dash–wedge notation is shown below. Draw its structure using dash–wedge notation after each of the indicated rotations has taken place.

(a)   180° rotation   **(b)**   180° rotation

D-**Glucose**   D-**Glucose**

2.36 Which ions or molecules would be attracted to CH_3^+? Indicate the type of intermolecular forces involved in the attraction.
   (a) H_2O **(b)** Na^+ **(c)** Cl^- **(d)** F^- **(e)** $H_2C{=}O$

2.37 Which pair of ions will attract each other most strongly? Explain.
   $Mg^{2+} \; O^{2-}$ $Al^{3+} \; O^{2-}$ $Na^+ \; O^{2-}$
       A   **B**   **C**

2.38 Which pair of ions will attract each other most strongly? Explain.
   $Mg^{2+} \; O^{2-}$ $Ca^{2+} \; O^{2-}$ $Sr^{2+} \; O^{2-}$ $Ba^{2+} \; O^{2-}$
     A   **B**   **C**   **D**

2.39 The dipole moment of HBr is 0.82 D and the internuclear distance is 141 pm. Estimate the amount of charge centered on each atom.

2.40 Rank the following molecules from least polar to most polar:
   BF_3 BF_2H BFH_2
    A   **B**   **C**

2.41 The dipole moments of a ketone and an ester are shown below. Why is the magnitude of the ester's dipole moment smaller?

3.0 D   **1.8 D**

2.42 Use the following electrostatic potential maps of a variety of uncharged molecules to determine which ones are polar (i.e., have nonzero molecular dipole moments) and which ones are nonpolar. For each one that is polar, indicate the direction of the net molecular dipole moment.

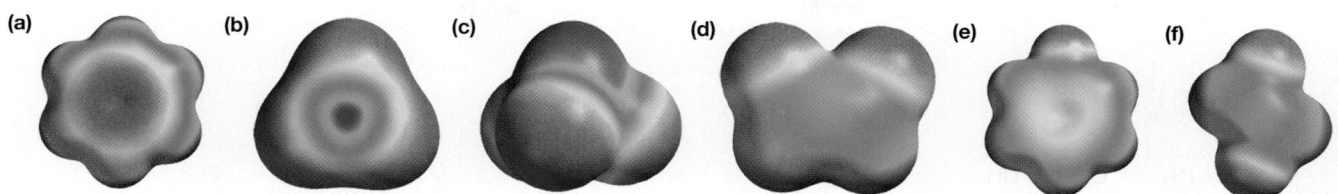

(a)　　　(b)　　　(c)　　　(d)　　　(e)　　　(f)

2.43 Determine whether or not each of the following molecules is polar. For those that are, indicate the direction of the net dipole moment.

(a)　　　(b)　　　(c) Cl——≡——Cl　　　(d) Cl——≡——CH₃　　　(e) Cl——≡——Br

(f)　　　(g)　　　(h)　　　(i)　　　(j)　　　(k)　　　(l)　　　(m)　　　(n)

2.44 In 1874, Dutch chemist Jacobus van't Hoff (1852–1911) and French chemist Joseph Le Bel (1847–1930) independently deduced that a carbon atom bonded to four atoms assumes a tetrahedral geometry. Prior to that time, it was believed that tetravalent carbons assumed a square planar geometry. One piece of evidence that can be used to support a tetrahedral geometry is the fact that molecules with the general formula CX_2Y_2 (where X and Y are either hydrogen or a halogen atom) are always polar. Explain how this supports a tetrahedral geometry and rules out a square planar geometry.

Tetrahedral geometry　　　**Square planar geometry**

2.45 In the following molecule, identify all H-bond donors and all H-bond acceptors.

NH_2

H_3C　　H　　C　　O
　　　C　　H　　C

CH_3　　OH

2.46 Rank the following molecules in order from lowest to highest boiling point.

A B C D

E F G H

2.47 Rank the following molecules in order from lowest to highest boiling point. Briefly rationalize your choice.

$CH_3CH_2OCH_2CH_3$ $CH_3CH_2CH_2CH_2CH_3$ $CH_3CH(OH)CH_2CH_3$ $FCH_2CH_2OCH_2CH_3$
  A   **B**   **C**   **D**

2.48 Rank the following molecules in order from lowest to highest boiling point. Briefly rationalize your choice.

$CH_3CH_2CH(OH)CH_2OH$ $CH_3CH_2CH_2CH_2OH$ $CH_3CH(OH)CH(OH)CH_2OH$ $FCH_2CH_2CH_2CH_3$
  A   **B**   **C**   **D**

2.49 The substance with the lowest known boiling point (4 K) is helium, an atomic element that has two electrons. Hydrogen is a diatomic molecule and also has two electrons, but its boiling point is significantly higher, at 20.65 K. (a) What is the dominant intermolecular force between a pair of helium atoms and a pair of H_2 molecules? Why do you think H_2 has a higher boiling point?

2.50 Propan-1-ol ($CH_3CH_2CH_2OH$) and propan-2-ol [$(CH_3)_2CHOH$] are both alcohols that have the same formula (C_3H_8O), but significantly different boiling points: 97.5 °C for propan-1-ol and 82 °C for propan-2-ol. Explain.

2.51 Which organic solvent of the four shown below would be most soluble in water? Explain your choice.

$CH_3CH_2CH_2CH_2CH_2CH_3$ $CH_3OCH_2CH_2CH_2CH_3$ $CH_3OCH_2CH_2OCH_3$ $CH_3CH_2CH_2CH_3$
  A   **B**   **C**   **D**

2.52 Recall from Section 2.10 that hard water ions like Ca^{2+} and Mg^{2+} bond more strongly to the carboxylate groups in soap than Na^+ and K^+ ions do. Why do you think this is so?

2.53 Recall from Section 2.10 that metal ions in hard water do not bind to the sulfate portion of detergents as strongly as they do to the carboxylate portion of soaps. Suggest a reason, based on intermolecular forces, for why this is so.

Alkyl sulfate = $R—OSO_3^-$ Carboxylate = $R—CO_2^-$

2.54 Would you expect 1,2-difluorobenzene or 1,3-difluorobenzene to have a higher boiling point? Explain.

1,2-Difluorobenzene

1,3-Difluorobenzene

2.55 The boiling points of several cyclic alkanes and ethers are listed in the table on the next page. For small rings, there is a significant difference in boiling points between the cyclic alkane and the cyclic ether (e.g., −76 °C vs. 10.7 °C), but for large rings, the boiling points are nearly the same (e.g., 80.7 °C vs. 88 °C). Explain.

Cyclic Alkane	Boiling Point (°C)	Cyclic Ether	Boiling Point (°C)
△	–76	△O	10.7
▢	–13	▢O	50
⬠	49	⬠O	66
⬡	80.7	⬡O	88

2.56 Naturally-occurring unsaturated fatty acids, such as **A** below, contain only cis double bonds. A corresponding saturated fatty acid (**B**) is also shown. **A** is a liquid at room temperature, whereas **B** is a solid. What does this suggest about the strength of the intermolecular forces in each substance? Can you explain? (*Hint:* Build a model and compare the surface area of contact for a pair of each molecule.)

2.57 In this chapter, we did not discuss the interaction that is primarily responsible for the attraction that exists between an anion like CH_3O^- and a nonpolar molecule like Br_2.
(a) What name would be ascribed to the strongest interaction that exists between those two species?
(b) For which pair of species would you expect that type of attractive interaction to be stronger: between CH_3O^- and Br_2 or between CH_3O^- and I_2? Why?

2.58 In which of the following alcohol solvents do you think NaCl would be the most soluble?
CH_3OH CH_3CH_2OH $CH_3CH_2CH_2OH$ $CH_3CH_2CH_2CH_2OH$
  A   **B**   **C**   **D**

2.59 Trimethylhexadecylammonium chloride, $CH_3(CH_2)_{15}N(CH_3)_3^+ Cl^-$, is one of a class of *cationic detergents*, commonly used in shampoos and as "clothes rinses."
(a) Identify the hydrophilic head group and the hydrophobic tail.
(b) Draw a depiction of a micelle that would form if this compound were dissolved in water.
(c) What are the intermolecular forces that are primarily responsible for the micelle's solubility in water?

2.60 Detergents need not be ionic. An example of a *nonionic detergent* is pentaerythrityl palmitate (shown below), which is used in dishwashing liquids.
(a) Identify the hydrophilic and hydrophobic portions of the molecule.
(b) Draw a depiction of a micelle that would form if this compound were dissolved in water.
(c) What intermolecular interactions are primarily responsible for the micelle's solubility in water?
(d) What advantages do nonionic detergents have over ionic detergents in hard water?

2.61 The following amines have the same molecular formula ($C_5H_{13}N$), but their boiling points are significantly different. Explain why.

2-Methylbutan-1-amine
Boiling point = 97 °C

N-Methylbutan-2-amine
Boiling point = 84 °C

N-Ethyl-N-methylethan-1-amine
Boiling point = 65 °C

2.62 Benzene and hexafluorobenzene have nearly identical boiling points, despite the fact that hexafluorobenzene has significantly more total electrons than benzene. Why do you think this is so?

Benzene
Boiling point = 80 °C

Hexafluorobenzene
Boiling point = 81 °C

2.63 An uncharged oxygen atom has two lone pairs of electrons, so a single oxygen atom can participate in two different hydrogen bonds (i.e., with two separate H-bond donors), as shown below. Estimate the angle between these two hydrogen bonds.

2.64 Would you expect the following compound to act as a soap? Why or why not? We explain in Chapter 3 that a chain composed of multiple, adjacent, single bonds is flexible. How does this new information affect your answer?

2.65 All of the following alcohols have the same molecular formula ($C_5H_{12}O$), but they have significantly different boiling points. Explain why.

Pentan-1-ol
Boiling point = 136–138 °C

Pentan-3-ol
Boiling point = 114–115 °C

2-Methylbutan-2-ol
Boiling point = 102 °C

2.66 Explain why 1,2-dihydroxybenzene and 1,3-dihydroxybenzene have such different boiling points.

1,2-Dihydroxybenzene
Boiling point = 245 °C

1,3-Dihydroxybenzene
Boiling point = 281 °C

2.67 Identify each compound below as either a *protic* solvent or an *aprotic* solvent.

Methanamide
(Formamide)

Ethanenitrile
(Acetonitrile)

Hexamethylphosphorotriamide
(HMPA)

Methanol

Ethane-1,2-diol
(Ethylene glycol)

2.68 The compound below is a terpenoid.
(a) Circle the individual isoprene units.
(b) Determine whether it is a monoterpene, sesquiterpene, diterpene, etc.

2.69 Draw a fat or oil molecule that is constructed from (a) three molecules of stearic acid and (b) two molecules of oleic acid and one molecule of linolenic acid.

2.70 Identify each of the following molecules as either a steroid, a fatty acid, or a wax.

(a)

Isolated from the seeds
of the jojoba plant

(b)

Medrogestone, a
synthetic drug

(c)

Erucic acid, isolated
from mustard seed

If your instructor assigns problems in smartwork, log in at **smartwork.wwnorton.com**.

Orbital Interactions 1

Hybridization and Two-Center Molecular Orbitals

In Chapters 1 and 2, we discussed Lewis structures and VSEPR theory, both of which are valuable tools for studying molecules. Lewis structures provide a model for the distribution of electrons and the formation of bonds within molecules; VSEPR theory allows us to understand and predict certain aspects of three-dimensional molecular structure.

Despite their many uses, both of these models have shortcomings. We saw in Chapter 1, for example, that Lewis structures cannot account for *electron delocalization* over multiple atoms. Hence the need for resonance theory (Section 1.10), which treats the true structure of a molecule as a weighted average of all resonance contributors.

VSEPR theory, moreover, is incapable of predicting a molecule's *extended* geometry—the relative positions of atoms separated by more than one atom. For example, we know from experiment that a molecule of ethene (Fig. 3-1), $H_2C\!\!=\!\!CH_2$, is entirely planar—all six atoms lie in the same plane—and we also know that there is no rotation of one CH_2 group relative to the other. Although VSEPR theory does successfully predict that the geometry about each individual carbon atom is trigonal planar, it cannot predict the absence of rotation about the $C\!\!=\!\!C$ bond.

In this chapter, we present a more powerful model of bonding that combines the concept of *hybridization* with *molecular orbital theory*. This model describes how *atomic orbitals*—the orbitals that are occupied by electrons in isolated atoms—combine to form *molecular orbitals*—the orbitals occupied by electrons in molecules. Together, hybridization and molecular orbit-

Ethene is a planar molecule; rotations do not occur about the $C\!\!=\!\!C$ bond.

Ethene

FIGURE 3-1 Ethene These characteristics of ethene can be understood by the orbital interaction discussed here in Chapter 3.

Very small particles such as electrons behave as waves, much like the ones on a water surface. As we will see in here Chapter 3, this wavelike behavior is what gives rise to orbitals and bonds.

al theory allow us to understand and make predictions about molecular geometry that span several atoms instead of just one. These concepts also provide substantially more information about the *energetics* involved in bonding (which we show in Chapter 7 is essential for accurately describing the dynamics of chemical reactions), as well as various aspects of spectroscopy, which is how molecules interact with light (an important subject with many applications that we discuss in Chapters 15 and 16).

To understand molecular orbital theory, we must first discuss some important results from *quantum mechanics*. Not only does quantum mechanics dictate the characteristics of atomic orbitals, it also provides the rules for constructing hybridized atomic orbitals and molecular orbitals.

3.1 Atomic Orbitals and the Wave Nature of Electrons

In Chapter 1, we briefly reviewed several aspects of quantum mechanics that you may have encountered in general chemistry, including the concept of an orbital and the meaning of quantum numbers. The central idea of **quantum mechanics** is that very small particles like electrons have traits that are characteristic of both particles and waves:

> Electrons exhibit wave–particle duality.

An electron has characteristics of a particle because it has mass and can undergo collisions with other forms of matter. It also has characteristics of a wave because it can exhibit phenomena such as *constructive interference* and *destructive interference*. We discuss these phenomena later in this chapter. Here in Section 3.1, we explore other outcomes of an electron's wavelike behavior, including the probability of finding an electron in space, electron density, and phase.

3.1a The 1s Orbital: Probability, Electron Density, and Phase

One crucial result of the electron's wave–particle duality is that we cannot know the exact position of an electron in an atom or molecule. This conclusion is the essence of the **Heisenberg uncertainty principle**, which states that the uncertainty in a measurement of an electron's position is inversely proportional to the uncertainty in a measurement of its momentum. In other words, the more precisely we know the electron's position, the less precisely we know where it is going, and vice versa. Thus,

> We cannot know the exact position of an electron in an **orbital**, but we can know the *probability* of finding the electron at a given location in space.

This probability of finding an electron is often referred to as **electron density**.

Upon completing Chapter 3 you should be able to:

- Explain how an electron's wave function relates to its phase, its probability of being found at a particular location in space, and the shape of its orbital.

- Draw *s* and *p* atomic orbitals, along with their relative phase and orientation.

- Explain the importance of hybridization and describe how hybridized atomic orbitals can be generated from pure *s* and *p* atomic orbitals via constructive interference and destructive interference.

- Specify the hybridization of an atom in a molecule, given only the molecule's Lewis structure.

- Explain the importance of molecular orbitals and describe how molecular orbitals can be generated from atomic orbitals via constructive interference and destructive interference.

- Identify a molecular orbital as having either σ or π symmetry and explain how σ and π symmetry relate to rotations about bonds.

- Determine the total number of σ and π bonds in a molecule, given only its Lewis structure.

- Construct the atomic orbital contribution picture of a molecule and its molecular orbital energy diagram.

- Determine a molecule's highest occupied molecular orbital and its lowest unoccupied molecular orbital.

- Specify when cis and trans configurations exist for a given double bond and identify whether a configuration is cis or trans.

- Explain how and why an atom's hybridization affects bond lengths and bond strengths, as well as how hybridization relates to the atom's effective electronegativity.

High electron density · · Low electron density

(a)

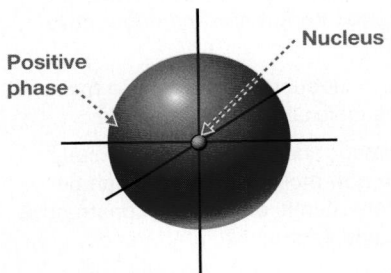

Positive phase · · Nucleus

An electron belonging to a 1s orbital has a 90% probability of being found inside this sphere.

(b)

FIGURE 3-2 **Representations of the 1s orbital** (a) Three-dimensional electron density picture of a 1s orbital. The greater the density of the dots, the higher the likelihood of finding a 1s electron. (b) The surface encompassing 90% of the electron's probability. A 1s orbital is shaped like a sphere.

The electron density of a particular electron is governed by the orbital to which the electron belongs. For an electron in a 1s orbital, electron density is highest at the nucleus and decreases continuously toward zero as the distance from the nucleus increases. This is shown in Figure 3-2a. In that representation, the more closely spaced the dots, the greater the probability of finding the electron.

Although the probability of finding a 1s electron is never zero, even at very large distances, it is convenient to draw a surface that encompasses 90% of the total probability, as shown in Figure 3-2b. This surface indicates the *shape* of the 1s orbital, which is a sphere.

The representation in Figure 3-2b depicts another important aspect of an electron, its relative **phase**. An electron's phase is not a measurable quantity, but it does play an important role in how orbitals interact with one another. It is therefore important to recognize whether two phases are the same or opposite. In this book, we will be consistent with the following convention:

> ▪ An electron will have the same phase in two different regions of space if the orbital to which the electron belongs is shaded in both regions, or is unshaded in both regions.
> ▪ The electron will have opposite phases in two regions of space if the orbital is shaded in one region, but unshaded in the other.

Notice that the entire orbital in Figure 3-2b is shaded, indicating that the electron's phase is the same everywhere.

The important points about a 1s orbital are as follows:

> ▪ A 1s orbital is spherical (Fig. 3-2).
> ▪ A 1s orbital has the same phase everywhere in space.

YOUR TURN 3.1

In Figure 3-2a, draw a circle around the dots so that the nucleus is at the center and the circle encompasses 90% of the dots.

Answers to Your Turns are in the back of the book.

3.1b The 2p Orbital

The three-dimensional electron density picture of a 2p orbital is depicted in Figure 3-3a. Figure 3-3b shows the surface that encompasses 90% of the total probability of an electron's location.

FIGURE 3-3 **Representations of a 2p orbital** (a) Electron density distribution of a 2p orbital. (b) A solid surface depicting a 2p orbital, indicating two lobes of opposite phase, separated by a nodal plane (left). A simplification of the 2p orbital (right).

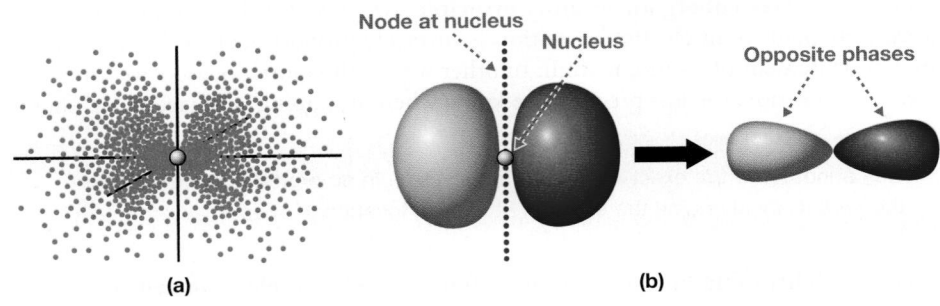

Node at nucleus · · Nucleus · · Opposite phases

(a) **(b)**

Quantum Teleportation

The teleportation of humans from one place to another is a popular plot device in science fiction movies. The basic idea is that a teleporter scans a person to extract all of his or her information, and that information is transmitted to a receiving location. The original person is disintegrated and, using the transmitted information, a perfect replica is constructed at the receiver. Is this possible?

Historically, teleportation has not been taken seriously, because extracting all of an object's information using a measuring or scanning device was thought to be a violation of the Heisenberg uncertainty principle (Section 3.1a). According to the uncertainty principle, the more accurately an object's information is measured, the more the object's state is disrupted. Thus, the object would be altered before all of its information could be extracted.

Although teleporting macroscopic objects is still science fiction, scientists have found a way to teleport information between "entangled" sub-atomic particles—particles that first interact physically and are then separated. If a third particle—a target particle—comes in contact with one of the entangled particles, that entangled particle can act as a scanner. That entangled particle and the target particle are disrupted, and the second entangled particle assumes the original state of the target. Quite impressively, this quantum teleportation has been achieved over distances as great at 89 miles.

Quantum teleportation is more than just a novelty; it has potential applications in quantum computing, a field that is in its infancy, but has the potential to revolutionize computing speed. In quantum computing, quantum particles are used to carry out operations. The resulting information, however, is too sensitive to be extracted by traditional means. Quantum teleportation may provide a mechanism by which to do so, thus helping to propel quantum computing out of its infancy.

From these plots, we can infer the following important points about a $2p$ orbital:

- A $2p$ orbital is composed of two *lobes* on opposite sides of the nucleus. The true shape of each lobe is a distorted sphere (Fig. 3-3b, left), but the pair can be depicted as a "dumbbell" (Fig. 3-3b, right).
- Electron density is zero at the nucleus.
- The two lobes are opposite in phase.

The location where electron density is zero is called a **node**. Thus,

An electron cannot be found at a node.

As shown in Figure 3-3b, the $2p$ orbital has a single **nodal plane** (indicated by the dotted line) that passes through the nucleus.

In Figure 3-3a, draw two lobes of equal size that together encompass about 90% of the dots.

3.2 YOUR TURN

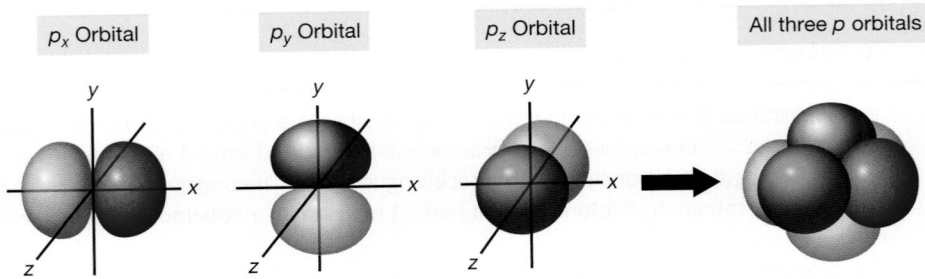

FIGURE 3-4 Orientations of the three 2p orbitals The first three images depict the p_x, p_y, and p_z orbitals individually, which are aligned along the x, y, and z axes, respectively. In the image at the far right, all three p orbitals are shown together.

Unlike the 1s orbital, the 2p orbital can be assigned an *orientation*. In a given atom there are three mutually perpendicular 2p orbitals, called $2p_x$, $2p_y$, and $2p_z$ (Fig. 3-4), aligned along the x, y, and z axes, respectively. Aside from their orientations in space, all three orbitals are exactly the same.

3.1c Other Orbitals

As we saw in Chapter 1, the 1s and the 2p orbitals are the lowest energy s and p orbitals, respectively. Higher-energy s and p orbitals (i.e., those belonging to higher shells) are larger, but the shape of an orbital is independent of the shell to which it belongs. That is, the 2s and 3s orbitals are both spherical, although the size of the orbitals increases in the order 1s < 2s < 3s. Similarly, the 3p orbital consists of two lobes, each larger than the lobes of the 2p orbital. As a result, all s orbitals behave the same qualitatively and all p orbitals behave the same qualitatively.

Recall from Chapter 1 that other atomic orbitals (AOs), such as d and f orbitals, are possible. We will not discuss those orbitals here, though, because the vast majority of organic chemistry involves only the interactions of s and p orbitals.

3.2 Interaction between Orbitals: Constructive and Destructive Interference

If the only orbitals available to electrons were the "pure" AOs we just examined (i.e., s, p, etc.), organic chemistry would be far less interesting than it is. Because AOs confine electrons to individual atoms instead of allowing them to be shared between atoms, covalent bonds—and therefore molecules—would not exist. But even if covalent bonding were possible, it would be difficult to imagine the diverse molecular geometries that we observe. How, for example, do we envision the carbon atom assuming tetrahedral geometries (with 109.5° bond angles) and trigonal planar geometries (with 120° bond angles), given that an s orbital has no specific orientation and the p_x, p_y, and p_z orbitals are all oriented 90° apart?

From an orbital perspective, both covalent bonding and the various molecular geometries can be described by *molecular orbitals* and *hybridized atomic orbitals*. Both are the product of orbital interactions involving two or more AOs.

To better understand such orbital interactions, consider the interaction between two waves generated by moving the ends of a rope back and forth rapidly (Fig. 3-5). In this case, notice that both waves are generated on the *same side* of the rope, which is analogous to two orbitals having the *same phase*—in this case, both positive. The waves will propagate toward each other (Fig. 3-5a and 3-5b), and when the centers of the two waves meet (Fig. 3-5c), the result is a new wave that *displaces the rope twice as much as each of the two individual waves*. That is,

Waves that overlap with the same phase undergo **constructive interference**.

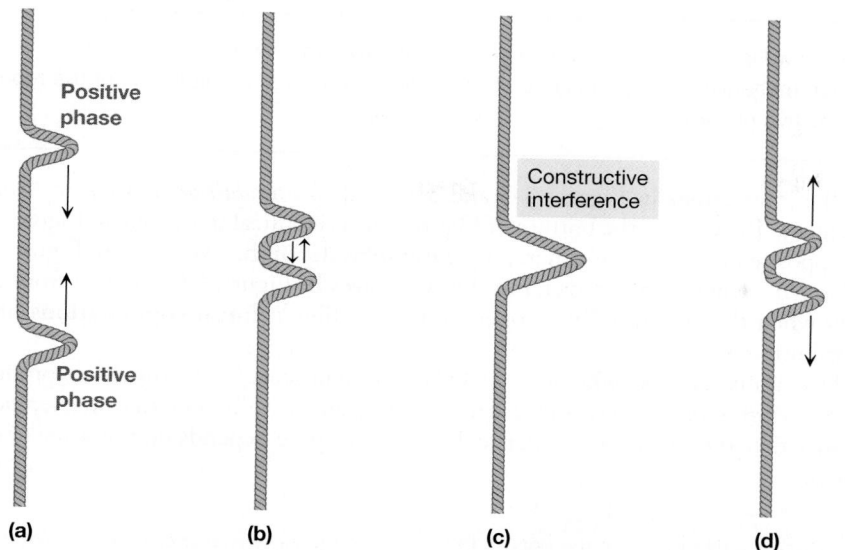

FIGURE 3-5 Constructive interference between two waves along a rope (a and b) Waves with the same phase (i.e., waves on the same side of the rope) propagate toward each other. (c) When the waves meet, they undergo constructive interference, temporarily creating a new wave with twice the amplitude of the original ones. (d) The waves continue to propagate toward opposite ends of the rope.

Constructive interference produces a new wave that has been *built up* by the addition of the amplitudes from the contributing waves.

Next, suppose that we generate waves at each end of the rope that have equal amplitudes but are *opposite in phase* (Fig. 3-6); that is, the sizes of the displacements are the same, but they are on opposite sides of the rope. This time, the two waves *cancel* each other when they meet, yielding zero displacement along the entire rope.

Waves that overlap with opposite phase undergo **destructive interference**.

Destructive interference produces a new wave with a reduced amplitude because the amplitude of one wave subtracts from the amplitude of the other.

Redraw Figure 3-5 with both initial waves generated on the left side of the rope to represent negative phases.

3.3 YOUR TURN

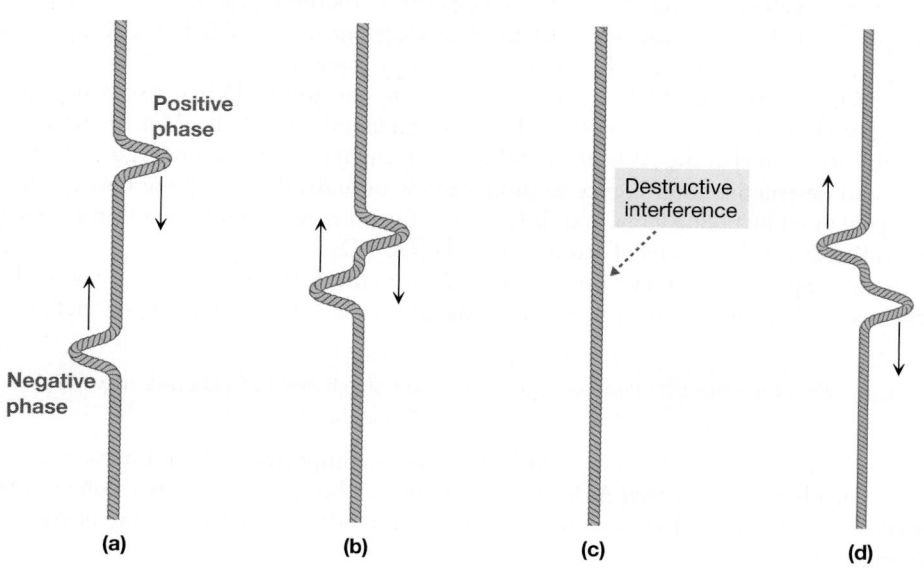

FIGURE 3-6 Destructive interference between waves generated along a rope (a and b) Waves with opposite phase (i.e., waves on opposite sides of the rope) propagate toward each other. (c) When the two waves meet, they undergo *destructive interference*, temporarily canceling each other. (d) The waves continue to propagate toward opposite ends of the rope.

YOUR TURN 3.4

Redraw Figure 3-5, but this time generate the top wave on the left side of the rope (negative phase) and generate the bottom wave on the right side of the rope (positive phase).

The interactions in Figures 3-5 and 3-6 result from *addition* and *subtraction*, respectively. The wave at the bottom of Figure 3-6a is identical to the one in Figure 3-5a but opposite in sign (phase). Therefore, the amplitudes of the two waves in Figure 3-5c *add together* when they interact, but the two waves in Figure 3-6c *subtract* from each other when they interact. These operations are different **linear combinations** of the contributing waves.

In a similar fashion, when two orbitals overlap in space, new orbitals are produced via *constructive interference* and *destructive interference*. Whether the interference is constructive or destructive at a particular point in space depends on the *phase* of each orbital.

- Constructive interference occurs in regions of space where two orbitals have the same phase.
- Destructive interference occurs where the respective orbitals have opposite phases.

As we will see in the sections that follow, these types of interactions can generate orbitals different from those that are interacting—namely, *molecular orbitals* and *hybridized atomic orbitals*.

3.3 An Introduction to Molecular Orbital Theory and σ Bonds: An Example with H$_2$

In Chapter 1, we saw that a covalent bond results from the sharing of electrons between atoms. AOs alone (i.e., *s*, *p*, etc.), however, cannot account for such electron sharing. To do so, we use *molecular orbital theory*.

The central concept of **molecular orbital theory** is that all electrons in a molecule can be thought of as occupying orbitals called **molecular orbitals** (**MOs**). Like AOs, MOs can each accommodate up to two electrons. Therefore, the two electrons that form a single bond occupy one MO, the four electrons of a double bond occupy two MOs, and the six electrons of a triple bond occupy three MOs.

MOs are constructed from the AOs of different atoms. When two atoms are brought close enough together (i.e., about a bond length apart), the AOs of one atom significantly overlap the AOs of the other atom, enabling them to undergo constructive and destructive interference to produce new orbitals. Because these new orbitals are generated by taking sums and differences of various AOs, MOs are often referred to as **linear combinations of atomic orbitals** (**LCAOs**).

Although several interactions among AOs from the different atoms are imaginable, there are relatively few that must be considered because of the following restriction:

Only AOs of roughly the same energy will interact significantly to generate new MOs.

Such a restriction on energy generally limits the important orbital interactions to those involving *valence shell* AOs of the two nuclei. This is consistent with one of the rules for constructing Lewis structures: Covalent bonds involve the sharing of *valence electrons* only.

This principle applies to the MO picture of the H_2 molecule, in which the two hydrogen atoms are held together by a single covalent bond:

H—H

Hydrogen molecule

In each *isolated* hydrogen atom, the valence shell is the $n = 1$ shell, which contains only a $1s$ orbital. Therefore, only the $1s$ orbitals of the hydrogen atoms are considered when constructing the MOs of H_2.

One way in which these $1s$ orbitals can mix is with the same relative phase, as shown in Figure 3-7a. In this case, both $1s$ orbitals are shaded, analogous to what we saw in Figure 3-5, with waves traveling along the same side of a rope. This will lead to *constructive interference*, resulting in an orbital that has been built up in the region of overlap.

A second way in which the two $1s$ orbitals can mix is with opposite phases, as shown in Figure 3-7b. In this case, the orbital on the left is shaded, and the one on the right is not, analogous to what we saw in Figure 3-6, in which waves are traveling on opposite sides of a rope. This will lead to destructive interference, resulting in an orbital that has been diminished in the region of overlap. More specifically, there is a plane between the two nuclei in which the orbitals completely cancel, producing a *nodal plane*.

Can still other orbitals be produced by mixing two $1s$ orbitals, using combinations of phases that are different from those in Figure 3-7? The answer is no. The molecular orbitals shown in Figure 3-7 are the only unique ones produced. This concept is explored in Solved Problem 3.1 and Problem 3.2.

FIGURE 3-7 MO formation from two 1s orbitals (a) The addition of two pure *s* AOs leads to constructive interference between the two nuclei. The result is an elliptical MO, which has been built up in the internuclear region. This MO is designated a σ MO because the AO overlap occurs along the internuclear axis. (b) The subtraction of one pure *s* AO from another leads to destructive interference between the two nuclei. The result is an MO that has been diminished in the internuclear region, and possesses a nodal plane between the two nuclei. This MO is designated σ*.

SOLVED problem 3.1 At the right is a representation of two 1s orbitals overlapping, each with a phase that is opposite the ones in Figure 3-7a. **(a)** Draw the MO that will result from this orbital interaction. **(b)** Is the resulting MO unique compared to the one shown at the right of Figure 3-7a?

Think In the region in which the orbitals overlap, are the phases the same or opposite? Will that lead to constructive interference or destructive interference?

Solve Both orbitals are unshaded, indicating that they have the same relative phase. Therefore, constructive interference will take place, and the resulting MO is one that has been built up in the overlap region.

The resulting MO is all a single phase (in this case, unshaded) and has precisely the same shape as the MO shown in Figure 3-7a. Because phase is not a measurable quantity, the two MOs are indistinguishable and thus are not unique.

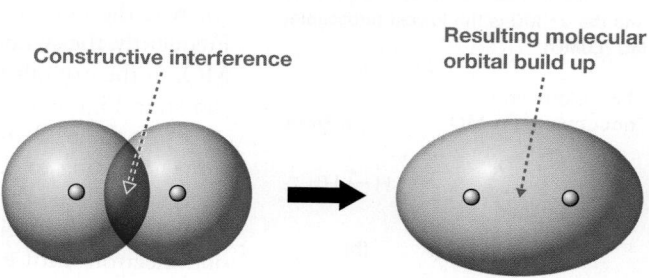

---

problem 3.2 To the right is a representation of two 1s orbitals overlapping, each with a phase that is opposite the ones in Figure 3-7b. **(a)** Draw the MO that will result from this orbital interaction. **(b)** Is the resulting MO unique compared to the one shown at the right of Figure 3-7b?

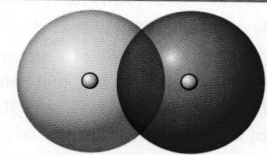

Looking back, we can see that when two 1*s* orbitals are mixed, two unique new orbitals are produced. This idea can be generalized as follows:

When *n* orbitals are mixed, *n* different orbitals must be produced.

We can think of this as an effective **conservation of number of orbitals**, a concept that will prove to be quite useful throughout the rest of this chapter and Chapter 14.

An important property of any MO is its *symmetry*, which is dictated by the locations in which the overlap of AOs takes place, relative to the bonding axis—the line connecting the two nuclei. The vast majority of MOs we will encounter in organic chemistry have either σ (sigma) or π (pi) symmetry.

In an MO of σ symmetry, overlap of the AOs takes place along a bonding axis.

FIGURE 3-8 **MO energy diagram of H_2** The overlap of two 1*s* AOs from separate H atoms results in the formation of a σ MO and a σ* MO. The individual AOs are on the left and the resulting MOs are on the right. The σ MO is lower in energy than the AOs by roughly the same amount that the σ* MO is higher in energy than the AOs. The σ MO is the highest occupied MO (HOMO) and the σ* MO is the lowest unoccupied MO (LUMO).

FIGURE 3-9 **Comparison between H_2's MO picture and its Lewis structure** (a) The MO picture of the H_2 molecule showing just the occupied bonding MO. (b) The Lewis structure of H_2, indicating that the two bonding electrons occupy a σ MO.

Later in this chapter, we will see that this is not the case for MOs of π symmetry. With this in mind, notice that in both Figure 3-7a and 3-7b, the 1*s* orbitals overlap along the bonding axis. Therefore, each of the MOs produced indeed has σ symmetry.

Not only do the MOs in Figure 3-7 differ in shape, but they also differ in energy, as shown in Figure 3-8. Notice that the MO that is produced from 1*s* AOs having the same phase is lower in energy (more stable) than a 1*s* AO. This is because that MO, having undergone constructive interference, gives an electron a greater probability of being found in the internuclear region, where it is simultaneously attracted to both nuclei. On the other hand, the MO produced from 1*s* AOs having opposite phase is higher in energy (less stable) than a 1*s* AO because the node largely excludes electrons from the internuclear region.

Based on the energy changes that accompany the formation of MOs from AOs, we can describe MOs as *bonding*, *antibonding*, or *nonbonding*.

- A **bonding MO** is lower in energy than its contributing AOs.
- An **antibonding MO** is higher in energy than its contributing AOs.
- A **nonbonding MO** has about the same energy as its contributing AOs.

Therefore, the MO that is produced from 1*s* AOs of the same phase is a bonding MO, whereas the one produced from 1*s* AOs of opposite phase is an antibonding MO. Frequently, the symbol "*" is used to distinguish an antibonding MO from a bonding MO, so the orbitals that are produced can simply be called σ and σ* (sigma-star), as shown in Figure 3-8.

Once the MOs have been established for a molecule, they can be filled with electrons according to the process described for AOs in Chapter 1. That is, the lowest-energy available orbital is filled first; each orbital can hold two electrons of opposite spin; and, when orbitals of equal energy are available, none of them receives a second electron until each of them already has one. Following this protocol, both of the electrons in H_2 will occupy the σ MO, leaving the higher-energy σ* MO empty (Fig. 3-8). Thus, the single bond we see in H_2's Lewis structure represents a pair of electrons in the σ MO and can therefore be called a **σ bond** (Fig. 3-9).

Two MOs, called the *highest occupied MO* (or *HOMO*) and the *lowest unoccupied MO* (or *LUMO*), are of particular interest in any molecule. As their names suggest, the **highest occupied molecular orbital** is the highest-energy MO that contains an electron, whereas the **lowest unoccupied molecular orbital** is the lowest-energy MO that is empty. In H_2, the HOMO is the σ MO and the LUMO is the σ* MO (Fig. 3-8).

SOLVED problem 3.3 In Chapter 15, we will learn that light can promote an electron to a higher energy MO. Suppose that light is used to promote one of the two electrons in the H_2 molecule from the σ MO to the σ* MO, as shown at the right. Is this state more stable, less stable, or about the same as the two separated hydrogen atoms?

Think Is an electron in the σ MO stabilized or destabilized compared to an electron in the 1s AO? Is an electron in the σ* MO stabilized or destabilized relative to an electron in the 1s AO?

Solve Compared to electrons in the 1s AO, one electron is stabilized in the σ MO, whereas the other is destabilized by about the same amount in the σ* MO. When the energies of the electrons are added up, there is essentially no difference between the energy of the H_2 molecule in the state described compared to the isolated atoms.

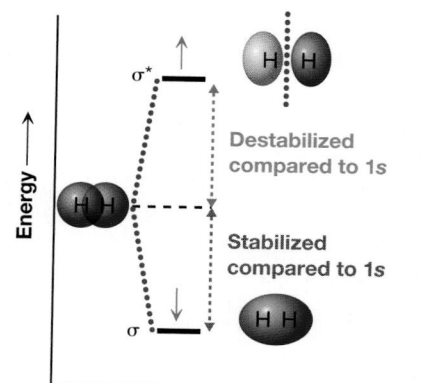

In the H_2 molecule described in Solved Problem 3.3, what is the HOMO?

3.5 **YOUR TURN**

---

problem 3.4 Suppose one electron is removed from an H_2 molecule, leaving H_2^+. Draw the energy diagram similar to Figure 3-8 for this species. Is H_2^+ more stable, less stable, or about the same as the isolated H atom and H^+ ion?

3.4 Hybridized Atomic Orbitals

The example of H_2 in the previous section provides insight into the role of orbitals in the formation of a covalent bond. That example involves only two atoms, so geometry is not a concern. However, as we saw in Chapter 2, the geometries about atoms commonly found in organic compounds are either linear, trigonal planar, or tetrahedral, forming bond angles of 180°, ~120°, and ~109.5°, respectively. How are such bond angles achieved, given that p orbitals are oriented 90° apart and s orbitals have no specific spatial orientation?

To answer this question, chemists often invoke the concept of *hybridization*, in which *hybridized* AOs are involved in bonding. As we will learn in greater detail in the sections that follow, these hybridized orbitals can be oriented about an atom in a linear, trigonal planar, or tetrahedral fashion. Therefore, invoking such hybridized orbitals makes it easier for us to envision molecular geometry from an orbital point of view.

A **hybridized atomic orbital** is a cross between two or more pure AOs from the valence shell of a single atom. Typically, the $2s$ and the $2p$ orbitals are the ones involved in hybridization, resulting in one of three types of hybridized AOs: sp, sp^2, or sp^3. Each of these hybridized orbitals has characteristics—such as its shape and its energy—that are intermediate between those of an s orbital and a p orbital. We will examine each type of hybridization in turn.

3.4a *sp* Hybridization

To account for linear geometries about atoms (180° bond angles), chemists invoke the concept of *sp hybridization*.

> *sp* **Hybridization** involves mixing the single s orbital and one of the three *p* orbitals—
> p_x, p_y, or p_z—from the valence shell to form two **sp-hybridized atomic orbitals.**

For elements in the second row of the periodic table, *sp* hybridization involves the 2*s* orbital and a 2*p* orbital. Each *sp*-hybridized orbital that is formed has 50% **s character** and 50% **p character**. In other words, an *sp*-hybridized orbital is distinct from either an *s* orbital or a *p* orbital, but its characteristics (e.g., shape and energy) resemble both orbitals equally.

Figure 3-10 depicts the result of the hybridization of an *s* orbital with a p_x orbital (chosen arbitrarily over the p_y and p_z orbitals). Because portions of these orbitals occupy the same space, they can be mixed to produce new orbitals. Notice that in *sp* hybridization, two pure AOs—one *s* and one *p*—are mixed to produce two hybridized orbitals. Thus, as expected, the number of orbitals is conserved in the mixing process.

Figure 3-10a represents the addition of the *s* and p_x orbitals, whereas Figure 3-10b represents the subtraction of one orbital from the other. In Figure 3-10b, the subtraction is indicated simply by reversing the phase (i.e., the shading) of the *p* orbital and adding the resulting orbitals together, analogous to what we saw in Figure 3-7 with two *s* orbitals.

On the left side of Figure 3-10a, notice that the right lobe of the contributing *p* orbital has the same phase as the contributing *s* orbital (both are shaded). Therefore, *constructive interference* takes place to the right of the nucleus. The resulting *sp*-hybridized orbital (middle) is built up to the right of the nucleus, denoted by the larger lobe. Notice, too, that the contributing *s* and *p* orbitals are opposite in phase to the left of the nucleus (one orbital is shaded but the other is not), so *destructive interference* takes place there. The resulting *sp*-hybridized orbital is diminished to the left of the nucleus.

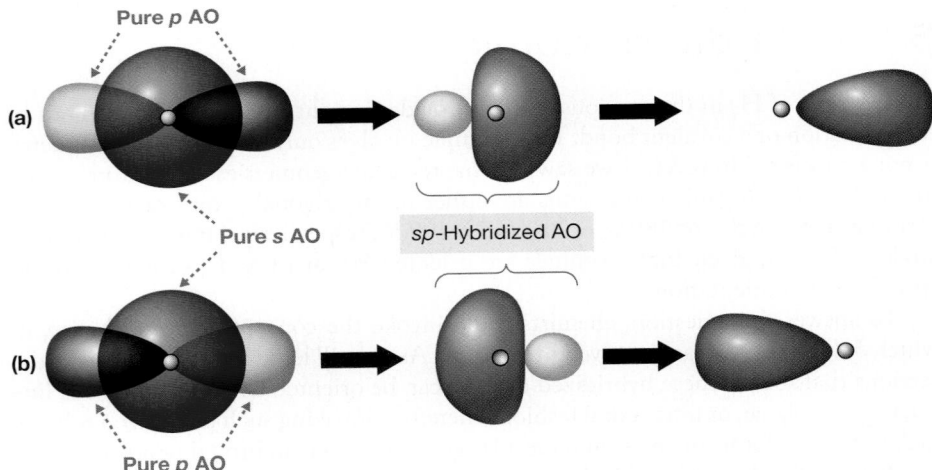

FIGURE 3-10 Generation of two sp-hybridized orbitals from pure AOs An *s* orbital and a *p* orbital of the same atom can mix in two different ways, forming two unique *sp*-hybridized AOs. (a) Left: Superposition of the *s* and *p* orbitals to be added. The phases of the orbitals are the same on the right side of the nucleus and opposite on the left side. Constructive interference therefore occurs to the right of the nucleus and destructive interference to the left. Middle: The *sp*-hybridized orbital that results is built up on the right and diminished on the left. Right: Often, as shown here, hybridized orbitals are depicted without the small lobe, and the shape of the large lobe is simplified to a distorted ellipsoid. (b) Left: The *p* orbital is subtracted from the *s* orbital. The phases of the *s* and *p* orbitals are the same on the left side of the nucleus and opposite on the right side of the nucleus. Middle: Constructive interference to the left of the nucleus and destructive interference to the right yields the *sp*-hybridized orbital shown. Right: The *sp*-hybridized orbital is shown without its small lobe.

In Figure 3-10a, circle and label the area of constructive interference and the area of destructive interference.

The *sp*-hybridized orbital that is generated from this process has two lobes that differ in size (one is much larger than the other) and are opposite in phase, and the larger lobe contains the nucleus. For convenience, however, the smaller lobe is often omitted (Fig. 3-10a, right) and the larger lobe is simplified to the distorted ellipsoid shown.

In Figure 3-10b, the left lobe of the contributing *p* orbital has the same phase as the contributing *s* orbital, whereas the right lobe has the opposite phase. Consequently, constructive interference takes place to the left of the nucleus and destructive interference to the right. As a result, the *sp*-hybridized orbital has the large lobe on the left and the small lobe on the right. Once again, for convenience the smaller lobe is not shown and the large lobe is simplified (Fig. 3-10b, right).

In Figure 3-10b, circle and label the area of constructive interference and the area of destructive interference.

The two *sp*-hybridized orbitals we have presented are the only unique orbitals that can be generated by mixing a 2*s* orbital with a 2p_x orbital. We can conceive of other linear combinations of the contributing *s* and *p* orbitals (see Problems 3.5 and 3.6), but the hybrid orbitals that are generated are redundant with the ones in Figure 3-10.

problem **3.5** Derive the hybridized orbital that would result from the interaction illustrated at the right, in which the phases of the s and p_x orbitals are opposite those in Figure 3-10a. Is the resulting orbital different from the ones in Figure 3-10? Explain.

problem **3.6** Derive the hybridized orbital that would result from the interaction illustrated at the right, in which the phases of the s and p_x orbitals are opposite those in Figure 3-10b. Is the resulting orbital different from the ones in Figure 3-10? Explain.

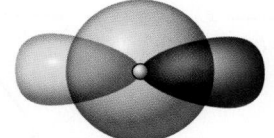

The only difference between the two *sp*-hybridized orbitals in Figure 3-10 is the direction the larger lobes point in space—they point 180° apart. When the p_x orbital is used for hybridization, the *sp*-hybridized orbitals are aligned along the x axis. The p_y and p_z orbitals that are left alone during the hybridization process remain perpendicular to the x axis, as well as perpendicular to each other, as shown in Figure 3-11a.

(a)

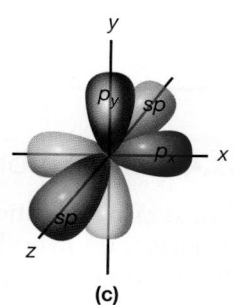

(b)

(c)

FIGURE 3-11 Three different but equivalent depictions of an *sp*-hybridized atom (a) An *sp*-hybridized atom has two *sp*-hybridized AOs and two unhybridized *p* orbitals. In this depiction the *sp*-hybridized orbitals are aligned along the x axis, indicating that the p_x orbital was used for hybridization. (b) In this depiction the p_y orbital was used for hybridization. (c) In this depiction the p_z orbital was used for hybridization.

In the previous treatment, the p_x orbital was *arbitrarily* chosen to be the p orbital used for sp hybridization. Figure 3-11b and 3-11c show what sp-hybridized atoms would look like if the $2p_y$ and $2p_z$ orbitals, respectively, were used for hybridization instead. The only difference among all three cases is the orientation of the orbitals. Therefore, we can make the following generalizations about an sp-hybridized atom:

- Any sp-hybridized atom has two sp-hybridized orbitals and two unhybridized p orbitals.
- The two sp-hybridized orbitals are aligned along the same axis and point in opposite directions.
- The unhybridized p orbitals are aligned along axes perpendicular to the axis containing the sp-hybridized orbitals.

SOLVED problem 3.7 Consider the pair of sp-hybridized orbitals at the right.

(a) Draw the orbital interaction that would be necessary to generate the orange sp-hybridized orbital from pure AOs. **(b)** Do the same for the green sp-hybridized orbital.

Think Along which axis are these sp-hybridized orbitals aligned? What orbitals must therefore interact? What is the phase of the large lobe of each hybridized orbital?

Solve Like any pair of sp-hybridized orbitals, the ones in the diagram must result from the interaction between an s and a p orbital. Because the sp-hybridized orbitals are aligned along the z axis, it must be the p_z orbital that has been used for hybridization, leaving the p_x and p_y orbitals unhybridized. The large lobe is shaded in both hybridized orbitals, so the s and p_z orbitals must both be shaded where the overlap occurs. The respective interactions are as follows:

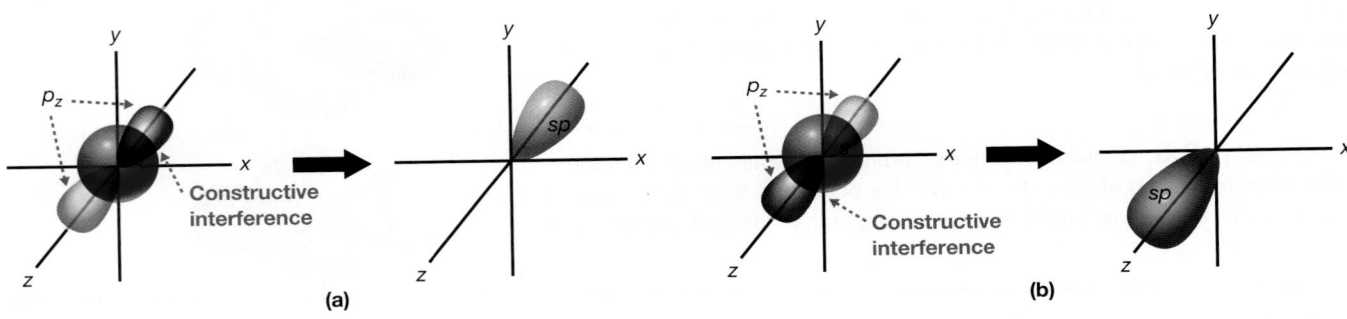

---

problem 3.8 Consider the pair of sp-hybridized orbitals at the right.

(a) Draw the orbital interaction that would generate the orange sp-hybridized orbital from pure AOs.
(b) Do the same for the green sp-hybridized orbital.

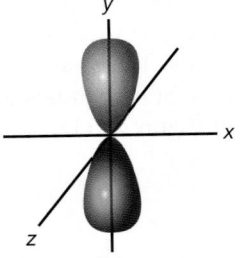

3.4b sp^2 Hybridization

Just as sp hybridization enables us to conceptualize the AOs of an atom with a linear geometry, **sp^2 hybridization** does so for an atom with a trigonal planar geometry. In sp^2 hybridization, three AOs—an s orbital and two of the three p orbitals from

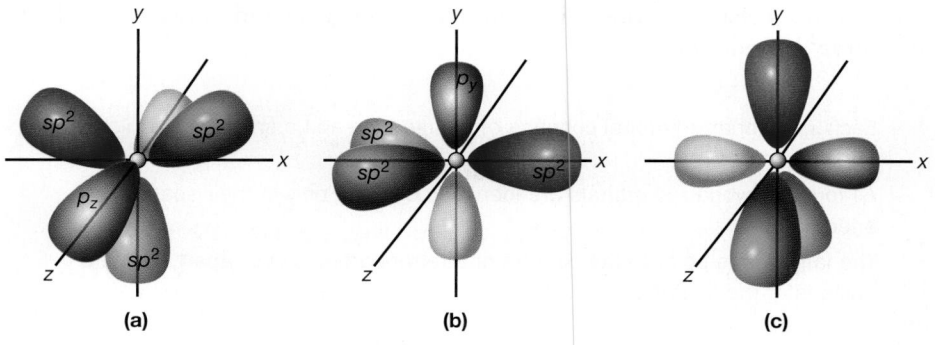

FIGURE 3-12 Three different but equivalent depictions of an sp^2-hybridized atom (a) The four valence AOs of an sp^2-hybridized atom in which the p_x and p_y orbitals were used for hybridization. The three sp^2-hybridized AOs occupy the xy plane, perpendicular to the unhybridized p_z AO. (b) An sp^2-hybridized atom in which the p_x and p_z orbitals were used for hybridization, leaving the p_y orbital unhybridized. (c) An sp^2-hybridized atom in which the p_y and p_z orbitals were used for hybridization, leaving the p_x orbital unhybridized.

the valence shell—are mixed. Three sp^2-**hybridized orbitals** result (because the total number of orbitals is conserved) and each has one-third s character and two-thirds p character. An sp^2 hybrid orbital, therefore, resembles a p orbital more than it does an s orbital.

sp^2 Hybridization involves constructive and destructive interference among three orbitals simultaneously, and are similar to sp-hybridized orbitals in the following ways:

- Each sp^2-hybridized orbital consists of a large lobe and a small lobe, opposite in phase to each other.
- All three sp^2 orbitals are identical to one another in shape, but have different orientations in space.

Whereas the two sp-hybridized orbitals point 180° apart, the three sp^2-hybridized orbitals point 120° apart and are all in the same plane. See Figure 3-12, which shows the three different possible combinations of s and p orbitals that can form an sp^2-hybridized atom.

Based on Figure 3-12, the following generalizations are true for an sp^2-hybridized atom:

- Any sp^2-hybridized atom possesses three sp^2-hybridized orbitals and one unhybridized p orbital.
- The three sp^2-hybridized orbitals are all in the same plane, pointing to the corners of a triangle; they are *trigonal planar*.
- The unhybridized p orbital is perpendicular to the *plane* that contains the hybridized orbitals.

In Figure 3-12c, label each orbital as either sp^2, p_x, p_y, or p_z.

3.8 YOUR TURN

3.4c sp^3 Hybridization

Whereas sp and sp^2 hybridization are invoked to explain atoms having linear and trigonal planar geometries, respectively, sp^3 hybridization is invoked to explain the bonding in atoms with tetrahedral geometries.

sp^3 **Hybridization** arises from the mixing of an s orbital and all three p orbitals from the valence shell. The result is four sp^3-**hybridized orbitals** (because the number of orbitals must be conserved), each of which has one-fourth s character and

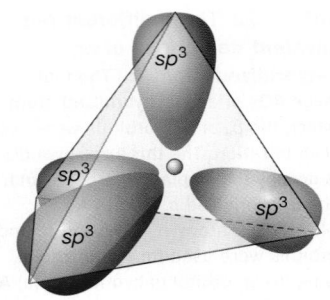

FIGURE 3-13 The four sp^3-hybridized orbitals of an sp^3-hybridized atom No unhybridized orbitals remain in the valence shell. The large lobes of the sp^3-hybridized orbitals point to the corners of a tetrahedron, so they are 109.5° apart.

three-fourths p character. The main characteristics of sp^3 hybridization are similar to those of sp^2 hybridization:

- Each sp^3-hybridized orbital consists of a large lobe and a small lobe, which are opposite in phase.
- All four sp^3-hybridized orbitals are identical, differing only in their spatial orientation.
- The large lobes point to the corners of a tetrahedron, 109.5° apart, with the nucleus at the center.

These features are illustrated in Figure 3-13, which depicts an sp^3-hybridized atom.

3.4d The Relationship between Hybridization and VSEPR Theory

Table 3-1 summarizes the relationship between hybridization, bond angles, and electron geometry. For example, an sp-hybridized atom has two hybrid orbitals pointing 180° apart, so they can account for a linear electron geometry about an atom. An sp^2-hybridized atom has three hybrid orbitals pointing 120° apart, which can therefore account for a trigonal planar electron geometry. And an sp^3-hybridized atom has four hybrid orbitals pointing 109.5° apart, which can account for a tetrahedral electron geometry.

TABLE 3-1 The Relationship between Hybridization and VSEPR theory

Hybridization	Orientation of Hybrid Orbitals	Bond Angles	Electron Geometry
sp		180°	Linear
sp^2		120°	Trigonal planar
sp^3		109.5°	Tetrahedral

Using VSEPR theory, then, we can deduce an atom's electron geometry from its total number of electron groups, and from its electron geometry we can infer its hybridization. This process is demonstrated in Solved Problem 3.9.

SOLVED problem 3.9 What is the hybridization of the N atom in the following molecule?

Think What is the electronic geometry about the N? What is the relationship between electron geometry and hybridization?

Solve According to VSEPR theory, the N atom is surrounded by two electron groups—a lone pair and a triple bond. The electronic geometry, therefore, is linear. According to Table 3-1, the atom must be *sp* hybridized.

---

problem **3.10** What is the hybridization of the atom indicated in each of the following molecules?

(a)   (b)   (c)   (d)   (e)

---

problem **3.11** **(a)** Draw the structure of a molecule with the formula C_5H_8 that has two *sp*-hybridized atoms. **(b)** Draw the structure of another molecule with that formula that has four sp^2-hybridized atoms.

3.5 Valence Bond Theory and Other Orbitals of σ Symmetry: An Example with Ethane (H_3C—CH_3)

The H—H bond discussed so extensively in Section 3.3 is not found in organic molecules, given that hydrogen forms only one bond. In Section 3.5, therefore, we extend our scope of bonding by examining the ethane molecule (H_3C—CH_3), which contains C—H single bonds and a C—C single bond, both of which are very common in organic molecules.

According to the following Lewis structure, ethane contains seven single bonds, representing a total of 14 valence electrons:

Ethane

In what kinds of orbitals do those electrons reside?

According to MO theory, the MOs would be derived by simultaneously mixing the valence AOs from all atoms: a $1s$ orbital from each hydrogen atom, and a $2s$, $2p_x$, $2p_y$, and $2p_z$ orbital from each carbon atom. In ethane, each of 14 different MOs would have simultaneous contributions from 14 different AOs. The result would be a picture that is much more complex than the one for H_2 shown previously in Figure 3-8.

The picture is simplified by applying **valence bond (VB) theory**, the basis of which can be stated as follows:

- Atoms that are bonded to two or more other atoms are considered to be hybridized.
- Hybridized atoms contribute their valence orbitals for mixing, which can include hybridized and unhybridized AOs.
- New orbitals are produced by mixing just two AOs at a time, one from each of two adjacent atoms.

sp³ Hybridized

Overlap occurs *along* the bonding axis.

(a)

(b)

FIGURE 3-14 AO overlap in ethane (a) Dash–wedge notation for a molecule of ethane, H_3C—CH_3, showing that each C atom is *sp³* hybridized. (b) The AOs contributed by each atom and their interaction in the ethane molecule. The seven pairs of interacting orbitals produce 14 MOs of σ symmetry: seven bonding and seven antibonding.

To apply these ideas to ethane, recall from VSEPR theory that each carbon atom has a tetrahedral electron geometry. According to Table 3-1, then, each carbon atom is *sp³* hybridized. Therefore, each carbon atom contributes four *sp³*-hybridized orbitals that are arranged tetrahedrally about the nucleus (Section 3.4c), as shown in Figure 3-14. Notice that there are seven orbital interactions involving AOs from adjacent nuclei. Six interactions involve the overlap between an *sp³*-hybridized orbital from C and a 1*s* orbital from H (purple with red), and one interaction involves the overlap between two *sp³*-hybridized C orbitals (purple with purple). All of these interactions will produce orbitals that have σ symmetry because all orbital overlap occurs *along* bonding axes, just as we saw previously with the σ and σ* MOs in H_2.

In all, the 14 AOs in Figure 3-14 will generate 14 new orbitals because the total number of orbitals must be conserved. Seven of the new orbitals are σ and seven are σ*, as illustrated in Figure 3-15. Similar to H_2 (Fig. 3-8),

> Each pair of AOs that overlap along a bonding axis adds both a σ and a σ* MO to the molecule.

The aufbau principle (Chapter 1) dictates that the 14 valence electrons (four from each of the C atoms and one from each of the H atoms) fill the seven lowest-energy orbitals available to them—the seven σ MOs (Fig. 3-15, right). The σ* MOs remain unoccupied. Therefore, the HOMO is the σ MO of highest energy, whereas the LUMO is the σ* MO of lowest energy.

As we can gather from Figures 3-14 and 3-15, there is a direct relationship between the number of occupied σ MOs and the number of single bonds that appear in the Lewis structure.

> The total number of single bonds in a Lewis structure is equal to the total number of filled σ MOs.

FIGURE 3-15 Energy diagram for the formation of ethane from its constituent atoms The individual AOs are shown on the left. Interaction among those orbitals generates 14 MOs of σ symmetry: seven bonding MOs and seven antibonding MOs (right).

Energy →

Out-of-phase overlap of AOs along bonding axes

Seven σ* MOs

All of these orbitals are empty, so one of these orbitals is the LUMO.

Fourteen electrons occupy these orbitals, so one of these orbitals is the HOMO.

14 AOs

In-phase overlap of AOs along bonding axes

Seven σ MOs

In ethane, in particular, there are seven total single bonds, and seven total σ MOs (Fig. 3-16). For this reason, it is common to refer to each single bond as a σ bond.

FIGURE 3-16 Bonds and Lewis structures In a Lewis structure, a single bond corresponds to a σ bond. The C—C bond and the six C—H bonds in ethane, therefore, are all σ bonds.

SOLVED problem **3.12** **(a)** Draw an orbital picture of methane (CH₄), similar to that in Figure 3-14, indicating the important overlap of AOs. **(b)** Draw an energy diagram for this molecule, similar to that in Figure 3-15.

Think What is the hybridization of C? What valence shell orbitals does it contribute? What orbitals do the H atoms contribute? How many total orbitals should be produced from AO mixing?

Solve The C atom is sp^3 hybridized, so it contributes four sp^3-hybridized orbitals in the valence shell. Each H atom contributes a 1s orbital. The overlap of AOs appears on the left in the following diagram:

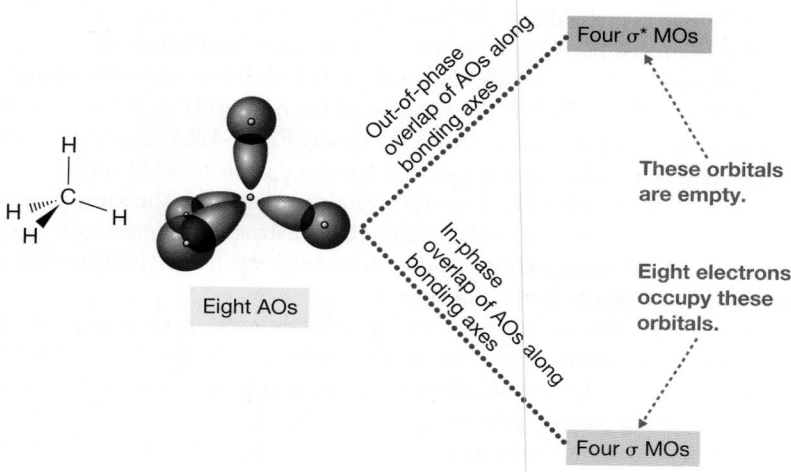

The energies of the MOs appear on the right. The eight AOs that overlap do so along the bonding axes, generating four σ and four σ* MOs. Eight valence electrons completely fill the σ MOs, leaving the σ* MOs empty.

problem **3.13** **(a)** Draw an orbital picture of the ammonium ion (NH₄⁺), similar to that in Figure 3-14, indicating the important overlap of AOs. **(b)** Draw an energy diagram for this species, similar to that in Figure 3-15.

problem **3.14** **(a)** Draw an orbital picture of the methylammonium ion (CH₃NH₃⁺), similar to that in Figure 3-14, indicating the important overlap of AOs. **(b)** Draw an energy diagram for this species, similar to that in Figure 3-15.

3.6 An Introduction to π Bonds: An Example with Ethene (H₂C═CH₂)

Ethene (H₂C═CH₂) contains a C═C double bond, which represents a total of four electrons. What kinds of orbitals do they occupy?

$$
\begin{array}{ccc}
H & & H \\
\diagdown & & \diagup \\
& C{=}C & \\
\diagup & & \diagdown \\
H & & H
\end{array}
$$

Ethene

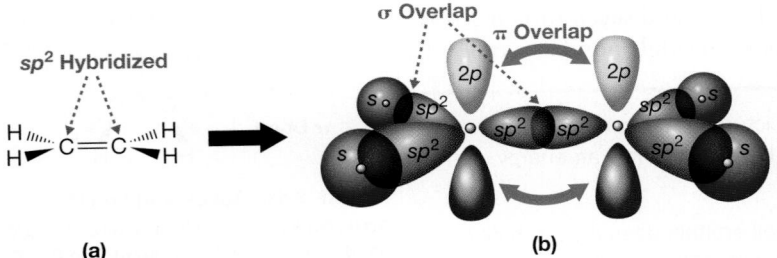

sp² Hybridized

σ Overlap
π Overlap

(a) **(b)**

FIGURE 3-17 AO overlap in ethene (a) Dash–wedge notation of a molecule of ethene (H_2C═CH_2), showing that it is entirely planar and that each C is sp² hybridized. (b) AOs that overlap to form the ethene molecule. Each carbon atom contributes three sp²-hybridized orbitals and one unhybridized p orbital. The hybridized orbitals and the s orbitals interact in a σ fashion, along the bonding axes. Although the p orbitals do not appear to physically overlap in this representation, they do, in fact, because each orbital extends beyond the surface indicated. The p orbitals overlap in a π manner on either side of the bonding axis.

We begin by identifying the hybridization of each C atom. Applying VSEPR theory, we predict that each C atom has a trigonal planar electron geometry, which, according to Table 3-1, makes it sp² hybridized. Thus, three hybridized AOs are arranged in a trigonal planar fashion about each C nucleus, as shown in purple in Figure 3-17. A single p orbital (blue) on each C atom is left unhybridized. Each is perpendicular to the plane defined by the three hybridized orbitals (recall Fig. 3-12).

Based on Figure 3-17b, there are five pairs of adjacent AOs that interact along the bonding axes, 10 orbitals in all (the six hybridized orbitals and the four s orbitals), which will give rise to 10 total orbitals of σ symmetry: five σ MOs and five σ* MOs. As discussed in Section 3.5, each of these interactions can be represented by a single bond in the Lewis structure, as shown in Figure 3-18.

This leaves two unhybridized p orbitals that are parallel to each other. Oriented as they are in Figure 3-18, these p orbitals overlap above and below the bonding axis connecting the two C atoms. As with any pair of AOs that mix, two new orbitals must be generated. One is generated by the addition of the orbitals (Fig. 3-18a); the other is generated by subtracting one orbital from the other (Fig. 3-18b). (Recall that subtraction is equivalent to reversing the phase of one orbital prior to addition.)

In Figure 3-18a, the lobes of the two p orbitals overlap with the same phase above and below the plane of the molecule, leading to constructive interference in each of those regions. The resulting orbital is, therefore, built up in the internuclear region above and below the plane of the molecule.

In Figure 3-18b, the lobes of the two p orbitals overlap with opposite phases above and below the plane of the molecule, leading to destructive interference in the internuclear region above and below the molecular plane. This leads to complete cancellation—a nodal plane—midway between the two nuclei, so the new orbital that is generated consists of four separate lobes, two above the molecular plane and two below.

These new orbitals do not have σ symmetry, because the overlap of AOs does not take place *along* the bonding axis. Instead, they are described as having *π symmetry*.

> An orbital of **π symmetry** is generated by AO overlap *on either side of* the bonding axis. The bonding axis itself contains a nodal plane.

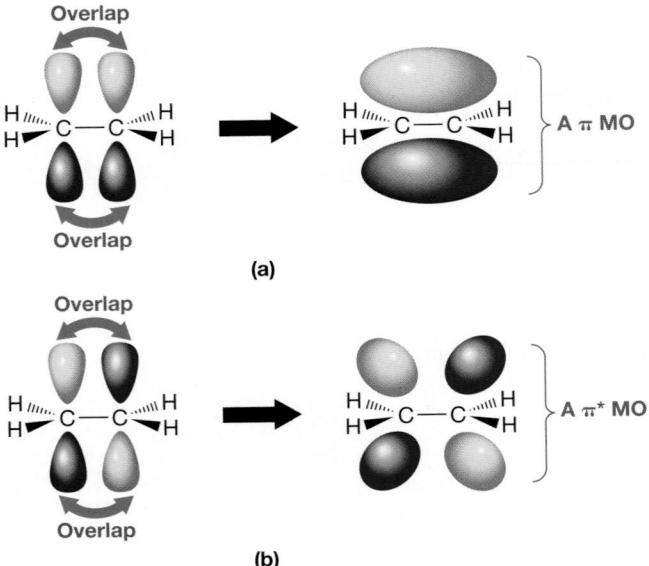

The new orbital in Figure 3-18a is a *π bonding* MO because the buildup of the orbital in the internuclear region lowers its energy relative to that of the p AOs from which it is generated (Fig. 3-19). The new orbital in Figure 3-18b is a *π* antibonding* MO. It has an additional nodal plane in the internuclear region that is *perpendicular* to the bonding axis. Just as in a σ* MO, this additional node largely excludes electrons from the internuclear region, which raises its energy relative to the p AOs from which it is generated.

FIGURE 3-18 MOs of π symmetry in ethene (a) The addition of two p AOs (left) results in constructive interference to generate a bonding MO of π symmetry, which has a lobe on either side of the bonding axis (right). (b) The subtraction of one p AO from the other can be accomplished by reversing the phase of one orbital before adding them (left). This results in destructive interference to generate an antibonding MO of π symmetry, which is composed of a total of four lobes (right).

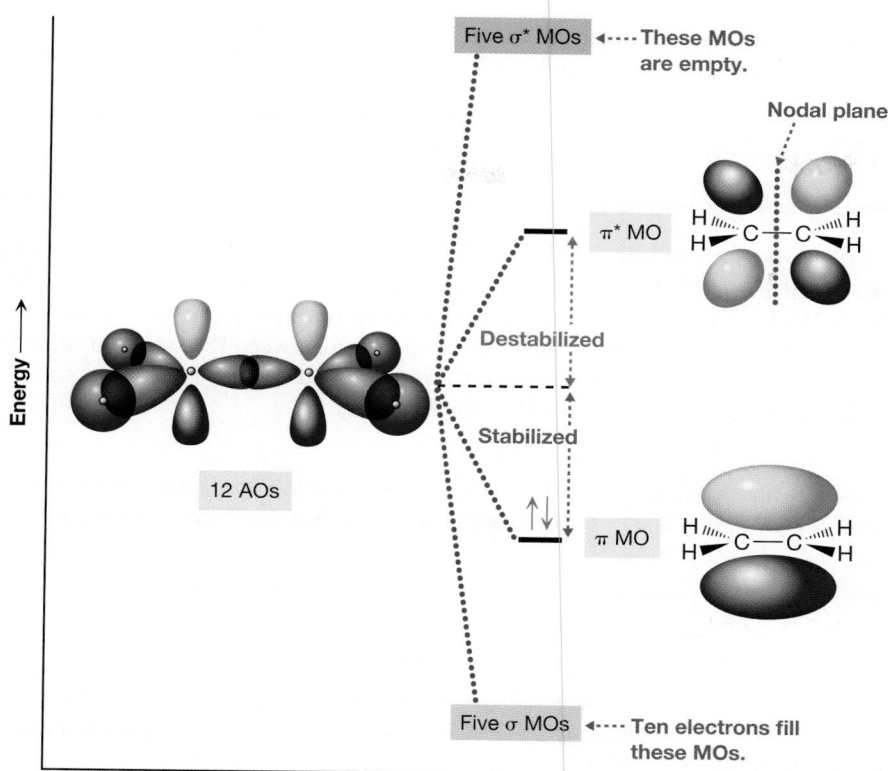

Five σ* MOs ◄---- These MOs are empty.

Nodal plane

π* MO

Destabilized

Stabilized

12 AOs

π MO

Five σ MOs ◄---- Ten electrons fill these MOs.

FIGURE 3-19 Energy diagram for the formation of ethene (H_2C=CH_2) from its constituent atoms The AOs are shown on the left and the resulting MOs are shown on the right. The energies of the five σ MOs are lower than that of the π MO, whereas the energies of the five σ* MOs are higher than that of the π* MO.

Ethene is shown twice below, with the molecular plane perpendicular to that in Figure 3-19. Identify the π and π* MOs, and in the π* MO, draw the nodal plane that is perpendicular to the bonding axis.

3.9 YOUR TURN

Based on the relative energies of the orbitals with σ and π symmetry shown in Figure 3-19,

- σ MOs are typically lower in energy than π MOs.
- σ* MOs are typically higher in energy than π* MOs.

The end-on overlap between orbitals that generates MOs with σ symmetry (Fig. 3-20a) is significantly more extensive than the side-by-side overlap of *p* orbitals that generates MOs with π symmetry (Fig. 3-20b). Consequently, there is more substantial interaction in σ and σ* orbitals than there is in π and π* orbitals.

With the relative energies of the orbitals shown in Figure 3-19, we can fill in the 12 valence electrons to generate ethene's electron configuration. The first

More extensive orbital overlap

Less extensive orbital overlap

(a)

(b)

FIGURE 3-20 Comparison of orbital overlap for orbitals of σ and π symmetry (a) Overlap of two hybrid AOs in the formation of an MO with σ symmetry. (b) Overlap of two *p* AOs in the formation of an MO with π symmetry. There is significantly more extensive orbital overlap in (a) than in (b).

10 valence electrons completely fill the σ MOs, and the last two fill the π MO. All of the antibonding MOs (i.e., σ* and π*) remain empty.

YOUR TURN 3.10   Label the HOMO and LUMO in Figure 3-19.

FIGURE 3-21 Double bonds and Lewis structures Two electrons of a double bond occupy a σ MO and the other two electrons occupy a π MO. A double bond, therefore, consists of a σ bond and a π bond.

Of the six total bonds in ethene's Lewis structure, five represent pairs of electrons occupying σ MOs and one represents a pair of electrons occupying the π MO (Fig. 3-21). In other words, ethene has five σ bonds (four joining C and H atoms and one between the two C atoms) and one π bond (also between the two C atoms). The C=C double bond, therefore, consists of both a σ bond and a π bond.

We will find that this is true of any double bond:

> A double bond consists of two electrons residing in a σ MO and two electrons residing in a π MO, so a double bond is said to consist of both a σ bond and a π bond.

SOLVED problem 3.15 How many π bonds are there in a molecule of cyclohexa-1,4-diene? How many σ bonds?

Think How many double bonds are there? What is each double bond composed of? How many single bonds are there? What is each single bond composed of?

Cyclohexa-1,4-diene

Solve There are two C=C double bonds in cyclohexa-1,4-diene. Each is composed of one π bond and one σ bond, for a total of two π bonds and two σ bonds. Additionally, there are 12 single bonds: four C—C single bonds and eight C—H single bonds. Each C—C and C—H single bond is a σ bond, giving a total of 14 σ bonds.

problem 3.16 How many π bonds and how many σ bonds are in each of the following species?

(a)   (b)   (c)   (d)

3.7 Nonbonding Orbitals: An Example with Formaldehyde (H₂C=O)

The AO contribution picture of methanal (formaldehyde, $H_2C{=}O$) shares a number of features in common with $H_2C{=}CH_2$ because both molecules possess a double bond involving a CH_2 group. The primary difference is that the double bond in ethene connects to a second CH_2 group, whereas the double bond in formaldehyde connects to an O atom that contains two lone pairs of electrons. Nevertheless, VSEPR theory suggests that the O atom in formaldehyde has a trigonal planar electron geometry, which means, according to Table 3-1, that it is sp^2 hybridized, similar to both C atoms in ethene.

With this in mind, we can construct the picture shown in Figure 3-22, which depicts the overlap of AOs in methanal. As we can see, there are three pairs of adjacent overlapping AOs (six orbitals in all) that are involved in significant overlap along bonding axes: the four hybridized AOs in purple and the two s orbitals in red. These

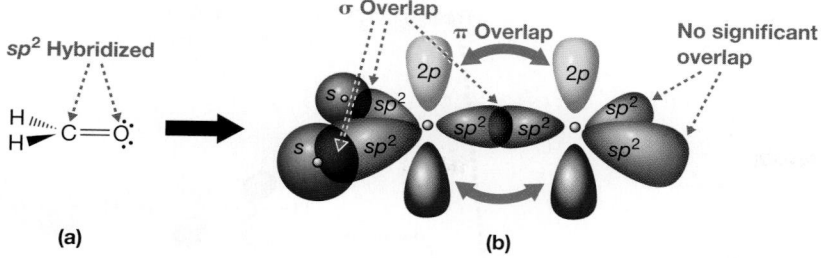

(a)

σ Overlap
π Overlap
No significant overlap

2p
2p

sp² Hybridized

s
sp²
sp²
sp²
sp²
sp²

(b)

FIGURE 3-22 Overlap of valence shell AOs in methanal (formaldehyde, $H_2C{=}O$) (a) Dash–wedge notation of a molecule of formaldehyde ($H_2C{=}O$), showing that it is entirely planar and that the O atom has two lone pairs. (b) The AOs in formaldehyde that overlap to produce new orbitals. The carbon and oxygen atoms each contribute three sp^2-hybridized orbitals and one unhybridized *p* orbital.

interactions give rise to six new MOs: three σ MOs and three σ* MOs. Also, similar to ethene, there are parallel adjacent *p* orbitals (one on C and one on O) that overlap on either side of a bonding axis, giving rise to one π MO and one π* MO.

Unlike the C atoms in ethene, however, the O atom in formaldehyde has two hybridized AOs (shown in orange in Fig. 3-22b) that do not overlap significantly with other AOs. Thus, there are two MOs that are relatively close in energy to the AOs from which they are derived. These are *nonbonding* MOs and they can be associated with the lone pairs of electrons in the Lewis structure shown in Figure 3-23.

σ Bond + π Bond

H

σ Bond

C ═ O

H

Nonbonding pairs

FIGURE 3-23 Lone pairs and Lewis structures Whereas single and double bonds in Lewis structures represent electron pairs occupying bonding MOs, lone pairs represent electrons occupying nonbonding MOs.

problem **3.17** Draw an orbital picture of $H_2C{=}NH$, similar to that in Figure 3-22. How many nonbonding MOs do you expect?

problem **3.18** For each of the following compounds, determine the number of σ bonds, the number of π bonds, and the number of electrons occupying nonbonding MOs.

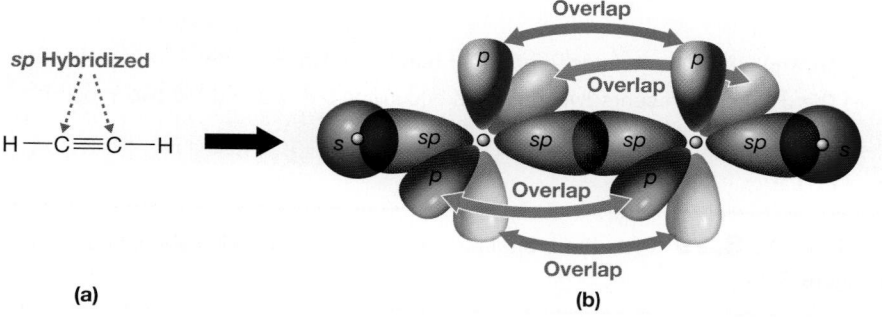

(a)   **(b)**   **(c)**   **(d)**   **(e)**

3.8 Triple Bonds: An Example with Ethyne (HC≡CH)

Ethyne (acetylene, HC≡CH) has a C≡C triple bond and, according to VSEPR theory, both C atoms have linear geometries. We can thus infer from Table 3-1 that both C atoms are *sp* hybridized. As a result, the valence shell AOs available for bonding include two *sp*-hybridized orbitals and two unhybridized *p* orbitals on each C atom. The hybridized orbitals form a linear arrangement, as shown in Figure 3-24.

sp Hybridized

H—C≡C—H

Overlap
Overlap

p   *p*
s *sp* *sp* *sp* *sp* s
p   *p*

Overlap
Overlap

(a)   **(b)**

FIGURE 3-24 The overlap of valence shell AOs in ethyne (acetylene) (a) The Lewis structure of ethyne (HC≡CH). (b) The *p* AOs in blue overlap each other, as do the *p* AOs in green. Two *sp*-hybridized orbitals (purple) overlap and each 1*s* orbital (red) from hydrogen overlaps an *sp*-hybridized orbital on carbon.

FIGURE 3-25 Energy diagram depicting the formation of MOs in ethyne, HC≡CH The AOs are on the left and the MOs are on the right. Six electrons occupy three σ MOs, and four electrons occupy two π MOs, thus accounting for all 10 valence electrons in ethyne.

Both of the p orbitals on each C are perpendicular to the line connecting the hybridized orbitals, and are perpendicular to each other.

There are six AOs that overlap along bonding axes, giving rise to three σ MOs and three σ* MOs (Fig. 3-25). Additionally, the C atoms contribute two pairs of overlapping p AOs, one shown in blue (p_y) and the other in green (p_z) in Figures 3-24 and 3-25. Each pair gives rise to a π MO and a π* MO.

The 10 valence electrons completely fill the three σ MOs and the two π MOs. The π* and σ* MOs remain empty.

YOUR TURN 3.11

Label the HOMO and the LUMO in Figure 3-25.

Since both π MOs between the two C atoms are filled, two of the three bonds in a C≡C triple bond are π bonds (Fig. 3-26). The remaining bond is a σ bond, representing one of the pairs of electrons occupying a σ MO in Figure 3-25. The two C—H bonds are also σ bonds, accounting for the remaining two pairs of electrons occupying σ MOs in Figure 3-25.

The conclusions we have drawn about the triple bond in ethyne can be extended to other triple bonds as well:

- Any triple bond is composed of one σ bond and two π bonds.
- The σ bond represents two electrons occupying a σ MO and the two π bonds represent four electrons occupying two π MOs.

FIGURE 3-26 Triple bonds and Lewis structures Two electrons from a triple bond occupy a σ MO, and the other four electrons occupy two π MOs. A triple bond, therefore, consists of one σ bond and two π bonds.

σ Bond

H—C≡C—H

σ Bond + two π bonds

problem **3.19** Draw the orbital picture of H—C≡N, which should be similar to Figure 3-24.

(a)   (b)   (c)   (d)

3.9 Bond Rotation about Single and Double Bonds: Cis-Trans Isomerism

There is a significant difference between single and double bonds that goes well beyond the differences in bond strength and bond length that we reviewed in Chapter 1:

> **Free rotation** can occur about single bonds but not about double bonds.

In a molecule of ethane ($H_3C—CH_3$), for example, the two CH_3 groups freely rotate relative to each other about the single bond that connects them (Fig. 3-27a). In a molecule of ethene ($H_2C=CH_2$), on the other hand, no rotation occurs. Instead, the entire molecule is planar, with the two CH_2 groups locked in place (Fig. 3-27b).

The CH_3 groups in ethane can rotate freely because, *during rotation, the σ bond that connects the two groups is unaffected* (Fig. 3-28a). Recall that a C—C σ bond results from the end-to-end overlap between the lobes of two hybridized orbitals. Because of the σ bond's symmetry, one CH_3 group can rotate relative to the other without affecting that overlap.

The picture is somewhat different with a molecule containing a double bond, which consists of one σ bond and one π bond. Imagine trying to rotate the CH_2 groups in $H_2C=CH_2$ relative to each other. The σ bond of the C=C double bond would be unaffected, but the π bond would be broken because rotation would destroy the overlap between the *p* orbitals (Fig. 3-28b). The substantial energy cost associated with breaking the π bond is what locks the CH_2 groups of ethene in place.

Free rotation occurs about single bonds.

Ethane
(a)

CH_2 groups locked in place.

Ethene
(b)

FIGURE 3-27 Bond rotation
Groups connected by a single bond (a) rotate freely relative to each other, but groups connected by a double bond (b) do not.

(a)

(b)

FIGURE 3-28 Orbital overlap during rotations about single and double bonds (a) Two different orientations of the CH_3 groups relative to each other during rotation about the C—C bond in $H_3C—CH_3$. The overlap of the hybridized orbitals is unaffected by the rotation, so this rotation is allowed in the molecule. (b) Two different orientations of the CH_2 groups relative to each other during rotation about the C=C bond in $H_2C=CH_2$. Although the overlap of the hybridized orbitals is unaffected, overlap between the *p* orbitals is destroyed, so the π bond must be broken for rotation to occur. Because breaking the π bond requires a significant input of energy, rotation about a double bond does not normally take place.

For the π bond in $H_2C\!=\!CH_2$ to remain intact, all six atoms must be in the same plane. In general,

> When two atoms are connected by a double bond, those atoms and any atoms to which they are directly bonded strongly prefer to lie in the same plane.

This principle is illustrated in Figure 3-29.

---

problem **3.21** In each of the following molecules, circle all of the atoms that are required to be in the same plane. (Because these are line structures, some atoms may need to be added back in.)

(a)   (b)   (c)

Because no rotation takes place about double bonds, some molecules containing a double bond can exist in either of two *configurations*. These two configurations differ only in the relative positions of the groups directly bonded to the doubly bonded atoms. In 1,2-dichloroethene, for example, the configuration in which the two Cl atoms are on the same side of the double bond is designated **cis**. The one in which the Cl atoms are on opposite sides of the double bond is designated **trans** (Fig. 3-30).

With no rotation about the double bond, the cis and trans forms do not interconvert and can therefore be isolated from each other. This makes it possible to study their properties separately. In the case of 1,2-dichloroethene, for example, the boiling point of the cis form is 60.3 °C, whereas that of the trans form is 47.5 °C.

---

problem **3.22** List all of the intermolecular interactions present in *cis*-1,2-dichloroethene and *trans*-1,2-dichloroethene. Using this information, explain the difference in their boiling points.

Not all molecules with double bonds have two configurations that differ by the hypothetical rotation about the double bond. Fluoroethene ($CH_2\!=\!CHF$), for example, does not (see Your Turn 3.12). A hypothetical 180° rotation of either group about the double bond in $CH_2\!=\!CHF$ returns the same molecule as before the rotation because one of the atoms of the double bond is bonded to two identical atoms or groups—in this case, hydrogen.

> A molecule with a double bond will have two distinct structures that differ by a hypothetical 180° rotation about the double bond only if *neither* of the atoms connected by the double bond is singly bonded to two identical atoms or groups.

These atoms must all lie in the same plane.

FIGURE 3-29 Planar geometry imposed by a double bond In a molecule of 2-methylhept-2-ene, the five C atoms and the H atom indicated must all lie in the same plane.

FIGURE 3-30 Cis and trans configurations These two molecules differ by a hypothetical 180° rotation about the C=C double bond, but the two configurations do not interconvert. The non-hydrogen substituents are on the same side of the double bond in the cis configuration and are on opposite sides in the trans configuration.

cis-1,2-Dichloroethene
Boiling point = 60.3 °C

trans-1,2-Dichloroethene
Boiling point = 47.5 °C

Using a molecular modeling kit, construct the following two molecules and try to line one up with the other. Are they the same or different? (Try all orientations of the two molecules.)

$$\begin{array}{ccc} H & & H \\ & C{=}C & \\ H & & F \end{array} \qquad \begin{array}{ccc} H & & F \\ & C{=}C & \\ H & & H \end{array}$$

SOLVED problem **3.23** Does the following molecule have two distinct configurations about the double bond?

$$\begin{array}{ccc} H & & Cl \\ & C{=}C & \\ Cl & & CH_3 \end{array}$$

Think Does a hypothetical rotation of 180° about the double bond give rise to a different molecule? How can you tell?

Solve One way to answer this question is to compare molecules before and after the hypothetical rotation.

$$\begin{array}{ccc} H & & Cl \\ & C{=}C & \\ Cl & & CH_3 \end{array} \xrightarrow[\text{bond rotation}]{\textbf{Hypothetical}} \begin{array}{ccc} H & & CH_3 \\ & C{=}C & \\ Cl & & Cl \end{array}$$

These molecules are different because the one on the left has the Cl atoms on opposite sides of the double bond, whereas the one on the right has the Cl atoms on the same side. A second way to answer this question is simply to look at the substituents (i.e., the atoms or groups) bonded to each atom joined by the double bond. Neither of them is singly bonded to two identical substituents, so two unique configurations about the double bond must exist.

problem **3.24** Does cyclooctene have two distinct configurations about its C=C double bond?

Cyclooctene

Some molecules have more than one C=C double bond, and each one can potentially exist in cis and trans forms. There are three double bonds, for example, in a molecule of α-linolenic acid, a natural fatty acid. Each double bond in α-linolenic acid is in the cis configuration, indicated in the following line structure by the fact that all of the hydrogen atoms (not shown) bonded to the C=C are on one side of the double bonds and all of the carbon atoms bonded to the C=C are on the other:

Three cis double bonds

α-Linolenic acid

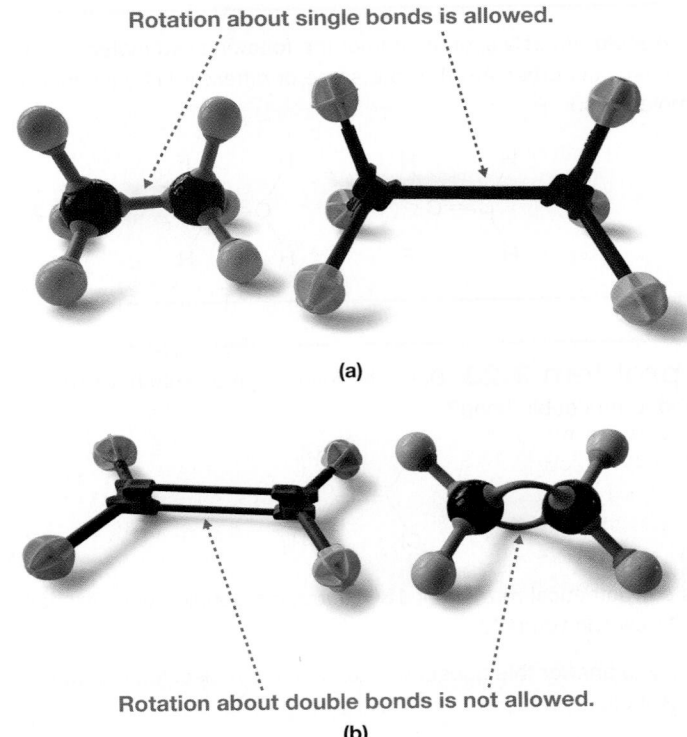

Rotation about single bonds is allowed.

(a)

Rotation about double bonds is not allowed.

(b)

FIGURE 3-31 Model kits and rotational characteristics of molecules (a) Molecular models of ethane (H_3C—CH_3) made with different model kits. In each case, one CH_3 group can rotate freely with respect to the second. (b) Molecular models of ethene (H_2C=CH_2) made with different model kits. In each case, the rotation of one CH_2 group relative to the other is restricted.

problem **3.25** Draw the form of α-linolenic acid in which all three double bonds are trans. How many unique structures can be made by changing the cis/trans configurations about the double bonds in α-linolenic acid?

3.10 Strategies for Success: Molecular Models and Extended Geometry about Single and Double Bonds

In Section 2.3, we saw that a molecular modeling kit can help you visualize three-dimensional molecules. Molecular models are particularly helpful for studying the *rotational* characteristics of the single and double bonds discussed in Section 3.9. Figure 3-31 shows molecular models of ethane (H_3C—CH_3) and ethene (H_2C=CH_2) created using different modeling kits. Free rotation of the CH_3 groups around the C—C single bond in ethane allows the hydrogens at one end of the molecule to assume any angle relative to those at the other end. In ethene, however, all of the hydrogens are locked into the same plane by the C=C bond.

Modeling kits, therefore, can show the *extended* geometries about double bonds—that is, the three-dimensional location of the atoms directly attached to each double bond. Because of the planar nature of double bonds, for example, it might be tempting to think that the entire propa-1,2-diene (allene, H_2C=C=CH_2) molecule is planar. Experiments show, however, that one CH_2 group is perpendicular to the other. As we can see in Figure 3-32, these are the same results we obtain from molecular modeling kits!

CH₂ groups perpendicular to each other

FIGURE 3-32 Molecular modeling kits and extended geometry A molecular model of propa-1,2-diene (allene, H_2C=C=CH_2) constructed with a modeling kit shows that the two CH_2 groups are perpendicular to each other, which agrees with experimental results.

(a) Using a molecular modeling kit, construct a molecule of $H_2C{=}C{=}C{=}CH_2$. Is the entire molecule planar? **(b)** Next, construct a molecule of $H_2C{=}C{=}C{=}C{=}CH_2$. Is that molecule entirely planar?

3.11 Hybridization, Bond Characteristics, and Effective Electronegativity

Thus far, we have not treated the various types of hybridized orbitals, or the MOs to which they give rise, as being significantly different. However, the type of hybridization an atom undergoes can significantly affect the structural and chemical properties of the molecule in which it is found. For example, the C—H bond in ethane ($H_3C{-}CH_3$) is longer and weaker than that in ethene ($H_2C{=}CH_2$), which is longer and weaker, in turn, than that in ethyne ($HC{\equiv}CH$) (Fig. 3-33). The hybridization of the carbon atoms in these compounds is sp^3, sp^2, and sp, respectively.

In addition to bond lengths and bond strengths, hybridization can also affect bond polarity. This can be seen in the electrostatic potential maps of ethane ($H_3C{-}CH_3$), ethene ($H_2C{=}CH_2$), and ethyne ($HC{\equiv}CH$) shown in Figure 3-34. Notice that as the carbon atom's hybridization goes from sp^3 (ethane) to sp^2 (ethene) to sp (ethyne), the concentration of positive charge (blue) on hydrogen increases. This is what we would expect if the carbon atom's electronegativity were increasing. However, because the nucleus is the same in each of these molecules (i.e., a carbon atom), the electronegativity is not changing. Instead, we say that the carbon atom's **effective electronegativity** increases.

We can generalize the preceding observations as follows:

As the hybridization of an atom goes from sp^3 to sp^2 to sp:

- Its bonds become shorter and stronger.
- Its effective electronegativity increases.

Increasing C—H bond length

Increasing C—H bond strength

108 pm
410 kJ/mol

107 pm
431 kJ/mol

105 pm
523 kJ/mol

sp^3

sp^2

sp

Increasing s character

FIGURE 3-33 Relationship between hybridization, bond length, and bond strength The bond distances and bond energies for the C—H bond in ethane, ethene, and ethyne (top). In all cases, the C—H bond involves the overlap of the 1s orbital of a H atom with a hybridized orbital from C. The s character of the C atom's hybridized orbital increases in the order $sp^3 < sp^2 < sp$ (bottom).

Lowest concentration
of positive charge
on hydrogen

Highest concentration
of positive charge
on hydrogen

Ethane
sp^3 Hybridized
(a)

Ethene
sp^2 Hybridized
(b)

Ethyne
sp Hybridized
(c)

Effective electronegativity of carbon

FIGURE 3-34 Hybridization and effective electronegativity Electrostatic potential maps of (a) ethane, H_3C—CH_3, (b) ethene, H_2C=CH_2, and (c) ethyne, HC≡CH. Ethyne has the highest concentration of positive charge on hydrogen. Therefore, as the hybridization of carbon goes from sp^3 to sp^2 to sp, the effective electronegativity of carbon increases.

Both of these observations can be explained by the fact that the *s* character of a hybridized orbital increases in going from sp^3 to sp^2 to sp, as we saw in Section 3.4. On average, a pure 2*s* orbital holds electrons closer to the nucleus than does a pure 2*p* orbital, as illustrated in Figure 3-35. Another way of saying this is that a 2*s* orbital is more *compact* than a 2*p* orbital. Therefore, as the *s* character of a hybridized orbital increases, its electrons lie closer (on average) to the nucleus, making the nucleus appear to have a greater electronegativity. And, as the hybridized orbital becomes more compact with greater *s* character, its overlap with an orbital from a second atom gives rise to a shorter bond.

YOUR TURN 3.14

To verify the statement that the s character of a hybridized orbital increases in going from sp^3 to sp^2 to sp, write the percent s character above each hybridized orbital in Figure 3-35.

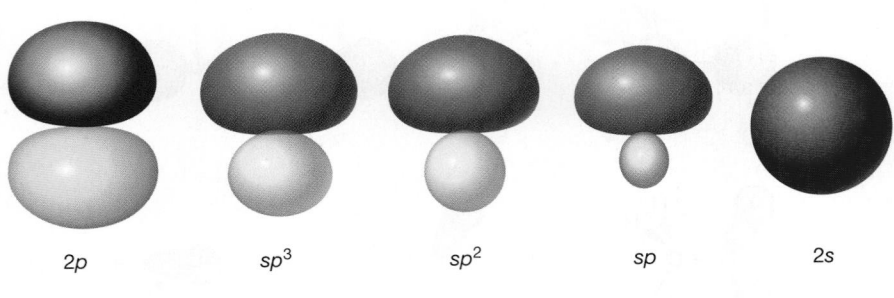

$2p$ sp^3 sp^2 sp $2s$

FIGURE 3-35 Comparison of pure and hybridized orbitals As s character increases, an orbital becomes more compact and therefore holds electrons closer to the nucleus, on average.

s Character increases

Orbital becomes more compact

Freezing Point Elevation?

Pure benzene (C_6H_6) melts at 5.5 °C, and pure hexafluorobenzene (C_6F_6) melts at 5.2 °C. Oddly, a 1:1 mixture of the two compounds melts at 23.7 °C.

Benzene
Melting point = 5.5 °C

Hexafluorobenzene
Melting point = 5.2 °C

Benzene:hexafluorobenzene
Melting point = 23.7 °C

In most cases, the melting point (i.e., the freezing point) of a compound is *lowered* when an impurity is introduced, a phenomenon known as freezing point depression. What makes the benzene:hexafluorobenzene mixture different?

The answer stems from the fact that benzene and hexafluorobenzene have essentially opposite charge distributions. As shown in its electrostatic potential map, benzene has excess negative charge in the center of the ring, owing to the six π electrons (Section 3.6), and excess positive charge on the H atoms, owing to the effective electronegativity of the sp^2-hybridized carbon atoms (Section 3.11). In hexafluorobenzene, on the other hand, the F atoms draw negative charge toward the periphery of the ring, leaving an excess of positive charge toward the center.

To maximize the attraction between oppositely charged regions in either pure benzene or pure hexafluorobenzene, the molecules adopt a T-shaped orientation. The contact surface area is minimized in these orientations, however, which compromises the dispersion forces. In the 1:1 mixture of the two compounds, on the other hand, maximum electrostatic interaction occurs when the two molecules are stacked on top of each other, taking full advantage of the complementary charge distribution. This stacked orientation maximizes the dispersion forces, too. Thus, the intermolecular interactions between the two different molecules are stronger than those between two of the same molecule, ultimately leading to a significantly higher melting point in the 1:1 mixture.

SOLVED problem 3.26 In which of the following molecules do you think the C=C distances are longer? In which molecule are the C=C bonds stronger?

A

B

Think What is the hybridization of each atom involved in the C=C bonds? How does that affect bond length and bond strength?

Solve In **A**, each C=C bond involves two sp^2-hybridized C atoms. In **B**, each C=C bond involves one sp^2-hybridized C atom and one sp-hybridized C atom. With greater s character, the hybridized orbitals are more compact, so the C=C bonds in **B** are shorter and stronger than the ones in **A**.

---

problem **3.27** In which molecule is the C—C bond shorter? In which molecule is it stronger? Explain.

$$H_3C—C\equiv N \qquad\qquad \overset{\displaystyle NH}{\underset{H_3C\qquad CH_3}{\|\\ C}}$$

3.12 Wrapping Up and Looking Ahead

This chapter extended our understanding of molecular structure by introducing hybridization and MO theory, both of which involve the mixing of orbitals to produce new orbitals. Hybridization enables us to account for electron geometry; MO theory accounts for the formation of covalent bonds.

We saw, in particular, that an atom's hybridization corresponds exactly to its electron geometry predicted by VSEPR theory: a linear atom is sp hybridized, a trigonal planar atom is sp^2 hybridized, and a tetrahedral atom is sp^3 hybridized. Further, the overlap of AOs from two separate atoms is what gives rise to MOs. The overlap of two AOs along an internuclear axis produces one σ MO and one σ* MO. Similarly, the overlap of two adjacent and parallel p orbitals produces one π MO and one π* MO. Using these ideas, we can construct MO energy diagrams for molecules, and determine which MOs are occupied and which remain empty.

With an understanding of MO theory, we can explain why single bonds undergo free rotation and double bonds do not, thus giving rise to cis–trans isomerism. Moreover, hybridization helps us explain trends in effective electronegativity, bond length, and bond strength. Specifically, with greater s character, an atom will have greater effective electronegativity, and will tend to be involved in shorter, stronger bonds.

The ideas we learned in this chapter will be very important in the chapters to come. In Chapters 4 and 5, for example, the rotational characteristics of single and double bonds will be explored in greater depth in the context of conformational isomers and diastereomers. In Chapter 7, we will begin to see that the symmetry of bonds (i.e., σ or π symmetry) that are formed and/or broken can have a major impact on the driving force for chemical reactions. In Chapter 14, we will delve even deeper into the ideas surrounding MO theory to explain resonance and aromaticity—two stabilizing phenomena. What we learn from that discussion will be applied toward aspects of spectroscopy (the interaction of electromagnetic radiation with matter) in Chapters 15 and 16.

Chapter Summary and Key Terms

- **Quantum mechanics** treats electrons as waves. It allows us to determine the probability of finding an electron at a certain location in space. (**Section 3.1**)

- The relative **phase** of an orbital in a region of space is indicated by whether the orbital is shaded or unshaded there. Phase is not a measurable quantity, but it determines how two or more orbitals interact. (**Section 3.1a**)

- The surface that encompasses 90% of an electron's probability defines the shape of the electron's orbital. The

volume inside that surface is commonly viewed as the **orbital**. The 1s orbital is spherical and is all one phase. (**Section 3.1a**)

- There are three 2p orbitals. Each is dumbbell shaped and consists of two lobes that are opposite in phase, separated by a **nodal plane** in which an electron has zero probability of being found. The three 2p orbitals (i.e., $2p_x$, $2p_y$, and $2p_z$) are perpendicular to one another. (**Section 3.1b**)

- Atomic s and p orbitals in higher shells are similar in shape and behavior to the 1s and 2p orbitals, respectively. (**Section 3.1c**)

- Orbitals interact through **constructive interference** and **destructive interference**. (Section 3.2)

- Electrons that bond atoms together occupy **molecular orbitals** (**MOs**). An MO arises from overlapping AOs of separate atoms. Each MO can accommodate up to two electrons. (Section 3.3)

- A **bonding MO** results from in-phase overlap in the bonding region and is lower in energy than the AOs from which it is constructed. An **antibonding MO** results from out-of-phase overlap in the bonding region and is higher in energy than the contributing AOs. A **nonbonding MO** is similar in energy to the contributing AOs. (Section 3.3)

- The highest-energy MO containing an electron is called the **highest occupied molecular orbital** (**HOMO**) and the lowest-energy empty MO is called the **lowest unoccupied molecular orbital** (**LUMO**). (Section 3.3)

- An MO of σ symmetry is generated by the interaction of AOs along bonding axes. Two electrons occupying a σ MO is called a σ **bond**. (Sections 3.3 and 3.5)

- When orbitals mix, the total number of orbitals is conserved: The interaction of n orbitals results in n unique new orbitals. (Section 3.3)

- In sp **hybridization**, one s and one p orbital from the valence shell mix to generate two sp**-hybridized orbitals**. Each has one large lobe and one small lobe that point in opposite directions. In an sp-hybridized atom, two p orbitals remain unhybridized; they are aligned perpendicular to each other and perpendicular to the sp orbitals. (Section 3.4a)

- In sp^2 **hybridization**, one s and two p orbitals interact to generate three sp^2**-hybridized orbitals**. All three hybridized orbitals lie in one plane and point to the corners of a triangle. In an sp^2-hybridized atom, there is one p orbital that remains unhybridized; it is perpendicular to the plane containing the hybridized orbitals. (Section 3.4b)

- In sp^3 **hybridization**, all three valence p orbitals interact with the valence s orbital to generate four sp^3**-hybridized orbitals**. The four hybridized orbitals point to the corners of a tetrahedron. (Section 3.4c)

- Hybridization is intimately related to VSEPR electron geometry (Table 3-1): sp-hybridized atoms have a linear geometry, sp^2-hybridized atoms are trigonal planar, and sp^3-hybridized atoms are tetrahedral. (Section 3.4d)

- An MO of π **symmetry** is generated by the interaction of adjacent parallel p orbitals, where the orbital overlap takes place on either side of the bonding axis. Two electrons occupying a π MO constitute a π bond. (Section 3.6)

- A double bond in a Lewis structure consists of one σ bond and one π bond. A triple bond consists of one σ bond and two π bonds. (Sections 3.6 and 3.8)

- A lone pair of electrons in a Lewis structure represents a pair of electrons occupying a nonbonding **MO**. (Section 3.7)

- **Free rotation** can occur about single bonds but not about double bonds. Rotation about a double bond can only occur if the π bond breaks. (Section 3.9)

- For a given double bond, **cis** and **trans** configurations are possible if an imaginary 180° rotation about the double bond results in a different molecule. Two groups are cis to each other if they are on the *same* side of a double bond and are trans to each other if they are on *opposite* sides. (Section 3.9)

- Molecular modeling kits display the rotational characteristics about single and double bonds and provide accurate representations of *extended* geometries involving double bonds. (Section 3.10)

- Single bonds involving a hybridized orbital become shorter and stronger as the *s* character of the hybridized orbital increases. **Effective electronegativity** also increases as the *s* character of the atom's hybridization increases. (Section 3.11)

Problems

3.28 For each of the following species, determine **(a)** the electron geometry and **(b)** the hybridization for all nonhydrogen atoms.

(i) CH_3NH_2 **(ii)** $CH_3N{=}O$ **(iii)** CH_2Cl_2 **(iv)** BrCN

3.29 For each of the following species, determine **(a)** the electron geometry and **(b)** the hybridization for all nonhydrogen atoms.

(i) $C_2H_5^+$ **(ii)** $C_2H_5^-$ **(iii)** $CH_2{=}OH^+$ **(iv)** CH_6N^+
(v) CH_5O^+ **(vi)** $C_3H_3^-$ (all hydrogens are on the same carbon)

3.30 Levomenol, a naturally occurring sesquiterpene alcohol, has a sweet-smelling aroma and has been used as a component in fragrances. It also is known to have antimicrobial and anti-inflammatory properties. Determine the electron geometry, hybridization, and molecular geometry for each non-hydrogen atom in levomenol.

Levomenol

3.31 Determine the total number of σ bonds and the total number of π bonds in levomenol (Problem 3.30).

3.32 Bombykol is a pheromone produced by silkworm moths.

Bombykol

Is the configuration around each C=C double bond cis or trans?

3.33 Norethynodrel was the synthetic hormone used in Enovid, the first oral contraceptive. **(a)** Determine the hybridization of each nonhydrogen atom. **(b)** How many total σ bonds and π bonds does norethynodrel have?

Norethynodrel

3.34 Suppose a linear molecule were constructed from three atoms, all of which are found in the second row of the periodic table. How many valence shell AOs would these three atoms contribute toward the production of MOs? How many MOs would be produced by the mixing of these valence shell AOs? (*Hint:* The answer is independent of which orbitals overlap.)

3.35 For the molecule described in Problem 3.34, suppose that two orbitals from the first atom were to mix with two orbitals from the second atom and that two other orbitals from the second atom were to mix with two orbitals from the third atom. In the resulting molecule, how many bonding MOs would there be in total? How many antibonding MOs? How many nonbonding MOs?

3.36 For which of the following molecules are there two unique configurations about the double bond? Explain.
(a) $(CH_3)_2C=CHCl$ **(b)** $H_2C=CHCH_2CH_3$
(c) $ClHC=CHBr$ **(d)** $HC≡CCH=CHCl$

3.37 How many total electrons reside in MOs of π symmetry in the following cation?

3.38 β-Carotene is the compound responsible for the orange color of carrots and is the precursor to vitamin A. Judging from the following Lewis structure, how many π bonds does β-carotene have?

β-Carotene

3.39 Which of the following molecules has a σ bond obtained from the overlap between an *sp*-hybridized orbital and an sp^2-hybridized orbital?

(a)   **(b)**   **(c)**   **(d)**

3.40 Adenine, cytosine, guanine, and thymine are the four nitrogenous bases found in DNA.

Adenine   **Cytosine**   **Guanine**   **Thymine**

For each molecule, identify all of the nonhydrogen atoms that are required to be in the same plane.

3.41 (a) Draw the orbital picture for :C≡O:, showing the explicit overlap of the contributing AOs. (b) How many MOs of π symmetry are there in total? (c) Draw the orbital energy diagram for CO and identify the HOMO and LUMO.

3.42 (a) Draw the molecular orbital picture for propa-1,2-diene, $H_2C=C=CH_2$. (*Hint:* The three-dimensional geometry is shown in the chapter.) (b) Draw the MO energy diagram for propa-1,2-diene. What is the HOMO? What is the LUMO?

3.43 Draw the MO picture for buta-1,2,3-triene, $H_2C=C=C=CH_2$. (*Hint:* See Your Turn 3.13.)

3.44 Do all of the atoms in buta-1,3-diene have to reside in the same plane? Why or why not?

Buta-1,3-diene

3.45 The boiling point of *cis*-but-2-ene is 3.7 °C, whereas that of *trans*-but-2-ene is 0.9 °C. Explain. (*Hint:* Identify which inter-molecular interaction is responsible for the difference in boiling points.)

cis-But-2-ene   **trans-But-2-ene**
Boiling point = 3.7 °C   Boiling point = 0.9 °C

3.46 Draw the AO contribution picture of $CH_3CH_2^+$ and the MO energy diagram. What is the HOMO? What is the LUMO?

3.47 Suppose that an electron were added to $CH_3CH_2^+$, the species in Problem 3.46, yielding uncharged CH_3CH_2. What is the HOMO? What is the LUMO?

3.48 One of the orbital interactions we did not consider in this chapter is that between an *s* AO from one atom and a *p* AO from another atom in the following fashion:

These orbitals will not interact with this orientation. Explain why.

3.49 The bond in H—Cl can be explained by the overlap between an *s* orbital from hydrogen and a *p* orbital from chlorine, as shown in the following diagram:

(a) Draw the bonding and antibonding MOs that would result from such an interaction. **(b)** What is the symmetry of each of these MOs?

3.50 The bond in Cl_2 can be explained by the end-on overlap between two *p* AOs, as shown in the following diagram:

(a) Draw the bonding and antibonding MOs that would result from such an interaction. **(b)** What is the symmetry of each of these MOs?

3.51 There are two C—C single bonds in penta-1,3-diyne.

Penta-1,3-diyne

(a) Which of those bonds would you expect to be stronger? **(b)** Which of those bonds would you expect to be shorter? **(c)** The molecule is moderately polar, with a dipole moment of 1.37 D. In which direction would you expect the dipole moment to point? Explain.

3.52 Consider the following two molecules:

A   **B**

(a) In which molecule would you expect the C—Cl bond to be stronger? **(b)** In which molecule would you expect the C—Cl bond to be shorter? **(c)** In which molecule would you expect there to be the greater concentration of negative charge on the Cl atom? Explain.

3.53 Octocrylene is an ingredient found in topical sunscreens. It is a water-resistant molecule that helps protect skin against harmful UVA and UVB radiation.

Octocrylene

(a) What is the hybridization of each nonhydrogen atom? **(b)** Circle all atoms bonded to the acyclic C=C double bond that are required to be in the same plane. **(c)** Are there two unique configurations possible about the acyclic C=C double bond? Explain. **(d)** Which of the two C—C single bonds indicated by arrows would you expect to be shorter? Explain.

3.54 In the chapter, we mentioned that ethene ($H_2C=CH_2$) does not undergo free rotation about the C=C double bond because the π bond must be broken to achieve the configuration in which the two CH_2 groups are perpendicular to each other (Fig. 3-28b). **(a)** Draw the MO energy diagram for this "twisted ethene" molecule and identify the HOMO and LUMO. **(b)** By comparing the diagram for this molecule to that in Figure 3-19, explain why the molecule is more stable when it is all planar.

3.55 An amide is typically drawn with a single bond connecting the carbonyl C atom to the N atom.

An amide

If this representation were accurate, we would expect the N atom to be pyramidal, and we would also expect C—N to undergo free rotation. In actuality, however, the N atom is rather planar, and the rotation is quite hindered—properties that give rise to important secondary structures of proteins, such as α helices and β sheets (Chapter 26). Explain these properties of the amide.

If your instructor assigns problems in smartwork, log in at **smartwork.wwnorton.com**.

Naming Alkenes, Alkynes, and Benzene Derivatives

Nomenclature 1 discussed how to name basic alkanes—hydrocarbons with only single bonds—as well as haloalkanes, nitroalkanes, and ethers. Nomenclature 2 extends what we have already learned, enabling us to name molecules that contain the *alkene* (C=C) and the *alkyne* (C≡C) functional groups. We also discuss how to name simple benzene derivatives (compounds that contain a benzene ring), which are related to alkenes because the Lewis structure for benzene (C_6H_6) contains three C=C bonds.

N2.1 Alkenes and Alkynes

When a new arrangement of atoms is encountered in a molecule, it frequently defines a particular class of compounds. We introduced these arrangements as *functional groups* in Chapter 1. Molecules with specific functional groups, such as alkenes and alkynes, are identified by modifying the root name.

- To name *alkenes* and *cycloalkenes*, which contain a C=C double bond, the root for the analogous alkane or cyclohexane is modified by replacing the suffix "ane" with "ene."
- To name *alkynes* and *cycloalkynes*, which contain a C≡C triple bond, the root for the analogous alkane or cyclohexane is modified by replacing the suffix "ane" with "yne."

For example, $CH_3CH_2CH_3$ is propane, so CH_2=$CHCH_3$ is propene and HC≡CCH_3 is propyne. Similarly, cyclooctane becomes cyclooctene if it contains a double bond, and cyclooctyne if it contains a triple bond:

The suffix "ene" indicates the presence of a C=C bond

The suffix "yne" indicates the presence of a C≡C bond

Cyclooctane   Cyclooctene   Cyclooctyne

problem **N2.1** Write the IUPAC name for each of the following molecules.

$$H_2C=CH_2$$

(a) (b) (c)

N2.1a The Numbering System for Simple Alkenes and Alkynes

If a carbon backbone is long enough, then a double or triple bond can appear at different locations, yielding *constitutional isomers* (Chapter 4). Pentene, for example, describes both $CH_2=CH-CH_2-CH_2-CH_3$ and $CH_3-CH=CH-CH_2-CH_3$. In the first molecule, the double bond involves a terminal carbon, whereas in the second molecule it is internal. Convince yourself that the same is true for pentyne by drawing the two structures.

Nomenclature must be able to distinguish one molecule from all others, so we use a numbering system to establish the location of the double or triple bond in compounds like these.

> ▪ Numbers are assigned to the carbons in the main chain so as to give the alkene or alkyne carbons (the carbons joined by a multiple bond) the *lowest possible numbers.*
>
> ▪ In naming the molecule, the lower of the two numbers designating the carbons of the multiple bond is placed immediately before the suffix.

Thus, as shown in Figure N2-1, $CH_2=CH-CH_2-CH_2-CH_3$ is pent-1-ene, because the double bond is at the end of the five-carbon chain (i.e., C1), whereas $CH_3-CH=CH-CH_2-CH_3$ is pent-2-ene, because the double bond begins at C2. Similarly, in pent-1-yne and pent-2-yne (Fig. N2-2), the triple bonds begin at C1 and C2, respectively.

It is important to know that in 1993, the IUPAC changed the rule that specifies the placement of the number locator in the name of an alkene or alkyne to what appears above. Prior to that, the number locator was required to appear immediately

FIGURE N2-1 The numbering system for an alkene In these cases, numbering begins from the left to give the doubly bonded C atoms the lowest possible numbers. The double bonds include C1 and C2 in the first molecule, and C2 and C3 in the second.

FIGURE N2-2 The numbering system for an alkyne In these cases, numbering begins from the right to give the triply bonded C atoms the lowest possible numbers. The triple bonds include C1 and C2 in the first molecule, and C2 and C3 in the second.

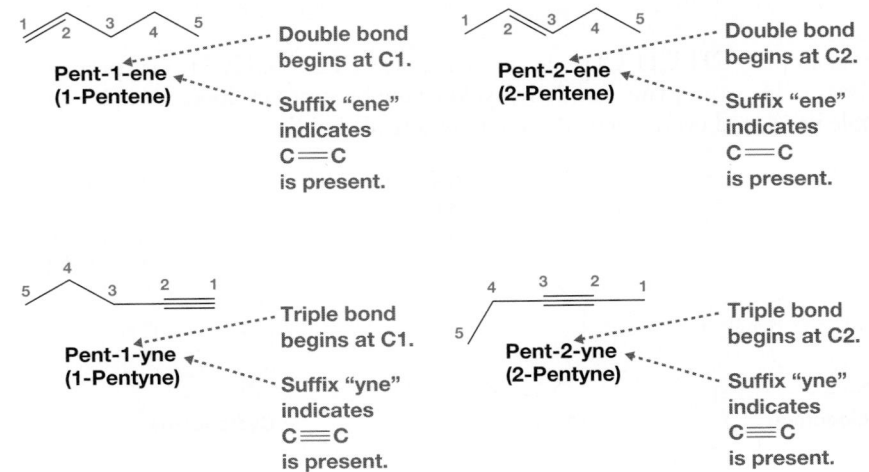

before the beginning of the root. Thus, applying this older rule, the previous four molecules would have been named 1-pentene, 2-pentene, 1-pentyne, and 2-pentyne. It is important to be aware of this change because you will likely encounter many examples of the older rule applied in chemical literature as well as on labels in the chemistry laboratory. In our discussions of nomenclature in this book, we will often provide the names derived using both the new and old rules, as is done in Figures N2-1 and N2-2. However, in other discussions throughout the book, we will be consistent with providing just the name derived using the new rule.

For cycloalkenes and cycloalkynes, assigning the lowest numbers to the alkene or alkyne carbons usually means numbering the carbons of the ring so that C1 and C2 are both part of the double or triple bond:

Cyclobutene
(Not cyclobut-1-ene)

Cyclopentene
(Not cyclopent-1-ene)

Cyclooctyne
(Not cyclooct-1-yne)

In other words, unless there are substituents on the ring (which we discuss shortly), it is redundant to include the number in the molecule's name. Thus, it is cyclobutene, not cyclobut-1-ene; cyclopentene, not cyclopent-1-ene; and cyclooctyne, not cylooct-1-yne.

---

problem **N2.2** What is the IUPAC name of each of the following compounds?

(a) (b) (c) (d) (e)

problem **N2.3** Draw the structures for **(a)** hex-2-ene (2-hexene), **(b)** hex-3-ene (3-hexene), **(c)** hept-1-ene (1-heptene), **(d)** oct-2-yne (2-octyne), and **(e)** cycloheptene.

N2.1b Substituents in Alkenes and Alkynes

Once the numbering system is established, substituents are appended to the name, as described in Sections N1.2 and N1.4:

Methyl group on C4

Chloro substituent on C3

Two methyl groups on C4

R

4-Methylpent-2-ene
(4-Methyl-2-pentene)

3-Chloropent-1-yne
(3-Chloro-1-pentyne)

4,4-Dimethylpent-2-yne
(4,4-Dimethyl-2-pentyne)

A generic 2-alkylbut-1-ene
(A generic 2-alkyl-1-butene)

Note in the fourth structure that R is used to indicate the presence of an alkyl group whose precise structure is not important to the current discussion (a convention introduced in Section N1.2).

In some cases, the alkene or alkyne carbons will have the same pair of numbers regardless of which end of the carbon chain (or ring) is designated as C1:

2-Methylhex-3-yne
(Not 5-methylhex-3-yne)

4-Methoxycyclohex-1-ene
(Not 5-methoxycyclohex-1-ene)

> If the locator numbers of the alkene or alkyne carbons are not affected by which carbon is assigned C1, we choose the system that also gives the *substituents* the lowest possible set of numbers.

If the numbering system in 2-methylhex-3-yne were to begin at the left, instead, then the name would be 5-methylhex-3-yne, which gives a higher number for the methyl substituent than it needs. Similarly, in 4-methoxycyclohex-1-ene, the carbon at the top of the ring could be designated C1, and the numbering could proceed clockwise to designate the second alkene carbon as C2. In this case, however, the OCH_3 substituent would be bonded to C5 instead of C4.

---

problem **N2.4** Name the following structures:

(a)   (b)   (c)

---

problem **N2.5** Provide correct names for the following compounds:

(a)   (b)   (c)

(d)   (e)   (f)

If the carbon backbone of an alkene or alkyne is branched, then the longest continuous chain of carbons might not include the double or triple bond. However, because functional groups are given priority when determining the names of compounds,

> The root name of an alkene or alkyne is derived from the longest continuous chain of carbon atoms *containing the C=C or C≡C bond.*

Longest continuous chain of
carbons containing the C=C

Longest continuous chain of
carbons containing the C=C

2-Ethylhex-1-ene
(2-Ethyl-1-hexene)

3-Propylhex-1-yne
(3-Propyl-1-hexyne)

FIGURE N2-3 Identifying the root
for an alkene and alkyne In both
molecules, the root derives from a six-
carbon chain because longest continuous
chain of C atoms entirely containing the
multiple bond has six C atoms. Each
molecule has a seven-carbon chain that
does *not* entirely contain the multiple
bond.

In 2-ethylhex-1-ene (Fig. N2-3), for example, the longest continuous chain contains seven C atoms, but the longest continuous chain that contains the C=C bond consists of six C atoms, so the compound is a hexene. Similarly, in 3-propylhex-1-yne, the longest continuous chain has seven C atoms, but the longest continuous chain of carbons that contains the C≡C bond consists of six C atoms. Therefore, this compound is a hexyne.

problem **N2.6** Provide the IUPAC names for the following compounds:

(a)

(b)

(c)

problem **N2.7** 2-Propylbut-1-ene is an incorrect name under the IUPAC system. What is the correct name for this compound? (*Hint:* Draw the structure based on the incorrect name and rename it.)

For some alkenes, it is impossible to name the molecule in a way that has the double bond entirely contained within the main chain or ring. In methylenecyclohexane (Fig. N2-4), for example, one alkene carbon is part of the ring and the other is not. This makes it impossible for the double bond to be entirely part of the ring or entirely part of the chain.

For these kinds of cases, the IUPAC system allows one end of the C=C double bond to be treated as a substituent. In the case in Figure N2-4, that substituent is $H_2C=$, called the **methylene substituent**. With the methylene group as a substituent, the root for the molecule becomes cyclohexane.

The methylene
substituent

CH_2

Methylenecyclohexane

FIGURE N2-4 The methylene
substituent The C=C double bond is
not entirely contained within a chain or
ring, so the $H_2C=$ group is treated as a
methylene substituent.

problem **N2.8** Draw the structures for **(a)** 2-chloro-1-methylenecyclopentane and **(b)** 1-methylene-2,4-dinitrocycloheptane.

N2.1c Multiple Double Bonds or Triple Bonds

Some compounds have more than one double bond or triple bond. For these compounds, the name must indicate how many double bonds or triple bonds are present.

- Prefixes (di, tri, etc.) specify how many double and/or triple bonds are present when there is more than one.
- Prior to this prefix, the letter "a" is appended.

Penta-1,3-diyne
(1,3-Pentadiyne)

Penta-1,4-diyne
(1,4-Pentadiyne)

Cyclohepta-1,3,5-triene
(1,3,5-Cycloheptatriene)

4,4-Dimethylpenta-1,2-diene
(4,4-Dimethyl-1,2-pentadiene)

FIGURE N2-5 Numbering systems with multiple C=C or C≡C bonds In the first molecule, two triple bonds begin at C1 and C3. In the second molecule, they begin at C1 and C4. In the third molecule, three double bonds begin at C1, C3, and C5. In the last molecule, two double bonds begin at C1 and C2.

Thus, the name propadiene identifies a chain of three carbons (propa-) containing two C=C bonds (-diene). Similarly, the name hexatriyne identifies a chain of six carbons (hexa-) containing three C≡C triple bonds (-triyne).

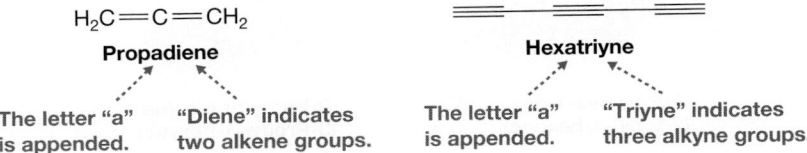

Numbers are unnecessary to indicate the positions of the double bonds or triple bonds in propadiene and hexatriyne because the only locations possible are the ones shown. When this is not the case, numbers must be included.

> A number is added to the name to specify the location of the first carbon in each double or triple bond. As usual, numbers are separated from each other by commas and from letters by hyphens.

As indicated in the examples in Figure N2-5, the numbers can be placed either between the root and the suffix or before the root.

problem **N2.9** In cyclohepta-1,3,5-triene and 4,4-dimethylpenta-1,2-diene in Figure N2-5, supply the numbering systems to the carbons.

problem **N2.10** Draw the structure of 1,4-cyclohexadiene. The "1,4" can appear in another place in the name. Write that name.

problem **N2.11** Draw the structures for penta-1,4-diene and cyclopenta-1,3-diene.

problem **N2.12** What is the name for ?

Occasionally, you will encounter molecules that contain both double and triple bonds such as the ones in Figure N2-6. For these molecules, apply the following additional rules:

> ■ If both double and triple bonds are present in the same chain or ring, both suffixes must be added to the root.
> ■ An "en" suffix (not "ene") appears before the "yne" suffix, and double bonds receive the lower numbers because they have a higher priority than triple bonds.

FIGURE N2-6 Molecules containing both C=C and C≡C bonds In the first molecule, the double bond begins at C1 and the triple bond begins at C4. In the second molecule, two double bonds begin at C1 and C3, and a triple bond begins at C6.

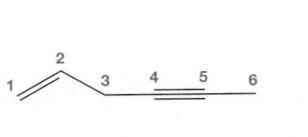

Hex-1-en-4-yne
(1-Hexen-4-yne)

Cycloocta-1,3-dien-6-yne
(1,3-Cyclooctadien-6-yne)

problem **N2.13** Draw the structures for pent-2-en-4-yne and
1,2-dimethylcycloocta-1,3-dien-6-yne.

problem **N2.14** What is the name of this
compound?

N2.1d Trivial Names of Alkenes and Alkynes

As with any class of organic compounds, alkenes and alkynes have trivial names that
are firmly entrenched in nomenclature. Some of the most common ones are shown in
Figure N2-7 and should be committed to memory.

$H_2C=CH_2$

Ethene
(Ethylene)

Propene
(Propylene)

Methylpropene
(Isobutylene)

$HC\equiv CH$

Ethyne
(Acetylene)

Propyne
(Methylacetylene)

2-Butyne
(Dimethylacetylene)

Trivial names are also commonly used for substituents containing double or triple
bonds:

Trivial names of common substituents
containing a multiple bond

Vinyl   **Allyl**   **Propargyl**

Some examples with these common names are as follows:

Cyclohexylethene
(Vinylcyclohexane)

Br

3-Bromoprop-1-ene
(Allyl bromide)

Cl

3-Chloroprop-1-yne
(Propargyl chloride)

Notice in these cases that the use of trivial names does *not* greatly simplify the
molecules' names. With more complex molecules, however, such as the following
substituted cyclopentene compound, the use of trivial names becomes advantageous.

3,5-Divinylcyclopentene

problem N2.15

Draw the structures of (a) 1,4-divinylcyclohexane, (b) allyl vinyl ether, and (c) 1-propargyl-3-chlorocyclopentane.

Work through the following problems to become as comfortable as possible with what you have learned about nomenclature so far.

problem N2.16

Draw the structures for each of the molecules.

(a) 2-Chloropropene

(b) 3-Methylbut-1-ene

(c) 2,3-Dimethyl-2-butene

(d) 2-Ethoxy-3,3-dimethylcyclohexene

(e) 3,4,5-Trimethoxycycloheptene

(f) 3-Bromo-2-methyl-4-nitrocyclopentene

(g) 3,3-Dibromo-4-methylcyclopentene

(h) 1,6-Dimethoxyhexa-1,5-diene

(i) 2-Methyl-1,3,5-hexatriene

(j) 4-Methyl-2-pentyne

problem N2.17

Give the IUPAC name for each of the molecules.

(a)   (b)   (c)   (d)

(e)   (f)   (g)   (h)   (i)

N2.2 Benzene and Benzene Derivatives

The Lewis structure of **benzene** (C_6H_6), shown in Figure N2-8, consists of three C=C double bonds that alternate with three C—C single bonds in a six-membered ring.

Based on the rules presented so far for naming alkenes, it might be tempting to rename benzene "cyclohexa-1,3,5-triene." This name is unacceptable, however, because benzene behaves significantly different than a typical alkene. Benzene is an *aromatic* compound (see Chapter 14 for an in-depth discussion) and, as its resonance hybrid indicates, its π electrons are fully delocalized around the ring. Thus, all six carbon atoms of benzene are identical, and we cannot formally assign any of the double bonds to a single pair of carbon atoms.

π Electrons delocalized around the ring

FIGURE N2-8 Benzene The Lewis structures of benzene are shown inside the brackets, and the resonance hybrid is shown on the right.

In recognition of the unique behavior of the benzene ring,

> Relatively simple compounds containing a benzene ring are usually named with "benzene" as the root.

Prefixes are appended to the name to indicate substituents, just as with cycloalkanes:

Chlorobenzene   **Nitrobenzene**   **Ethylbenzene**   **(*tert*-Butylbenzene)**   **Propoxybenzene**

problem **N2.18** Draw the structures for hexylbenzene and bromobenzene.

problem **N2.19** What is the name of this compound?

N2.2a The Numbering System for the Benzene Ring: *Ortho*, *Meta*, and *Para* Designations

A number is not included in the name of monosubstituted benzenes to indicate the position of the substituent because all six carbon atoms of benzene are equivalent. Therefore, the substituent can always be viewed as being bonded to C1. The story is different, however, if there are two or more substituents.

> ■ When a benzene ring has two or more substituents, a numbering system is used to identify their locations on the ring.
> ■ The ring is numbered to give the substituents the lowest possible set of numbers.
> ■ If there is a tie between two numbering systems, choose the one that gives the lowest number to the substituent that comes first alphabetically.
> ■ Prefixes (di, tri, etc.) indicate the number of each type of substituent.

Because of the cyclic nature of benzene, one of the carbon atoms to which a substituent is attached will be designated C1.

1-Bromo-3-chlorobenzene   **1,2,4-Trinitrobenzene**   **A generic 1,4-dialkylbenzene**   **A generic 2,3,4-trialkylnitrobenzene**

Notice in 1-bromo-3-chlorobenzene that C1 is attached to Br, because alphabetically "bromo" comes before "chloro."

---

problem **N2.20** Draw the structures of 1,2,3-trimethylbenzene and 4-bromo-2-chloro-1-nitrobenzene.

---

problem **N2.21** What are the IUPAC names for the following molecules?

(a)   (b)

For benzene rings that have two substituents, that is, **disubstituted benzenes**, the relative positions of the substituents can be designated using the non-numerical prefixes "*ortho*," "*meta*," or "*para*," which can be abbreviated as *o*, *m*, or *p*, respectively.

> ■ *ortho* = *o* = 1,2-positioning
> ■ *meta* = *m* = 1,3-positioning
> ■ *para* = *p* = 1,4-positioning
>
> (*Note*: When more than two substituents are attached to an aromatic ring, the *ortho*, *meta*, and *para* prefix system is never used.)

The following disubstituted benzenes show how to use the *ortho*, *meta*, and *para* notation.

1-Bromo-2-chlorobenzene	1,3-Dibromobenzene	A generic 1,4-dialkylbenzene
ortho-Bromochlorobenzene	*meta*-Dibromobenzene	A generic *para*-dialkylbenzene
o-Bromochlorobenzene	*m*-Dibromobenzene	A generic *p*-dialkylbenzene

---

problem **N2.22** Draw the structures and give the IUPAC names for *p*-dichlorobenzene and *m*-bromoethoxybenzene.

N2.2b Trivial Names Involving the Benzene Ring

Because of the rich history of aromatic compounds in chemistry, several trivial names are in widespread use. The following examples should be committed to memory.

Methylbenzene or toluene   **Methoxybenzene or anisole**   **Isopropylbenzene (Cumene)**   **Vinylbenzene (Styrene)**

A handful of these trivial names, such as toluene and anisole, have been adopted by the IUPAC system, which is why those two names are not in parentheses above (we will encounter other examples in Nomenclature 4). In such cases, the substituted benzene becomes the root and the numbering is established by the position of the substituent on which the name is based (methyl in the case of toluene, and methoxy in the case of anisole).

2-Ethyl-4-pentyltoluene   **2-Chlorotoluene** *o*-Chlorotoluene   **2,4,6-Trinitrotoluene (TNT)**   **3-Nitroanisole** *m*-Nitroanisole   **4-Propoxyanisole** *p*-Propoxyanisole

In the first three molecules, C1 is the carbon to which the methyl group is attached, because the root is toluene. In the last two molecules, C1 is the carbon to which the methoxy group is attached, because the root is anisole.

---

problem **N2.23** Draw the structures and give the IUPAC names for
(a) *m*-bromotoluene, **(b)** 2,5-dinitrotoluene, and **(c)** *p*-chloroanisole.

Some trivial names describe two substituents on benzene, not just one. The most common of these is xylene, the trivial name for dimethylbenzene, which has *ortho*, *meta*, and *para* isomers.

1,2-Dimethylbenzene *ortho*-Dimethylbenzene *o*-Dimethylbenzene (*o*-Xylene)   **1,3-Dimethylbenzene** *meta*-Dimethylbenzene *m*-Dimethylbenzene (*m*-Xylene)   **1,4-Dimethylbenzene** *para*-Dimethylbenzene *p*-Dimethylbenzene (*p*-Xylene)

N2.2c Substituents Containing the Benzene Ring

If a substituent on a benzene ring is sufficiently complicated, the molecule is probably easier to name if we treat the benzene ring as a substituent instead.

As a substituent, C_6H_5 is called the **phenyl** group, pronounced the same as the spice "fennel." It is often abbreviated as Ph.

Phenyl is abbreviated
C_6H_5 or simply Ph

The phenyl group

1-Phenyl-4-chlorohexane

2,3-Diphenylpent-1-ene

or Ph_3CH or $(C_6H_5)_3CH$

Triphenylmethane

or C_6H_5 or   Ph

1,2-Diphenylbutane

problem **N2.24** Draw the structures for **(a)** 2-phenyl-1-hexene and **(b)** 1,5-diphenylpentane.

problem **N2.25** What is the name of this compound?

$$Ph-\underset{\underset{Ph}{|}}{\overset{\overset{Ph}{|}}{C}}-\underset{\underset{Ph}{|}}{\overset{\overset{Ph}{|}}{C}}-Ph$$

A closely related substituent is $C_6H_5CH_2$—, which is called the **phenylmethyl** group or the **benzyl** group. It can be abbreviated as $PhCH_2$— or Bn—.

The benzyl group

or   Bn

3-Benzylhexane

or

1,4-Dibenzylcyclohexane

Finally, if we want to indicate the presence of an aromatic substituent, but are unconcerned with its specific structure, we can use the general abbreviation Ar, which stands for **aryl**. For example, a generic aryl chloride might be represented as Ar—Cl.

Work through the following practice problems to become as comfortable as possible with what you have learned about nomenclature so far.

problem **N2.27** Draw the structures for each of the following molecules.

(a) Fluorobenzene
(c) 1-Iodo-4-nitrobenzene
(e) 2-Fluorotoluene
(g) 2-Ethoxyanisole
(i) 1,3-Diphenylheptane

(b) 1-Chloro-2-fluorobenzene
(d) 1,3-Dibromobenzene
(f) 4-Ethoxytoluene
(h) 2,3-Dimethyl-1-cyclopentylbenzene
(j) 4,4-Diphenyl-1-octene

problem **N2.28** Give the IUPAC name for each of the following molecules. Which of these compounds could be named using *ortho-*, *meta-*, or *para-* as a prefix?

(a)　　　　(b)　　　　(c)　　　　(d)　　　　(e)

(f)　　　　(g)　　　　(h)　　　　(i)

Isomerism 1

Conformational and Constitutional Isomers

There are tens of millions of known organic compounds, and more are being isolated, synthesized, and identified every year! Fortunately, the chemical behavior of every compound is *not* entirely unique. Instead, *molecules that are similar in composition and structure tend to have similar reactivity*. We first encountered this principle in Chapter 1, where we saw that identical functional groups in different molecules tend to react similarly due to the specific arrangement of their atoms.

Here in Chapter 4 and later in Chapter 5, we focus more closely on the relationships between similar molecules, in particular those with the same molecular formulas: *isomers*. In Chapter 4 we concentrate on the distinctive characteristics of two classes of isomers: *conformational isomers* and *constitutional isomers*. In Chapter 5 we explore two other types of isomers: *enantiomers* and *diastereomers*.

When a rollercoaster car travels along its rails, it gains and loses potential energy, much like a molecule's energy changes as it rotates about its single bonds. These energy changes within a molecule give rise to conformational isomers, a topic of Chapter 4.

4.1 Isomerism: A Relationship

Two molecules are said to be **isomers** of each other if they have the same molecular formula, but they are different in some way.

There are a variety of ways in which two molecules with the same molecular formula can differ from each other. Their atoms can be *connected in different ways*; they can differ as a result of *rotations about single bonds*; or they can be *mirror images* of each other. These differences give rise to the different types of *isomerism* shown in the flow chart in Figure 4-1. Constitutional isomers and conformational isomers (i.e., the categories highlighted in red in Fig. 4-1) are examined in depth here in Chapter 4; configurational isomers, diastereomers, and enantiomers are discussed in Chapter 5.

Each type of **isomerism** is a *specific relationship* between molecules, so the term "isomer" cannot be applied to a single molecule. It is a mistake, for example, to refer to a single molecule as a constitutional isomer, a diastereomer, or an enantiomer without reference to a second molecule. This is no different from saying: "John is a cousin." A cousin of whom? We *can* say, however, that two molecules are constitutional isomers *of each other*, or that one molecule is an enantiomer *of a second*. Similarly, we can say, "John is Jane's cousin," or "John and Jane are cousins."

4.2 Conformational Isomers: Rotational Conformations, Newman Projections, and Dihedral Angles

According to the flow chart in Figure 4-1, a pair of **conformational isomers**, also called **conformers**, are **stereoisomers** of each other that differ only by rotations about single bonds. That is, conformational isomers have the same **connectivity**, meaning

FIGURE 4-1 Flow chart for categorizing types of isomers Constitutional isomers and conformational isomers (red) are discussed in depth here in Chapter 4. Diastereomers and enantiomers are discussed in Chapter 5.

Bond not visible in Newman projection

Back carbon

90° Rotation

Front carbon

FIGURE 4-2 Interpretation of a Newman projection The generic molecule on the left can be depicted by the Newman projection on the right. Each carbon has three substituents pointing outward.

that the same atoms in both species are bonded together by the same types of bonds. This must be the case because, as we saw in Chapter 3, no bonds are broken or formed in the rotation about single bonds. (We examine aspects of connectivity in greater detail in Sections 4.11–4.13, when we discuss *constitutional isomers*.)

To more easily study conformational isomers, it is helpful to have a systematic way to depict molecules having different angles of rotation about single bonds—that is, different **rotational conformations**. One convenient way to illustrate rotational conformations is with a **Newman projection**, which is a two-dimensional representation of a molecule *viewed down the bond of interest* (Fig. 4-2). The following conventions are used when drawing Newman projections:

- In a Newman projection, the two atoms directly connected by the bond of interest are shown explicitly.
  - The nearer atom is depicted as a point.
  - The more distant atom is depicted as a circle.
- Bonds to the front atom converge at the point, whereas bonds to the back atom connect to the circle.

YOUR TURN 4.1

Based on the Newman projection given, label the "front carbon" and the "back carbon" in the structure on the left.

90° Rotation

Answers to Your Turns are in the back of the book.

The Newman projection in Figure 4-2 shows that *the bond of interest is not visible; instead, it must be imagined as connecting the front and back carbons*. A Newman projection does, however, highlight the other bonds to the front and back atoms.

YOUR TURN 4.2

Construct a molecule of 1,1,1-tribromoethane (H_3C—CBr_3) using a molecular modeling kit. Rotate the CH_3 and CBr_3 groups relative to each other until the molecule looks like the structure shown in dash–wedge notation on the left side of Figure 4-2. Look down the C—C bonding axis and fill in the missing atoms in the incomplete Newman projection provided here.

Figure 4-3 shows the Newman projections for a series of rotations about the C—C bond of a generic molecule XCH_2—CH_2Y. In the figure, we have arbitrarily chosen to leave the C atom in front (the CH_2Y group) frozen in place and to rotate the back C (the CH_2X group).

Each angle of rotation defines a particular **dihedral angle**, θ, corresponding to the angle between the C—X and C—Y bonds as they appear in the Newman projection. By convention, θ can assume values between $-180°$ and $+180°$, where the conformations at $-180°$ and $+180°$ are exactly the same because they are 360° apart.

Dihedral angle, θ	−180°	−120°	−60°	0°	+60°	+120°	+180°
View of C—C bond from the side							
View down the C—C bond							
Newman projection							

FIGURE 4-3 Newman projections and bond rotations A generic molecule of XCH_2—CH_2Y is shown at various angles of rotation (dihedral angles, θ) about the C—C bond. (Top) View of the molecule from the side. (Middle) View of the molecule down the C—C bonding axis (from the right side). (Bottom) Newman projection of the molecule that corresponds to the view down the C—C bond.

4.3 **YOUR TURN**

Add the substituents to the incomplete Newman projections to represent the molecule at the left after +60° and +120° rotations of the back carbon.

0°   +60°   +120°

SOLVED problem 4.1 Draw the Lewis structure of the molecule whose Newman projection is as follows:

Think Which atoms are connected by the bond not observable in the Newman projection? What atoms or groups are attached to those atoms?

Solve The bond not shown is a C—C single bond. The C atom in the front (represented by the point) is bonded to two H atoms and a CH_3 group, whereas the C in the back (represented by a circle) is bonded to three H atoms. Thus, the structure consists of three C atoms bonded together. The compound is propane.

···· Front carbon of Newman projection

···· Back carbon of Newman projection

Propane

problem **4.2** Draw the Lewis structure that corresponds to each of the following Newman projections:

(a)   (b)   (c)

problem **4.3** The Newman projection at the right is of the same generic molecule used in Figure 4-3. Taking this dihedral angle to be −180°, draw Newman projections for each 60° rotation about the C—C single bond, from −180° to +180°, in which *the back carbon remains frozen in place*.

4.3 Conformational Isomers: Energy Changes and Conformational Analysis

To better understand the nature of a rotation about a given bond, we can perform a **conformational analysis**, which is a plot of a molecule's energy as a function of that bond's dihedral angle. The energy values for various angles of rotation are relative to that of the lowest-energy conformation, and may be obtained from experiments in spectroscopy (see Chapters 15 and 16), or they may be estimated from computer calculations or various other means.

4.3a Conformational Analysis of Ethane: Torsional Strain, Eclipsed Conformations, and Staggered Conformations

Figure 4-4 depicts the conformational analysis of ethane (H_3C—CH_3) about the C—C bond. There are three rotational conformations in Figure 4-4 in which ethane's energy is at a *maximum* (i.e., θ = 0°, +120°, and −120°). These are called **eclipsed conformations** because the C—H bonds on the front carbon atom cover, or "eclipse," those on the rear carbon atom in the Newman projections. Because the hydrogen substituents are all identical, these three conformations are indistinguishable.

There are also three rotational conformations of H_3C—CH_3 in which the energy is at a minimum (i.e., θ = ±180°, −60°, and +60°). In the Newman projections of these conformations, each C—H bond on the front carbon atom bisects a pair of C—H bonds on the rear carbon, so the bonds to the front and rear carbon atoms alternate around the circle. These are therefore called **staggered conformations**.

The eclipsed and staggered conformations differ only by a rotation about the C—C single bond. However, the eclipsed conformations do not exist for significant amounts of time because they represent energy maxima (see Fig. 4-4). Instead, essentially all molecules of ethane exist in staggered conformations, which appear at energy minima.

YOUR TURN 4.4

In Figure 4-4, label each of the unlabeled conformations as either "eclipsed" or "staggered."

FIGURE 4-4 Conformational analysis of ethane, H_3C—CH_3 The plot shows the relative energy of ethane as a function of the C—C dihedral angle. Energies are relative to the lowest-energy conformation. The double-headed arrow represents the energy barrier for rotation about that bond.

The difference in energy between a staggered conformation and an eclipsed conformation of H_3C—CH_3 is a result of *torsional strain*. In general,

Torsional strain is an increase in energy (i.e., decrease in stability) that arises in an eclipsed conformation.

In ethane, that torsional strain is brought about by *electron repulsion between eclipsed bonds* (Fig. 4-5). In a staggered conformation of ethane, the distance between C—H bonds on adjacent carbon atoms is at a maximum, so the resulting electron repulsion is at a minimum (Fig. 4-5a). In an eclipsed conformation, however, the distance between those bonds is at a minimum, so the electron repulsion is at a maximum (Fig. 4-5b).

4.5 YOUR TURN

Construct a molecule of ethane (H_3C—CH_3) using a molecular modeling kit. Rotate the molecule so that the C—C bond is in an eclipsed conformation. Now rotate the C—C bond so that it is in a staggered conformation. Are the C—H bonds closer in the eclipsed conformation or in the staggered conformation?

More stable because electron repulsion is at a minimum

Less stable because electron repulsion is at a maximum

Staggered

Eclipsed

(a)

(b)

FIGURE 4-5 Origin of torsional strain (a) In the staggered conformation, repulsion among the electrons in the C—H bonds is at a minimum, making this the more stable conformation. (b) In the eclipsed conformation, that repulsion is at a maximum, making it the less stable conformation.

Use Figure 4-4 to estimate the torsional strain in a molecule of ethane in an eclipsed conformation. _____

Although essentially all ethane molecules exist in staggered conformations, a molecule of $H_3C—CH_3$ does not exist in one staggered conformation indefinitely. Instead,

At room temperature, staggered conformations constantly interconvert through rotation about the C—C single bond.

Ethane's staggered conformations rapidly interconvert because the *thermal energy* in the surroundings is comparable to (i.e., is of the same order of magnitude as) the molecule's *rotational energy barrier*. The **rotational energy barrier** is the amount of energy needed to convert one conformational isomer into another via rotation about a specified bond. In the case of ethane, this is simply the 12 kJ/mol (2.9 kcal/mol) difference in energy between the staggered and eclipsed conformations (Fig. 4-4). **Thermal energy** is the average energy available through molecular collisions, and it increases as temperature increases. The average thermal energy is calculated as the product RT, where R is the universal gas constant (8.314 J/mol · K or 1.987 cal/mol · K) and T is the temperature in kelvins. If T is 298 K, then the average thermal energy is ~2.5 kJ/mol (~0.6 kcal/mol).

It may seem peculiar at first that 2.5 kJ/mol (0.6 kcal/mol) of thermal energy is sufficient to surmount ethane's 12 kJ/mol (2.9 kcal/mol) rotational energy barrier. Keep in mind, however, that thermal energy is an *average*, so some molecules possess more energy than that value and some possess less. The graph in Figure 4-6 may help you understand this concept better. For a given value of energy on the x axis, the percentage of molecules that possess *at least* that much energy at room temperature is given on the y axis. Notice that roughly 2% of ethane molecules at any moment in time possess enough energy to surmount ethane's rotational energy barrier, which, in a mole of ethane, is approximately 1.2×10^{22} molecules!

FIGURE 4-6 Distribution of thermal energy in molecules at 298 K For each energy plotted on the x axis, the percentage of molecules possessing *at least* that much energy at 298 K is plotted on the y axis. At any given time, about 2% of molecules possess enough energy to surmount an energy barrier of 12 kJ/mol (2.9 kcal/mol), the rotational energy barrier for ethane.

About 2% of molecules possess at least 12 kJ/mol from thermal energy.

4.3b Conformational Analysis of 1,2-Dibromoethane: Steric Strain, Gauche Conformations, and Anti Conformations

If one H atom of each CH_3 group in ethane ($H_3C—CH_3$) is replaced by another substituent, the energies of the three staggered conformations are no longer the same. Neither are the energies of the three eclipsed conformations. This can be seen in the conformational analysis of 1,2-dibromoethane ($Br—CH_2—CH_2—Br$) shown in Figure 4-7.

Label each unlabeled structure in Figure 4-7 as either "eclipsed" or "staggered."

Use Figure 4-7 to estimate how much higher in energy one eclipsed conformation of 1,2-dibromoethane ($BrCH_2—CH_2Br$) is than the other two.

_____ kJ/mol   _____ kcal/mol

Additionally, estimate how much lower in energy one staggered conformation is than the other two.

_____ kJ/mol   _____ kcal/mol

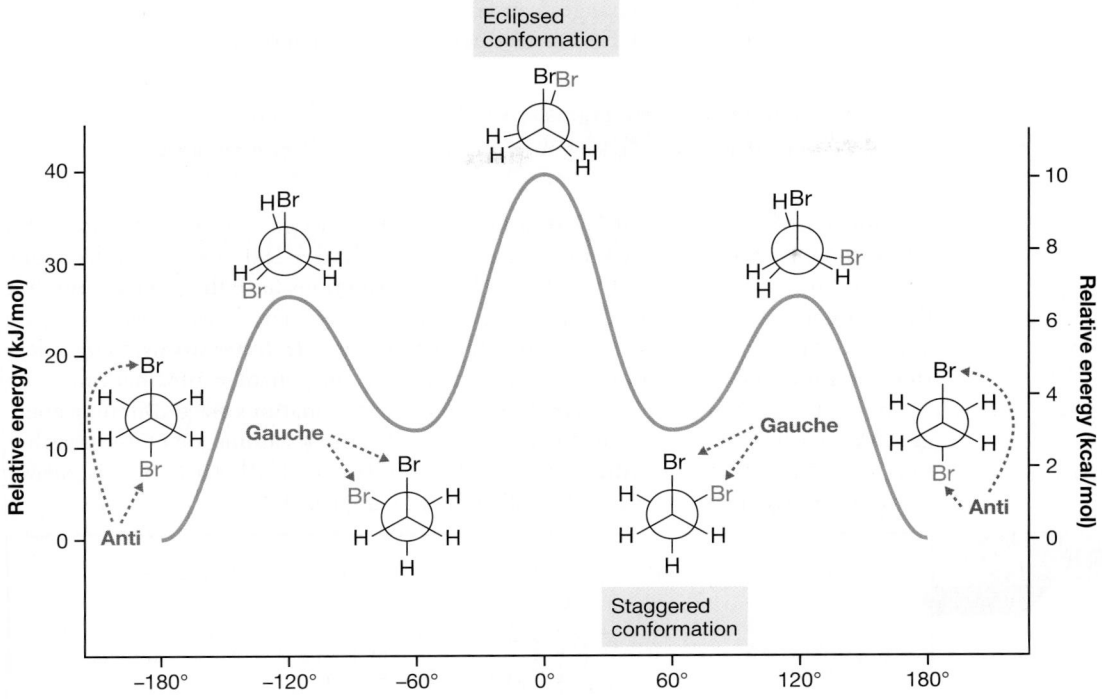

FIGURE 4-7 **Conformational analysis of 1,2-dibromoethane, Br—CH$_2$—CH$_2$—Br** The energy of 1,2-dibromoethane is plotted as a function of the Br—C—C—Br dihedral angle. Energies are relative to that of the most stable conformation. The gauche and anti conformations are labeled.

The differences in energy that you estimated in Your Turn 4.8 are mainly due to the much larger size of a bromine atom compared to a hydrogen atom. As the Br atoms are brought closer together through rotation about the C—C bond, they begin to crash into each other, and their electrons, forced to occupy the same space, repel one another. This is a form of strain called *steric strain* and is depicted in Figure 4-8.

More generally,

Steric strain is an increase in energy that results from electron repulsion between atoms or groups of atoms that are not directly bonded together but occupy the same space.

(a)

(b)

FIGURE 4-8 **Steric strain in 1,2-dibromoethane, BrCH$_2$CH$_2$Br** (a) 1,2-Dibromoethane in its gauche conformation, shown as a Newman projection (left), a ball-and-stick model (middle), and a space-filling model (right). Because the two Br atoms are quite large, their electrons begin to occupy the same space, causing steric strain. (b) 1,2-Dibromoethane in its anti conformation, shown as a Newman projection (left), a ball-and-stick model (middle), and a space-filling model (right). Because the Br atoms are 180° apart, this conformation does not possess steric strain.

4.3 Conformational Isomers: Energy Changes and Conformational Analysis / **183**

Because of the energy changes caused by bulky groups like bromine atoms, the staggered conformations are further distinguished as anti and gauche:

- **Anti conformation:** Bulky groups are 180° apart in a Newman projection.
- **Gauche conformation:** Bulky groups are 60° apart in a Newman projection.

Notice in Figure 4-7 that 1,2-dibromoethane has two gauche conformations and one anti conformation (the conformations at −180° and +180° are identical, so they count as one conformation). The two gauche conformations have the same energy, but the anti conformation is lower in energy than either of them because the bulky Br atoms are farther apart in space in the anti conformation. In other words, there is less steric strain in the anti conformation than there is in the gauche conformation.

Similar arguments explain why the eclipsed conformations have different energies. With a Br—C—C—Br dihedral angle of 0°, the Br atoms are closer together in space than when the dihedral angle is either +120° or −120°. Thus, there is greater steric strain when the Br—C—C—Br dihedral angle is 0°.

YOUR TURN 4.9

Identify which of the following conformations is anti and which is gauche.

The three staggered (i.e., the two gauche and one anti) conformations of Br—CH$_2$—CH$_2$—Br are stable relative to the eclipsed conformations. Being staggered, the gauche and anti conformations occur at energy minima in Figure 4-7, whereas the eclipsed conformations appear at energy maxima. Consequently, gauche and anti conformations are considered to be *isomers* of each other because they have the same molecular formula, yet are not exactly the same. More specifically, they differ only by the rotation about a single bond within the molecule. According to Figure 4-1, this means that

Gauche and anti conformations are *conformational isomers* of each other.

Like those in ethane, all three staggered conformations of 1,2-dibromoethane interconvert rapidly at room temperature, because the rotational energy barrier between the anti and gauche conformations (see Your Turn 4.10) is similar in magnitude to the thermal energy available at room temperature (2.5 kJ/mol; 0.60 kcal/mol). Consequently, just as in ethane, a significant (but smaller) percentage of 1,2-dibromoethane molecules possess sufficient energy to surmount that barrier.

YOUR TURN 4.10

In Figure 4-7, draw a double-headed arrow (↔) to indicate the rotational energy barrier upon going from the anti conformation to one of the gauche conformations. Estimate that energy barrier here: _____ kJ/mol; _____ kcal/mol.

Even though the gauche and anti conformations of 1,2-dibromoethane rapidly interconvert, they do not exist in equal abundance.

An anti conformation is more stable than the corresponding gauche conformation, so an anti conformation is preferred.

In fact, Br—CH$_2$—CH$_2$—Br molecules at room temperature are 98% anti and 2% gauche.

---

problem **4.4** Perform a conformational analysis of 1,2-dichloroethane (ClCH$_2$—CH$_2$Cl) by sketching its energy as a function of the dihedral angle about the C—C bond. Make sure the *relative* energies are correct, but do not concern yourself with the exact values. Draw Newman projections for each staggered and eclipsed conformation, and identify the anti and gauche conformations.

---

problem **4.5** Which molecule do you think has a larger rotational energy barrier about the C—C bond: 1,2-dibromoethane or 1,2-difluoroethane? Why?

4.3c Longer Molecules and the Zigzag Conformation

For longer molecules, such as hexane (CH$_3$CH$_2$CH$_2$CH$_2$CH$_2$CH$_3$), there are several single bonds about which rotation can occur, giving rise to many possible conformations. Each of the three interior C—C bonds in hexane has the form R—CH$_2$—CH$_2$—R, in which R is an alkyl group. As a result, each of these bonds can exist in either a gauche or an anti conformation. Just as we saw with 1,2-dibromoethane, anti conformations are favored over gauche, so the most stable conformation of hexane is the **all-anti conformation**, sometimes called the **zigzag conformation** (Fig. 4-9a). This is the basis for the zigzag convention we use to depict alkyl chains in line structures (Fig. 4-9b).

In a hexane chain, the all-anti conformation is most stable, giving rise to a zigzag structure.

(a)

(b)

FIGURE 4-9 The all-anti conformation of hexane (a) Ball-and-stick model of hexane. (b) Line structure of hexane.

4.4 Conformational Isomers: Cyclic Alkanes and Ring Strain

Thus far, we have limited our conformational analyses to single bonds in acyclic compounds. Single bonds, however, may also be part of ring structures, so ring structures can also have conformations associated with them, too (Fig. 4-10). We examine these conformations in Sections 4.5 and 4.6. Here, we introduce ring structures in general and discuss their relative stabilities.

Cyclopropane **Cyclobutane** **Cyclopentane** **Cyclohexane** **Cycloheptane** **Cyclooctane**

FIGURE 4-10 Cycloalkanes Because their rings consist of single bonds, some cycloalkanes can attain different conformations.

Ring structures consisting only of single bonds are particularly abundant in nature (Fig. 4-11). Menthol, for example, is a natural oil that contains a six-membered ring. The basic structure of all steroids, such as androsterone, consists of three six-membered rings and one five-membered ring. Diamond, one of the hardest substances known, is an extended network of six-membered rings. And a

FIGURE 4-11 Rings in nature Menthol, androsterone, and diamond have rings made of only carbon. Ribose and deoxyribose have rings that consist of carbon and oxygen.

Menthol   **Androsterone**   **Diamond**   **Ribose**   **Deoxyribose**

An All-Gauche Alkane

How can a linear alkane be made to adopt an all-gauche conformation? The answer is to trap it inside a molecular capsule whose cavity is shorter than the stretched alkane in its zigzag conformation. Recall that a linear alkane, free of any external influences, prefers the all-anti conformation. But in 2004, Julius Rebek, Jr., at the Scripps Research Institute, trapped linear tetradecane ($C_{14}H_{30}$) inside a capsule that self-assembles via hydrogen bonds between its two halves.

R = $C_{11}H_{23}$

To fit inside the cavity, the molecule must twist into a helical structure to become shorter. In that helical structure, all C—C bonds adopt a gauche conformation. Each gauche conformation raises the energy of, and thus destabilizes, the alkane, but that destabilization is more than compensated by the hydrogen bonding between the two halves of the capsule and by the favorable interactions between the alkane and the capsule walls.

Studies such as this are more than just a novelty. They can, in fact, help us better understand the way that some medicines function. Medicines are typically small molecules that operate inside a relatively small and confining cavity of a protein or nucleic acid. And it is known that the severe confinement of a molecule can have a dramatic effect on its chemical behavior.

principal component of RNA and DNA are five-membered rings called ribose and deoxyribose, respectively.

Although compounds can exist with rings of any size, five- and six-membered rings are most prevalent in nature, suggesting that rings of these sizes are particularly stable. We attribute the instability of other ring sizes to **ring strain**, the increase in energy due to geometric constraints on the ring.

Ring strain can be quantified using **heats of combustion**, the energy given off in the form of heat ($\Delta H°$) during a combustion reaction. The heats of combustion for some cyclic alkanes are listed in Table 4-1.

We can extract information about strain from heats of combustion because the *balanced* chemical equation for the combustion of each of these compounds has the same form; all that varies is n, the number of carbon atoms:

$$(CH_2)_n + (3n/2)\, O_2 \rightarrow n\, CO_2 + n\, H_2O + \text{heat} \qquad (4\text{-}1)$$

The number of moles of each combustion product is also n, so the amount of product depends upon the size of the ring. Therefore, to compare heats of combustion, we must divide each term in Equation 4-1 by n, as shown in Equation 4-2a:

$$(CH_2)_n/n + (3n/2)/n\ O_2 \rightarrow n/n\ CO_2 + n/n\ H_2O + \text{heat}/n \qquad (4\text{-}2a)$$

This reduces to Equation 4-2b, and yields the ring's heat of combustion *per CH_2 group*: heat/n. Values of heat/n are listed for a variety of rings in Table 4-1.

$$\text{"}(CH_2)\text{"} + (3/2)\ O_2 \rightarrow CO_2 + H_2O + \text{heat}/n \qquad (4\text{-}2b)$$

Because the products of combustion are otherwise identical, we may assume the following for a cyclic alkane:

- Any difference in the heat of combustion per CH_2 group for two cyclic alkanes reflects a difference in *ring strain* per CH_2 group.
- The total ring strain is obtained by multiplying the strain per CH_2 group by the number of CH_2 groups, n.

According to Table 4-1, ring strain is smallest for cyclohexane, the six-membered ring. In fact, as we will see in Section 4.5, cyclohexane is considered to have no ring strain at all. Cyclopentane (the five-membered ring) and cycloheptane (the

TABLE 4-1 Heats of Combustion and Ring Strain for Cyclic Alkanes

Cycloalkane	Number of CH_2 Units	Heat of Combustion, $\Delta H°$		Heat of Combustion, $\Delta H°$, per CH_2 Unit		Ring Strain per CH_2 Unit, Relative to Cyclohexane		Total Ring Strain Relative to Cyclohexane	
		kJ/mol	kcal/mol	kJ/mol	kcal/mol	kJ/mol	kcal/mol	kJ/mol	kcal/mol
Cyclopropane	3	1961.0	468.7	653.7	156.2	653.7 − 615.1 = 38.6	156.2 − 147.0 = 9.2	3(38.6) = 115.8	3(9.2) = 27.6
Cyclobutane	4	2570.2	614.3	642.6	153.6	642.6 − 615.1 = 27.5	153.6 − 147.0 = 6.6	4(27.5) = 110.0	4(6.6) = 26.4
Cyclopentane	5	3102.5	741.5	620.5	148.3	620.5 − 615.1 = 5.4	148.3 − 147.0 = 1.3	5(5.4) = 27.0	5(1.3) = 6.5
Cyclohexane	6	3690.7	882.1	615.1	147.0	615.1 − 615.1 = 0	147.0 − 147.0 = 0	6(0) = 0	6(0) = 0
Cycloheptane	7	4332.1	1035.4	618.9	147.9	618.9 − 615.1 = 3.8	147.9 − 147.0 = 0.9	7(3.8) = 26.6	7(0.9) = 6.3

seven-membered ring), on the other hand, both have small amounts of ring strain. Cyclopropane (the three-membered ring) and cyclobutane (the four-membered ring) are highly strained.

YOUR TURN 4.11

The heat of combustion of cyclooctane is 4962.2 kJ/mol (1186.0 kcal/mol).

Cyclooctane

How many CH_2 groups compose cyclooctane? _____

Compute the heat of combustion per CH_2 group. _____

Compute the ring strain per CH_2 group by subtracting cyclohexane's heat of combustion per CH_2 group from that determined for cyclooctane. _____

Compute the total ring strain for cyclooctane by taking into account the total number of CH_2 groups in the molecule. _____

How do these values compare to those for the other molecules in Table 4-1?

SOLVED problem 4.6 Rank the following cycloalkanes in order from lowest heat of combustion to highest.

(a)   (b)   (c)   (d)

Think What is the same among these molecules? What is different? How do those differences translate into heats of combustion?

Solve All four of these compounds have the molecular formula C_7H_{14} and contain only C—C and C—H single bonds. Differences in heats of combustion, therefore, will largely reflect differences in ring strain. The three-membered ring (**D**) is the most strained, so it will give off the most heat during combustion—it will have the highest heat of combustion. Next comes the four-membered ring (**A**), followed by the seven-membered ring (**C**). The six-membered ring (**B**) is the least strained, so it will have the lowest heat of combustion. In order, then, the heats of combustion are: **B < C < A < D.**

problem 4.7 Rank the following cycloalkanes in order from lowest heat of combustion to highest.

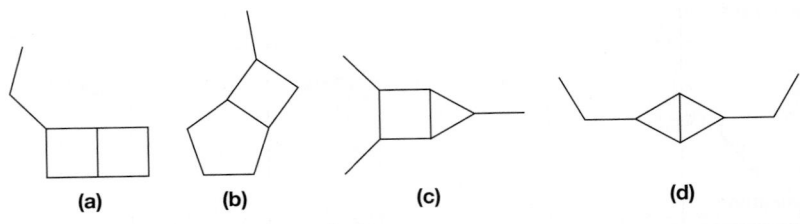

(a)   (b)   (c)   (d)

4.5 Conformational Isomers: The Most Stable Conformations of Cyclohexane, Cyclopentane, Cyclobutane, and Cyclopropane

In Section 4.4, we saw that rings of different size possess different amounts of *ring strain*. What structural factors give rise to ring strain?

Angle strain, which we first encountered in Chapter 2, makes a significant contribution to ring strain. In the cyclic alkanes discussed so far, the atoms of the ring are sp^3 hybridized, with an *ideal* bond angle of 109.5°. Geometric constraints imposed by the ring, however, can force significant deviations in these bond angles, thereby causing an increase in energy. Additionally, *torsional strain* and *steric strain* can contribute to ring strain. This should come as no surprise, because torsional strain and steric strain can exist for any single bond, whether or not the bond is part of a ring structure.

Here in Section 4.5, we study these contributions toward ring strain in rings of various sizes.

4.5a Cyclohexane

According to Table 4-1, cyclohexane has less ring strain than other cycloalkanes. In fact, cyclohexane has no ring strain at all. This can be better understood by examining its geometry. Cyclohexane is not a planar molecule; instead, its lowest energy conformation resembles a chair (Fig. 4-12a) and is therefore called a **chair conformation**. In this conformation, all bond angles of the ring are about 111° (Fig. 4-12b), which is very close to the ideal tetrahedral angle of 109.5° (Chapter 2). Consequently, the six-membered ring of cyclohexane has essentially no angle strain.

FIGURE 4-12 Various representations of cyclohexane (a) Ball-and-stick model of cyclohexane viewed from the side, illustrating its chair conformation. (b) Ball-and-stick model of cyclohexane viewed from the top. All C—C—C bond angles are about 111°. (c) Ball-and-stick model of cyclohexane viewed down two C—C bonds, illustrating that those bonds are in staggered conformations. (d) Newman projection of cyclohexane.

The most stable conformation of cyclohexane resembles a chair.

Essentially no angle strain

111°

Essentially no torsional strain

(a)　　(b)　　(c)　　(d)

It also has little to no torsional strain (Section 4.3a), because all of the rotational conformations about the C—C bonds are staggered. This can be seen in both the ball-and-stick model in Figure 4-12c and the Newman projection in Figure 4-12d.

If the six carbon atoms of cyclohexane were co-planar, as shown in Figure 4-13, then there would be considerable strain in the ring. One contributing factor would

120° = Angle strain

90° Rotation

Steric strain　Torsional strain

Eclipsed conformations

(a)　　(b)　　(c)

FIGURE 4-13 Strain in planar cyclohexane (a) Ball-and-stick representation of planar cyclohexane viewed from the top, illustrating that its C—C—C bond angles are 120°. This deviation from the ideal angle of 109.5° adds angle strain to the ring. (b) Ball-and-stick representation and (c) Newman projection of planar cyclohexane viewed from the side, showing two C—C bonds in their eclipsed conformations. These conformations add both steric strain and torsional strain to the ring.

be the significantly increased angle strain, given that a regular hexagon has an interior angle of 120°—an angle that is quite different from the ideal angle of 109.5°. Two other reasons stem from the fact that all six C—C bonds would be in eclipsed conformations. As with any eclipsed conformation, this adds torsional strain, which we can see explicitly for two of the six C—C bonds in the Newman projection in Figure 4-13. Additionally, the Newman projection shows that these eclipsed conformations bring nonbonded CH_2 groups closer together, which adds steric strain—that is, the electron clouds on those CH_2 groups begin to crash together, causing electron repulsion.

YOUR TURN 4.12

Using a molecular modeling kit, build a molecule of cyclohexane in its chair conformation, then rotate it in space until it appears as in Figure 4-12c. Observe the staggered conformation about each C—C bond. Next force all of the carbon atoms to be planar as in Figure 4-13 and observe the eclipsed conformation about each C—C bond.

4.5b Cyclopentane

The lowest-energy conformation of cyclopentane is shown in Figure 4-14. Four of its five carbon atoms lie essentially in one plane, with the fifth carbon outside of that plane. If you imagine the carbon atoms located at the five corners of the envelope in Figure 4.14c, you can see why this geometry is referred to as an **envelope conformation**.

As in cyclohexane, strain in cyclopentane is minimized by adopting a nonplanar configuration. In a hypothetical planar structure (Fig. 4-15a), cyclopentane would have each of its C—C bonds in an eclipsed conformation, just as in the hypothetical planar structure of cyclohexane (Fig. 4-13a). Lifting one carbon atom out of the plane, however, allows four of the five bonds in the ring to become more staggered. For those bonds, the C—C—C—C dihedral angles range between about 28° and 45°, compared to 60° for an ideal staggered conformation (Fig. 4-15b). Torsional strain, therefore, is relieved to some extent.

(a)

Bond in front of plane of paper and parallel to paper

(b)

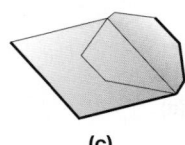

(c)

FIGURE 4-14 The envelope conformation of cyclopentane (a) Ball-and-stick representation of the envelope conformation of cyclopentane. (b) Line structure of the envelope conformation with dash–wedge notation. The thicker bond on the bottom indicates that it is in front of, and parallel to, the plane of the paper. (c) The lowest-energy conformation of cyclopentane resembles an envelope.

(a)   **(b)**

FIGURE 4-15 Comparison of the planar and envelope conformations of cyclopentane (a) Ball-and-stick model (bottom) and Newman projection (top) of planar cyclopentane, showing the eclipsed conformation at one of the C—C bonds. (b) Ball-and-stick model (bottom) and Newman projection (top) of cyclopentane in its envelope conformation, showing how the C—C bond is partly staggered, thereby relieving some torsional strain.

Despite relaxing to a nonplanar geometry, cyclopentane still has more ring strain than cyclohexane, in large part because torsional strain remains. One of the C—C bonds is essentially eclipsed (see Fig. 4-14), whereas the other four are intermediate between eclipsed and staggered.

4.13 YOUR TURN

Identify the C—C bond in Figure 4-14 that is in the eclipsed conformation. It may help if you build a model of cyclopentane in its envelope conformation using a modeling kit.

Cyclopentane also possesses some angle strain. Its bond angles, which range from 102° to 106° in the envelope conformation, are somewhat farther from the ideal tetrahedral bond angle of 109.5° than the 111° angles of cyclohexane.

4.5c Cyclobutane and Cyclopropane

The most stable conformation for cyclobutane is close to square, but not exactly; instead, the ring is slightly puckered, with interior angles of about 88° (Fig. 4-16). As with cyclohexane and cyclopentane, puckering of the cyclobutane ring relieves some torsional strain. If all its carbon atoms were coplanar, cyclobutane would have four perfectly eclipsed bonds (Fig. 4-17a) and the interior angle would be exactly 90°. In the puckered ring, on the other hand, the conformations of those bonds are slightly staggered (Fig. 4-17b), with C—C—C—C dihedral angles of about 16°. The puckered ring actually has more angle strain than the planar ring, but the relief in torsional strain compensates.

For cyclopropane, there is no alternative to having all three carbon atoms in the same plane, because three points define a plane. All three angles of the ring are exactly 60° and all three C—C bonds are the same length, so the ring forms an equilateral triangle. As a result, cyclopropane has more angle strain than the other cyclic molecules we have examined. Furthermore, the conformation at all of the C—C bonds is fully eclipsed, as shown in Figure 4-18, resulting in significant torsional strain.

The cyclobutane ring puckers slightly to relieve torsional strain.

(a)   (b)

FIGURE 4-16 Puckered conformation of cyclobutane The most stable conformation of cyclobutane has a slightly puckered ring. (a) Ball-and-stick representation of cyclobutane. (b) Line structure.

FIGURE 4-17 Planar versus puckered cyclobutane (a) Ball-and-stick model (bottom) and Newman projection (top) of planar cyclobutane, illustrating the eclipsed conformation of the C—C bonds. (b) Ball-and-stick model (bottom) and Newman projection (top) of puckered cyclobutane, illustrating the slight staggering of the C—C bond.

FIGURE 4-18 Cyclopropane (a) Ball-and-stick model (bottom) and Newman projection (top) of cyclopropane, showing that the C—C bonds are eclipsed. (b) Line structure of cyclopropane.

4.6 Conformational Isomers: Cyclopentane, Cyclohexane, Pseudorotation, and Chair Flips

In the envelope conformation of cyclopentane (Fig. 4-14), four of the carbon atoms lie in the same plane, with the remaining carbon outside that plane. It appears, therefore, that one carbon atom is distinct from the other four. Experiments cannot distinguish among the five carbons, however, so the five carbon atoms are said to be *equivalent*.

How can we explain this discrepancy? The answer is **pseudorotation**, which enables all five possible envelope conformations—each with a different carbon atom out of the plane—to interconvert. The energy barrier for pseudorotation is very small, which means that these interconversions take place constantly and very rapidly. It is impossible, therefore, to isolate any one of the envelope conformations from the others.

The mechanism by which pseudorotation occurs in cyclopentane is shown in Figure 4-19. The out-of-plane carbon atom (orange) moves into the plane, whereas one of the carbon atoms in the plane (green) moves out of plane. Although it may be difficult to see from the figure,

Pseudorotation occurs by *partial rotations* about the C—C bonds.

This explains, at least in part, why the energy barrier for the interconversion is so small—namely, covalent bonds are never broken. You may find a molecular modeling kit useful in illustrating these bond rotations (see Your Turn 4.14).

YOUR TURN 4.14

Use a molecular modeling kit to build a molecule of cyclopentane in its envelope conformation. View it so that it appears as shown on the left in Figure 4-19 and move the atoms as indicated in the figure. Identify which C—C bonds rotate and circle those bonds on the drawing on the left in Figure 4-19.

All six carbon atoms in the chair conformation of cyclohexane (Fig. 4-20) are identical, making it impossible to distinguish among them. To see this more clearly, carry out the molecular rotations indicated in Figure 4-20. Notice that the green carbon atom can be made to occupy the position of any of the other five.

The carbon atoms in cyclohexane may be identical, but there are two different types of hydrogen atoms, as shown in Figure 4-21. Six hydrogen atoms occupy *equatorial* positions and six occupy *axial* positions. Each carbon atom in cyclohexane is bonded to one of each.

- **Equatorial** bonds lie almost in the plane that is roughly defined by the ring (i.e., the *equator* of the molecule; see Fig. 4-21) and point outward from the center of the ring.
- **Axial** bonds are perpendicular to the plane.

FIGURE 4-19 Pseudorotation in cyclopentane The envelope conformation of cyclopentane on the left is converted into the one on the right by moving the highlighted carbon atoms in the direction of the red arrows.

Pseudorotation

FIGURE 4-20 All six carbon atoms in cyclohexane are equivalent The green C atom in cyclohexane is tracked through several rotations of the molecule. The rotations that transform (a) into (b), (b) into (c), (d) into (e), and (e) into (f) are each 120° about the vertical axis (indicated by the solid blue line). The rotation that transforms (a) into (d) is 180° about the diagonal axis represented by the solid blue line. Because the green C atom can be transformed into any of the other five C atoms by rotation of the entire molecule in space, all six C atoms are identical.

In most experiments, however, the axial and equatorial hydrogens in cyclohexane are indistinguishable. This is because there are two chair conformations that interconvert very rapidly—on the order of millions of times per second at room temperature—through a process called a **chair flip** (Fig. 4-22).

A chair flip converts axial hydrogens into equatorial hydrogens, and vice versa.

The rate at which the two chair conformations interconvert is so high because the energy barrier between the two chair conformations is quite low. The energy barrier is low because a chair flip does not involve breaking or forming bonds. Instead, like pseudorotation in cyclopentane, a chair flip in cyclohexane involves only rotations about bonds that make up the ring.

FIGURE 4-21 The axial and equatorial hydrogens in cyclohexane This chair conformation of cyclohexane shows that there are six axial H atoms (orange) and six equatorial H atoms (green). Bonds to axial H atoms are perpendicular to the plane indicated, whereas bonds to equatorial H atoms are nearly in the plane indicated.

FIGURE 4-22 Chair flip of a cyclohexane ring The chair conformation on the left can be converted into the chair conformation on the right by moving the two specified C atoms in the directions indicated by the red arrows. Notice that all the equatorial positions in the structure on the left (green) become axial after the chair flip, and all the axial positions (orange) become equatorial.

Cubane: A Useful "Impossible" Compound?

Cubane, C_8H_8, is an exotic molecule in which the eight carbon atoms are located at the corners of a cube. It was once thought to be an "impossible" compound, unable to exist due to excessive strain. In 1964, however, Philip Eaton and Thomas Cole, Jr., at the University of Chicago, successfully carried out its synthesis.

Cubane   **A dipivaloylcubane**   **A cubane-based liquid crystal**

At the time, such a synthesis was more a novelty than anything else. But since then, cubane and its derivatives have been finding widespread potential applications. One of the longest-studied applications is its potential use as a high-energy fuel. With the large amount of strain it has, combined with its relatively high density (nearly twice that of gasoline), cubane can store energy more efficiently than conventional fuels.

Derivatives of cubane also have potential use in medicine, owing to the rigid, lipophilic framework to which up to eight, independent, functional groups can be attached at eight specific locations. Consider, for example, the dipivaloylcubane shown above, which has been shown to exhibit moderate activity against the human immunodeficiency virus (HIV).

Cubane derivatives also have applications in materials science. For example, liquid crystals have been synthesized with cubane as the central structural component, and individual cubane units have been linked together to form polymers with interesting properties. With each passing year, more applications of cubane and its derivatives are sure to be found.

YOUR TURN 4.15

Use a molecular modeling kit to construct a model of cyclohexane and view it so that it parallels the structure on the left in Figure 4-22. Perform a chair flip by moving the two carbon atoms indicated by the red arrows to obtain the conformation on the right in the figure. Reverse and repeat this procedure several more times. As you flip the chair back and forth, identify the bonds of the ring that rotate and the direction in which they rotate. Circle and label those bonds in the drawing on the left in Figure 4-22.

Even though a chair flip interconverts axial and equatorial positions on a cyclohexane ring, it does *not* allow substituents to switch sides of the ring's plane. During a chair flip, a hydrogen atom in an axial position on one side of the ring's plane becomes an equatorial hydrogen on the same side. Likewise, an equatorial hydrogen on one side of the plane becomes an axial hydrogen on that same side. Viewing the cyclohexane ring from the side, as in Figure 4-22:

■ A chair flip converts "axial-up" to "equatorial-up" (and vice versa).
■ A chair flip converts "axial-down" to "equatorial-down" (and vice versa).

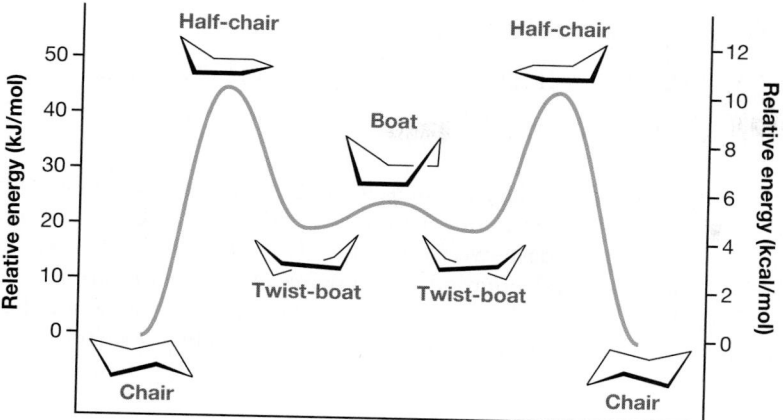

FIGURE 4-23 Energy diagram of a chair flip A chair flip that converts the conformation on the left into the one on the right goes through several key conformations, including the "half-chair," the "twist-boat," and the "boat." Energies are relative to the chair conformation.

Rather than occurring in a single step, a cyclohexane chair flip involves the multiple independent steps shown in Figure 4-23. Throughout such a chair flip, cyclohexane assumes key conformations known as the "half-chair," the "twist-boat," and the "boat."

As can be seen from Figure 4-23,

> The half-chair, twist-boat, and boat conformations in a chair flip are higher in energy than the chair conformation itself, due to added ring strain.

In the boat conformation (Fig. 4-24), for example, two of the C—C bonds are in an eclipsed conformation. In addition, there is a **flagpole interaction** between hydrogen atoms on an opposing pair of carbon atoms. This is a form of steric strain that is absent in the chair conformation.

4.16 **YOUR TURN**

Use a molecular modeling kit to build the following half-chair conformation:

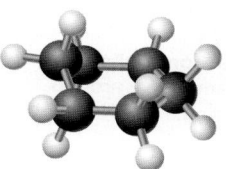

Examine the model from different points of view, and note the different types of strain. In the figure provided here, indicate where the strain exists, as well as the type of strain (i.e., angle, steric, or torsional). With these observations, can you justify why the half-chair conformation is so high in energy?

"Flagpole" interaction

Eclipsed conformations

(a)

(b)

FIGURE 4-24 Strain in the boat conformation of cyclohexane
(a) The boat conformation viewed from the side, showing the flagpole interaction.
(b) The boat conformation viewed from one end, showing the two eclipsed conformations.

4.7 Strategies for Success: Drawing Chair Conformations of Cyclohexane

Given the abundance of cyclohexane rings, it would soon become cumbersome if we always had to represent chair conformations three-dimensionally as ball-and-stick models (Fig. 4-25a) or in dash–wedge notation (Fig. 4-25b). Chemists, therefore, have devised the shorthand notation for drawing chair conformations shown in Figure 4-25c. Working with these structures is not trivial, so we devote the rest of Section 4.7 to practicing both drawing and interpreting chair structures using this shorthand method.

FIGURE 4-25 Various representations of the chair conformation of cyclohexane
(a) Ball-and-stick model. (b) Dash–wedge notation. (c) Shorthand notation.

(a) (b) (c)

Before we begin drawing chair conformations, please take note of the following important features of the shorthand notation:

■ By convention, the C—C bonds toward the bottom of the ring are interpreted as being in front of the plane of the paper (Fig. 4-26a).
■ When the ring is oriented as it is in Figure 4-25, all of the axial bonds (red) are perfectly vertical, *alternating up and down* around the ring (Fig. 4-26b).
■ All equatorial bonds (blue) are either slightly up or slightly down, *alternating around the ring*; on a carbon where the axial bond is down, there is an equatorial bond slightly up, and vice versa (Fig. 4-26c).

FIGURE 4-26 Interpreting the shorthand notation for the chair conformation of cyclohexane (a) By convention, the bonds on the bottom are in front of the plane of the paper. (b) Axial bonds (red) are vertical and alternate up and down around the ring. (c) Equatorial bonds (blue) alternate slightly up and slightly down around the ring.

Axial bonds alternate up and down around the ring.

Equatorial bonds alternate slightly up and down around the ring.

These bonds are in front.

(a) (b) (c)

FIGURE 4-27 Parallel bonds in the shorthand notation for cyclohexane The chair conformation of cyclohexane is color-coded to show its four sets of parallel bonds. Each C—C bond is parallel to the one opposite it in the ring and to a pair of equatorial C—H bonds. The six axial C—H bonds are also parallel to one another.

Figure 4-27 uses color coding to illustrate which sets of bonds are parallel in the shorthand notation for cyclohexane. All six axial C—H bonds (red) are parallel to one another, since they are all drawn vertically. Each equatorial C—H bond (brown, purple, or green) is parallel to one equatorial C—H bond on the opposite side of the ring and to two C—C bonds that are part of the ring. *Paying attention to these sets of parallel lines when you draw chair structures will help with the accuracy of your drawings.*

Begin by drawing the two brown C—C bonds from Figure 4-27, as shown in Figure 4-28a. Next, close the left side of the ring by drawing a V, which points slightly down (Fig. 4-28b). The right side of the ring is closed in the same fashion, but its V must point in the opposite direction of the first (Fig. 4-28c).

(a) (b) (c) (d) (e) (f) (g)

FIGURE 4-28 The progression in drawing a chair conformation of cyclohexane The steps are described in the text. The lines that are added in each step are indicated in red.

Next, draw the six axial bonds, all of which must point either directly up or directly down. On the leftmost carbon, draw an axial bond pointing directly down, and on the rightmost carbon, draw one pointing directly up (Fig. 4-28d). To help you remember this convention, the axial bonds on these two carbons point in roughly the same direction as their respective Vs: The V on the left points down, and the one on the right points up. Once these axial bonds are drawn, the remaining ones can be added by alternating the vertical lines up and down as you go around the ring (Fig. 4-28e).

Finally, the equatorial bonds are drawn (Fig. 4-28f and 4-28g). Wherever there is an axial bond pointing down, there must be an equatorial bond pointing slightly up, and wherever there is an axial bond pointing up, there must be an equatorial bond pointing slightly down. Remember to draw each equatorial bond parallel to the appropriate C—C bonds of the ring, as shown in Figure 4-27.

4.17 YOUR TURN

Draw the chair conformation of cyclohexane in Figure 4-28 using the steps described above. Practice doing so until you can draw it without having to refer to the figure.

4.18 YOUR TURN

The following chair conformation of cyclohexane is obtained after the one in Figure 4-28 has undergone a chair flip:

Use the steps in Figure 4-28 to practice drawing this chair conformation until you can draw it without having to refer to the figure.

4.8 Conformational Isomers: Monosubstituted Cyclohexane

If one of the hydrogen atoms in cyclohexane is replaced by a substituent such as CH_3, the result is a **monosubstituted cyclohexane**.

The two chair conformations of a monosubstituted cyclohexane are *not* equivalent.

For example, the CH_3 group of methylcyclohexane is axial in one chair form (Fig. 4-29a), whereas it is equatorial in the other (Fig. 4-29b).

FIGURE 4-29 The two chair conformations of methylcyclohexane The two chair conformations of monosubstituted cyclohexanes are not equivalent. (a) The methyl group occupies an axial position. (b) The methyl group occupies an equatorial position.

(a)   (b)

YOUR TURN 4.19

In Figure 4-29a, label each of the 11 hydrogen atoms bonded to the ring as either axial or equatorial. Do the same for Figure 4-29b.

Because they have the same molecular formulas, but they differ due to rotations about single bonds,

> Two nonequivalent chair forms are *conformational isomers* of each other.

The two chair conformations of methylcyclohexane rapidly interconvert, but they are not equally favored. At any given time, about 95% of the molecules exist in the form with an equatorial CH_3 group, and the remaining 5% exist in the form with an axial CH_3 group. Other monosubstituted cyclohexanes exhibit the same trend:

> A monosubstituted cyclohexane is more stable when the substituent occupies an equatorial position.

The equatorial conformer is more stable than the axial conformer because there is *more room for the substituent in the equatorial position.* To see why this is so, compare methylcyclohexane with an axial CH_3 group (Fig. 4-30a) to methylcyclohexane with an equatorial CH_3 group (Fig. 4-30b).

FIGURE 4-30 Strain from axial versus equatorial substituents on cyclohexane (a) The methyl group is in an axial position. A Newman projection looking down the C1—C2 and C5—C4 bonds (top) shows strain from a gauche interaction between the CH_3 substituent on the ring and a CH_2 group of the ring. A chair conformation (bottom) shows strain from 1,3-diaxial interactions. Ball-and-stick models are also provided. (b) The methyl group is in an equatorial position. A Newman projection looking down the C1—C2 and C5—C4 bonds (top) shows that the CH_3 substituent on the ring is anti to the CH_2 group of the ring. A chair conformation (bottom) shows that no 1,3-diaxial interactions are present. Ball-and-stick models are also provided.

(a)   (b)

In the axial position, the CH_3 group experiences significant steric strain from gauche interactions, one of which is illustrated in the Newman projection in Figure 4-30a. If we view the CH_2 group containing C3 of the ring as a substituent on C2, then that CH_2 group is gauche to the CH_3 substituent bonded to C1. Repulsion between the electrons from the respective CH_3 and CH_2 groups gives rise to the strain.

A major contribution to these gauche interactions results specifically from repulsion between the electrons on the axial CH_3 group bonded to C1 and those on the axial H atoms two positions away on the ring (i.e., at positions 3 and 5). There are two of these **1,3-diaxial interactions** (the "1,3" identifies the *relative* positions on the ring for the interacting substituents), as indicated in Figure 4-30a.

No such steric strain exists when the CH_3 group is in the equatorial position (Fig. 4-30b). In the equatorial position, the CH_2 group at position 3 on the ring is anti to the CH_3 group, so no steric strain arises from their interaction. More specifically, the 1,3-diaxial interactions are eliminated because the CH_3 group is relatively far away from the axial H atoms at positions 3 and 5.

4.20 **YOUR TURN**

The model of methylcyclohexane below corresponds to the top structure in Figure 4-30a. It shows two hydrogen atoms on the ring involved in 1,3-diaxial interactions with the CH_3 group. Identify these two hydrogens. Also, *two* different CH_2 groups are gauche to the CH_3 group. One is indicated in Figure 4-30a. Identify the second one in the structure below. (*Hint:* Build a molecular model of methylcyclohexane.)

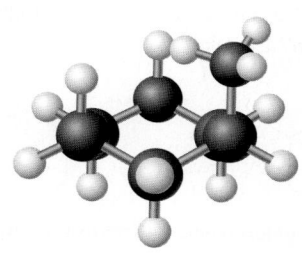

SOLVED problem 4.8 Which species below is more stable?

Think How are the two molecules related? How are they different? How do those differences translate into stabilities?

Solve The two species are related by a simple chair flip:

The Br atom occupies an equatorial position in **A**, whereas it occupies an axial position in **B**. Bromine is much larger than any of the 11 H atoms also bonded to the cyclohexane ring, so it requires more room. Because the equatorial position provides more room than the axial position, conformation **A** is more stable.

problem **4.9** Draw both chair conformations of cyclohexane-d_1, in which a hydrogen atom on cyclohexane has been replaced by a deuterium atom. Which conformation would you expect to be in greater abundance, if any? Explain. (*Hint:* Deuterium is an isotope of hydrogen, possessing one more neutron than hydrogen. Recall from Chapter 1 that the size of an atom is dictated by the size of its electron cloud.)

Cyclohexane-d_1

Even though all nonhydrogen groups prefer the equatorial position over the axial position, some groups have a stronger preference than others. This can be seen in Table 4-2, which shows the relative percentages of the two chair conformations of various alkylcyclohexanes.

As can be seen from Table 4-2,

> The bulkier the substituent on a cyclohexane ring, the more favored is the chair conformation with the substituent in the equatorial position.

Notice, especially, that the *tert*-butyl group $[C(CH_3)_3]$ is sufficiently bulky that only a very small percentage of molecules exist with the group in the axial position.

problem **4.10** For which compound—triiodomethylcyclohexane or trifluoromethyl-cyclohexane—would you expect to find a greater percentage of molecules that have the substituent in the axial position? Explain.

Triiodomethylcyclohexane   **Trifluoromethylcyclohexane**

TABLE 4-2 Relative Percentages of Chair Conformations of Alkylcyclohexanes

Substituted Cyclohexane	Percent Axial	Percent Equatorial
R = H	50.0	50.0
R = CH$_3$	5.3	94.7
R = CH$_2$CH$_3$	4.5	95.5
R = CH(CH$_3$)$_2$	2.8	97.2
R = C(CH$_3$)$_3$	0.02	99.98

4.9 Conformational Isomers: Disubstituted Cyclohexanes, Cis and Trans Isomers, and Haworth Projections

With **disubstituted** cyclohexanes (i.e., those with two substituents), we have to take into account the relationship of each substituent to the plane of the ring. Are they on the same side of the ring (i.e., **cis** to each other) or are they on opposite sides (i.e., **trans**)? A chair flip does *not* switch a substituent from one side of the plane to the other, so *the cis–trans relationship between any pair of substituents on a cyclohexane ring is independent of the particular chair conformation the species is in.*

> Substituents that are cis to each other on a cyclohexane ring remain cis after a chair flip; substituents that are trans remain trans.

Because the cis–trans relationship of any pair of substituents is unaffected by a chair flip, chemists often find it more convenient to represent substituted cyclohexanes (and rings of other size, for that matter) using *Haworth projections*. In a **Haworth projection**, the ring is depicted as being planar and substituents are drawn perpendicular to that plane (Fig. 4-31). Despite their convenience, Haworth projections are inaccurate representations of the true structure and should be used with caution. In particular, they are incapable of portraying axial versus equatorial positions, so they do not accurately represent steric interactions.

> Draw each line structure as a Haworth projection and each Haworth projection as a line structure including dash–wedge notation.
>
> (a) (b) (c) (d)

Because each substituent on a cyclohexane ring is more stable in an equatorial position than in an axial position, we can reasonably predict the more stable chair conformation of a number of disubstituted cyclohexanes. In *cis*-1,3-dimethylcyclohexane (Fig. 4-32), for example, one chair conformation has two equatorial CH_3 groups, whereas the other has both groups axial. The diequatorial conformer is favored over the diaxial conformer.

As indicated in Figure 4-33, both chair conformations of *trans*-1,3-dimethylcyclohexane are equally favored because both have one axial and one equatorial CH_3 group. With a chair flip, the axial CH_3 group becomes equatorial and the equatorial group becomes axial, thus yielding a conformation identical to the first.

Cis isomer

Same side of the ring's plane

Trans isomer

Opposite sides of the ring's plane

Haworth projection

(a) (b)

FIGURE 4-31 Haworth projections of disubstituted cyclohexanes (a) A cis isomer of a generic 1,2-disubstituted cyclohexane shown in its dash–wedge notation (top) and as a Haworth projection (bottom). (b) A trans isomer of a generic 1,2-disubstituted cyclohexane shown in its dash–wedge notation (top) and as a Haworth projection (bottom).

4.21 YOUR TURN

cis-1,3-Dimethylcyclohexane

(a)

Favored chair conformation

(b)

FIGURE 4-32 Relative stabilities of *cis*-1,3-dimethylcyclohexane chair conformations (a) Haworth projection of *cis*-1,3-dimethylcyclohexane. (b) The two chair conformations of *cis*-1,3-dimethylcyclohexane. The one on the left has both methyl groups in equatorial positions and the one on the right has them both axial. The conformation on the left is favored over the one on the right.

FIGURE 4-33 Relative stabilities of trans-1,3-dimethylcyclohexane chair conformations (a) Haworth projection of *trans*-1,3-dimethylcyclohexane. (b) The two chair conformations of *trans*-1,3-dimethylcyclohexane. Both chair conformations have one methyl group in an equatorial position and one in an axial position, so both conformations are favored equally.

YOUR TURN 4.22

Use a molecular modeling kit to build a model of *trans*-1,3-dimethylcyclohexane so that it looks exactly like the chair conformation shown on the left in Figure 4-33b. Without flipping the chair, simply rotate the molecule again until it looks like the chair conformation on the right.

SOLVED problem 4.11 Draw the more stable chair conformation of *cis*-1,2-dimethylcyclohexane.

cis-1,2-Dimethylcyclohexane

Think Which substituents require the most room? Which position, axial or equatorial, offers more room? Can both substituents achieve that position?

Solve The two methyl groups are the largest substituents on the ring and each is more stable in an equatorial position. To determine whether they can both occupy equatorial positions, begin by drawing a chair conformation with one equatorial methyl group.

For the two groups to be cis, the other methyl group must be in the less favorable axial position.

If we perform a chair flip, the equatorial methyl group becomes axial and vice versa, yielding the following conformation.

This chair has precisely the same stability as the first one, so both conformations are favored equally.

problem 4.12 Draw the most stable conformation of *trans*-1,2-dimethylcyclohexane.

problem 4.13 Draw the most stable conformation of the following molecule.

Attach CH₃ groups on same side of ring's plane

Axial "up"

Chair flip

Axial "up"

Equatorial "up"

Equatorial "up"

(a)

CH_3
CH_3
(b)

H_3C
CH_3
(c)

FIGURE 4-34 **Model kits and chair flips** (a) A cyclohexane ring without hydrogen atoms. Adjacent bonds on the same side of the ring's plane are identified. (b) CH_3 groups have been added. One is axial and the other equatorial. Rotating the C atoms on the left and right sides according to the red arrows flips the chair. (c) The model after the chair flip. Both CH_3 groups remain up, with one axial and the other equatorial.

problem **4.14** Draw the most stable conformation of *cis*-1-methyl-4-trichloromethylcyclohexane.

4.10 Strategies for Success: Molecular Modeling Kits and Chair Flips

Molecular modeling kits can be really useful because they help us "see" molecules in three dimensions from different vantage points and they accurately portray the rotational characteristics of single bonds. Because chair flips affect the three-dimensional arrangement of atoms in space and involve only rotations about single bonds, molecular modeling kits can be *extremely* helpful in problems that ask you to compare chair conformations.

For example, instead of working Solved Problem 4.11 entirely on paper, you can simplify the problem by making the modeling kit do much of the work for you. First, build a cyclohexane ring in its chair conformation, temporarily leaving off all of the hydrogen atoms, as shown in Figure 4-34a. Being able to see all of the bonding positions on the ring that are available, you can then attach two CH_3 groups on adjacent carbon atoms, on the same side of the plane of the ring. As shown in Figure 4-34b, one position is axial, pointing straight up, and the other is equatorial, pointing slightly up. After adding the remaining hydrogen atoms, one of the two chair conformations of *cis*-1,2-dimethylcyclohexane is complete. Flipping the chair as indicated in Figure 4-34b, you arrive at the second chair conformation in Figure 4-34c, which also has one axial and one equatorial CH_3 group. Hence, the two conformations are equivalent.

4.11 Constitutional Isomerism: Identifying Constitutional Isomers

Constitutional isomers are the second type of isomers we discuss in this chapter. As can be seen from the portion of the Figure 4-1 flow chart reproduced in Figure 4-35,

Constitutional isomers, also called **structural isomers**, share the same molecular formula but *differ in their connectivity*.

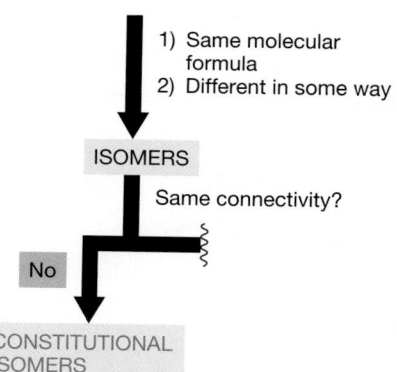

1) Same molecular formula
2) Different in some way

ISOMERS

Same connectivity?

No

CONSTITUTIONAL ISOMERS

FIGURE 4-35 **Constitutional isomers** This is part of the flow chart in Figure 4-1, showing that constitutional isomers have the same formula but different connectivity.

Unlike conformational isomers, constitutional isomers generally do not interconvert and can be separated from each other. Recall from Chapter 1 that the **connectivity** of a molecule describes its *bonding scheme*, including which atoms are bonded together and by what type of bond (e.g., single, double, or triple). Cyclobutane and but-1-ene are constitutional isomers, for example, because both have the molecular formula C_4H_8, but their connectivities are quite different. Cyclobutane has only single bonds, whereas but-1-ene contains a $C=C$ double bond. Furthermore, the $C-C$ single bonds in cyclobutane form a ring, whereas but-1-ene is acyclic.

**These are constitutional isomers
because they have different connectivity.**

Cyclobutane   **But-1-ene**

Both of the following structures have the molecular formula C_8H_{18}, but are they constitutional isomers?

These two molecules are not constitutional isomers of each other!

Although they are drawn differently, closer inspection reveals that they have the same connectivity, so they are *not* constitutional isomers of each other. How can we tell?

Perhaps the most straightforward method to determine whether two molecules with the same formula are constitutional isomers is one that takes advantage of some aspects of nomenclature that were discussed in the first two nomenclature units.

1. For each molecule, identify the longest continuous chain or ring of carbons that contains any $C=C$ double and $C\equiv C$ triple bonds.
2. Number the carbons in the chain or ring sequentially so that:
   - The carbon atoms involved in the double or triple bonds receive the lowest numbers, or, if there are no such multiple bonds,
   - The first substituent is attached to the lowest-numbered carbon.
3. If there is a difference in any of the following, the molecules must have different connectivity, and are therefore constitutional isomers:
   - The size of the longest continuous chain or ring.
   - The numbers assigned to the carbons involved in the multiple bonds.
   - The numbers assigned to the carbons to which any substituent is attached.
4. Otherwise, the molecules have the same connectivity and are not constitutional isomers.

Let's now apply this method to the two molecules above. For Step 1, the longest continuous chain of carbon atoms in each molecule has six carbons. For Step 2, the carbons are numbered 1 through 6 so that the first methyl group is encountered on C2. Because the methyl groups are attached to C2 and C4 in both cases, the molecules have the same connectivity and therefore are not constitutional isomers.

Longest continuous chain has six carbon atoms.

Methyl groups on C2 and C4

Longest continuous chain has six carbon atoms.

Methyl groups on C2 and C4

4.23 YOUR TURN

Using the above method, show that the following molecule is not a constitutional isomer of the previous two molecules.

SOLVED problem 4.15 Are the two molecules at the right constitutional isomers of each other?

Think Do both compounds have the same molecular formula? Is the largest continuous chain or ring in each molecule the same? Are the C atoms that are involved in the double bonds assigned the same numbers in each molecule? In each molecule, is the C atom to which the methyl group is attached assigned the same number?

Solve Both compounds have a molecular formula of C_6H_{10}. In each molecule, the largest continuous ring has five C atoms. As shown at the right, the numbers assigned to the doubly bonded C atoms are the same in each molecule. However, the molecules differ in the number assigned to the C atom to which the methyl group is attached. In the first molecule, the methyl group is attached to C3, whereas in the second molecule, it is attached to C4. Thus, the molecules have different connectivity, making them constitutional isomers.

Methyl groups attached to different carbons

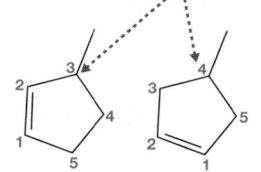

---

problem 4.16 For each pair of molecules, determine whether they are constitutional isomers.

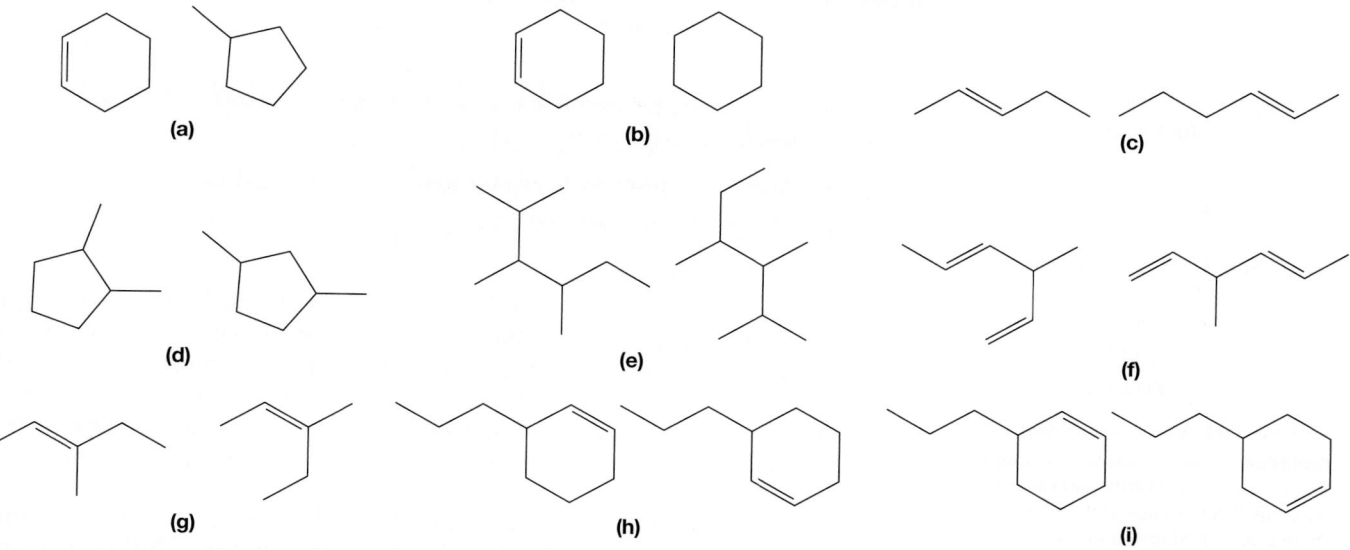

(a)

(b)

(c)

(d)

(e)

(f)

(g)

(h)

(i)

(a)

(b)

(c)

(d)

(e)

(f)

(g)

(a) Methane

(b) Ethane

(c) Ethanol

FIGURE 4-36 Some saturated molecules These molecules contain the most number of hydrogen atoms possible, given (a) one carbon atom, (b) two carbon atoms, and (c) two carbon atoms and one oxygen atom.

4.12 Constitutional Isomers: Index of Hydrogen Deficiency (Degree of Unsaturation)

Section 4.13 is devoted to drawing constitutional isomers of a given formula. There are a variety of contexts where being able to do so is useful, including in *spectroscopy*, which we discuss in Chapters 15 and 16.

To help you address these kinds of issues, we introduce here in Section 4.12 a concept called the **index of hydrogen deficiency** (**IHD**), also known as **degree of unsaturation**. Before we examine IHD, however, you must understand what it means for a molecule to be *saturated*.

A molecule is said to be **saturated** if it *has the maximum number of hydrogen atoms possible*, consistent with:

- The number and type of each nonhydrogen atom in the molecule
- The octet rule and the duet rule

Each of the molecules in Figure 4-36, for example, is saturated. Methane (CH_4) is saturated because a single carbon atom can bond at most to four other atoms; in this case, they are all hydrogen atoms. Ethane (CH_3CH_3) is also saturated because both carbon atoms are bonded to four other atoms (i.e., one carbon and three hydrogens). Ethanol (CH_3CH_2OH) is saturated because both carbon atoms are bonded to four other atoms, and the oxygen atom, being uncharged, is bonded to its maximum of two other atoms.

Molecules that contain double bonds, triple bonds, or rings are said to be **un-saturated** because they have fewer than the maximum number of hydrogen atoms

FIGURE 4-37 Some unsaturated molecules Ethene (a) and ethyne (b) are unsaturated because they have fewer hydrogen atoms than ethane. Ethanal (c) and oxirane (d) are unsaturated because they have fewer hydrogen atoms than ethanol.

possible (Fig. 4-37). Ethene (ethylene, $H_2C{=}CH_2$), for example, has the same number of carbon atoms as ethane, but has two fewer hydrogen atoms. Ethyne (acetylene, $HC{\equiv}CH$) has four fewer hydrogen atoms than ethane. Ethanal (acetaldehyde, $CH_3CH{=}O$) and oxirane each have the same number of carbon and oxygen atoms as ethanol, but have two fewer hydrogen atoms.

Draw the structure of a *saturated* molecule that corresponds to each of the unsaturated molecules in Figure 4-37.

4.24 YOUR TURN

The *extent* to which a molecule is unsaturated is indicated by its index of hydrogen deficiency:

A molecule's **index of hydrogen deficiency** (**IHD**), or **degree of unsaturation**, is defined as *half* the number of hydrogen atoms missing from that molecule compared to an analogous, completely saturated molecule.

Alternatively, it is the number of H_2 *molecules* that are missing from the molecule compared to an analogous saturated structure. Based on this definition, ethene, ethanal, and oxirane all have an IHD of 1, whereas ethyne has an IHD of 2. In other words,

- Each double bond contributes 1 to a molecule's IHD.
- Each triple bond contributes 2 to a molecule's IHD.
- Each ring contributes 1 to a molecule's IHD.

Benzene (C_6H_6), therefore, has an IHD of 4, because it has three double bonds and one ring, each of which contributes 1 to its IHD (Fig. 4-38).

IHD = 4

Three double bonds and one ring

C_6H_6
Benzene

FIGURE 4-38 The IHD of benzene Each of the three double bonds contributes one to the IHD, and the ring also contributes one to the IHD, for a total of four.

SOLVED problem 4.18 What is the IHD of naphthalene?

Naphthalene

Think How many double bonds are there? How many triple bonds are there? How many rings are there? How much does each contribute to the overall IHD?

Solve There are five double bonds, each contributing 1 to the IHD, and there are two rings, each contributing 1 to the IHD. There are no triple bonds. The total IHD, therefore, is $5(1) + 2(1) = 7$.

---

problem **4.19** Determine the IHD for each of the following molecules.

(a) (b) (c) (d) (e)

problem **4.20** Identify all of the functional groups listed in Table 1-6 that contribute **(a)** an IHD of 1, **(b)** an IHD of 2, **(c)** an IHD of 3, and **(d)** an IHD of 4.

We can determine the IHD of a compound from its molecular formula without knowing its structure if we know the formula for an analogous saturated compound. *The saturated compound must contain the same number of each nonhydrogen atom, but its specific connectivity does not matter.*

Suppose, for example, that we want to determine the IHD for a molecule with the formula C_5H_6. We first draw *any* completely saturated molecule with five carbon atoms, possessing no double bonds, triple bonds, or rings. We may choose to connect all five C atoms in a linear fashion, as with pentane (Fig. 4-39), or we may choose a branched structure such as dimethylpropane. In both cases, the resulting molecular formula is C_5H_{12}. Compared to this, the formula we are given (C_5H_6) is missing six hydrogen atoms (or three H_2 molecules), for an IHD of 3.

Pentane and dimethylpropane have the same molecular formula, but different connectivities, so they are constitutional isomers of each other. In general:

> The IHD for any pair of isomers must be the same.

Therefore, when constructing a saturated molecule to compare against a molecule of interest, that saturated compound can have *any* connectivity, and need not be real or stable. Each atom, however, should have its "normal" number of bonds and lone pairs to avoid formal charges.

The procedure for finding the IHD is no different if the molecular formula given contains atoms other than carbon and hydrogen, as shown in Solved Problem 4.21.

C_5H_{12}

Pentane

C_5H_{12}

Dimethylpropane

FIGURE 4-39 Completely saturated molecules with five carbon atoms Both pentane and dimethylpropane are saturated molecules because a molecule with five carbon atoms can be bonded to at most 12 hydrogen atoms. The two molecules are, furthermore, isomers of each other.

SOLVED problem 4.21 Determine the IHD for a compound whose molecular formula is $C_7H_7NO_2$.

Think What is the formula for an analogous saturated compound? How many hydrogen atoms are missing from the formula we are given?

Solve We can construct a saturated compound containing seven carbon atoms, one nitrogen atom, and two oxygen atoms simply by connecting all of these atoms in a row using single bonds only.

A saturated compound

$$H_3C-\overset{H_2}{C}-\overset{H_2}{C}-\overset{H_2}{C}-\overset{H_2}{C}-\overset{H_2}{C}-\overset{H_2}{C}-\overset{H}{N}-O-OH$$
$$C_7H_{17}NO_2$$

It takes 17 hydrogen atoms in total to saturate each carbon, nitrogen, and oxygen in this compound. Thus, the formula we were given is missing 10 H atoms, for an IHD of 5.

problem **4.22** Compute the IHD for a compound whose molecular formula is $C_4N_2OH_7F$.

Computing a formula's IHD can be sped up considerably by realizing that a saturated hydrocarbon with n carbon atoms has $2n + 2$ hydrogen atoms; that is, its formula is C_nH_{2n+2}. See Table 4-3, which lists a variety of saturated hydrocarbons.

Problems 4.23 and 4.24 show that adjustments can be made to the generic formula of C_nH_{2n+2} for saturated compounds containing atoms such as nitrogen, oxygen, and halogens.

problem **4.23** If a saturated compound contains four carbon atoms as the only nonhydrogen atoms, how many hydrogen atoms does it contain? How many hydrogen atoms are in a saturated compound containing four carbon atoms and one oxygen atom? How many hydrogen atoms are in a saturated compound containing four carbon atoms and two oxygen atoms? What can you conclude about the effect that each oxygen atom has on the formula for a completely saturated molecule?

problem **4.24** (a) Repeat Problem 4.23, adding nitrogen atoms instead of oxygen atoms. (b) Repeat it once again, adding fluorine atoms instead of oxygen atoms.

TABLE 4-3 The Number of Hydrogen Atoms in a Saturated Hydrocarbon with n Carbon Atoms

Hydrocarbon	Number of H Atoms	$2n + 2$
CH_4 ($n = 1$)	4	4
CH_3CH_3 ($n = 2$)	6	6
$CH_3CH_2CH_3$ ($n = 3$)	8	8
$CH_3(CH_2)_2CH_3$ ($n = 4$)	10	10
$CH_3(CH_2)_3CH_3$ ($n = 5$)	12	12
$CH_3(CH_2)_4CH_3$ ($n = 6$)	14	14

4.13 Strategies for Success: Drawing All Constitutional Isomers of a Given Formula

Being able to draw all constitutional isomers of a given molecular formula can be useful, especially when trying to determine a compound's structure using results from spectroscopy (see Chapters 15 and 16). More immediately, however, the time you spend drawing constitutional isomers will deepen your understanding of the relationships between them.

It helps to have a systematic method to tackle these kinds of problems. Here we present one method, and you may even develop your own.

1. Determine the formula's IHD. This will tell you the possible combinations of double bonds, triple bonds, and rings required in each isomer you draw.
2. Draw all possible isomers that omit double bonds, triple bonds, and halogen atoms. (It is most convenient to work with line structures so that the H atoms are accounted for appropriately when features are added in Steps 3 and 4.)
   - Double bonds, triple bonds, and halogen atoms will be added later.
   - Include rings. The number of rings must not exceed the IHD computed from Step 1.
3. For each structure generated in Step 2, add double and/or a triple bonds to satisfy the total IHD calculated in Step 1. Try to add these double/triple bonds at various locations to generate as many unique connectivities as possible.
4. For each structure generated in Step 3, add halogen atoms at various locations to generate as many unique connectivities as possible.

How we apply these steps depends specifically on the nature of the formula we are given. We present two examples.

4.13a Drawing All Constitutional Isomers of $C_4H_8F_2$

Let's draw all possible constitutional isomers having the formula $C_4H_8F_2$. According to Step 1, we first determine the IHD for this formula. To do so, we construct a saturated molecule having four C atoms and two F atoms:

Saturated

$C_4H_8F_2$

- Four C atoms
- No double bonds, triple bonds, or halogens
- No rings

Because this compound has eight H atoms, *the formula we are given has an IHD of 0*. That means that every constitutional isomer we draw must have *no double bonds, no triple bonds,* and *no rings*.

For Step 2, we draw all isomers that have four C atoms only and just single bonds. Because the IHD is 0, these structures also should contain no rings. There are only two such structures, **A** and **B**, shown in Figure 4-40.

Normally, for Step 3, we would add double or triple bonds to satisfy the IHD. We skip Step 3 in this case, however, because the IHD = 0.

For Step 4, we have two F atoms to add. Let's add one F atom at a time to produce different connectivities. By adding the first F atom (in red) to structure **A**, we obtain two structures with different connectivities, **C** and **D**, shown in Figure 4-41a. Structure **C** is obtained by adding the F atom to C1, and structure **D** is obtained by adding it to C2. We do not add the F atom to C3 or C4 because the resulting structures would be redundant to **D** and **C**, respectively. If we add the first F atom to structure **B** instead, we arrive at structures **E** and **F**, shown in Figure 4-41b. Notice that those structures were obtained by adding the F atom to C1 and C2, respectively. Once again, adding that F atom to either of the other two C atoms would result in redundant structures.

We can now add the second F atom separately to structures **C–F**. First let's do so for structures **C** and **D**, as shown in Figure 4-42. In Figure 4-42a, the second F atom (blue) has been added to each of the four different C atoms in Structure **C** to yield Structures **G–J**, which have the desired formula of $C_4H_8F_2$. In Figure 4-42b, the same has been done for Structure **D** to yield Structures **K–N**. Notice, however, that Structures **K** and **N** are redundant, because they have the same connectivity as molecules **H** and **I**, respectively.

Now let's add the second F atom to structures **E** and **F**, as shown in Figure 4-43a and 4-43b, respectively. This yields Structures **O–U**, each of which has the

FIGURE 4-40 Four-C isomers (Step 2) These isomers have four C atoms, the same as the formula we were given: $C_4H_8F_2$. Step 2 requires them to have no double bonds, triple bonds, or halogen atoms. There are no rings because the IHD = 0.

FIGURE 4-41 Isomers resulting from the addition of the first F atom (Step 4) (a) Structures **C** and **D** were produced by adding F to C1 and C2, respectively, of Structure **A**. (b) Structures **E** and **F** were produced by adding F to C1 and C2, respectively, of Structure **B**.

An F atom was added to structure A.

(a)

An F atom was added to structure B.

(b)

A second F atom was
added to structure C.

G **H** **I** **J**

(a)

A second F atom was
added to structure D.

K
Redundant to H

L

M

N
Redundant to I

(b)

FIGURE 4-42 Isomers resulting from
the addition of the second F atom
to Structures C and D (Step 4)
(a) The second F atom (blue) has been
added to each of the four C atoms in
Structure **C**, producing Structures **G–J**.
(b) The second F atom (blue) has been
added to each of the four C atoms in
Structure **D**, producing Structures **K–N**.
Structures **K** and **N** are redundant
to Structures **H** and **I**, respectively.

desired formula $C_4H_8F_2$. Of these, only molecules **O**, **P**, and **Q** have unique connectivities. Therefore, there are nine total constitutional isomers: **G, H, I, J, L, M, O, P,** and **Q**.

A second F atom was
added to structure E.

O **P** **Q** **R**
Redundant to Q

(a)

A second F atom was
added to structure F.

S
Redundant to P

T
Redundant to P

U
Redundant to P

(b)

FIGURE 4-43 Isomers resulting from
the addition of the second F atom
to Structures E and F (Step 4)
(a) The second F atom (blue) has been
added to each of the four C atoms
in Structure **E**, producing Structures
O–R. (b) The second F atom (blue) has
been added to each of three C atoms
in Structure **F**, producing Structures
S–U. Structure **R** is redundant to **Q**, and
Structures **S–U** are redundant to **P**.

4.13b Drawing All Constitutional Isomers of C_3H_6O

Let us now consider the formula C_3H_6O. Using the same systematic approach, we first determine the IHD. We begin by comparing the formula we are given with a completely saturated molecule containing three C atoms and one O atom. One possibility

FIGURE 4-44 Isomers containing three C atoms and one O atom (Step 2) These isomers have three C atoms and one O atom, the same as the formula we were given: C_3H_6O. Step 2 requires them to have no double bonds or triple bonds. These isomers can contain up to one ring (Structures A–C) because the formula we were given corresponds to an IHD of 1.

Contain 1 ring

A   B   C

Contain no rings

D   E   F

is $CH_3CH_2CH_2OH$, which has a formula C_3H_8O. Consequently, our formula has an IHD of 1, which can be due to either a double bond or a ring.

In Step 2, we draw all unique line structures containing three C atoms and one O atom, leaving out double bonds and triple bonds. As shown in Figure 4-44, there are six such structures (**A–F**), which can contain up to one ring without exceeding an IHD of 1.

For Step 3, we add double and/or triple bonds to obtain a total IHD of 1 (calculated in Step 1). We cannot add double or triple bonds to structures **A–C**, however, because with the ring present, their IHD is already 1. For structures **D–F**, on the other hand, we must add a double bond to various locations, as shown in Figure 4-45. Notice that a double bond can be added to structure **D** at three different locations to yield structures **G, H**, and **I**. A double bond can be added to only one location in structure **E**, however, yielding structure **J**; adding the double bond to any other location would introduce a formal charge on O. A double bond also can be added to three locations in structure **F** to yield structures **K, L**, and **M**, but, as indicated, **L** is redundant to **K**.

Normally, Step 4 would have us add halogens, but because the formula we were given has no halogen atoms, the structures we have constructed thus far all have the formula C_3H_6O. Therefore, there are nine constitutional isomers of C_3H_6O: **A, B, C, G, H, I, J, K,** and **M.**

YOUR TURN 4.25

Identify the functional groups present in *each* molecule in Figure 4-45.

Even though we can draw these nine constitutional isomers, molecules **H** and **K** are unstable. As a class of compounds, they are called **enols**: They have a carbon that is bonded to an OH group and is also part of a C=C double bond. We explain in Section 7.9 that these two molecules will undergo a rearrangement reaction in solution to form the more stable isomers **I** and **M**, respectively.

A double bond was added to structure D.

A double bond was added to structure E.

A double bond was added to structure F.

G   H   I   J   K   L   M

L is redundant to K.

FIGURE 4-45 Isomers resulting from the addition of a double bond to Structures D–F (Step 3) Structures **G–I** were produced by adding a double bond to various locations in Structure D. Structure **J** was produced by adding a double bond to Structure E. Structures **K–M** were produced by adding a double bond to various locations in Structure F. Structure **L** is redundant to **K**.

4.14 Wrapping Up and Looking Ahead

This chapter began by introducing the concept of isomerism in general: relationships among two species that have the same molecular formula, but differ in some way. There are a variety of ways in which isomers can differ, but this chapter focused only on conformational isomers and constitutional isomers. Conformational isomers differ only by rotations about single bonds, whereas constitutional isomers differ in their connectivities.

In the discussion on conformational isomers, we saw that energy changes during the course of a rotation about a single bond. Thus, different conformational isomers can have different stabilities and we can make predictions of the most stable conformation by identifying various kinds of strain that are present. This is particularly useful when determining the relative stabilities of chair conformations of various substituted cyclohexanes.

In our discussion of constitutional isomers, we learned how to determine whether two molecules have different connectivities. We further learned how to determine the index of hydrogen deficiency for a species, and took advantage of the IHD when drawing all constitutional isomers of a given molecular formula.

This was the first of two chapters devoted to isomerism. The second is Chapter 5, which deals with enantiomers and diastereomers—isomers that do not relate by free rotations about single bonds or by differences in connectivity. As we will see in Chapter 5, knowing the specific type of isomeric relationship between two molecules helps us determine whether they should have the same or different properties. We will begin to use this kind of information in Chapter 8 to make predictions about relative amounts of products made in competing reactions.

THE ORGANIC CHEMISTRY OF BIOMOLECULES

4.15 Constitutional Isomers and Biomolecules: Amino Acids and Monosaccharides

Leucine and isoleucine (Fig. 4-46), two naturally occurring amino acids, are constitutional isomers because they have the same molecular formula, but differ in their connectivity. (The prefix "iso," in fact, stands for "isomer.") Notice, in particular, that the only difference between the two molecules is in the location of a methyl group.

FIGURE 4-46 Isomeric amino acids Leucine and isoleucine are naturally occurring amino acids that have the same formula but different connectivity, so they are constitutional isomers.

Monosaccharides provide many more examples of constitutional isomers. The cyclic and acyclic forms of ribose are constitutional isomers, for instance, as are the cyclic and acyclic forms of glucose (Fig. 4-47).

Constitutional isomers of ribose

Constitutional isomers of glucose

FIGURE 4-47 Acyclic and cyclic forms of sugars as constitutional isomers Acyclic and cyclic ribose have the same formula ($C_5H_{10}O_5$) but different connectivities. Acyclic and cyclic glucose have the same formula ($C_6H_{12}O_6$) but different connectivities.

Among the acyclic forms of monosaccharides, ribose and ribulose are constitutional isomers, as are glucose and fructose (Fig. 4-48).

Ribose
An *aldopentose*

Ribulose
A *ketopentose*

Glucose
An *aldohexose*

Fructose
A *ketohexose*

FIGURE 4-48 Acyclic sugars as constitutional isomers Ribose and ribulose have the same formula ($C_5H_{10}O_5$) but different connectivities. Glucose and fructose have the same formula ($C_6H_{12}O_6$) but different connectivities.

These isomers differ by the location of the carbonyl group. In ribose, for example, the carbonyl group is part of an aldehyde, whereas in ribulose, it is part of a ketone. Thus, ribose is classified as an **aldose**, whereas ribulose is a **ketose**. Glucose is similarly classified as an aldose, whereas fructose is a ketose.

To further distinguish sugars on the basis of their carbon atoms, ribose and ribulose are **pentoses**, because they both contain five carbons, whereas glucose and fructose are **hexoses**, because they both contain six carbons. Combining these terminologies,

Butyric acid, C₄H₈O₂
(Found in rancid butter)
Melting point = –7.9 °C

Stearic acid, C₁₈H₃₆O₂
(Major constituent of beef fat)
Melting point = 70 °C

FIGURE 4-49 Saturated fatty acids Butyric acid and stearic acid are saturated fatty acids because they contain the maximum possible number of hydrogen atoms in their carbon chains.

we say that ribose is an **aldopentose**, ribulose is a **ketopentose**, glucose is an **aldohexose**, and fructose is a **ketohexose**.

---

problem **4.25** Draw the Lewis structure of each of the following.

(a) An aldotetrose   **(b)** A ketotetrose   **(c)** An aldotriose
(d) A ketotriose   **(e)** A ketohexose different from fructose

---

4.16 Saturation and Unsaturation in Fats and Oils

Diets high in saturated fats, typically of animal origin, are unhealthy because they increase the risk of coronary heart disease. Conversely, unsaturated fats, typically derived from plants, are a healthy part of our diets.

The concepts of saturation and unsaturation apply to fats and oils in the same way they do to other organic compounds (see Section 4.13). That is, each C=C double bond that is present adds one unit of unsaturation (or IHD) to the molecule. Thus, fats that contain fatty acids such as butyric acid or stearic acid (Fig. 4-49) are *saturated*, because they contain no C=C double bonds, whereas fats that contain oleic, linoleic, or linolenic acids (Fig. 4-50) are *unsaturated*, because these fatty acids have one or more C=C double bonds.

Oleic acid, C₁₈H₃₄O₂
(Major constituent of olive oil)
Melting point = 16 °C

Linoleic acid, C₁₈H₃₂O₂
(Major constituent of safflower oil)
Melting point = –5 °C

Essential fatty acids

Linolenic acid, C₁₈H₃₀O₂
(Major constituent of flaxseed oil)
Melting point = –11 °C

FIGURE 4-50 Unsaturated fatty acids Oleic acid, linoleic acid, and linolenic acid are unsaturated fatty acids because they contain units of unsaturation (i.e., C=C double bonds) in their carbon chains.

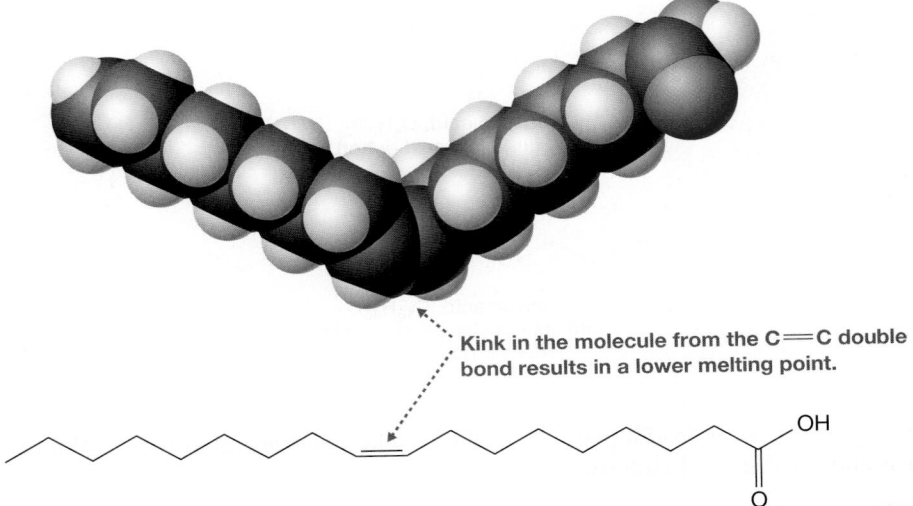

FIGURE 4-51 Space-filling model of oleic acid A cis double bond introduces a "kink" in the carbon chain of a naturally occurring fatty acid such as oleic acid. This decreases the contact surface area among separate fatty acid molecules, which reduces the strength of their intermolecular interactions, and thus decreases the melting point.

Kink in the molecule from the C=C double bond results in a lower melting point.

We can also distinguish the extent of unsaturation on the basis of the number of C=C double bonds present. For example, oleic acid is a **monounsaturated** fatty acid, because it has just one C=C double bond, whereas linolenic acid is a **polyunsaturated** fatty acid, because it has three.

Linoleic and linolenic acids are further classified as **essential fatty acids** because these are the only naturally occurring fatty acids that cannot be synthesized in the human body by any known chemical pathways. Instead, they must be consumed.

The different health effects of saturated and unsaturated fats seem to correlate with their different physical properties. Namely,

> Each cis C=C double bond that is present in a fatty acid lowers the melting point.

Notice, in particular, that stearic acid, a saturated fatty acid, melts well *above* room temperature and is therefore a solid under normal conditions. By contrast, oleic, linoleic, and linolenic acids melt well *below* room temperature and are therefore liquids, despite having the same number of carbon atoms as stearic acid.

The number of double bonds affects the melting points of these compounds because all C=C double bonds found in naturally occurring fatty acids are cis. Thus, each double bond introduces a "kink" in the carbon chain, as shown for oleic acid in Figure 4-51.

Such a kink makes it more difficult for separate fatty acid molecules to align, so the contact surface area among the molecules is decreased. In turn, this weakens the induced dipole–induced dipole interactions (Section 2.6), resulting in a lower melting point.

Chapter Summary and Key Terms

- **Isomerism** is a relationship between two or more molecular species. Molecules are **isomers** of each other if they have the same molecular formula, but are different in some way. **(Section 4.1)**

- **Constitutional isomers** (Section 4.11) have different connectivity, whereas **conformational isomers** differ by rotations about single bonds. **(Section 4.2)**

- **Newman projections** are used to show conformations about bonds. They depict the view down a bonding axis. The atom

in front is represented as a dot and the atom in back is represented as a circle. **(Section 4.2)**

- In a **conformational analysis** of ethane, energy is plotted as a function of the H—C—C—H **dihedral angle (θ)**. In a 360° rotation, we observe three equivalent **staggered conformations**, each at an energy minimum, and three equivalent **eclipsed conformations**, each at an energy maximum. **(Section 4.3a)**

- **Torsional strain** is the energy increase that appears in an eclipsed conformation. **(Section 4.3a)**

- Staggered conformations in ethane rapidly interconvert because the **rotational energy barrier** is comparable to the **thermal energy** available. **(Section 4.3a)**
- Based on the conformational analysis of CH_2Br—CH_2Br, one staggered conformation, called the **anti conformation**, is lower in energy than the other two, called **gauche conformations**. This is because there is less **steric strain** between the two Br atoms in the anti conformation. Similarly, one eclipsed conformation is higher in energy than the other two. **(Section 4.3b)**
- Although the anti conformation is more stable than the gauche, the two conformational isomers rapidly interconvert because the energy barrier between them is comparable to the available thermal energy. **(Section 4.3b)**
- The lowest energy conformation of an alkyl chain is the **zigzag**, which has the anti conformation at each C—C bond. **(Section 4.3c)**
- **Heats of combustion** provide insight into **ring strain** for rings of various sizes. Cyclohexane (a six-membered ring) has essentially no ring strain. Cyclopentane and cycloheptane (five- and seven-membered rings, respectively) have mild ring strain. Cyclopropane and cyclobutane (three- and four-membered rings, respectively) are highly strained. **(Section 4.4)**
- Cyclohexane has no ring strain because it adopts a **chair conformation**, in which all angles are about 111° (close to the ideal tetrahedral angle of 109.5°) and all C—C bonds are staggered. Cyclopentane adopts an **envelope conformation**, which has slightly more angle strain and torsional strain. Cyclobutane and cyclopropane rings are highly strained due to substantial torsional and angle strain. **(Section 4.5)**
- Envelope conformations of cyclopentane rapidly interconvert via **pseudorotation**—slight rotations about the single bonds that compose the ring. Similarly, chair conformations interconvert via **chair flips. (Section 4.6)**

- In each chair conformation of cyclohexane, one hydrogen atom on each carbon is in an **axial position** and one is in an **equatorial position**. The two positions interconvert during a chair flip. **(Section 4.6)**
- The two chair conformations of a **monosubstituted cyclohexane** are no longer equivalent. Because the substituent is larger than the hydrogen atoms, the cyclohexane ring is more stable with the substituent in an equatorial position, where there is more room. **(Section 4.8)**
- The bulkier a substituent, the greater its tendency to occupy an equatorial position. A *tert*-butyl group, $C(CH_3)_3$, is so bulky that it is almost exclusively found in an equatorial position. **(Section 4.8)**
- **Disubstituted** cyclohexanes introduce **cis** and **trans** relationships relative to the plane of the ring. **Haworth projections** illustrate cis and trans relationships well, but they do not accurately convey three-dimensional relationships or steric strain because they portray the cyclohexane ring as planar. **(Section 4.9)**
- The most stable chair conformation of a disubstituted cyclohexane is the one in which the substituents experience the least amount of strain. This usually calls for the larger substituent to occupy an equatorial position. **(Section 4.9)**
- Molecular modeling kits effectively demonstrate changes that occur during a chair flip. **(Section 4.10)**
- A molecule's **index of hydrogen deficiency**, also called its **degree of unsaturation**, is half the number of hydrogen atoms missing from that molecule compared to an analogous completely **saturated** molecule. A saturated molecule has the most possible hydrogen atoms given the nonhydrogen atoms it contains; its IHD is 0. **(Section 4.12)**
- Each double bond a molecule possesses contributes 1 to its IHD. Each triple bond contributes 2, and each ring contributes 1. **(Section 4.12)**

Problems

4.26 Determine the IHD for each of the following compounds.

(a)

(b)

(c)

(d)

(e)

(f)

(g)

(h)

4.27 Determine the number of hydrogen atoms in each compound, given the number and type of nonhydrogen atoms it contains and its IHD.

(a) Four carbon atoms; IHD = 0.

(b) Four carbon atoms; IHD = 2.

(c) Three carbon atoms; two oxygen atoms; IHD = 1.

(d) Five carbon atoms; two chlorine atoms; one nitrogen atom; IHD = 3.

(e) One carbon atom; one nitrogen atom; IHD = 2.

(f) Six carbon atoms; two nitrogen atoms; one oxygen atom; three fluorine atoms; IHD = 4.

4.28 Calculate the IHD for each of the following molecular formulas.

(a) C_6H_6 (b) $C_6H_5NO_2$ (c) $C_8H_{13}NOF_2$ (d) $C_4H_{12}Si$ (e) $C_6H_5O^-$ (f) $C_4H_6O_3S$

4.29 Draw the more stable chair conformation of each of the following molecules.

(a)   (b)   (c)   (d)   (e)   (f)   (g)

4.30 Draw the Newman projection for each of the following species, looking down the bond that is indicated in red.

(a) (b) (c) (d)

(e) (f) (g) (h)

4.31 (a) Draw the Newman projection for each molecule shown below, looking down the C—C bond indicated by the arrow.

(b) Which configuration do you think is more stable? Explain.

cis-1,2-Dimethylcyclopropane   **trans-1,2-Dimethylcyclopropane**

4.32 Draw the corresponding dash–wedge structure for each of the following Newman projections.

(a) (b) (c) (d)

4.33 Rank the following conformations in order from least stable to most stable.

A

B

C

D

E

F

4.34 Identify which C—H bonds are axial and which are equatorial in the following Newman projection for cyclohexane.

4.35 For each of the following substituted cyclohexane rings, **(a)** draw the corresponding dash–wedge structure, **(b)** draw the corresponding Haworth projection, **(c)** determine whether the given conformation is the most stable one, and **(d)** if it is *not* the most stable conformation, draw the Newman projection of the one that is.

(i)

(ii)

(iii)

4.36 Perform a conformational analysis of propane, $CH_3CH_2CH_3$. Pay attention to the relative energies of the various conformations, but do not concern yourself with the actual energy values.

4.37 Perform a conformational analysis of butane, $CH_3CH_2CH_2CH_3$, looking down the C2—C3 bond. Pay attention to the relative energies of the various conformations, but do not concern yourself with the actual energy values.

4.38 Perform a conformational analysis of 1-bromo-2-chloroethane, $BrCH_2CH_2Cl$. Pay attention to the relative energies of the various conformations, but do not concern yourself with the actual energy values.

4.39 Perform a conformational analysis of 2-methylbutane, $(CH_3)_2CHCH_2CH_3$, looking down the C2—C3 bond. Pay attention to the relative energies of the various conformations, but do not concern yourself with the actual energy values.

4.40 Perform a conformational analysis of 1,2-dibromo-1-fluoroethane, $CHBrF—CH_2Br$. Pay attention to the relative energies of the various conformations, but do not concern yourself with the actual energy values.

4.41 The heat of combustion of cyclononane is 5586 kJ/mol (1335 kcal/mol). Calculate the ring strain per CH_2 group and the total ring strain of cyclononane. Which compound has more ring strain, cyclononane or cycloheptane?

4.42 Draw a chair conformation of the following molecule with **(a)** all CH_3 groups in axial positions and **(b)** all CH_3 groups in equatorial positions.

4.43 Identify each of the following disubstituted cyclohexanes as either a cis or a trans isomer.

(a)

(b)

(c)

(d)

(e)

(f)

4.44 For each of the following disubstituted cyclohexanes, determine whether the cis or trans isomer is more stable.

(a)

(b)

(c)

4.45 Rank the following compounds in order from smallest heat of combustion to largest heat of combustion.

A   B   C   D

4.46 There are three distinct chair conformations for cyclohexylcyclohexane. **(a)** Draw all three and **(b)** determine which one is the most stable.

Cyclohexylcyclohexane

4.47 Draw all constitutional isomers that have the formula $C_5H_{11}Br$.

4.48 Draw all constitutional isomers that have the formula C_3H_5N. In each isomer, identify any functional groups that are listed in Table 1-6.

4.49 Draw all constitutional isomers that have the molecular formula $C_3H_6OF_2$, in which the oxygen is bonded to only one carbon atom. (There are 14 isomers.)

4.50 Draw all constitutional isomers that have the molecular formula $C_3H_6OF_2$, in which the oxygen is bonded to two carbon atoms.

4.51 Draw all constitutional isomers that have the molecular formula C_4H_6. (There are nine isomers.)

4.52 Draw all constitutional isomers of C_9H_{12} that contain a benzene ring.

4.53 Draw all constitutional isomers of $C_5H_8O_2$ that contain a carboxylic acid functional group.

4.54 Draw all constitutional isomers of C_5H_8O that contain a ketone functional group, but *not* an alkene.

4.55 The two compounds shown below are each in their more stable chair conformation.

(a) Which occupies more space, a lone pair of electrons or a N—H bond?
(b) Which occupies more space, a lone pair of electrons or a CH_3 group?

4.56 Both cis and trans isomers exist for the following molecule. Which one is more stable?

4.57 Draw the more stable chair conformation for each of the following disubstituted cyclohexanes.

(a) OH (b) OH (c) CBr₃ / CCl₃ (d) CBr₃ / CCl₃ (e) D / F (f) CCl₃ / Cl

4.58 Draw all constitutional isomers of all-cis ethylmethylisopropylcyclohexane—that is, in which a methyl group (CH_3), an ethyl group (CH_2CH_3), and an isopropyl group [$CH(CH_3)_2$] are all bonded to a cyclohexane ring on the same side of the ring's plane. Which of those isomers do you think is the most stable? Explain.

4.59 For which isomer would you expect a greater equilibrium percentage of molecules with the alkyl group in the axial position, isopropylcyclohexane or propylcyclohexane? Explain.

Isopropylcyclohexane   **Propylcyclohexane**

4.60 Glucose, a monosaccharide, has both acyclic and cyclic forms. One of the cyclic forms possible, called β-D-glucopyranose, is shown below. Draw the more stable chair conformation of β-D-glucopyranose.

HO
HO‚ᵢᵢᵢ
HO   O
         OH
    OH
β-D-Glucopyranose

4.61 In addition to β-D-glucopyranose (see Problem 4.60), glucose can exist in another cyclic form, called β-D-glucofuranose, shown below. Which form is more stable, β-D-glucopyranose or β-D-glucofuranose? Explain.

β-D-Glucofuranose

4.62 Draw the more stable chair conformation of each of the following compounds in which an sp^3-hybridized heteroatom is part of the ring.

(a)　　　　　(b)　　　　　(c)

4.63 Draw the more stable chair conformation of each of the following compounds in which an sp^2-hybridized carbon atom is part of the ring.

(a)　　　(b)　　　(c)　　　(d)　　　(e)

4.64 Which monosaccharide has a greater heat of combustion, β-D-glucopyranose or β-D-allopyranose? Explain.

β-D-Glucopyranose　　　　　**β-D-Allopyranose**

4.65 Rank the following compounds in order from smallest heat of combustion to largest heat of combustion.

A　　　　　B　　　　　C　　　　　D　　　　　E

4.66 Behenic acid and erucic acid are two fatty acids isolated from rapeseed oil. Which fatty acid has a higher melting point?

Behenic acid

Erucic acid

4.67 Even though an iodine atom is larger in size than a bromine atom, both bromocyclohexane and iodocyclohexane exist with 31% of molecules having the halogen atom in the axial position. Explain why.

4.68 5-Hydroxy-1,3-dioxane is more stable with the OH group in the axial position than in the equatorial position. Explain why.

5-Hydroxy-1,3-dioxane

If your instructor assigns problems in smartwork, log in at **smartwork.wwnorton.com**.

Isomerism 2

Chirality, Enantiomers, and Diastereomers

I n Chapter 4, we examined *constitutional isomers* and *conformational isomers* in detail. Recall that constitutional isomers differ in their connectivity, whereas conformational isomers have the same connectivity and are thus a type of *stereoisomer*. Specifically, conformational isomers differ only by rotations about single bonds. Since these rotations typically occur freely, conformational isomers interconvert rapidly, making it difficult or impossible to isolate one from another.

Here in Chapter 5, we examine *configurational isomers* in detail. As another type of stereoisomer, configurational isomers can be further categorized as either *enantiomers* or *diastereomers*. We will study in depth the structural relationships among these types of stereoisomers, and we will begin to examine how these relationships affect their respective physical and chemical behaviors.

When objects are reflected through a mirror, some are identical to their mirror image and others are not. This is the basis for what is called *chirality*, a major topic of this chapter.

5.1 Defining Configurational Isomers, Enantiomers, and Diastereomers

To formally define *configurational isomers, enantiomers,* and *diastereomers,* review the flow chart first shown in Figure 4-1, which is shown once again in Figure 5-1. The types of isomers we are interested in here are highlighted in red. According to the flow chart,

> **Configurational isomers** have the same connectivity but differ in a way *other* than by rotations about single bonds. They include two types:
>
> - **Enantiomers:** Configurational isomers that are mirror images of each other.
> - **Diastereomers:** Configurational isomers that are *not* mirror images of each other.

Whereas conformational isomers are related solely by *rotations* about single bonds, converting from one configurational isomer to another usually requires the breaking of a covalent bond, something that generally does not readily take place at room temperature. Therefore, configurational isomers can usually be isolated from one another.

Even though the structural differences between a pair of configurational isomers may seem subtle, the differences in their behavior may not be. For example, one of two enantiomers of the drug thalidomide acts as a sedative and antinausea medication for pregnant women suffering from morning sickness, whereas the other enantiomer causes terrible birth defects in newborns. Widely marketed in Europe during the late 1950s and early 1960s, thalidomide is now used to treat multiple myeloma (a kind of cancer) and erythema nodosum leprosum (an inflammation of fat cells under the skin). We will have more to say about this phenomenon later in the chapter.

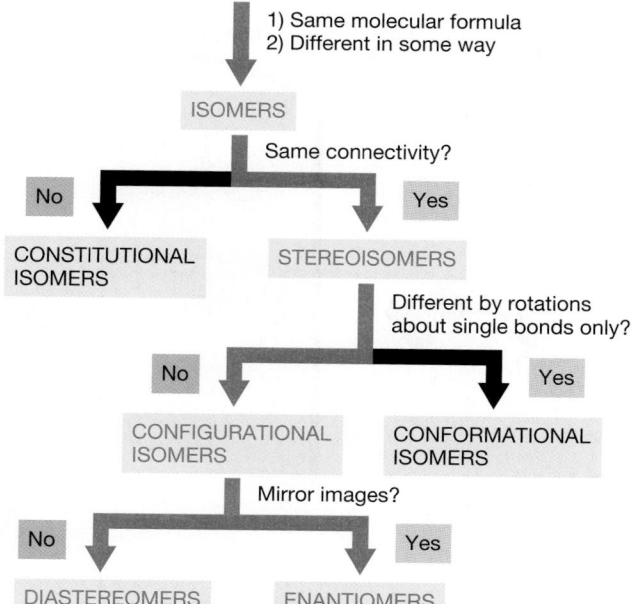

FIGURE 5-1 Flow chart illustrating the subcategories of isomers Those in red are examined in detail in this chapter.

5.2 Enantiomers, Mirror Images, and Superimposability

Enantiomers are mirror images of each other. To be *isomers* of each other, however, they must be *different* in some way. Enantiomers, therefore, are *nonsuperimposable mirror images*.

> Molecules are **nonsuperimposable** if there is no orientation in which *all* atoms of both molecules can be superimposed (i.e., lined up perfectly).

The two molecules of CHFClBr in Figure 5-2a are enantiomers. They are mirror images of each other—that is, if we looked at the reflection in the mirror of the molecule on the right (Fig. 5-2b), we would see an image resembling the molecule on the left. There is no orientation of the two molecules, however, that allows *all* of the atoms to be superimposed (see Your Turn 5.1).

YOUR TURN 5.1

The molecule on the left in Figure 5-2a is redrawn on the left below. Only part of the other molecule is redrawn on the right. Using a molecular modeling kit, construct the molecule on the right in Figure 5-2a and orient it so that the C, Br, and F atoms occupy the positions shown on the right below. Complete the drawing of the molecule by adding the H and Cl atoms in the positions where they appear in your model. Comparing these two structures, are the molecules superimposable?

Molecule on the left
in Figure 5-2a

Molecule on the right
in Figure 5-2a

Answers to Your Turns are in the back of the book.

Unlike CHFClBr, a molecule of CH_2FCl does not have an enantiomer. CH_2FCl has a mirror image (Fig. 5-3a), but its mirror image is exactly the same as itself

FIGURE 5-2 Enantiomers The two molecules in (a) are enantiomers because they are nonsuperimposable mirror images of each other (b). Ball-and-stick models are shown below each dash–wedge structure.

(a)

(b)

FIGURE 5-3 Superimposable mirror images CH_2FCl does not have an enantiomer because the molecule and its mirror image (a) are superimposable. If the image in the mirror is rotated by 109.5° (b), then the molecule and its mirror image are exactly the same. Ball-and-stick models are shown below each dash–wedge structure.

(a)   **(b)**

(Fig. 5-3b). If the image in the mirror is rotated, it can be superimposed on the original molecule (see Your Turn 5.2). Remember,

> Every molecule has a mirror image, but not every molecule has a *nonsuperimposable* mirror image.

5.2 YOUR TURN

Use a molecular modeling kit to construct both the original molecule and the mirror image shown in Figure 5-3a. Rotate the mirror image structure as indicated in Figure 5-3b to verify that the two molecules are superimposable.

problem **5.1** For each pair of molecules given, determine whether they are *superimposable* or *nonsuperimposable*. (*Hint:* It may help to build a model of each molecule.)

5.3 Strategies for Success: Drawing Mirror Images

Learning to draw a molecule's mirror image quickly and correctly is an essential skill in this chapter. The key to doing this is drawing the mirror image of each atom one at a time, following these three guiding principles:

- An atom and its mirror image are directly opposite each other on opposite sides of the mirror.
- An atom and its mirror image are identical distances *away from* the mirror.
- Dash–wedge notation in the mirror image is identical to that in the original molecule.

Let's practice by drawing the mirror image of the molecule on the left in Figure 5-2, one atom at a time. This step-by-step process is shown in Figure 5-4.

FIGURE 5-4 Progression of drawing the mirror image of CHFClBr
In each frame, the mirror is represented by the blue dotted line, the original molecule is on the left, and the mirror image is on the right. The atoms and bonds added to the mirror image of the dash–wedge structure are highlighted in red. A ball-and-stick model is shown below each dash–wedge structure.

SOLVED problem 5.2 Given the following Newman projection of butane in a gauche conformation, draw the Newman projection of its mirror image. The blue dotted line next to the molecule represents a mirror.

Think For each atom or group in the original molecule, where should its mirror image be with respect to the mirror?

Solve Six substituents must be added to the Newman projection on the right to complete the mirror image—two CH_3 groups and four H atoms. Begin by drawing the mirror image of the topmost CH_3 group, which must be directly opposite the original CH_3 group and must also be the same distance from the mirror.

Next, add the second CH_3 group to the mirror image so that it is directly opposite the second CH_3 group in the original molecule and the two CH_3 groups are the same distance from the mirror.

To complete the mirror image, add the remaining four H atoms.

Notice that each H atom in the mirror image is directly opposite its original H atom and is the same distance from the mirror.

---

problem 5.3 Draw the mirror image of each of the following molecules.

(a) (b) (c) (d)

problem 5.4 For each molecule in Problem 5.3, determine whether the mirror image is superimposable on the original molecule.

FIGURE 5-5 Familiar objects that are chiral (a) Hands. (b) A corkscrew. (c) A sea shell.

(a)   (b)   (c)

problem **5.5** Determine which pairs of molecules are mirror images.

and

(a)

and

(b)

and

(c)

5.4 Chirality

Whether a molecule has an enantiomer can significantly affect its chemical reactions, as well as some of its physical properties.

> ■ A molecule is **chiral** (kai-rul) if it *has an enantiomer*.
> ■ A molecule is **achiral** if it *does not have an enantiomer*.

The word *chiral* is derived from Latin for "hand," given that your hands are chiral. In fact, the enantiomer of your left hand is your right hand, and vice versa, because your left and right hands are mirror images of each other but they are not superimposable (Fig. 5-5a). Other objects with handedness, or chirality, include corkscrews, and certain sea shells (Fig. 5-5b and 5-5c). Biomolecules such as amino acids and sugars are chiral, too (see Section 5.13).

Of the molecules we have examined thus far, CHFClBr (Fig. 5-2) is chiral, whereas CH$_2$FCl is not (Fig. 5-3). Both molecules have mirror images, but only CHFClBr has a mirror image that is not superimposable.

SOLVED problem **5.6** Determine whether *trans*-1,2-dichlorocyclopropane is chiral or achiral.

Cl
⌲····ıllıCl

***trans*-1,2-Dichlorocyclopropane**

Think Are *trans*-1,2-dichlorocylopropane and its mirror image superimposable or non-superimposable? How can you tell?

Solve First draw the mirror image (shown below in red).

It then becomes a matter of lining up the two molecules to see if they are superimposable:

In neither of these two cases is the mirror image superimposable on the original molecule. We can see that the Cl atoms, in particular, are in different locations with respect to the plane of the page. There are other orientations you can try, but you will find that none of them allows the mirror image to be superimposable on the original molecule. Therefore, *trans*-1,2-dichlorocyclopropane is chiral.

Determining whether a molecule is superimposable onto its mirror image is a non-trivial task. If you cannot determine on paper whether two molecules are superimposable, construct a model of each molecule and then rotate each one in space until you are convinced one way or the other (see Your Turn 5.3).

5.3 YOUR TURN

Build a molecular model of both the original molecule and the mirror image in Solved Problem 5.6. Verify that the two molecules are nonsuperimposable by performing the rotations indicated in Solved Problem 5.6. Are there other rotations you can imagine to determine whether the two molecules line up perfectly?

5.4 YOUR TURN

Is the mirror image of the molecule in Solved Problem 5.6 (i.e., the molecule in red) chiral or achiral?

problem **5.7** Is *trans*-1,2-dichloroethene chiral or achiral?

trans-**1,2-Dichloroethene**

5.4a Chirality and Conformational Isomers

When determining whether a molecule is chiral or achiral, keep in mind rotations about single bonds.

If a molecule and its mirror image are rapidly interconverting conformational isomers, then the molecule is effectively *achiral*.

For example, the mirror image of 1,2-dibromoethane in one of its gauche conformations is simply the second gauche conformation (Fig. 5-6a). These two gauche conformations are nonsuperimposable (see Your Turn 5.5), which suggests that 1,2-dibromoethane is chiral. Recall from Chapter 4, however, that the

120° rotation about C–C bond

The original molecule and its mirror image are superimposable after the C—C bond rotation.

Gauche 1,2-dibromoethane (a)

(b)

(c)

FIGURE 5-6 **Bond rotations and chirality** (a) Newman projection of 1,2-dibromoethane in a gauche conformation (black), along with its mirror image (red). The mirror is indicated by the blue dotted line. Ball-and-stick models are provided next to the corresponding Newman projections. (b) The mirror image after a 120° rotation of the rear CH_2Br group about the C—C single bond. (c) Overlaying the conformer from (b) with the original molecule from (a) shows that the two are superimposable.

two conformations rapidly interconvert via rotation about the C—C single bond (Fig. 5-6b), which makes the original molecule and its mirror image superimposable (Fig. 5-6c). 1,2-Dibromoethane, then, is achiral.

YOUR TURN 5.5

Build molecular models of the gauche-1,2-dibromoethane shown on the left in Figure 5-6a and its mirror image. By overlaying the two molecules, verify that they are nonsuperimposable. Next, rotate the bond as in Figure 5-6b and show that the resulting molecules are superimposable.

problem **5.8** Determine whether each of the following molecules is chiral or achiral.

(a)   (b)   (c)

Figure 5-7a and 5-7b show that one chair conformation of *cis*-1,2-difluorocyclohexane is nonsuperimposable on its mirror image. However, as shown in Figure 5-7c, the mirror image is simply the *second* chair conformation. And, because the two chair conformations rapidly interconvert, *cis*-1,2-difluorocyclohexane is *achiral*.

YOUR TURN 5.6

Build molecular models of *cis*-1,2-difluorocyclohexane and its mirror image (as indicated in Fig. 5-7a). Rotate the two molecules in space to convince yourself that they are *not* superimposable. Then perform a chair flip on the mirror image and again rotate the molecules in space to convince yourself that the resulting structure *is* superimposable with the original molecule.

The preceding example suggests that we do not need to take into account the specific chair conformation assumed by the molecule when evaluating the chirality of molecules with cyclohexane rings that undergo rapid chair flips. Rather, the chirality of these molecules is governed only by the *relative* positioning of the substituents with respect to the ring (i.e., cis or trans). Therefore, we may simplify the problem by working with Haworth projections.

(a)

(b)

(c)

FIGURE 5-7 Chair conformations and chirality (a) The chair conformation of *cis*-1,2-difluorocyclohexane (black) and its mirror image (red), along with their respective ball-and-stick models. The mirror is represented by the blue dotted line. (b) The mirror image (red) is rotated 180° in an attempt to superimpose it upon the original molecule (black). (The C atoms in the ring are numbered to keep track of them during rotation.) The two are nonsuperimposable. (c) The mirror image (red) undergoes a chair flip, followed by a rotation of 120°. The resulting orientation is superimposable upon the original molecule (black). Therefore, *cis*-1,2-difluorocyclohexane is achiral.

For example, the Haworth projection of *cis*-1,2-difluorocyclohexane has a superimposable mirror image (Fig. 5-8a), so it is achiral, whereas the Haworth projection of *trans*-1,2-difluorocyclohexane does *not* have a superimposable mirror image (Fig. 5-8b), so it is chiral.

(a)

(b)

FIGURE 5-8 Haworth projections and chirality Haworth projections accurately depict the chirality of a substituted cyclohexane ring. (a) The Haworth projection of *cis*-1,2-difluorocyclohexane has a superimposable mirror image, so it is achiral. (b) The Haworth projection of *trans*-1,2-difluorocyclohexane has a nonsuperimposable mirror image, so it is chiral.

Use molecular models to prove to yourself that *trans*-1,2-difluorocyclohexane is chiral, as predicted using Haworth projections in Figure 5-8. The molecule and its mirror image are as follows:

trans-1,2-Difluorocyclohexane

(a) Build a model of the original molecule and a model of its mirror image.
(b) Rotate the two molecules in space to show that they are *not* superimposable.
(c) Carry out a chair flip on the model of the mirror image and once again rotate the molecules in space to show that they are *not* superimposable.

---

problem **5.9** Determine whether each of the following molecules is chiral or achiral.

   (a)   (b)   (c)   (d)   (e)

5.4b The Plane of Symmetry Test for Chirality

A convenient way to determine whether or not a molecule is chiral is to search for a *plane of symmetry*. A molecule has a **plane of symmetry** if it can be bisected in such a way that *one half of the molecule is the mirror image of the other half.* If a molecule has at least one plane of symmetry, then its mirror image is the same as itself. Thus,

> A molecule that possesses a plane of symmetry must be achiral.

Most molecules that do not possess a plane of symmetry are chiral, though there are some exceptions. (See Problem 5.70 at the end of the chapter.)

CH$_2$FCl is achiral (Fig. 5-3), and Figure 5-9a shows that there is a plane of symmetry that bisects the H—C—H bond angle. CHFClBr is chiral (Fig. 5-2), on the other hand, so it must *not* have a plane of symmetry. Notice that what was a plane of symmetry in CH$_2$FCl (Fig. 5-9a) is no longer one in CHFClBr (Fig. 5-9b).

Recall from Figure 5-8 that *cis*-1,2-difluorocyclohexane is achiral. Find the plane of symmetry in the following Haworth projection and indicate where it is, using a dashed line.

cis-1,2-Difluorocyclohexane

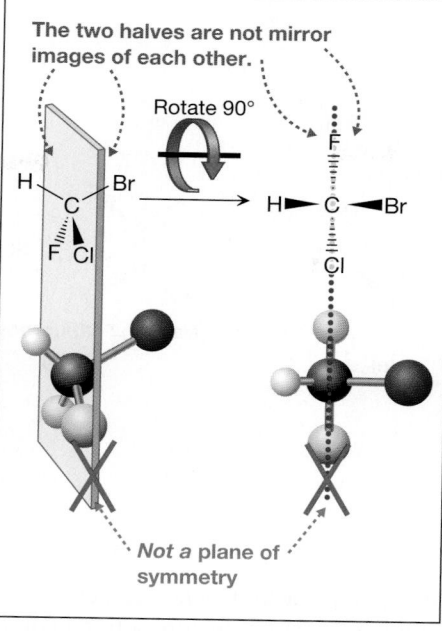

FIGURE 5-9 Plane of symmetry test for chirality (a) Two orientations of CH₂FCl. To obtain the orientation on the right, the molecule on the left is rotated 90° about the axis indicated. In both orientations, the molecule possesses a plane of symmetry (indicated in blue) that bisects the H—C—H angle. As a result, CH₂FCl is achiral. (b) Two orientations of CHFClBr that differ by a rotation of 90° about the axis are indicated. Replacing one H atom from CH₂FCl by a Br atom destroys the plane of symmetry that exists in CH₂FCl. As a result, CHFClBr is chiral.

(a)   **(b)**

5.9 **YOUR TURN**

Recall, too, that *trans*-1,2-difluorocyclohexane is chiral, in which case it does *not* have a plane of symmetry. Convince yourself that each plane in the following diagrams is *not* a plane of symmetry. Write an "O" at the position where each F's mirror image would be. Is there an F already at each "O" you drew?

trans-**1,2-Difluorocyclohexane**

---

problem **5.10** Determine whether each of the following molecules possesses a plane of symmetry. If it does, indicate that plane of symmetry using a dashed line.

(a)   **(b)**   **(c)**   **(d)**   **(e)**   **(f)**

5.4c Stereocenters and Stereochemical Configurations

Thus far, we have shown that a molecule's chirality is governed by a particular structural feature—the plane of symmetry. Specifically, a molecule that has a plane of

FIGURE 5-10 Tetrahedral
stereocenters (a) The C atom is
a tetrahedral stereocenter because
it is bonded to four different atoms.
(b) C1 and C2 are both tetrahedral
stereocenters (left) because each is
bonded to four different groups.
For each stereocenter, the groups
that make up the ring are shown
to be different (right).

trans-1,2-Difluorocyclohexane

symmetry is achiral, whereas a chiral molecule has no planes of symmetry. However,
another structural feature, a tetrahedral stereocenter, is also related to chirality.

> ■ A **stereocenter** is an atom with the property that interchanging any two of its
> attached groups produces a different stereoisomer.
> ■ A tetrahedral stereocenter is bonded to four *different groups*.

Each of the molecules we have identified thus far as being chiral possesses at least
one tetrahedral stereocenter. The C atom in CHFClBr, for example, is a tetrahedral
stereocenter, because it is bonded to four different atoms (Fig. 5-10a). *trans*-1,2-
Difluorocyclohexane contains two tetrahedral stereocenters—namely, each C that is
bonded to a F (Fig. 5-10b). While it should be obvious that both of those C atoms
are tetrahedral, it may be less obvious that both are also bonded to four *different*
groups. Two of the groups are H and F. The other two groups, which are part of
the ring, are not only different from H and F, but also are different from each other
because they have *different connectivity*. Specifically, one side of the ring is bonded
to a given stereocenter by a CH$_2$ group and the other side of the ring is bonded to
that stereocenter by a CHF group.

In general, to determine whether groups that are part of a ring are different
from each other, we can use a *plane of symmetry test*. For a given potential stereocenter, draw a plane that bisects the angle made by that atom's bonds to the ring. Ignoring dash–wedge notation, if one half of the ring is the mirror image of the other
half, then the groups are the same. Otherwise, they are different. This is shown in
Figure 5-11 for both stereocenters in *trans*-1,2-difluorocyclohexane.

FIGURE 5-11 Plane of symmetry
test for stereocenters (a) The plane
indicated by the dotted line is *not* a plane
of symmetry. Therefore, the indicated
groups that make up the ring, which
are bonded to the labeled C atom, are
different from each other. (b) The groups
indicated are different from each other
for the same reason as in (a).

SOLVED problem **5.11** How many stereocenters are in the following
molecule?

Think Which atoms are bonded to four *different* groups? How can you determine
whether groups that are part of a ring are different from each other?

Solve The three C atoms indicated by an asterisk are bonded to four different groups and are, therefore, stereocenters.

The right-most C is bonded to H, OH, CH₃, and the ring. Each of the other stereocenters is bonded to H, a group that is not part of the ring, and two groups that are part of the ring. For those C atoms, we can apply the plane of symmetry test to determine whether the groups that are part of the ring are different from each other:

In each case, the plane indicated by the dotted line is *not* a plane of symmetry, so the two groups that are part of the ring are, indeed, different from each other.

---

problem 5.12 Determine how many stereocenters exist in each of the following molecules.

(a)   (b)   (c)   (d)   (e)

Although tetrahedral stereocenters are related to the concept of chirality,

> The presence of tetrahedral stereocenters does not guarantee chirality.

For example, *cis*-1,2-difluorocyclohexane contains two stereocenters, but it is achiral because there is a plane of symmetry between the two stereocenters (Fig. 5-12).

> A molecule is **meso** if it contains at least two tetrahedral stereocenters but has a plane of symmetry that makes it achiral *overall*.

The term comes from the Greek word for "middle," and refers to the fact that the molecule reflects about its middle—its plane of symmetry.

cis-1,2-Difluorocyclohexane
(*meso*-1,2-Difluorocyclohexane)

FIGURE 5-12 A meso compound
This molecule is meso because it contains tetrahedral stereocenters but, due to the plane of symmetry, is achiral.

problem **5.13** Which of the following molecules are *meso*? (*Hint:* Consider rotations about single bonds.)

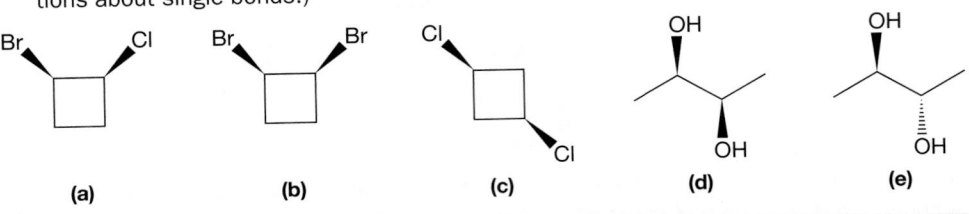

In most cases, a molecule (e.g., CH_2FCl) is achiral if it does not have a tetrahedral stereocenter. However,

> The absence of a tetrahedral stereocenter does not guarantee that a molecule is achiral.

Consider, for example, 1,3-difluoropropa-1,2-diene (HFC=C=CHF; Fig. 5-13). It has no tetrahedral stereocenters (it has no tetrahedral atoms), yet it is chiral.

Nonsuperimposable

Compare

FIGURE 5-13 A chiral molecule with no tetrahedral stereocenters HFC=C=CHF contains no tetrahedral stereocenters, yet it is chiral. It does not have a plane of symmetry.

No tetrahedral stereocenters

There is, however, one strict rule that relates the number of tetrahedral stereocenters and chirality:

> Any molecule that contains exactly one tetrahedral stereocenter must be chiral.

We have previously seen (Fig. 5-2) that CHFClBr has a single stereocenter and is chiral. The same would be true if H, F, Cl, and Br were replaced by any four distinct groups. Examples include butan-2-ol and 3-chloro-1,1-dimethylcyclohexane (Fig. 5-14).

YOUR TURN 5.10

Place an asterisk at the stereocenter in butan-2-ol and 3-chloro-1,1-dimethylcyclohexane in Figure 5-14.

YOUR TURN 5.11

Prove to yourself that butan-2-ol is chiral by (1) drawing the mirror image, (2) building models of both the original molecule and the mirror image, and (3) lining the models up to determine whether they are superimposable or nonsuperimposable.

FIGURE 5-14 Tetrahedral stereocenters and chirality Each molecule has exactly one tetrahedral stereocenter and is therefore chiral.

OH
Butan-2-ol

Cl
3-Chloro-1,1-dimethylcyclohexane

Figure at top (5-15 illustration).

FIGURE 5-15 Two ways to invert stereochemical configurations (a) Reflection of the stereocenter through a mirror. (b) Exchanging two groups on the stereocenter. Reflection through a mirror (red structure) and the exchange of any two groups (blue structure) give exactly the same stereochemical configuration, which is opposite that in the original (black) structure.

Reflect through a mirror — Nonsuperimposable — Compare — Identical stereochemical configurations — (a)

Exchange two groups — Compare — (b) — Nonsuperimposable

There are two possible **stereochemical configurations** for any given tetrahedral stereocenter. That is, there are two possible ways in which to arrange the four atoms or groups about the stereocenter in a tetrahedral fashion. One configuration is called "*R*" and the other is "*S*." The details of assigning these configurations, however, are not essential to understanding the aspects of stereochemistry discussed here in Chapter 5.

> The convention for naming a stereochemical configuration as *R* or *S* is discussed in Section N3.1 in Nomenclature 3, which immediately follows this chapter.

What is critical to this chapter is understanding that for any tetrahedral stereocenter, one configuration is the *opposite* or *inverse* of the other, and that the two configurations are related as follows:

- One configuration of a tetrahedral stereocenter is the mirror image of the other.
- Exchanging any two of the four groups on a tetrahedral stereocenter gives the opposite stereochemical configuration.

These relationships are illustrated in Figure 5-15. The black structures in Figure 5-15a and 5-15b are identical; hence, the stereochemical configuration at the C atom (the stereocenter) is identical in both cases. In Figure 5-15a, the structure is reflected through a mirror plane, giving the structure shown in red. In Figure 5-15b, groups W and X in the black structure are interchanged, giving the structure shown in blue. Accordingly, the stereochemical configuration about the C atom in the red and blue structures is opposite that in the black structure, but the red and blue structures have the same configuration.

problem **5.14** Figure 5-15b shows the result of exchanging groups W and X on the black structure, giving the blue structure. Repeat this exercise, but instead, exchange groups W and Y on the black structure and determine whether or not the resulting structure is superimposable on the blue structure in Figure 5-15.

5.4d Stereocenters Other Than Carbon

Although most of the tetrahedral stereocenters you will encounter in organic chemistry will be carbon atoms, other tetrahedral atoms can also be stereocenters. Nitrogen atoms, for example, can be sp^3 hybridized, and can therefore have a tetrahedral electron geometry. Consequently, a N atom bonded to four different substituents, such as in a quaternary ammonium ion, is a stereocenter (Fig. 5-16).

CH_3 ---- Stereocenter

C_6H_5 ⁞⁞⁞⁞ N

H_3CH_2C ⊕ $CH_2CH_2CH_3$

A quaternary ammonium ion

FIGURE 5-16 A nitrogen stereocenter The nitrogen atom is tetrahedral and is bonded to four different groups, so it is a stereocenter.

FIGURE 5-17 Stereocenters and uncharged nitrogen atoms Counting the lone pair as a group, methylethylamine has a tetrahedral nitrogen atom with four different groups. As a result, the mirror image is nonsuperimposable on the original molecule, suggesting that methylethylamine should be chiral.

A N atom singly bonded to three different substituents, as in *N*-methylethanamine (methylethylamine; Fig. 5-17) has a lone pair of electrons, so the N atom is still surrounded by four different groups in a tetrahedral arrangement. (Recall from Section 2.1 that a lone pair is counted as a group in VSEPR theory.) In other words, the nitrogen atom in NRR′R″ could be a stereocenter and the molecule could be chiral. It turns out, however, that the two methylethylamine species rapidly interconvert and cannot be isolated, so methylethylamine is achiral and its nitrogen atom is not a stereocenter.

The process by which NRR′R″ species interconvert, called **nitrogen inversion** (Fig. 5-18), proceeds through a structure in which the nitrogen atom's electron geometry is planar. In that geometry, the nitrogen atom is sp^2 hybridized (not sp^3) and possesses an unhybridized p orbital in which the lone pair of electrons is located. In other words, the nitrogen atom is temporarily rehybridized from sp^3 to sp^2 back to sp^3 during the transformation from one tetrahedral structure to the other.

> Uncharged nitrogen atoms that undergo *nitrogen inversion* are not stereocenters.

FIGURE 5-18 Nitrogen inversion *N*-Methylethanamine is achiral because the two sp^3-hybridized mirror images rapidly interconvert via a planar sp^2-hybridized intermediate. Consequently, the uncharged nitrogen atom is not a stereocenter.

Nanocars

Can you imagine an electric car about 10,000 times smaller than the thickness of a human hair? Dr. Ben Feringa could, and in 2011, he and his research team at the University of Groningen (Netherlands) constructed one! It is a single molecule, roughly 1 nm long, as shown below on the left. Each of the four "wheels" comprises a planar three-ring system.

Rotation is induced about the double bonds screened in red

Meso-(R,S-R,S) isomer

(1)

(2)

(3)

(4)

After the car is sublimed onto a copper surface, it is powered by electrons from a scanning tunneling microscope (STM). Each of the wheels functions independently of the others, making the nanocar analogous to a four-wheel-drive vehicle. As shown above on the right, a full rotation of each wheel consists of four separate steps. First, excitation by a pulse from the STM provides energy to temporarily break the π bond of the double bond that connects the wheel. This isomerizes the double bond, resulting in a partial turn. The wheel is left in a sterically strained conformation as a result, and in Step 2 it relieves that strain by relaxing into a more stable conformation. Steps 3 and 4 are the same as Steps 1 and 2.

The direction in which each wheel rotates is governed by the specific configuration of its four stereocenters. With the configurations shown, the four wheels work in concert to propel the nanocar forward. Other configurations can lead to the wheels working against each other, resulting in either no net motion, or in the car turning.

Powering a nanocar in a single direction like this provides a proof of concept for more advanced nanomachines that could carry out specific tasks within our bodies. One task that scientists envision is targeting and killing cancer cells.

Not all nitrogen atoms undergo nitrogen inversion. One example, as mentioned previously, is a *positively* charged nitrogen atom. Unlike an uncharged nitrogen atom, a positively charged nitrogen cannot rehybridize from sp^3 to sp^2, because it is bonded to four different substituents. Another example is one in which geometrical constraints prevent inversion at nitrogen (see Problem 5.69).

(a) (b) (c) (d)

5.5 Diasteromers

Recall that **diastereomers** are *stereoisomers that are not mirror images of each other* (review the flow chart in Fig. 5-1). To be stereoisomers, they must have the same molecular formula, be different molecules, and have the same connectivity.

Cis and trans alkenes, such as *cis*-1,2-dichloroethene and *trans*-1,2-dichloroethene, are diastereomers. They have the same molecular formula and the same connectivity, but they are different molecules, given that one has chlorine atoms on the same side of the double bond, and the other has chlorine atoms on opposite sides. They are, furthermore, not mirror images of each other.

Cis and trans isomers with respect to a double bond are diastereomers of each other.

cis-1,2-Dichloroethene   *trans*-1,2-Dichloroethene

The compounds *cis*- and *trans*-cyclohexane-1,3-diol are diastereomers, too. Unlike the previous example, however, *cis*- and *trans*-cyclohexane-1,3-diol do not differ by a 180° rotation about a double bond. In fact, neither compound has a double bond. Nevertheless, the two have the same molecular formula and the same connectivity, but they are different molecules and are not mirror images of each other.

Cis and trans isomers with respect to a ring are diastereomers of each other.

cis-Cyclohexane-1,3-diol   *trans*-Cyclohexane-1,3-diol

Diastereomers need not be cis–trans pairs:

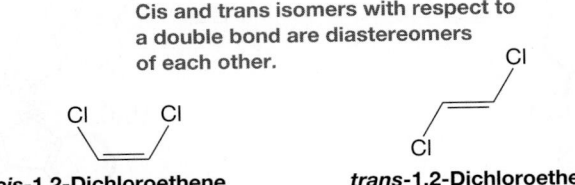

Same stereochemical configuration

Diastereomers

Opposite stereochemical configurations

As long as they are different molecules, but not mirror images of each other, with the same molecular formula and the same connectivity, then they are diastereomers (see Your Turn 5.12).

Confirm that the following compounds are different by building a model of each, and rotating one to determine if it is superimposable on the other.

It can be quite helpful to use the relative configurations at the tetrahedral stereo-centers to determine whether two molecules are diastereomers or enantiomers:

- If two isomers are related by the inversion of some, but not all, tetrahedral stereocenters, then the two molecules are *diastereomers* of each other.
- If two isomers are related by the inversion of *all* tetrahedral stereocenters, then the two molecules are *enantiomers* of each other.

These criteria apply regardless of the number of stereocenters a molecule contains. For example, if two molecules have opposite configurations at two out of three tetrahedral stereocenters, then they are *diastereomers* (Fig. 5-19a). On the other hand, if they have opposite configurations at all three stereocenters, then they are *enantiomers* (Fig. 5-19b).

FIGURE 5-19 Stereoisomers and relative configurations about tetrahedral stereocenters (a) Two of the three stereocenters in the first molecule have configurations that are opposite those in the second molecule, so the molecules are diastereomers. (b) All three stereocenters in the first molecule have configurations that are opposite those in the second molecule, so the molecules are enantiomers.

SOLVED problem 5.16 Given the following configurational isomer of 1-chloro-2,3-dimethylcyclopentane, draw its enantiomer and one of its diastereomers.

Think How many tetrahedral stereocenters exist? To obtain its enantiomer, how many of those configurations must be reversed? To obtain one of its diastereomers, how many of those configurations must be reversed?

Solve There are three tetrahedral stereocenters, indicated by asterisks below. To obtain its enantiomer, we must reverse *all* three of those configurations. To obtain a diastereomer, we must reverse at least one configuration, but not all of them. The enantiomer and one diastereomer are shown.

problem **5.17** Draw two more diastereomers of the molecule given in Solved Problem 5.16.

Because a new configurational isomer can be obtained for *each* inversion of a stereocenter's configuration, the number of configurational isomers that exist can double with the addition of each stereocenter. Therefore,

The maximum number of configurational isomers that can exist for a molecule with n stereocenters is 2^n.

2,3-Dibromo-4-methylhexane, for example, has three stereocenters, so it has the $2^3 = 8$ configurational isomers shown in Figure 5-20.

YOUR TURN 5.13

In molecule **A** in Figure 5-20, indicate each stereocenter with an asterisk.

problem **5.18** Which molecules in Figure 5-20 are enantiomers of molecule **A**? Which molecules are diastereomers? (*Hint:* How many stereocenters are there? In each structure, how many stereocenters are inverted relative to those in molecule **A**?)

FIGURE 5-20 All possible configurational isomers of 2,3-dibromo-4-methylhexane
2,3-Dibromo-4-methylhexane has three stereocenters, giving rise to a total of $2^3 = 8$ configurational isomers.

It is important to realize that 2^n represents a *maximum* number.

> Fewer than 2^n configurational isomers can exist when at least one of the isomers is *meso*.

This is illustrated in Solved Problem 5.19.

SOLVED problem **5.19** Draw all configurational isomers of cyclohexane-1,3-diol. Does the total number of configurational isomers equal 2^n? Explain.

Cyclohexane-1,3-diol

Think How many stereocenters are there? How can we change each stereocenter to convert from one configurational isomer to another?

Solve There are two stereocenters, indicated by the asterisks.

To obtain all possible configurational isomers, we can systematically reverse the configurations at the different stereocenters. Begin with the isomer in which both OH groups point toward us (Structure **A**):

Structures A and D are the same molecule.

A   **B**   **C**   **D**

It may appear that there are four total isomers, which does equal 2^n with n (the number of stereocenters) being two. Structures **A** and **D** are exactly the same, however, so there are only three total isomers. There are fewer than 2^n isomers because one of the isomers is meso. Although it has two stereocenters, it has a plane of symmetry, too, and is therefore achiral—its mirror image is no different from itself.

Plane of symmetry

problem **5.20** Draw all possible configurational isomers of each molecule. How many stereocenters are there? Does the number of configurational isomers equal 2^n? Explain.

(a)　　　　　(b)

5.6 Fischer Projections and Stereochemistry

In his study of simple sugars (saccharides) in the late nineteenth century, Emil Fischer found himself working with several tetrahedral stereocenters in a given molecule and several molecules at a time. This led Fischer to develop a quicker and more convenient way to depict configurations about these stereocenters, now known as **Fischer projections**. There are two conventions for drawing Fischer projections (Fig. 5-21):

■ The intersection of a horizontal line and a vertical line indicates a carbon atom—typically a tetrahedral carbon stereocenter.

■ The substituents on the horizontal bonds are understood to point toward you, reminiscent of a bow tie, whereas the substituents on the vertical bonds are understood to point away from you.

YOUR TURN 5.14

Build a model of the molecule represented by the Fischer projection in Figure 5-21 (shown again on the left below). Use different colored balls to represent the four different substituents. View the molecule from the vantage point indicated on the right below and fill in the atoms in the boxes provided.

FIGURE 5-21 Fischer projections (Left) A generic Fischer projection of a carbon stereocenter, where W, X, Y, and Z are different substituents. (Middle) The dash–wedge structure that corresponds to the Fischer projection. The substituents on the horizontal point toward you, reminiscent of a bow tie, whereas the substituents on the vertical point away from you. (Right) A ball-and-stick model of the molecule.

FIGURE 5-22 Rotation of a Fischer projection by 90° (a) (Top left) A generic Fischer projection and its dash–wedge representation. (Bottom left) Rotation of the Fischer projection by 90° and the resulting dash–wedge structure. (Right) The two structures are nonsuperimposable, so rotation of a Fischer projection by 90° gives the opposite configuration at the stereocenter. (b) Three successive exchanges of substituents about a stereocenter in a Fischer projection is equivalent to a 90° rotation of the Fischer projection. Each exchange of substituents gives the opposite stereochemical configuration. Overall, three exchanges gives the opposite configuration.

To work comfortably with Fischer projections, we must know how certain manipulations affect the configurations about their tetrahedral stereocenters.

- Exchanging any two substituents on a carbon stereocenter in a Fischer projection gives the opposite stereochemical configuration.
- Taking the mirror image of a Fischer projection gives the opposite stereochemical configuration.
- Rotating a Fischer projection 90° in the plane of the paper gives the opposite stereochemical configuration.

The first two also are true of dash–wedge representations, whereas the third is peculiar to Fischer projections. When a Fischer projection is rotated 90°, the bonds that were pointing toward us instead point away from us, and vice versa. As illustrated in Figure 5-22a, this results in a configuration that is nonsuperimposable on the original molecule.

To better understand this, we can view a 90° rotation as the equivalent of three successive exchanges of pairs of substituents, as illustrated in Figure 5-22b. Each exchange reverses the configuration, and three exchanges result in an overall inversion of configuration.

Using the same logic, *a 180° rotation of a Fischer projection results in no change in the configuration*, because this is the same as two 90° rotations. The first 90° rotation inverts the configurations of all stereocenters and the second 90° rotation inverts them again, thus restoring them to their original configurations.

Just as with any chiral molecule, the mirror image of a Fischer projection gives the enantiomer because the configuration at each stereocenter is opposite that in the original molecule. Figure 5-23 illustrates why this is so. The mirror reflection taken of the Fischer projection in Figure 5-23a is equivalent to the exchange of the X and

FIGURE 5-23 Reflection of a Fischer projection gives its enantiomer (a) Reflection of a Fischer projection through a mirror. (b) Exchanging two groups on the Fischer projection gives the same result as a reflection. Because exchange of any two groups on a stereocenter gives the opposite configuration, taking a mirror reflection of a Fischer projection must also give the opposite stereochemical configuration.

FIGURE 5-24 Fischer projections with multiple stereocenters Each Fischer projection represents four tetrahedral stereocenters. D-Allose and D-glucose differ in the configuration at one of these stereocenters, so they are diastereomers. D-Glucose and L-glucose are nonsuperimposable mirror images, so they are enantiomers.

Z groups in Figure 5-23b. And, as we just learned, exchange of any two groups in a Fischer projection gives the opposite configuration about a stereocenter.

The convenience of Fischer projections is fully realized when multiple carbon stereocenters exist in the same molecule, as in the case of simple sugars. D-Allose, for example, is a molecule with six adjacent C atoms, four of which are stereocenters (Fig. 5.24). Notice that the two C atoms that are not stereocenters are *not* represented by the intersection of perpendicular lines.

D-Glucose has the same molecular formula and the same connectivity as D-allose. Based on their Fischer projections, however, they are diastereomers of each other (not enantiomers) because their stereochemical configurations are opposite at only one of the four C stereocenters. Similarly, we can see from their Fischer projections that L-glucose and D-glucose are enantiomers of each other: They are mirror images that are nonsuperimposable.

YOUR TURN 5.15

Identify the C atoms at which the stereochemical configurations are different in D-allose and D-glucose.

problem **5.21** What is the stereochemical relationship between D-allose and L-glucose? Explain.

problem **5.22** Draw the enantiomer of D-allose as a Fischer projection.

5.7 Strategies for Success: Converting between Fischer Projections and Zigzag Conformations

A molecule containing a chain of carbon atoms is frequently represented in its zigzag conformation. However, if it contains multiple tetrahedral stereocenters, it may be more convenient to work with its Fischer projection. How does one convert from a zigzag conformation to a Fischer projection?

Consider, for example, the following molecule in its zigzag conformation:

It consists of a chain of seven C atoms, each of which is numbered. C2 through C6 are stereocenters. We can therefore begin to draw the Fischer projection with the following framework, which has five adjacent tetrahedral C stereocenters:

C1 is part of the CO_2H group and C7 is part of the CH_2OH group, so neither is a stereocenter. All that's left to do is to place an H and an OH on each of the stereocenters. But, for each stereocenter, which substituent should be placed on the left and which should be placed on the right? This choice dictates each stereocenter's specific configuration.

Build a molecular model of the molecule in its zigzag conformation using the given dash–wedge structure as your guide (Figure 5-25a). Orient the model vertically so that the carbon chain matches the Fischer projection (i.e., C1 on top and C7 on the bottom).

Notice that there is a discrepancy between the zigzag conformation and the Fischer projection. The Fischer projection assumes that the horizontal bonds on *each* stereocenter point toward us, but in the zigzag conformation, it is every *second* C atom that has its horizontal bonds pointed toward us, and the remaining C atoms have their horizontal bonds pointed away from us. This is shown on the left and right sides of Figure 5-25b. In other words, *a Fischer projection implies a conformation of the carbon chain that is not zigzag.* (In fact, if the molecule shown in Fig. 5-25 actually adopted the conformation implied by the Fisher projection, it would curve around on itself in a loop.)

To account for these different conformations, we can complete the Fischer projection by doing the following:

1. For each stereocenter with the horizontal bonds pointing toward us, simply add the substituents on the left and right of the Fischer projection as they appear in the molecular model.
2. Turn the molecule over so that the remaining stereocenters have their horizontal bonds pointing toward us, and repeat Step 1.

(a)

(b)

FIGURE 5-25 Converting a zigzag conformation to a Fischer projection (a) Ball-and-stick model of the molecule in question in its zigzag conformation. (b) On the left, the molecule is oriented such that the stereocenters at C2, C4, and C6 have the horizontal bonds pointing toward us. This allows us to complete the Fischer projection at C2, C4, and C6. On the right, the molecule is oriented such that the horizontal substituents at C3 and C5 are pointing toward us, allowing us to complete the Fischer projection at C3 and C5.

Therefore, holding the molecule as it appears on the left in Figure 5-25b, we can complete the Fischer projection at C2, C4, and C6. Then, turning it over and holding the molecule as it appears on the right in Figure 5-25b, we can complete the Fischer projection at C3 and C5.

YOUR TURN 5.16

Complete the Fischer projection in Figure 5-25b by placing the H and OH substituents appropriately on C5 and C6.

Just as it is important to be able to convert from a zigzag conformation to a Fischer projection, it is also important to be able to convert from a Fischer projection to a zigzag conformation. Suppose, for example, that we want to convert the Fischer projection below, which contains stereocenters at C2 through C5, into a zigzag formula. C1 (i.e., CN) is not a stereocenter because it is *sp* hybridized and linear, and C6 (i.e., CH_2OH) is not a stereocenter because it has only three different substituents.

$$
\begin{array}{c}
\text{OH} \\
\text{H}\overset{2}{-\!\!-\!\!-}\overset{1}{\text{CN}} \\
\text{HO}\overset{3}{-\!\!-\!\!-}\text{H} \\
\text{H}\overset{4}{-\!\!-\!\!-}\text{OH} \\
\text{H}\overset{5}{-\!\!-\!\!-}\text{OH} \\
\overset{6}{\text{CH}_2\text{OH}}
\end{array}
$$

Begin by building a molecular model of a chain of five tetrahedral carbons in a zigzag conformation, representing C2 through C6, as shown in Figure 5-26a. Then, orient the model so that the chain of stereocenters is vertical, as in Figure 5-26b. Recalling, once again, that Fischer projections, by convention, have their horizontal bonds pointing toward us, complete the molecular model as follows:

1. For each tetrahedral stereocenter in the model in which the horizontal bonds are pointing toward us, attach the substituents on the left and right as they appear in the Fischer projection.
2. Turn the molecule over so that the remaining stereocenters have their horizontal bonds pointing toward us and repeat Step 1.

Therefore, orienting the model as shown on the left in Figure 5-26b allows us to attach substituents at C2 and C4, while orienting it as shown on the right in Figure 5-26b allows us to attach substituents at C3 and C5.

YOUR TURN 5.17

In Figure 5-26b, write the atoms in the boxes provided to match the stereochemical configurations indicated at C4 and C5 in the Fischer projection.

FIGURE 5-26 Converting a Fischer projection to a zigzag conformation (a) The backbone of a molecular model in the zigzag conformation showing just the carbon stereocenters and the terminal OH groups. (b) On the left, the backbone is oriented so that the substituents on the horizontal at C2 and C4 are pointed toward us. On the right, the backbone is oriented so that the substituents on C3 and C5 are pointed toward us.

(a)   (b)

problem **5.23** Draw each of the following molecules as Fischer projections so that the carbon indicated is at the top.

(a)

(b)

problem **5.24** Draw the following molecules in their zigzag conformations.

(a)

(b)

5.8 Physical and Chemical Properties of Isomers

One benefit of knowing the specific relationship between two isomers is the insight it provides into their relative behavior, both their chemical behavior (e.g., the products, reaction rates, and equilibrium constants of their various reactions) and their physical properties (e.g., their boiling points, melting points, solubilities). Do the two isomers behave identically or differently? If differently, to what extent?

5.8a Constitutional Isomers

A pair of constitutional isomers must have different connectivities, so they must have some difference in bonding, too. For example, one of two constitutional isomers may have a C=O double bond, whereas the second molecule may have a C=C double bond. Or one constitutional isomer may contain a ring and the other may not. These differences in connectivity lead to differences in polarities and bond energies. As a result:

Constitutional isomers must have different physical properties and chemical properties.

How differently constitutional isomers behave depends largely on how different their connectivities are. For example, pent-1-ene and *trans*-pent-2-ene are two constitutional isomers of C_5H_{10}. They both contain one C=C double bond in addition to C—C and C—H single bonds. Although they have different connectivities, the differences are not great—the double bonds are simply found at different locations within the molecules. As a result, these two molecules behave similarly, both in their physical and chemical characteristics.

Similar connectivities give similar properties.

Pent-1-ene
Melting point = –165 °C
Boiling point = 30 °C

***trans*-Pent-2-ene**
Melting point = –140 °C
Boiling point = 37 °C

On the other hand, butanal and tetrahydrofuran are two constitutional isomers of C_4H_8O that have quite different physical and chemical behavior because one has a highly polar $C=O$ double bond and the other does not.

Different connectivities give different properties.

Butanal
Melting point = –8 °C
Boiling point = 163.5 °C

Tetrahydrofuran
Melting point = –108 °C
Boiling point = 66 °C

5.8b Enantiomers

Enantiomers are mirror images of each other, so they have exactly the same connectivity and precisely the same polarities. For these reasons, it might seem that enantiomers should behave identically. Indeed, both enantiomers of butan-2-ol boil at 99 °C.

Mirror images ➡ identical properties

OH

OH

Boiling point = 99 °C

Boiling point = 99 °C

Enantiomers of butan-2-ol

In general, however, whether enantiomers have identical properties depends on if they are in a *chiral environment* or an *achiral environment*.

- A **chiral environment** is one that is nonsuperimposable upon its mirror image.
  - Chiral species must be present, other than the enantiomers of interest.
  - At least one of those chiral species must be in unequal proportions of its enantiomers.
- An **achiral environment** is one that *is* superimposable upon its mirror image.
  - One way this can occur is if no chiral species are present other than the enantiomers of interest.
  - A second way is if chiral species (other than the enantiomers of interest) are present in equal proportions of their enantiomers.

Most environments we encounter in the laboratory are *achiral*. This is the case, for example, when a pure enantiomer boils. Furthermore, most solvents, solutes, and reactants an enantiomer might encounter are achiral.

Plane of symmetry

(a) (b) (c)

FIGURE 5-27 Chiral and achiral environments (a) A sock has a plane of symmetry and is therefore achiral. (b) Because a sock is an "achiral environment" for feet, socks fit both the left and right feet equally well. (c) A shoe is chiral and is therefore a "chiral environment" for feet. This is why a left shoe fits a left foot far better than a right shoe and why a right shoe fits a right foot far better than a left shoe.

Chiral environments can occur in a variety of ways. In chromatography, for example, a chiral stationary phase can be used. In chemical reactions, we can use chiral catalysts. Perhaps more importantly, biological systems are chiral environments because biomolecules such as proteins, sugars, and DNA are chiral and, in the body, are each present in exclusively one of their enantiomers.

With an understanding of chiral and achiral environments, we can now be more precise in our statement regarding the relative behavior of enantiomers.

- In an *achiral environment*, enantiomers have exactly the same physical and chemical properties.
- In a *chiral environment*, enantiomers must have different physical and chemical properties. Depending on the specific situation, the behavior of the enantiomers can be slightly different or dramatically different.

To better understand the behavior of enantiomers in chiral versus achiral environments, we present an analogy using feet, socks, and shoes. Your left and right feet are chiral—they are enantiomers of each other. A sock, however, is achiral, with a plane of symmetry along its length (Fig. 5-27a). Thus, both a left foot and right foot should "behave" identically when they interact with the achiral environment of a sock. From experience we know this to be the case; a sock will fit either foot equally well (Fig. 5-27b).

Shoes, however, are chiral objects—left and right shoes are enantiomers of each other. Left and right feet, therefore, should "behave" differently in the presence of the chiral environment of a given shoe. Indeed, a right shoe fits the right foot better than the left (Fig. 5-27c) and vice versa.

5.8c Diastereomers

Like enantiomers, a pair of diastereomers have the same connectivity. They are *not* mirror images of each other, however, so they must behave differently.

Diastereomers must have different physical and chemical properties.

This can be illustrated most straightforwardly with a pair of cis and trans isomers, which, as we learned in Section 5.5, constitute one category of diastereomers.

trans-1,2-Dichloroethene, for example, is a nonpolar molecule; *cis*-1,2-dichloroethene, its diastereomer, has a substantial dipole moment. This difference in polarity results in physical properties that can be quite different.

Polar	Nonpolar

cis-1,2-Dichloroethene
Melting point = −85 °C
Boiling point = 60 °C

trans-1,2-Dichloroethene
Melting point = −50 °C
Boiling point = 47.5 °C

The difference in behavior between diastereomers isn't always so dramatic, such as with the diastereomers of butane-2,3-diol.

(2S,3S)-Butane-2,3-diol
Melting point = 25 °C
Boiling point = 179–182 °C

meso-Butane-2,3-diol
Melting point = 32–34 °C
Boiling point = 183–184 °C

SOLVED problem 5.25 In Chapter 8, you will learn that 2-bromo-4-methyl-hexane (molecule **Z**) can undergo the following substitution reaction when treated with NaCl.

NaCl

Z

The following molecules, **A–E**, undergo a similar substitution reaction with NaCl.

A

B

C

D

E

For which of these molecules will the *rate* of the reaction be precisely the same as that of the reaction involving **Z**? Explain.

Think How is each molecule **A** through **E** related to **Z**? How do those relationships translate into relative behavior?

Solve **A** and **B** are unrelated to **Z** because they have different molecular formulas. **C**, **D**, and **E** are all isomers of **Z**. **C** is its enantiomer, **D** is a diastereomer of it, and **E** is one of its constitutional isomers. Only enantiomers have precisely the same behavior, so the correct answer is **C**.

As discussed in Chapter 6, the carboxylic acid functional group (CO_2H) in the following molecule, **Y**, is moderately acidic.

For which of the following molecules, **A–E**, must the carboxylic acid functional group have an acidity that is *different* from that of **Y**? Explain.

| A | B | C | D | E |

5.9 Stability of Double Bonds and Chemical Properties of Isomers

Recall from Section 5.8 that constitutional isomers and diastereomers *must* have different chemical properties. This can be seen in the heats of combustion of the five C_6H_{12} isomeric alkenes listed in Table 5-1.

The particular isomers in Table 5-1 are chosen because their C=C double bonds have different *alkyl substitutions*. **Alkyl substitution** is the number of alkyl groups bonded to the alkene carbon atoms. 2,3-Dimethylbut-2-ene, for example, has a **tetrasubstituted** double bond because the carbons of the C=C double bond are connected to four alkyl groups (in this case, all CH_3 groups). On the other hand, the double bond in hex-1-ene is **monosubstituted** because the carbons of the C=C double bond are connected to one alkyl group (in this case, a butyl group, $CH_2CH_2CH_2CH_3$).

Because all of the molecules in Table 5-1 are isomers of one another, their combustion products are identical. That is, each mole of C_6H_{12} that undergoes combustion produces 6 moles of CO_2 and 6 moles of H_2O, as shown in Equation 5-1.

$$C_6H_{12} + 9\ O_2 \rightarrow 6\ CO_2 + 6\ H_2O \qquad (5\text{-}1)$$

As a result, any difference in the heat of combustion must be attributed to differences in stabilities among the reactant alkenes. Because combustion releases heat, *the compound with the smallest heat of combustion is the most stable and vice versa.* The stability of a C=C double bond increases as follows in going from monosubstituted to tetrasubstituted, where R represents a generic alkyl group:

Increasing stability of the double bond

| Tetrasubstituted | Trisubstituted | Disubstituted | Monosubstituted |

Alkene (C_6H_{12})	Type of Alkene	Heat of Combustion		Difference	
		kJ/mol	kcal/mol	kJ/mol	kcal/mol
2,3-Dimethylbut-2-ene	Tetrasubstituted	–3741.5	–894.2	–	–
3-Methylpent-2-ene	Trisubstituted	–3749.9	–896.2	8.4	2.0
2-Ethylbut-1-ene	Disubstituted	–3727.1	–890.8	14.4	3.4
trans-Hex-3-ene	Disubstituted	–3762.6	–899.3	21.1	5.1
cis-Hex-3-ene	Disubstituted	–3766.3	–900.2	24.8	6.0
Hex-1-ene	Monosubstituted	–3769.7	–901.0	28.2	6.8

Steric repulsion

Cis isomer
Less stable

Trans isomer
More stable

FIGURE 5-28 Steric strain in cis alkenes Steric repulsion between the alkyl groups causes the cis isomer of an alkene to be less stable than the trans isomer.

Two features of this sequence are particularly worth noting:

- Double bond stability increases as the amount of *alkyl substitution* increases.
- Trans alkenes are more stable than cis alkenes.

Trans isomers are more stable than cis isomers due to *steric strain* (Chapter 4). The *bulkiness* of the two alkyl substituents causes electrons from those groups to occupy the same space in the cis isomer (Fig. 5-28), whereas they are on opposite sides of the double bond in the trans isomer. Repulsion of those electrons decreases the stability of the cis isomer.

Double bonds become more stable as the number of R groups increases, due to **hyperconjugation**. A detailed discussion of hyperconjugation is inappropriate here. Briefly, however, the electrons in the single bonds of an R group can interact with the empty π^* MO of the double bond. As a result, those electrons are *delocalized* to some extent, similar to what we observe in resonance. *Hyperconjugation leads, therefore, to an energy lowering (i.e., to stabilization), and each additional alkyl group present leads to additional hyperconjugation.*

SOLVED problem 5.27 Which of the following molecules do you think will have the smaller heat of combustion? Explain.

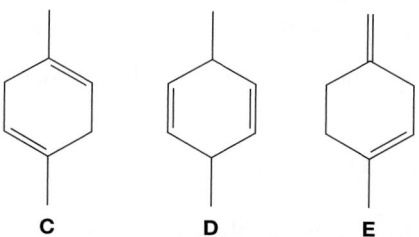

A   **B**

Think Which of the C=C double bonds is more highly alkyl substituted? More stable? What is the relationship between relative stability and relative heats of combustion?

Solve The two molecules are isomers of each other, each with the formula C_8H_{14}. As a result, they have precisely the same combustion products, in which case any difference in the heat of combustion must come from differences in the stability of the reactants. The C=C double bond in molecule **A** is trisubstituted (one CH_3 group and two alkyl groups that are part of the ring), whereas the C=C double bond in molecule **B** is tetra-substituted (two CH_3 groups and two alkyl groups that are part of the ring). Thus, **B** is more stable and will have a *smaller* heat of combustion (i.e., less heat is released).

---

problem 5.28 Which of the following do you think will have the greatest heat of combustion? The smallest? Explain.

C   **D**   **E**

5.10 Separating Configurational Isomers

As we discuss in Chapter 8, if a chemical reaction forms a chiral product, it usually forms a mixture of stereoisomers. How, then, do we separate stereoisomers from one another?

Recall that diastereomers have different physical properties, whereas enantiomers have identical properties in achiral environments. Consequently,

- Diastereomers often can be separated by common laboratory techniques such as fractional distillation, crystallization, and simple chromatography.
- Enantiomers generally cannot be separated by these methods.

Louis Pasteur was the first to isolate a pair of enantiomers from each other. The enantiomers he separated were those of sodium ammonium tartrate, an ionic compound that forms crystals. As Pasteur noted, the crystals appeared to grow in one of

"Left-handed" crystal

"Right-handed" crystal

Sodium ammonium tartrate

FIGURE 5-29 Separation of sodium ammonium tartrate enantiomers
Right- and left-handed crystals of sodium ammonium tartrate that Louis Pasteur separated by hand are depicted on the top. The two crystals are mirror images of each other

two varieties—left-handed crystals and right-handed crystals—that are mirror images of each other (Fig. 5-29). Using nothing more than a microscope and a pair of tweezers, he physically separated the two types of crystals.

Today, other techniques are used to separate enantiomers. Chromatography, for example, can exploit the fact that *enantiomers have different physical properties in a chiral environment*. A sample containing a mixture of enantiomers is passed through a chiral stationary medium, for which the enantiomers have different affinities. Traveling through the chiral medium at different rates allows them to be collected separately.

A second method of separating enantiomers takes advantage of the fact that diastereomers are readily separable, as mentioned previously. The key to this method involves:

1. Temporarily converting the enantiomers into a pair of diastereomers.
2. Separating those diastereomers from each other by exploiting their different physical and chemical properties.
3. Regenerating the enantiomers from the separated diastereomers.

See Problem 5.71 for a specific example of how this method is applied.

5.11 Optical Activity

Although enantiomers have identical physical and chemical properties in an achiral environment, they behave differently in a chiral environment (Section 5.8b). They also interact differently with **plane-polarized light**.

Light can be regarded as both a particle and a wave. When treated as a particle, we think of light as consisting of **photons**, each of which carries a specific quantity of energy that can be associated with its frequency and wavelength (this is discussed further in Chapters 15 and 16). When treated as a wave, we think of light as consisting of oscillating electric and magnetic fields. The frequency of oscillation of those fields is what defines the frequency of light.

FIGURE 5-30 Plane-polarized light (Left) The electric field (red) oscillates in a vertical direction and the magnetic field (black) oscillates in the plane perpendicular to the page as the light wave travels to the right. (Right) A double-headed arrow represents the electric field's plane of oscillation.

Electric field

Light traveling in this direction

Electric field polarized in vertical direction

Magnetic field

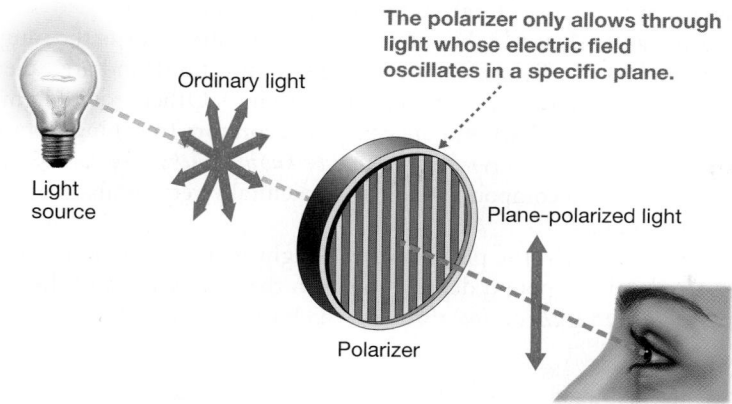

FIGURE 5-31 Function of a polarizer
(Left) Light emitted from most sources is unpolarized. The electric-field vectors of its photons oscillate in all planes perpendicular to the direction of travel. (Middle) A polarizer effectively filters out photons whose electric-field vector does not oscillate in the specified plane (in this case, the vertical direction). (Right) The light that passes through the polarizer is plane polarized.

As a ray of light travels through space, its electric and magnetic fields oscillate in planes perpendicular to each other and also perpendicular to the direction of motion (Fig. 5-30). If all photons from a light source have their electric fields oscillating in the same plane, then the light is plane polarized. In these cases, we can represent that plane of polarization using a double-headed arrow, as shown on the right in Figure 5-30. Note that for simplicity, the plane in which the magnetic field oscillates is not shown in that representation; it is understood, though, to be perpendicular to the electric field's plane of oscillation.

Most light sources emit light that is **unpolarized**. That is, if we could view all of the photons traveling in the same direction, we would see each one's electric field oscillating in a different plane. A **polarizer** (Fig. 5-31) generates plane-polarized light by allowing through only those photons whose electric field is oscillating in a specified plane, effectively filtering out light whose electric field oscillates in any other plane.

If plane-polarized light passes through a sample of a compound (Fig. 5-32), the plane in which the light is polarized can change, depending upon whether the compound is chiral or achiral.

- Enantiomerically pure chiral compounds are said to be **optically active** because they rotate the plane of polarization.
- Achiral compounds are said to be **optically inactive** because they leave the plane of polarization unchanged.

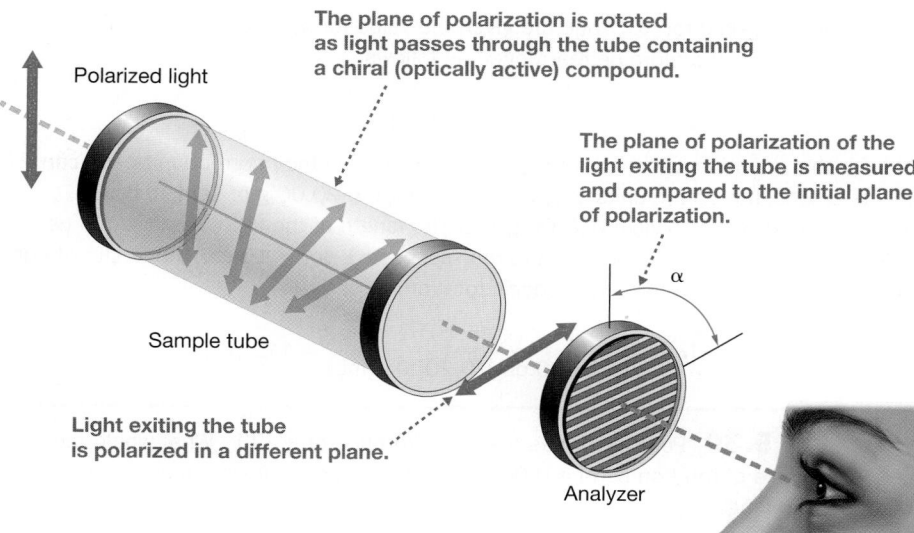

FIGURE 5-32 Optical activity
(Left) Plane-polarized light enters a tube containing a solution of a compound being studied. If the compound is optically active, the plane of polarization is rotated as the light passes through the tube. (Right) The angle of rotation, α, is measured using an analyzer.

The angle by which a chiral compound rotates plane-polarized light can be measured using an analyzer. Light enters the analyzer after it exits the sample. Some chiral compounds rotate light clockwise (in the + direction), and are called **dextrorotatory** (from Latin, meaning "rotating to the right"). Others rotate light counterclockwise (in the − direction), and are called **levorotatory** (meaning "rotating to the left"). *The direction of rotation generally cannot be known without performing the experiment.* In fact, two chiral compounds that are structurally very similar may rotate light in opposite directions!

The amount by which the plane-polarized light is rotated upon passing through a sample of a chiral compound depends on both the *concentration* of the chiral compound and the *length of the sample* through which the light travels:

> ■ As the concentration of a chiral compound increases, so does the angle of rotation.
> ■ As the length of the sample tube increases, so does the angle of rotation.

This should make sense, because the number of molecules the light encounters increases with increasing concentration or length of sample tube.

The **measured angle of rotation**, α, can thus be expressed by Equation 5-2, where c is the concentration of the sample in units of g/mL and l is the length of the sample tube in units of decimeters (dm) (1 dm = 1/10 meter).

$$\alpha = ([\alpha]_\lambda^T)(l)(c) \tag{5-2}$$

The $[\alpha]_\lambda^T$ term is the **specific rotation**, and it is a constant that is *unique for a given chiral compound*. It is the angle of rotation of light of a given wavelength in nanometers, λ, if it passes through a sample whose concentration is 1 g/mL, whose length is 1 dm, and whose temperature is T°C. Most often, the light used for measurement is the "sodium D line" (589.6 nm), abbreviated simply D, and the temperature is 20 °C. Therefore, the specific rotation is usually reported as $[\alpha]_D^{20}$.

SOLVED problem **5.29** Suppose that 20.00 g of a chiral compound is dissolved in 0.1000 L of solution and is placed in a tube that is 20.00 cm long. What is its specific rotation of light if the observed rotation is determined experimentally to be +45.00°?

Think Which variable in Equation 5-2 are we solving for? The equation calls for concentration—how do we calculate it? Are the units correct for the length of the tube, l?

Solve We are asked for $[\alpha]_D^{20}$ and are given the value for α, which is +45.00°. Equation 5-2 must therefore be rearranged as follows:

$$[\alpha]_D^{20} = \frac{\alpha}{(l)(c)}$$

To solve this equation, we must have values for c and l in their correct units. We can calculate concentration in units of g/mL by dividing the 20.00 g of sample by the 100.0 mL of solution, to yield 0.2000 g/mL. The length of the tube, l, is given to us as 20.00 cm, but must be converted to dm. Because 1 dm = 10 cm, the length of our tube is 2.000 dm. Therefore, the specific rotation is:

$$[\alpha]_D^{20} = \frac{(+45.00°)}{(2.000\ \text{dm})(0.2000\ \text{g/mL})} = 112.5°$$

---

problem **5.30** Penicillin V has a specific rotation of +223°. What would the measured angle of rotation be of a 0.00300 g/mL solution, if it were measured in a tube 10.0 cm long?

Recall that a pair of enantiomers interact differently with plane-polarized light. In fact,

> A pair of enantiomers have equal but opposite specific rotations.

Just as enantiomers are mirror images of each other, the mirror image of a rotation in the clockwise direction is an identical rotation in the counterclockwise direction. For this reason, one enantiomer can always be designated as the (+) enantiomer and the other as the (−) enantiomer.

A **racemic mixture** contains equal amounts of the (+) and (−) enantiomers of a chiral molecule. That is, light traveling through a racemic mixture encounters an equal number of molecules of each enantiomer. Therefore, the tendency of one enantiomer to rotate the light in one direction is exactly balanced by the tendency of the other enantiomer to rotate the light in the opposite direction. The net result is zero rotation of the light. Consequently,

> A racemic mixture of enantiomers is optically inactive, despite being made up of chiral molecules.

If a mixture of enantiomers is not racemic, then it will be optically active, but it will not rotate light as much as one of the pure enantiomers. Such a mixture can be viewed as being a certain percentage racemic, with the remaining percentage, called the **enantiomeric excess (ee)**, viewed as being composed of one of the pure enantiomers. The percentage that is racemic will not contribute toward the rotation of plane-polarized light, but the enantiomeric excess will. This idea is summarized in Equation 5-3.

$$\% \text{ ee} = \frac{(\text{specific rotation of mixture})}{(\text{specific rotation of pure enantiomer})} \times 100 \qquad (5\text{-}3)$$

SOLVED problem 5.31 Suppose a solution of a pure chiral molecule has a specific rotation of −32°. What is the specific rotation of a solution that is 90% of the (+) enantiomer and 10% of the (−) enantiomer?

Think Which enantiomer is in excess and what is its ee? How does that govern the mixture's ability to rotate plane-polarized light?

Solve Because the solution is excess in the (+) enantiomer, the specific rotation of the mixture should be in the (+) direction. Realize that the 10% that is the (−) enantiomer can be combined with 10% of the solution that is the (+) enantiomer, such that 20% of the solution is effectively racemic and the other 80% is excess in the (+) enantiomer. Thus, 20% of the solution does not contribute toward the rotation of plane-polarized light and 80% contributes toward a specific rotation of +32° [the value for the pure (+) enantiomer]. This can be summarized by rearranging Equation 5-3 and substituting the appropriate numbers.

$$(\text{specific rotation of mixture}) = (\% \text{ ee})(\text{specific rotation of pure enantiomer})/(100)$$
$$= (80)(+32°)/(100)$$
$$= 25.6°$$

---

problem 5.32 Suppose that a pure compound has a specific rotation of +49°. In the laboratory, a solution in which the compound is mixed with its enantiomer is found to have a specific rotation of +12°. What is the ee of the mixture? What percentage of the mixture is the (+) enantiomer and what percentage is the (−) enantiomer?

Proteins can take on a wide variety of three-dimensional shapes—and it is that shape that gives a protein its highly specialized function. At first glance, a protein might seem to be folded in upon itself randomly, but careful examination reveals that when a protein is folded properly, portions of the main chain typically assume regular structural patterns, or *secondary structures* (see Chapter 26). Examples of secondary structures include α helices (red) and β sheets (gold), as shown below.

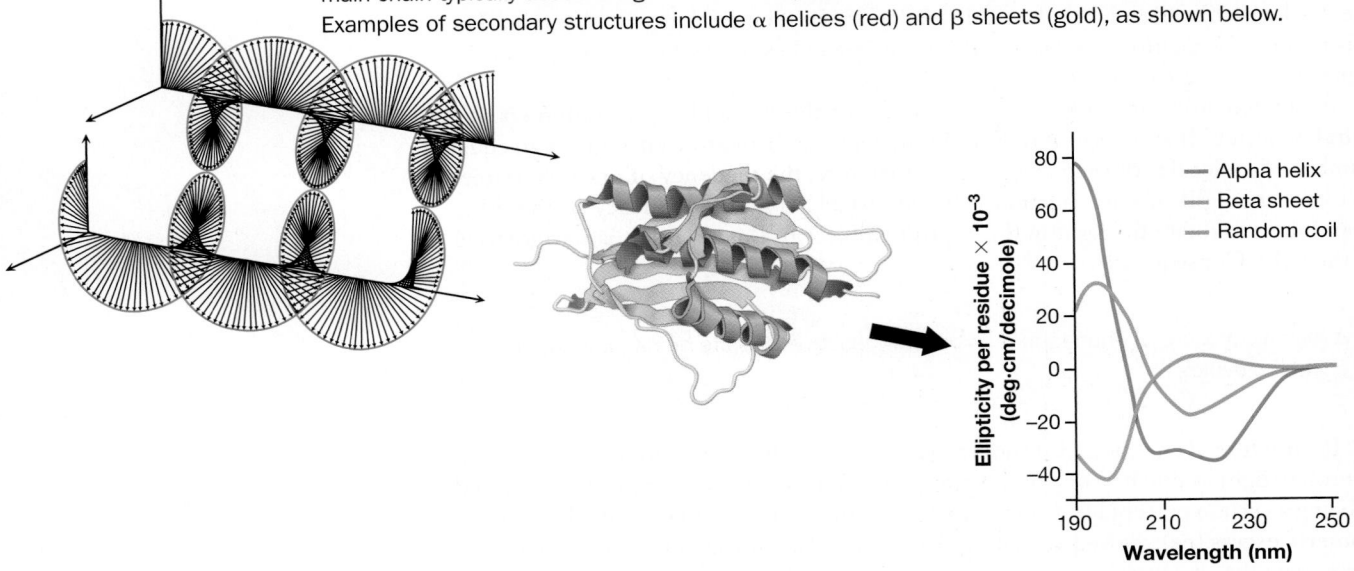

The existence of such secondary structures can be detected by a technique called circular dichroism (CD) spectroscopy. CD spectroscopy is related to polarimetry, which, as we have seen, can be used to detect chiral compounds. Unlike standard polarimetry, whose focus is on the interaction of plane-polarized light with matter, CD spectroscopy involves circularly polarized light. Recall that the electric and magnetic vectors in plane-polarized light each oscillate in a fixed plane perpendicular to the direction in which the light propagates. In circularly polarized light, the plane of polarization coils around the axis of propagation. As shown above on the left, the vectors can coil in a clockwise or counterclockwise direction, giving rise to right-handed and left-handed circularly polarized light.

CD spectroscopy works by measuring how much of each type of circularly polarized light a sample absorbs at various wavelengths. That difference can be plotted as a function of wavelength to produce a CD spectrum. Regular patterns in a protein (i.e., secondary structures) give rise to characteristic features in such a spectrum.

One of the main applications of CD spectroscopy involves studying the dynamics of the unfolding and folding of proteins in solution. By monitoring features in a CD spectrum over time, the kinetics and thermodynamics of these processes can be determined. This kind of information can help us solve problems related to mad cow disease, Alzheimer's disease, and Parkinson's disease, which arise from the unfolding and misfolding of proteins.

5.12 Wrapping Up and Looking Ahead

This was the second of two chapters devoted to isomerism. The first, Chapter 4, dealt with conformational isomers and constitutional isomers; Chapter 5 dealt with enantiomers and diastereomers. Both enantiomers and diastereomers are types of configurational isomers, meaning that they must have the same connectivity. Enantiomers are mirror images of each other; diastereomers are not.

Configurational isomers commonly differ in their configurations about tetrahedral stereocenters, which are bonded to four different groups. Usually, tetrahedral stereocenters are carbon atoms, but other atoms, such as positively charged nitrogen atoms, can be, too. Uncharged nitrogen atoms that undergo nitrogen inversion are not. Because of the relevance of configurations about tetrahedral stereocenters, we saw that it is important to be able to depict such configurations accurately, using dash–wedge notation, Haworth projections, and Fischer projections.

Intimately related to the discussion of configurational isomers is the notion of chirality: the property of having a nonsuperimposable mirror image. Species that have a single tetrahedral stereocenter must be chiral; species that have two or more tetrahedral stereocenters may or may not be, depending upon their symmetry. Enantiomers that rapidly interconvert via bond rotations or nitrogen inversion are effectively achiral. Chirality also relates to optical activity—pure chiral compounds are optically active, meaning that they rotate plane-polarized light. Thus, using polarimetry, we can quantify the amount of a chiral compound present in solution.

The ideas presented in this chapter will be very important as we move into discussions surrounding specific reactions. Probably the most important ideas that relate to reactions come from our discussion of the physical and chemical properties of isomers. Namely, enantiomers must have identical properties in an achiral environment, whereas diastereomers must have different properties. Therefore, as we will see in Chapter 8 (and will apply throughout the rest of the book), when a reaction produces a mixture of configurational isomers, we will be able to determine whether they will be produced in equal or unequal amounts.

THE ORGANIC CHEMISTRY OF BIOMOLECULES

5.13 The Chirality of Biomolecules

The tragedy of thalidomide was mentioned briefly in Section 5.1. In the 1950s and 1960s, thalidomide was prescribed as an antinausea medication for pregnant women with morning sickness. Unfortunately, thalidomide is teratogenic—that is, it causes birth defects. As a direct result of prescribing the drug, it is estimated that more than 10,000 children worldwide were born with deformed or missing limbs.

Like many drugs, thalidomide is chiral and was sold as a *racemic mixture* of its two enantiomers:

This enantiomer suppresses nausea. **This enantiomer causes birth defects.**

Enantiomers of thalidomide

Later testing on mice showed that the enantiomer on the left is primarily responsible for suppressing nausea, whereas the one on the right is primarily responsible for the teratogenic properties. (It turns out, however, that administering only the enantiomer on the left would not have solved the problem because the two enantiomers interconvert in the body.)

How can enantiomers—molecules that are mirror images of each other—behave so differently? Recall from Section 5.8 that *enantiomers have different physical and chemical properties in a chiral environment*. In other words,

The body acts as a chiral environment.

Most biomolecules, including those encountered in previous chapters, are chiral. For example, as shown in Figure 5-33, a typical amino acid has a single stereocenter (marked by an asterisk), and is thus chiral. There is one exception (see Problem 5.72).

FIGURE 5-33 Stereocenters in biomolecules In each of these biomolecules, tetrahedral stereocenters are marked by asterisks.

Glucose, in its acyclic form, has four stereocenters—each C atom that is bonded to an H and an OH group. A nucleotide in DNA has three stereocenters, and testosterone, a steroid, has six stereocenters.

Despite the presence of these chiral compounds, the body would remain an achiral environment if each pair of enantiomers were present in equal amounts. (This is analogous to a racemic mixture being optically inactive, as discussed in Section 5.11.) Instead,

> Natural amino acids and monosaccharides appear in the body exclusively in one enantiomeric form.

For amino acids, that form is the L enantiomer, and for monosaccharides, it is the D enantiomer. (These designations are discussed in greater detail in Section 5.14.) Reasons why these compounds appear exclusively in these forms are not known and are the subject of debate.

5.14 The D/L System for Classifying Monosaccharides and Amino Acids

Each chiral amino acid and monosaccharide has two enantiomers, specified using the D/L system. The system was established around 1910, before the advent of the IUPAC system of nomenclature and before the technology existed to determine the specific location of atoms in three-dimensional space.

The basis of the D/L system is the optical rotation of glyceraldehyde. Glyceraldehyde is an *aldotriose*, a three-carbon monosaccharide possessing an aldehyde group (see Section 4.15). It has a single tetrahedral stereocenter, and thus has enantiomers that rotate plane-polarized light in equal but opposite directions (Section 5.11). The enantiomer that rotates plane-polarized light in the clockwise direction was designated as D-glyceraldehyde, because it is dextrorotatory. (Recall that dextrorotatory derives from Latin, and means "rotating to the right.") The other enantiomer, which rotates plane-polarized light in the counterclockwise direction, was designated as L-glyceraldehyde, because it is levorotatory. (Levorotatory means "rotating to the left.")

At the time the D/L system was established, other sugars could be synthesized from glyceraldehyde by lengthening the molecule in the direction of the aldehyde group, while leaving unchanged the configuration of glyceraldehyde's stereocenter. As a result, sugars synthesized this way from D-glyceraldehyde were designated as D sugars and sugars synthesized from L-glyceraldehyde were designated as L sugars.

D-Glyceraldehyde rotates plane-polarized light in the clockwise direction, and is thus *dextrorotatory*.

D-Glyceraldehyde

Add *n* H–C–OH groups →

A D sugar other than glyceraldehyde can be either *dextrorotatory* or *levorotatory*.

A generic D sugar

L-Glyceraldehyde rotates plane-polarized light in the counterclockwise direction, and is thus *levorotatory*.

L-Glyceraldehyde

Add *n* H–C–OH groups →

An L sugar other than glyceraldehyde can be either *dextrorotatory* or *levorotatory*.

A generic L sugar

Based on their origins, the D/L designation for a sugar other than glyceraldehyde is simply part of the *name*, and does not have any connection to the direction in which the sugar rotates plane-polarized light.

- Some D sugars rotate plane-polarized light in the clockwise direction, and others rotate it in the counterclockwise direction.
- Some L sugars rotate plane-polarized light in the clockwise direction, and others rotate it in the counterclockwise direction.

The D and L designations for amino acids are assigned by analogy. The second conformation of each glyceraldehyde enantiomer below resembles the enantiomer of the amino acid next to it—specifically, glyceraldehyde's aldehyde, OH, and CH_2OH groups are analogous to the amino acid's carboxyl, NH_2, and R groups. Thus, the top amino acid is the D enantiomer, because its R group points toward you, just as D-glyceraldehyde's CH_2OH group does. For similar reasons, the bottom amino acid is the L enantiomer.

Rotation about this bond

D-Glyceraldehyde

Analogous to glyceraldehyde's OH

Analogous to glyceraldehyde's C═O

Analogous to glyceraldehyde's CH_2OH

A D amino acid

Rotation about this bond

L-Glyceraldehyde

An L amino acid

5.15 The D Family of Aldoses

Each H—C—OH group added to glyceraldehyde introduces a new stereocenter, which can have either of two stereochemical configurations. Thus, D-glyceraldehyde is the only D aldotriose that is possible (shown in Fig. 5-34 in its Fischer projection), but two D aldotetroses are possible—namely, D-erythrose and D-threose. Two more D aldoses can be produced from each of those sugars upon the addition of another H—C—OH group, giving rise to four possible D aldopentoses—namely, D-ribose, D-arabinose, D-xylose, and D-lyxose. And, with yet another H—C—OH group, there

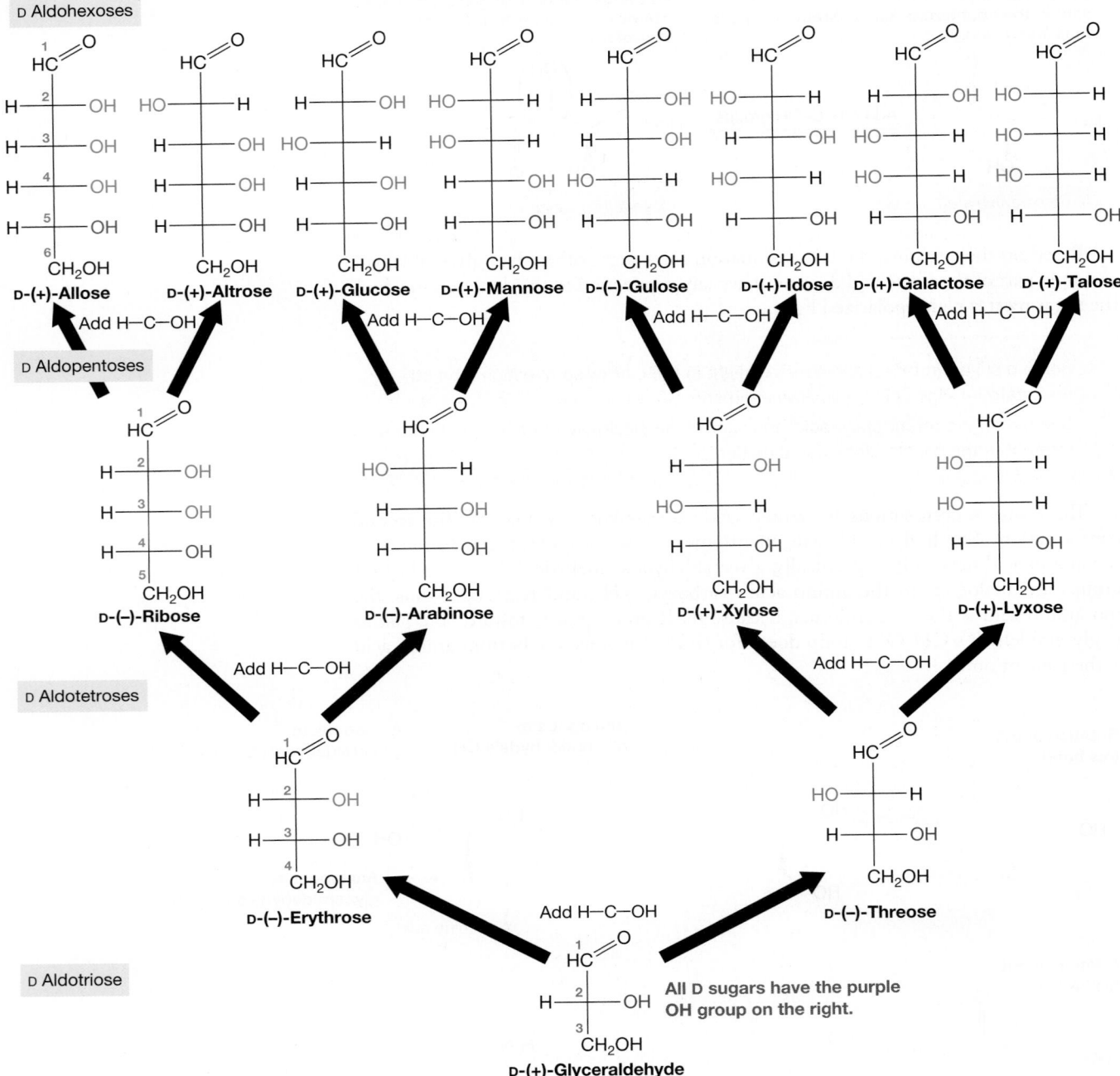

FIGURE 5-34 **The D family of aldoses** In the Fischer projection of any D aldose, C1 is an aldehyde carbon and the OH group of the highest numbered carbon stereocenter (colored purple) appears on the right side. The sugars are distinguished by the number of carbon atoms (bottom to top) and the stereochemical configuration at each carbon stereocenter (left to right).

are eight possible D aldohexoses—namely, D-allose, D-altrose, D-glucose, D-mannose, D-gulose, D-idose, D-galactose, and D-talose.

Notice that in the Fischer projection of D-glyceraldehyde, the OH group attached to the stereocenter (purple) appears on the right. Similarly,

> Any D sugar is distinguished by having the OH group of the highest-numbered carbon stereocenter appear on the right in its Fischer projection.

Notice, too, that D-glyceraldehyde rotates plane-polarized light in the clockwise direction, denoted by (+), but this is not true of all D sugars. For example, D-erythrose rotates plane-polarized light in the counterclockwise direction, denoted by (−). Finally, sugars in the same row all have the same connectivity, but are not mirror images. Thus,

> Any two sugars in the same row of Figure 5-34 are diastereomers.

Recall from Section 5.8 that diastereomers have different physical and chemical properties, which is why no two sugars in the same row have the same name.

Among the various diastereomers in a particular row in Figure 5-34, some differ only by the stereochemical configuration at one carbon atom. Compounds with this specific relationship are called **epimers**. For example, D-allose and D-altrose are epimers, differing only in the configuration at C2; they are called C2 epimers. D-Allose and D-glucose are C3 epimers.

---

problem **5.33** Name the sugar that fits each of the following descriptions.
(a) The C2 epimer of D-glucose. **(b)** The C3 epimer of D-glucose. **(c)** The C4 epimer of D-talose. **(d)** The C3 epimer of D-xylose.

The mirror image of any D sugar would result in a Fischer projection in which the OH group of the highest-numbered carbon stereocenter appears on the left. This is shown below for D-glucose:

D-Glucose and L-glucose are enantiomers.

This OH group appears on the right in a D sugar, and on the left in an L sugar.

D-(+)-Glucose   L-(−)-Glucose

As a result, the enantiomer of any D sugar must not be a D sugar. Instead,

> The enantiomer of a D sugar is designated as an L sugar of the same name, and in its Fischer projection, the OH group of the highest-numbered carbon stereocenter appears on the left.

Thus, the enantiomer of D-glucose is L-glucose.

---

problem **5.34** Draw the Fischer projection of each of the following sugars:
(a) L-mannose; **(b)** L-arabinose; **(c)** L-threose; **(d)** the C2 epimer of L-arabinose.

Chapter Summary and Key Terms

- **Configurational isomers** are isomers that have the same connectivity but are **nonsuperimposable**. **Enantiomers** and **diastereomers** are types of configurational isomers; enantiomers are mirror images of each other, whereas diastereomers are not. (**Sections 5.1 and 5.2**)

- A molecule is **chiral** if it has an enantiomer. Otherwise it is **achiral**. (**Section 5.4**)

- If a molecule and its mirror image rapidly interconvert via rotations about single bonds, the molecule is achiral. (**Section 5.4a**)

- If a molecule and its mirror image rapidly interconvert via a chair flip of a cyclohexane ring, the molecule is achiral. As a result, Haworth projections can be used to evaluate the chirality of a cyclohexane ring. (**Section 5.4a**)

- A molecule that has at least one **plane of symmetry** must be achiral. A molecule that is chiral must have no plane of symmetry. (**Section 5.4b**)

- A tetrahedral **stereocenter** is bonded to four *different* substituents. Every tetrahedral stereocenter has two different configurations possible, related to each other by either (1) reflection through a mirror or (2) the interchange of any two groups. (**Section 5.4c**)

- A molecule that contains exactly one tetrahedral stereocenter must be chiral. (**Section 5.4c**)

- A molecule with no tetrahedral stereocenters may or may not be chiral. Similarly, a molecule with two or more tetrahedral stereocenters may or may not be chiral. (**Section 5.4c**)

- A molecule with at least two tetrahedral stereocenters is **meso** if *overall* it is achiral. (**Section 5.4c**)

- A nitrogen atom may be a stereocenter if it is bonded to four different groups (and thus bears a +1 formal charge). An uncharged nitrogen atom that undergoes **nitrogen inversion** is not a stereocenter. (**Section 5.4d**)

- Cis–trans pairs are **diastereomers**. If a molecule contains two or more tetrahedral stereocenters, then inversion of some, but not all, of their configurations gives a different diastereomer. (**Section 5.5**)

- Inversion of configuration at all tetrahedral stereocenters in a chiral molecule gives the molecule's enantiomer. (**Section 5.5**)

- For a molecule that contains n tetrahedral stereocenters, there are, at most, 2^n configurational isomers. (**Section 5.5**)

- **Fischer projections** are shorthand notations used to represent the configurations about tetrahedral carbon stereocenters in a molecule. Perpendicular intersecting lines represent a carbon atom—typically a stereocenter. Horizontal bonds point toward the viewer, whereas vertical bonds point away from the viewer. (**Section 5.6**)

- Rotation of a Fischer projection 90° in the plane of the page gives the opposite configuration at all tetrahedral stereocenters. Rotation of 180° leaves all stereochemical configurations unchanged. Taking a mirror image of a chiral Fischer projection gives the molecule's enantiomer. (**Section 5.6**)

- Constitutional isomers have different chemical and physical properties. Similarly, diastereomers have different chemical and physical properties. In an **achiral environment**, enantiomers have *identical* physical and chemical properties. In a **chiral environment**, enantiomers have *different* properties. (**Section 5.8**)

- Heats of combustion can be used to determine the relative stabilities of isomeric alkenes. The stability of C=C double bonds increases as the amount of **alkyl substitution** increases; trans alkenes are more stable than cis alkenes. (**Section 5.9**)

- Diastereomers can be separated based on their physical properties. Enantiomers, however, can be separated only in a chiral environment, such as a chromatography column containing a chiral stationary phase. (**Section 5.10**)

- Chiral compounds rotate **plane-polarized light**, and are therefore **optically active**. Achiral compounds are **optically inactive**. (**Section 5.11**)

- A compound's **specific rotation** (i.e., $[\alpha]_D^{20}$) reflects its ability to rotate plane-polarized light and is a constant that is unique to every chiral compound. (**Section 5.11**)

- Enantiomers have equal but opposite specific rotations. A **racemic mixture** of enantiomers is optically inactive. (**Section 5.11**)

- The **enantiomeric excess** (**ee**) is the fraction of a mixture that is not racemic. It is the fraction of a mixture that contributes to the rotation of plane-polarized light. (**Section 5.11**)

Problems

5.35 If molecules **A** and **B** are isomers of each other, then what kinds of isomers could they be (i.e., *enantiomers, diastereomers,* or *constitutional isomers*) under each of the following conditions?
- **(a)** Both molecules have the same IHD.
- **(b)** Molecule **A** has a ring but molecule **B** does not.
- **(c)** Molecules **A** and **B** contain different functional groups.
- **(d)** Molecules **A** and **B** share exactly the same functional groups.
- **(e)** Molecule **A** has a plane of symmetry but molecule **B** does not.

5.36 Determine whether each of the following objects is chiral or achiral. (Assume that there are no graphics on any of these objects.)
- **(a)** A coffee mug
- **(b)** Your ears
- **(c)** A bowling ball
- **(d)** An automobile
- **(e)** A pair of scissors
- **(f)** A t-shirt
- **(g)** Eyeglasses
- **(h)** A piano
- **(i)** Golf clubs
- **(j)** A tennis racquet

5.37 Draw the Newman projection of 1,2-dichloroethane in its *anti* conformation and in each of its *gauche* conformations. Determine which of these conformational isomers possess a plane of symmetry. For those that do, indicate the plane of symmetry using a dashed line.

5.38 In which molecules below does the stereocenter have the same configuration as the molecule at the right?

(a) (b) (c) (d) CH₃ (e) CH₃

(f) OH (g) OH (h) OH

5.39 Identify the specific type of relationship between each of the following pairs of molecules (i.e., either *same molecules*, *constitutional isomers*, *enantiomers*, *diastereomers*, *conformational isomers*, or *unrelated*).

(a) (b) (c)

(d) (e)

(f) (g) (h)

5.40 Identify the specific relationship between each of the following pairs of molecules (i.e., either *same molecules*, *constitutional isomers*, *enantiomers*, *diastereomers*, *conformational isomers*, or *unrelated*).

(a) Cl OH (b) Br CH₃ (c) Br CH₃

(d) CH₃ CH₃ (e) CH₃ CH₃ (f) Br OCH₃

5.41 Identify the specific relationship between each of the following pairs of molecules (i.e., either *same molecules, constitutional isomers, enantiomers, diastereomers, conformational isomers,* or *unrelated*).

5.42 Identify all of the following species that are chiral.

5.43 For each molecule in Problem 5.42, identify all of the tetrahedral stereocenters that exist. Which of those molecules, if any, are meso?

5.44 Which of the following species are chiral?

5.45 For each molecule in Problem 5.44, identify all stereocenters that exist. Which of those molecules, if any, are meso?

5.46 Consider the following molecule.

(a) How many stereocenters does it have?

(b) How many total configurational isomers are possible? (*Hint:* Determine whether it is possible for any of the configurational isomers to be meso.)

5.47 Which of the following molecules are chiral? (*Hint:* Do these Lewis structures accurately depict the three-dimensional geometry?)

(a) H₃C, H, C=C=C, Cl, H **(b)** H₃C, Cl, C=C=C, H, H **(c)** H₃C, CH₃, C=C=C, H, H **(d)** Cl, Br, C=C=C, Br, Cl **(e)** Cl, Br, C=C=C, Cl, Br

5.48 Is it possible for a meso compound to contain three tetrahedral stereocenters? Why or why not?

5.49 Draw each of the following molecules in a zigzag conformation.

(a)
```
        CH₂OH
   HO ──┼── H
    H ──┼── OH
   HO ──┼── H
        CO₂H
```

(b)
```
         CH₃
   HO ──┼── H
    H ──┼── Cl
   HO ──┼── CH₃
        CH₂OH
```

(c)
```
      HC⟍O
    H ──┼── OH
    H ──┼── OH
        CH₂OH
```

(d)
```
         CN
    H ──┼── OH
    H ──┼── OH
   HO ──┼── H
   HO ──┼── H
        CH₂Cl
```

5.50 Draw a Fischer projection for each of the following molecules shown in its zigzag conformation.

(a)
H, O, OH, OH, HO, C₆H₅, OH

(b)
OH, HO, CN, OH

(c)
O, OH, Cl, HO, OH, OH, OH

5.51 How many tetrahedral stereocenters are in each of the following molecules? Mark each one with an asterisk.

5.52 Draw all possible configurational isomers of the following molecule. Which ones are meso?

5.53 Draw all possible configurational isomers of the following molecule. Which ones are optically active?

5.54 Draw all configurational isomers of $C_4H_{11}N$ that are optically active.

5.55 Butanal (C_4H_8O) has eight H atoms.

Butanal

Suppose that any of these H atoms can be replaced by a Cl atom to yield a molecule with the formula C_4H_7ClO.
(a) Identify two H atoms where, if this substitution takes place, it would yield *constitutional isomers* of C_4H_7ClO.
(b) Identify two H atoms where, if this substitution takes place, it would yield *enantiomers* of C_4H_7ClO.
(c) Identify two H atoms where, if this substitution takes place, it would yield *conformational isomers* of C_4H_7ClO.

5.56 2,3-Dibromoprop-1-ene ($C_3H_4Br_2$) has four H atoms.

2,3-Dibromoprop-1-ene

Suppose that any of these H atoms can be replaced by a Cl atom to yield a molecule with the formula $C_3H_3Br_2Cl$.
(a) Identify two H atoms where, if this substitution takes place, it would yield *constitutional isomers* of $C_3H_3Br_2Cl$.
(b) Identify two H atoms where, if this substitution takes place, it would yield *enantiomers* of $C_3H_3Br_2Cl$.
(c) Identify two H atoms where, if this substitution takes place, it would yield *diastereomers* of $C_3H_3Br_2Cl$.

5.57 For each of the following pairs, determine if the compounds have the same boiling point or different boiling points.

(a)

(b)

(c)

(d)

```
      CH₂OH              CH₂OH
   H——OH            H——OH
   H——OH            HO——H
   H——OH            H——OH
      CH₂OH              CH₂OH
```

(e)

5.58 An unknown compound is optically active and has the molecular formula C_6H_{12}. Draw all possible isomers of the compound.

5.59 Aldosterone, a steroid involved in regulating blood pressure, is shown below without dash–wedge notation. How many stereocenters does aldosterone have?

Aldosterone

5.60 Taxol, an anticancer drug, is shown below without dash–wedge notation. How many stereocenters does Taxol have?

Taxol

5.61 If Taxol (see Problem 5.60) has a specific rotation of −49°, then what is the specific rotation of its enantiomer?

5.62 Consider a mixture that is 60% (+) enantiomer and 40% (−) enantiomer. In which direction will the mixture rotate plane-polarized light? What is the enantiomeric excess of the mixture?

5.63 (S)-2-Bromobutane has a specific rotation of +23.1°. Suppose that a solution is made by dissolving 20.00 g of (S)-2-bromobutane in 1.00 L total solution. What would be the measured rotation of plane-polarized light sent through a 10.00-cm tube containing that solution?

5.64 Ibuprofen is a drug used to manage mild pain, fever, and inflammation. It is a chiral drug, and only the (S) enantiomer, whose specific rotation is +25.0°, is effective. The (R) enantiomer exhibits no biological activity. If a particular process is capable of producing ibuprofen in 84% enantiomeric excess of the (S) enantiomer, then what is the specific rotation of that mixture? What is the percentage (S) enantiomer in that mixture?

5.65 Which of the following would you expect to have the greatest heat of combustion? The smallest? Explain.

A   B   C

5.66 Which of the following isomers would you expect to have the greater heat of combustion? Explain.

A   B

5.67 The heat of combustion of either but-1-yne or but-2-yne is −2577 kJ/mol (−615.9 kcal/mol), whereas the other's is −2597 kJ/mol (−620.7 kcal/mol). Match each compound to its heat of combustion.

But-1-yne   **But-2-yne**

5.68 1,1′-Bi-2-naphthol is a chiral compound, and its enantiomers can be separated from each other. The enantiomer shown here has a specific rotation of −32.70° in tetrahydrofuran. (Notice that the two fused-ring systems are not in the same plane. In the top ring system, the left side is in front of the plane of the page, whereas in the bottom ring system, the right side is in front.)

1,1′-Bi-2-naphthol

(a) How many tetrahedral stereocenters does 1,1′-bi-2-naphthol have?
(b) Draw the enantiomer of the molecule shown. What is its specific rotation?
(c) Aside from being mirror images of each other, how else are the two enantiomers related? Why do you think they do not readily interconvert? (*Hint:* It may help to construct a molecular model.)

5.69 Although we learned in Section 5.4 that uncharged nitrogen atoms generally cannot be stereocenters (due to nitrogen inversion), an exception is Tröger's base.

Tröger's base has two enantiomers that can be separated from each other. They are different in their configurations at the N atoms. One enantiomer is shown on the right.
(a) Draw the second enantiomer of Tröger's base.
(b) Explain why the two enantiomers do not interconvert.

5.70 In this chapter, you learned that a molecule that possesses a plane of symmetry must be achiral. It is possible, however, for a molecule to be achiral without possessing a plane of symmetry, as long as it possesses a *point of symmetry*. If a point of symmetry exists for a molecule, then each atom can be reflected through that point (rather than a plane) to arrive at an identical atom. The following molecule is achiral, yet it possesses no plane of symmetry.
(a) Convince yourself that the molecule has no plane of symmetry.
(b) Draw the molecule's mirror image and show that the molecule and its mirror image are superimposable.
(c) Where is the point of symmetry located?

5.71 Molecule **A** is an acid that has a single tetrahedral stereocenter and is therefore chiral. As with any chiral compound, the enantiomers of **A** have identical physical and chemical properties and therefore cannot be separated in an achiral environment. However, when a racemic mixture of **A** is reacted with an enantiomerically pure chiral base such as **B**, then two salts of the form **C** are produced.

The two product salts have *different* physical and chemical properties, which allow them to be separated in an achiral environment.
(a) Draw the two salts that are produced, including their complete dash–wedge notations.
(b) Explain why the two salts have different physical and chemical properties.

5.72 In this chapter, we saw that amino acids typically have a single tetrahedral stereocenter and are therefore chiral. One amino acid, however, is achiral. Which one? Explain. (*Hint:* Review Table 1-7.)

If your instructor assigns problems in smartwork, log in at **smartwork.wwnorton.com**.

Considerations of Stereochemistry

R and S Configurations about Tetrahedral Stereocenters and Z and E Configurations about Double Bonds

In Chapter 5, we saw that different configurations can exist about a tetrahedral stereocenter or a double bond, giving rise to different configurational isomers—namely, enantiomers and diastereomers. Here in Nomenclature 3, we explain how to assign a specific designation to each type of configuration and to incorporate these designations into the molecule's IUPAC name. The collection of rules used to assign these configurations is called the **Cahn–Ingold–Prelog convention**.[1,2]

The basis of this convention involves first assigning priorities to the various substituents attached to the tetrahedral stereocenter or double bond and then examining the spatial arrangement of a subset of those substituents. In Section N3.1, we show how the Cahn–Ingold–Prelog rules are used to determine the configuration about a tetrahedral stereocenter, and in Section N3.2, we do the same for the configuration about a double bond.

N3.1 Priority of Substituents and Stereochemical Configurations at Tetrahedral Centers: R/S Designations

The IUPAC rules for assigning the configuration at a particular tetrahedral stereocenter involve three basic steps:

> **Step 1.** Assign a priority, 1 through 4 (where 1 is the highest and 4 is the lowest), to each of the four substituents bonded to the stereocenter.
>
> **Step 2.** Orient the molecule so that *the lowest-priority substituent is pointed away from you*.

[1]Cahn, R. S.; Ingold, C. K.; Prelog, V. *Angew. Chem.* **1966**, 78, 413–447.
[2]Prelog, V.; Helmchen, G. *Angew. Chem. Int. Ed.*, **1982**, 21, 567–583.

Step 3. If the substituents having the first, second, and third priorities are arranged *clockwise*, then the stereocenter is assigned the *R* configuration (Fig. N3-1a). If they are arranged *counterclockwise*, then the stereocenter is assigned the *S* configuration (Fig. N3-1b). (The *R* and *S* designations derive from Latin—*rectus* means right and *sinister* means left.)

As you can see, determining whether a tetrahedral stereocenter's configuration is *R* or *S* depends on how the priorities are assigned to its four substituents. Under the Cahn–Ingold–Prelog convention, a set of rules is applied, which is a system of tiebreakers based primarily on atomic numbers. Because of the complexity introduced when substituents contain double or triple bonds, we first present the basic system involving substituents that contain only single bonds (Section N3.1a). We then look at the modifications required to deal with multiple bonds in a substituent (Section N3.1b).

N3.1a Substituents Involving Only Single Bonds

The basic system for establishing relative priorities of substituents under the Cahn–Ingold–Prelog system are as follows:

The Cahn–Ingold–Prelog Rules for Assigning Substituent Priorities for Stereochemical Configurations

Rule 1. If the atoms at the substituents' points of attachment are different, then the relative priorities of the substituents are assigned as follows:
  a. The substituent whose atom has the higher atomic number has the higher priority.
  b. If atoms have the same atomic number but are isotopes of each other (i.e., have the same number of protons but different numbers of neutrons), then the substituent with the heavier isotope has the higher priority.

Rule 2. If the atoms at the points of attachment are identical, then the priorities of the substituents are established by the *sets of atoms one bond away* from the respective points of attachment:
  a. Order the atoms in each set from highest priority to lowest, as described in Rule 1.
  b. Compare each set's highest-priority atom. If they are different, then the atom that has the higher priority corresponds to the higher-priority substituent.
  c. If the highest-priority atoms in the two sets are identical, then compare each set's second-highest-priority atom to break the tie.
  d. If the second-highest-priority atoms in the two sets are identical, then compare each set's lowest-priority atom to break the tie.

Rule 3. If sets of atoms one bond away from the point of attachment are identical, then Rule 2 is applied to the sets of atoms *one additional bond away* from the point of attachment. Continue to move farther away from the point of attachment until there is a difference.
  a. If the substituent's backbone is branched, then follow the chain on which the higher-priority atoms are encountered first.
  b. If two atoms being compared are the same and one of them is a substituent's terminal atom (i.e., the atom farthest from the point of attachment), then the substituent to which it belongs has the lower priority.
  c. If a point of difference is never encountered, then the substituents must be identical.

Turning a steering wheel in the *clockwise* direction turns a car to the *right*.

R configuration

(a)

Turning a steering wheel in the *counterclockwise* direction turns a car to the *left*.

S configuration

(b)

FIGURE N3-1 ***R* and *S* configurations** With the lowest-priority substituent pointed away, a clockwise arrangement of the top-three-priority substituents defines an *R* configuration (a) and a counterclockwise arrangement defines an *S* configuration (b).

Let's now apply these rules to determine the configuration about the stereocenter at C1 in the following isomer of 1-bromo-1-chloroethane:

1-Bromo-1-chloroethane

First we must assign priorities to each of the four substituents that are attached to the stereocenter: CH_3, H, Cl, and Br. According to Rule 1, we identify the atoms at the points of attachment as C, H, Cl, and Br. The atomic numbers of these atoms decrease in the order: Br > Cl > C > H. Therefore, the substituent priorities are assigned as follows:

1st priority ·····▶Br

3rd priority·.....·····.C·······''Cl◀····· 2nd priority

H₃C   H ◀····· 4th priority

Having established the priorities of all four substituents, we next orient the molecule so that the lowest-priority substituent (in this case, H) is pointed away. In the structure above, however, it is pointing toward us, so the molecule must be reoriented as shown in Figure N3-2 (you will find a molecular modeling kit helpful for doing this).

With the lowest-priority substituent (H) pointing away from us, the first (Br), second (Cl), and third (CH_3) priority substituents are arranged counterclockwise (as indicated by the red arrow), so the stereocenter has the S configuration.

If you have difficulty reorienting a molecule to place the lowest-priority substituent pointing away from you, then consider one of the following alternatives:

▪ If the lowest-priority substituent points *toward* you, determine the arrangement of the first-, second-, and third-priority substituents (i.e., clockwise or counterclockwise) in the orientation that is given, and then reverse that arrangement prior to assigning the configuration.

▪ If the lowest-priority substituent lies in the plane of the page, exchange that group with another substituent that is *not* in the plane of the page, and then determine the configuration as usual. The true configuration will be the opposite of what you derive, because exchanging any two groups reverses the configuration (Section 5.4c).

FIGURE N3-2 Assigning R and S designations To assign the configuration of this tetrahedral stereocenter, the molecule is reoriented so the fourth-priority substituent (in this case, H) that points toward you (left) instead points away (right). The resulting counterclockwise arrangement of the top-three-priority substituents designates the configuration as S.

The fourth-priority substituent points towards you.

Rotate

The fourth-priority substituent points away.

1st priority

2nd priority

3rd priority

(S)-1-Bromo-1-chloroethane

Consider $CH_3CHClBr$ again in the orientation that was initially given.

1st priority ······▶ Br
3rd priority ····
2nd priority
H_3C — C ···· Cl
H ···· 4th priority

Reverse the arrangement

Counterclockwise = S

First-, second-, and third-priority substituents are arranged clockwise, but the fourth-priority substituent points toward you, not away.

In that orientation, the first-, second-, and third-priority substituents are arranged clockwise. Because the fourth-priority substituent (H) points toward us, we reverse that arrangement to counterclockwise, which corresponds to an S configuration, the same that we determined previously.

As a final step, the name of the molecule is altered to identify the configuration of each stereocenter.

- For a molecule with one stereocenter, a single designation of (R) or (S) is added to the molecule's name.
- That designation should appear before any number in the name that identifies the location of a substituent attached to the stereocenter.

For relatively simple molecules, that often means that the $(R)/(S)$ designation appears first in the name. Therefore, the name of the preceding molecule becomes (S)-1-bromo-1-chloroethane. For molecules with more than one stereocenter, there is some flexibility as to the location of these designations (see Section N3.1c).

---

problem **N3.1** The atoms in each of the following sets are arranged alphabetically. Reorder the substituents in each set from *highest* to *lowest* priority.

(a) Br, CH_3, Cl, F   (b) Br, CH_3, H, I   (c) CH_3, F, O, N

---

problem **N3.2** Designate the configuration of each stereocenter in the following molecules as R or S. What is the complete IUPAC name for each molecule?

(a)   (b)   (c)   (d)

---

problem **N3.3** Draw the specific stereoisomer that corresponds to each name.

(a) (S)-1-Methoxy-1-nitrobutane
(b) (R)-1-Methoxy-1-nitrobutane
(c) (R)-1-Fluoro-1-methoxy-1-nitropropane
(d) (S)-3,3-Dichloro-1-ethoxy-1-fluorohexane

The following stereoisomer illustrates how to deal with isotopes:

With the fourth-priority substituent pointing away, the first-, second-, and third-priority substituents are clockwise, so the configuration is *R*.

The two CH_3 substituents have a higher priority than deuterium (D) or H because C has a higher atomic number. ^{13}C is heavier than ^{12}C, however, so the CH_3 substituent with the ^{13}C isotope is assigned first priority, and the one with the ^{12}C isotope is assigned second priority. Similarly, D and H have the same atomic number, but D, being the heavier isotope (i.e., ^{2}H vs. ^{1}H), is assigned a higher priority than H; thus, D is assigned third priority and H is assigned fourth. Looking back at the structure, the fourth-priority substituent (H) is pointing toward us. When the molecule is reoriented to have that substituent pointing away, the first-, second-, and third-priority substituents are arranged clockwise, so the configuration is *R*.

---

problem **N3.4** Designate the configuration of each stereocenter in the following molecules as *R* or *S*.

(a)   (b)   (c)

In the following stereoisomer of 2-chlorobutane, the substituents bonded to the stereocenter are CH_3, CH_2CH_3, Cl, and H:

1st priority ------- Cl

4th priority

Rule 1 results in a tie because both substituents are attached by C.

By considering the atoms at the points of attachment, we can immediately assign Cl as the first-priority substituent and H as the fourth-priority substituent. The priorities of the CH_3 and CH_2CH_3 substituents, however, remain tied.

To break the tie, we proceed to Rule 2, which has us examine the set of atoms one bond away from the point of attachment. For the CH_2CH_3 substituent, that set of atoms is {C,H,H} and for the CH_3 group it is {H,H,H}:

The set of atoms one bond
away from the point of
attachment is {C,H,H}.

2nd priority

1st priority

4th priority

3rd priority

(S)-2-Chlorobutane

The set of atoms one bond
away from the point of
attachment is {H,H,H}.

Comparing the highest-priority atoms in each set (C vs. H), we assign the higher priority to the CH_2CH_3 substituent, so CH_2CH_3 is second priority and CH_3 is third. Notice that the fourth-priority substituent (H) is pointing away and the first-, second-, and third-priority substituents are arranged counterclockwise, so the configuration is (S). The complete name of the molecule is therefore (S)-2-chlorobutane.

You must remember to compare substituents one atom at a time, not entire groups. The following example demonstrates why.

1st priority

4th priority

Rule 1 results in a tie because both
substituents are attached by C.

Once again, by examining the atoms at the points of attachment, we can apply Rule 1 to assign Cl as the first-priority substituent and H as the fourth-priority substituent. The remaining two substituents are attached by C, however, so Rule 1 results in a tie. To apply Rule 2, we identify the sets of atoms one bond away from the points of attachment. For both substituents, the set of atoms is {C,H,H}, as shown on the left below, resulting in a tie once again.

The set of atoms one bond
away from the point of
attachment is {C,H,H}.

The set of atoms two bonds
away from the point of
attachment is {C,H,H}.

1st priority

4th priority

The set of atoms one bond
away from the point of
attachment is {C,H,H}.

3rd priority

2nd priority

The set of atoms two bonds
away from the point of
attachment is {Cl,H,H}.

We must therefore apply Rule 3 by examining the sets of atoms one bond farther away from the points of attachment. As shown on the right above, that set of atoms for the Br-containing substituent is {C,H,H} and that set of atoms for the Cl-containing substituent is {Cl,H,H}. Because the atomic number of Cl is higher than that of C, the Cl-containing substituent wins. Therefore, the Cl-containing substituent is assigned second priority and the Br-containing substituent is assigned third priority.

Notice in this case that the Br atom, whose atomic number is higher than any other atom in the molecule, never figured in the evaluation of assignments.

Now that the priorities of the four substituents are established, we can assign the configuration of the stereocenter and write the complete name of the molecule. With the fourth-priority substituent pointing away, the first-, second-, and third-priority substituents are arranged clockwise, so the configuration is *R*. Thus, the complete IUPAC name of this molecule is (*R*)-1,3-dichloro-6-bromohexane.

---

problem **N3.5** What is the complete IUPAC name for each molecule?

(a)   (b)   (c)   (d)

---

problem **N3.6** Draw the specific stereoisomer that corresponds to each name.

(a) (*R*)-2-Bromohexane
(b) (*S*)-4-Ethoxy-1,1,1,2-tetrafluorobutane
(c) (*R*)-1,4-Dinitropentane
(d) (*S*)-2-Ethoxy-2-methoxypentane

N3.1b Substituents with Double or Triple Bonds

Substituents with double or triple bonds are treated somewhat differently from substituents that contain only single bonds.

> ▪ If atoms X and Y in a substituent are connected by a *double bond*, then we treat X as being singly bonded to two Y atoms. Likewise, we treat Y as being singly bonded to two X atoms.
>
> ▪ If atoms X and Y in a substituent are connected by a *triple bond*, then we treat X as being singly bonded to three Y atoms, and we treat Y as being singly bonded to three X atoms.

To apply these rules, it is helpful to consider adding *imaginary* atoms when replacing double and triple bonds with single bonds. This is shown in Figure N3-3 for a variety of cases, where the atoms in red are the imaginary atoms that have been added.

FIGURE N3-3 Priorities of substituents with double and triple bonds (a) An atom that is doubly bonded to another atom is treated as having two single bonds to the atom—one real atom (black) and one imaginary (red). (b) An atom that is triply bonded to another atom is treated as having three single bonds to the atom—one real atom (black) and two imaginary (red).

With these rules in mind, let's determine the configuration of the stereocenter in the following isomer of 3-methylhex-1-ene:

The set of atoms one bond away from the point of attachment is {C,H,H}.
2nd priority

The set of atoms one bond away from the point of attachment is {C,C,H}.
1st priority

Stereocenter

Replace the double bond with single bonds.

The set of atoms one bond away from the point of attachment is {H,H,H}.

3-Methylhex-1-ene

The set of atoms one bond away from the point of attachment is {H,H,H}.
3rd priority

In the structure on the right above, the C=C double bond has been replaced by a single bond, and single bonds to two imaginary C atoms (red) have been added. The H atom is assigned fourth priority because C atoms attach the other three substituents; C has a higher atomic number than H. The priorities of the remaining substituents are determined by the sets of atoms one bond away from the respective points of attachment: {H,H,H} for the CH_3 substituent, {C,H,H} for the $CH_2CH_2CH_3$ substituent, and {C,C,H} for the $CH=CH_2$ substituent. Therefore, $CH=CH_2$ is first priority, $CH_2CH_2CH_3$ is second priority, and CH_3 is third priority. Given that these substituents are arranged counterclockwise with the fourth-priority substituent pointing away, the configuration is assigned S and the complete name of the molecule is (S)-3-methylhex-1-ene.

problem **N3.7** What is the configuration, R or S, of the tetrahedral stereocenter in each of the following molecules?

(a)   (b)   (c)   (d)   (e)

problem **N3.8** Designate the configuration of each tetrahedral stereocenter in the following molecules as R or S.

(a)   (b)   (c)   (d)

N3.1c Molecules with More Than One Stereocenter

If a molecule has two or more stereocenters, then it must have a unique name that accounts for this.

- If multiple stereocenters exist in a molecule, each stereocenter receives its own (R) or (S) designation.
- In the name, each designation appears before any number that identifies the location of a substituent on the corresponding stereocenter.

For example, both C2 and C3 in 3-bromo-2-chloropentane are stereocenters.

3-Bromo-2-chloropentane

The configuration of each stereocenter must be determined independently of the other, as shown in Figure N3-4.

The priorities of the groups attached to C2 (Fig. N3-4a), from highest to lowest, are: Cl, CHBrCH$_2$CH$_3$, CH$_3$, and H. The first-, second-, and third-priority groups are arranged counterclockwise with the fourth-priority group pointing away, so the configuration is S. At C3 (Fig. N3-4b), the priorities of the groups in decreasing order are: Br, CHClCH$_3$, CH$_2$CH$_3$, and H. The first-, second-, and third-priority groups are also arranged counterclockwise, but in this case, the fourth-priority substituent is pointed toward you, not away, so the direction is reversed and the configuration is R.

FIGURE N3-4 Compounds with multiple tetrahedral stereocenters This compound has two tetrahedral stereocenters, so each is designated R or S. (a) The stereocenter (red) has its first-, second-, and third-priority substituents arranged counterclockwise with the fourth-priority substituent pointing away, so the configuration is S. (b) The stereocenter (red) has its first-, second-, and third-priority substituents arranged counterclockwise, but the fourth-priority substituent points toward us, so the direction is reversed and the configuration is R.

We can now append these designations to the molecule's name, 3-bromo-2-chloropentane. The configuration for C2 (which is S) can be placed before the "2" because that is the number that locates the Cl substituent on C2. The configuration for C3 (which is R) can be placed before the "3" because that number locates the Br substituent on C3. Thus, the molecule can be named (R)-3-bromo-(S)-2-chloropentane.

An alternative is to group all of the R and S designations at the beginning of the name and use numbers to specify which stereocenter each designation corresponds to. For the preceding example, the name would be (2S,3R)-3-bromo-2-chloropentane. Notice that this format repeats the number locators—once for the location of the stereocenter and once for the location of the substituent.

problem **N3.9** Determine the IUPAC name for each of the following molecules.

(a)   (b)   (c)   (d)

problem **N3.10** Draw the structure of each of the following molecules.

(a) (S)-2-Chloro-(S)-3-ethoxypentane
(b) (4R,5S)-2,4-Dimethyl-5-nitrohex-2-ene
(c) (4R,5R,6S)-4,5,6-Trichloro-2-methyl-3-phenylhept-2-ene

N3.1d Stereocenters Incorporated into Rings

When a stereocenter is part of a ring, the protocol for assigning R and S configurations does not change. In 1,1-dibromo-(S)-3-chloro-2,2,4,4-tetramethylcyclopentane, the ring carbon bonded to Cl is a stereocenter (red).

1,1-Dibromo-(S)-3-chloro-2,2,4,4-tetramethylcyclopentane

The Cl substituent has the highest priority and H has the lowest priority. The two substituents involved in the ring, however, are attached by C, resulting in a tie.

When we examine each of those substituents' atoms one bond farther away, we find that both are bonded to a {C, C, C} set of atoms (i.e., another tie). The tie is broken, however, when we reach the atoms that are two bonds away from the point of attachment. The set of atoms is {C, H, H} for the substituent that makes up the top half of the ring and {Br, Br, C} for the substituent that makes up the bottom half of the ring, so the latter substituent has the higher priority. The configuration is therefore S, because, with the H atom pointing away, the first-, second-, and third-priority substituents are arranged counterclockwise.

If a ring contains two or more stereocenters, then each configuration should be specified in the name in precisely the same way we learned for acyclic molecules. In some cases, however, the cis/trans convention for rings (Section 4.9) is sufficient to describe the stereochemical configurations.

> For a ring that contains exactly two stereocenters, each of which is bonded to one H atom, the cis/trans convention is sufficient to unambiguously describe the stereochemical configurations only if the species is achiral.

Consider, for example, the following stereoisomer of 1,3-dibromocyclohexane.

(1R,3S)-1,3-Dibromocyclohexane
or
cis-1,3-Dibromocyclohexane

Because one stereocenter is R and the other is S, the compound can be called (1R,3S)-1,3-dibromocyclohexane. Alternatively, the molecule is achiral (notice that it has a plane of symmetry), so we can call it *cis*-1,3-dibromocyclohexane because both Br substituents appear on the same side of the ring's plane.

The species must be achiral to use the cis/trans convention so that the name of the molecule unambiguously describes one stereoisomer. If the molecule is chiral, then two enantiomers exist, in which case the cis/trans convention is ambiguous and insufficient to distinguish one enantiomer from the other. In both of the following enantiomers, for example, the Br substituents are on opposite sides of the roughly defined by the ring plane, so both would be called trans isomers.

Enantiomers of *trans*-1,3-dibromocyclohexane

(1*R*,3*R*)-1,3-Dibromocyclohexane

(1*S*,3*S*)-1,3-Dibromocyclohexane

---

problem **N3.11** Determine the IUPAC name of each of the following compounds.

(a)   (b)   (c)   (d)

---

problem **N3.12** Draw the structure of each molecule.

(a) (*S*)-1-Chloro-2,2-dimethyl-1-phenylcyclopentane
(b) (1*R*,2*S*)-1-Methyl-1,2-dinitrocyclopropane
(c) (*R*)-4-Ethoxycyclohexene
(d) (3*S*,4*S*)-3-Chloro-4-fluoro-2-methylhepta-1,6-diene

---

problem **N3.13** Determine whether each of the following names describes a single stereoisomer unambiguously. For each one that does, rewrite the name using the *R* and *S* designations for the stereocenters if appropriate.

(a) *cis*-1,2-Difluorocyclohexane   (b) *trans*-1,2-Difluorocyclohexane
(c) *trans*-1,4-Difluorocyclohexane   (d) *cis*-1-Chloro-2-fluorocyclohexane
(e) *trans*-1,4-Dimethylcycloheptane

N3.1e Fischer Projections and the *R/S* Designations

If you have learned how to use Fischer projections, then the assignment of the stereocenters can be quite straightforward. Recall that *vertical* bonds in a Fischer projection point *away* from you and horizontal bonds point *toward* you. Thus, if the lowest-priority group is on a vertical bond, then the projection represents a properly oriented model. (It doesn't matter if the bond points up or down on the page, it still points *away* from you.)

Consider, for example, the Fischer projection of 3-bromo-1,2-dichloro-3-methylbutane.

First-, second-, and third-priority groups arranged counterclockwise with the fourth-priority group pointing away

is equivalent to ... **S configuration**

(S)-3-Bromo-1,2-dichloro-3-methylbutane

On the right, the Fischer projection has been converted into a dash–wedge formula, where we can more clearly see that the fourth-priority substituent (H) points away from us. In this orientation, the first-, second-, and third-priority substituents are arranged counterclockwise, so the stereocenter has the S configuration and the molecule is named (S)-3-bromo-1,2-dichloro-3-methylbutane.

If the lowest-priority substituent is on a horizontal bond, instead, then it points *toward* us rather than away. Therefore, prior to assigning the configuration, we can simply reverse the order in which the first-, second-, and third-priority substituents are arranged. As an example, both stereocenters in the following Fischer projection have the fourth-priority substituent (H) on a horizontal bond.

Clockwise, but H points toward us instead of away

Reverse direction ... **S configuration**

Clockwise, but H points toward us instead of away

Reverse direction ... **S configuration**

In the Fischer projection itself, each stereocenter has its first-, second-, and third-priority groups arranged clockwise. But, because the fourth-priority substituent is pointing *toward* us, we reverse those directions, yielding a counterclockwise arrangement for each stereocenter. Thus, each stereocenter has an S configuration.

problem **N3.14** Determine the IUPAC name of each of the following compounds.

(a)   (b)   (c)

N3.1f Using the *R/S* Designations to Identify Enantiomers and Diastereomers

The R/S designation in the names of molecules can be used to determine the specific relationship between a pair of configurational isomers—that is, whether they are enantiomers or diastereomers of each other. This is done by applying the following rules we learned in Section 5.5.

> - If two molecules are related by the inversion of some, but not all tetrahedral stereocenters, then they are *diastereomers* of each other.
> - If two molecules are related by the inversion of *all* tetrahedral stereocenters, then they are *enantiomers* of each other.

For example, (2*R*,3*S*,4*S*)-2,3,4-trichlorohexane and (2*S*,3*R*,4*R*)-2,3,4-trichlorohexane are enantiomers of each other because each of the stereocenters in one is the reverse of the corresponding stereocenter in the other. In other words, wherever there is an *R* configuration in one, there is an *S* configuration in the other.

(2R,3S,4S)-2,3,4-Trichlorohexane   **(2S,3R,4R)-2,3,4-Trichlorohexane**

Configurations are different
at all three stereocenters.

Are (2*R*,3*S*,4*S*)-2,3,4-trichlorohexane and (2*R*,3*R*,4*R*)-2,3,4-trichlorohexane enantiomers or diastereomers of each other? The configuration is the same at C2 in both molecules (i.e., *R*), but opposite at C3 and C4, so they are diastereomers.

(2R,3S,4S)-2,3,4-Trichlorohexane   **(2R,3R,4R)-2,3,4-Trichlorohexane**

Configurations are different at
two of the three stereocenters.

---

problem **N3.15** If the tetrahedral carbon stereocenters in a molecule have been designated (2*R*,3*S*,5*R*), what will the designations be in the molecule's enantiomer? What will the designations be in *one* of its diastereomers?

---

problem **N3.16** Considering only the IUPAC names, determine whether each of the following pairs of molecules are enantiomers, diastereomers, or neither.

(a) (1*R*,3*S*)-1-Methyl-3-nitrocyclohexane   (1*S*,3*R*)-1-Methyl-3-nitrocyclohexane
(b) (1*R*,3*S*)-1-Methyl-3-nitrocyclohexane   (1*S*,3*S*)-1-Methyl-3-nitrocyclohexane
(c) (1*R*,3*S*)-1-Methyl-3-nitrocyclohexane   (1*S*,2*R*)-1-Methyl-2-nitrocyclohexane
(d) *cis*-1,2-Dimethylcyclobutane   *trans*-1,2-Dimethylcyclobutane

---

Work through the following practice problems to become as comfortable as possible with what you have learned about nomenclature so far.

problem N3.17 Draw the structure of each of the following molecules.

(a) (R)-1-Chloro-1-fluorobutane
(d) (R)-2,2,3-Trichlorobutane
(g) (R)-3-Chloropent-1-ene
(j) (S)-3-Chlorocyclohexene

(b) (S)-2-Chloropentane
(e) (S)-3-Methylhexane
(h) (2S,3S)-2-Bromo-3-chloropentane
(k) (1R,2S)-1,2-Dibromocyclopentane

(c) (R)-2-Chloro-2-methoxypentane
(f) (S)-2-Bromo-1-nitropentane
(i) (R)-1-Bromo-(R)-2-iodocyclopentane

problem N3.18 Determine the IUPAC name of each of the following molecules.

(a)   (b)   (c)   (d)   (e)

(f)   (g)   (h)   (i)

N3.2 Stereochemical Configurations of Alkenes: Z/E Designations

Recall from Chapter 5 that a C=C double bond can have two configurations that differ by an imaginary 180° rotation about the double bond. However, the rules we have learned thus far for naming alkenes do not take these kinds of stereochemical configurations into account. In this section, we learn how to make the configuration at each double bond an explicit and unambiguous part of the compound's name.

N3.2a The Limitations of Cis and Trans Designations for Alkenes

Recall from Chapter 3 that two substituents are cis to each other if they are on the same side of the double bond, and they are trans to each other if they are on opposite sides of the double bond. These terms can be used to describe the configuration about a double bond if the groups to which we are referring are unambiguous. Generally, this limits us to cases in which each alkene carbon has one hydrogen atom and one nonhydrogen substituent.

> ■ A cis alkene describes two nonhydrogen substituents that are on the same side of the double bond.
>
> ■ A trans alkene describes two nonhydrogen substituents that are on opposite sides of the double bond.

These definitions make it possible to identify the cis and trans forms of an alkene such as pent-2-ene.

Nonhydrogen substituents on same side of double bond

cis-Pent-2-ene

Nonhydrogen substituents on opposite sides of double bond

trans-Pent-2-ene

A problem arises, however, when there is more than one nonhydrogen substituent bonded to one or both alkene carbons. Consider, for example, the two stereoisomers of 1-chloro-1-fluoro-2-nitro-2-phenylethene:

Which is the cis isomer? Which is the trans isomer?

1-Chloro-1-fluoro-2-nitro-2-phenylethene

It is unclear whether the cis and trans designations should refer to the relationship between the F and NO_2 groups, the F and C_6H_5 groups, the Cl and NO_2 groups, or the Cl and C_6H_5 groups. To avoid this ambiguity, we turn to a method for assigning configurations about a double bond known as the *Z/E system*.

N3.2b The *Z/E* System for Double Bond Configurations

We determine the configuration about a double bond by first determining *which of the two substituents on each alkene carbon has the higher priority*. The rules for determining these substituent priorities are the same Cahn–Ingold–Prelog conventions discussed previously in Section N3.1a.

Once these relative priorities have been established for the substituents on each alkene carbon, the configuration can be designated as either *Z* or *E*, depending upon the orientation of the two higher-priority substituents with respect to the double bond.

> ■ If the two higher-priority groups are on the same side of the double bond, then the alkene is assigned the *Z* configuration, and the molecule's name is given the prefix (*Z*).
>
> ■ If the two higher-priority groups are on opposite sides of the double bond, then the alkene is assigned the *E* configuration, and the molecule's name is given the prefix (*E*).

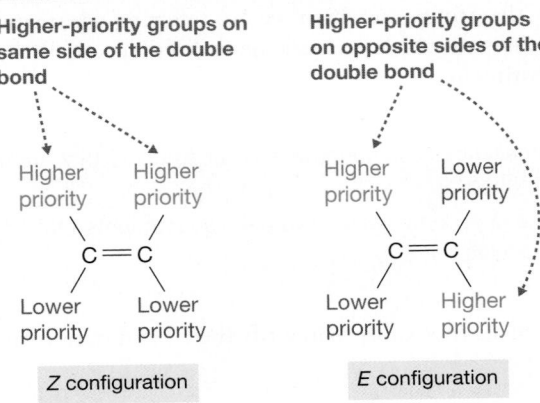

Higher-priority groups on same side of the double bond

Z configuration

Higher-priority groups on opposite sides of the double bond

E configuration

The Z/E notation derives from German, in which *zusammen* means "together" and *entgegen* means "opposed." Alternatively, you can note whether the two higher-priority groups are on the "Zame" side of the double bond or on "Epposite" sides.

Let's see how this system is applied to the two isomers of pent-2-ene below.

Higher priority substituents are on the same side.

Higher priority substituents are on opposite sides.

C wins over H

C wins over H

(Z)-Pent-2-ene

(E)-Pent-2-ene

Notice that C2 and C3 are the atoms connected by the double bond. The substituents on C2 are H and CH_3, and CH_3 beats H because its atom at the point of attachment, C, has the higher atomic number. Similarly, on C3, the CH_2CH_3 substituent beats H. Notice in the isomer above on the left that the two higher-priority substituents are on the same side of the double bond, so the configuration is Z. Thus, the name of the molecule is (Z)-pent-2-ene. In the isomer on the right, the two higher-priority substituents are on opposite sides of the double bond, so the configuration is E and the name of the molecule is (E)-pent-2-ene.

---

problem **N3.19** What is the configuration, Z or E, of each of the following double bonds?

(a)   **(b)**   **(c)**   **(d)**

As another example, consider the following isomer of 6-bromo-3-(1-chloroethyl) hept-2-ene:

The set of atoms one bond away from the point of attachment is {C,H,H}.

Lower priority

Higher priority

The set of atoms one bond away from the point of attachment is {Cl,C,H}.

Lower priority   Higher priority

(E)-6-Bromo-3-(1-chloroethyl)hept-2-ene

On C2, the CH_3 group beats H because its atom at the point of attachment (C) has the higher atomic number. On C3, the two substituents are both attached by C, so to break the tie, we examine the set of atoms one bond farther away. For the Cl-containing substituent, that set of atoms is {Cl,C,H} and for the Br-containing substituent it is {C,H,H}. Because Cl has a higher priority than C, the Cl-containing substituent is assigned the higher priority on C3. Thus, the higher priority substituents on C2 and

C3 are on opposite sides of the double bond, making the configuration E. The complete name of the molecule is (E)-6-bromo-3-(1-chloroethyl)hept-2-ene.

problem **N3.20** What is the configuration, Z or E, of each of the following double bonds?

(a)   (b)   (c)   (d)

problem **N3.21** What is the configuration, Z or E, of each of the following double bonds?

(a)   (b)

N3.2c Assigning Configurations to Molecules with More Than One Alkene Group

When a molecule has two or more alkene groups that can be assigned either an E or a Z configuration, the name of the molecule must do so unambiguously.

> For each C=C double bond that can be assigned either an E or Z configuration, the corresponding designation must appear as a prefix in the name, following the number that corresponds to the location of the double bond.

In the penta-1,3-diene isomer on the left below, there are two double bonds but only one E/Z configuration is assigned. The double bond at C1 cannot have an E/Z configuration because two H atoms are bonded to C1. In the 1-chloropenta-1,3-diene isomer below on the right, however, a Z/E designation is required for both double bonds. In this case, the double bond involving C1 has the E configuration and the one involving C3 has the Z configuration.

C3 is E; C1 is neither.   C1 is E; C3 is Z.

(E)-Penta-1,3-diene $(1E,3Z)$-1-Chloropenta-1,3-diene

Note that in the name of the previous molecule on the right, each number appears in the prefix, even though they are duplicated later in the name.

---

problem **N3.22** What is the complete IUPAC name for each of the following compounds?

(a)   (b)   (c)

N3.2d Configurations of Alkenes as Part of Rings

A double bond that is part of a ring, such as that in cyclodecene, can have both *Z* and *E* configurations.

Double bond is in front of the single bond

(Z)-Cyclodecene   **(E)-Cyclodecene**

For rings consisting of seven or fewer carbons, however, the *E* configuration about the double bond is unstable due to significant ring strain (Section 4.4). Consequently, the *Z/E* convention for small rings is usually omitted; it is understood that the double bond is in the configuration in which the ring carbons are cis to each other.

---

problem **N3.23** Given each of the following names, draw the corresponding structure.

(a) 3,3-Dichlorocyclohexene   **(d)** (*R*)-3-Fluorocycloheptene
(b) 1-Chlorocyclohexene   **(e)** Cyclohepta-1,3-diene
(c) (*E*)-4,4-Dinitrocyclodecene

---

problem **N3.24** What is the complete IUPAC name for each of the following molecules?

(a)   (b)   (c)

Work through the following practice problems to become as comfortable as possible with what you have learned about nomenclature so far.

problem **N3.25** Given each of the following names, draw the corresponding structure.

(a) (Z)-2-Methoxypent-2-ene

(b) (E)-3-Methylpent-2-ene

(c) (Z)-1-Chloro-2-methylpent-1-ene

(d) (E)-2-Chloro-3-methoxybut-2-ene

(e) (Z)-1-Bromo-1-chloropent-1-ene

(f) (Z)-3-Methylpent-2-ene

(g) (Z)-3-Phenylhex-2-ene

(h) 1,2-Dichlorocyclopentene

(i) (2E,4E)-2-Ethoxyhexa-2,4-diene

(j) (1E,3E,5E)-1,3,4,6-Tetrachlorohexa-1,3,5-triene

problem **N3.26** What is the complete IUPAC name for each of the following molecules?

(a)

(b)

(c)

(d)

(e)

(f)

(g)

(h)

(i)

Note: Make a model; the diagonal bond is *behind* the double bond.

CHAPTER 6

The Proton Transfer Reaction

An Introduction to Mechanisms, Thermodynamics, and Charge Stability

In Chapters 1 through 5, we focused primarily on structural aspects of atoms, molecules, and ions. We asked questions about the nature of the chemical bond, about how electrons are distributed in a given molecule or ion, and about the similarities and differences among the various types of isomers. Here in Chapter 6, we shift our focus to *chemical reactions*. A **chemical reaction** is the transformation of one substance (called a reactant) into another substance (called a product), typically through changes in chemical bonds. Although this definition is quite broad, in this book the term *reaction* will usually be applied only when the starting material is converted into another substance that can be isolated.

When we watch a chemical reaction take place in the laboratory, it may seem to proceed smoothly, with the reactants disappearing and the products appearing in one continuous process. If we could observe the reaction on the molecular level, however, we would see that it actually takes place in distinct events called **elementary steps,** each of which involves the breaking and/or formation of specific bonds. The precise sequence of elementary steps that results in the conversion of the original reactants to the final products is called the **mechanism**

This statue of George Washington (Washington Square Park, New York City) exhibits damage from acid rain. Sulfuric acid from acid rain reacts with calcium carbonate in the stone via a proton transfer reaction, the same type of reaction that is the primary focus of Chapter 6.

Upon completing Chapter 6 you should be able to:

- Draw the curved arrow notation for a proton transfer reaction, given the reactants and products.

- Write the expression for the acidity constant, K_a, for any acid, and determine the relative strengths of acids based on their K_a or pK_a values.

- Determine which side of a proton transfer reaction is favored given only the pK_a values of the acids involved.

- Estimate an acid's pK_a value by comparing its structure to that of an acid with known pK_a.

- Calculate the percent dissociation of an acid at various solution pH values given the acid's pK_a.

- Calculate ΔG_{rxn}° for a reaction from its K_a value and explain how ΔG_{rxn}° relates to the relative stability of products versus reactants.

- Interpret a free energy diagram to identify the locations of reactants, products, and transition states and infer relative values for ΔG_{rxn}° and $\Delta G^{\circ\ddagger}$.

- Determine relative stabilities of ions from the pK_a values of acids.

- Explain how the stability of an ion is governed by the identity and hybridization of the atom bearing the charge, the extent of charge delocalization via resonance, and inductive effects from nearby electron-donating and electron-withdrawing substituents.

- Predict relative acid and base strengths based on charge stability.

- Predict the relative contribution of a resonance structure toward the resonance hybrid on the basis of charge stability.

of the reaction. Depending on the nature of the particular reaction, the mechanism may consist of a single step (i.e., Reactants → Products) or multiple steps (e.g., Reactants → A → B → C → Products).

As a result, a reaction mechanism can be used to illustrate *how* a reaction takes place. Much more importantly, however, mechanisms can help us understand *why* a reaction takes place, and ultimately can help us make *predictions* about reactions we may not have seen before. Therefore, each time a new reaction is introduced, we will focus on its mechanism.

Another important aspect of reaction mechanisms is the way in which they simplify organic chemistry. As we will see throughout the rest of the book, there are many seemingly different organic reactions. However, the mechanisms for most of those reactions are constructed from just a few types of elementary steps. Here in Chapter 6, we examine one such elementary step—the *proton transfer*—and in Chapter 7 we examine others.

We begin with the proton transfer reaction because it is one of the simplest reactions that you will encounter in organic chemistry and you are likely already familiar with it from general chemistry. Many of the concepts we learn here will be applied to the other elementary steps discussed in Chapter 7.

6.1 An Introduction to Reaction Mechanisms: The Proton Transfer Reaction and Curved Arrow Notation

A **proton transfer reaction** (or a **Brønsted–Lowry acid–base reaction**) is one in which a *Brønsted–Lowry base* reacts with a *Brønsted–Lowry acid*:

Base	Acid		Conjugate acid	Conjugate base

$$HO:^{\ominus} \;+\; H\!-\!\ddot{C}l: \longrightarrow H\ddot{O}\!-\!H \;+\; :\ddot{C}l:^{\ominus} \qquad (6\text{-}1)$$

Hydroxide anion   **Hydrochloric acid**

In Equation 6-1, HO^- is the **Brønsted–Lowry base** because it *accepts a proton* (H^+), and HCl is the **Brønsted–Lowry acid** because it *donates a proton*. Overall, a proton (H^+), highlighted in blue, is transferred from the acid to the base. Notice that the **conjugate acid** is the species that the base becomes after picking up a proton, and the **conjugate base** is the species that the acid becomes after losing a proton.

Proton transfer reactions take place as a single event, which is to say:

> A proton transfer reaction consists of a single **elementary step**.

As a result, all of the changes that occur in the reaction at the molecular level do so simultaneously.

Organic chemists follow these changes by keeping track of the arrangement of the electrons. As shown in Equation 6-2, a single bond between H and Cl in the reactants is broken, while a single bond between H and O in the products is formed:

Bond is broken.   **Bond is formed.**

$$HO^{\ominus} \;+\; H-Cl: \longrightarrow HO-H \;+\; :Cl^{\ominus} \qquad \text{(6-2)}$$

Moreover, organic chemists use **curved arrow notation** (also called **arrow pushing**) to keep track of the valence electrons as they move throughout a mechanism. There are essentially three rules to using curved arrow notation:

1. A curved arrow represents the movement of *valence electrons*, not *atoms*.
2. Each *double-barbed curved arrow* (⌒) represents the movement of two valence electrons.
3. To show bond breaking, the tail of the arrow originates from the center of a bond. To show bond formation, the head of the arrow points to the region where the bond is formed.

The curved arrow notation for the reaction in Equation 6-1 is illustrated in Equation 6-3.

$$HO^{\ominus} \;+\; H-Cl: \longrightarrow HO-H \;+\; :Cl^{\ominus} \qquad \text{(6-3)}$$

Two curved arrows are required because the reaction involves a total of four electrons. The curved arrow on the left uses a lone pair on O to form an O—H bond. The curved arrow on the right represents the breaking of the H—Cl bond. The pair of electrons from that bond ends up as an additional lone pair on Cl. You will learn in Solved Problem 6.1 that this set of curved arrows can be used to describe other proton transfer reactions as well.

6.1 YOUR TURN

In Equation 6-3, circle all of the electrons that are involved in the chemical reaction. Label each curved arrow as either "bond breaking" or "bond formation."

Answers to Your Turns are in the back of the book.

SOLVED problem 6.1 Draw the curved arrow notation for the proton transfer between ammonia (NH_3) and water, where water acts as the acid and ammonia acts as the base.

Think What does it mean to be an acid? A base? What are the important electrons to keep track of during the course of the reaction? What bonds are broken? What bonds are formed?

Solve Because water acts as an acid (a proton donor) and ammonia acts as a base (a proton acceptor), we may write the reaction as follows:

Acid = H⁺ donor   **Conjugate base**
  Base = H⁺ acceptor   **Conjugate acid**

$$H_2O: \;+\; NH_3 \longrightarrow HO:^{\ominus} \;+\; {}^{\oplus}NH_4$$

Bond breaking and bond formation involve only valence electrons, so draw all valence electrons in the two reactants. That way, we can clearly see which electrons are involved in the reaction, both from the reactants and from the products.

It then becomes a matter of using curved arrows to show the appropriate movement of those electrons. One curved arrow is drawn from the lone pair on N to the H on water to illustrate the formation of the H—N bond. A second curved arrow originates from the center of the O—H bond to illustrate the breaking of that bond, and points to the O atom, showing that those electrons end up as a lone pair on O.

problem **6.2** Draw the curved arrow notation for the reverse of the reaction in Solved Problem 6.1.

problem **6.3** Draw the curved arrow notation for the proton transfer reaction between NH_3 and H_2O, in which NH_3 acts as the acid and H_2O acts as the base.

6.2 Chemical Equilibrium and the Equilibrium Constant, K_{eq}

Even though a proton transfer reaction can be *written* for any combination of acid and base, not every reaction will actually occur to a significant extent. A proton transfer reaction between HCl and HO^- will take place readily, for example, producing Cl^- and H_2O, but essentially no proton transfer will occur between HO^- and NH_3.

A reaction's tendency to form products is described by its **equilibrium constant, K_{eq}**. An equilibrium constant for a given reaction can be obtained experimentally by allowing that reaction to come to equilibrium (the point at which the concentrations of reactants and products no longer change) and then measuring the concentrations of all products and all reactants. Those equilibrium concentrations are then substituted into the reaction's equilibrium constant expression, which is shown for a generic reaction in Equation 6-4.

$$a\,A + b\,B + c\,C + \cdots \rightleftharpoons w\,W + x\,X + y\,Y + \cdots$$

$$K_{eq} = \frac{[W]_{eq}^{w}[X]_{eq}^{x}[Y]_{eq}^{y} \cdots}{[A]_{eq}^{a}[B]_{eq}^{b}[C]_{eq}^{c} \cdots} \tag{6-4}$$

In this reaction, the uppercase letters are the reactants and products, and the lowercase letters are the coefficients used to balance the equation. The numerator of the K_{eq} expression contains the equilibrium concentrations of the products, whereas the denominator contains the equilibrium concentrations of the reactants. Each concentration is raised to a power specified by the exponent corresponding to its coefficient (i.e., a, b, c... and w, x, y...).

For a proton transfer reaction between HA (the acid) and B^- (the base), the equilibrium constant expression is written as in Equation 6-5.

$$HA + B^- \rightleftharpoons A^- + HB \qquad K_{eq} = \frac{[A^-]_{eq}[HB]_{eq}}{[HA]_{eq}[B^-]_{eq}} \qquad (6\text{-}5)$$

If the equilibrium concentrations of the products are high (and hence the equilibrium concentrations of the reactants are low), then the value of the numerator will be large and the value of the denominator will be small. The converse is true, as well. Thus,

- A very large K_{eq} value (e.g., 10^{10}) heavily favors products.
- A very small K_{eq} value (e.g., 10^{-10}) heavily favors reactants.
- A K_{eq} value close to 1 favors significant concentrations of both reactants and products.

6.2 YOUR TURN

Add the curved arrow notation to each of the following reactions. Based on the K_{eq} values, which reaction tends to form more products at equilibrium?

$K_{eq} = 7.1 \times 10^{-8}$

$K_{eq} = 4.0 \times 10^{-3}$

6.2a Acid Strengths: K_a and pK_a

The strength of an acid is defined as its *tendency to donate a proton*. Acid strengths can be obtained experimentally via the equilibrium that is established between a particular acid (HA) and water, as shown in Equation 6-6. This equilibrium is convenient because water acts as the base *and* the solvent.

$$HA(aq) + H_2O(l) \rightleftharpoons A^-(aq) + H{-}OH_2^+(aq) \qquad K_{eq} = \frac{[A^-]_{eq}[H_3O^+]_{eq}}{[HA]_{eq}[H_2O]_{eq}} \qquad (6\text{-}6)$$

Water is in such a huge excess in this equilibrium that its concentration is effectively constant, at 55.5 mol/L. Therefore, $[H_2O]_{eq}$ can be eliminated from the equilibrium expression without any loss of information, yielding what is called the **acidity constant**, K_a.

$$K_a = \frac{[A^-]_{eq}[H_3O^+]_{eq}}{[HA]_{eq}}$$

As a result, K_{eq} and K_a for any acid are different by a factor of 55.5, the molarity of pure water, as indicated by Equation 6-7.

$$K_a = K_{eq}[H_2O]_{eq} = K_{eq}(55.5 \text{ mol/L}) \tag{6-7}$$

Because all K_a values are obtained with water acting as the base, any difference in K_a for two compounds reflects a difference in acid strength.

> When comparing two acids, the one with the larger K_a value is the stronger acid.

Values of K_a for several common acids encountered in organic chemistry are listed in Table 6-1. (A more comprehensive table can be found in Appendix A.) Notice the incredibly wide range of values—from about 10^{-50} for ethane (CH_3CH_3) to about 10^{13} for trifluoromethanesulfonic acid (CF_3SO_3H). In other words, CF_3SO_3H is 10^{63} times stronger than CH_3CH_3—a number that dwarfs even Avogadro's number (6.02×10^{23})!

Due to the immensely large range of values for K_a, chemists frequently work with values of **pK_a**, which is related to K_a through Equation 6-8.

$$pK_a = -\log(K_a) \tag{6-8}$$

The log function can make very large and very small numbers easier to work with. For example, the pK_a values for the acids in Table 6-1 range from 50 to -13, a difference of only 63.

The negative sign in front of the log function in Equation 6-8 has the effect of reversing the relative values of K_a and pK_a. That is, as the value of K_a increases (becomes more positive), the value of the corresponding pK_a decreases (becomes more negative). Therefore, when comparing pK_a values of two different acids,

> - The acid with the less positive (or more negative) pK_a value is the stronger acid.
> - Each difference of 1 in pK_a values represents a factor of 10 difference in acid strength.

SOLVED problem 6.4 Which of the following is a stronger acid? By what factor?

4-Methylphenol
$pK_a = 10.26$

4-Chlorophenol
$pK_a = 9.43$

Think Is the stronger acid the one with the more positive pK_a or the less positive pK_a? What is the difference in their pK_a values and how can that difference be used to calculate the difference in their acid strengths?

Solve The compound with the less positive pK_a value, 4-chlorophenol, is the stronger acid. The difference in pK_a values is $10.26 - 9.43 = 0.83$, which corresponds to a difference in acid strength of $10^{0.83} = 6.8$. Thus, 4-chlorophenol is 6.8 times stronger an acid than 4-methylphenol.

TABLE 6-1 Values of K_a and pK_a for Various Acids[a]

Acid	Conjugate Base	K_a	pK_a	Acid	Conjugate Base	K_a	pK_a
F₃C—S(=O)(=O)—OH Trifluoromethane-sulfonic acid	F₃C—S(=O)(=O)—O⁻	1×10^{13}	−13	CH₃OH Methanol	CH₃O⁻	3.2×10^{-16}	15.5
HO—S(=O)(=O)—OH Sulfuric acid	HO—S(=O)(=O)—O⁻	1×10^{9}	−9	H₂O Water	HO⁻	2×10^{-16}	15.7
HCl Hydrochloric acid	Cl⁻	1×10^{7}	−7	⌐OH Ethanol	⌐O⁻	1×10^{-16}	16
H₃O⁺ Hydronium ion	H₂O	55	−1.7	⌐OH Propan-2-ol (Isopropyl alcohol)	⌐O⁻	3.2×10^{-17}	16.5
Cl₃C—C(=O)—OH Trichloroethanoic acid (Trichloroacetic acid)	Cl₃C—C(=O)—O⁻	0.17	0.77	⌐OH Methylpropan-2-ol (tert-Butyl alcohol)	⌐O⁻	1×10^{-19}	19
HF Hydrofluoric acid	F⁻	6.3×10^{-4}	3.2	(CH₃)(C=O) Propanone (Acetone)	(CH₂⁻)(C=O)	1×10^{-20}	20
Benzoic acid (C₆H₅COOH)	C₆H₅COO⁻	6.3×10^{-5}	4.2	HC≡CH Ethyne (Acetylene)	HC≡C⁻	1×10^{-25}	25
Ethanoic acid (Acetic acid) CH₃COOH	CH₃COO⁻	1.8×10^{-5}	4.75	C₆H₅NH₂ Aniline (Phenylamine)	C₆H₅NH⁻	1×10^{-27}	27
H₂S Hydrogen sulfide	HS⁻	6.3×10^{-8}	7.2	H₂ Hydrogen gas	H⁻	1×10^{-35}	35
H₄N⁺ Ammonium ion	NH₃	4×10^{-10}	9.4	H₃C—NH—CH₃ N-Methylmethanamine (Dimethylamine)	H₃C—N⁻—CH₃	1×10^{-38}	38
Phenol (C₆H₅OH)	C₆H₅O⁻	1×10^{-10}	10.0	H₂C=CH₂ Ethene (Ethylene)	H₂C=CH⁻	1×10^{-44}	44
H₃C—NH₃⁺ Methylammonium ion	H₃C—NH₂	2.3×10^{-11}	10.63	CH₃CH₂—O—CH₂—CH₃... Ethoxyethane (Diethyl ether)	...O—C⁻H...	$\sim 1 \times 10^{-45}$	~45
F₃C—CH₂—OH 2,2,2-Trifluoroethanol	F₃C—CH₂—O⁻	4×10^{-13}	12.4	CH₄ Methane	H₃C⁻	1×10^{-48}	48
Cl—CH₂CH₂—OH 2-Chloroethanol	Cl—CH₂CH₂—O⁻	1.3×10^{-13}	12.9	CH₃CH₃ Ethane	CH₃CH₂⁻	1×10^{-50}	50

[a] pK_a = −logK_a. The less positive (or more negative) the pK_a value, the stronger the acid relative to another acid.

problem **6.5** Which is a stronger acid, phenol or 4-methylphenol? By what factor? (*Hint:* Consult Table 6-1 and Solved Problem 6.4.)

Phenol   **4-Methylphenol**

problem **6.6** Recall from general chemistry that water undergoes *autoionization*, a reaction in which water acts as both an acid and a base:

$$H_2O(\ell) + H_2O(\ell) \rightleftharpoons H_3O^+(aq) + HO^-(aq)$$

The expression for the autoionization constant for water, K_w, is the product $K_w = [H_3O^+(aq)]_{eq}[HO^-(aq)]_{eq}$. Using water's pK_a value, calculate K_w.

problem **6.7** In the following proton transfer reaction, the hydronium ion (H_3O^+) acts as the acid and water acts as the base.

(a) Write the expression for the equilibrium constant, K_{eq}, according to Equation 6-6 and evaluate it.

(b) Next, evaluate K_a according to Equation 6-7. Convert this value of K_a to pK_a and compare your answer to the value in Table 6-1.

To determine relative *base* strengths, compare the pK_a values of the bases' respective *conjugate acids* and apply the following rule:

The strength of a base decreases as the strength of its conjugate acid increases.

For example, what are the relative base strengths of the hydroxide anion (HO^-) and ammonia (H_3N)? The conjugate acids of these bases are H_2O and H_4N^+, whose pK_a values are 15.7 and 9.4, respectively (Table 6-1). Because H_4N^+ is the stronger of these two acids (it has the less positive pK_a value), H_3N must be a weaker base than HO^-. Furthermore, because the difference in those pK_a values is $15.7 - 9.4 = 6.3$, H_3N is weaker by a factor of $10^{6.3} = 2.0 \times 10^6$.

problem **6.8** Which is a stronger base, Cl^- or the phenoxide anion ($C_6H_5O^-$)? By what factor? Explain. (*Hint:* Consult Table 6-1.)

6.2b Predicting the Outcome of a Proton Transfer Reaction Using pK_a Values

Although pK_a values can be found for a wide variety of acids, many proton transfer reactions do not involve water acting as a base. In these cases, pK_a values can still be used to predict the outcome of a proton transfer reaction—that is, to determine which side of the reaction is favored at equilibrium and by how much.

To make use of pK_a values in this way, keep in mind that there are two acids involved in an equilibrium for a proton transfer reaction: the acid on the left side of the equation (a reactant) and the conjugate acid of the base on the right side (a product) (Equation 6-9).

HA drives the reaction to the right. → ← HB drives the reaction to the left.

$$B:^{\ominus} + \quad H{-}A \quad \rightleftharpoons \quad B{-}H \quad + \quad :A^{\ominus} \qquad (6\text{-}9)$$

Base   Acid   Conjugate acid   Conjugate base

The two acids can be thought of as being in competition with each other. The acid on the left drives the reaction to the right, while the acid on the right drives the reaction to the left. The outcome of this competition is dictated by the stronger acid:

Proton transfer reactions favor the side *opposite* the stronger acid.

For example, consider the reaction in which CH_3CO_2H donates a proton to CH_3NH_2 (Equation 6-10).

Stronger acid

Weaker acid

$pK_a = 4.75$

$pK_a = 10.63$

$K_{eq} = 7.6 \times 10^5$ (6-10)

Equilibrium favors the side opposite the stronger acid.

The conjugate acid on the product side is $CH_3NH_3^+$ and the conjugate base is $CH_3CO_2^-$. The pK_a of CH_3CO_2H is 4.75 and the pK_a of $CH_3NH_3^+$ is 10.63 (Table 6-1). Because the pK_a value of CH_3CO_2H is less positive, it is a stronger acid than $CH_3NH_3^+$. Thus, the product side of this proton transfer reaction is favored over the reactant side. This is signified in Equation 6-10 with an equilibrium arrow that is longer in the forward direction than in the reverse direction ($\rightleftharpoons$).

The extent to which a proton transfer reaction is favored—that is, the equilibrium constant for the proton transfer reaction—is related to the *difference* in pK_a values of the two acids involved in the reaction (see Problem 6.11):

$$K_{eq}(\text{proton transfer}) = 10^{[pK_a\,(\text{product acid})\,-\,pK_a\,(\text{reactant acid})]} = 10^{(\Delta pK_a)} \qquad (6\text{-}11)$$

The difference in pK_a values between the product and reactant acids (ΔpK_a) appears in the exponent because, as we learned earlier, every difference of 1 in the value of pK_a corresponds to a factor of 10 in acid strength. For the reaction in Equation 6-10, for example, $\Delta pK_a = 10.63 - 4.75 = 5.88$. Thus, the equilibrium constant for the reaction, K_{eq}, is $10^{5.88}$, or 7.6×10^5. In other words, the proton transfer reaction in Equation 6-10 favors products over reactants by a factor of 7.6×10^5.

SOLVED problem 6.9 Predict which side of the following reaction is favored. To what extent is that side favored?

Think What acid is present on each side of the reaction? Which one is stronger? What is the difference in their pK_a values?

Solve The acid on the left is propanone (acetone), $(CH_3)_2C$=O, whose pK_a value is 20. The conjugate acid on the right is methylpropan-2-ol (*tert*-butyl alcohol), $(CH_3)_3COH$, whose pK_a is 19. Therefore, the acid on the right is stronger (lower pK_a), making the left side of the equilibrium favored. Using Equation 6-11, $K_{eq} = 10^{(19-20)} = 10^{-1} = 0.1$. Thus, the left side of the equation is favored over the right by a factor of 10.

---

problem **6.10** Repeat Solved Problem 6.9 using $(CH_3)_2N^-$ as the base.

---

problem **6.11** Derive the result in Equation 6-11. (*Hint:* The expression for K_{eq} is obtained from the reaction in Equation 6-9. You will need to use Equation 6-8 to substitute for pK_a. You will also need to use the following two properties: $10^{\log x} = x$ and $\log x + \log y = \log xy$.)

6.2c The Leveling Effect

Even if a solvent does not act as an acid or a base in an *intended* proton transfer reaction, it can limit the existence of certain acids and bases in solution. These restrictions are the result of what is called the **leveling effect**.

To understand the leveling effect, let's examine the proton transfer equilibria in Equations 6-12 and 6-13, which involve water, a solvent that can act as either an acid or a base.

(6-12)

(6-13)

In Equation 6-12, HCl acts as an acid and water acts as a base in the forward direction, whereas H_3O^+ is the acid and Cl^- is the base in the reverse direction. HCl is a stronger acid than H_3O^+ (see Table 6-1), so the right side of the equation is favored. In other words, HCl cannot exist to a significant extent in water. Instead, HCl is consumed by its reaction with water and produces H_3O^+.

> **YOUR TURN 6.3**
>
> What are the pK_a values of HCl and H_3O^+ in Equation 6-12, and do they verify that HCl is a stronger acid than H_3O^+?

In Equation 6-13, water is the acid in the forward direction and $(CH_3)_2NH$ is the acid in the reverse direction. According to Table 6-1, water is a stronger acid than $(CH_3)_2NH$, [making $(CH_3)_2N^-$ the stronger base] so the right side of the reaction is again favored. We conclude, therefore, that $(CH_3)_2N^-$ cannot exist in water to any appreciable concentration. It is consumed by its reaction with water to produce OH^- as the base in solution.

> **YOUR TURN 6.4**
>
> What are the pK_a values of H_2O and $(CH_3)_2NH$ in Equation 6-13, and do they verify that H_2O is a stronger acid than $(CH_3)_2NH$?

These two examples demonstrate water's *leveling effect* as follows:

> In water, no acid stronger than H_3O^+ and no base stronger than HO^- can exist to any appreciable extent.

The leveling effect applies in other solvents as well:

> - The strongest acid that can exist in solution to any appreciable concentration is the protonated solvent.
> - The strongest base that can exist in solution is the deprotonated solvent.

If an acid stronger than the protonated solvent is dissolved, then it will be consumed by its reaction with the solvent acting as the base, generating the protonated solvent as the predominant acid in solution. Similarly, if a base stronger than the deprotonated solvent is dissolved in solution, then it will be consumed by its reaction with the solvent acting as an acid, generating the deprotonated solvent as the predominant base in solution.

The take-home lesson here is that a solvent for a reaction must be chosen wisely. For example, if we want to use $(CH_3)_2N^-$ as a reagent for a given reaction, then water is a poor choice as a solvent: water will consume $(CH_3)_2N^-$ according to Equation 6-13. A better choice of solvent would be a compound that is much less acidic, such as diethyl ether ($CH_3CH_2OCH_2CH_3$).

6.5 YOUR TURN

Verify the preceding statement that diethyl ether would be a suitable solvent for $(CH_3)_2N^-$. To do so, use Table 6-1 to fill in the boxes below with the appropriate pK_a values and label which one is the stronger acid. Indicate which side of the reaction is favored at equilibrium. Is this the same side that is favored in Equation 6-13?

problem **6.12** With respect to the leveling effect, determine whether each of the following solvents would be suitable for a reaction involving $HC\equiv C:^-$ as a reactant? (*Hint:* Is it desirable for HCC^- to react with the solvent?)

H_2O (a)   (b)   (c)   (d)   (e)

6.2d Le Châtelier's Principle, pH, and Percent Dissociation

A chemical equilibrium is a *stationary state* of a reaction, in which the concentrations of reactants and products do not change with time. Even after equilibrium has been established, however, the reaction can be driven in the forward or reverse direction by altering the concentrations of the species involved in the reaction. This response by a chemical equilibrium is one consequence of **Le Châtelier's principle**, which states that *if a reaction at equilibrium experiences a change in reaction conditions (e.g., concentrations, temperature, pressure, or volume), then the equilibrium will shift to counteract that change.* Specifically,

- A reaction at equilibrium can be shifted in the forward direction (i.e., to form more products) by the addition of reactants or the removal of products.
- A reaction at equilibrium can be shifted in the reverse direction (i.e., to form more reactants) by the addition of products or the removal of reactants.

In an acid's equilibrium with water (Equation 6-6), for example, the concentration of $H_3O^+(aq)$—a product of the reaction—directly affects the relative amounts of HA and A^- at equilibrium. Increasing the concentration of H_3O^+ (e.g., by adding a strong acid) will cause the equilibrium to shift toward reactants. At the new equilibrium, there will be more HA and less A^-, so the **percent dissociation** of HA (Equation 6-14) is said to decrease.

$$\% \text{ Dissociation} = \frac{[A^-]_{eq}}{[HA]_{initial}} \times 100 = \frac{[A^-]_{eq}}{[HA]_{eq} + [A^-]_{eq}} \times 100 \qquad (6\text{-}14)$$

Likewise, decreasing the concentration of H_3O^+ (e.g., by adding a strong base) will cause the equilibrium to shift toward products, leading to an increase in the percent dissociation of HA.

The concentration of H_3O^+ in solution is commonly reported as pH:

$$\text{pH} = -\log[H_3O^+(aq)] \qquad (6\text{-}15)$$

A solution is considered *acidic* if its pH is <7, *basic* if its pH is >7, and *neutral* if its pH equals 7. Furthermore, pH increases as the concentration of H_3O^+ decreases. Thus,

In the equilibrium between an acid (HA) and water, increasing the solution's pH causes an increase in the percent dissociation of the acid into its conjugate base (A^-).

This can be seen graphically in Figure 6-1.

FIGURE 6-1 Percent dissociation of an acid with pK_a = 5 as a function of pH As the pH of the solution increases (i.e., as the solution becomes more basic), the percent dissociation of the acid (HA) increases. When the pH of the solution is equal to the pK_a of the acid, the acid has dissociated 50%.

Caution: Hydrofluoric Acid Is a Weak Acid

With a pK_a of -7, hydrochloric acid (HCl) is a strong acid and will protonate many substances with which it comes in contact, including your skin. As long as the exposure is minor enough, however, damage will be limited primarily to the skin's outer layers. HCl's cousin, hydrofluoric acid (HF), has a pK_a of 3.2, so it is a weak acid. You might think, therefore, that HF is less dangerous than HCl, but you would be mistaken: Concentrated HF is more hazardous than concentrated HCl, and, ironically, its weak acidity is partly what makes HF so dangerous!

Because HF is a weak acid, it will remain largely unreacted when it comes in contact with your skin. Those unreacted HF molecules will form rather stable dimers, $(HF)_2$, which are lipophilic. This allows HF to penetrate your skin and enter the bloodstream somewhat easily. Once in the blood, HF is a potent poison. It not only destroys tissue, but it also decalcifies bone and damages the central nervous system. Because of this mode of attack, exposure of concentrated HF to as little as 2% of the body's surface area has been known to cause death.

Despite HF's immense health hazards, it remains an integral part of some highly valuable industries—most notably, the semiconductor industry. HF has the unusual ability to etch silicon dioxide (SiO_2) according to the following reaction.

$$SiO_2(s) + 6\ HF(aq) \longrightarrow H_2SiF_6(aq) + 2\ H_2O(\ell)$$

Thus, it is used in the process of fabricating intricate nanoscale structures on silicon wafers to produce powerful integrated circuits.

Of particular significance in Figure 6-1 is the pH at which the acid dissociates 50%. This significance can best be seen with the Henderson–Hasselbalch equation (Equation 6-16), which is commonly applied in general chemistry toward problems dealing with buffers:

Henderson–Hasselbalch equation

$$pH = pK_a + \log\left[\frac{[A^-(aq)]_{eq}}{[HA(aq)]_{eq}}\right] \qquad (6\text{-}16)$$

At 50% dissociation of the acid, the equilibrium concentrations of $A^-(aq)$ and $HA(aq)$ are equal. This makes the second term in Equation 6-16 equal to zero, in which case $pH = pK_a$.

> An acid (HA) dissolved in water dissociates 50% (into A^-) when the solution's pH is equal to the acid's pK_a.

Thus, the graph in Figure 6-1 represents an acid whose pK_a is 5. We will find in Section 6.11 that this general relationship is useful in *gel electrophoresis*, a method used to analyze mixtures of amino acids and proteins.

6.6 YOUR TURN

In Figure 6-1, how many pH units above the acid's pK_a must the solution be to cause the acid to dissociate nearly 100%? _____
How many pH units below the acid's pK_a must the solution be for the acid to be nearly 100% associated? _____

In Figure 6-1, sketch a plot of the percent dissociation of an acid with $pK_a = 9$ as a function of pH.

6.3 Thermodynamics and Gibbs Free Energy

In Section 6.2, we saw that the value of a reaction's equilibrium constant (K_{eq}) increases as more products are formed at equilibrium. As you may recall from general chemistry, the equilibrium constant is also related to the **standard Gibbs free energy difference** between reactants and products, ΔG°_{rxn}, as shown in Equation 6-17.

$$G^\circ_{products} - G^\circ_{reactants} = \Delta G^\circ_{rxn} = -RT \ln K_{eq} \qquad (6\text{-}17)$$

Here R is the universal gas constant (8.314 J/mol · K; 1.987 cal/mol · K), T is the temperature in kelvins, and K_{eq} is the equilibrium constant calculated from the expression in Equation 6-4. The naught (°) signifies *standard conditions*, in which all pure substances are in their most stable states at 298 K, the partial pressures of all gases are 1 atm, and the concentrations of all solutions are 1 mol/L.

With the relationship in Equation 6-17, we can associate ΔG°_{rxn} with the reaction's tendency to favor products at equilibrium (i.e., to be *spontaneous*).

- A reaction tends to be spontaneous if $\Delta G^\circ_{rxn} < 0$. Such a reaction is said to be **exergonic** (energy out), because it releases free energy.
- A reaction tends to be nonspontaneous (i.e., the reverse reaction is spontaneous) if $\Delta G^\circ_{rxn} > 0$. Such a reaction is said to be **endergonic** (energy in), because it absorbs free energy.

Because R and T are both positive, a negative value for ΔG°_{rxn} results in $K_{eq} > 1$, in which case products are favored over reactants. On the other hand, a positive value for ΔG°_{rxn} means that $K_{eq} < 1$, in which case reactants are favored over products.

Knowing the value of ΔG°_{rxn} for a given reaction can be helpful further, because it conveys information about the *stability* of the products relative to the reactants. If $\Delta G^\circ_{rxn} < 0$, then the standard Gibbs free energy of the products ($G^\circ_{products}$) must be lower than that of the reactants ($G^\circ_{reactants}$). Conversely, if $\Delta G^\circ_{rxn} > 0$, then $G^\circ_{products}$ is higher than $G^\circ_{reactants}$. Thus,

- If $\Delta G^\circ_{rxn} < 0$, then products are more stable than reactants.
- If $\Delta G^\circ_{rxn} > 0$, then products are less stable than reactants.

6.3a Enthalpy and Entropy

The ΔG°_{rxn} can be expressed in terms of the reaction's *enthalpy change* and *entropy change*, as follows:

$$\Delta G^\circ_{rxn} = \Delta H^\circ_{rxn} - T\Delta S^\circ_{rxn} \qquad (6\text{-}18)$$

The ΔH°_{rxn} term in Equation 6-18 is the **standard enthalpy difference** between the reactants and products. At constant pressure, ΔH°_{rxn} equals the heat absorbed or released by the reaction. If ΔH°_{rxn} is positive, then the reaction absorbs heat and is said to be **endothermic**; if ΔH°_{rxn} is negative, then the reaction releases heat and is **exothermic**.

In the second term on the right side of Equation 6-18, T is the temperature in units of kelvin and ΔS°_{rxn} is the **standard entropy difference** between the reactants and products. As mentioned in Section 2.7, entropy is related to the number of different states available to a system and is often thought of as a "measure of disorder."

For a proton transfer reaction, the entropy term in Equation 6-18 is usually quite small. This is because the most significant factor that governs ΔS°_{rxn} is the number of independent species on each side of the reaction, and, as we can see from the general proton transfer reaction below, both the reactant side and the product side contain two separate species. Therefore, the enthalpy term in Equation 6-18 generally dominates and $\Delta G^\circ_{rxn} \approx \Delta H^\circ_{rxn}$.

ΔS°_{rxn} is small because both reactants and products have the same number of species.

$$B:\frown H-A \rightleftharpoons B-H \ + \ :A^\ominus \qquad \Delta G^\circ_{rxn} \approx \Delta H^\circ_{rxn}$$

For most other reactions we will encounter, ΔH°_{rxn} also governs whether ΔG°_{rxn} is positive or negative. In situations where ΔS°_{rxn} is significant, however, the sign of ΔG°_{rxn} depends upon the signs of both ΔH°_{rxn} and ΔS°_{rxn}, as well as the temperature. We will discuss this idea further in the context of the reactions to which it is pertinent.

6.3b The Reaction Free Energy Diagram

So far, we have considered only the end points of the proton transfer reaction—namely, the reactants and the products. However, the changes that the reactants undergo to become products are continuous, not instantaneous. To help us discuss these changes, chemists often use free energy diagrams.

In a **reaction free energy diagram**, Gibbs free energy is plotted as a function of the *reaction coordinate*. A **reaction coordinate** is a variable that corresponds to the changes in geometry, on a molecular level, as reactants are converted into products. The way in which a reaction coordinate is quantified is unique to each individual reaction and can be a complicated function of bond angles and bond lengths. In general, however, we may say the following:

As the reaction coordinate increases, the geometries of the species involved in the reaction increasingly resemble those of the products.

This becomes clearer if we examine the specific examples in Figure 6-2. Figure 6-2a is the free energy diagram for the proton transfer between HO^- and HCl, whereas Figure 6-2b is the diagram for the reverse reaction, between Cl^- and H_2O.

FIGURE 6-2 Free energy diagrams (a) Free energy diagram for the proton transfer between HCl and HO^-. The reaction is exergonic ($\Delta G^\circ_{rxn} < 0$), but it still has an energy barrier. (b) Free energy diagram for the proton transfer between Cl^- and H_2O, an endergonic reaction ($\Delta G^\circ_{rxn} > 0$).

For the reaction in Figure 6-2a, an H—Cl bond is broken during the course of the reaction and an O—H bond is formed. Therefore, we can say that as the reaction coordinate increases, the distance between the H and Cl atoms increases and the distance between the H and O atoms decreases.

YOUR TURN 6.8

In Figure 6-2b, as the reaction coordinate increases, does the distance between Cl and H increase or does it decrease? _____ Does the distance between the O and H increase or decrease? _____

One of the main benefits of free energy diagrams is that they allow us to see quickly whether the products are more stable or less stable than the reactants. In Figure 6-2a, for example, the products are lower in energy than the reactants (the reaction is exergonic), signifying that the products are more stable; this difference in energy, ΔG°_{rxn}, is indicated by the blue double-headed arrow. For the reverse reaction (Fig. 6-2b), the reactants are more stable than the products.

Notice in Figure 6-2a and 6-2b that there is an *energy maximum along the reaction coordinate*. The structure that corresponds to this energy maximum is the elementary step's **transition state**. Because there is a maximum in energy between the reactants and products, an *energy barrier* must be surmounted to form products. *This is true regardless of whether the elementary step is exergonic (Fig. 6–2a) or endergonic (Fig. 6–2b).* This energy barrier, called the **free energy of activation** ($\Delta G^{\circ\ddagger}$, "delta-*G*-naught-double-dagger"), is the difference in standard free energy between the reactants and the transition state. As we discuss in greater detail in Chapter 9, $\Delta G^{\circ\ddagger}$ is an important factor that governs reaction rates.

YOUR TURN 6.9

Indicate ΔG°_{rxn} and $\Delta G^{\circ\ddagger}$ in Figure 6-2b the way it is done in Figure 6-2a. Which reaction has a greater $\Delta G^{\circ\ddagger}$, the one in Figure 6-2a or the one in Figure 6-2b? _____

6.4 Strategies for Success: Functional Groups and Acidity

Although the pK_a values for several different compounds are listed in Table 6-1, they represent only a small fraction of the millions of compounds known. There is a good chance, then, that the compound for which you need to know the pK_a value is not included in the table. How can we obtain the pK_a values for those other compounds?

One way is to *estimate* the value based on structural similarities to a compound that is listed in the table. This may seem to be a formidable task at first, due to the many structural variations of the compounds listed in Table 6-1. Recall from Section 1.13, however, that the chemical behavior of a compound is governed by the functional groups it possesses. The functional group's behavior is relatively independent of the overall structure of the molecule in which it is found. This applies to proton transfer reactions as follows:

The acidity (pK_a) of a compound is governed largely by the functional group on which the acidic proton is found.

For example, the pK_a of ethanol (CH$_3$CH$_2$OH) is 16 and the pK_a of propan-2-ol [isopropyl alcohol, (CH$_3$)$_2$CHOH] is 16.5 (Fig. 6-3a). Their pK_a values are similar because their acidic protons are both part of *alcohol* functional groups, R—OH. Likewise, ethanoic acid (acetic acid, CH$_3$CO$_2$H) and phenylmethanoic acid (benzoic acid, C$_6$H$_5$CO$_2$H) have similar pK_a values, at 4.75 and 4.20, respectively, because their acidic protons are both part of *carboxylic acid* functional groups, R—CO$_2$H (Fig. 6-3b).

Carboxylic acids are substantially more acidic than alcohols, even though both have their acidic protons on an oxygen atom. Ethanoic acid (acetic acid, CH$_3$CO$_2$H) has a pK_a of 4.75, whereas ethanol (CH$_3$CH$_2$OH) has a pK_a of 16—a difference in acid strength of $>10^{11}$. Thus, the presence of a carbonyl group (C=O) significantly increases the acidity of an adjacent OH group.

Although acidity is governed primarily by which functional group the acidic proton is on, some nearby structural features, such as a highly electronegative substituent, can alter the acidity significantly. Trichloroacetic acid (Cl$_3$CCO$_2$H; pK_a = 0.77) is a stronger acid than acetic acid (H$_3$CCO$_2$H; pK_a = 4.75) by nearly four pK_a units—a factor of 10^4 in acid strength (Fig. 6-4a). The presence of a double bond adjacent to the atom containing the acidic proton can significantly increase the acidity, too. Phenol (C$_6$H$_5$OH; pK_a = 10), for example, is more acidic than ethanol (CH$_3$CH$_2$OH; pK_a = 16) by a factor of about 10^6 (Fig. 6-4b).

The reasons for these structural effects on pK_a are discussed in greater detail in Sections 6.5 and 6.6. Based on these few examples, however, we can quickly recognize when *not* to expect acidic protons on the same functional group to have similar acidities.

Both molecules are alcohols, R—OH.

pK_a = 16   pK_a = 16.5

Ethanol   **2-Propanol (Isopropyl alcohol)**

(a)

Both molecules are carboxylic acids, R—CO$_2$H.

pK_a = 4.75   pK_a = 4.20

Ethanoic acid (Acetic acid)   **Phenylmethanoic acid (Benzoic acid)**

(b)

FIGURE 6-3 Functional groups and pK_a (a) The acidic proton (in blue) in both molecules is part of an alcohol functional group, and the pK_a values are similar. (b) The acidic proton (in blue) in both molecules is part of a carboxylic acid functional, and the pK_a values are similar. The pK_a value of a carboxylic acid, however, is significantly different from that of an alcohol.

SOLVED problem 6.13 Using Table 6-1, estimate the pK_a for the NH$_2$ protons in cyclohexylamine.

Cyclohexylamine

Think On what functional group do the H atoms appear? What molecule(s) in Table 6-1 have the same functional group? Are there any nearby electronegative atoms or adjacent double bonds?

Solve The H atom in question is part of an amine functional group. According to Table 6-1, the pK_a of (CH$_3$)$_2$NH, another amine, is 38. Because cyclohexylamine does not have any nearby electronegative atoms or adjacent double bonds, we estimate its pK_a to be ~38, too.

Nearby electronegative atoms significantly affect pK_a.

pK_a = 4.75   pK_a = 0.77

Ethanoic acid (Acetic acid)   **Trichloroethanoic acid (Trichloroacetic acid)**

(a)

Adjacent double bond significantly affects pK_a.

pK_a = 16   pK_a = 10

Ethanol   **Phenol**

(b)

FIGURE 6-4 Effects on pK_a from nearby structural features (a) The nearby electronegative Cl atoms significantly lower the pK_a of the proton in blue. (b) The adjacent double bond significantly lowers the pK_a of the proton in blue.

problem **6.14** Using Table 6-1 and/or Appendix A, estimate the pK_a for the proton explicitly shown on each of the following compounds.

(a)　　　　　(b)　　　　　(c)　　　　　(d)　　　　　(e)

6.5 Relative Strengths of Charged and Uncharged Acids: The Reactivity of Charged Species

In Table 6-1, we saw that acid strengths can vary by as much as a factor of 10^{63}. There are a number of clear trends in these values, some of which were mentioned in Section 6.4. Here in Section 6.5 and in Section 6.6, we discuss these trends in greater detail, as well as the reasons behind them.

Perhaps the most conspicuous trend from Table 6-1 is that involving the charge on the atom to which a proton is attached.

> A proton is significantly more acidic when the atom to which it is attached is positively charged than when that atom is uncharged.

For example, the pK_a of H_3O^+ is -1.7, whereas that of H_2O is 15.7. Additionally, the pK_a of H_4N^+ is 9.4, whereas that of H_3N is 36. Why should the presence of a charge have this effect?

To begin to answer this question, consider Equation 6-19, the *autoionization* equilibrium for water.

$$H_2\overset{..}{O}:(aq) \; + \; H\overset{.}{-}\overset{..}{O}H(aq) \; \longrightarrow \; H_2\overset{\oplus}{O}\text{—}H(aq) \; + \; \overset{\ominus}{\overset{..}{:}}\overset{..}{O}H(aq) \qquad \begin{array}{l} K_{eq} \approx 10^{-14} \\ \Delta G^\circ_{rxn} \approx +80 \text{ kJ/mol} \end{array} \qquad \text{(6-19)}$$

The equilibrium constant for the autoionization of water is very small ($K_{eq} \approx 10^{-14}$), which corresponds (via Equation 6-17) to a substantially positive change in Gibbs free energy ($\Delta G^\circ_{rxn} \approx +80$ kJ/mol). See the free energy diagram in Figure 6-5.

The charged products are significantly less stable than the uncharged reactants.

$H_2\overset{\oplus}{O}\text{—}H \; + \; \overset{\ominus}{:}\overset{..}{O}H$

$\Delta G^\circ_{rxn} = +80$ kJ/mol

$H_2\overset{..}{O}: \; + \; H\text{—}\overset{..}{O}H$

Gibbs free energy →

Reaction coordinate →

FIGURE 6-5 Free energy diagram for the autoionization of H_2O (Equation 6-19) The charged products are much higher in energy than the uncharged reactants, suggesting that the charged species are significantly less stable.

The H_3O^+ and HO^- products, which are charged, are less stable than the uncharged reactants by 80 kJ/mol.

> Generally speaking, a charged species is significantly higher in energy (and thus less stable) than its uncharged counterpart.

With this in mind, we can understand the trend among charged and uncharged acids, mentioned previously. Consider the reactions shown in Equation 6-20a and 6-20b, in which water deprotonates NH_3 and H_4N^+, respectively.

The reactants in Equation 6-20a are both uncharged, whereas the products are both charged. Thus, this reaction has a significantly positive ΔG_{rxn}°, similar to the reaction in Equation 6-19. In Equation 6-20b, on the other hand, one reactant and one product are charged, so no additional charges are introduced during the course of the reaction. For this reason, the deprotonation of H_4N^+ (Equation 6-20b) is not as difficult energetically as the deprotonation of NH_3 (Equation 6-20a), so H_4N^+ is more acidic than NH_3.

Figure 6-6 compares the reaction free energy diagrams for the deprotonation of H_4N^+ and NH_3. The reactants from Equation 6-20b (the blue curve in Fig. 6-6) are higher in energy than the reactants from Equation 6-20a (the red curve in Fig. 6-6), because H_4N^+, which carries a +1 charge, is less stable than H_3N, which is uncharged. Moreover, the products of Equation 6-20a (red curve) are higher in energy than those of Equation 6-20b (blue curve), because NH_2^-, a charged species, is less stable than

FIGURE 6-6 Energy diagrams for the deprotonation of NH_3 (red) and H_4N^+ (blue) by water The reactants in Equation 6-20a ($NH_3 + H_2O$; red curve) are lower in energy than the reactants in Equation 6-20b ($H_4N^+ + H_2O$; blue curve) because they have fewer charges. Similarly, the products in Equation 6-20b ($NH_3 + H_3O^+$; blue curve) are lower in energy than the products in Equation 6-20a ($H_2N^- + H_3O^+$; red curve) because they have fewer charges. For the red curve, the products are much less stable than the reactants because the products have two additional charges. For the blue curve, the reactants and products have more similar energies because they each have one total charge. The deprotonation of H_4N^+ is more energetically favorable than the deprotonation of NH_3, so H_4N^+ is a stronger acid than NH_3.

H_3N. Finally, the products for Equation 6-20a (red curve) carry two total charges and appear much higher in energy than the reactants, which are both uncharged. For Equation 6-20b (blue curve), on the other hand, the reactants and products are more similar in energy, because they each carry one total charge. The energy diagrams in Figure 6-6 show, therefore, that the deprotonation of H_4N^+ (blue curve) is much more energetically favorable than the deprotonation of NH_3 (red curve).

problem **6.15** Draw an energy diagram similar to Figure 6-6, using H_3O^+ and H_2O as the acids. Based on the energy diagram, which acid is predicted to be stronger? Is this consistent with their relative pK_a values?

6.6 Relative Acidities of Protons on Atoms with Like Charges

The analysis in Section 6.5 explains the greater acidity of a positively charged species relative to a comparable uncharged molecule, but we have yet to explain the relative acidities of species bearing the *same* charge. As we explain here in Section 6.6, there are a variety of trends to examine. In Sections 6.6a and 6.6b, for example, we discuss how the identity of an atom affects the acidity of protons attached to it; in Section 6.6c, we describe how hybridization affects acidity; and in Sections 6.6d and 6.6e, we explain the effects of nearby π bonds and nearby atoms.

6.6a Protons on Different Atoms in the Same Row of the Periodic Table

The acidic protons in CH_4, H_3N, H_2O, and HF are bonded to different atoms in the second row of the periodic table—namely, C, N, O, and F, respectively. The pK_a values of these acids are 48, 36, 15.7, and 3.2, respectively.

> The farther to the right an atom appears in the periodic table, the more acidic the protons that are attached to it.

To better understand the basis for such a trend, consider the equilibria in which H_3N and CH_4 are deprotonated by water.

$$H_2N{-}H \quad + \quad H_2\ddot{O}: \quad \rightleftharpoons \quad H_2\ddot{N}^{\ominus} \quad + \quad H_3O^{\oplus} \qquad \text{(6-21a)}$$

$$H_3C{-}H \quad + \quad H_2\ddot{O}: \quad \rightleftharpoons \quad H_3\ddot{C}^{\ominus} \quad + \quad H_3O^{\oplus} \qquad \text{(6-21b)}$$

The free energy diagrams for these two reactions are shown in Figure 6-7. Just as we saw in Figure 6-6, the deprotonation of H_3N by water in Figure 6-7 (Equation 6-21a, red curve) is energetically unfavorable due to the addition of two charges during the course of the reaction. The same is true for the deprotonation of CH_4 by water (Equation 6-21b, blue curve). However, because H_3N is a stronger acid than CH_4, the deprotonation of H_3N is more favorable than the deprotonation of CH_4.

H_2O and H_3O^+ have no impact on why the deprotonation of H_3N is more energetically favorable than the deprotonation of CH_4, because H_2O is a reactant and H_3O^+ is a product in both reactions. As a result, their contributions to both reactions are the same.

FIGURE 6-7 Relative stabilities of H_3C^- and H_2N^- Because H_3N is a stronger acid than CH_4, the reaction in Equation 6-21a (the red curve) is more favorable than the reaction in Equation 6-21b (the blue curve). Thus, H_2N^- is more stable than H_3C^-.

Factors that affect stability have a substantially greater influence on a charged species than they do on an uncharged species. In other words, compared to charged species, structural variations in an uncharged species should lead to relatively small changes in stability. Carrying this further,

> When constructing free energy diagrams for reactions that involve both charged and uncharged species, assume that the various uncharged species have similar stabilities.

Applying this idea to Figure 6-7, notice that the difference between the two sets of reactants is in the acids themselves, H_3N and CH_4. Because both are uncharged, both sets of reactants are placed at essentially the same energy in the diagram. Thus, for the deprotonation of H_3N to be more energetically favorable than the deprotonation of CH_4, the products of the deprotonation of H_3N must be more stable than the products of the deprotonation of CH_4. This is indicated in Figure 6-7 by the fact that the products of Equation 6-21a (red curve) appear below the products of Equation 6-21b (blue curve). We can therefore conclude that H_2N^- is more stable than H_3C^-.

Why should H_2N^- be more stable than H_3C^-? The answer lies in the difference between the atoms on which the charge resides. As we know from Section 1.7, N has a higher electronegativity than C, meaning that N has a stronger attraction for electrons (i.e., for negative charge). Thus, whereas both species are destabilized by the addition of a negative charge, the destabilization of N is not as great as it is of C.

A similar analysis shows that H_2N^- is less stable than HO^-, which, in turn, is less stable than F^-. Thus,

> The stability of an anion increases as the electronegativity of the atom bearing the negative charge increases.

Negative charges on atoms in the same row of the periodic table

Increasing electronegativity of atom bearing the negative charge

→

H_3C^- H_2N^- HO^- F^-

Increasing stability of the anion

The same trend is observed among acids with acidic protons bonded to atoms in other rows of the periodic table. For example, HCl is a stronger acid than H_2S because Cl is more electronegative than S, allowing Cl to accommodate a negative charge better.

> Verify that HCl is a stronger acid than H_2S by looking up their pK_a values in Table 6-1. HCl _____ H_2S _____

6.10 **YOUR TURN**

The preceding analysis involving uncharged acids applies equally well to positively charged acids. H_3O^+, for example, is a stronger acid than H_4N^+.

YOUR TURN 6.11

Verify that H_3O^+ is a stronger acid than H_4N^+ by looking up their pK_a values in Table 6-1. H_3O^+ _____ H_4N^+ _____

The proton transfer equilibria for H_3O^+ and H_4N^+ with water are shown in Equation 6-22a and 6-22b.

Stronger acid $H_2O^+\!-\!H$ + $H_2\ddot{O}:$ ⇌ $H_2\ddot{O}:$ + $H_3O^{\oplus}$ (6-22a)

Weaker acid $H_3N^+\!-\!H$ + $H_2\ddot{O}:$ ⇌ $\ddot{N}H_3$ + $H_3O^{\oplus}$ (6-22b)

Because H_3O^+ is more acidic than H_4N^+, we know that the reaction in Equation 6-22a is more favorable than the reaction in Equation 6-22b, as shown in the free energy diagrams in Figure 6-8. Notice that Figure 6-8 differs somewhat from Figure 6-7. In Figure 6-8, the acids—H_3O^+ and H_4N^+—are charged, so they have different stabilities. The products, on the other hand, appear at essentially the same energy because the difference between the two sets of products is in the conjugate bases—H_3N and H_2O—both of which are uncharged (and are therefore assumed to have similar stabilities). On the reactant side, therefore, H_4N^+ is more stable than H_3O^+.

The difference in stability between H_4N^+ and H_3O^+ can be attributed, once again, to a difference in electronegativity. The O atom bearing the positive charge in H_3O^+ is more electronegative than the N atom bearing the positive charge in H_4N^+. Thus, for the same reason that O can accommodate a *negative* charge better than N, N can accommodate a *positive* charge better than O.

The relative stabilities of H_3O^+ and H_4N^+ are consistent with the following periodic trend for the stabilities of positively charged ions:

Positive charges on atoms in the same row of the periodic table

Increasing electronegativity of atom bearing the positive charge

$H_4N^{\oplus}$ $H_3O^{\oplus}$ $H_2F^{\oplus}$

Increasing stability of the cation

The stability of a cation decreases as the electronegativity of the atom bearing the positive charge increases.

FIGURE 6-8 Relative stabilities of H_3O^+ and H_4N^+ H_3O^+ is a stronger acid than H_4N^+, so the reaction in Equation 6-22a (the red curve) is more favorable than the reaction in Equation 6-22b (the blue curve). Thus, H_4N^+ is more stable than H_3O^+.

6.6b Protons on Different Atoms in the Same Column of the Periodic Table

HCl ($pK_a = -7$) is a *stronger* acid than HF ($pK_a = 3.2$), which means that Cl⁻ is *more* stable than F⁻. F is *more* electronegative than Cl, however, in which case you might expect that F could accommodate a negative charge better than Cl.

To untangle this apparent discrepancy, recall that *Cl is a substantially larger atom than F*, because Cl is one row below F in the periodic table. A negative charge on Cl is therefore less concentrated than a negative charge on F because the same -1 charge is spread out over a larger volume in Cl. This lower concentration of charge contributes to the greater stability of Cl⁻ than F⁻.

The electrostatic potential maps of F⁻ and Cl⁻, shown in Figure 6-9, are consistent with this analysis. The electron cloud of F⁻ is a deep red, whereas the electron cloud of Cl⁻ is orange, which suggests that the concentration of charge is higher in F⁻ than in Cl⁻.

This trend continues down group 17 of the periodic table, because HBr ($pK_a = -9$) is a stronger acid than HCl ($pK_a = -7$). Br⁻ is more stable than Cl⁻, because Br is larger than Cl, so the negative charge on Br⁻ is more spread out than it is on Cl⁻. The following general trend applies to other columns of the periodic table as well, as shown in Solved Problem 6.16:

> The stability of an anion tends to increase when the negative charge is on an atom farther down a column of the periodic table.

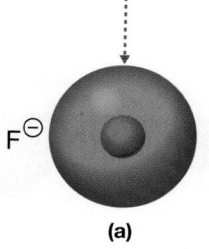

- Smaller volume
- More concentrated negative charge
- Less stable

F⊖

(a)

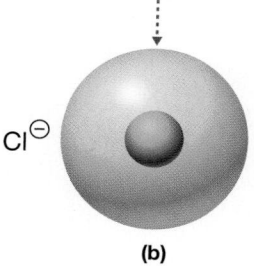

- Larger volume
- Less concentrated negative charge
- More stable

Cl⊖

(b)

FIGURE 6-9 Atomic size and concentration of negative charge The electrostatic potential map of F⁻ is shown in (a) and that of Cl⁻ is shown in (b). Because Cl is larger than F, the negative charge on Cl⁻ is less concentrated, which is why it appears less red than F⁻. This lower concentration of charge helps make Cl⁻ more stable than F⁻.

SOLVED problem 6.16 Predict which has a lower pK_a: CH_4 or SiH_4.

Think Which is more stable, H_3C^- or H_3Si^-? Based on their relative stabilities, which anion's conjugate acid is deprotonated more favorably? How does that correspond to the relative pK_a values of the uncharged acids?

Solve The negative charges in H_3C^- and H_3Si^- appear on C and Si, respectively, which are different atoms in the same column of the periodic table. Because Si is significantly larger than C, H_3Si^- is more stable than H_3C^-. Thus, the products of the second reaction below are more stable than the products of the first.

Consequently, SiH_4 is deprotonated more favorably than CH_4, making SiH_4 the stronger acid. SiH_4 has a lower pK_a than CH_4.

---

problem **6.17** Which is a stronger acid, HBr or HI? Explain.

problem **6.18** Which is a stronger acid, CH_4 or PH_3? Explain.

6.6c Hybridization of the Atom to Which the Proton Is Attached

According to Table 6-1, $H_3C—CH_3$ is an extremely weak acid (among the weakest known), $H_2C=CH_2$ is somewhat stronger, and $HC\equiv CH$ is nearly as acidic as some alcohols (R—OH).

YOUR TURN 6.12

Verify the relative acidities of ethane, ethene, and ethyne by looking up their pK_a values in Table 6-1.

pK_a: $H_3C—CH_3$ _____ $H_2C=CH_2$ _____ $HC\equiv CH$ _____

Because ethane, ethene, and ethyne are uncharged acids, their differences in acidity must be caused by differences in the stability of their conjugate bases. In other words, $HC\equiv C^-$ must be more stable than $H_2C=CH^-$, which, in turn, must be more stable than $H_3C—CH_2^-$.

These differences in stability cannot come from differences in the size of the charged atom, because the negative charge is on carbon in each conjugate base. Instead, they arise from differences in the carbon atom's **effective electronegativity** in each of these species. As we saw in Section 3.11, the effective electronegativity of an atom depends on its hybridization, increasing in the order: $sp^3 < sp^2 < sp$. The effective electronegativity has the same effect on charge stability that true electronegativity does:

> The stability of a charged species increases as the *effective electronegativity* of an atom bearing a *negative* charge increases.

The C atom bearing the negative charge in $HC\equiv C^-$ is sp hybridized (highest effective electronegativity = greatest stability); that in $H_2C=CH^-$ is sp^2 hybridized (lower effective electronegativity = less stability); and that in $H_3C—CH_2^-$ is sp^3 hybridized (lowest effective electronegativity = least stability).

As you might expect, the effective electronegativity has the opposite effect on the stability of positively charged species.

> The stability of a charged species decreases as the *effective electronegativity* of an atom bearing a *positive* charge increases.

All else being equal, a positive charge on an sp^2-hybridized atom is *less* stable than a positive charge on an sp^3-hybridized atom (see Solved Problem 6.19). The greater effective electronegativity of the sp^2-hybridized atom makes it less able to accommodate a positive charge.

SOLVED problem 6.19 Predict which acid is stronger: $(CH_3)_2C=OH^+$ or $(CH_3)_2CH—OH_2^+$.

Think Which atom bears the positive charge in each acid? What is the hybridization of each of those atoms, and which atom has a greater effective electronegativity? How do their relative effective electronegativities correspond to the relative stabilities of the positively charged species? Based on those stabilities, which acid undergoes deprotonation more favorably?

Solve Each acid is positively charged, with the charge appearing on O. The O atom in $(CH_3)_2C=OH^+$ is sp^2 hybridized, whereas the O atom in $(CH_3)_2CH—OH_2^+$ is sp^3 hybridized, giving the O atom in $(CH_3)_2C=OH^+$ a higher effective electronegativity. Consequently, $(CH_3)_2C=OH^+$ cannot accommodate a positive charge on O as well as $(CH_3)_2CH—OH_2^+$ can, making $(CH_3)_2C=OH^+$ less stable than $(CH_3)_2CH—OH_2^+$.

To see how this translates into relative acidities, we can compare their free energy diagrams, which are similar to those in Figure 6-8.

$(CH_3)_2CH—OH_2^+$ (the blue ion) is more stable than $(CH_3)_2C=OH^+$ (the red ion), so the reaction depicted by the red curve is more energetically favorable than the one depicted by the blue curve. In other words, $(CH_3)_2C=OH^+$ is the stronger acid because its acidic proton is on an sp^2 hybridized O atom, whereas the acidic proton in $(CH_3)_2CH—OH_2^+$ is on an sp^3 hybridized O atom.

---

problem **6.20** Which is a stronger acid: $CH_3—NH_3^+$ or $HC≡NH^+$?

6.6d Effects from Adjacent Double and Triple Bonds: Resonance Effects

Recall from Section 6.4 that the pK_a of an acid can be altered significantly by the presence of a double or triple bond *adjacent* to the atom bonded to the acidic proton. For example, ethanoic acid (CH_3CO_2H), which has a C=O bond adjacent to the acidic OH group, is more acidic than ethanol (CH₃CH₂OH) by about 11 pK_a units.

Adjacent double bond increases acid strength

Ethanoic acid (Acetic acid)   Ethanol

YOUR TURN 6.13

Use Table 6-1 to look up the pK_a values of ethanoic acid and ethanol.

pK_a: Ethanoic acid _____   Ethanol _____

One of the main reasons ethanoic acid is so much more acidic than ethanol has to do with **resonance effects**. Both of these acids are uncharged (of the type in Equation 6-21), so the difference in their acidities is largely due to differences in the stability of their conjugate bases, $CH_3CO_2^-$ and $CH_3CH_2O^-$ (Equation 6-23a and 6-23b). Because ethanoic acid (acetic acid) is more acidic, we can say that $CH_3CO_2^-$ is more stable than $CH_3CH_2O^-$ (review Fig. 6-7).

More stable anion

Negative charge delocalized over two O atoms

Ethanoic acid (Acetic acid)   Ethanoate anion (Acetate anion)   (6-23a)

Negative charge localized on one O atom

Ethanol   Ethoxide anion   (6-23b)

As we can see in Equation 6-23a, ethanoic acid's conjugate base, the ethanoate anion, has two resonance structures, each of which has the negative charge on a different O atom. We say that the negative charge is **delocalized** over these two O atoms.

On the other hand, the negative charge is **localized** on a single O atom in ethanol's conjugate base (the ethoxide anion), shown in Equation 6-23b. In other words, the negative charge is *less concentrated* in the ethanoate anion than it is in the ethoxide anion. This is confirmed in the electrostatic potential maps of the two anions in Figure 6-10. With lower concentration of charge comes greater charge stability.

Delocalizing a charge via resonance lowers the *concentration* of the charge and increases the stability of the charged species.

Negative charge concentrated on one O atom

More stable anion

Negative charge delocalized over two O atoms

Ethoxide anion

(a)

Ethanoate anion (Acetate anion)

(b)

FIGURE 6-10 Resonance delocalization of a negative charge (a) An electrostatic potential map of the ethoxide anion ($CH_3CH_2O^-$) shows that the negative charge is localized on just the one O atom. (b) An electrostatic potential map of the ethanoate anion ($CH_3CO_2^-$) shows that the negative charge is delocalized onto both O atoms, consistent with its resonance hybrid. The concentration of negative charge is lower (less red) in the ethanoate anion, so the ethanoate anion is more stable than the ethoxide anion.

In the space provided here, draw the curved arrow notation for the conversion of one of the ethanoate anion's resonance structures into the other and draw the resonance hybrid. (You may want to review Section 1.10.)

6.14 **YOUR TURN**

Resonance hybrid

SOLVED problem 6.21 Predict which of the following species is a stronger acid:

A   **B**

Think Write reactions that depict each acid being deprotonated by a base. Do you expect a significant difference in energy between the reactants of one and the reactants of the other? Between the products of one and the products of the other? Are the charge-bearing atoms different in these two acids? Do they have different effective electronegativities? Do the ions differ in resonance delocalization of the charge?

Solve The acids can be deprotonated by water according to the following reactions:

More stable cation

H_3C OH **A** $+$ H_2O $\rightleftharpoons$ H_3C OH $+$ H_3O^+

Less stable cation

H_3C CH_3 **B** $+$ H_2O $\rightleftharpoons$ H_3C CH_3 $+$ H_3O^+

Being positively charged, these acids have the same form as H_3O^+ and H_4N^+ in Equations 6-22a and 6-22b. So, as we learned from Figure 6-8, the energetically more favorable deprotonation will involve the less stable reactant ion. Thus, the stronger acid is the one that is *less stable*. In both of these acids, the acidic proton is attached to a positively charged O atom. Both of those O atoms are sp^2 hybridized, so effective electronegativity does not play a role. However, *acid A has two resonance structures*, which means the positive charge can be delocalized over two O atoms, as shown below.

Acid **B**, by contrast, has no resonance structures, so the positive charge is *localized* on one oxygen atom. As a result, acid **B** is the stronger acid because it is less stable than the first.

problem **6.22** Which of the following compounds is a stronger acid? Explain.

The number of atoms over which a charge is delocalized by resonance affects charge stability.

> Stabilization via resonance generally increases as the number of atoms over which a charge is delocalized increases.

For example, sulfuric acid (H_2SO_4; $pK_a = -9$) is a stronger acid than ethanoic acid (CH_3CO_2H; $pK_a = 4.75$). Both acids are uncharged, so the conjugate base of the stronger acid must be more stable than that of the weaker acid (Equation 6-21 and Fig. 6-7). In HSO_4^-, the negative charge is delocalized over three O atoms, whereas in $CH_3CO_2^-$, the negative charge is delocalized over only two. Delocalizing over more atoms in HSO_4^- means the negative charge is less concentrated, leading to greater stability. We can see that the negative charge is more delocalized in HSO_4^- than in $CH_3CO_2^-$ by comparing their electrostatic potential maps in Figure 6-11.

FIGURE 6-11 Resonance delocalization of a negative charge onto different numbers of atoms (a) An electrostatic potential map of the ethanoate anion ($CH_3CO_2^-$) shows that the negative charge is delocalized by resonance over two O atoms. (b) An electrostatic potential map of the HSO_4^- anion shows that the negative charge is delocalized over three O atoms. The concentration of negative charge is lower in HSO_4^- (is less red around the O atoms), so HSO_4^- is more stable than $CH_3CO_2^-$.

Negative charge delocalized over two O atoms

More stable anion

Negative charge delocalized over three O atoms

$CH_3CO_2^{\ominus}$ anion
(a)

$HSO_4^{\ominus}$ anion
(b)

In the space provided, draw the two remaining resonance structures of HSO_4^- that illustrate the sharing of its negative charge. Be sure to include the appropriate curved arrows. Then, draw the corresponding resonance hybrid.

Resonance hybrid

SOLVED problem 6.23 Deprotonation in pentane-2,4-dione may occur at a terminal C atom or at the central C atom. Which site is more acidic? Explain.

Pentane-2,4-dione

Think Does charge stability play a role in the acid or in the conjugate bases? What, specifically, leads to differences in charge stability? The type of atom? Effective electronegativity? Resonance delocalization of the charge? How are the answers to these questions related to the relative strengths of the acids?

Solve The acid is uncharged, so we must look for differences in charge stability in the possible conjugate bases. Since both acidic protons are attached to sp^3-hybridized C atoms, neither the type of atom nor effective electronegativity should play a role. We do find, however, that there is a difference in resonance stabilization of the resulting charge.

If deprotonation occurs at a terminal C, then the resulting negative charge in the conjugate base is delocalized over the C atom and an O atom via resonance. On the other hand, if deprotonation occurs at the central C, then the negative charge is delocalized over the C atom and *two* O atoms. As a result, the negative charge that develops is less concentrated, so the conjugate base is more stable. This makes the central C atom more acidic than a terminal C atom.

Add the appropriate curved arrows to the resonance structures in Solved Problem 6.23 to show how each is transformed into the one on its right.

More stable anion

Some negative
charge has been
removed.

Ethoxide anion

2-Chloroethoxide anion

**FIGURE 6-12 Delocalization of a
negative charge via induction**
Electrostatic potential maps of the
ethoxide anion (CH₃CH₂O⁻, left) and the
2-chloroethoxide anion (ClCH₂CH₂O⁻,
right). The presence of the Cl atom
in place of an H atom removes some
electron density (red) from the negatively
charged O atom. This decrease in charge
concentration stabilizes the anion.

problem 6.24 Based on differences in resonance delocalization, predict whether HNO_3 or CH_3CO_2H is the stronger acid. Explain. (*Hint*: In this case, you can ignore the fact that the O atoms are bonded to different atoms [i.e., N vs. C].)

6.6e Effects from Nearby Atoms: Inductive Effects

Recall from Section 6.4 that another of the structural differences responsible for differences in acid strength is *the presence of electronegative atoms near the acidic proton*. For example, 2-chloroethanol ($ClCH_2CH_2OH$; $pK_a = 12.9$) is more acidic than ethanol (CH_3CH_2OH; $pK_a = 16$), by about three pK_a units. Both acids are uncharged, so the conjugate base of 2-chloroethanol ($ClCH_2CH_2O^-$) must be more stable than that of ethanol ($CH_3CH_2O^-$).

In both conjugate bases, the negative charge is on the same type of atom (oxygen) with the same hybridization (sp^3). Furthermore, neither of the conjugate bases has any resonance structures that place the negative charge on other atoms. Therefore, the difference in charge stability must come from another factor.

The electrostatic potential maps of the two anions are shown in Figure 6-12. Notice that some of the negative charge (red) surrounding the O atom has been removed when the Cl atom is present. In other words, the O atom bears less of a negative charge in $ClCH_2CH_2O^-$ than it does in $CH_3CH_2O^-$. Just as with charge delocalization via resonance, this decrease in the concentration of charge increases the stability of the anion.

Why does the presence of the Cl atom remove electron density from the nearby O atom? Cl is more electronegative than H, so Cl is **electron withdrawing** relative to H. Therefore, if the left-most H atom in $CH_3CH_2O^-$ were replaced by Cl, then electron density on the left-most C atom would be shifted toward the Cl atom, along the C—Cl bond. As a result, the left-most C atom would develop a deficiency of electrons, and to compensate, it would draw electron density away from other atoms to which it is bonded. This effect would be repeated down the chain until, ultimately, electron density was removed from the O atom bearing the negative charge:

More stable anion

Electron density is removed
from the O, thus stabilizing
the negative charge.

Cl is more electronegative
than H and is thus electron
withdrawing.

In general, anions are *stabilized* by electron-withdrawing groups near the negative charge.

This phenomenon is called **induction**, and it can be defined as the distortion of electron density along covalent bonds, brought about by the replacement of an H atom with another substituent. The effect that induction has on stability is called an **inductive effect**. In $ClCH_2CH_2O^-$, for example, the Cl atom is *inductively stabilizing*.

An electron-withdrawing inductive effect does not always lead to stabilization. If electron density (i.e., negative charge) is drawn away from a *positively* charged atom, then the concentration of positive charge *increases*, which is destabilizing. This is what we observe in $ClCH_2CH_2OH_2^+$, the conjugate acid of 2-chloroethanol. In this case, the Cl atom is *inductively destabilizing*.

Less stable cation

Cl is more electronegative than H and is thus *electron withdrawing*.

Electron density is removed from the O, thus *destabilizing the positive charge*.

In general, cations are *destabilized* by electron-withdrawing groups near the positive charge.

problem **6.25** Predict which of the following protonated alcohols is the stronger acid.

A   B

Most substituents, like chlorine, are inductively electron-withdrawing groups because most atoms common in organic compounds are more electronegative than hydrogen. However, there are a handful of substituents that are inductively **electron donating**. A silicon atom, for example, is electron donating relative to hydrogen because silicon is less electronegative than hydrogen.

The most common electron-donating groups in organic chemistry are *alkyl groups*. $(CH_3)_3COH$ ($pK_a = 19$), for example, is about three pK_a units *less acidic* than CH_3CH_2OH ($pK_a = 16$), suggesting that $(CH_3)_3CO^-$ is less stable than $CH_3CH_2O^-$. $(CH_3)_3CO^-$ is less stable because the electron-donating ability of the CH_3 groups compared to H atoms increases the concentration of negative charge on O.

Less stable anion

Alkyl groups are *electron donating* compared to H.

An increase in concentration of charge on O leads to destabilization.

In general, anions are *destabilized* by electron-donating groups near the negative charge.

The opposite is true for cations. $CH_3NH_3^+$ ($pK_a = 10.6$), for example, is a weaker acid than NH_4^+ ($pK_a = 9.4$), suggesting that $CH_3NH_3^+$ is the more stable cation. $CH_3NH_3^+$ is more stable because the electron-donating ability of the CH_3 group compared to H decreases the concentration of positive charge on N.

More stable cation

$$H\overset{\oplus}{-}NH_3 \qquad H_3C\overset{\oplus}{-}NH_3$$

$$pK_a = 9.4 \qquad pK_a = 10.6$$

In general, cations are *stabilized* by electron-donating groups near the positive charge.

YOUR TURN 6.17

Draw the arrow in $CH_3NH_3^+$ in the previous graphic to represent the inductive effect of the CH_3 group.

problem **6.26** Predict which of these compounds is a stronger acid. Explain.

A   SH   H₂S

A   B

This effect is particularly important for **carbocations**—species that contain a positively charged carbon atom (C^+)—which are key reactive intermediates in a variety of chemical reactions. Carbocation stability generally increases with each additional alkyl group adjacent to the positively charged carbon atom, in part because the electron-donating ability of each alkyl group helps to diminish the concentration of positive charge on C^+.[1]

Carbocation stability

Methyl   Primary (1°)   Secondary (2°)   Tertiary (3°)

Decreasing concentration of positive charge

Increasing stability

---

[1]Each alkyl group also stabilizes the carbocation via *hyperconjugation*, in which electron density from adjacent σ bonds is donated into the empty *p* orbital on the positively charged carbon.

Carbocations are therefore distinguished based on the number of alkyl groups surrounding C^+. A **methyl cation** (H_3C^+) has no adjacent alkyl groups; a **primary (1°) carbocation**, RCH_2^+, has one alkyl group adjacent to C^+; a **secondary (2°) carbocation**, R_2CH^+, has two alkyl groups; and a **tertiary (3°) carbocation**, R_3C^+, has three alkyl groups. Carbocation stability increases in the following order: methyl < 1° < 2° < 3°.

The difference in electronegativity between carbon and hydrogen is quite small—2.5 for carbon compared to 2.2 for hydrogen—but it is enough to make alkyl groups electron donating relative to hydrogen, even though they are made up of only C—H and C—C single bonds. The reason becomes clearer if we closely examine CH_3, the simplest alkyl group (Fig. 6-13). In each of the C—H bonds, a small bond dipole points toward the C atom, giving the carbon atom a partial negative charge (δ^-). This buildup of negative charge enables that carbon atom to donate some electron density to a group to which it is bonded.

Alkyl groups of different size have different electron-donating capabilities, but the difference is typically insubstantial. For example, the alkyl group attached to the acidic OH in ethanol (CH_3CH_2OH) is twice the size of that in methanol (CH_3OH), but both have pK_a values of about 16.

Thus far, we have discussed inductive effects on ion stability only in a completely qualitative sense—that is, electron-withdrawing groups stabilize nearby negative charges and destabilize nearby positive charges, whereas electron-donating groups destabilize nearby negative charges and stabilize nearby positive charges. However, we can be somewhat quantitative as well, by considering a few straightforward principles.

FIGURE 6-13 The electron-donating ability of an alkyl group (left) C is slightly more electronegative than H, so each C—H bond dipole points toward C. (right) The electron density (δ^-) built up on C can be donated to adjacent atoms.

1. The inductive electron-withdrawing ability of an uncharged atom increases as its electronegativity increases, whereas an atom's electron-donating ability increases as its electronegativity decreases. For example,
   - Fluorine is more electronegative than chlorine, so fluorine is more inductively electron withdrawing than chlorine.
   - Silicon is less electronegative than carbon, so the H_3Si group is more electron donating than the H_3C group.

These relationships occur because inductive effects arise from a distortion of electron density along covalent bonds. Thus, there is a greater shift of electron density toward the more electronegative group and away from the less electronegative group.

2. Charged substituents have more pronounced inductive effects than uncharged substituents. For example,
   - $(CH_3)_3N^+$ is more electron withdrawing than fluorine.
   - CO_2^- is more electron donating than CH_3.

A full positive charge signifies a substituent that is very highly electron deficient. A full negative charge, on the other hand, signifies a large excess of electron density.

3. Inductive effects are additive. For example,
   - $CF_3CH_2^+$ is *less stable* than $CH_2FCH_2^+$.
   - $(CH_3CH_2)_2NH_2^+$ is *more stable* than $CH_3CH_2NH_3^+$.

In $CF_3CH_2^+$, three F atoms withdraw electron density from the positively charged C atom, whereas only one F atom does in $CH_2FCH_2^+$. As a result, there is a greater

Sulfuric acid (H_2SO_4) is arguably the most important industrial compound—nearly 200 million tons of it are produced each year, about 30% in the United States alone. It is so important, in part, because it is a strong acid ($pK_a = -9$), and strong acids catalyze numerous chemical reactions.

According to Table 6-1, however, sulfuric acid is not the strongest acid that exists. Trifluoromethanesulfonic acid (CF_3SO_3H; $pK_a = -13$), for example, is about four pK_a units (a factor of 10,000) stronger. As a result, CF_3SO_3H is one of a variety of acids classified as a "superacid." A few others are shown below, along with their pK_a values. The acid at the far right—the strongest acid known—is some 14 pK_a units lower than sulfuric acid, making it 100 trillion times stronger!

$pK_a = -10.7$ $pK_a = -13.3$ $pK_a = -18$ $pK_a = -25$

What all of these superacids have in common is very heavy stabilization of the negative charge that appears in the conjugate base. That stabilization results from extensive resonance delocalization of the charge, multiple inductively electron-withdrawing groups, or both.

What is the purpose of superacids, beyond simply pushing the limits of chemistry? One application is to make acid catalysis "greener" by carrying out certain reactions using smaller quantities of strong acid at lower temperatures. A second use is as a medium in which to generate and stabilize highly reactive species, such as carbocations, so they can be studied. And a third application, used in the petroleum industry, is to convert long-chain hydrocarbons into smaller ones for the production of high-octane fuels.

---

concentration of positive charge in $CF_3CH_2^+$ than in $CH_2FCH_2^+$. In $(CH_3CH_2)_2NH_2^+$, two alkyl groups donate electron density to the positively charged N atom, whereas only one alkyl group does in $CH_3CH_2NH_3^+$. Thus, there is a smaller concentration of charge in $(CH_3CH_2)_2NH_2^+$.

4. Inductive effects fall off very quickly with distance. For example, $CH_3CH_2CHFCH_2NH^-$ is *more stable* than $CH_3CHFCH_2CH_2NH^-$.

The F atom is separated from the negatively charged N by one fewer bond in $CH_3CH_2CHFCH_2NH^-$ than in $CH_3CHFCH_2CH_2NH^-$. In $CH_3CH_2CHFCH_2NH^-$, therefore, the electron-withdrawing effect of F is transmitted more efficiently to the negatively charged N, resulting in a lower concentration of negative charge on N.

SOLVED problem 6.27 Predict which of the following carboxylic acids is more acidic.

or

Think For each acid, does the stability of the acid or the stability of the conjugate base dictate pK_a? Do electron-donating or electron-withdrawing effects stabilize those species? Are the Br and I substituents electron donating or electron withdrawing? Which substituent invokes stronger inductive effects?

Solve These are uncharged acids, so their pK_a values are dictated by the stability of their negatively charged conjugate bases. Both Br and I are electron-withdrawing substituents, so they stabilize negatively charged species. Br is more electronegative than I, so Br better stabilizes the conjugate base, in which case the acid on the left is stronger.

problem **6.28** Which acid is stronger, $O_2NCH_2CH_2OH$ or $H_2NCH_2CH_2OH$? Explain. (*Hint:* Draw out the complete Lewis structure for each.)

problem **6.29** Which carboxylic acid functional group has the more acidic proton? Explain.

6.7 Strategies for Success: Ranking Acid and Base Strengths—The Relative Importance of Effects on Charge

Thus far in Chapter 6, we have analyzed relative acid strengths by considering only one charge-stability factor at a time. Often in organic chemistry, however, we must consider multiple factors simultaneously. To do so, we need to understand the relative importance of each factor in evaluating the stability of a particular species.

With relatively few exceptions, the order of priority for these factors is as follows: whether the acid or the conjugate base bears a charge is more important than the type of atom on which the charge appears, which is more important than the extent of charge delocalization via resonance, which is more important than inductive effects. When evaluating the relative stabilities of two species, therefore, ask the following questions in order. The first question to which the answer is "yes" corresponds to the factor that most influences the species' relative stabilities.

1. Do the species have different charges?

A charged species is generally more reactive than one that bears no formal charge.

2. Do the charges appear on different atoms?

The size, electronegativity, and hybridization of the atom on which the charge is located affects stability.

All else being equal, stability increases as the number of atoms over which a charge is shared increases.

Electron-withdrawing groups stabilize nearby negative charges but destabilize nearby positive charges. Electron-donating groups stabilize nearby positive charges but destabilize nearby negative charges.

The relative importance of these factors is reflected in the magnitude of their effect on pK_a. The appearance of a charge has quite a significant effect. A protonated amine ($R—NH_3^+$), for example, is nearly 30 pK_a units more acidic than a comparable uncharged amine ($R—NH_2$). Somewhat less dramatically, a protonated alcohol ($R—OH_2^+$) is about 18 pK_a units more acidic than an uncharged alcohol ($R—OH$).

YOUR TURN 6.18

Verify the differences in pK_a between the protonated and uncharged amines and alcohols just mentioned by estimating the pK_a values of the generic species shown below based on actual compounds in Table 6-1. Write them in the spaces provided here.

pK_a: $R—NH_3^+$ _____ $R—NH_2$ _____ $R—OH_2^+$ _____ $R—OH$ _____

The type of atom on which an acidic proton is found also has a dramatic effect on pK_a. A typical alcohol ($R—OH$), for example, is about 20 pK_a units more acidic than a typical amine ($R—NH_2$). The main difference is that the negative charge resides on an oxygen atom in the conjugate base $R—O^-$, and on the nitrogen atom in the $R—NH^-$ conjugate base. When we consider differences in type of atom, we should also consider effective electronegativity because an acidic hydrogen on an sp-hybridized carbon atom is about 23 pK_a units more acidic than a comparable hydrogen on an sp^3-hybridized carbon atom.

YOUR TURN 6.19

Verify the preceding statements by estimating the pK_a values of the generic alcohol, amine, alkane, and terminal alkyne species shown below based on actual compounds in Table 6-1. Write them in the spaces provided here.

pK_a: $R—OH$ _____ $R—NH_2$ _____ $R—CH_3$ _____ $RC\equiv CH$ _____

Resonance effects have less of an impact on pK_a than the type of atom on which the acidic hydrogen is found. Nevertheless, they are quite important because a carboxylic acid ($R—CO_2H$) is about 11 pK_a units more acidic than an alcohol ($R—OH$). Similarly, the negative charge on N in the conjugate base of aniline ($C_6H_5NH^-$) can be delocalized by resonance, as shown below, making aniline about 11 pK_a units more acidic than a typical amine ($R—NH_2$).

In the preceding graphic, draw the curved arrows needed to show how each resonance structure of $C_6H_5NH^-$ is converted to the one on its right. Then, draw the
resonance hybrid to the right of the resonance structures.

Inductive effects generally have the least effect on pK_a values. For example, the
pK_a of 2,2,2-trifluoroethanol (F_3CCH_2OH) is 12.4, whereas the pK_a of ethanol
(CH_3CH_2OH) is 16. With its three F atoms near the acidic alcohol group,
2,2,2-trifluoroethanol exhibits relatively strong inductive effects, yet it is only about
four pK_a units more acidic than ethanol.

Let's now apply this knowledge toward ranking the strengths of the following
four acids:

Because pK_a is defined in terms of each acid's reaction with water, we begin by
writing out their reactions, as shown in Equation 6-24.

(6-24a)

(6-24b)

(6-24c)

(6-24d)

To help determine how favorable each reaction is (the more favorable the reaction,
the stronger the acid), construct a free energy diagram of all four reactions on a single
plot (see Fig. 6-14). We do so by determining the relative stabilities of the reactant
and product species. H_2O (a reactant) can be ignored because it appears in all cases.
The remaining reactants can be evaluated using the tie-breaking questions. Do the remaining reactants have different charges? The acid in Equation 6-24c (red) is charged,
whereas those in Equation 6-24a, 6-24b, and 6-24d are uncharged. In Figure 6-14,
therefore, the reactants in Equation 6-24a, 6-24b, and 6-24d are lower in energy than
the reactants in Equation 6-24c. Because the acids in Equation 6-24a, 6-24b, and
6-24d are all uncharged, the remaining three tie-breaking questions cannot be used to
distinguish their relative stabilities; their stabilities are assumed to be similar.

Moving to the product side, notice first that H_3O^+ can be ignored in each case,
so the tie-breaking questions should be applied just to the conjugate bases. Do the
species have different charges? The conjugate bases in Equation 6-24a, 6-24b, and
6-24d are all negatively charged, whereas the one in Equation 6-24c is uncharged
(red). Therefore, the products of Equation 6-24c are more stable than the ones from
the other reactions.

FIGURE 6-14 Free energy diagrams for the proton transfer reactions shown in Equation 6-24 The reactants of the red curve are highest in energy because the acid in that reaction is charged, whereas the ones in the other reactions are uncharged. The products of the red curve are lowest in energy because the conjugate base in that reaction is uncharged; the conjugate bases in the other reactions are charged. The products of the purple curve are next lowest in energy because the negative charge on the conjugate base is resonance delocalized. The products of the green curve are lower in energy than the products of the blue curve due to inductive effects. In the green conjugate base, F is closer to the negative charge, which more substantially decreases the concentration of negative charge on O.

For the charged conjugate bases in Equation 6-24a, 6-24b, and 6-24d, do the charges appear on different atoms? The −1 charge appears on O in all three cases, so the answer is no. Are the charges delocalized differently via resonance? The conjugate base in Equation 6-24d (purple) has two resonance structures that delocalize the negative charge over the two O atoms, making it more stable than the negatively charged conjugate bases in Equation 6-24a (blue) and 6-24b (green).

Are there differences in inductive effects for the conjugate bases in Equation 6-24a and 6-24b? Both have a highly electron-withdrawing F atom that inductively stabilizes the negative charge, but the one in Equation 6-24b (green) is closer to the negative charge. Inductive effects decrease with distance, so the conjugate base in Equation 6-24b is more stable than the one in Equation 6-24a.

By establishing the relative stabilities of all four sets of products, we can see in Figure 6-14 that deprotonation of the acids becomes more favorable (less positive ΔG°_{rxn}) in the order $\mathbf{A} < \mathbf{B} < \mathbf{D} < \mathbf{C}$. This, then, is the order of increasing acid strength.

If you know how to rank species according to acid strength, then you can rank species according to base strength by applying the following concept from Section 6.2a:

> The stronger a base's conjugate acid, the weaker the base.

Thus, the conjugate bases in Equation 6-24 increase in base strength in the following order:

Weakest base   Strongest base

The molecule on the left is the weakest base because it is the conjugate base of the strongest acid in Equation 6-24. The species on the right is the strongest base because its conjugate acid is the weakest acid in Equation 6-24.

6.8 Strategies for Success: Determining Relative Contributions by Resonance Structures

Our knowledge of charge stability is applicable beyond simply predicting the outcomes of proton transfer reactions. In this section, we discuss how charge stability can be used to determine the relative contributions by resonance structures. In Chapter 7, we explain how charge stability can be applied to elementary steps other than proton transfers.

Although the resonance hybrid is an *average* of all the resonance contributors, not all resonance contributors are weighted equally. As discussed in Section 1.10,

> The resonance hybrid looks most like the lowest-energy (most stable) resonance structure.

We also learned in Chapter 1 that a variety of factors govern the relative stabilities of resonance structures. In particular, a resonance structure is more stable with:

- a greater number of atoms with octets
- more covalent bonds
- fewer atoms with nonzero formal charges

Based on what you have learned here in Chapter 6, a resonance structure is more stable with fewer atoms possessing nonzero formal charges because charged species are inherently less stable than the corresponding uncharged molecules. Consider, for example, the following resonance contributors to $H_2C{=}CH{-}OCH_3$:

Greater contribution

No nonzero
formal charges

Two nonzero
formal charges

No atoms in the structure on the left possess a nonzero formal charge, but the O atom has a +1 formal charge and one C atom has a −1 formal charge in the structure on the right. The structure on the right is less stable, therefore, so it makes a smaller contribution to the resonance hybrid. Thus, a molecule of $H_2C{=}CH{-}OCH_3$ looks more like the resonance contributor on the left.

---

problem 6.30

(a) Draw the curved arrows necessary to convert the resonance structure on the left into the one on the right.

(b) Which resonance contributes more to the hybrid? Explain.

$$\left[H_3C{-}C{\equiv}C{-}Cl \quad \longleftrightarrow \quad H_3C{-}\overset{\ominus}{C}{=}C{=}\overset{\oplus}{Cl} \right]$$

Frequently, the relative stabilities of resonance structures cannot be determined on the basis of the octet rule, the total number of bonds, or the number of atoms bearing a formal charge, but they can be determined on the basis of charge stability. In these cases, first examine the atoms on which the formal charges appear (the more important factor) and then examine any inductive effects. Because we are considering a single resonance contributor only, we do not consider resonance effects.

In both resonance contributors to an enolate anion, for example, each atom has its octet or duet, both resonance structures have the same total number of bonds, and each has a single atom with a −1 formal charge.

Greater contribution

Negative charge ····▶ on O

Negative charge on C

The main difference between these structures is the atom on which the formal charge appears—namely, C in the structure on the left and O in the structure on the right. O is more electronegative than C, so O can better accommodate the negative charge than C. The structure on the right, therefore, is more stable and makes the greater contribution to the resonance hybrid.

---

problem 6.31 Which resonance structure contributes more to the hybrid? Explain.

A   B

(a) Explain why this is counterintuitive based on charge stability.

(b) Why does the structure on the right contribute more?

Greater contribution

In the following two resonance contributors, there is a $+1$ formal charge on a different C atom in each.

Greater contribution

Alkyl group is electron donating

In the structure on the left, the C atom bearing the positive charge is bonded to two H atoms and a C atom that is part of the double bond. In the structure on the right, the positively charged C atom is bonded to one H atom, a CH_3 group, and a C atom that is part of the double bond. Because the CH_3 group is electron donating compared to H, the positive charge on the C atom is reduced in the structure on the right, making this structure more stable and a more significant resonance contributor.

problem **6.33** Predict which of the following is the more important resonance contributor. Explain.

A   B

6.9 Wrapping Up and Looking Ahead

The focus of Chapter 6 has been the proton transfer (or Brønsted–Lowry acid–base) reaction—a reaction that is among the simplest you will encounter. We began by introducing curved arrow notation so you could learn how chemists keep track of valence electrons throughout a reaction. Next, we discussed aspects of chemical equilibrium and the equilibrium constant, K_{eq}. To describe the deprotonation of an acid by water, chemists use a related constant, called the acidity constant, K_a, and its negative log, pK_a. One of the most convenient uses of pK_a values is to predict K_{eq} for a proton transfer reaction in which water is *not* the base.

We also saw how a reaction's equilibrium constant is related to the difference in standard Gibbs free energy between products and reactants, $\Delta G°_{rxn}$, and that $\Delta G°_{rxn}$ has both an enthalpy term and an entropy term. Knowing ΔG°_{rxn} for a reaction makes it possible to determine product stability relative to reactant stability; that is, $\Delta G^\circ_{rxn} > 0$ means that the products are higher in energy (and thus less stable) than the reactants, whereas $\Delta G^\circ_{rxn} < 0$ means the opposite.

We then focused on the reasons for the dramatically different pK_a values that acids can exhibit. It turns out that charged species are generally much less stable than the corresponding uncharged molecules, and their relative stabilities are sensitive to a variety of factors. The size and electronegativity of the atom bearing the charge tend to have the greatest influence, followed by resonance effects, and finally inductive effects.

As we continue on to other chapters, pay attention to how often the material from this chapter reappears. You will see, for example, that proton transfer reactions appear as elementary steps in mechanisms more often than any other elementary step. Moreover, charge stability is a major driving force for many different elementary steps, just as we saw it is for proton transfers. For these reasons, the greater depth to which you understand proton transfer reactions, the greater your success will be with other reactions we study.

THE ORGANIC CHEMISTRY OF BIOMOLECULES

6.10 The Structure of Amino Acids in Solution as a Function of pH

In Section 1.12, the following structure was introduced as the general form of an α-amino acid, in which both an amino group (NH_2) and a carboxyl group (CO_2H) are bonded to the same C atom.

An α-amino acid

It turns out, however, that this form never dominates in aqueous solution because the carboxyl group is weakly acidic and the amine group is weakly basic.

What form, then, does an α-amino acid take in aqueous solution? The answer depends upon the pH of the solution. Under highly acidic (pH < 2) conditions, the weakly basic N atom is protonated. The resulting species, which bears an overall charge of +1, is shown on the left in Equation 6-25.

This is a zwitterion, because it has a separated positive and negative charge, but its net charge is zero.

(6-25)

As the solution becomes more basic (i.e., as the pH of the solution increases), a proton can be removed. It does *not* come from the protonated amine, however, because the most acidic proton in the cationic species is the one that belongs to the carboxylic acid; its pK_a is around 2, whereas that of the protonated amine group is around 9 to 10. (The exact pK_a values depend upon the specific amino acid.) Therefore, when the pH of the solution has risen significantly above 2, the OH proton is lost and the dominant form is the middle species in Equation 6-25. This species, called a **zwitterion** (pronounced zvitter-ion), has both a positive and a negative formal charge, but a net charge of zero.

As the solution becomes more basic still, the proton of the protonated amine group is lost, yielding the species on the right in Equation 6-25. This second deprotonation takes place when the pH of the solution is significantly above 9 or 10, the pK_a of the protonated amine.

Table 6-2 lists the pK_a values associated with the 20 naturally occurring amino acids. For the majority of these amino acids, only two pK_a values are listed—one for the carboxylic acid group, and one for the protonated amine. For seven of them, however, a third pK_a value is given. Those amino acids have side chains that are **ionizable**, meaning that an atom in the side chain can gain or lose a proton to become charged.

---

problem **6.34** Draw the dominant form of alanine in solutions whose pH values are 1, 4, 8, and 11.

Aspartic acid is one of the seven amino acids with an ionizable side chain. Its side chain (shown below in blue) contains a second carboxylic acid.

Aspartic acid

(6-26)

At low pH (i.e., strongly acidic conditions), aspartic acid is in its fully protonated form, in which both carboxylic acid groups are uncharged and the amine is positively charged. As the pH is increased, the first proton that is lost comes from the carboxylic acid group that is part of the amino acid backbone (pK_a = 2.10), thus yielding the zwitterion. Increasing the pH further causes the carboxylic acid on the side chain (pK_a = 3.86) to lose its proton. Under more strongly basic conditions, the ammonium group (pK_a = 9.82) loses its proton.

---

problem **6.35** Draw the structure of the most abundant form of glutamic acid in solutions whose pH values are 1, 3, 5, and 11. Do the same for cysteine in solutions whose pH values are 1, 4, 6, 9, and 11.

Lysine is another amino acid that has an ionizable side chain, though it has a N atom that can be protonated, rather than a carboxylic acid that can be deprotonated. As indicated in Equation 6-27, the carboxylic acid group is uncharged and both N atoms are positively charged (i.e., are protonated) at very low pH. As the pH of the solution is increased, the first group to lose its proton is the carboxylic acid (pK_a = 2.18).

TABLE 6-2 pK$_a$ Values of the 20 Naturally Occurring Amino Acids

Amino Acid	Side Chain	pK$_a$ of -CO$_2$H	pK$_a$ of -NH$_3^+$	pK$_a$ of Side Chain	Amino Acid	Side Chain	pK$_a$ of -CO$_2$H	pK$_a$ of -NH$_3^+$	pK$_a$ of Side Chain
Alanine	—CH$_3$	2.35	9.87	N/A	Leucine		2.33	9.74	N/A
Arginine		2.01	9.04	12.48	Lysine		2.18	8.95	10.53
Aspargine		2.02	8.80	N/A	Methionine		2.28	9.21	N/A
Aspartic acid		2.10	9.82	3.86	Phenylalanine		2.58	9.24	N/A
Cysteine		2.05	10.25	8.00	Proline[a]		2.00	10.60	N/A
Glutamic acid		2.10	9.47	4.07	Serine		2.21	9.15	N/A
Glutamine		2.17	9.13	N/A	Threonine		2.09	9.10	N/A
Glycine	—H	2.35	9.78	N/A	Tryptophan		2.38	9.39	N/A
Histidine		1.77	9.18	6.10	Tyrosine		2.20	9.11	10.07
Isoleucine		2.32	9.76	N/A	Valine		2.29	9.72	N/A

[a] For proline, the side chain is shown in black.

Next is the protonated amine group of the amino acid backbone ($pK_a = 8.95$), and last is the protonated amine group of the side chain ($pK_a = 10.53$). With lysine, the zwitterion is the product of the second deprotonation, not the first.

problem **6.36** Draw the structure of the most abundant form of arginine in solutions whose pH values are 1, 4, 8, 10, and 14. Do the same for histidine in solutions whose pH values are 1, 3, 5, 7, and 11.

6.11 Electrophoresis and Isoelectric Focusing

Knowing the pK_a values of amino acids is quite helpful in **electrophoresis**, one of the most common ways of separating a mixture of amino acids or proteins. In electrophoresis, the mixture is spotted on a gel or strip of paper that has been buffered to a specific pH (Fig. 6-15). A high voltage (50–1000 V) is applied by positively and negatively charged poles—the **anode** and **cathode**, respectively—situated at opposite ends of the apparatus. Any ion that has a net positive charge at that pH will migrate toward the negatively charged terminal, whereas any ion having a net negative charge will migrate toward the positively charged terminal. If the net charge is zero (as in

FIGURE 6-15 Paper electrophoresis Any substance carrying a positive charge will migrate toward the negatively charged terminal (the cathode) and any substance carrying a negative charge will migrate toward the positively charged terminal (the anode). An uncharged substance will not migrate.

a zwitterion), then the species will not move. To help you remember this, use the following mnemonic:

- **Cat**ions migrate toward the **cat**hode.
- **An**ions migrate toward the **an**ode.

An amino acid's **isoelectric pH**, or **isoelectric point (pI)**, is *the pH at which the substance has a charge of zero*, and it is unique for each of the 20 amino acids. If the pH of the solution is higher (i.e., more basic) than the pI, then the substance will, on average, be deprotonated to a greater extent and will carry a net negative charge. Conversely, if the pH is lower (i.e., more acidic) than the pI, then the substance will, on average, be protonated to a greater extent and will carry a net positive charge.

An amino acid's pI can be computed straightforwardly from its pK_a values:

An amino acid's pI is the average of its two pK_a values associated with the zwitterion—namely, the pK_a value corresponding to the proton transfer reaction in which the zwitterion is a product and the pK_a value corresponding to the proton transfer reaction in which the zwitterion is a reactant.

For an amino acid with only two ionizable groups (i.e., no ionizable side chains), there are only two pK_a values in all. As shown previously in Equation 6-25, the first pK_a corresponds to the proton transfer reaction in which the zwitterion is produced, and the second corresponds to the proton transfer reaction in which the zwitterion reacts. Therefore, the pI is simply the average of those two pK_a values. For example, the pI of glycine is: pI = (2.35 + 9.78)/2 = 6.07.

problem **6.37** In an electrophoresis experiment where the pH of the gel is 7, will glycine migrate toward the anode or the cathode? Explain.

problem **6.38** Calculate the pI of alanine. In an electrophoresis experiment where the pH of the gel is 4, will alanine migrate toward the anode or the cathode? Explain.

If an amino acid has an ionizable side chain, then there are three ionizable groups in all and three pK_a values to consider. The way in which pI is computed does not change, however, because it is still the average of the two pK_a values associated with the reactions that involve the zwitterion. Which two pK_a values those are, however, depends on whether the side chain is acidic or basic.

Aspartic acid, for example, has an ionizable side chain that contains a carboxylic acid and is therefore acidic. According to Equation 6-26, aspartic acid's zwitterion is a product of the first deprotonation (pK_a = 2.10) and a reactant in the second (pK_a = 3.86). Therefore, aspartic acid's pI is (2.10 + 3.86)/2 = 2.98.

Lysine, on the other hand, has a side chain that contains an amine and is therefore basic. According to Equation 6-27, lysine's zwitterion is a product in the second deprotonation (pK_a = 8.95) and a reactant in the third (pK_a = 10.53). As a result, the pI for lysine is (8.95 + 10.53)/2 = 9.74.

The three examples we have discussed show how an amino acid's pI depends on the nature of the side chain.

- Amino acids with *acidic* side chains have pI values that lie between about 2 and 3.
- Amino acids with *neutral* side chains have pI values that lie between about 5 and 6.
- Amino acids with *basic* side chains have pI values >7.

problem **6.39** Compute the pI values of glutamic acid and tyrosine. If the gel in an electrophoresis experiment is at pH 7, in which direction will each of these amino acids migrate, toward the cathode or toward the anode? Explain.

Could you use simple electrophoresis to separate a solution containing glycine (pI = 6.07), aspartic acid (pI = 2.98), lysine (pI = 9.74), and asparagine (pI = 5.41)? If the pH of the gel is buffered to 5.41, then asparagine will not migrate (because pH = pI), aspartic acid will migrate toward the positively charged terminal (because the gel is more basic than its pI), but both glycine and lysine will migrate toward the negatively charged terminal (because the gel is more acidic than their pIs). It is difficult, therefore, to separate glycine from lysine in this manner.

In cases like this, chemists often turn to **isoelectric focusing**, a particularly elegant way of separating amino acids or proteins. Isoelectric focusing is similar to the electrophoresis experiments described previously, except that a **pH gradient** is used. That is, the pH increases toward the cathode along the paper or gel (Fig. 6-16), rather than being constant throughout. If an amino acid is at a location where the pH is higher (more basic) than its pI value, then it will have a net negative charge and will move away from the cathode, which is in the direction of lower pH. It will continue to migrate until pH = pI, at which point the amino acid will no longer migrate.

Another species at a location where the pH is lower than its pI will have a net positive charge, and will move toward the cathode, which is in the direction of increasing pH. It, too, will continue to migrate until pH = pI. In this way it is possible to separate amino acids or proteins whose isoelectric points differ by only a few hundredths of a pH unit.

problem **6.40** The pI for lysine is 9.74. In an isoelectric focusing experiment, toward which terminal will lysine migrate if it is placed on the gel at a pH of 12? If it is placed on the gel at a pH of 1? If it is placed on the gel at a pH of 7? At what pH will lysine remain stationary?

Positively charged species migrate toward the cathode.

Component is focused at its isoelectric point.

Negatively charged species migrate toward the anode.

⊕ Anode

⊖ Cathode

Lower pH

pH Gradient

Higher pH

Electrodes

FIGURE 6-16 Isoelectric focusing of peptides and proteins The paper or gel has a pH gradient in which pH increases toward the cathode and decreases toward the anode. Each peptide "seeks" the point on the paper or gel that corresponds to its pI.

Chapter Summary and Key Terms

- In a **proton transfer reaction**, a proton is transferred from a **Brønsted–Lowry acid** to a **Brønsted–Lowry base** in a single **elementary step**; that is, one bond is broken and another is formed simultaneously. (**Section 6.1**)

- **Curved arrow notation** describes the movement of electrons in an elementary step of a mechanism, showing explicitly the breaking and/or forming of bonds. A *double-barbed* curved arrow (⌒⟶) represents the movement of a pair of valence electrons. (**Section 6.1**)

- A reaction's **equilibrium constant**, K_{eq}, reflects the tendency of that reaction to form products. A reaction with $K_{eq} > 1$ favors products and one with $K_{eq} < 1$ favors reactants. (**Section 6.2**)

- The equilibrium between an acid (HA) and water is described by the **acidity constant**, K_a, where $K_a = [A^-]_{eq}[H_3O^+]_{eq}/[HA]_{eq}$. A compound's K_a reflects its strength as an acid; for convenience, acid strength is usually reported as $\mathbf{p}K_a$, where $pK_a = -\log K_a$. (**Section 6.2a**)

- In a proton transfer reaction, the equilibrium favors the side *opposite* the stronger acid. (**Section 6.2b**)

- A solvent's **leveling effect** dictates the maximum strength of an acid or a base that can exist in solution. The strongest acid that can exist is the protonated solvent; the strongest base that can exist is the deprotonated solvent. (**Section 6.2c**)

- The **percent dissociation** of an acid (HA) in water increases as the pH of the solution increases (i.e., as the concentration of H_3O^+ decreases). This observation is embedded in the Henderson–Hasselbalch equation: $pH = pK_a + \log([A^-]/[HA])$. When the pH of the solution equals the acid's pK_a, the acid is 50% dissociated. (**Section 6.2d**)

- A compound's pK_a is governed primarily by the functional group on which the acidic proton resides. The pK_a value can be affected by nearby electronegative atoms or adjacent double bonds. (**Section 6.3**)

- A reaction tends to be spontaneous if its change in **standard Gibbs free energy**, $\Delta G°_{rxn}$, is negative. If $\Delta G°_{rxn}$ is positive, then the reverse reaction tends to be spontaneous. (**Section 6.3**)

- $\Delta G°_{rxn}$ consists of an **enthalpy** term and an **entropy** term: $\Delta G°_{rxn} = \Delta H°_{rxn} - T\Delta S°_{rxn}$. For many reactions, including the proton transfer reaction, $\Delta G°_{rxn} \approx \Delta H°_{rxn}$ because the change in entropy is so small. (**Section 6.3a**)

- A reaction's **free energy diagram** plots its free energy as a function of the **reaction coordinate**: a measure of geometric changes of the species involved in a reaction as reactants are transformed into products. The maximum in free energy between reactants and products, which corresponds to the **transition state**, creates an energy barrier for the reaction called the **free energy of activation**, $\Delta G°^{\ddagger}$. (**Section 6.3b**)

- A charged species is generally high in energy, and therefore tends to be unstable and reactive. Thus, positively charged acids tend to be more acidic than similar uncharged acids. (**Section 6.5**)

- Relative pK_a values for acids reflect the relative *charge stability* of the reactants and products. For an uncharged acid (HA), the stability of the conjugate base (A^-) increases as the pK_a decreases. For a positively charged acid (HA^+), the stability of the acid decreases as the pK_a decreases. (**Section 6.6**)

- For two ions in which the formal charge is on a different atom in the same row of the periodic table, the electronegativity of the atom governs the stability of the species. A negative charge is energetically favored on the more electronegative atom, whereas a positive charge is energetically favored on the less electronegative atom. (**Section 6.6a**)

- For two ions in which the formal charge is on a different atom in the same column of the periodic table, the size of the atom governs stability. The charge is favored on the atom that is larger, and hence in a lower row of the periodic table. (**Section 6.6b**)

- For two ions in which the formal charge is on an atom of the same element, hybridization governs stability. A negative charge is energetically favored on the atom with the greater **effective electronegativity** (i.e., $sp^3 < sp^2 < sp$). A positive charge is energetically favored on the atom with the lower effective electronegativity. (**Section 6.6c**)

- **Resonance effects** can stabilize a charged species. A species in which a charge is **delocalized** by resonance is more stable than one in which the charge is **localized**. All else being equal, the stability of the charged species increases as the number of atoms over which the charge is delocalized increases. (**Section 6.6d**)

- **Inductive effects** can affect the stability of a charged species by shifting electron density through covalent bonds. An atom that is more electronegative than hydrogen is considered to be **electron withdrawing**, so it stabilizes a nearby negative charge but destabilizes a nearby positive charge. An **electron-donating** group, such as an alkyl group or an atom that is less electronegative than hydrogen, stabilizes a nearby positive charge but destabilizes a nearby negative charge. (**Section 6.6e**)

- Inductive effects are additive. The greater the number of groups that contribute to an inductive effect, the greater the effect. (**Section 6.6e**)

- Inductive effects fall off quickly with distance. (**Section 6.6e**)

- In general, the order of importance for factors affecting charge stability is (**Section 6.7**):
  1. the presence of charge,
  2. the type of atom on which the charge resides,
  3. resonance effects, and
  4. inductive effects.

- Charge stability is an excellent predictor of a resonance structure's relative contribution to the overall resonance hybrid. All else being equal, the contribution of a resonance structure increases as the stability of the charge increases. (**Section 6.8**)

6.41 Given the curved arrow notation for each of the following proton transfer reactions, draw the appropriate products.

(a)

→ ?

(b)

→ ?

(c)

→ ?

(d)

→ ?

6.42 Given the reactants and products of each of the following proton transfer reactions, supply the missing curved arrows. Add relevant electrons if they are not shown.

(a)

(b)

(c)

(d)

6.43 In each of the following species, identify the most acidic proton and estimate its pK_a.

(a)

(b)

(c)

(d)

(e)

(f)

(g)

(h)

6.44 For each of the following species, identify the most basic site.

6.45 For each pair of molecules, predict which is the stronger acid and explain your reasoning.

(a)

H₃C–C(=O)–CH₃   or   F₃C–C(=O)–CH₃

(b)

[two benzene rings joined by CH₂]   or   [benzene ring with CH₃]

6.46 For each pair of species, predict which is the stronger base? Explain.

(a) [propanol with OH] or [propoxide with O⁻]

(b) [propanol with OH] or [propyl oxonium OH₂⁺]

(c) [2-butene with ⊖] or [1-butyne with ⊖]

(d) [tetrahydropyran, O] or [piperidine, N–H]

(e) [piperidide, N⁻] or [phosphorinanide, P⁻]

(f) [cyclopentadienyl anion ⊖] or [cyclopentyl anion ⊖]

(g) [cyclopentadienyl anion ⊖] or [fluorocyclopentyl anion ⊖, F]

(h) [cyclopentadienyl anion ⊖] or [pentafluorocyclopentadienyl anion, F substituted ⊖]

(i) [cyclopentadienyl anion ⊖] or [cyclohexadienyl anion ⊖]

6.47 Sulfuric acid (H_2SO_4) is called a *diprotic acid* because it has two acidic protons. The pK_a for the first deprotonation is −9, whereas the pK_a for the second deprotonation is 2. Explain these relative acid strengths.

6.48 Based on the pK_a values of the following substituted acetic acids, which is a stronger electron-withdrawing group, CO_2H or NO_2? Can you explain why? (*Hint:* Write out the complete Lewis structures.)

$pK_a = 1.68$ $pK_a = 2.83$

6.49 Use the electrostatic potential maps provided to predict whether C≡N or CF₃ is a stronger electron-withdrawing substituent. Explain.

6.50 An important step in one synthesis of carboxylic acids is the deprotonation of diethyl malonate and its alkyl substituted derivative:

Diethyl malonate

Alkyl substituted diethyl malonate

NaOH can deprotonate diethyl malonate effectively, but NaOC(CH₃)₃ is typically used to deprotonate the alkyl substituted derivative. Explain why.

6.51 The pK_a of phenol (C₆H₅OH) is 10.0. When a nitro group (NO₂) is attached to the ring, the pK_a decreases, as shown for the ortho, meta, and para isomers:

Phenol

pK_a = 10.0   7.23   8.35   7.14

(a) Explain why the pK_a values of all three isomers are lower than the pK_a of phenol itself.
(b) Explain why the meta isomer has the highest pK_a of the three isomers.

6.52 The pK_a of phenol (C₆H₅OH) is 10.0. When a methyl (CH₃) group is attached to the ring, the pK_a increases, as shown for the ortho, meta, and para isomers:

Phenol

pK_a = 10.0   10.3   10.1   10.2

(a) Explain why the pK_a values of all three isomers are higher than the pK_a of phenol itself.

(b) Explain why the meta isomer has the lowest pK_a of the three isomers.

6.53 The pK_a values of the following three acids are 9.0, 9.1, and 9.2. Match each of these values with its structure.

A   B   C

6.54 Which of the following resonance structures has the greatest contribution to the hybrid? Explain.

D   E   F   G

6.55 Draw all resonance structures of the following ion and determine which is the strongest contributor to its resonance hybrid.

6.56 Based on the following pK_a values, which do you think is more important in determining inductive effects: electronegativity or distance from the reaction center? Explain.

$pK_a = 2.68$ $pK_a = 3.85$ $pK_a = 2.95$ $pK_a = 3.97$

6.57 Students are often taught in general chemistry that HCl, HBr, and HI are all "strong acids" and no distinction is made among them. Based on charge stability, rank these acids from least acidic to most acidic.

6.58 Based on charge stability, rank the following species from weakest base to strongest base.

A   B   C   D

E   F   G

6.59 *cis*-Cyclohexane-1,2-diol is more acidic than *trans*-Cyclohexane-1,2-diol. Based on charge stability, explain why this is the case.

cis-Cyclohexane-1,2-diol   **trans-Cyclohexane-1,2-diol**

6.60 The pK_a of a typical ketone is 20, whereas the pK_a of a typical ester is 25. Explain why.

pK_a = 20   pK_a = 25

6.61 Explain why phenol (C_6H_5OH) is substantially more acidic than methanol (CH_3OH), but benzoic acid ($C_6H_5CO_2H$) is not much more acidic than acetic acid (CH_3CO_2H).

pK_a = 16   pK_a = 10   pK_a = 4.75   pK_a = 4.20

6.62 Which do you predict will be a stronger acid, *m*-hydroxybenzaldehyde or *p*-hydroxybenzaldehyde? Explain.

m-Hydroxybenzaldehyde   **p-Hydroxybenzaldehyde**

6.63 Which of the following compounds do you expect to be the stronger base? Explain.

or

6.64 Which do you expect to be the stronger base: HCN or HNC? Explain. (*Hint:* Draw out the complete Lewis structure for each molecule.)

6.65 Which do you expect to be the stronger acid: CH_3CN or CH_3NC? Explain. (*Hint:* Draw out the complete Lewis structure for each molecule.)

6.66 The pK_a of formaldehyde ($H_2C=O$) is not listed in Table 6-1.
   (a) Based on your understanding of charge stability, which of the following compounds would you expect to have a pK_a most similar to that of formaldehyde?

   (b) Using Table 6-1, as well as your knowledge of the factors that affect charge stability, estimate the pK_a of formaldehyde.

6.67 The acid-catalyzed hydrolysis of an ester converts an ester into a carboxylic acid:

| Ester | | Carboxylic acid | Alcohol |

Although there are two O atoms that can be protonated, the first step in the mechanism is believed to be protonation of the oxygen in the C=O group. Based on charge stability, why is it favorable to protonate that oxygen? (*Hint:* Draw out the products of each protonation.)

6.68 For which of the following molecules do you expect the pK_a value of the OH proton to be the most similar to that of cyclohexanol? Which do you expect to be the most different? Explain.

Cyclohexanol   A   B   C   D

6.69 If 0.100 mol of phenol, C_6H_5OH, were dissolved in pure water to make 1.000 L of total solution, what is the concentration of $C_6H_5O^-$ at equilibrium? What is the acid's percent dissociation?

6.70 For each of the following proton transfer reactions,
(a) draw the products,
(b) determine which side of the equilibrium is favored, and
(c) compute the equilibrium constant for the reaction. (*Hint:* For acids not in Table 6-1 or in Appendix A, you will need to estimate pK_a values based on charge stability and the functional group in which the acidic proton appears.)

(i)

(ii)

(iii)

(iv)

(v)

(vi)

(vii)

6.71 Draw the curved arrow notation for the proton transfer reaction between the hydride anion (H⁻) and ethanol (CH_3CH_2OH).

Hydride anion   Ethanol

Using the pK_a values listed in Table 6-1, predict which side of this reaction is favored. What is the equilibrium constant for this reaction?

6.72 Keeping in mind the leveling effect, can the following species be used as a reactant in ethanol?

(a)   (b)   (c)   (d)   (e)   (f)

6.73 Keeping in mind the leveling effect, can the following species be used as a reactant in ethanamine ($CH_3CH_2NH_2$)?

(a) $Cl^{\ominus}$ (b) HCl (c) $^{\ominus}CH_3$ (d) $^{\ominus}NH_2$ (e) $^{\oplus}NH_4$ (f) $^{\ominus}OH$

6.74 At what pH will Cl_3CCO_2H dissociate 50% in water? At what pH will it dissociate 90%? At what pH will it dissociate 10%?

6.75 The protonated form of aniline has a pK_a of about 10.7. At what pH would you expect the species to exist predominantly in the protonated (cationic) form? At what pH would you expect it to exist predominantly in the uncharged form? At what pH would you expect an equal mixture of the two forms?

$pK_a = 10.7$

6.76 (a) Draw the products of the following proton transfer reaction.

$+ \quad ^{\ominus}NH_2 \longrightarrow$

(b) Draw a free energy diagram for this reaction, indicating whether it is endothermic or exothermic.

6.77 (a) Draw the products of the following proton transfer reaction.

$+ \quad \text{OH} \longrightarrow$

(b) Draw a free energy diagram for this reaction, indicating whether it is endothermic or exothermic.

6.78 Two possible proton transfer reactions can take place between the following reactants.

$\text{OH} \quad + \quad \overset{H}{N} \longrightarrow$

(a) Write the products of each possible proton transfer reaction.
(b) Determine which reaction is more energetically favorable.

6.79 A reactant **W** can undergo two separate reactions to yield either **X** or **Y**, as indicated in the following free energy diagram. Which product is in greater abundance at equilibrium?

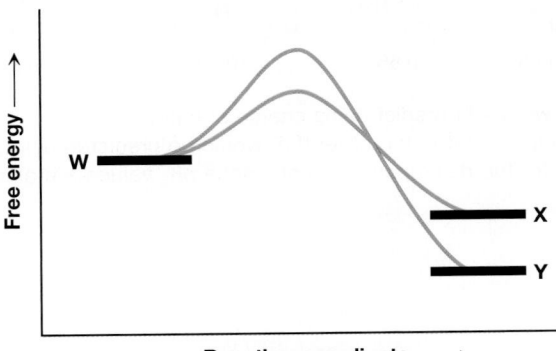

Free energy

W

X

Y

Reaction coordinate ⟶

6.80 Repeat Problem 6.79 for the following free energy diagram.

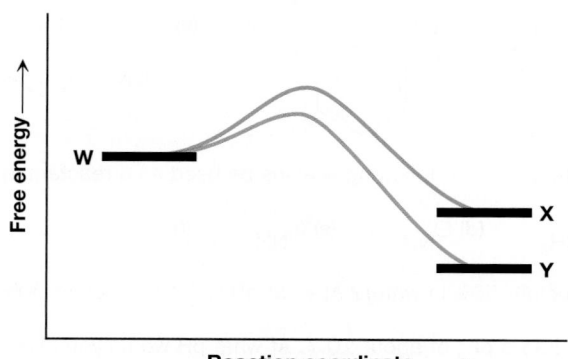

Reaction coordinate ⟶

6.81 Deuterium (D) is an isotope of H. Both D and H have one proton and one electron; H has no neutrons and D has one. Consequently, D and H have nearly identical behavior, but they can be distinguished from each other experimentally due to their different masses. Therefore, replacing an H with a D in a molecule—*deuterium isotope labeling*—can provide valuable information about a mechanism. With this in mind, how would you synthesize each of the following deuterium-labeled compounds from the analogous unlabeled compound, using D_2O as your only source of deuterium? (*Hint:* You will have to use two separate proton transfer reactions to synthesize each one.)

(a)

(b)

(c)

6.82 When water is the solvent, the pK_a of acetic acid (CH_3CO_2H) is 4.75, but when DMSO is the solvent, the pK_a is 12.6. Explain why. (*Hint:* Review Section 2.9 and consider the ability of each solvent to solvate cations and anions.)

6.83 When water is the solvent, the pK_a of NH_4^+ is 9.4, but when DMSO is the solvent, the pK_a is 10.5. Explain why the acid strength of NH_4^+ is similar in the two solvents, but, as shown in Problem 6.82, the acid strength of acetic acid is much different in the two solvents. (*Hint:* Review Section 2.9 and consider the ability of each solvent to solvate cations and anions.)

6.84 In which of the following solvents will CH_3NH_2 be the *weakest* base? Explain.

A   B   C   D CCl_4
                                                E

6.85 The pK_a value for an ammonium ion (R_3NH^+) depends on the number of alkyl groups attached to N:

$\overset{\oplus}{N}H_4$ $H_3C\overset{\oplus}{N}H_3$ $H_3C\overset{\oplus}{\underset{H_2}{N}}CH_3$ $H_3C\overset{CH_3}{\underset{H}{\overset{\oplus}{N}}}CH_3$

$pK_a = 9.25$ 10.66   10.73   9.81

This order disagrees with what we would predict using charge stability.
(a) From least acidic to most acidic, what is the order that would be predicted using charge stability?
(b) Can you explain the reason for the discrepancy? (*Hint:* These pK_a values are all measured in water.)

If your instructor assigns problems in smartwork, log in at **smartwork.wwnorton.com**.

CHAPTER 7

An Overview of the Most Common Elementary Steps

In Chapter 6, we learned that a proton transfer occurs as a single event, and is thus a type of *elementary step*. If it takes place in isolation from other steps, as we saw in the examples in Chapter 6, then it constitutes an *overall reaction*. Frequently, however, a proton transfer makes up an individual step of a multistep mechanism—something we will begin to see in Chapter 8.

There are a handful of other quite common elementary steps as well, which can be combined in various ways to produce mechanisms for numerous reactions. Chapter 7 provides an overview of nine of these elementary steps:

1. Bimolecular nucleophilic substitution (S_N2, Section 7.2)
2. Coordination (Section 7.3)
3. Heterolysis (Section 7.3)
4. Bimolecular elimination (E2, Section 7.4)
5. Nucleophilic addition (Section 7.5)
6. Nucleophile elimination involving a polar π bond (Section 7.5)
7. Electrophilic addition (Section 7.6)
8. Electrophile elimination involving a nonpolar π bond (Section 7.6)
9. Carbocation rearrangements (Section 7.7)

These steps, along with proton transfer steps (10 in all), compose nearly all of the reaction mechanisms you will encounter through Chapter 23.

Coordination and heterolysis (Section 7.3), nucleophilic addition and nucleophile elimination (Section 7.5), and electrophilic addition and electrophile elimination (Section 7.6) are discussed in pairs in their respective sections because one is simply the reverse of the other. This means both are governed by the same factors.

Pollen sticks to this bee due to electrostatic attraction. When the bee flies, it loses electrons and builds up a positive electrical charge. The flower's pollen, on the other hand, is negatively charged, so when the bee lands, the pollen experiences a driving force from the flower to the bee. Similarly, the elementary steps we examine here in Chapter 7 are driven by the flow of electrons from a site with excess negative charge toward a site with excess positive charge.

Upon completing Chapter 7 you should be able to:

- Describe how the curved arrow notation for a proton transfer step reflects the flow of electrons from an *electron-rich* site to an *electron-poor* site.

- Identify the following common elementary steps in reaction mechanisms:
    - Proton transfer
    - Bimolecular nucleophilic substitution (S_N2)
    - Coordination
    - Heterolysis
    - Bimolecular elimination (E2)
    - Nucleophilic addition
    - Nucleophile elimination
    - Electrophilic addition
    - Electrophile elimination
    - Carbocation rearrangements

- Describe the electron flow in each of the above steps in terms of electron-rich and electron-poor sites.

- Rationalize the driving force for each of the above elementary steps in terms of charge stability and bond energies.

- Draw the keto and enol forms of a keto–enol tautomerization, and explain the driving force for such a reaction.

Each of these nine elementary steps can be depicted using curved arrow notation, much as proton transfer steps were depicted in Chapter 6. However, beyond simply describing *how* each step takes place, we need to know *why* it would (or would not) take place. Therefore, we devote Section 7.8 to the *driving force* for elementary steps. Understanding the driving force for each step will ultimately enable us to use mechanisms to make *predictions* about the outcomes of reactions.

7.1 Mechanisms as Predictive Tools: The Proton Transfer Step Revisited

A reaction mechanism can be very helpful as a *predictive* tool. In this section we will begin to see how, by revisiting *curved arrow notation* for a proton transfer step and examining specifically how it can be used to make predictions about bond formation and bond breaking.

Even though our discussion here is in the context of proton transfer steps, we can apply the same logic to other elementary steps as well. Effectively, then, this section establishes a process by which to *analyze* an elementary step. The utility of the process becomes apparent when we discuss new elementary steps later in this chapter.

7.1a Curved Arrow Notation: Electron Rich to Electron Poor

Curved arrow notation was introduced in Section 6.1 solely as a means for keeping track of valence electrons in an elementary step. It can be far more powerful than that, though, if we recall two concepts:

1. *Opposite charges attract; like charges repel.*
2. *Atoms in the first and second rows of the periodic table must obey the duet and octet rules, respectively.*

Equation 7-1 shows the appropriate curved arrows for the proton transfer between HCl and HO⁻.

(7-1)

Notice that HO⁻ bears a full negative charge. This excess electron density on O means the electrons on O are not particularly stable due to their mutual repulsion. The proton on HCl bears a partial positive charge, δ^+, because Cl is more electronegative than H. As a result, the electrons on O are attracted to the proton on HCl. This simultaneous charge repulsion among the electrons on O and their attraction to H facilitates the flow of electrons from O to H and results in the formation of the new O—H bond.

This example illustrates one of the most important guidelines for drawing curved arrows:

> In an elementary step, electrons tend to flow from an *electron-rich* site to an *electron-poor* site.

In Equation 7-1, the HO^- anion is relatively electron rich and the H atom on HCl is relatively electron poor, denoted by the light red screen behind the negatively charged O atom and the light blue screen behind the H bearing a partial positive charge. (This is the same color scheme used in electrostatic potential maps to represent areas of more and less electron density, respectively.)

This kind of electron flow can also be explained in terms of *Lewis bases* and *Lewis acids*. A **Lewis base** is an *electron-pair donor*, whereas a **Lewis acid** is an *electron-pair acceptor*. For the proton transfer in Equation 7-1, HO^- is the Lewis base, because it donates a lone pair of electrons to form a bond to H, whereas HCl is the Lewis acid, because its proton accepts that pair of electrons. Thus,

> In an elementary step, a *Lewis base* tends to form a bond with a *Lewis acid*.

The curved arrow in Equation 7-1 that goes from the H—Cl bond to the Cl atom is needed to maintain H's duet of electrons. Without that curved arrow—and hence without breaking that bond—H would end up with two bonds, which is one too many.

7.1 YOUR TURN

In the following proton transfer step, identify the electron-rich and electron-poor sites, and label the curved arrow that connects the two as "electron rich to electron poor." Which reactant is the Lewis acid and which is the Lewis base?

$$HS^{\ominus} \quad + \quad H—Br: \quad \longrightarrow \quad HS—H \quad + \quad :Br^{\ominus}$$

Answers to Your Turns are in the back of the book.

7.2 YOUR TURN

Use the box provided to draw the product suggested by the faulty curved arrow notation in the following chemical equation. What is unacceptable about the product you drew?

$$HO^{\ominus} \quad + \quad H—Cl: \quad \longrightarrow \quad \boxed{}$$

SOLVED problem 7.1 Identify the electron-poor H atom in methanol. Draw the mechanism by which methanol acts as an acid in a proton transfer reaction with H_2N^-.

$$H—\overset{\displaystyle H}{\underset{\displaystyle H}{C}}—\overset{..}{\underset{..}{O}}—H$$

Methanol

Think What kinds of charges characterize an electron-poor atom? Are there any formal charges present? Any strong partial charges? When H_2N^- and CH_3OH are combined, what curved arrow can we draw to depict the flow of electrons from an electron-rich site to an electron-poor site?

Solve CH_3OH does not bear any full charges, but the highly electronegative O places a strong partial positive charge on the H to which it is bonded. That H is therefore electron poor, and a blue screen is placed behind it as a reminder. H_2N^- bears a full negative charge and is therefore electron rich. As a reminder, a red screen is placed behind N. To represent the flow of electrons from the electron-rich site (H_2N^-) to the electron-poor site (CH_3OH), draw a curved arrow from a lone pair on N to the H atom on O. A second curved arrow is needed to make sure H has only one bond, not two.

Electron rich to electron poor

---

problem **7.2** Identify the electron-rich and electron-poor sites in each of the following reactant molecules:

Draw the curved arrows and the products for the proton transfer between these two molecules and label the curved arrow that represents the flow of electrons from an electron-rich site to an electron-poor site.

7.1b Simplifying Assumptions Regarding Electron-Rich and Electron-Poor Species

Contrary to what is suggested in Equation 7-1, we cannot simply add hydroxide anion (HO^-) to HCl to initiate a proton transfer reaction, because anions do not exist in the solid or liquid phase without the presence of cations, and vice versa. We can, however, add a *source* of HO^-, such as NaOH; NaOH is an ionic compound, so in solution it dissolves as Na^+ and HO^-.

This initially may seem to complicate the picture, because Na^+ is electron poor. We can thus envision Na^+ reacting with an electron-rich site of another species. However, Na^+ is relatively inert in solution and tends not to react. It behaves instead as a **spectator ion**, which is generally true of group IA metal cations (i.e., Li^+, Na^+, and K^+). Consequently, when we envision the flow of electrons from an electron-rich site to an electron-poor site, we can disregard such metal cations.

SOLVED problem **7.3** Draw the necessary curved arrows for the proton transfer between $KOCH_3$ and HCN in solution.

Think What are the electron-rich and electron-poor sites in each compound? Are there any simplifying assumptions we can make?

Solve $KOCH_3$ is an ionic compound that dissolves in solution as K^+ and CH_3O^-. We can therefore treat K^+ as a spectator ion and CH_3O^- as an electron-rich Lewis base. HCN has an electron-poor H atom due to the high effective electronegativity of the sp-hybridized C and N atoms. Thus, HCN acts as a Lewis acid. A curved arrow is drawn from the electron-rich O to the electron-poor H to initiate the proton transfer. A second curved arrow is drawn to break the H—C bond to avoid two bonds to H.

---

problem 7.4 Use curved arrow notation to indicate the proton transfer between NaSH and CH_3CO_2H.

Organometallic compounds contain *a metal atom bonded directly to a carbon atom*. Examples include: alkyllithium (R—Li); alkylmagnesium halide (R—MgX, where X = Cl, Br, or I), also called a **Grignard reagent**; and lithium dialkyl cuprate $[Li^+(R—Cu—R)^-]$. These kinds of organometallic compounds are useful reagents for forming new carbon–carbon bonds (see Chapter 10).

Although the carbon–metal bond in an organometallic compound joins a non-metal and a metal, these bonds exhibit significant covalent character. Nevertheless, the carbon–metal bonds are highly polar due to the large difference in electronegativity between carbon (2.55) and the metal (Li = 0.98, Mg = 1.31, and Cu = 1.90). As a result, there is a large partial negative charge on carbon (making it electron rich) and a large partial positive charge on the metal atom (making it electron poor; see Fig. 7-1).

Just as we did with the group IA metal cations, we can often treat the metal-containing portion of an organometallic compound as a spectator and disregard it in the context of electron flow from an electron-rich site to an electron-poor site. Therefore, as shown on the right in Figure 7-1, we can often treat organometallic compounds simply as electron-rich **carbanions**—compounds in which a negative formal charge appears on C. Applying this simplifying assumption allows us, for example, to treat CH_3CH_2Li as a source of $CH_3CH_2^-$, C_6H_5MgBr as a source of $C_6H_5^-$, and $(CH_3)_2CuLi$ as a source of CH_3^-, as shown in Figure. 7-2.

FIGURE 7-1 Simplifying assumptions in organometallic compounds (Left) Because of the high polarity in a carbon–metal bond, organometallic compounds contain an electron-rich site on C and an electron-poor site on the metal. (Right) We can usually ignore the reactivity of the metal-containing portion and treat the organometallic compound simply as a carbanion, which is electron-rich. The quotation marks remind us that the carbanion does not actually exist in solution.

(a) Li—CH_2CH_3 ⟶ " $:CH_2CH_3^{\ominus}$ "

(b) [benzene ring]—MgBr ⟶ " [benzene ring with $\ominus$] " — These three C atoms are electron rich.

(c) H_3C—Cu(Li)—CH_3 ⟶ " $H_3C:^{\ominus}$ "

FIGURE 7-2 Simplifying some specific organometallic compounds
(a) $LiCH_2CH_3$ is simplified to $^-CH_2CH_3$.
(b) C_6H_5MgBr is simplified to $C_6H_5^-$.
(c) $(CH_3)_2CuLi$ is simplified to H_3C^-.

SOLVED problem 7.5 What are the products of the proton transfer step between C_6H_5MgBr and H_2O?

[benzene ring]—MgBr + H_2O ⟶ ?

Think What are the electron-rich sites? What are the electron-poor sites? What can be disregarded?

Solve In H_2O, O is electron rich, and both H atoms are electron poor. In C_6H_5MgBr, we can disregard the MgBr portion because it contains the metal, and treat the reactive species as $C_6H_5^-$. Hence, the proton transfer is initiated by drawing a curved arrow from the electron-rich C on $C_6H_5^-$ to an electron-poor H on H_2O.

Electron rich to electron poor

[benzene ring with $\ominus$] + H—Ö—H ⟶ [benzene ring]—H + $:\overset{\ominus}{Ö}$—H

---

problem 7.6 Use curved arrow notation to show the proton transfer step that occurs between CH_3Li and CH_3OH. Predict the products of this reaction.

FIGURE 7-3 Simplifying assumptions in hydride reagents (a) Lithium aluminum hydride, LiAlH₄, consists of Li⁺ and AlH₄⁻ ions. The reactive species is AlH₄⁻ (middle), which can be treated simply as H⁻. The quotation marks indicate that H⁻ does not actually exist in solution. (b) Sodium borohydride, NaBH₄, consists of Na⁺ and BH₄⁻ ions. The reactive species is BH₄⁻ (middle), which can be treated as H⁻, too.

Hydride reagents, which include lithium aluminum hydride ($LiAlH_4$) and sodium borohydride ($NaBH_4$), are commonly used as *reducing agents* (see Chapter 17). $LiAlH_4$ consists of Li^+ ions and AlH_4^- ions, as shown on the left in Figure 7-3a. We can treat Li^+ as a spectator ion, leaving AlH_4^- as the reactive species (Fig. 7-3a, middle). According to the Lewis structure of AlH_4^-, the H atoms (electronegativity [EN] = 2.20) are bonded to an Al metal atom (EN = 1.61). The excess negative charge built up on each H atom means that we can treat $LiAlH_4$ as a source of hydride, H^-, as indicated on the right in Figure 7-3a. Using similar arguments, we can treat $NaBH_4$ as a source of H^-, too.

---

problem **7.7** Use curved arrow notation to show the proton transfer step that takes place between $LiAlH_4$ and water. Predict the products of the reaction. Do the same for the proton transfer step between $NaBH_4$ and phenol (C_6H_5OH).

Later, we will need to be careful when making these simplifying assumptions, particularly when treating organometallic compounds as carbanions, R^-, and hydride reagents as hydride anions, H^-. Experimentally, significant differences in reactivity are observed among the different organometallic compounds and among the different hydride reagents. These differences are discussed more extensively in the context of nucleophilic addition reactions in Chapters 17 and 18.

7.2 Bimolecular Nucleophilic Substitution (S$_N$2) Steps

In a **bimolecular nucleophilic substitution (S$_N$2) step**, a molecular species, called a **substrate**, undergoes substitution in which one atom or group of atoms is replaced by another. Examples are shown in Equations 7-2 and 7-3.

HO^- substitutes for Cl^- in Equation 7-2, and H_3N substitutes for $CH_3SO_3^-$ in Equation 7-3.

Reactions involving S_N2 steps are really important in organic chemistry, especially in *organic synthesis* (Chapter 13). Sometimes the product of an S_N2 reaction is the compound you want to synthesize. In other cases, an S_N2 reaction, by substituting one atom or group for another, can be used to alter the reactivity of a molecule in ways that make further reactions possible. Because of this central role in synthesis, we revisit S_N2 reactions in Chapters 8–10.

During the course of the S_N2 steps in Equations 7-2 and 7-3, a **nucleophile** forms a bond to the substrate at the same time that a bond is broken between the substrate and a **leaving group**. The step is said to be "bimolecular" because it contains two separate reacting species in an elementary step. In other words, the step's **molecularity** is 2. It is called "nucleophilic" because a nucleophile is the species that reacts with the substrate.

Equations 7-2 and 7-3 also show that a leaving group often comes off in the form of a negatively charged species. *Common leaving groups are relatively stable with a negative charge.* Applying what we learned in Sections 6.6 and 6.7, we can say that

Leaving groups are typically *conjugate bases of strong acids.*

Leaving groups thus include Cl^-, Br^-, and I^- (conjugate bases of the strong acids HCl, HBr, and HI, respectively) because, in each case, the negative charge is on a relatively large and/or electronegative atom. They also include anions in which there is substantial resonance and/or inductive stabilization, such as an alkylsulfonate anion, RSO_3^- (the conjugate base of RSO_3H). Water and alcohols (ROH) are also leaving groups because they, too, are the conjugate bases of strong acids (H_3O^+ and ROH_2^+, respectively).

A nucleophile tends to be attracted by and form a bond to an atom that bears a partial or full positive charge.

The term "nucleophile" literally means *nucleus loving.* It is given this name because the nucleus of an atom bears a positive charge, the type of charge to which a nucleophile is attracted.

Species that act as nucleophiles generally have the following two attributes:

1. A nucleophile has an atom that carries a full negative charge or a partial negative charge. The charge is necessary for it to be attracted to an atom bearing a positive charge.
2. The atom with the negative charge on the nucleophile has a pair of electrons that can be used to form a bond to an atom in the substrate. As shown in Equations 7-2 and 7-3, those electrons are usually lone pairs.

In the nucleophile in Equation 7-2 (HO^-), the O atom possesses a full negative charge and three lone pairs of electrons, as indicated below. In Equation 7-3, on the other hand, the nucleophile (H_3N) has a N atom bearing a *partial* negative charge and one lone pair of electrons.

Other common, negatively charged nucleophiles include CH_3O^-, Cl^-, Br^-, I^-, H_2N^-, CH_3NH^-, HS^-, CH_3S^-, $N\equiv C^-$, and N_3^-. Other common uncharged nucleophiles include H_2O, CH_3OH, CH_3NH_2, H_2S, and CH_3SH.

Draw the complete Lewis structures for three of the *negatively charged* nucleophiles listed on the previous page (other than HO^-) and for three of the uncharged nucleophiles (other than H_3N). Include all lone pairs. For each of the uncharged nucleophiles, write "δ^-" next to the atom bearing a partial negative charge.

SOLVED problem 7.8 Should CH_4 act as a nucleophile? Why or why not?

Think Does CH_4 have an atom that carries a partial or full negative charge? Does that atom have a pair of electrons that can be used to form a bond to another atom?

Solve The complete Lewis structure for CH_4 is as follows:

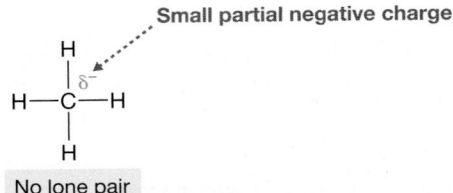

Small partial negative charge

No lone pair

The C atom has a small partial negative charge (because C is slightly more electronegative than H), but it does not possess a lone pair of electrons that can be used to form a bond with another atom. Thus, CH_4 should not act as a nucleophile.

---

problem 7.9 Which of the following species can behave as a nucleophile? Explain.

(a) SiH_4 **(b)** NaSCN **(c)** NH_4^+ **(d)** CH_3Li

Recall from Section 7.1a that the electrons in an elementary step tend to flow from an *electron–rich* site to an *electron–poor* site. In the proton transfer in Equation 7-1 (shown again in Equation 7-4a), for example, a curved arrow is drawn from the O in HO^- to the H in HCl, and a second curved arrow is drawn to show the initial bond to H being broken, with its electrons becoming a lone pair on Cl^-.

Proton transfer

Electron rich   Electron poor

$$HO^- + H-Cl: \longrightarrow HO-H + :Cl:^- \qquad (7\text{-}4a)$$

S_N2

Electron rich   Electron poor

$$HO^- + H_3C-Cl: \longrightarrow HO-CH_3 + :Cl:^- \qquad (7\text{-}4b)$$

A similar thing happens in the S_N2 step shown in Equation 7-4b. In this case, HO^- is still relatively electron rich, but the C atom of CH_3Cl is relatively electron poor because it carries a partial positive charge. Thus, a curved arrow is drawn from a pair of electrons on O to the C atom to signify bond formation, and to avoid five bonds to C (which would exceed the C atom's octet), a second curved arrow shows that the pair of electrons initially composing the C—Cl bond becomes a lone pair on Cl^-.

An S_N2 reaction can also be viewed as a Lewis acid–Lewis base reaction. In Equation 7-4b, the nucleophile (HO^-) behaves as a Lewis base (because it donates the lone pair of electrons used to form the bond to the substrate), whereas the substrate behaves as a Lewis acid (because it accepts the pair of electrons from the nucleophile).

The following S_N2 step is similar to the one in Equation 7-4b:

$$:\overset{..}{\underset{..}{Cl}}:^{\ominus} \ + \ H_3C-\overset{..}{\underset{..}{Br}}: \ \longrightarrow \ :\overset{..}{\underset{..}{Cl}}-CH_3 \ + \ :\overset{..}{\underset{..}{Br}}:^{\ominus}$$

Label the appropriate reacting species as "electron rich" or "electron poor" and draw in the correct curved arrows.

SOLVED problem 7.10 Draw the S_N2 step that would occur between $C_6H_5CH_2I$ and CH_3SNa.

Think Which species is the nucleophile? Which is the substrate? What do we do with the metal atom? Which species is electron rich? Electron poor?

Solve $C_6H_5CH_2I$ will behave as the substrate because it possesses as I, a good leaving group that departs as I^-. The conjugate acid of I^-, HI, is a very strong acid. CH_3SNa has a metal atom that can be treated as a spectator ion and thus ignored. The nucleophile is therefore CH_3S^-. In an S_N2 step, a curved arrow is drawn from the lone pair of electrons on the electron-rich S atom to the electron-poor C atom bonded to I. A second curved arrow must be drawn to indicate that the C—I bond is broken (otherwise that C would have five bonds).

---

problem 7.11 Draw the S_N2 step that would occur between the two compounds at the right.

7.3 Bond-Formation (Coordination) and Bond-Breaking (Heterolysis) Steps

In both the proton transfer and the S_N2 steps we have examined so far, a bond is formed and a separate bond is broken simultaneously. It is possible, however, for bond formation and bond breaking to occur as independent steps. Equations 7-5 and 7-6, for example, show steps in which only a single covalent bond is formed. These are called **coordination steps**.

An elementary step can also occur in which only a single bond is broken and *both electrons from that bond end up on one of the atoms initially involved in the bond*, as shown in Equations 7-7 and 7-8. These are called **heterolytic bond dissociation steps**, or **heterolysis** (*hetero* = different; *lysis* = break) **steps**, to emphasize that, once the bond is broken, the two electrons are not distributed equally to the atoms initially involved in the bond. *Heterolysis steps are the reverse of coordination steps.*

Heterolysis step

$$\text{cyclohexyl–}\ddot{\text{I}}\!: \;\rightleftharpoons\; \text{cyclohexyl}^{+} \;+\; :\!\ddot{\text{I}}\!:^{\ominus} \qquad (7\text{-}7)$$

Heterolysis step

$$H_3C-\underset{\overset{|}{\overset{\oplus}{:}\ddot{C}l-\overset{\ominus}{Al}(Cl)_2Cl}}{C}\!\!=\!\!O \;\rightleftharpoons\; H_3C-\overset{\oplus}{C}\!\!=\!\!O \;+\; :\ddot{C}l-\overset{\ominus}{Al}(Cl)_2Cl \qquad (7\text{-}8)$$

Unlike proton transfer and S_N2 steps, coordination and heterolysis generally do not take place in isolation. This is because one of the reactant or product species lacks an octet and is therefore highly unstable and reactive. Instead, coordination and heterolysis steps usually compose one step of a mechanism involving two or more elementary steps—a so-called *multistep mechanism* (Chapter 8). We will see, for example, that coordination and heterolysis occur in mechanisms for S_N1 and E1 reactions (Chapter 8), electrophilic addition reactions (Chapter 11), and Friedel–Crafts reactions (Chapter 22).

YOUR TURN 7.5

In Equations 7-5 through 7-8, identify all atoms lacking an octet.

In a coordination step, electrons flow directly from an *electron-rich* site to an *electron-poor* site. In Equation 7-5, Br^- is negatively charged and hence electron rich. $(CH_3)_3C^+$, on the other hand, has a C atom that is positively charged and has only six shared electrons, so it is electron poor. Br^-, therefore, is a Lewis base and $(CH_3)_3C^+$ is a Lewis acid.

In Equation 7-6, CH_3COCl has an electron-rich Cl atom that bears a partial negative charge, whereas the Al atom in $AlCl_3$ is electron poor, largely because it is short of an octet. Moreover, the surrounding Cl atoms on $AlCl_3$ are electron withdrawing via induction (Section 6.6e), which makes Al even more electron deficient.

YOUR TURN 7.6

Label each reactant in Equation 7-6 as either a Lewis acid or a Lewis base.

problem 7.12

(a) Draw the appropriate curved arrows for the coordination step between $FeCl_3$ and Cl^-. Draw the reaction products. Identify which species acts as the Lewis acid and which acts as the Lewis base.

(b) Use curved arrow notation to show the product from part (a) undergoing heterolysis to regenerate $FeCl_3$ and Cl^-.

7.4 Bimolecular Elimination (E2) Steps

The elementary steps we have examined so far involve the breaking and/or formation of single bonds only. A **bimolecular elimination (E2) step**, however, forms a double bond or triple bond in the products, as shown in Equations 7-9 through 7-11.

$$(7\text{-}9)$$

$$(7\text{-}10)$$

$$(7\text{-}11)$$

An E2 step can take place when a strong base is in the presence of a substrate in which *a leaving group (L) and a hydrogen atom are on adjacent carbon atoms*. That is,

> In an E2 step, the substrate generally has the form H—C—C—L or H—C=C—L.

It is called an elimination step because both the H atom and the leaving group (L) are *eliminated* from the substrate. It is a *bimolecular* step because there are two reactant species in an elementary step—the base and the substrate.

Equations 7-9, 7-10, and 7-11 illustrate the diversity of reactants and products that can be involved in E2 steps. The base is HO⁻ in both Equations 7-9 and 7-10, whereas it is H_2N^- in Equation 7-11. The leaving group can come off as a negatively charged species, as in Equations 7-9 and 7-11, or it can be uncharged, as in Equation 7-10. And the C atoms containing the H atom and the leaving group in the substrate may be joined by either a single bond (Equations 7-9 and 7-10) or a double bond (Equation 7-11).

The primary importance of E2 steps is the incorporation of a carbon–carbon multiple bond into a molecule at a particular site. In Equations 7-9 and 7-10, for example, a C=C double bond is generated. And, in Equation 7-11, a C≡C triple bond is formed. As we discuss later in this chapter, as well as in Chapters 11 and 12, the reactivity of carbon–carbon double and triple bonds is quite important in organic chemistry. We therefore revisit E2 steps in greater detail in Chapters 8, 9, and 10.

The base in an E2 step is the electron-rich species, but the hydrogen atom that the base attacks is not particularly electron poor; instead, the electron-poor atom is the carbon atom bonded to the leaving group. Thus, the movement of electrons from the *electron-rich* site to the *electron-poor* site is depicted with two curved arrows (Equation 7-12). One curved arrow is drawn from the negatively charged atom in the base to the proton in the substrate. The second curved arrow is then drawn from the C—H bond to the bonding region between the two C atoms. Overall, the electron-poor C atom gains a share of electrons from the C—H bond. Furthermore, that C atom is no longer electron poor in the products because it is no longer bonded to the electronegative leaving group.

In Equation 7-12, the strong base (B:$^-$) is the Lewis base because its lone pair of electrons is used to form a bond to the substrate. The substrate, on the other hand, is the Lewis acid because it receives the pair of electrons from the base.

---

problem **7.13** Supply the appropriate curved arrows and the products for each of the following E2 steps.

7.5 Nucleophilic Addition and Nucleophile Elimination Steps

In Section 7.2, we introduced the S$_N$2 step, in which a nucleophile bonds to an atom containing a suitable leaving group. A nucleophile can also bond to an atom that is involved in a *polar π bond*—that is, a π bond that is part of a double or triple bond connecting atoms with significantly different electronegativities, such as those in carbonyl groups (C=O), imine groups (C=N), and cyano groups (C≡N). Specifically, as shown in Equations 7-13 to 7-15, the nucleophile forms a bond to the less electronegative atom and the π bond breaks, becoming a lone pair on the more electronegative atom. A nucleophile adds to the polar π bond in these steps, so they are called **nucleophilic addition steps**.

Nucleophilic addition step

$$(7\text{-}14)$$

Nucleophilic addition step

$$(7\text{-}15)$$

The reverse of nucleophilic addition (Equations 7-16 and 7-17) is also commonly encountered in organic chemistry.

Nucleophile elimination step

The leaving group becomes nucleophilic

$$(7\text{-}16)$$

Nucleophile elimination step

The leaving group becomes nucleophilic

$$(7\text{-}17)$$

In both examples, a lone pair of electrons from a more electronegative atom forms a π bond with a less electronegative atom. A *leaving group* is simultaneously expelled to avoid exceeding an octet on the less electronegative atom. In the products, the leaving group ends up with a lone pair of electrons and an excess of negative charge, both of which are characteristics of a nucleophile. As a result, these are called **nucleophile elimination steps**.

Nucleophilic addition and nucleophile elimination steps are important in organic chemistry for three main reasons:

1. They can cause the number of bonds between two atoms of significantly different polarity to change, thus leading to *reduction* and *oxidation* reactions. We discuss these kinds of reactions in greater detail in Chapter 17.
2. They allow two relatively large organic species to be joined to make one even larger organic species, thus providing significant flexibility in designing molecular backbones for organic synthesis. We discuss these kinds of reactions in Chapters 18 and 21.
3. When nucleophilic addition and nucleophile elimination steps occur sequentially, *substitution* can take place at a polar π bond. We discuss these kinds of reactions in Chapters 20 and 21.

The *electron-rich to electron-poor* nature of a nucleophilic addition step is fairly straightforward. The nucleophile in a nucleophilic addition step, which has an excess of negative charge, is relatively electron rich, and the less electronegative atom of the polar π bond is relatively electron poor. Thus, the curved arrow drawn from the nucleophile to the polar π bond represents the flow of electrons from an electron-rich site to an electron-poor site:

In this sense, the nucleophile behaves as the Lewis base, whereas the species containing the polar π bond behaves as the Lewis acid. The second curved arrow, drawn from the center of the double (or triple) bond to the electronegative atom (X), is necessary to avoid exceeding an octet on the less electronegative atom (C).

YOUR TURN 7.7

For the following nucleophilic addition step, label the pertinent electron-rich and electron-poor sites. Add the appropriate curved arrows and draw the product. Identify the curved arrow that is drawn from the electron-rich site to the electron-poor site.

In nucleophile elimination (the reverse step), the more electronegative atom (X) is relatively electron rich: it bears either a full negative charge or a partial negative charge. The less electronegative atom (typically C) is relatively electron poor. Thus, the curved arrow that originates from a lone pair of electrons on X and points to the bonding region between C and X represents the flow of electrons from an electron-rich site to an electron-poor site:

Once again, a second curved arrow (to illustrate the breaking of the C—L bond) is necessary to avoid exceeding an octet on the less electronegative atom (C).

YOUR TURN 7.8

For the following nucleophile elimination step, label the pertinent electron-rich and electron-poor sites. Add the appropriate curved arrows and identify the curved arrow that is drawn from the electron-rich site to the electron-poor site.

(a)   (b)

problem **7.15** Draw the appropriate curved arrows necessary for each of the following nucleophile elimination steps to produce a ketone, and draw the resulting ketone.

(a)   (b)

7.6 Electrophilic Addition and Electrophile Elimination Steps

An **electrophilic addition** occurs when a species containing a nonpolar π bond (as part of a double or triple bond) approaches a strongly electron-deficient species, called an **electrophile**, and a bond forms between an atom of the π bond and the electrophile (Equations 7-18 and 7-19).

Electrophilic addition step

Nonpolar π bond   Electrophile

Electrophilic addition step

Nonpolar π bond   Electrophile

The nonpolar π bonds involved in electrophilic addition steps are typically ones that join a pair of carbon atoms. Additionally, the electrophile can be a strong acid, as in HCl (where the H atom is highly electron deficient, carrying a strong partial positive charge), or it can be a species that bears a full positive charge, such as NO_2^+.

The product of each of these electrophilic addition steps is a *carbocation*, which is highly unstable and will react further because it has a positive charge and lacks an octet. Electrophilic additions, therefore, are generally part of multistep mechanisms. Equation 7-18, for example, is the first step in the electrophilic addition of an acid across a multiple bond (Chapters 11 and 12). Equation 7-19, on the other hand, is the first step in electrophilic aromatic substitution (Chapters 22 and 23).

Add the appropriate curved arrows for the following electrophilic addition step.

Carbocations are typically quite unstable, so the reverse of electrophilic addition is also a common elementary step in organic reactions. In the reverse step, called **electrophile elimination**, an electrophile is *eliminated* from the carbocation, generating a stable, uncharged, organic species. Equations 7-20 and 7-21 show examples in which H^+ is the electrophile that is eliminated.

Electrophile elimination

(7-20)

Electrophile elimination

(7-21)

Equation 7-20 is the second step of an E1 reaction (Chapter 8) and Equation 7-21 is the second step of an electrophilic aromatic substitution reaction (Chapters 22 and 23).

Add the appropriate curved arrow(s) to the following electrophile elimination step, which is the reverse of the addition step in Your Turn 7.9.

Even though Equations 7-20 and 7-21 show that H^+ is the electrophile that is eliminated, *a proton does not exist on its own in solution*. Rather, it must be associated with a base. Any base that is present in solution will therefore assist in the removal of a proton in an electrophile elimination step. If water is present, for example, then Equation 7-20 would more appropriately be written as follows in Equation 7-22:

Electrophile elimination

(7-22)

As shown in Equation 7-23, the electrophile (E^+) in an electrophilic addition step is relatively electron poor because it either carries a full positive charge (Equation 7-19) or has an atom with a significant partial positive charge (Equation 7-18). The double or triple bond, on the other hand, is relatively electron rich. In a C=C double bond, for example, four electrons are localized in the region between two atoms, and in a C≡C triple bond, six electrons are localized between two atoms. Therefore, the movement of electrons from *electron rich to electron poor* is indicated by a curved arrow that originates from the center of the multiple bond (the Lewis base) and terminates at the electrophile (the Lewis acid).

In electrophile elimination (Equation 7-24), the positively charged C atom is relatively electron poor, whereas the C—E single bond is relatively electron rich. Therefore, the curved arrow originates from the C—E single bond and points to the bond between C and C^+.

Electrophile elimination

(7-24)

In both Your Turn 7.9 and 7.10, label the pertinent *electron-rich* and *electron-poor* sites.

7.11 **YOUR TURN**

---

problem **7.16** Supply the appropriate curved arrows and draw the product of each of the following electrophilic addition steps.

(a)   (b)

+ HBr ⟶ ?

---

problem **7.17** Supply the appropriate curved arrows and draw the product of each of the following electrophile elimination steps. (You may assume that a weak base is present.)

(a)   (b)

7.7 Carbocation Rearrangements: 1,2-Hydride Shifts and 1,2-Alkyl Shifts

Carbocations typically appear as intermediates in multistep mechanisms. They are usually too unstable to exist for a prolonged period of time because they are extremely electron deficient due to (1) the carbon atom's formal positive charge and (2) its lack of an octet. As we saw in Section 7.3, carbocations commonly behave as Lewis acids to form a bond with an electron-rich Lewis base. Carbocations can also eliminate H^+ (Section 7.6) to yield an alkene or alkyne. Given a chance, however, a carbocation can also undergo a **rearrangement** before it takes part in one of these steps with another species.

Equations 7-25 and 7-26 show two carbocation rearrangements. Equation 7-25 is a **1,2-hydride shift**. A *hydride anion* (H^-) is said to shift because a hydrogen atom migrates along with the pair of electrons initially making up the C—H bond. The numbering system denotes the number of atoms over which the hydride anion migrates; the atom to which the hydrogen is initially bonded is designated as the number 1 atom and any adjacent atom can be designated as a number 2 atom. Thus, a "1,2" shift refers to the hydrogen migrating to an adjacent atom.

A 1,2-hydride shift

H migrates with two electrons.

(7-25)

A 1,2-methyl shift

CH_3 migrates with two electrons.

(7-26)

Equation 7-26 shows a **1,2-alkyl shift**; more specifically, a methyl group is transferred, so this rearrangement is called a **1,2-methyl shift**. The numbering system is no different from that of the hydride shift because the migrating group—the methyl group—is transferred to an adjacent atom.

Both Equations 7-25 and 7-26 share identical curved arrow notation. In both cases, a single curved arrow indicates that the initial bond between the C atom and the migrating group is broken, and those electrons are used to form a bond to the adjacent C. Meanwhile, the formal positive charge is also shifted over one atom to the atom that was initially bonded to the migrating group.

Carbocation rearrangements are important to consider whenever carbocations are formed in a particular reaction. These reactions include S_N1 and E1 reactions (Chapters 8 and 9) and electrophilic addition reactions (Chapter 11).

In a carbocation rearrangement, the positively charged carbon atom of the carbocation is very electron poor because it carries a full positive charge and it has less

"Watching" a Bond Break

An elementary step can be extremely fast—on the order of picoseconds (1 ps = 10^{-12} s) or femtoseconds (1 fs = 10^{-15} s). Is it possible, then, to actually observe one taking place? A few decades ago, the answer would have been no, but now we have the ability to do so, at least with some types of reactions. How is it done? In essence, snapshots of a reaction are taken using a really fast "camera." But not a camera in the traditional sense. In this case, the "camera" is constructed from lasers capable of producing light pulses lasting <1000 fs. Such a femtosecond laser is shown at the right.

The decomposition of gaseous ICN into I + CN, carried out in 1988 by Stewart O. Williams and Dan G. Imre of the University of Washington, has been studied using this technology. Two lasers were used—one to provide a ~125-fs pulse of 306-nm light to break the I—C bond and a second to provide a ~125-fs pulse of ~389-nm to 433-nm light at various delay times after the first pulse. The second pulse excited the newly forming CN species, causing it to emit fluorescence that could be detected and correlated with the I—C distance over time. They found that the I—C bond was nearly completely broken at ~60 fs.

Even though studies such as this one focus on specific chemical reactions in the gas phase, the knowledge we gain can be used to develop a deeper understanding of reaction dynamics in general.

---

than an octet of electrons. On the other hand, a single bond to hydrogen or carbon on an adjacent atom is relatively electron rich because two electrons are localized in the bonding region. Therefore, the single curved arrow that is used to depict a carbocation rearrangement in Equation 7-27 represents the flow of electrons from an electron-rich site to an electron-poor site.

$$(7\text{-}27)$$

7.12 YOUR TURN

Supply the curved arrow notation for the following carbocation rearrangement.

---

problem **7.18** Supply the appropriate curved arrows and draw the product for the following carbocation undergoing **(a)** a 1,2-hydride shift and **(b)** a 1,2-methyl shift.

7.8 The Driving Force for Chemical Reactions

To this point in Chapter 7, our focus on elementary steps has been limited to their curved arrow notation. Namely, when a particular elementary step takes place, the dynamics of the reaction can be described in terms of the flow of electrons from an electron-rich site to an electron-poor site. But why should these elementary steps occur in the first place? That is, what is their *driving force*?

The **driving force** for a reaction reflects the extent to which the reaction favors products over reactants, and that tendency increases with increasing stability of the products relative to the reactants. Thus, to understand a reaction's driving force, it is a matter of identifying and evaluating the factors that help stabilize the products, destabilize the reactants, or both.

Although there are various factors that dictate the stability of a species, we can often make reasonable predictions about a reaction's driving force by examining just two factors. One is charge stability, which, as we saw in Chapter 6, can vary dramatically from one species to another. The other is total bond energy. As we saw in Chapter 1, each covalent bond provides stability to a molecule, and the amount of stabilization depends heavily on the type of bond that is formed—that is, which atoms are bound together, and whether it is by a single, double, or triple bond.

A reaction's driving force generally increases with:

- Greater charge stabilization in the products relative to the reactants.
- Greater total bond energy in the products relative to the reactants.

With this in mind, we can see that a coordination step such as the one in Equation 7-28 is unambiguously driven toward products.

$$(7\text{-}28)$$

Charge stability heavily favors products because, although there are two formal charges in the reactants, there are no formal charges in the products. Furthermore, total bond energy favors products because, during the course of the reaction, one covalent bond is formed, giving carbon an octet, and none are broken.

In other elementary steps, the driving force may not immediately be as clear cut. For the proton transfer in Equation 7-29, the charge in the products is located on a different atom than it is in the reactants. Furthermore, during the course of the reaction, one covalent bond is broken and another is formed.

The H——F bond formed is stronger than the H——Cl bond broken.

The negative charge is better stabilized on Cl than F.

$$K_{eq} = 1.6 \times 10^{10} \qquad (7\text{-}29)$$

Therefore, we need to consider both charge stability and total bond energy.

In this case, however, both factors work in the same direction. The negative charge is better accommodated on Cl in the products than on F in the reactants because Cl is the larger atom. Furthermore, the H—F bond that is formed (569 kJ/mol; 136 kcal/mol) is much stronger than the H—Cl bond that is broken (431 kJ/mol; 103 kcal/mol). Thus, the products are unambiguously favored in this reaction, consistent with the fact that HCl is a much stronger acid than HF, resulting in a K_{eq} of 1.6×10^{10}.

7.13 YOUR TURN

Consult Table 6-1 to verify that HCl is a stronger acid than HF. Write the pK_a values of the two acids below. With those pK_a's, use Equation 6-11 to compute the value of K_{eq} for the proton transfer reaction in Equation 7-29.

pK_a of HCl _____ pK_a of HF _____

The situation is similar with the S_N2 reaction in Equation 7-30.

The C—Cl bond formed is stronger than the C—I bond broken.

The negative charge is better stabilized on I than Cl.

(7-30)

Charge stability favors products because the negative charge is better accommodated on I than on Cl due to the fact that I is the larger atom. Bond energy also heavily favors products. The average bond energy of the C—Cl bond that is formed is 331 kJ/mol (79 kcal/mol), whereas the average bond energy of the C—I bond that is broken is 238 kJ/mol (57 kcal/mol). Thus, once again, the products are unambiguously favored.

Charge stability and bond energy do not always work in the same direction, such as in the proton transfer step in Equation 7-31.

The H—Cl bond broken is stronger than the H—N bond formed.

The negative charge is better accommodated on Cl than N.

$K_{eq} = 1 \times 10^{43}$ (7-31)

In this case, charge stability favors products because the negative charge is better accommodated on Cl than on N. Bond energy, however, favors the reactants because the H—Cl bond that is broken (431 kJ/mol; 103 kcal/mol) is stronger than the H—N bond that is formed (389 kJ/mol; 93 kcal/mol). Despite the disagreement, notice that the products are heavily favored; HCl is a much stronger acid than NH_3, giving rise to a very large K_{eq} of 1×10^{43}.

7.14 YOUR TURN

Consult Table 6-1 to verify that HCl is a stronger acid than NH_3. Write the pK_a values of the two acids below. With those pK_a's, use Equation 6-11 to compute the value of K_{eq} for the proton transfer reaction in Equation 7-31.

pK_a of HCl _____ pK_a of NH_3 _____

The previous example leads to the following general rule:

When charge stability and bond energy favor opposite sides of a chemical reaction, charge stability usually wins.

This idea is very useful when considering the E2 step in Equation 7-32.

σ Bonds broken   σ Bond formed   π Bond formed   **The negative charge is better stabilized on Cl than O.**

$$HO^{\ominus} \quad H_2C - CH_2 \xrightarrow{E2} HO - H + H_2C = CH_2 + :Cl:^{\ominus} \tag{7-32}$$

Charge stability favors the products because the negative charge is better accommodated on Cl than on O. Bond energy, however, favors the reactants because, during the course of the reaction, a σ and a π bond are formed, while two σ bonds are broken. Effectively, there is a net gain of one π bond and a net loss of one σ bond and, as we learned in Chapter 3, a σ bond is generally stronger. Experimentally, we know that the product side of the reaction is heavily favored, in agreement with the prediction obtained using charge stability.

SOLVED problem **7.19** Determine which side of this carbocation rearrangement is favored.

Think On the two sides of the reaction, is there a difference in charge stability? Is there a difference in total bond energy?

Solve The positive charge is on a tertiary carbon in the reactant and is on a secondary carbon in the product. Because a tertiary carbocation is more stable (Section 6.6e), charge stability favors the reactant side. Total bond energy is not a significant factor in this reaction because a C—C σ bond is broken in the reactant and another one is formed in the product. Overall, then, this reaction will favor the reactant side.

problem **7.20** Determine which side of each of the following carbocation rearrangements is favored.

(a)   (b)

Caution must be taken when using charge stability and total bond energy to predict the driving force for an elementary step. One reason is that a reaction that is product favored is not guaranteed to occur at a rate that is practical. For example, both charge stability and total bond energy heavily favor the products of the S_N2 reaction in Equation 7-33, but essentially no reaction occurs.

The negative charge is more stable on Br than Cl.

$$:Cl:^{\ominus} + \quad \diagup\diagdown Br: \quad \xcancel{\quad} \quad \diagup\diagdown Cl: + :Br:^{\ominus} \tag{7-33}$$

Understanding the reason for this requires a discussion of reaction kinetics, a topic discussed in Chapter 9.

A second reason to be cautious is that other factors can come into play. Consider, for example, the heterolysis step in Equation 7-34, which is the reverse of the coordination step in Equation 7-28.

Bond broken | **Two formal charges**

(7-34)

It might seem that this step shouldn't take place at all because both charge stability and total bond energy heavily disfavor the products. However, heterolysis steps such as this one often appear in multistep mechanisms. As it turns out, factors that help contribute to the driving force include effects from the solvent, as well as entropy—topics that we will discuss in Chapter 9.

By and large, charge stability and total bond energy are the dominant contributors to the driving force of elementary steps in most situations. These are the ones you should focus on the most. We will discuss other factors in the context of the reactions in which they are relevant.

7.9 Keto–Enol Tautomerization

In Section 7.8, we saw that charge stability and bond energies can contribute significantly to the driving force for a chemical reaction. However, charge stability does not always dictate a chemical reaction's outcome. Consider, for example, the competing proton transfers shown in Equations 7-35a and 7-35b.

The base, called an **enolate anion,** has two resonance structures that delocalize the negative charge onto a C atom and an O atom. As a result, an acid (HA) can protonate either the O atom (Equation 7-35a) or the C atom (Equation 7-35b). Depending upon which atom is protonated, one of two constitutional isomers can be generated. One is an **enol,** in which an alk*ene* functional group (C=C) and an alcoh*ol* functional group (ROH) are on the same C. The other is a compound containing a carbonyl group (C=O)—either a ketone or an aldehyde—and is referred to as the **keto form**.

For most enolate anions, protonation leads predominantly to the keto form, as shown in Table 7-1. The enol form is present only in trace amounts, suggesting that the keto form is significantly more stable. In other words, the driving force for the step leading to the keto form is greater than the driving force leading to the enol form.

Because the charged species in Equation 7-35a are precisely the same as those in Equation 7-35b, charge stability must have the same contribution to both reactions. Thus, the fact that the keto form is favored must not stem from a difference in charge

TABLE 7-1 Relative Percentages of Keto and Enol Forms

Tautomerization Reaction	Percent Keto Form	Percent Enol Form
	99.99994%	0.00006%
	99.986%	0.014%
	99.9999995%	0.0000005%
	99.9999988%	0.0000012%
	99.99996%	0.00004%

stability, but rather is an outcome of a greater total bond energy in the keto form than in the enol form. This is shown explicitly in Figure 7-4, which tallies the energies of the bonds that appear in one form but not the other. As we can see, the total bond energy of the keto form is roughly 47 kJ/mol (11 kcal/mol) greater than that of the enol form, making the keto form more stable.

FIGURE 7-4 Relative stabilities of enol and keto forms (a) Energies of the bonds that appear in the enol form but not in the keto form. The sum of the energies is 1430 kJ/mol (342 kcal/mol). (b) Energies of the bonds that appear in the keto form but not in the enol form. The sum of the energies is 1477 kJ/mol (353 kcal/mol). Because of its greater total bond energy, the keto form is more stable than the enol form.

In Figure 7-4, there is a similar bond in red in the keto form for each bond in blue in the enol form. Therefore, we can pair these bonds as follows:

Bond in Keto Form	Bond in Enol Form	Difference in Bond Energy
C=O	C=C	
C—C	C—O	
C—H	O—H	

For each pair, compute the *difference* in bond energy and enter it into the table. Based on these data, which bond is most responsible for the additional stability of the keto form? _____

Because keto and enol forms are related by the different placement of a proton in competing proton transfer reactions (Equation 7-35), the two forms will generally equilibrate. Such an equilibrium, called **keto–enol tautomerization**, is shown in Equation 7-36.

The keto form generally dominates this equilibrium.

$$\text{Tautomerization} \qquad (7\text{-}36)$$

Enol form   Keto form

In later chapters, we will see that this equilibrium has important consequences in a variety of chemical reactions. One consequence is that when an enol is produced in a reaction, it will spontaneously rearrange to the more stable keto form (see Problem 7.21). A second consequence is that when a ketone or aldehyde is present in a chemical reaction, a small fraction of it will be available in the enol form, which has somewhat different reactivity than the keto form.

problem **7.21** Decarboxylation (i.e., elimination of CO_2) occurs when a β-ketoacid is heated under acidic conditions.

The immediate product of decarboxylation is an enol, which quickly rearranges. Draw the overall product of the rearrangement.

7.10 Wrapping Up and Looking Ahead

This chapter began by revisiting proton transfer steps in order to identify the tendency of electrons to flow in an elementary step. We saw that electrons tend to flow from an electron-rich site to an electron-poor site, and this flow of electrons is reflected in the step's curved arrow notation. Throughout much of the rest of the chapter, we

Sugar Transformers

A sugar inside a cell can be different from the one that might be needed for a particular purpose, but the body has developed an elegant way to deal with this: It can transform one sugar into another! It does so using keto–enol tautomerization reactions (Section 7.9). This is exemplified in the reaction scheme below, which is a key part of **glycolysis**, a metabolic pathway that breaks down simple carbohydrates for their energy.

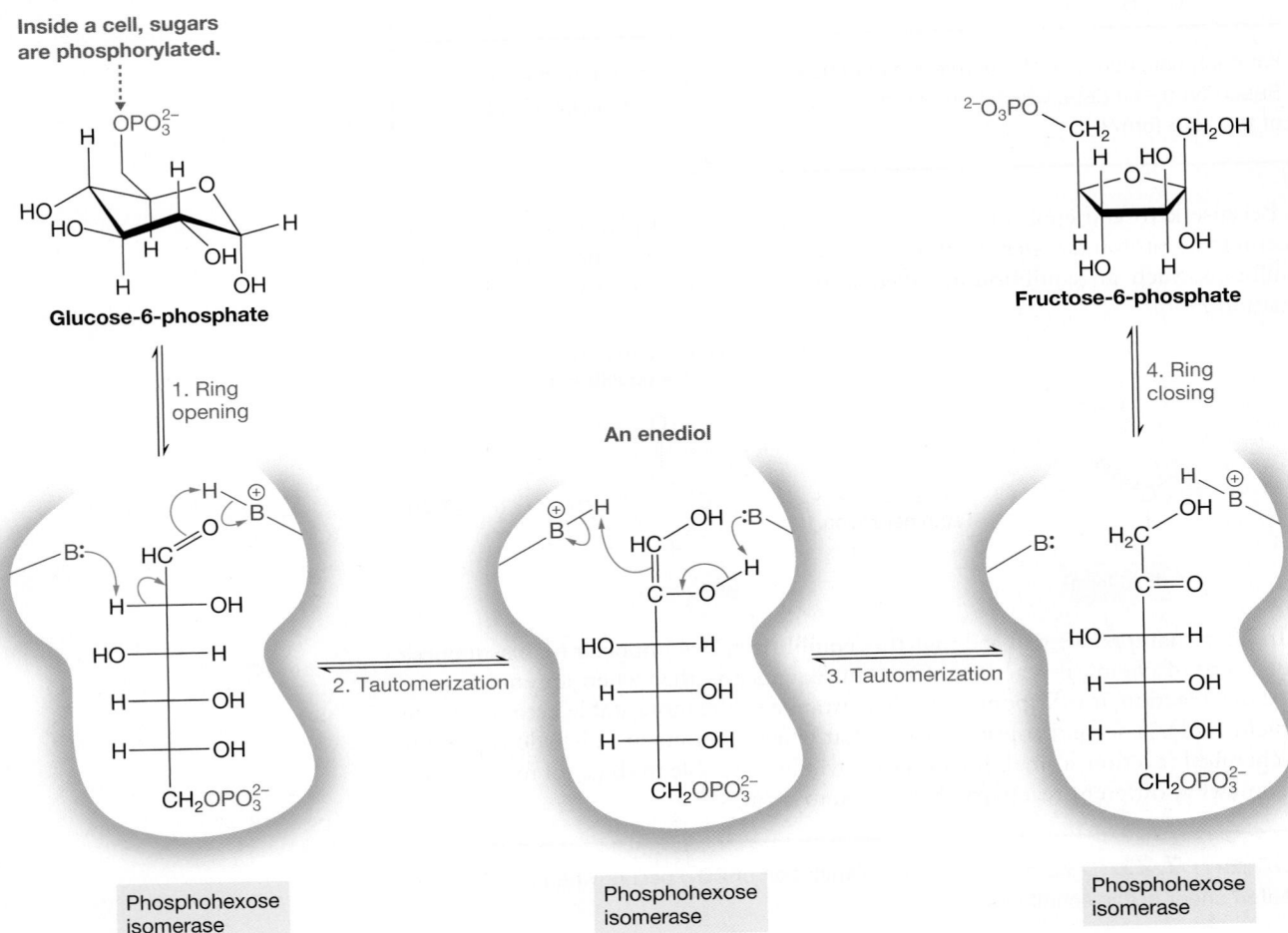

When D-glucose enters a cell, it is *phosphorylated* to become glucose-6-phosphate (shown above at left), in which the hydroxyl group on C6 (the bottom-most carbon in the Fischer projection) has been replaced by a phosphate (PO_4^{2-}) group. Before it can be broken down, however, glucose-6-phosphate must be converted into fructose-6-phosphate (shown at right). This conversion involves back-to-back tautomerization reactions catalyzed by the enzyme *phosphohexose isomerase* (the active site is shown in purple). The enzyme supplies the basic (B:) and acidic sites (B^+—H) necessary for the proton transfer steps. The first tautomerization produces an enol that is part of two different OH groups, so it is more precisely called an *enediol*. The second tautomerization converts the enediol back into a keto form, but upon doing so, the C=O bond is part of a different carbon atom than in the initial sugar. The result is a different phosphorylated sugar, fructose-6-phosphate.

This process of transforming one sugar into another is not limited to just six-carbon sugars. At a later stage in glycolysis, an enzyme called *phosphotriose isomerase* converts one three-carbon sugar into another. These kinds of processes truly are a testament to how efficient biological organisms are.

introduced nine more of the most common elementary steps in organic chemistry. For each step, we first provided the curved arrow notation, along with specific examples. Then, we analyzed how the curved arrow notation reflects the flow of electrons from *electron rich to electron poor*.

We then considered how charge stability and bond energies contribute to each step's driving force. Often, charge stability and bond energies work to drive the reaction in the same direction, making it relatively straightforward to predict which side of the reaction is favored. When these factors disagree in an elementary step that involves charges, the outcome of the reaction is usually governed by charge stability.

Finally, keto–enol tautomerization was introduced as a reaction whose outcome does not depend on charge stability, but rather is governed largely by a difference in total bond energy. With a substantially greater total bond energy, the keto form is generally favored heavily over its enol form.

By definition, all the changes that the valence electrons undergo in an elementary step to form products occur simultaneously. Most organic reactions, however, have mechanisms consisting of two or more steps, something we will begin to study in Chapter 8. Despite their additional complexity, the vast majority of these mechanisms are constructed from the steps we have seen here in Chapter 7. Therefore, understanding *how* and *why* each elementary step takes place will provide you with the tools necessary to understand and make predictions about reactions with more complex mechanisms.

Chapter Summary and Key Terms

- The curved arrow notation for an elementary step reflects the flow of electrons from an *electron-rich* site to an *electron-poor* site. **(Section 7.1a)**

- Metal cations from group IA of the periodic table behave as spectator ions, so they can be disregarded when identifying electron-rich sites in an elementary step. **(Section 7.1b)**

- **Organometallic** reagents, such as alkyllithium reagents (RLi), Grignard reagents (RMgBr), and lithium dialkylcuprates (R_2CuLi), can be treated as sources of R^-. Hydride reagents, such as lithium aluminum hydride ($LiAlH_4$) and sodium borohydride ($NaBH_4$), can be treated as sources of H^-. **(Section 7.1b)**

- Charge stability and total bond energy are two major factors that contribute to a reaction's **driving force**. For a reaction or elementary step involving both ions and uncharged molecules, the side that is favored generally exhibits greater charge stability. **(Section 7.8)**

- For a reaction or elementary step involving only uncharged species, the side that is favored generally has the greater bond energies. **(Section 7.9)**

- In a **keto–enol tautomerization**, the **keto form** is in equilibrium with its **enol form** via proton transfer steps. In general, the keto form is much more stable than, and therefore favored over, the enol form because it has a greater total bond energy. **(Section 7.9)**

- **Bimolecular nucleophilic substitution (S_N2) steps.** **(Section 7.2)**

- A **substrate** (R–L) contains a **leaving group** (L). Good leaving groups have strong conjugate acids. A **nucleophile** contains an atom that has a full or partial negative charge and possesses a lone pair of electrons.

- The nucleophile is relatively *electron rich* and the atom attached to the leaving group is relatively *electron poor*.

- **Coordination steps** and **heterolytic bond dissociation (heterolysis) steps. (Section 7.3)**

- In a coordination step, the Lewis acid is usually deficient of an octet, and the Lewis base has an atom with a partial or full negative charge and a lone pair of electrons.

- The Lewis base is relatively *electron rich*, whereas the Lewis acid is relatively *electron poor*.

- In a heterolysis step, the bond to the substrate is broken and the bonding electrons become a lone pair on the leaving group.

- **Bimolecular elimination (E2) steps.** (Section 7.4)

- In an E2 step, a base deprotonates an atom on the substrate at the same time that a leaving group is expelled, forming an additional π bond between the atoms to which the hydrogen and the leaving group were initially bonded.
- The base is relatively *electron rich*; the carbon atom bonded to the leaving group is relatively *electron poor*.

- **Nucleophilic addition steps** and **nucleophile elimination steps.** (Section 7.5)

- A nucleophile forms a bond to the positive end of a polar C—X multiple bond, forcing a pair of electrons from a π bond onto X.
- The nucleophile is relatively *electron rich*, and the atom at the positive end of the polar C—X multiple bond is relatively *electron poor*.

- In a nucleophile elimination step a new C—X π bond is formed at the same time that a leaving group is expelled.

- The X atom of the C—X bond is relatively *electron rich*, whereas the C atom is relatively *electron poor*.
- **Electrophilic addition steps** and **electrophile elimination steps.** (Section 7.6)

- In electrophilic addition, a pair of electrons from a nonpolar π bond forms a bond to an **electrophile**, an electron-deficient species.
- The nonpolar π bond is relatively *electron rich*, whereas the electrophile is relatively *electron poor*.

- In an electrophile elimination step, an electrophile is eliminated from a carbocation species and a nonpolar π bond is formed simultaneously.
- The C—E single bond is relatively *electron rich*; the positively charged C atom is relatively *electron poor*.
- **Carbocation rearrangements.** (Section 7.7)

- In a **1,2-hydride shift** or **1,2-alkyl shift**, a C—H or C—C bond adjacent to a carbocation is broken, and the bond is reformed to the C atom initially with the positive charge. The positive charge moves to the C atom whose bond is broken.
- The positively charged C atom is relatively *electron poor*; the C—H or C—C single bond that is broken is relatively *electron rich*.

Problems

7.22 (a) Draw the appropriate curved arrows and products for each set of reactants undergoing a coordination step. Identify each reactant species as either a Lewis acid or Lewis base.
(b) Use curved arrow notation to show each product undergoing heterolysis to regenerate reactants.

7.23 The following reaction, which is discussed in Chapter 8, is an example of a unimolecular nucleophilic substitution (S_N1) reaction. It consists of the four elementary steps shown here. For each step (i–iv),
(a) identify all electron-rich sites and all electron-poor sites,
(b) draw in the appropriate curved arrows to show the bond formation and bond breaking that occur, and
(c) name the elementary step.

(i)

(ii)

(iii)

(iv)

7.24 The following reaction, which is discussed in Chapter 8, is an example of a unimolecular elimination (E1) reaction and consists of the three elementary steps shown. For each step (i–iii),
(a) identify all electron-rich sites and all electron-poor sites,
(b) draw in the appropriate curved arrows to show the bond formation and bond breaking that occur, and
(c) name the elementary step.

(i)

(ii)

(iii)

7.25 The following reaction, which converts an epoxide into a bromohydrin, is discussed in Chapter 10. It consists of the two elementary steps shown. For each step (i and ii),
(a) identify all electron-rich sites and all electron-poor sites,

(b) draw in the appropriate curved arrows to show the bond formation and bond breaking that occur, and

(c) name the elementary step.

(i)

(ii)

7.26 The following reaction, which is discussed in Chapter 10, consists of the two elementary steps shown. For each step (i and ii),

(a) identify all electron-rich sites and all electron-poor sites,

(b) draw in the appropriate curved arrows to show the bond formation and bond breaking that occur, and

(c) name the elementary step.

(i)

(ii)

7.27 The following reaction, which converts a cyclic ether into a diol, is discussed in Chapter 8. It consists of the three elementary steps shown. For each step (i–iii),

(a) identify all electron-rich sites and all electron-poor sites,

(b) draw in the appropriate curved arrows to show the bond formation and bond breaking that occur, and

(c) name the elementary step.

(i)

(ii)

(iii)

$HO\text{—}\cdots\text{—}\overset{\overset{H}{|}}{\underset{\overset{|}{H}}{O^{\oplus}}} + H_2O \longrightarrow HO\text{—}\cdots\text{—}OH + H_3O^{\oplus}$

7.28 The following reaction, which is discussed in Chapter 21, consists of the three elementary steps shown. For each step (i–iii),

(a) identify all electron-rich sites and all electron-poor sites,

(b) draw in the appropriate curved arrows to show the bond formation and bond breaking that occur, and

(c) name the elementary step.

(i)

(ii)

(iii)

7.29 The following reaction, which is discussed in Chapter 17, consists of the two elementary steps shown. For each step (i and ii),

(a) identify all electron-rich sites and all electron-poor sites,

(b) draw in the appropriate curved arrows to show the bond formation and bond breaking that occur, and

(c) name the elementary step.

(i)

(ii)

7.30 The following is an example of a *Fischer esterification* reaction, which is discussed in Chapter 21. The mechanism consists of the six elementary steps shown. For each step (i–vi),

(a) identify all electron-rich sites and all electron-poor sites,

(b) draw in the appropriate curved arrows to show the bond formation and bond breaking that occur, and

(c) name the elementary step.

(i)

(ii)

(iii)

(iv)

(v)

(vi)

7.31 The following reaction is an example of an *acid-catalyzed hydrolysis* of an amide bond. This reaction, which is the same one that breaks down proteins into amino acids in your stomach, is discussed in Chapter 21. The mechanism consists of the six elementary steps shown. For each step (i–vi),
(a) identify all electron-rich sites and all electron-poor sites,
(b) draw in the appropriate curved arrows to show the bond formation and bond breaking that occur, and
(c) name the elementary step.

(i)

(ii)

$$H_3C-CH_2-\overset{\overset{\displaystyle H-\overset{\oplus}{O}}{\|}}{C}-\overset{\displaystyle CH_3}{\underset{\displaystyle CH_3}{N}} \quad + \quad H_2O \quad \longrightarrow \quad H_3C-CH_2-\overset{\overset{\displaystyle H-O}{|}}{\underset{\underset{\displaystyle H_3C}{N-CH_3}}{C}}-\overset{\oplus}{O}H_2$$

(iii)

$$H_3C-CH_2-\overset{\overset{\displaystyle H-O}{|}}{\underset{\underset{\displaystyle H_3C}{N-CH_3}}{C}}-\overset{\oplus}{O}\overset{H}{H} \quad + \quad H_2O \quad \longrightarrow \quad H_3C-CH_2-\overset{\overset{\displaystyle H-O}{|}}{\underset{\underset{\displaystyle H_3C}{N-CH_3}}{C}}-OH \quad + \quad H_3O^{\oplus}$$

(iv)

$$H_3C-CH_2-\overset{\overset{\displaystyle H-O}{|}}{\underset{\underset{\displaystyle H_3C}{N-CH_3}}{C}}-OH \quad + \quad H-\overset{\oplus}{O}H_2 \quad \longrightarrow \quad H_3C-CH_2-\overset{\overset{\displaystyle H-O}{|}}{\underset{\underset{\displaystyle H_3C}{\overset{\oplus}{N}-H}\ \ CH_3}{C}}-OH \quad + \quad H_2O$$

(v)

$$H_3C-CH_2-\overset{\overset{\displaystyle H-O}{|}}{\underset{\underset{\displaystyle H_3C\ \ CH_3}{\overset{\oplus}{N}-H}}{C}}-OH \quad \longrightarrow \quad H_3C-CH_2-\overset{\overset{\displaystyle H-\overset{\oplus}{O}}{\|}}{C}-OH \quad + \quad H_3C-\overset{\displaystyle H}{\underset{}{N}}-CH_3$$

(vi)

$$H_3C-CH_2-\overset{\overset{\displaystyle H-\overset{\oplus}{O}}{\|}}{C}-OH \quad + \quad H_3C-\overset{\displaystyle H}{\underset{}{N}}-CH_3 \quad \longrightarrow \quad H_3C-CH_2-\overset{\overset{\displaystyle O}{\|}}{C}-OH \quad + \quad H_3C-\overset{\overset{\displaystyle H_2}{\oplus}}{N}-CH_3$$

7.32 The following is an example of an *electrophilic aromatic substitution* reaction, which we examine in Chapter 22. The mechanism consists of the four elementary steps shown. For each step (i–iv),
(a) identify all electron-rich sites and all electron-poor sites,
(b) draw in the appropriate curved arrows to show the bond formation and bond breaking that occur, and
(c) name the elementary step.

(i)

$$Cl-Cl \quad + \quad \overset{\overset{\displaystyle Cl}{|}}{\underset{\underset{\displaystyle Cl}{}}{Cl-Fe-Cl}} \quad \longrightarrow \quad Cl-Cl\overset{\oplus}{-}\overset{\overset{\displaystyle Cl}{|}}{\underset{\underset{\displaystyle Cl}{}}{\overset{\ominus}{Fe}-Cl}}$$

(ii)

$$Cl-Cl\overset{\oplus}{-}\overset{\overset{\displaystyle Cl}{|}}{\underset{\underset{\displaystyle Cl}{}}{\overset{\ominus}{Fe}-Cl}} \quad \longrightarrow \quad Cl^{\oplus} \quad + \quad \overset{\overset{\displaystyle Cl}{|}}{\underset{\underset{\displaystyle Cl}{}}{Cl-\overset{\ominus}{Fe}-Cl}}$$

(iii)

(iv)

7.33 Determine whether each of the following elementary steps is acceptable. For those that are, draw the products. For those that are not, explain why. (*Hint:* Explaining why may involve drawing the products.)

(a)

(b)

(c)

(d)

(e)

7.34 Each heterolysis step on the left does not readily occur, but the corresponding one on the right does. Explain why. (*Hint:* Draw the products of each heterolysis and determine the contributions to their driving force.)

(a)

(b)

(c)

(d)

7.35 If the H colored red below is eliminated as H^+, then two possible diastereomers can form. Draw the curved arrows for this electrophile elimination step, and draw each of the diastereomeric products.

$$H_2O \quad + \quad H_3C{-}\underset{\underset{H}{|}}{\overset{\overset{H}{|}}{C}}{-}\overset{\overset{H\oplus}{|}}{C}{-}CH_3 \longrightarrow$$

7.36 There are two possible products in the electrophilic addition step shown here. Draw both possible products and predict which one is more stable.

7.37 If H^+ is eliminated from the following carbocation in an electrophilic elimination step, then three possible constitutional isomers can form. Draw the mechanism for the formation of all three of those products.

$$H_2O \quad +$$

7.38 In the following electrophilic addition step, where NO_2^+ adds to phenol, the electrophile can add either ortho, meta, or para to the OH substituent on the ring. Draw the curved arrows and products of each electrophilic addition step.

7.39 If the anionic species shown were to eliminate a leaving group, the three possibilities would be H_3C^-, Cl^-, or CH_3O^-. Draw the curved arrow notation and the products for each of these elimination steps. Which is the major product? Why?

7.40 Which of the following carbocations will rearrange to produce a more stable carbocation? For each one, draw the curved arrows and products for the rearrangement.

(a)

(b)

(c)

(d)

(e)

(f)

7.41 Draw the curved arrows and the product for each of the following S_N2 steps.

(a)

$$+ \quad NaSH \longrightarrow \quad ?$$

(b)

$$+ \quad KCN \longrightarrow \quad ?$$

7.42 Draw the curved arrows and the product for each of the following nucleophilic addition steps.

(a)

[structure of a ketone] + CH₃OK ⟶ ?

(b)

[structure of cyclohexanone] + CH₃Li ⟶ ?

(c)

[structure of a nitrile] + [phenyl MgBr] ⟶ ?

(d)

[structure of acyl chloride] + NaBH₄ ⟶ ?

(e)

[structure of ester] + NaOH ⟶ ?

(f)

[structure of two acyl groups] ⟶ ?

7.43 For each of the steps in Problem 7.42, determine whether the product can eliminate a leaving group to produce a compound that is different from the reactants. For those that can, draw the appropriate curved arrows and the new product that forms.

7.44 Predict the product of the reaction between phenol (C₆H₅OH) and each of the following compounds. (*Hint:* First determine which elementary step is likely to occur.)

(a)

CH₃OK

(b)

CH₃Li

(c)

[phenyl MgBr]

(d)

LiAlH₄

7.45 A proton (H⁺) from CH₃OH₂⁺ can add to the alkyne shown to yield two different carbocation products.
(a) Draw the mechanism for each of these steps, along with the corresponding products.
(b) Which carbocation is more stable?

[structure: H₃C–CH₂–CH₂–CH₂–C≡CH] + CH₃OH₂⁺ ⟶ ?

7.46 The carbocation shown below is formed in one step of an electrophilic aromatic substitution reaction (discussed in Chapter 22).
(a) Draw the curved arrow notation and the product for the elimination of H⁺.
(b) Do the same for the elimination of SO₃.

[structure of sulfonate carbocation intermediate] →H₂O→ ?

7.47 According to Table 7-1, the equilibrium percentage of in its keto form is lower than that of [structure]. Explain why.
(*Hint:* What do you know about the stability of C=C double bonds?)

7.48 According to Table 7-1, the equilibrium percentage of in its keto form is lower than that of the cyclopentanone structure. Explain why.

7.49 Recall from Section 7.9 that most ketones and aldehydes exist primarily in their keto form, as shown in Table 7-1. Propanedial, however, exists primarily (>99%) in its enol form. Explain why. (*Hint:* Examine its structure in the enol form.)

Propanedial

7.50 A tautomerization reaction can occur with an imine in a way analogous to that of a ketone or aldehyde. Using the appropriate bond energies from Section 1.2, estimate ΔH°_{rxn} for this reaction and determine which form is more stable, the imine or enamine.

Imine   **Enamine**

7.51 The first of the two heterolysis reactions below takes place readily, but the second one does not. Explain why.

If your instructor assigns problems in **smartwork**, log in at **smartwork.wwnorton.com**.

Problems / **387**

Molecular Orbital Theory and Chemical Reactions

Chapters 6 and 7 introduced 10 of the most common elementary steps that make up organic reaction mechanisms. For each step, we examined the curved arrow notation that enables us to depict the changes that take place among valence electrons throughout the course of the reaction. Furthermore, we discussed how such changes in valence electrons tend to represent a flow of electrons from an electron-rich site to an electron-poor site. Thus far, however, these aspects have been discussed only in terms of the Lewis structure model—there has been no mention of orbitals.

Here in Interchapter 1, each of the elementary steps we have previously learned is presented again from the perspective of molecular orbital (MO) theory. The reason, as we saw in Chapter 3, is that MO theory can provide deeper insights than Lewis structures can. In Chapter 3, those insights led to a deeper understanding of molecular structure and stability. As we will see here in Interchapter 1, the application of MO theory to chemical reactions provides a deeper understanding of reaction dynamics as well.

To gain an understanding of elementary steps from the perspective of MOs, we first give an overview of what is called **frontier molecular orbital (FMO) theory**. Once FMO theory has been developed, it is applied to each of the elementary steps discussed in Chapter 7. In so doing, we will be able to see how FMO theory accounts for the flow of electrons from an electron-rich site to an electron-poor one. Additionally, we will see how FMO theory justifies the *stereochemistry* of certain elementary steps, a topic discussed more fully in Chapter 8.

IC1.1 An Overview of Frontier Molecular Orbital Theory

The application of frontier molecular orbital (FMO) theory to chemical reactions begins with the notion that the reactants in an elementary step must typically surmount a significant energy barrier—the free energy of activation—to proceed to products. The energy barrier arises from the relative instability of the transition state of the reaction relative to the reactants. Without a source of significant stabilization of that transition state (red curve, Fig. IC1-1), many elementary steps would have very large energy barriers, making their rates excessively slow. Conversely, stabilization of the transition state can lower the energy barrier to allow the elementary step to take place at a reasonable rate (blue curve, Fig. IC1-1).

What source of stabilization in the transition state can bring about such a lowering of the energy barrier? One of the main contributions involves what are called the **frontier molecular orbitals** of the reacting species, which are defined as the species' *highest occupied* and *lowest unoccupied* MOs (i.e., the HOMO and LUMO, respectively; review Section 3.4, if necessary).

The energy barrier can be substantially lowered to make an elementary step feasible.

Without a source of stabilization in the transition state, the energy barrier can be excessively large.

Reaction coordinate ⟶

FIGURE IC1-1 Energy barriers and transition-state stabilization The red curve represents an elementary step in which the transition state is not significantly stabilized (large energy barrier), making the reaction excessively slow. When the transition state is significantly stabilized (small energy barrier), as indicated in the blue curve, the reaction rate is substantially increased.

- If the HOMO of one reactant and the LUMO of another have the appropriate symmetry to interact in the transition state, then the transition state tends to be significantly stabilized, and the reaction of interest is said to be "allowed."

- Otherwise, the transition state tends not to be sufficiently stabilized, and the reaction of interest is said to be "forbidden."

The frontier orbitals of the reactants are the focus for two reasons. One is that, of all imaginable interactions of a filled orbital with an empty orbital, the HOMO–LUMO interaction is generally the one that involves orbitals that are closest in energy (Fig. IC1-2a). This maximizes any interaction that takes place between the orbitals (Section 3.3).

The second reason we focus on the frontier orbitals is that if the HOMO and LUMO orbitals can interact, then the interaction *must* stabilize the entire species, as shown in Figure IC1-2b. This interaction must be stabilizing because, by definition, the HOMO contains electrons and the LUMO is empty. When the orbitals mix (via constructive and destructive interference [Section 3.2]), therefore, two new orbitals must be produced—one that has been lowered in energy and one that has been raised in energy—and the electrons from the HOMO will end up in the lower of the two orbitals in the transition state.

With what we have seen so far, we can make a very useful association between FMO interactions and the tendency of electrons to flow from an electron-rich site to

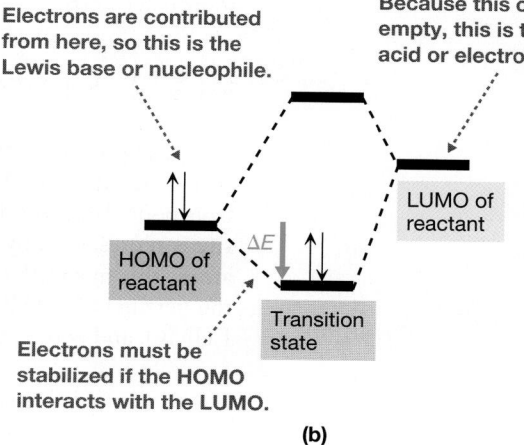

FIGURE IC1-2 HOMO–LUMO interactions (a) The similar energies of the HOMO and LUMO maximize any interaction between the two orbitals. (b) If an interaction takes place between the HOMO of one reactant and the LUMO of another, then stabilization is guaranteed. When the two orbitals interact, two new orbitals in the transition state (center) are produced—one that is lower in energy than either the HOMO or LUMO, and one that is higher in energy. The electrons contributed by the HOMO end up in the lower of the two orbitals.

an electron-poor one. Notice in Figure IC1-2b that the pertinent electrons originate from the HOMO of one species and interact with the empty LUMO of the other. Thus, there is an effective flow of electrons from the HOMO to the LUMO. Consequently,

> ■ The species that contributes the pertinent HOMO in the FMO interaction is relatively electron rich; it is the Lewis base or the nucleophile.
> ■ The species that contributes the pertinent LUMO is relatively electron poor; it is the Lewis acid or the electrophile.

One of the important messages from Chapter 7 is that some elementary steps might simply be the reverse of each other, but receive different names. Examples include coordination and heterolysis (Section 7.3), nucleophilic addition and nucleophile elimination (Section 7.5), and electrophilic addition and electrophile elimination (Section 7.6). In such cases, we can imagine applying FMO theory separately to the reactions in each pair, and indeed this can be done. However, it is much more convenient to take advantage of the following result.

> ■ If a reaction is allowed in the forward direction, then it will be allowed in the reverse direction as well.
> ■ If a reaction is forbidden in the forward direction, then it will be forbidden in the reverse direction as well.

This result is related to the **principle of microscopic reversibility**, which demands that if one reaction is the reverse of another, then the two reactions must proceed along the same path in the energy diagram and, therefore, must proceed through the same transition state. Thus, the FMO interactions that might contribute to the stabilization of the transition state in the forward reaction will be identical to the ones for the reaction in the reverse direction.

Keep in mind that FMO theory focuses only on the energy barrier of a reaction—that is, kinetics. Even if a reaction is allowed, thermodynamics (i.e., the relative energies of the reactants and products) will still govern whether the reaction is favorable at equilibrium.

With all of this in mind, let's now apply FMO theory toward the various elementary steps that we encountered in Chapter 7. Doing so entails identifying the FMOs in the reacting species and determining whether they have the appropriate symmetries to interact in the transition state.

IC1.2 Proton Transfer Steps

A typical proton transfer step, including curved arrow notation, is shown in Equation IC1-1. In this case, HO^- is electron rich and the H atom of HCl is relatively electron poor. Thus, the flow of electrons from an electron-rich site to an electron-poor site is represented by the curved arrow drawn from a pair of electrons on O to the H atom. This means that HO^- is the Lewis base, and is thus the species that contributes the pertinent HOMO to the FMO interaction. HCl, then, contributes the pertinent LUMO, and is thus the Lewis acid.

The HOMO and LUMO have appropriate symmetries to interact, so this reaction is *allowed*.

HOMO   LUMO

(a)

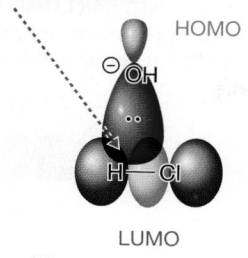

The HOMO and LUMO do *not* have appropriate symmetries to interact, so this reaction is *forbidden*.

HOMO

LUMO

(b)

FIGURE IC1-3 Frontier orbital interactions for the proton transfer involving HO^- and HCl (a) The approach of HO^- from the side opposite the Cl atom. As indicated, the HOMO and LUMO have the appropriate symmetries to interact. The specific phases shown would lead primarily to constructive interference, and reversing the phase of one of the orbitals would lead primarily to destructive interference. (b) The approach of HO^- from the side of the H—Cl bond. In this transition state, the HOMO and LUMO do *not* have a significant net interaction. Whereas the left side in this depiction (shaded red and shaded blue) indicates constructive interference, the right side (shaded red and unshaded blue) contributes to destructive interference.

It is now a matter of identifying the HOMO of HO^- and the LUMO of HCl and determining whether they have the appropriate symmetries to interact in the transition state. In the two-center MO approach we used in Chapter 3, the HOMO of HO^- is a nonbonding pair of electrons, as shown in Figure IC1-3, and the LUMO of HCl is a σ^* antibonding MO. When HO^- approaches HCl from the side opposite the Cl atom (Fig. IC1-3a), the HOMO and LUMO will undergo substantial interaction in the transition state. The specific phases shown in the figure will lead to a net constructive interference, and destructive interference would result from reversing the phase of one of the orbitals. As we learned in the previous section (Fig. IC1-2b), this kind of HOMO–LUMO interaction will stabilize the transition state, making the reaction allowed.

Construct the atomic orbital overlap and MO energy diagrams separately for HO^- and HCl (review Figs. 3-13 through 3-21). From these diagrams, determine the HOMO of HO^- and the LUMO of HCl. Do they agree with the ones shown in Figure IC1-3?

Answers to Your Turns are in the back of the book.

IC1.1 YOUR TURN

In Figure IC1-3a and IC1-3b, label every region of constructive interference and destructive interference between the HOMO and LUMO.

IC1.2 YOUR TURN

What would give rise to a transition state in which these frontier orbitals do not undergo significant interaction? An example is shown in Figure IC1-3b, in which HO^- approaches HCl from the side of the H—Cl bond. In that transition state, one portion of the orbital overlap leads to constructive interference (in this case, the left side), whereas the other portion (the right side) leads to destructive interference. Thus, there is very little net interaction, so the transition state remains unstabilized and the reaction is forbidden.

IC1.3 Bimolecular Nucleophilic Substitution (S_N2) Steps

The S_N2 step we first encountered in Chapter 7 is shown once again in Equation IC1-2.

S_N2 step

Leaving group

Nucleophile   **Substrate**

(IC1-2)

The curved arrow notation for this reaction is essentially the same as we saw for the previous proton transfer step in Equation IC1-1, so it is reasonable to expect the

FIGURE IC1-4 Frontier orbital interactions for the S_N2 step involving HO^- and CH_3Cl (a) The approach of HO^- from opposite the Cl leaving group. As indicated, the HOMO and LUMO have the appropriate symmetries to interact. (b) The approach of HO^- from the side of the C—Cl bond. In this transition state, the HOMO and LUMO do *not* have a significant net interaction. Whereas the right side in this depiction (shaded red and shaded blue) indicates constructive interference, the left side (shaded red and unshaded blue) contributes to destructive interference.

The HOMO and LUMO have appropriate symmetries to interact, so this reaction is *allowed*.

(a)

The HOMO and LUMO do *not* have appropriate symmetries to interact, so this reaction is *forbidden*.

(b)

FMO picture to be similar. Indeed, HO^- is once again the electron-rich species, making it the Lewis base or nucleophile, so it will contribute the relevant HOMO to the FMO interaction. CH_3Cl has a relatively electron-poor C atom, so it will act as the Lewis acid or electrophile and will contribute the relevant LUMO.

Just as we saw in Figure IC1-3, the HOMO that is contributed by HO^- in Equation IC1-2 contains the nonbonded electrons. The LUMO of CH_3Cl, once again, is a σ^* MO, this time of the C—Cl bond. These orbitals are shown in Figure IC1-4. If the HO^- nucleophile approaches the CH_3Cl substrate from the side opposite the Cl leaving group (Fig. IC1-4a), then the resulting overlap of the HOMO and LUMO orbitals will lead to significant interaction, the transition state will be stabilized, and the reaction will be allowed. Also, similar to what we saw in Figure IC1-3b, if the approach of the HO^- nucleophile is from the side of the C—Cl bond, then there will be no significant interaction between the FMOs, making the reaction forbidden.

YOUR TURN IC1.3

Construct the atomic orbital overlap and MO energy diagrams for CH_3Cl (review Figs. 3-13 through 3-21) and use these diagrams to determine its LUMO. Does your answer agree with the one shown in Figure IC1-4?

YOUR TURN IC1.4

In Figure IC1-4a and IC1-4b, label every region of constructive interference and destructive interference between the HOMO and LUMO.

Important consequences arise from the fact that whether an S_N2 reaction is allowed or forbidden depends on the particular approach of the nucleophile. Such a dependency gives rise to what is called the *stereochemistry* of the reaction—a topic that is more completely discussed in Chapter 8. In short, in the case of S_N2 reactions, if the carbon atom bonded to the leaving group is a stereocenter, then its stereochemical configuration in the product will be the reverse of what it initially was in the reactant.

IC1.4 Bond-Formation (Coordination) and Bond-Breaking (Heterolysis) Steps

The coordination step depicted in Equation IC1-3 was shown previously in Section 7.3.

Coordination step

$$\begin{array}{ccc} \text{:}\overset{\ominus}{\text{Br}}\text{:} & + & \text{Lewis acid} \\ \text{Lewis base} & & \text{or electrophile} \\ \text{or nucleophile} & & \end{array} \longrightarrow \quad (\text{IC1-3})$$

Br⁻ is relatively electron rich and is the Lewis base or nucleophile, so it will contribute a nonbonding orbital as the relevant HOMO to the FMO interaction. By contrast, the carbocation is electron poor, so it serves as the Lewis acid or electrophile in this example, and it will contribute the relevant LUMO (the empty p orbital in Fig. IC1-5) to the FMO interaction.

Figure IC1-5a shows that the HOMO and LUMO have the appropriate symmetries to interact when the Br⁻ nucleophile approaches from the left face of the planar C atom. Figure IC1-5b shows that the same is true for the approach of the nucleophile from the right face of the planar C atom, too. Thus, coordination is allowed from either face.

The HOMO and LUMO have appropriate symmetries to interact, so this reaction is *allowed*.

HOMO LUMO
(a)

The HOMO and LUMO have appropriate symmetries to interact, so this reaction is *allowed*.

LUMO HOMO
(b)

FIGURE IC1-5 Frontier orbital interactions for the coordination of Br⁻ and $(CH_3)_3C^+$ (a) The approach of Br⁻ from one face of the planar C atom. (b) The approach of Br⁻ from the other face of the planar C atom. As indicated, the HOMO and LUMO for both approaches have the appropriate symmetries to interact, so both approaches are allowed.

Construct the atomic orbital overlap and MO energy diagrams for $(CH_3)_3C^+$ (review Figs. 3-13 through 3-21) and use these diagrams to determine its LUMO. Does your answer agree with the one shown in Figure IC1-5?

IC1.5 YOUR TURN

In Figure IC1-5a and IC1-5b, label every region of constructive interference and destructive interference between the HOMO and LUMO.

IC1.6 YOUR TURN

Let's now turn our attention to heterolysis, the reverse of coordination. An example that we previously encountered in Chapter 7 is shown again in Equation IC1-4.

Heterolysis step

(IC1-4)

We don't need to begin an FMO analysis of such a reaction from scratch to determine whether it is allowed or forbidden, however, because we can use the principle of microscopic reversibility (Section IC1.1) instead. Because coordination steps are allowed, so, too, are heterolysis steps.

IC1.5 Bimolecular Elimination (E2) Steps

In Section 7.4, we learned that a bimolecular elimination (E2) reaction involves a base and a substrate. As shown again in Equation IC1-5, the base removes a proton at the same time a leaving group is displaced from an adjacent C atom, thus producing a new π bond.

E2 step

(IC1-5)

Lewis base or nucleophile   Lewis acid or electrophile

FIGURE IC1-6 The first part of the frontier orbital interactions of an E2 step The nonbonding orbital of the HO⁻ base (red; HOMO) has the appropriate symmetry to interact with the σ* orbital of the H—C bond (blue; LUMO), so this part of the E2 step is allowed.

Unlike the previous examples, an E2 step involves the formation of two separate bonds (in this case, the HO—H bond and the π bond) and the breaking of another two bonds (in this case, the H—C and C—Br bonds). To simplify the picture using FMO theory, we can treat the entire process as two separate parts that take place at essentially the same time—one that forms the HO—H bond and breaks the H—C bond, and the second that forms the π bond and breaks the C—Br bond.

In the first of these parts of the E2 step—namely, forming the HO—H bond and breaking the H—C bond—the HO⁻ base is electron rich and the H—C bond of the substrate is relatively electron poor. Thus, HO⁻ will contribute its nonbonding MO as the HOMO, and the substrate will contribute its σ* orbital of the H—C bond as the LUMO, as shown in Figure IC1-6. Similar to what we saw previously in Figure IC1-3, the HOMO and LUMO have the appropriate symmetries to interact.

In the second part of the E2 step—namely, forming the π bond and breaking the C—Br bond—the electrons originate from the H—C bond that is undergoing bond breaking. Thus, the H—C bond is treated as electron rich, while the carbon of the C—Br bond is electron poor. As such, the HOMO that is contributed to the FMO interaction is the σ bonding orbital of the H—C bond, and the LUMO is the σ* orbital of the C—Br bond. This is shown in Figure IC1-7.

Notice, however, that there are two orientations in Figure IC1-7 in which the HOMO and LUMO have appropriate symmetries to interact: One has the H—C and C—Br bonds anti to each other (Fig. IC1-7a) and the other has the two bonds syn to each other (Fig. IC1-7b), in which case they eclipse each other. Thus, an E2 reaction is allowed in both of these conformations. The HOMO and LUMO overlap to a greater extent, however, when the two bonds are anti to each other. As a result, an E2 reaction tends to proceed faster when the hydrogen and the leaving group on the adjacent carbon atom in the substrate are anti to each other than when they are syn. This kind of preference for the particular conformation of the substrate has significant impact on the *stereochemistry* of the reaction, often favoring one diastereomeric product over another (see Chapter 8).

YOUR TURN IC1.7

Construct the atomic orbital overlap and MO energy diagrams for CH₃CH₂Br (review Figs. 3-13 through 3-21) and use these diagrams to determine its HOMO and LUMO. Do your answers agree with the ones shown in Figure IC1-7a?

YOUR TURN IC1.8

In Figure IC1-7a and IC1-7b, label every region of constructive interference and destructive interference between the HOMO and LUMO.

FIGURE IC1-7 The second part of the frontier orbital interactions of an E2 step In both the (a) anti and (b) syn conformations, the filled C—H σ orbital (red; HOMO) has the appropriate symmetry to interact with the empty C—Br σ* orbital (blue; LUMO), so both reactions are allowed. The anti conformation is favored, however, due to the greater extent of orbital overlap.

Anti conformation

(a)

Syn conformation

(b)

IC1.6 Nucleophilic Addition and Nucleophile Elimination Steps

Recall from Section 7.5 that a nucleophilic addition step involves the formation of a bond between a nucleophile and the atom at the positive end of a polar π bond. The π bond breaks during such a step, and its electrons end up as a lone pair on the other atom that was initially part of the polar π bond. An example is shown in Equation IC1-6, in which the HO⁻ nucleophile adds to the C atom of a C=O group.

Nucleophilic addition

(IC1-6)

Lewis base or nucleophile   **Lewis acid or electrophile**

Once again, the HO⁻ nucleophile is relatively electron rich, making it the Lewis base or nucleophile, and the C atom of the C=O group is relatively electron poor, making it the Lewis acid or electrophile. As a result, the relevant HOMO will be contributed by HO⁻, and the relevant LUMO will be contributed by the C=O group. As shown in Figure IC1-8, the HOMO of HO⁻ is a nonbonding orbital, and the LUMO of the C=O group is a π* MO.

IC1.9 **YOUR TURN**

Construct the atomic orbital overlap and MO energy diagrams for CH_3COCl (review Figs. 3-13 through 3-21) and use these diagrams to determine its LUMO. Does your answer agree with the one shown in Figure IC1-8a?

IC1.10 **YOUR TURN**

In Figure IC1-8a and IC1-8b, label every region of constructive interference and destructive interference between the HOMO and LUMO.

Figure IC1-8 shows the approach of the nucleophile from either of the two faces of the planar carbon. Regardless of the nucleophile's approach, the HOMO and LUMO have the appropriate symmetries to interact, so either approach leads to an allowed nucleophilic addition.

The reverse of nucleophilic addition is nucleophile elimination. An example is shown in Equation IC1-7, in which Cl⁻ is eliminated from a reactant species.

The **HOMO and LUMO have the appropriate symmetries to interact.**

The HOMO is a nonbonding orbital from the nucleophile.

The LUMO is the π* MO from the C=O bond.

(a)

The LUMO is the π* MO from the C=O bond.

The **HOMO and LUMO have the appropriate symmetries to interact.**

The HOMO is a nonbonding orbital from the nucleophile.

(b)

FIGURE IC1-8 Frontier orbital interaction for the nucleophilic addition of HO⁻ to CH_3COCl The nonbonding orbital of the HO⁻ nucleophile (red; HOMO) has the appropriate symmetry to interact with the π* orbital of the C=O bond (blue; LUMO), so these nucleophilic additions are allowed. (Note: The π* MO on O is smaller than on C because, being more electronegative, O contributes more of its p AO to the π MO.)

(IC1-7)

According to the principle of microscopic reversibility, nucleophile elimination reactions are allowed because nucleophilic addition reactions are allowed.

IC1.7 Electrophilic Addition and Electrophile Elimination Steps

Electrophilic addition was first discussed in Section 7.6. In a typical electrophilic addition step, a carbon atom of an alkene or alkyne group forms a new σ bond to an electrophile. An example is shown in Equation IC1-8, in which a C atom of a C=C bond forms a new bond to the proton of H—Cl.

Electrophilic addition

(IC1-8)

Lewis base or nucleophile   **Lewis acid or electrophile**

Recall that the C=C double bond of an alkene is relatively electron rich, making it the Lewis base or nucleophile, whereas the electrophile (the proton of HCl) is electron poor, making it the Lewis acid or electrophile. Therefore, the alkene donates the relevant HOMO to the FMO interaction, and the electrophile donates the relevant LUMO. According to Figure IC1-9, the HOMO of the alkene is a π MO, and the LUMO of H—Cl is a σ* orbital.

The LUMO of HCl is a σ* MO.

The HOMO of the alkene and the LUMO of HCl have the appropriate symmetries to interact.

The HOMO of an alkene is a π MO.

FIGURE IC1-9 Frontier orbital interaction for the electrophilic addition involving an alkene and HCl The π MO of the alkene (red; HOMO) has the appropriate symmetry to interact with the σ* orbital of the H—Cl bond (blue; LUMO), so this electrophilic addition step is allowed.

YOUR TURN IC1.11

Construct the atomic orbital overlap and MO energy diagrams for CH_3CH=$CHCH_3$ (review Figs. 3-13 through 3-21) and use these diagrams to determine its HOMO. Does your answer agree with the one shown in Figure IC1-9?

YOUR TURN IC1.12

In Figure IC1-9, label every region of constructive interference and destructive interference between the HOMO and LUMO.

Electrophile elimination is the reverse of electrophilic addition, so electrophile elimination is allowed, too, according to the principle of microscopic reversibility. A typical electrophile elimination step is shown in Equation IC1-9, in which an H^+ electrophile is eliminated from a carbocation species, producing a new π bond. Simultaneously, a base forms a bond to the departing H^+ electrophile.

Electrophile elimination

(IC1-9)

IC1.8 Carbocation Rearrangements

Carbocation rearrangements were discussed in Section 7.7. Recall that a hydrogen or methyl group can migrate from one carbon atom to an adjacent one to increase carbocation stability. An example of a 1,2-hydride shift is shown in Equation IC1-10.

1,2-Hydride shift

(IC1-10)

This reaction involves just one species, so the carbocation must provide both the HOMO and the LUMO to the FMO interaction. As shown in Figure IC1-10, the HOMO is the σ orbital that connects to the migrating group, and the LUMO is the empty p orbital of the positively charged C atom.

Figure IC1-10 shows that the HOMO and LUMO do, indeed, have the appropriate symmetries to interact, so these carbocation rearrangements are allowed.

The HOMO and LUMO have the appropriate symmetries to interact, so this reaction is *allowed*.

The HOMO is the bonding σ orbital of the C—H bond.

The LUMO is the empty p orbital of the positively charged C atom.

FIGURE IC1-10 The frontier orbital interaction for a 1,2-hydride shift The σ orbital of the C—H bond (red; HOMO) has the appropriate symmetry to interact with the empty p orbital of the positively charged carbon atom (blue; LUMO), so this carbocation rearrangement step is allowed.

IC1.13 YOUR TURN

Construct the atomic orbital overlap and MO energy diagrams for $(CH_3)_2CHC^+HCH_3$ (review Figs. 3-13 through 3-21) and use these diagrams to determine its HOMO and LUMO. Do your answers agree with the ones shown in Figure IC1-10?

IC1.14 YOUR TURN

In Figure IC1-10, label every region of constructive interference and destructive interference between the HOMO and LUMO.

Problems

IC1.1 Would the reaction between HO^- and HCl be allowed if HO^- were to approach HCl from the same side as Cl (i.e., directly opposite the H)? If your answer is yes, would you expect that reaction to take place? Why or why not?

IC1.2 Would the reaction between HO^- and CH_3Cl be allowed if HO^- were to approach CH_3Cl from the same side as Cl (i.e., directly opposite the CH_3)? If your answer is yes, would you expect that reaction to take place? Why or why not?

IC1.3 Would the coordination step between Br^- and $(CH_3)_3C^+$ be allowed if Br^- were to approach from within the plane of the central C? Explain.

IC1.4 Would the nucleophilic addition step between HO^- and CH_3COCl be allowed if HO^- were to approach the carbonyl C atom directly along the C=O bond? Explain.

IC1.5 Show that the electrophilic addition involving $CH_3CH{=}CHCH_3$ and $(CH_3)_3C^+$ is allowed.

IC1.6 Show that a 1,2-methyl shift involving $(CH_3)_3CC^+HCH_3$ is allowed.

IC1.7 Determine whether the addition of a nucleophile to the O atom of a C=O group is allowed or forbidden. If you determine that it is allowed, would you expect that reaction to take place? Why or why not?

IC1.8 Using FMO theory, determine whether a carbanion rearrangement, analogous to the 1,2-hydride shift in Equation IC1-10, is allowed or forbidden.

Naming Compounds with Common Functional Groups

Alcohols, Amines, Ketones, Aldehydes, Carboxylic Acids, Acid Halides, Acid Anhydrides, Nitriles, and Esters

In previous nomenclature units, we have seen that the presence of a functional group invariably leads to a very specific modification to the name. The presence of an alkene or alkyne functional group, for example, is reflected by a modification to the molecule's root—specifically, the "ane" portion of the corresponding alkane's root is changed to "ene" or "yne," respectively. Groups such as halogen atoms, nitro groups, and ethers, on the other hand, are indicated by appropriate prefixes.

As we will see in this nomenclature unit, a *suffix* must be added to the appropriate root if functional groups other than the ones just mentioned are present in a molecule. In Section N4.1, the basic system for naming such compounds is presented. Sections N4.2, N4.3, and N4.4 then discuss how these new rules are combined with ones presented in earlier nomenclature sections pertaining to modifying the root when alkene or alkyne groups are present, adding prefixes to identify substituents, and designating stereochemical configurations. In Section N4.5, we discuss the hierarchy of functional groups, which is necessary to establish the suffix when more than one suffix is possible. Finally, in Section N4.6, we examine trivial names for a variety of compounds encountered throughout this nomenclature unit.

N4.1 The Basic System for Naming Carboxylic Acids, Acid Anydrides, Esters, Acid Chlorides, Amides, Nitriles, Aldehydes, Ketones, Alcohols, and Amines

The name of a compound requires a suffix to be added to the root if it contains something other than halo, nitro, ether, alkene, or alkyne groups. For such compounds, the general form of the name is as follows:

The root indicates the longest chain or ring of carbons, and the presence of alkenes and alkynes.

Prefixes indicate the presence and locations of substituents.

The suffix indicates the presence of a group other than halo, nitro, ether, alkene, or alkyne groups.

prefixesrootsuffix

Table N4-1 lists the specific ways in which suffixes are added for a variety of the most commonly occurring functional groups. The discussion of esters is postponed until Section N4.1a, because naming esters involves rules that do not apply to the functional groups in Table N4-1.

Adding a suffix generally requires first dropping the "e" from the root. To name $CH_3CH_2CO_2H$, for example, first recognize that the root is propane, because the longest continuous chain has three C atoms, all connected by single bonds. The presence of the carboxylic acid (CO_2H) requires the suffix "oic acid." According to Table N4-1, however, the "e" at the end of propane must first be removed, so the IUPAC name is propanoic acid.

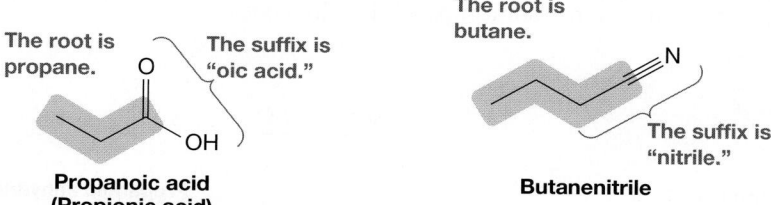

The root is propane.
The suffix is "oic acid."
Propanoic acid (Propionic acid)

The root is butane.
The suffix is "nitrile."
Butanenitrile

Nitriles are an exception. The longest continuous chain in $CH_3CH_2CH_2CN$ has four C atoms, so its root is butane. According to Table N4-1, the presence of the CN

TABLE N4-1 Common Functional Groups and Their Corresponding Suffixes

Functional Group	Drop "e" from Root?	Add	Example	Functional Group	Drop "e" from Root?	Add	Example
Carboxylic acid	Yes	-oic acid	Propanoic acid	Nitrile	No	-nitrile	Butanenitrile
Acid anhydride	Yes	-oic anhydride	Propanoic anhydride	Aldehyde	Yes	-al	Pentanal
Ester	See Section N4.1a			Ketone	Yes	-one	Pentan-2-one
Acid chloride	Yes	-oyl chloride	Butanoyl chloride	Alcohol	Yes	-ol	Propan-1-ol
Amide	Yes	-amide	Hexanamide	Amine	Yes	-amine	Butan-2-amine

group is indicated by the suffix "nitrile" without dropping the "e" at the end of the root name. Thus, the IUPAC name is butanenitrile, not butannitrile.

Acid anhydrides deserve additional discussion because they involve two separate carbon chains. Those carbon chains can be the same, as in the example in Table N4-1, or they can be different from each other. We focus only on the cases in which the carbon chains are the same, because anhydrides with different carbon chains are not as common.

When its carbon chains are the same, an acid anhydride can be prepared from an analogous carboxylic acid by a *dehydration reaction*, which serves to remove water. (See Chapter 21 for a discussion of this reaction.) Propanoic anhydride, for example, can be prepared from propanoic acid. Thus, even though the acid anhydride has two separate carbon chains, the root (propane) describes both.

Propanoic acid   **Propanoic anhydride**

problem **N4.1** Provide the IUPAC name for each of the following compounds.

problem **N4.2** Draw the structure that corresponds to each of the following IUPAC names.

(a) Pentanoic acid   (b) Propanenitrile   (c) Hexanal
(d) Hexanoyl chloride   (e) Butanamide   (f) Butanoic anhydride

Notice in Table N4-1 that number locators appear in some of the names (e.g., propan-1-ol). Specifically,

> - When the suffix corresponds to a ketone, alcohol, or amine functional group, a number is generally used to locate that group along the carbon chain.
> - The carbon chain is numbered such that the carbon atom involved in one of these functional groups is assigned the lowest possible number.

A number locator must be included in their names because ketone, alcohol, and amine functional groups can appear at different locations along a carbon chain, thus leading to constitutional isomers. Pentan-2-one and pentan-3-one are isomers, for example, because they differ only in the location of the ketone group along the five-carbon chain (Fig. N4-1a). Propan-1-ol and propan-2-ol are isomers of each other (Fig. N4-1b), too, as are butan-1-amine and butan-2-amine (Fig. N4-1c). Notice in each of these examples that the ketone, alcohol, and amine functional groups are assigned the lowest possible number. In pentan-2-one, numbering begins at the right-most C atom so that the ketone is located at C2. If numbering were to begin at the left, the ketone would instead be located at C4. In propan-1-ol, numbering begins on the left so that the C atom to which the OH is attached receives the lowest possible number. And, to assign the lowest possible number to the C atoms attached to the amine groups in both butan-1-amine and butan-2-amine, numbering begins on the right.

In contrast to ketones, alcohols, and amines, the other names in Table N4-1 do not have number locators. Specifically,

- When the suffix corresponds to a carboxylic acid, acid anhydride, acid chloride, amide, nitrile, or aldehyde functional group, no locator is required.
- In acyclic compounds, the carbon atom involved in one of these functional groups is designated C1.

Ketone group located on C2

Pentan-2-one

Ketone group located on C3

Pentan-3-one

(a)

Alcohol group located on C1

Propan-1-ol
(n-Propanol)

Alcohol group located on C2

Propan-2-ol
(Isopropanol)
(Isopropyl alcohol)

(b)

Amine group located on C2

Butan-2-amine
(sec-Butylamine)

Amine group located on C1

Butan-1-amine
(n-Butylamine)

(c)

FIGURE N4-1 Suffixes requiring number locators. Functional groups such as (a) ketones, (b) alcohols, and (c) amines require number locators for their suffixes because these functional groups can appear at different locations along the main chain.

These functional groups need no number locator because they cannot be located at various positions along a carbon chain. Rather, the carbon atom in one of these functional groups must appear at one end of the chain. As a result, that carbon atom can always be designated as C1.

The above rules are modified only slightly if the molecule contains multiple functional groups of the type corresponding to the suffix.

- To indicate two, three, or more functional groups of the type corresponding to the suffix, "di," "tri," and so forth must appear directly before the suffix.
- The "e" at the end of the root is *not* removed prior to adding the suffix.
- Numbers must be used to locate each of these functional groups unless the name, without such locators, corresponds to exactly one molecule.

The following examples illustrate these points:

Propanedioic acid

Two carboxylic acid groups

Butanedinitrile

Two nitrile groups

Pentanedial

Two aldehyde groups

Pentane-2,4-dione

Two ketone groups located at C2 and C4

Cyclohexane-1,3,5-trione

Three ketone groups located at C1, C3, and C5

Cyclopentane-1,2,3,4,5-pentaol

Five alcohol groups located at C1, C2, C3, C4, and C5

Hexane-1,6-diamine

Two amine groups located at C1 and C6

In propanedioic acid, butanedinitrile, and pentanedial, "di" indicates the presence of two functional groups corresponding to the suffix—namely, carboxylic acids, nitriles, and aldehydes, respectively. No number locators are used in these names because all three of these types of functional groups must be located at the ends of carbon chains. This is not true for the ketones, the alcohols, and the amines, however, so locators must be used. In pentane-2,4-dione, "di" indicates the presence of two ketone groups, and the locators 2 and 4 describe where they are along the five-carbon chain. The name cyclohexane-1,3,5-trione indicates that three ketone groups are present, located at C1, C3, and C5 of a cyclohexane ring. Similarly, the name cyclopentane-1,2,3,4,5-pentaol indicates that five alcohol groups are attached to a cyclopentane ring, one alcohol group at each C atom. And, in hexane-1,6-diamine, two amine groups are present, one attached to C1 and one to C6.

problem **N4.3** Provide the IUPAC name for each of the following compounds.

(a)   (b)   (c)   (d)

(e)   (f)   (g)   (h)

problem **N4.4** Draw the molecule that corresponds to each of the following IUPAC names.

(a) Hexanedinitrile   (b) Hexane-2,3-dione   (c) Heptane-2,4-diol
(d) Propane-1,2,3-triamine   (e) Octane-2,5-dione   (f) Propanedioyl chloride
(g) Hexanediamide

N4.1a Nomenclature Rules for Esters

Esters are common functional groups that, as indicated in Table N4-1, require the name to have a suffix. The general rules for naming esters are somewhat different from those discussed so far, however, because an ester functional group consists of two separate alkyl groups, and it is quite common for those alkyl groups to be different from each other.

These alkyl groups can consist of any number of carbons.

Alkanoate group

Alkyl group

An ester

Because the two alkyl groups (R and R′) can consist of any number of carbon atoms, the name of the ester must accurately describe both. We begin by dividing the molecule into two parts, as shown above on the right. One part is simply the R′ alkyl group bonded to the O atom. The other is the RCO_2 group, called the

alkanoate group, which contains the carbonyl (C=O) group and the singly bonded O atom. Esters are then named as follows:

- The name of an ester has the general form alkyl alkanoate.
- The alkyl group is named precisely as described in Section N1.2a (i.e., methyl, ethyl, propyl, etc.)
- The alkanoate portion derives from the analogous alkane having the same number of carbons.

In ethyl butanoate, for example, an ethyl group is attached to the singly bonded O atom. The alkanoate group is named butanoate because it is made up of a four-carbon chain. In propyl ethanoate (propyl acetate), the alkyl group is a propyl group and the alkanoate group has two C atoms, so it is the ethanoate group (acetate group).

problem **N4.5** What is the IUPAC name for each of the following molecules?

(a)   (b)   (c)

problem **N4.6** Draw the molecules that correspond to the following IUPAC names.

(a) Pentyl pentanoate   (b) Propyl butanoate   (c) Ethyl methanoate

N4.2 Substituents in Compounds with Functional Groups That Call for a Suffix

Recall from Nomenclature 1 that the longest continuous chain or ring of carbon atoms establishes the molecule's root. Groups other than hydrogen that are attached to that chain or ring are designated as *substituents*. These include alkyl (R) groups, halogen atoms, nitro (NO_2) groups, and alkoxy (OR) groups, and are indicated in the molecule's name by appropriate prefixes before the root. (We discuss other types of substituents in Section N4.5.)

The same basic principles are applied when substituents appear in compounds containing a functional group that calls for a suffix, with one additional rule:

When the name of a molecule requires a suffix to be appended after the root, the root must derive from the longest continuous chain of carbon atoms that contains the functional group corresponding to the suffix.

Consider, for example, the following molecules:

The longest chain of carbon atoms containing the carboxylic acid group is five carbons long.

2-Ethylpentanoic acid

The longest chain of carbon atoms containing the alcohol group is six carbons long.

2-Ethyl-2-methoxyhexan-1-ol

The longest chain of carbon atoms containing the amine group is three carbons long.

1-Cyclopentylpropan-2-amine

The longest chain of carbon atoms containing the alkanoate group is six carbons long.

Ethyl 3-methyl-5-nitrohexanoate

The longest continuous chain in 2-ethylpentanoic acid has six carbons, but the root derives from pentane because the longest chain of carbons containing the carboxylic acid group has only five carbons. In 2-ethyl-2-methoxyhexan-1-ol, the root derives from hexane because the longest chain of carbons containing the alcohol group has six carbons. In 1-cyclopentylpropan-2-amine, the five-membered ring is treated as a substituent because the amine group is part of the three-carbon chain. And, in ethyl 3-methyl-5-nitrohexanoate, the longest carbon chain containing the alkanoate group has six carbons.

The preceding rule can be extended to account for multiple functional groups corresponding to the suffix. In 2-ethyl-3-propylbutanedinitrile, for example, there are several different carbon chains: two that are four carbons long, two that are five carbons long, one that is six carbons long, and one that is seven carbons long. The root derives from butane, however, because the longest of these chains containing *both* nitrile groups has four carbons. Similarly, in 2-butyl-3-ethyl-3-fluoropentane-1,5-diol, there are several chains ranging in length from five to eight carbon atoms. However, the root derives from pentane because the longest continuous chain containing *both* alcohol groups has five carbon atoms.

The longest chain of carbon atoms containing *both* nitrile groups is four carbons long.

2-Ethyl-3-propylbutanedinitrile

The longest chain of carbon atoms containing *both* alcohol groups is five carbons long.

2-Butyl-3-ethyl-3-fluoropentane-1,5-diol

problem **N4.7** What is the IUPAC name for each of the following compounds?

(a)

(b)

(c)

(d)

(e)

(f)

(g)

problem **N4.8** Draw the structure of the molecule that corresponds to each of the following IUPAC names. For each structure, determine the number of carbon atoms composing the longest continuous carbon chain.

(a) 3,3-Dibutylpentane-2,4-dione

(b) 2-Nitro-2-pentylpropanedial

(c) 2,2-Dichloro-3-hexylbutanedioic acid

(d) 3-Butyl-2-heptanamine

(e) 3,4-Dimethylhexanamide

(f) Propyl 5-ethoxypentanoate

N4.2a Alkyl Groups Bonded to Nitrogen in Amines and Amides

The nitrogen atom of an amine can be bonded to two, one, or zero hydrogen atoms (Table N4-1), giving rise to structures of the form RNH_2, R_2NH, and R_3N, respectively. The same is true of the nitrogen atom of an amide, so amides can have the form $RCONH_2$, $RCONHR$, or $RCONR_2$. However, the rules we have learned thus far that pertain to amines and amides only take into account structures of the form RNH_2 and $RCONH_2$, respectively.

The following rules are needed to name compounds of the form R_2NH, R_3N, $RCONHR$, and $RCONR_2$:

- Alkyl groups attached to the nitrogen atom of an amine or amide are treated as substituents if they are not part of the root, and therefore appear as prefixes in the name.
- To avoid confusion with alkyl groups attached to the main chain of an amine or amide, "*N*" must precede the name of each alkyl group attached to nitrogen.
- Substituents should appear alphabetically in the name, regardless of whether they are attached to nitrogen or the main chain of the molecule.

The longest carbon chain attached to N has five C atoms.

Ethyl substitutent attached to N

N-Ethylpentan-1-amine

The longest carbon chain attached to N has six C atoms.

N,N-Diethylhexan-3-amine

The first molecule on the right is a pentan-1-amine because the longest continuous carbon chain attached to the N atom consists of five C atoms, and the N atom is attached at C1. The *N*-ethyl prefix indicates that an ethyl substituent is attached to N.

The second molecule on the previous page is a hexan-3-amine because the longest carbon chain to which the N atom is attached consists of six C atoms, and N is attached to C3. The two ethyl groups attached to N are indicated by the *N,N*-diethyl prefix.

The following molecules show how substituents attached to N are indicated in amides.

The amide chain has three C atoms.   **A propyl substituent attached to N**

The amide chain has four C atoms.   **A cyclopentyl substituent attached to N**

A methyl substituent attached to N

2,2-Dichloro-*N*-propylpropanamide   **N-Cyclopentyl-*N*,3-dimethylbutanamide**

The molecule on the left is a propanamide because the carbon chain that contains the amide group consists of three C atoms. The 2,2-dichloro prefix indicates the two Cl atoms attached to the main chain, whereas the *N*-propyl identifies the propyl group attached to N. The molecule on the right is a butanamide because the longest carbon chain containing the amide group has four C atoms. The *N*-cyclopentyl prefix identifies the cyclopentyl group attached to N, and the *N*,3-dimethyl prefix indicates that there are two methyl groups—one attached to N and the other attached to C3 of the main chain.

---

problem **N4.9** What is the IUPAC name for each of the following compounds?

(a)   (b)   (c)   (d)

---

problem **N4.10** Draw the structure that corresponds to each IUPAC name given.

(a) *N*-Cyclohexyl-3,3-dimethylpentanamide   **(b)** *N,N*-Diethylethanamine

(c) 4,4,4-Trichloro-*N*-methyl-*N*-propylbutanamide   **(d)** *N*-Butyl-3-methoxypentanamide

---

N4.3 Naming Compounds That Have an Alkene or Alkyne Group as Well as a Functional Group That Calls for a Suffix

Recall from Nomenclature 2 that the root must be modified slightly to name molecules containing alkene or alkyne functional groups. Specifically, "ane" in the analogous alkane or cycloalkane is replaced by "ene" or "yne," respectively. This rule remains unchanged for molecules containing a functional group that calls for a suffix—that is, a functional group from Table N4-1. Two examples are as follows:

Propenoic acid   **Propynenitrile**

Indicates the presence of a C═C bond   **Indicates the presence of a C≡C bond**

In propenoic acid, the longest carbon chain has three atoms. If those carbons were connected by single bonds only, the root would have derived from propane. The presence of the double bond, however, requires the root to be changed to propene. With the root established, the suffix is then added as we saw in Table N4-1, to indicate the presence of the carboxylic acid. Namely, the final "e" from propene is dropped, and the suffix "oic acid" is added, yielding propenoic acid as the IUPAC name.

In propynenitrile, the root of a three-carbon chain possessing a C≡C triple bond is propyne. To indicate the presence of the nitrile functional group, the "e" in propyne remains, and we simply add "nitrile" as the suffix. Thus, the IUPAC name is propynenitrile.

Neither propenoic acid nor propynenitrile have number locators because each name already corresponds to an unambiguous structure. A three-carbon chain that has an alkene or alkyne group must involve a terminal carbon, and both carboxylic acid and nitrile functional groups are always located at the end of a carbon chain. However, when longer carbon chains and other functional groups are involved, numbers must be used to locate each such group.

This requirement leads to a contradiction between two previous rules regarding the numbering system of carbon atoms. In Section N2.1a, we learned that for compounds containing an alkene or alkyne group, the carbon atoms of the main carbon chain are numbered so as to give the carbon atoms involving the alkene or alkyne group the lowest possible numbers. Earlier in Nomenclature 4, however, we learned that when a molecule contains a functional group that calls for a suffix, the carbon atom involved with that functional group should receive the lowest possible number. The following rule resolves this conflict:

> In establishing numbering systems for naming molecules, a functional group that requires a suffix (Table N4-1) is given precedence over an alkene or alkyne group.

In other words, when a molecule contains an alkene or alkyne group as well as another functional group requiring a suffix, the carbon atom involved with the latter receives the lowest possible number. The following examples illustrate this point:

The ketone carbon receives a lower number than the alkene carbons.

4-Methylpent-3-en-2-one

The alcohol carbon receives a lower number than the alkyne carbons.

4-Nitrohex-5-yn-3-ol

The nitrile carbons are numbered 1 and 6 irrespective of the numbering system, so the carbons are numbered to give the alkyne carbons the lowest numbers.

5,5-Dichlorohex-2-ynedinitrile

Both "en" and "yn" make up the root.

4-Ethoxy-5-ethylhept-4-en-2-ynal

In 4-methylpent-3-en-2-one, numbering begins from the right to give the ketone carbon the lowest possible number. If numbering were to begin from the left instead, then the first alkene carbon would be C2 and the ketone carbon would be C4. Similarly, in 4-nitrohex-5-yn-3-ol, numbering begins from the right to give the alcohol carbon the lowest possible number. In 5,5-dichlorohex-2-ynedinitrile, the nitrile carbons would be C1 and C6 regardless of which end of the chain numbering begins. Therefore, we begin the numbering from the left so the alkyne carbons have the lowest possible numbers. Finally, 4-ethoxy-5-ethylhept-4-en-2-ynal contains both an alkene and an alkyne group. As a result, the root is modified to include both "en" and "yn" in that order.

problem **N4.11** Provide the IUPAC name for each of the following molecules.

(a)

(b)

(c)

(d)

(e)

(f)

problem **N4.12** Draw the structure that corresponds to each of the following IUPAC names.

(a) But-2-yn-1,4-diamine
(b) 4,4-Dinitrobut-3-enoic acid
(c) 2,3,5,6-Tetramethylcyclohex-2,5-dien-1,4-diol

problem **N4.13** Provide the correct IUPAC name for each of the following molecules, which were named using an incorrect numbering system. (*Hint:* Draw each molecule before you name it.)

(a) Hept-4-yndioic acid
(b) 2,2-Dinitrocyclopent-4-enone
(c) *N*,3-Dimethylhex-1-en-4-amine
(d) 4-Methoxy-2,3-dimethylpent-2-en-5-ol

N4.4 Stereochemistry and Functional Groups That Require a Suffix

Nomenclature 3 discussed how to incorporate stereochemical configurations into a molecule's name. Section N3.1 discussed the R/S system for designating configurations about tetrahedral stereocenters, and Section N3.2 discussed the Z/E system for designating configurations about double bonds. These systems can also be invoked to describe stereochemical configurations in molecules that contain functional groups requiring suffixes.

Consider, for example, the isomer of 3-methoxybut-2-enal shown on the next page. The C=C double bond has both Z and E configurations possible because each C atom is singly bonded to two different groups. On C3, those groups are H_3C- and CH_3O-, and on C2 those groups are $H-$ and the aldehyde group. The CH_3O- group is the higher-priority group on C3, and the aldehyde group is the higher-priority group on C2. Because these two groups are on the same side of the double bond, the configuration is Z. This is indicated in the name by writing (Z) in front, so the molecule is (Z)-3-methoxybut-2-enal.

OCH₃ has a higher priority than CH₃.

The two higher-priority substituents are on the same side of the double bond, defining a **Z** configuration.

In each case, the 1ˢᵗ, 2ⁿᵈ, and 3ʳᵈ priority substituents are arranged clockwise with the 4ᵗʰ priority substituent pointing away, so the configuration is **R**.

HC=O has a higher priority than H.

(Z)-3-Methoxybut-2-enal

(2R,5R)-2-Methylhexane-1,5-diol

There are two stereocenters in the isomer of 2-methylhexane-1,5-diol shown above on the right, at C2 and C5. Thus, each stereocenter must be designated as either *R* or *S*. The four groups attached to C2, from highest to lowest priority, are HOCH₂—, the four-carbon group, CH₃—, and H—. Because the fourth priority substituent (H) is pointing away and the first, second, and third priority substituents are arranged clockwise, the stereocenter is designated *R*. At C5, the attached groups, in order of decreasing priority, are HO—, the five-carbon group, CH₃—, and H—. Once again, the fourth priority substituent (H) is pointing away and the first, second, and third priority substituents are arranged clockwise, so the stereocenter is designated *R*.

As we saw in Section N3.2, these stereochemical designations should appear at the front of the name, so the molecule is (2*R*,5*R*)-2-methylhexane-1,5-diol. Note that the number that locates the position of each stereocenter along the main chain appears before its corresponding stereochemical designation.

---

problem **N4.14** Provide the IUPAC name for each of the following molecules.

(a)

(b)

(c)

(d)

(e)

---

problem **N4.15** Draw the structure that corresponds to each of the following IUPAC names.

(a) (1*R*,3*S*)-Cyclopent-4-en-1,3-diol

(b) (2*E*,4*Z*)-Hepta-2,4,6-trienoic acid

(c) (2*S*,5*S*)-2,5-Diethoxyhexane-3,4-dione

(d) (2*R*,3*Z*)-3-Chloro-2-phenylhex-3-enal

N4.5 The Hierarchy of Functional Groups

In Section N4.1, we saw that if a molecule contains a functional group other than a halo, nitro, ether, alkene, or alkyne, then the name must have a suffix appended to the name after the root. From Table N4-1, we learned how to append such suffixes for a variety of functional groups.

Thus far, the examples here in Nomenclature 4 have each contained only one type of functional group that requires adding a suffix. Many molecules, however, contain more than one type of these functional groups, so the IUPAC has established a hierarchy of functional groups to eliminate any potential conflicts. The relative priorities of some of the most common functional groups are listed in Table N4-2: the lower the number, the higher the priority.

> - If a molecule contains two or more types of functional groups associated with different suffixes, the suffix assigned to the molecule is the one corresponding to the highest-priority functional group.
> - The highest priority functional group establishes the numbering system for the molecule.
> - Prefixes are used to indicate the presence of the other functional groups.

TABLE N4-2 Hierarchy of Functional Groups in Nomenclature

Functional Group	Relative Priority	Number of Carbon–Heteroatom Bonds	Prefix	Functional Group	Relative Priority	Number of Carbon–Heteroatom Bonds	Prefix
Carboxylic acid	1	3	carboxy-	Nitrile	6	3	cyano-
Acid anhydride	2	3	—	Aldehyde	7	2	oxo-
Ester	3	3	—	Ketone	8	2	oxo-
Acid chloride	4	3	chlorocarbonyl-	Alcohol	9	1	hydroxy-
Amide	5	3	carbamoyl-	Amine	10	1	amino-

The functional groups not listed in Table N4-2 will not be discussed because we will not encounter many molecules that require prefixes corresponding to those groups.

The following examples illustrate how to apply the preceding rules.

4-Hydroxybutanal

OH group indicated by a prefix

5-Cyanopentanoic acid

CN group indicated by a prefix

2-Amino-5-ethylcyclohexane-1,3-dione

NH$_2$ group indicated by a prefix

(E)-2,2-Dibromo-6-oxohept-4-enal

C═O group indicated by a prefix

(3R,4S)-3-Hydroxy-6-oxo-4-phenylhexanoic acid

OH group indicated by a prefix

C═O group indicated by a prefix

4-Hydroxybutanal contains two functional groups that are associated with suffixes—an alcohol group and an aldehyde group. According to Table N4-2, the aldehyde group has a higher priority than the alcohol, so the suffix corresponds to the aldehyde only. The alcohol group is treated as a substituent attached to C4, indicated by the prefix "hydroxy." In 5-cyanopentanoic acid, there are two functional groups that have suffixes associated with them—the nitrile group and the carboxylic acid group. According to Table N4-2, the carboxylic acid group has the higher priority, which is why it is the group that corresponds to the suffix. The nitrile group is treated as a substituent attached to C5, indicated by the prefix "cyano." In 2-amino-5-ethylcyclohexane-1,3-dione, the NH$_2$ group is treated as a substituent, indicated by the prefix "amino." The name (E)-2,2-dibromo-6-oxohept-4-enal illustrates how the prefix "oxo" is used to indicate the presence of the ketone group at C6. In (3R,4S)-3-hydroxy-6-oxo-4-phenylhexanoic acid, both the alcohol and aldehyde groups are indicated by prefixes—"hydroxy" for the alcohol group at C3, and "oxo" for the aldehyde group at C6.

In light of these examples, you must commit to memory the relative priorities of functional groups in Table N4-2. In doing so, it helps to know that there is a correlation between a functional group's relative priority and the number of bonds to heteroatoms involving the carbon atom of that functional group.

In general, the greater the number of bonds to heteroatoms involving the carbon atom of a particular functional group, the higher the functional group's relative priority in nomenclature.

The number of carbon–heteroatom bonds for such carbon atoms are listed in Table N4-2. As we can see, the carbon atoms of carboxylic acids, acid anhydrides, esters, acid chlorides, amides, and nitriles have the greatest number of carbon–heteroatom bonds (three), consistent with the fact that those functional groups have the highest priorities in the table. Aldehydes and ketones have the next highest priorities because they each have two carbon–heteroatom bonds. And the functional groups whose carbon atoms have the fewest bonds to heteroatoms are alcohols and amines (one), consistent with those functional groups having lower priorities than the others.

---

problem **N4.16** Provide the IUPAC name for each of the following compounds.

(a)   (b)   (c)   (d)

---

problem **N4.17** For each IUPAC name provided, draw the corresponding structure.

(a) 4,4-Dinitro-3-oxohexanoic acid
(c) 2,3-Dihydroxypropanenitrile

(b) Methyl 2-cyanopentanoate
(d) 4-Hydroxy-N,N-dimethylpentanamide

N4.6 Trivial Names Involving Functional Groups That Call for a Suffix

In Section N1.3, we learned that many trivial names of alkanes and ethers remain in common use. In Sections N2.1d and N2.2b, we encountered a variety of these trivial names involving alkenes and alkynes, and substituted benzenes, respectively. Here in Section N4.6, we will examine some important aspects of trivial names involving functional groups we have encountered throughout this nomenclature unit. Related functional groups will be discussed together.

N4.6a Trivial Names of Carboxylic Acids

Trivial names of some common carboxylic acids are shown below. They include mono-carboxylic acids as well as dicarboxylic acids.

Monocarboxylic acids

Methanoic acid
(Formic acid)

Ethanoic acid
(Acetic acid)

Propanoic acid
(Propionic acid)

Butanoic acid
(Butyric acid)

Pentanoic acid
(Valeric acid)

Hexanoic acid
(Caproic acid)

Propenoic acid
(Acrylic acid)

Phenylmethanoic acid
(Benzoic acid)

Dicarboxylic acids

Ethanedioic acid
(Oxalic acid)

Propanedioic acid
(Malonic acid)

Butanedioic acid
(Succinic acid)

(Z)-Butenedioic acid
(Maleic acid)

(E)-Butenedioic acid
(Fumaric acid)

2-Carboxybenzoic acid
(Phthalic acid)

Several of the preceding trivial names are used almost to the exclusion of their IUPAC names. Acetic acid, the principle component of vinegar, is one example. Another is formic acid, a main constituent of ant venom.

---

problem **N4.18** For each trivial name, draw the complete structure and provide the IUPAC name.

(a) Trichloroacetic acid   **(b)** 2,2-Dimethylbutyric acid   **(c)** 2-Aminopropionic acid
(d) Dimethylmaleic acid   **(e)** Diethylmalonic acid

With trivial names, a system of Greek letters is used to locate substituents relative to the carboxylic acid group. More specifically, the carbon atom adjacent to the carboxylic acid group is designated α (alpha), and the atoms that are two, three, four, and five carbons away from the carboxylic acid's carbon atom are designated β (beta), γ (gamma), δ (delta), and ε (epsilon), respectively.

An alpha amino acid   A beta keto acid

Thus, a compound that has an amino group attached to a carbon atom adjacent to the carboxylic acid group is called an alpha amino acid, and a compound in which a ketone group is two carbons removed from the carboxylic acid is called a beta keto acid.

N4.6b Trivial Names of Carboxylic Acid Derivatives: Acid Anhydrides, Esters, Acid Chlorides, Amides, and Nitriles

Acid anhydrides, esters, acid chlorides, amides, and nitriles are collectively known as *carboxylic acid derivatives* because, as we will see in Chapters 20 and 21, they can be synthesized relatively easily from analogous carboxylic acids. Due to their chemical relationship to carboxylic acids, the trivial names of many carboxylic acid derivatives derive from the trivial names of the analogous carboxylic acids.

Acid anhydrides

Ethanoic anhydride
(Acetic anhydride)

2-Benzofuran-1,3-dione
(Phthalic anhydride)

Esters

Methyl methanoate
(Methyl formate)

Ethyl ethanoate
(Ethyl acetate)

Ethyl phenylmethanoate
(Ethyl benzoate)

Acid chlorides

Ethanoyl chloride
(Acetyl chloride)

Phenylmethanoyl chloride
(Benzoyl chloride)

Amides

Methanamide
(Formamide)

N,N-Dimethylmethanamide
(N,N-Dimethylformamide, DMF)

Ethanamide
(Acetamide)

Benzamide

Nitriles

Ethanenitrile
(Acetonitrile)

Benzonitrile

Notice, in particular, the prevalence of the roots "form," "acet," and "benz," which derive from formic acid, acetic acid, and benzoic acid, respectively.

---

problem **N4.19** For each trivial name, draw the complete structure and provide the IUPAC name.

(a) *N,N*-Dimethylacetamide **(b)** Methoxyacetonitrile **(c)** Ethyl trichloroacetate

(d) Isopropyl formate **(e)** *N,N*-Diphenylbenzamide

N4.6c Trivial Names of Aldehydes

Many trivial names for aldehydes are derived from trivial names for the analogous carboxylic acids in much the same way as we just learned for carboxylic acid derivatives. For example, the trivial name for methanal, which has one carbon atom, is formaldehyde. This is analogous to formic acid, the carboxylic acid that contains just one carbon atom. Similarly acetaldehyde, the trivial name for the two-carbon aldehyde, derives from acetic acid, the two-carbon carboxylic acid.

Aldehydes

Methanal (Formaldehyde) **Ethanal** (Acetaldehyde) **Propanal** (Propionaldehyde) **Butanal** (Butyraldehyde) **Pentanal** (Valeraldehyde) **Hexanal** (Capronaldehyde) **Benzaldehyde**

problem **N4.20** For each trivial name, draw the complete structure and provide the IUPAC name.

(a) Phenylacetaldehyde **(b)** 2,3-Dichloropropionaldehyde **(c)** 4-Nitrobutyraldehyde

N4.6d Trivial Names of Ketones

Because ketones consist of two alkyl groups attached to a carbonyl (C=O) group, their trivial names consist of identifying the alkyl groups and appending the suffix "ketone." Thus, the trivial name for propanone, $(CH_3)_2C$=O, is dimethyl ketone, the trivial name for 3-pentanone is diethyl ketone, and the trivial name for diphenylmethanone, $(C_6H_5)_2C$=O, is diphenyl ketone.

Ketones

Propanone (Dimethyl ketone) (Acetone) **Pentan-3-one** (Diethyl ketone) **Diphenylmethanone** (Diphenyl ketone) (Benzophenone) **Butanone** (Methyl ethyl ketone)

Phenylethanone (Methyl phenyl ketone) (Acetophenone) **Butenone** (Methyl vinyl ketone) **3-Methylbutanone** (Methyl isopropyl ketone)

This system can also be used when the two alkyl groups are different, as in the case of methyl ethyl ketone, the trivial name of butanone. Other examples include methyl phenyl ketone, methyl vinyl ketone, and methyl isopropyl ketone. Notice in the last two examples that the trivial names vinyl and isopropyl are used to identify the alkyl groups.

Some ketones have more than one trivial name—for example, propanone is also known as dimethyl ketone and acetone. Diphenylmethanone (trivial name: diphenyl ketone) is also known as benzophenone, and phenylethanone (trivial name: methyl phenyl ketone) is also known as acetophenone.

---

problem **N4.21** For each trivial name, draw the complete structure and provide the IUPAC name.

(a) Divinyl ketone   **(b)** Benzyl isopropyl ketone   **(c)** Cyclohexyl methyl ketone
(d) Diisopropyl ketone   **(e)** Isobutyl phenyl ketone

N4.6e Trivial Names of Alcohols

Because alcohols consist of an alkyl group attached to a hydroxyl (OH) group, many trivial names of alcohols are derived simply by identifying the specific alkyl group present and adding the second word "alcohol."

Primary (1°) alcohols

Secondary (2°) alcohol

Tertiary (3°) alcohol

H_3C—OH
Methanol
(Methyl alcohol)
(Wood alcohol)

Ethanol
(Ethyl alcohol)
(Grain alcohol)

Propan-1-ol
(Propyl alcohol)

Propan-2-ol
(Isopropyl alcohol)
(Rubbing alcohol)

Methylpropan-2-ol
(*tert*-Butyl alcohol)

Phenol

Phenylmethanol
(Benzyl alcohol)

For example, methanol (sometimes referred to as wood alcohol because it was first isolated from the pyrolysis of wood) consists of a methyl group attached to OH, so its trivial name is methyl alcohol. Similarly, the trivial name of ethanol (sometimes called grain alcohol because of its production during the fermentation of grains) is ethyl alcohol, and the trivial name of propan-1-ol is propyl alcohol. Examples that incorporate trivial names for the alkyl group include isopropyl alcohol (also known as rubbing alcohol) and *tert*-butyl alcohol. Phenol and benzyl alcohol represent examples that contain a phenyl ring.

Alcohols are often classified as either primary (1°), secondary (2°), or tertiary (3°), depending on the degree of alkyl substitution of the carbon to which the OH group is attached.

> ▪ In a *primary* (1°) alcohol, the OH group is attached to a primary carbon.
> ▪ In a *secondary* (2°) alcohol, the OH group is attached to a secondary carbon.
> ▪ In a *tertiary* (3°) alcohol, the OH group is attached to a tertiary carbon.

Recall from Section N1.3 that a carbon's degree of substitution is defined by the number of alkyl groups to which it is bonded. A primary carbon is bonded to one alkyl group, a secondary carbon is bonded to two alkyl groups, and a tertiary carbon is bonded to three alkyl groups. Therefore, ethanol (ethyl alcohol) and propan-1-ol (propyl alcohol) are primary alcohols, whereas propan-2-ol (isopropyl alcohol) and methylpropan-2-ol (*tert*-butyl alcohol) are secondary and tertiary alcohols, respectively.

problem **N4.22** For each trivial name, draw the complete structure and provide the IUPAC name.

(a) Trichloromethyl alcohol   **(b)** Isobutyl alcohol
(c) Pentyl alcohol   **(d)** sec-Butyl alcohol

problem **N4.23** Determine whether each alcohol in Problem N4.22 is a primary, secondary, or tertiary alcohol.

N4.6f Trivial Names of Amines

The trivial names for amines are constructed in much the same way as they are for alcohols. The alkyl groups involved with the amine are named first, followed by the suffix "amine."

Primary (1°) amines

H₃C—NH₂
Methanamine
(Methylamine)

Ethanamine
(Ethylamine)

Butan-2-amine
(sec-Butylamine)

Secondary (2°) amines

N-Methylmethanamine
(Dimethylamine)

N-Ethylcyclohexanamine
(Ethylcyclohexylamine)

Tertiary (3°) amine

N,N-Dimethylmethanamine
(Trimethylamine)

Aniline

Phenylmethanamine
(Benzylamine)

The main difference between naming amines and naming alcohols is in the number of alkyl groups that must be identified. Whereas an alcohol oxygen is bonded to only one alkyl group, the nitrogen atom of an amine can be bonded to one, two, or three alkyl groups.

In methylamine, for example, the N atom is bonded to just a methyl group, so the other two bonds to N are N—H bonds. The names ethylamine and *sec*-butylamine indicate that the N atom is attached to an ethyl group and a *sec*-butyl group, respectively. Dimethylamine and ethylcyclohexylamine are examples in which the N atom is bonded to two alkyl groups, and trimethylamine is an example in which the N atom is bonded to three alkyl groups. Aniline and benzylamine involve a phenyl ring.

Amines are classified as either primary (1°), secondary (2°), or tertiary (3°). Although this terminology is identical to that used for classifying alcohols (Section N4.6e), the way in which amines are classified is different from that for alcohols. Whereas alcohols are classified by the number of alkyl groups bonded to the C to which the OH group is attached, an amine is classified by the number of alkyl groups to which the N atom is attached.

> ■ In a *primary* (1°) amine, the N atom is bonded to one alkyl group.
>
> ■ In a *secondary* (2°) amine, the N atom is bonded to two alkyl groups.
>
> ■ In a *tertiary* (3°) amine, the N atom is bonded to three alkyl groups.

Thus, methylamine, ethylamine, and *sec*-butylamine are all primary amines, whereas dimethylamine and ethylcyclohexylamine are secondary amines, and trimethylamine is a tertiary amine.

---

problem **N4.24** For each trivial name, draw the complete structure and provide the IUPAC name.

(a) Diisopropylamine
(b) sec-Butylisopropylamine
(c) *tert*-Butyldimethylamine
(d) Triethylamine
(e) Diphenylamine

---

problem **N4.25** Identify each molecule in Problem N4.24 as either a primary, secondary, or tertiary amine.

N4.6g Trivial Names of Compounds in Which a Functional Group That Calls for a Suffix Is Attached to a Phenyl Ring

Throughout Section N4.6, we have encountered trivial names of several compounds in which a phenyl ring is attached to a single substituent. The ones shown below have, in fact, been incorporated into the IUPAC system, and therefore should be committed to memory.

Benzoic acid   **An alkyl benzoate**   **Benzamide**   **Benzonitrile**

Benzaldehyde   **Phenol**   **Aniline**

As a result, the root names of many compounds containing a phenyl ring are derived from one of the above names, similar to what we saw for toluene (C_6H_5—CH_3) and anisole (C_6H_5—OCH_3) in Section N2.2a.

> ▪ If two or more substituents are attached to a phenyl ring and one of them is a carboxylic acid, ester, amide, nitrile, aldehyde, alcohol, or amine, then the compound is named as a benzoic acid, alkyl benzoate, benzonitrile, benzaldehyde, phenol, or aniline, respectively.
>
> ▪ If two or more substituents attached to a phenyl ring are among the functional groups listed above, the highest-priority functional group (Table N4-2) establishes the root, as well as C1 of the phenyl ring.

Consider, for example, the following compounds:

2,4,6-Trimethylphenol

4-Nitroaniline
p-Nitroaniline

3-Hydroxybenzoic acid
m-Hydroxybenzoic acid

2,4-Dimethoxy-*N*,*N*-dimethylbenzamide

Phenyl 2,3-diaminobenzoate

3,4-Diaminobenzonitrile

2,4,6-Trimethylphenol is a phenol because OH is the highest-priority functional group attached to the phenyl ring. 4-Nitroaniline is named as an aniline because of the NH₂ group attached to the phenyl ring. In 3-hydroxybenzoic acid, the highest-priority functional group attached to the phenyl ring is a carboxylic acid. 2,4-Dimethoxy-*N*,*N*-dimethylbenzamide is an example in which an amide is the highest-priority functional group attached to the phenyl ring. Notice also that the "*N*,*N*" designation identifies the two methyl groups as being attached to the nitrogen atom rather than to the phenyl ring. Phenyl 2,3-diaminobenzoate is an ester, and 3,4-diaminobenzonitrile is a nitrile.

---

problem **N4.26** Provide the IUPAC name for each of the following compounds.

(a) (b) (c) (d)

problem **N4.27** Draw the structure that corresponds to each of the following names.

(a) 2,3-Diaminoaniline (b) 2,4-Dinitrophenol (c) 2,4,5-Trihydroxybenzoic acid
(d) 3-Hydroxybenzonitrile (e) 4-Cyanobenzonitrile (f) Ethyl 4-methylbenzoate

An Introduction to Multistep Mechanisms

S$_N$1 and E1 Reactions

Many organic reactions take place by mechanisms that involve several steps. Fortunately, such **multistep mechanisms** are typically built up from just a handful of different *elementary steps*. We have already studied 10 of these elementary steps in Chapters 6 and 7. Just as letters of the alphabet can be combined in various ways to create a large number of different words, these elementary steps can be combined in various ways to create a large number of different reaction mechanisms.

Here in Chapter 8, we will introduce two of the simplest multistep mechanisms: the unimolecular nucleophilic substitution (S$_N$1) reaction and the unimolecular elimination (E1) reaction. Each of these reactions consists of just two steps, but by studying them we can learn much about

Eventually, all of these dominos will fall, but they must do so in a particular order. Analogously, the overall reactions we will learn in this chapter—S$_N$1 and E1 reactions—take place via multistep mechanisms constructed from precise sequences of elementary steps.

multistep mechanisms in general, including those with three, four, five, or more steps.

We begin Chapter 8 by examining key aspects of these two reactions—namely, their curved arrow notation, reaction energy diagrams, chemical kinetics (reaction rates), and stereoselectivity (the tendency to form one stereoisomer over another). We then present general guidelines for reasonable multistep mechanisms, which are intended to help you develop a certain level of "chemical intuition" that can be applied to other reaction mechanisms in subsequent chapters.

8.1 The Unimolecular Nucleophilic Substitution (S$_N$1) Reaction

In Section 7.2, we saw that a nucleophile (Nu$^-$) can replace a leaving group (L) on a substrate (R—L) in a single step via a bimolecular nucleophilic substitution (S$_N$2) reaction (Equation 8-1). However, we can also envision a nucleophilic substitution reaction taking place in two steps, as shown in Equation 8-2. This is an example of a **unimolecular nucleophilic substitution** (**S$_N$1**) mechanism. (Note that the numbers in the names of the mechanisms do *not* correspond to the number of steps in the mechanism.)

S$_N$2 mechanism

$$\text{Nu:}^{\ominus} + \overset{|}{\underset{|}{-\text{C}}}-\text{L} \longrightarrow \text{Nu}-\overset{|}{\underset{|}{\text{C}}}- + \text{:L}^{\ominus} \qquad (8\text{-}1)$$

Nucleophile   Substrate

S$_N$1 mechanism

Leaving group leaves

$$\overset{|}{\underset{|}{-\text{C}}}-\text{L} \longrightarrow \text{C}^{\oplus} + \text{:L}^{\ominus} \qquad (8\text{-}2a)$$

Nucleophile attacks

$$\text{Nu:}^{\ominus} + \text{C}^{\oplus} \longrightarrow \text{Nu}-\overset{|}{\underset{|}{\text{C}}}- \qquad (8\text{-}2b)$$

Equation 8-2a shows that the leaving group (L) simply leaves in the first step of an S$_N$1 mechanism, yielding L$^-$ and a carbocation. In the second step (Equation 8-2b), the nucleophile (Nu⁻) forms a bond to the carbocation (which is very reactive) to complete the reaction. Even though this mechanism is different from what we have seen before, it is composed of elementary steps with which we are familiar (see Your Turn 8.1).

YOUR TURN 8.1

Under each arrow in Equation 8-2, write the name of the elementary step that is occurring. (*Hint:* See Chapter 7.)

Answers to Your Turns are in the back of the book.

Although a nucleophile replaces a leaving group in both the S_N1 and S_N2 mechanisms, there are important differences in their respective outcomes. We will see, for example, that S_N2 reactions tend to produce a single stereoisomer, whereas S_N1 reactions tend to produce mixtures of stereoisomers. We will also see that the carbon backbone can rearrange in an S_N1 mechanism, but cannot in an S_N2 mechanism.

---

problem **8.1** Using curved arrow notation, draw **(a)** an S_N2 mechanism and **(b)** an S_N1 mechanism for the substitution reaction at the right.

8.1a Overall Reactants, Overall Products, and Intermediates

In any reaction that takes place in multiple steps, we must be able to distinguish the *overall reaction* from the elementary steps of its mechanism. The **overall reaction** (or **net reaction**) describes the *net changes* that occur, and can be obtained by simply adding together all of the elementary steps.

An S_N1 mechanism

When we add together the steps of an S_N1 mechanism in Equation 8-2, for example, we cancel the carbocation species because it is a *product* in Equation 8-2a and a *reactant* in Equation 8-2b. In other words, there is no net consumption or net production of the carbocation. The species that remain appear in the overall reaction, as shown in Equation 8-3.

We must also be able to distinguish between *overall reactants*, *overall products*, and *intermediates*.

> ▪ A species is an **overall reactant** or an **overall product** if it appears in the overall (i.e., net) reaction.
>
> ▪ A species is an **intermediate** if it appears in a mechanism but does *not* appear in the overall reaction.

The overall reactants in Equation 8-3 are R—L and Nu⁻, and the overall products are Nu—R and L⁻. The carbocation (R⁺) that appears in the reaction mechanism (Equation 8-2) is the only intermediate.

8.2 YOUR TURN

In the following S_N1 mechanism, **(a)** identify each species as an *overall reactant*, *overall product*, or *intermediate*, and **(b)** sum the steps to yield the overall reaction.

We need to distinguish intermediates from overall reactants and products because *intermediates usually cannot be isolated.* They are intermediates in the first place because they tend to be highly reactive: as soon as one is formed as a product in one step, it is consumed as a reactant in the next step. On the other hand, *overall reactants and products tend to be stable and can be isolated.* Overall reactants are typically the compounds that are physically added together to initiate a chemical reaction, and overall products are the newly formed compounds remaining in the reaction mixture when the reaction has come to completion.

---

problem **8.2** For the S_N1 reaction in Problem 8.1, identify the overall reactants, overall products, and any intermediates.

8.1b Free Energy Diagram of an S_N1 Reaction

The free energy diagram for the S_N1 mechanism in Equation 8-2 is shown in Figure 8-1. Just as in any reaction free energy diagram, the Gibbs free energy of the species involved in the reaction is plotted as a function of the *reaction coordinate* (Section 6.3b). As the reaction coordinate increases from left to right, the species that are involved in the reaction less closely resemble the overall reactants and more closely resemble the intermediate or overall products. That is, the *overall reactants* are located on the far left and the *overall products* are located on the far right.

FIGURE 8-1 Free energy diagram for an S$_N$1 reaction The overall reactants are on the far left, the overall products are on the far right, and any intermediates are at local energy minima in between.

Unlike energy diagrams we have seen before, the one in Figure 8-1 has two humps connecting reactants to products, not one. This is because there are two separate elementary steps for the S$_N$1 mechanism. Each step is an individual reaction that has its own reactants and products and proceeds through a *transition state*; each transition state occurs at a local energy maximum along the reaction coordinate.

There are two characteristics of the intermediate worth noting:

1. The energy of the intermediate is higher than that of the overall reactants and the overall products.
2. The intermediate occurs at a **local minimum** in energy along the reaction coordinate, so energy increases when going either forward or backward from the intermediate along the reaction coordinate.

The energy of the reaction intermediate in Equation 8-2 is high because two additional charges are generated and because the C atom loses its octet. Although the intermediate is itself highly unstable, even more destabilization is created by the partial formation or breaking of bonds as the reaction proceeds either forward or backward from the intermediate along the reaction coordinate.

YOUR TURN 8.3

In Figure 8-1, label the transition state for the first step of the mechanism "transition state 1" and label the one for the second step "transition state 2." Also, label the intermediate.

Because they are situated at local energy minima, intermediates can *theoretically* be isolated. We can think of the intermediate as "trapped" between the two energy barriers on either side. As noted previously, however, isolating an intermediate is very difficult because, compared to the energy difference between the intermediate and either the overall reactants or overall products, those energy barriers are typically quite small.

These observations can be generalized for any mechanism involving any number of steps.

For a mechanism that contains n total steps, there must be n total transition states.

Because an intermediate is located in between two transition states, there must be $n - 1$ total intermediates.

How many elementary steps are there in the mechanism represented by the following free energy diagram? How many intermediates are there? Mark the locations of the reactants (R), intermediates (I), transition states (TS), and products (P) on the diagram.

problem 8.3 Draw a detailed free energy diagram for the S_N1 reaction in Your Turn 8.2. Include and label the overall reactants, the overall products, the intermediate(s), and the transition state(s).

8.2 The Unimolecular Elimination (E1) Reaction

In Section 7.4, we saw that a leaving group and a proton (H^+) are eliminated from adjacent carbon atoms in a *bimolecular elimination (E2) reaction*, resulting in an additional π bond between those two carbon atoms. Equation 8-4 shows an E2 reaction between a generic base (B^-) and a generic substrate ($R—L$).

E2 mechanism

$$B:^{\ominus} + \quad H—C—C—L \longrightarrow B—H + \quad \diagup C=C \diagdown + \quad :L^{\ominus} \quad (8\text{-}4)$$

Base   Substrate

Elimination reactions can also take place in two steps, via a **unimolecular elimination (E1) reaction**, as shown in Equation 8-5. As with S_N1 and S_N2 reactions, the numbers in the names of the elimination reactions do *not* refer to the number of steps in each mechanism.

E1 mechanism

$$H—C—C—L \xrightarrow{\text{Leaving group leaves}} H—C—C^{\oplus} + :L^{\ominus} \quad (8\text{-}5a)$$

$$B:^{\ominus} + \quad H—C—C^{\oplus} \xrightarrow{\text{Base attacks}} B—H + \quad \diagup C=C \diagdown \quad (8\text{-}5b)$$

Notice that the first step in an E1 reaction (Equation 8-5a) is precisely the same as the first step in an S_N1 reaction (Equation 8-2a). The difference between the S_N1 and E1 mechanisms is in the second step. Whereas the second step of an S_N1 mechanism is a *coordination* step (Equation 8-2b), H^+ is eliminated in the second step of an E1 mechanism (Equation 8-5b), thus forming a π bond.

YOUR TURN 8.5

Under each arrow in Equation 8-5, name the elementary step that is occurring. (*Hint:* Review Chapter 7.)

Just as we did with the S_N1 mechanism, we can obtain the overall reactants and products of an E1 mechanism by simply adding together the steps in Equation 8-5, yielding Equation 8-6. Notice once again that the carbocation is an intermediate in the E1 mechanism because it does not appear in the overall reaction.

Overall reaction for the E1 mechanism in Equation 8-5

$$ B:^{\ominus} \ + \ H-\overset{|}{\underset{|}{C}}-\overset{|}{\underset{|}{C}}-L \ \longrightarrow \ B-H \ + \ \overset{}{\underset{}{C}}=\overset{}{\underset{}{C} } \ + \ :L^{\ominus} \quad (8\text{-}6) $$

Base   Substrate

YOUR TURN 8.6

Label any intermediates in Equations 8-5a and 8-5b.

problem **8.4** Draw the mechanism for the E1 reaction that would take place between the two species at the right. Label each species in the mechanism as an *overall reactant*, an *overall product*, or an *intermediate*.

$$ \text{(cyclohexane ring with } C(CH_3)_3 \text{ and } Br) \ + \ HOCH_3 \ \longrightarrow \ ? $$

Figure 8-2 illustrates the free energy diagram for an E1 reaction. It is strikingly similar to the one in Figure 8-1 for an S_N1 reaction. Notice that there are two transition states (energy maxima), consistent with the fact that two separate steps compose the E1 mechanism. Furthermore, there is a single intermediate, appearing at a *local energy minimum* between the two transition states. Finally, the energy of the intermediate

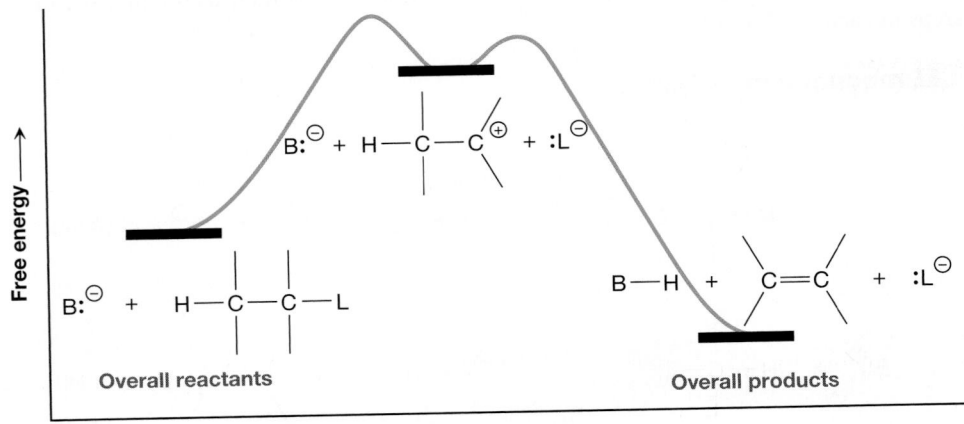

FIGURE 8-2 Free energy diagram for an E1 reaction The overall reactants are on the far left, the overall products are on the far right, and any intermediates are at local energy minima in between.

is substantially higher than either the reactants or products due to the appearance of charges and the loss of an octet on carbon.

8.7 YOUR TURN

In Figure 8-2, label the transition state for the first step of the mechanism "transition state 1" and label the one for the second step "transition state 2." Also, label the intermediate.

---

problem **8.5** Draw the energy diagram for the E1 reaction in Problem 8.4. Include and label the *overall reactants*, the *overall products*, the *transition state(s)*, and the *intermediate(s)*.

8.3 Direct Experimental Evidence for Reaction Mechanisms

So far in this chapter we have seen that S_N1 products are similar to S_N2 products and that E1 products are similar to E2 products. What evidence, then, do we have for each mechanism? How do we know that the mechanisms we have presented are "correct"?

Perhaps the most powerful evidence for the correctness of a reaction mechanism is the direct observation of the intermediates that are proposed. For both S_N1 and E1 reactions, for example, we would try to directly observe the carbocation that is presumed to be an intermediate. As mentioned previously, however, intermediates (e.g., carbocations) tend to be very reactive and thus have very short lifetimes. The direct observation of intermediates, therefore, is a formidable task and quite rare.

In 1962, however, George A. Olah, a Hungarian-born American chemist, directly observed the *tert*-butyl carbocation, $(CH_3)_3C^+$, using *nuclear magnetic resonance spectroscopy* (an important technique that we discuss in Chapter 16). Olah dissolved 2-fluoro-2-methylpropane, $(CH_3)_3C$—F, in a "superacid," which provided the extreme conditions necessary for the carbocation to be long-lived enough for observation.

Key intermediates in a few other reactions have been directly observed as well. Most intermediates, however, do not lend themselves to direct observation, so:

In general, mechanisms cannot be proved.

Consequently, other pieces of evidence, such as those described in Sections 8.4 and 8.5, become critical. This kind of evidence can be used either to *support* a mechanism or to *disprove* it.

8.4 The Kinetics of S_N2, S_N1, E2, and E1 Reactions

Some of the most important evidence supporting the different mechanisms proposed for S_N1 and S_N2 nucleophilic substitution reactions, as well as for the E1 and E2 reactions, comes from *reaction kinetics*, the study of *reaction rates*. Here in Section 8.4, we discuss differences in the *rate laws* for the different reactions, which, as we will see, correspond to the differences in their respective free energy diagrams.

8.4a Empirical and Theoretical Rate Laws for the S_N2, S_N1, E2, and E1 Reactions

The rates of S_N1 and S_N2 reactions differ in the way they depend on reactant concentrations. We can see this difference from the experimentally determined rate laws, or **empirical rate laws**, for the two reactions, shown in Equations 8-7 and 8-8.

$$\text{Rate } (S_N2) = k_{S_N2}[\text{Nu}^{\ominus}][R-L] \tag{8-7}$$

$$\text{Rate } (S_N1) = k_{S_N1}[R-L] \tag{8-8}$$

Notice in Equation 8-7 that the rate of an S_N2 reaction is directly proportional to both the nucleophile ($[\text{Nu}^-]$) and substrate ($[R-L]$) concentrations. The rate of an S_N1 reaction, on the other hand, is directly proportional only to the substrate concentration ($[R-L]$). In other words,

> The rate of an S_N2 reaction is *directly proportional to* the concentration of the nucleophile, whereas the rate of an S_N1 reaction is *independent of* the concentration of the nucleophile!

The proportionality constant in Equation 8-7, k_{S_N2}, is the **rate constant** for the S_N2 reaction. Similarly, the rate constant for the S_N1 reaction is denoted k_{S_N1} in Equation 8-8.

SOLVED problem 8.6 A chemist wants to determine whether the following reaction proceeds via an S_N1 or S_N2 reaction.

She therefore decided to carry out kinetics experiments, measuring the dependence of the reaction rate on the initial concentrations of each reactant. Her results are provided in the following table:

Trial Number	$[C_6H_5CH_2Cl]$	$[HO^-]$	Rate (M/s)
1	0.10 M	0.10 M	2.6×10^{-6}
2	0.10 M	0.20 M	5.1×10^{-6}
3	0.20 M	0.20 M	1.0×10^{-5}

Based on these results, does the reaction proceed by the S_N1 or S_N2 mechanism?

Think What is the dependence of the reaction rate on each of the reactants? What is the corresponding empirical rate law? Is it consistent with the rate law for an S_N1 or an S_N2 reaction?

Solve In going from Trial 1 to Trial 2, the concentration of HO^- doubles but the concentration of $C_6H_5CH_2Cl$ remains the same. This causes the reaction rate to double—from 2.6×10^{-6} M/s to 5.1×10^{-6} M/s—suggesting that the rate is directly proportional to the concentration of HO^-, the nucleophile. Similarly, in going from Trial 2 to Trial 3, the concentration of $C_6H_5CH_2Cl$, the substrate, is doubled, but the concentration of HO^- remains the same, and the rate also doubles. Thus, the rate is directly proportional to the concentration of the substrate as well. These results indicate an empirical rate law of the form rate $= k[C_6H_5CH_2Cl][HO^-]$, which is consistent with an S_N2 reaction.

Rate data for the following substitution reaction are presented in the accompanying table.

Trial Number	[R—Br]	[HO⁻]	Rate (M/s)
1	0.10 M	0.10 M	5.5×10^{-7}
2	0.10 M	0.05 M	5.5×10^{-7}
3	0.20 M	0.10 M	1.0×10^{-6}

Do the data suggest an S_N1 reaction or an S_N2 reaction?

Why do the rates of the S_N2 and S_N1 reactions have different dependences on reactant concentrations? To help answer this question, we introduce one of the foundational principles of chemical kinetics:

For an *elementary step*, $aA + bB \rightarrow$ products, where A and B are reactants and a and b are the coefficients required to balance the chemical equation, the **theoretical rate law** is

$$\text{Rate}_{\text{elementary step}} = k_{\text{elementary step}}[A]^a[B]^b \qquad (8\text{-}9)$$

The quantities [A] and [B] are the concentrations of the reactants, and their exponents are the **orders** of the reaction with respect to the reactants A and B. That is, a is the order of the reaction with respect to A, and b is the order of the reaction with respect to B. Importantly, a and b are the same as the coefficients used to balance the elementary step. This is because for an elementary step to take place, the reactants must collide at effectively the same time, and the frequency with which that happens is proportional to $[A]^a[B]^b$. Note that Equation 8-9 is referred to as a *theoretical* rate law (not an *empirical* rate law) because it is derived from a proposed elementary step, not from an experiment.

For an S_N2 reaction, which is an elementary step itself, there are two reactant species—the nucleophile and the substrate. According to Equation 8-9, therefore, the concentrations of both reactants should appear in the theoretical rate law. Moreover, the exponent for each reactant concentration should be 1, the coefficient required to balance the corresponding species in the reaction. Thus, the theoretical rate law should appear as in Equation 8-10, which matches the rate law obtained experimentally (Equation 8-7).

S_N2 **Because this is an elementary step, both reactants appear in the rate law.**

                               Theoretical rate law from (Eq. 8-9)

$$\text{Nu:}^{\ominus} + \text{R—L} \longrightarrow \text{Nu—R} + \text{:L}^{\ominus} \qquad \text{Rate} = k_{S_N2}[\text{Nu}^{\ominus}][\text{R—L}] \quad (8\text{-}10)$$

For an S_N1 reaction, Equation 8-9 can be applied to each of the two elementary steps, as shown in Equations 8-11a and 8-11b.

Atom loses its octet

S_N1

Slow

Theoretical rate law for each elemtary step (from Eq. 8-9)

$R\text{—}L \xrightarrow{\text{Slow}} R^{\oplus} + :L^{\ominus}$ Rate $= k_{1,S_N1} [R\text{—}L]$ (8-11a)

$Nu:^{\ominus} + R^{\oplus} \xrightarrow{\text{Fast}} Nu\text{—}R$ Rate $= k_{2,S_N1} [Nu^{\ominus}][R^{\oplus}]$ (8-11b)

Atom gains an octet

The first step is **unimolecular** (its *molecularity* is 1) because only one reactant species appears—the substrate, R—L. According to Equation 8-9, the rate for that step should therefore depend on the concentration of R—L only, as indicated by its theoretical rate law in Equation 8-11a. The second step, on the other hand, is bimolecular (its molecularity is 2), because it involves the nucleophile and the newly formed carbocation as reactants. Consequently, the rate for that step depends on the concentrations of both species, as we can see from its theoretical rate law in Equation 8-11b.

Notice that the theoretical rate law for the first step of an S_N1 reaction (Equation 8-11a) is essentially identical to the empirical rate law, shown previously in Equation 8-8. In other words,

> The first step of an S_N1 reaction establishes the rate of the overall reaction, and is thus the **rate-determining step**.

In fact, the overall reaction is called a *unimolecular* nucleophilic substitution reaction because the rate-determining step is unimolecular.

The first step of the S_N1 mechanism is rate determining because it is intrinsically much slower than the second step, as indicated in Equation 8-11. In large part, this is because an atom *loses* its octet in the first step, whereas an atom *gains* an octet in the second step. (The reason for these relative rates will become clearer in Section 8.4b, where we discuss reaction energy barriers.)

We can better understand the rate-determining step by making an analogy to the flow of water through a system of pipes. If a narrower pipe is connected to a wider pipe, as shown in Figure 8-3, then it is the width of the narrower pipe that dictates the rate at which water can flow through the entire system. This is true regardless of the width of the wider pipe. Similarly, the slowest step of a mechanism governs the rate of the overall reaction.

Like S_N2 and S_N1 reactions, E2 and E1 reactions differ in their empirical rate laws. An E2 reaction is first order with respect to both the base and the substrate (Equation 8-12). By contrast, an E1 reaction is first order with respect to the substrate only (Equation 8-13).

$$\text{Rate}_{E2} = k_{E2}[\text{Base}][R\text{—}L] \tag{8-12}$$

$$\text{Rate}_{E1} = k_{E1}[R\text{—}L] \tag{8-13}$$

In other words,

> Although the rate of an E2 reaction is directly proportional to the concentration of the base, the rate of an E1 reaction is independent of the concentration of the base.

The rate law for the E2 reaction in Equation 8-12 is consistent with a single-step mechanism involving both the base and the substrate, as indicated in Equation 8-14.

Narrower pipe

Wider pipe

The rate of water flow is determined by the width of the narrower pipe.

FIGURE 8-3 Rate-determining steps The narrower and wider pipes represent slower and faster elementary steps, respectively. Just as the rate that water can flow through the system of pipes is governed by the narrower pipe, the rate of an overall reaction is governed by its slowest step—that is, the rate-determining step.

According to Equation 8-9, the rate should be directly proportional to the concentration of each reactant.

E2   **Because an E2 reaction is an elementary step, both reactants appear in the rate law.**

Theoretical rate law (from Eq. 8-9)

$$\text{Rate} = k_{E2}\,[\text{B}^{\ominus}][\text{R}{-}\text{L}] \qquad (8\text{-}14)$$

The rate law for the E1 reaction in Equation 8-13 is consistent with the two-step mechanism shown in Equation 8-15, where the first step is substantially slower than the second. That is,

The first step of an E1 reaction is the rate-determining step.

E1 mechanism   **Atom loses its octet**

Theoretical rate law for each elementary step (from Eq. 8-9)

$$\text{Rate} = k_{1,E1}\,[\text{R}{-}\text{L}] \qquad (8\text{-}15a)$$

$$\text{Rate} = k_{2,E1}\,[\text{B}^{\ominus}][\text{R}^{\oplus}] \qquad (8\text{-}15b)$$

Atom gains an octet

As with the S_N1 reaction, the E1 reaction is said to be *unimolecular* because its rate-determining step is unimolecular.

SOLVED problem 8.8   Suppose that the first step of an S_N1 reaction were much faster than the second step. Would the resulting theoretical rate law for the overall reaction agree or disagree with the observed dependence of the S_N1 reaction on the concentration of the nucleophile?

Think   What would the rate-determining step be under this assumption? What is the theoretical rate law for that step?

Solve   If the first step of an S_N1 reaction (Equation 8-11a) were much faster than the second step (Equation 8-11b), then the second step would be the rate-determining step for the reaction. The theoretical rate law for the overall reaction would simply be that for the second step—namely,

$$\text{Rate} = k_{2,S_N1}\,[\text{Nu}^{\ominus}][\text{R}^{\oplus}].$$

Thus, the reaction rate would be directly proportional to $[\text{Nu}^{\ominus}]$. This contradicts the observed S_N1 reaction rate, which is independent of $[\text{Nu}^{\ominus}]$.

---

problem 8.9   Suppose that the first step of an E1 reaction were much faster than the second step. Would the resulting theoretical rate law for the overall reaction agree or disagree with the observed dependence of the E1 reaction on the base concentration? Explain.

8.4b Energy Barriers and Rate Constants: Transition State Theory

In Section 8.4a, we learned that the first step of an S_N1 reaction tends to be much slower than the second. Likewise, the first step of an E1 reaction tends to be the slow step.

The reason for these different reaction rates can also be seen in each reaction's free energy diagram, shown previously in Figures 8-1 and 8-2. Both diagrams resemble Figure 8-4, which depicts two separate steps, the first of which leads to the formation of a high-energy intermediate.

Each step must overcome an energy barrier, called the *standard free energy of activation*, $\Delta G^{\circ\ddagger}$ (Section 6.3b). Notice that the energy barrier for the first step, $\Delta G^{\circ\ddagger}(1)$, is much larger than the energy barrier for the second step, $\Delta G^{\circ\ddagger}(2)$. Thus, the first step tends to be much more sluggish than the second.

The quantitative relationship between reaction rates and free energies of activation is given by Equation 8-16, which is a result from **transition state theory**.

$$k_{\text{elementary step}} = CTe^{-(\Delta G^{\circ\ddagger}/RT)} \qquad (8\text{-}16)$$

Here, $k_{\text{elementary step}}$ is the rate constant for an elementary step, C is a constant, T is the absolute temperature (in kelvin), $\Delta G^{\circ\ddagger}$ is the standard Gibbs free energy of activation, and R is the universal gas constant.

Because $\Delta G^{\circ\ddagger}$ appears in the numerator of a negative exponent in Equation 8-16,

> The $k_{\text{elementary step}}$ decreases as $\Delta G^{\circ\ddagger}$ increases, so the corresponding reaction rate decreases.

Temperature (T), on the other hand, appears in the denominator of the negative exponent and as part of the *pre-exponential factor*—the collection of terms that multiply the exponential factor. Consequently,

> The $k_{\text{elementary step}}$ increases as T increases, so the reaction rate increases.

Both of these relationships stem from the fact that, at a particular temperature, only a certain percentage of molecules possess enough energy to surmount an energy barrier (Section 4.3). Figure 8-5 shows how this percentage of molecules changes with the size of the energy barrier for two different temperatures—namely, 0 °C for the blue curve and 100 °C for the red curve.

FIGURE 8-4 Dependence of k on $\Delta G^{\circ\ddagger}$ The free-energy diagram for an S_N1 or E1 reaction shows that the standard free energy of activation for the first step, $\Delta G^{\circ\ddagger}(1)$, is much larger than the free energy of activation for the second step, $\Delta G^{\circ\ddagger}(2)$. Thus, the rate constant for the first step is much smaller than the rate constant for the second step and the rate of the first step is much slower than the rate of the second step.

FIGURE 8-5 Percentage of molecules able to surmount an energy barrier ($\Delta G^{\circ\ddagger}$) as a function of the size of the energy barrier and the temperature The percentage of molecules able to surmount an energy barrier is plotted against the size of the energy barrier for two different temperatures: 0 °C (blue curve) and 100 °C (red curve). The region inside the box is expanded in the inset. Notice that as the size of the energy barrier increases, the percentage of molecules having enough energy to react decreases dramatically. Notice also that for a given energy barrier, increasing the temperature significantly increases the percentage of molecules able to react.

At each temperature, the percentage of molecules able to surmount the energy barrier, and thus form products, decreases as the size of the barrier increases. Thus, the reaction rate decreases as $\Delta G^{\circ\ddagger}$ increases.

At the higher temperature (red curve), moreover, a greater percentage of molecules can surmount the energy barrier. This is because the reactants have, on average, more kinetic energy and will be moving faster as T increases. With higher molecular speeds, each collision will be more likely to have enough energy to surmount the $\Delta G^{\circ\ddagger}$ barrier. Thus, the reaction rate increases.

Using Figure 8-5, estimate the percentage of molecules at 100 °C having enough energy to surmount an energy barrier of **(a)** 15 kJ/mol and **(b)** 25 kJ/mol.

8.8 YOUR TURN

Using Figure 8-5, estimate the percentage of molecules having enough energy to surmount an energy barrier of 20 kJ/mol **(a)** at 0 °C and **(b)** at 100 °C.

8.9 YOUR TURN

8.5 Stereochemistry of Nucleophilic Substitution and Elimination Reactions

From Equations 8-1 and 8-2, it may appear that both the S_N2 and S_N1 mechanisms yield the same products. Likewise, from Equations 8-4 and 8-5, it may appear that E2 and E1 reactions yield the same products. However, depending upon the particular nature of the reaction, there can be subtle but important differences. One way in which the reaction products can differ is in their stereochemical configurations. Thus, the **stereochemistry** of an S_N1 reaction differs from that of an S_N2 reaction, and the stereochemistry of an E1 reaction differs from that of an E2 reaction.

8.5a Stereochemistry of an S_N1 Reaction

Stereochemistry is pertinent to an S_N1 reaction when the atom that is attached to the leaving group is a *stereocenter*, because the breaking and formation of bonds to the stereocenter can impact the stereochemical configuration of that atom.

If an S_N1 reaction is carried out on a stereochemically pure substrate, such as the one shown in Equation 8-17, then a *mixture* of both the R and S enantiomers is produced.

Mixture of R and S configurations

(8-17)

(S)-2-Chloro-2-phenylbutane   **(S)-2-Iodo-2-phenylbutane**   **(R)-2-Iodo-2-phenylbutane**

Why does the reaction produce both configurations of the stereocenter?

To answer this question, let's look at the mechanism, shown in Equation 8-18.

Mechanism for Equation 8-17

$I^{\ominus}$ can attack from in front of or from behind the plane of C.

(8-18a)

Planar C

(8-18b)

In the first step, Cl^- simply departs, leaving behind a planar carbocation. The C atom initially bonded to Cl in the carbocation is no longer a stereocenter (i.e., the stereochemistry of that C atom has been lost). In the second step, in which I^- forms a bond to the carbocation, that same C atom becomes a stereocenter once again.

YOUR TURN 8.10

Label each organic species in Equation 8-18 as either *chiral* or *achiral*.

In Figure 8-6, we focus on the second step of the mechanism. The carbocation intermediate is shown at the left in Figures 8-6a and 8-6b, and the plane that is indicated is the one defined by the bonds to the positively charged C. When I^- attacks, it can do so from either side of that plane. The approach of I^- from one side of the plane leads to the formation of the R enantiomer (Fig. 8-6a), whereas approach from the other side of the plane leads to the formation of the S enantiomer (Fig. 8-6b). This idea can be summarized as follows:

If a tetrahedral stereocenter is generated in a single elementary step from an atom that is not a stereocenter, then both R and S configurations will be produced.

You might expect the enantiomeric products of the reaction in Equation 8-17 to be produced in equal amounts—that is, as a *racemic mixture.* The carbocation shown in Figure 8-6 has a plane of symmetry and is therefore achiral, in which case the attack of the I^- nucleophile would appear to be equally likely from either side of the

(a)

C—I bond formed on the right

(R)-2-Iodo-2-phenylbutane

C—I bond formed on the left

(S)-2-Iodo-2-phenylbutane

(b)

FIGURE 8-6 Generating a mixture of stereoisomers in an S$_N$1 reaction The specific stereoisomer that is formed in the reaction in Equation 8-17 is determined by the approach of the I$^-$ nucleophile. (a) I$^-$ approaches from the right, generating the new C—I bond shown, thus producing the *R* configuration. (b) I$^-$ approaches from the left, generating the new C—I bond shown, thus producing the *S* enantiomer.

carbocation's plane. Indeed, if the nucleophile were to attack the free carbocation, this would be the case, according to the following general rules:

- If a new tetrahedral stereocenter is produced in a step in which the reactants and the environment are achiral, then the *R* and S configurations of the stereocenter are produced in *equal* amounts.
- Otherwise, the *R* and S configurations are produced in *unequal* amounts.

However, the nucleophile does not attack the free carbocation in S$_N$1 reactions like this one. After the bond to the leaving group is broken in the first step, the leaving group remains associated with the side of the carbon from which it was initially attached, forming what is called an **ion pair**. The ion pair produced in the reaction in Equation 8-17 retains some of the chiral character from the original substrate, so effectively there is no plane of symmetry when the nucleophile attacks. This results in an *unequal* likelihood of nucleophilic attack from the two sides of the C atom, producing a mixture that is close to, but not completely racemic.

The S$_N$1 reaction in Equation 8-19 is more clear cut with regard to stereochemistry.

I$^-$ can attack from in front of or from behind the plane of C.

Planar C

Chiral carbocation

Unequal mixture of diastereomers

(2S,3S)-3-Iodo-3-methyl-3-phenylpropan-2-ol

(2S,3R)-3-Iodo-3-methyl-3-phenylpropan-2-ol

(8-19a)

(8-19b)

8.5 Stereochemistry of Nucleophilic Substitution and Elimination Reactions / **435**

FIGURE 8-7 Formation of an unequal mixture of diastereomers in an S_N1 reaction The OH group is present only on one side of the plane, so whether it interacts with an approaching I^- depends on the direction from which the ion arrives. (a) If the I^- nucleophile comes from the right, it encounters the OH group. (b) If the I^- nucleophile comes from the left—the side opposite the OH group—it avoids the OH group. The two possible resulting reactions thus occur with different probabilities, producing an unequal mixture of diastereomers.

Here, the overall reactant has two stereocenters: the C atom bonded to the Cl leaving group and the C atom bonded to OH. Notice that the stereochemical configuration of the C bonded to OH remains unchanged because the reaction takes place only at the C atom bonded to Cl. Just as in Equation 8-18, though, the C atom bonded to Cl becomes planar when Cl^- leaves, so that C atom loses its stereochemistry. When I^- forms a bond in the second step, the C atom once again becomes a stereocenter, but, because I^- can approach from either side of the C atom's plane, both the R and S configurations can be produced. As indicated in Equation 8-19, the result is a mixture of two *diastereomers*.

Unlike the previous situation, the carbocation intermediate shown in Equation 8-19 has a single tetrahedral stereocenter, so it is unambiguously chiral and has no plane of symmetry. Without a plane of symmetry in the carbocation, the nucleophile will be influenced differently depending upon which side of the carbocation it approaches. For the conformation shown in Figure 8-7, the nucleophile would encounter the OH group on the left side but not the right, so the likelihood of each approach would be unequal and the diastereomeric products would be produced in *unequal* amounts.

You may be tempted to try to predict which configuration is produced in greater abundance. We caution against it, however, because the factors that favor the production of one configuration over another can be subtle, involving the interplay between steric effects and electronic effects (such as hydrogen bonding, inductive effects, and polarizability). As a result, we limit such discussions to reactions in which these factors are clear cut.

SOLVED problem 8.10 Draw the complete, detailed mechanism for the reaction at the right, assuming that it takes place via the S_N1 mechanism. Will the reaction produce a single stereoisomer, or will it produce a mixture of stereoisomers? If it produces a mixture, then will the isomers be produced in equal amounts?

Think What can act as the leaving group? What can act as the nucleophile? What are the steps that compose an S_N1 mechanism? Does the reaction take place at a stereocenter?

Solve The Br substituent can act as a leaving group, coming off as Br⁻, which is a relatively stable species. The HS⁻ ion can act as a nucleophile. The S_N1 mechanism takes place in two steps. First, the leaving group leaves via heterolysis, yielding a planar carbocation, then HS⁻ attacks C⁺ via a coordination step, yielding the overall products.

Only this stereoisomer is formed.

Notice in the alkyl halide reactant that there is a single stereocenter, but no bonds to it are affected throughout the course of the reaction. Therefore, its stereochemical configuration in the product is the same as in the reactant. Notice also in the second step of the reaction that the planar C⁺ becomes tetrahedral, but it does *not* become a stereocenter—it is bonded to two CH_3 groups. Therefore, stereochemistry is not a concern at that C, and the only stereoisomer formed is the one shown.

---

problem **8.11** Draw the complete mechanism and products for an S_N1 reaction between the reactants at the right. Will the reaction produce a single stereoisomer, or will it produce a mixture of stereoisomers? If it produces a mixture, will those isomers be produced in equal amounts?

8.5b Stereospecificity of an S_N2 Reaction

Previously in Section 8.5 we said that the S_N1 and S_N2 reactions have different stereochemistry. This is evident in Equation 8-20, which shows the outcome of an S_N2 reaction between I⁻ and (*S*)-1-chloro-1-phenylethane.

Only this stereoisomer is formed.

(S)-1-Chloro-1-phenylethane   **(R)-1-Iodo-1-phenylethane**

Whereas the S_N1 mechanism involving similar reactants (Equation 8-17) produces a mixture of enantiomers, the S_N2 mechanism yields only one of those enantiomers. In general:

> A stereochemically pure substrate that undergoes an S_N2 reaction yields only one stereoisomer, which depends on the specific configuration of the substrate, so S_N2 reactions are **stereospecific**.

More specifically, as shown in Equation 8-20, the C atom's bond to the nucleophile in the products is on the side of the substrate *opposite* the initial bond to the leaving group in the reactants.

The S_N2 reaction takes place in a single step, so this stereospecificity suggests that the nucleophile attacks the substrate exclusively from the side opposite the leaving group (Equation 8-21a). In other words,

> S_N2 reactions require **backside attack** of the substrate by the nucleophile.

Backside attack

The three R groups must flip over to the other side.

This stereoisomer is formed exclusively.

$$\text{:}\overset{\ominus}{\underset{..}{\text{I}}}\text{:} \quad + \quad \left[\overset{\delta^-}{\underset{H_3C}{\text{:}\underset{..}{\text{I}}}} - - - \overset{H}{\underset{C_6H_5}{\text{C}}} - - - \overset{\delta^-}{\underset{..}{\text{Cl}}}\text{:} \right]^{\ddagger} \longrightarrow \text{:}\underset{..}{\text{I}}\text{—}\overset{H}{\underset{CH_3}{\text{C}}}\text{\\C_6H_5} \quad + \quad \text{:}\overset{\ominus}{\underset{..}{\text{Cl}}}\text{:} \qquad \text{(8-21a)}$$

The stereoisomer that is *not* produced would be generated by attack of the nucleophile on the *same* side as the leaving group (Equation 8-21b), called **frontside attack**.

Frontside attack

This stereoisomer is *not* formed.

$$\text{:}\overset{\ominus}{\underset{..}{\text{I}}}\text{:} \quad \underset{C_6H_5}{\overset{H}{H_3C\text{''''}}}\text{C}\text{—}\underset{..}{\text{Cl}}\text{:} \quad \times \quad \left[\overset{H}{\underset{C_6H_5}{H_3C\text{''''}}}\text{C} \overset{\delta^-}{\underset{\underset{\delta^-}{\text{Cl}}}{\text{I}}} \right]^{\ddagger} \longrightarrow \overset{H}{\underset{C_6H_5}{H_3C\text{''''}}}\text{C}\text{—}\underset{..}{\text{I}}\text{:} \quad + \quad \text{:}\overset{\ominus}{\underset{..}{\text{Cl}}}\text{:} \qquad \text{(8-21b)}$$

As shown in Equation 8-21a, the backside attack requires the remaining three groups of the substrate to "flip over" to the other side. This is known as a **Walden inversion**, and is analogous to an umbrella inverting in a windstorm. In a frontside attack, the three groups would remain on the same side in the products.

One factor that contributes to the stereospecificity of an S_N2 reaction is **steric hindrance**. The leaving group is often large (otherwise it could not accommodate the negative charge with which it leaves), so it would crash into any nucleophile approaching in a frontside attack.

Frontside attack is disfavored, too, due to *charge repulsion* (Fig. 8-8). The atom of the leaving group bonded to the substrate is typically highly electronegative, and therefore usually bears a significant partial negative charge. The nucleophile, which itself bears either a partial or a full negative charge, is thus repelled from that side of the substrate.

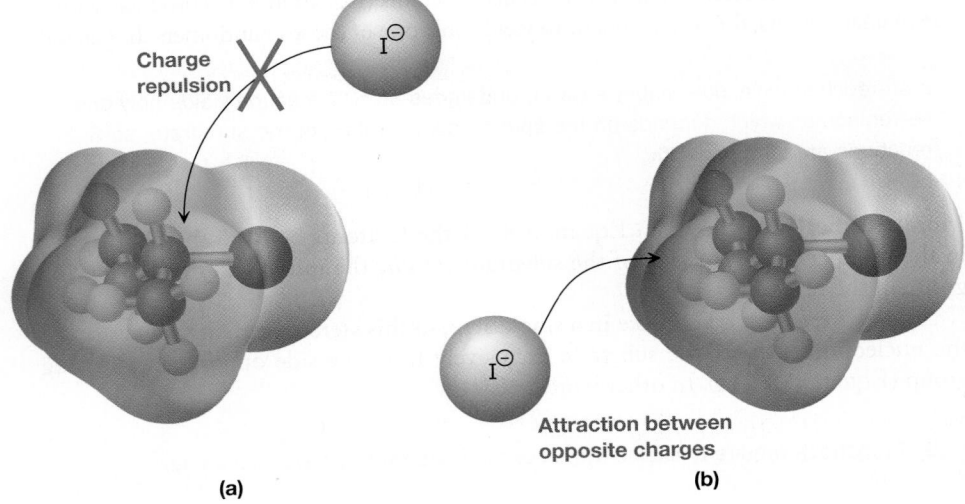

FIGURE 8-8 Stereospecificity of an S_N2 reaction (a) *Frontside attack* of a nucleophile on a substrate is disfavored due to charge repulsion between the incoming nucleophile and the leaving group. (b) *Backside attack* of a nucleophile on a substrate is favored because the negatively charged nucleophile is attracted to the partially positively charged carbon that has the leaving group.

Charge repulsion

Attraction between opposite charges

(a)

(b)

The stereospecificity of an S_N2 reaction can be understood instead from a molecular orbital point of view. For such a discussion, see Section IC1.3 in the interchapter just after Chapter 7.

SOLVED problem 8.12 Draw the complete, detailed mechanism for the following reaction, assuming that it takes place via an S_N2 mechanism. Pay attention to stereochemistry.

Think What can act as the leaving group? What can act as the nucleophile? How many steps make up an S_N2 mechanism? How does the nucleophile approach the substrate during attack?

Solve The leaving group is Br^- and the nucleophile is HS^-. In an S_N2 reaction, the HS^- nucleophile should attack from the side opposite the C—Br bond—in this case, from behind the plane of the page. Thus, the new C—S bond remains behind the plane of the page.

The nucleophile attacks from behind the plane of the page.

The new C—S bond remains behind the plane of the page.

---

problem 8.13 Draw the complete mechanism and the products for each of the following S_N2 reactions, paying close attention to stereochemistry.

(a)

(b)

(c)

8.5c Stereochemistry of an E1 Reaction

When an E1 reaction produces a new double bond, stereochemistry is an issue if both E and Z configurations about the double bond exist. An example is shown in Equation 8-22.

E1

(1S,2R)-1-Bromo-1,2-diphenylpropane → CH₃CH₂OH → **(E)-1,2-Diphenylpropene** (Major product) + **(Z)-1,2-Diphenylpropene** (Minor product) + HBr   (8-22)

As indicated, both of these diastereomers are produced in the reaction. This is a general result for all E1 reactions:

> An E1 reaction produces a mixture of the *E* and *Z* configurations about a double bond formed in the products.

To understand why both stereoisomers are produced, you must understand the mechanism of the E1 reaction, shown in Equation 8-23.

Rotation about C—C single bond

Elimination of H⁺   (8-23a)

Loss of Br⁻

Conformational isomers are in equilibrium.

A mixture of *E* and *Z* isomers is produced.

Rotation about C—C single bond

Elimination of H⁺   (8-23b)

• Steric strain
• Minor product

In the carbocation intermediate that is formed, a single bond connects the C atoms that were initially bonded to the leaving group and the H$^+$ undergoing elimination. Thus, an equilibrium is established among the various conformational isomers about that single bond. Depending upon the specific conformation of the carbocation when the second step of the mechanism occurs, the product can be either the E isomer (Equation 8-23a) or the Z isomer (Equation 8-23b).

SOLVED problem 8.14 Draw the complete, detailed mechanism for the reaction at the right, assuming it proceeds via an E1 mechanism. Pay attention to stereochemistry.

Think What can act as the leaving group? What can act as the base? What proton can be removed in the second step of an E1 mechanism? Do E and Z configurations exist for the product? If so, can both be formed?

Solve As shown below, Br$^-$ can act as the leaving group in the first step, thus generating a carbocation intermediate. In the second step, HCO$_3^-$ can act as a base to remove a proton from the adjacent carbon and produce the alkene product. Both E and Z configurations exist about that double bond, and because the bond indicated in the carbocation can undergo rotation, both the E and Z configurations are formed.

Mixture of E and Z configurations

Rotation takes place about this C—C bond.

---

problem 8.15 Draw the complete, detailed mechanism for each of the following reactions, assuming they proceed via E1 mechanisms. Pay attention to stereochemistry.

(a)   (b)

Notice that the E isomer in Equation 8-22 is favored over the Z isomer. This is because bulky phenyl rings cause steric strain. Thus, the conformational isomer from which the E isomer is produced (Equation 8-23a) will be in greater abundance than the one from which the Z isomer is produced (Equation 8-23b). This idea can be generalized for other E1 reactions as well.

> If an E1 reaction produces both E and Z isomers, the isomer with less steric strain will generally be favored.

---

problem 8.16 For each reaction in Problem 8.15 that produces a mixture of diastereomers, predict which diastereomer will be produced in greater abundance.

8.5d Stereospecificity of an E2 Reaction

Unlike the E1 reaction in Equation 8-22, which yields a mixture of both the *E* and *Z* alkene products, an E2 reaction between the same reactants yields only the diastereomer shown in Equation 8-24, making the E2 reaction *stereospecific*.

Only this diastereomer is formed.

E2

(1S,2R)-1-Bromo-1,2-diphenylpropane   (E)-1,2-Diphenylpropene

$$+ \quad {}^{\ominus}\ddot{\text{O}}\text{H} \longrightarrow \quad + \quad \text{H—OH} \quad + \quad \text{Br}^{\ominus} \qquad (8\text{-}24)$$

The stereospecificity of the E2 reaction can be explained as follows:

> E2 reactions are favored by the substrate conformation in which the leaving group and the hydrogen atom that are eliminated are anti to each other.

Because a single bond joins the C atoms that are bonded to the H and the leaving group, the stable conformations are those in which the H atom and the leaving group are either gauche to each other or anti to each other (see Section 4.9):

Rotate molecule 90°   Rotation about C—C bond   Rotation about C—C bond

Anti   Gauche   Gauche

E2 elimination is favored by the anti conformation.

The different conformations interconvert via rotation about that C—C bond.

In the conformation in which the H and the leaving group are anti to each other, the H, the leaving group, and the C atoms to which they are attached all reside in a single plane (Fig. 8-9). This conformation is therefore sometimes referred to as **anticoplanar** or **antiperiplanar**.

With the substrate in the anticoplanar conformation, the stereoisomer that is formed is determined by the location of the other substituents on the carbon atoms of the substrate. Notice in Figure 8-9 that the CH_3 and C_6H_5 substituents are on the same side of the plane in the substrate, so they are on the same side of the double bond in the product, too. Similarly, the H and C_6H_5 substituents are on the same side of the double bond in the product because they began on the same side of the plane in the substrate.

---

problem **8.17** Draw the substrate that would undergo an E2 reaction to yield the diastereomer of the product alkene in Equation 8-24. Be sure to include an accurate dash–wedge structure.

FIGURE 8-9 Stereospecificity in E2 reactions An E2 reaction is favored when the substrate is in the *anticoplanar* conformation (left), in which the H atom and the leaving group (in this case, Br) on adjacent C atoms are anti to each other and in the same plane. Because the CH_3 and the C_6H_5 groups are on the same side of the plane in the substrate, they end up on the same side of the double bond in the product. Similarly, the H and C_6H_5 groups are on the same side of the plane in the substrate and are on the same side of the double bond in the product.

If there are two H atoms that can be eliminated from the same C atom, then a *mixture* of diastereomers can form in an E2 reaction (see Equation 8-25).

Each of the products in Equation 8-25 is the result of elimination from the substrate in an anticoplanar conformation. Figure 8-10 shows that there are two different anticoplanar conformations, which differ by rotation about the C—C single bond. One of them leads to the *E* configuration about the C=C double bond (Fig. 8-10a) and the other leads to the *Z* configuration (Fig. 8-10b).

SOLVED problem 8.18   Draw the products of E2 elimination involving (a) the D atom and (b) the H atom indicated. Pay attention to stereochemistry, and note that D is an isotope of H, so the two atoms exhibit nearly identical chemical behavior.

Think What conformations of the substrate facilitate an E2 reaction? In those conformations, what dictates which atoms/groups are on the same side of the double bond in the products?

Solve An E2 reaction is facilitated by an anticoplanar conformation involving H (or D), Br, and the two C atoms to which they are bonded. There are two such conformations, as shown on the left in each of the following reactions: one involves D and the other involves H.

(a)

(b)

In the first reaction, D is eliminated along with Br. Because H and a CH$_3$ group appear on each side of the D—C—C—Br plane in the substrate, H and a CH$_3$ group appear on each side of the double bond in the products. In the second reaction, H is eliminated along with Br. Because two CH$_3$ groups appear on one side of the H—C—C—Br plane in the substrate, two CH$_3$ groups appear on the same side of the double bond in the products. Similarly, because H and D appear on the same side of that plane in the substrate, they also appear on the same side of the double bond in the product.

(a)

Rotate about
C—C bond

(b)

FIGURE 8-10 Formation of diastereomers in an E2 reaction
The substrate in Equation 8-25 (left) has two different conformations in which an H and the Br are anti. The E diastereomer is formed from the conformation shown in (a), whereas the Z diastereomer is formed from the conformation shown in (b).

Phosphorylation: An Enzyme's On/Off Switch

Phosphorylation is a process that regulates the function of certain enzymes, such as glycogen phosphorylase, which catalyzes the breaking down of glycogen. The process is essentially a nucleophilic substitution reaction, and is facilitated by another enzyme called a *kinase*, whereby adenosine triphosphate (ATP, abbreviated $^{2-}O_3P{-}ADP$), acting as the substrate, undergoes nucleophilic attack to produce the phosphorylated product and adenosine diphosphate (ADP).

In glycogen phosphorylase, in particular, a serine amino acid in the enzyme's active site is subject to phosphorylation, owing to the nucleophilic character of the OH group in serine's side group. The introduction of the negatively charged phosphate group onto that residue significantly changes the interactions with other amino acid residues, causing a change in the conformation of the enzyme. Such a change in conformation, which can be seen above (particularly the region highlighted in red), causes the enzyme's activity to increase by roughly 25%. For this and other enzymes, therefore, phosphorylation can be thought of as a convenient "on/off" switch.

Unphosphorylated

Phosphorylated

---

problem **8.19** Draw the products of E2 elimination involving **(a)** the H atom and **(b)** the D atom indicated. Pay attention to stereochemistry.

---

problem **8.20** Which of these compounds would you expect to undergo E2 elimination more readily? Why? (*Hint:* Can the H and leaving group attain an anticoplanar conformation in each structure?)

Looking back at Equation 8-25, notice that the *E* isomer is favored over the *Z* isomer. Once again, this is primarily due to steric strain. In the anticoplanar conformation from which the *Z* isomer is produced (Fig. 8-10b), the bulky CH_3 groups are gauche to each other, whereas in the anticoplanar conformation from which the *E* isomer is produced (Fig. 8-10a), those CH_3 groups are anti to each other. Thus, the conformation that produced the *E* conformation is more abundant.

> In general, if an E2 reaction produces a mixture of diastereomers, the diastereomer that is favored will be the one that is produced from the anticoplanar conformation that is more stable.

Negative charge

Partial positive charge

HO⊖

(a)

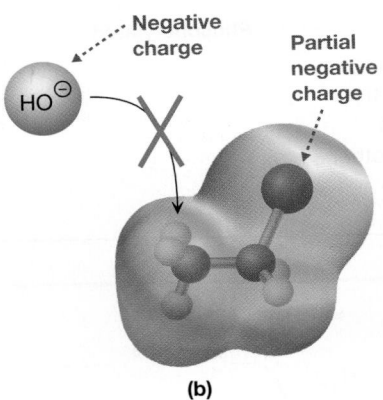

Repulsion between like charges

Negative charge

Partial negative charge

HO⊖

(b)

FIGURE 8-11 Stereospecificity in E2 reactions E2 reactions with the substrate in different conformations. (a) The H and leaving group are anti to each other. In this conformation, the incoming base is attracted by the positively charged end of the substrate. (b) The H and leaving group are gauche to each other. In this conformation, the base is repelled by the partial negative charge on the leaving group.

Why is it that the E2 reaction is favored by the anticoplanar orientation of the substrate? One important factor has to do with electrostatic attraction and repulsion. As in the S_N2 reaction, the bond between the leaving group and the C atom to which it is attached is polar—the atom bonded to C bears a partial negative charge, whereas the C atom bears a partial positive charge. Therefore, if the strong base (which is negatively charged) attacks the H atom anti to the leaving group, then it would be attracted to the partial positive charge on C (Fig. 8-11a). On the other hand, if the base attacks the H atom that is gauche to the leaving group, then the incoming base would be repelled by the partial negative charge on the leaving group (Fig. 8-11b).

Just as with S_N2 reactions, the stereospecificity of an E2 reaction can be understood instead from a molecular orbital point of view. For this discussion, see Section IC1.5 in the interchapter just after Chapter 7.

8.6 Proton Transfers and Carbocation Rearrangements as Part of Multistep Mechanisms: The Reasonableness of a Mechanism

So far, we have considered just the simplest of nucleophilic substitution and elimination reactions. Each of the S_N2 and E2 mechanisms that have been presented consists of precisely one step, and each of the S_N1 and E1 reactions that have been presented consists of precisely two steps. However, you will encounter many instances where other elementary steps are incorporated into these rudimentary mechanisms, making the mechanisms slightly longer and more complex. The most common of these steps are *proton transfer reactions* and *carbocation rearrangements*.

Because of the greater complexity resulting from including these additional steps, it is important for you to gain a sense of how they can be incorporated into a mechanism in a *reasonable* way. To help you acquire this "chemical intuition," we present four general rules. The first three pertain to proton transfer reactions and are discussed in Sections 8.6a–8.6c. The fourth rule, discussed in Section 8.6d, pertains to carbocation rearrangements.

Although these rules are introduced in the context of S_N2, S_N1, E2, and E1 reactions, they apply generally to all reaction mechanisms. Therefore, you should try to apply the lessons you learn here each time you encounter a new reaction mechanism.

8.6a Acidic and Basic Conditions: Proton Transfer Reactions Are Fast

The first general rule addresses the conditions under which a reaction takes place:

Proton transfer reactions should be incorporated in such a way as to avoid the appearance of species that are incompatible with the conditions of the solution. Specifically,

- Strong acids are compatible with acidic and neutral conditions, but are incompatible with basic conditions.
- Strong bases are compatible with basic conditions and neutral conditions, but are incompatible with acidic conditions.

These ideas stem from the fact that under basic conditions, the equilibrium concentration of strongly acidic species is extremely small, and under acidic conditions, the equilibrium concentration of strongly basic species is extremely small.

Proton transfer reactions are fast, so they can be incorporated before or after another elementary step as necessary. Recall, for example, the acid–base titrations you have carried out in lab. Changes in the pH of a solution occur almost instantaneously upon the addition of acid or base. On the other hand, many organic reactions require hours to reach completion, even at an elevated temperature.

To apply this rule correctly, we must be able to recognize strong acids and strong bases. As we learned in Chapter 6, many strongly basic species bear a negative charge, including HO^-, H_2N^-, and H_3C^-. However, some anions are reasonable to include in such mechanisms because they are weak bases and thus can be found in high concentration in acidic media. These include the halide anions Cl^-, Br^-, and I^-, as well as anions whose charge is heavily stabilized by resonance, such as HSO_4^-.

Many strong acids, on the other hand, are positively charged, including H_3O^+ and $CH_3OH_2^+$. However, some positively charged species are reasonable to include under basic conditions because they are weak acids. Examples include H_4N^+ and protonated amines, $R—NH_3^+$.

Consider the substitution reaction in Equation 8-26, in which phenylmethanol is converted to methoxyphenylmethane under *basic* conditions (indicated by the presence of KOH).

Phenylmethanol　　　　　　　　　　　**Methoxyphenylmethane**
75%

Basic conditions

(8-26)

In the proposed mechanism in Equation 8-27, the alcohol acts as a nucleophile in an S_N2 reaction, followed by a proton transfer step.

Proposed mechanism for Equation 8-26

Unreasonable

A strong acid is incompatible with basic conditions.

1. S_N2　　　2. Proton transfer

(8-27)

This mechanism, however, is unreasonable because the product of the first step has a highly acidic proton, which is incompatible with the basic conditions of the reaction.

A more reasonable mechanism is shown in Equation 8-28, in which HO^- first deprotonates phenylmethanol to produce an alkoxide anion, RO^-. The alkoxide anion then acts as the nucleophile to displace I^-. In contrast to Equation 8-27, no strongly acidic species appear in this mechanism.

Proposed mechanism for Equation 8-26

Reasonable　　　　　　　　　　No strong acids appear.

1. Proton transfer　　　2. S_N2

(8-28)

Consider the following reaction, which takes place under *basic* conditions.

In the following mechanism, label the incompatible species that makes the mechanism unreasonable.

1. S$_N$2

2. Proton transfer

---

problem **8.21** Propose a reasonable mechanism for the reaction in Your Turn 8.11.

The substitution reaction in Equation 8-29 takes place under *acidic* conditions.

2-Methylpropan-2-ol

HBr

H$_2$SO$_4$, 20 °C
30 min.

2-Methyl-2-bromopropane
85%

+ H$_2$O (8-29)

Acidic conditions

The S$_N$1 mechanism shown in Equation 8-30 accounts for the formation of the product, but it is unreasonable. Notice that the product of the first step is HO$^-$, which is a strong base and is thus incompatible with the acidic conditions of the reaction. (In Chapter 9, we will also learn that this mechanism is unreasonable because strong bases are poor *leaving groups*.)

Proposed mechanism for Equation 8-29

Unreasonable

A strong base is incompatible with the acidic conditions.

1. Heterolysis

2. Proton transfer (8-30)

3. Coordination

A more reasonable mechanism is shown in Equation 8-31, in which the OH group is protonated prior to the S_N1 mechanism. Thus, the leaving group departs as H_2O instead of HO^-, and no strongly basic species appear at any stage of the mechanism.

Proposed mechanism for Equation 8-29

Reasonable

(8-31)

1. Proton transfer

2. Heterolysis

3. Coordination

8.12 YOUR TURN

The reaction at the right takes place under *acidic* conditions.

In the following proposed mechanism, label the incompatible species that makes the mechanism unreasonable.

1. Heterolysis

2. Proton transfer

3. Coordination

4. Proton transfer

---

problem **8.22** Propose a reasonable mechanism for the reaction in Your Turn 8.12.

SOLVED problem **8.23** Draw a reasonable mechanism for the elimination reaction at the right, assuming it takes place via an E1 mechanism.

Think What are the first and second steps of a typical two-step E1 mechanism? What incompatible species appear in such a mechanism for this reaction? How can you incorporate a proton transfer reaction in a reasonable way to avoid the formation of such a species?

Solve The normal two-step E1 mechanism includes a heterolysis step, followed by elimination of H⁺ to form the double bond.

A strong base is incompatible
with acidic conditions.

This requires the formation of CH_3O^-, however, which is a strong base and is thus incompatible with the acidic conditions given. To avoid the formation of this species, the strong acid can protonate oxygen prior to the heterolysis step (remember, proton transfer reactions are *fast*). Therefore, in the step where the C—O bond is broken, a weakly basic CH_3OH molecule is formed instead.

1. Proton
transfer

2. Heterolysis

3. Elimination
of H⁺

problem **8.24** Draw a reasonable mechanism for the reaction in Solved Problem 8.23, assuming that it proceeds via an E2 mechanism.

8.6b Intramolecular versus Solvent-Mediated Proton Transfer Reactions

Some mechanisms must account for the removal of a proton at one site within a molecular species and the addition of a proton at another site within the *same* species. Consider, for example, the reaction shown in Equation 8-32, in which ammonia attacks oxirane to produce 2-aminoethanol. (We discuss this type of ring-opening reaction in greater detail in Chapter 10.)

Oxirane
(Ethylene oxide)

2-Aminoethanol
70%

(8-32)

A reasonable first step is shown in Equation 8-33, in which an S_N2 reaction opens the ring to produce a species with a positively charged ammonium ion ($R—NH_3^+$) and a negatively charged alkoxide anion (RO^-). From there, the N atom must be deprotonated and the negatively charged O atom must be protonated to arrive at the final, uncharged product. But how does this happen?

A proton must
be added.

A proton must
be removed.

S_N2

(8-33)

We could imagine that the proton is transferred directly from the N to the O, through an **intramolecular proton transfer** reaction. The curved arrow notation for that process is shown in Equation 8-34.

Proposed mechanism for Equation 8-33

Unreasonable

An *intramolecular* proton transfer is generally not reasonable.

(8-34)

In general, however,

> Intramolecular proton transfer reactions are unreasonable.

This is because there are typically several solvent molecules that reside between the acidic and basic sites at any given time, making it difficult for the *direct* transfer of the proton from one site to the other.

Instead, if these solvent molecules are sufficiently acidic or basic, they invariably participate in the transfer of a proton from one site to another in a particular species, via a **solvent-mediated proton transfer**. In Equation 8-32, notice that the solvent, water, is both weakly acidic and weakly basic. Therefore, the solvent-mediated proton transfer in Equation 8-35 is reasonable.

Proposed mechanism for Equation 8-33

Reasonable

Water assists the transfer of a proton from N to O via a *solvent-mediated proton transfer*.

Proton transfer

Proton transfer

(8-35)

8.13 YOUR TURN

Consider the following reaction, in which methanethiol attacks β-propiolactone to open the ring.

β-Propiolactone

3-Methylthiopropanoic acid

The following mechanism for this reaction is unreasonable. Label the step that is unreasonable and explain why.

Using Proton Transfer Reactions to Discover New Drugs

Proton transfer reactions are among the simplest of reactions, but they can be a powerful tool in our never-ending quest to discover new drugs. The key, as we learned here in Chapter 8, is that proton transfer reactions tend to be quite fast, and there are several mildly acidic protons throughout the structure of a protein, both in the amide groups that make up the protein's back-bone and in the side groups of certain amino acids. If a protein is dissolved in deuterated water (D_2O), these protons can exchange with the D atoms of the solvent via simple proton transfer reactions. The rate of this H/D exchange can be monitored with mass spectrometry (see Chapter 16), because the atomic mass of D is greater than that of H.

How can this help us discover new drugs? The answer lies in the fact that drugs are typically designed to bind to target proteins that are in their *folded* state, as shown below. A potentially viable drug, therefore, will help keep the protein folded, preventing D_2O from exchanging with protons on the interior of the protein. Overall, then, the rate of H/D exchange will be slowed.

A drug bound to a protein stabilizes the protein in its folded state.

This technique is especially attractive because it requires only picomole amounts of protein, can be carried out even in the presence of impurities, and can be automated. As many as 10,000 potential drugs can be tested in a single day!

---

problem **8.25** Draw a reasonable mechanism for the reaction provided in Your Turn 8.13.

8.6c Molecularity

A third general rule addresses the number of reactant species in a particular elementary step. Equation 8-36, for example, accomplishes the same proton transfers we saw previously in Equation 8-35 but is unreasonable.

Proposed mechanism for Equation 8-33

Unreasonable

A *termolecular step* is, in general, highly unlikely.

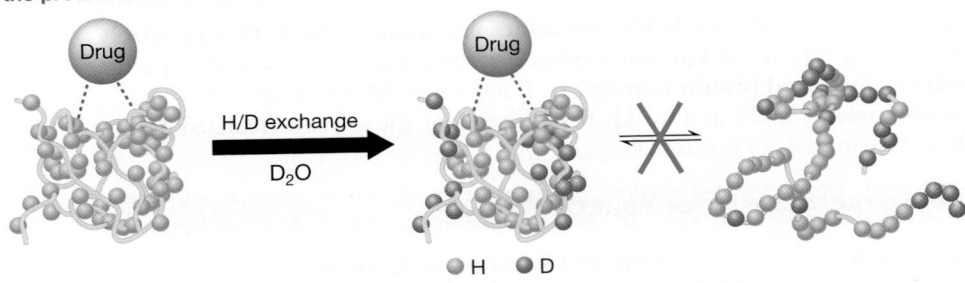

(8-36)

Neither of the rules we have encountered so far is violated in Equation 8-36—namely, there are no incompatible species together and the proton transfers are solvent-mediated, not intramolecular. Why, then, is Equation 8-36 unreasonable?

Equation 8-36 is unreasonable because it is a **termolecular** elementary step—that is, it involves three reactant species simultaneously.

> In general, termolecular steps (and steps of higher molecularity) are unreasonable.

By definition, an elementary step takes place in a *single event*—that is, the breaking and/or forming of bonds occur simultaneously. Thus, a termolecular step would require the collision of all three reactant species at precisely the same moment, the probability of which is vanishingly small. By contrast, bimolecular and unimolecular steps are reasonable because a bimolecular step requires the collision of just two species, and a unimolecular step is a spontaneous transformation that does not require a collision with another reactant species at all.

SOLVED problem 8.26 Consider this overall reaction.

Why is the following mechanism unreasonable? Suggest an alternate mechanism that would be more reasonable.

Think Are all of the species compatible with the acidic conditions under which the reaction takes place? Does an intramolecular proton transfer appear in the mechanism? Does a termolecular step appear?

Solve The reaction takes place under acidic conditions. The species that appear are either strongly acidic or neutral, so they are compatible with the reaction conditions. No intramolecular proton transfer appears, but the step shown is termolecular, because it involves the alcohol, H_2O, and H_3PO_4. To avoid this, the single step can be split into two separate steps.

---

problem 8.27 Suggest why the following S_N2 step is unreasonable and provide an alternate mechanism that is reasonable.

8.6d Carbocation Rearrangements

Carbocations, because of their net positive charge and lack of an octet, are inherently quite reactive (Chapter 6). We saw in Section 8.1 that they can achieve greater stability by undergoing a coordination reaction, thus forming S_N1 products. As we saw in Section 8.2, they can also increase their stability by eliminating H^+, thus forming E1 products. Furthermore, as we learned in Section 7.7, they can undergo **carbocation rearrangements**, in which both the reactant and the product carbocations are isomers of each other.

S_N1 reactions provide clear evidence of these carbocation rearrangements. Consider the S_N1 reaction in Equation 8-37 between 2-iodo-3-methylbutane and water. It *appears* that substitution occurs at a C atom without the leaving group!

2-Iodo-3-methylbutane   **2-Methylbutan-2-ol**

The mechanism in Equation 8-38 shows that a carbocation rearrangement takes place. In Step 1, the leaving group leaves as in the usual S_N1 reaction, yielding I^- and a secondary carbocation intermediate. In Step 2, a 1,2-hydride shift converts the secondary carbocation into the tertiary carbocation. In Step 3, the normal second step of an S_N1 reaction—a coordination step—takes place between H_2O and the carbocation. A final proton transfer yields the uncharged alcohol product.

Mechanism for Equation 8-37

Why should this carbocation rearrangement take place? Recall from Section 7.8 that there are two major factors that can contribute to the driving force for an elementary step: charge stability and bond energies. In this case, bond energies will not contribute significantly because one σ bond is broken and a second σ bond is formed. Charge stability, however, significantly favors the tertiary carbocation over the secondary one, because, as we learned in Section 6.6e, the additional alkyl group stabilizes the positive charge.

Carbocation rearrangements that result in significantly greater stability tend to be quite rapid. George Olah and Joachim Lukas experimentally measured the rate constant for these steps to be on the order of $10^4 \, s^{-1}$ or faster, which, under normal conditions, makes them faster than most other elementary steps we encounter. Therefore,

> If an energetically favorable 1,2-hydride shift or 1,2-methyl shift competes with another possible elementary step, the carbocation rearrangement will usually win.

In the mechanism in Equation 8-38, for example, such a carbocation rearrangement beats out the coordination step that would otherwise take place. (Recall from Section 8.6a that proton transfer reactions are fast, too, but we seldom have to consider competition between a simple proton transfer and a carbocation rearrangement.)

Because carbocation rearrangements tend to be so fast, it is important that you pay attention whenever you see a carbocation produced in a mechanism. In such cases, you should consider every possible 1,2-hydride shift and 1,2-methyl shift. Consider, for example, an S_N1 reaction involving the substrate and nucleophile in Equation 8-39.

S_N1

(8-39)

Once the carbocation is produced, there are two possible carbocation rearrangements to consider. Equation 8-40 shows a 1,2-hydride shift, and Equation 8-41 shows a 1,2-methyl shift.

This rearrangement doesn't occur because the carbocation does not gain stability.

Tertiary carbocation   Primary carbocation

(8-40)

The carbocation gains significant stability from resonance delocalization of the charge.

Charge is localized.   1,2-Methyl shift   Charge is resonance delocalized.   Coordination   Proton transfer

(8-41)

As indicated, the 1,2-hydride shift does not take place because a more stable tertiary carbocation is converted to a less stable primary one. On the other hand, the 1,2-methyl shift is expected to take place rapidly because it produces a carbocation that has gained significant stability via resonance delocalization of the charge. From there, a coordination step and a proton transfer step complete the S_N1 reaction.

E1 reactions are also susceptible to carbocation rearrangements, as illustrated in Solved Problem 8.28. Later in this book, we discuss other mechanisms that involve carbocations. Keep in mind the possibility of carbocation rearrangements in those reactions as well.

SOLVED problem 8.28 Predict the products of the reaction at the right, which takes place via an E1 mechanism.

Think What are the normal steps of an E1 mechanism? Is a carbocation generated as an intermediate? If so, can it undergo a 1,2-hydride shift or a 1,2-methyl shift to become more stable?

Solve The normal E1 reaction takes place in two steps. The leaving group leaves in the first step, and a proton is eliminated with the help of a base in the second step.

Notice, however, that there is a secondary carbocation intermediate that can rearrange to become more stable via a 1,2-methyl shift, as shown below. After this rearrangement, elimination of an H⁺ takes place to complete the E1 mechanism.

problem **8.29** Draw the complete, detailed mechanism for the S_N1 reaction between the same reactants as in Solved Problem 8.28.

8.7 Wrapping Up and Looking Ahead

Chapter 8 introduced two of the simplest multistep mechanisms: the S_N1 and E1 mechanisms. Even though these *overall* reactions were new to you, the elementary steps of which they are composed were introduced in Chapter 7. The S_N1 mechanism consists of a *heterolysis* step followed by a *coordination* step, and the E1 reaction consists of a *heterolysis* step followed by *elimination of an H⁺ electrophile*. This general idea—that multistep mechanisms are constructed from a handful of elementary steps—is one that can greatly simplify learning organic chemistry, and thus will be revisited frequently in the chapters to come.

Chapter 8 also discussed various types of experiments used to gain insight into the mechanism of a reaction. The experiments that provide the strongest support for a mechanism—namely, direct evidence of specific intermediates—are also the rarest, given how unstable intermediates tend to be. Therefore, most evidence used to provide support for a mechanism comes from other experimental results, such as the *chemical kinetics* and *stereochemistry* of the reaction in question. We saw here in Chapter 8 how such results are applied specifically to nucleophilic substitution and elimination reactions.

Finally, we saw that in addition to being consistent with experimental results, a mechanism should be *reasonable*. We introduced four general rules to help determine whether a given mechanism is reasonable. Three of these rules pertained to how proton transfer reactions should be incorporated into a multistep mechanism, and one rule addressed the role of carbocation rearrangements in multistep mechanisms.

What has been presented here in Chapter 8 will be integral in Chapter 9, where we discuss the competition that exists between S_N1, S_N2, E1, and E2 reactions. To predict the outcome of such a competition, we will examine how the rate of each reaction is governed by specific structural characteristics of the reactants, as well as by the reaction conditions. In turn, to understand the effects that those factors have on reaction rate, we must carry forward a keen understanding of each reaction's mechanism.

Chapter Summary and Key Terms

- A **unimolecular nucleophilic substitution (S_N1) reaction** consists of two steps. First, the leaving group leaves in a *heterolysis* step, yielding a carbocation **intermediate**, then a nucleophile attacks the carbocation in a *coordination* step. **(Section 8.1)**

- An **overall reaction** is obtained by summing all of the steps in a mechanism. Intermediates do not appear in the overall reaction—just **overall reactants** and **overall products**. **(Section 8.1a)**

- The reaction free energy diagram of an S_N1 reaction (Fig. 8-1) contains two transition states—one for each step. Between the transition states is the intermediate, which occurs at a **local energy minimum**. **(Section 8.1b)**

- The **unimolecular elimination (E1) reaction** also consists of two steps. First the leaving group leaves, generating a carbocation intermediate, then H^+ is eliminated with the aid of a base to yield a double bond. **(Section 8.2)**

- The free energy diagram of an E1 reaction (Fig. 8-2), like that of an S_N1 reaction, shows two transition states flanking the local energy minimum, which represents the carbocation intermediate. **(Section 8.2)**

- Direct observation of an intermediate is powerful evidence for a mechanism, but it is rare because intermediates tend to be unstable. Thus, a mechanism generally cannot be proved. **(Section 8.3)**

- Most of what we know about mechanisms comes from *reaction kinetics*. A *proposed mechanism* yields a **theoretical rate law**. Agreement between the theoretical rate law and the **empirical rate law** lends support to a mechanism. **(Section 8.4)**

- S_N2 and E2 reactions are both *second-order* reactions. In each case, the reaction rate is directly proportional to the concentration of both the nucleophile/base and the substrate. **(Section 8.4)**

- S_N1 and E1 reactions are both *first order*. In each case, the reaction rate depends only on the concentration of the

substrate; it is independent of the concentration of the nucleophile or the base, respectively. **(Section 8.4)**

- For S_N1 and E1 reactions to be first order, the first step of each one—that is, the departure of the leaving group—must be the **rate-determining step** of the mechanism. As a result, the rate of the entire reaction is essentially the rate of the first step. **(Section 8.4)**

- If an S_N1 reaction takes place at a tetrahedral stereocenter, the products contain a mixture of both stereochemical configurations. **(Section 8.5a)**

- In an S_N2 reaction, the nucleophile attacks the substrate only from the side opposite the leaving group—so-called **backside attack**. The substituents that remain on the atom being attacked undergo **Walden inversion**, and a single stereoisomer is produced, making the S_N2 reaction **stereospecific**. **(Section 8.5b)**

- Because an E1 reaction takes place in two steps, both E and Z configurations about the double bond are produced. **(Section 8.5c)**

- An E2 reaction is favored by the **anticoplanar** conformation of the substrate, in which the H atom and the leaving group that are eliminated are anti to each other about the C—C bond involved in the reaction. Thus, the E2 reaction is stereospecific. **(Section 8.5d)**

- Under acidic conditions, strong bases should not appear in a mechanism and, under basic conditions, strong acids should not appear. **(Section 8.6a)**

- **Intramolecular proton transfer** reactions are generally *unreasonable*. Instead, a proton is usually transferred from one part of a molecule to another by a **solvent-mediated proton transfer**. **(Section 8.6b)**

- **Termolecular** elementary steps are generally *unreasonable* because they require three species to collide at precisely the same time. **(Section 8.6c)**

- **Carbocation rearrangements** are fast. If a 1,2-hydride shift or a 1,2-methyl shift is energetically favorable, it will generally occur before any other step. **(Section 8.6d)**

Problems

8.30 Draw the complete, detailed mechanism (including curved arrows) for each of the following reactions occurring via **(a)** an S_N2 mechanism and **(b)** an S_N1 mechanism. Pay attention to stereochemistry.

(i)

+ NaOH $\longrightarrow$?

(ii)

+ NaOH $\longrightarrow$?

(iii)

+ NaOH → ?

(iv)

+ KBr → ?

(v)

+ NaOCH$_3$ → ?

8.31 Draw the complete, detailed mechanism (including curved arrows) for each of the following reactions occurring via **(a)** an E2 mechanism and **(b)** an E1 mechanism. If more than one possible product can be produced from the same type of mechanism, draw the complete mechanism that leads to each one. Pay attention to stereochemistry.

(i)

+ NaOH → ?

(ii)

+ NaOH → ?

(iii)

+ KOC(CH$_3$)$_3$ → ?

(iv)

+ NaOCH$_3$ → ?

(v)

+ KOH → ?

(vi)

+ KOH → ?

8.32 The cis isomer of 4-*tert*-butyl-1-bromocyclohexane undergoes E2 elimination about 1,000 times faster than the trans isomer. Explain why the cis isomer reacts faster. (*Hint:* It is *not* because of steric hindrance.)

+ NaOH —Fast→

+ NaOH —Slow→

8.33 For each of the following substrates, predict whether a carbocation rearrangement will take place in an S$_N$1 or E1 mechanism? Explain. Draw the curved arrow notation illustrating the carbocation rearrangement that is likely to occur.

(a)

(b)

(c)

(d)

(e)

(f)

(g)

(h) H$_3$CH$_2$C CH$_2$CH$_3$

(i)

8.34 Consider the following *overall* reaction, which will be discussed in Chapter 20.

$$H_3C—C≡N \xrightarrow[\text{HO}^{\ominus}]{\text{H}_2\text{O}} \underset{H_3C}{}\overset{O}{\underset{}{||}}C—O^{\ominus} + NH_3$$

Below is a proposed mechanism for this reaction. Use the rules we learned in this chapter to evaluate whether or not each step in the proposed mechanism is reasonable. For each step that is *not* reasonable, explain why.

(i)

(ii)

(iii)

(iv)

(v)

(vi)

8.35 Consider the following *overall* reaction, which will be discussed in Chapter 20.

Below is a proposed mechanism. Evaluate whether each step of the mechanism is reasonable or not. For each step that is *not* reasonable, explain why.

(i)

(ii)

(iii)

8.36 Racemization occurs when (S)-3,3-dimethylcyclohexanol is dissolved in dilute acid. Draw a mechanism to account for this, including curved arrows.

(S)-3,3-Dimethylcyclohexanol

8.37 Draw a reasonable, detailed mechanism that shows the following racemization at the alpha carbon. (*Note:* The reaction takes place under basic conditions.)

Racemic mixture

8.38 Draw a reasonable, detailed mechanism that shows the following racemization at the alpha carbon. (*Note:* The reaction takes place under acidic conditions.)

Racemic mixture

8.39 The following elimination reaction yields the same alkene product independent of whether it proceeds by the E2 or E1 mechanism.

The mechanism by which the reaction proceeds could be determined if the following D-labeled substrate were used instead. (*Note:* D is an isotope of H. They both have one proton and one electron, so they have nearly identical chemical properties, but D, having one additional neutron, is heavier by 1 amu.)

(a) Draw the complete mechanism for an E2 reaction involving the D-labeled substrate and predict the major product(s).
(b) Draw the complete mechanism for an E1 reaction involving the D-labeled substrate and predict the major product(s).
(c) What is the molar mass of each product from (a) and (b)?

8.40 Consider the following E1 reaction.

(a) Draw a complete, detailed mechanism for this reaction.
(b) Draw a reaction free energy diagram that agrees with that mechanism, labeling overall reactants, overall products, all transition states, and all intermediates.

8.41 According to the rules for reasonable mechanisms, the following E1 reaction should undergo a 1,2-hydride shift. However, the same product is produced regardless of whether the rearrangement occurs.

(a) Draw the mechanism that includes that carbocation rearrangement.
(b) Draw the mechanism that does not include the rearrangement.
(c) Experimentally, how can we use ^{13}C isotope labeling to determine whether the rearrangement occurs? In other words, can a ^{12}C atom in the substrate be replaced by a ^{13}C atom so that the E1 products would depend on whether the rearrangement takes place?
(d) How can we use deuterium isotope labeling to determine whether the rearrangement occurs?

8.42 Consider the following *intramolecular* nucleophilic substitution reaction.

Considering stereochemistry, does this suggest an S_N1 or S_N2 mechanism? Draw the complete mechanism for this reaction, including curved arrows.

8.43 Consider the following nucleophilic substitution reaction, which yields a mixture of constitutional isomers.

(a) Does this occur via an S_N1 or an S_N2 mechanism? How do you know?
(b) Propose a mechanism that accounts for the formation of *each* product.

8.44 Consider the following elimination reaction, which produces a mixture of diastereomers.

Based on the stereochemistry of this reaction alone, is it possible to tell whether the reaction takes place via an E1 or an E2 reaction? Explain.

8.45 Consider the following nucleophilic substitution reaction.

(a) Argue whether this reaction takes place via an S_N1 or an S_N2 reaction.
(b) Draw the complete mechanism (including curved arrows) for this reaction.

8.46 The following reaction yields three different nucleophilic substitution products that are constitutional isomers of one another.

(a) Does this suggest an S_N1 or S_N2 mechanism?
(b) Draw the mechanism for the formation of each of these products.

8.47 When benzyl bromide is treated separately with KI and CH_3OH, the substitution products are different but the reaction rates are about the same.

(a) What does this suggest about the mechanism—is it S_N1 or S_N2? Explain.
(b) Draw the complete mechanism (including curved arrows) for the formation of each product.
(c) If the concentration of KI were doubled, what would happen to the rate of the substitution reaction?

8.48 The initial rates for the following elimination reaction were measured under different concentrations of the substrate and base (water); the data are tabulated at the right.

Trial Number	[R—OCH₃]	[H₂O]	Rate (M/s)
1	0.010 M	0.45 M	9.50×10^{-4}
2	0.020 M	0.45 M	1.85×10^{-3}
3	0.020 M	0.22 M	1.85×10^{-3}

Do the data suggest an E1 reaction or an E2 reaction?

8.49 Draw the reaction free energy diagram for the reaction in Problem 8.48. Include and label the overall reactants, overall products, all transition states, and all intermediates.

8.50 The initial rates for the following elimination reaction were measured under different concentrations of the substrate and base; the data are tabulated at the right.

Trial Number	[R—Br]	[KOCH₂CH₃]	Rate (M/s)
1	1.0 M	1.0 M	2.35×10^{-6}
2	0.50 M	0.50 M	5.9×10^{-7}
3	0.50 M	1.0 M	1.20×10^{-6}

Do the data suggest an E1 reaction or an E2 reaction?

8.51 Draw the reaction free energy diagram for the reaction in Problem 8.50. Include and label the overall reactants, overall products, all transition states, and all intermediates.

8.52 The following reaction proceeds via a carbocation rearrangement.

Draw a complete, detailed mechanism to account for the product. Explain why the carbocation rearrangement is favorable.

8.53 One way to synthesize diethyl ether is to heat ethanol in the presence of a strong acid:

Draw a complete, detailed mechanism for this reaction.

8.54 Consider the following nucleophilic substitution reaction.

Based on the stereochemistry, does it proceed by an S_N1 or S_N2 mechanism? Explain.

8.55 Suggest a reasonable mechanism for the following reaction.

8.56 (S)-Adenosyl methionine (SAM) is a cosubstrate that is involved in biological methyl group transfers. SAM is believed to be produced by an S_N2 type of reaction between methionine and ATP, as shown below.

Methionine **ATP** **SAM**

(a) Draw the appropriate curved arrows for this reaction.
(b) Suggest why the reaction takes place with S as the nucleophilic atom instead of one of the negatively charged O atoms on methionine.

8.57 Creatine is a naturally-occurring compound that helps provide energy to cells in the body, especially muscle cells. Draw an S_N2 mechanism that shows how creatine is produced from guanidoacetate and SAM (see Problem 8.56). Pay attention to stereochemistry.

Guanidoacetate **Creatine**

8.58 One of the ways in which an L α-amino acid can be synthesized is to carry out an S_N2 reaction between an α-bromo acid and ammonia. (The wavy line indicates that the bond could be a dash or a wedge.)

An α-bromo acid   **An L α-amino acid**

Draw the stereoisomer of the α-bromo acid that would be necessary to produce L-alanine, in which the R group is CH_3.

8.59 (*E*)-Anethole is the major component of anise oil, which is used as artificial licorice flavoring and has potential antimicrobial and antifungal properties. It can be synthesized from an alkyl halide precursor, as follows:

(E)-Anethole

(a) Will an E2 reaction produce (*E*)-anethole exclusively, or will the reaction produce a mixture of stereoisomers?
(b) Will an E1 reaction produce the pure stereoisomer or a mixture?
(c) For each of these reactions that produces a mixture, which stereoisomer will be produced in greater abundance?

8.60 Structures **A** and **B** are intermediates in the biosynthesis of steroids. **(a)** Draw the mechanism (including curved arrows) that shows how **A** can be converted to **B** through two 1,2-hydride shifts followed by two 1,2-methyl shifts. **(b)** Draw the mechanism (including curved arrows) that shows how **B** is converted to lanosterol.

Intermediate **A**   Intermediate **B**   **Lanosterol**

8.61 The following reaction is called the pinacol rearrangement.

A carbocation rearrangement is believed to be involved. **(a)** Propose a reasonable mechanism for this reaction.
(b) Suggest why the carbocation rearrangement is favorable.

8.62 Propose a mechanism for the following reaction, which produces 1,4-dioxane.

1,4-Dioxane

8.63 The specific angle of rotation of (*R*)-2-bromobutane is −23.1°. Treatment of (*R*)-2-bromobutane with potassium bromide produces a racemic mixture of (*R*)- and (*S*)-2-bromobutane, which is optically inactive. The rate at which the product's angle of rotation decreases is directly proportional to the concentration of KBr. Does this suggest that the reaction takes place by an S_N1 or an S_N2 mechanism? Explain.

8.64 Propose a mechanism for the following reaction, which takes place under conditions that favor an S_N1 reaction.

8.65 Propose a mechanism for the following reaction, which takes place under conditions that favor an S_N1 reaction.

Based on the mechanism, do you think that the products will be formed in a mixture of stereoisomers?

8.66 The following nucleophilic substitution reaction is monitored by measuring the optical rotation of the solution as a function of time. Based on the results graphed on the right, suggest whether the reaction takes place by the S_N1 or S_N2 mechanism. (*Hint:* Review optical rotation in Chapter 5.)

NaBr → CH(Br)CH₃

8.67 A chemist proposes that the following reaction proceeds by an S_N2 mechanism.

She carries out the reaction with oxygen-18 (^{18}O)-labeled hydroxide in ^{18}O-labeled water (i.e., $^{18}OH^-/H_2{}^{18}O$). When analyzing the products, she finds that the ^{18}O isotope appeared only in $CH_3CO_2^-$, and not in CH_3OH. What does this result suggest about her hypothesis?

8.68 A chemist proposes that the following reaction occurs via an S_N2 mechanism.

Upon carrying out the reaction using ^{18}O-labeled hydroxide anion in ^{18}O-labeled water, she finds that ^{18}O-labeled methanol is produced. What does this suggest about her hypothesis?

If your instructor assigns problems in smartwork, log in at **smartwork.wwnorton.com**.

CHAPTER 10

Nucleophilic Substitution and Elimination Reactions 1

Competition among S_N2, S_N1, E2, and E1 Reactions

We have studied nucleophilic substitution and elimination reactions extensively in the last two chapters: S_N2 and E2 in Chapter 7, and S_N1 and E1 in Chapter 8. Up to this point, we have considered these four reactions as if they were independent of one another. Here in Chapter 9, however, we will see that they are generally in competition, so under most circumstances, *if one reaction is feasible, we must consider all four of them.* As a result, one of the main goals of Chapter 9

In this children's game, players compete to accumulate the most marbles in their respective bins. Analogously, S_N2, S_N1, E2, and E1 reactions compete to establish the major product.

is to provide you with a method of predicting the major product in such a competition. To achieve this goal, we will first examine how a number of factors affect each of the four reactions, then we will see how we can use that knowledge to predict reaction products.

Sometimes, a competition can take place between reactions that proceed by the same mechanism. In the two examples discussed here in Chapter 9, one involves an aspect of reactions known as *regioselectivity*—the tendency of a particular reaction to be favored at one site within a molecule over another. In the second, an *intermolecular reaction* (involving functional groups on separate reactant species) competes with an *intramolecular reaction* (involving functional groups on the same reactant).

Many of the reactions we will encounter in subsequent chapters also involve competitions. Thus, the ideas we learn in Chapter 9 pertaining to nucleophilic substitution and elimination reactions will be applied throughout the book.

9.1 The Competition among S_N2, S_N1, E2, and E1 Reactions

A competition usually exists among S_N2, S_N1, E2, and E1 reactions. In the four reactions shown in Equations 9-1 through 9-4 (p. 468), for example, the reactants are identical, but the mechanisms are different.

The products of all four mechanisms are different, too, particularly between the substitution products and the elimination products. Furthermore, the *stereochemistry* of each unimolecular reaction (i.e., S_N1 and E1) is different from that of the corresponding bimolecular reaction (i.e., S_N2 and E2, respectively). Whereas each bimolecular reaction is stereospecific, each unimolecular reaction leads to a mixture of stereoisomers.

There are essentially two reasons that this competition occurs:

1. S_N2, S_N1, E2, and E1 reactions all involve a substrate containing a *leaving group*.
2. Any species that can act as a nucleophile also has the potential to act as a base, and vice versa. Such a species is called an **attacking species**.

Acting as a nucleophile, an attacking species uses a lone pair of electrons to form a bond to an electron-poor *nonhydrogen atom*. Acting as a base, on the other hand, an attacking species forms a bond to a *hydrogen atom*. Notice that $CH_3CO_2^-$ acts as a nucleophile in Equations 9-1 and 9-2, but acts as a base in Equations 9-3 and 9-4.

CHAPTER OBJECTIVES

Upon completing Chapter 9 you should be able to:

- Recognize suitable *substrates* for nucleophilic substitution and elimination reactions and draw the S_N2, S_N1, E2, and E1 mechanisms that compete with each other when a substrate is treated with a nucleophile and/or base.

- Determine the strength of an attacking species as a nucleophile and as a base, and on that basis predict whether an S_N2, S_N1, E2, or E1 reaction is favored.

- Determine which reactions—S_N2, S_N1, E2, or E1—are favored by high concentration of the attacking species and which are favored by low concentration of the attacking species.

- Distinguish among good, moderate, and poor leaving groups, and specify which substitution and elimination reactions each type of leaving group favors.

- Establish whether or not nucleophilic substitution and elimination reactions are feasible based on the hybridization of the carbon atom bonded to the leaving group.

- Classify the carbon atom bonded to the leaving group as either primary (1°), secondary (2°), or tertiary (3°), and on that basis predict which substitution and elimination reactions are favored.

- Identify a solvent as either protic or aprotic, and on that basis determine which reactions—S_N2, S_N1, E2, or E1—are favored.

- Explain the role of heat in nucleophilic substitution and elimination reactions.

- Predict the major product(s) of a given nucleophilic substitution or elimination reaction by systematically evaluating the factors affecting the reaction.

- Recognize when an elimination reaction can produce two or more alkene products, and predict which is the major product.

- Predict the major product of competing intermolecular and intramolecular substitution reactions.

S_N2 mechanism

The attacking species acts as a nucleophile.

(9-1)

S_N1 mechanism

The attacking species acts as a nucleophile.

(9-2)

E2 mechanism

The attacking species acts as a base.

(9-3)

E1 mechanism

The attacking species acts as a base.

(9-4)

Which of the following reactions, **A** or **B**, shows NH_3 acting as a base and which shows NH_3 acting as a nucleophile?

| A | B |

Answers to Your Turns are in the back of the book.

problem **9.1** Draw the complete, detailed mechanisms for the S_N2, S_N1, E2, and E1 reactions between iodocyclohexane and ammonia. Include the necessary curved arrows and draw the products.

Because S_N2, S_N1, E2, and E1 reactions are in competition and can therefore lead to a mixture of products, it might seem that such reactions would not be very useful to synthesize specific compounds. Frequently, however, one of these reactions dominates, though which reaction that is can depend upon a variety of factors. Therefore, we devote much of the rest of this chapter to discussing those factors. With an understanding of how such factors govern the various reactions, we can predict the major product of an $S_N2/S_N1/E2/E1$ competition under certain conditions. Moreover, we can often adjust reaction conditions to favor the desired reaction, and hence the desired product.

Before we can discuss how to predict the major product of any competition, however, we must know whether the competition takes place under *kinetic control* or *thermodynamic control*.

■ In a competition that takes place under **kinetic control**, the major product is the one that is produced the fastest.
■ Under **thermodynamic control**, the major product is the one that is the most stable (i.e., the lowest energy).

For nucleophilic substitution and elimination reactions,

The $S_N2/S_N1/E2/E1$ competition usually takes place under *kinetic control*.

The reasons kinetics usually governs these reactions are explained in greater detail in Section 9.12, after the conditions that favor each reaction have been more fully discussed.

9.2 Rate-Determining Steps Revisited: Simplified Pictures of the S_N2, S_N1, E2, and E1 Reactions

Because substitution and elimination reactions generally take place under kinetic control, predicting the outcome of an $S_N2/S_N1/E2/E1$ competition means we have to know how to predict the relative rates of the competing reactions. To begin that process, we first review each reaction's rate-determining step to gain a sense of the kinds of factors to which each reaction is sensitive.

As we learned in Section 8.4, the rate-determining step of a reaction dictates the rate of the overall reaction. The S_N2 mechanism consists of just a single step, for example, so that single step (shown in Equation 9-5) must be rate determining. The S_N1 reaction, on the other hand, takes place in two steps. As indicated in Equation 9-6, the first step is rate determining because formation of the carbocation is much slower than formation of the final product.

S_N2 mechanism

Rate-determining step

$$Nu:^{\ominus} \; + \; R-L \longrightarrow Nu-R \; + \; :L^{\ominus} \qquad (9\text{-}5)$$

S_N1 mechanism

Rate-determining step

$$R-L \xrightarrow[\text{Slow}]{} R^{\oplus} \; + \; :L^{\ominus} \quad \xrightarrow[\text{Fast}]{:Nu^{\ominus}} \quad Nu-R \qquad (9\text{-}6)$$

In the rate-determining step of an S_N2 reaction, the nucleophile forms a bond to the substrate at the same time the leaving group leaves. In the rate-determining step of an S_N1 reaction, on the other hand, the leaving group departs only. The nucleophile is present in solution throughout the reaction, despite the fact that it does not enter the mechanism until the second step. Therefore, the role of the nucleophile differs in S_N2 and S_N1 reactions:

- In an S_N2 reaction, the nucleophile forces the leaving group out.
- In an S_N1 reaction, the nucleophile waits until the leaving group has left.

As a result, the rate of an S_N2 reaction is highly sensitive to factors that affect the nucleophile's ability to attack the substrate and to factors that affect the ability of the leaving group to depart. The rate of an S_N1 reaction, on the other hand, is highly sensitive only to factors that help the leaving group depart.

We can differentiate the E1 and E2 mechanisms in a similar way (see Equations 9-7 and 9-8). Like an S_N2 reaction, an E2 reaction consists of a single step that must be rate determining. And, as in an S_N1 reaction, the rate-determining step of an E1 reaction is the first of its two steps.

E2 mechanism

Rate-determining step

$$Base:^{\ominus} \; + \; \overset{H}{\underset{L}{C-C}} \longrightarrow Base-H \; + \; C=C \; + \; :L^{\ominus} \qquad (9\text{-}7)$$

E1 mechanism

Rate-determining step

$$\overset{H}{\underset{L}{C-C}} \xrightarrow[\text{Slow}]{} \overset{H}{C-C^{\oplus}} \; + \; :L^{\ominus} \quad \xrightarrow[\text{Fast}]{:Base^{\ominus}} \quad C=C \; + \; Base-H \qquad (9\text{-}8)$$

In the E2 reaction, the substrate is deprotonated at the same time the leaving group leaves. In an E1 reaction, the base does not enter the picture until the second step, even though it is in solution throughout the reaction. That is,

- In an E2 reaction, the base pulls off the proton, thus forcing the leaving group to leave.
- In an E1 reaction, the base waits until the leaving group has left.

Consequently, the rate of an E2 reaction is highly sensitive to factors that affect the ability of the base to pull off the proton and to factors that affect the ability of the leaving group to leave. And, like an S_N1 reaction, the rate of an E1 reaction is highly sensitive only to factors that help the leaving group to depart.

We are now ready to examine in detail how each reaction rate is affected by six different factors. In Sections 9.3 and 9.4, we discuss two factors related to the attacking species. The nature of the leaving group is discussed in Section 9.5, and the type of carbon to which the leaving group is attached is discussed in Section 9.6. Section 9.7 deals with the nature of the solvent, and Section 9.8 discusses the impact of heat on the reactions. Following our discussions of these six factors, Section 9.9 presents a strategy that can be used to predict the major product of an $S_N2/S_N1/E2/E1$ competition by bringing together all of these factors in a systematic way.

9.3 Factor 1: Strength of the Attacking Species

Because the S_N2, S_N1, E2, and E1 reactions are sensitive to the attacking species in different ways, the *identity* of the attacking species can play a major role in the outcome of the $S_N2/S_N1/E2/E1$ competition. Here in Section 9.3, we examine how the attacking species affects each of these reactions differently. First we discuss the nature of the attacking species in S_N2 and S_N1 reactions, in which it behaves as a nucleophile. Then we turn our attention to E2 and E1 reactions, in which the attacking species behaves as a base.

9.3a The Nucleophile Strength in S_N2 and S_N1 Reactions: An Introduction to the Hammond Postulate

Table 9-1 lists the S_N2 reaction rates for various nucleophiles attacking CH_3I in the solvent *N,N*-dimethylformamide (DMF). The rate of the S_N2 reaction depends very heavily upon the identity of the nucleophile. Notice, for example, that Cl^- undergoes

TABLE 9-1 S_N2 Reaction Rate in DMF of the Reaction:

$$Nu^{\ominus} + CH_3I \longrightarrow NuCH_3 + I^{\ominus}$$

Nucleophile	H_2O	(pyridine)	$NCS^{\ominus}$	$Br^{\ominus}$	$Cl^{\ominus}$	$N_3^{\ominus}$	$C_6H_5S^{\ominus}$	$CH_3CO_2^{\ominus}$	$NC^{\ominus}$
Relative Reaction Rate	~0	~0.0005	0.06	1	2	3	12	15	250

the S_N2 reaction about twice as fast as Br^-, and the S_N2 reaction involving the cyanide anion (NC^-) is about 250 times faster than that involving Br^-. These rate differences reflect differences in *nucleophile strength*, or **nucleophilicity**.

> The stronger nucleophile promotes a faster S_N2 reaction.

Thus, in DMF, the relative nucleophilicities of Br^-, Cl^-, and NC^- are 1, 2, and 250, respectively.

As we learned in Section 8.4, such differences in rate reflect differences in the sizes of the energy barriers that must be traversed for reactants to become products: The smaller the energy barrier, the faster the reaction. In turn, the size of the energy barrier is determined by the energy of the transition state relative to that of the reactants. So, one way to understand the variations in reaction rate is to evaluate the relative energies of transition states. This can be a formidable task, however, because it involves carefully considering partial bonds and partial charges.

Instead, the problem can be greatly simplified by applying the **Hammond postulate**, proposed in 1955 by the American chemist George S. Hammond: "If two states . . . occur consecutively during a reaction process and have nearly the same energy content, their interconversion will involve only a small reorganization of the molecular structures." In other words,

> If, in a reaction free energy diagram, two species lie near each other along the reaction coordinate (the x axis) and are similar in energy, then they will have very similar structures.

The Hammond postulate can be used to provide insight into the structure and energy of a transition state by considering the reactants that immediately precede it and the products that immediately follow it.

> - For a reaction whose ΔG°_{rxn} is negative (i.e., an *exergonic* reaction), the transition state resembles the reactants more than it does the products, both in structure and energy.
> - For a reaction whose ΔG°_{rxn} is positive (i.e., an *endergonic* reaction), the transition state resembles the products more than it does the reactants, both in structure and energy.

As shown in Figure 9-1, these two points stem from the fact that a transition state lies higher in energy than either the reactants or products. If ΔG°_{rxn} is negative (Fig. 9-1a), then the reactants are higher in energy than the products, so the energy of the transition state must be closer to that of the reactants than the products. Consequently, according to the Hammond postulate, the transition-state structure must resemble the reactants more than the products, and this is consistent with the shorter distance along the reaction coordinate between the transition state and reactants than between the transition state and products. By contrast, if ΔG°_{rxn} is positive (Fig. 9-1b), then the energy of the transition state is closer to that of the products than reactants, and the same is true of its structure.

We can extend the Hammond postulate to say that, *all else being equal*, the more negative (less positive) the value of ΔG°_{rxn}, the more the transition state tends to resemble the reactants in both structure and energy. In turn, as the transition state's energy becomes closer to that of the reactants, the energy barrier decreases. Thus, we

An exergonic reaction

The transition state lies closer to the reactants than to the products, both along the energy axis and along the reaction coordinate.

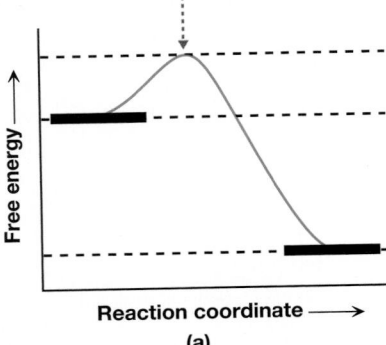

(a)

An endergonic reaction

The transition state lies closer to the products than to the reactants, both along the energy axis and along the reaction coordinate.

(b)

FIGURE 9-1 The Hammond postulate and ΔG°_{rxn} (a) Free energy diagram for an exergonic reaction, showing that the transition state lies closer in energy to the reactants than to the products, and that its structure resembles reactants more than it does products. (b) Free energy diagram for an endergonic reaction, showing that the transition state lies closer in energy to the products than to the reactants, and that its structure resembles products more than it does reactants.

The transition state is closer in energy to the reactants when Cl$^{\ominus}$ is the nucleophile.

The energy barrier is smaller, so the rate is faster, making the nucleophile stronger.

Cl$^{\ominus}$ + H$_3$C—I

Br$^{\ominus}$ is more stable than Cl$^{\ominus}$.

Br$^{\ominus}$ + H$_3$C—I

Cl—CH$_3$ + I$^{\ominus}$

Br—CH$_3$ + I$^{\ominus}$

Uncharged products are assumed to have similar stabilities.

Free energy →

Reaction coordinate →

FIGURE 9-2 ΔG°_{rxn} and S$_N$2 energy barriers Free energy diagrams for the S$_N$2 reactions of Cl$^-$ (red) and Br$^-$ (blue) with CH$_3$I. With Cl$^-$ as the nucleophile, the reaction is energetically more favorable. Consequently, the transition state involving Cl$^-$ as the nucleophile lies closer in energy to the reactants than does the transition state involving Br$^-$ as the nucleophile. This corresponds to a smaller energy barrier, and thus a faster reaction, when Cl$^-$ is the nucleophile.

arrive at a very useful rule that allows us to understand and make predictions about relative reaction rates:

> For two reactions that proceed by the same mechanism, the one with the more negative (less positive) value of ΔG°_{rxn} tends to have the smaller energy barrier, and thus tends to be faster.

To see how this rule applies to S$_N$2 reactions, consider the reactions in Equations 9-9 and 9-10, involving Cl$^-$ and Br$^-$ as nucleophiles.

$$:\!\overset{\ominus}{\underset{..}{\text{Cl}}}\!: \quad \text{H}_3\text{C}—\overset{..}{\underset{..}{\text{I}}}\!: \longrightarrow :\!\overset{..}{\underset{..}{\text{Cl}}}\!—\text{CH}_3 \ + \ :\!\overset{..}{\underset{..}{\text{I}}}\!:^{\ominus} \qquad (9\text{-}9)$$

$$:\!\overset{\ominus}{\underset{..}{\text{Br}}}\!: \quad \text{H}_3\text{C}—\overset{..}{\underset{..}{\text{I}}}\!: \longrightarrow :\!\overset{..}{\underset{..}{\text{Br}}}\!—\text{CH}_3 \ + \ :\!\overset{..}{\underset{..}{\text{I}}}\!:^{\ominus} \qquad (9\text{-}10)$$

The free energy diagrams for these two reactions are shown together in Figure 9-2.

As we learned in Section 7.8, the relative energies of the reactants and products are governed largely by charge stability. In both cases, ΔG°_{rxn} is negative because I$^-$ is more stable than either Br$^-$ or Cl$^-$. (Recall from Section 6.6b that larger atoms can accommodate charges better.) Moreover, Br$^-$ is more stable than Cl$^-$ (Br is larger), making the set of reactants belonging to the blue curve lower in energy than the set of reactants belonging to the red curve. Thus, the reaction involving Cl$^-$ as the nucleophile (the red curve) has the more negative value for ΔG°_{rxn}. According to the rule above, this corresponds to a smaller energy barrier—and thus a faster rate—for the Cl$^-$ reaction, consistent with the results in Table 9-1.

problem **9.2** Based on the values in Table 9-1, draw a free energy diagram similar to that in Figure 9-2, comparing the S$_N$2 reactions of NC$^-$ and Br$^-$ with CH$_3$I.

The principles we have learned in this section are quite helpful in predicting the relative strengths of nucleophiles *not* listed in Table 9-1, including uncharged nucleophiles (see Solved Problem 9.3). Moreover, as we will see on several occasions throughout this book, these principles represent powerful tools that can help us understand the relative reactivities of species in other reactions.

SOLVED problem **9.3** Which nucleophile will react faster with CH_3I in an S_N2 reaction: H_2O or H_2S? What does this indicate about their relative nucleophilicities?

Think What are the products of the two reactions? Are the products higher or lower in energy than the reactants? Do the reactants of the two reactions differ significantly in energy? The products? Which reaction has a more negative (less positive) value for $\Delta G°_{rxn}$?

Solve The two reactions are as follows:

The free energy diagrams for these reactions are as follows:

The transition state is closer to the products when H_2O is the nucleophile.

The energy barrier is smaller, so the reaction is faster, and H_2S is the stronger nucleophile.

Reaction of H_2O

Reaction of H_2S

H_2O + H_3C—I

H_2S + H_3C—I

$H_2O^{\oplus}$—CH_3 + $I^{\ominus}$

$H_2S^{\oplus}$—CH_3 + $I^{\ominus}$

The positive charge is more stable on S than on O.

Free energy ⟶

Reaction coordinate ⟶

Both reactions are endergonic because two new charges are produced. Both sets of reactants are uncharged, so they appear at the same energy. The set of products from the H_2S reaction (the blue curve) is lower in energy than the set of products from the H_2O reaction (the red curve) because a positive charge is more stable on the larger S atom than it is on the smaller O atom. Thus, the reaction in which H_2S is the nucleophile has a less positive value for $\Delta G°_{rxn}$ and, according to the Hammond postulate, should have a smaller activation energy. In other words, H_2S is a stronger nucleophile than H_2O.

---

problem **9.4**

(a) For each pair of nucleophiles, predict which will react faster with CH_3I in an S_N2 reaction: (i) CH_3O^- or $CH_3CO_2^-$; (ii) H_3N or H_3P.

(b) Which of each pair is the stronger nucleophile?

How, then, does the nucleophile influence the rate of an S_N1 reaction, such as those shown in Equations 9-11a and 9-11b? The substrate in both reactions is chlorodiphenylmethane, but the nucleophiles are the thiocyanate ion (NCS^-) in Equation 9-11a and the azide anion (N_3^-) in Equation 9-11b.

$$\text{(9-11a)}$$

The rates of both reactions are nearly the same.

$$\text{(9-11b)}$$

Recall from Table 9-1 that N_3^- is a stronger nucleophile than NCS^- by a factor of roughly 50. The rates of the two reactions in Equation 9-11, however, are about the same. It turns out that

> The rate of an S_N1 reaction is essentially independent of the identity of the nucleophile.

Recall from Section 9.2 that the rate of an S_N1 reaction is sensitive only to factors that affect the ability of the leaving group to leave. The nucleophile is not involved in an S_N1 reaction until after the leaving group has left, so the specific identity of the nucleophile should have little impact on the overall rate.

What happens to the rates of S_N2 and S_N1 reactions as the strength of the nucleophile changes? The rate of an S_N2 reaction increases as the strength of the nucleophile increases, whereas the rate of an S_N1 reaction remains essentially unchanged. If the nucleophile is strong enough, the S_N2 reaction becomes faster than the S_N1 reaction. Conversely, as the nucleophile becomes weaker, the S_N2 reaction is slowed, but the S_N1 reaction is not. And, if the nucleophile is weak enough, the S_N2 reaction becomes slower than the S_N1 reaction. Stated another way,

> - Strong nucleophiles tend to favor S_N2 reactions.
> - Weak nucleophiles tend to favor S_N1 reactions.

The distinction between "strong" and "weak" nucleophiles is purely empirical. In practice, with all else being equal, S_N2 reactions are usually favored by nucleophiles bearing a total -1 charge, whereas S_N1 reactions are usually favored by uncharged nucleophiles. Thus,

> - Strong nucleophiles tend to have a total -1 charge.
> - Weak nucleophiles tend to be uncharged.

Examples of strong nucleophiles include the halide anions Cl^-, Br^-, and I^-, as well as HO^-, alkoxide anions (RO^-), thiolate anions (RS^-), and deprotonated amines (R_2N^-) (Fig. 9-3). Organometallic reagents, which behave as R^-, are strong nucleophiles if the

FIGURE 9-3 Intrinsic nucleophile strength Uncharged nucleophiles tend to be intrinsically weak, whereas negatively charged nucleophiles tend to be strong.

Weak　　　　　　　　　　　Strong

$$H_2O, ROH < H_3N, R_2NH < I^{\ominus} < Br^{\ominus} < Cl^{\ominus} < HS^{\ominus}, RS^{\ominus} \approx RCO_2^{\ominus} < HO^{\ominus}, RO^{\ominus} < H_2N^{\ominus}, R_2N^{\ominus} < H_3C^{\ominus}, R^{\ominus}$$

Intrinsic nucleophile strength increases

Stability of negative charge decreases

partial negative charge on carbon is sufficiently strong. These include alkyllithium (R—Li) and Grignard (R—MgX) reagents. Examples of weak nucleophiles include H_2O, alcohols (ROH), and amines (R_2NH).

---

problem **9.5** Is H_2P^- a strong nucleophile or a weak nucleophile? Will it favor the S_N2 or the S_N1 mechanism? Explain.

9.3b Generating Carbon Nucleophiles

As we will see in Chapter 13, nucleophilic carbon atoms are often important in synthesis, especially in the formation of carbon–carbon bonds. In uncharged molecules, however, carbon atoms are typically non-nucleophilic for two reasons: (1) They do not possess a lone pair of electrons, and (2) they are rarely considered electron rich because they are generally bonded to atoms with electronegativities comparable to their own (e.g., hydrogen and other carbon atoms), if not greater (e.g., oxygen, nitrogen, and halogens).

A carbon atom is quite nucleophilic, however, when it bears a formal negative charge—that is, when it is a **carbanion** (Fig. 9-4). These carbon atoms not only are electron rich, but also possess a lone pair of electrons that can be used to form a bond.

It is often beneficial, then, to generate carbanions from uncharged carbon atoms. The simplest way to do so would be to deprotonate the uncharged carbon, but carbon atoms typically do not possess acidic hydrogens. Alkanes, for example, have pK_a values around 50, so they are such weak acids that deprotonation is unfeasible.

Some carbon atoms, on the other hand, are deprotonated much more readily. A terminal alkyne (RC≡C—H), for example, is weakly acidic ($pK_a \approx 25$) because the alkyne C atom is sp hybridized. Recall from Section 3.11 that an sp-hybridized atom has a greater effective electronegativity than its sp^3-hybridized counterpart, so it can better stabilize a negative charge.

Once deprotonated (Equation 9-12a), the resulting **acetylide anion** (RC≡C⁻) can behave as a strong nucleophile, as shown in Equation 9-12b. This reaction forms a new C—C bond, as noted previously.

Not nucleophilic   **A strong nucleophile**

FIGURE 9-4 Carbon nucleophiles An uncharged, tetrahedral C atom (left) is not nucleophilic. A carbanion (right) is a strong nucleophile because of the negative charge and the lone pair of electrons.

$$R—C{\equiv}CH \ + \ NaH \ \longrightarrow \ R—C{\equiv}C{:}^{\ominus} \ + \ H—H \ + \ Na^{\oplus} \tag{9-12a}$$

$$R—C{\equiv}C{:}^{\ominus} \ + \ R'—Br \ \xrightarrow{\ S_N2\ } \ R—C{\equiv}C—R' \ + \ Br^{\ominus} \tag{9-12b}$$

Strong nucleophile   **New C—C bond**

Other functional groups have mildly acidic protons on carbon atoms, including ketones (R—CR=O), aldehydes (R—CH=O), and nitriles (R—C≡N). Thus, compounds containing these functional groups can be converted into strong carbon nucleophiles as well. This is discussed in greater detail in Section 10.3.

YOUR TURN 9.2

Hydrocyanic acid (HCN), like a terminal alkyne, can be converted into a carbon nucleophile by treatment with a sufficiently strong base:

$$N{\equiv}C—H \ + \ {:}\ddot{O}H^{\ominus} \ \longrightarrow$$

Hydrocyanic acid

Draw the products of this reaction and label the nucleophilic atom that is produced.

9.3c The Base Strength in E2 and E1 Reactions

Table 9-2 lists the relative rates of E2 reactions involving 1,2-dichloroethane as the substrate. These data show that the relative rates of E2 reactions depend on the identity of the base—more specifically, on the strength of the base.

> The rate of an E2 reaction generally increases as the strength of the base increases (i.e., as the pK_a of the base's conjugate acid increases).

The base in an E2 reaction pulls off the proton, thereby forcing the leaving group to leave. The stronger the base, the faster it can pull off the proton, and the faster the reaction occurs.

Reaction free energy diagrams can provide insight into how the rate of an E2 reaction depends on the strength of the base. Consider the E2 reactions in Equations 9-13 and 9-14, in which 1,2-dichloroethane is the substrate, and the acetate anion $(CH_3CO_2^-)$ and hydroxide anion (HO^-), respectively, are the bases.

$$\text{(9-13)}$$

$$\text{(9-14)}$$

The free energy diagrams for these two reactions are shown in Figure 9-5. For both reactions, ΔG°_{rxn} is negative because a negative charge is better stabilized on Cl than it is on O. Also, ΔG°_{rxn} is more negative when HO^- is the base than when $CH_3CO_2^-$ is, because the negative charge is resonance delocalized in $CH_3CO_2^-$. Thus, as we learned in Section 9.3a using the Hammond postulate, the reaction involving HO^- as the base has a smaller energy barrier and proceeds faster, consistent with the data in Table 9-2.

TABLE 9-2 E2 Reaction Rates in H_2O for the Reaction:

Base					
Relative Reaction Rate	1	3	40	60	353
pK_a of Base–H	4.75	5.2	10.0	9.8	15.7

The reaction is faster when the energy barrier is smaller.

- $CH_3CO_2^{\ominus}$ is the more stable base.
- ΔG_{rxn}° is more negative for the reaction involving $HO^{\ominus}$.

FIGURE 9-5 ΔG_{rxn}° **and E2 energy barriers** Free energy diagrams for the E2 reactions of HO^- (red) and $CH_3CO_2^-$ (blue) with $CIH_2C—CH_2CI$. With HO^- as the base (i.e., with the stronger base), the reaction is energetically more favorable, has a smaller energy barrier, and proceeds faster.

Similar to what we saw with S_N2 and S_N1 reactions, there is a stark contrast between the E2 and E1 reactions with regard to the nature of the attacking species.

> The rate of an E2 reaction depends on the strength of the base, whereas the rate of an E1 reaction is essentially independent of the identity of the base.

The base does not participate in an E1 reaction until the leaving group has left (i.e., until after the rate-determining step), at which point it removes the proton on the substrate. Thus, because the base does *not* help the leaving group to leave, the specific identity of the base has little effect on the reaction rate.

SOLVED problem 9.7 F. G. Bordwell and S. R. Mrozack carried out the following E2 elimination reaction in DMSO, using a variety of bases.

Two of the bases they used are shown below.

$pK_a(Base—H) = 17.1$ $pK_a(Base—H) = 16.2$

A   **B**

The rate constants for these reactions are $55.0 \times 10^{-3} \ M^{-1}s^{-1}$ and $16.0 \times 10^{-3} \ M^{-1}s^{-1}$. Based on the pK_a values given for each base's conjugate acid (measured in DMSO), match each rate constant to the appropriate base.

Think How does the strength of a base correspond to the strength of its conjugate acid? How does the strength of a base affect the rate of an E2 reaction?

478 / CHAPTER 9 Nucleophilic Substitution and Elimination Reactions 1

Solve The conjugate acid of anion **A** is weaker (has a higher pK_a) than that of anion **B**, so **A** is the stronger base. Stronger bases promote faster E2 reactions, so the rate constant involving **A** should be 55.0×10^{-3} M^{-1}s^{-1} and the rate constant involving **B** should be 16.0×10^{-3} M^{-1}s^{-1}.

problem 9.8 Which promotes a faster E2 reaction with bromocyclohexane:

(a) F$^-$ or HO$^-$?

(b) $CH_3CH_2O^-$ or $CF_3CH_2O^-$? Explain your answers.

The E2 rate is highly sensitive to base strength, whereas the E1 rate is not, but what impact does that have on the competition between the two reactions? With a sufficiently strong base, the E2 reaction becomes faster than the corresponding E1 reaction, and with a sufficiently weak base, the E1 reaction becomes faster than the corresponding E2 reaction:

▪ Strong bases tend to favor E2 reactions.
▪ Weak bases tend to favor E1 reactions.

To use these concepts to predict products, we must be able to distinguish strong bases from weak bases. In practice, with all else being equal, E2 reactions are usually favored by bases that are as strong or stronger than HO$^-$. Otherwise, E1 reactions tend to be favored. Thus,

▪ Strong bases are at least as strong as HO$^-$.
▪ Weak bases are weaker than HO$^-$.

Alkoxides (RO$^-$), then, are strong bases, as are deprotonated amines (R$_2$N$^-$) and organometallic reagents such as alkyllithium (R—Li) and Grignard (R—MgX) reagents (Fig. 9-6). Weak bases include halide anions (i.e., F$^-$, Cl$^-$, Br$^-$, and I$^-$), thiolate anions (RS$^-$), and carboxylate anions (RCO$_2^-$). Similarly, uncharged species like H$_2$O, alcohols (ROH), and amines (R$_2$NH) are weak bases.

SOLVED problem 9.9 Is the phenoxide anion, $C_6H_5O^-$, a strong base or a weak base? Will it favor the E1 or the E2 mechanism?

Think What is the pK_a of C_6H_5OH, the conjugate acid of $C_6H_5O^-$? How does this compare to that of H$_2$O, the conjugate acid of HO$^-$? Based on these relative pK_as, is $C_6H_5O^-$ a stronger or weaker base than HO$^-$?

Solve The pK_a values of C_6H_5OH and H$_2$O are 10.0 and 15.7, respectively. C_6H_5OH, therefore, is a stronger acid than H$_2$O, making $C_6H_5O^-$ a weaker base than HO$^-$. Thus, $C_6H_5O^-$ is a weak base and weak bases favor the E1 mechanism over the E2.

problem 9.10 Is NC$^-$ a strong base or a weak base? Will it favor the E1 or the E2 mechanism? Explain. (*Hint:* Consult Table 6-1.)

FIGURE 9-6 Intrinsic base strength Strong bases are at least as strong as HO$^-$. Weak bases are significantly weaker than HO$^-$.

Weak   Strong

$$H_2O, ROH < H_3N, R_2NH < I^{\ominus} < Br^{\ominus} < Cl^{\ominus} < F^{\ominus} < RCO_2^{\ominus} < HS^{\ominus}, RS^{\ominus} < HO^{\ominus}, RO^{\ominus} < H_2N^{\ominus}, R_2N^{\ominus} < H_3C^{\ominus}, R^{\ominus}$$

Base strength generally increases

Stability of negative charge decreases

Difficult for the O atom to approach the C to form a bond

(a)

No significant steric hinderance in a proton transfer

Fast

(b)

FIGURE 9-7 Steric repulsion from the *tert*-butoxide anion (a) Steric hindrance by the CH₃ groups diminishes the nucleophilicity of (CH₃)₃CO⁻. (b) The basicity of (CH₃)₃CO⁻ is unaffected because steric hindrance by the CH₃ groups does not come into play—protons are very small and usually well exposed.

9.3d Strong, Bulky Bases

Based on the guidelines just described, the *tert*-butoxide anion, $(CH_3)_3CO^-$, should be both a strong nucleophile (because it has a full negative charge) and a strong base (because it is stronger than HO^-). Thus, it should favor both S_N2 and E2 reactions. In actuality, in the competition between substitution and elimination, the *tert*-butoxide anion usually favors just E2 products because *$(CH_3)_3CO^-$ is a much weaker nucleophile than we would expect based on charge stability*.

Figure 9-7a shows that the diminished nucleophilicity of the *tert*-butoxide anion stems from the bulkiness of the methyl groups surrounding the nucleophilic O atom. They are so bulky that it is difficult for the O atom to form a bond to a C atom and to displace a leaving group, which slows the reaction. Thus, the S_N2 reaction suffers from **steric hindrance** caused by the methyl groups. The basicity of the anion, however, is essentially unaffected by steric hindrance because protons are very small and are usually well exposed. Consequently, protons are not encumbered by the methyl groups surrounding the O atom (Fig. 9-7b) and can therefore be attacked by O easily.

> Strong, bulky bases (such as the *tert*-butoxide anion) favor E2 reactions over S_N2 reactions.

In addition to the *tert*-butoxide anion, other common, strong, bulky bases include the neopentoxide anion and the diisopropylamide anion:

Strong, bulky bases

The *tert*-butoxide anion   **The neopentoxide anion**   **Lithium diisopropylamide (LDA)**

YOUR TURN 9.3

In the three bases above, identify the groups that contribute to significant steric hindrance.

problem **9.11** Rank the following alkoxide nucleophiles in order of increasing S_N2 reaction rate.

A   **B**   **C**

9.4 Factor 2: Concentration of the Attacking Species

The dependence of each reaction upon the concentration of the substrate and the attacking species is summarized in its respective *empirical rate law* (i.e., its experimentally derived rate law). These were first introduced in Chapter 8 and are shown again in Equations 9-15 through 9-18.

$$\text{Rate}_{S_N2} = k_{S_N2}[\text{Att}^-][\text{R—L}] \qquad (9\text{-}15)$$

$$\text{Rate}_{E2} = k_{E2}[\text{Att}^-][\text{R—L}] \qquad (9\text{-}16)$$

$$\text{Rate}_{S_N1} = k_{S_N1}[\text{R—L}] \qquad (9\text{-}17)$$

$$\text{Rate}_{E1} = k_{E1}[\text{R—L}] \qquad (9\text{-}18)$$

[Att$^-$] represents the concentration of the attacking species, whether it is acting as a nucleophile (in the substitution reactions) or a base (in the elimination reactions). [R—L] is the substrate concentration. The rate constants for the various reactions are different and are therefore denoted as k_{S_N2}, k_{E2}, k_{S_N1}, and k_{E1}.

Note that the S_N2 and E2 reaction rates depend on the concentration of the attacking species, [Att$^-$], whereas the S_N1 and E1 reactions do not. S_N2 and E2 reactions proceed faster with higher concentration of the attacking species because the role of the attacking species is to *force off the leaving group*. In both the S_N1 and E1 mechanisms, by contrast, the attacking species does not participate in the reaction until the leaving group has departed. Therefore, a higher concentration of the attacking species simply means more attacking species waiting for the leaving group to come off, but the reaction rate does not change.

Because the S_N2, S_N1, E2, and E1 reaction rates have different dependencies on the concentration of the attacking species, these concentrations can be used to control the outcome of the competition as follows:

- A high concentration of the attacking species tends to favor both S_N2 and E2 reactions over S_N1 and E1 reactions.
- A low concentration of the attacking species tends to favor both the S_N1 and E1 reactions over S_N2 and E2 reactions.

In practice, *we may assume that a high concentration of the attacking species is present unless a special effort has been made to ensure that it is dilute*. If the concentration is indeed dilute, it should be indicated in the reaction conditions—for example, $\xrightarrow{\text{dil. Br}^{\ominus}}$.

We must be careful when drawing these conclusions about high and low concentrations of the attacking species. If the attacking species is a weak nucleophile, for example, then it is essentially incapable of forcing off the leaving group, regardless of the number of nucleophile molecules or ions present in solution. Therefore, *a high concentration of a weak nucleophile does not substantially promote an S_N2 reaction*. By the same token, *a high concentration of a weak base does not substantially promote an E2 reaction*.

SOLVED problem 9.12 State which reaction(s)—S_N2, S_N1, E2, or E1—are favored by:

(a) A high concentration of HS$^-$
(b) A low concentration of HS$^-$

Think Is HS$^-$ a strong nucleophile or a weak nucleophile? A strong base or a weak base? What relative concentrations favor each reaction? What are the exceptions?

Solve HS$^-$ is a strong nucleophile because it possesses a -1 charge. Therefore, at a high concentration of HS$^-$, S_N2 is favored over S_N1, and at low concentrations, S_N1 is favored over S_N2. HS$^-$ is a weak base because it is weaker than HO$^-$. Therefore, at both low and high concentrations of HS$^-$, E1 is favored over E2.

---

problem 9.13 For each of the following conditions, state whether an S_N1 or S_N2 reaction is favored and whether an E1 or E2 reaction is favored.

(a) A high concentration of HO$^-$
(b) A low concentration of HO$^-$
(c) A high concentration of Br$^-$
(d) A low concentration of Br$^-$
(e) A high concentration of $(CH_3)_3CO^-$
(f) A low concentration of $(CH_3)_3CO^-$

An alkyl tosylate (R—OTs)

FIGURE 9-8 The tosylate leaving group A tosylate group (red), abbreviated OTs, has an excellent leaving group ability.

9.5 Factor 3: Leaving Group Ability

A leaving group must leave whether the reaction is S_N2, S_N1, E2, or E1. Not surprisingly, then, the identity of the leaving group has an effect on the rate of each reaction. As shown in Table 9-3, for example, an alkyl tosylate (R—OTs) (Fig. 9-8) undergoes an S_N2 reaction about 300 times faster than a comparable alkyl chloride (R—Cl). We say, therefore, that the **leaving group ability** of the tosylate anion (TsO^-) in an S_N2 reaction is 300 times greater than that of the chloride anion. More dramatically, Table 9-4 shows that the leaving group ability of TsO^- in an S_N1 reaction is about 10,000 times greater than that of Cl^-!

> S_N1 reactions are more sensitive to leaving group ability than S_N2 reactions are.

What are the reasons for these relative leaving group abilities? And why should the reactions have different sensitivities to leaving group ability? We will answer these questions here in Section 9.5.

TABLE 9-3 Effects of Leaving Group Ability on the Relative Rate of the S_N2 Reaction:

Leaving Group ($L^\ominus$)	$HO^\ominus$, $H_2N^\ominus$, $RO^\ominus$	$F^\ominus$	$Cl^\ominus$	$Br^\ominus$	$I^\ominus$	H_3C—⬡—S—$O^\ominus$ ($TsO^\ominus$)
Relative S_N2 Reaction Rate	~0	0.005	1	50	150	300
pK_a of Conjugate Acid	> ~16	3.2	–7	–9	–10	–2.8

TABLE 9-4 Effects of Leaving Group Ability on the Relative Rate of the S_N1 Reaction:

Leaving Group ($L^\ominus$)	$HO^\ominus$, $RO^\ominus$, $H_2N^\ominus$	$Cl^\ominus$	$Br^\ominus$	H_2O	$I^\ominus$	H_3C—⬡—S—$O^\ominus$ ($TsO^\ominus$)	F_3C—S—$O^\ominus$ ($TfO^\ominus$)
Relative S_N1 Reaction Rate	~0	1	~10	~10	~100	~10,000	~100,000,000
pK_a of Conjugate Acid	> ~16	–7	–9	–1.7	–10	–2.8	–13

9.5a Leaving Group Ability, Charge Stability, and Base Strength

Relative leaving group abilities are governed largely by the stabilities of the leaving groups in the form in which they have come off the substrate.

> The more stable the leaving group is in the form in which it has left, the better its leaving group ability.

For example, H_2O is a much better leaving group than HO^- because water is uncharged. (Recall from Section 6.5 that negatively charged species are intrinsically less stable than analogous uncharged ones.) Similarly, Br^- is a much better leaving group than Cl^- because the larger Br atom can better stabilize the negative charge.

Leaving group ability increases as the stability of the leaving group increases because of the relationship between $\Delta G°_{rxn}$ and the size of the energy barrier, discussed in Section 9.3a. As the leaving group becomes more stable in the form in which it departs, $\Delta G°_{rxn}$ becomes more favorable (less positive). Thus, $\Delta G°^{\ddagger}$ decreases and the rate increases.

Conveniently, relative leaving group abilities correlate rather well with relative base strengths—namely,

> The weaker the leaving group is as a base, the better it tends to be as a leaving group.

Br^-, for example, is a weaker base than Cl^-, and is also a better leaving group.

The correlation between leaving group ability and base strength is particularly useful when it comes to understanding why the tosylate anion (TsO^-), the mesylate anion (MsO^-), and the triflate anion (TfO^-) are among the best leaving groups (Fig. 9-9). Structurally, they are very similar to the bisulfate anion (HSO_4^-), which is the conjugate base of sulfuric acid. Sulfuric acid is among the strongest acids known, in which case the HSO_4^- anion is among the weakest bases.

> Sulfonate anions owe much of their stability to resonance delocalization of the negative charge. Draw all of the resonance structures of the MsO^- anion.

We can apply this reasoning to understand why the rate of nucleophilic substitution and elimination is essentially zero when the leaving group is HO^-. HO^- is a strong base (it has a weak conjugate acid, H_2O), so it is a very poor leaving group. Stronger bases, such as RO^-, H_2N^-, H^-, and R^-, are even poorer leaving groups. H_2O, on the other hand, is an excellent leaving group, because it is a weak base (it has a strong conjugate acid, H_3O^+).

---

problem **9.14** Which is a better leaving group, HCO_2^- or $C_6H_5O^-$? Explain.

As mentioned previously, S_N1 reactions are more sensitive to leaving group ability than S_N2 reactions. You can understand this observation in terms of their respective mechanisms. In the S_N1 reaction, the rate is governed entirely by the step in which the leaving group departs. In the S_N2 reaction, on the other hand, the departure of the leaving group is greatly assisted by the attacking nucleophile.

The effects of leaving group abilities on E2 and E1 reaction rates parallel those of S_N2 and S_N1 reactions. Both E2 and E1 reaction rates are enhanced by better leaving group abilities, but E1 reactions are much more sensitive to leaving group ability than E2 reactions are. (See Solved Problem 9.15.)

Bisulfate anion

Methyl sulfonate (Mesylate) ($MsO^{\ominus}$)

p-Toluene sulfonate (Tosylate) ($TsO^{\ominus}$)

Trifluoromethyl sulfonate (Triflate) ($TfO^{\ominus}$)

FIGURE 9-9 **Sulfonate leaving groups** The excellent leaving group abilities of the mesylate, tosylate, and triflate groups derive from the same type of stability exhibited by HSO_4^-.

SOLVED problem **9.15** Which of the following reactions will proceed *fastest* by the E1 mechanism? Which will proceed *slowest* by the E2 mechanism?

| A | B | C |

Think What are the relative strengths of the bases in **A–C**, and how are the E1 and E2 reactions affected by base strength? What are the relative leaving group abilities of the leaving groups that appear in **A–C**, and how are the E1 and E2 reactions affected by leaving group ability?

Solve The E1 reaction is sensitive to leaving group ability, but is relatively insensitive to base strength. Because TsO⁻ is a better leaving group than Br⁻, **A** proceeds by the E1 mechanism faster than either **B** or **C**. The E2 reaction is sensitive to both leaving group ability and base strength. Because Br⁻ is a worse leaving group than TsO⁻ and NH_3 is a weaker base than HO⁻, **C** proceeds slower by the E2 mechanism than either **A** or **B**.

problem **9.16** Which of the following reactions will proceed the *slowest* by the E1 mechanism? Which will proceed the *fastest* by the E2 mechanism?

| D | E | F |

Given their greater sensitivities to leaving group ability,

> Excellent leaving groups favor S_N1 and E1 reactions over the corresponding S_N2 and E2 reactions.

Br⁻ is generally a very good leaving group, whereas Cl⁻ is a moderate leaving group. On the other hand, F⁻ is a poor leaving group, along with HO⁻, RO⁻, and H_2N^- (Fig. 9-10).

| Poor | Moderate | Good |

$$H^\ominus, H_3C^\ominus < H_2N^\ominus < HO^\ominus, RO^\ominus < F^\ominus < RCO_2^\ominus < Cl^\ominus < Br^\ominus, H_2O < I^\ominus < MsO^\ominus < TsO^\ominus < TfO^\ominus$$

Leaving group ability increases

→

Charge stability generally increases

Poor leaving groups favor S_N2 and E2 reactions over S_N1 and E1 reactions. According to Tables 9-3 and 9-4, however,

> The rates of nucleophilic substitution and elimination reactions under normal conditions are essentially zero unless the leaving group is at least as stable as F⁻.

9.5b Converting a Bad Leaving Group into a Good Leaving Group

Frequently, a desired nucleophilic substitution or elimination reaction is unfeasible because the leaving group is unsuitable—that is, it is not stable enough in the form in which it leaves. Consider, for example, Equation 9-19, which shows that no reaction takes place when Br^- is added to butan-1-ol under normal conditions.

$$\text{OH} + \text{NaBr} \xrightarrow{H_2O} \boxed{\text{No reaction}} \qquad (9\text{-}19)$$

This result should not be surprising, though, because the leaving group in this case is HO^-, which is a very poor leaving group in substitution and elimination reactions. Under acidic conditions, however, a reaction does take place, as shown in Equation 9-20.

$$\text{OH} + \text{HBr} \xrightarrow[\substack{H_2O, H_2SO_4 \\ \text{Reflux 90 min.}}]{} \quad \text{Br} + H_2O \quad (9\text{-}20)$$

The acidic conditions in Equation 9-20 facilitate the substitution reaction because the O atom on the OH group is weakly basic and therefore becomes protonated (remember, proton transfer reactions are *fast*), as shown in Equation 9-21. The leaving group then leaves as H_2O, not HO^-. Being uncharged, H_2O is much more stable than HO^-, and is a *good* leaving group (Fig. 9-10). This is consistent with the fact that H_2O is a much weaker base than HO^-: the pK_a of H_3O^+ is -1.7, whereas the pK_a of H_2O is 15.7.

Mechanism for Equation 9-20

H_2O is a good leaving group.

$$(9\text{-}21)$$

Circle the potential leaving group in both the reactant and product of the following proton transfer reaction:

Next to each, write the form of the leaving group in which it would leave, and label it as either a good leaving group or a bad leaving group.

The protonation of ethers works in much the same way. Under normal conditions (Equation 9-22), methoxybenzene ($C_6H_5OCH_3$) does not undergo nucleophilic substitution or elimination reactions because the leaving group would be an RO^- anion, which is a relatively poor leaving group. A reaction does take place, however, under strongly acidic conditions, as shown in Equation 9-23.

(9-22)

(9-23)

Just as we saw in Equation 9-21, protonation of the O atom generates an excellent leaving group: an uncharged and very weakly basic molecule, C_6H_5OH (Equation 9-24).

Mechanism for Equation 9-23

C_6H_5OH is a good leaving group.

(9-24)

Generating a good leaving group in this fashion can also facilitate elimination reactions, as shown in the **dehydration** reaction in Equation 9-25.

A dehydration reaction

The substrate undergoes a net loss of H_2O.

(9-25)

HO^- would be the leaving group under neutral conditions, but acidic conditions generate a water leaving group, as shown in Equation 9-26.

Mechanism for Equation 9-25

(9-26)

Like alcohols and ethers, amines tend *not* to undergo nucleophilic substitution or elimination reactions under normal conditions (Equation 9-27). To occur, these reactions would require the departure of a very poor leaving group—namely, H_2N^-, HRN^-, or R_2N^-. Protonation of the mildly basic nitrogen of the amine group would make the leaving group better, but *even under acidic conditions, amines tend not to act as substrates in nucleophilic substitution or elimination reactions.* Instead, the reaction stops at the formation of the ammonium ion under normal conditions (Equation 9-28).

$$\text{CH}_3\text{CH}_2\text{CH}_2\text{CH}_2\text{NH}_2 + \text{NaBr} \longrightarrow \boxed{\text{No reaction}} \qquad (9\text{-}27)$$

$$\text{CH}_3\text{CH}_2\text{CH}_2\text{CH}_2\text{NH}_2 + \text{HBr} \longrightarrow \text{CH}_3\text{CH}_2\text{CH}_2\text{CH}_2\overset{\oplus}{\text{NH}_2}\text{Br}^{\ominus} \qquad (9\text{-}28)$$

Ammonium salt
(No S$_N$2)

Why do protonated alcohols tend to undergo nucleophilic substitution and elimination reactions but protonated amines do not? Evidently, the leaving group in a protonated alcohol (i.e., H_2O) is much better than the leaving group in a protonated amine (i.e., R_2NH). This can be explained by the much greater basicity of NH_3 relative to that of H_2O (the pK_a of NH_4^+ is about 10, and the pK_a of H_3O^+ is -1.7).

9.6 Factor 4: Type of Carbon Bonded to the Leaving Group

Although any substrate with a carbon atom bonded to a good leaving group can *theoretically* participate in a nucleophilic substitution or elimination reaction, not all such reactions are *practical* under normal conditions. The nature of the carbon atom bonded to the leaving group can dramatically influence the outcome of these reactions. The relevant factors include the particular hybridization of the carbon, the number of alkyl groups to which it is bonded, and the proximity of double or triple bonds.

9.6a Hybridization of the Carbon Atom Bonded to the Leaving Group

An important characteristic of the carbon atom bonded to the leaving group is its hybridization.

> S$_N$2, S$_N$1, E2, and E1 reactions generally do *not* occur unless the carbon atom bonded to the leaving group is sp^3 hybridized.

(As we will see in Chapter 10, however, substrates in which the leaving group is bonded to an sp^2-hybridized carbon can undergo E2 reactions under extreme conditions.)

One of the main reasons the leaving group needs to be bonded to an sp^3-hybridized carbon atom is that sp^2- and sp-hybridized carbons form stronger σ bonds than sp^3-hybridized carbons do, thereby making it more difficult for the leaving group to leave. This effect stems from the greater s character in the sp^2- and sp-hybridized atomic orbitals (see Section 3.11).

9.6 YOUR TURN

To verify that bond strength increases as the percent s character in the hybridized orbital used to form the bond increases, find the following C—H bond energies in Section 3.11 and write them in the boxes provided here.

$$\underset{sp^3}{\text{C}}\text{—H} \qquad \underset{sp^2}{\text{C}}\text{—H} \qquad \underset{sp}{\text{C}}\text{—H}$$

The π electrons in the C═C and C≡C bonds repel negatively charged nucleophiles.

FIGURE 9-11 Electrostatic repulsion between π electrons and a nucleophile When a leaving group is bonded to an alkene or an alkyne carbon, S_N2 reactions are hindered by repulsion between the negatively charged π electron cloud and the incoming nucleophile.

When the leaving group is on an *sp-* or *sp²*-hybridized carbon atom, S_N1 and E1 reactions are further hindered by the carbon atom's effective electronegativity (Section 3.11). In the rate-determining step of such proposed reactions, departure of the leaving group produces a carbocation (Equations 9-29a and 9-29b).

$$\text{(9-29a)}$$

$$\text{(9-29b)}$$

Unstable carbocations

Because *sp-* and *sp²*-hybridized carbon atoms possess greater *s* character than an *sp³*-hybridized carbon does, these atoms have a higher effective electronegativity, which makes the resulting positive charge less stable. This makes $\Delta G°_{rxn}$ less favorable (more positive) for the rate-determining step, which, as we saw in Section 9.3a, makes the energy barrier larger and slows the rate.

With the leaving group on an *sp-* or *sp²*-hybridized carbon, S_N2 reactions are further hindered by electrostatic repulsion. As we learned in Section 7.6, C═C and C≡C bonds are relatively electron rich, so they repel an incoming negatively charged nucleophile (Fig. 9-11).

problem **9.18** Which carbon atom in the molecule at the right would undergo nucleophilic substitution most readily? Explain.

9.6b The Number of Alkyl Groups on the Carbon Bonded to the Leaving Group

Even with a good leaving group on an *sp³*-hybridized carbon atom, nucleophilic substitution and elimination reactions can be strongly influenced by the number of alkyl groups bonded to that carbon. Recall from Nomenclature 1 that a carbon atom bonded to three alkyl groups is called a *tertiary (3°) carbon*; if it is bonded to two alkyl groups, then it is called a *secondary (2°) carbon*; and if it is bonded to one alkyl group, then it is called a *primary (1°) carbon*. Alternatively, if the carbon is bonded only to hydrogen atoms, then it is called a *methyl carbon* and the substrate takes the form CH_3—L.

The data in Tables 9-5 and 9-6 show that both the S_N1 and S_N2 reaction rates are very sensitive to the type of carbon atom bonded to the leaving group. Specifically,

> As the number of alkyl groups on the carbon atom to which the leaving group is bonded increases, the S_N1 reaction rate sharply *increases*, whereas the S_N2 reaction rate sharply *decreases*.

To understand why the number of alkyl groups affects reaction rates in these ways, we must revisit their mechanisms. For an S_N2 reaction, recall that the nucleophile *forces off the leaving group*. Each alkyl group surrounding the carbon atom bonded to the leaving group adds *steric hindrance*, similar to what we saw with the *tert*-butoxide anion, $(CH_3)_3CO^-$, in Section 9.3d. Very little steric hindrance exists with methyl

TABLE 9-5 Relative Reaction Rates in the S_N1 Reaction:

$$H_2O \;+\; \overset{|}{\underset{|}{C}}-Br \longrightarrow HO-\overset{|}{\underset{|}{C}}- \;+\; HBr$$

Substrate	H₃C—Br	H₃C–CH₂–Br	H₃C–CH(CH₃)–Br	H₃C–C(CH₃)(CH₃)–Br
Type of Carbon	Methyl	1°	2°	3°
Relative S_N1 Reaction Rate	~0	0.08	1	100,000

TABLE 9-6 Relative Reaction Rates in the S_N2 Reaction:

$$Cl^{\ominus} \;+\; \overset{|}{\underset{|}{C}}-Br \longrightarrow Cl-\overset{|}{\underset{|}{C}}- \;+\; Br^{\ominus}$$

Substrate	H₃C—Br	H₃C–CH₂–Br	H₃C–CH(CH₃)–Br	H₃C–C(CH₃)(CH₃)–Br
Type of Carbon	Methyl	1°	2°	3°
Relative S_N2 Reaction Rate	4,000	80	1	~0

and primary carbons, so it is relatively easy for the nucleophile to form a bond to the carbon attached to the leaving group. With secondary carbons, the increased steric hindrance causes S_N2 reactions to proceed somewhat more slowly. With tertiary carbons, S_N2 reactions are essentially nonexistent because the excessive steric hindrance from the alkyl groups makes the carbon atom bonded to the leaving group inaccessible (Fig. 9-12).

In an S_N1 reaction, the reaction rate is dictated by the ability of the leaving group to leave, so steric hindrance does not come into play. Instead, *each additional alkyl group helps*

Excessive steric hindrance blocks the attacking nucleophile.

Methyl   1°   2°   3°

Steric hindrance increases

S_N2 reaction rate decreases

FIGURE 9-12 How steric hindrance affects the rate of an S_N2 reaction Steric hindrance of the nucleophile in an S_N2 reaction increases as the number of alkyl groups surrounding the C atom bonded to the leaving group increases. As steric hindrance increases, the S_N2 reaction rate decreases.

FIGURE 9-13 How S_N1 and E1 reaction rates depend on carbocation stability Carbocation stability increases as the number of alkyl groups bonded to the positively charged C of a carbocation increases. In the substrate, therefore, each alkyl group surrounding the C atom bonded to the leaving group increases the rate of S_N1 and E1 reactions.

Methyl substrate | Primary substrate | Secondary substrate | Tertiary substrate

Inductive stabilization from one R group | Inductive stabilization from two R groups | Inductive stabilization from three R groups

Stability of carbocation increases

Rate of S_N1 and E1 reactions increases

the leaving group leave by inductively stabilizing the carbocation intermediate (Fig. 9-13). A tertiary carbocation is the most stable because the maximum number of alkyl groups is bonded to the carbon atom bearing the formal positive charge. Methyl and primary carbocations are the least stable. Consequently, S_N1 reactions occur most readily when the leaving group is bonded to a tertiary carbon and do not readily occur when the leaving group is bonded to a methyl or primary carbon. S_N1 reactions can occur when the leaving group is bonded to a secondary carbon, but they tend to proceed slowly.

E1 reaction rates depend on the number of alkyl groups in the same way as S_N1 reaction rates:

> The rate of an E1 reaction increases as the number of alkyl groups on the carbon bonded to the leaving group increases.

E1 and S_N1 reaction rates depend on carbocation stability in the same way because they both have *exactly* the same rate-determining step. (See Solved Problem 9.19.)

SOLVED problem 9.19 Which of the substrates at the right undergoes an E1 reaction faster?

Think Which rate-determining step produces a more stable carbocation? How does this affect the E1 reaction rate?

Solve The two rate-determining steps are as follows:

A **B**

A

B

More stable carbocation

The carbocation indicated is more stable, because its additional CH_3 group stabilizes the electron-deficient C^+. Thus, substrate **B** reacts faster via the E1 mechanism.

Protons are typically well exposed on 1°, 2°, and 3° substrates

FIGURE 9-14 Accessibility of the proton in E2 reactions Unlike in S_N2 reactions, the addition of alkyl groups to the carbon bonded to the leaving group does not provide much steric hindrance in E2 reactions. Even with tertiary C atoms, the protons on adjacent C atoms remain well exposed to the base.

problem **9.20** Which substrate, **C** or **D**, undergoes an E1 reaction faster? Explain.

Unlike S_N2, S_N1, and E1 reactions, *E2 reactions are relatively insensitive to the number of alkyl groups on the carbon bonded to the leaving group.* That is,

> Substrates with leaving groups on primary, secondary, and tertiary carbons can usually undergo E2 reactions quickly.

This may be surprising because the attacking species in an E2 reaction *must force off the leaving group*, just as in an S_N2 reaction. In an E2 reaction, however, the attacking species does so by pulling off a proton from a carbon *adjacent* to the one bonded to the leaving group. Because a proton is small and generally well exposed, *steric hindrance tends to be a minor factor in E2 reactions*, even with the leaving group on a 3° carbon (Fig. 9-14).

Given that S_N2, S_N1, E2, and E1 reaction rates depend differently on the number of alkyl groups bonded to the carbon with the leaving group, the type of substrate (i.e., methyl, primary, secondary, or tertiary) can dramatically influence the outcome of the competition. Specifically, as summarized in Table 9-7, certain reactions are feasible for some substrates and others are not. Notice, for example, that S_N2 reactions are generally feasible for methyl, primary, and secondary substrates, because they have little steric hindrance surrounding the carbon bonded to the leaving group. By contrast, S_N2 reactions are *not* feasible for tertiary substrates because of the excessive steric hindrance. S_N1 and E1 reactions are feasible for secondary and tertiary substrates, which form relatively stable carbocations upon the departure of the leaving group, but they are *not* feasible for methyl or primary substrates, which form relatively unstable carbocations. E2 reactions are feasible for all of the substrates except those in which the leaving group is bonded to a methyl group.

TABLE 9-7 Feasibility of S_N2, S_N1, E2, and E1 Reactions for Each Type of Substrate

Type of Substrate (R—L)	S_N2	S_N1	E2	E1
CH₃—L (methyl)	✓			
RCH₂—L (1°)	✓		✓	
R₂CH—L (2°)	✓	✓	✓	✓
R₃C—L (3°)		✓	✓	✓

problem **9.21** Why are E2 reactions feasible for substrates in which the leaving group is bonded to a primary, secondary, or tertiary carbon, but not for substrates in which the leaving group is bonded to a methyl carbon?

problem **9.22** Which substrate, **X**, **Y**, or **Z**, will undergo an S_N2 reaction the fastest? Which will undergo an S_N1 reaction the fastest? Which will undergo an E1 reaction the fastest? Explain.

X   Y   Z

Exceptions arise when the leaving group is bonded to an **allyl** (CH_2=CH—CH_2—L) or **benzyl** (C_6H_5—CH_2—L) C atom, as shown in Equations 9-30 and 9-31.

1° Allyl carbon

Resonance–stabilized charge

Allyl cation

(9–30)

Resonance-stabilized charge

1° Benzyl carbon   **Benzyl cation**

(9–31)

If the leaving group is bonded to a primary allyl or benzyl carbon, then both S_N2 and S_N1 reactions are feasible.

Normally, S_N1 mechanisms are unavailable to substrates in which the leaving group is bonded to a primary carbon because the resulting primary carbocation is too unstable. However, if the leaving group is bonded to a primary allyl or benzyl carbon, then the resulting carbocation is heavily stabilized by resonance with the adjacent π electrons.

YOUR TURN 9.7

Draw all possible resonance structures for the benzyl cation to illustrate how resonance stabilizes the charge.

SOLVED problem 9.23   Which substrate, **A** or **B**, would undergo an S_N1 reaction more rapidly?

A   B

Think What is the carbocation intermediate involved in each S_N1 mechanism? Which one is more stable? How does this affect the S_N1 reaction rate?

Solve In each case, the carbocation is generated by loss of the leaving group, Cl^-.

Strong resonance contributor

A

No resonance stabilization

B

The carbocation from **A** is resonance stabilized by a lone pair of electrons on the neighboring O atom, yielding a strong resonance contributor with all atoms having their octet. The carbocation from **B** is not resonance stabilized, so it is less stable. Thus, the carbocation from **A** is generated faster than the carbocation from **B**. Since these are the rate-determining steps of the respective S_N1 reactions, **A** undergoes the S_N1 reaction faster than **B**.

Rotaxanes: Exploiting Steric Hindrance

As we have just seen, steric hindrance is a major factor that helps govern the outcome of an $S_N2/S_N1/E2/E1$ competition. Steric hindrance generally has a negative connotation, though, because the role of steric hindrance is to *prevent* a process from occurring. However, steric hindrance can, in fact, be exploited to make possible the fabrication of new molecules such as *rotaxanes*. **Rotaxanes** are a class of compounds in which a dumbbell-shaped molecule is threaded through a large cyclic molecule called a *macrocycle*, as shown schematically on the right. The graphic on the bottom is the crystal structure of an actual rotaxane.

Dumbbell-shaped molecule

Macrocycle

Once this particular rotaxane is formed, the dumbbell-shaped molecule can rotate about its axis relative to the macrocycle, much like a wheel and axle. The macrocycle can also slide back and forth along the axis of the dumbbell. The dumbbell-shaped molecule doesn't slip out of the macrocycle entirely, however, because of the steric hindrance that is induced by the bulky end groups.

By applying an external stimulus (e.g., a voltage), the geometries of some rotaxanes can be manipulated in specific ways, giving rise to a variety of interesting potential applications. These include molecular switches, memory storage devices, and molecular muscles.

The rotaxane motif has also been shown to exist in biological systems. One example is microcin J25, which is produced by *Escherichia coli* during periods of nutrient depletion. Microcin J25 belongs to a class of peptides called "lasso peptides," in which a C-terminal tail is threaded through an N-terminal cyclized ring.

---

problem **9.24** Will substrate **C** or **D** undergo an E1 reaction faster? Explain.

9.7 Factor 5: Solvent Effects

There are two types of solvent in which S_N2, S_N1, E2, and E1 reactions can take place: *polar protic solvents* and *polar aprotic solvents*. The solvent must be polar to dissolve the reactants efficiently, given that the reactants are typically polar or ionic. As we learned in Section 2.9, protic solvents (e.g., water and alcohols) possess hydrogen-bond donors, whereas aprotic solvents (e.g., dimethyl sulfoxide [DMSO], *N,N*-dimethylformamide [DMF], and acetone) do not. Here in Section 9.7 we will examine the impact that the choice of solvent has on the competition between S_N2, S_N1, E2, and E1 reactions, as well as on relative nucleophile strengths.

9.7a Protic Solvents, Aprotic Solvents, and the $S_N2/S_N1/E2/E1$ Competition

Experimentally, the choice of solvent can have a significant influence on the outcome of nucleophilic substitution and elimination reactions.

- Polar *aprotic* solvents tend to favor S_N2 and E2 reactions.
- Polar *protic* solvents tend to favor S_N1 and E1 reactions.

These effects are illustrated with the data listed in Tables 9-8 and 9-9.

TABLE 9-8 Reaction Rates in Various Solvents for the S$_N$2 Reaction:

Solvent	Type of Solvent	Relative S$_N$2 Reaction Rate
CH$_3$OH	Protic	1
H$_2$O	Protic	7
(CH$_3$)$_2$S=O, DMSO	Aprotic	1,300
HCON(CH$_3$)$_2$, DMF	Aprotic	2,800
CH$_3$CN	Aprotic	5,000

Anion very strongly solvated (heavily stabilized)

Protic solvent

(a)

TABLE 9-9 Reaction Rates in Various Solvents for the S$_N$1 Reaction:

Solvent	Type of Solvent	Relative S$_N$1 Reaction Rate
CH$_3$OH	Protic	7,400,000
HCONH$_2$	Protic	1,200,000
HCON(CH$_3$)$_2$, DMF	Aprotic	12.5
CH$_3$CON(CH$_3$)$_2$	Aprotic	1

Anion weakly solvated (not heavily stabilized)

Aprotic solvent

(b)

FIGURE 9-15 Comparing solvation of anions in protic and aprotic solvents (a) Solvation of an anion by water, a polar protic solvent. The large concentration of positive charge on a well-exposed hydrogen atom enables water to strongly solvate negative charges. (b) Solvation of an anion by acetone, a polar aprotic solvent. The partial positive end of the net dipole is buried inside the solvent molecule, which severely decreases acetone's ability to solvate negative charges.

These solvent effects are due to two factors: solvation of the *attacking species* and solvation of the *leaving group*. As we learned in Section 2.9, *solvation of anions is very strong in protic solvents but is much weaker in aprotic solvents* (Fig. 9-15). Solvation occurs when solvent molecules bind to solute molecules through the various intermolecular interactions. Solvation of negative ions by protic solvents is dominated by strong ion–dipole interactions between the solvent and anion (Fig. 9-15a), which substantially stabilizes the nucleophile. As we learned in Section 9.3, an increase in the stability of a nucleophile slows an S$_N$2 reaction, but essentially leaves the S$_N$1 rate unchanged. With aprotic solvents, however, these ion–dipole interactions are much weaker because the positive end of the net dipole is typically buried inside the solvent molecule (Fig. 9-15b). *Steric hindrance* keeps the anion away from that partial positive charge. As a result, aprotic solvents do not stabilize nucleophiles as much as polar protic solvents do, allowing S_N2 reactions to remain faster than S_N1 reactions. Putting these ideas together:

> Protic solvents weaken nucleophiles substantially via *solvation*; aprotic solvents do not.

The strong solvation by polar protic solvents also serves to weaken nucleophiles as a result of the "cage" that is formed by the solvent molecules around the nucleophile. To form a bond with a substrate, the nucleophile must first shed some of those solvent molecules.

---

problem **9.25** In which solvent—dimethyl sulfoxide or ethanol—does this reaction proceed faster via an S_N1 mechanism? In which solvent does it proceed faster via an S_N2 mechanism? Explain.

The second reason for the solvent effects shown in Tables 9-8 and 9-9 derives from the different abilities of the solvent to solvate the *leaving group*. Recall that the S_N1 reaction rate is dictated entirely by the departure of the leaving group. That step often produces an anion along with the carbocation. In a protic solvent, the anion is stabilized heavily by the strong solvation that ensues, but in an aprotic solvent it is not. Therefore,

> A protic solvent will dramatically speed up the rate-determining step of an S_N1 reaction by solvating the leaving group; an aprotic solvent will not.

S_N2 reaction rates, on the other hand, are not affected as greatly, due to the assistance provided by the attacking species. In S_N2 reactions, recall that the attacking species is viewed as forcing the leaving group to leave.

---

problem **9.26** In which solvent—ethanol or acetone—does this reaction proceed faster by an S_N2 mechanism? Explain.

Similar reasons explain why protic solvents tend to favor E1 reactions over E2 reactions, whereas aprotic solvents tend to favor E2 over E1. Strong solvation of the base by a protic solvent weakens the base, thereby slowing the E2 reaction rate. At the same time, protic solvents enhance the ability of the leaving group to leave in the rate-determining step of an E1 reaction. Conversely, aprotic solvents favor E2 reactions over E1 because the base is not weakened as much. The leaving group is slow to depart on its own, moreover, because it is so poorly solvated by the aprotic solvent.

---

problem **9.27** In which solvent, **Y** or **Z**, does the reaction in Problem 9.26 proceed faster by the E1 mechanism? Explain.

TABLE 9-10 S_N2 Reaction Rate in Ethanol for the Reaction:

$$\text{Nu:}^{\ominus} \quad H_3C{-}Br \longrightarrow Nu{-}CH_3 + Br^{\ominus}$$

Nucleophile	CH_3OH	$F^{\ominus}$	$CH_3CO_2^{\ominus}$	$Cl^{\ominus}$	NH_3	$N_3^{\ominus}$
Relative Reaction Rate	1	500	20,000	23,000	320,000	600,000
Nucleophile	$Br^{\ominus}$	$CH_3O^{\ominus}$	$NC^{\ominus}$	Ph_3P	$HS^{\ominus}$	
Relative Reaction Rate	620,000	2,000,000	5,000,000	10,000,000	100,000,000	

9.7b Relative Nucleophilicities in Protic and Aprotic Solvents

We have just seen that solvation in protic solvents has the effect of weakening nucleophiles substantially. The solvation of anions by protic solvents is so dramatic that it can *reverse* the relative strengths of some nucleophiles. This can be seen in Table 9-10, which lists the S_N2 reaction rates of various nucleophiles in ethanol (a protic solvent).

Note, for example, that the nucleophilicity of Br^- in ethanol is about 30 times greater than that of Cl^- (i.e., 620,000 vs. 23,000). This is contrary to DMF, an aprotic solvent (Table 9-1), in which Cl^- is a stronger nucleophile than Br^- (i.e., 2 vs. 1). This reversal occurs in protic solvents because Cl^- is substantially smaller than Br^- (Cl is above Br in the periodic table), which means the negative charge is more concentrated when it is on Cl^-. With a higher concentration of negative charge, Cl^- is solvated much more strongly in a protic solvent and is thus more weakened as a nucleophile.

YOUR TURN 9.8

Compare Tables 9-1 and 9-10 to find another pair of nucleophiles (other than Cl^- and Br^-) whose *relative* nucleophilicities in ethanol are the reverse of what they are in DMF. What does this observation suggest about which of the two species is more strongly solvated in ethanol?

In general,

We see a reversal of nucleophilicity in protic versus aprotic solvents when we compare two nucleophiles that have negative charges localized on atoms in different rows of the periodic table.

This is illustrated further in Solved Problem 9.28.

SOLVED problem 9.28 Is HS^- or HO^- a stronger nucleophile in water?

Think Which is a stronger nucleophile *intrinsically* (i.e., without considerations of solvation)? Is water a protic or an aprotic solvent? Is the solvation of HO^- in water much different from the solvation of HS^-?

Solve Without considering solvation, HO^- is a stronger nucleophile because O does not stabilize the negative charge as well as S. Water is a *protic solvent*, however, so solvation is very important. The solvation of HO^- is much stronger than that of HS^- because the atoms on which the negative charge is localized (O and S) are in different rows of the periodic table. We therefore expect a reversal of the nucleophile strengths in a protic solvent versus an aprotic solvent. Thus, HS^- is a stronger nucleophile than HO^- in water.

You Can't Always Want What You Get: How an Enzyme Can Manipulate the Reactivity of a Substrate

We have just seen how we can exploit solvent effects to manipulate the outcome of an $S_N2/S_N1/E2/E1$ competition. Similarly, enzymes can facilitate reactions that are otherwise difficult or impossible under physiological conditions, such as the following isomerization of 5-androstene-3,17-dione:

These hydrogen bonds stabilize the enolate anion intermediate.

Normally, this reaction would require a strong base such as HO^- in the first step because the pK_a of 5-androstene-3,17-dione is ~20. But soil bacteria such as *Pseudomonas putida* possess an enzyme called ketosteroid isomerase, which allows this transformation to be carried out under mild conditions, using the weakly basic CO_2^- group of the aspartate-40 (Asp40) residue. The key is that ketosteroid isomerase enhances the acidity of 5-androstene-3,17-dione by stabilizing its conjugate base via two hydrogen bonds in the enzyme's active site involving the tyrosine-16 (Tyr16) and aspartic acid-103 (Asp103) amino acid residues. These hydrogen bonds can be seen at the right in the crystal structure of the complex between ketosteroid isomerase and a model substrate, equilenin.

Importantly, the environment within the active site also serves to diminish the acidity of the Asp103 residue; otherwise, the acidic H would be absent, making it impossible for the residue to supply the necessary hydrogen-bond donor. With more than 20 nonpolar amino acid residues in the active site, water is largely excluded, so the aspartate anion is not heavily solvated.

The ways in which ketosteroid isomerase uses hydrogen bonding and solvation (or lack thereof) in its active site to manipulate the reactivity of a substrate are important to understand because these modes of action are exploited by other enzymes as well. As a result, studies of ketosteroid isomerase allow us to have a deeper understanding of how enzymes function in general.

---

problem **9.29** Is F^- or I^- a stronger nucleophile in ethanol? Explain.

The tremendous solvation that occurs in protic solvents causes some negatively charged nucleophiles to be weaker than some uncharged nucleophiles. Notice in Table 9-10, for example, that NH_3 is more than 10 times stronger than Cl^- in ethanol (i.e., 320,000 vs. 23,000), making NH_3 a moderately strong nucleophile. More impressively, triphenylphosphine (Ph_3P) is more than 400 times stronger than Cl^- (i.e., 10,000,000 vs. 23,000).

If we compare nucleophiles with localized negative charges on atoms in the *same row of the periodic table*, then protic solvents do *not* reverse nucleophilicities. In both protic and aprotic solvents, for example, nucleophilicities increase in the order $F^- < CH_3O^- < NC^-$. This can be explained, once again, by solvation factors. The atoms upon which the negative charge resides in these three ions—F, O, and C—are not very different in size because they appear in the same row of the periodic table. Consequently, the *concentration* of charge on the respective atoms is roughly the same. Therefore, even though each ion is solvated quite strongly, they are not solvated very differently from each other—certainly not enough to reverse their nucleophilicities.

---

problem **9.30** Would you expect H_2S to be a stronger or weaker nucleophile than H_3N in ethanol? Explain.

9.8 Factor 6: Heat

Products from an S_N2 reaction are often formed along with abundant E2 products. Similarly, S_N1 and E1 reaction products are often formed together. This occurs because conditions that favor S_N2 reactions generally favor E2 reactions as well—for example, a high concentration of a strong attacking species, a leaving group bonded to a primary carbon, and a polar aprotic solvent. S_N1 and E1 reactions are favored by the same conditions, too—namely, a weak attacking species (or a low concentration of a strong attacking species); a good leaving group; a leaving group on a tertiary, allylic, or benzylic carbon; and a polar protic solvent.

When substitution and elimination reactions are both favored under a specific set of conditions, it is often possible to influence the outcome by changing the temperature under which the reactions take place. Specifically,

> Increasing the temperature of the reaction generally tips the balance in the favor of elimination.

This effect can be seen in Equations 9-32 and 9-33, in which 2-bromopropane reacts with hydroxide anion at 45 °C and 100 °C, respectively. Substitution and elimination products are formed in roughly equal amounts at the lower temperature; elimination is favored at the higher temperature.

This temperature effect can be understood by considering *entropy*, which, as we discussed in Section 2.7, is often thought of as a measure of disorder. If we examine an *overall* nucleophilic substitution reaction alongside an *overall* elimination reaction, as in Equations 9-34a and 9-34b, then we can see that their reactants are identical, but they differ in the number of product species: There are two product species in a substitution reaction and three in an elimination reaction. As a result, *there is significantly more disorder in the products of an elimination reaction than in the products of a competing substitution reaction.* In other words, the standard change in entropy of a reaction, ΔS°_{rxn}, is more positive for an elimination reaction than for a substitution reaction.

Two product species

Substitution (9-34a)

Br
2-Bromopropane

$^{\ominus}$OH

Elimination (9-34b)

Three product species
= greater entropy

For the reaction below, identify the products that have greater entropy. Label the products of substitution and the products of elimination.

Recall from Section 6.3 (Equation 6-18) that $\Delta G^{\circ}_{rxn} = \Delta H^{\circ}_{rxn} - T\Delta S^{\circ}_{rxn}$. Because ΔS°_{rxn} is more positive for elimination than substitution, an increase in T causes a larger *decrease* in ΔG°_{rxn} for an elimination reaction. The Hammond postulate (Section 9.3a) would then suggest that $\Delta G^{\circ\ddagger}$ is lowered—and reaction rate is increased—more for elimination than for substitution.

There are a variety of ways to indicate that a reaction is run at high temperatures. One is simply to write the temperature at which the reaction is run underneath the reaction arrow (e.g., $\xrightarrow{90\ °C}$). Presumably that temperature will be significantly above room temperature. Alternatively, you can indicate that heat has been added to the reaction mixture by writing either "heat" or its shorthand notation, Δ, underneath the reaction arrow (e.g., $\xrightarrow{\text{heat}}$ or $\xrightarrow{\Delta}$).

problem **9.31** Suppose this reaction produces a substantial amount of both products at room temperature. If the temperature at which the reaction is run is raised, then which product will be favored? If the temperature is lowered, then which product will be favored?

9.9 Predicting the Outcome of Nucleophilic Substitution and Elimination Reactions

If all of the factors we have discussed were to favor the same reaction, then predicting the outcome of the competition between the S_N2, S_N1, E2, and E1 reactions would be pretty easy. Generally speaking, however, those factors do *not* all pull in the same direction, so predicting the major products is not always clear cut. In many cases, definitively knowing the major product requires carrying out the reaction in question and measuring the relative concentrations of each product. However, by systematically evaluating all of the factors we have examined thus far, we can often reasonably *predict* the major products.

Step 1: Does the substrate have a suitable leaving group?

None of the four reactions can take place readily unless the leaving group is at least as stable as F^-. If the potential leaving group is HO^-, RO^-, H_2N^-, H^-, or R^-, for example, then no practical nucleophilic substitution or elimination reaction can occur under normal conditions (Tables 9-3 and 9-4). If it does have a suitable leaving group, then proceed to Step 2.

Step 2: How many alkyl groups are attached to the carbon bonded to the leaving group?

The type of carbon—methyl, primary, secondary, or tertiary—has a great influence on which of the four reactions dominates. As we saw in Table 9-7, *if the leaving group is on a primary carbon, then the S_N1 and E1 reactions are not feasible*, regardless of which reactions the other factors favor. Not only would the S_N1 and E1 mechanisms proceed through a very unstable primary carbocation intermediate, but the S_N2 mechanism would be favored by very little steric hindrance. Exceptions arise for allylic and benzylic substrates, whose corresponding carbocations are resonance stabilized. *If the leaving group is on a tertiary carbon, then the S_N2 reaction is not feasible*, due to the excessive steric hindrance introduced by the three alkyl groups. Finally, *if the leaving group is on a secondary carbon, then all four reactions must be considered*.

Step 3: Examine the influences of the four factors:

- the strength of the attacking species as a nucleophile and as a base,
- the concentration of the attacking species,
- the leaving group ability, and
- the effects of the solvent.

Temporarily ignoring the influence of heat, we essentially give equal weight to each of those factors and tally up a score to determine the "winner" of the competition.

Step 4: If both substitution and elimination reactions appear to be favored, then factor in the influence of heat.

High temperatures tend to favor elimination, whereas low temperatures tend to favor substitution.

To illustrate this systematic method, let's first try to predict the outcome of the reaction in Equation 9-35.

(S)-2-Chloropentane $+$ NaBr $\xrightarrow{\text{DMSO}}$ (9-35)

The attacking species is Br^-, the substrate is (*S*)-2-chloropentane, the solvent is DMSO, and the leaving group would be Cl^-.

In Step 1, we need to determine whether the substrate has a suitable leaving group. Cl$^-$ is more stable than F$^-$, so Cl$^-$ has adequate leaving group ability (Tables 9-3 and 9-4).

In Step 2, we need to determine how many alkyl groups in the substrate are attached to the C atom bonded to the leaving group. The leaving group in (S)-2-chloropentane is on a 2° carbon, so we must consider all four mechanisms (Table 9-7).

For Step 3, we can construct a table (Table 9-11) to keep track of the reactions favored by each factor. We begin by considering the strength of the attacking species, Br$^-$. Because it has a full negative charge, Br$^-$ is a strong nucleophile, thereby favoring S$_N$2 reactions over S$_N$1 reactions. This is recorded in Table 9-11 with a check mark. On the other hand, Br$^-$ is a weak base because it is substantially weaker than HO$^-$. Therefore, the E1 reaction is favored over E2. Again, this is recorded in Table 9-11 with a check mark.

Next, what influence does the attacking species' concentration have on the competing reactions? Because we are not explicitly told that Br$^-$ is dilute, we may assume that it is concentrated. Given that Br$^-$ is a strong nucleophile, this high concentration will favor the S$_N$2 reaction over the S$_N$1, and we record this in Table 9-11 with another check mark. On the other hand, because Br$^-$ is a weak base, a high concentration does *not* favor E2; we can remind ourselves of this by placing a dash in the E2 column in Table 9-11.

The leaving group ability of Cl$^-$ is considered next. Cl$^-$ is a moderate leaving group (Tables 9-3 and 9-4), so it does not discriminate greatly between S$_N$1 and S$_N$2 or between E1 and E2. We therefore record a check in all four columns in Table 9-11 (alternatively, we could leave all four columns blank).

Finally, what are the solvent effects? DMSO is a polar *aprotic* solvent, so it favors S$_N$2 and E2 reactions over their corresponding S$_N$1 and E1 reactions. Once again, this is recorded in Table 9-11.

We essentially give equal weight to these four factors, so tallying up the number of factors that favor each of the four reactions gives four "votes" to the S$_N$2 reaction, and the other three reactions receive two or fewer. As a result, the S$_N$2 reaction is heavily favored, which means we can skip Step 4. The effect of heat must be considered only if there is a tie between substitution and elimination.

To predict the major product, enter the specific reactants into the S$_N$2 mechanism, as is shown in Equation 9-36. Notice that *stereochemistry is important* here, given that the C atom bonded to the leaving group is a stereocenter. Backside attack of the nucleophile ensures that the only stereoisomer produced is the one shown, which is (R)-2-bromopentane.

TABLE 9-11 Summary Table for the Reaction in Equation 9-35

Factor	S$_N$1	S$_N$2	E1	E2
Strength		✓	✓	
Concentration		✓		–
Leaving group	✓	✓	✓	✓
Solvent		✓		✓
Total	1	4	2	2

Mechanism for Equation 9-35

(S)-2-Chloropentane   →(S$_N$2)→   (R)-2-Bromopentane   + :Cl:$^{\ominus}$ (9-36)

Now, predict the major product(s) of the reaction in Equation 9-37.

3-Bromo-3-ethylpentane   —CH$_3$OH, Δ→   (9-37)

TABLE 9-12 Summary Table for the Reaction in Equation 9-37

Factor	S_N1	S_N2	E1	E2
Strength	✓		✓	
Concentration				–
Leaving group	✓		✓	
Solvent	✓		✓	
Total	3		3	0

In this reaction, 3-bromo-3-ethylpentane is the substrate and Br^- is the leaving group. The substrate is dissolved in methanol, so methanol must be the attacking species as well as the solvent. This type of reaction is called **solvolysis** because the solvent acts as a reactant.

For Step 1, Br^- is a better leaving group than F^-, so it is indeed a suitable leaving group for nucleophilic substitution and elimination. For Step 2, the leaving group is on a tertiary carbon, so S_N2 reactions can be ruled out (Table 9-7). The S_N2 column in Table 9-12 is therefore grayed out.

For Step 3, we determine which reactions are favored by the remaining factors (except heat), and enter the results into Table 9-12.

The attacking species, CH_3OH, is a weak nucleophile and a weak base, so S_N1 and E1 are favored over S_N2 and E2. The concentration of the nucleophile appears to be high (it is, after all, the solvent), which normally would favor S_N2 and E2. However, the attacking species is weak, both as a nucleophile and as a base, so CH_3OH cannot force off the leaving group, regardless of concentration (Section 9.4). As a reminder, we enter a dash in Table 9-12. Next, Br^- is a good leaving group, so this factor favors S_N1 and E1 over S_N2 and E2. Finally, methanol is a polar protic solvent, which favors the S_N1 and E1 reactions over the S_N2 and E2.

The final tally of the results in Table 9-12 shows that both the S_N1 and E1 reactions are heavily favored. This suggests that both the S_N1 and E1 products should be formed in significant amounts. To break the tie, we must consider the effect of heat (Step 4). The Δ symbol beneath the reaction arrow indicates that the reaction is heated, which tips the balance in the favor of elimination over substitution (Section 9.8). We therefore conclude that the E1 reaction (Equation 9-38) leads to the major product, and the S_N1 reaction (Equation 9-39) leads to a minor product.

E1 mechanism for Equation 9-37

(9-38)

Major product

3-Ethylpent-2-ene

S_N1 mechanism for Equation 9-37

(9-39)

Minor product

3-Methoxy-3-ethylpentane

A proton can be removed from one of three possible carbons of 3-bromo-3-ethylpentane in an E1 reaction. In this case, however, those carbons are chemically equivalent, so removal of any one of them leads to the same alkene product (Equation 9-38). (In Section 9.10 we discuss how to predict the major elimination product when such protons are chemically distinct.)

Equation 9-39 shows that there are three steps for this S_N1 mechanism instead of two. The nucleophile that attacks in the second step of the S_N1 reaction, CH_3OH, is uncharged, so it develops a positive charge on oxygen. That positive charge is removed in the third step, which is a simple proton transfer involving the solvent.

What is the major product of the reaction in Equation 9-40? The substrate contains a tosylate (OTs) leaving group, the attacking species is the *tert*-butoxide anion, $(CH_3)_3CO^-$, and the solvent is *N,N*-dimethylformamide (DMF).

(9-40)

For Step 1, the TsO leaving group is suitable for nucleophilic substitution and elimination; in fact, it is among the best leaving groups possible. For Step 2, the leaving group is bonded to a 2° carbon, so all four reactions must be considered.

For Step 3, consider all of the remaining factors except heat. The attacking species, $(CH_3)_3CO^-$, is a strong base, so it favors the E2 reaction over the E1. We would normally expect alkoxides (RO^-) to favor S_N2 reactions as well, but because of the excessive bulkiness of $(CH_3)_3CO^-$, it favors only the E2 reaction. The attacking species' concentration is assumed to be high because we are not told otherwise. This favors the E2 reaction because $(CH_3)_3CO^-$ is a strong base, but it does *not* favor the S_N2 reaction because of the diminished nucleophilicity of $(CH_3)_3CO^-$. TsO$^-$ is an excellent leaving group, so it favors the S_N1 reaction over the S_N2 and the E1 reaction over the E2. Finally, DMF is a polar aprotic solvent, which favors the S_N2 and E2 reactions over the S_N1 and E1 reactions. With these factors tallied at the bottom of Table 9-13, we can conclude that the E2 reaction will predominate.

TABLE 9-13 Summary Table for the Reaction in Equation 9-40

Factor	S_N1	S_N2	E1	E2
Strength				
Concentration				
Leaving group				
Solvent				
Total	1	1	1	3

9.10 YOUR TURN

Complete Table 9-13 by placing check marks in the appropriate boxes, using the information provided above.

Notice in the substrate that there is a H atom on each C adjacent to the C atom bonded to the leaving group. Thus, it may appear that there are two possible E2 products, depending upon which of those protons is removed. The E2 reaction is *stereospecific*, however, favoring the anti-coplanar conformation of the substrate (Section 8.5d). In this substrate, only the proton indicated in Equation 9-41 can be anti to the leaving group, giving rise to only one E2 product.

Mechanisms for Equation 9-40

This H is anti to the leaving group.

This H is *not* anti to the leaving group.

E2 Reaction

(9-41)

Major product

SOLVED problem **9.32** Draw the complete mechanism and predict the products for the reaction at the right. In this reaction, (R)-(bromomethyl-d)-benzene is dissolved in methanol.

(R)-(Bromomethyl-d)-benzene

Think Is there a suitable leaving group? Does the type of carbon bonded to the leaving group rule out any of the four reactions? What is the attacking species? What is the solvent? Which reaction does each of the factors favor?

Solve The substrate is (R)-(bromomethyl-d)-benzene. This is a solvolysis reaction because methanol is both the attacking species and the solvent. The leaving group, Br^-, is suitable for nucleophilic substitution and elimination (it has much better leaving group ability than F^-). The leaving group is on a 1° carbon, suggesting that we might not have to consider the S_N1 and E1 options. The leaving group is bonded to a *benzylic carbon*, however, which makes the S_N1 and E1 reactions possible. Thus, all four reactions must be considered. Looking ahead, though, we can omit both the E1 and E2 options because we do not have a hydrogen and a suitable leaving group on *adjacent* carbons. That leaves only S_N1 and S_N2 to consider.

The attacking species (CH_3OH) is a weak nucleophile, favoring S_N1 reactions. Although the concentration of the attacking species is high, it does not favor the S_N2 mechanism because the attacking species is a weak nucleophile. Br^- is a very good leaving group, favoring S_N1. Finally, the solvent (CH_3OH) is polar protic, which also favors S_N1. Tallying up the scores using the table in the margin, the S_N1 mechanism should be the predominant mechanism.

Factor	S_N1	S_N2	E1	E2
Strength	✓			
Concentration		–		
Leaving group	✓			
Solvent	✓			
Total	3	0		

The mechanism is as follows:

The loss of the leaving group produces an achiral carbocation, so the attack of the nucleophile produces a mixture of enantiomers. The S_N1 product is then deprotonated to yield an uncharged ether.

---

problem **9.33** Predict the major product of the following reaction:

In this reaction, the alkyl bromide is treated with sodium methoxide in acetone.

9.10 Regioselectivity in Elimination Reactions: Zaitsev's Rule

A substrate can possess two or more distinct hydrogen atoms that can be removed as protons in an elimination reaction, leading to two or more possible alkene products. This is the case for 2-iodohexane, shown in Equation 9-42. Removing a proton from C3 produces hex-2-ene, whereas removing a proton from C1 produces hex-1-ene. With CH_3O^- as the base, the major product is hex-2-ene.

The major product is the
more highly substituted alkene.

$$CH_3CH_2CH_2CH_2CHCH_3 \quad \xrightarrow[\substack{CH_3OH \\ 100\ °C}]{CH_3ONa} \quad CH_3CH_2CH_2CH{=}CHCH_3 \quad + \quad CH_3CH_2CH_2CH_2CH{=}CH_2 \qquad (9\text{-}42)$$

2-Iodohexane

68%
Hex-2-ene

16%
Hex-1-ene

This elimination reaction exhibits **regioselectivity**, which is the tendency of a reaction to take place at one site within a molecule over another.

Using observations such as this, Alexander M. Zaitsev, a Russian chemist, summarized the regioselectivity of elimination reactions as follows: *The major elimination product is the one produced by deprotonating the carbon atom initially attached to the fewest hydrogen atoms.* In other words,

> Elimination usually takes place so as to produce the most highly alkyl-substituted alkene.

This empirical rule came to be known as **Zaitsev's rule**, and the most highly substituted alkene product is called the **Zaitsev product**.

The conditions in Equation 9-42 favor E2 reactions, so each product is formed from a competing E2 reaction. The reaction free energy diagram for each of these reactions is shown in Figure 9-16.

Hex-2-ene is more stable than hex-1-ene because it is more highly alkyl substituted (Section 5.9), so the reaction that leads to the formation of hex-2-ene has a more negative $\Delta G°_{rxn}$ than that leading to the formation of hex-1-ene. Thus, as we learned in Section 9.3a, the reaction that produces hex-2-ene should have a lower energy barrier and proceed faster than the reaction that produces hex-1-ene.

Zaitsev's rule also holds for E1 reactions, such as those in Equation 9-43. A proton on 2-methylbutan-2-ol can be removed from C1 to produce 2-methylbut-1-ene, or from C3 to produce 2-methylbut-2-ene. The major product is 2-methylbut-2-ene, because, once again, it is the more highly substituted alkene product.

The more highly
substituted alkene
product

(9-43)

2-Methylbutan-2-ol

2-Methylbut-1-ene
21%

2-Methylbut-2-ene
78%

Smaller energy barrier when
deprotonation takes place at C3

Gibbs free energy

$CH_3CH_2CH_2CH_2CHCH_3$

$CH_3CH_2CH_2CH_2CH{=}CH_2$
Hex-1-ene

$CH_3CH_2CH_2CH{=}CHCH_3$
Hex-2-ene

The more highly substituted
alkene product is more stable.

Reaction coordinate ⟶

FIGURE 9-16 Zaitsev's rule The free energy diagram is shown for the reaction leading to each product in Equation 9-42. The more stable alkene product is hex-2-ene because it is more highly substituted. According to the Hammond postulate, then, the transition state leading to hex-2-ene is lower in energy than that leading to hex-1-ene, so the energy barrier leading to hex-2-ene is smaller, too.

One noteworthy exception to Zaitsev's rule involves a strong, bulky base, such as the *tert*-butoxide ion, $(CH_3)_3CO^-$. When 2-iodohexane is treated with potassium *tert*-butoxide, the major product is hex-1-ene (Equation 9-44), which is the *less* highly substituted alkene product—that is, the **anti-Zaitsev product**.

There is less steric hindrance surrounding these H atoms.

Strong bulky base

The major product is the less highly substituted alkene.

$$CH_3CH_2CH_2CH_2CHCH_3 \xrightarrow[\substack{(CH_3)_3COH \\ 99\,°C}]{(CH_3)_3COK} CH_3CH_2CH_2CH=CHCH_3 \ + \ CH_3CH_2CH_2CH_2CH=CH_2 \qquad (9\text{-}44)$$

$\ddot{\underset{\cdot\cdot}{I}}\colon$

2-Iodohexane   27%   60%

Hex-2-ene   **Hex-1-ene**

Although the H atoms at both C1 and C3 are relatively well exposed to the base (Section 9.6b), there is more steric hindrance surrounding those at C3. Thus, the base, which itself is quite bulky, favors abstraction of the proton at C1.

SOLVED problem **9.34** Predict the major product of this E2 reaction.

Think What are the possible alkene products? Which is the most stable?

$\xrightarrow[CH_3CH_2OH,\ \Delta]{NaOCH_2CH_3}$

Solve The possible E2 products are obtained by eliminating the leaving group (Br⁻) and a proton on a carbon *adjacent* to the one bonded to the leaving group. These protons are highlighted on the reactant side below and the corresponding products are shown on the right.

$\xrightarrow[CH_3CH_2OH,\ \Delta]{NaOCH_2CH_3}$

A

More highly substituted alkene product

B

Major product

Alkene **B** is more highly substituted than alkene **A**, so **B** is the more stable alkene and is thus the major product.

---

problem **9.35** For each of the following substrates, draw *all* possible E2 products using CH_3ONa as the base and determine which of them is the major product.

(a)   (b)   (c)

9.11 Intermolecular Reactions versus Intramolecular Cyclizations

We generally think of a chemical reaction being between two *separate species*—a so-called **inter**molecular (between molecule) reaction. A reaction typically requires two separate *functional groups*, but those functional groups need not be on separate molecules. Instead, they can be part of the *same molecule*, attached at different places on the molecule's backbone. In that case, the reaction is said to be **intra**molecular (within the same molecule).

Equation 9-45 shows an intramolecular S_N2 reaction. It is intramolecular because the nucleophile (the negatively charged O atom) and the leaving group (the Cl atom) are on the same molecule, resulting in the formation of a ring.

An *intramolecular* reaction

Whenever an intramolecular reaction can take place, an intermolecular reaction is also possible—that is, the two reactions compete with each other. This can be seen in Equation 9-46, in which the reactants are the same as those in Equation 9-45. Which is the predominant reaction?

An *intermolecular* reaction

$$(9\text{-}46)$$

Experimentally,

> An *intramolecular* reaction typically "wins out" over its competing *inter*molecular reaction when the formation of a five- or six-membered ring is possible.

This is an outcome of the interplay between the changes in entropy and strain during the formation of a ring. As a ring forms, entropy decreases (resulting in more order), which is unfavorable. And, as the size of the ring increases, that decrease in entropy becomes more substantial; it becomes less likely for the ends of the ring to meet. On the other hand, recall from Section 4.4 that ring strain decreases sharply to zero upon going from a three-membered ring to a six-membered ring, and larger rings have a small amount of ring strain. Therefore, as the size of the ring being formed increases from three to six members, the process of forming the ring becomes easier energetically. These opposing factors—entropy and strain energy—reach an optimum upon the formation of a five- or six-membered ring.

It is possible to affect the competition between intermolecular and intramolecular reactions by changing the concentrations of the reactants.

> High concentrations tend to favor the intermolecular reaction, whereas low concentrations tend to favor the intramolecular reaction.

An intermolecular reaction requires two separate reactant species to collide to form products. At low concentrations, these collisions become less probable. The reacting species in an intramolecular reaction, on the other hand, are tethered together, so the likelihood that they interact is unaffected by low concentrations.

SOLVED problem 9.36 Two different intramolecular nucleophilic substitution reactions are possible with the substrate at the right. Draw the complete, detailed mechanism that leads to each product. Predict which one is the major product and explain why.

Think Under what conditions does the reaction take place? For an intramolecular nucleophilic substitution reaction under these conditions, what can act as the nucleophile? What can act as the leaving group? For each intramolecular nucleophilic substitution reaction, what is the size of the ring that is formed?

Solve The reaction takes place under basic conditions, so deprotonation of an OH group on the organic species would generate a strongly nucleophilic O⁻ site. The leaving group is Br. The two possible intramolecular nucleophilic substitution reactions that can take place are as follows:

The first reaction yields a seven-membered ring, whereas the second reaction yields a five-membered ring. Because intramolecular reactions favor the formation of five- and six-membered rings, the second reaction yields the major product.

problem 9.37 What is the major product of this following nucleophilic substitution reaction?

9.12 Kinetic Control, Thermodynamic Control, and Reversibility

Recall from Section 9.1 that competing reactions can take place under **kinetic control**, in which case the major product is the one that forms the fastest, or under **thermodynamic control**, in which case the major product is the one that is most stable. Recall, too, that nucleophilic substitution and elimination reactions generally compete under kinetic control. Why?

Kinetic control and thermodynamic control differ depending on whether the products formed from each of the competing reactions are in equilibrium with each other. If an equilibrium involving the various products is established, then, as with any equilibrium, the species with the lowest free energy *must* be present in the greatest amount (Section 6.3). This describes thermodynamic control for the competing reactions.

For products from competing reactions to be in equilibrium, there must be a way that those products interconvert. One typical scheme, shown in Equation 9-47, is **reversible** competing reactions. That is, as suggested by the **reversible reaction arrows** (⇌), each reaction can take place readily in both the forward and reverse directions.

Because of this reversibility, a molecule of Product 1, once produced, can react in the reverse direction to regenerate reactants. Subsequently, the regenerated reactants can undergo reaction again to produce Product 2.

Conversely, if competing reactions are **irreversible**, in which case they do *not* take place readily in the reverse direction, then equilibrium is *not* established between the products from the respective reactions. This is illustrated in Equation 9-48, using **irreversible reaction arrows** (⟶) to connect reactants to products.

If the respective products are not in equilibrium, then the major product is *not* necessarily the most stable product. This is often the case with competing reactions that take place under kinetic control.

These ideas can be summarized as follows:

- Reversible reactions tend to take place under thermodynamic control.
- Irreversible reactions tend to take place under kinetic control.

Thus, competing S_N2, S_N1, E2, and E1 reactions usually take place under kinetic control because *these reactions tend to be irreversible.*

Whether a reaction is reversible or irreversible can often be determined by carefully examining its free energy diagram. For example, consider a typical S_N2 reaction, such as the one in Equation 9-49.

S_N2

2-Bromopropane   2-Ethoxypropane   (9-49)

In its free energy diagram in Figure 9-17, the products are much lower in energy than the reactants, making ΔG°_{rxn} substantially negative. This is mainly due to the greater charge stability on the product side; the Br atom, because it is substantially larger,

FIGURE 9-17 Standard Gibbs free energy change, ΔG°_{rxn}, and irreversible reactions The free energy diagram for the substitution reaction in Equation 9-49. The products are much more stable than the reactants, so the reaction has a large negative value for ΔG°_{rxn}. As a result, the energy barrier in the reverse direction, $\Delta G^{\circ\ddagger}_{reverse}$, is large. This makes the reverse reaction very slow, in which case the reaction is *irreversible*.

can better accommodate the negative charge than the O atom can (Section 6.6b). Consequently, the sizes of the energy barriers in the forward and reverse directions—$\Delta G^{\circ\ddagger}_{forward}$ and $\Delta G^{\circ\ddagger}_{reverse}$, respectively—differ significantly. Specifically, $\Delta G^{\circ\ddagger}_{reverse}$ is quite large, so the reaction in the reverse direction is very slow, thus making the reaction virtually irreversible.

Some S_N2 reactions, including that in Equation 9-50, are reversible.

A *reversible* S_N2 reaction

$$:\overset{\ominus}{\underset{..}{I}}: \;+\; H_3C-\overset{..}{\underset{..}{Br}}: \;\rightleftharpoons\; :\overset{..}{\underset{..}{I}}-CH_3 \;+\; :\overset{..}{\underset{..}{Br}}:^{\ominus} \qquad (9\text{-}50)$$

Notice in the free energy diagram for Equation 9-50 (Fig. 9-18) that ΔG°_{rxn} is *not* substantially negative, in contrast to what we saw in Figure 9-17 for the irreversible reaction in Equation 9-49. In fact, the products are somewhat higher in energy than the reactants, making ΔG°_{rxn} somewhat positive. This is primarily because Br^- is less stable than I^-. Consequently, $\Delta G^{\circ\ddagger}_{reverse}$ is not very large, allowing the reaction to proceed in the reverse direction at a signifant rate.

The connection between ΔG°_{rxn} and reversibility extends to other reactions as well.

■ A reaction tends to be *irreversible* if its ΔG°_{rxn} is substantially negative (i.e., if the products are much more stable than the reactants).

■ Otherwise, the reaction tends to be *reversible*.

FIGURE 9-18 Standard Gibbs free energy change, ΔG°_{rxn}, and reversible reactions The free energy diagram for the substitution reaction in Equation 9-50. The products are *not* dramatically more stable than the reactants, so the energy barrier in the reverse direction, $\Delta G^{\circ\ddagger}_{reverse}$, is comparable to that in the forward direction, $\Delta G^{\circ\ddagger}_{forward}$. Thus, the rate in the reverse direction is significant, and the reaction is *reversible*.

How, then, can S_N1 and E1 reactions, such as those in Equations 9-51 and 9-52, participate in the $S_N2/S_N1/E2/E1$ competition under kinetic control? Kinetic control of these reactions might seem counterintuitive because both of them *directly* produce charged products, which are higher in energy than the uncharged reactants.

This product is irreversibly removed from the equilibrium, thus driving the equilibrium more toward products.

(9-51)

This product can be distilled away from the reaction mixture as the reaction progresses, thus driving the equilibrium more toward products.

(9-52)

Indeed, as indicated, the reactions leading to the immediate substitution and elimination products are intrinsically reversible. However, typical conditions for nucleophilic substitution and elimination reactions quite often prevent such reactions from reaching equilibrium. For example, the substitution reaction in Equation 9-51 is a solvolysis reaction, with water as the solvent. Because water is present in such a large amount, Le Châtelier's principle dictates that the deprotonation of the $C_6H_{11}OH_2^+$ product is shifted heavily to the right. Once $C_6H_{11}OH_2^+$ is formed, therefore, it is rapidly deprotonated and removed from the equilibrium. Without the necessary products present for the reaction to take place in the reverse direction, the S_N1 reaction is effectively *irreversible*.

A similar phenomenon occurs with the elimination reaction in Equation 9-52. As we saw in Section 9.8, elimination reactions such as this are heated substantially. The relatively low-boiling product, cyclohexene, can thus be distilled away from the product mixture as the reaction progresses. Once again, without this product present, the reverse reaction cannot take place, effectively making the E1 reaction *irreversible*.

---

problem **9.38** The E2 reaction at the right competes with the S_N2 reaction from Equation 9-49. Is this reaction *reversible* or *irreversible*? Explain.

9.13 Wrapping Up and Looking Ahead

Chapter 9 has dealt primarily with the competition among S_N2, S_N1, E2, and E1 reactions. This competition arises because all four reactions require a substrate and because any species that can act as a nucleophile can also act as a base. These reactions tend to proceed under kinetic control, so the outcome of this competition is governed by their relative rates. To predict the major product, therefore, we examined six different factors that affect the rate of each reaction: the strength of the attacking species as a nucleophile and as a base, the concentration of the attacking species, the leaving group ability of the leaving group, the type of carbon atom bonded to the leaving group, the type of solvent used, and the effect of heat.

To understand the specific effect a given factor has on the relative rate of each reaction, we examined the role of the attacking species in each mechanism. In S_N2 and E2 reactions, the attacking species forces off the leaving group; the nucleophile does so by forming a bond to the atom attached to the leaving group in an S_N2 reaction, whereas the base does so by deprotonating an adjacent atom in an E2 reaction. In S_N1 and E1 reactions, on the other hand, the attacking species does not participate in the reaction until the leaving group has departed. With this in mind, we were able to determine which mechanism is favored by each factor under a given set of conditions, and we saw how to tabulate this information systematically to arrive at a reasonable *prediction* of the predominant mechanism.

We also discussed two other aspects of the competition among nucleophilic substitution and elimination reactions—namely, regioselectivity in elimination reactions and intermolecular versus intramolecular reactions in nucleophilic substitution. When two or more distinct protons can be removed in an elimination reaction, the major alkene product is usually the one that is the most stable—that is, the Zaitsev product. Intramolecular substitution reactions that form five- or six-membered rings are favored over corresponding intermolecular substitution reactions.

Finally, we learned that nucleophilic substitution and elimination reactions tend to take place under kinetic control because they tend to be irreversible under normal conditions. S_N2 and E2 reactions tend to be irreversible because their products are usually much more stable than their reactants. Although the immediate S_N1 and E1 products are usually *not* substantially more stable than the reactants, a product is often removed as the reaction progresses, which makes the reverse reaction unfeasible.

Our study of nucleophilic substitution and elimination reactions is not yet complete. In Chapter 10, we will explore a variety of new, useful reactions that proceed by the same nucleophilic substitution and elimination mechanisms we have already seen. This will allow us to continue to focus on reaction mechanisms as we encounter new reactions.

Two sugar units of a complex carbohydrate, such as starch or cellulose

(a)

Acetal carbon

RO⋯OR
C
(H or) R⟍⟋R (or H)

A generic acetal

(b)

FIGURE 9-19 Acetal groups in carbohydrates (a) In complex carbohydrates, one cyclic sugar unit is connected to another via an acetal functional group, highlighted in red. (b) An acetal carbon is bonded to two RO— groups.

9.14 Nucleophilic Substitution Reactions and Monosaccharides: The Formation and Hydrolysis of Glycosides

As we saw in Sections 1.15b, 4.15, and 5.15, monosaccharides are relatively small carbohydrates and are often shown in their acyclic form. In nature, however, carbohydrates largely exist as more complex structures, such as starch and cellulose, in which numerous simple sugars (frequently many thousands) in their ring forms are connected together. A portion of one such carbohydrate is shown in Figure 9-19a. (We describe these macroscopic structures in greater detail in Chapter 26.)

As indicated in red, an *acetal* functional group connects the sugar units together. First introduced in Table 1-6, an acetal group (shown in its generic form in Fig. 9-19b) is characterized by a C atom bonded to two alkyl (or H) groups and two alkyoxy (RO—) groups. In the case of sugar units connected together, the alkoxy groups belong to different sugar units.

An acetal group involving a sugar can be produced in the laboratory simply by treating a monosaccharide with an alcohol under acidic conditions. An example is shown in Equation 9-53, in which β-D-glucopyranose is treated with methanol and hydrochloric acid. In this case, only one alkoxy group in the product belongs to a sugar unit. The second alkoxy group is simply a methoxy (CH_3O—) group from the alcohol used.

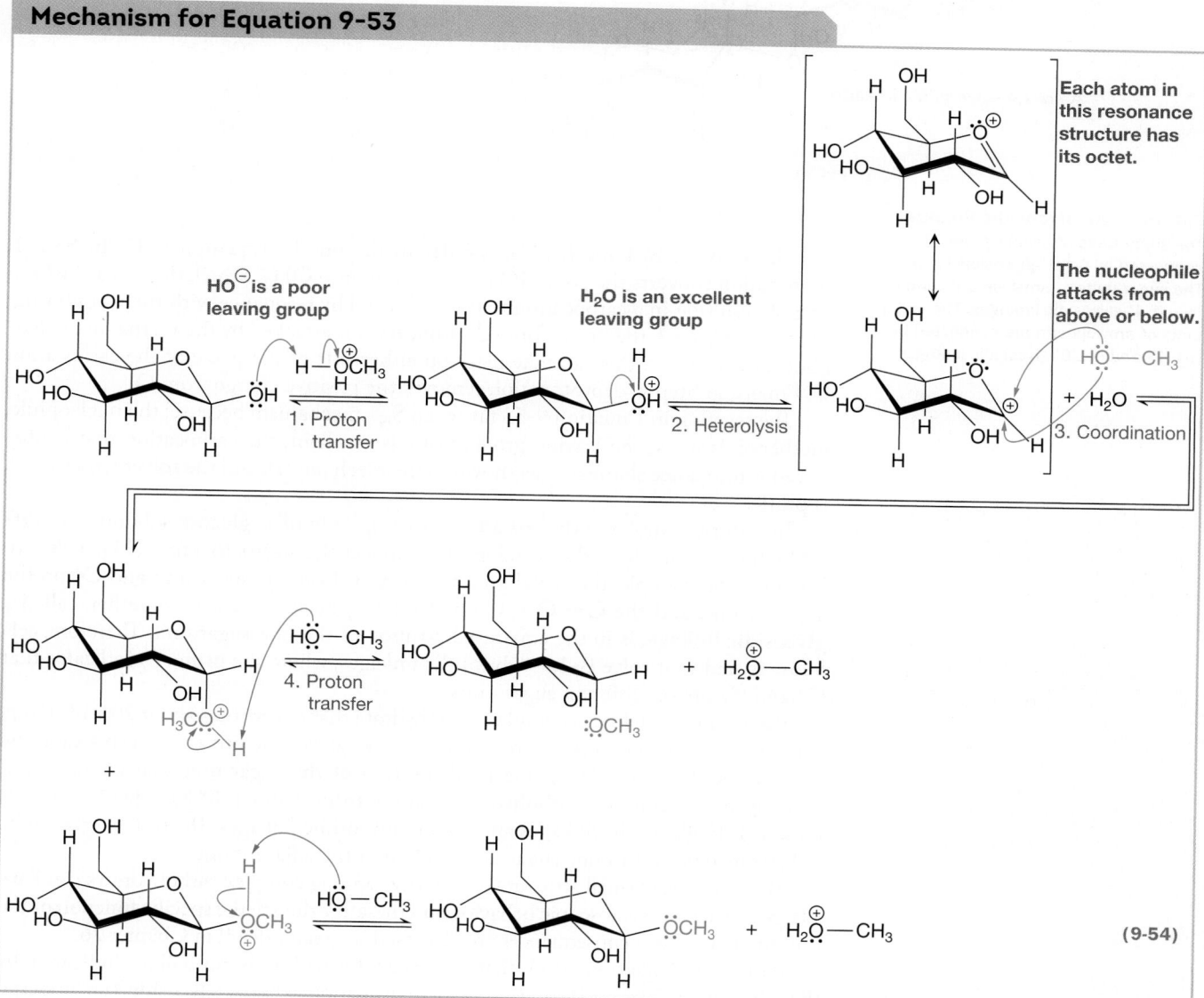

The products in Equation 9-53 are called **glycosides**, and they are produced as a mixture of diastereomers. The first diastereomer, in which the alkoxy group is in the axial position, is designated as α, and the second one, in which the alkoxy group is in the equatorial position, is designated as β. (The α and β designations are explained in greater detail in Section 18.14.)

To understand why a mixture of diastereomers is produced, study the mechanism of the reaction, which is shown in Equation 9-54.

FIGURE 9-20 Glycosidic linkages
The sugar units of cellulose are connected by β-1,4'-glycosidic linkages. The sugar units of amylose are connected by α-1,4'-glycosidic linkages. The sugar units of amylopectin are connected by α-1,4'- and α-1,6'-glycosidic linkages.

This is really an S_N1 mechanism, similar to the one in Equation 9-37. In Step 1, protonation converts the poor HO leaving group into OH_2^+, which departs as H_2O in Step 2, resulting in a carbocation that is stabilized by resonance with the neighboring O atom. In Step 3, the newly formed carbocation is attacked by the methanol nucleophile. As indicated, this can take place on either side of the plane of the carbocation C. Finally, in Step 4, deprotonation removes the positive charge from O.

The reaction in Equation 9-53 favors an S_N1 mechanism because: the nucleophile, methanol, is weak; the leaving group, H_2O, is excellent; the carbocation that is produced is resonance stabilized (as shown in the mechanism); and the solvent, methanol, is protic.

In nature, complex carbohydrates consisting only of D-glucose subunits can differ by the way in which the acetal groups connect the sugars together (Fig. 9-20). In cellulose, for example, the acetal group involves C1 on one sugar unit and C4 on the adjacent one, and the C—O—C bond that connects the sugars together, called a **glycosidic linkage**, is in the equatorial position of the first sugar unit. Therefore, cellulose is said to involve β-1,4'-glycosidic linkages, where the prime (') indicates that C4 and C1 are on different sugar units.

By contrast, amylose, a complex carbohydrate that constitutes about 20% of starch, consists of D-glucose units connected by α-1,4'-glycosidic linkages. In this case, the C—O—C bond resides in the axial position of the sugar unit that contains C1. Amylopectin, a complex carbohydrate that constitutes about 80% of starch, contains both α-1,4'-glycosidic linkages and α-1,6'-glycosidic linkages. The α-1,6'-glycosidic linkage involves C1 of one sugar unit and C6 of the adjacent one.

These relatively small structural differences among complex carbohydrates can have a large impact on macroscopic behavior, and therefore dictate the specific function of the carbohydrate. These consequences are discussed in greater detail in Chapter 26.

Notice in Equation 9-53 that glycoside formation is reversible, indicated by the equilibrium arrow. Thus, as shown in Equation 9-55, a glycoside can undergo acid-catalyzed hydrolysis to produce the monosaccharide. The mechanism for this reaction is identical to the one in Equation 9-54, except the roles of water and methanol are reversed—namely, water is the nucleophile and the alcohol is the leaving group in the hydrolysis of a glycoside.

Glycoside hydrolysis

Methyl β-D-glucopyranoside $\rightleftharpoons$ (H$_2$O / HCl) **α-D-Glucopyranose** + **β-D-Glucopyranose** (9-55)

The stomach provides an acidic environment, so it might seem plausible that the breakdown of complex carbohydrates into their monosaccharide subunits takes place in the stomach. Acid hydrolysis is too slow, however, for the body to make use of complex carbohydrates as quick sources of energy. Instead, hydrolysis depends on various enzymes located in the saliva as well as in the small intestine. These enzymes are highly specialized, and target only the α-1,4′-glycosidic linkages. Thus, humans can metabolize starches, but not cellulose.

Chapter Summary and Key Terms

- S_N2, S_N1, E2, and E1 reactions generally compete with one another under **kinetic control**. The fastest reaction yields the major product. (**Section 9.1**)

- S_N2 reaction rates are highly sensitive to **nucleophilicity**, whereas S_N1 reactions are not. Thus, strong nucleophiles favor S_N2 reactions and weak nucleophiles favor S_N1 reactions. (**Section 9.3a**)

- According to the **Hammond postulate**, the transition state for a reaction resembles reactants more than products if ΔG°_{rxn} is negative, and it resembles products more than reactants if ΔG°_{rxn} is positive. If two reactions proceed by the same mechanism, then the one with the more negative (less positive) value for ΔG°_{rxn} tends to be faster. (**Section 9.3a**)

- E2 reaction rates are highly sensitive to the strength of the attacking base, whereas E1 reactions are not. Thus, strong bases favor E2 reactions and weak bases favor E1 reactions. (**Section 9.3c**)

- Nucleophilicity can be weakened significantly by bulky alkyl groups surrounding the nucleophilic site. Base strength, however, is not significantly affected. Thus, strong, bulky bases favor E2 over S_N2 reactions. (**Section 9.3d**)

- If a nucleophile is strong, then high concentration favors S_N2 and low concentration favors S_N1. If a base is strong, then high concentration favors E2 and low concentration favors E1. (**Section 9.4**)

- All four reaction rates are affected by **leaving group ability**. However, S_N1 and E1 reactions are more sensitive to this factor than S_N2 and E2 reactions are. Excellent leaving groups favor S_N1 and E1, whereas poor leaving groups favor S_N2 and E2. (**Section 9.5**)

- Leaving group ability is determined largely by charge stability. The stronger the leaving group's conjugate acid, the better the leaving group. (**Section 9.5**)

- Substrates with very poor leaving groups, such as HO^-, RO^-, H_2N^-, H^-, and R^-, generally cannot undergo any of the four reactions. (**Section 9.5**)

- Nucleophilic substitution and elimination reactions generally do not occur when leaving groups are on sp^2- or sp-hybridized carbon atoms. (**Section 9.6a**)

- S_N2, S_N1, and E1 reactions are highly sensitive to the type of carbon to which the leaving group is bonded (i.e., 1°, 2°, or 3°), whereas E2 reactions are not. S_N2 reactions are feasible when the leaving group is on a primary carbon because steric hindrance is minimized. S_N1 and E1 reactions are feasible when the leaving group is on a tertiary, allylic, or benzylic carbon, because the carbocation intermediate is stabilized inductively or by resonance. E2 reactions are feasible for all types of carbons because hydrogen atoms are well exposed. (**Section 9.6b**)

- *Aprotic solvents* favor S_N2 and E2; *protic solvents* favor S_N1 and E1. (**Section 9.7**)

- Nucleophilicity in a protic solvent is reversed from that in an aprotic solvent when the nucleophilic atoms are negatively charged and are in the same column, but different rows, of the periodic table. These reversals are due to dramatic differences in solvation. (**Section 9.7**)

- Heat favors elimination over substitution due to the greater *entropy* in the elimination products. (Section 9.8)

- When multiple elimination products are possible, the major product is usually the one that is the most stable, in accordance with **Zaitsev's rule**. The **anti-Zaitsev product** can be obtained using a strong, bulky base like $(CH_3)_3CO^-$. (Section 9.10)

- **Intramolecular cyclization reactions** are favored over their corresponding **intermolecular reactions** when a five- or six-membered ring can be formed. (Section 9.11)

- A reaction with a substantially negative ΔG°_{rxn} tends to be **irreversible** and takes place under **kinetic control**. Otherwise, the reaction tends to be **reversible** and takes place under **thermodynamic control**. (Section 9.12)

Reaction Tables

This is the first of several end-of-chapter sections in which the reactions from the chapter are summarized in tabular form. For each reaction, the starting functional group, the functional group formed, and the typical reagents and reaction conditions are provided. The sections in which the reaction is discussed are also listed. Similar tables will be provided at the ends of future chapters in which new reactions are encountered.

There are two other important features of these tables to note. First, each entry identifies the key electron-rich and electron-poor species that appear in the mechanism of the reaction (recall from Section 7.1a that these species tend to react with each other). Knowing key electron-rich and electron-poor species for various reactions can help you focus on the mechanisms as you learn new reactions. Second, we have collected the reactions into two tables (Tables 9-14 and 9-15), depending upon whether the reaction leads to the formation and/or breaking of a carbon–carbon σ bond. Reactions that do *not* are designated as *functional group transformations*, whereas reactions that do are said to alter the *carbon skeleton*. We discuss the importance of these designations in Chapter 13 in the context of *organic synthesis*.

TABLE 9-14 Functional Group Transformations[a]

	Starting Functional Group	Typical Reagents and Reaction Conditions	Functional Group Formed	Key Electron–Rich Species	Key Electron–Poor Species	Comments	Discussed in Section(s)
(1)	R–CH₂–X, Primary alkyl halide	NaOH	R–CH₂–OH, 1° Alcohol	$HO^{\ominus}$	R–CH₂–X (δ^+)	S_N2 reaction	7.2, 8.4, 8.5, 9.9
(2)	R–CR₂–X (R on all sides), Tertiary alkyl halide	H₂O	R–CR₂–OH, 3° Alcohol	H_2O	$R_2C^{\oplus}R$ carbocation	S_N1 reaction	7.3, 8.1, 8.4, 8.5, 9.5a, 9.9
(3)	R–O–R, Ether	H₂O, H⁺	R–OH, Alcohol	H_2O	$R^{\oplus}$ or $ROH_2^{\oplus}$	S_N1 reaction	7.3, 8.1, 8.4, 8.5, 9.5a, 9.9
(4)	R–CH₂–X, Primary alkyl halide	R'ONa	R–CH₂–O–R', Ether	$R'O^{\ominus}$	R–CH₂–X (δ^+)	S_N2 reaction	7.2, 8.4, 8.5, 9.9
(5)	R–OH, Alcohol	H₃PO₄ or H₂SO₄, Δ	R–O–R, Ether	ROH	$R^{\oplus}$ or $ROH_2^{\oplus}$	S_N1 reaction	7.3, 8.1, 8.4, 8.5, 9.5a, 9.9
(6)	R–L, L = Cl, Br, I, OTs, OMs, or OTf	NaX	R–X, Alkyl halide	$X^{\ominus}$	R–CH₂–L (δ^+) or $R^{\oplus}$	S_N1 or S_N2 reaction	7.2, 7.3, 8.1, 8.4, 8.5, 9.9

[a] X = Cl, Br, I

TABLE 9-14 Functional Group Transformations[a] (continued)

Starting Functional Group	Typical Reagents and Reaction Conditions	Functional Group Formed	Key Electron–Rich Species	Key Electron–Poor Species	Comments	Discussed in Section(s)
(7) Alcohol	HX	Alkyl halide	X⊖	R⊕ or ROH₂⊕	S_N1 or S_N2 reaction	7.2, 7.3, 8.1, 8.4, 8.5, 9.5a, 9.9
(8) Alkyl halide	(CH₃)₃CONa, Δ	Alkene	(CH₃)₃CO⊖		E2 reaction	7.4, 8.4, 8.5, 9.3d, 9.9
(9) Alcohol	H₃PO₄ or H₂SO₄, Δ	Alkene	H₂O		E1 reaction	7.3, 7.6, 8.2, 8.4, 8.5, 9.5a, 9.9

[a] X = Cl, Br, I

TABLE 9-15 Reactions That Alter the Carbon Skeleton[a]

Reactant	Typical Reagents and Reaction Conditions	Product Formed	Key Electron–Rich Species	Key Electron–Poor Species	Comments	Discussed in Section(s)
(1) Alkyne	1. NaH 2. R′–X	Alkyne	Alkynide anion	Alkyl halide	S_N2 reaction	7.2, 8.4, 8.5, 9.3b, 9.9
(2) Alkyl halide	NaCN	Nitrile	Cyanide anion	Alkyl halide	S_N2 reaction	7.2, 8.4, 8.5, 9.9

[a] X = Cl, Br, I

Problems

9.39 Rank the following substrates in order from slowest E1 reaction rate to fastest.

9.40 Rank the following substrates in order from slowest S_N2 reaction rate to fastest.

9.41 Rank the following bases in order from slowest E2 reaction rate to fastest.

| A | B | C | D | E | F |

9.42 Rank the following substrates in order from slowest S_N1 reaction rate to fastest.

| A | B | C | D |

9.43 Suggest an alkyl bromide that can be treated with CH_3SNa to form each of the following products exclusively. By what mechanism should the respective reactions proceed?

(a)

R—Br $\xrightarrow{CH_3SNa}$

(b) R—Br $\xrightarrow{CH_3SNa}$

9.44 Both of the following reactions will give the same S_N2 product.

(a) $H_3C—Cl \xrightarrow{(CH_3)_2CHONa}$?

(b) $H_3C—ONa \xrightarrow{(CH_3)_2CHBr}$?

Draw the mechanism for each of these reactions and show the product. Which reaction is more efficient?

9.45 For each of the following pairs of species, which is the stronger nucleophile in acetone? Explain.

(a) $H_3C—OH$ or $H_3C—O^{\ominus}$

(b) $H_3C—O^{\ominus}$ or $H_3C—\overset{\oplus}{O}H_2$

(c) $H_3C—OH$ or $H_3C—NH_2$

(d) or

(e) or

(f) or

(g) $F^{\ominus}$ or

(h) or

(i) $CH_3S^{\ominus}$ or $CH_3Se^{\ominus}$

(j) $CH_3Se^{\ominus}$ or $Br^{\ominus}$

9.46 For each of the following pairs of species, which is the stronger nucleophile in ethanol? Explain.

(a) $H_3C—OH$ or $H_3C—O^{\ominus}$

(b) $H_3C—OH$ or $H_3C—NH_2$

(c) or

(d) or

(e) $CH_3S^{\ominus}$ or $CH_3Se^{\ominus}$

(f) $CH_3Se^{\ominus}$ or $Br^{\ominus}$

9.47 Suggest how each of the following reactions could be carried out, focusing in particular on the identity of the nucleophile and the choice of solvent.

(a)

(b)

Racemic

9.48 Draw the complete, detailed mechanism for the following reaction and predict the major products, including stereochemistry.

9.49 Show how pent-2-yne can be made from two different alkyl bromides.

9.50 Did the following overall reaction occur by an S_N1, S_N2, E1, or E2 mechanism?

How do you know? Draw the complete, detailed mechanism to account for the formation of both products.

9.51 Did the following overall reaction occur by an S_N1, S_N2, E1, or E2 mechanism?

How do you know? Draw the complete, detailed mechanism to account for the formation of both products.

9.52 In which of the following reactions would you expect a carbocation rearrangement? Explain. (*Hint:* You will need to figure out whether the predominant mechanism is S_N1, S_N2, E1, or E2.)

(a)

$\xrightarrow{H^+}{\Delta}$

(b)

$\xrightarrow{H^+}{\Delta}$

(c)

$\xrightarrow{H^+}{\Delta}$

(d)

$\xrightarrow{CH_3OH}$

(e)

$\xrightarrow[\text{DMSO}]{\text{NaBr}}$

(f)

$\xrightarrow[\text{(CH}_3\text{)}_3\text{COH}]{\text{dilute (CH}_3\text{)}_3\text{COK}}$

(g)

$\xrightarrow{\text{Ethanol}}$

9.53 Determine the major product of each reaction in Problem 9.52 and draw the complete, detailed mechanism. Pay attention to stereochemistry where appropriate.

9.54 The following compound is highly unreactive under conditions that favor E2 reactions.

Explain why. (*Hint:* It may help to build a model of this compound.)

9.55 Provide a complete, detailed mechanism for the following reaction.

9.56 For each of the following reactions, provide a complete, detailed mechanism and predict the products, including stereochemistry where appropriate. Determine whether the reaction will yield exclusively one product or a mixture of products. For each reaction that yields a mixture, determine which is the major product.

(a)
(S)-3-Chlorooctane $\xrightarrow[\text{DMSO}]{\text{CH}_3\text{CH}_2\text{OK}}$

(b)
(S)-3-Methyl-3-chlorooctane $\xrightarrow[\text{CH}_3\text{CH}_2\text{OH}]{\text{CH}_3\text{CH}_2\text{OK}}$

(c)
$\xrightarrow[\Delta]{\text{conc. H}_3\text{PO}_4}$

(d)
$\xrightarrow[\text{DMF}]{(\text{CH}_3)_3\text{CONa}}$

(e)
$\xrightarrow[\text{Ethanol}]{\text{NaN}_3}$

(f)
$\xrightarrow{\text{CH}_3\text{CH}_2\text{OH}}$

9.57 For each of the following reactions, provide a complete, detailed mechanism and predict the products, including stereochemistry where appropriate. Determine whether the reaction will yield exclusively one product or a mixture of products. For each reaction that yields a mixture, determine which is the major product.

(a)
$\xrightarrow{\text{CH}_3\text{CH}_2\text{OH}}$

(b)
$\xrightarrow[\text{DMF}]{\text{NaCN}}$

(c)
$\xrightarrow[\text{DMSO}]{(\text{CH}_3)_3\text{CONa}}$

(d)
$\xrightarrow[\text{CH}_3\text{CH}_2\text{OH, }\Delta]{\text{Na}_2\text{CO}_3}$

(e)
$\xrightarrow{\text{CH}_3\text{CH}_2\text{OH, }\Delta}$

(f)
$\xrightarrow[\text{DMSO, 95 °C}]{\text{NaOCH}_2\text{CH}_3}$

9.58 Predict the major products of each of the following reactions.

(a)
$\xrightarrow[\text{DMSO}]{\text{NaBr}}$

(b)
$\xrightarrow[\text{DMF}]{(\text{CH}_3)_3\text{COK}}$

9.59 The following isomers react separately with sodium hydroxide to give different products with the formulas shown.

(a) Draw the structure of each product.
(b) Draw the mechanism that accounts for the formation of each of those products.
(c) Explain why the isomeric reactants lead to different products.

9.60 The following reaction produces two isomers with the formula $C_7H_{11}Br$.

(a) Draw each product.
(b) Draw the mechanism that accounts for the formation of each product.

9.61 When the reaction mixture in Problem 9.60 is heated, the following compounds are produced.

Draw the complete, detailed mechanism that accounts for the formation of each of these products.

9.62 Given the following reaction sequence, determine the structures of **A** and **B**, including proper stereochemistry.

9.63 When acetic acid is treated with a strong base, followed by benzyl bromide, a compound is formed whose formula is $C_9H_{10}O_2$. Draw the structure of this product, and draw the mechanism leading to its formation.

9.64 When the following deuterium-labeled compound is treated with potassium *tert*-butoxide in *N,N*-dimethylformamide, a single product is observed.

When the same substrate is heated in the presence of dilute potassium ethoxide in ethanol, a mixture of two products is formed. Provide the complete, detailed mechanism for each reaction and explain these results.

9.65 The formula of the precursor is given for each of the following reactions. Draw its structure, paying attention to stereochemistry, if appropriate.

(a)

(b)

9.66 Suggest why each of the following reactions will not occur as indicated.

(a)

conc. H₃PO₄, Δ

(b)

NaSH, DMSO

(c)

1. NaH 2.

(d)

Ethanol

9.67 Determine whether each of the following S$_N$2 reactions is reversible or irreversible.

(a)

NaSCH₃ / EtOH

(b)

KBr / EtOH

(c)

NaCN / EtOH

9.68 In a Finkelstein reaction, an alkyl chloride (R-Cl) or alkyl bromide (R-Br) is treated with NaI in acetone to produce an alkyl iodide (R-I). Whereas NaI is soluble in acetone, NaCl and NaBr are not. With this understanding, argue whether a Finkelstein reaction is reversible.

9.69 In Chapter 9, reversibility was discussed only in terms of S$_N$2, S$_N$1, E2, and E1 reactions, but the ideas apply to other reactions as well. Considering charge stability, determine whether each of the following elementary steps is reversible or irreversible.

(a)

(b)

(c)

9.70 2,5-Dimethylfuran is a liquid biofuel that can be synthesized from 5-hydroxymethylfurfural (HMF). As shown below, HMF can be synthesized from D-fructose by treatment with sulfuric acid.

D-Fructose → H₂SO₄ / DMSO / 48 h, 110 °C → **5-Hydroxymethylfurfural (HFM)** 68% → **2,5-Dimethylfuran**

The mechanism for the formation of HMF from D-fructose is believed to involve the following series of dehydration reactions:

5-Hydroxymethylfurfural

Draw the complete, detailed mechanism for each of these dehydration reactions.

9.71 Draw the complete, detailed mechanism and the products for the following glycoside formation. Pay attention to stereochemistry.

Galactose

$$\xrightarrow[\text{H}^+]{\text{CH}_3\text{CH}_2\text{OH}}$$

9.72 Lactose is a disaccharide in which a glycosidic linkage connects the monosaccharides galactose and glucose.

Lactose

(a) Identify the glycosidic linkage and the acetal carbon in lactose.
(b) What type of glycosidic linkage does lactose have (i.e., is it 1,1′-, 1,2′-, etc., and is it α or β)?
(c) People who are lactose intolerant are deficient in the enzyme lactase, and therefore cannot efficiently break down the disaccharide into its monosaccharides. When lactose is treated with aqueous acid, however, this hydrolysis can take place, though relatively slowly. Draw the complete, detailed mechanism and the products of the acid-catalyzed hydrolysis of lactose.

9.73 DNA is damaged when a base from the DNA chain is removed after an alkylation has occurred. In a "depurination" reaction, the nitrogen purine base is displaced from its sugar as shown in the following reaction:

Draw the mechanism for this reaction and suggest a reason why it occurs so easily.

If your instructor assigns problems in smartwork, log in at **smartwork.wwnorton.com**.

Nucleophilic Substitution and Elimination Reactions 2

Reactions That Are Useful for Synthesis

In the last few chapters we have focused a great deal on S_N2, S_N1, E2, and E1 reactions. In Chapters 7 and 8, the mechanisms of these four reactions were introduced, including their stereochemistries. In Chapter 9, we learned that all four reactions are generally in competition with one another, and we analyzed six main factors that can be used to predict the outcome of that competition.

Thus far, however, our focus on such reactions has primarily been on their fundamental mechanisms, which has limited us to a relatively narrow range of reactions. Here in Chapter 10, we learn how to use nucleophilic substitution and elimination reactions to carry out specific transformations that are useful in synthesis (the topic of Chapter 13). In doing so, we hope to illustrate the importance of these reactions throughout organic chemistry and to reinforce important aspects of nucleophilic substitution and elimination in general.

Creating a plastinate, such as this one, for the Body Worlds exhibition requires a plastination process that was invented by Dr. Gunther von Hagens. One step in plastination can include treatment with an epoxy resin, which produces a polymer via the opening of epoxide rings—a reaction that we will study here in Chapter 10. (Photo credit: Gunther von Hagens' BODY WORLDS and the Institute for Plastination, www.bodyworlds.com)

Because both nucleophilic substitution and elimination reactions are presented in this chapter, we have grouped the reactions according to their mechanisms. Sections 10.1 through 10.7 discuss reactions involving nucleophilic substitution, then Sections 10.8 and 10.9 discuss reactions involving elimination.

10.1 Nucleophilic Substitution: Converting Alcohols to Alkyl Halides Using PBr₃ and PCl₃

In Section 9.5, we saw that under normal conditions, alcohols (R—OH) tend not to undergo nucleophilic substitution and elimination reactions, but alkyl chlorides (R—Cl) and alkyl bromides (R—Br) do so quite readily. This is because HO^- is a rather poor leaving group for nucleophilic substitution and elimination reactions, whereas halides such as Br^- and Cl^- are quite good. (Recall that Cl and Br are substantially larger atoms than O and can better stabilize a negative charge.) Thus, if we wish to carry out a nucleophilic substitution or elimination reaction at a carbon atom that is attached to an OH group, it is extremely useful to be able to convert that alcohol into an alkyl chloride or alkyl bromide.

In Section 9.5b, we learned that one way of doing so is to treat an alcohol with a strong acid like HBr or HCl (Equation 10-1).

Alcohol   Alkyl halide

$$ (10\text{-}1) $$

Under these acidic conditions, a poor leaving group (HO^-) is converted to a very good leaving group (H_2O), as shown in the mechanism in Equation 10-2. The Br^- that is generated in the process can then act as a nucleophile to displace H_2O. If the substrate is a primary alcohol, as in Equation 10-1, then this displacement takes place via an S_N2 reaction.

Mechanism for Equation 10-1

$$ (10\text{-}2) $$

However, a variety of problems are often associated with this kind of reaction. For example, there may be other functional groups in the substrate (not shown in Equation 10-1) that are sensitive to strongly acidic conditions, leading to unwanted side reactions. If the substrate is a secondary or tertiary alcohol, moreover, then these

CHAPTER OBJECTIVES

Upon completing Chapter 10 you should be able to:

● Draw the products of the reaction that takes place when an alcohol is treated with PBr₃ or PCl₃, and explain the role of the S_N2 mechanism in these reactions, including aspects of stereochemistry.

● Show how ammonia and amines can be alkylated by treatment with an alkyl halide, and explain the potential problem of *polyalkylation*.

● Show how a ketone or aldehyde can be alkylated at an alpha carbon by treatment with an alkyl halide under basic conditions, and explain how the regioselectivity can be governed by the choice of base.

● Draw the products of *halogenation* at an α carbon of a ketone or aldehyde, and explain the effect that acidic or basic conditions have on *polyhalogenation*.

● Show how *diazomethane* converts a carboxylic acid into a methyl ester via an S_N2 reaction, and explain the role of the nitrogen leaving group.

● Show how an ether can be synthesized under basic conditions (via the *Williamson ether synthesis*) and under acidic conditions, and explain the role of the S_N2 and S_N1 mechanisms in these reactions.

● Recognize *epoxides* and *oxetanes* as suitable substrates in S_N2 reactions and explain how the regioselectivity depends upon whether the reaction conditions are acidic.

● Explain how to convert a vinyl halide into an alkyne via an E2 reaction, and describe the role of the base in governing the regioselectivity.

● Predict the products of a *Hofmann elimination* reaction, including regioselectivity, and explain the role of both the S_N2 and E2 mechanisms in this reaction.

reaction conditions would favor both S_N1 and E1 mechanisms (see Section 9.6b). Because of the formation of elimination products, the yield of the intended substitution product would be compromised. Also, S_N1 reactions proceed through a planar *carbocation intermediate*, so any stereochemistry that might exist at the carbon atom bonded to the leaving group would be lost, as shown in Equations 10-3 and 10-4 (see Problem 10.1).

Mixture of stereoisomers

(10-3)

(10-4)

---

problem **10.1** Draw the complete, detailed mechanism for the reaction in **(a)** Equation 10-3 and **(b)** Equation 10-4.

Finally, generating a carbocation means a carbocation rearrangement may be possible (see Section 8.6d), as shown in Equation 10-5 (see also Problem 10.2).

Carbocation rearrangement

(10-5)

---

problem **10.2** Draw the complete detailed mechanism for the reaction in Equation 10-5.

A much better way to convert an alcohol to an alkyl halide is to use phosphorus tribromide, PBr_3 (Equation 10-6), or phosphorus trichloride, PCl_3 (Equation 10-7). (Another reagent, $SOCl_2$, is even better than PCl_3, as we discuss in Chapter 20.)

Reaction is *stereospecific*
(inversion of configuration)

(10-6)

No carbocation rearrangement

(10-7)

These reagents have none of the disadvantages that arise from using HBr or HCl:

- PBr$_3$ and PCl$_3$ convert an alcohol to an alkyl halide under relatively mild conditions.
- No elimination products are generated.
- The conversion is *stereospecific*. It takes place with *inversion of configuration* at the alcohol carbon atom (see Equation 10-6).
- No carbocation rearrangements occur (see Equation 10-7).

These observations are explained by the two-step mechanism shown in Equation 10-8, which consists of back-to-back S$_N$2 steps.

Mechanism for Equation 10-6

In Step 1, the O atom of the alcohol is electron rich and acts as the nucleophile, whereas the P atom of PBr$_3$ is electron poor and acts as the substrate. A Br$^-$ anion is the leaving group and is therefore liberated. In Step 2, the Br$^-$ anion generated in Step 1 acts as the nucleophile and the phosphorus-containing species acts as the substrate. The leaving group is HOPBr$_2$.

The phosphorus-containing species in the second step of the mechanism resembles a protonated alcohol, ROH$_2^+$. The leaving group (HOPBr$_2$) resembles a stable, uncharged H$_2$O molecule, which is an excellent leaving group (Fig. 10-1).

Repeat the mechanism shown in Equation 10-8, using PCl$_3$ instead of PBr$_3$.

Answers to Your Turns are in the back of the book.

10.1 YOUR TURN

How does the mechanism in Equation 10-8 account for the stereospecificity of the reaction? Notice that the S$_N$2 reaction in Step 1 does not involve the formation or the breaking of any bonds at the C atom attached to the OH group, so the configuration at that C atom is unaffected. In Step 2, that same C atom undergoes S$_N$2 attack,

FIGURE 10-1 The HOPBr$_2$ leaving group The HOPBr$_2$ leaving group in Step 2 of Equation 10-8 is highlighted in red in the first structure. This leaving group resembles the excellent H$_2$O leaving group in a protonated alcohol, the second structure.

10.1 Nucleophilic Substitution: Converting Alcohols to Alkyl Halides Using PBr$_3$ and PCl$_3$ / **527**

which inverts the stereochemical configuration. There is no carbocation rearrangement, moreover, because no carbocations are ever formed!

---

problem **10.3** Draw the complete, detailed mechanism for each of the following reactions and predict the product.

(a)

(b)

Recall from Section 9.6a that nucleophilic substitution reactions generally do *not* occur when a leaving group is on an sp^2- or sp-hybridized carbon atom. The same is true of these reactions involving phosphorus trihalides:

> The conversion of an alcohol to an alkyl halide using PBr_3 or PCl_3 occurs only when the OH group is bonded to an sp^3-hybridized carbon atom.

The OH group in Equation 10-9 is on an sp^2-hybridized C, for example, so no reaction takes place.

OH group on sp^2-hybridized C

No reaction   (10-9)

Recall from Section 9.6b, too, that S_N2 reactions effectively do not occur when the leaving group is bonded to a tertiary carbon (i.e., a carbon bonded to three R groups), due to excessive steric hindrance. Thus,

> PBr_3 or PCl_3 cannot convert a tertiary alcohol (R_3COH) into an alkyl halide.

For example, no reaction occurs when the tertiary alcohol in Equation 10-10 is treated with PBr_3.

Excessive steric hindrance in a tertiary alcohol

No reaction   (10-10)

As we can see in Solved Problem 10.4, these restrictions on the type of carbon atom bonded to OH make it possible to carry out a selective conversion to the alkyl halide.

SOLVED problem **10.4** Predict the products of this reaction and draw its complete, detailed mechanism. Pay attention to stereochemistry.

Think Which OH group is more susceptible to reaction with PBr_3? What type of functional group is produced? How is the stereochemistry of the molecule affected?

Solve PBr_3 converts an OH group into a Br substituent. Of the two OH groups in the molecule given, the one on the right is part of a secondary alcohol, and the one on the left is part of a tertiary alcohol. Therefore, the OH group on the right will be converted.

According to the mechanism shown, inversion of configuration at the C atom bonded to OH results in a C—Br bond that points above the plane of the page.

problem **10.5** Which of the following compounds will react with PCl_3? For those that will, **(a)** draw the products, including stereochemistry, and **(b)** draw the complete, detailed mechanism.

A   B   C   D

10.2 Nucleophilic Substitution: Alkylation of Ammonia and Amines

Although ammonia (NH_3) is uncharged, it is a moderately strong nucleophile (Section 9.7b). Therefore, when an alkyl halide such as bromoethane is treated with ammonia, an S_N2 reaction can ensue, producing a protonated amine (Equation 10-11). Ammonia is also mildly basic, so it can deprotonate the S_N2 product to yield an uncharged amine. Overall, ammonia has undergone a single **alkylation**, a reaction in which a hydrogen atom is replaced by an alkyl group.

(10-11)

This reaction is *not* an effective method of synthesizing primary amines ($R-NH_2$), however, due to the mixture of other products formed (Equation 10-12). (We present a more efficient synthesis of primary amines in Chapter 21.)

$$\text{Br} \xrightarrow[\text{(1 equiv.)}]{NH_3} \quad NH_2 \quad + \quad NH \quad + \quad N \quad + \quad \overset{\oplus}{N}$$

1° amine   2° amine   3° amine   4° ammonium ion
57%   24%   3%   Trace amount

(10-12)

In other words,

> If an alkyl halide is treated with an equal molar amount of ammonia (i.e., if there is one equivalent of each), then the 1° amine (RNH_2) is formed in a mixture that also contains a 2° amine (R_2NH), a 3° amine (R_3N), and a 4° ammonium ion (R_4N^+).

The primary amine that is formed from the first alkylation is also nucleophilic at the N atom. It can therefore react with any unreacted alkyl halide species remaining in solution (Equation 10-13), leading to further alkylations.

Partial Mechanism for Equation 10-12

1° Amine product from first alkylation (Steps 1 and 2)

A 2° amine

Unreacted alkyl halide

3. S_N2

4. Proton transfer

(10-13)

The same process—an S_N2 reaction followed by deprotonation—can occur a third time to yield the tertiary amine, $(CH_3CH_2)_3N$ (see Your Turn 10.3). The tertiary amine is nucleophilic, too, so it can react in a fourth S_N2 reaction to yield the quaternary ammonium ion, $(CH_3CH_2)_4N^+$. Unlike the first three S_N2 reactions, the quaternary ammonium ion cannot be deprotonated because there is no acidic hydrogen left on the N atom.

YOUR TURN 10.3

The following reaction scheme shows the mechanism for the formation of the quaternary ammonium ion in Equation 10-12 from the secondary amine shown in Equation 10-13. Draw in the necessary curved arrows, and below each reaction arrow label each step as either "proton transfer" or "S_N2."

Alkylation does not stop at the formation of the primary amine because *the primary amine that is formed is a stronger nucleophile than NH_3.* The additional alkyl group in the primary amine is electron donating and thus helps to stabilize the positive charge in the S_N2 product [i.e., in $(CH_3CH_2)_2NH_2^+$]. Consequently, the S_N2 reaction in Equation 10-13 is *faster* than that in Equation 10-11. Similar arguments explain the subsequent alkylations as well.

YOUR TURN 10.4

In Your Turn 10.3, write "more stable" and "less stable" under the appropriate ammonium ions.

Although the inductive stabilization provided by each additional alkyl group makes the alkylation of ammonia a poor method of synthesizing primary amines, those inductive effects do facilitate the formation of the quaternary ammonium ion. As shown in Equation 10-14, this can be done simply by adding a large excess of the alkyl halide.

Large excess of alkyl halide

The 4° ammonium ion is the major product.

$$\text{(10-14)}$$

87%

If desired, the ammonium ion can be isolated with an anion as a **quaternary ammonium salt**.

This alkylation process is not limited to just ammonia. As shown in Solved Problem 10.6, other amines can be alkylated as well, which is particularly useful in Hofmann elimination reactions (discussed in Section 10.9).

SOLVED problem 10.6 Draw the complete, detailed mechanism for the reaction at the right and predict the major product.

Think Which species is electron rich? Which is electron poor? What kind of reaction will take place? Can that reaction take place a second time?

excess CH$_3$Br

Solve The amine is electron rich at the N atom, and CH$_3$Br is electron poor at the C atom. The resulting reaction is an S$_N$2, yielding a protonated tertiary amine. The product can be deprotonated by another molecule of the amine to yield an uncharged tertiary amine. Because there is excess CH$_3$Br, a further alkylation generates a quaternary ammonium ion as the major product. No further reaction can take place because there are no acidic H atoms on the ammonium ion.

---

problem 10.7 Draw the complete, detailed mechanism for the reaction at the right and predict the major product.

excess CH$_3$Br

10.3 Nucleophilic Substitution: Alkylation of α Carbons

No reaction occurs when an aldehyde (RCH=O) or ketone (R$_2$C=O) is treated with an alkyl halide alone:

$$\text{(10-15)}$$

$$\text{(10-16)}$$

When a strong base like sodium hydride (NaH) or sodium amide (NaNH₂) is included, however, *alkylation* can take place at the *α carbon* (i.e., at the C atom adjacent to the C=O group). Examples are shown in Equations 10-17 and 10-18. (In Chapter 21, we show that the same is true for other carbonyl-containing functional groups, such as esters, although the base must be chosen carefully to avoid unwanted side reactions.)

1. NaH, THF
2. CH₃I

(10-17)

77%

The α carbon is alkylated.

1. NaNH₂
2. Br⌁⌁⌁

(10-18)

72%

The mechanism for these reactions, shown in Equation 10-19, is essentially a proton transfer step followed by an S_N2 step.

Mechanism for Equations 10-17 and 10-18

Weakly acidic α hydrogen

Nucleophilic carbon atom

:base⊖

1. Proton transfer

Enolate anion

+ H—base

2. S_N2

+ :Br:⊖ (10-19)

A base is required because an α carbon is non-nucleophilic in the uncharged form of a ketone or aldehyde. It is non-nucleophilic because the α carbon does not have a lone pair of electrons, nor does it bear a significant partial or full negative charge (see Section 9.3b). As we learned in Chapter 6, however, a hydrogen atom on such a carbon is weakly acidic (pK_a ≈ 20), so it can be deprotonated by a sufficiently strong base to generate an *enolate anion*. In the first resonance structure in Equation 10-19, the α carbon possesses a lone pair and a −1 formal charge. Thus,

An enolate anion is nucleophilic at the α carbon.

This enables the nucleophilic substitution to take place in the second step.

YOUR TURN 10.5

The mechanism for the reaction in Equation 10-17 is as follows:

:H⊖

+ H₂

CH₃—I:

Provide the appropriate curved arrows, label the enolate anion, and write "proton transfer" or "S_N2" below each reaction arrow.

10.3a Regioselectivity in α Alkylations

Alkylation at an α carbon is rather straightforward if there is only one distinct type of α carbon to consider. This is the case for the aldehyde in Equation 10-17, which contains only one α carbon altogether. It is also the case for the ketone in Equation 10-18, because the two α carbons are indistinguishable (Fig. 10-2).

For many ketones, however, such as 2-methylcyclohexanone, the α carbons are distinct, so alkylation of the different C atoms leads to different products (Fig. 10-3). *Regioselectivity*, therefore, is a concern for the alkylation of ketones that have distinct α carbons, but regioselectivity can be controlled by the choice of base, as shown in Equation 10-20.

FIGURE 10-2 Species with one distinct α carbon Regiochemistry is not an issue in the α alkylation of these species because each has just one distinct α carbon.

The only α carbon

Indistinguishable α carbons

From Equation 10-17

From Equation 10-18

Lithium diisopropylamide (LDA)

1. $(C_3H_7)_2N^-Li^+$
   THF/pentane

2. ═══ Br

α Carbons are distinct

Alkylation at the *less* substituted α carbon

(10-20a)

═══ Br

$(CH_3)_3COK$, $(CH_3)_3COH$

Alkylation at the *more* substituted α carbon

(10-20b)

- Lithium diisopropylamide (LDA) causes alkylation to occur at the *less* substituted α carbon of a ketone.
- Alkoxide bases (RO^-) cause alkylation to occur at the *more* highly substituted α carbon of a ketone.

The regioselectivity demonstrated by LDA versus RO^- is dictated by whether deprotonation at the α carbon takes place *reversibly* or *irreversibly*. When LDA

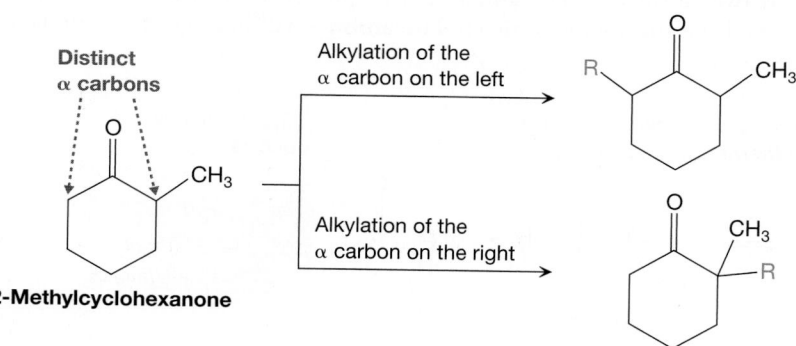

Distinct α carbons

Alkylation of the α carbon on the left

Alkylation of the α carbon on the right

2-Methylcyclohexanone

FIGURE 10-3 Regiochemistry in α alkylation 2-Methylcyclohexanone has two distinct α carbons, so α alkylation can produce different constitutional isomers.

is used as the base (Equation 10-21), K_{eq} is $\gg 1$, making ΔG°_{rxn} substantially negative.

Deprotonation is *irreversible*, so it takes place under *kinetic control*.

ΔG°_{rxn} is very negative.

$$(C_3H_7)_2\overset{\ominus}{\underset{..}{N}}: + H-\overset{O}{\overset{\|}{\underset{|}{C}}}-C \longrightarrow (C_3H_7)_2\overset{..}{N}-H + \overset{\ominus}{\underset{..}{C}}-\overset{O}{\overset{\|}{C}}$$

$pK_a \approx 20 \qquad pK_a = 38$

$K_{eq} \approx 10^{18}$
$\Delta G^{\circ}_{rxn} \approx -103$ kJ/mol
≈ -25 kcal/mol

(10-21)

Therefore, applying what we learned from Section 9.12, we can say that deprotonation by LDA is *irreversible* and hence takes place under *kinetic control*. In other words, LDA will predominantly deprotonate the α hydrogen that can be removed the fastest, yielding the **kinetic enolate anion**. As shown in Equation 10-22, that is the proton on the less substituted α carbon, because LDA, a very bulky base, will encounter less steric hindrance there.

Mechanism for Equation 10-20a

The kinetic enolate

(10-22)

When $(CH_3)_3CO^-$ is used as the base, K_{eq} for the proton transfer step is somewhat <1, making ΔG°_{rxn} slightly positive (Equation 10-23). The reaction is *reversible*, therefore, so it takes place under *thermodynamic control* and an *equilibrium* is established. If two distinct enolate anions can be produced, then the more stable of the two—called the **thermodynamic enolate anion**—will have greater abundance prior to the S_N2 step.

Deprotonation is *reversible*, so it takes place under *thermodynamic control*.

ΔG°_{rxn} is slightly positive.

$$\overset{\ominus}{\underset{..}{\overset{..}{O}}}: + H-\overset{O}{\overset{\|}{\underset{|}{C}}}-C \rightleftharpoons \overset{..}{\underset{..}{O}}-H + \overset{\ominus}{\underset{..}{C}}-\overset{O}{\overset{\|}{C}}$$

$pK_a \approx 20 \qquad pK_a = 19$

$K_{eq} \approx 10^{-1}$
$\Delta G^{\circ}_{rxn} \approx +6$ kJ/mol
$\approx +1.4$ kcal/mol

(10-23)

To identify the thermodynamic enolate anion, let's examine both enolate anions that can be formed.

Dialkyl substituted alkene = less stable

(10-24a)

Trialkyl substituted alkene = more stable

(10-24b)

Thermodynamic enolate anion

Both enolate anions stabilize the negative charge roughly equally because the negative charge in each case is resonance-delocalized over a C atom and an O atom. The difference is in the stability contributed by the C=C double bond in the respective resonance forms in Equation 10-24. The C=C double bond in Equation 10-24a is dialkyl substituted, whereas that in Equation 10-24b is trialkyl substituted, so the enolate anion in Equation 10-24b is more stable (Section 5.9), making it the *thermodynamic enolate anion*.

Equation 10-25 shows the mechanism for the reaction in Equation 10-20b. In Step 1, $(CH_3)_3CO^-$ deprotonates the ketone reversibly to produce the thermodynamic enolate anion. In Step 2, the thermodynamic enolate anion displaces Br^- from the alkyl halide in an S_N2 reaction.

Mechanism for Equation 10-20b

The thermodynamic enolate

Alkyl substitution stabilizes C — C bond

(10-25)

SOLVED problem **10.9** **(a)** Predict the major product for the reaction at the right and draw the complete, detailed mechanism. **(b)** What would the major product be if the base were NaOC(CH₃)₃ instead?

Think Are the two α carbons distinct? How does LDA differentiate between α carbons? How does NaOC(CH₃)₃ differentiate between those carbons?

Solve

(a) The two α carbons are indeed distinct. The one on the left is part of a CH₃ group, and the one on the right is part of a CH₂ group. LDA is a very strong, bulky base, so it will deprotonate the less sterically hindered α carbon under kinetic control. That α carbon is the one on the left in the given ketone because it has fewer alkyl groups attached. The enolate anion that is formed then attacks benzyl chloride in an S_N2 reaction, yielding an alkylated ketone.

Kinetic enolate

(b) If NaOC(CH₃)₃ were used as the base instead, then the α carbon on the right would be deprotonated reversibly to yield the more stable (thermodynamic) enolate anion. Therefore, alkylation would take place at the α carbon on the right, yielding the product at the right.

problem **10.10**

(a) Show the complete mechanism for the reaction at the right and predict the major product.
(b) Do the same for the reaction in which KOC(CH₃)₃ is used as the base instead of LDA.

10.4 Nucleophilic Substitution: Halogenation of α Carbons

When a ketone or aldehyde with an α hydrogen is treated with a strong base in the presence of a molecular halogen, such as Cl_2, Br_2, or I_2, a halogen atom replaces the α hydrogen (Equation 10-26).

The α carbon has been halogenated.

α Hydrogen

(10-26)

The mechanism for this **halogenation** reaction is shown in Equation 10-27.

Mechanism for Equation 10-26

The mechanism for halogenation at the α carbon is essentially the same as that for *alkylation* at an α carbon (Section 10.3). In Step 1, the base deprotonates the α carbon, producing a nucleophilic enolate anion. In Step 2, the enolate anion attacks the molecular halogen in an S_N2 reaction.

The following reaction is nearly identical to that in Equation 10-27. The only difference is that I_2 is the molecular halogen instead of Br_2.

Draw its complete, detailed mechanism, including any curved arrows.

10.6 **YOUR TURN**

Why can a molecular halogen species act as a substrate in a nucleophilic substitution reaction, given that it does not appear to possess an electron-poor site that a nucleophilic enolate anion would seek? The halogen molecule is nonpolar, and the three lone pairs of electrons surrounding each halogen atom may even suggest that the substrate is electron rich. However,

Halogens like Cl_2, Br_2, and I_2 can behave as substrates in S_N2 reactions in part because of their *polarizability*.

Cl_2, Br_2, and I_2 are relatively highly polarizable due to their abundance of electrons, especially nonbonded electrons (Section 2.6d). Consequently, when an electron-rich species like the enolate anion approaches, the electrons on the halogen molecule are repelled quite readily, generating a substantial *induced dipole* (Fig. 10-4). The halogen atom that is

FIGURE 10-4 Molecular halogens as substrates in nucleophilic substitution (a) An isolated bromine molecule is nonpolar. (b) As the nucleophile approaches, electron repulsion forces the bulk of electron density around Br_2 to the opposite side, generating an induced dipole moment. The near side, therefore, becomes electron poor.

An isolated molecule of Br_2 is nonpolar.

(a)

The negative charge repels electrons on Br_2.

Electron rich to electron poor

This end develops a δ⁺ and is thus electron poor.

(b)

nearer the nucleophile, therefore, becomes electron poor. As a result, a curved arrow can be drawn from the electron-rich nucleophile to the electron-poor halogen atom.

Notice in the halogenation shown in Equation 10-27 that the ketone has only a single α hydrogen. The ketone in Equation 10-28 contains two α hydrogens, and *both* α hydrogens are replaced.

(10-28)

2,2-Dimethylpentan-3-one

Both α hydrogens were replaced.

Under *basic* conditions, **polyhalogenation** generally occurs, in which every α hydrogen is replaced by a halogen atom.

In the mechanism for Equation 10-28, which is shown in Equation 10-29, the mechanism for a single halogenation (Equation 10-27) essentially occurs twice.

Mechanism for Equation 10-28

(10-29)

Enolate anion

Halogen atom increases acidity.

Br stabilizes the negative charge on the enolate anion.

Enolate anion

Under each *reaction arrow* in the mechanism in Equation 10-29, label the step as either "proton transfer" or "S_N2."

10.7 YOUR TURN

Halogenation does not stop after a single substitution because the product after the first halogenation is more acidic at the α carbon than the original ketone was. As indicated in Equation 10-29, this greater acidity is due to the increased stability of the halogenated ketone's enolate anion, which, in turn, is a result of the halogen atom's ability to withdraw electron density from the negative charge. As a result, the second deprotonation is more energetically favorable than the first.

Which of these ketones will undergo chlorination faster under basic conditions?

10.8 YOUR TURN

A   B

SOLVED problem 10.11 Draw the complete, detailed mechanism for the chlorination of butanal.

Think In the presence of a strong base, what reaction takes place at an α carbon? How does this affect the chemical properties at that carbon? What species will be attacked as a result? How many times will such a reaction take place at that α carbon?

Butanal

Solve In the presence of a strong base, an α carbon can be deprotonated to generate a nucleophilic C atom. Subsequently, Cl_2 will be attacked in an S_N2 reaction. Because this reaction takes place under basic conditions, all α hydrogens will be replaced. In this case, there are two.

Draw the complete mechanism for the iodination of 2,6-dimethylcyclohexanone, which takes place under basic conditions, and predict the major product.

2,6-Dimethylcyclohexanone

If polyhalogenation is *not* desired, then we can change the chemical environment in which the reaction is carried out.

> Under *acidic conditions*, only a single halogenation takes place.

This is exemplified in Equation 10-30, in which the ketone from Equation 10-28 is halogenated in the presence of acetic acid (CH_3CO_2H, HOAc).

Under acidic conditions, halogenation occurs only once.

(10-30)

Under these conditions, a negatively charged enolate anion cannot exist at any substantial concentration because it is strongly basic (the pK_a of a ketone or aldehyde is ~20). However, the α carbon can still become electron rich via acid-catalyzed keto–enol tautomerization, as shown in Equation 10-31.

Mechanism for Equation 10-30

(10-31)

In the enol form, the α carbon is electron rich, due to the small contribution by the resonance structure in which a negative charge and a lone pair of electrons are located on the α carbon. As before, one end of the halogen molecule becomes electron poor upon approach of the enol, thus setting up the *electron-rich to electron-poor* driving force for the ensuing nucleophilic substitution.

10.9 **YOUR TURN**

The following reaction is identical to the one in Equation 10-30, except Br_2 takes the place of Cl_2.

Draw the complete, detailed mechanism for this reaction.

Why does halogenation take place only once under acidic conditions? In contrast to what occurs under basic conditions, generation of the nucleophilic carbon becomes *more difficult* with each additional halogen. To initiate the second halogenation, the monohalogenated product must be protonated. As indicated in Equation 10-31, however, the resulting positive charge on O is inductively *destabilized* by the presence of the halogen atom.

10.10 **YOUR TURN**

Which of the following ketones will undergo bromination faster under acidic conditions?

A   **B**

problem **10.13** Draw the complete, detailed mechanism for the following iodination under acidic conditions and predict the major product.

I_2 / Acetic acid

10.5 Nucleophilic Substitution: Diazomethane Formation of Methyl Esters

Treatment of a carboxylic acid with **diazomethane** (CH_2N_2) yields a methyl ester, in which the acidic H atom is replaced by a CH_3 group:

$\xrightarrow[\text{Et}_2\text{O, CH}_2\text{Cl}_2]{CH_2N_2}$
30 min

+ $N_2(g)$ (10-32)

90%

The mechanism for this reaction consists of a proton transfer followed by a nucleophilic substitution, as shown in Equation 10-33.

Mechanism for Equation 10-32

(10-33)

In Step 1, the C atom of diazomethane is protonated by the acidic proton of the carboxylic acid. The C atom of diazomethane is electron rich, indicated by the negative charge that appears on C in one of the resonance structures.

In Step 2 of the mechanism, the carboxylate anion (RCO_2^-) is electron rich, and acts as the nucleophile in the ensuing S_N2 reaction. The protonated form of diazomethane is electron poor and acts as the substrate. The leaving group is nitrogen gas, which is an *excellent* leaving group for two reasons:

1. $N_2(g)$ is extremely stable—one of the most inert compounds known.
2. It is a gas, so it bubbles out of solution as it is formed. The nitrogen product is removed permanently. In fact, these characteristics of the $N_2(g)$ leaving group make the reaction *irreversible*.

YOUR TURN 10.11

Diazomethane can be used to convert acetic acid into methyl acetate:

**Ethanoic acid
(Acetic acid)**   **Methyl ethanoate
(Methyl acetate)**

The mechanism for this reaction is as follows, with the curved arrows omitted:

Complete the mechanism by drawing the curved arrows, and below the reaction arrows, label each step as either "proton transfer" or "S_N2."

problem **10.14** Predict the products of each of the following reactions and provide complete, detailed mechanisms.

(a)   (b)

Although turning a carboxylic acid into a methyl ester with CH_2N_2 is a very clean reaction on paper, care must be taken whenever using diazomethane in the lab because it is toxic and explosive! As a result, other methods for forming methyl esters are often much more attractive. Trimethylsilyldiazomethane [$(CH_3)_3SiCHN_2$], for example, is a less explosive alternative to CH_2N_2. And the *Fischer esterification reaction*, presented in detail in Chapter 21, is a much milder, safer, and altogether different reaction that allows us to produce a wide variety of esters from carboxylic acids.

10.6 Nucleophilic Substitution: Formation of Ethers and Epoxides

One of the most convenient ways of forming an ether (R—O—R′) is via the **Williamson ether synthesis,** in which an alkyl halide is treated with a salt of an alkoxide anion. Examples are shown in Equations 10-34 and 10-35.

Williamson ether synthesis

(10-34)

Ethoxyethane
(Diethyl ether)

A *symmetric* ether

(10-35)

Methoxybenzene
or anisole

An *unsymmetric* ether

The Williamson ether synthesis makes it possible to synthesize a wide variety of ethers:

The *Williamson ether synthesis* can be used to synthesize both *symmetric* ethers (in which the alkyl groups bonded to O are the same) and *unsymmetric* ethers (in which the alkyl groups bonded to O are different).

In solution, the alkoxide salts dissolve to form alkoxide anions (RO⁻), which are strong nucleophiles. In the presence of an alkyl halide, then, an S_N2 reaction takes place, as shown in Equation 10-36.

Mechanism for Equations 10-34 and 10-35

$$X = Cl, Br, I$$

(10-36)

An ether

As in any S_N2 reaction, steric hindrance must be taken into consideration. For example, reaction of sodium ethoxide with 2-bromo-2-methylpropane (Equation 10-37) yields essentially no ether product. The C atom bonded to the leaving group is tertiary, and, as we learned in Section 9.6, tertiary carbon atoms are not susceptible to nucleophilic attack in an S_N2 reaction. Instead, the major product is methylpropene, which is the product of E2 elimination. (Recall from Section 9.6 that steric hindrance plays a much smaller role in E2 reactions.)

(10-37)

Methylpropene

E2 product only   **No S_N2 product**

As we learned in Section 9.11, an S_N2 reaction between an alkoxide anion and an alkyl halide can also take place *intramolecularly*, yielding cyclic ethers. For example, treating 4-bromobutan-1-ol with a strong base produces tetrahydrofuran (THF) (Equation 10-38):

4-Bromobutan-1-ol   **Tetrahydrofuran (THF)**

(10-38)

Equation 10-38 proceeds by a two-step mechanism, as shown in Equation 10-39. In Step 1, the alkoxide nucleophile is generated by a proton transfer. Step 2 is an intramolecular S_N2 reaction. In this case, the reaction is favored in large part by the formation of a five-membered ring, which, as we learned in Chapter 4, tends to be quite stable.

Mechanism for Equation 10-38

The intramolecular reaction is favored by formation of a 5-membered ring.

1. Proton transfer   2. S_N2

(10-39)

YOUR TURN 10.12

Tetrahydropyran (THP) can be made with the reaction at the right, which is analogous to Equation 10-38. If the mechanism mimics the one in Equation 10-39, draw the complete, detailed mechanism for this reaction.

Tetrahydropyran (THP)

problem **10.15** Draw the complete, detailed mechanism for this reaction and predict the major product.

Epoxides—three-membered ring ethers—can be made in a similar fashion using **halohydrins** (Equation 10-40).

A halohydrin   **An epoxide**   **(10-40)**

In a halohydrin, a halogen atom and an alcohol group are on adjacent carbon atoms. (Halohydrins are typically made via *electrophilic addition* to an alkene, described in Chapter 12.)

Draw the complete, detailed mechanism for the reaction in Equation 10-40.

10.13 **YOUR TURN**

Whereas the Williamson ether synthesis takes place under *basic* conditions (indicated by the presence of alkoxide anions, RO⁻), ether formation can also take place via nucleophilic substitution under *acidic conditions*. Diisopropyl ether, for example, is formed upon heating propan-2-ol in the presence of HCl (Equation 10-41).

Propan-2-ol   **(Diisopropyl ether)** $+ \ H_2O$ **(10-41)**

Equation 10-41 is an example of a *dehydration* reaction (see Section 9.5b), so called because water is lost in the process.

The mechanism for this reaction, which is essentially an S_N1, is shown in Equation 10-42. In Step 1, the OH group is protonated to become a very good leaving group (i.e., H_2O). Steps 2 and 3 make up the two-step S_N1 reaction—namely, water leaves in a heterolysis step, followed by attack of the nucleophile, propan-2-ol, in a coordination step. In Step 4, another proton transfer produces the uncharged ether.

Mechanism for Equation 10-41

Propan-1-ol also forms an ether under acidic conditions (Equation 10-43), but the reaction proceeds through an S_N2 reaction instead of an S_N1 (see Your Turn 10.14).

$$\text{Propan-1-ol} \xrightarrow[\Delta]{\text{dilute HCl}} \text{Propoxypropane (Dipropyl ether)}$$

(10-43)

Despite the very good leaving group (H_2O), primary carbocations are too unstable for the leaving group to leave spontaneously.

YOUR TURN 10.14

The mechanism for the reaction in Equation 10-43 is as follows:

Supply the appropriate curved arrows, and under each reaction arrow, write "proton transfer" or "S_N2."

---

problem **10.16** Draw the complete, detailed mechanism for the reaction at the right and predict the major product.

$$\xrightarrow[\Delta]{\text{dilute HCl}}$$

Whereas the Williamson ether synthesis can be used to make either symmetric or unsymmetric ethers,

> The synthesis of an ether via the dehydration of an alcohol is useful only when the target is a *symmetric* ether.

If you try to synthesize an unsymmetric ether via dehydration, a mixture of ethers is produced, since the reactants provide two different potential nucleophiles and two different potential substrates. For example, if propan-1-ol is treated with propan-2-ol under acidic conditions, the three ethers shown in Equation 10-44 are produced.

(10-44)

---

problem **10.17** Mechanisms for the formation of the two symmetric ethers in Equation 10-44 (i.e., the second and third products) have been shown previously. Draw a complete, detailed mechanism for the formation of the unsymmetric ether.

SOLVED problem 10.18 Predict the ether that would be formed from the reaction at the right, and then draw the complete, detailed mechanism.

HO~~~~OH $\xrightarrow[\Delta]{\text{dilute HCl}}$?

Think In a substitution reaction, what could act as the nucleophile? What could act as the substrate? Is an S_N1 or S_N2 mechanism favored? Can an intramolecular reaction take place? If so, is it more favorable or less favorable than the corresponding intermolecular reaction?

Solve The acidic conditions cause an OH group to be protonated, generating a good leaving group (H_2O). The second OH group in the molecule can act as a nucleophile in an intramolecular S_N2 reaction. The intramolecular reaction is favored over the intermolecular one, due to the formation of the six-membered ring.

problem 10.19 Predict the product of the reaction at the right and draw the complete, detailed mechanism.

$\xrightarrow[\Delta]{\text{dilute HCl}}$?

10.7 Nucleophilic Substitution: Epoxides and Oxetanes as Substrates

In Section 9.5, we learned that ethers (R—O—R′) are resistant to nucleophilic substitution and elimination reactions under normal conditions because the leaving group would have to be an alkoxide anion (RO⁻), a very poor leaving group. Thus, if diethyl ether is treated with a strong nucleophile such as the methoxide anion (CH_3O^-), then no reaction takes place, as shown in Equation 10-45.

Under normal conditions, ethers tend not to react with nucleophiles.

~~~O~~~ $\xrightarrow{\text{NaOCH}_3}$ No reaction   (10-45)

An exception arises with ethers that are part of small rings, such as oxirane and other **epoxides** (three-membered ring ethers), as well as **oxetanes** (four-membered ring ethers) (Fig. 10-5). Because oxiranes have the smaller ring size, they are less stable and therefore react with a greater variety of nucleophiles than oxetanes do. Thus, we will spend the bulk of this section discussing the reactivity of epoxides. We first discuss their reactivity under neutral and basic conditions (Section 10.7a), and then under acidic conditions (Section 10.7b). In each of these sections, we examine not only the mechanisms by which these types of compounds react, but also aspects of *stereochemistry* and *regiochemistry*. Finally, we conclude this section by introducing some of the reactions that oxetanes undergo.

Oxirane (An epoxide)   **Oxetane**

FIGURE 10-5 Epoxides and oxetanes as S_N2 substrates Epoxides and oxetanes can behave as substrates in S_N2 reactions despite having RO⁻ leaving groups, owing to their ring strain.

10.7a Reactions of Epoxides under Neutral and Basic Conditions

Equations 10-46 and 10-47 show that oxirane reacts readily under neutral and basic conditions, respectively.

Neutral conditions

$$
\text{Oxirane (An epoxide)} \xrightarrow[\text{CH}_3\text{OH}]{\text{NaCl}} \text{Cl}\overset{\frown}{}\text{OH} \tag{10-46}
$$

Basic conditions

$$
\xrightarrow[\text{CH}_3\text{OH}]{\text{NaOCH}_3} \text{H}_3\text{CO}\overset{\frown}{}\text{OH} \tag{10-47}
$$

The mechanism for the reaction in Equation 10-47, an S_N2 step followed by a proton transfer, is shown in Equation 10-48.

Mechanism for Equation 10-47

(10-48)

> **YOUR TURN 10.15**
>
> Draw the complete, detailed mechanism for the reaction in Equation 10-46, which is nearly identical to that of Equation 10-47.

As we can see in the mechanism, the S_N2 reaction causes the highly strained epoxide ring to open, which compensates for the poor leaving group ability of alkoxides (RO^-).

> Epoxides can undergo S_N2 reactions in neutral or basic solution due to the relief of ring strain.

An S_N2 reaction takes place in Step 1, rather than an S_N1 reaction, because the carbon atom that is attacked is primary. Substrates in which the leaving group is attached to a primary carbon effectively cannot undergo S_N1 reactions due to the highly unstable primary carbocation that would be formed. Additionally, the leaving group on that carbon (which leaves with a negative charge on the oxygen atom) is very poor, and poor leaving groups favor S_N2 reactions over S_N1.

The ring opening of epoxides can be carried out with a variety of different nucleophiles. Equations 10-49 and 10-50 show examples of treating an epoxide with an alkyllithium reagent ($R-Li$) and a Grignard reagent ($R-MgX$), respectively.

New C—C bond

$$
\xrightarrow[\text{2. H}_3\text{O}^{\oplus}]{\text{1. CH}_3\text{CH}_2\text{Li}} \text{H}_3\text{CH}_2\text{C}\overset{\frown}{}\text{OH} \tag{10-49}
$$

Acid workup

$$(10\text{-}50)$$

75%

Treatment of an epoxide with R—Li or R—MgX results in the formation of a new C—C bond.

The general mechanism for these types of reactions is shown in Equation 10-51. Notice that R—Li and R—MgX are treated as strong R$^-$ nucleophiles, just as we learned in Section 7.1b.

Mechanism for Equations 10-49 and 10-50

$$(10\text{-}51)$$

Alkyllithium and Grignard reagents are extremely strong bases, so Step 1 must take place in the absence of any acidic species, including protic solvents such as water. As shown in Equations 10-49 and 10-50, the acid must be added in Step 2, only after the S$_N$2 reaction is complete, in what is called an **acid workup**.

Other carbon nucleophiles can also react with epoxides to form a new C—C bond. Examples include the cyanide anion (NC$^-$; Equation 10-52) and alkynide anions (RC≡C$^-$; Equation 10-53).

$$(10\text{-}52)$$

38%

$$(10\text{-}53)$$

In a similar fashion, *hydride nucleophiles* can open an epoxide ring via an S$_N$2 reaction. Equation 10-54, for example, shows that LiAlH$_4$ or NaBH$_4$ can behave as H$^-$ (see Section 7.1b) to convert oxirane into ethanol.

$$(10\text{-}54)$$

With oxirane as the substrate in these ring-opening reactions, we do *not* need to consider *stereochemistry* or *regiochemistry*. Stereochemistry is not an issue because the C atom that is attacked in oxirane is not a stereocenter. Regiochemistry is not a concern because the C atoms attached to the ether O are indistinguishable. Therefore, the product is the same, regardless of which C atom is attacked.

This is not the case for (2*S*,3*R*)-2-ethyl-2,3-dimethyloxirane (Equation 10-55).

(2S,3R)-2-Ethyl-2,3-dimethyloxirane (10-55)

Not only are the C atoms distinct from each other, but each one is also a stereocenter. Thus, different constitutional isomers are possible, depending upon which C atom is attacked. Furthermore, because the reaction involves attack at a stereocenter, different stereoisomers are imaginable. As shown in Equation 10-55, however, only one substitution product is formed.

These results are consistent with the following rule:

> Under neutral or basic conditions, a nucleophile attacks an epoxide at the *less highly alkyl-substituted C atom*, from the side opposite the O atom.

The mechanism in Equation 10-56 shows how this rule applies to the reaction in Equation 10-55. Specifically, the C atom on the left is attacked because it has one fewer alkyl group than the one on the right.

Mechanism for Equation 10-55

(10-56)

These observations are consistent with aspects of the S$_N$2 reaction we have encountered previously. In Section 9.6b, we saw that with less alkyl substitution on the carbon bonded to the leaving group, there is less *steric hindrance*, which enables an S$_N$2 reaction to proceed faster. Also, as we learned in Section 8.5b, backside attack is required of all S$_N$2 reactions, leading to *inversion* of stereochemical configuration at the carbon atom that was attacked.

SOLVED problem **10.21** Predict the major product of this reaction.

Think What is the nucleophile? Which C atom of the epoxide ring will it attack? What is the stereochemistry of such a reaction?

(2S,3S)-2-Ethyl-2,3-dimethyloxirane

Solve The nucleophile is the ethoxide anion, $CH_3CH_2O^-$. It will undergo an S_N2 reaction with the epoxide ring, attacking the less sterically hindered C, which is the one on the left. The nucleophile attacks from the bottom of the epoxide (i.e., opposite the ether O atom), forcing the H and CH_3 substituents upward. Notice that because the epoxide here differs from that in Equation 10-55 only by the configuration at the C atom that is attacked, the products also differ only by the configuration at that C atom.

problem **10.22** Predict the major product for each of the following reactions, and draw the complete, detailed mechanisms.

(a)

(b)

10.7b Reactions of Epoxides under Acidic Conditions

The ring opening of epoxides occurs not only under neutral or basic conditions, but under acidic conditions as well. For example, (R)-2-methyloxirane reacts with HBr to form the bromoalcohol shown in Equation 10-57.

Acidic conditions → HBr →

Nucleophilic attack at the more highly alkyl-substituted C

(10-57)

(R)-2-Methyloxirane

Note, however, that the *regiochemistry* is different from that observed under neutral or basic conditions.

> Under *acidic conditions*, a nucleophile attacks an epoxide at the *more highly alkyl-substituted carbon*.

This regiochemistry is governed by the mechanism illustrated in Equation 10-58, which is different from that in Equation 10-56 for neutral or basic conditions.

Mechanism for Equation 10-57

- Weaker bond to O
- Larger δ+

1. Proton transfer

2. S_N2

(10-58)

Prior to the attack of the nucleophile, a fast proton transfer occurs, generating an intermediate with a positive charge on the O atom in the ring.

Longer and
weaker bond
to the more highly
alkyl-substituted C

147 pm
147 pm
154 pm
159 pm

$\delta+$
Nu:$^{\ominus}$
$\delta+$

Larger partial
positive charge

The C—O bonds are longer
in the protonated form.

Because of the strain in the ring, both C—O bonds *partially* break in the intermediate, indicated by the dashed bonds in Equation 10-58. This allows the formal positive charge on O to be shared on the two C atoms as well, and thus helps to stabilize the intermediate. As shown in Figure 10-6, however, the C—O bond involving the more highly alkyl-substituted C is weakened more. Thus, a larger partial positive charge appears on the side of the ring with the greater alkyl substitution, which attracts the nucleophile more strongly in the subsequent S$_N$2 step.

This particular C—O bond is preferentially weakened because the additional alkyl group is electron donating, so it stabilizes the partial positive charge that develops on the carbon atom of the ring. The greater stabilization of the charge on this C facilitates the weakening of the C—O bond in which it is involved.

FIGURE 10-6 Effect of protonation on an epoxide's C—O bond lengths
The C—O bonds in an epoxide (left) partially break upon protonation (right), as shown by the lengthening of the C—O bonds. As a result, both C atoms bear a partial positive charge, but a larger partial positive charge appears on the side of the ring with greater alkyl substitution, drawing the incoming nucleophile to that side.

SOLVED problem 10.23 (S)-2-Methyloxirane undergoes a chemical reaction when it is dissolved in ethanol. However, what products are formed depends on whether the reaction takes place under acidic or basic conditions. Draw the complete, detailed mechanism for each reaction and predict the products.

(S)-2-Methyloxirane

dilute HCl
CH$_3$CH$_2$OH
?

CH$_3$CH$_2$ONa
CH$_3$CH$_2$OH
?

Think What types of species are allowed under acidic conditions and what types are not allowed? Under basic conditions? Under each set of conditions, what is the substrate being attacked? What is the nucleophile?

Solve Under basic conditions, the nucleophile is CH$_3$CH$_2$O$^-$ and the substrate is the uncharged epoxide:

Less sterically
hindered C

ÖCH$_2$CH$_3$
H
Ö:$^\ominus$
H$_3$CH$_2$CÖ:
CH$_3$CH$_2$Ö:$^\ominus$

$^\ominus$ÖCH$_2$CH$_3$
:ÖH
H$_3$CH$_2$CÖ:

The mechanism follows the one in Equation 10-56. Because the epoxide being attacked is uncharged, steric hindrance guides the nucleophile to the less-substituted carbon of the ring.

Under acidic conditions, however, strong bases are not allowed, so we may *not* include species with a localized negative charge on O. Instead, the mechanism in Equation 10-58 takes place, as shown:

HÖCH$_2$CH$_3$
H
:Ö:
CH$_3$CH$_2$ÖH

• Weaker bond to O
• Larger $\delta+$

H
$^\oplus$Ö:
:ÖCH$_2$CH$_3$
H
CH$_3$CH$_2$ÖH

HÖ:
:ÖCH$_2$CH$_3$

HÖ:
:ÖCH$_2$CH$_3$
CH$_3$CH$_2$ÖH$_2$$^\oplus$

A protonated epoxide is generated in the first step, which is attacked by an uncharged molecule of ethanol in the second step. The nucleophile attacks the more highly substituted C on the ring because that side of the ring bears a larger partial positive charge than the other side.

---

problem **10.24** Draw the complete, detailed mechanism for each of the following reactions and predict the products.

(a)   (b)

10.7c Reactions of Oxetanes and Cyclic Ethers That Are Part of Larger Rings

Oxetanes, which are four-membered ring ethers, can also open via nucleophilic attack under neutral or basic conditions, aided by the relief of ring strain. Oxetanes are larger rings than epoxides, however, so they have less ring strain, giving these reactions less of a driving force.

Consequently, reactions that open oxetanes are generally limited to ones involving very strong nucleophiles, such as alkyllithium reagents (R—Li) and Grignard reagents (R—MgBr). As shown in Equation 10-59, these reactions result in the formation of a new C—C bond, just as we saw with the analogous reactions involving epoxides (Equations 10-49 and 10-50).

Ethers that are part of larger rings do not readily undergo ring opening under neutral or basic conditions, because five-membered rings and larger have little or no ring strain, and thus behave much like acyclic ethers. As shown in Equations 10-60 and 10-61, for example, treating tetrahydrofuran (THF) or tetrahydropyran (THP) with a very strong R⁻ nucleophile results in no reaction. (In fact, because of this lack of reactivity, THF and THP are excellent solvents for reactions that involve such strong R⁻ nucleophiles.)

DNA Alkylation: Cancer Causing and Cancer Curing

DNA bases (adenine, guanine, cytosine, and thymine) contain nucleophilic nitrogen atoms, which is why many halogenated compounds are carcinogenic. The moderate-to-good leaving group abilities of halogen atoms facilitate a nucleophilic substitution reaction, leaving the DNA base *alkylated*, much like the way in which ammonia and amines become alkylated upon treatment with an alkyl halide (Section 10.2). Alkylated DNA can still function in its process of replication, though it will do so abnormally, resulting in mutations in the DNA and, ultimately, cancerous cells.

Guanine

Damaging DNA via alkylation can also be used to *treat* cancer. The key is that cancer cells grow and divide more rapidly than normal cells, and thus are more susceptible to mechanisms that damage DNA and impair its function. A number of successful chemotherapeutic drugs, such as mechlorethamine, have a bis(2-chloroethyl)amino motif to carry out these alkylations.

Mechlorethamine

With a leaving group at two locations, a single molecule of the drug will alkylate two DNA bases and tether them together, disrupting DNA function even more severely. Each alkylation by mechlorethamine involves the formation of an aziridinium ring followed by ring opening, analogous to the formation and opening of an epoxide ring that we learned about in Sections 10.6 and 10.7, respectively.

Unfortunately, alkylating agents like mechlorethamine cannot differentiate the DNA in cancer cells from DNA in normal cells. Other cells that grow and divide rapidly, such as cells that grow hair and cells in the epithelial lining of the gastrointestinal tract, will also be targeted. This accounts for some of the side effects associated with chemotherapy, such as hair loss, nausea and vomiting, and diarrhea.

10.8 Elimination: Generating Alkynes via Elimination Reactions

Although most of the instances in which we discuss elimination reactions lead to the formation of alkenes, it is also possible to generate an alkyne from an elimination reaction (Equation 10-62). As with the other E2 reactions we have examined, a proton and a leaving group are required on adjacent atoms.

$$(10\text{-}62)$$

The leaving group is in a **vinyl** position (i.e., attached to a C=C double bond), so these substrates are quite resistant to nucleophilic substitution and elimination reactions. As we learned in Section 9.6a, the C—L bond strength is significantly greater with an sp^2-hybridized carbon atom than with an sp^3-hybridized carbon atom. Furthermore, because the four electrons of the C=C bond establish a region that is relatively electron rich, bases, which are themselves electron rich, tend to be repelled. Under normal conditions, therefore, even a strong base such as HO⁻ does not facilitate such a reaction. As a result, these reactions are carried out under more extreme conditions, often at temperatures >200 °C. An example is shown in Equation 10-63 with a bromoalkene.

$$(10\text{-}63)$$

Depending upon the nature of the substrate, the choice of base may have a dramatic effect on the outcome of this reaction. For example, 1-bromopent-1-ene leads to pent-1-yne as the major product when $NaNH_2$ is used as the base (Equation 10-64a), whereas pent-2-yne is the major product when NaOH is used as the base (Equation 10-64b).

The mechanism with H_2N^- as the base is shown in Equation 10-65.

Mechanism for Equation 10-64a

This proton transfer step is *irreversible*.

(10-65)

1. E2

2. Proton transfer

add H—OH₂
3. Proton transfer

The reaction stops here before adding H₃O⊕.

Step 1 is the E2 elimination of HBr, which *initially* forms pent-1-yne. In Step 2, this terminal alkyne is *irreversibly* deprotonated because the pK_a of NH_3 is substantially higher than that of a terminal alkyne (see Your Turn 10.16). The reaction stops there, so an *acid workup* is necessary to replace that proton on the terminal C atom and recover the final uncharged product.

YOUR TURN 10.16

Look up or estimate the pK_a values of NH_3 and a terminal alkyne to verify the preceding statement. NH_3 _____ R—C≡C—H _____

With NaOH as the base, on the other hand, the mechanism is distinctly different (Equation 10-66).

Mechanism for Equation 10-64b

(10-66)

1. E2

Pent-1-yne

2. Proton transfer

3. Proton transfer

4. Proton transfer

5. Proton transfer

Pent-2-yne

Step 1 is precisely the same E2 step as in Equation 10-65, which initially forms pent-1-yne. The remaining four steps are proton transfer reactions, which serve to transfer two protons from C3 to C1.

The deprotonation at C1 in Equation 10-65 (i.e., Step 2) does not appear in the mechanism in Equation 10-66. Whereas that proton transfer step is *irreversible* with H_2N^- as the base, it is *reversible* with HO^- as the base, since the pK_a of a terminal alkyne is significantly higher than that of H_2O (see Your Turn 10.17). Consequently, even though that step can take place rapidly, it does not lead to products and thus is omitted from the mechanism.

Look up or estimate the pK_a values of H_2O and a terminal alkyne to verify the preceding statement. H_2O _____ $R—C{\equiv}C—H$ _____

10.17 YOUR TURN

Because all of the proton transfer steps in Equation 10-66 are reversible, *pent–1-yne and pent-2-yne are in equilibrium with each other.* Given that the triple bond is more highly alkyl substituted in pent-2-yne than in pent-1-yne, pent-2-yne is more stable and is therefore the major product.

These elimination reactions are particularly useful when we begin with a dihalide such as 1,2-dibromo-1-phenylethane (Equation 10-67) or 1,1-dichloropentane (Equation 10-68).

1,2-Dibromo-1-phenylethane

$\xrightarrow[\text{CH}_3\text{OH, }\Delta,\text{ 1h}]{\text{KOH}}$

(10-67)

**Phenylethyne
(Phenylacetylene)**
40–50%

1,1-Dichloropentane

$\xrightarrow[\text{2. H}_3\text{O}^\oplus]{\text{1. 3 equiv. NaNH}_2,\ \Delta}$

(10-68)

Pent-1-yne

Three equivalents of H_2N^- are used in Equation 10-68 to convert the dihalide into an alkyne. The first two equivalents bring about separate E2 reactions, removing H and Cl each time. The third equivalent of base deprotonates the acidic proton on the newly formed terminal alkyne. *Acid workup* replaces that proton.

---

problem **10.25** Draw the complete, detailed mechanisms for the reactions in Equations 10-67 and 10-68.

---

problem **10.26** Predict the major product of each of the following reactions, and draw their complete, detailed mechanisms.

$\xrightarrow[\text{2. H}_3\text{O}^\oplus]{\text{1. 3 equiv. NaNH}_2}$?

(a)

$\xrightarrow[\text{2. H}_3\text{O}^\oplus]{\text{1. 3 equiv. NaNH}_2}$?

(b)

10.9 Elimination: Hofmann Elimination

But-1-ene is produced when butan-2-amine is treated with excess bromomethane in the presence of sodium hydroxide:

Butan-2-amine → CH₃Br (excess), NaOH, Δ → **But-1-ene** (*Anti-Zaitsev elimination product*) + (CH₃)₃N   (10-69)

Equation 10-69 is an example of a **Hofmann elimination reaction**. The formation of a C=C double bond strongly suggests that an elimination reaction has occurred in which the leaving group contains the N atom. Two problems, however, must be reconciled: (1) amino groups are terrible leaving groups and therefore do not participate as substrates in elimination reactions, and (2) the major product is a terminal alkene, which is the less alkyl substituted, and therefore the less stable, of two possible elimination products (Section 9.10). In other words,

> The major product of Hofmann elimination is the *anti-Zaitsev* (or Hofmann) product.

The more stable elimination product (i.e., the Zaitsev product) would be but-2-ene, in which the alkene group is more highly alkyl substituted.

The presence of CH_3Br ensures that the leaving group is not simply H_2N^-. As we saw in Section 10.2, an amine reacts with multiple equivalents of CH_3Br to yield the quaternary ammonium ion (Equation 10-70). The leaving group is therefore a stable, uncharged amine, $N(CH_3)_3$.

Mechanism for Equation 10-69

The last step in Equation 10-70 is an E2 elimination reaction, with HO^- as the base. Notice that elimination of the proton can occur at two places: at C1 or at C3. Because the major product is but-1-ene, HO^- must deprotonate at C1. This is peculiar because, as we learned in Section 9.10, a small, strong base like HO^- typically leads to the more highly alkyl substituted elimination product. Here, however, it leads to the less substituted product, called the **Hofmann product**.

This regioselectivity of the Hofmann elimination reaction can be explained by the steric bulk of the amine leaving group. Recall from Section 8.5d that an E2 reaction favors an *anticoplanar* conformation of the proton and leaving group that are eliminated. However, with a proton at C3 in an anticoplanar arrangement with the leaving group (Fig. 10-7a), substantial *steric strain* exists because the C4 methyl group is *gauche* to the leaving group. Thus, the more stable conformation is the one with the C4 methyl group *anti* to the leaving group (Fig. 10-7b). In that conformation, both H atoms on C3 are *gauche* to the leaving group, so elimination involving either of those H atoms is *not* favored.

As shown in Figure 10-7c, on the other hand, a H atom on C1 can be *anti* to the leaving group without causing substantial steric strain. In this case, elimination involving the H atom is favored.

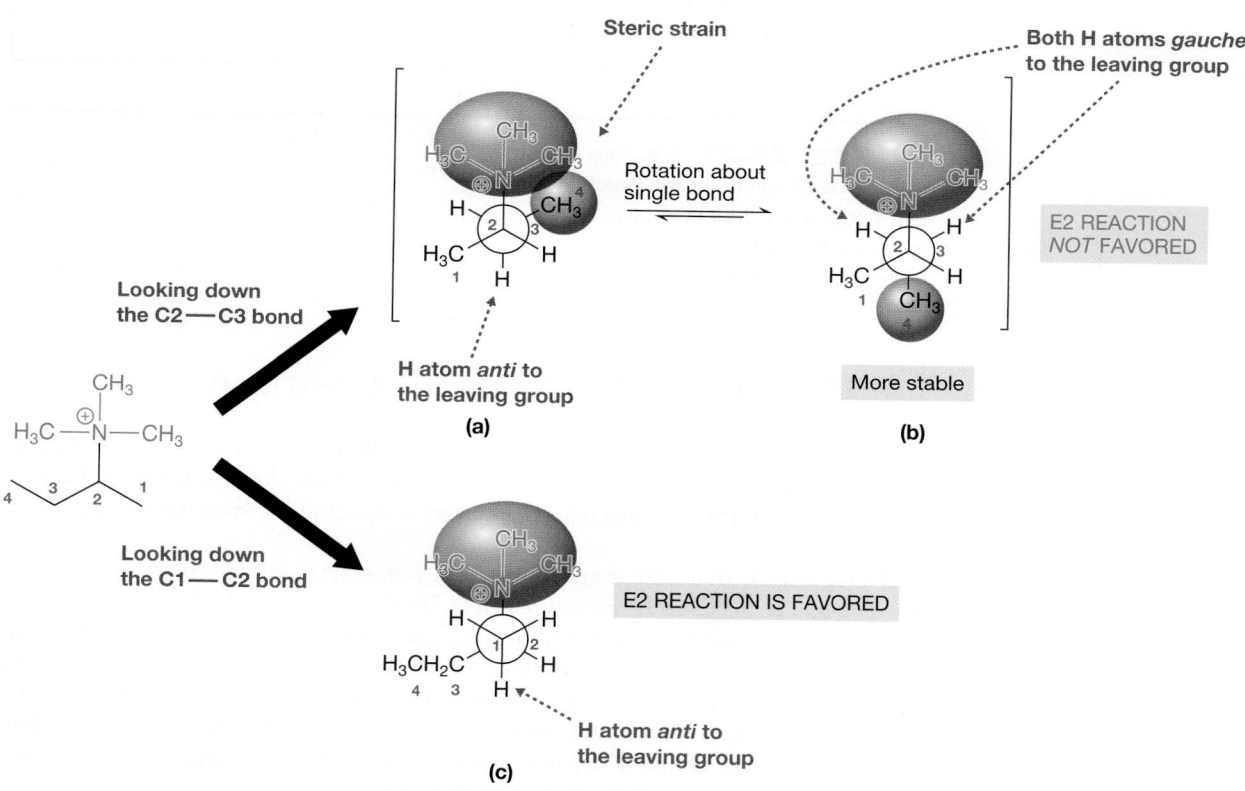

FIGURE 10-7 Regioselectivity in Hofmann elimination reactions Newman projections looking down the C2—C3 bond (top) and the C1—C2 bond (bottom) of the tetraalkyl ammonium ion shown at the left. (a) Conformation necessary for formation of the Zaitsev product, where the H on C3 is *anti* to the amine leaving group. In this conformation, there is substantial steric strain between the C4 methyl group and the leaving group. (b) The more stable conformation is the one in which the C4 methyl group is *anti* to the leaving group. This conformation does not favor the E2 reaction because the H and the amine leaving group are not in the anticoplanar conformation. (c) A H atom is *anti* to the leaving group, and no substantial steric strain exists, so an E2 reaction favors elimination of a proton on C1.

2-Methylpentan-3-amine undergoes complete alkylation to give the quaternary ammonium ion shown below in the middle.

As indicated in the graphic, the leaving group on C3 can be *anticoplanar* with the H atom on C2 (shown in the Newman projection in **A**), as well as with a H atom on C4 (shown in **B**). In the Newman projections, label the H atoms that are in the *anticoplanar* conformation with the leaving group. Also, identify and label all sources of steric strain along the C2—C3 bond in **A** and the C3—C4 bond in **B**. Label the conformation with the least steric strain.

---

problem **10.27** Draw the complete, detailed mechanism for the reaction at the right and predict the major product. (*Hint:* See Your Turn 10.18.)

excess CH$_3$Br

HO$^{\ominus}$, Δ

10.10 Wrapping Up and Looking Ahead

Here in Chapter 10, we examined several reactions that involve a wide variety of functional groups. They are closely related through their mechanisms, however, because the core of every reaction mechanism is either a nucleophilic substitution step or an elimination step.

Nevertheless, there are some significant differences among these mechanisms that ultimately make each reaction distinct. Those differences include where in the mechanism such steps appear, the number of times they appear, and the role of proton transfer reactions. In turn, these differences in the respective mechanisms arise from variations in the structures of the reactants and reagents, or from variations in the reaction conditions. What is the strength of the attacking nucleophile or base? How good is the leaving group? Does the reaction take place under acidic or basic conditions? Are there acidic protons that come into play?

The reactions here in Chapter 10 will form the basis of much of our discussion in Chapter 13, in which organic synthesis is introduced. There, we will begin to see how reactions such as these can be used strategically to alter the functional group at a particular site within a molecule, or to construct new C—C bonds. We will see how, for some reactions, the specific choice of reagents can be chosen to manipulate regiochemistry, and even stereochemistry.

Be aware that the reactions we learned in this chapter are not the only ones that involve nucleophilic substitution or elimination steps. We will encounter others in subsequent chapters, with mechanisms that involve other critical steps as well.

gem-Dichloropropanated indene unit

Cross-linked poly(methyl acrylate)

When you bend or pull on a piece of plastic, it can crack or tear. And for typical plastics, that damage is essentially permanent, without some outside intervention. But imagine if that plastic could, instead, repair the damage on its own. This might seem like fiction, but is, in fact, possible. Plastics consist of very long-chain molecules called polymers, and a crack

Pressure

+ HCl

or tear in a plastic represents bonds in polymer having been severed. For the polymer to "heal" itself—and thus for the plastic to be repaired—a chemical reaction might be initiated automatically, which would reconnect the broken bonds.

How can such a reaction be initiated? Jeffrey Moore at the University of Illinois at Urbana-Champaign envisions that the physical stress placed on the polymer, which would have caused the polymer's damage in the first place, would produce a significant amount of acid at the location where the damage occurred. The acid could behave, in turn, as a catalyst to start a reaction that repairs the polymer molecule.

Moore developed a polymer system that produces this acid when pressure is applied—so-called mechanically generated acid. The key was to covalently bond gem-dichloropropanated indene units into cross-linked polymer chains of poly(methyl acrylate).

When pressure is applied, the induced strain causes HCl to be eliminated, analogous to the elimination reactions we have seen here in Chapter 10 and in Chapter 9. This reaction is driven by the additional stabilization gained by the opening of three-membered ring and the resonance stabilization involving the newly formed double bonds.

Chapter Summary and Key Terms

- Phosphorus tribromide (PBr_3) and phosphorus trichloride (PCl_3) convert primary and secondary alcohols to alkyl halides *stereospecifically* via back-to-back S_N2 reactions. The configuration of the alkyl halide is opposite that of the initial alcohol. **(Section 10.1)**

- Ammonia and amines undergo **alkylation** when treated with an alkyl halide that has a good leaving group. Multiple alkylations occur because the product amine is often more nucleophilic than the reactant. Alkylation of ammonia is a poor method to synthesize a primary amine, but it is an effective way of generating a **quaternary ammonium salt**. **(Section 10.2)**

- Alkylation can take place at an α carbon of a ketone or aldehyde. Under basic conditions, the α carbon is deprotonated and becomes nucleophilic. The resulting enolate anion may

then attack an alkyl halide in an S_N2 reaction. The choice of base can dictate the regiochemistry in the alkylation of a ketone. Reaction with LDA leads to alkylation at the *less* highly alkyl substituted α carbon, whereas reaction with an RO^- base leads to alkylation at the *more* highly alkyl substituted carbon. **(Section 10.3)**

- **Halogenation** may occur at an α carbon under either acidic or basic conditions. The reaction can be stopped after just a single halogenation under acidic conditions, whereas multiple halogenations occur under basic conditions. **(Section 10.4)**

- **Diazomethane** (CH_2N_2) converts a carboxylic acid into a methyl ester (RCO_2CH_3). After an initial proton transfer, an S_N2 reaction occurs, displacing $N_2(g)$ as the leaving group. **(Section 10.5)**

- The **Williamson ether synthesis** can be used to synthesize either symmetric or unsymmetric ethers via an S_N2 reaction between an alkoxide anion (RO^-) and an alkyl halide ($R'X$). The Williamson synthesis takes place under basic conditions. Under acidic conditions, a dehydration reaction can be used to synthesize a symmetric ether. **(Section 10.6)**

- **Epoxides** and **oxetanes** are three- and four-membered ring ethers, respectively, that can be opened in an S_N2 reaction that relieves their substantial ring strain. Under neutral or basic conditions, a nucleophile attacks the less alkyl-substituted carbon of the epoxide ring. Under acidic conditions, the nucleophile attacks the more highly alkyl-substituted carbon of the ring. **(Section 10.7)**

- A **vinyl** halide can undergo an E2 reaction to produce an alkyne. Because bonds to an sp^2-hybridized carbon atom are quite strong, and because alkenes are electron rich, these reactions require strong bases and often elevated temperatures. A very strong base such as H_2N^- has the tendency to produce a terminal alkyne, whereas a less strong base such as HO^- has the tendency to produce an internal alkyne. **(Section 10.8)**

- **Hofmann elimination** leads to an *anti-Zaitsev* elimination product. A poor amine leaving group is converted to a tetra alkyl ammonium ion using excess of an alkyl halide like CH_3Br, which has a relatively good amine leaving group. The bulkiness of the leaving group causes the anti-Zaitsev regioselectivity. **(Section 10.9)**

Reaction Tables

The functional group transformations from this chapter have been collected in Table 10-1; reactions that alter the carbon skeleton are listed in Table 10-2.

TABLE 10-1 Functions Group Transformations[a]

	Starting Functional Group	Typical Reagents and Reaction Conditions	Functional Group Formed	Key Electron–Rich Species	Key Electron–Poor Species	Comments	Discussed in Section
(1)	1° or 2° alcohol	PBr_3	1° or 2° alkyl halide	$Br^{\ominus}$		Back-to-back S_N2 reactions	10.1
(2)	1° alkyl halide	1 eq. NH_3	1° amine + other amines	NH_3		S_N2 reaction (Not useful in synthesis)	10.2
(3)	Amine	excess CH_3Br	Quaternary ammonium salt			Multiple S_N2 reactions	10.2
(4)	Ketone or aldehyde	X_2 / NaOH	α-Halogenated ketone or aldehyde	Enolate anion		Multiple S_N2 reactions	10.4
(5)	Ketone or aldehyde	X_2 / acid	α-Halogenated ketone or aldehyde	Enol		Single S_N2 reaction	10.4
(6)	Carboxylic acid	CH_2N_2	Methyl ester	Carboxylate anion		S_N2	10.5

[a] X = Cl, Br, I

TABLE 10-1 Functions Group Transformations[a] (continued)

Starting Functional Group	Typical Reagents and Reaction Conditions	Functional Group Formed	Key Electron–Rich Species	Key Electron–Poor Species	Comments	Discussed in Section
(7) R—X Alkyl halide	NaOR' →	R—O—R' Ether (symmetric or unsymmetric)	R'O⁻	δ+ R—X	S_N2	10.6
(8) Halohydrin (Br, R, R', OH)	NaOH	Epoxide (R, O, R')	Br δ+ R—R' ⊖O	Intramolecular S_N2	10.6
(9) R—OH Alcohol	dilute HCl / Δ →	R—O—R Ether (symmetric)	R—OH	R—⊕OH₂	S_N1 or S_N2 (dehydration)	10.6
(10) Epoxide (R, O, R')	:Nu⁻ neutral or basic →	OH R—R' Nu Alcohol (2-substituted)	:Nu⁻	R—R' δ+ (epoxide)	S_N2	10.7a
(11) Epoxide (R, O, R')	H—Nu acidic →	OH R—R' Nu Alcohol (2-substituted)	:Nu⁻	⊕O—H R—R' δ+	S_N2	10.7b
(12) Vinyl halide (R, X, H, H)	1. NaNH₂ / 2. H₃O⁺ →	R—C≡CH Terminal alkyne	⊖NH₂	R δ+ H C=C X H	E2	10.8
(13) Vinyl halide (RCH₂, X, H, H)	NaOH / Δ →	R—C≡C—CH₃ Internal alkyne	⊖OH	R δ+ H C=C X H	E2	10.8
(14) Amine (NH₂, R, R')	CH₃Br (excess) / NaOH, Δ →	R—R' Alkene	⊖OH	⊕N(CH₃)₃ R—R'	E2	10.9

[a] X = Cl, Br, I

TABLE 10-2 Reactions That Alter the Carbon Skeleton[a]

Reactant	Typical Reagents and Reaction Conditions	Functional Group Formed	Key Electron–Rich Species	Key Electron–Poor Species	Comments	Discussed in Section
(1) (H or) R—C(=O)—CH₂ Ketone or aldehyde	1. Base⁻ / 2. R''—X →	(H or) R—C(=O)—R' R'' α-Alkylated ketone or aldehyde	(H or) R—C(=O)—⊖R' :: Enolate anion	δ+ R''—X Alkyl halide	S_N2	10.3

[a] X = Cl, Br, I

(continued)

TABLE 10-2 Reactions That Alter the Carbon Skeleton[a] (continued)

Reactant	Typical Reagents and Reaction Conditions	Functional Group Formed	Key Electron–Rich Species	Key Electron–Poor Species	Comments	Discussed in Section
(2) Epoxide	1. R″—Li or R″—MgBr 2. H₃O⁺	Alcohol (2-alkyl substituted)	R″⊖		S_N2	10.7a
(3) Epoxide	NaCN / H₂O	Nitrile (β-hydroxy)	NC⊖		S_N2	10.7a
(4) Alkyne (terminal)	1. NaH 2. 3. H₃O⁺	Alcohol (3-Alkyn-1-ol)	R″C≡C⊖		S_N2	10.7a
(5) Oxetane	1. R″—Li or R″—MgBr 2. H₃O⁺	Alcohol	R″⊖		S_N2	10.7a

[a] X = Cl, Br, I

Problems

10.28 Draw the complete, detailed mechanism for each of the following reactions and predict the major product(s).

(a) PBr₃ → ? → NaCN → ?

(b) Br₂/HO⊖ / H₂O → ?

(c) NaOH, Δ; NaCl; HCl → ?

(d) Br₂/H⁺ / H₂O → ?

(e) excess CH₃Br / HO⊖ → ?

10.29 Predict the products for each of the following reactions and draw the complete, detailed mechanism.

(a)

$$\xrightarrow[\text{CH}_3\text{OCH}_3, \ 0 \ °\text{C}]{\text{CH}_2\text{N}_2} \quad ?$$

(b)

$$\xrightarrow[\text{CH}_3\text{OH, benzene, } 0 \ °\text{C}]{\text{CH}_2\text{N}_2} \quad ?$$

10.30 Draw the structure of the carboxylic acid that can be reacted with diazomethane to form each of the following compounds.

(a)

(b)

(c)

10.31 (a) Draw the alcohol that would be required to form the following alkyl chloride using PCl$_3$.
(b) Draw the complete, detailed mechanism by which this transformation would occur.

10.32 Draw a complete, detailed mechanism for the following reaction.

$$\xrightarrow[\text{H}^{\oplus}]{\text{CH}_3\text{CH}_2\text{OH}}$$

10.33 When pentane-2,4-dione is treated with one molar equivalent of sodium carbonate and bromoethane, 3-ethylpentane-2,4-dione is the major product. If NaNH$_2$ is used as the base, however, then heptane-2,4-dione is the major product. Explain these observations.

1. Na$_2$CO$_3$ (1 equiv.)
2. CH$_3$CH$_2$Br

1. NaNH$_2$ (1 equiv.)
2. CH$_3$CH$_2$Br

70%

10.34 Propose how you would carry out the following transformation. (*Hint:* It may take more than a single reaction.)

10.35 In this chapter, we showed that an α carbon can be alkylated or halogenated under basic conditions. Recall that multiple halogenations take place under these conditions, whereas polyalkylation is generally not a concern. Suggest why.

10.36 In the elimination of a vicinal dihalide to yield an internal alkyne, two equivalents of $NaNH_2$ are required—one for each equivalent of HX that is eliminated. When the product is a terminal alkyne, however, three equivalents of $NaNH_2$ are required. Explain why.

10.37 When oxirane is treated with NaOH, an S_N2 reaction predominantly occurs, thus opening the ring. Given that conditions that favor an S_N2 reaction generally favor an E2 reaction, too, we can also write an E2 mechanism that opens the ring. The E2 reaction, however, does *not* occur readily. Why not? (*Hint:* It may help to build a molecular model of the epoxide.)

10.38 Elimination occurs when (Z)-3-bromohex-3-ene is treated with $NaNH_2$. Under the same conditions, 1-bromocyclohexene undergoes elimination much more sluggishly. Explain why.

10.39 Draw a mechanism to account for the formation of each product in the following reaction.

10.40 The following Hofmann elimination reaction is significantly slower than one involving pentan-3-amine. Explain why. (*Hint:* You may want to build a molecular model of this molecule and the intermediates formed in the mechanism.)

10.41 A student wanted to prepare 3-cyclopentoxypentane by heating cyclopentanol and pentan-3-ol under acidic conditions. Upon carrying out the reaction, however, she found that there were three different ethers produced, each containing 10 carbon atoms. Draw the structure of each of these ethers, and draw the complete, detailed mechanisms leading to their formation.

3-Cyclopentoxypentane

10.42 Devise a synthesis that would yield 3-cyclopentoxypentane as the exclusive ether, beginning with the same two alcohols given in the Problem 10.41.

10.43 When the following 2,3-epoxyalcohol is dissolved in aqueous base, an equal mixture of the two compounds shown is produced. Draw a complete, detailed mechanism for this reaction.

10.44 One stereoisomer of 2,6-dibromocyclohexanol is labeled with a ^{13}C isotope (indicated by an asterisk) at one of the C atoms bonded to Br. When this compound is treated with a strong base and heated, only the product shown is formed. Draw the stereoisomer of 2,6-dibromocyclohexanol that is consistent with these results. Explain.

10.45 Draw the complete, detailed mechanism for the following reaction.

10.46 Halogenation readily takes place at an α carbon under basic conditions if the halogen is Cl_2, Br_2, or I_2. However, it does *not* readily take place with F_2. Explain.

10.47 Draw and name the stereoisomer of 3-bromobutan-2-ol that, upon heating in the presence of sodium hydroxide, will produce an epoxide that has no optical activity.

3-Bromobutan-2-ol

10.48 Draw the product of each of the following reactions.

(a)

$$\xrightarrow[H_2O]{NaN_3} \quad ?$$

(b)

$$\xrightarrow[HOCH_2CH_2OH]{KCN} \quad ?$$

(c)

$$\xrightarrow[HOCH_2CH_2OH]{HCN} \quad ?$$

(d)

$$\xrightarrow[2.\ H_3O^{\oplus}]{1.\ LiAlH_4} \quad ?$$

(e)

$$\xrightarrow[H_2SO_4]{CH_3OH} \quad ?$$

(f)

$$\xrightarrow[CH_3OH]{CH_3ONa} \quad ?$$

(g)

$$\xrightarrow[CH_3OH]{C_6H_5CH_2NH_2} \quad ?$$

(h)

$$\xrightarrow[2.\ H_3O^{\oplus}]{1.\ CH_3Li} \quad ?$$

10.49 Draw the complete, detailed mechanism for the following reaction, and explain why nucleophilic attack takes place predominantly at the epoxide carbon that is attached to the vinyl group.

63%

10.50 A student attempted to synthesize an epoxide according to the following reaction scheme, but no epoxide was formed. Explain why.

NaOH → No epoxide

10.51 Suggest a reagent that can be used to carry out the following transformation, and draw the complete, detailed mechanism for the reaction.

10.52 An *aziridine* is a compound that contains a three-membered ring consisting of two carbon atoms and a nitrogen atom. Because of the strain in the ring, it behaves similarly to an epoxide. Predict the product of the following reaction and draw the complete, detailed mechanism.

H_2S → ?

10.53 Show how you would carry out the following syntheses.

(a)

(b)

10.54 Draw the complete, detailed mechanism for the following reaction.

CH_3I (excess)
NaOH, Δ

10.55 Under conditions that favor Hofmann elimination, *N*-ethylhexan-3-amine can lead to three different alkene products. Draw the complete mechanism that leads to each alkene product, and predict the major product.

1. CH_3I (excess)
2. NaOH, Δ → ?

N-Ethylhexan-3-amine

10.56 An amine, whose formula is $C_9H_{13}N$, was treated with excess bromomethane and subsequently heated under basic conditions. When the product mixture was analyzed, two isomeric alkenes were found, along with trimethylamine, as indicated below. Draw the structure of the initial amine.

$$C_9H_{13}N \xrightarrow[\text{2. NaOH, } \Delta]{\text{1. CH}_3\text{Br (excess)}}$$

Minor + **Major** + $N(CH_3)_3$

10.57 Given the following information, determine the structure of **A** and **B**.

$$\xrightarrow[\text{2. NaOH, } \Delta]{\text{1. CH}_3\text{I}} \quad \mathbf{A} \quad \xrightarrow[\text{2. NaOH, } \Delta]{\text{1. CH}_3\text{I}} \quad \mathbf{B} \quad + \quad N(CH_3)_3$$
$$(C_7H_{15}N) \qquad (C_5H_8)$$

10.58 In the following protonated epoxide, which C—O bond would you expect to be longer? Why?

$$\longrightarrow \quad ?$$

10.59 Given the following information, determine the structures of compounds **A**, **B**, **C**, and **D**.

$$\xrightarrow[\text{2. H}_3\text{O}^{\oplus}]{\text{1.}} \quad \mathbf{A} \quad \xrightarrow{\text{PBr}_3} \quad \mathbf{B}$$

$$\xrightarrow[\text{Compound B}]{\text{NaOC(CH}_3)_3} \quad \mathbf{C} \quad \xrightarrow[\text{acetic acid}]{\text{Br}_2} \quad \mathbf{D} \quad \xrightarrow{\text{H}_3\text{C—C}\equiv\text{CLi}}$$

10.60 Amides are moderately acidic at the N atom, so they can be alkylated in a fashion that is quite similar to the alkylation of ketones and aldehydes. Predict the product of the following reaction, and draw its complete, detailed mechanism.

$$\xrightarrow[\text{2.}]{\text{1. NaH}} \quad ?$$

If your instructor assigns problems in smartwork, log in at **smartwork.wwnorton.com**.

Electrophilic Addition to Nonpolar π Bonds 1

Addition of a Brønsted Acid

In the last several chapters, we have seen that bond formation is frequently driven by the flow of electrons from an electron-rich species to an electron-poor species. In the reactions we have examined in depth thus far, the electron-rich species, acting as a nucleophile or base, typically possesses an atom that has not only a partial or full negative charge, but also a lone pair of electrons. In such cases, those electrons are the ones used to form the new bond.

Here in Chapter 11 and in Chapter 12, we will discuss **electrophilic addition reactions**, whose mechanisms contain an *electrophilic addition step* (or a variation of it). Recall from Section 7.6 that the electrons used to form a new bond in an electrophilic addition step originate from a nonpolar π bond, such as the C=C double bond of an alkene or the C≡C triple bond of an alkyne. These bonds are relatively electron rich because there are multiple electrons confined to the region between two atoms (Fig. 11-1a). Thus, alkenes and alkynes tend to react with **electrophiles** (i.e., with electron-poor species).

Mycobacterium tuberculosis is the bacterium species responsible for most cases of tuberculosis, which attacks the lungs. These bacteria use an electrophilic addition reaction to produce branched-chain fatty acids, which become incorporated into their cell walls.

Two other factors contribute to the reactivity of the π electrons of a nonpolar double or triple bond compared to an analogous single bond. One is the fact that π electrons are, on average, located farther away from a molecule's nuclei than σ electrons are, as shown in Figure 11-1b. Thus, π electrons are more accessible spatially. The second factor is the relative energy of the π electrons. As indicated by Figure 11-1c, π electrons are higher in energy, and thus less stable than σ electrons. Consequently,

It is easier to break the π bond of a double bond than the σ bond of an analogous single bond.

There are a variety of electrophilic addition reactions we will examine here in Chapter 11 and in Chapter 12. The complexity of these reactions can vary greatly. Some are stepwise additions, whose mechanisms consist of two or more steps, while others are concerted, taking place in a single step. Some even result in the formation of rings. Realizing the complexity that electrophilic addition reactions can have, we begin with the addition of a Brønsted acid (HA) to an alkene, a prototype of electrophilic addition reactions. We then explore the intricacies of the other electrophilic addition mechanisms.

11.1 The General Electrophilic Addition Mechanism: Addition of a Strong Brønsted Acid to an Alkene

Equation 11-1 shows a prototypical electrophilic addition reaction, in which cyclohexene reacts with HCl, a strong Brønsted acid, to produce chlorocyclohexane, the **adduct** (i.e., addition product). As we can see, one alkene carbon forms a new bond to H, and the other one forms a new bond to Cl. The HCl molecule is said to *add across*

FIGURE 11-1 Relative reactivity of π electrons (a) The electrons that make up a nonpolar π bond are relatively electron rich because there are multiple electrons confined to the region between two atoms. In general, π electrons are more reactive than σ electrons because π electrons are (b) more easily accessed spatially, and (c) higher in energy.

An alkene is relatively electron rich because 4 electrons are confined between 2 atoms.

The π electrons reside farther away from the nuclei than the σ electrons and are thus more accessible.

The π electrons are higher in energy and are thus more reactive than the σ electrons.

(a)

(b)

(c)

the C=C double bond. Similarly, in the reaction in Equation 11-2, H_2SO_4 adds across the C=C double bond in 2,3-dimethylbut-2-ene.

Cyclohexene → (HCl) → **Chlorocyclohexane** 95%

HCl has added across the double bond. (11-1)

2,3-Dimethylbut-2-ene → (conc. H_2SO_4) →

H_2SO_4 has added across the double bond. (11-2)

YOUR TURN 11.1

Describe what takes place in the following reaction:

(E)-But-2-ene + HBr → **2-Bromobutane**

Answers to Your Turns are in the back of the book.

These reactions take place via the two-step mechanism shown in Equation 11-3 for the addition of HCl.

Mechanism for Equation 11-1

The electron-rich alkene attacks the electron-poor H.

The electron-rich chloride anion attacks the electron-poor carbocation.

Slow
1. Electrophilic addition

Fast
2. Coordination

(11-3)

Step 1 is an electrophilic addition step, in which a pair of electrons from the electron-rich π bond forms a bond to the acid's electron-poor H atom. This leaves one of the initial alkene C atoms with only three bonds, giving it a +1 formal charge. Cl⁻ is produced as a result of the concurrent breaking of the H—Cl bond. Step 2 is a coordination step, whereby Cl⁻ forms a bond to the carbocation. It is identical to the second step of the S_N1 mechanism discussed in Chapter 8.

YOUR TURN 11.2

The mechanism for the addition of HBr across the double bond in cyclohexene is as follows, but the curved arrows have been omitted:

+ H—Br: → + :Br:⁻ →

The mechanism is essentially identical to the one in Equation 11-3. Supply the missing curved arrows and write the name of each elementary step below the appropriate reaction arrow. Also, label the appropriate electron-rich and electron-poor sites in each step.

problem **11.1** Draw the complete, detailed mechanism and predict the major product for each of the following reactions. (Neglect stereochemistry.)

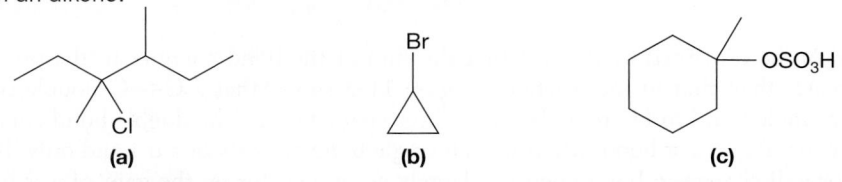

(a)

(b)

(c)

(d)

problem **11.2** Show how each of the following compounds can be produced from an alkene.

(a)

(b)

(c)

Step 1 in Equation 11-3 is much slower than Step 2, so

In an electrophilic addition of a Brønsted acid to an alkene, the rate-determining step is Step 1: addition of the H^+ electrophile.

Step 1 is much slower because the intermediates are substantially higher in energy than either the reactants or products, as can be seen from the free energy diagram shown in Figure 11-2. Thus, the first step has a much larger energy barrier than the

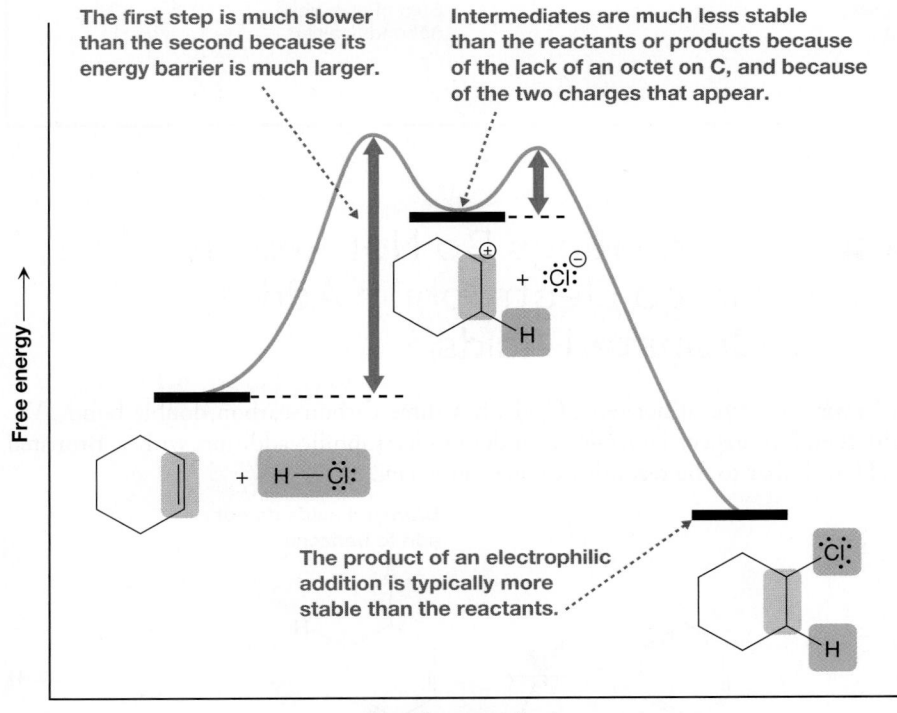

The first step is much slower than the second because its energy barrier is much larger.

Intermediates are much less stable than the reactants or products because of the lack of an octet on C, and because of the two charges that appear.

The product of an electrophilic addition is typically more stable than the reactants.

Reaction coordinate ⟶

FIGURE 11-2 Reaction energy diagram for the electrophilic addition of a strong acid to an alkene The intermediates are much higher in energy than the reactants or products, due to the lack of an octet on C and the presence of two charges. As a result, the first step has a much larger energy barrier, and is thus much slower than the second step. The first step, consequently, is rate determining.

σ Bond + π Bond   σ Bond   σ Bond   σ Bond   σ Bond

A σ bond is stronger than a π bond, so the total bond strength is greater in the product than in the reactants.

FIGURE 11-3 Driving force for electrophilic addition Two σ bonds and a π bond are lost (red) and three σ bonds are gained (blue). Because of the greater strength of a σ bond, the products are lower in energy.

second step. The high energy of the intermediates, in turn, is due to both the lack of an octet on C and the creation of two charges.

As indicated in Figure 11-2, the product of an electrophilic addition reaction is typically more stable than the reactants. Thus,

> In general, electrophilic addition reactions are energetically quite favorable.

This can be explained by the fact that the sum of the bond energies in the product is greater than that in the reactants. Figure 11-3 shows that a C=C double bond and a single bond to H are replaced by three single bonds. The double bond consists of a σ bond and a π bond, whereas each single bond consists of a σ bond only. Thus, the overall change in bond energy is largely accounted for by the gain of a σ bond at the expense of the loss of a π bond. Because a σ bond is stronger than a π bond (Section 3.6), the total bond strength increases throughout the course of the reaction.

YOUR TURN 11.3

Verify that the addition of HCl across a C=C double bond is energetically favorable by using the table of bond energies from Chapter 1 to estimate ΔH°_{rxn}.

ΔH°_{rxn} = (_____ kJ/mol + _____ kJ/mol) − (_____ kJ/mol + _____ kJ/mol + _____ kJ/mol)

C=C   H—Cl   C—C   H—C   C—Cl

Sum of energies of bonds lost   Sum of energies of bonds gained

= _____ kJ/mol

11.2 Benzene Rings Do Not Readily Undergo Electrophilic Addition of Brønsted Acids

The Lewis structure of benzene, C_6H_6, has three carbon–carbon double bonds. We might therefore expect benzene to undergo electrophilic addition with a Brønsted acid, HA, similar to the reactions we have just seen.

Brønsted acids do not add to benzene.

(11-4)

However, as indicated in Equation 11-4, these reactions do not take place.

In general, benzene does *not* undergo addition reactions involving Brønsted acids.

The π electrons that would be used to form the new bond to H are too heavily stabilized.

The source of this stability is a phenomenon called *aromaticity*, a topic of Chapter 14. Although we will not fully discuss aromaticity here, it is helpful to know that benzene's aromaticity results in the six π electrons being completely delocalized around the ring, just as we would expect in the hybrid of benzene's two resonance structures:

The π electrons in benzene are heavily stabilized by electron delocalization.

If HA were to add across one of the C=C double bonds, as shown previously in Equation 11-4, then this cyclic delocalization of the π electrons would cease to exist, and the resulting stabilization would be destroyed.

If a species contains both a benzene ring and a separate C=C, then treatment with a Brønsted acid will lead to a reaction at the separate C=C double bond, as shown in Equation 11-5.

HBr adds across this C=C double bond.

(11-5)

1,2-Diphenylethene

11.3 Regiochemistry: Production of the More Stable Carbocation and Markovnikov's Rule

In the addition of a Brønsted acid across a double bond, the proton can bond to one of two possible carbon atoms. In each of the reactions we have examined thus far, the alkene is *symmetric* (i.e., the groups attached to one of the alkene carbons are identical to the groups attached to the other), so the same adduct is produced regardless of which alkene carbon gains the proton. With an *asymmetric* alkene, on the other hand, two constitutional isomers can be produced. For example, Equation 11-6 shows that propene reacts with HCl to produce 1-chloropropane and 2-chloropropane.

Minor product   Major product

Propene $\xrightarrow{\text{HCl}}$ **1-Chloropropane**   **2-Chloropropane**
                                                                              61%

The H⁺ (proton) has bonded to the alkene C that was initially bonded to the greater number of H atoms.   (11-6)

2-Chloropropane is the major product, which is consistent with the generalization proposed in 1865 by the Russian chemist Vladimir Markovnikov (1838–1904), and which is now known as **Markovnikov's rule:**

The addition of a hydrogen halide to an alkene favors the product in which the proton adds to the alkene carbon that is initially bonded to the greater number of hydrogen atoms.

Thus, the H^+ forms a bond to the terminal alkene C when HCl adds to propene; the terminal alkene C atom is initially bonded to two H atoms, whereas the central C atom is initially bonded to only one. This type of regioselectivity is called **Markovnikov addition**. In Chapter 12, we present examples of electrophilic addition with the opposite regioselectivity—so-called **anti-Markovnikov addition**.

Although Markovnikov's rule resulted from observations made solely on hydrogen halides, it holds for Brønsted acids in general. For example, as shown in the addition of sulfuric acid across the C=C double bond in dodec-1-ene, H adds primarily to C1 instead of C2. C1 is initially bonded to two H atoms, whereas C2 is initially bonded to just one (Equation 11-7).

The H⁺ adds primarily to the alkene C atom initially bonded to the greater number of H atoms.

Dodec-1-ene $\xrightarrow{\text{H}_2\text{SO}_4}$ 78%   (11-7)

Markovnikov's work was carried out without knowing the mechanism of the reaction, so he did not know the reason for the regioselectivity he observed. By understanding the mechanisms for these reactions (Equation 11-3), however, we can now rationalize his observations.

To begin, recall from Section 11.1 that the overall product of an electrophilic addition reaction is significantly more stable than the overall reactants. Thus, applying what we learned from Section 9.12,

Electrophilic addition reactions tend to be irreversible and generally take place under *kinetic control.*

As a result, the major product is the one that is produced the *fastest*.

Also recall from Section 11.1 that the first step of the mechanism—formation of the carbocation intermediate—is the rate-determining step of the overall reaction. Thus, the most rapidly produced adduct derives from the carbocation intermediate that is produced most rapidly.

With this in mind, let's examine Figure 11-4, the free-energy diagrams for the formation of 1-chloropropane (minor product; red curve) and 2-chloropropane (major product; blue curve) from propene. The carbocation that is produced by the addition of H^+ to the terminal C (blue curve) is lower in energy than the one that is produced by the addition of H^+ to the central C (red curve). This is because adding H^+ to the terminal C forms a secondary carbocation, whereas adding H^+ to the central C forms a primary one.

According to the Hammond postulate (Section 9.3a), production of the secondary carbocation involves a smaller energy barrier than production of the primary carbocation does, so the secondary carbocation is produced faster than the primary one. In other words, 2-chloropropane (the adduct that derives from the secondary

The energy barrier is smaller when H⁺ adds to the terminal C.

The 1° carbocation is less stable.

The 2° carbocation is more stable.

2-Chloropropane

1-Chloropropane

Free energy →

Reaction coordinate →

FIGURE 11-4 Markovnikov's rule and relative energy barriers in electrophilic addition Electrophilic addition of HCl to propene to form 2-chloropropane is represented by the blue curve, whereas formation of 1-chloropropane via the same reaction is represented by the red curve. The faster rate of the blue reaction is consistent with a smaller energy barrier, which, in turn, is consistent with a more stable carbocation intermediate.

carbocation) is produced faster than 1-chloropropane (the one that derives from the primary carbocation), making 2-chloropropane the major product in these kinetically controlled reactions.

11.4 YOUR TURN

The free energy diagrams for the reactions that produce the isomeric adducts in Equation 11-7 are shown in the figure below. Complete the diagram by drawing the carbocation intermediates in the appropriate boxes provided.

$$\left(\!\!\!\!\!\!\!\!\!\right)_9 + H_2SO_4$$

OSO₃H

H

OSO₃H

Free energy →

Reaction coordinate →

We can now rewrite Markovnikov's rule from the perspective of the mechanism:

The major product of an electrophilic addition of a Brønsted acid to an alkene is the one that proceeds through the more stable carbocation intermediate.

Although this form of Markovnikov's rule may at first seem to be redundant to the one presented earlier, it is, in fact, much more powerful as a predictive tool. This is because the degree of alkyl substitution is not the only factor that affects carbocation stability, and is frequently not the most important factor. This is illustrated in Solved Problem 11.3.

SOLVED problem 11.3 Predict the major product of this reaction.

$\xrightarrow{\text{HCl}}$?

Think Which C=C double bond will undergo electrophilic addition? What are the *possible* products and the corresponding carbocation intermediates from which they derive? Which carbocation intermediate is more stable?

Solve The right-most C=C double bond is the one that will undergo electrophilic addition. The others make up a phenyl ring, and are thus much too stable to react under these conditions. The two possible products of HCl addition differ by which C atom gains the H$^+$ and which gains the Cl$^-$:

The benzylic carbocation is resonance stabilized.

A

B

Major product

In this case, both alkene C atoms are initially bonded to one H atom, so the original form of Markovnikov's rule is not applicable. However, we can predict the major product by identifying the more stable carbocation intermediate. As indicated in the reaction scheme, carbocation **B** is benzylic and therefore is resonance stabilized, giving it significantly greater stability than carbocation **A**. Thus, the product that derives from **B** is the major product.

problem 11.4 Draw the detailed mechanism for the reaction of each of the following with HCl and predict the major product.

(a)   (b)   (c)   (d)

problem 11.5 Show how each of the following compounds can be synthesized from two different alkenes.

(a)   (b)   (c)

The laser printer you use to print a document onto paper is based on a technology that relies on a photoconducting material such as the polymer poly(N-vinylcarbazole), or PVK. Like all photoconducting materials, PVK's electrical conductivity is altered when it absorbs light. PVK, in particular, is an organic photoconductor (OPC), which is synthesized using a variation of the electrophilic addition mechanism that is presented here in Chapter 11. When vinylcarbazole is treated with a strong acid, H^+ adds to the $C=C$ bond to produce a carbocation, R_1^+. An unreacted molecule of N-vinylcarbazole then undergoes electrophilic addition of R_1^+ to produce a larger carbocation, R_2^+. This process is repeated thousands of times, resulting in PVK.

In a laser printer, an imaging drum coated with an OPC picks up charged particles when it is exposed to a high voltage. As the drum spins, an image to be printed is scanned onto the drum using a light source such as a laser or an array of light-emitting diodes. Because of the OPC's properties, areas of the drum that are exposed to that image of light eject their charged particles. The surface of the drum is then treated with toner (dry ink), which adheres to the remaining charged particles. As paper is fed through the printer and comes in contact with the drum's surface, the toner is deposited and burned onto the paper. Once the drum's surface is cleaned, the cycle repeats.

11.4 Carbocation Rearrangements

When 3-methylbut-1-ene is treated with hydrochloric acid (Equation 11-8), significant amounts of both 2-chloro-3-methylbutane and 2-chloro-2-methylbutane are produced.

The production of this adduct involves a carbocation rearrangement.

3-Methylbut-1-ene → (HCl) → 2-Chloro-3-methylbutane (45%) + 2-Chloro-2-methylbutane (45%)   (11-8)

2-Chloro-3-methylbutane is the product of a normal Markovnikov addition of HCl across the C=C double bond. Specifically, H⁺ adds to the terminal alkene C and Cl⁻ adds to the adjacent, secondary alkene C. By contrast, 2-chloro-2-methylbutane does not appear to be the product of addition of HCl across the double bond because the C atom to which the Cl⁻ attaches is not initially part of the double bond.

As with any mechanism that proceeds through a carbocation intermediate,

> Electrophilic addition of a Brønsted acid across a C=C double bond is suscep-
> tible to carbocation rearrangements.

Recall from Chapter 8 that carbocation rearrangements are fast! In this case, a 1,2-hydride shift can account for the formation of 2-chloro-2-methylbutane, as shown in the mechanism in Equation 11-9. This 1,2-hydride shift transforms a less stable secondary carbocation to a more stable tertiary carbocation prior to the attack of the Cl⁻ nucleophile.

Mechanism for Equation 11-8

1,2-Methyl shifts also take place quickly, so they can appear in the mechanism of an electrophilic addition reaction, too. An example is illustrated in Solved Problem 11.6.

SOLVED problem 11.6 Draw the complete, detailed mechanism and predict the major product of the reaction at the right.

Think What carbocation intermediate is produced upon addition of H⁺? Can that carbocation intermediate undergo a 1,2-hydride shift or a 1,2-methyl shift to attain greater stability?

Solve In Step 1 of this electrophilic addition reaction, H⁺ adds to the terminal alkene C to produce a secondary carbocation intermediate, as shown below. (Addition of H⁺ to the internal alkene C would instead produce a less stable primary carbocation intermediate.)

The secondary carbocation intermediate rapidly converts to a more stable tertiary one via a 1,2-methyl shift in Step 2. After coordination of Br⁻ in Step 3, the reaction is complete.

(a)   (b)   (c)

problem **11.8** Each of the following deuterated bromoalkanes can be produced by treating an alkene with deuterium bromide (DBr). Draw the corresponding alkenes.

(a)   (b)   (c)   (d)

11.5 Stereochemistry

In an electrophilic addition reaction involving an alkene, both alkene carbons, which are initially planar, become tetrahedral in the product. It is possible, therefore, for new stereocenters to be produced during the course of the reaction. Depending upon the symmetry of the product, stereochemistry may or may not be an issue.

In the reaction shown previously in Equation 11-1 (shown again in Equation 11-10), stereochemistry is *not* a concern because neither alkene C becomes a stereocenter, and the product is achiral; in the product, none of the C atoms is bonded to four *different* groups.

$$\text{Cyclohexene} \xrightarrow{\text{HCl}} \text{Chlorocyclohexane}$$

Neither C becomes a stereocenter.   (11-10)

By contrast, in the reaction shown previously in Equation 11-5 (shown again in Equation 11-11), a single tetrahedral stereocenter is formed, so each product molecule is chiral. Because the starting material and the conditions under which the reaction takes place are achiral, a racemic mixture of enantiomers is produced.

A new stereocenter is produced.

1,2-Diphenylethene   →   HBr   →   (11-11)

Racemic mixture of products

When DBr (whose reactivity is nearly identical to that of HBr) reacts with an achiral alkene such as cyclohexene (Equation 11-12), the product molecules are chiral. Unlike the ones in Equation 11-11, however, these product molecules contain two stereocenters, so there are $2^n = 2^2 = 4$ stereoisomers that exist. Which of these stereoisomers are produced, and what are their relative abundances?

Two stereocenters (*)
are produced.

$$D—Br \qquad (11\text{-}12)$$

To answer these questions, we turn to the mechanism. The addition of DBr takes place in two steps, as shown in Equation 11-13, similar to the general mechanism in Equation 11-3. In the first step, D^+ (just like H^+) forms a bond to an alkene C, producing a carbocation intermediate. In the second step, Br^- forms a bond to the positively charged C atom.

Mechanism for Equation 11-12

In Step 1, D^+ can form a bond to the alkene C from either side of that C atom's plane, producing one of two enantiomeric carbocations, **X** or **Y**. In Step 2, Br^- can attack the positively charged C atom from either side of the plane—either syn to (i.e., on the same side as) or anti to (i.e., on the opposite side from) the D atom. This gives rise to two cis (**A** and **D**) and two trans (**B** and **C**) isomers. Thus, the reaction produces a mixture of all four of the aforementioned stereoisomers.

What is the relative abundance of these stereoisomers? The two cis isomers are enantiomers, just as the two trans isomers are. Because the starting material and the environment are achiral, each of these enantiomeric pairs (i.e., **A/D** and **B/C**) is produced as a racemic mixture. By contrast, cis and trans isomers (e.g., **A/B, A/C, B/D,** or **C/D**) are diastereomers, so they must be produced in unequal amounts. (In this case, because deuterium and hydrogen atoms are nearly equal in size, the cis and trans products will be nearly equal in abundance.)

---

problem **11.9** Draw the complete mechanism for the formation of all products formed in this reaction. Which products, if any, will be formed in equal amounts?

$$\text{(structure with D substituents)} \xrightarrow{\text{HCl}} ?$$

11.6 Addition of a Weak Acid: Acid Catalysis

Thus far, we have examined the addition of only strong acids to alkenes. What happens when weak acids are added to alkenes?

As indicated in Equation 11-14, effectively no reaction occurs if an alkene is treated with water—a weak acid—under neutral conditions.

$$\text{(alkene)} \xrightarrow{H_2O} \boxed{\text{No reaction}} \qquad (11\text{-}14)$$

With water as the acid, the first step of the general electrophilic addition mechanism (Equation 11-3) is too unfavorable to take place at a reasonable rate. Not only would the first step produce a carbocation, it would also produce HO^-, in which the negative charge is relatively poorly stabilized. (See Your Turn 11.6.)

11.6 YOUR TURN

Draw the hypothetical two-step mechanism just mentioned, in which water adds across the double bond in Equation 11-14 according to the mechanism in Equation 11-3. Identify the HO^- intermediate.

Equation 11-15 shows, on the other hand, that

The addition of water to an alkene takes place readily in the presence of a strong acid, such as sulfuric acid, producing an alcohol.

Equation 11-15 is an example of an **acid-catalyzed hydration reaction**.

Water has added across the C—C double bond.

Acid-catalyzed hydration

$$\text{But-1-ene} \xrightarrow[H_2SO_4]{H_2O} \text{Butan-2-ol} \qquad (11\text{-}15)$$
$$\text{OH}$$
$$90\%$$

The mechanism for the hydration reaction is shown in Equation 11-16. Due to the leveling effect (Section 6.2C), H_3O^+ is the strongest acid that can exist in significant amounts in water. Being a strong Brønsted acid, H_3O^+ can protonate the alkene to produce a carbocation intermediate, as illustrated in Step 1. Notice that Step 1 adheres to Markovnikov's rule: The proton adds to the terminal C to yield the more stable carbocation intermediate. In Step 2, that carbocation intermediate is attacked by a weak H_2O nucleophile. A final deprotonation in Step 3 removes the positive charge from the adduct, yielding an uncharged product.

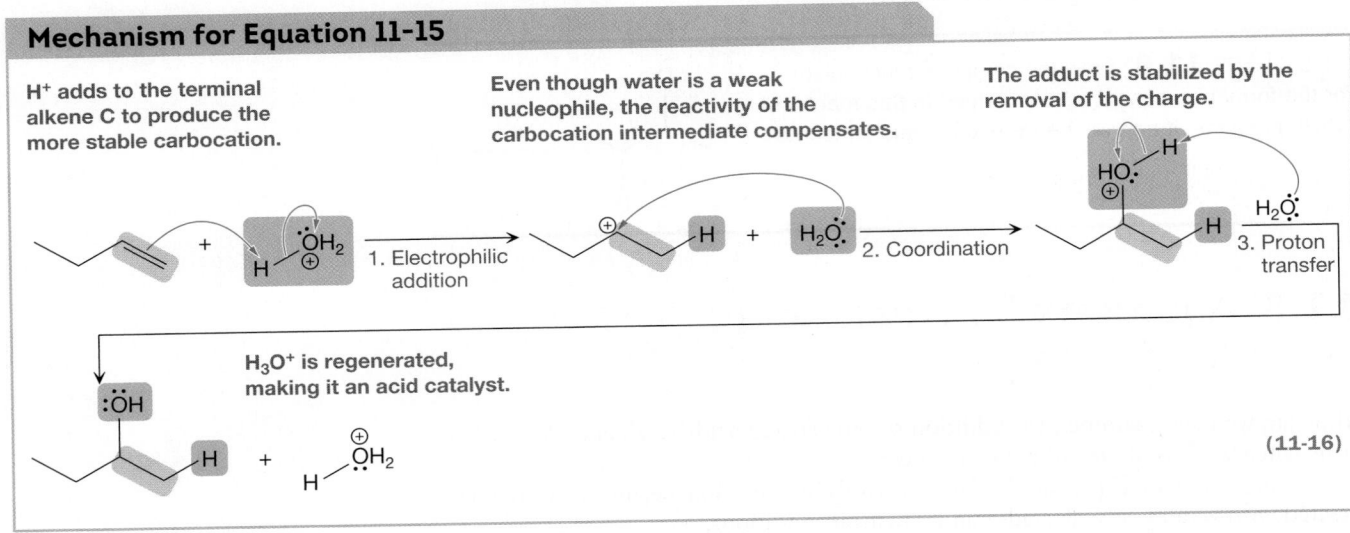

Mechanism for Equation 11-15

H⁺ adds to the terminal alkene C to produce the more stable carbocation.

Even though water is a weak nucleophile, the reactivity of the carbocation intermediate compensates.

The adduct is stabilized by the removal of the charge.

1. Electrophilic addition

2. Coordination

3. Proton transfer

H_3O^+ is regenerated, making it an acid catalyst.

(11-16)

Water is not the only weak nucleophile that can add across a C=C double bond. When an alcohol adds, as shown in Solved Problem 11.10, it is called an **acid-catalyzed alkoxylation reaction**.

SOLVED problem 11.10 Draw the complete, detailed mechanism for this reaction and predict the products.

$$\xrightarrow[\text{H}_2\text{SO}_4]{\text{CH}_3\text{OH}} \quad ?$$

Think What is the electrophile? What is the nucleophile? To which alkene carbon does the electrophile add?

Solve The strongest acid that can exist in methanol is $CH_3OH_2^+$, which can react with the alkene in an electrophilic addition step. Formally, H⁺ from this acid adds to the terminal C in Step 1, producing a secondary, resonance-stabilized carbocation intermediate. (This is consistent with Markovnikov's rule.)

1. Electrophilic addition

2. Coordination

3. Proton transfer

In Step 2, CH_3OH can behave as a nucleophile, forming a bond to the positively charged C. In Step 3, methanol acts as a base to remove a proton from the adduct, yielding an uncharged product. Overall, this is the same mechanism as in Equation 11-16. Unlike Equation 11-16, however, the product here is an *ether*, not an alcohol.

problem **11.11** Draw the complete, detailed mechanism for each of the following reactions and predict the products.

(a)

(b)

(c)

(d)

problem **11.12** Show how each of the following compounds can be produced from an alkene.

(a)

(b)

(c)

11.7 Electrophilic Addition of a Strong Brønsted Acid to an Alkyne

Like the C=C double bond of an alkene, the C≡C triple bond of an alkyne is relatively electron rich—that is, the six electrons of the C≡C triple bond reside largely in the region between the two carbon atoms, and the π electrons are relatively easily accessible. We might expect, therefore, that alkynes undergo electrophilic addition with strong Brønsted acids in much the same way alkenes do.

When hex-1-yne is treated with HBr, for example, some vinylic bromide is produced (Equation 11-17), resulting from a single addition of HBr across the triple bond.

A single addition of HBr across the triple bond produces a vinylic bromide.

(11-17)

Hex-1-yne

2-Bromohex-1-ene

< 8%

The mechanism for this reaction is shown in Equation 11-18.

Mechanism for Equation 11-17

Markovnikov addition proceeds through the more stable carbocation intermediate.

1. Electrophilic addition

2. Coordination

(11-18)

As in electrophilic addition to an alkene, a hydrogen halide adds to an alkyne in a Markovnikov fashion, proceeding through the more stable of two possible carbocation intermediates. In this case, H^+ adds to the terminal C in Step 1 to produce a vinylic carbocation that is stabilized by the attached alkyl group. In Step 2, Br^- attacks the positively charged C in a coordination step.

YOUR TURN 11.7

The mechanism for the addition of HCl to hex-1-yne is as follows, but the curved arrows have been omitted:

Supply the missing curved arrows and, under each reaction arrow, write the appropriate name of the elementary step taking place.

Equation 11-17 shows that

> Electrophilic addition of a single equivalent of a strong Brønsted acid to an alkyne generally gives a poor yield, so it is not very useful in synthesis.

The yield is poor because the positive charge in the carbocation intermediate is on a vinylic C. That C is *sp* hybridized (notice that only two electron groups are attached to it), so it has a relatively high effective electronegativity (Section 3.11). Consequently, the positive charge is rather poorly stabilized, which makes the reaction proceed slowly.

Stereochemistry becomes an issue in these reactions when one equivalent of a hydrogen halide adds to an internal alkyne, as in Equation 11-19. Because the product is an internal alkene, both *E* and *Z* configurations exist. Typically, however, the major product is produced from *anti* addition across the alkyne (in this case, it is the *Z* isomer). Why this is so is not completely understood.

The major product is from *anti* addition.

But-2-yne   →HBr   (*Z*)-2-Bromobut-2-ene

(11-19)

The product of a single addition of a hydrogen halide to an alkyne has a C=C double bond, so a second addition can take place with excess HX to produce a dihaloalkane. An example is shown in Equation 11-20, in which propyne is treated with excess HBr and is allowed to react for four days.

Addition of two equivalents of HBr to an alkyne produces a *geminal* dibromide, in which both Br atoms are on the same C atom.

(11-20)

Propyne

2,2-Dibromopropane
100%

These reactions are highly regioselective, as can be seen from Equation 11-20. Specifically,

When two equivalents of a hydrogen halide add to an alkyne, both halogen atoms end up on the same carbon atom in the product, producing what is called a **geminal dihalide**.

The reason for this regioselectivity is explained by the mechanism, shown in Equation 11-21.

Mechanism for Equation 11-20

(11-21)

The first two steps of the mechanism are identical to those in Equation 11-18. In Step 1, H^+ adds to the terminal C, thus producing the more stable carbocation intermediate (i.e., Markovnikov addition), and Step 2 is coordination involving the newly formed carbocation and bromide ions. In Step 3, H^+ adds to the terminal C of the alkene group, which allows the carbocation that is produced to be stabilized by resonance involving a lone pair of electrons from Br; no such resonance stabilization would be possible if the H$^+$ were to add to the central C atom. Finally, in Step 4, a second Br$^-$ anion attacks the positively charged C to produce the uncharged geminal dibromide.

The mechanism for the addition of HCl to propyne is as follows, but the curved arrows have been omitted:

Supply the missing curved arrows and, under each reaction arrow, write the appropriate name of the elementary step taking place. Also, draw the missing resonance structure in the box provided.

SOLVED problem 11.13 Predict the major product of the following reaction and draw its complete, detailed mechanism.

Think Based on the electron-rich and electron-poor sites in the reactants, what type of reaction will take place? What aspects of regiochemistry should be considered?

Solve The alkyne is relatively electron rich and the H atom of HCl is relatively electron poor. So, the conditions favor an electrophilic addition reaction in which HCl adds across the multiple bond. The product of that addition to the alkyne is a vinyl halide, which, in the presence of excess HCl, can undergo a second electrophilic addition.

Resonance-stabilized carbocation

In the first addition, regiochemistry is not an issue because the alkyne is symmetric—the same vinyl halide is produced regardless of which C forms the new bond to H$^+$. In the second addition, however, regiochemistry is a concern because the vinyl halide is not symmetric. Addition of the H$^+$ is favored at the C atom that is *not* bonded to Cl, because that produces a resonance-stabilized carbocation intermediate.

problem **11.14** For each of the following reactions, draw the complete, detailed mechanism, predict the major product, and state whether the product will be produced in high yield.

(a)   **(b)**

problem **11.15** Show how to synthesize each of the following compounds from an alkyne and state which one(s) will be produced in high yield.

(a)   **(b)**

11.8 Acid-Catalyzed Hydration of an Alkyne: Synthesis of a Ketone

Recall from Section 11.6 that under acidic conditions, water adds across the double bond of an alkene to produce an alcohol. It makes sense, therefore, that an alkyne would react in a similar way under the same conditions. Indeed, Equation 11-22 shows that a reaction does take place, but

Acid-catalyzed hydration of an alkyne produces a ketone, not an alcohol.

Acid-catalyzed hydration of an alkyne produces a ketone.

Ethynylbenzene   **Phenylethanone**   (11-22)
                            (trace amount)

The mechanism for Equation 11-22 is shown in Equation 11-23.

The vinylic carbocation intermediate is resonance stabilized.

1. Electrophilic addition

2. Coordination

(11-23)

3. Proton transfer

4. Electrophilic addition

The enol rapidly converts to the more stable keto form via *tautomerization*.

Resonance stabilization

5. Proton transfer

The more stable keto form

Steps 1–3 are the same as the ones that make up the mechanism for the acid-catalyzed hydration of an alkene, shown previously in Equation 11-16. Step 1 is electrophilic addition of H$^+$, which obeys Markovnikov's rule and produces a relatively unstable vinylic carbocation intermediate. In Step 2, water behaves as a nucleophile, attacking the positively charged C of that intermediate. Deprotonation in Step 3 produces the uncharged alcohol group, which is attached to an alkene C. The product of Step 3 is an *enol*.

Under these acidic conditions, the enol rapidly tautomerizes (Section 7.9) to the more stable *keto* form in two steps. First, in Step 4, a proton adds to the terminal C atom. Then, in Step 5, the O atom is deprotonated, yielding the final uncharged ketone.

There are two important aspects of this reaction that you need to understand. First, the product is a ketone, not an aldehyde. This is an outcome of Markovnikov's rule, which dictates that H$^+$ should add preferentially to the terminal C in Step 1 to give the more stable carbocation intermediate. This forces water to add to the internal C atom in the second step. An aldehyde could form if the proton added to the internal C atom in the first step, but that would produce the less stable carbocation intermediate (see Your Turn 11.9).

YOUR TURN 11.9

Draw the carbocation intermediate that must be generated to produce an aldehyde from the acid-catalyzed hydration of ethynylbenzene. Explain why it is less stable than the carbocation intermediate shown in Equation 11-23.

The second aspect of this reaction to take note of is that the product yield of the reaction in Equation 11-22 is very low; in fact, only a trace amount of ketone is produced, so this reaction is not very useful synthetically. The reason for such a low yield is that the addition of H^+ in the first step of the mechanism produces a vinylic carbocation intermediate, which, as mentioned in Section 11.7, is generally highly unstable. The first step of the mechanism, therefore, is excessively unfavorable.

This problem can be circumvented by using a Brønsted acid that is significantly stronger than sulfuric acid, thus making the first step of the mechanism more favorable. Examples of such acids include trifluoromethanesulfonic acid, CF_3SO_3H (TfOH), and trifluoromethanesulfonamide, $(CF_3SO_2)_2NH$ (Tf_2NH). As shown in Equations 11-24 and 11-25, a relatively high yield of the ketone product can be obtained by adding a catalytic amount of one of these acids.

Acid-catalyzed hydration of an alkyne is facilitated by acids that are significantly stronger than H_2SO_4.

$$\text{Ethynylbenzene} \xrightarrow[\text{catalytic TfOH, 48 h}]{H_2O} \text{Phenylethanone } 93\%$$

(11-24)

$$\text{Oct-4-yne} \xrightarrow[\text{catalytic Tf}_2\text{NH, 48 h}]{H_2O} \text{Octan-4-one } 64\%$$

(11-25)

11.10 YOUR TURN

Verify that trifluoromethanesulfonic acid is a significantly stronger acid than sulfuric acid by looking up their pK_a values and writing them in the spaces provided here.

H_2SO_4 _____   TfOH _____

Another way to circumvent the problem associated with the acid-catalyzed hydration of an alkyne is to use a mercury(II) catalyst, Hg^{2+}, as shown in Equation 11-26. The presence of a mercury(II) catalyst changes the mechanism, which is discussed in Chapter 12.

Mercury(II) catalyst

$$\text{Hex-1-yne} \xrightarrow[\substack{H_2SO_4,\ HgSO_4 \\ \text{acetic acid} \\ 60\ °C,\ 4\ h}]{H_2O} \text{Hexan-2-one } 80\%$$

(11-26)

problem **11.16** Draw the complete, detailed mechanism for the following reaction, and predict the major product.

$$\xrightarrow[\text{catalytic Tf}_2\text{NH}]{H_2O} \quad ?$$

11.9 Electrophilic Addition of a Brønsted Acid to a Conjugated Diene: 1,2-Addition and 1,4-Addition

A molecule such as buta-1,3-diene is said to have **conjugated** double bonds because the two double bonds are separated by another bond (in this case, a single bond). A conjugated diene such as buta-1,3-diene is electron rich, much like the previous alkenes we have studied in this chapter, so it undergoes electrophilic addition with Brønsted acids (an example is shown in Equation 11-27 with HCl as the acid). Notice that this reaction yields a mixture of isomeric products. One product, 3-chlorobut-1-ene, appears to be as expected because the H⁺ and Cl⁻ have added across one of the double bonds with Markovnikov regiochemistry. The other product, 1-chlorobut-2-ene, cannot be produced simply by the addition of HCl across one of the double bonds, because the double bond in the product is in a different location than either of the double bonds in the reactant species.

The location of the double bond differs from those in the starting material.

Buta-1,3-diene → HCl → **3-Chlorobut-1-ene** + **1-Chlorobut-2-ene** (11-27)

Both reaction products are produced from the same general mechanism, as shown in Equation 11-28. In Step 1, H⁺ adds to a π bond, and in Step 2, the Cl⁻ anion attacks the newly formed carbocation intermediate in a coordination step.

Mechanism for Equation 11-27

Buta-1,3-diene → 1. Electrophilic addition → 2. Coordination → **3-Chlorobut-1-ene** — Product of 1,2-addition (11-28a)

2. Coordination → **1-Chlorobut-2-ene** — Product of 1,4-addition (11-28b)

Resonance hybrid

Both products are produced from the *same carbocation intermediate*. As indicated in Equation 11-28, this intermediate has two resonance structures—one with the positive charge on C1 and one with the positive charge on C3. In the resonance hybrid, therefore, each of those two C atoms bears a partial positive charge, so each can be attacked by Cl⁻. Attack on one of those C atoms (Equation 11-28a) yields 3-chlorobut-1-ene, whereas attack on the other carbon atom (Equation 11-28b) yields 1-chlorobut-2-ene.

The H⁺ and Cl⁻ that added to the diene to produce 3-chlorobut-1-ene are separated by two C atoms, making it the product of **1,2-addition**. On the other hand, 1-chlorobut-2-ene is the product of **1,4-addition** because the H⁺ and Cl⁻ that added to the diene are separated by four C atoms.

11.11 YOUR TURN

Which of the following products is produced from 1,2-addition, and which is produced from 1,4-addition?

Addition of the proton in the first step of the mechanism occurs to give *the more stable carbocation intermediate*, in accord with Markovnikov's rule, just as we saw previously in the electrophilic addition to propene (Equation 11-6). The carbocation intermediate shown in Equation 11-28, which is produced by the addition of H⁺ to a terminal C, is *resonance stabilized*. If one of the internal C atoms were to gain the H⁺, on the other hand, then the positive charge in the resulting carbocation would be *localized* on a primary C, as shown in Equation 11-29.

This carbocation intermediate is not as stable as the one in Eq. 11-28.

(11-29)

SOLVED problem 11.17 Draw the major 1,2-addition and 1,4-addition products of this reaction.

Think How many distinct carbocation intermediates are possible from protonation of a double bond? Which one is the most stable? What are the species that can be produced upon nucleophilic attack of that carbocation intermediate?

Solve Each of the alkene groups can gain a proton at either of its C atoms, giving rise to four possible carbocation intermediates.

Intermediates **A** and **D** are more stable than **B** and **C** due to resonance delocalization of the positive charge. Intermediate **A** is more stable than **D** (and is the most stable of all four) because its positive charge is shared on a tertiary C atom.

With the most stable intermediate identified, the 1,2- and 1,4-addition products are obtained by attack of the nucleophile, H_2O, on the carbon atoms sharing the positive charge.

problem **11.18** Draw the complete, detailed mechanism for the formation of the major 1,2- and 1,4-addition products of this reaction.

problem **11.19** When an unknown conjugated diene is treated with HCl, this mixture of two chloroalkenes is produced:

Draw the conjugated diene that was used as the reactant and draw the complete, detailed mechanism that leads to the formation of each product.

11.10 Kinetic versus Thermodynamic Control in Electrophilic Addition to a Conjugated Diene

As we discussed in Section 11.9, electrophilic addition to buta-1,3-diene yields a mixture of addition products—a 1,2-adduct and a 1,4-adduct—from the same carbocation intermediate. Which adduct is the major product? The answer to that question can depend upon the *temperature* at which the reaction is carried out. If the electrophilic addition of HCl to buta-1,3-diene is carried out at room temperature, for example, then the 1,4-adduct is the major product (Equation 11-30).

Major product at warm temperatures

Buta-1,3-diene → (HCl, 20 °C)

3-Chlorobut-1-ene
25% of adduct mixture

1-Chlorobut-2-ene
75% of adduct mixture

(11-30)

If the reaction is carried out at cold temperatures, however, then the 1,2-adduct is the major product (Equation 11-31).

Major product at cold temperatures

Buta-1,3-diene → (HCl, −80 °C)

3-Chlorobut-1-ene
80% of adduct mixture

1-Chlorobut-2-ene
20% of adduct mixture

(11-31)

We observe these temperature effects because the temperature at which the reaction is run governs whether the reaction is *reversible* or *irreversible*. At low temperatures, the product molecules do not possess enough energy to climb over the energy barrier in the reverse direction to reform reactants at a significant rate (review Fig. 11-2), effectively making electrophilic addition to the diene irreversible. At high temperatures, on the other hand, the product molecules do possess enough energy to make the reaction reversible. Combining this information with what we learned in Section 9.12:

- At low temperatures, electrophilic addition to a conjugated diene takes place under *kinetic control*, so the major product is the one that is produced most rapidly.
- At high temperatures, electrophilic addition to a conjugated diene takes place under *thermodynamic control*, so the major product is the one that is most stable.

Based on the temperatures reported in Equations 11-30 and 11-31, therefore, the 1,4-adduct must be the thermodynamic product and the 1,2-adduct must the kinetic product.

In Equations 11-30 and 11-31, label each product as either the "kinetic product" or the "thermodynamic product."

11.12 YOUR TURN

We can rationalize the 1,4-adduct being the thermodynamic product by the stability of the alkene group (Fig. 11-5). Notice that the alkene group in the 1,4-adduct is *disubstituted* (i.e., bonded to two alkyl groups), whereas the alkene group in the 1,2-adduct is only *monosubstituted*. Then recall from Section 5.9 that the more highly alkyl substituted an alkene is, the more stable it is.

YOUR TURN 11.13

Identify the thermodynamic product in the following electrophilic addition reaction.

A monosubstituted alkene
(a)

The disubstituted alkene is more stable.

A disubstituted alkene
(b)

FIGURE 11-5 1,2- and 1,4-adduct stability (a) The 1,2-adduct and (b) the 1,4-adduct of Equations 11-30 and 11-31. The 1,4-adduct is more stable because its double bond is more highly substituted.

The 1,2-adduct is the kinetic product simply as a result of the location of Cl⁻ after completion of the first step of the mechanism. Upon addition of the H^+ to the diene, Cl⁻ is closer to C2 than it is to C4 (Equation 11-32).

Cl⁻ is closer to this C atom.

The 1,2-adduct is formed faster. (11-32)

Therefore, Cl⁻ can attack C2 more quickly than it can attack C4.

In the case of electrophilic addition to buta-1,3-diene, the kinetic and thermodynamic products are different—the 1,2-adduct is the kinetic product and the 1,4-adduct is the thermodynamic product. This is not always the case, however, as shown in Solved Problem 11.20.

SOLVED problem 11.20 Determine the major thermodynamic and kinetic products in this reaction and draw the complete, detailed mechanism for the formation of each.

Think What are the *possible* carbocation intermediates? What is the most stable carbocation intermediate? What are the adducts that can be produced upon nucleophilic attack of that carbocation intermediate? Which of those adducts is produced the fastest? Which is the more stable adduct?

Solve Two carbocations can be produced if the π bond at the top is used to form a bond to H^+.

A   **B**

The more stable carbocation intermediate is **A**, because it is resonance stabilized:

The 1,2-adduct is formed faster, and is also the more stable alkene product.

The 1,4-adduct

A

Adding I⁻ yields the 1,2- and 1,4-adducts. As usual, the 1,2-adduct is the kinetic product. It is also the more stable alkene product because it is more highly alkyl substituted. Therefore, the 1,2-adduct is the thermodynamic product, too.

---

problem **11.21** Determine the major thermodynamic and kinetic products in the following reaction and draw the complete, detailed mechanism for the formation of each.

11.11 Wrapping Up and Looking Ahead

Chapter 11 introduced electrophilic addition reactions, using the addition of a hydrogen halide to an alkene as the prototype reaction. This reaction proceeds by a mechanism that consists of two steps: addition of a proton to the alkene, which produces a carbocation intermediate, followed by coordination of a nucleophile to the newly formed carbocation intermediate.

The appearance of a carbocation intermediate in the mechanism explains two important aspects of the reaction. One is the regioselectivity that is described by *Markovnikov's rule*—namely, the mechanism proceeds through the most stable carbocation intermediate. The second is the rearrangement of the carbon skeleton that is occasionally observed and is expected of carbocations that can become more stable via a 1,2-hydride or 1,2-methyl shift.

With a solid foundation in this prototypical electrophilic addition reaction, we subsequently explored electrophilic addition reactions involving other Brønsted acids and other functional groups containing nonpolar π bonds. We saw, for example, that the addition of a weak acid, such as water or an alcohol, can take place via acid catalysis, and that the π bond that undergoes electrophilic addition can be part of an alkyne or a diene.

Special attention must be paid when an alkyne or conjugated diene undergoes electrophilic addition. When water adds to an alkyne, for example, a ketone or an aldehyde is produced via an unstable enol intermediate. And electrophilic addition

Mad cow disease and its human form, Creutzfeldt–Jakob disease (CJD), are neurologic disorders that are incurable, degenerative, and fatal. They are classified as prion (short for "protein infec-tion") diseases because of the mode by which they are transmit-ted and proliferate: A protein in its native conformation (i.e., the conformation in which the protein functions normally) becomes mis-folded, and that diseased protein promotes the refolding of other native proteins into their diseased states. In the case of CJD, the protein that undergoes this refold-ing (i.e., the prion protein) is water

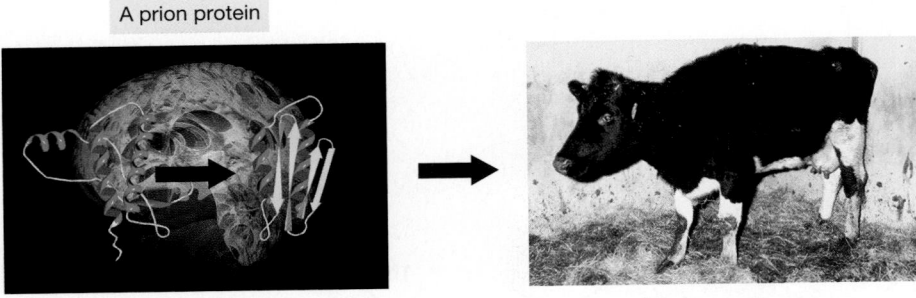

A prion protein

Native state   Diseased state

Mad cow disease

soluble and is believed to be responsible for transmembrane transport or cell signaling when in its native state. In its diseased state, that prion protein forms aggregates and becomes water *insoluble*.

The mechanism by which prion proteins refold into their diseased states is not exactly clear, but a key piece of the puzzle has to do with kinetic control versus thermodynamic control in the protein folding process—the same type of phenomenon that governs electrophilic addition to conjugated dienes that we have studied here in Chapter 11. In their native states, many proteins are believed to adopt their most energetically stable conformations—that is, they fold under ther-modynamic control. But F. E. Cohen and coworkers at the University of California, San Francisco, showed that the native state of the mouse recombinant prion protein is *not* in the most stable conformation; it folds under kinetic control. Understanding the specific factors that lead to such a kinetically controlled process could someday lead to treatments or cures for prion diseases.

to a conjugated diene can take place via 1,2- or 1,4-addition, with the major product frequently governed by the temperature at which the reaction is carried out.

This was the first of two chapters dealing with electrophilic addition reactions. In Chapter 12, our focus turns toward electrophiles that are *not* Brønsted acids. Interestingly, many of those reactions proceed through mechanisms that involve a cyclic transition state, and have regioselectivity or stereoselectivity that make them quite useful for synthesis.

THE ORGANIC CHEMISTRY OF BIOMOLECULES

11.12 Terpene Biosynthesis: Carbocation Chemistry in Nature

Recall from Section 2.10c that many natural products are *terpenes* or *terpenoids*, including essential oils from plants and steroids. More importantly, recall that ter-penes and terpenoids are distinguished from other natural products by the structure of their carbon backbone—namely, the carbon backbone of a terpene or terpenoid consists of multiple *isoprene units* linked by their terminal carbons (Fig. 11-6). One of the keys to terpene synthesis in nature is isopentenyl pyrophosphate, which itself is produced in several steps from acetic acid (Equation 11-33). As you can see, isopen-tenyl pyrophosphate has the same carbon structure as the fundamental isoprene unit.

Isoprene unit

FIGURE 11-6 An isoprene unit Four carbons make up a chain and a fifth carbon is connected to C2 of that chain.

Acetic acid → **Mevalonic acid** → **Isopentenyl pyrophosphate** (11-33)

The pyrophosphate group

Some fraction of isopentenyl pyrophosphate reacts with an enzyme (a protein catalyst), isomerizing to dimethylallyl pyrophosphate. As shown in Equation 11-34, the first step is electrophilic addition of a proton, and the second step is the elimination of a different proton.

Addition of H⁺ produces the more stable carbocation (Markovnikov's rule).

Elimination of H⁺ produces the more substituted alkene (Zaitsev's rule).

Isopentenyl pyrophosphate → Electrophilic addition → Electrophile elimination → **Dimethylallyl pyrophosphate**

(11-34)

Notice that the electrophilic addition step produces the more stable carbocation intermediate, thus adhering to Markovnikov's rule (Section 11.3), and that the elimination step produces the more highly substituted alkene product, consistent with Zaitsev's rule (Section 9.10).

Dimethylallyl pyrophosphate then loses its pyrophosphate group (a rather good leaving group) in a heterolysis step, as shown in Equation 11-35, resulting in an allylic carbocation.

Allylic carbocation

Dimethylallyl pyrophosphate → Heterolysis → + **Isopentenyl pyrophosphate** → Electrophilic addition

(11-35)

More stable carbocation intermediate → Electrophile elimination → **Geranyl pyrophosphate** + **Neryl pyrophosphate**

The allylic carbocation then reacts with isopentenyl pyrophosphate in an electrophilic addition step, producing another carbocation intermediate that contains 10 carbon atoms. Subsequent elimination of H⁺ produces geranyl pyrophosphate and neryl pyrophosphate, which are *E/Z* isomers.

Neryl pyrophosphate can then react in a number of ways to produce various terpenes and terpenoids containing 10 carbons. For example, Equation 11-36 shows it can undergo an intramolecular electrophilic addition reaction to produce a carbocation intermediate with a six-membered ring.

Electrophilic addition produces the more stable carbocation intermediate.

Neryl pyrophosphate → Heterolysis → **Neryl cation** → Electrophilic addition → → :Enzyme / Electrophile elimination → **Limonene**

(11-36)

Subsequent elimination of H^+ yields limonene, a monoterpene that is a constituent of citrus oils. Alternatively, geranyl pyrophosphate can undergo nucleophilic substitution with water to produce geraniol (Equation 11-37), found in oils of the rose and geranium plants.

Geranyl pyrophosphate → Heterolysis → **Geranyl cation** → H_2O: / Coordination (11-37)

→ H_2O: / Proton transfer → **Geraniol**

---

problem **11.22** Neryl pyrophosphate can react with water to produce α-terpineol and terpin hydrate, terpenoids found in natural oils. Draw the mechanism for each of these reactions.

α-Terpineol   **Terpin hydrate**

Geranyl pyrophosphate can be used to further grow the carbon chain in the synthesis of terpenes and terpenoids. For example, after the pyrophosphate group has left, the geranyl cation can react with isopentenyl pyrophosphate to produce a carbocation intermediate containing 15 carbon atoms, and subsequent elimination of a proton produces farnesyl pyrophosphate, as shown in Equation 11-38.

(11-38)

Farnesyl pyrophosphate can be hydrolyzed with water to produce farnesol (Equation 11-39)—a pheromone for some insects, and an oil used to enhance the aroma of perfumes for some humans.

(11-39)

Alternatively, two farnesyl pyrophosphate molecules can couple in a "tail-to-tail" fashion to produce squalene, a triterpene (Equation 11-40).

(11-40)

Squalene is an important terpene because it is the precursor to cholesterol, from which all other steroid hormones are biosynthesized. Equation 11-41 shows that squalene is oxidized by the enzyme squalene epoxidase in the presence of oxygen, producing squalene oxide.

Under acid catalysis, the epoxide ring in squalene oxide is electrophilic at the tertiary C that is indicated, which facilitates an electrophilic addition step that involves three alkene groups simultaneously. Three new C—C bonds are formed, resulting in three new rings: two that are six membered, and one that is five membered. When each π bond is broken, notice that the pair of electrons remains associated with the *less* alkyl-substituted alkene carbon, which is what we would expect in a Markovnikov addition. Thus, in the carbocation that is produced, the positive charge is on a tertiary carbon.

In the next step, a 1,2-alkyl shift converts the five-membered ring into an additional six-membered ring, and the positive charge moves to a secondary carbon. Subsequently, another electrophilic addition takes place, yielding the protosterol cation; in that step, the final ring is formed. Lanosterol is produced when a proton is eliminated, which also causes the shifting of two methyl groups and two other protons. From lanosterol, several additional steps are required to produce cholesterol.

Chapter Summary and Key Terms

- A strong Brønsted acid, such as HCl or HBr, has an electron-poor proton, thus making it an **electrophile**. These acids can therefore react with the nonpolar π bond of an alkene, which is relatively electron rich. An **electrophilic addition reaction** ensues, in which the Brønsted acid adds across a C=C double bond: the proton to one carbon, and the conjugate base to the other. **(Section 11.1)**

- The rate-determining step in the addition of a Brønsted acid to an alkene is the addition of the proton, which is Step 1 of the general mechanism. **(Section 11.1)**

- Benzene does not undergo electrophilic addition with Brønsted acids, because doing so would destroy the aromaticity of the ring. **(Section 11.2)**

- The *regiochemistry* of the addition of a Brønsted acid across a C=C double bond is described by **Markovnikov's rule**. Markovnikov's rule derives, in turn, from the fact that the addition of a proton to an alkene produces the most stable carbocation intermediate. **(Section 11.3)**

- Because the mechanism for the addition of a Brønsted acid to an alkene proceeds through a carbocation intermediate, these reactions are susceptible to carbocation rearrangements. **(Section 11.4)**

- Stereochemistry is an issue in electrophilic addition reactions involving alkenes if one of the alkene carbons becomes a tetrahedral stereocenter in the product. If the starting alkene is achiral, then any chiral products are typically produced as a racemic mixture of enantiomers. **(Section 11.5)**

- Weak Brønsted acids, such as water and alcohols, can add across the double bond of an alkene under acid catalysis.

When water adds, the reaction is called an **acid-catalyzed hydration reaction**. **(Section 11.6)**

- An alkyne can undergo electrophilic addition of a Brønsted acid to produce a vinyl-substituted alkene. As with electrophilic addition to an alkene, the regiochemistry of the addition to an alkyne is described by Markovnikov's rule. **(Section 11.7)**

- When two equivalents of a hydrogen halide add to a C≡C triple bond, the major product is a **geminal dihalide**. This is an extension of Markovnikov's rule: During the addition of the second equivalent of the acid, the carbocation intermediate that is produced is resonance stabilized if the positive charge appears on the carbon atom already attached to the halogen atom. **(Section 11.7)**

- The acid-catalyzed hydration of an alkyne produces an enol, which quickly tautomerizes to a ketone. **(Section 11.8)**

- A conjugated diene undergoes electrophilic addition via **1,2-addition** and **1,4-addition**, producing a mixture of products. Both mechanisms involve precisely the same carbocation intermediate, in which the positive charge is resonance delocalized over two carbon atoms. **(Section 11.9)**

- Electrophilic addition to a conjugated diene takes place under kinetic control at cold temperatures, and under thermodynamic control at warm temperatures. The product of 1,2-addition is typically the kinetic product. The thermodynamic product is the one in which the alkene product is best stabilized. **(Section 11.10)**

Reaction Table

TABLE 11-1 Functional Group Transformations[a]

	Starting Functional Group	Typical Reagent Required	Functional Group Formed	Key Electron-Rich Species	Key Electron-Poor Species	Comments	Discussed in Section(s)
(1)	Alkene	HX	Alkyl halide			Markovnikov addition	11.1
(2)	Alkene	H_2O / $H^{\oplus}$	Alcohol			Markovnikov addition, acid catalysis	11.6
(3)	Alkene	ROH / $H^{\oplus}$	Ether			Markovnikov addition, acid catalysis	11.6
(4)	Alkyne	HX (1 equiv)	Vinyl halide			Markovnikov addition, predominantly trans; not useful for synthesis	11.7
(5)	Alkyne	HX (2 equiv)	Geminal dihalide			Markovnikov addition twice	11.7
(6)	Alkyne	H_2O / TfOH or Tf_2NH	Ketone			Markovnikov addition of H_2O, keto–enol tautomerization, acid catalyst significantly stronger than H_2SO_4	11.8
(7)	Conjugated diene	HX (1 equiv), cold	1,2-adduct			Kinetic control	11.9; 11.10
(8)	Conjugated diene	HX (1 equiv), warm	1,4-adduct			Thermodynamic control	11.9; 11.10

[a] X = Cl, Br, I

11.23 Cyclohexene can react with hydrogen halides, HX, to yield the various halocyclohexanes, $C_6H_{11}X$. Rank HF, HCl, HBr, and HI in order from slowest reaction rate to fastest. Explain. (*Hint:* What is the rate-determining step?)

11.24 Rank alkenes **A–C** in order from slowest rate of electrophilic addition of HCl to fastest. Explain. (*Hint:* What is the rate-determining step?)

A   **B**   **C**

11.25 Which of alkenes **D–H** will produce 2-chloro-3-methyl-2-phenylbutane as the major product when treated with HCl? Explain.

D   **E**   **F**   **G**   **H**

11.26 Predict the major product(s) for each of the following reactions.

(a) 4-Chlorobut-1-ene + HBr $\longrightarrow$?   **(b)** 1-Chlorobut-1-ene + HBr $\longrightarrow$?

(c) 4,4-Dimethylcyclopentene + H_2O, H^+ $\longrightarrow$?   **(d)** Propyne + 2 HCl $\longrightarrow$?

(e) Cyclopentylethene $\xrightarrow{H_3O^+}$?

11.27 Each of the following compounds can be produced from an alkene, using a single electrophilic addition reaction. Write that reaction and draw its complete, detailed mechanism.
(a) 4-Chloro-1,2-dimethylcyclohexane   **(b)** 1-Chloro-1,2-dimethylcyclohexane
(c) 1-Bromo-1,1-diphenylbutane   **(d)** 2,2-Dichloropentane

11.28 Each of the following compounds can be produced from an alkene, using a single electrophilic addition reaction. Write that reaction and draw its complete, detailed mechanism.
(a) Pentan-2-ol   **(b)** 3-Methylpentan-3-ol
(c) 1-Methoxy-1,4-dimethylcyclohexane   **(d)** Cyclopentyldimethoxyphenylmethane

11.29 Each of the following compounds can be produced from an alkyne, via electrophilic addition. Write the reactions and draw complete, detailed mechanisms.

(a) 1,1-Dichloro-2-cyclopentyl-1-phenylethane   **(b)** 3,3-Dibromohexane

(c)   **(d)**

11.30 Each of the following compounds can be produced from an alkyne, using a single electrophilic addition reaction. Write the reactions and draw complete, detailed mechanisms.

(a)   **(b)**

11.31 Recall that electrophilic addition to an internal alkyne generally leads to a *mixture* of isomeric adducts. The acid-catalyzed hydration of the following internal alkyne leads, however, to only a single adduct. Draw the product and explain why a mixture of adducts is not produced.

$$H_3O^{\oplus} \longrightarrow \quad ?$$

11.32 Show how 1-methylcyclohexanol can be produced from two *different* alkenes.

11.33 Show how you would carry out each of the following transformations. (*Hint:* Two or more separate reactions may be required.)

(a)   **(b)**

11.34 Treatment of (R)-4-chlorocyclohexene with HCl produces a mixture of four products.

HCl ⟶ [Four products]

Draw the mechanism that accounts for the formation of each product, and identify which products are optically active.

11.35 Consider the following addition of HBr.

+ HBr ⟶ ?

(a) Draw all four carbocation intermediates possible upon protonation of the diene and identify the most stable one.
(b) Draw both halogenated products formed by attack of Br⁻ on that carbocation.
(c) Which of those products would you expect to be formed in the greatest amount at low temperatures?
(d) Which would you expect to be formed in the greatest amount at high temperatures?

11.36 Consider the following addition of HBr.

HBr ⟶ ?

(a) There are three carbocation intermediates possible from the protonation of this triene. Draw all three of them and identify the most stable one.
(b) Draw all halogenated products formed by attack of Br⁻ on the most stable carbocation.
(c) Which of those products would you expect to be formed in the greatest amount at low temperatures?
(d) Which would you expect to be formed in the greatest amount at high temperatures?

11.37 Recall that the addition of HBr to buta-1,3-diene results in 1,2-addition at cold temperatures and 1,4-addition at warm temperatures.

1,2-Adduct
Major product
at –78 °C

1,4-Adduct
Major product
at +50 °C

If the 1,2-adduct is first formed at cold temperatures and then warmed up, the 1,4-adduct is formed, as shown below. Draw a mechanism for this isomerization.

heat

11.38 Draw the complete, detailed mechanism for the addition of HCl to dihydropyran and predict the major product.

+ HCl ⟶ ?

11.39 Draw the complete, detailed mechanism for the addition of hexan-3-ol to dihydropyran and predict the major product.

11.40 Determine the structures of compounds **A** through **D** in the following reaction scheme:

$\xrightarrow[H^{\oplus}]{H_2O}$ **A** $\xrightarrow[2.]{1.\ NaH}$ **B** $\xrightarrow{PBr_3}$ **C** $\xrightarrow[\Delta]{NaOH}$ **D**

3. H⊕

11.41 Determine the structures of compounds **E** through **I** in the following reaction scheme:

$\xrightarrow[Tf_2NH]{H_2O}$ **E** $\xrightarrow[2.]{1.\ NaNH_2}$ **F** $\xrightarrow[H^{\oplus}]{CH_3OH}$ **G** $\xrightarrow[NaOH]{Br_2}$ **H** $\xrightarrow[\Delta]{NaOH}$ **I**

11.42 Determine the structures of compounds **J** through **N** in the following reaction scheme.

$\xrightarrow[2.\ H_2O]{1.\ 2\ equiv.\ NaNH_2}$ **J** $\xrightarrow[2.]{1.\ NaNH_2}$ **K** $\xrightarrow{PBr_3}$ **L** $\xrightarrow[Tf_2NH]{H_2O}$ **M** $\xrightarrow{NaCN}$ **N**

3. H₂O

11.43 As discussed in greater detail in Chapter 22, $AlCl_3$ is a powerful Lewis acid that effectively catalyzes the dissociation of an alkyl chloride, RCl, into its respective ions, R^+ and Cl^-.

(a) Propose a mechanism for the addition of RCl across a double bond, as shown in the following reaction:

(b) Using that mechanism, what are the two possible isomers that can be formed when 2-methylpropene is treated with 2-chloropropane in the presence of $AlCl_3$? Which one is the major product? Explain.

11.44 As discussed in Chapters 25 and 26, a polymer is a very large molecule that contains many repeating units called monomers. The following reaction shows, for example, how styrene reacts to form polystyrene.

Styrene　　　　　**Polystyrene**

The reaction is *initiated* by the electrophilic addition of H^+ from an acid like sulfuric acid, which generates an initial carbocation. Afterwards, that carbocation behaves as an electrophile in the presence of another molecule of styrene, resulting in yet another carbocation. This reaction can repeat many thousands of times to build up the polymer. With this in mind, draw the detailed mechanism that illustrates the initiation of the polymerization reaction and the addition of the first two monomers, as shown in the following:

11.45 The treatment of but-1-en-3-yne with HBr produces 4-bromobuta-1,2-diene, which is an allene. Draw the complete, detailed mechanism for this reaction.

But-1-en-3-yne　　　　　**4-Bromobuta-1,2-diene**

11.46 Propose a mechanism for the following reaction, in which HCl adds to hepta-1,6-diene.

11.47 When penta-1,4-diene is heated in the presence of HCl, the major product is 1-chloropent-2-ene. Draw the complete, detailed mechanism that accounts for this reaction.

Penta-1,4-diene　　　　　**1-Chloropent-2-ene**

11.48 Propose a reasonable mechanism that would account for the following reaction.

11.49 The regiochemistry in the following electrophilic addition reaction does not adhere to the original generalization put forth by Markovnikov.

Draw the complete mechanism for this reaction and explain its regiochemistry.

11.50 In the biosynthesis of aromatic amino acids, erythrose-4-phosphate undergoes electrophilic addition to phosphoenolpyruvate (PEP).

Phosphoenolpyruvate (PEP)   **Erythrose-4-phosphate**

Draw the products of this step, paying particular attention to regiochemistry.

11.51 α-Terpineol, a naturally occurring monoterpene alcohol, isomerizes to 1,8-cineole and 1,4-cineole when treated with acid.

α-Terpineol   **1,8-Cineole**   **1,4-Cineole**

Draw a complete, detailed mechanism to account for the formation of each product.

If your instructor assigns problems in **smartwork**, log in at **smartwork.wwnorton.com**.

CHAPTER 12

Electrophilic Addition to Nonpolar π Bonds 2

Reactions Involving Cyclic Transition States

A pyrethrin

C hapter 11 discussed reactions in which a Brønsted acid adds to a C=C double bond of an alkene or a C≡C triple bond of an alkyne. The π bond of the alkene or alkyne is relatively electron rich in those reactions, whereas the proton (H^+) from the Brønsted acid is electron poor. Thus, the proton acts as an *electrophile* and, as it adds to the alkene or alkyne, a carbocation intermediate is produced. Subsequently, that carbocation intermediate is attacked by a nucleophile.

Here in Chapter 12, we will see that species other than protons can act as electrophiles, and thus can also be added to an alkene or alkyne. Examples include molecular halogens like Cl_2 and Br_2, peroxyacids (RCO_3H), carbenes ($R_2C\colon$), and borane (BH_3).

A pyrethrin, which has a characteristic cyclopropane ring, is a natural, potent insecticide that can be isolated from chrysanthemum flowers. Pyrethroids are synthetic variations of pyrethrin and constitute the bulk of household insecticides. Many of these pyrethroids are synthesized using a cyclopropanation reaction analogous to the one we present here in Chapter 12.

Unlike the electrophilic addition of a Brønsted acid, all of these reaction mechanisms involve a step whose transition state is cyclic. This prevents the formation of a carbocation intermediate, and therefore has significant consequences on the outcomes of the reactions, on both regiochemistry and stereochemistry.

We begin Chapter 12 with a brief look at a key elementary step that is involved in several of the above reactions: electrophilic addition to form a three-membered ring. We then examine the details of several reactions that proceed by such a step, including carbene addition, halogenation, oxymercuration, and epoxide formation. Following that, we examine hydroboration, which proceeds through a four-membered-ring transition state.

12.1 Electrophilic Addition via a Three-Membered Ring: The General Mechanism

The mechanisms of all of the electrophilic addition reactions we saw in Chapter 11 include a step in which a carbocation is produced. Recall that this step is generally highly unfavorable, due in large part to the loss of an octet on a C atom, as indicated in Equation 12-1.

$$(12\text{-}1)$$

However, if the electrophilic atom on the electrophile has a lone pair of electrons, then addition can take place in a way that avoids losing an octet. This is illustrated in Equation 12-2, in which "E:" represents a generic electrophile that possesses a lone pair of electrons.

$$(12\text{-}2)$$

In this step, two new E—C sigma bonds are formed simultaneously to produce a three-membered ring. One of those bonds is formed by the electrons from the initial carbon–carbon π bond, and the other is formed by the lone pair of electrons from the electrophile. Each of the carbon atoms that forms a new bond to the electrophile maintains its octet throughout the course of the reaction step because, as one of those carbon atoms loses its share of the two π electrons, it gains a share of two electrons in the new σ bond to the electrophile. The electrophile, on the other hand, gains a share of two electrons, but its formal charge does not change because it is assigned the same number of electrons from the lone pair as it is from two covalent bonds.

Notice in Equation 12-2 that as the alkene reacts, two C atoms re-hybridize from sp^2 to sp^3, so *stereochemistry* can be an issue. When it is, stereochemistry is governed by the following rules.

- The cis/trans relationship is conserved for the groups attached to the alkene carbons in Equation 12-2. Groups that are on the same side of the double bond in the reactant must end up on the same side of the plane of the three-membered ring in the product (and vice versa).

- If the cyclic product in Equation 12-2 is chiral, then a mixture of stereoisomers is produced—a racemic mixture of enantiomers if the original alkene and other reagents are achiral, or an unequal mixture of diastereomers if otherwise.

These rules are exemplified by Equation 12-3.

A mixture of stereoisomers is produced because the electrophile can add to either face of the initial C=C double bond.

(12-3)

R_2 and R_4 are on the same side of the C=C bond.

R_2 and R_4 are on the same side of the ring.

Stereochemistry is conserved in such reactions because both C—E bonds are formed simultaneously. Normally, if a C=C double bond is converted into a C—C single bond, rotation about that single bond can lead to scrambling of the cis/trans relationship among the groups attached to those carbon atoms. In this case, however, the C—C single bond that remains is part of a ring, so free rotation cannot take place.

The reaction produces a mixture of stereoisomers because an alkene functional group is planar, and the electrophile can add to either face of the alkene. One stereoisomer is produced by the addition of the electrophile to one face, and the second is produced by addition to the other face. If the initial alkene is achiral (e.g., by possessing a plane of symmetry), then addition to each face of the alkene is equally likely. If the initial alkene is chiral, however, then it possesses no plane of symmetry, making electrophilic addition to one face more likely than addition to the other.

Having examined this elementary step in a generic fashion, let's now turn our attention toward four specific reaction mechanisms in which it appears. Section 12.2 discusses the electrophilic addition of a carbene to an alkene; Section 12.3 discusses electrophilic addition involving molecular halogens; Section 12.4 discusses oxymercuration–reduction; and Section 12.5 discusses epoxide formation.

12.2 Electrophilic Addition of Carbenes: Formation of Cyclopropane Rings

Mechanistically, the simplest of the reactions we discuss here in Chapter 12 occurs between an alkene and a **carbene**. A carbene is a species containing a carbon atom that possesses two bonds and a lone pair of electrons, as shown in Figure 12-1. Like a carbocation, a carbene has a C atom that lacks an octet (it is surrounded by only six electrons: two single bonds and one lone pair of electrons) and is generally *highly* electron poor. Unlike a carbocation, however, the electron-poor C atom of a carbene has a formal charge of 0.

No octet = highly electron deficient

A carbene

FIGURE 12-1 A generic carbene A carbene is characterized by a C atom with two bonds and a lone pair of electrons. Even though the C atom is uncharged, carbenes are typically highly electrophilic due to the lack of an octet.

Most carbenes are highly reactive, so they typically have very short lifetimes and cannot be isolated. In these cases, *we cannot simply add a carbene as a reagent*. Instead,

> Highly reactive carbenes must be generated in situ (i.e., "on site") from precursors that can be added as reagents.

Diazomethane, CH_2N_2, is one precursor from which a carbene can be made. An example of how it is used is shown in Equation 12-4. Notice that it produces a new cyclopropane ring.

New cyclopropane ring

80%

[1] Indicates light (h = Planck's constant; ν = photon frequency)

The complete mechanism for this reaction is shown in Equation 12-5.

Mechanism for Equation 12-4

Heat or light causes the weak C—N bond to break.

Carbene

Nitrogen gas escapes the solution irreversibly.

1. Heterolysis

2. Electrophilic addition

Two new C—C bonds are formed simultaneously.

(12-5)

In Step 1, the C—N bond in diazomethane is broken, which produces $H_2C:$. This step is helped by the production of $N_2(g)$, which is a very stable leaving group and a gas, so it escapes the reaction mixture *irreversibly*. In Step 2, $H_2C:$ reacts with the alkene to generate the three-membered ring via the same mechanism as in Equation 12-2.

---

problem **12.1** Draw the complete, detailed mechanism for the reaction at the right and predict the major products.

$$\xrightarrow[\Delta]{CH_2N_2}$$

problem **12.2** Show how each of the following compounds can be produced from an alkene.

(a) (b) (c) (d)

Although the formation of carbene from diazomethane can lead to a good yield of the cyclopropane-containing product, the reaction is of rather limited use in synthesis. One of the main reasons is that diazomethane is explosive and requires extreme caution in the laboratory. A safer way to produce a cyclopropane ring from an alkene is with the Simmons–Smith reaction (see Problem 12.46).

Chloroform ($CHCl_3$) is another precursor that can be used to generate a carbene. Equation 12-6 provides an example in which chloroform is treated with a strong base in the presence of cyclohexene, once again producing a cyclopropane ring. Based on Equation 12-6, the carbene that is generated from chloroform must be dichloromethylene ($Cl_2C:$), also called **dichlorocarbene**, which is different from the $H_2C:$ generated from diazomethane.

$$+ \; CHCl_3 \xrightarrow[\substack{CH_2Cl_2 \\ 45\ h,\ 40-45\ °C}]{NaOH}$$

52%

(12-6)

As shown in the mechanism in Equation 12-7, dichlorocarbene is produced in two steps via **α-elimination**. The base first deprotonates CHCl$_3$, which is mildly acidic (pK_a = 24). In Step 2, Cl$^-$ departs to generate the Cl$_2$C: carbene. Although Cl$^-$ is a good leaving group, this step occurs quite slowly because the carbene that is produced is very unstable. In Step 3, Cl$_2$C: adds to the alkene, producing the cyclopropane ring.

Mechanism for Equation 12-6

12.3 **YOUR TURN**

The following reaction is similar to the one in Equation 12-6.

The mechanism for this reaction is given below, but the curved arrows have been omitted. Complete the mechanism by adding the curved arrows, and write the name of each elementary step below the appropriate reaction arrow.

problem **12.3** Draw the complete, detailed mechanism for the reaction at the right. Using the mechanism, predict the products, paying attention to stereochemistry.

$$\xrightarrow[\text{NaOC(CH}_3)_3]{\text{CHBr}_3} \quad ?$$

problem **12.4** Provide two different ways of synthesizing the compound at the right via carbene addition, each using a different starting alkene.

$$? \longrightarrow$$

12.3 Electrophilic Addition Involving Molecular Halogens: Synthesis of 1,2-Dihalides and Halohydrins

Section 12.2 highlighted the reactions of carbenes with alkenes. The highly electrophilic carbon atom of the carbene possesses a lone pair of electrons, which is what facilitates the formation of a three-membered ring. Carbenes are not the only species in which an electrophilic atom possesses a lone pair of electrons. As we will see here in Section 12.3, molecular halogens, such as Cl_2 and Br_2, share that feature with carbenes, and can therefore react with an alkene to produce a three-membered ring. Unlike what is observed with carbenes, however, the three-membered ring that is produced from a molecular halogen is an unstable intermediate and reacts further to produce relatively stable products such as a *1,2-dihalide* (i.e., a *vicinal dihalide*) or a *halohydrin*.

12.3a Synthesis of 1,2-Dihalides

When cyclohexene is treated with molecular bromine (Br_2) in tetrachloromethane (CCl_4), also called carbon tetrachloride, a racemic mixture of *trans*-1,2-dibromocyclohexane is produced, as shown in Equation 12-8.

Products of *anti* addition only

(12-8)

Cyclohexene   ***trans*-1,2-Dibromocyclohexane**
                              59%

In other words,

> Molecular bromine undergoes anti addition to a C=C double bond.

Although Br_2 does not appear to possess an electron-poor atom, recall that we have seen Br_2 behave as an electrophile once before—in the S_N2 bromination of enols and enolate anions (Section 10.4). As the electron-rich π bond approaches Br_2, Br_2 becomes *polarized*, temporarily making one of the Br atoms electron poor (Fig. 12-2).

Electron-rich π bond

FIGURE 12-2 The electrophilic nature of a molecular halogen in the presence of an alkene (a) When it is isolated from other species, Br_2 is not electron poor. **(b)** As the electron-rich alkene approaches the Br_2 molecule, however, electrons on Br_2 are repelled to the side opposite the alkene, temporarily generating an electron-poor site on the Br atom closer to the alkene. Subsequent flow of electrons takes place between the electron-rich alkene and the electron-poor Br atom.

This Br atom becomes electron poor.

Electrons are pushed to the side opposite the alkene.

(a)   (b)

To account for the stereochemistry of the reaction in Equation 12-8, the mechanism must *not* proceed through a carbocation intermediate; otherwise, both syn and anti addition would take place (review Section 11.5). Instead, as shown in Equation 12-9, the mechanism proceeds through a **bromonium ion intermediate** (possessing a positively charged bromine atom), which is produced in Step 1.

Mechanism for Equation 12-8

The curved arrow from the π bond to Br represents the flow of electrons from an electron-rich site to an electron-poor site. Similar to the reaction in Equation 12-2, a lone pair of electrons on Br forms a bond back to one of the alkene C atoms to avoid breaking the C atom's octet, resulting in a three-membered ring. In this case, the ring consists of two C atoms and one Br atom. Simultaneously, the weak Br—Br bond breaks, and one of the atoms leaves as Br⁻.

Step 2 of the mechanism is an S_N2 reaction: The Br⁻ ion produced in Step 1 acts as the nucleophile, and the positively charged Br atom in the ring acts as the leaving group. This is exactly the same step we saw in the opening of a protonated epoxide ring in Section 10.7b, where a positively charged oxygen atom behaves as the leaving group.

Notice in Equation 12-9 that the S_N2 reaction in Step 2 is precisely what requires the two Br atoms to be anti to each other in the product. Recall from Section 8.5b that the nucleophile in an S_N2 reaction must attack from the side *opposite the leaving group*. The only way to achieve syn addition would be for the Br⁻ nucleophile to attack from the same side on which the leaving group leaves—something that is forbidden in an S_N2 reaction.

Notice also in Step 2 of Equation 12-9 that nucleophilic attack is shown to occur at just the left-hand C atom of the three-membered ring. As shown in Equation 12-10, nucleophilic attack can also occur at the other C atom of the three-membered ring, resulting in the enantiomer of the product shown in Equation 12-9. Because this particular bromonium ion is achiral, the mixture must be *racemic*.

SOLVED problem 12.5 Predict the products of the reaction at the right. Do you expect the product mixture to be optically active? Why or why not?

Think What reaction takes place between Br_2 and an alkene? Are the product molecules chiral? How does chirality relate to optical activity?

Solve Br_2 undergoes anti addition across the double bond, yielding the following mixture of products.

Each product molecule is chiral. Whereas the products in Equation 12-8 are enantiomers, here the two products are diastereomers. They are formed in unequal amounts, therefore, and do not have equal and opposite values of $[\alpha]_D^{20}$. As a result, the product solution will be optically active.

---

problem **12.6** Draw the product(s) that would be produced from reacting each of the following compounds with molecular bromine in carbon tetrachloride. Will the product mixture be optically active?

(a)   (b)   (c)   (d)

Cl_2 also undergoes addition to alkenes to produce vicinal dichlorides (Equation 12-11). As in the addition of Br_2, Cl_2 adds in an anti fashion to the double bond. The mechanism is similar, proceeding through a **chloronium ion intermediate** to avoid breaking the C atom's octet (see Your Turn 12.4).

Anti addition of Cl_2

trans-But-2-ene   (2R,3S)-2,3-Dichlorobutane
73%

(12-11)

The mechanism for the reaction in Equation 12-11 is as follows, but the curved arrows have been omitted.

Complete the mechanism by adding the curved arrows and write the name of each elementary step under the appropriate reaction arrow. Label the chloronium ion intermediate.

Molecular fluorine (F_2) and iodine (I_2) react with alkenes as well. These reactions are not as useful in organic synthesis, however, because F_2 reacts explosively with alkenes, whereas I_2 reacts too slowly. Therefore, we will not discuss these reactions any further.

Treatment of an alkene with molecular chlorine in carbon tetrachloride yields a racemic mixture of (*R*,*R*)-3,4-dichloro-2-methylhexane and its enantiomer. Draw the structure of the initial alkene.

Br$_2$ and Cl$_2$ add to alkynes in much the same way they add to alkenes. For example, oct-1-yne reacts with *one equivalent* of Br$_2$ to yield 1,2-dibromooct-1-ene (Equation 12-12) and with *excess* Br$_2$ to yield the tetrabromide (Equation 12-13).

Br$_2$ adds both syn and anti to the C≡C bond.

Oct-1-yne
Br$_2$ (1 eq.)
CH$_2$Cl$_2$,
16 h, 25 °C

(E)-1,2-Dibromooct-1-ene
65%

+

(Z)-1,2-Dibromooct-1-ene
19%

(12-12)

Two equivalents of Br$_2$ add to an alkyne.

Oct-1-yne
Br$_2$ (excess)

1,1,2-Tetrabromooctane

(12-13)

Notice in Equation 12-12 that a mixture of both *E* and *Z* isomers is formed. In other words,

> Br$_2$ adds to an alkyne via both syn and anti addition.

This means that the reaction must not proceed cleanly through a bromonium ion intermediate—otherwise, only anti addition would take place, producing the *E* isomer exclusively. Why is this so?

Essentially, it is because of ring strain (Equation 12-14). When a bromonium ion intermediate is produced from an alkyne, the resulting three-membered ring consists of two sp^2-hybridized C atoms, whose ideal angles are 120°. By contrast, when a bromonium ion intermediate is produced from an alkene, those two C atoms are sp^3 hybridized, and their ideal angles are 109.5°. As a result, the sp^3 C atoms from the alkene reaction can better handle the small angles required for the three-membered ring, which are near 60°.

A bromonium ion is not produced when Br$_2$ reacts with an alkyne, because the sp^2-hybridized C atoms would be too highly strained in the ring.

(12-14)

Because of this ring strain, chemists believe that the reaction proceeds through a carbocation-like intermediate instead, much as in the addition of a Brønsted acid. As we saw in Section 11.5, nucleophilic attack of a carbocation intermediate is generally not stereoselective.

12.3b Synthesis of Halohydrins

In Section 12.3a (Equation 12-8), we saw that a vicinal dibromide is produced if an alkene reacts with Br_2 in carbon tetrachloride. If the same reaction is carried out in water, however, as shown in Equation 12-15, then a **bromohydrin** is produced instead. We can view the formation of a bromohydrin as the *net* addition of HO—Br across the double bond. Similarly, a **chlorohydrin** is produced when an alkene is treated with Cl_2 in water (Equation 12-16).

Water is the solvent.   **Anti addition of HO and Br**

$$\xrightarrow[H_2O]{Br_2}$$

Cyclohexene   **+**   (12-15)

trans-**2-Bromocyclohexanol**

A bromohydrin

Water is the solvent.   **Anti addition of HO and Cl**

$$\xrightarrow[H_2O]{Cl_2}$$

(*E*)-**But-2-ene**   **+**   (12-16)

(**2S,3R**)- and (**2R,3S**)-**3-Chlorobutan-2-ol**

A chlorohydrin

These two reactions are *stereospecific*—namely,

> In the formation of a **halohydrin** from an alkene, the OH group and the halogen atom add anti to each other.

This stereochemistry is the same as in the addition of Br_2 or Cl_2 across a double bond (recall Equations 12-8 and 12-11), strongly suggesting that halohydrin formation, too, proceeds through a *halonium ion intermediate*. This is illustrated in the mechanism in Equation 12-17.

Mechanism for Equation 12-15

The first two steps are essentially the same as the ones in Equation 12-9, in which Br_2 adds to an alkene. Step 1 is electrophilic addition that produces the bromonium ion intermediate, and Step 2 is an S_N2 reaction that produces a racemic mixture of enantiomers—the enantiomers are produced from attack of the two different C atoms bonded to the Br leaving group. In contrast to the mechanism in Equation 12-9, however, the ring is opened in Step 2 via nucleophilic attack by

H_2O instead of attack by Br^-. Finally, in Step 3, a proton transfer produces the uncharged bromohydrin.

It might seem counterintuitive that H_2O acts as the nucleophile in Step 2, even though the much stronger Br^- nucleophile is present in solution. However, keep in mind the relative concentrations of the two nucleophiles. Br^- is produced in Step 1, which is a slow step because of the relatively unstable bromonium ion intermediate that is also produced. Thus, the concentration of Br^- is maintained at a very low concentration throughout the reaction. By contrast, H_2O is the solvent, and is present in very high concentrations (the concentration of pure water is 55 mol/L). Therefore, the bromonium ion intermediate is much more likely to encounter a molecule of H_2O than Br^-.

12.5 YOUR TURN

The mechanism for the reaction in Equation 12-16 is as follows, but the curved arrows have been omitted.

Complete the mechanism by drawing in the curved arrows. Also, write the name of each elementary step below the appropriate reaction arrow.

---

problem **12.8** Draw the complete, detailed mechanism leading to the formation of the second enantiomer shown in Equation 12-15, which is not shown explicitly in Equation 12-17.

SOLVED problem **12.9** When the reaction in Equation 12-15 is carried out in an aqueous NaCl solution instead of pure water, racemic *trans*-1-bromo-2-chlorocyclohexane is produced along with the bromohydrin. Draw a complete mechanism that accounts for the formation of the dihalo compound.

Think How can you account for the stereochemistry of the reaction—that is, the formation of only the trans isomer? What nucleophiles are present in solution?

Solve To account for the production of the trans isomers exclusively, the reaction must proceed through a bromonium ion intermediate. To open the ring and form a C—Cl bond, Cl^- in solution must act as the nucleophile.

Cyclohexene

trans-1-Bromo-2-chlorocyclohexane

---

problem **12.10** Draw the complete, detailed mechanism of the reaction in Equation 12-15, assuming it was carried out in ethanol instead of water.

Regiochemistry becomes an issue with halohydrin formation if the alkene reactant is *unsymmetrical*. With distinct alkene carbon atoms, two possible constitutional isomers can be produced, depending upon which carbon atom ends up bonded to the halogen atom, and which carbon atom ends up bonded to the OH group. This is the case, for example, with the reaction in Equation 12-18.

Phenylethene

Major product

2-Bromo-1-phenylethan-1-ol
78%

2-Bromo-2-phenylethan-1-ol

(12-18)

These C atoms are distinct.

Why is the first of these products the major product? The answer can be understood by studying the mechanism, which is presented in Equation 12-19.

Mechanism for Equation 12-18

This side of the ring acquires more of Br's positive charge than does the other side of the ring because this C atom is benzylic.

H_2O attacks the side of the ring that acquires the greater amount of positive charge.

1. Electrophilic addition

2. S_N2

3. Proton transfer

(12-19)

The specific isomer that is formed is dictated by Step 2 of the mechanism—namely, nucleophilic attack by water. Although attack can occur at either C atom of the three-membered ring, attack is favored at C1 instead of C2. Recall that we saw a very similar mechanistic step in Equation 10-58 of Section 10.7b, in which a protonated epoxide acts as a substrate in an S_N2 reaction. Even though the positive charge is formally represented on the heteroatom—in this case, Br—it is actually shared among the three atoms of the ring. Given that C1 is benzylic, it can better handle a positive charge than C2, an isolated primary C atom. Therefore, the side of the ring that contains C1 acquires more of that positive charge than does the side of the ring that contains C2. As a result, when the H_2O nucleophile approaches the ring, it is more attracted to the side that contains C1 than it is to the side that contains C2, despite the fact that C2 is less sterically hindered.

problem **12.11** Draw a complete, detailed mechanism for the reaction at the right and predict the major product, paying attention to both regiochemistry and stereochemistry.

$\xrightarrow[H_2O]{Cl_2}$?

Halogenated Metabolites: True Sea Treasures

Natural products isolated from marine organisms have a wide range of potential applications in medicine and beyond. Of particular interest are halogenated compounds that serve as metabolites for many of these organisms. Compounds **A–D** below, for example, are some of the halogenated compounds that have been isolated from marine red algae. Compound **A** has antitumor and cytotoxic properties; **B** has anthelminthic properties (i.e., it expels parasitic worms from the body); **C** has antifouling properties; and **D** has antimicrobial properties.

Marine red algae

A **B** **C** **D**

A bromonium ion intermediate

J. N. Carter-Franklin and A. Butler, from the University of California, Santa Barbara, have shown that the biosynthesis of these halogenated metabolites probably involves a bromonium ion intermediate, much like we saw in the bromination of alkenes here in Chapter 12. Production of the bromonium ion intermediate in these biosynthetic pathways, however, does not involve molecular Br_2. Rather, an alkene is believed to react with Br^+, which is produced from Br^- (an abundant ion in seawater) in the presence of hydrogen peroxide (H_2O_2), catalyzed by the enzyme vanadium bromoperoxidase (V—BrPO). Br^- undergoes a two-electron oxidation in the active site of the enzyme to make Br^+ (or its equivalent), which then adds to the C=C double bond. In many cases, the opening of the three-membered ring results in a new carbocycle, as shown with (*E*)-(+)-nerolidol in the example above.

12.4 Oxymercuration–Reduction: Addition of Water

Recall from Chapter 11 that water can undergo an acid-catalyzed addition to an alkene or alkyne, producing an alcohol or ketone, respectively. Because the reaction proceeds through a carbocation intermediate, we observe *Markovnikov addition* of water. One drawback of these reactions, however, is that carbocation rearrangements are also possible. An example is shown in Equation 12-20.

The product of a carbocation rearrangement

OH

3-Methylbut-1-ene $\xrightarrow[\text{HCl}]{\text{H}_2\text{O}}$ 2-Methylbutan-2-ol

(12-20)

YOUR TURN 12.6

Write the mechanism that accounts for the formation of the product in Equation 12-20.

An alternate method to add water across a double bond is **oxymercuration–reduction** (also called **oxymercuration–demercuration**), an example of which is shown in Equation 12-21. The alkene is first treated with mercury(II) acetate, $Hg(OAc)_2$, in a water–tetrahydrofuran (THF) solution, and that is followed by reduction with sodium borohydride.

Oxymercuration–reduction

Water undergoes Markovnikov addition across the C=C bond with no rearrangement.

OH

3-Methylbut-1-ene $\xrightarrow[\text{2. NaBH}_4, \text{ ethanol}]{\text{1. Hg(OAc)}_2, \text{ H}_2\text{O/THF}}$ 3-Methylbutan-2-ol

(12-21)

There are two aspects of this reaction to be aware of:

- The product of oxymercuration–reduction is the one expected from Markovnikov's rule—that is, the OH group forms a bond to the carbon atom that can better stabilize a positive charge.
- No rearrangement takes place with oxymercuration–reduction.

To better understand the reasons for these outcomes, study the mechanism shown in Equation 12-22.

Mechanism for Equation 12-21

(12-22)

1. Electrophilic addition

This side of the ring bears the larger positive charge.

2. S_N2

3. Proton transfer

Sodium borohydride reduces the C atom, replacing the Hg group with an H atom.

add NaBH₄

The first three steps of the mechanism are identical to those in the formation of a halohydrin, shown previously in Equation 12-19. In Step 1, the Hg atom is electron poor, given that it is bonded to two highly electronegative O atoms. It is therefore attacked by the electron-rich double bond, and simultaneously a lone pair of electrons on Hg forms a bond to C to produce the three-membered ring. The result is a **mercurinium ion intermediate**, which is analogous to the bromonium (or chloronium) ion intermediate we encountered previously. In Step 2 of the mechanism, H_2O acts as a nucleophile to open the three-membered ring, and in Step 3, the positively charged O atom is deprotonated.

Subsequent reduction with sodium borohydride ($NaBH_4$) replaces the Hg-containing substituent with H. Although this may appear to be a simple nucleophilic substitution reaction with H^- as the nucleophile, it actually proceeds through a more complex mechanism that is believed to involve free radicals (species with unpaired electrons; see Chapter 25). Consequently, even though oxymercuration takes place with anti addition, any stereochemistry set up by oxymercuration is scrambled during the reduction step, giving a mixture of both syn and anti addition of water. This is exemplified in the oxymercuration–reduction of methyl (E)-2-methylbut-2-enoate shown in Equation 12-23.

Reduction with NaBH₄ scrambles the stereochemistry.

Syn addition

Anti addition

(12-23)

Methyl (E)-2-methylbut-2-enoate

+ Enantiomer
43% of product

+ Enantiomer
57% of product

---

problem **12.12** Draw a detailed mechanism for each of the following reactions and identify the major products.

(a)

(b)

The addition of H_2O in Step 2 of Equation 12-22 follows Markovnikov's rule, because the C atom that is attacked is on the side of the ring that can accommodate more of the positive charge that appears on the Hg atom—in this case, the secondary C atom instead of the primary one. This is the same C atom that would bear the positive charge in the carbocation intermediate produced when a proton adds to the alkene, and thus is the same carbon atom that would be attacked by water in an acid-catalyzed hydration.

The mechanism also shows why rearrangement does not take place in oxymercuration–reduction, in contrast to acid-catalyzed hydration. In the acid-catalyzed hydration of an alkene, rearrangement takes place when a carbocation intermediate can undergo a 1,2-hydride shift or a 1,2-alkyl shift to produce a more stable carbocation. In oxymercuration–reduction, no carbocation intermediate is ever formed!

---

problem **12.13** Draw two possible alkenes that can undergo oxymercuration–reduction to yield 2-methylpentan-2-ol as the major product.

---

problem **12.14** Propose how to carry out the following transformation.

Alkynes can also undergo Markovnikov addition of water via oxymercuration. However, just as we saw with the acid-catalyzed hydration of an alkyne (Section 11.8), an unstable enol is produced initially. Subsequent tautomerization converts the enol into the more stable keto form, as shown in Equation 12-24.

Hydration of an alkyne leads to an initial enol, which tautomerizes into the more stable keto form.

(12-24)

The oxymercuration reaction in Equation 12-24 does not require reduction with NaBH$_4$ to remove the mercury(II) substituent. Instead, as shown in Equation 12-25, the mercurinium ion intermediate opens to produce a mercuric enol, which, after tautomerizing to a mercuric ketone, is hydrolyzed by water to produce the enol from Equation 12-24.

(12-25)

Frequently, other mercury(II) species are used to facilitate the addition of water to an alkyne. Examples include HgCl$_2$ and HgSO$_4$, as shown in Equations 12-26 and 12-27.

(12-26)

(12-27)

Equations 12-26 and 12-27 show the overall products upon hydration of the alkynes. For each reaction, draw the enol that is produced prior to tautomerization to the more stable keto form.

Internal alkynes can also be converted to ketones. A mixture of isomeric ketones will be produced, however, as we saw in Chapter 11, unless the alkyne is symmetric. This is shown in Equation 12-28.

Hydration of an internal alkyne leads to a mixture of isomeric ketones.

Hept-2-yne

H$_2$O

HgO, H$_2$SO$_4$
methanol

Heptan-2-one
~67%

+

Heptan-3-one
~33%

(12-28)

SOLVED problem 12.15 In an **alkoxymercuration–reduction** reaction, an alcohol is used as the solvent in the first step instead of water. Draw the complete, detailed mechanism for this reaction and predict the major product.

1. Hg(OAc)$_2$, EtOH/THF
2. NaBH$_4$

?

Think How does the substitution of EtOH for water change the mechanism in Equation 12-22? What considerations should be made about *regiochemistry*? What considerations should be made about rearrangements?

Solve The mechanism still proceeds through a cyclic mercurinium ion intermediate. Normally, H$_2$O acts as the nucleophile to open the ring. Here, however, this is the role of ethanol. Because the alkene is not symmetric, the alcohol can attack either C atom of the ring, but predominantly attacks the more highly substituted one because it can accommodate more positive charge. Finally, because no carbocations are formed as intermediates, carbocation rearrangements are not an issue.

---

problem 12.16 Predict the product of the reaction at the right.

1. Hg(OAc)$_2$, EtOH/THF
2. NaBH$_4$

?

12.5 Epoxide Formation Using Peroxyacids

Recall from Section 10.7 that epoxides can be useful in synthesis. In particular, an epoxide can undergo nucleophilic attack to produce a 2-substituted alcohol. To produce such an epoxide, recall from Section 10.6 that a halohydrin can be treated with a strong base. An example of this kind of reaction is shown in Equation 12-29.

(12-29)

Unfortunately, synthesizing an epoxide in this way is often impractical, in part because HO⁻, a strong base and strong nucleophile, could lead to unwanted side reactions. A second reason is that halohydrins are not very common, so they may not be readily available to be used as precursors.

Fortunately, an epoxide can be produced from an alkene using a **peroxyacid** (RCO_3H), also called a peracid, such as ***meta*-chloroperbenzoic acid** (**MCPBA**) as shown in the **epoxidation reactions** in Equations 12-30 and 12-31.

(12-30)

(12-31)

Synthesizing epoxides in this way is advantageous because the reaction conditions are relatively mild and many alkenes are common or easily synthesized.

The mechanism for these epoxidation reactions is shown in Equation 12-32. Notice that it takes place in a single step—that is, it is *concerted*.

Mechanism for Equation 12-31

(12-32)

The O atom of the OH group is the one that is transferred to the alkene because both of its original covalent bonds are relatively weak. The O—O single bond is inherently very weak. On the other hand, an O—H bond is normally quite strong, but in this case the O's bond to H is weakened by the internal hydrogen bond involving the OH and the carbonyl O atom.

> Verify that the O—O single bond is weak by looking up its average value (Table 1-2). For comparison, do the same for an average C—C single bond.
>
> O—O _____ kJ/mol   C—C _____ kJ/mol

With its weakened bonds, the O atom of the OH group is highly reactive and is attacked by the π electrons from the alkene. To avoid generating a carbocation on one of the C atoms from the initial alkene, the pair of electrons from the OH bond is used to form an O—C bond, thereby completing the three-membered epoxide ring. The carbonyl's O atom acquires the H, and the pair of electrons from the initial O—O bond goes to make a C=O bond.

Because epoxidation takes place in a single step, the stereochemical requirements presented in Section 12.1 apply. As shown previously in Equation 12-31, for example, stereochemistry is conserved, so the trans relationship of the CH₃ groups about the C=C double bond in the reactants establishes a trans relationship about the plane of the epoxide ring in the products. Furthermore, because the products are chiral, a mixture of stereoisomers is produced.

---

problem **12.17** Draw the complete, detailed mechanism for the reaction at the right and predict the major product(s), paying particular attention to the stereochemistry.

MCPBA → ?

---

problem **12.18** What alkene can be epoxidized using MCPBA to yield the following compound?

Because an epoxide readily undergoes ring opening (Section 10.7), the product of an epoxidation reaction is often used as a reactant in a subsequent reaction. This is exemplified in Solved Problem 12.19.

SOLVED problem **12.19** Identify the missing compounds **A–C**.

Think In the first reaction, what is the nature of the leaving group? The attacking species? The solvent? What type of reaction does heat promote? For the second reaction, what type of reaction takes place between an alkene and a peroxyacid? In the third reaction, what acts as the nucleophile? The substrate?

Solve The missing products are as follows:

The first reaction is an acid-catalyzed dehydration, which is an E1 reaction. Under the acidic conditions, the leaving group is H_2O, the attacking species is weak, and the solvent is protic. Heat favors elimination over substitution. The second reaction involves an alkene and MCPBA, which is a peroxyacid, leading to the formation of an epoxide. In the third reaction, HO^- acts as the nucleophile to open the epoxide in an S_N2 reaction. The trans stereochemistry is governed by the attack of the nucleophile from the side of the ring opposite the epoxide O.

problem **12.20** Identify the missing compounds **D–F**.

12.6 Hydroboration–Oxidation: Anti-Markovnikov Syn Addition of Water to an Alkene

So far, we have seen two reactions that serve to add water across a carbon–carbon double or triple bond. One is *acid-catalyzed hydration* (Chapter 11), which proceeds through a carbocation intermediate. The second is *oxymercuration–reduction* (Section 12.4), which proceeds through a cyclic mercurinium ion intermediate, thereby avoiding carbocation rearrangements. Both of these reactions add water in a *Markovnikov* fashion, in which the carbon atom that gains the OH group is the one that is better able to stabilize a positive charge. Also, neither of these reactions is stereoselective—a mixture of stereoisomers from both syn addition and anti addition is produced.

Hydroboration–oxidation provides a third way to add water across a nonpolar π bond, as shown in Equations 12-33 and 12-34. In each case, the alkene first undergoes **hydroboration**, in which borane, BH_3 (from either $BH_3 \cdot THF$ or B_2H_6), adds across the double bond. The product is then *oxidized* with a basic solution of hydrogen peroxide, H_2O_2.

(12-33)

(12-34)

Notice the regiochemistry that is exhibited by these reactions. Specifically, in both Equations 12-33 and 12-34, addition of OH is favored at the alkene C atom that can *least* stabilize a positive charge—namely, a primary C atom in Equation 12-33 and a secondary C atom in Equation 12-34. This is in contrast to the Markovnikov

Benzo[a]pyrene: Smoking, Epoxidation, and Cancer

Benzo[a]pyrene is found in smoke that is produced from the combustion of organic compounds (such as the tobacco in cigarettes), and several different studies have linked benzo[a]pyrene to various forms of cancer. Benzo[a]pyrene is classified as a procarcinogen because it is a *precursor* to the specific compound that is directly responsible for causing cancer. In this case, in the detoxification process in the liver, benzo[a]pyrene reacts with molecular oxygen in the presence of the enzymes cytochrome P450 1A1 (CYP1A1) and cytochrome P450 1B1 (CYP1B1). This is an epoxidation of the C=C bond between C7 and C8, analogous to the MCPBA epoxidation of alkenes we have studied here in Chapter 12, resulting in (+)-benzo[a]pyrene-7,8-epoxide (**A**).

The epoxide in **A** then undergoes ring opening with H_2O in the presence of the enzyme epoxide hydrolase, which produces a trans diol, much like we saw in Section 10.7. A second epoxidation targets the C=C bond between C9 and C10, resulting in (+)-benzo[a]pyrene-7,8-dihydrodiol-9,10-epoxide (**B**). **B** can intercalate into the DNA double helix, where it covalently bonds to a nucleophilic guanine nitrogenous base. As you can see above on the right, this adduct distorts the double helix, which interferes with the normal process by which DNA is copied, thus inducing mutations.

---

regiochemistry exhibited by acid-catalyzed hydration and oxymercuration–reduction. Thus,

> Hydroboration–oxidation adds H and OH to an alkene with **anti-Markovnikov regiochemistry**.

Notice also that hydroboration–oxidation is stereoselective. In Equation 12-34, for example, both the H and OH groups add to the same face of the planar alkene functional group. In other words,

> In hydroboration–oxidation, an alkene undergoes syn addition of H and OH.

To understand these aspects of hydroboration–oxidation reactions, we must understand their mechanisms. First, we study the mechanism of hydroboration in Section 12.6a, then we examine the mechanism of the subsequent oxidation in Section 12.6b.

Diborane, B$_2$H$_6$

(a)

Coordination of THF to BH$_3$ provides some stability by giving B an octet.

Borane•THF complex (BH$_3$•THF)

(b)

FIGURE 12-3 Sources of BH$_3$ (a) Diborane, B$_2$H$_6$, is a dimer of BH$_3$, formed by two separate, three-center, two-electron bonds indicated by the dashed lines. (b) In BH$_3$•THF, all atoms have an octet.

12.6a Hydroboration: Addition of BH$_3$ across a Carbon–Carbon Double Bond

Hydroboration is a very useful reaction because it is the starting point for a variety of synthetic sequences. It is the first of two sequential reactions in the anti-Markovnikov hydration of alkenes, for example, and it can be used as a starting point for hydrogenating alkenes and alkynes, thereby reducing them to alkanes and alkenes, respectively. For these reasons, Herbert C. Brown (1912–2004), who discovered hydroboration, shared the 1979 Nobel Prize in Chemistry.

Borane (BH$_3$) is the reactive species in the hydroboration of an alkene. Borane is highly unstable, however, because the central B atom does not have an octet. Therefore, BH$_3$ cannot be isolated. Instead, in its pure form it exists as a gaseous dimer, **diborane** (B$_2$H$_6$), in which two H atoms constitute a bridge between the B atoms (shown in Fig. 12-3a). Those H atoms are involved in what are called **three-center, two-electron bonds**. Although these bonds provide some stability, B$_2$H$_6$ is highly reactive, and thus is both *toxic* and *explosive*. A more stable variation of BH$_3$ is sold commercially as a one-to-one complex with tetrahydrofuran (THF; shown in Fig. 12-3b), where THF acts as a Lewis base to give the B atom its octet.

The mechanism for the hydroboration reaction is shown in Equation 12-35.

Partial mechanism for Equation 12-34

In the transition state, a partial positive charge is better stabilized on this C atom because it is more highly alkyl substituted.

H and BH$_2$ add to the double bond in a syn fashion.

(12-35)

It is driven primarily by the flow of electrons from the electron-rich π bond to the electron-poor boron atom of BH$_3$. Similar to the mechanisms we have seen previously in this chapter, this reaction avoids the formation of a highly unstable carbocation by simultaneously forming a bond back to the second C atom of the double bond. Whereas a lone pair was used in previous reactions to form that second bond to C (Section 12.1), in hydroboration that bond comes from electrons originally part of a B—H bond in BH$_3$. Overall, then, two bonds are broken and two bonds are formed in a *concerted* fashion—that is, without the formation of intermediates.

According to Equation 12-35,

Hydroboration is *stereospecific*, with the H and BH$_2$ groups adding to the alkene in a syn fashion.

Hydroboration is also *regioselective*. The H atom primarily adds to the C with the greater number of alkyl groups, whereas the BH$_2$ group adds to the C with the fewer number of alkyl groups. This, too, can be explained by the fact that the reaction is concerted. With no intermediates formed along the reaction coordinate, no formal charge is generated. Charge stability, therefore, does not provide as much of a driving force as it does in other reactions we have seen in this chapter and Chapter 11, allowing *steric hindrance* to play a more significant role.

> When BH_3 adds to an alkene, *steric repulsion* directs the larger group, BH_2, to the carbon atom with the fewer number of alkyl groups.

Equation 12-36 shows the steric repulsion that would ensue if the BH_2 group were to add to the more highly substituted alkene carbon.

(12-36)

Even though no formal charges appear, charge stability is also a factor in the regioselectivity of hydroboration. In the transition states shown in Equations 12-35 and 12-36, a *partial* positive charge appears on the C atom that has a partial bond to H. This is because the step is driven by the formation of a bond between the electron-rich alkene and the electron-poor B atom, so the C—B bond is formed to a greater extent in the transition state than the C—H bond. The transition state is more stable when the BH_2 group adds to the C atom with the fewer number of alkyl groups because that allows the partial positive charge that is generated to appear on the C atom with the greater number of alkyl groups. The additional alkyl groups inductively stabilize that partial positive charge. Overall, then, charge stability and steric hindrance promote the same regiochemistry.

The product of hydroboration is an **alkylborane**, R—BH_2 (Equation 12-37a), which has two B—H bonds remaining. That alkylborane can therefore react with an additional unreacted alkene, as shown in Equation 12-37b. That is, RHB—H adds across the double bond of the alkene to produce a **dialkylborane**, R_2B—H. In turn, the dialkylborane product adds across yet another equivalent of the unreacted alkene to produce a **trialkylborane**, R_3B, as shown in Equation 21-37c. As with the addition of BH_3, *each of these is a syn addition, and the bulkier B-containing portion adds preferentially to the less sterically hindered alkene C atom.*

The bulkier B-containing group adds to this C atom.

Syn addition of H and BH_2

H_2B—H

An alkylborane

(12-37a)

The bulkier B-containing group adds to this C atom.

H_3C H   H B H

CH_3

H_3C H   BH

Syn addition of H and BHR

(12-37b)

A dialkylborane

The bulkier B-containing group adds to this C atom.

CH_3

H

H_3C B H

CH_3

CH_3

H

H_3C B

Syn addition of H and BR_2

H

H_3C

(12-37c)

A trialkylborane

problem **12.21** Draw the detailed mechanism for the formation of the monoalkylborane in the reaction at the right. Also draw the trialkylborane that is ultimately produced.

$BH_3 \cdot THF$?

12.6b Oxidation of the Trialkylborane: Formation of the Alcohol

Equations 12-33 and 12-34 show that after an alkene has undergone hydroboration, treatment with a basic solution of H_2O_2 produces the alcohol. We can now see from Equation 12-37 that the actual species that undergoes oxidation is a trialkylborane. The net reaction is shown in Equation 12-38.

Each C—B bond is replaced by a C—OH bond.

Three equivalents of the alcohol are produced.

CH_3

H

H_3C H   B

$\xrightarrow[\text{H}_2\text{O}]{\text{H}_2\text{O}_2,\ \text{NaOH}}$

3

HO

H

H_3C

+ $^{\ominus}B(OH)_4$ (12-38)

H

H_3C

A trialkylborane

Note the following features of this oxidation:

- Each B-containing group attached to C is replaced by an OH group, producing three equivalents of the alcohol.
- Each OH group is syn to the hydrogen atom added from the previous hydroboration reaction. Thus, these substitutions take place with *retention of stereochemistry* at each C atom bonded to B.

These aspects of the oxidation reaction can be better understood by studying the mechanism, which is shown in Equation 12-39.

Mechanism for Equation 12-38

Under basic conditions, there is an equilibrium amount of the deprotonated peroxide, HOO^-, called the **hydroperoxide ion**. The mechanism begins with coordination of HOO^- to the electron-deficient B atom of the trialkylborane, thereby producing an unstable tetrahedral intermediate. In Step 2, the breaking of a weak peroxide (O—O) bond drives a 1,2-alkyl shift that yields a **borate ester**. This pair of steps occurs twice more (Steps 3–6), resulting in a **trialkylborate ester**. In Step 7, hydroxide anion (HO^-) coordinates to the B atom of the trialkylborate ester, and in Step 8, heterolysis occurs to liberate an alkoxide anion (RO^-) leaving group. In Step 9, the strongly basic alkoxide anion gains a proton from H_2O to produce the first equivalent of the final alcohol. This trio of steps is then repeated twice (Steps 10–15).

With this mechanism, we can now see how each B-containing group is replaced by OH with retention of stereochemistry. The critical step is Step 2, where a C—B bond breaks at the same time a C—O bond forms. Because of geometric constraints during this *concerted* process, the position that the O-containing group assumes about the C atom must be the same as that originally occupied by the B-containing group. Thus, the configuration about that C atom must not change.

YOUR TURN 12.10

This trialkylborane is an intermediate in a hydroboration–oxidation reaction. Draw the alcohol that is produced upon treatment with a basic solution of H_2O_2.

H_2O_2, NaOH
H_2O

---

problem **12.22** Draw the complete detailed mechanism of the reaction that takes place when the product of the reaction in Problem 12.21 is treated with a basic solution of hydrogen peroxide.

SOLVED problem **12.23** Show how to carry out each of the following transformations.

(a)   (b)

Think In each case, there is a net addition of what molecule across the double bond? What is the regiochemistry required in each reaction? Are carbocation rearrangements a concern?

Solve In each case, H_2O is being added across the double bond. In **(a)**, the addition takes place with Markovnikov regiochemistry, so there is a choice between acid-catalyzed hydration and oxymercuration–reduction. In this case, the carbocation that would be produced under an acid-catalyzed hydration would undergo a carbocation rearrangement, which is undesirable. Therefore, oxymercuration–reduction would be the better choice. In **(b)**, the regiochemistry is anti-Markovnikov, which requires hydroboration–oxidation.

1. Hg(OAc)$_2$, H_2O
2. NaBH$_4$

1. BH$_3$•THF
2. H_2O_2, NaOH, H_2O

(a)   (b)

(a)

(b)

12.7 Hydroboration–Oxidation of Alkynes

The addition of BH_3 to an alkyne takes place in much the same way as it does to an alkene, with BH_2 adding to the less sterically hindered carbon and H adding to the more sterically hindered one (Equation 12-40). An alkene is produced, however, and if that alkene is terminal a second hydroboration takes place readily. As a result, such a reaction would lead to a mixture of products, making it a fairly useless reaction.

(12-40)

Chemists avoid this problem by using a bulky dialkyl borane, such as **disiamyl-borane**, instead of BH_3, as shown in Equation 12-41. Disiamylborane reacts with an alkyne in the usual way, with the R_2B group adding to the less sterically hindered C atom. With the bulky R_2B group attached to the alkene, *a second addition of disiamylborane does not occur*, so hydroboration stops at the alkene stage.

(12-41)

Subsequent oxidation with a basic solution of H_2O_2 converts the R_2B substituent on the alkene into an OH group, similar to Equation 12-38. In this case, however, the product is an *enol*, which tautomerizes into the more stable keto form.

As we can see from Equation 12-41, hydroboration–oxidation is a useful way to convert a *terminal alkyne* into an aldehyde. This reaction can also be used to convert an internal alkyne into a ketone, but a mixture of products results unless the alkyne is symmetric.

---

problem **12.25** Alkyne **A** is treated with disiamylborane followed by a basic solution of H_2O_2. The overall product is an aldehyde:

(a) Draw the structure of the initial alkyne **A**.
(b) Draw the structure of intermediate **B**.

12.8 Wrapping Up and Looking Ahead

The focus of Chapter 12 has been on reactions in which an electrophile adds to an alkene or alkyne, without proceeding through a carbocation intermediate. The reactions proceed, instead, through a cyclic transition state, ultimately leading to two new bonds that join both of the alkene or alkyne carbons to one or more atoms of the electrophile. One of those new bonds uses a pair of electrons that is originally a π bond of the alkene or alkyne. The other new bond uses a pair of electrons originally from the electrophile.

One of the main differences among the reactions discussed here in Chapter 12 is the number of atoms that compose the ring of the cyclic transition state in the electrophilic addition step. The first several reactions we examined proceed through a three-membered ring transition state. Those include the addition of a carbene to produce a cyclopropane ring (Section 12.2), the addition of molecular halogens to produce 1,2-dihalides (Section 12.3), oxymercuration–reduction to produce an alcohol (Section 12.4), and epoxide formation using a peroxyacid (Section 12.5). By contrast, hydroboration–oxidation (Sections 12.6 and 12.7) proceeds through a four-membered-ring cyclic transition state.

A striking feature of the reactions in this chapter is their tendency to be stereospecific. For example, the formation of cyclopropane rings and epoxides conserves the stereochemistry about a double bond; groups that are on the same side of the initial alkene end up on the same side of the ring in the product. Halogenation takes place with anti addition, whereas hydroboration–oxidation take place with syn addition.

Some of the reactions in this chapter are also highly regioselective. In oxymercuration–reduction, for example, water adds across an alkene or alkyne in a Markovnikov fashion. By contrast, in hydroboration–oxidation, water adds in an anti-Markovnikov fashion. Finally, in the conversion of an alkene to a halohydrin, the halogen atom adds preferentially to the less alkyl-substituted carbon, whereas the OH group adds to the more alkyl-substituted one.

This was the second of two chapters dealing with electrophilic addition reactions. However, it is not the last chapter that deals with reactions involving electrophiles and nonpolar π bonds. Chapters 22 and 23 discuss electrophilic reactions involving benzene and other aromatic compounds, and Chapter 24 deals with the Diels–Alder reaction and other pericyclic reactions.

Before those chapters, however, we need to turn our attention to other concepts and other types of reactions. In Chapters 13 and 19, we examine aspects of organic synthesis; Chapters 14–16 deal with aspects of structure and the determination of structure using spectroscopy; and Chapters 17, 18, 20, and 21 discuss reactions that involve nucleophilic addition to polar π bonds.

Chapter Summary and Key Terms

- Addition of an electrophile to an alkene or alkyne tends to proceed through a three-membered ring transition state if the electron-poor atom of the electrophile possesses a lone pair of electrons. Thus, no carbocation is produced. **(Section 12.1)**

- A **carbene** is characterized by an uncharged carbon atom that possesses a lone pair of electrons, and is typically produced in situ from an appropriate precursor. A carbene adds to an alkene, via a three-membered-ring transition state, to produce a cyclopropane ring. This reaction preserves the cis/trans relationships of the groups attached to the alkene carbons. **(Section 12.2)**

- In a non-nucleophilic solvent (such as carbon tetrachloride), Br_2 or Cl_2 adds to an alkene to produce a 1,2-dihalide. This reaction takes place with anti addition of the halogens (i. e., a vicinal dihalide), signifying the initial production of a **bromonium** or **chloronium ion intermediate**. **(Section 12.3a)**

- In water, treatment of an alkene with a molecular halogen produces a **halohydrin**, in which a halogen atom and an OH group add in an anti fashion to the alkene carbon atoms. **(Section 12.3b)**

- **Oxymercuration–reduction** proceeds through a **mercurinium ion intermediate** and results in the addition of water to an alkene or alkyne with Markovnikov regiochemistry. **(Section 12.4)**

- Treatment of an alkene with a **peroxyacid** (or peracid) produces an epoxide. This reaction preserves the cis/trans relationships of the groups attached to the alkene. **(Section 12.5)**

- In a **hydroboration–oxidation reaction**, water adds to an alkene in a syn fashion, with **anti-Markovnikov regiochemistry**, to produce an alcohol. **(Section 12.6)**

  - The hydroboration step involves the addition of BH_3 to the alkene to produce an **alkylborane** and proceeds through a four-membered-ring transition state. **(Section 12.6a)**

  - The oxidation step converts the alkylborane to the alcohol, with retention of stereochemistry. **(Section 12.6b)**

- A single hydroboration–oxidation of an alkyne produces an enol that tautomerizes into a ketone or an aldehyde. To ensure that an alkyne undergoes only a single hydroboration, a bulky dialkylborane is used, such as **disiamylborane**. **(Section 12.7)**

Reaction Tables

TABLE 12-1 Functional Group Transformations[a]

Starting Functional Group	Typical Reagent Required	Functional Group Formed	Key Electron-Rich Species	Key Electron-Poor Species	Comments	Discussed in Section
Alkene	X_2 / CCl_4	Vicinal dihalide	:Ẍ:⁻		Anti addition	12.3a
Alkyne	X_2 2 equiv / CCl_4	1,1,2,2-Tetrahalide	:Ẍ:⁻		2 equivalents of halogen	12.3a
Alkene	X_2 / H_2O	Halohydrin	$H_2\ddot{O}$:		OH ends up on more highly substituted carbon; anti addition	12.3b
Alkene	1. $Hg(OAc)_2$, H_2O 2. $NaBH_4$	Alcohol	$H_2\ddot{O}$:		Markovnikov addition of water; no carbocation rearrangements	12.4

[a] X = Cl, Br

(continued)

TABLE 12-1 Functional Group Transformations[a] (continued)

Starting Functional Group	Typical Reagent Required	Functional Group Formed	Key Electron-Rich Species	Key Electron-Poor Species	Comments	Discussed in Section
Alkene (C=C)	1. B_2H_6 or BH_3·THF 2. $OH^\ominus$, H_2O_2	Alcohol (H—C—C—OH)	C=C	BH_3	Anti-Markovnikov addition of water	12.6
Alkene (C=C)	R—C(=O)—O—OH	Epoxide	C=C	R—C(=O)—O—O—H	Conservation of cis/trans configuration	12.5
Alkyne (C≡C)	$Hg(OAc)_2$, H_2O	Ketone	$H_2\ddot{O}:$	$\overset{\oplus}{Hg}$—OAc on C=C	Markovnikov addition of water	12.4
Alkyne (C≡C)	1. $(C_5H_{11})_2BH$ 2. $OH^\ominus$, H_2O_2	Ketone or aldehyde	C≡C	$(C_5H_{11})_2BH$	Anti-Markovnikov addition of water	12.7

[a] X = Cl, Br

TABLE 12-2 Reactions That Alter the Carbon Skeleton

Starting Functional Group	Typical Reagent Required	Functional Group Formed	Key Electron-Rich Species	Key Electron-Poor Species	Comments	Discussed in Section
Alkene (C=C)	CH_2N_2 Δ or hv	Cyclopropyl ring	C=C	$:CH_2$	Syn addition; retention of cis/trans configuration; not very useful in synthesis	12.2
Alkene (C=C)	$CHCl_3$ $(CH_3)_3CONa$	Dichlorocyclopropyl ring	C=C	$:CCl_2$	Syn addition; retention of cis/trans configuration	12.2

Problems

12.26 Br_2 undergoes electrophilic addition to maleic anhydride as follows:

Explain why this reaction is much slower than the analogous one with cyclopentene.

12.27 The electrophilic addition of Br_2 to several alkenes was examined. Explain why the relative reaction rates are as follows:

Increasing reaction rate with Br_2

12.28 Predict the major product(s) for each of the following reactions.

(a) 4-Chlorobut-1-ene + HBr ⟶ ?

(b) 1-Chlorobut-1-ene + HBr ⟶ ?

(c) 4,4-Dimethylcyclopentene + H_2O, H^+ ⟶ ?

(d) (R)-1,6-Dimethylcyclohexene + diazomethane, $h\nu$ ⟶ ?

(e) Hexa-1,5-diene + 2 Cl_2 ⟶ ?

(f) 2-Methylbut-2-ene + Br_2, H_2O ⟶ ?

(g) Propyne + 2 HCl ⟶ ?

(h) 3-Ethylpent-1-yne + 2 Br_2 ⟶ ?

(i) Cyclopentylethene $\xrightarrow{H_3O^+}$?

(j) Cyclopentylethene $\xrightarrow[\text{2. } NaBH_4/OH^-]{\text{1. } Hg(OAc)_2/H_2O}$?

(k) Cyclopentylethene $\xrightarrow[\text{2. } NaOH/H_2O_2, H_2O]{\text{1. } BH_3 \cdot THF}$?

12.29 For each of the following reactions, draw the complete mechanism and use the mechanism to predict the major product(s). Include stereochemistry where appropriate.

(a) $\xrightarrow{HCl}$?

(b) $\xrightarrow[\text{2. } NaBH_4/OH^-]{\text{1. } Hg(OAc)_2/H_2O}$?

(c) $\xrightarrow{H_3O^\oplus}$?

(d) $\xrightarrow{Br_2}{CH_3OH}$?

(e) $\xrightarrow[\text{2. } OH^-/H_2O_2, H_2O]{\text{1. } B_2H_6}$?

(f) $\xrightarrow{CH_2N_2}{h\nu}$?

(g) $\xrightarrow[NaOCH_3]{CHBr_3}$?

(h) $\xrightarrow{MCPBA}$?

(i) $\xrightarrow{Br_2}{CCl_4}$?

(j) $\xrightarrow{HBr}{\Delta}$?

(k) $\xrightarrow{HBr}{-78\ ^\circ C}$?

(l) $\xrightarrow{H_3O^\oplus}$?

(m) $\xrightarrow[\text{2. } NaOH, H_2O_2, H_2O]{\text{1. Disiamylborane, THF}}$?

(n) $\xrightarrow{HBr\ (excess)}$?

12.30 Show how each of the following can be synthesized from an alkene.

(a)

(b)

(c)

(d)

OH

(e)

(f)

(g)

(h)

OH

Br

12.31 Show how each of the following can be synthesized from an alkyne.

(a)

O

(b)

O

(c)

O

O

(d)

(e)

Br

Br

(f)

Br   Br

Br   Br

12.32 Supply the missing compounds in the following sequence of reactions.

$\xrightarrow{\text{MCPBA}}$ **A** $\xrightarrow[\text{2. H}_3\text{O}^{\oplus}]{\text{1. CH}_3\text{CH}_2\text{Li}}$ **B** $\xrightarrow[\Delta]{\text{H}_3\text{O}^{\oplus}}$ **C** $\xrightarrow{\text{MCPBA}}$ **D** $\xrightarrow[\text{CH}_3\text{CH}_2\text{OH}]{\text{NaBr}}$ **E** $\xrightarrow[\Delta]{\text{NaOH}}$ **F** $\xrightarrow[h\nu]{\text{CH}_2\text{N}_2}$ **G**

12.33 Supply the missing compounds in the following synthesis scheme.

$\xrightarrow[\text{CCl}_4]{\text{Br}_2}$ **H** $\xrightarrow[\text{2. H}_2\text{O}]{\text{1. NaNH}_2 \text{ (3 equiv)}}$ **I** $\xrightarrow[\text{2. NaBH}_4]{\text{1. Hg(OAc)}_2, \text{H}_2\text{O}}$ **J** $\xrightarrow[\substack{\text{Acetic} \\ \text{acid}}]{\text{Br}_2}$ **K** $\xrightarrow[\Delta]{\text{NaOH}}$ **L** $\xrightarrow[\text{H}_2\text{O}]{\text{Br}_2}$ **M**

12.34 For each of the following reactions, draw a complete, detailed mechanism and predict the major products, paying attention to regiochemistry and stereochemistry.

(a)

$\xrightarrow[\text{H}_2\text{O}]{\text{Br}_2}$?

(b)

$\xrightarrow[\text{H}_2\text{O}]{\text{Cl}_2}$?

12.35 Predict the product of the following reaction.

Buta-1,3-diene $\xrightarrow[\text{2. NaOH, H}_2\text{O}_2, \text{H}_2\text{O}]{\text{1. BH}_3\text{•THF (excess)}}$?

12.36 Show two different ways to convert 2-methylbut-2-ene to 3-bromo-2-methylbutan-2-ol.

$\xrightarrow{?}$

Br

OH

12.37 In the following oxymercuration–reduction reaction, H adds to the more highly substituted C atom, which might appear to violate Markovnikov's rule. Explain why this reaction exhibits this regiochemistry.

Methyl (E)-2-methylbut-2-enoate

1. Hg(OAc)$_2$, H$_2$O/THF
2. NaBH$_4$, ethanol

+ Enantiomer   + Enantiomer

12.38 Bromination can occur in a 1,4 fashion across conjugated double bonds, as shown for cyclohexa-1,3-diene:

$$\xrightarrow[\text{CCl}_4]{\text{Br}_2}$$

One mechanism that has been proposed involves a five-membered ring, bromonium ion intermediate.

(a) According to this mechanism, what should the stereochemistry be for the products—namely, all cis, all trans, or a mixture of both?

(b) Observations from experiment show that both cis and trans products are formed. Does this support or discredit the proposed mechanism shown here?

12.39 Propose a mechanism for the following reaction that accounts for the observed stereochemistry.

+ Enantiomer

12.40 When cyclohexene is treated with m-chloroperbenzoic acid and H$_2$O, trans-cyclohexane-1,2-diol is produced. Propose a mechanism for this reaction, accounting for the observed stereochemistry. (*Hint:* Recall what a peroxyacid does to an alkene.)

$$\xrightarrow[\text{H}_2\text{O}]{\text{MCPBA}}$$

+ Enantiomer

12.41 The high reactivity of carbenes can facilitate the synthesis of some unusual compounds. Show how each of the following can be synthesized from acyclic compounds.

(a)   **(b)**

12.42 Dichloromethane (CH_2Cl_2) can be treated with butyllithium, $CH_3CH_2CH_2CH_2$—Li, to make a carbene in situ, analogous to the way a carbene is generated from trichloromethane ($CHCl_3$) using HO^-.
(a) Show the mechanism for the generation of a carbene from CH_2Cl_2.
(b) Why is butyllithium used instead of HO^-?
(c) The following reaction leads to a mixture of four products. Draw all four products.

12.43 Propose how to convert hex-1-yne into (a) 2,2-dibromohexane and (b) 1,2-dibromohexane.

12.44 A student attempted to brominate the double bond in pent-4-en-1-ol, but ended up with the following cyclic ether instead. Propose a mechanism for the formation of this product.

12.45 Iodine monochloride (ICl) is a mixed halogen that adds to an alkene via the same mechanism by which bromination takes place. With that in mind, propose a mechanism for the following reaction, and use that mechanism to predict the products, paying attention to both *regiochemistry* and *stereochemistry*. (*Hint:* In ICl, one atom is more electrophilic than the other.)

12.46 In Section 12.2, we learned that diazomethane (CH_2N_2) can be used as a precursor to generate carbene, which subsequently adds to an alkene to produce a cyclopropane ring. Diazomethane is explosive, however, so it is not very useful synthetically. A safer way to produce a cyclopropane ring from an alkene is with the Simmons–Smith reaction, which occurs in a single elementary step:

Simmons—Smith reagent

The Simmons–Smith reagent (ICH_2ZnI) is produced by treating CH_2I_2 with a zinc–copper couple. Complete this mechanism by adding the necessary curved arrows. (*Hint:* The curved arrow notation is identical to that for epoxidation involving a peroxyacid.)

12.47 When benzene is treated with diazomethane and irradiated with light, cyclohepta-1,3,5-triene is produced. Propose a mechanism for this reaction.

If your instructor assigns problems in smartwork, log in at **smartwork.wwnorton.com**.

CHAPTER 13

Organic Synthesis 1

Beginning Concepts

Before chemistry developed into a scientific discipline, alchemists sought various *elixirs* to cure diseases and lengthen life. These elixirs were mixtures of compounds, often produced by carrying out extractions or distillations on samples of plant or animal material, and they contained what was called the "essence" of that material. Examples include peppermint oil and clove oil, produced by the steam distillation of peppermint and cloves, respectively. As the discipline of chemistry became more sophisticated throughout the 19th and 20th centuries, scientists realized that the bioactivity of a particular elixir was typically due to a single compound, now called a **natural product**. Menthol and eugenol, for example, are natural products found in peppermint oil and clove oil, respectively.

Because of the potential benefits they offer, many natural products are in high demand today. Whenever we find a use for a natural product, however, demand almost invariably exceeds nature's supply. One notable example of this is taxol.

Originally isolated from the bark of the Pacific yew tree (*Taxus brevifolia*) in 1966, taxol is an effective anticancer drug. Unfortunately, the amount of taxol available from the bark of one tree can provide only one 300-mg dose for one person. Even worse, isolating

Taxol

Taxol, used in cancer chemotherapy, was first isolated from the bark of the Pacific yew tree. For many years, taxol was produced industrially from readily available starting compounds via a synthesis, prior to the development of the modern process involving plant cell fermentation and extraction. We discuss some beginning concepts of organic synthesis here in Chapter 13.

Upon completing Chapter 13 you should be able to:

- Specify the difference between a *synthesis* and a *mechanism* and distinguish the ways in which each is reported.

- Define a *transform* and use transforms to carry out a *retrosynthetic analysis*.

- Identify *synthetic traps* and explain some of the ways to avoid them.

- Distinguish between *functional group transformations* and reactions that alter the *carbon skeleton*, and explain the importance of each.

- Design a synthesis that incorporates stereo-specific reactions.

- Explain how the overall yield in a synthesis is affected by the total number of steps, the yield of each step, and whether the synthesis is *linear* or *convergent*.

taxol requires harvesting the bark, which kills the tree. Thus, many people became concerned about the possible extinction of the Pacific yew.

Organic chemists sought a *synthesis* of taxol. A **synthesis** is a specific sequence of chemical reactions that converts **starting materials** into the desired compound, called the **target** of the synthesis (or the **synthetic target**). In 1994, Robert A. Holton (b. 1944) of Florida State University devised the first route for synthesizing taxol from starting materials that were all commercially available. This synthesis, which was the culmination of 12 years of work, consisted of 37 separate reactions, called *synthetic steps*. Since then, other schemes to synthesize taxol have been designed with somewhat fewer steps.

Taxol is but one of many compounds whose natural abundance is insufficient to meet human demand. As a result, the field of organic synthesis is immense, making it impossible to present in a single chapter all that organic synthesis entails. Instead, Chapter 13 is devoted to *introductory* aspects of organic synthesis, focusing on targets that are much simpler than taxol. First, we present proper ways to write the reactions of a synthesis. Then we explore the basic thinking that goes into planning an efficient synthesis of an organic molecule, including *retrosynthetic analysis*, a powerful tool that helps organize the search for synthetic pathways.

In Chapter 19, we tackle some of the more complex aspects of organic synthesis, such as the proper placement of functional groups within a molecule, and how to use selective reagents and protecting groups.

13.1 Writing the Reactions of an Organic Synthesis

Although formalizing an organic synthesis on paper is typically the last step in designing a synthesis scheme, we begin by examining some of the conventions used to do so. These conventions are not new to you; they have been used consistently in the last several chapters when reactions have been introduced and discussed. We take the time to review them here, however, so that they are at the forefront of your mind when you write out a synthesis. These conventions help with effective communication, ensuring that a synthesis scheme you have devised is interpreted by others in the way that you intend.

There are essentially three main conventions routinely used in writing a synthesis scheme. The first stems from the fact that a synthesis is an abbreviated *recipe*. As such,

For each **synthetic step**, provide just the overall reactants, the reagents added, the reaction conditions, and the structures of the overall products.

In particular, *do not include the details of any reaction mechanism.* The mechanisms that you have seen throughout the last several chapters enable you to understand *how* and *why* each reaction takes place as it does, but when conveying information about a synthesis to others, the focus changes to just the sequence of actions that must be done in the laboratory to produce the target from the starting material.

Analogously, when writing a recipe for chocolate chip cookies, it is not important to explain the details of the chemical processes that take place when an egg is added; it is important, however, to know *that* an egg is needed, as well as *when* and *how* to add the egg.

For example, a synthetic step showing how to convert 2-phenyl-2-tosylpropane into 2-bromo-2-phenylpropane might be written according to Equation 13-1:

(13-1)

Overall reactants — 2-Phenyl-2-tosylpropane + KBr

Overall product — 2-Bromo-2-phenylpropane

This synthetic step implies that when 2-phenyl-2-tosylpropane and potassium bromide are combined, 2-bromo-2-phenylpropane is generated as an overall product. By contrast, the mechanism for this reaction, which is an S_N1, would be written as follows:

Mechanism for Equation 13-1

(13-2)

Even though KBr dissolves to form K^+ and Br^-, and even though Br^- is the active nucleophile in the mechanism in Equation 13-2, it is inappropriate to include Br^- in a synthetic step. Br^- cannot exist in a pure form, due to the charge that it carries, so the synthetic step proposed in Equation 13-3 is technically incorrect.

Not an appropriate synthetic step because Br$^\ominus$ does not exist in pure form

(13-3)

In general,

> Reagents must be written in the form in which they can be added, not as they appear in the mechanism.

For this reason, Equation 13-4 is also unacceptable as a synthetic step.

Not an appropriate synthetic step because :CH₂ does not exist in pure form

$$+ \quad :CH_2 \quad \longrightarrow \qquad \text{(13-4)}$$

Recall from Section 12.2 that $:CH_2$ is a very highly reactive species and does not exist in pure form. Rather, $:CH_2$ is produced in situ from a precursor such as diazomethane (CH_2N_2), which can exist in pure form. Thus, Equation 13-5 would be an acceptable synthetic step to accomplish the electrophilic addition of carbene to cyclohexene.

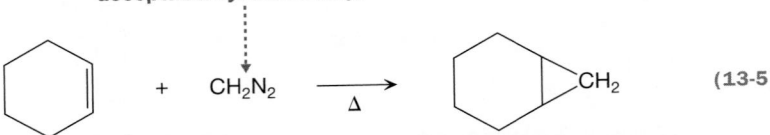

CH_2N_2 does exist in pure form and is a precursor of :CH₂, so this is an acceptable synthetic step.

$$+ \quad CH_2N_2 \quad \xrightarrow{\Delta} \qquad \text{(13-5)}$$

Notice in Equations 13-1 and 13-5 that not all of the products are included in the synthesis scheme—the tosylate anion (TsO^-), in particular, has been omitted from Equation 13-1 and $N_2(g)$ has been omitted from Equation 13-5. This simply suggests that TsO^- and N_2 are by-products that we are not interested in collecting. Whereas organic products are usually of interest,

> Inorganic by-products and leaving groups are often irrelevant to the synthesis and are omitted.

YOUR TURN 13.1

(a) Write the synthetic step that shows 2-phenyl-2-tosylpropane reacting with NaCl to produce 2-chloro-2-phenylpropane.

(b) Draw the mechanism for this reaction.

Answers to Your Turns are in the back of the book.

The second convention for writing synthesis schemes relates to how the information in a chemical reaction is presented:

> Reagents added in a particular synthetic step can be written *above* the reaction arrow (⟶) that connects reactants to products, whereas reaction conditions (including solvent, temperature, pH, time of reaction, etc.) are usually written *below* the arrow.

For example, the conversion of 7-chlorohept-1-ene to 7-iodohept-1-ene is shown in Equation 13-6.

Reagent

7-Chlorohept-1-ene $\xrightarrow[\text{Acetone, 40 °C, 3 days}]{\text{NaI}}$ 7-Iodohept-1-ene (13-6)

Reaction conditions

Sodium iodide is added to 7-chlorohept-1-ene in this synthetic step. The reaction takes place in acetone at an elevated temperature (40 °C) for 3 days, producing 7-iodohept-1-ene as the overall product.

Complete the following synthetic step by indicating that water is used as the solvent and that the reaction is carried out at 70 °C.

The third convention provides some flexibility in how we can write a synthesis that has multiple steps.

> Synthetic steps that are customarily carried out sequentially may be combined, using numbers above and below the reaction arrow to distinguish the steps.

An example is shown in Equation 13-7, which shows how hex-2-yne can be synthesized from pent-1-yne in a two-step process.

The numbers indicate that the reaction with NaH is carried out to completion before CH_3I is added.

1. NaH
2. CH_3I (13-7)

Pent-1-yne   **Hex-2-yne**

According to this scheme, NaH is added directly to pent-1-yne in the first reaction. Once that reaction has come to completion, CH_3I is added, and the subsequent reaction is allowed to go to completion, yielding hex-2-yne.

A critical part of this convention is that each reaction that is denoted by a number is allowed to go to completion before the next reagent is added. Therefore, in the case of the reaction sequence in Equation 13-7, it is possible to carry out the pair of reactions without the two reagents—NaH and CH_3I—coming into contact with each other.

Describe in words the sequence of reactions depicted in the following chemical equation.

1. NaOH
2. CH_3CH_2Br

Butanoic acid   **Ethyl butanoate**

Using this convention for sequential steps, reaction conditions can be written after the reagent for each numbered step. The reaction conditions are typically separated from the reactant or reagent by either a comma or a slash. For example, in the two-step synthesis shown in Equation 13-8, 1-bromohexane is first treated with $(CH_3)_3COK$ using DMSO as the solvent. In the second step, the resulting product is treated with Br_2, using H_2O as the solvent.

1-Bromohexane

$$\xrightarrow[\text{2. Br}_2 / \text{H}_2\text{O}]{\text{1. (CH}_3\text{)}_3\text{COK / DMSO}}$$

Reagent ⋯ Conditions (top)

Reagent ⋯ Conditions (bottom)

(13-8)

YOUR TURN 13.4

Using the number convention for common sequential steps, show how the following synthetic step would be represented: (Z)-hex-3-ene is treated with mercury(II) acetate in water, and upon completion of that reaction, the product is treated with sodium borohydride to yield hexan-3-ol as the overall product.

13.2 Cataloging Reactions: Functional Group Transformations and Carbon–Carbon Bond Formation/Breaking Reactions

How challenging it is to design a practical synthesis partly depends on the nature of the target molecule and on the types of compounds available as starting materials. It also depends on how familiar the chemist is with the wide variety of existing reactions, knowing specifically what each reaction can accomplish. Therefore, it becomes important to *catalog* reactions you have encountered according to their utility in a synthesis. Specifically,

For each reaction you encounter, you should ask yourself the following questions:

- What functional groups in the reactants are involved?
- What functional groups are produced in the products?
- What are the major structural changes that occur?

To help you answer these questions quickly, reaction summary tables are included at the end of every chapter in which new reactions are introduced, beginning with Chapter 9. As shown in the following example, each entry in these tables provides the reactants, typical reagents and reaction conditions, products, and the section in which the reaction is discussed. Also, to better enable you to apply your knowledge of mechanisms in synthesis, each entry provides key electron-rich and electron-poor species, along with the reaction type.

	Reactant	Typical Reagent and Reaction Conditions	Product Formed	Key Electron-Rich Species	Key Electron-Poor Species	Comments	Discussed in Section
(1)	(H or) R—C(=O)—CH₂—R' Ketone or aldehyde	1. Base⁻ 2. R''—X (X = Cl, Br, I)	(H or) R—C(=O)—CR''—R' α-Alkylated ketone or aldehyde	(H or) R—C(=O)—C⁻—R' Enolate anion	δ^+ R''—X Alkyl halide	S_N2	10.3

In general, you will find two of these tables at the end of a chapter. One contains **functional group transformations** or **functional group conversions**, which, as their name suggests, simply transform one functional group into another. They do so by leaving alone the **carbon skeleton** (i.e., the bonding arrangement of the carbon atoms).

The second of these tables contains reactions that do, indeed, bring about changes in the carbon skeleton. These reactions *require that carbon–carbon σ bonds be broken and/or formed.* (The breaking and/or formation of a carbon–carbon π bond alone is not considered to be among these reactions, because the σ bond between those atoms remains intact, thus preserving the carbon skeleton.)

Your ability to design a synthesis depends heavily on your familiarity with a variety of reactions. Therefore, take the time now to assess how familiar you are with the reactions we have learned so far, which appear in the reaction tables at the ends of Chapters 9–12 (Table 9-14, p. 516; Table 9-15, p. 517; Table 10-1, p. 562; Table 10-2, p. 563; Table 11-1, p. 604; Table 12-1, p. 639; and Table 12-2, p. 640). Go through the tables first to see if you can determine the typical reagents and reaction conditions for each reaction by examining just the reactants and products. Then, go through the reaction tables again to see if you can determine the reactants for each reaction by examining just the products and the typical reagents and reaction conditions. If you encounter difficulty with a particular reaction, make sure to review the sections where the reaction is discussed and study the mechanism.

13.5 YOUR TURN

Equation 13-9 shows the various ways in which the two types of reactions are used in synthesis.

(13-9)

Synthetic Steps **A** and **C** are functional group conversions, whereas Steps **B** and **D** are carbon–carbon bond-formation reactions that alter the carbon skeleton.

Notice in Equation 13-9 that functional group conversions do not necessarily produce a functional group that appears in the overall product. In Synthetic Step **A**, for example, the epoxide that is produced sets up the Grignard reaction in the subsequent step. And the reaction in Synthetic Step **C** converts a bad HO⁻ leaving group into a good Br⁻ one, and therefore sets up the S$_N$2 reaction in Step **D**.

Although Equation 13-9 illustrates how reactions can be used to alter a carbon skeleton and to convert one functional group into another, our ability to use these schemes is quite limited because we have not learned many reactions so far. As you encounter more reactions in future chapters, you will acquire a better feel for how these types of reactions can and should be used in synthesis.

Label each step in the following synthesis as either a functional group conversion or a reaction that alters the carbon skeleton.

problem **13.1** Supply the missing reagents for each step in Your Turn 13.6.

13.3 Retrosynthetic Analysis: Thinking Backward to Go Forward

When presented with a synthesis problem, the natural tendency is to think in the *forward* direction, first deciding which compound(s) to use as the starting material, and then determining the specific sequence of reactions to carry out. Solving the problem in this way can be straightforward for syntheses that require only one or two synthetic steps, but it becomes impractical as the number of steps in an organic synthesis increases. In many cases, it is not obvious what the starting material should be, especially since the starting material does not need to resemble the target. Moreover, even if a viable starting compound is established, you will find that there are many dozens or even hundreds of different reactions that can be carried out for each synthetic step. Therefore, even for a three-step synthesis, there might be more than a million different imaginable schemes to consider!

In the 1960s, this problem was greatly simplified when Elias J. Corey (b. 1928) of Harvard University pioneered a new method of designing a synthesis scheme. This method, called **retrosynthetic analysis**, changed the face of organic synthesis and helped earn Professor Corey the 1990 Nobel Prize in Chemistry.

The basis of retrosynthetic analysis is the **transform**, which is the *proposed undoing* of a single reaction or set of reactions. To begin the design of a synthesis scheme, the chemist carries out a transform on the target molecule to dissect it into smaller and/or simpler **precursors**, often without consideration of the starting material. An open arrow ($\Longrightarrow$), called a **retrosynthetic arrow**, is the convention used to indicate a transform, and is drawn from the target to the precursor. Therefore,

When considering a reaction or set of reactions that would produce a target A from a precursor B, a transform can be written as follows:

Target $\cdots$ Precursor

A $\Longrightarrow$ B

Each precursor is then dissected in the same way, and the cycle is repeated until the chemist arrives at compounds that are either readily available or easy to produce.

To see how retrosynthetic analysis can be used, let's design a synthesis of 1-methoxypent-2-yne, limiting ourselves to starting compounds that contain three or fewer carbons.

1-Methoxypent-2-yne

Because the target contains a continuous chain of five C atoms, and the compounds we may use as our starting material can contain at most three C atoms, the synthesis of 1-methoxypent-2-yne must contain at least one reaction that forms a C—C bond.

There are essentially nine reactions we have encountered so far that produce a new C—C bond, but only two leave a C≡C bond in the product, as shown in Table 13-1 (Entry 1 is taken from Table 9-15 and entry 2 is taken from Table 10-2.) We should therefore consider using one of these reactions.

Of these two reactions, the first (entry 1 of Table 13-1) is the better choice because the product of the second reaction contains an alcohol functional group on the second carbon away from the alkyne group, and no such carbon has a functional group in our target molecule. We can therefore apply a transform to our target molecule that undoes the reaction, so that it *disconnects* the single bond between C3 and C4. This is indicated in Equation 13-10 by the wavy line.

Undo the C—C bond formation reaction in entry 1 of Table 13-1. **Contains 2 carbon atoms** **Contains 4 carbon atoms**

(13-10)

Bromoethane   **3-Methoxyprop-1-yne**

According to Table 13-1, the two necessary precursors for such a reaction are an alkyl halide and a terminal alkyne. In this case, the alkyl halide precursor contains two C atoms, and the other precursor, 3-methoxyprop-1-yne, contains the terminal alkyne.

TABLE 13-1 Reactions That Alter the Carbon Skeleton and Leave a C≡C Bond

	Reactant	Typical Reagent and Reaction Conditions	Product Formed	Key Electron-Rich Species	Key Electron-Poor Species	Comments	Discussed in Section
(1)	RC≡CH Alkyne (terminal)	1. NaH 2. R′—X (X = Cl, Br, I)	RC≡C—R′ Alkyne	RC≡C:$^{\ominus}$	R′—X	S_N2	9.3b
(2)	RC≡CH Alkyne (terminal)	1. NaH 2. (epoxide) R′, R″ 3. $H_3O^{\oplus}$	OH R′ — C — R″ ‖ CR Alcohol (3-alkyn-1-ol)	RC≡C:$^{\ominus}$	R′ — (epoxide) — R″	S_N2	10.7a

Draw the appropriate precursors indicated by the transform at the right, similar to what was done in Equation 13-10.

$$H_2C \overset{\xi}{-} C \equiv C - CH_2 \quad \Longrightarrow$$

problem **13.2** Draw the appropriate precursors to the target at the right by applying a transform that undoes the reaction in entry 2 of Table 13-1. (*Hint*: Which C—C bond should be disconnected?)

OH

Of those two precursors, only bromoethane is acceptable for our starting material, because it contains three or fewer C atoms. 3-Methoxyprop-1-yne contains four C atoms, however, so it cannot be used as starting material. Therefore, we must apply a transform to dissect it into smaller precursors. 3-Methoxyprop-1-yne contains an ether functional group, so we can apply a transform that undoes an ether-forming reaction. According to Table 13-2 (taken from Table 10-1), one such reaction—the Williamson ether synthesis—requires an alkyl halide (R—X) and a sodium alkoxide (NaOR) as precursors.

The resulting transform might appear as follows:

Undo the ether formation reaction in Table 13-2 (a Williamson ether synthesis).

Contains 3 carbon atoms   **Contains 1 carbon atom**

$$\Longrightarrow \qquad \qquad + \quad NaOCH_3 \qquad (13\text{-}11)$$

Both of these precursors now contain three or fewer carbons and can be used as starting materials.

What remains to complete the synthesis is to reverse the transforms and to include the necessary reagents and conditions that will accomplish each reaction, as shown in Equation 13-12. In other words, write the synthesis in the *forward* direction.

$$\xrightarrow{NaOCH_3} \qquad \xrightarrow[\text{2. } CH_3CH_2Br]{\text{1. NaH}} \qquad (13\text{-}12)$$

TABLE 13-2 Functional Group Transformation

Starting Functional Group	Typical Reagent and Reaction Conditions	Functional Group Formed	Key Electron-Rich Species	Key Electron-Poor Species	Comments	Discussed in Section
R—X (X = Cl, Br, I) Alkyl halide	$\xrightarrow{NaOR'}$	R—O—R' Ether (symmetric or unsymmetric)	R'O$^{\ominus}$	R—X δ^+	Williamson ether synthesis, S$_N$2	10.6

Notice in each of the previous transforms that the reactions can be undone in a different way from the one that was shown. For example, instead of disconnecting the C3—C4 bond to the left of the triple bond in the target (Equation 13-10), we can begin by disconnecting the C1—C2 bond to the right of the triple bond, as shown in Equation 13-13.

Undo the C—C bond formation reaction in entry 1 of Table 13-1.

Undo the C—C bond formation reaction in entry 1 of Table 13-1.

(13-13)

The alkyl halide precursor contains two C atoms, so it is acceptable as starting material. The alkyne-containing precursor, however, contains four C atoms, but another transform that disconnects the C—C bond to the left of the triple bond results in two precursors that each have two C atoms, and are therefore acceptable as starting material.

The synthesis can then be written as shown in Equation 13-14.

(13-14)

The previous example illustrates an important point:

> A synthesis can usually be designed in multiple different ways.

There are a variety of factors that can make one synthesis better than another, and we will examine some of these factors throughout the rest of our discussion of synthesis here in Chapter 13 and in Chapter 19.

13.8 YOUR TURN

Show how to carry out the transform in Equation 13-11 by disconnecting the other O—C bond, and modify the synthesis in Equation 13-12 accordingly.

SOLVED problem 13.3 Propose a way to carry out the synthesis at the right.

Think Is there a reaction we have encountered that can transform an alcohol group into a nitrile group directly? If not, then what other precursor can be transformed into a nitrile? Can that precursor be made directly from the alcohol?

Solve There are no reactions we have learned that convert an alcohol into a nitrile directly. Notice, however, that a C—C single bond must be formed to carry out this synthesis. Only the S_N2 reaction at the right (taken from entry 2 in Table 9-15) accomplishes this while leaving a nitrile group in the product.

We can thus undo this reaction, as shown in the first transform.

Undo the S_N2 reaction in entry 2 of Table 9-15.

Undo the PBr₃ bromination in entry 1 of Table 10-1.

This yields a primary alkyl halide as a precursor, which can be made directly from an alcohol using PBr₃ (entry 1 of Table 10-1), as shown in the second step of the transform. To complete the synthesis, we reverse the transforms and apply the appropriate reagents.

---

problem **13.4** Propose a synthesis of pent-2-yne from compounds containing two or fewer carbons.

Pent-2-yne

---

problem **13.5** Propose a synthesis of 2-benzyl-3-phenylpropanal from bromophenylmethane (benzyl bromide), using any reagents containing two or fewer carbon atoms.

Bromophenylmethane
(Benzyl bromide)

2-Benzyl-3-phenylpropanal

13.4 Synthetic Traps

When we undo a particular reaction by applying a transform, we tend to consider only the major structural changes the reaction brings about, and to focus mainly on the specific location in our target molecule where we want those changes to occur. However, if we restrict ourselves in these ways while considering various transforms, we may encounter a **synthetic trap**, in which some factor we have overlooked prevents us from executing the reaction in the *forward* direction as planned.

When we encounter a synthetic trap, we must alter our synthesis in some way. Sometimes a synthetic trap can be circumvented with only minor changes to the synthesis scheme, perhaps simply by reordering the steps. At other times, the synthesis might require a specialized reaction, many of which we will learn in the chapters to come. Occasionally, however, there will be no apparent way around a synthetic trap, forcing us to search for an altogether different synthetic route. As we gain more experience with organic synthesis, and as we learn more reactions in subsequent chapters, we will be better equipped to make these kinds of decisions.

13.4a Reactants with More Than One Reactive Functional Group

Among the most conspicuous synthetic traps is one in which there are two or more functional groups on a precursor that can react in a particular synthetic step. Consider, for example, the transform proposed in Equation 13-15, which undoes a C—C bond formation reaction from entry 1 of Table 13-1.

Undo the C—C bond formation reaction in entry 1 of Table 13-1.

Hex-4-yn-1-ol   **Propyne**   **3-Bromopropan-1-ol** (13-15)

In the forward direction, this synthetic step would appear as shown in Equation 13-16.

$$\text{(13-16)}$$

In the first step, the terminal alkyne is deprotonated to generate the acetylide anion, $CH_3C{\equiv}C^-$. A problem arises in the second of these sequential reactions, however, because the acetylide anion is *intended* to behave as a nucleophile and displace the Br leaving group, as shown in Equation 13-17a. Unfortunately, it can also behave as a strong base, deprotonating the OH group, as shown in Equation 13-17b. Because proton transfer reactions tend to be much faster than nucleophilic substitution reactions (Section 8.6a), the reaction in Equation 13-17b dominates, and the synthesis does not proceed as intended.

The desired reaction does not take place.

$$\text{(13-17a)}$$

There are two possible reaction sites that can involve the acetylide ion.

Proton transfer

$$\text{(13-17b)}$$

This undesired reaction takes place instead.

One way to circumvent this kind of problem involves what are called *protecting groups*. In this case, the OH group can be protected to prevent the proton transfer reaction from occurring, leaving the S$_N$2 reaction as the only possibility. We discuss the use of protecting groups in synthesis in greater detail in Chapter 19. For now, we will simply focus on identifying synthetic traps so that we can avoid them.

problem **13.6** For each of the following proposed transforms, determine whether it leads to a synthetic trap (i.e., it will not proceed as planned in the forward direction). Explain your reasoning.

(a)

(b)

(c)

(d)

(e)

(f)

13.4b Synthetic Traps and Regiochemistry

In Section 13.4a, we learned that a synthetic trap can arise when a reactant has two or more reactive functional groups. Even when there is only one reactive functional group, however, a synthetic trap can still arise if that functional group has two or more reactive sites—that is, if *regiochemistry* is an issue. Thus, running such a reaction in the forward direction might take place at a different site than we envisioned when we undid the reaction in a transform.

Consider, for example, a transform that undoes the hydrobromination of an alkene, which we learned in Section 11.1 (Equation 13-18):

$$\text{(13-18)}$$

In the forward direction, the synthetic step would call for the addition of HBr, as shown in Equation 13-19.

$$\text{(13-19)}$$

However, the major product of this reaction is not the desired one because, as we saw in Section 11.3, the reaction takes place with Markovnikov regiochemistry.

One way to solve this problem is to consider a different reaction to obtain the desired regiochemistry. We have not yet encountered a reaction that directly adds HBr to an alkene with anti-Markovnikov regiochemistry, but we learned in Section 12.6 that hydroboration–oxidation results in the anti-Markovnikov addition of H_2O. As shown in Equation 13-20, the alcohol product from this reaction can then be converted to the bromide using PBr_3, resulting in the desired product in Equation 13-18.

$$\text{(13-20)}$$

Because of the importance of regiochemistry in synthesis, it is helpful to review the reactions we have encountered in Chapters 9–12 that enable us to carry out particular transformations with different regiochemistry. We can often choose one of these reactions to provide the regiochemistry that a synthesis might call for.

For example, Equations 13-21 through 13-23 are three different reactions that add H_2O across a C=C double bond. Equation 13-21 is the hydroboration–oxidation reaction we just reviewed, which takes place with anti-Markovnikov regiochemistry. Equations 13-22 and 13-23, on the other hand, take place with Markovnikov regiochemistry.

$$\text{(13-21)}$$

Markovnikov addition of H$_2$O

$$\text{(13-22)}$$

Markovnikov addition of H$_2$O

$$\text{(13-23)}$$

SOLVED problem 13.7 Show how to carry out the synthesis at the right.

Think What precursor can be used to produce the alkyl bromide target? Is regiochemistry a concern? If so, can the alkyl bromide be produced directly with the desired regiochemistry? If not, what other reaction can be carried out to give the desired regiochemistry?

Solve As shown in the transform below, the alkyl bromide target can be produced from a hydrobromination of an alkene.

Undo an HBr addition.

However, the HBr addition in the forward direction has Markovnikov regiochemistry, which would produce an undesired product.

Br is not in the desired location.

Instead, the alkyl bromide target can be produced from the corresponding alcohol, which can be produced from an alkene, in turn, by an anti-Markovnikov addition of H$_2$O. The alkene can be produced from the starting compound by dehydration.

Undo a bromination. **Undo an H$_2$O addition.** **Undo a dehydration.**

The synthesis in the forward direction would then be written as follows:

conc. H$_3$PO$_4$, Δ   1. BH$_3$·THF 2. H$_2$O$_2$, NaOH, H$_2$O   PBr$_3$

---

problem 13.8 Show how to carry out the synthesis at the right.

The alkylation of an α carbon (Section 10.3) is another example of a reaction whose regiochemistry we can control, as shown in Equation 13-24.

Alkylation at the less substituted α C

The reaction can take place at either α C

(13-24a)

(13-24b)

Alkylation at the more substituted α C

In this case, the choice of base can dictate which α carbon undergoes reaction. A very strong base such as LDA leads to alkylation at the less substituted α C (Equation 13-24a), whereas a moderately strong base such as the *tert*-butoxide anion leads to alkylation at the more highly substituted α C (Equation 13-24b).

---

problem **13.9** Show how each of the following transforms would appear in a synthesis.

(a)

(b)

Epoxide ring opening (Equation 13-25) is another reaction whose regiochemistry can be controlled.

$Br^{\ominus}$ attacks the more substituted C.

(13-25a)

$Br^{\ominus}$ attacks the less substituted C.

(13-25b)

As we learned in Section 10.7, the nucleophile tends to attack the more highly substituted C atom under acidic conditions (Equation 13-25a), and tends to attack the less highly substituted C under neutral or basic conditions (Equation 13-25b).

Show how each of the following transforms would appear in a synthesis.

(a)   (b)

Regiochemistry can also be controlled in elimination reactions, as shown in Equations 13-26 and 13-27.

Zaitsev product

$$\text{NaOCH}_3, \Delta$$

(13-26)

Hofmann product

1. CH_3I
(excess)

2. NaOH, Δ

(13-27)

As we learned in Section 9.10, the elimination reaction in Equation 13-26 tends to produce the most highly substituted alkene product, or the Zaitsev product, because it is the most stable. On the other hand, as we saw in Section 10.9, the elimination reaction in Equation 13-27, a Hofmann elimination reaction, tends to produce the less substituted alkene product.

SOLVED problem 13.11 The alkene in Equation 13-27 can also be synthesized from a precursor in which the leaving group is on C2 of the propane chain, as shown at the right. What leaving group would you choose, and how would you carry out the transformation?

H and L have been added to the alkene C's

Undo an elimination.

1-Cyclohexylpropene

(L = leaving group)

Think Which protons can be eliminated along with the leaving group? What would be the elimination product in each case? Which is the Zaitsev product, which is the Hofmann product, and which corresponds to the target?

Solve The protons that can be eliminated are on C1 and C3 of the propane chain.

Deprotonation at C1

Zaitsev product

Deprotonation at C3

Hofmann product

Deprotonation at C1 gives the desired product, which, in this case, is the Zaitsev product. This deprotonation can be achieved by choosing a good leaving group and a strong base, neither of which is bulky. The leaving group can be a halide anion and the base can be $CH_3CH_2O^-$. Thus, the synthesis can be written as follows:

problem **13.12** Show how you can synthesize 3,4-dimethylhex-2-ene via two different elimination reactions—one in which the leaving group in the precursor is on C2 and the other in which the leaving group is on C3. (Do not concern yourself with stereochemistry.)

3,4-Dimethylhex-2-ene

Finally, in the electrophilic addition to a conjugated diene (Section 11.9), regiochemistry can be governed by whether the conditions lead to kinetic or thermodynamic control of the reaction. For example, HCl adds to buta-1,3-diene in a 1,2 fashion at cold temperatures (Equation 13-28), but in a 1,4 fashion at warm temperatures (Equation 13-29).

1,2-Addition

(13-28)

1,4-Addition

(13-29)

problem **13.13** Show how to produce each of the following compounds beginning with buta-1,3-diene. (Do not concern yourself with stereochemistry.)

(a)　　　　(b)　　　　(c)　　　　(d)

13.5 Choice of the Solvent

With a variety of factors to consider in devising a synthesis, it is easy to overlook the importance of the solvent. However, the nature of the solvent can have a dramatic effect on the outcome of a reaction. For example, consider the transform shown in Equation 13-30, which undoes the bromination of an alkene.

Undo a bromination.

(13-30)

To carry out the reaction in the forward direction, Br_2 must add across the double bond as discussed in Section 12.3a. If we choose water as the solvent, however, then the unwanted bromohydrin shown in Equation 13-31 would be produced, as we learned in Section 12.3b.

The solvent participates as a reactant to produce an unwanted bromohydrin.

$$\text{(13-31)}$$

To produce the desired 1,2-dibromide instead, the solvent must be non-nucleophilic, such as CCl_4. This is shown in Equation 13-32.

$$\text{(13-32)}$$

As another example, consider the transform shown in Equation 13-33, which undoes a nucleophilic substitution reaction.

Undo a nucleophilic substitution.

$$\text{(13-33)}$$

To carry out the reaction in the forward direction, we might choose $^-OCH(CH_3)_2$ as the nucleophile, with water as the solvent, so that the synthetic step would be as follows (Equation 13-34):

$$\text{(13-34)}$$

Keep in mind, however, that water has a mildly acidic hydrogen that can be deprotonated by $^-OCH(CH_3)_2$, according to Equation 13-35.

This side of the reaction is favored.

$$\text{(13-35)}$$

pK$_a$ = 15.7   pK$_a$ = 16.5

This proton transfer reaction effectively removes $^-OCH(CH_3)_2$ and replaces it with HO^-, so the *actual* nucleophile is primarily HO^-. Thus, even though the *intended* product is an ether, the major product of the reaction would be an alcohol, as shown in Equation 13-36.

Unwanted side reaction

$$\text{(13-36)}$$

One way to solve this problem is to choose an aprotic solvent whose protons are not acidic enough to be deprotonated, such as DMSO or DMF. A second way is to use a solvent whose conjugate base is the same as the intended nucleophile. In the example at hand, an appropriate solvent would be propan-2-ol, $(CH_3)_2CHOH$. Thus, deprotonation of the solvent by the nucleophile (Equation 13-37) does not result in any new species.

The reactants and products are identical.

(13-37)

YOUR TURN 13.9

If $CH_3CH_2O^-$ is to be used as a nucleophile, then what alcohol would be an appropriate solvent?

SOLVED problem 13.14 The transform at the right undoes a nucleophilic substitution reaction. How would you carry out the reaction in the forward direction using a *protic* solvent?

Undo a nucleophilic substitution.

Think Should the reaction be S_N1 or S_N2? To ensure that this reaction takes place, should you use a strong nucleophile or a weak nucleophile? What protic solvent could be used that would not react with that nucleophile to form a new nucleophile?

Solve Because there is inversion of stereochemical configuration at the carbon atom where substitution must take place, we want to carry out an S_N2 reaction, not an S_N1. To do so, we should choose a strong nucleophile, $CH_3CH_2CH_2O^-$. With this nucleophile, the proper choice of protic solvent would be $CH_3CH_2CH_2OH$, because deprotonating it would yield our desired nucleophile. Therefore, the synthetic step would be written as follows.

$NaOCH_2CH_2CH_3$

$CH_3CH_2CH_2OH$

---

problem 13.15 The following transform undoes a nucleophilic substitution reaction. What nucleophile would you use to carry out this reaction in the forward direction? What *protic* solvent would you use?

Undo a nucleophilic substitution.

13.6 Considerations of Stereochemistry in Synthesis

It is common for a target molecule to be one of multiple stereoisomers, each with different behavior. This is particularly true with biologically active compounds such as *naproxen*, a drug (marketed under the trade name Aleve) that is used for pain relief and the reduction of fever and inflammation.

Prescription drug that relieves pain, fever, and inflammation

(S)-Naproxen

Causes liver damage

(R)-Naproxen

Naproxen has one tetrahedral stereocenter and is therefore chiral—its two enantiomers are designated as (R) and (S). Whereas the (S) enantiomer is relatively safe to use, the (R) enantiomer causes liver damage! It would therefore be extremely useful to devise a synthesis in which the *desired* (S) stereoisomer was predominantly produced. Syntheses that accomplish this are called **asymmetric syntheses**, and they constitute the subject of a major area of ongoing research in organic synthesis.

A key to asymmetric synthesis is using a reaction that favors one type of product stereochemistry at the stage at which it is possible to generate more than one stereoisomer. Recall that *stereospecific* reactions accomplish this by producing one type of configuration exclusively, which depends on the reactant configuration. The S_N2 reaction in Equation 13-38, for example, is stereospecific. As we learned in Section 8.5b, S_N2 reactions result in the inversion of stereochemical configuration because the nucleophile must attack from the side opposite the leaving group.

An S_N2 reaction reverses the configuration.

(13-38)

A particularly useful stereospecific reaction is the conversion of an alcohol to an alkyl halide using PBr_3 or PCl_3, an example of which is shown in Equation 13-39.

Opposite configuration

(13-39)

As we learned in Section 10.1, the stereospecificity of this reaction is accounted for by the mechanism, which consists of back-to-back S_N2 steps. Only the second of those two steps involves the C atom bonded to the OH group, so the reaction leads to inversion of configuration at that C.

To successfully implement an S_N2 reaction when designing an organic synthesis, we must take stereochemistry into account in a retrosynthetic analysis. Suppose, for example, that we want to synthesize (S)-2-ethylpentanenitrile via an S_N2 reaction. Because S_N2 reactions cause inversion of configuration, the precursor should have the opposite configuration at the atom bonded to the leaving group. In this case, the stereospecific transform might be written as shown in Equation 13-40.

Opposite configuration

(S)-2-Ethylpentanenitrile   **Undo an S_N2.**   **(R)-3-Chlorohexane**   (13-40)

problem 13.16 The transform at the right, which is intended to undo an S_N2 reaction, does not take stereochemistry into consideration. Draw the precursor with the appropriate stereochemical configuration.

An E2 reaction is also stereospecific. As we learned in Section 8.5d, an E2 reaction favors the anticoplanar conformation of the substrate, in which a hydrogen atom and a leaving group on adjacent carbon atoms are anti to each other (Equation 13-41):

Only this configuration of the alkene is produced.

$$\xrightarrow[\Delta]{\text{NaOH}}$$

(13-41)

To take stereochemistry into account in a transform that undoes an E2 reaction, the hydrogen atom and the leaving group must be anti to each other before they are eliminated from the substrate. Thus,

> ■ The precursor of an E2 reaction is constructed from an alkene by adding the hydrogen and leaving group (L) substituents to the alkene carbon atoms in an anti fashion.
>
> ■ Groups attached to the same side of the C=C bond in the target should appear on the same side of the H—C—C—L plane in the precursor.

An example of such a transform is shown in Equation 13-42, in which (E)-3-methylpent-2-ene is the target.

CH₃ groups on same side of C=C bond

H, C, C, and Cl atoms are anticoplanar.

CH₃ groups on same side of H—C—C—Cl plane

Undo an E2.

(13-42)

(E)-3-Methylpent-2-ene

(2R,3R)-2-Chloro-3-methylpentane

In applying the transform, H and Cl atoms have been added to the alkene C atoms to arrive at (2R,3R)-2-chloro-3-methylpentane. The H and Cl atoms that are added are in an anti conformation about the C—C single bond indicated in the precursor. Furthermore, the two CH₃ groups are on the same side of the H—C—C—Cl plane in the precursor (i.e., behind the plane of the paper), just as they are on the same side of the C=C double bond in the target.

problem 13.17 The transform at the right shows another way in which H and Cl atoms can be added to (E)-3-methylpent-2-ene to undo an E2 reaction. The H and Cl atoms are in an anticoplanar conformation about the indicated C—C bond. Supply the three substituents that are missing.

(E)-3-Methylpent-2-ene

Undo an E2.

Electrophilic addition reactions can be stereospecific, taking place in a syn or anti fashion. As we saw in Chapter 12, carbene addition, epoxidation, and hydroboration–oxidation reactions (Equations 13-43 through 13-45) result in syn addition.

Syn addition

$$\text{(13-43)}$$

with CH₂N₂, hν

$$\text{(13-44)}$$

with MCPBA

$$\text{(13-45)}$$

1. BH₃·THF
2. H₂O₂, NaOH, H₂O → + Enantiomer

By contrast, the formation of 1,2-dihalides and halohydrins (Equations 13-46 and 13-47) results in anti addition.

Anti addition

$$\text{(13-46)}$$

Br₂ / CCl₄ → + Enantiomer

$$\text{(13-47)}$$

Br₂ / H₂O → + Enantiomer

Recall that when chirality is introduced in the products (e.g., Equations 13-45 through 13-47), these reactions result in a *mixture* of stereoisomers.

To take stereochemistry into account when incorporating these reactions in a synthesis, we can take advantage of the fact that in a syn addition, the cis/trans relationships are conserved among the groups attached to the alkene carbons. In an anti addition, these relationships are inverted. With this in mind, suppose that we want to apply a transform to (2R,3R)-3-phenylbutan-2-ol that undoes a hydroboration–oxidation reaction, as shown in Equation 13-48.

In this conformation, H and OH are syn.

The cis/trans relationships among the groups attached to the alkene are conserved.

Undo a hydroboration–oxidation.

$$\text{(13-48)}$$

(2R,3R)-3-Phenylbutan-2-ol   **(Z)-2-Phenylbut-2-ene**

Ensuring that the H and OH groups in the target are syn to each other, we can determine the precursor by removing those substituents and adding a π bond between the two C atoms to which they are initially attached. The cis/trans relationships of the groups attached to the alkene C atoms in the precursor remain the same as those in the target.

problem **13.18** Apply a transform to (2S,3R)-3-phenylbutan-2-ol that undoes a hydroboration–oxidation reaction.

Undo a hydroboration–oxidation.

(2S,3R)-3-Phenylbutan-2-ol

Even though a transform like the one in Equation 13-48 yields a single precursor, keep in mind that the reaction in the forward direction produces a racemic mixture of enantiomers because both the reactants and the environment are achiral. It is often possible, however, to carry out these kinds of reactions in a way that would favor one enantiomer over the other—a so-called **enantioselective synthesis**. Commonly, this is done by carrying out the reaction in the presence of a chiral catalyst. We will not discuss these techniques in depth here, but if you are interested in learning more about this topic, see I. Ojima's *Catalytic Asymmetric Synthesis*, 3rd ed. (Wiley, 2010).

The reactions that we have revisited here in Section 13.6 are not the only reactions with characteristic stereochemistry that we will encounter in this book. When you encounter other such reactions in subsequent chapters, be sure to pay particular attention to the aspects of their mechanisms that dictate their stereochemistry. Having a keen sense of the mechanisms of stereospecific reactions will better equip you to design more efficient syntheses.

13.7 Percent Yield

For any given target, there may be several different syntheses involving different synthetic intermediates, and perhaps different starting materials. What makes one synthesis better or worse than another? Much has to do with costs that would be incurred by carrying out the synthesis, including monetary costs of the materials, costs associated with time, and costs associated with environmental hazards. It also has to do with how "green" (i.e., environmentally friendly) the synthesis is. To minimize these costs and to help make the synthesis as green as possible, the *percent yield* of the target should be maximized. For this reason, there are two basic rules that should always be applied when designing a synthesis.

> 1. The number of steps in a synthesis scheme should be minimized.
> 2. Each step in the synthesis should proceed with the highest possible percent yield.

These rules are essentially an outcome of how percent yield is computed for a **linear synthesis** (i.e., a synthesis composed of sequential steps): *For a linear synthesis, the overall percent yield is equal to the product of the yields of the individual steps.* For example, compare a three-step linear synthesis to one that is six steps long, and suppose that each individual step in both syntheses proceeds with an 80% yield of product (which is often considered quite good). The three-step synthesis will have an *overall* yield of $(0.80) \times (0.80) \times (0.80) = (0.80)^3 = 0.51$, or 51%, whereas the six-step synthesis will have an overall yield of 26%. Thus, the synthesis with the fewer number of steps has the greater yield, consistent with the first rule.

YOUR TURN 13.10

Show how 26% was obtained as the overall yield for the six-step synthesis.

Now consider another pair of syntheses, each of which consists of three steps. Suppose that each step in the first synthesis proceeds with a 90% yield, to give an overall yield of $(0.90) \times (0.90) \times (0.90) = 0.73$, or 73%. Suppose, in the second synthesis, that each step proceeds with a 70% yield, so the overall yield is $(0.70) \times (0.70) \times (0.70) = 0.34$, or 34%. Thus, as suggested in rule 2, the first synthesis has the higher overall yield because each of its steps has a higher yield than the individual steps in the second synthesis.

Table 13-3 illustrates how dramatically the overall yield of a synthesis can change with the number of steps it has and the yield of each step. It is common to see the published synthesis of a natural product include 15 to 30 steps, with an overall yield of <5%!

---

problem **13.19** Calculate the percent yield of a synthesis consisting of eight steps, each of which proceeds with 75% yield.

---

problem **13.20** Calculate the percent yield of the following seven-step synthesis that converts the starting material **S** into the target **T**.

$$\textbf{S} \xrightarrow{90\%} \textbf{A} \xrightarrow{81\%} \textbf{B} \xrightarrow{85\%} \textbf{C} \xrightarrow{98\%} \textbf{D} \xrightarrow{82\%} \textbf{E} \xrightarrow{72\%} \textbf{F} \xrightarrow{94\%} \textbf{T}$$

Thus far, we have considered percent yield in a *linear synthesis* only. The yield can generally be improved, however, if the same target can be produced from a *convergent synthesis*. In a **convergent synthesis**, portions of a target molecule are synthesized separately and are assembled together at a later stage.

In an overly simplified example, suppose our target contains 16 carbons, and our only carbon source contains two carbons. We could then envision a *linear synthesis* that involved adding two carbons at a time. According to Figure 13-1a, this would require seven steps. If we assume an 80% yield for each step, then the overall yield would be $(0.80)^7 \times 100\% = 21\%$. In a convergent synthesis (Fig. 13-1b), however, we might connect two eight-carbon pieces together, which were constructed separately from our two-carbon source. Reaching the final target still requires seven separate steps, but the longest branch of the synthesis is only four steps, so the overall yield would be $(0.80)^4 \times 100\% = 41\%$. This is roughly twice that from the linear synthesis!

The better yield often obtained from a convergent synthesis leads to the following general rule:

When possible, a synthesis should be designed to be *convergent* rather than *linear*.

TABLE 13-3 How Overall Yield in a Synthesis Is Affected by the Number and Yield of Individual Steps

Number of Steps in a Synthesis	Percent Yield per Step	Overall Percent Yield
5	90%	59%
5	80%	33%
5	70%	17%
10	90%	35%
10	80%	11%
10	70%	2.8%
20	90%	12%
20	80%	1.1%
20	70%	0.08%

FIGURE 13-1 Linear versus convergent synthesis schemes Hypothetical synthesis of a 16-carbon target from two-carbon pieces, using (a) a linear synthesis scheme and (b) a convergent synthesis scheme. Assuming each step has an 80% yield, the overall yield of the linear synthesis is 21%, whereas that of the convergent synthesis is 41%.

Although these rules that maximize percent yield in a synthesis are of paramount importance in the laboratory, their importance in the classroom can vary significantly from one instructor to the next. If it has not already been made clear to you by your instructor, you should ask whether the application of these rules will impact the credit you are awarded for a particular synthesis problem.

13.8 Wrapping Up and Looking Ahead

Chapter 13 has introduced some of the basic concepts that underlie organic synthesis. The goal of organic synthesis is to design a scheme of reactions that enables us to produce a target molecule from available starting material. First, we reviewed some of the conventions chemists use to write individual reactions of a synthesis so that we can effectively communicate synthesis schemes. Then we saw the importance of cataloging reactions, so that we can become as familiar as possible with the utility of each reaction. In particular, we distinguished between functional group conversions and reactions that alter the carbon skeleton. Next, retrosynthetic analysis, in which we start with the target molecule and work backward, was introduced as an aid in designing a synthesis. With this technique, we apply transforms (i.e., proposed reaction reversals) to dissect the target molecule into smaller and/or simpler precursors, ultimately working back to compounds that can be used as starting material.

After carrying out a retrosynthetic analysis, there are often other matters to consider to arrive at a viable synthesis. We learned to avoid synthetic traps, where some transforms cannot be carried out in the forward direction as planned. Stereochemistry can be an issue, too, so a stereospecific or stereoselective synthesis may be necessary. And, for the most efficient synthesis, we need to maximize the percent yield. This often calls for a minimum number of steps in the synthesis and, if possible, a convergent synthesis in place of a linear synthesis.

As we said at the beginning of this chapter, the scope of organic synthesis is immense. We have only begun to explore what the field has to offer, in part because we have had relatively little exposure to reactions. We will encounter new reactions in subsequent chapters, giving us additional opportunities to hone the skills developed here in Chapter 13, and to learn new ones.

Chapter Summary and Key Terms

- A **synthesis** scheme is an abbreviated recipe that tells how **starting material** is converted into a particular **target**. **(Introduction)**

- In each **synthetic step**, only the reactants, reagents added, reaction conditions, and products are included. Reagents are usually written above the reaction arrow, whereas reaction conditions are usually written below. **(Section 13.1)**

- Reactions that form and/or break C—C σ bonds alter the **carbon skeleton**, whereas **functional group conversions** do not. **(Section 13.2)**

- **Retrosynthetic analysis** streamlines the process of designing an organic synthesis. **Transforms** are applied to the target molecule to arrive at smaller or simpler **precursors**, and the cycle is repeated on the precursors until a suitable starting material is reached. **(Section 13.3)**

- Some transforms cannot be carried out in the forward direction. These **synthetic traps** can occur if a precursor has multiple reactive functional groups or if one reactive functional group has two reactive sites. **(Section 13.4)**

- Often the choice of solvent can have a dramatic impact on the outcome of a reaction. **(Section 13.5)**

- When incorporating a stereospecific reaction into a synthesis, the stereochemistry of the reaction should be considered when applying a transform. **(Section 13.6)**

- For the most efficient synthesis, the percent yield should be maximized. This often calls for minimizing the number of steps and using a **convergent synthesis** scheme instead of a **linear** one. **(Section 13.7)**

13.21 Determine whether each of the following transformations requires a reaction that alters the carbon skeleton.

(a)

(b)

(c)

(d)

(e)

(f) H₃CO

(g) H₃CO

(h)

13.22 Each of the following is a set of directions for carrying out a reaction or sequence of reactions. Rewrite each set of directions in the form of a synthesis.

(a) To 2-ethylcyclohexanone, add lithium diisopropylamide, and when that reaction is complete, add bromoethane to yield 2,6-diethylcyclohexanone.

(b) Add molecular bromine to 2,2-dimethylcyclohexanone in the presence of acetic acid to yield 2,2-dimethyl-6-bromocyclohexanone. To the resulting mixture, add sodium cyanide to yield 2,2-dimethyl-6-cyanocyclohexanone.

(c) Treat pent-4-ynoic acid with diazomethane to produce methyl pent-4-ynoate. Next, add sodium hydride, followed by bromophenylmethane, to yield methyl 6-phenylhex-4-ynoate.

13.23 Describe how each of the following reactions or sequence of reactions is to be carried out in the laboratory.

(a)

(b)

(c)

13.24 Show how a retrosynthetic analysis might be constructed for each synthesis in Problems 13.22 and 13.23.

13.25 For each of the following proposed synthetic steps, determine whether or not the solvent is appropriate. For those that are inappropriate, explain why and suggest another solvent that could be used instead.

(a)

(b)

(c)

(d)

13.26 Determine whether each of the following proposed transforms represents a synthetic trap (i.e., it will not proceed as planned in the forward direction). Explain your reasoning.

(a)

(b)

(c)

(d)

HO⌐...O⌐cyclohexane ⟹ HO...Br + cyclohexane-ONa

(e)

13.27 Show how pent-2-yne can be made from two different alkyl bromides.

13.28 Show how propoxybenzene can be synthesized using phenol and alcohols as your only source of carbon atoms that appear in the target.

13.29 Show how you would synthesize each of the following compounds from the materials specified.

(a)

From reagents containing
4 or fewer carbons

(b)

From any alcohols

(c)

From reagents containing
5 or fewer carbons

13.30 For each of the following alkenes, provide an alkyl halide that can be used to synthesize it exclusively via an E2 reaction. Pay attention to stereochemistry.

(a)

(b)

(c)   **(d)**   **(e)**

13.31 Show how you would carry out the synthesis at the right from the starting material given, using any other compound containing, at most, one carbon atom.

13.32 Show how you would carry out the synthesis at the right using the starting material given, plus propyne, and any other inorganic reagents (i.e., reagents containing no carbon).

13.33 Suggest how you would carry out each of the syntheses below.

(a)

(b)

13.34 Compute the *overall* percent yield of the following proposed synthesis. The percent yield of each synthetic step or sequence of steps is provided above the appropriate reaction arrow.

13.35 The proposed synthesis in Problem 13.34 is a *linear* scheme. Using the same starting compounds and the same types of reactions shown in the proposed synthesis, construct a *convergent* synthesis that might produce the same target in a higher yield.

13.36 Propose how to carry out the synthesis at the right using any reagents necessary.

+ Enantiomer

13.37 Propose how to carry out the synthesis at the right using any reagents necessary.

13.38 Propose how to carry out the following synthesis using any reagents necessary.

+ Enantiomer

Pay attention to the stereochemistry in the target to determine the required stereochemistry of the starting material.

If your instructor assigns problems in smartwork, log in at **smartwork.wwnorton.com**.

Orbital Interactions 2

Extended π Systems, Conjugation, and Aromaticity

In Chapter 3, we were first introduced to the concept of *molecular orbitals (MOs)*. There, we saw that atomic orbitals (AOs) from separate atoms can overlap in space to yield new orbitals. Regions of *constructive interference* lower the energy of the resulting MO, whereas regions of *destructive interference* raise the energy.

The discussion in Chapter 3 dealt only with the MOs generated by the interaction of two AOs—one orbital from each of two atoms. These kinds of MOs are called *two-center molecular orbitals*, and include the σ bonding and σ* antibonding MOs produced by the interaction of two s orbitals, as well as the π bonding and π* antibonding MOs produced by the interaction of two p orbitals.

Most MOs, however, are not two-center MOs, but instead result from the simultaneous interaction of AOs from three or more atoms. These are called *multiple-center molecular orbitals*. In these situations, we learned that valence bond (VB) theory (Section 3.5) can simplify the picture by considering the interactions of just two AOs at a time—that is, one from each of two adjacent atoms. VB theory can account for a number of important aspects of molecular structure, but it does have significant shortcomings. Therefore, here in Chapter 14, we extend our discussion of MO theory by examining multiple-center MOs.

Graphene aerogel—approximately seven times lighter than air—rests on the petals of a flower. The material, whose fundamental structural motif is graphene (molecular chicken wire made entirely of carbon; see p. 5), is mechanically strong and electrically conductive, giving it a wide range of potential applications. What makes graphene conductive is the conjugation of its p orbitals, a topic discussed here in Chapter 14.

Although multiple-center MOs can have σ or π symmetry (or even other symmetries), we restrict the scope of our discussion solely to π MOs. We do so because of their importance in *conjugation* and *aromaticity*, concepts that are key to our understanding of chemical structure, reactivity, and energetics.

We begin our discussion by outlining some specific shortcomings of VB theory, demonstrating the need for a better model. We then show how multiple-center MOs can provide more accurate pictures and energy diagrams of two relatively simple species—namely, the allyl cation (CH_2=CH—CH_2^+) and buta-1,3-diene (CH_2=CH—CH=CH_2). These species serve as models for the concept of *conjugation*, which we subsequently use to develop the general idea of *aromaticity*. Together, these two concepts are crucial to our understanding of molecular structure.

What we learn about multiple-center MOs here is quite useful in the chapters that follow. Important concepts of spectroscopy (the study of how molecules interact with light, discussed in Chapters 15 and 16) can be better understood with a solid foundation in multiple-center MOs. Moreover, the energetic consequences of conjugation and aromaticity can significantly affect the reactivity of particular species and the outcome of chemical reactions in which they take part. We introduce examples of this beginning in Chapter 17.

14.1 The Shortcomings of VB Theory

Some chemical and structural features of molecules are unaccounted for by VB theory—most notably *electron delocalization*, or *resonance* (Section 1.10), over three or more nuclei. Here in Section 14.1, we examine these issues, using as examples the allyl cation (CH_2=CH—CH_2^+) and buta-1,3-diene (H_2C=CH—CH=CH_2).

14.1a The Allyl Cation

The allyl cation (CH_2=CH—CH_2^+) has the following two resonance structures:

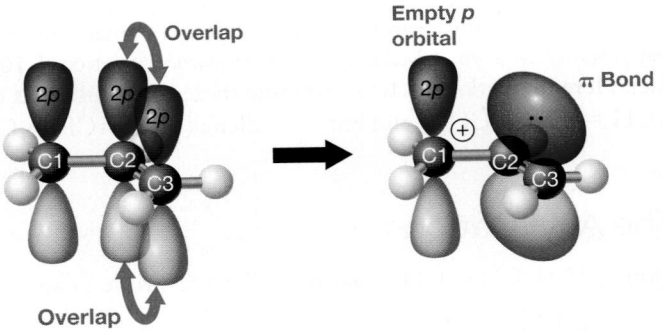

FIGURE 14-1 VB theory and the allyl cation (a) Atomic *p* orbitals on the C atoms in the allyl cation. (b) A π bond has been generated by the interaction between the *p* AOs on C1 and C2. This VB theory picture represents the resonance structure shown above it, which is an inaccurate description of the species.

The one true structure is better represented by the *resonance hybrid*. The hybrid is entirely planar, with all three C atoms sp^2 hybridized, identical C—C bonds that are intermediate in character between single and double bonds, and identical partial positive charges on the terminal C atoms.

VB theory does not generate this picture of the allyl cation, however, as shown in Figure 14-1.

Each C atom is sp^2 hybridized, so each has a single valence *p* AO. If we suppose that the *p* orbital on C1 interacts with the one on C2, then the result is an isolated π bond between C1 and C2. This suggests that there should be a formal double bond connecting C1 and C2, and there should be an empty *p* orbital on C3, giving C3 a +1 formal charge. As shown in Figure 14-1, this VB theory picture is analogous to the first of the two resonance structures shown previously, not the resonance hybrid. A similar picture emerges if we suppose that the *p* orbitals on C2 and C3 interact instead (see Your Turn 14.1). In general:

> The VB theory picture of a species that has resonance is inaccurate because it represents a single resonance structure, not the hybrid.

YOUR TURN 14.1

In the allyl cation, the *p* orbitals on C2 and C3 can interact, as shown on the left.

The resulting VB picture is shown on the right. Draw the resonance structure that corresponds to this VB picture.

Answers to Your Turns are in the back of the book.

problem **14.1** Draw a VB picture for the π orbitals of the allyl anion, CH_2=CH—CH_2^- (all C atoms are sp^2 hybridized). Draw the resonance hybrid of the allyl anion. Do the two pictures agree or disagree? Explain.

14.1b Buta-1,3-diene

This shortcoming of VB theory is not limited to just the allyl cation. Like the allyl cation, an uncharged molecule such as buta-1,3-diene ($H_2C=CH—CH=CH_2$) has structural features that VB theory does not account for. First, although the Lewis structure of buta-1,3-diene indicates a formal single bond between C2 and C3, rotation about that bond is more difficult than would be expected of a regular C—C single bond. Rotation does, in fact, occur, but the molecule favors an all-planar conformation. Furthermore, as shown in Figure 14-2, the formal double bonds at either end of the molecule are found experimentally to be slightly longer than the C=C double bond in ethene ($H_2C=CH_2$), whereas the C—C single bond in the middle of the molecule is found to be substantially shorter than that in ethane ($H_3C—CH_3$).

Both of these phenomena can be explained by taking into account the following weak resonance contributors:

Their contribution gives rise to a resonance hybrid in which the central single bond has some double-bond character and the terminal double bonds have some single-bond character. (These resonance contributors are weak because each has a C atom that lacks an octet and carries both a positive and a negative charge.) The single-bond character lengthens the double bonds, and the double-bond character shortens the single bonds. It is this double-bond character in the C2—C3 bond that causes the molecule to favor an all-planar conformation (Section 3.9).

Figure 14-3 shows, however, that a VB picture fails to account for these observations. Each of the C atoms in Figure 14-3a, which are sp^2 hybridized, has one unhybridized p orbital. Interaction between the p orbitals on C1 and C2 results in a π bond between those two carbons. A second π bond results from the interaction between the p orbitals on C3 and C4. Thus, the bond between C2 and C3 remains a single bond, and the VB picture resembles the resonance structure above on the left, not the resonance hybrid.

132 pm Ethene

154 pm Ethane

133.8 pm **135.4 pm** Buta-1,3-diene

FIGURE 14-2 Bond lengths in buta-1,3-diene The formal C—C single bond in buta-1,3-diene is significantly shorter than the one in ethane. Each of the formal C=C double bonds in buta-1,3-diene is slightly longer than the one in ethene.

FIGURE 14-3 VB theory and buta-1,3-diene (a) The atomic *p* orbitals that contribute to the MOs. (b) The interaction between the *p* orbitals on C1 and C2 results in a π bond between C1 and C2. The interaction between the *p* orbitals on C3 and C4 results in a π bond between C3 and C4. The VB picture shown corresponds to the Lewis structure above it.

14.2 Multiple-Center MOs

To accurately describe a species like the allyl cation or buta-1,3-diene, we must use **multiple-center molecular orbitals**: MOs that span *multiple atoms* and involve the simultaneous interaction of *multiple AOs*.

To begin to understand the nature of these orbitals, we can apply some general results from our treatment of two-center MOs previously discussed in Section 3.3:

> 1. MOs arise from interacting AOs on different atoms.
> 2. The number of MOs formed must equal the number of AOs that contribute to their creation (the *conservation of number of orbitals*).
> 3. Differences in the resulting MOs are due to the different ways in which the AOs contribute—specifically, to the different phase relationships of the contributing AOs.
> 4. The energy of the MO increases as the number of nodes perpendicular to the bonding axis increases.
> 5. An orbital must have opposite phases on either side of a nodal surface.

With these rules in mind, let us see how the MOs of π symmetry are constructed for the allyl cation and buta-1,3-diene.

14.2a The Allyl Cation ($H_2C=CH-CH_2^+$)

Three conjugated atomic *p* orbitals

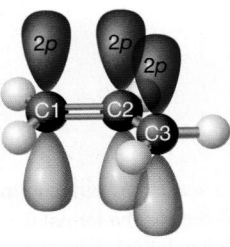

FIGURE 14-4 Conjugation in the allyl system These three *p* orbitals are conjugated because they are adjacent and parallel.

There are three unhybridized *p* orbitals in the allyl cation that are **conjugated**, meaning that they are *adjacent* and *overlap in a side-by-side fashion* (Fig. 14-4). From the simultaneous interaction of these three orbitals, *three MOs must be formed*. What do those MOs look like, and how do the *p* AOs mix to form them?

The following rule from quantum mechanics helps us answer these questions:

> For a linear system of conjugated *p* orbitals, all nodal planes in a resulting π MO must be positioned symmetrically about the center of that set of *p* orbitals.

The lowest-energy π MO, called π_1, has no nodal planes perpendicular to the bonding axes, so all three *p* AOs must mix together with the same phase, as shown on the left in the following graphic.

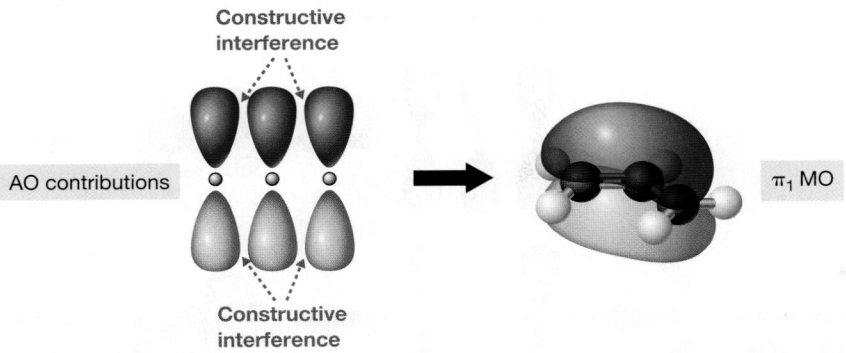

Notice that only constructive interference takes place above and below the bonding axes. This results in one large lobe above the bonding axes and one large lobe below, as indicated by the picture of the MO above on the right.

The second π MO, called π_2, has one nodal plane perpendicular to the bonding axes, so π_2 will appear at a higher energy than π_1. The nodal plane must appear on the central C atom because it must be positioned symmetrically about the center of the set of p orbitals:

The presence of the nodal plane precisely at the central C atom means that its atomic p orbital must make no contribution to the MO. This ensures that an electron in that MO has zero probability of being found in that plane. The remaining two p orbitals contribute with opposite phase, as they appear on opposite sides of the nodal plane.

The third π MO, called π_3, has two nodal planes perpendicular to the bonding axes, so π_3 has a higher energy than either π_1 or π_2. To be symmetric, the nodal planes must appear on either side of the central C atom, as shown below on the left:

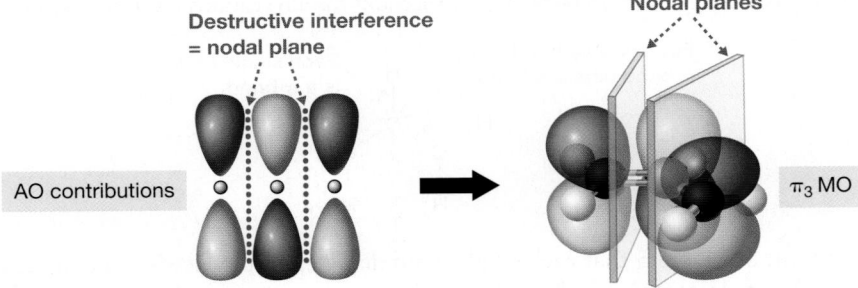

All three p AOs, therefore, contribute to the MO. Because the phases change going across each nodal plane, each interaction among adjacent p orbitals results in destructive interference.

For comparison, Figure 14-5 shows all three π MOs of the allyl cation in a single MO energy diagram.

14.2 **YOUR TURN**

In Figure 14-5, indicate the location of every nodal plane in each MO. (Try to do so without looking at the previous diagrams.)

Antibonding interactions

Energy

π_3 **Antibonding MO**

π_2 **Nonbonding MO**

Bonding interactions

π_1 **Bonding MO**

Noninteracting AOs

AO contributions

MOs

FIGURE 14-5 π MOs of the allyl cation The energy of the three noninteracting *p* AOs is indicated on the left. The phase relationships of those AOs that give rise to the three MOs are shown in the center, and the MOs themselves (designated π_1, π_2, and π_3), with their respective energies, are shown on the right.

We can also identify each of these π MOs as *bonding, nonbonding,* or *antibonding,* based on its energy relative to that of the isolated *p* AOs (represented by the horizontal dashed line in the middle of Fig. 14-5). The π_1 MO is lower in energy than the isolated *p* AOs because each pair of adjacent *p* AOs results in constructive interference (and thus a decrease in energy), so it is a **bonding MO** (or, more simply, a π MO). By contrast, π_3 is an **antibonding MO** (or, more simply, a π* MO), because each pair of adjacent *p* AOs results in destructive interference (and thus an increase in energy). Finally, the energy of π_2 is roughly the same as that of the noninteracting *p* AOs, because the *p* AOs shown are not adjacent, so they essentially do not interact. This makes π_2 a **nonbonding MO**.

Having established the relative energies of the π MOs of the allyl cation, we can complete the MO energy diagram by taking into account the allyl cation's MOs of σ symmetry.

Each double bond is a σ bond plus a π bond.

Each single bond is a σ bond.

Recall from Section 3.5 that each single bond depicted in the Lewis structure represents a pair of electrons in a σ MO. The double bond, moreover, which consists of one σ bond and one π bond, represents another pair of electrons occupying a σ MO. Thus, there are seven σ MOs in all. As we learned in Section 3.6, these σ MOs should be lower in energy than any π MOs, as shown in Figure 14-6.

As we learned in Section 3.5, too, the number of σ* MOs should equal the number of σ MOs. The σ* MOs should appear higher in energy than any MO of π symmetry.

The 16 valence electrons (depicted as eight total bonds in the Lewis structure) occupy these MOs according to the aufbau principle. The first 14 electrons completely fill all seven of the σ MOs. The remaining two electrons are placed into π_1, leaving the π_2 and π_3 MOs empty. As a result, π_1 is the HOMO and π_2 is the LUMO.

7 σ* MOs ◄············ These MOs are empty.

π_3

π_2 —— LUMO

π_1 —— HOMO

7 σ MOs ◄············ 14 electrons occupy these MOs.

Energy

FIGURE 14-6 Complete orbital energy diagram of the allyl cation The σ MOs are lower in energy than the MOs of π symmetry, whereas the σ* MOs are higher in energy. The 16 total valence electrons fill the eight MOs that are lowest in energy, making π_1 the HOMO and π_2 the LUMO.

SOLVED problem 14.2 Draw the complete MO picture and energy diagram for the allyl anion ($CH_2=CH-CH_2^-$), which is similar to that shown in Figure 14-6. Identify the HOMO and the LUMO. (All carbon atoms are sp^2 hybridized.)

Think How many p orbitals are conjugated? How many π MOs should result? How do those π MOs differ from each other? How many total σ and σ* MOs should there be? How many total valence electrons are there, and how should they be arranged in the orbitals?

Solve Each sp^2-hybridized carbon contributes one p orbital, so there are three p orbitals that are conjugated. Their simultaneous overlap forms three π MOs, the same as those in the allyl cation. Additionally, as we can see from their Lewis structures, the allyl anion has the same types of σ MOs as the allyl cation. The difference between the two species is that the allyl anion has two more electrons than the allyl cation. In this case, the last two electrons are placed in π_2, making π_2 the HOMO and π_3 the LUMO.

7 σ* MOs ◄············ These MOs are empty.

π_3 —— LUMO

π_2 —— HOMO

π_1

Energy

7 σ MOs ◄············ 14 electrons occupy these MOs.

problem 14.3 Draw the complete MO picture and energy diagram for the allyl *radical* ($CH_2=CH-CH_2^{\cdot}$), which has one unpaired electron. Identify the HOMO and the LUMO. (All carbon atoms are sp^2 hybridized.)

One bonding interaction

Two bonding interactions

This *p* AO is not conjugated to the other two.

π_1

Less stabilization

FIGURE 14-7 The planarity of the allyl cation When the allyl cation is entirely planar (left), two bonding interactions stabilize the electrons in π_1. When the terminal CH_2 group is rotated 90° (right), only one bonding interaction remains, which raises the energy of the orbital and its electrons.

Half of a π bond   Half of a π bond

The π_1 MO extends over all three C atoms.

FIGURE 14-8 Symmetric carbon–carbon bonds in the allyl cation The π electrons that occupy the π_1 MO are symmetrically distributed over the two carbon–carbon bonding regions. Because these two electrons account for one bond total, each pair of C atoms effectively receives half of a π bond.

How does this MO treatment account for the characteristics of the allyl cation that VB theory could not? That is, how does MO theory account for the allyl cation's planarity, its symmetric carbon–carbon bonds, and the identical partial positive charges on the terminal carbons?

The planarity of the allyl cation is largely due to π_1, which contains both π electrons. Recall that π_1 is the result of two bonding interactions among the three contributing *p* AOs (shown again in Fig. 14-7, left), and each bonding interaction is stabilizing. When a terminal CH_2 group is rotated 90° (Fig. 14-7, right), its *p* AO is no longer conjugated to the other two *p* AOs, which leaves only one bonding interaction. This raises the energy of the π MO, and of the π electrons, too.

The π_1 MO also accounts for the equivalent carbon–carbon bonds. As shown in Figure 14-8, the electrons in π_1 are distributed symmetrically over both the C1—C2 and the C2—C3 internuclear regions. These two electrons correspond to one π bond in total, so C1—C2 and C2—C3 each receive half of a π bond. Taking into account the C—C σ bonds, there are effectively one and one-half bonds connecting each pair of C atoms, consistent with the resonance hybrid.

Finally, to understand why each terminal carbon bears the same partial positive charge, examine the orbitals in Figure 14-9. Figure 14-9a shows the empty *p* orbital from the VB theory picture (Fig. 14-1), which we associate with a full positive charge. The analogous orbital in MO theory is the empty π_2 MO, which is equally shared between C1 and C3 (Fig. 14-9b). This results in an equal sharing of that positive charge on C1 and C3, or +½ on each.

VB theory

Full positive charge

The empty *p* orbital is localized on C3.

(a)

MO theory

Half of a positive charge   Half of a positive charge

The empty π_2 MO is shared equally on C1 and C3.

(b)

FIGURE 14-9 Equivalent partial positive charges in the allyl cation (a) In the VB picture the localized positive charge is associated with an empty *p* AO. (b) The corresponding MO is symmetrically distributed over C1 and C3, so the positive charge is equally shared by those two atoms.

SOLVED problem 14.4 Using the MO picture of the allyl anion ($H_2C=CH-CH_2^-$), explain why each terminal carbon atom bears the same partial negative charge. (If necessary, review Solved Problem 14.2.)

VB theory

MO theory

Think There are two more electrons in the allyl anion than in the allyl cation. Which orbital do those electrons occupy in the VB theory picture? What kind of charge do we associate with an occupied orbital like that? Which orbital do those two electrons occupy in the MO picture? How is the distribution of charge affected when that multiple-center MO is occupied?

Solve In the VB theory picture, the two additional electrons constitute a lone pair of electrons occupying a p orbital, which we associate with a full negative charge. In the MO theory picture, those electrons occupy the π_2 MO, which delocalizes the electrons onto C1 and C3. Thus, the negative charge is shared equally on C1 and C3, for a charge of $-\frac{1}{2}$ on each.

The p orbital localizes the nonbonding electrons on C3.

The π_2 MO delocalizes the nonbonding electrons onto C1 and C3.

problem 14.5 Suppose that one electron from the π_1 MO in the allyl cation were promoted to the π_2 MO. How would this affect the C—C bond lengths? How would it affect the total partial charges on the terminal carbon atoms?

14.2b Buta-1,3-diene ($H_2C=CH-CH=CH_2$)

To arrive at the MO picture and electron configuration of buta-1,3-diene, we can apply the same principles we just used for the allyl cation. With an all-planar conformation of the molecule, the four p AOs are parallel and, therefore, *conjugated* to one another (Fig. 14-10). As a result, they interact simultaneously to form *four MOs of* π *symmetry*.

The π MOs of buta-1,3-diene are shown in Figure 14-11. As with the allyl cation, the π MOs of buta-1,3-diene differ in the number of nodal planes perpendicular to the bonding axes. These differences are due, in turn, to the various unique ways in which the atomic p orbitals can contribute.

In the lowest-energy π MO, π_1, all of the p AOs have the same phase, so there are no nodal planes perpendicular to the C—C bonding axes. The next MO, π_2, contains one nodal plane, so it is higher in energy than π_1. Just as we saw with π_2 of the allyl cation, that single nodal plane must appear at the center of the system of p orbitals. The π_3 MO has two nodal planes and is thus higher in energy than π_2. Notice that those two nodal planes are positioned symmetrically about the center of the molecule—one between C1 and C2, and the other between C3 and C4. Finally, π_4, the highest-energy MO, contains three nodal planes positioned symmetrically about the center of the molecule.

Unlike the allyl cation, buta-1,3-diene has no nonbonding π MOs. This can be explained by the phases of p AOs contributed to each π MO. As in the allyl cation, the lowest-energy π MO in buta-1,3-diene (π_1) is the one in which all of the atomic p orbitals are in phase, resulting in three separate bonding contributions to the MO.

Each bonding contribution lowers the energy of a π MO relative to the p AOs.

Thus, π_1 is a bonding MO (or, more simply, a π MO). Similarly, the highest-energy π MO in buta-1,3-diene (π_4) is the one in which each pair of adjacent atomic p orbitals is out of phase, resulting in a total of three antibonding contributions to the MO.

Four conjugated p AOs

FIGURE 14-10 Conjugation in buta-1,3-diene These four p orbitals are conjugated because they are adjacent and parallel.

Nodal planes

π_4 ___
Antibonding MO

π_3 ___
Antibonding MO

π_2 ___
Bonding MO

π_1 ___
Bonding MO

Energy

Four noninteracting *p* AOs

AO contributions

MOs

FIGURE 14-11 π MOs of buta-1,3-diene The energy of the four noninteracting *p* orbitals is indicated on the left. The phase relationships of those AOs that give rise to the four MOs are shown in the center, and the MOs themselves (designated as π_1, π_2, π_3, and π_4), with their respective energies, are shown on the right. The vertical blue dotted lines represent nodal planes.

Each antibonding contribution raises the energy of a π MO relative to the *p* AOs.

Overall, then, π_4 is an antibonding MO (or, more simply, a π* MO). In the π_2 MO of buta-1,3-diene, there are two bonding contributions and one antibonding contribution (indicated by the nodal plane), resulting in a net lowering of the energy relative to the noninteracting atomic *p* orbitals. Overall, then, π_2 is a bonding MO. The π_3 MO, on the other hand, is an antibonding MO, because the overlap of the *p* AOs gives rise to one bonding contribution and two antibonding contributions (indicated by the two nodal planes), resulting in a net increase in the energy.

YOUR TURN 14.3

For each MO in Figure 14-11, label each region that gives rise to a bonding contribution and each region that gives rise to an antibonding contribution.

problem **14.6** The drawing at the right indicates the *p* AO contribution to a single MO of π symmetry. Locate the regions that give rise to bonding contributions and to antibonding contributions. How many are there of each? Will the contributions of the *p* AOs in this fashion produce a bonding, nonbonding, or antibonding MO? Explain.

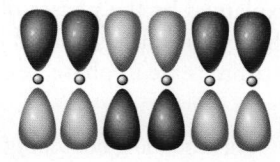

The complete orbital energy diagram of buta-1,3-diene is shown in Figure 14-12. There are nine total σ MOs, one for each of the seven single bonds shown in the

9 σ* MOs ←----- These MOs are empty.

π_4

π_3 LUMO

π_2 HOMO

π_1

Energy

9 σ MOs ←----- 18 electrons occupy
these MOs.

FIGURE 14-12 Complete orbital energy diagram of buta-1,3-diene The σ MOs are lower in energy than the MOs of π symmetry, whereas the σ* MOs are higher in energy. The 22 total valence electrons occupy the 11 MOs that are lowest in energy, making π_2 the HOMO and π_3 the LUMO.

Lewis structure and one for each of the two double bonds. There are also nine total σ* MOs, one for each bonding MO. Just as with the allyl cation, the σ MOs are lower in energy than the π MOs, and the σ* MOs are higher in energy than the π* MOs.

Looking again at the Lewis structure of buta-1,3-diene, notice that there are 22 total valence electrons. The first 18 of those electrons occupy the nine σ MOs. The remaining four electrons occupy π_1 and π_2, the two lowest-energy π MOs. The remaining two MOs of π symmetry, π_3 and π_4, remain empty. Therefore, π_2 is the HOMO, and π_3 is the LUMO.

---

problem **14.7** Draw the complete MO energy diagram for the butadienyl dication, $[H_2C—CH=CH—CH_2]^{2+}$, similar to the one in Figure 14-12. Draw in all of the valence electrons and identify the HOMO and the LUMO. (*Note:* All carbons are sp^2 hybridized.)

Now that we have developed the MO picture of buta-1,3-diene, we can see why the molecule favors the all-planar conformation about the C2—C3 bond, and why the C1—C2 and C3—C4 bonds are longer than normal, whereas the C2—C3 bond is unusually short. Buta-1,3-diene favors planarity because it allows the four *p* AOs to be parallel, which gives rise to three bonding interactions for the π_1 MO, as shown in Figure 14-13a. When the C2—C3 bond is rotated 90°, however, as shown in Figure 14-13b, there is no longer a π bonding interaction between C2 and C3, which results in an increase in the energy of two π electrons.

The π_1 MO also accounts for the unusual carbon–carbon bond lengths. As shown in Figure 14-14, the two electrons in π_1 are delocalized over the entire molecule. Thus, the electrons that occupy that MO give the C2—C3 bond some π bond character, at the expense of the π bond character of the C1—C2 and the C3—C4 bonds.

Three bonding interactions

This is no longer a π bonding interaction, so (b) is not as stable as (a).

π_1

Rotate C2—C3 bond 90°

Buta-1,3-diene
(a)

(b)

FIGURE 14-13 Planarity of buta-1,3-diene (a) With buta-1,3-diene in an all-planar conformation, the four atomic *p* orbitals contribute to the π_1 MO to give rise to three bonding interactions. (b) Rotating 90° about the C2—C3 bond destroys one of the bonding interactions, resulting in a less stable molecule.

MO theory

π Bond character

The π_1 MO delocalizes electrons over the C1—C2, C2—C3, and C3—C4 bonding regions.

FIGURE 14-14 The unusual carbon–carbon bond lengths of buta-1,3-diene The π_1 MO is delocalized over the entire molecule, which gives the C2—C3 bond some π bond character at the expense of some π bond character from the C1—C2 and C3—C4 bonds. This shortens the C2—C3 bond and lengthens the other two.

Effectively, then, the central bond is slightly more than a formal single bond, and is therefore slightly stronger and shorter than a normal C—C single bond. By the same token, the C1—C2 and C3—C4 bonds are each slightly less than a formal double bond, such that each is weaker and longer than a normal C=C double bond.

14.3 Aromaticity and Hückel's Rules

Sections 14.1 and 14.2 discussed aspects of conjugation pertaining to the allyl cation and buta-1,3-diene. Conjugation is also apparent in benzene (C_6H_6), a cyclic, six-carbon molecule whose Lewis structure depicts alternating single and double bonds. Based on the Lewis structure, benzene should have alternating shorter and longer carbon–carbon bonds. Experimentally, however, all six of them are identical in length, intermediate between that of a formal single bond and a formal double bond, as shown in Figure 14-15a. Furthermore, the entire molecule is perfectly planar and all six C—C—C bond angles are 120°. This high degree of symmetry is consistent with benzene's resonance hybrid, shown in Figure 14-15b.

Identical C—C bond lengths

139 pm

139 pm

(a)   (b)

FIGURE 14-15 The structure of benzene (a) The carbon–carbon bond lengths in benzene are all identical. (b) Benzene has equivalent resonance structures (left), the hybrid of which (right) suggests a symmetric molecule.

Perhaps more importantly, from a chemical standpoint,

Benzene's system of π electrons is *unusually stable*.

For example, whereas Br_2 adds to a normal alkene under relatively mild conditions (Equation 14-1; Section 12.3), no reaction occurs with benzene under these same conditions (Equation 14-2).

$$\text{(diagram)} \xrightarrow{Br_2} \text{(diagram)} \quad + \quad \text{Enantiomer} \qquad \text{(14-1)}$$

$$\text{(diagram)} \xrightarrow{Br_2} \text{No reaction} \qquad \text{(14-2)}$$

If a strong Lewis acid catalyst such as $FeBr_3$ is present (Equation 14-3), then a reaction does take place, but rather than addition, benzene undergoes substitution to preserve the π system. (We discuss these kinds of reactions in greater detail in Chapters 22 and 23.)

Benzene undergoes substitution rather than addition.

$$\text{(diagram)} \xrightarrow[FeBr_3]{Br_2} \text{(diagram)} \quad Br \qquad \text{(14-3)}$$

Conjugated Linoleic Acids

Unsaturated and polyunsaturated fatty acids (PUFAs) have received a lot of attention over the years for their health benefits, including lowering the risk of heart attacks and cardiovascular disease. Some of these fatty acids—oleic acid, linoleic acid, and linolenic acid—were examined in Section 4.16. In the biosynthesis of these fatty acids, the cis configurations of the double bonds are produced almost exclusively, and the double bonds themselves are separated by a single CH_2 group. These double bonds, therefore, are isolated from each other.

In the stomach of a ruminant animal (such as cattle), certain bacteria carry out biohydrogenation to convert dietary linoleic acid to stearic acid, its saturated form. Conjugated linoleic acids (CLAs), which are isomers of linoleic acid, are intermediates in this process, and the predominant CLA is the *cis*-9-*trans*-11 isomer.

Linoleic acid, $C_{18}H_{32}O_2$

cis-9-*trans*-11 CLA

These CLAs have significantly different bioactivity than PUFAs, and some studies suggest that CLAs exhibit remarkable health benefits of their own. CLAs have been purported to be an effective anticarcinogen, they are used to treat inflammatory bowel disease and Crohn's disease, and they are even thought to be effective weight-loss supplements.

It is not entirely clear what gives rise to the unique bioactivity of these CLAs, but at least in some cases, the additional rigidity brought about by the conjugation of the double bonds is believed to play a role. Just as we saw with buta-1,3-diene here in Chapter 14, the conjugation in *cis*-9-*trans*-11 CLA favors a coplanar arrangement of the double bonds, and this rigidity can affect the ability of the fatty acid to dock with specific enzyme active sites or receptors.

We can gain a sense of benzene's stability from **heat of hydrogenation** data. As Equation 14-4 shows, 1 mol of benzene reacts with 3 mol of H_2 in the presence of a catalyst, under high temperature and pressure, to produce 1 mol of cyclohexane (this reaction is discussed in greater detail in Chapter 19).

$$\Delta H^\circ_{rxn} = -208 \text{ kJ } (-49.7 \text{ kcal})$$ (14-4)

Similarly, 3 mol of cyclohexene will undergo hydrogenation in the presence of a catalyst (Equation 14-5) to produce 3 mol of cyclohexane, but under much milder

$$\Delta H^\circ_{rxn} = -360 \text{ kJ } (-86.0 \text{ kcal})$$ (14-5)

Benzene's π system is more stable, so less heat is released upon hydrogenation.

$\Delta H^{\circ}_{rxn} = -208$ kJ (−49.7 kcal)

$\Delta H^{\circ}_{rxn} = -360$ kJ (−86.0 kcal)

FIGURE 14-16 The stability of benzene The hydrogenation of one mole of benzene (left) and three moles of cyclohexene (right) are both exothermic. The π system of benzene is stabilized by aromaticity, however, so it releases significantly less heat upon hydrogenation.

conditions. Both of these reactions can be simplified to Equation 14-6, in which 3 mol of C=C double bonds react with 3 mol of H_2 to produce 3 mol of C—C single bonds and 6 mol of C—H bonds.

$$3\ C{=}C \quad + \quad 3\ H_2 \quad \longrightarrow \quad 3\ C{-}C \quad + \quad 6\ C{-}H \qquad \text{(14-6)}$$

However, ΔH°_{rxn} for the two reactions differs significantly. Whereas the hydrogenation of 1 mol of benzene (Equation 14-4) releases 208 kJ (49.7 kcal), the hydrogenation of 3 mol of cyclohexene (Equation 14-5) releases 360 kJ (86.0 kcal). Thus, as shown in the energy diagram in Figure 14-16, benzene's π system is more stable by 360 kJ − 208 kJ = 152 kJ (36.3 kcal). This "extra" stability is attributed to **aromaticity** and is sometimes referred to as benzene's **resonance energy**.

YOUR TURN 14.4

What fraction of an average C—C single bond energy is benzene's resonance energy? (*Hint:* You may need to review Section 1.4.)

SOLVED problem **14.8** Based on its heat of hydrogenation, is cycloocta-1,3,6-triene aromatic? Explain.

Cycloocta-1,3,6-triene

$\Delta H^{\circ}_{hyd} = -334$ kJ/mol (−79.8 kcal/mol)

Think How many moles of H_2 are required to completely hydrogenate cycloocta-1,3,6-triene? How much heat is released when that number of moles of hydrogen reacts with cyclohexene instead? How does that quantity compare to the heat of hydrogenation for cycloocta-1,3,6-triene?

Solve Complete hydrogenation of cycloocta-1,3,6-triene requires 3 mol of H_2—one for each mole of C=C double bonds. If 3 mol of H_2 were to react with cyclohexene instead, then (3)(120 kJ/mol) = 360 kJ/mol (86.0 kcal/mol) of heat would be released. This is 26 kJ/mol (6.2 kcal/mol) more heat than is released when hydrogenating cycloocta-1,3,6-triene, suggesting that the three double bonds in cycloocta-1,3,6-triene are about 26 kJ/mol (6.2 kcal/mol) more stable than in cyclohexene. This is much smaller than the 152 kJ/mol (36.3 kcal/mol) of stabilization provided by the three double bonds in benzene, so cycloocta-1,3,6-triene is *not* aromatic.

---

problem **14.9** Based on their heats of hydrogenation, predict whether or not each of these compounds is aromatic.

Cyclopenta-1,3-diene

$\Delta H^{\circ}_{hyd} = -211$ kJ/mol (−50 kcal/mol)

Naphthalene

$\Delta H^{\circ}_{hyd} = -318$ kJ/mol (−76 kcal/mol)

132 pm

←----160 pm

(a)

Not resonance structures
because atoms moved

(b)

FIGURE 14-17 The structure of
cyclobutadiene (a) Cyclobutadiene
has two different bond lengths, one that
is characteristic of a C=C double bond
and one that is characteristic of a C—C
single bond. (b) Cyclobutadiene does not
enjoy the same resonance stabilization
that benzene does.

The unusual stability of benzene's π system can be explained, in part, by electron delocalization through *resonance* (Chapter 1). The two resonance structures of benzene, called **Kekulé structures** after the German chemist, August Kekulé (1829–1896), who proposed them, are shown in Figure 14-15b (p. 688). Because the resonance structures are *equivalent*, they should contribute equally to the hybrid structure, allowing the six π electrons to be delocalized completely around the ring.

If resonance fully explained this phenomenon, however, then we should expect a similar stability in the π system of cyclobutadiene. Just like benzene, cyclobutadiene can be represented as a fully conjugated ring that has two equivalent resonance structures (Fig. 14-17). Unlike benzene, however,

Cyclobutadiene's system of π electrons is *unusually unstable*.

Cyclobutadiene's π system is so unstable, in fact, that it has only been isolated and studied in an argon matrix at extremely low temperatures. Moreover, those studies reveal that cyclobutadiene consists of *two different carbon–carbon bonds*: one that is consistent with a C=C double bond and one that is consistent with a C—C single bond; the molecule is rectangular! In other words, cyclobutadiene does not exist as a resonance hybrid, but rather equilibrates between two different structures.

Compare the carbon–carbon bond lengths in ethane and ethene to those in cyclobutadiene. (See Section 14.1b.)

14.5 YOUR TURN

Benzene and cyclobutadiene are members of two separate *classes* of compounds that exhibit different characteristic behavior. Benzene is classified as *aromatic*, whereas cyclobutadiene is *antiaromatic*. In general,

- **Aromatic** compounds have cyclic π systems that are unusually *stable*.
- **Antiaromatic** compounds have cyclic π systems that are unusually *unstable*.
- All other compounds are classified as **nonaromatic** (i.e., as neither unusually stable nor unusually unstable).

From his observations on the structural similarities of compounds in each of these classes, Erich Hückel (1896–1980), a German physicist and physical chemist, proposed what are now known as **Hückel's rules** for aromaticity:

- If a species possesses a π system of MOs constructed from atomic *p* orbitals that are fully conjugated around a ring, then the species is:
  - *Aromatic* if the number of electrons in that cyclic π system is either 2, 6, 10, 14, 18, and so on. (These are called **Hückel numbers**.)
  - *Antiaromatic* if the number of electrons in that cyclic π system is either 4, 8, 12, 16, 20, and so on. (These are called **anti-Hückel numbers**.)
- All other species are *nonaromatic*.

Benzene

Fully conjugated ring of *p* orbitals

2 π electrons

2 π electrons

2 π electrons

(a)

Cyclobutadiene

Fully conjugated ring of *p* orbitals

2 π electrons

2 π electrons

(b)

FIGURE 14-18 Benzene, cyclobutadiene, and Hückel's rules (a) Benzene has a cyclic π system that is fully conjugated (left), which is occupied by six electrons (right). (b) Cyclobutadiene has a cyclic π system that is fully conjugated (left), which is occupied by four electrons (right).

Each Hückel number is an *odd number of pairs* (e.g., 6 is the same as three pairs), whereas each anti-Hückel number is an *even number of pairs* (e.g., 4 is the same as two pairs). Alternatively, the Hückel numbers correspond to $4n + 2$, where *n* is any integer ≥ 0, and the anti-Hückel numbers correspond to $4n$, where *n* is any integer ≥ 1.

Benzene satisfies the criteria for an aromatic species, as shown in Figure 14-18a. Each of its six C atoms has an unhybridized atomic *p* orbital (because each is sp^2 hybridized). The six *p* orbitals are all adjacent and parallel, making them conjugated in a complete ring. Additionally, there are a total of six π electrons, which is a Hückel number.

Cyclobutadiene (Fig. 14-18b) also has a fully conjugated ring of atomic *p* orbitals because all of its C atoms, too, are sp^2 hybridized. Cyclobutadiene's π system contains four electrons, however, which is an anti-Hückel number.

SOLVED problem 14.10 Assuming the molecule is entirely planar, would cycloocta-1,3,5,7-tetraene be aromatic, antiaromatic, or nonaromatic? Explain your reasoning.

Think Does planar cycloocta-1,3,5,7-tetraene contain a set of atomic *p* orbitals that are fully conjugated around a ring? If so, does it contain a Hückel or anti-Hückel number of π electrons?

Solve All eight C atoms are sp^2 hybridized, so each contributes an atomic *p* orbital. Thus, if the molecule is planar, then the eight *p* orbitals are fully conjugated around the ring. In that π system, there are eight electrons (two from each of the four double bonds), which is an anti-Hückel number. According to Hückel's rules, this compound should be antiaromatic. (In actuality, cycloocta-1,3,5,7-tetraene is *not* planar, as discussed later in Section 14.6.)

---

problem **14.11** Predict whether each of the following compounds is aromatic, antiaromatic, or nonaromatic.

(a) **(b)** **(c)** **(d)** **(e)** **(f)** **(g)**

14.4 The MO Picture of Benzene: Why It's Aromatic

Given the difficulties of resonance theory in explaining the unusual stability of benzene, as well as the unusual instability of cyclobutadiene, we must turn to MO theory. In benzene, there are six unhybridized *p* orbitals fully conjugated around the ring (Fig. 14-19a), because each C atom is sp^2 hybridized. The simultaneous interaction among benzene's atomic *p* orbitals must yield six MOs of π symmetry.

(a)

(b)

6 p AOs   6 π MOs

(c)

AO energies

3 H₂C═CH₂

FIGURE 14-19 Energy diagram for the π MOs of benzene (a) The six atomic p orbitals of benzene. (b) The energies of the six MOs of benzene. (c) Three molecules of ethene. The dashed line indicates the energy of the noninteracting atomic p orbitals.

We can derive the relative energies of benzene's π MOs using the **Frost method**, a shortcut developed by the American chemist Arthur A. Frost (1909–2002) that can be applied to any species with a fully conjugated, cyclic π system:

1. Draw the polygon that represents the cyclic compound's line structure, with one of the vertices pointed directly downward.
2. Place the energy of the noninteracting atomic p orbitals at the center of the polygon.
3. Place the energies of the π MOs at the vertices of the polygon.

Thus, as shown in Figure 14-19b, benzene has a single π MO (π_1) that is lowest in energy. At a somewhat higher energy, there are the π_2 and π_3 MOs—they have identical energies and are thus called **degenerate orbitals**. A second pair of degenerate orbitals, π_4 and π_5, are found at a higher energy still, and π_6 has the highest energy. Benzene's six π electrons fill the lowest-energy π MOs first, and therefore occupy π_1, π_2, and π_3.

In Figure 14-19b, label each of the six MOs as either "bonding," "nonbonding," or "antibonding." Also, identify each *pair* of degenerate orbitals.

14.6 YOUR TURN

---

problem 14.12 Draw the complete MO energy diagram for benzene, including the MOs of σ symmetry, similar to Figures 14-6 and 14-12.

For comparison, the energy of the six π electrons in three molecules of $H_2C═CH_2$, whose double bond is isolated, is shown in Figure 14-19c. Whereas benzene's π_2 and π_3 MOs are similar in energy to the π MO in $H_2C═CH_2$, benzene's π_1 MO is significantly lower in energy. This provides insight into why benzene's π system of electrons is so stable.

Figure 14-20 shows how each of benzene's π MOs is produced from the contributing p AOs. As we have seen before, one way in which these MOs differ is in the phases of the contributing p orbitals. Some also differ in the number of nodal planes perpendicular to the bonding axes.

- Each pair of degenerate MOs has the same number of nodal planes perpendicular to the bonding axes.
- The energy of each MO rises with each additional nodal plane.

In particular, there are zero nodes for π_1, one for π_2 and π_3, and two for π_4 and π_5.

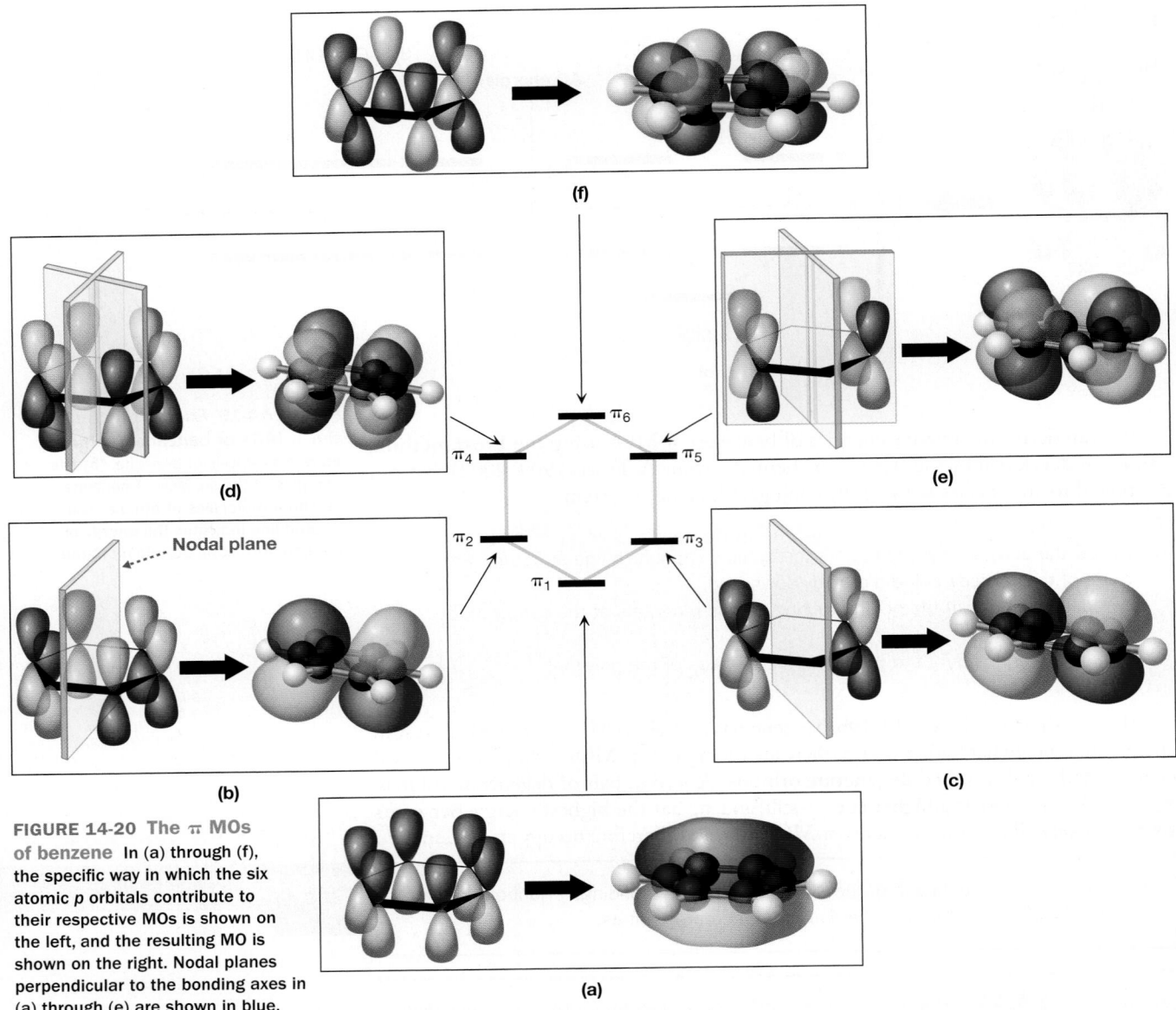

FIGURE 14-20 The π MOs of benzene In (a) through (f), the specific way in which the six atomic *p* orbitals contribute to their respective MOs is shown on the left, and the resulting MO is shown on the right. Nodal planes perpendicular to the bonding axes in (a) through (e) are shown in blue.

YOUR TURN 14.7

In Figure 14-20, how many nodal planes perpendicular to the bonding axes should π_6 have? Can you locate them?

14.5 The MO Picture of Cyclobutadiene: Why It's Antiaromatic

Cyclobutadiene presents a somewhat different situation. In cyclobutadiene, there must be four MOs of π symmetry generated from the four unhybridized *p* AOs that are fully conjugated around the ring. To obtain their relative energies, we can use the Frost method (Section 14.4). We orient a square such that one vertex is pointed downward (Fig. 14-21). Each of the four vertices then represents the energy of a π MO (π_1, π_2, π_3, and π_4), and the center of the square represents the energy of each unhybridized *p* AO.

YOUR TURN 14.8

In Figure 14-21, label each MO as either bonding, antibonding, or nonbonding.

(a) Four *p* AOs

(b) Four π MOs

FIGURE 14-21 **Energy diagram for the π MOs of square cyclobutadiene** (a) Four unhybridized atomic *p* orbitals on the C atoms of cyclobutadiene. (b) Energy diagram of the four π MOs of cyclobutadiene. The dashed line represents the energy of the unhybridized *p* AOs.

Figure 14-21 shows that the four π electrons in cyclobutadiene occupy π_1, π_2, and π_3. The π_1 MO is completely filled, whereas π_2 and π_3 contain only one electron each. This is in accordance with *Hund's rule*—namely, orbitals of the same energy are not doubly occupied unless it is absolutely necessary.

---

problem **14.13** Draw the complete MO energy diagram for cyclobutadiene, including the MOs of σ symmetry, similar to Figures 14-6 and 14-12.

We can see why cyclobutadiene's π system is so unstable by comparing the MO energy diagram of square cyclobutadiene (Fig. 14-21) to that of benzene (Fig. 14-19). In benzene, all valence electrons occupy bonding MOs; they are highly stabilized relative to their energies in unhybridized AOs. In cyclobutadiene, on the other hand, only two of the four π electrons are found in a bonding MO; the other two are in higher-energy nonbonding MOs. Furthermore, whereas all of benzene's π electrons are paired, cyclobutadiene has two *unpaired* electrons. As we discuss in greater detail in Chapter 25, species with unpaired electrons, known as *free radicals,* tend to be quite unstable.

Figure 14-22 shows how each of cyclobutadiene's π MOs is produced from the contributing *p* AOs. As with the other π systems we have seen, the lowest-energy MO

FIGURE 14-22 **The π MOs of cyclobutadiene** In (a) through (d), the specific way in which the four atomic *p* orbitals contribute to their respective MOs is shown on the left, and the resulting MO is shown on the right. Nodal planes perpendicular to the bonding axes are shown in blue.

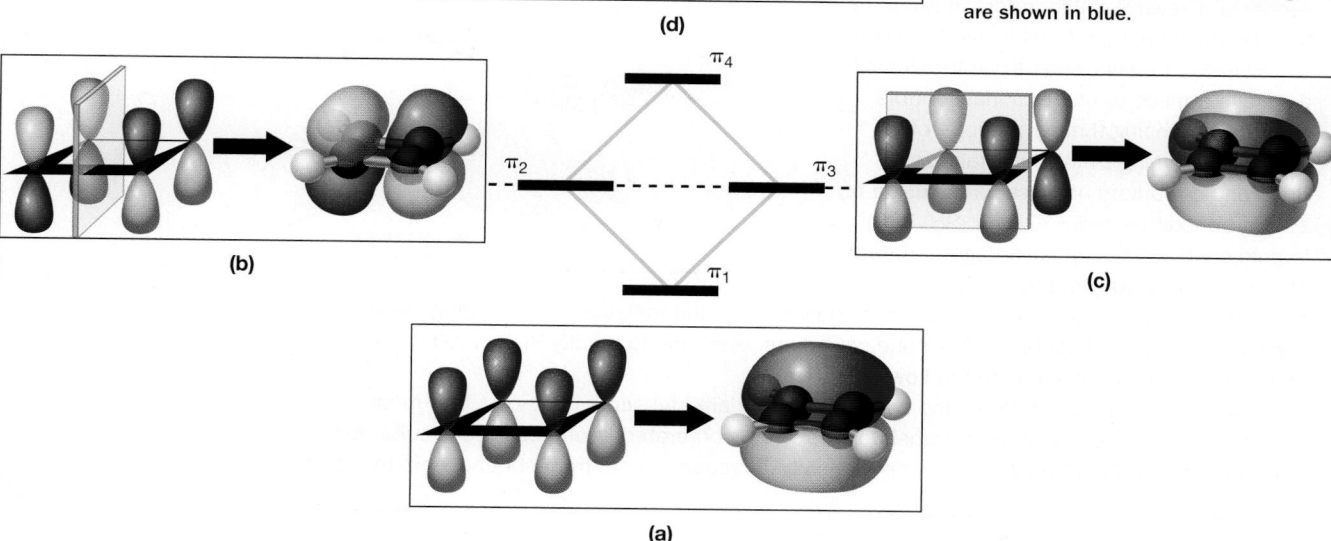

(π_1) has no nodal planes perpendicular to the bonding axes, and each additional nodal plane raises the energy of the other MOs.

SOLVED problem 14.14 Based on the characteristics of its MO diagram, would you expect the cyclobutadienyl dication to be aromatic, antiaromatic, or nonaromatic? Explain. (*Hint:* How would Fig. 14-21 change upon going from the uncharged molecule to this species with a +2 charge?)

The cyclobutadienyl dication

Think Are the *p* AOs in the cyclobutadienyl dication fully conjugated in a ring? What does the Frost method suggest for the relative energies of the π MOs? How many π electrons are there? Are they all paired? Do they all reside in bonding MOs?

Solve Each of the C atoms is sp^2 hybridized and all four *p* orbitals are conjugated, just as in cyclobutadiene. The Frost method yields four π MOs with the same relative energies as in cyclobutadiene— one bonding, two nonbonding, and one antibonding.

There is a total of two π electrons. The +2 charge indicates that there are two fewer electrons than in the uncharged cyclobutadiene molecule. Both electrons are paired and occupy the bonding MO only, so this species should be aromatic.

All π electrons are paired and reside in a bonding MO.

Benzene and Molecular Transistors

Over the last 40 years, computing speed has doubled roughly every 18 months—an idea related to a prediction in 1965 by Gordon Moore (cofounder of Intel Corporation) that is now called Moore's law. Much of this trend has to do with our ability to fabricate transistors, which are key electronic components of integrated circuits. Quite simply, as the number of transistors that can fit in an integrated circuit increases, so does computing speed. Smaller transistors, therefore, means faster processing speed.

Many people believe that Moore's law will come to an end within the next decade because we are fast approaching a fundamental barrier in miniaturization—namely, the size of the atom. In fact, in 2009, M. A. Reed at Yale University and several collaborators in South Korea engineered the first functional single-molecule transistor.

A typical transistor has three terminals; a voltage applied to one terminal controls the current running through the other two. To make their single-molecule transistor, Reed and coworkers anchored a molecule of 1,4-benzenedithiol (HS—C_6H_4—SH, BDT) to gold contacts, taking advantage of the relatively strong sulfur–gold bond. The conjugation within BDT's π system allows current to flow through the molecule. By changing the voltage applied from the gold contacts, Reed and coworkers were able to modify BDT's orbital energies, and so were able to manipulate the flow of current.

Electron flow

Despite this triumph by Reed and coworkers, there are still some obstacles that must be overcome before these molecular transistors can be incorporated into integrated circuits. But when they are, today's transistors will dramatically decrease in size, from about 20 nm to <1 nm.

problem **14.15** Based on the characteristics of its MO diagram, do you think the benzene dication should be aromatic, antiaromatic, or nonaromatic? What would you expect for the benzene dianion? Explain.

The benzene dication

The benzene dianion

14.6 Aromaticity in Larger Rings: [*n*]Annulenes

The two species we have examined thus far with cyclic π systems—benzene and cyclobutadiene—are *monocyclic* (one ring), fully conjugated molecules that contain only carbon and hydrogen. These molecules belong to a class of compounds collectively called **annulenes**. They are commonly named "[*n*]annulenes," where *n* is the number of carbon atoms in the ring. Therefore, cyclobutadiene is [4]annulene and benzene is [6]annulene.

Cyclobutadiene and benzene are not the only [*n*]annulenes. For example, cycloocta-1,3,5,7-tetraene (Fig. 14-23a) is [8]annulene. According to Hückel's rules, [8]annulene should be antiaromatic if it is planar. This would give it a fully conjugated, cyclic π system, and it has eight π electrons, which is an anti-Hückel number. Because of the instability associated with antiaromaticity, however, [8]annulene resists planarity. Its most stable conformation is "tub" shaped (Fig. 14-23b), in which the *p* AOs are not all parallel to one another (Fig. 14-23c), and are therefore not fully conjugated. Thus, [8]annulene is better characterized as nonaromatic.

[10]Annulene (Fig. 14-24a) should be aromatic according to Hückel's rules, if it is planar. Each C atom contributes a *p* AO to make a complete ring, and there are 10 π electrons (a Hückel number). The molecule does not attain planarity, however, because a planar [10]annulene would have excessive strain. With all of the double bonds in a cis configuration, there would be excessive *angle strain*; the average interior angle would have to be 144°, compared to the ideal angle of 120° for an sp^2-hybridized carbon atom. The angle strain could be eliminated by introducing two trans double bonds into the ring, as shown in Figure 14-24b, but that molecule would have excessive *steric strain*, due to the H atoms crashing into each other in the center of the ring. Because [10]annulene cannot achieve planarity, there is no experimental evidence that it is aromatic.

As shown in Figure 14-24c, [14]annulene also suffers from steric strain on the inside of the ring. That strain, however, is significantly less than in [10]annulene, due to the larger ring size. Thus, [14]annulene is stable and is aromatic.

[18]Annulene (Fig. 14-24d) is large enough to accommodate six trans double bonds without introducing substantial steric strain. Because it has a Hückel number of π electrons, it is aromatic.

Tub-shaped conformation

Cycloocta-1,3,5,7-tetraene ([8]Annulene)
(a)

(b)

The *p* AOs are *not* fully conjugated around the ring.

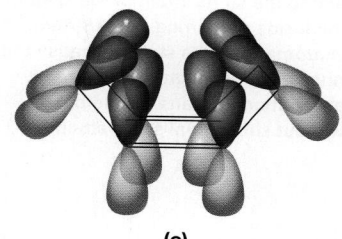

(c)

FIGURE 14-23 Cycloocta-1,3,5,7-tetraene ([8]annulene) (a) Lewis structure and **(b)** ball-and-stick model of [8]annulene. **(c)** In its tub shape, the *p* AOs do not make a fully conjugated ring, so the compound is nonaromatic.

In the molecule of [18]annulene shown in Figure 14-24d, draw bonds to the H atoms on the inside of the ring. Is there significant steric strain?

14.9 YOUR TURN

---

problem **14.16** Draw the most stable configuration of [16]annulene and suggest whether it should be aromatic, antiaromatic, or nonaromatic. Explain.

UNSTABLE

Angle strain

144°

[10]Annulene
(a)

UNSTABLE

Excessive steric strain

H
H

[10]Annulene
(b)

STABLE

Moderate steric strain

H H

H H

[14]Annulene
(c)

STABLE

[18]Annulene
(d)

FIGURE 14-24 Aromaticity in [10]annulene, [14]annulene, and [18]annulene [10]Annulene, (a) and (b), should be aromatic according to Hückel's rule, but it is unstable. The all-cis configuration (a) has too much angle strain, whereas the configuration with two trans double bonds (b) has too much steric strain. (c) [14]Annulene is aromatic, but is less stable than benzene due to the steric strain of the H atoms on the inside of the ring. (d) [18]Annulene is aromatic and is stable because the ring is large enough to comfortably accommodate trans double bonds, without substantial steric strain.

14.7 Aromaticity and Multiple Rings

Compounds with two or more rings can also be aromatic. Biphenyl (Fig. 14-25a) is a somewhat trivial example, consisting of two separate benzene rings connected by a single bond. Although the two rings are conjugated to each other, the π systems form *two separate rings*, each with six π electrons. Therefore, each essentially behaves as an independent aromatic system.

Naphthalene (Fig. 14-25b) and anthracene (Fig. 14-25c) contain two and three **fused rings**, respectively, so called because *the rings have bonds in common*. These molecules are aromatic, and as a class are called **polycyclic aromatic hydrocarbons** (**PAHs**). Like benzene, each of these molecules possesses a system of conjugated *p* AOs that form a single loop and contain a Hückel number of π electrons. Respectively, they contain 10 and 14 π electrons.

The Lewis structures of PAHs can be misleading. In the Lewis structures of naphthalene and anthracene, for example, the bond lines might make it appear that there are separate cyclic π systems. The atomic *p* orbitals do indeed make a single loop, however, so they can be viewed as a single aromatic π system.

YOUR TURN 14.10

In Figures 14-25b and 14-25c, connect all of the shaded lobes on the *p* orbitals to illustrate the single loop that exists in each.

YOUR TURN 14.11

In the resonance structures of naphthalene and anthracene in Figure 14-25, the double bonds appear to make one ring around the periphery of the molecules. Draw a resonance structure of each compound that shows the double bonds forming *two* separate rings. Do these resonance structures have an effect on the locations of the *p* AOs in Figure 14-25?

Two separate aromatic systems

The *p* AOs form a single loop around the periphery.

Biphenyl
(a)

Naphthalene
(b)

Anthracene
(c)

FIGURE 14-25 Aromaticity in polycyclic compounds (a) Biphenyl can be viewed as two independent aromatic systems, given that the atomic *p* orbitals form two different closed loops. (b) Naphthalene and (c) anthracene, on the other hand, have *p* orbitals that form one complete ring around the periphery.

SOLVED problem 14.17 Do you think that the compound at the right is aromatic, antiaromatic, or nonaromatic? Explain.

Think Does the molecule contain a π system formed from a ring of fully conjugated p orbitals? How many electrons are in that π system?

Solve Because the C atoms are all sp^2 hybridized and lie in the same plane, there is a fully conjugated system of p AOs around the periphery of the molecule, as shown by the blue dashed line at the right. In that π system, there are 10 π electrons (indicated by the five double bonds), which is a Hückel number (five pairs), so the molecule should be aromatic.

problem 14.18 Are the compounds at the right aromatic, antiaromatic, or nonaromatic? Explain.

(a)   (b)   (c)

14.8 Heterocyclic Aromatic Compounds

Thus far, we have only studied examples of aromatic hydrocarbons. There are many aromatic compounds, however, in which atoms other than carbon (i.e, heteroatoms) are part of the aromatic ring. Examples of these **heterocyclic aromatic compounds** include pyridine, pyrrole, and furan (Fig. 14-26).

Pyridine has a six-membered ring containing one N atom that resembles benzene. The main difference is that the N atom possesses a lone pair of electrons instead of a bond to H.

According to Hückel's rules, pyridine will be aromatic if there is a complete ring of conjugated p AOs possessing a Hückel number of electrons. Just as in benzene, all six atoms in pyridine (the N atom included) are sp^2 hybridized, so each possesses an unhybridized p AO. Furthermore, pyridine contains a total of six π electrons, as shown by the three formal double bonds in the Lewis structure.

This must mean that the lone pair of electrons on the N atom in pyridine is not part of the π system. If it were, then that system would contain eight total π electrons, making it *antiaromatic*. Instead, that lone pair resides in a MO that has contribution from the sp^2-hybridized AO on N, as shown in Figure 14-27a. *Because that hybridized orbital and the π system are perpendicular to each other, the two cannot interact.*

Pyrrole is slightly more complex. According to VSEPR theory, the N atom should be sp^3 hybridized, but it is actually sp^2 hybridized. Although this imposes some angle

Pyridine   Pyrrole   Furan

FIGURE 14-26 Heterocyclic aromatic compounds In each compound, a noncarbon atom contributes a p AO to the aromatic π system.

The lone pair is perpendicular to the p AOs, so it is not part of the π system.

The lone pair resides in a p AO and is therefore part of the π system.

This lone pair is part of the π system.

This lone pair is isolated from the π system.

(a)   (b)   (c)

FIGURE 14-27 Aromaticity in heterocyclic compounds (a) The orbital picture of pyridine, showing that the lone pair of electrons on N is isolated from the aromatic π system. (b) The orbital picture of pyrrole, showing that the lone pair is part of the π system. (c) The orbital picture of furan, showing that one lone pair is part of the aromatic π system and the other is not.

strain on the N atom, the sp^2 hybridization allows the molecule to get back much more stability by becoming aromatic. Notice in Figure 14-27b that the N atom contributes its unhybridized p orbital to the π system of the ring and the lone pair of electrons on N is part of that π system. Consequently, there is a fully conjugated ring of p orbitals containing six electrons—a Hückel number.

Looking back, we can see that the lone pair on N in pyridine is not part of the aromatic π system, but it is in pyrrole. In general,

> A heteroatom can contribute a lone pair of electrons to the π system to attain aromaticity.

This is evident in furan, whose O atom has two lone pairs of electrons. Only one of those lone pairs, however, is part of the π system, giving furan a total of six π electrons: the two electrons from the lone pair and the four electrons from the two formal π bonds. If both lone pairs were part of the π system, then there would be eight total π electrons, in which case furan would be antiaromatic. As shown in Figure 14-27c, this means that the O atom must be sp^2 hybridized with one lone pair residing in an atomic p orbital and the other lone pair *isolated* from the aromatic π system.

SOLVED problem 14.19 Do you think this compound is aromatic, antiaromatic, or nonaromatic? Explain.

Think How many lone pairs from S are part of the π system? How many total π electrons are there?

Solve Just like the O atom in furan, the S atom here can contribute one of its two lone pairs (a total of two electrons) to the π system involving the double bonds. Including the four π electrons from the two double bonds, this makes a total of six electrons (a Hückel number) in a completely conjugated, cyclic π system. The compound is aromatic.

---

problem 14.20 Identify each of the following compounds as aromatic, antiaromatic, or nonaromatic.

The cyclopentadienyl anion

The cyclopropenyl anion

The cycloheptatrienyl cation (The tropylium ion)

The cyclopentadienyl cation

FIGURE 14-28 Aromatic ions In each ion, the π system is made of a fully conjugated ring of p AOs and contains a Hückel number of electrons.

FIGURE 14-29 Antiaromatic ions In each ion, the π system is made of a fully conjugated ring of p AOs and contains a Hückel number of electrons.

14.9 Aromatic Ions

Aromaticity is not limited to uncharged molecules, because several molecular *ions* that conform to Hückel's rules are also aromatic. The cyclopentadienyl anion and the cycloheptatrienyl cation (also called the tropylium ion), are two common examples (Fig. 14-28).

The cyclopentadienyl anion is similar to pyrrole because it is a five-membered ring with two conjugated double bonds and a lone pair of electrons. Therefore, it has six π electrons. The cycloheptatrienyl cation, on the other hand, is a seven-membered ring that does not possess any lone pairs of electrons. Each C atom is sp^2 hybridized, and the three conjugated double bonds provide the six π electrons.

Antiaromatic ions can also exist (Fig. 14-29). One example is the cyclopropenyl anion, which contains a fully conjugated ring of atomic p orbitals with a total of four π electrons—an anti-Hückel number. Additionally, the cyclopentadienyl cation contains a total of four π electrons, making it antiaromatic as well.

problem **14.21** Based on Hückel's rules, determine whether each of the species at the right is aromatic, antiaromatic, or nonaromatic. Explain.

(a)　　　　(b)

14.10 Strategies for Success: Counting π Systems and π Electrons Using the Lewis Structure

Determining the number of π electrons in a given π system is important, especially when it comes to predicting whether or not a species is aromatic. This kind of accounting can be straightforward for any species using the MO approach used throughout this chapter, but can also be somewhat involved, even for relatively small molecules. Fortunately, we can arrive at the same result more quickly and easily using just the species' Lewis structure.

The basis for this method is the close connection between *resonance* and *conjugation*, which we learned in Sections 14.1 and 14.2. For a particular set of electrons to be delocalized over multiple atoms via resonance, there must be a single π system of MOs that encompasses those atoms. That system originates from a set of conjugated *p* AOs, one contributed by each atom. With such a connection between resonance and conjugation, we can make the following simplifications:

- A single π system of MOs encompasses the atoms over which electrons can be delocalized via resonance.
- The number of electrons in that π system is *the number of electrons that can be shifted via resonance*, either from one end of the π system to the other (if it is acyclic), or once around the ring (if it is cyclic).

We can show how this applies to the following ion, which resembles the allyl cation.

π System

2 electrons are shifted.

The π system encompasses the three C atoms indicated by the blue screen, because those are the atoms over which electrons can be resonance delocalized. Furthermore, two electrons can be shifted from the right side of that π system to the left, so the π system contains two electrons, the same as we saw in the allyl cation in Section 14.2a.

In the following allylic anion, there is a single π system that encompasses three C atoms:

π System

4 electrons are shifted.

In this case, however, four electrons can be shifted to the right via resonance, so the π system contains four total electrons.

We can also apply this method to hexa-2,4-diene, which resembles buta-1,3-diene.

π System

Hexa-2,4-diene

4 electrons are shifted.

There is a single π system that encompasses the four C atoms indicated. And, because four electrons can be shifted toward the right via resonance, that system contains a total of four electrons, the same as we obtained for buta-1,3-diene (Section 14.2b).

This method can also be applied to both hexa-1,3,5-triene and benzene.

Hexa-1,3,5-triene

6 electrons are shifted.

Benzene

6 electrons are shifted.

Each compound has a single π system that encompasses six C atoms and contains six electrons. In the case of hexa-1,3,5-triene, six electrons can be shifted to the right. In the case of benzene, six electrons can be shifted cyclically around the ring.

From all of these examples, we can make some very useful generalizations.

- A single π system cannot encompass sp^3-hybridized carbon atoms.
- Two double bonds are part of the same π system if they are separated by exactly one bond. These kinds of double bonds are said to be *conjugated*.

The first generalization arises because sp^3-hybridized carbon atoms cannot be involved in resonance (see Section 1.10). This makes sense from the MO point of view, too, because an sp^3-hybridized atom has no p orbital to contribute to a π system. The conditions in the second generalization enable the electrons from one double bond to participate in resonance with the electrons from the other double bond. This, too, makes sense from the MO point of view. As we saw in Section 14.2b, such an arrangement allows the p AOs of one double bond to be both parallel and adjacent to those from the other double bond. Thus, all the p AOs can interact.

With these generalizations in mind, notice that octa-1,3,6-triene has six total π electrons in two separate π systems:

The sp^3-hybridized C isolates one π system from the other.

π System with 4 electrons

π System with 2 electrons

Octa-1,3,6-triene

The sp^3-hybridized C prevents the ring from being fully conjugated.

π System with 4 electrons

Cyclopenta-1,3-diene

One π system consists of two conjugated double bonds, whereas the other is a lone double bond. The two systems are isolated from each other by the sp^3-hybridized C that separates them. Notice, too, that the π system of cyclopenta-1,3-diene is *not* fully conjugated around the ring. Although the two double bonds are conjugated to each other, the sp^3-hybridized C atom isolates one end of that π system from the other.

Moreover, the top two C—C bonds are both single bonds, so the ring does not consist *entirely* of alternating double and single bonds.

problem **14.22** How many π systems does β-carotene contain? How many electrons are in each?

β-Carotene

Using this method of counting π systems and π electrons is particularly helpful when triple bonds are involved, as shown with octa-3,5-dien-1-yne:

According to the drawing above on the left, there are eight total π electrons in the molecule. They reside in two different π systems, however—namely, one that contains six electrons and one that contains two. This is because, at any given time, only two of the four π electrons from the triple bond can participate in resonance with the four electrons from the conjugated double bonds. If *both* pairs of π electrons from the triple bond were to be involved in resonance at the same time, on the other hand, then we would arrive at a resonance structure in which a carbon atom's octet is exceeded:

From an MO point of view, this restriction arises because the two π bonds of a triple bond are perpendicular to each other (Section 3.8), so only one of those π bonds can be conjugated with a neighboring π system.

A similar situation arises in penta-1,2,3-triene:

Although the C1—C2 and C3—C4 double bonds are conjugated to each other, neither is conjugated to the C2—C3 double bond. No resonance structure can be drawn in which electrons from the C2—C3 bond have been shifted, because doing so would require breaking an atom's octet. MO theory accounts for this result by the fact that the central π bond is perpendicular to the other two.

The previous two examples can be generalized as follows:

> Two π bonds to the same atom must belong to different π systems.

SOLVED problem 14.23 If it is planar, should the compound at the right be aromatic, antiaromatic, or nonaromatic? Explain.

Think Can we draw a resonance structure by shifting electrons fully around the ring? If so, how many electrons must be shifted?

Solve We can shift electrons fully around the ring via resonance:

12 electrons are shifted.

Doing so involves two electrons from each of the double bonds, as well as two electrons from the triple bond, for a total of 12 electrons. Thus, there is a fully conjugated cyclic π system that contains 12 electrons (an anti-Hückel number), so the compound should be antiaromatic if it is planar. Notice that one pair of electrons of the triple bond is not shifted, indicating that those two electrons are in a separate π system.

problem 14.24 If it is planar, should the compound at the right be aromatic, antiaromatic, or nonaromatic? Explain.

The lone pair is part of the π system.

6 π electrons are shifted.

Pyrrole

Finally, using resonance structures can be quite useful when counting electrons in a π system that involves atoms with lone pairs. Consider pyrrole, which we examined in Section 14.8. As shown at the left, a resonance structure can be drawn by shifting six electrons around the ring. Thus, the π system contains six electrons, including the lone pair on N.

In pyridine, however, the lone pair on N is *not* part of the cyclic π system (Section 14.8). This is consistent with the fact that the lone pair is not involved in resonance with the double bonds of the ring, as shown on the left below. The π system thus contains a total of six electrons—namely, those that are part of the double bonds only.

The lone pair is *not* part of the π system.

6 π electrons are shifted.

The lone pair is unaffected.

Pyridine

Notice that, as shown in parentheses on the previous page, an attempt can be made to include N's lone pair in resonance. In the resulting resonance structure, however, the N atom retains its lone pair, so that lone pair is effectively unchanged.

A slightly different situation arises with furan, which we also examined in Section 14.8. The O atom has two lone pairs, but, as shown below, only one pair at a time can be involved in resonance with the double bonds. Thus, only one of those lone pairs is part of the π system that includes the double bonds, giving the system a total of six electrons.

Only one of the two lone pairs can be part of the π system.

6 π electrons are shifted.

Furan

What we observe with the lone pairs in furan can be generalized as follows:

> A single π system can include at most one lone pair of electrons on a particular atom.

---

problem **14.25** Identify each of the compounds at the right as aromatic, antiaromatic, or nonaromatic.

(a)   (b)   (c)

14.11 Wrapping Up and Looking Ahead

At the beginning of the chapter, we cited situations in which the two-center orbital picture of VB theory fails to explain the structural characteristics that arise from resonance delocalization. Thus, there was a need for multiple-center MOs. Although these orbitals can be of other symmetries, our focus here in Chapter 14 was on multiple-center MOs of π symmetry, generated by the simultaneous interaction of adjacent and parallel atomic p orbitals. These multiple-center π MOs are the ones that provide insight into resonance and aromaticity.

Multiple-center π MOs are constructed using many of the rules that we learned in Chapter 3. Among the most important rules are (1) the number of orbitals must be conserved, and (2) MOs differ in the phase relationships of the contributing AOs. With these and other rules, we were able to construct the ground-state electron configurations for the allyl cation and buta-1,3-diene—two species that serve as models for electron delocalization. Thus, we saw how the characteristics that each species derived from the multiple-center MO picture were consistent with those suggested by their respective resonance hybrids.

We also learned that certain compounds, like benzene, have unusually stable π systems and are classified as aromatic. Other compounds, like cyclobutadiene, have unusually unstable π systems and are classified as antiaromatic. Whereas resonance theory fails to account for both of these results, they can be explained using multiple-center π MOs.

Finally, using the intimate connection between resonance delocalization and conjugation, we can determine the number of electrons in a particular π system solely from a species' Lewis structure. This method is particularly convenient when establishing whether a species is aromatic or antiaromatic.

What we have learned here has important implications for aspects of chemistry we will encounter in the chapters to come. In Chapter 15, for example, we will see that multiple-center π MOs are the basis for *ultraviolet–visible spectroscopy*. In Chapter 16, we will see that aromaticity, in particular, plays a key role in *nuclear magnetic resonance spectroscopy*. In Chapters 17 and 20, we will see how the conjugation of double bonds can affect the outcome of chemical reactions. And, in Chapter 22, we will learn about the unique reactivity of benzene and other aromatic compounds.

THE ORGANIC CHEMISTRY OF BIOMOLECULES

14.12 Aromaticity and DNA

Although the discussion of aromaticity here in Chapter 14 predominantly focused on prototypical molecules, the phenomenon is observed in biological molecules as well. Aromaticity, for example, affects the structure and properties of DNA. Recall from Section 1.15c that a nucleic acid is a long chain of nucleotides and that each nucleotide has three components: a sugar group, a phosphate group, and a nitrogenous base. Furthermore, a nucleic acid's backbone consists of alternating sugar and phosphate groups, and the nitrogenous bases are bonded specifically to the sugar units of the backbone.

One nucleic acid is distinguished from another by the specific sequence of nitrogenous bases attached to the sugar–phosphate backbone. The four nitrogenous bases found in DNA are guanine (G), adenine (A), cytosine (C), and thymine (T).

Each of these ring systems consists of atoms that have planar geometries and are sp^2 hybridized. Thus, each nitrogenous base has a fully conjugated, cyclic system of *p* AOs. Additionally, each of these π systems contains a Hückel number of electrons—10 π electrons for guanine and adenine, and six for cytosine and thymine—which are highlighted in red in the preceding structures. According to Hückel's rules (Section 14.3), then,

All nitrogenous bases in nucleic acids are aromatic.

To understand the implications of the aromatic nature of these nitrogenous bases, we must consider the general structure of DNA shown in Figure 14-30.

DNA consists of two nucleic acid strands, wrapped around each other to form a "double helix." The nitrogenous bases from each nucleic acid point toward the center of the cylinder that is created by the two sugar–phosphate backbones. Moreover, the specific sequence of nitrogenous bases in one nucleic acid strand dictates the sequence in the other, according to the following rules for base pairing:

- Guanine in one strand of DNA is matched with cytosine in the other.
- Adenine in one strand of DNA is matched with thymine in the other.

Thus, the strands are said to be **complementary**.

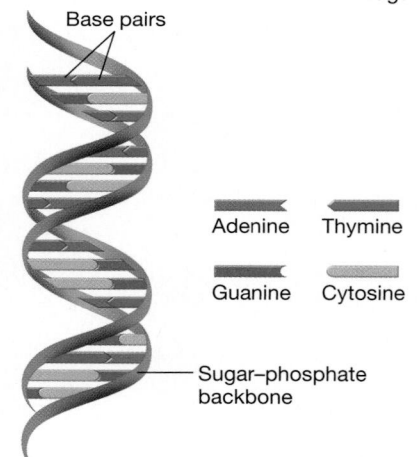

FIGURE 14-30 The general structure of DNA DNA consists of two nucleic acid strands that are wrapped around each other in a double-helix structure. The sugar–phosphate backbones of the strands define a cylinder, inside of which the nitrogenous bases are located. The strands have *complementary* sequences of nitrogenous bases—that is, adenine in one strand is matched with thymine in the other, and guanine in one strand is matched with cytosine in the other.

The G-C/A-T base pairing is heavily favored because the two strands are held together relatively strongly by hydrogen bonding among the nitrogenous bases, as shown in Figure 14-31, and this hydrogen bonding is maximized when the strands are complementary.

As shown Figure 14-31a, guanine has two hydrogen-bond donors and one hydrogen-bond acceptor that are ideally located to establish hydrogen bonds with two acceptors and one donor on cytosine. Thus, guanine and cytosine form three hydrogen bonds. Similarly, adenine and thymine form two hydrogen bonds, as shown in Figure 14-31b.

Considering again the double-helix structure of DNA, we can see how the aromaticity of the nitrogenous bases is advantageous. First, it allows each base pair to be planar, as shown in Figure 14-32a. As we learned earlier in Chapter 14, aromatic rings favor a planar structure to maximize the overlap of the p orbitals that make up the π system of MOs. Furthermore, the hydrogen bonding essentially locks complementary bases in the same plane. Thus, as shown in Figure 14-32, the DNA base pairs are "stacked" on the inside of DNA, much like sheets of paper, which keeps steric crowding to a minimum. As a result, the nitrogenous bases are packed in a highly efficient way, allowing DNA to store a large amount of genetic information in relatively little space.

The second way in which the aromaticity of the nitrogenous bases is advantageous is in the stability it provides to the double-helix structure. Because nitrogenous bases are stacked on top of one another, the p AOs from the aromatic system of one nitrogenous base overlap with the p AOs from another nitrogenous base attached to an adjacent sugar. This is a stabilizing interaction called π **stacking**, and is illustrated in Figure 14-32a. Individually, these interactions are relatively weak, but, viewing DNA from the top (Fig. 14-32b), notice that there is a high density of these interactions inside the double helix. The sum of these interactions along the entire length of DNA can provide significant stability.

The fact that the nitrogenous bases in DNA are aromatic delayed the elucidation of its structure, which was eventually published in 1953 by James Watson (b. 1928),

The G and C bases are paired because they form three hydrogen bonds.

Guanine   **Cytosine**

(a)

The A and T bases are paired because they form two hydrogen bonds.

Adenine   **Thymine**

(b)

FIGURE 14-31 Complementarity among nitrogenous bases (a) G and C bases are complementary and (b) A and T bases are complementary because each hydrogen-bond donor in one nitrogenous base is aligned with a hydrogen-bond acceptor in the other.

π Stacking takes place inside the double helix, whereby the orbitals from the π system of one base pair overlap with those from another.

Each base pair is essentially planar, making for efficient packing inside the double helix.

(a)   (b)

FIGURE 14-32 Consequences of aromaticity on the DNA structure (a) The DNA double helix shown from the side. Each base pair is essentially planar, stemming from the aromaticity of each nitrogenous base and the hydrogen bonding among complementary nitrogenous bases. This allows the nitrogenous bases to pack efficiently inside the double helix, and also facilitates π stacking, which arises from the overlap among the p AOs. (b) The DNA double helix shown from the top. This view shows the nitrogenous bases on top of one another, which enhances the stability from the π stacking that takes place.

With guanine and thymine in their enol forms, fewer hydrogen bonds are possible in the pairing of G with C and A with T.

FIGURE 14-33 The enol forms of nitrogenous bases In their enol forms (a) G and C form only two hydrogen bonds, and (b) A and T form only one hydrogen bond.

Keto form

Enol form

Guanine   Cytosine   Adenine   Thymine

an American biologist, and Francis Crick (1916–2004), a British physicist. Watson and Crick had two important pieces of information with which to work. One was the parity relationship among the nitrogenous bases, discovered by the Austrian chemist Erwin Chargaff (1905–2002), which stated that the amounts of guanine and cytosine in DNA are equal, as are the amounts of adenine and thymine. The second piece of information was the X-ray diffraction images of DNA taken by Rosalind Franklin (1920–1958), a British biophysicist, which suggested that the structure was helical.

Watson and Crick used physical models to determine how the base pairing could occur. They were working with the incorrect structures for guanine and thymine, however, because these bases were assumed to be more stable in their enol forms (Section 7.9) instead of their keto forms.

This assumption is understandable, because the guanine and thymine rings in their enol forms (Fig. 14-33) have alternating single and double bonds, which resemble a benzene ring. In their enol forms, however, guanine and cytosine cannot participate in as many hydrogen bonds with their complementary bases. The enol form of guanine can participate in only two hydrogen bonds with cytosine, and the enol form of thymine can participate in only one hydrogen bond with adenine. Thus, Crick and Watson could not devise a model that was consistent with the complementarity among the nitrogenous bases.

Fortunately, Jerry Donohue (1920–1985), an American chemist, was studying at Cambridge on a six-month grant, and was sharing an office with Crick and Watson. Donohue had expertise with small organic molecules, and suggested that guanine and thymine were more stable in their keto forms. This suggestion turned out to be the final piece of the puzzle that Crick and Watson needed to elucidate the structure of DNA.

Chapter Summary and Key Terms

- Electron delocalization via resonance cannot be accounted for by the two-center orbital picture of VB theory. The orbital picture of resonance requires **multiple-center MOs**, which span multiple adjacent atoms. **(Sections 14.1 and 14.2)**

- MOs are created by the interaction of different phases of the contributing AOs, the total number of orbitals must be conserved, and additional nodes increase energy. **(Section 14.2)**

- Atomic *p* orbitals are **conjugated** to each other if they are *adjacent* and *parallel*. **(Section 14.2a)**

- Three conjugated *p* AOs in the allyl cation yield three π MOs: one bonding, one nonbonding, and one antibonding. **(Section 14.2a)**

- Buta-1,3-diene has four conjugated *p* AOs that give rise to four π MOs: two bonding and two antibonding. **(Section 14.2b)**

- The allyl cation and buta-1,3-diene favor an all-planar conformation to allow all of their *p* AOs to be conjugated. **(Sections 14.2a and 14.2b)**

- Benzene (C_6H_6) is symmetric, with six identical carbon–carbon bonds, and it has an unusually stable π system of electrons. These characteristics are attributable to **aromaticity**. **Heat of hydrogenation** data can be used to determine the stabilities of π systems. From these experiments, we know that aromaticity provides benzene with ~150 kJ/mol (~36 kcal/mol) of *extra* stability. **(Section 14.3)**

- Cyclobutadiene (C_4H_4) is highly unstable and is classified as **antiaromatic**. **(Section 14.3)**

- **Hückel's rules** are *empirical* rules for aromaticity. A species is **aromatic** if it has a cyclic system of conjugated *p* orbitals that contains an odd number of pairs of electrons ($4n + 2$ π electrons). If it contains an even number of pairs ($4n$ π elec-

trons), then it is antiaromatic. Otherwise, it is **nonaromatic**. **(Section 14.3)**

- Benzene has six π MOs. Its six π electrons are all paired and reside in bonding MOs, and all π MOs are filled, giving benzene its significant stability. **(Section 14.4)**

- Cyclobutadiene has four π MOs. Only two of its four π electrons are paired in a bonding MO. The other two are *unpaired* and reside in nonbonding MOs. This distribution of electrons makes cyclobutadiene very unstable. **(Section 14.5)**

- Hückel's rules apply to **[*n*]annulenes**. However, some [*n*]annulenes that should be aromatic according to the rules are, in fact, not, due to excessive angle strain and/or steric strain. **(Section 14.6)**

- **Polycyclic aromatic hydrocarbons (PAHs)** generally have a single π system, although their Lewis structures show them as separate rings. **(Section 14.7)**

- Heteroatoms can be part of an aromatic ring, giving rise to **heterocyclic aromatic compounds**. In these rings, the heteroatoms can contribute a lone pair of electrons to achieve a Hückel number of π electrons. **(Section 14.8)**

- Ions follow the same rules that Hückel outlined, and can therefore be aromatic, antiaromatic, or nonaromatic. **(Section 14.9)**

Problems

14.26 The following are three π MOs for the heptatrienyl cation:

For each MO,
(a) determine the number of nodal planes perpendicular to the bonding axes and rank them in order of increasing energy;
(b) draw the *p* AO contributions on each C atom that would give rise to the MO;
(c) identify each internuclear region as having either a *bonding* or an *antibonding* type of interaction; and
(d) based on your answer to part (c), determine whether each MO is overall *bonding*, *nonbonding*, or *antibonding*.

14.27 The following are three π MOs of octa-1,3,5,7-tetraene:

Repeat Problem 14.26 for these orbitals.

14.28 Buta-1,3-diene has one strong and two weak resonance contributors:

The valence bond picture that describes the strong resonance contributor was previously shown in Figure 14-3. Draw the valence bond picture that describes each of the two weak contributors.

14.29 An "excited state" of buta-1,3-diene is achieved when one electron from π_2 has been promoted to π_3. How do you think this electronic transition affects the bond length between C1 and C2? Does it increase, decrease, or stay about the same? Explain. What about the C2—C3 bond length?

14.30 For the penta-1,3-dienyl anion (H_2C=CH—CH=CH—CH_2^-), draw the π MOs and the MO energy diagram similar to what is shown in Figures 14-5 and 14-6 for the allyl cation. Identify each MO as either bonding, nonbonding, or antibonding. Label the HOMO and the LUMO.

14.31 For the penta-1,3-dienyl cation (H_2C=CH—CH=CH—CH_2^+), draw the π MOs and the MO energy diagram similar to what is shown in Figures 14-5 and 14-6 for the allyl cation. Identify each MO as either bonding, nonbonding, or antibonding. Label the HOMO and the LUMO.

14.32 For hexa-1,3,5-triene (H_2C=CH—CH=CH—CH=CH_2), draw the π MOs and the MO energy diagram similar to what is shown in Figures 14-5 and 14-6 for the allyl cation. Identify each MO as either bonding, nonbonding, or antibonding. Label the HOMO and the LUMO.

14.33 The bond length of the C—C single bond in hexa-1,3,5-triene is about 146 pm, which is considerably shorter than that in ethane, H_3C—CH_3.

146 pm

(a) How much shorter is it? (*Hint:* See Fig. 14-2.)
(b) Which occupied π MOs (obtained in Problem 14.32) help decrease this bond length?
(c) If an electron were promoted from the HOMO to the LUMO, would the length of the C—C single bond increase, decrease, or stay roughly the same? Explain.

14.34 How many total π electrons are in the following molecule? In how many separate π systems do they reside? How many electrons are in each π system?

14.35 Although the hypothetical molecule at the right has alternating single and double bonds, those double bonds are not considered to be conjugated. Why not?

14.36 Pyrene has been determined experimentally to be aromatic. At first glance, however, its structure appears to break Hückel's rule. How so? Can you explain why pyrene exhibits aromaticity? (*Hint:* What are the characteristics of the π system on the periphery of the molecule?)

Pyrene

14.37 Draw the π MO energy diagram for [10]annulene, similar to those in Figures 14-19 and 14-21. Fill up the orbitals with the appropriate number of π electrons. Based on this diagram, should [10]annulene be aromatic or antiaromatic? Explain.

14.38 Repeat Problem 14.37 for [8]annulene.

14.39 Repeat Problem 14.37 for the following species:

(a)   (b)   (c)   (d)   (e)   (f)

14.40 The following molecule has been synthesized. Should it be aromatic, antiaromatic, or nonaromatic? Explain.

14.41 Each N atom in the molecule in Problem 14.40 contains a lone pair of electrons. In what kind of orbital does each one reside?

14.42 Determine whether each of the following species is aromatic, antiaromatic, or nonaromatic. (*Hint:* Don't forget the lone pairs.) Explain.

(a)   (b)   (c)   (d)   (e)   (f)   (g)

14.43 Rank the following in order of increasingly exothermic heat of hydrogenation. Explain.

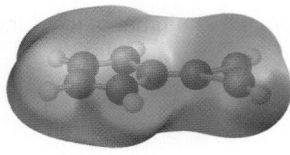

A   **B**   **C**   **D**

14.44 Which of the following ketones, **E** or **F**, do you think has the more stable π system? Why? (*Hint:* Consider various resonance structures of each species.)

E   **F**

14.45 The molecule shown has quite a large dipole, as indicated in its electrostatic potential map. Explain why. (*Hint:* Consider various resonance structures.)

14.46 Based on your answer to Problem 14.45, do you think the following compound should have a significant dipole moment? If so, in which direction does it point?

14.47 Using your knowledge of aromaticity, explain the difference in the calculated dipole moments of compounds **W** and **X**. (*Hint:* Consider various resonance structures.)

4.87 D   3.41 D

W   **X**

14.48 Molecules **Y** and **Z** are isomers of each other, and they have the same number of C=C double bonds. Which one has the more exothermic heat of hydrogenation? Why?

Y   **Z**

14.49 Coronene is a polycyclic aromatic hydrocarbon.

Coronene

(a) How many total π electrons does it have?
(b) Is that number consistent with the fact that the compound is aromatic? Explain.

14.50 Bromobenzene is insoluble in water, but 7-bromocyclohepta-1,3,5-triene is water soluble. Explain why. (*Hint:* To have dramatic differences in solubility, do you think the type of intermolecular forces between each compound and water can be the same?)

Insoluble   Soluble

14.51 As we saw here in Chapter 14, the tropylium ion is aromatic. Ion **X**, however, is not. Explain.

The tropylium ion   **X**

14.52 Most dianions of hydrocarbons are very unstable. However, if cyclooctatetraene is treated with potassium, the cyclooctatetraenyl dianion is readily formed via a redox reaction.

$$+ \ 2 \ K(s) \longrightarrow C_8H_8^{2-} \ + \ 2 \ K^+$$

(a) Explain why $C_8H_8^{2-}$ is easy to make.

(b) As we saw in the chapter, cyclooctatetraene is tub shaped. What do you think the geometry of $C_8H_8^{2-}$ is? Explain.

14.53 Rank the following compounds in order of decreasing acid strength, from lowest pK_a to highest pK_a. Explain.

14.54 Draw the complete, detailed mechanism for the following reaction and predict the major product. Explain.

14.55 Which of the following cations do you think is a stronger acid? Why?

14.56 Which O atom of 4-pyrone do you think is more *basic*? Explain.

4-Pyrone

14.57 For each of the following pairs of molecules, which do you think should have the more acidic α hydrogen? Explain why.

14.58 Which do you think is a stronger nucleophile in an S_N2 reaction—pyridine or pyrrole? Explain.

Pyridine   **Pyrrole**

14.59 For each of the following pairs of substrates, which do you think will undergo an S_N1 reaction faster? Explain.

14.60 Which of the following, **U** or **V**, do you think undergoes dehydration more quickly? Why?

14.61 Treatment of propadiene (an allene) with hydrogen bromide produces 2-bromopropene as the major product. This suggests that the more stable carbocation intermediate is produced by the addition of a proton to a terminal carbon rather than to the central carbon.

(a) Draw both carbocation intermediates that can be produced by the addition of a proton to the allene.

(b) Explain the relative stabilities of those intermediates. (*Hint:* Draw the orbital picture of the intermediates and consider whether the CH_2 groups in propadiene are in the same plane.)

14.62 Alkynes behave quite like alkenes when it comes to carbene addition. For example, a carbene can add to an alkyne to yield a cyclopropene:

Alkynes behave differently from alkenes, however, when treated with a peroxyacid. Whereas an alkene would be converted into an epoxide by the addition of an oxygen atom, this addition product is not observed for alkynes under normal conditions. Suggest why. (*Hint:* Pay special attention to the lone pairs.)

If your instructor assigns problems in smartwork, log in at **smartwork.wwnorton.com**.

Structure Determination 1

Ultraviolet–Visible and Infrared Spectroscopies

By now you have seen the structures of many hundreds of different molecules throughout this book. What assurance do you have that those structures are accurate? When organic chemistry was a relatively immature field, chemists would derive molecular structure by measuring a compound's physical properties, and by carrying out different reactions designed to provide information about the presence and relative positions of specific functional groups. As you might imagine, these processes were quite painstaking and often unreliable.

Today, much of our knowledge about chemical structure comes from *spectroscopy*, the study of how electromagnetic radiation (such as visible light) interacts with matter. In a spectroscopic experiment, we *shine light (or other radiation) on molecules and carefully observe the outcome* using specialized instruments. There are many different types of spectroscopy, but the ones we will focus on here in Chapter 15 and in Chapter 16 are *ultraviolet–visible (UV–vis), infrared (IR)*, and *nuclear magnetic resonance (NMR)*. Because molecules interact with different kinds of electromagnetic radiation in characteristic ways, each type of spectroscopy reveals different aspects of how a molecule is put together.

Chloroplasts

Plants harness the sun's energy to convert carbon dioxide and water into sugar for storable energy—a process called photosynthesis that takes place inside chloroplasts. Chlorophyll, which gives plants their green color, is the molecule responsible for absorbing visible light from the sun. As we will see here in Chapter 15, this type of interaction with light is the basis for spectroscopy—a laboratory technique that provides information about a molecule's structure.

A plant cell

Upon completing Chapter 15 you should be able to:

- Define *spectroscopy* and specify the general features of a *spectrum*.

- Describe how a *UV–vis spectrum* is acquired.

- Explain what a *photon* is and how its energy is related to its frequency and wavelength.

- Identify the longest-wavelength λ_{max} in a UV–vis spectrum and relate it to the energy difference between a molecule's HOMO and LUMO.

- Determine when the longest-wavelength λ_{max} corresponds to a *π-to-π* transition* and when it corresponds to an *n-to-π* transition*, and explain why the former generally occurs at shorter wavelength than the latter.

- Describe the relationship between the longest-wavelength λ_{max} and the extent of conjugation in a molecule.

- Relate the frequency of light that is absorbed in IR spectroscopy to the frequency of a molecule's normal mode of vibration.

- Identify the general regions in an IR spectrum where stretching modes and bending modes of vibration appear.

- Specify the approximate frequencies of IR absorptions for stretching modes of a bond between hydrogen and a heavy atom, and explain why these stretching modes correspond to the highest-frequency IR absorptions.

- Outline the factors that cause variations in the stretching frequency of a bond between hydrogen and a heavy atom.

- Explain why stretching frequencies decrease in the following order: triple bond > double bond > single bond.

- Specify the factors that dictate the intensity of a stretching-mode absorption band.

- Obtain structural information about molecules from their UV–vis and IR spectra.

Another important tool for elucidating molecular structure (which we discuss in Chapter 16) is *mass spectrometry*. Rather than using electromagnetic radiation, it uses various means to ionize molecules and even shatter them into fragments. A detector then measures the masses of the resulting charged species.

Nowadays, spectroscopy and mass spectrometry are used quite heavily in the area of synthesis. They often help us to identify newly discovered compounds in nature (so-called natural products) so that we can synthesize them from simpler compounds in the laboratory (e.g., as described in Chapters 13 and 19). Additionally, we routinely rely on these techniques to ensure that the products we make in the laboratory are in fact the ones we intended to make.

Spectroscopy and mass spectrometry even have many everyday applications. In criminal investigations, for example, an unknown substance encountered at a crime scene can be brought to a lab and analyzed via spectroscopy or mass spectrometry. Often the identity of the compound (or at least certain important characteristics) can be revealed in seconds or minutes.

Here in Chapter 15, we discuss UV–vis and IR spectroscopy. For each type of spectroscopy, we examine the characteristic changes that take place within a molecule when it interacts with electromagnetic radiation, then we explain how to interpret the results of each type of spectroscopic experiment so that we can use spectroscopy to glean important information about molecular structure. In Chapter 16, we discuss NMR spectroscopy and provide a brief introduction to mass spectrometry.

15.1 An Overview of Ultraviolet–Visible Spectroscopy

Spectroscopy is the study of the interaction of *electromagnetic radiation* with matter. As shown in Figure 15-1, electromagnetic radiation is categorized according to its wavelength, from very-short-wavelength gamma rays to relatively long-wavelength radio waves.

Ultraviolet–visible spectroscopy (UV-vis), in particular, uses light from the UV and visible regions of the spectrum, which are probably more familiar to you than the other regions. The visible region includes wavelengths we associate with colors we can see with our eye, ranging from red to violet. UV light has shorter wavelengths than visible light and contains the harmful radiation associated with sunburns and skin cancer.

In a typical UV-vis spectroscopy experiment (Fig. 15-2a), a range of wavelengths (usually 200–800 nm) is sent through an **analyte**—a sample we wish to analyze. At each wavelength, the intensity of light that reaches the detector ($I_{detected}$) is measured,

FIGURE 15-1 The electromagnetic spectrum Different types of spectroscopy use different wavelengths of electromagnetic radiation.

and so is the intensity of light from the source (I_{source}). **Transmittance** ($\%T$) is computed from these values according to Equation 15-1.

$$\%T = (I_{detected}/I_{source}) \times 100 \quad \text{(15-1)}$$

Although intensity is a quantity that has units such as watts (W), these units cancel when $I_{detected}$ is divided by I_{source}.

Equation 15-2 is used to convert transmittance to **absorbance** (A), which ranges from 0 (no absorption) to infinity (complete absorption).

$$A = \log(I_{detected}/I_{source}) = 2 - \log(\%T) \quad \text{(15-2)}$$

Finally, a **UV-vis spectrum** is produced by plotting absorbance against wavelength, as shown for a generic case in Figure 15-2b. Features of the spectrum where absorbance is high are called **absorption bands** or **peaks**.

15.1 YOUR TURN

Which scenario corresponds to a greater proportion of light absorbed by a sample: **(a)** $\%T$ = 20% or 40%? **(b)** A = 0.500 or 0.750?

Answers to Your Turns are in the back of the book.

---

problem **15.1** What is the percent transmittance if the intensity from a radiation source is measured to be 1.0×10^{-4} W and the intensity measured after sending the radiation through a sample is 2.5×10^{-5} W? What is the absorbance of the sample?

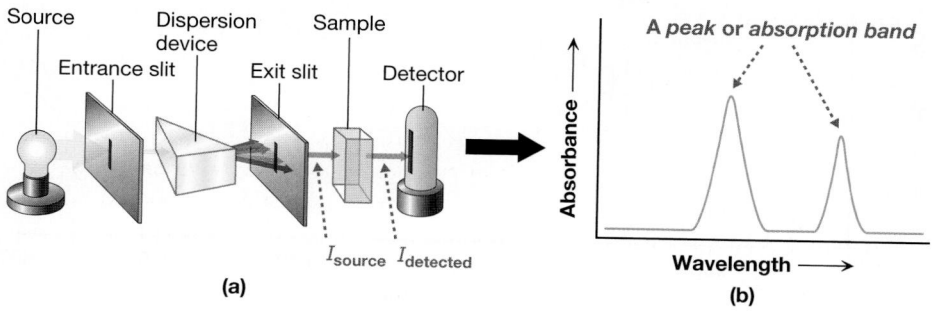

FIGURE 15-2 UV-vis spectroscopy (a) A general schematic of a UV-vis spectrophotometer. Specific wavelengths of light are selected to enter a sample. The intensity of light entering the sample is I_{source}, whereas the intensity of light that exits the sample and is detected by the detector is $I_{detected}$. (b) A generic UV-vis spectrum in which wavelength is plotted on the x axis and absorbance is plotted on the y axis.

The magnitude of absorbance is governed by the **Beer–Lambert law** (Equation 15-3), which you may have encountered in general chemistry:

$$A = \varepsilon \cdot l \cdot C \tag{15-3}$$

This law, which applies to *all* forms of spectroscopy, states that absorbance is directly proportional to three different variables: (1) the concentration, C, of the species responsible for absorbing light (often called the **chromophore**); (2) the length, l, of the sample through which the light travels; and (3) the **molar absorptivity**, ε (the Greek letter epsilon), an experimentally derived quantity that is characteristic of a given species at a given wavelength of radiation.

The Beer–Lambert law takes the form it does because the amount of light that a sample absorbs is proportional to the number of chromomphores the light encounters as it travels through the sample; the number of chromophores, in turn, increases with both C and l. The molar absorptivity reflects *the probability that light of a given wavelength will be absorbed when it encounters a chromophore*. Later here in Chapter 15 and in Chapter 16, we discuss some of the factors that govern the relative magnitude of ε for various types of spectroscopy. This makes it possible to better understand certain characteristics of the different spectra we encounter.

YOUR TURN 15.2

If increasing the wavelength of light causes the molar absorptivity of a sample to increase by a factor of three, what happens to the measured absorbance?

problem **15.2** Calculate the molar absorptivity if a sample's absorbance is 0.78, the concentration of the sample is 0.600×10^{-5} M, and the length of the sample the light travels through is 1.00 cm. What are the units?

15.2 The UV–vis Spectrum: Photon Absorption and Electron Transitions

The UV–vis absorption spectrum of 2-methylbuta-1,3-diene (isoprene) is shown in Figure 15-3. Notice that there is a peak in the spectrum, centered at 224 nm. This is called λ_{max} for that peak, because it is *the wavelength of light at which absorbance is a local maximum*. Notice, too, that there are no absorptions at longer wavelength.

To understand why the spectrum of 2-methylbuta-1,3-diene has these characteristics, we must first understand that light (and electromagnetic radiation in general) has a dual nature, behaving both as a wave and a particle. As a wave, light can be assigned a *wavelength* and a *frequency*. The **wavelength**, λ (the Greek letter lambda), is

FIGURE 15-3 UV–vis spectrum of 2-methylbuta-1,3-diene (isoprene) Relative absorbance is plotted against wavelength. The wavelength of maximum absorption, λ_{max}, is 224 nm.

the distance between two successive maxima or minima in the wave's oscillating amplitude. **Frequency**, ν (the Greek letter nu), on the other hand, is the number of complete oscillations that occur in a given period of time. It is given in units of hertz (Hz), cycles/second, or s^{-1}, all of which are equivalent. Wavelength and frequency are related by Equation 15-4:

$$c = \lambda\nu \quad \text{or} \quad \nu = c/\lambda \tag{15-4}$$

Here c stands for the speed of light, a universal constant: $c = 2.9979 \times 10^8$ m/s. According to this equation, *wavelength and frequency are inversely related*, so as one increases, the other must decrease.

Behaving as a particle, electromagnetic radiation exists as **photons**. Each photon possesses a characteristic energy that depends only on its frequency (or wavelength), according to Equation 15-5:

$$E_{photon} = h\nu_{photon} = hc/\lambda_{photon} \tag{15-5}$$

In this equation, E_{photon} is the energy of a *single* photon, and h is **Planck's constant** $(6.626 \times 10^{-34}$ J · s). Thus, *the energy of a given photon is directly proportional to its frequency*: As one increases, so does the other. On the other hand, *a photon's energy and wavelength are inversely proportional.*

> Which wavelength of light corresponds to photons with a greater energy: 375 nm or 530 nm?

15.3 YOUR TURN

According to the **law of conservation of energy**, energy cannot be created or destroyed, so when a molecule absorbs a photon, it acquires the photon's energy, too. More specifically,

> The absorption of a UV–vis photon causes an electron in the molecule to be promoted to a higher-energy MO—a so-called **electron transition**.

The blue arrows in Figure 15-4 depict examples of these kinds of transitions. Typically, many are possible, even for simple molecules.

> In Figure 15-4, draw arrows to represent five more electron transitions involving the six MOs that are given.

15.4 YOUR TURN

FIGURE 15-4 Electron transitions in UV–vis spectroscopy In UV–vis spectroscopy, the absorption of a photon typically corresponds to the promotion of an electron from an occupied MO to an unoccupied one. These electron transitions are represented by the blue arrows.

Which MOs, in particular, are involved in the transition that corresponds to the absorption band at 224 nm for 2-methylbuta-1,3-diene? Because no absorption bands appear at longer wavelength in the UV–vis spectrum, the electron transition must be the one that requires the *least* amount of energy, which, as indicated in Figure 15-4, is the one that promotes an electron from the HOMO to the LUMO. Thus, in general,

> The longest-wavelength UV–vis absorption band corresponds to the HOMO–LUMO transition of the analyte.

YOUR TURN 15.5

How do the energies required for the transitions you drew in Your Turn 15.4 compare to the energy required for the HOMO–LUMO transition?

SOLVED problem **15.3** Given the following UV–vis spectrum, calculate the approximate energy difference between the HOMO and LUMO.

Think Which is the longest-wavelength absorption? What is its λ_{max}? What energy does a photon with that wavelength possess?

Solve Although there may be a peak that is partially shown around 230 nm, the longest-wavelength absorption has a λ_{max} of 310 nm. This corresponds to an energy of:

$$E = h\nu = h(c/\lambda) = (6.626 \times 10^{-34} \text{ J} \cdot \text{s})(2.9979 \times 10^8 \text{ m/s})/(310 \times 10^{-9} \text{ m})$$

$$= 6.41 \times 10^{-19} \text{ J}$$

problem **15.4** Given the following UV–vis spectrum, calculate the approximate energy difference between the HOMO and LUMO of the unspecified organic compound.

FIGURE 15-5 Absorption of a UV–vis photon in buta-1,3-diene (a) Ground-state electron configuration of buta-1,3-diene. The red wavy arrow represents a photon with energy $h\nu$. (b) Electron configuration of the same molecule after absorption of the photon. Notice that an electron has been promoted from the HOMO (π_2) to the LUMO (π_3).

To gain a better feel for the HOMO–LUMO transition in 2-methylbuta-1,3-diene, consider Figure 15-5a, which shows the MO energy diagram of buta-1,3-diene (shown previously in Fig. 14-12). This diagram is very similar to the MO energy diagram of 2-methylbuta-1,3-diene because of the pair of conjugated double bonds that the two molecules have in common. In both species, the valence electrons fill the σ bonding MOs and the first two of the four MOs of π symmetry. As a result, the HOMO is π_2, which is a bonding MO, and the LUMO is π_3, an antibonding MO. When an incoming photon with the appropriate energy (represented by the wavy red arrow) is absorbed, the molecule undergoes what is called a $\boldsymbol{\pi \rightarrow \pi^*}$ ("pi-to-pi-star") **transition**.

In Figure 15-5a, circle the electron that gains the energy of the incoming photon and draw an arrow from it to the orbital to which it is promoted. Label each of the π MOs as either bonding, nonbonding, or antibonding.

15.6 YOUR TURN

problem **15.5** Construct a diagram, similar to Figure 15-5, that depicts the longest-wavelength UV–vis absorption for ethene ($H_2C{=}CH_2$).

15.3 Effects of Structure on λ_{max}

As we saw in Section 15.2, λ_{max} for the longest-wavelength UV–vis absorption of 2-methylbuta-1,3-diene is 224 nm (Fig. 15-3). The specific structure of the molecule, however, can have significant effects on the λ_{max} for these kinds of absorption bands, as shown in Table 15-1. The range is wide, from $\sim$161 nm (in the UV region) to >450 nm (in the visible region). These values account for the colors of the compounds, or lack thereof. Cyclohexene, for example, is a clear, *colorless* liquid because it absorbs entirely

TABLE 15-1 λ_{max} for the Longest-Wavelength UV–vis Absorptions of a Variety of Organic Compounds

Compound	λ_{max} (nm)	Compound	λ_{max} (nm)	Compound	λ_{max} (nm)
Alkanes and cycloalkanes	<150				
Ethene	161	Buta-1,3-diene	217	Cyclohexa-1,3-diene	256
Hex-1-ene	177	cis-Penta-1,3-diene	223	Hexa-1,3,5-triene	274
Penta-1,4-diene	178	trans-Penta-1,3-diene	223.5	Methanal (Formaldehyde)	280
Cyclohexene	182	2-Methylbuta-1,3-diene (Isoprene)	224	Octa-1,3,5,7-tetraene	290
Hex-1-yne	185	Cyclopentadiene	239	Propenal (Acrolein)	340

β-Carotene

Absorbs blue light, so β-carotene appears red-orange.   455

FIGURE 15-6 UV-vis absorption and β-carotene Absorption by β-carotene in the UV–vis portion of the electromagnetic spectrum gives carrots their characteristic color.

in the UV region. No wavelengths from the visible region are absorbed; all are allowed to pass through. β-Carotene, on the other hand, absorbs strongly at 455 nm, which is essentially blue. The remaining wavelengths that reach our eye appear deep red-orange, characteristic of vegetables like carrots and sweet potatoes (Fig. 15-6).

Table 15-1 shows that λ_{max} for the longest-wavelength absorption depends on the nature of bonding within the molecule.

- Molecules that contain only σ bonds (e.g., alkanes and cycloalkanes) do not absorb in the UV or visible regions—their λ_{max} is much too short.
- Molecules that contain at least one π bond (e.g., ethene or hex-1-ene) have a λ_{max} greater than about ~160 nm.

Of the molecules containing π bonds, λ_{max} *depends highly on the extent of conjugation.* For example, the λ_{max} of $H_2C{=}CH_2$ and cyclohexene, each of which contains an isolated double bond, are similar—161 nm and 182 nm, respectively. The values for λ_{max} of *cis*-penta-1,3-diene and cyclopentadiene, each of which contains two conjugated double bonds, are similar, too, but they are at 223 nm and 239 nm, respectively. In general,

> The value of the longest-wavelength λ_{max} increases as the extent of conjugation increases.

15.7 YOUR TURN

Examine the compounds in Table 15-1. For each one that contains only carbon and hydrogen, determine the number of conjugated π bonds that make up the largest π system and compare that number to the compound's longest-wavelength λ_{max}.

Notice, too, from Table 15-1 that molecules with a C=O bond tend to absorb at wavelengths longer than analogous molecules with only C=C bonds. For instance, $\lambda_{max} = 280$ nm for formaldehyde ($H_2C{=}O$), whereas $\lambda_{max} = 161$ nm for ethene. Similarly, $\lambda_{max} \approx 340$ nm for propenal, but only 217 nm for buta-1,3-diene.

problem **15.6** Considering the trend observed in Table 15-1, estimate λ_{max} for the longest-wavelength absorption of deca-1,3,5,7,9-pentaene ($C_{10}H_{12}$).

problem **15.7** Considering the trend observed in Table 15-1, estimate λ_{max} for the longest-wavelength absorption of penta-2,4-dienal ($H_2C{=}CH{-}CH{=}CH{-}CH{=}O$).

15.3a Effect of Conjugation on λ_{max}

As conjugation increases, λ_{max} for the longest-wavelength absorption increases, too. We can see why from Figure 15-7, which shows the specific electron transitions that occur when a photon is absorbed by ethene, buta-1,3-diene, and hexa-1,3,5-triene—

FIGURE 15-7 Conjugation and HOMO–LUMO transitions MO energy diagram for the π MOs of (a) ethene, (b) buta-1,3-diene, and (c) hexa-1,3,5-triene. As conjugation increases, the HOMO–LUMO energy difference ($\Delta E_{HOMO-LUMO}$) decreases, so the value of λ_{max} must increase.

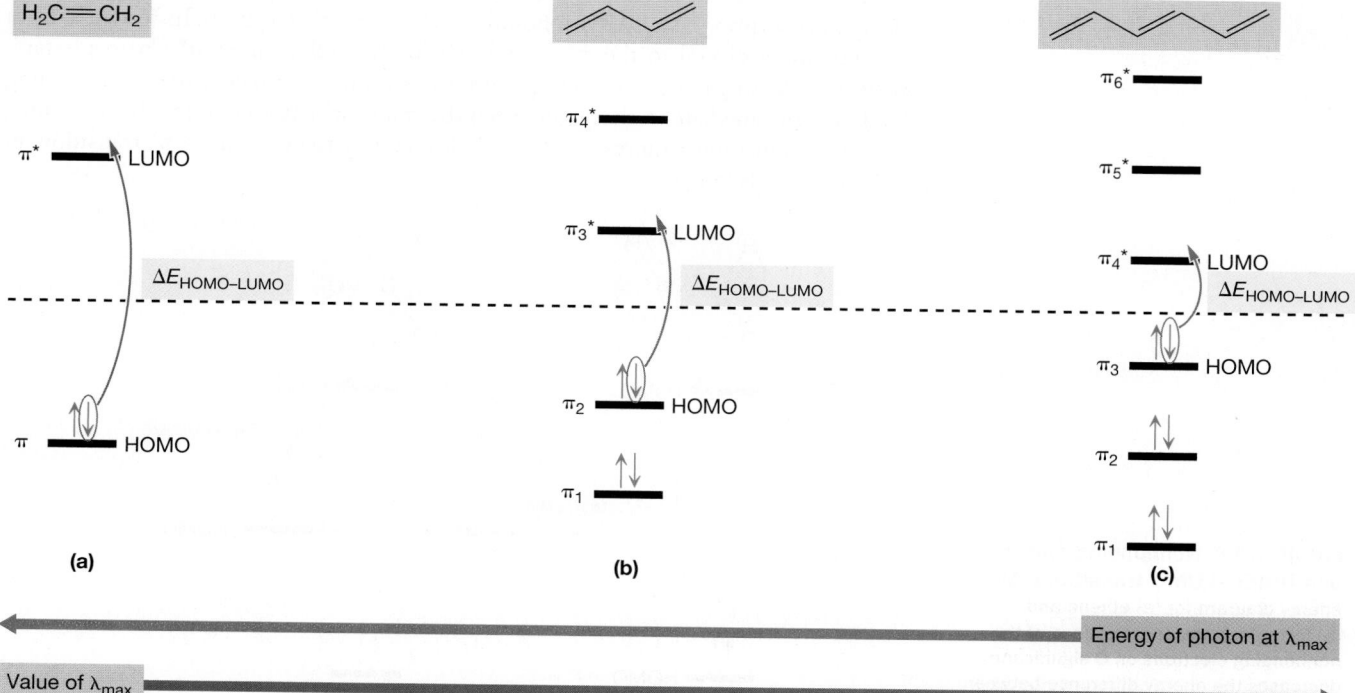

molecules that differ in the extent of the conjugation of their π systems. In all three cases, the lower-energy σ MOs are all filled, so the HOMO–LUMO transition is from a π MO to a π* MO (a π → π* transition).

According to Figure 15-7, the HOMO–LUMO energy difference ($\Delta E_{HOMO-LUMO}$) decreases as the extent of conjugation increases. Because this energy difference corresponds to the energy of the photon that is absorbed, the photon's energy must also decrease. And, according to Equation 15-5, that means the photon's wavelength must *increase*.

SOLVED problem 15.8 Both *trans*-penta-1,3-diene and penta-1,4-diene have a total of four MOs of π symmetry, yet penta-1,3-diene absorbs at a much longer wavelength. Explain why.

Think In each compound, are the double bonds conjugated or isolated? What do the resulting MO energy diagrams look like? What are the relative energy differences between the HOMO and LUMO in each molecule?

Solve The structures of the compounds are as follows: ***trans*-Penta-1,3-diene**   **Penta-1,4-diene**

The double bonds in *trans*-penta-1,3-diene are conjugated, so all four π electrons are in the same π system, resembling that shown in Figure 15-7b. In penta-1,4-diene, however, the double bonds are isolated from each other, so each π system looks like that in Figure 15-7a. The HOMO–LUMO energy difference is smaller in Figure 15-7b, so the λ_{max} of the corresponding absorption is longer.

---

problem 15.9 Which of the compounds at the right do you expect to have the longer λ_{max}? Explain.

15.3b Effect of Lone Pairs on λ_{max}

A molecule such as formaldehyde, $H_2C=O$, does not at first seem to be consistent with the dependence of λ_{max} on the extent of conjugation. Its λ_{max} is 280 nm, which is significantly longer than that of its carbon analog, $H_2C=CH_2$ ($\lambda_{max} = 161$ nm). In fact, the λ_{max} of formaldehyde is about the same as that of hexa-1,3,5-triene ($\lambda_{max} = 274$ nm), which has three conjugated double bonds!

In formaldehyde, the HOMO is a nonbonding MO (i.e., *not* a π MO), which holds a lone pair of electrons on O. As a result, the transition that occurs in formaldehyde upon absorption of a UV–vis photon is *not* a π → π* transition. Instead, it is from the nonbonding orbital to the π* orbital, which is called an **n → π*** ("n-to-pi-star") **transitions**. A simple valence bond picture suggests that formaldehyde's nonbonding orbitals are intermediate in energy between the π and π* MOs (Fig. 15-8b). Therefore, the n → π* transition requires substantially less energy than the π → π* transition, in which case λ_{max} is longer.

FIGURE 15-8 Nonbonding electrons and HOMO–LUMO transitions MO energy diagram for (a) ethene and (b) formaldehyde. The presence of the nonbonding electrons on O significantly decreases the energy difference between the HOMO and LUMO ($\Delta E_{HOMO-LUMO}$).

The Chemistry of Vision

The apparent color of a substance depends on the wavelengths of light that it absorbs. β-Carotene, for example, has a λ_{max} of 455 nm, which is in the blue part of the visible spectrum. When white light shines upon a sample of β-carotene, therefore, blue light is absorbed and the remaining wavelengths that reach our eye are interpreted by our brain as orange. But how does our eye detect light in the first place?

The answer has to do with a cis–trans isomerization involving 11-*cis*-retinal, which has several C=C bonds conjugated in a chain; the one at C11 is cis, and all the others are trans.

The retina at the back of the eye is lined with >100 million rods and >5 million cones, which are photoreceptor cells. Certain proteins in rods and cones are designed to bind 11-*cis*-retinal. In rods, that protein is opsin, and when 11-*cis*-retinal is bound to it, it is called rhodopsin. A visible photon that impinges upon rhodopsin causes a π-to-π* transition in the system of conjugated double bonds, which effectively breaks the C11–C12 π bond to allow free rotation. Within nanoseconds, the conjugated chain isomerizes to the all-trans configuration. The longer all-trans configuration does not fit as well into the protein's active site, which triggers conformational changes in the protein and ultimately causes retinal to be released. When this happens, a cascade of events leads to the closing of Na$^+$ channels, which produces an electrical signal that is sent to the brain.

SOLVED problem 15.10 Which of the following has a longer λ_{max}: $H_2C=CH_2$ or $H_2C=NH$? Explain.

Think For each compound, does the HOMO–LUMO transition correspond to a π → π* transition or an n → π* transition? In general, what are the relative energies of these types of transitions?

Solve The HOMO–LUMO transition for $H_2C=CH_2$ is π → π*, as shown in Figure 15-8a. The transition for $H_2C=NH$, however, is n → π*, similar to Figure 15-8b, given the presence of both a lone pair on N and the π bond. Accordingly, the HOMO–LUMO transition for $H_2C=NH$ should require less energy, and will thus appear at a longer wavelength.

---

problem 15.11 Which of the following do you think has a longer λ_{max}: $H_2C=NH_2^+$ or $H_2C=OH^+$? Explain.

15.4 IR Spectroscopy

In IR spectroscopy, radiation from the IR region of the electromagnetic spectrum is sent through the analyte, similar to what we saw with UV–vis spectroscopy (Fig. 15-2). IR radiation, which includes wavelengths from ~800 nm to ~10^6 nm (1 mm), is invisible to our eyes, but we can feel it—it feels warm. It is the radiation that emanates from lamps used at fast-food restaurants to keep your food warm, and it is responsible for the heat you can feel when you stand several feet away from a campfire, or when sunlight shines on your face. IR radiation feels warm to us because of the molecular property that is affected:

> IR absorption causes excitations in the vibrational motions of molecules.

That includes molecules in our skin. IR spectroscopy takes advantage of this phenomenon.

15.4a General Theory of IR Spectroscopy

Recall from Chapter 1 that chemical bonds behave like springs. Thus, the atoms of a molecule, like masses connected by springs, can undergo different types of regular oscillations in which interatomic distances and bond angles vary rhythmically. There are two basic, independent types of vibrational motion, called **normal modes of vibration**: **stretching** and **bending**. In stretching, the *distance* between two atoms in a chemical bond grows longer and shorter; in bending, an *angle*—either a bond angle or a dihedral angle—becomes larger and smaller.

Figure 15-9 shows several modes of vibration that take place simultaneously in a molecule of formaldehyde ($H_2C=O$). The carbonyl (C=O) group undergoes a

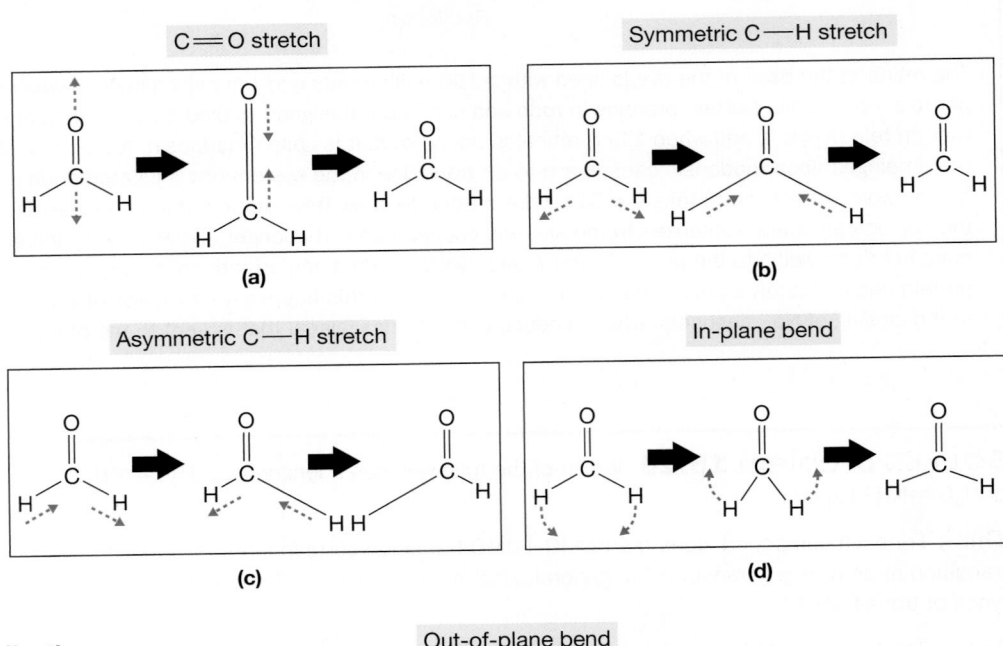

FIGURE 15-9 Normal modes of vibration
Various normal modes of vibration are shown for a molecule of formaldehyde. (a) The C=O bond stretches and compresses. (b) Both C—H bonds stretch and compress in unison. (c) While one C—H bond stretches, the other compresses. (d) The H—C—H bond angle closes and opens. (e) The two H atoms and the O atom shift downward and upward. (*Note*: The magnitudes of displacement are exaggerated for illustration.)

stretching vibration, as indicated in Figure 15-9a. The C—H bonds can also undergo stretching vibrations, but not independently of each other. They undergo a **symmetric stretch** (Fig. 15-9b) and an **asymmetric stretch** (Fig. 15-9c).

Two types of bending modes are also shown for formaldehyde—namely, an **in-plane bending vibration** (Fig. 15-9d) and an **out-of-plane bending vibration** (Fig. 15-9e).

---

problem **15.12** At the right is one more normal mode of vibration that formaldehyde undergoes, but is not shown in Figure 15-9. Classify it as either a *symmetric stretch*, an *asymmetric stretch*, an *in-plane bend*, or an *out-of-plane bend*.

Quantum mechanics dictates that *each normal mode of vibration is quantized*. For each normal mode, therefore, only certain energy levels are allowed, so only photons of certain energies can be absorbed. Molecules at room temperature, in particular, tend to absorb an IR photon when the photon's frequency equals the frequency of vibration for a particular normal mode. Therefore,

> The frequency at which a peak appears in an IR spectrum is often the same as the frequency of the vibration that is responsible for the absorption of the photon.

A typical IR spectrum is shown in Figure 15-10, in which percent transmittance is plotted against the frequency of light in units of cm^{-1}, called **inverse centimeters** or **wavenumbers**, $\bar{\nu}$. This unit, the reciprocal of the wavelength, is calculated by dividing 1 by the wavelength in cm (Equation 15-6). Physically, it corresponds to the number of waves that fit in 1 cm. The two common units of frequency—cm^{-1} and Hz—are related by the speed of light, as shown in Equation 15-7. (The factor of 100 accounts for different units of length used for wavenumbers and the speed of light: cm for wave numbers and m for the speed of light.)

$$\bar{\nu} \text{ (in } cm^{-1}) = 1/\text{wavelength (in cm)} \tag{15-6}$$

$$\bar{\nu} \text{ (in } cm^{-1}) = \nu \text{ (in Hz)}/(100\ c) \tag{15-7}$$

For historical reasons, frequency increases from right to left along the x axis. Commonly, the frequency range in an IR spectrum is from $\sim 400\ cm^{-1}$ to $\sim 4000\ cm^{-1}$.

Because percent transmittance is plotted on the y axis, the baseline (no absorption) is at 100% and *peaks in the spectrum appear as dips in percent transmittance*. Therefore, a **strong absorption** is one whose percent transmittance is near zero (near the bottom of the spectrum), and a **weak absorption** is one whose percent transmittance is near 100% (near the top of the spectrum). In Figure 15-10, for example, we can identify peaks at 2961, 1717, and 1167 cm^{-1}. Applying what we learned earlier, we can say

FIGURE 15-10 A typical infrared spectrum *Percent transmittance* is plotted on the *y* axis and frequency of light in *wavenumbers* (cm^{-1}) is plotted on the *x* axis. Each peak drops down from the top and represents the frequency at which various vibrations occur within the molecule.

that the molecule that produces this IR spectrum has normal modes of vibration with these same frequencies.

YOUR TURN 15.8

Choose a fourth peak in Figure 15-10 and annotate it with the frequency of the normal mode of vibration it represents.

problem **15.13** Calculate the wavelengths of the photons required to excite the three vibrations indicated in Figure 15-10.

15.4b Location of Peaks in an Infrared Spectrum

One of the greatest advantages of IR spectroscopy is that the frequency of a particular normal mode of vibration is relatively independent of the functional group with which the vibration is associated. Thus,

Molecules with similar bonds have characteristic IR absorptions that appear at similar frequencies.

For example, compare the IR spectra of hexan-2-one and 2-ethylhexanal, shown in Figure 15-11. Both compounds contain a C=O bond—hexan-2-one as part of a ketone and 2-ethylhexanal as part of an aldehyde. Despite the different functional groups, the absorptions corresponding to the C=O stretching frequencies are quite similar, appearing at 1717 cm^{-1} and 1723 cm^{-1}, respectively.

The fact that absorptions by certain vibrations appear with characteristic frequency ranges is what enables us to use IR spectroscopy to obtain structural information about a molecule. For example, based on the spectra in Figure 15-11, we can safely

FIGURE 15-11 Characteristic absorption frequency of the C=O stretching mode of vibration The IR spectra of hexan-2-one (a) and 2-ethylhexanal (b) are shown. Although the C=O bonds are part of different functional groups—hexan-2-one is a ketone and 2-ethylhexanal is an aldehyde—the absorptions corresponding to the C=O stretching modes are quite similar.

say that the appearance of a strong, sharp peak around 1720 cm^{-1} in the IR spectrum of an unknown compound strongly suggests that the compound contains a carbonyl (C=O) group.

15.9 **YOUR TURN**

Benzaldehyde, whose IR spectrum is shown below, contains a C=O bond. Identify the absorption that corresponds to the C=O bond.

Absorptions for other types of vibrational modes appear within characteristic frequency ranges, too, as shown in Table 15-2. These are all for stretching modes of vibration (we examine some for bending modes of vibration in Section 15.5d). Notice that these stretching modes have frequencies from ~1000 cm^{-1} to a few thousand cm^{-1} (bending modes generally occur from ~500 cm^{-1} to ~1000 cm^{-1}).

Although the absorptions for a particular mode of vibration appear within a relatively narrow range of frequencies, small differences in frequency can provide valuable information about the functional group involved. For example, the C=O absorptions for ketones, aldehydes, esters, carboxylic acids, and amides all appear between ~1630 and ~1780 cm^{-1}. However, as we will see in Section 15.5b, an amide's C=O absorption tends to appear at the low end of that range, whereas that for a carboxylic acid tends to appear at the high end.

TABLE 15-2 Characteristic Frequencies of Absorption in IR Spectroscopy: Stretching Modes of Vibration

Type of Bond	Functional Group	Frequency Range (cm^{-1})	Appearance	Type of Bond	Functional Group	Frequency Range (cm^{-1})	Appearance
O—H	Alcohol and phenol	3200–3600	Broad, strong	C≡N	Nitrile	2210–2260	Medium
	Carboxylic acid	2500–3000	Broad, strong	C≡C	Alkyne	2100–2260	Variable
N—H	Amine	3300–3500	Medium	C=O	Ketones/ aldehydes	1680–1750	Strong
	Amide	3350–3500	Medium		Esters	1730–1750	Strong
C—H	Alkane	2800–3000	Variable		Carboxylic acid	1710–1780	Strong
	Alkene	3000–3100	Weak		Amide	1630–1690	Strong
	Alkyne	~3300	Strong	C=C	Alkene	1620–1680	Variable
	Aldehyde	2820 and 2720	Strong		Aromatic	1450–1550	Variable
				C—O	Alcohol, ester, ether	1050–1150	Medium

SOLVED problem **15.14** Use Table 15-2 to estimate the frequencies of the stretching vibrations indicated in the molecule at the right.

Think What type of bond does each arrow indicate? Is there more than one type of functional group to which each of those bonds could belong?

Solve The C=C bond belongs to a simple alkene, not an aromatic ring, so its stretching frequency is ~1620 to 1680 cm^{-1}. The C=O bond belongs to an ester, so its stretching frequency is ~1730 to 1750 cm^{-1}. The C—O bond's stretching frequency will be ~1100 cm^{-1}. And the C—H bond most closely resembles that from an alkane (rather than an alkene, alkyne, or aldehyde), so its frequency is ~2800 to 3000 cm^{-1}.

problem **15.15** For each of the following molecules, use Table 15-2 to estimate the frequencies of the stretching vibrations indicated.

In Table 15-2, there are a number of trends worth noting. The first defines five major regions in which absorptions appear.

- Stretching vibrations of M—H bonds (where M is a "heavy atom" such as C, N, and O) occur in the region between ~2500 and 4000 cm^{-1}.
- Absorptions by triple bonds (C≡C or C≡N) appear between 2000 and 2500 cm^{-1}.
- Double bonds appear between ~1500 and 2000 cm^{-1}.
- Single bonds between two heavy atoms appear between ~1000 and 1500 cm^{-1}.
- The frequencies of bending vibrational modes appear below 1000 cm^{-1}.

These trends are illustrated in Figure 15-12.

The region below ~1400 cm^{-1} is called the **fingerprint region**. Numerous peaks typically appear in this region, even for relatively simple molecules, and many of those peaks overlap. Therefore, we will focus primarily on identifying absorptions at greater than ~1400 cm^{-1}.

FIGURE 15-12 The five major regions of absorption in IR spectroscopy Absorptions corresponding to the stretching frequencies of O—H, N—H, and C—H bonds appear between 2500 and 4000 cm^{-1}. Those for triple bonds generally appear between 2000 and 2500 cm^{-1}. Those for double bonds generally appear between 1500 and 2000 cm^{-1}. Those for single bonds not involving hydrogen appear between 1000 and 1500 cm^{-1}. Absorptions corresponding to bending modes appear below 1000 cm^{-1}.

Identify the two regions in Figure 15-12 that encompass the fingerprint region.

15.10 **YOUR TURN**

Two other trends that are useful to know involve bonds between hydrogen and a heavy atom—that is, absorptions that appear in the region above 2500 cm^{-1}. First, a periodic-table trend is in effect for these stretching frequencies:

- C—H stretches appear between ~2700 and 3300 cm^{-1}.
- N—H stretches appear around 3300 to 3500 cm^{-1}.
- Alcohol O—H stretches creep up a bit higher to ~3600 cm^{-1}.

Second, there is a trend involving hybridization for C—H stretches:

- Alkane (sp^3) C—H stretches appear below 3000 cm^{-1}.
- Alkene (sp^2) C—H stretches appear between ~3000 and 3100 cm^{-1}.
- Alkyne (sp) C—H stretches appear at ~3300 cm^{-1}.

Reasons for these trends are discussed next in Section 15.4c.

15.4c The Ball-and-Spring Model for Explaining Peak Location

IR spectroscopy can be a valuable tool for quickly determining which functional groups are present in a molecule. To use it properly, however, you must know which peaks correspond to which modes of vibration. Table 15-2 is quite useful for this purpose, but this information may not be provided on an exam, in which case you will have to commit certain frequencies to memory. To help with this task, recall that *each IR peak corresponds to the frequency of the vibration that is responsible for photon absorption.* Therefore, if we can understand the factors governing the frequencies of molecular vibrations, we can begin to make sense of the trends in absorption frequencies in IR spectroscopy.

The picture of bond vibration is simplified greatly if we view each bond as a spring connecting two atoms, as shown in Figure 15-13a. If the mass of an atom decreases (Fig. 15-3b), the spring can move the atom more easily—that is, *faster*. Thus, all else being equal, the vibrational frequency of the spring increases as the atoms attached to the spring become lighter. This explains why C—H, N—H, and O—H stretches have such high frequencies: The H atom is *very light*.

Moreover, if the *spring stiffness* increases (Fig. 15-13c), a greater force is exerted on the attached atoms. Thus, for a given pair of bonded atoms, vibrational frequency increases as the spring stiffness increases. This explains why vibrational frequency decreases in the order: triple bonds > double bonds > single bonds. Triple bonds tend to be stronger and stiffer than double bonds, just as double bonds tend to be stronger and stiffer than single bonds.

The generalization about spring stiffness also accounts for the relationship between the C—H stretching frequency and the hybridization of C. We learned in Chapter 3 that the C—H bond energy increases as the *s* character of the carbon atom increases. Thus, a bond

FIGURE 15-13 Vibrational frequency, mass, and spring stiffness (a) A mass connected to a spring undergoes vibration. (b) With a lighter mass, the vibrational frequency increases. (c) With a stiffer spring, the vibrational frequency increases, too.

Time

(a)

Decrease mass

(b)

Vibrational frequency has increased.

Increase spring stiffness

(c)

Vibrational frequency has increased.

between hydrogen and an *sp*-hybridized carbon atom is stronger and stiffer than one with an *sp*^{2}-hybridized carbon, which is stronger and stiffer, in turn, than one involving an *sp*^{3}-hybridized carbon. For this reason, the C—H stretching frequency decreases in the same order: alkyne C—H > alkene C—H > alkane C—H.

SOLVED problem 15.16 Deuterium (D), an isotope of hydrogen, has roughly twice the mass of hydrogen, but the two have nearly identical chemical properties. A C—H bond, therefore, has essentially the same strength as a C—D bond. Which vibrational mode, a C—H stretch or a C—D stretch, absorbs at a higher frequency in the IR region? Explain.

Think Considering the model of masses connected by a spring, are the masses the same? Is the stiffness of each spring the same? How do these factors govern vibrational frequency? How does vibrational frequency affect IR absorption frequencies?

Solve We are told that C—H and C—D bonds are essentially identical in strength, so their spring stiffnesses must be about the same. Spring stiffness, therefore, does not come into play, but the greater mass of D contributes to a lower frequency of vibration. Because the frequency of vibration is the same as the photon frequency that can be absorbed, a C—D stretch must absorb lower-frequency photons than a C—H stretch.

---

problem **15.17** For each of the following pairs of compounds, indicate which C—N stretching mode absorbs the higher-frequency IR photons. Explain.

(a)   (b)   (c)

15.4d Amplitudes of Peaks in an IR Spectrum

As shown in Table 15-2, some absorptions have characteristically strong intensities, whereas others are typically weak. Still others are variable, depending upon the specific molecules in which they are found. Why do these absorption peaks have the intensities they do? The answer has to do with a result from quantum mechanics, which states that a photon is more likely to be absorbed if a particular vibrational mode causes oscillations in the molecule's dipole moment. Therefore,

> ■ A stretching mode involving a bond connecting atoms with substantially different electronegativities tends to have a strong IR absorption.
>
> ■ A stretching mode involving a bond connecting atoms with similar electronegativities tends to have a weak (or nonexistent) IR absorption unless the two portions of the molecule connected by that bond are highly dissimilar.

C=O and O—H bonds, for example, are highly polar, so their stretching modes tend to have strong absorptions. The intensity of a C=C stretch, on the other hand, is variable, as shown in the IR spectra of hept-1-ene and hept-3-ene in Figure 15-14.

In the IR spectrum of hept-1-ene (Fig. 15-14a), the C=C stretch band at 1643 cm^{-1} is moderately intense because the two portions of the molecule connected by the C=C bond (i.e., the CH$_2$ and C$_6$H$_{12}$ portions) are rather dissimilar. The C=C stretch is essentially absent in the spectrum of hept-3-ene (Fig. 15-14b), however, because of the much greater symmetry about the C=C bond.

FIGURE 15-14 Molecular symmetry and intensity of stretching absorptions (a) IR spectrum of hept-1-ene, illustrating the presence of the C=C stretching mode. Because of the considerable asymmetry about the C=C double bond, that absorption is moderately intense. (b) IR spectrum of hept-3-ene. The C=C stretch band is essentially absent because of the greater symmetry about the C=C double bond.

15.11 **YOUR TURN**

In the spectrum of hept-3-ene in Figure 15-14b, indicate where you would expect an absorption corresponding to a C=C stretch.

problem **15.18** Based on the intensities of the C=C stretching bands, match each of the following compounds with its IR spectrum.

15.5 A Closer Look at Some Important Absorption Bands

When you encounter an IR spectrum of an unknown compound, it may seem daunting at first. Even a simple organic molecule can have dozens of peaks in its IR spectrum. As you gain experience interpreting these spectra, though, these problems will become less and less intimidating. But where do we begin?

The best way to begin is by examining in detail the regions of the spectrum above ~1400 cm^{-1}, where peaks tend to be well separated, are relatively easy to spot, and provide unambiguous information. Then look for the presence or absence of specific peaks in the fingerprint region (less than ~1400 cm^{-1}) if the information provided by the spectrum >1400 cm^{-1} needs to be clarified.

The sections that follow will help you identify and interpret several of these absorptions, which, in turn, will help you determine aspects of a molecule's structure.

15.5a The O—H Stretch

Among the easiest IR absorption bands to spot are those associated with the O—H stretch.

> OH stretch bands are intense and are usually rather broad, centered above ~3000 cm^{-1}.

Such absorption bands are evident in the IR spectra of cyclohexanol (Fig. 15-15) and 3,3-dimethylbutanoic acid (Fig. 15-16).

FIGURE 15-15 IR spectrum of cyclohexanol Notice the broad OH stretch band centered at ~3350 cm^{-1}.

FIGURE 15-16 IR spectrum of 3,3-dimethylbutanoic acid Notice the broad OH stretch band centered at ~3000 cm^{-1}.

Identify the peaks in Figures 15-15 and 15-16 that correspond to the O—H stretches.

The broadening of an OH stretch band is due to hydrogen bonding involving the OH groups of the sample. A network of hydrogen bonding in a sample facilitates a rapid exchange of hydrogen atoms from one OH group to another, since each O—H hydrogen atom is already partially bonded to two different oxygen atoms.

This process makes each O—H bond less well defined, and gives rise to a wide range of O—H stretching frequencies that overlap in the spectrum, resulting in what appears to us as a single broad peak.

Notice that the OH stretch bands in Figures 15-15 and 15-16 are markedly different.

> ■ An alcohol O—H stretch is typically centered near 3300 cm^{-1} and is usually well resolved from the alkane C—H stretch bands between 2800 and 3000 cm^{-1}.
>
> ■ A carboxylic acid O—H stretch is centered at around 3000 cm^{-1} and can extend down to ~2500 cm^{-1}, thus overlapping the alkane C—H stretches.

A carboxylic acid O—H stretch is generally lower in frequency and much broader than that of an alcohol, because carboxylic acids can form dimers in which there is more extensive hydrogen bonding than is normally found in alcohols.

---

problem **15.19** Which of the following isomers, **A** or **B**, is consistent with the IR spectrum provided? Explain.

An O—H stretching band in an IR spectrum can also indicate the presence of an impurity such as water, which is often difficult to eliminate from a sample. This is

FIGURE 15-17 IR spectrum of 2,5-dimethyltetrahydrofuran
Because the compound itself does not contain an OH group, the weak, broad absorption near 3500 cm^{-1} must indicate the presence of an impurity, such as water.

the case with the IR spectrum of 2,5-dimethyltetrahydrofuran shown in Figure 15-17. No OH group appears in the molecule, but a weak, broad absorption appears around 3500 cm^{-1}. If the compound itself were to contain an OH group, that peak would be *much* more intense, as we saw in Figures 15-15 and 15-16.

YOUR TURN 15.13

The IR spectrum of hept-3-ene shown previously in Figure 15-14b exhibits a water impurity. Identify that peak.

15.5b The Carbonyl (C=O) Stretch

The carbonyl group (C=O) is found in quite a few classes of compounds, including simple ketones and aldehydes, carboxylic acids, esters, and amides. A carbonyl-stretch absorption band is quite strong and relatively narrow. Given the variety of functional groups that contain a carbonyl group, we find a moderately large range of frequencies at which the carbonyl stretch absorption can appear—generally between 1600 and 1800 cm^{-1}.

Within that range, however, the C=O stretch absorption tends to appear at frequencies that are somewhat characteristic of the functional group to which the C=O group belongs.

For each of the following functional groups, the C=O stretch typically appears in the corresponding range:

Ketone   = 1710–1730 cm^{-1}
Aldehyde = 1720–1740 cm^{-1}
Amide   = 1630–1690 cm^{-1}
Ester   = 1730–1750 cm^{-1}

Moreover, conjugation can have a significant impact on the location of a C=O stretch.

When a C=O group is conjugated to a C=C or C≡C bond, the frequency of the C=O absorption is lowered by as much as 30–40 cm^{-1}.

The C=O stretch absorption for 4-methylpentan-2-one, for example, appears at 1721 cm^{-1}. For 4-methylpent-3-en-2-one, however, the conjugated C=C double bond lowers the frequency to 1690 cm^{-1}.

1721 cm^{-1} 1690 cm^{-1}

4-Methylpentan-2-one   **4-Methylpent-3-en-2-one**

These differences in absorption frequencies reflect differences in *stiffness* of the C=O bond (Section 15.4c). Specifically, the stiffness of the C=O bond in general increases in the order: amide < ketone < aldehyde < ester. Furthermore, conjugation decreases the stiffness of the C=O bond. The reasons for this are explored in Problems 15.41–15.44 at the end of the chapter.

problem 15.20 The IR spectrum of an unknown compound, whose formula is $C_9H_{10}O$, has a strong absorption band at 1686 cm^{-1}. Which of these compounds is consistent with that spectrum? Explain.

A   B

15.5c Alkyne (C≡C) and Nitrile (C≡N) Stretches

Triple bonds are generally among the easiest to identify in an IR spectrum, largely because no other absorption bands appear where they do. Unfortunately, however, C≡C and C≡N bonds both appear as sharp bands at roughly the same position, with C≡C between 2100 and 2260 cm^{-1} and C≡N between 2210 and 2260 cm^{-1}. Without additional information, therefore, such as whether or not the compound contains nitrogen, it can be difficult to decide between these two functional groups based solely on the frequency of the triple-bond absorption band.

Probably the best way to distinguish between the two functional groups is to search for an alkyne C—H peak at ~3300 cm^{-1}. The presence of one, which appears as a *sharp* peak (in contrast to a broad peak representative of an N—H or O—H stretch), is consistent with a terminal alkyne (C≡C—H). The absence of one, however, may still leave us the choice between a C≡N or an internal alkyne (R—C≡C—R).

In these cases, the intensity of the absorption band for the triple-bond stretch can help you distinguish between functional groups.

Whereas RC≡CR stretches typically have weak absorption intensities, RC≡N absorptions have moderate intensities.

This is because internal alkynes tend to have significant symmetry about the triple bond, whereas the C≡N group has a significant bond dipole (Section 15.4d).

15.14 YOUR TURN

In each of the following spectra, identify the band corresponding to a triple-bond stretching mode.

Spectrum 1
% Transmittance
Wavenumbers (cm^{-1})

Spectrum 2
% Transmittance
Wavenumbers (cm^{-1})

problem **15.21** Which of the two spectra in Your Turn 15.14 most likely corresponds to a nitrile? Explain.

15.5d The C=C Stretch and C—H Bends

Carbon–carbon double bonds are common in organic molecules, but verifying their presence from an IR spectrum is not quite as easy as detecting the signs of the other functional groups discussed so far. One reason is that the C=C stretching bands occur in the same region (1620–1680 cm^{-1}) as some C=O stretches. The C=C absorption bands in simple alkenes are sharper and frequently less intense, but it is not unusual for them to be obscured by the carbonyl absorption bands.

These absorptions can be difficult to spot, too, because they can vary greatly in intensity, just as we saw with the C≡C stretch. In some highly symmetric alkenes, such as *trans*-hept-3-ene, the C=C stretch has very little absorption (see Fig. 15-14b). An asymmetric alkene, however, can exhibit an absorption band that is of moderate intensity (Fig. 15-14a).

Aromatic rings also exhibit C=C stretching modes in their IR spectra, but these modes appear somewhat lower in frequency than isolated C=C stretches:

> Stretching bands between ~1450 and 1600 cm^{-1} usually indicate an aromatic ring.

There can be up to three of these stretching modes—one around 1600 cm^{-1}, one around 1500 cm^{-1}, and one slightly <1500 cm^{-1}. The latter two are usually the more intense peaks, as shown in Figure 15-18. In Figure 15-18a, all three peaks are present; only two of the peaks are present in Figure 15-18b.

FIGURE 15-18 Aromatic C=C stretches Bands that appear around 1600, 1500, and just below 1500 cm^{-1} are characteristic of aromatic C=C stretches. All three are present in the spectrum of *sec*-butylbenzene (a), but only two of the three are present in the spectrum of 1,4-di-*tert*-butylbenzene (b).

problem **15.22** Match isomers **C** and **D** with the correct IR spectrum. Explain.

C   **D**

The presence of a C=C stretch in an IR spectrum can make it worthwhile to examine the region below 1000 cm^{-1} (Table 15-3). A simple alkene can often be identified as either cis or trans, depending upon the frequencies of specific bending modes present. Similarly, substituted benzene rings can be identified as mono- or disubstituted based on the bending modes present, and if they are disubstituted, it is often possible to distinguish between ortho, meta, and para substitution. For example, in the spectrum of *sec*-butylbenzene (Fig. 15-18a), relatively strong absorptions appear at 759 and 703 cm^{-1}, representing the bending modes of a monosubstituted benzene. In the spectrum of di-*tert*-butylbenzene (Fig. 15-18b), only one bending mode appears at 833 cm^{-1}, indicating a para-disubstituted benzene.

TABLE 15-3 Characteristic Frequencies of Absorption in IR Spectroscopy: C—H Bending Modes of Vibration

Type of Bond	Number of Bands	Frequency Range (cm^{-1})	Appearance
R—CH=CH$_2$	2	910 and 990 (two peaks)	Strong
R$_2$C=CH$_2$	1	890	Strong
RCH=CHR (cis)	1	660–730	Strong
RCH=CHR (trans)	1	970	Strong
R$_2$C=CHR	1	815	Moderate
Aromatic (monosubstituted)	2	700 and 750	Strong
Aromatic (ortho)	1	750	Strong
Aromatic (meta)	2	780 and 880	Strong
Aromatic (para)	1	830	Strong

15.5e The C—H Stretch

All C—H stretches appear between 2700 and ~3300 cm^{-1} (Table 15-2). Alkane C—H stretches, in particular, appear between 2800 and 3000 cm^{-1}. They are so commonplace in organic molecules, though, that these peaks are generally *not* useful in structure elucidation. Instead, they serve as benchmarks. Trying to use these bands to elucidate a structure is like trying to identify a car based on its tires.

The alkane C—H stretching band is perhaps most useful when it is low in intensity or absent altogether. This is because:

> The relative intensity of the alkane C—H band decreases as the relative number of alkane C—H bonds in the molecule decreases.

In the structure of 2-ethylhexan-1-ol, for example, there are 17 C—H bonds, making the C—H stretching band even more intense than the O—H stretch (Fig. 15-19a).

FIGURE 15-19 **Relative intensities of the alkane C—H bands** IR spectra of (a) 2-ethylhexan-1-ol, (b) ethanol, and (c) diphenylmethanone are shown. As the number of alkane C—H bonds decreases, so does the intensity of the alkane C—H absorption band.

In ethanol's IR spectrum (Fig. 15-19b), on the other hand, the alkane C—H stretching band is less intense than the O—H stretch because there are only five alkane C—H bonds. Diphenylmethanone (benzophenone) has no alkane C—H bonds at all. Its IR spectrum (Fig. 15-19c), therefore, has no absorption band between 2800 and 3000 cm^{-1}.

15.15 **YOUR TURN**

Identify the alkane C—H stretching absorption bands in Figure 15-19a and 15-19b. In Figure 15-19c, indicate where alkane C—H stretching bands would normally appear.

problem **15.23** Based on the relative intensities of the alkane C—H absorption bands, match compounds **E–G** with spectra 5–7.

Other types of C—H stretches, however, can be quite useful. Alkyne C—H stretches appear as sharp bands of moderate intensity at around 3300 cm^{-1}. The appearance of such a band indicates that a C≡C stretch must also appear between 2100 and 2260 cm^{-1}.

Alkene C—H stretches usually appear as small bands around 3050 cm^{-1}. This is true for both simple alkenes and aromatic rings. Given the difficulty one can encounter in identifying a C═C stretch, especially in highly symmetric alkenes, these C—H stretching bands can be important in classifying a compound as an alkene.

The aldehyde C—H stretch appears as two bands—one at ~2820 cm^{-1} and the other at ~2720 cm^{-1}. The band at ~2720 cm^{-1} is generally quite easy to spot (Fig. 15-20), given that all other C—H stretching bands appear at frequencies higher than 2800 cm^{-1}. These bands are particularly useful in distinguishing between aldehydes and ketones.

FIGURE 15-20 IR spectrum of heptanal Notice the aldehyde C—H stretch at ~2720 cm^{-1}.

YOUR TURN 15.16

Identify the band in Figure 15-20 that corresponds to the aldehyde C—H stretch.

15.5f The N—H Stretch

Bands representing N—H stretches share many similarities with those of O—H stretches, including being generally easy to identify. N—H stretches appear between 3300 and 3500 cm^{-1}. Hydrogen bonding between amines usually causes these bands to be rather broad, although typically not as broad as those for an analogous O—H stretch. Because of the smaller difference in electronegativity between N and H, an N—H bond dipole is smaller than that of an O—H bond dipole. Consequently,

The intensity of an N—H stretch is usually less than that of an O—H stretch.

The number of N—H peaks that appear can be used to distinguish between various types of amines or amides.

- A primary amine (RNH$_2$) or amide (RCONH$_2$) exhibits two separate (but closely spaced) N—H stretching bands (Fig. 15-21a).
- A secondary amine (R$_2$NH) or amide (RCONHR) exhibits one N—H stretching band (Fig. 15-21b).
- A tertiary amine (R$_3$N) or amide (RCONR$_2$) exhibits no N—H stretching bands (Fig. 15-21c).

Identify *all* of the peaks in Figure 15-21 that correspond to N—H stretching modes of vibration.

15.17 **YOUR TURN**

Two N—H stretching bands appear for a primary amine or amide because an NH$_2$ group has two different stretching modes of vibration: a symmetric stretch (both N—H bonds stretch simultaneously) and an asymmetric stretch (one N—H bond

FIGURE 15-21 N—H stretching bands in IR spectra IR spectra of (a) 2-methylpropanamide, a primary amide; (b) *N*-methyl-2,2-dimethylpropanamide, a secondary amide; and (c) *N,N*-dibutylformamide, a tertiary amide, are shown.

15.5 A Closer Look at Some Important Absorption Bands / **743**

IR Spectroscopy and the Search for Extraterrestrial Life

Our culture is obsessed with extraterrestrial life. Even though we have yet to find any hard evidence that extraterrestrial life exists, we continue to invest significant time and resources looking for it. One way this is done by the SETI (Search for Extraterrestrial Intelligence) Institute is to use radio telescopes to detect radio waves emanating from deep space. The data collected are then analyzed for patterns that might suggest the radio waves were produced by intelligent life.

But radio telescopes are not the only tool used to search for life on other planets; IR spectroscopy is a pivotal tool in this quest as well. Researchers are able to record an IR spectrum produced by the atmosphere of a distant planet, and with such a spectrum in hand, they look for absorbances that are characteristic of gases generally believed to be conducive to life. Some of these gases include H_2O, CO_2, CH_4, and O_3 (ozone), as can be seen in the IR spectrum of Earth's atmosphere.

Of these gases, researchers are particularly interested in the existence of O_3. Not only is O_3 produced from atmospheric O_2 (which sustains life), but it also provides an important layer of protection from harmful UV radiation.

Whereas an IR spectrum can indicate whether an atmosphere might be able to support life, researchers believe that the detection of IR radiation in the range of 750–1000 nm (13,000–10,000 cm^{-1}) would provide more direct evidence of life. That's because on Earth, photosynthetic plants tend to reflect IR radiation in that range with relatively high efficiency.

---

lengthens while the other shortens). A secondary amine or amide has just one N—H bond, and a tertiary amine or amide has none.

YOUR TURN 15.18

In the generic primary amine on the left, add arrows to illustrate the motion of the *symmetric* N—H stretch. Illustrate the motion of the *asymmetric* N—H stretch in the molecule on the right. (*Hint:* Review Fig. 15-9.)

Symmetric stretch   Asymmetric stretch

15.6 Structure Elucidation Using IR Spectroscopy

Interpreting an IR spectrum requires practice. For this reason, the IR spectra of three unknown compounds are presented here in Section 15.6. The interpretations of the spectra apply the concepts presented in Sections 15.4 and 15.5. In some cases, there are features in the spectra that are quite obvious. In others, they are more subtle. In still other cases, the *absence* of specific absorption bands can provide useful structural information.

When you approach an IR spectrum of an unknown compound, it is generally a good idea to consider the six types of absorption bands examined in detail in Section 15.5: (1) the O—H stretch, (2) the C=O stretch, (3) triple bonds (C≡C and C≡N), (4) C=C double bonds, (5) C—H stretches, and (6) N—H stretches. The order in which you examine these bands is not critical.

15.6a Unknown 1

The IR spectrum of Unknown 1 is shown in Figure 15-22. Its molecular formula is C_7H_8O, so its index of hydrogen deficiency (IHD; Section 4.12) is 4, because a completely saturated molecule with 7 C atoms and 1 O atom would have 16 H atoms, not just 8. Unknown 1, therefore, could contain up to four rings, four double bonds, or two triple bonds.

FIGURE 15-22 IR spectrum of Unknown 1, C_7H_8O

The first striking feature of the spectrum in Figure 15-22 is the large, broad peak centered at ~3300 cm^{-1}. This is an O—H stretch. The O atom is *not* part of a C=O bond because there is no intense band around 1700 cm^{-1}.

> 15.19 **YOUR TURN**
>
> Identify the O—H stretch in Figure 15-22 and indicate where a C=O stretch would normally appear.

Another striking feature of Figure 15-22 is the absence of an intense alkane C—H band between 2800 and 3000 cm^{-1}. There are alkane C—H peaks present, but their intensity is low (less than the O—H band). This suggests that there are relatively few alkane C—H bonds.

> 15.20 **YOUR TURN**
>
> Identify the alkane C—H stretching band in Figure 15-22.

There are alkene C—H bonds present in Unknown 1, based on the C—H stretching band at ~3050 cm^{-1}. The presence of a C=C double bond is supported by the combination of an intense peak just above 1500 cm^{-1} and a moderate peak near 1600 cm^{-1}, which together are consistent with an aromatic ring.

> 15.21 **YOUR TURN**
>
> In Figure 15-22, identify the band corresponding to an alkene C—H stretch and the bands corresponding to C=C stretches.

A substituted benzene ring would account for the IHD of 4, given its three formal double bonds and a ring. If true, then there should be no other double bonds, triple bonds, or rings aside from the aromatic ring.

Because Unknown 1 likely contains a substituted benzene ring, we can look in the region below 1000 cm^{-1} for characteristic bending modes. There is a relatively intense peak around 830 cm^{-1}, suggesting that the benzene ring is para disubstituted (Table 15-3).

YOUR TURN 15.22

Identify the C—H bending mode in Figure 15-22 that indicates a para-disubstituted benzene ring.

At this stage, we know that the compound has the general form M—C$_6$H$_4$—Q, where the C$_6$H$_4$ portion is the para-disubstituted benzene ring. Given that the molecular formula is C$_7$H$_8$O, the M and Q portions must contain a total of one carbon, four hydrogens, and one oxygen. We also know that there is an OH group, leaving one carbon and three hydrogens yet to be accounted for. That could simply be a CH$_3$ group, in which case the unknown would be CH$_3$—C$_6$H$_4$—OH:

Unknown 1

15.6b Unknown 2

The IR spectrum of Unknown 2 (molecular formula C$_8$H$_{14}$O) is shown in Figure 15-23. We begin again with determining the compound's IHD. A completely saturated molecule with 8 C and 1 O would have 18 H, not 14, so Unknown 2 has an IHD of 2. Thus, Unknown 2 could have one triple bond, two double bonds, two rings, or a double bond and a ring.

FIGURE 15-23 IR spectrum of Unknown 2, C$_8$H$_{14}$O

The most conspicuous absorption band in Figure 15-23 is the intense one at ~1690 cm^{-1}, which is indicative of a carbonyl (C=O) group. Recall, however, that "normal" carbonyl absorptions belonging to ketones and aldehydes appear above 1700 cm^{-1}, so this carbonyl is either conjugated to a double bond (Section 15.5b) or belongs to an amide. There are no N atoms in Unknown 2, however, so we can conclude that the C=O bond is conjugated, appearing as C=C—C=O.

YOUR TURN 15.23

Identify the C=O stretch in Figure 15-23.

The IHD of 2 for Unknown 2 is accounted for by these two double bonds, so the rest of the molecule cannot contain any multiple bonds or rings.

Because we believe that a C=C double bond is present, we can look for confirmation by searching for both a C=C stretch and an alkene C—H stretch. Indeed, there is a sharp band that appears around 1650 cm^{-1}, characteristic of a C=C stretch.

Additionally, there is evidence of absorption at ~3050 cm^{-1}, suggesting an alkene C—H stretch.

In Figure 15-23, identify the bands corresponding to the C=C stretch and the alkene C—H stretch.

15.24 **YOUR TURN**

Because there is at least one alkene C—H, we can turn to the fingerprint region to search for C—H bending modes that would help us determine what type of C=C bond is present (Table 15-3). There appear to be two such bands, around 920 and 980 cm^{-1}, indicative of a terminal alkene of the form R—CH=CH$_2$.

Identify the C—H bending bands in Figure 15-23 that indicate the presence of a terminal alkene.

15.25 **YOUR TURN**

We have sorted out quite a bit of the structure of Unknown 2, but is it a ketone or aldehyde? That is, is there an aldehyde C—H stretch around 2720 cm^{-1}? Seeing nothing significant there, we can conclude that Unknown 2 is a ketone, having the general structure C—CO—CH=CH$_2$.

Indicate in Figure 15-23 where an aldehyde C—H stretch would typically appear.

15.26 **YOUR TURN**

Four carbon atoms and 11 hydrogen atoms still have not been accounted for. The remaining carbon atoms are sp^3 hybridized, but lacking any other information we cannot determine definitively which of multiple possible isomers our unknown compound is. Two possibilities are shown at the right.

The IR spectrum of Unknown 2 is an excellent example of the limitations of IR spectroscopy. Specific absorption bands in an IR spectrum provide structural information about only *pieces* of a molecule. Frequently, there is insufficient explicit information to allow you to converge on a single structure.

Two possibilities for Unknown 2

SOLVED problem 15.25 Draw a third possible structure that is consistent with the IR spectrum in Figure 15-23.

Think What is the basic structural unit that we derived from the IR spectrum? How can the remaining atoms be attached without disturbing that basic structure? What kinds of bonds must connect those atoms?

Solve The first structure at the right is the basic structural unit we derived from the spectrum, which accounts for four C, three H, and one O atom. That leaves 4 C and 11 H atoms yet to add, all of which can be attached only by single bonds—otherwise, we would exceed the compound's IHD of 2. To preserve the basic structure, we must add C$_4$H$_{11}$ to the left-hand C atom. One way to do so, which does not repeat either of the previous two structures, is as shown in the second structure at the right

---

problem 15.26 Draw two additional structures consistent with the IR spectrum in Figure 15-23, each of which is different from the previous three.

15.6c Unknown 3

The IR spectrum for Unknown 3 (molecular formula C$_6$H$_7$N) is shown in Figure 15-24. Suppose, too, that the longest-wavelength absorption in its UV–vis spectrum has λ_{max} around 190 nm.

FIGURE 15-24 IR spectrum of Unknown 3, C_6H_7N

Once again, we begin by determining the IHD, which for Unknown 3 is 4. This corresponds to a variety of combinations of double bonds, triple bonds, and rings (including an aromatic ring).

Probably the most noticeable peak in the IR spectrum is the one centered at ~ 3300 cm^{-1}. It is somewhat broad, suggesting either an O—H stretch or an N—H stretch. Given the absence of any O atoms, though, it must be an N—H stretch, belonging to an amine. An O—H stretch, moreover, would be significantly broader.

Unknown 3 is most likely a secondary amine of the form R—NH—R because there is only one N—H band. A primary amine would have two separate N—H stretching bands, and a tertiary amine would have no N—H peaks.

How do we account for the IHD of 4? A conspicuous feature of the spectrum is the small, sharp peak around 2100 cm^{-1}, indicative of a triple bond. This must be a C≡C triple bond, not a nitrile (C≡N) group, because the one nitrogen atom in Unknown 3 has already been associated with an amine. Furthermore, as discussed in Section 15.5c, a nitrile stretch generally has a much more intense peak than the one shown here.

To confirm that Unknown 3 has a C≡C triple bond, we can look for a terminal alkyne C—H stretch around 3300 cm^{-1}. The presence of one is difficult to see, given that the N—H stretch appears there as well. However, there does appear to be some evidence of one. Normally, N—H stretches are fairly broad. In this case, however, the absorption band becomes somewhat "pointy" around 3300 cm^{-1}. This is what we would expect if the alkyne C—H stretch overlapped with the N—H stretch as follows:

N—H stretch

C≡C—H stretch

Identify the band in Figure 15-24 that corresponds to an alkyne C—H stretch.

The presence of the C≡C triple bond now accounts for an IHD of 2. What about the other IHD of 2? It could be another C≡C triple bond or two rings. It is unlikely to be from a C=C double bond, because there are no sharp peaks near 1650 cm⁻¹.

If Unknown 3 has a second C≡C triple bond, then the two must not be conjugated to each other. According to Table 15-1, a λ_{max} of ~190 nm in a UV–vis spectrum suggests an isolated π bond. One possibility that fits all of these criteria is shown in the structure at the right.

Possible structure for Unknown 3

problem 15.27 Draw another molecule that is consistent with the IR spectrum in Figure 15-24.

15.7 Wrapping Up and Looking Ahead

UV–vis and IR spectroscopy, the two types of spectroscopy introduced here in Chapter 15, provide different types of structural information about molecules. In UV–vis spectroscopy, absorption of a photon promotes an electron from a lower-energy MO to a higher-energy MO. UV–vis spectra, therefore, give us information about the electronic structure of the molecule, particularly the extent of conjugation within the molecule's π system. In IR spectroscopy, absorption of a photon excites one of a molecule's normal modes of vibration. An IR spectrum thus provides information about which atoms are bonded together, and by what types of bonds—single, double, or triple.

Interpreting a spectrum requires familiarity with the types of peaks that can appear. In UV–vis spectroscopy, we often look for the longest-wavelength absorption, which generally corresponds to the HOMO–LUMO electron transition. In IR spectroscopy, we look for characteristic peaks that appear within small ranges of frequency. Among the most evident peaks are the C=O stretch, appearing as a strong, sharp band centered near 1700 cm⁻¹, and the O—H stretch, appearing as a strong, broad band centered above 3000 cm⁻¹.

Although UV–vis and IR spectroscopy can be quite useful in determining (or verifying) aspects of a compound's structure, we discuss two other methods of structural elucidation in Chapter 16: NMR spectroscopy and mass spectrometry (MS). All four—UV–vis, IR, NMR, and MS—provide different information about a molecule, and for a particular problem at hand, one might be more useful than another. Moreover, combining the information from multiple methods often leads to a complete elucidation of a molecule's structure.

Chapter Summary and Key Terms

- In **UV–vis spectroscopy**, a range of wavelengths of UV and visible light from a source is sent through an **analyte**, and at each wavelength, the intensity of light that is transmitted through the analyte is compared to the intensity of light from the source. **(Section 15.1)**

- A **UV–vis spectrum** is obtained by plotting **absorbance** against the wavelength of light. **(Section 15.1)**

- Electromagnetic radiation has a dual nature, behaving both as a wave and as a particle. As a wave, we can assign it a wavelength and frequency. As a particle, or **photon**, radiation carries only certain, discrete amounts of energy. **(Section 15.2)**

- In UV–vis spectroscopy, absorption of a photon promotes an electron from a lower-energy to a higher-energy MO. The

λ_{max} for the longest-wavelength (lowest-energy) absorption in the spectrum generally corresponds to the HOMO–LUMO transition. **(Section 15.2)**

- In a $\pi \rightarrow \pi^*$ ("pi-to-pi-star") **transition**, an electron from a π MO is promoted to a π* MO. **(Section 15.2)**

- In an $n \rightarrow \pi^*$ ("n-to-pi-star") **transition**, an electron from a nonbonding MO is promoted to a π* MO. **(Section 15.3b)**

- Conjugation decreases the energy difference between the HOMO and LUMO, and thus increases the wavelengths of the photons absorbed. **(Section 15.3a)**

- Compounds with lone pairs tend to have longer-wavelength absorptions than analogous compounds without lone pairs. **(Section 15.3b)**

- Absorption of an IR photon excites a molecule's **normal mode of vibration**. Absorption occurs when the frequency of the photon equals the frequency of that vibration. (Section 15.4a)

- In an IR spectrum, transmittance is plotted against IR frequency, in units of **wavenumbers** (cm^{-1}). Each peak in the spectrum corresponds to a vibrational frequency of the molecule. (Section 15.4a)

- The highest-frequency vibrations in an organic molecule are those of M—H bonds (where M is a heavy atom like C, N, or O) because H is very light. (Sections 15.4b and 15.4c)

- For a bond between two heavy atoms, the vibrational frequency depends on the strength and stiffness of the bond. Thus, vibrational frequency decreases in the order: triple bond > double bond > single bond. Vibrational frequency also decreases in the order: alkyne C—H > alkene C—H > alkane C—H. (Sections 15.4b and 15.4c)

- Bending modes of vibration are low in frequency, appearing below 1000 cm^{-1}. (Section 15.4b)

- The intensity of an IR stretching absorption increases, in general, as the magnitude of the dipole undergoing vibration increases. (Section 15.4d)

- The O—H stretch typically appears as a broad absorption due to extensive hydrogen bonding. (Section 15.5a)

- The C=O stretch of an amide is lower in frequency than that of a ketone or aldehyde, whereas an ester's C=O stretch is at a higher frequency than that of a ketone or aldehyde. Conjugation of a C=O bond tends to lower the absorption frequency by up to 30–40 cm^{-1}. (Section 15.5b)

- The nitrile (C≡N) stretch is usually more intense than that of an alkyne (C≡C), due to the greater bond dipole of the nitrile. (Section 15.5c)

- The intensity of a C—H stretching band is indicative of the number of C—H bonds that contribute to it. (Section 15.5e)

- The N—H stretch appears as two peaks for a primary amide or amine and one peak for a secondary amide or amine. No N—H peaks appear for a tertiary amide or amine. (Section 15.5f)

Problems

15.28 Naturally occurring carotene primarily exists in two different forms, called α-carotene and β-carotene.

α-Carotene

β-Carotene

Which has the longer-wavelength UV–vis absorption? Explain.

15.29 In the UV–vis spectrum of buta-1,2-diene ($CH_3CH=C=CH_2$), the longest-wavelength absorption appears at 178 nm. Compare this to the longest-wavelength absorption in buta-1,3-diene (see Table 15-1) and explain the significant difference.

15.30 An unknown compound has the formula C_5H_6. Its longest-wavelength UV–vis absorption is centered at 215 nm. Draw four isomers that are consistent with these results.

15.31 A compound, whose formula is C_7H_{12}, is known to have a six-membered ring. In its UV–vis spectrum, the longest-wavelength λ_{max} appears at 191 nm. Draw four isomers that are consistent with these results.

15.32 Phenolphthalein is often used as an indicator in acid–base titration experiments because its color depends upon the pH of the solution. When the solution is acidic or near neutral (pH < 8), it is colorless. Under mildly basic conditions (pH 9–13), the solution is red. Under strongly basic conditions (pH > 14), the solution is colorless again. Given the following structures of phenolphthalein under the various pH conditions indicated, explain the color dependence on pH.

Phenolphthalein
pH < 8

pH 9–13

pH > 14

15.33 Which compound has the longer-wavelength λ_{max}: propyne or acetonitrile ($CH_3C\equiv N$)? Explain.

15.34 Suggest how UV–vis spectroscopy could be used to determine whether each of the following reactions actually took place. Explain your reasoning.

(a)

Base

(b)

1. LiAlH$_4$
2. H$_3$O$^+$

(c)

Li
NH$_3$/EtOH

(d)

Fe
HCl

(e)

AlCl$_3$

15.35 According to Table 15-1, propenal (acrolein) has a λ_{max} at 340 nm. In the UV–vis spectrum shown below, the region that contains that absorption peak is magnified and shown in the inset. A more intense absorption appears at 202 nm. What kind of electron transition does that absorption correspond to? How do you know?

15.36 In the UV–vis spectrum of benzene, the absorption that corresponds to the HOMO–LUMO transition occurs at 184 nm. How does this compare to the corresponding electron transition in hexa-1,3,5-triene (see Table 15-1)? Explain why there is a significant difference.

15.37 For each of the following compounds, estimate the IR stretching frequencies for the bonds indicated by the arrows.

15.38 What differences in the IR spectra of the reactant and product would enable you to tell that each of the following reactions took place?

(a)

85% H₃PO₄ / Δ

(b)

CH₂N₂

(c)

(d)

CH₃Br

(e)

NaOH / H₂O

(f)

H₂ / Pd

(g)

Li / NH₃/EtOH

(h)

NaOH / Δ

(i)

1. LiAlH₄
2. H₃O⁺

15.39 How could you use IR spectroscopy to distinguish between the two compounds at the right?

A   B

15.40 IR spectroscopy does not distinguish very well between the two isomers shown.

C   D

Explain why. How could UV–vis spectroscopy be used to distinguish between them?

15.41 In Section 15.5b, we learned that the frequency of the C=O stretch for an amide is lower than it is for a ketone, suggesting that the C=O bond is weaker in an amide. This weakening of the C=O bond can be explained by the significant contribution from one of its resonance structures toward the resonance hybrid. Draw that resonance structure for N,N-dimethylacetamide [CH₃CON(CH₃)₂] and explain how it accounts for the weakening of the C=O bond.

15.42 Problem 15.41 calls attention to a resonance structure of an amide that accounts for the lowering of the C=O stretch frequency. An analogous resonance structure can be drawn for an ester.
(a) Draw that resonance structure for methyl acetate (CH₃CO₂CH₃).
(b) Based on the fact that an ester's C=O stretch frequency is higher than an amide's, does that resonance structure have a larger contribution toward the resonance hybrid of an ester or an amide? Using arguments of charge stability, explain why that should be so.

15.43 The carbonyl stretch of acetyl chloride (CH₃COCl) is found at 1806 cm⁻¹. Is the C=O bond stronger or weaker than the one in an amide? What does this suggest about the resonance contribution of a lone pair on Cl in an acid chloride compared to that of the lone pair on N in an amide? Explain. (*Hint:* See Problems 15.41 and 15.42.)

15.44 In Section 15.5b, we learned that the frequency of a C=O stretch decreases when the C=O bond is conjugated to a C=C bond. Draw the pertinent resonance contributor of a conjugated carbonyl (C=C—C=O) and, based on the resulting resonance hybrid, explain why the frequency decreases.

15.45 A student acquires the IR spectra of two isomeric compounds—namely, cyclohepta-1,3-diene and 3-methylcyclohexa-1,4-diene. One has an absorption band at 1648 cm⁻¹ and the other at 1618 cm⁻¹. Which spectrum belongs to which compound?

15.46 The C=C stretch frequency of an isolated C=C bond is normally ~1620 cm⁻¹. The C=C stretch frequencies for an aromatic ring are typically found between 1450 and 1600 cm⁻¹. Explain why the frequencies are lower in an aromatic ring.

15.47 In an IR spectrum, the C—O stretch for an alcohol (ROH) or an ether (ROR) appears near 1050 cm⁻¹, but for a carboxylic acid (RCO₂H) or an ester (RCO₂R), the C—O stretch appears near 1250 cm⁻¹. Explain why.

15.48 The C=O stretching frequency for butanal is centered at 1746 cm⁻¹, whereas the one for butan-2-one is centered at 1735 cm⁻¹. Explain the difference.

1746 cm⁻¹   1735 cm⁻¹

15.49 A compound whose molecular formula is C₆H₇N has the IR spectrum shown. Suggest a reasonable structure for this compound.

15.50 A compound whose molecular formula is C₈H₈O has the IR spectrum shown. Suggest a reasonable structure for this compound.

15.51 A compound whose molecular formula is $C_3H_2O_2$ has the IR spectrum shown. Propose a structure for this compound and estimate its pK_a.

15.52 Match each IR spectrum provided with one of the following three compounds.

X   Y   Z

15.53 In which compound, **A** or **B**, would you expect the C—O stretch band to be higher in frequency? Explain.

A   B

15.54 In which compound, **C** or **D**, would you expect the C=C stretch absorption to be higher in intensity? Explain.

C   D

15.55 The IR spectra for cyclohexanone and cyclobutanone are shown. In which compound is the C=O bond stronger? How do you know?

15.56 Based on your answer to Problem 15.55, in which compound—cyclohexanone or cyclobutanone—is there greater s character in the σ bond between C and O? Explain your answer.

15.57 In which compound, **E** or **F**, do you expect that the C=O stretch has a higher frequency? Explain.

15.58 Sodium acetate has a strong, sharp IR peak appearing at 1569 cm^{-1}, as shown in the following spectrum. To what kind of stretching mode does this band correspond? Why is its frequency so different from that of an ester?

15.59 A student ran the reaction at the right separately using two different bases: once with NaOH and again with LDA.

When NaOH was used, the organic product's UV–vis spectrum had a λ_{max} of ~220 nm. When LDA was used, the product's spectrum had a λ_{max} of ~180 nm. In both cases, the formula of the organic product was C_7H_{10}. Explain.

15.60 A compound whose molecular formula is $C_5H_{10}O$ has the IR spectrum shown. The compound's pK_a is >40. Propose a structure for this compound that is consistent with these data.

15.61 When 1,2-dibromo-1-phenylethane is heated with sodium hydroxide, a compound with the IR spectrum provided is produced. Propose a structure for this product.

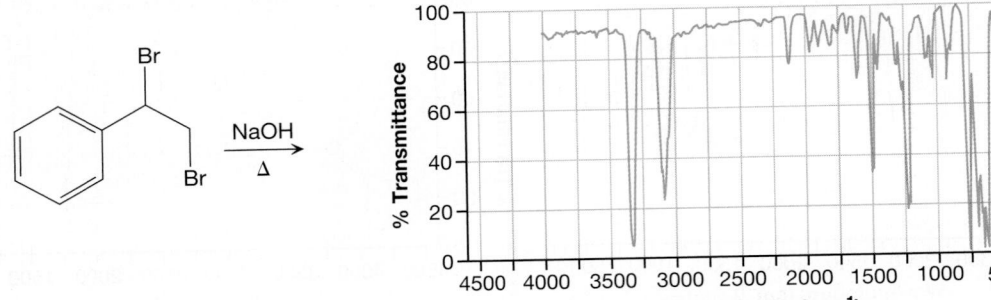

15.62 The product obtained from the reaction in Problem 15.61 is then treated with sodium hydride, followed by treatment with bromomethane. The IR spectrum of the resulting compound is provided. Propose a structure for this product.

15.63 The IR spectrum of the product of the following reaction is shown below. Propose a structure for this compound.

15.64 The IR spectrum of the product of the following reaction is shown below. Propose a structure for this compound, paying particular attention to its stereochemistry.

If your instructor assigns problems in smartwork, log in at **smartwork.wwnorton.com**.

Structure Determination 2

Nuclear Magnetic Resonance Spectroscopy and Mass Spectrometry

Chapter 15 discussed UV–vis and IR spectroscopy. Here in Chapter 16, we continue the discussion of spectroscopy, focusing on **nuclear magnetic resonance (NMR) spectroscopy.** Toward the end of this chapter, we will also provide an introduction to *mass spectrometry*.

Because of the kind of information it provides, NMR spectroscopy is generally considered the single most powerful tool in organic chemistry for the elucidation of molecular structure. NMR spectroscopy can tell us about the number of distinct types of hydrogen and carbon atoms in a given molecule, as well as each atom's specific environment. This, in turn, allows us to determine the *connectivity* of atoms within the molecule. By comparison, IR spectroscopy provides relatively little information about connectivity, and UV–vis spectroscopy provides even less.

Understanding NMR spectroscopy, however, requires a grasp of principles that are somewhat more complex than those underlying UV–vis or IR spectroscopy. Moreover, an NMR spectrum often has more features to analyze than a typical UV–vis or IR spectrum, so interpreting an NMR spectrum is slightly more involved. For these reasons, we devote nearly an entire chapter to NMR spectroscopy.

A magnetic resonance imaging (MRI) scanner, such as this one, is a medical instrument used to analyze soft tissue, such as in the heart or brain. The basis of MRI is essentially the same as that of nuclear magnetic resonance (NMR) spectroscopy, a topic discussed here in Chapter 16.

Upon completing Chapter 16 you should be able to:

- Describe how an NMR spectrum is generated.

- Identify the region of the electromagnetic spectrum containing wavelengths that are absorbed by nuclei, and specify which quantum states are affected by such an absorption.

- Define *chemical shift* and estimate the chemical shift of each hydrogen and carbon nucleus in a molecule, taking into account the roles played by *inductive effects* and *magnetic anisotropy*.

- Determine the number of chemically distinct hydrogen atoms and carbon atoms in a compound from its proton NMR spectrum and its carbon NMR spectrum, respectively, and establish whether a molecule's structure is consistent with that information.

- Define the *integration* of a signal, and use a proton spectrum's integral trace to determine the relative number of protons that each signal represents.

- Identify protons in a molecule that exhibit *spin–spin coupling*, and explain how this coupling affects *signal splitting*.

- Interpret results from *distortionless enhancement by polarization transfer (DEPT) spectroscopy* to determine the type of carbon atom giving rise to a particular carbon signal.

- Describe the features of a *mass spectrum*, and explain how one is generated.

- Identify the *molecular ion* (M^+) peak in a mass spectrum, and explain the origin of the M + 1 peak and fragment peaks.

- Exploit the relative intensities of isotope peaks to determine the number of carbon atoms present in a molecule, and to determine whether a molecule contains bromine, chlorine, or sulfur.

Mass spectrometry is another very powerful tool for structural elucidation. Unlike UV–vis, IR, and NMR spectroscopy, however, mass spectrometry does not involve the interaction of radiation with molecules. Rather, it allows us to measure the masses of the molecule itself and the fragments that are created when it is split apart. Nevertheless, we group mass spectrometry with the three spectroscopies because all four complement one another in the determination of molecular structure.

16.1 NMR Spectroscopy: An Overview

Recall from Chapter 15 that UV–vis and IR spectroscopies are characterized by the kind of radiation that they use. UV–vis spectroscopy uses radiation from the ultraviolet and visible regions of the electromagnetic spectrum, whereas IR spectroscopy uses infrared radiation. NMR spectroscopy, on the other hand, uses radiation from the **radio frequency (RF)** region of the spectrum, also called *radio waves* (review Fig. 15.1).

As its name suggests, RF radiation is largely used for communication purposes, including AM and FM radio signals and wireless network connections for computers. Radiation in this region is less energetic than that in the UV–vis or IR regions, encompassing frequencies between roughly 3 Hz and 3×10^{11} Hz (i.e., wavelengths between 1 mm and 100,000 km). Most modern NMR spectrometers, however, work in the relatively narrow region between about 10^8 and 10^9 Hz.

A typical setup for an NMR experiment is shown in Figure 16-1a. The sample is placed in the hollow bore of a superconducting magnet and is irradiated with a pulse of RF radiation emitted from an NMR probe. As in any spectroscopic experiment, we want to obtain a *spectrum* that can tell us which frequencies are absorbed and how strong each absorption is.

The general setup for NMR spectroscopy, however, differs from the typical setup for UV–vis or IR spectroscopy in two major ways. The first is the use of a superconducting magnet, which subjects the sample to a very high **external magnetic field**, abbreviated as B_{ext}. The second is in the way that each absorbed frequency is detected. In UV–vis and IR spectroscopy, absorbance at a particular frequency is obtained by comparing the intensity of radiation that has passed through the sample, $I_{detected}$, to the intensity of radiation from a source, I_{source} (review Equations 15-1 and 15-2).

By contrast, most NMR spectrometers take advantage of the fact that a portion of the RF radiation absorbed by the sample at each frequency is *re-emitted* at the same frequency. Thus, each of these re-emitted frequencies, or **signals**, corresponds to a frequency that has been absorbed by the sample. Moreover, the *amplitude* of each signal is proportional to the amount of radiation that was originally absorbed. This allows the data from each recorded signal to be converted automatically into a spectrum analogous to other spectra we have seen in which relative absorbance is plotted against a characteristic of radiation (Fig. 16-1b). In this case, the *x* axis of the spectrum is *chemical shift*, which we can think of for now as representing *relative frequency*. (We discuss chemical shift in greater detail in Section 16.6.)

(a)

(b)

FIGURE 16-1 Typical nuclear magnetic resonance experiment
(a) A modern superconducting NMR spectrometer. A sample is placed in the hollow bore of a superconducting magnet. An NMR probe irradiates the sample with RF radiation, and detects RF signals emitted by the sample. (b) The RF radiation emitted by the sample is recorded and converted into an NMR spectrum, in which amplitude is plotted against chemical shift.

With this overview of NMR spectroscopy in mind, we ask the following questions:

- Why must a large magnetic field be applied to the sample?
- Why are only certain frequencies of RF radiation absorbed by a given sample?
- How can an NMR spectrum be used to interpret molecular structure?

We answer these questions in the sections that follow.

16.2 Nuclear Spin and the NMR Signal

The sample is placed in a strong magnetic field in an NMR experiment because *no RF radiation can be absorbed by the sample in the absence of such a field, and no NMR spectrum can be recorded.* Why?

The answer stems from the fact that some atomic nuclei possess a property called **nuclear spin**. The most common example is the nucleus of a hydrogen atom (i.e., a proton), whose spin is analogous to that of an electron. Recall from Section 1.3c that electrons can assume one of two spin states. A proton's spin is quantized in the same way.

> A proton has a spin of $+\frac{1}{2}$ or $-\frac{1}{2}$ a.u., which corresponds to the **α spin state** or the **β spin state**, respectively.

Nuclear spin generates a small magnetic field, an outcome expected of a charge in motion. Consequently, protons (and other nuclei with spin) have quantized magnetic dipoles, and thus behave as tiny bar magnets, with a north and a south pole. Nuclei are thus often represented as arrows, indicating the direction in which their magnetic dipoles point (Fig. 16-2).

Outside of an external magnetic field, there is no energy difference between the α and β states, so there is an equal probability of a proton being in either state (Fig. 16-3, left). Moreover, the magnetic dipoles of the protons point in random directions. The situation is similar to what we would observe if we were to grab several bar magnets and toss them onto the floor. Unless two magnets lie close enough to affect each other, they will end up with random orientations.

(a) **(b)**

FIGURE 16-2 Quantized magnetic dipoles of the hydrogen nucleus
(a) The α spin or spin-up state. (b) The β spin or spin-down state.

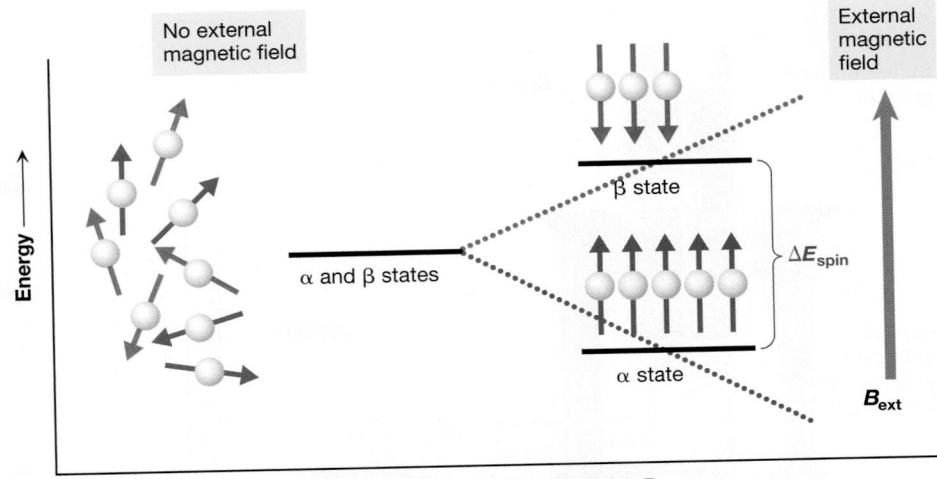

FIGURE 16-3 Effect of an external magnetic field on nuclear spin states In the absence of an external magnetic field, the α and β states are at exactly the same energy, and the nuclear magnetic dipoles are oriented in random directions (left). In an external magnetic field, represented by the green arrow, the α spins align with the external magnetic field and the β spins align against it (right). This makes the energy of the α state slightly lower than that of the β state, by a difference ΔE_{spin}.

In a B_{ext}, however, the situation is different. Each nucleus with spin has its average magnetic dipole aligned parallel to B_{ext}. As shown on the right in Figure 16-3, the dipoles of nuclei in the α spin state become aligned *with* B_{ext}, whereas the dipoles of nuclei in the β spin state become aligned *against* it. In these orientations, the energy of the α state is slightly lower than the energy of the β state. (It's like being in a canoe on a river—paddling with the current requires less energy than paddling against it.) At equilibrium, therefore, there is a small, but significant, excess of nuclei in the α state.

Equation 16-1 shows that the difference in energy between the two spin states of a proton, ΔE_{spin}, is directly proportional to B_{ext}.

$$\Delta E_{spin} = \gamma h B_{ext}/2\pi = \text{(constant)} \times B_{ext} \qquad \text{(16-1)}$$

Here, B_{ext} is in units of **tesla (T)**, h is Planck's constant (6.626×10^{-34} J · s), and the Greek letter γ (gamma) is the **gyromagnetic ratio**. Every nucleus that has spin has a characteristic value for the gyromagnetic ratio; for a proton, its value is 2.67512×10^8 T^{-1} s^{-1}.

In a strong magnetic field, typical values of ΔE_{spin} correspond to photon energies from the RF region of the electromagnetic spectrum. Therefore, a nucleus in the α state can absorb such a photon and be promoted to the β state—a so-called **spin flip** (Fig. 16-4). A signal is produced when a nucleus in the β state relaxes back to the α state.

We can now understand why no NMR spectrum can be acquired in the absence of a B_{ext}. As we saw in Figure 16-3, the α and β states of a proton have exactly the

FIGURE 16-4 A spin flip in NMR spectroscopy (a) Several protons in an external magnetic field. An incoming photon is absorbed by one of the nuclei in the α state, promoting it to the β state. When a nucleus relaxes back to the α state, a signal is produced. (b) A spectrum is obtained by plotting signal amplitude versus the frequency of radiation detected.

(a)

(b)

same energy in the absence of a B_{ext}. Under these circumstances, there is no higher-energy spin state to which a nucleus can be promoted, so no RF photon can be absorbed.

Most of this chapter focuses on **proton NMR (or ^{1}H NMR) spectroscopy**, in which the RF radiation used causes spin flips in the nuclei of hydrogen atoms. NMR spectroscopy, however, can also be used to probe other nuclei that possess spin. In general,

A nucleus possesses spin and thus can be studied with NMR spectroscopy if it has an odd number of protons, an odd number of neutrons, or both.

Thus, the carbon-13 (^{13}C) nucleus, which has six protons (an even number) and seven neutrons (an odd number), can be studied with NMR spectroscopy (Sections 16.13 and 16.14), but the carbon-12 (^{12}C) nucleus, which has six protons (even) and six neutrons (even), cannot.

problem **16.1** At which magnetic field experienced by a hydrogen nucleus is the energy difference between the α and β states greater: 0.5 T or 5.0 T? Explain.

SOLVED problem **16.2** A common magnetic field strength of modern NMR instruments is 7.064 T. In this magnetic field, calculate the energy difference between the α and β spin states of a proton.

Think In Equation 16-1, what value should be substituted for γ? For h? For B_{ext}?

Solve As mentioned in the text, γ for a proton is 2.67512×10^8 T^{-1} s^{-1} and h is 6.626×10^{-34} J · s. In the problem, the value for B_{ext} is given as 7.064 T. Therefore,

$$\Delta E_{spin} = \gamma h B_{ext}/2\pi$$
$$= (2.67512 \times 10^8 \text{ T}^{-1} \text{ s}^{-1})(6.626 \times 10^{-34} \text{ J} \cdot \text{s})(7.064 \text{ T})/2\pi$$
$$= 1.993 \times 10^{-25} \text{ J}$$

The units s and T both cancel, leaving energy in units of J.

problem **16.3** Calculate the energy difference between the α and β spin states of a proton in a magnetic field of 2.2 T.

16.3 Shielding, Chemical Distinction, and the Number of NMR Signals

All protons in a molecule typically do *not* absorb at the same frequency. Instead, protons surrounded by different electron distributions—that is, ones that reside in different **chemical environments**—absorb at different frequencies. These protons are called **heterotopic protons**, or **chemically distinct protons**. By the same token, protons that reside in identical chemical environments, called **homotopic protons**, or **chemically equivalent protons**, absorb at the same RF frequency.

Why should a proton's chemical environment govern its absorption frequency? When subjected to B_{ext}, the electrons that surround a proton create a **local magnetic field** (B_{loc}) that opposes B_{ext}. The magnetic field that is "felt" by the nucleus, called the **effective magnetic field** (B_{eff}), is the sum of B_{ext} and B_{loc}:

$$B_{eff} = B_{ext} + B_{loc} \qquad \text{(16-2)}$$

Because B_{loc} opposes B_{ext}, B_{eff} is somewhat less than B_{ext}, so protons in a molecule are said to be **shielded**.

Chemically distinct protons, which are surrounded by different electron distributions, are subjected to different local magnetic fields, and hence differ in their value of B_{eff}; they are *shielded* to different extents. According to Figure 16-3, these protons will have different values of ΔE_{spin}, so they will absorb at different frequencies.

YOUR TURN 16.2

The following diagram represents the spin states of two different protons (proton 1 and proton 2) exposed to the same B_{ext}.

B_{eff} for one proton differs from B_{eff} of the other. Identify the spin states of the proton that is shielded to a greater extent.

Because a proton's absorption frequency depends on its chemical environment, we arrive at one of the most important rules for ^1H NMR spectroscopy:

Each chemically distinct type of hydrogen atom in a molecule gives rise to a signal in a proton NMR spectrum.

In benzene, for example, the electron distribution about each proton is identical, so all six H atoms are chemically equivalent. This gives rise to a single ^1H NMR signal, which appears at 7.3 ppm in the NMR spectrum (Fig. 16-5).

FIGURE 16-5 ^1H **NMR signal of benzene** The six H atoms of benzene are chemically equivalent, giving rise to a single ^1H NMR signal at 7.3 ppm. The small signal at 0 ppm arises from tetramethylsilane (TMS) that is added to provide a reference signal.

The second signal in the spectrum, which appears at 0 ppm, is due to a small amount of **tetramethylsilane** (**TMS**), $(CH_3)_4Si$, which can be added to samples to provide a reference signal.

---

problem **16.4** Based on the positions of the two signals in Figure 16-5, do you think that the protons in benzene are chemically equivalent to those in TMS? Explain.

Propynoic acid ($HC\equiv C-CO_2H$) gives rise to two ^1H NMR signals, as shown in the spectrum in Figure 16-6. (Once again, the signal at 0 ppm corresponds to TMS that is added to the sample.) Thus, there are two chemically distinct H atoms in $HC\equiv C-CO_2H$. This should not be surprising, because one proton is bonded to an alkyne C, whereas the other is bonded to the O atom of a carboxyl group.

FIGURE 16-6 ^1H NMR spectrum of propynoic acid The two signals on the left correspond to the protons in propynoic acid, indicating that propynoic acid has two chemically distinct protons. The signal at 0 ppm corresponds to TMS, the reference compound.

SOLVED problem **16.5** From the ^1H NMR spectrum given, determine the number of chemically distinct protons in dichloroacetic acid.

Think How many signals appear in the proton NMR spectrum? Of those signals, which are generated by dichloroacetic acid?

Solve Although there are three peaks in the ^1H NMR spectrum, only the ones at 6.0 ppm and 11.8 ppm correspond to protons in dichloroacetic acid. The peak at 0 ppm corresponds to added TMS. Thus, there are only two chemically distinct types of protons in dichloroacetic acid.

---

problem **16.6** From the ^1H NMR spectrum given, determine the number of chemically distinct protons in *trans*-1,2-dichloroethene.

16.4 Strategies for Success: The Chemical Distinction Test and Molecular Symmetry

As discussed in Section 16.3, a proton NMR spectrum can tell us the number of chemically distinct hydrogen atoms in a particular compound. To make use of this information, however, it is important to be able to determine the number of chemically distinct hydrogen atoms solely from a molecule's structure. The ability to do so helps us to establish whether or not a compound is consistent with the NMR spectrum.

To determine the number of chemically distinct hydrogen atoms in a molecule, we use a protocol called the **chemical distinction test**.

The Chemical Distinction Test for Hydrogen Atoms

1. For *each* hydrogen atom in question, draw the complete structure of the molecule in which *just that hydrogen atom* is replaced by an imaginary atom, "X." There should be one X-substituted molecule for each hydrogen atom being tested.
2. Determine which of those X-substituted molecules are either *identical* or *enantiomers*. The corresponding hydrogen atoms in the original molecule are *chemically equivalent*.
3. Determine which of those X-substituted molecules are either *constitutional isomers* or *diastereomers*. The corresponding hydrogen atoms in the original molecule are *chemically distinct*.

If Step 1 yields *enantiomers*, then the corresponding hydrogen atoms are said to be **enantiotopic**. If Step 1 yields *diastereomers*, then the corresponding hydrogen atoms are said to be **diastereotopic**.

Let's apply the chemical distinction test to the six H atoms in benzene. Step 1 results in the following six molecules shown on the right:

Replace each H with an imaginary "X."

These are *identical* molecules, so all H atoms are chemically equivalent, or homotopic.

All six of the X-substituted molecules are identical to each other, so, according to Step 2, all six H atoms must be chemically equivalent. This conclusion is consistent with the fact that benzene gives rise to just one signal in its ^{1}H NMR spectrum (Fig. 16-5).

Now consider propynoic acid, $HC \equiv C-CO_2H$. When each H atom is separately replaced by "X," we get the following two structures shown on the right:

These are *constitutional isomers*, so the two H atoms are chemically distinct.

Replace each H atom with "X."

The resulting molecules are *constitutional isomers* because they have the same molecular formula but different connectivities. Thus, the two H atoms are chemically distinct, consistent with the fact that the compound gives rise to two signals in its ^1H NMR spectrum (Fig. 16-6).

SOLVED problem **16.7** How many chemically distinct H atoms are there in chloroethene, $H_2C = CHCl$?

Think What molecules are generated by substituting each H atom separately by "X"? Which of those molecules are identical? Enantiomers? Diastereomers? Constitutional isomers?

Solve We can draw three X-substituted molecules as follows:

Diastereomers   **Constitutional isomers**

Replace each H atom with "X."

Molecules **A** and **B** are diastereomers—they have the same connectivity but are not mirror images. Molecules **B** and **C** are constitutional isomers. Therefore, according to Step 3 in the chemical distinction test, all three H atoms are chemically distinct.

---

problem **16.8** How many chemically distinct H atoms are in each of the following molecules?

(a)   (b)   (c)   (d)   (e)   (f)

---

problem **16.9** A compound gives rise to two signals in its ^1H NMR spectrum. Which of the following compounds can it be? Explain.

D   E   F   G

H_A and H_B are mirror images of
each other, so they are equivalent.

This 180° rotation of the molecule exchanges
H_C and H_D, but the molecule otherwise appears
unchanged, so they are equivalent.

Rotate 180°

1-Bromobutane

(a)

(1*R*,2*R*)-1,2-Dimethylcyclobutane

(b)

FIGURE 16-7 Molecular symmetry and chemical equivalence (a) H_A and H_B are mirror images of each other through the molecule's plane of symmetry, so the two H atoms are equivalent. (b) The 180° rotation of the molecule causes H_C and H_D to exchange locations but the molecule otherwise appears unchanged, so the two H atoms are equivalent.

After you apply the chemical distinction test to a variety of different molecules, you will notice that molecules with greater symmetry tend to have fewer chemically distinct hydrogens. That's because a molecule's symmetry can place demands on the chemical equivalence of certain atoms.

Two atoms must be chemically equivalent if:

- The atoms are mirror images with respect to a plane of symmetry of the molecule, or
- A rotation of the entire molecule causes the atoms to exchange locations, but the molecule otherwise appears unchanged.

These ideas can help you identify equivalent atoms quickly. In 1-bromobutane (Fig. 16-7a), for example, H_A and H_B are mirror images with respect to the plane of symmetry, so they are equivalent. In (1*R*,2*R*)-1,2-dimethylcyclobutane (Fig. 16-7b), the 180° rotation of the molecule results in the exchange of H_C and H_D, but the molecule otherwise appears unchanged, so the two H atoms are equivalent.

YOUR TURN 16.3

In benzene (Fig. 16-5), all six H atoms are chemically equivalent.
(a) Can you identify a plane of symmetry through which one H atom is the mirror image of another?
(b) Can benzene be rotated in such a way as to exchange two H atoms, but otherwise leave the molecule apparently unchanged?

16.5 The Time Scale of NMR Spectroscopy

If we apply the chemical distinction test to the conformation of acetic acid (CH₃CO₂H) shown on the left of the following graphic, then we obtain the four X-substituted molecules shown on the right (under each one is the corresponding Newman projection looking down the C—C bond):

Because structure **1** is a constitutional isomer of structures **2–4**, the chemical distinction test predicts that the H atoms of CH₃ should be chemically distinct from the carboxyl H atom. The test also predicts that two H atoms of CH₃ should be chemically distinct from each other given that, if frozen in their respective conformations, structures **2** and **3** are diastereomers (i.e., they have the same connectivity but are not mirror images). Thus, we might expect at least three chemically distinct H atoms, which should give a ¹H NMR spectrum with at least three peaks. The spectrum in Figure 16-8, however, shows only *two* peaks, so there are only *two* chemically distinct types of protons.

In other words,

> All three hydrogen atoms of a CH₃ group give rise to a single proton signal.

It turns out that the CH₃ hydrogens rapidly interchange environments via rotation about the C—C bond. More specifically, *the time it takes for a typical free rotation about a single bond (~10⁻¹⁰ s) is much shorter than the time it takes to acquire an NMR spectrum (~1 s).* Thus, the different environments of the interchanging protons in acetic acid are blurred, in much the same way that a camera with a slow shutter speed blurs objects in motion. The NMR signal, as a result, reflects the *average* proton environment of each proton during its 360° rotation about the C—C single bond.

FIGURE 16-8 ¹H NMR spectrum of acetic acid The three H atoms of CH₃ give rise to the same signal because they rapidly interconvert via rotation of the CH₃ group.

Axial and equatorial
H atoms rapidly interconvert.

Chemical shift (ppm)

FIGURE 16-9 ^{1}H NMR spectrum of cyclohexane The axial and equatorial H atoms give rise to the same signal because they rapidly interconvert via a chair flip.

This phenomenon is not limited to just CH_3 groups. In general,

> If protons in different chemical environments interchange rapidly, then they give rise to the same NMR signal that reflects the *average* of those environments.

This is why, under normal conditions, cyclohexane's ^{1}H NMR spectrum (Fig. 16-9) shows only one signal despite having two types of H atoms: axial and equatorial. (This is explored further in Problem 16.68.)

16.6 Chemical Shift

Chemical shift, often abbreviated as the Greek letter δ (delta), is a measure of the extent to which a signal's frequency differs from that of a reference compound. The reference compound used for ^{1}H NMR spectroscopy is TMS. Formally, **chemical shift**, in units of *parts per million (ppm)*, is defined by the mathematical expression in Equation 16-3:

$$\text{Chemical shift (ppm)} = [(\nu_{sample} - \nu_{TMS})/\nu_{op}] \times 10^6 \qquad \text{(16-3)}$$

Here, ν_{sample} is the frequency of the signal of interest, ν_{TMS} is the frequency of the signal from the protons in TMS, and ν_{op} is the *operating frequency* of the NMR spectrometer, all of which are in units of Hz.

The spectrometer's **operating frequency** is the frequency at which a bare proton absorbs radiation when subjected to the spectrometer's magnetic field (Equation 16-4).

$$\nu_{op} = \gamma B_{ext}/2\pi \qquad \text{(16-4)}$$

Because the gyromagnetic ratio (γ) is a constant, the operating frequency is directly proportional to B_{ext}. Thus, the operating frequency is constant for a given magnetic field strength. If the strength of the spectrometer's magnet is 7.046 T, for example, then the operating frequency is calculated as follows:

$$\nu_{op} = (2.67512 \times 10^8 \text{ T}^{-1} \text{ s}^{-1})(7.046 \text{ T})/2(3.14159)$$

$$= 3.00 \times 10^8 \text{ s}^{-1}$$

Alternatively, because the unit s^{-1} is equivalent to the unit Hz, and because 1 MHz (i.e., megahertz) equals 10^6 Hz, a spectrometer with a 7.046 T magnet has an operating frequency of 300 MHz. We say that the spectrometer is a 300-MHz instrument.

problem **16.10** What is the operating frequency of an NMR spectrometer using a magnet whose strength is 11.74 T? Give your answer in units of MHz.

Chemical shift is plotted on the x axis instead of frequency, because signal frequency depends on B_{ext} (see Equation 16-1). Therefore, signal frequency can vary from one NMR instrument to the next, depending on the strength of its magnet. By contrast, a proton's *chemical shift is independent of* B_{ext} because the operating frequency that appears in the denominator of Equation 16-3 has the same dependence on magnetic field strength as the frequency terms in the numerator. The magnetic field dependence thus cancels out, so a proton's chemical shift is the same regardless of the instrument used to measure it.

---

problem 16.11 Suppose that the signal of a specific hydrogen nucleus appears 2200 Hz higher than that of the protons in TMS. If the NMR spectrometer uses a 7.046 T magnet, what is the chemical shift of that proton?

---

problem 16.12 Suppose that a proton's chemical shift is 2.4 ppm in a 300-MHz NMR spectrometer. What is the signal frequency of that proton relative to the signal frequency of a proton in TMS?

By convention, chemical shift increases from right to left in an NMR spectrum. The chemical shift of the protons in TMS always appears at 0 ppm. (We can solve for this value by substituting ν_{TMS} for ν_{sample} in Equation 16-3, which makes the numerator zero.) The farther to the left an absorption peak appears in a spectrum (i.e., the more positive the chemical shift), the more it is said to be shifted **downfield**. By the same token, the farther to the right an absorption peak appears, the more it is said to be shifted **upfield**. (These terms originate from older types of NMR spectrometers, in which a sample was irradiated with a fixed frequency of RF radiation, and the magnetic field was adjusted to make ΔE_{spin} equal E_{photon}. The more positive a proton's chemical shift, the stronger the magnetic field that was required, and vice versa.)

16.7 Characteristic Chemical Shifts, Inductive Effects, and Magnetic Anisotropy

Recall from Section 16.3 that protons in different chemical environments give rise to different signals in an NMR spectrum. Table 16-1, which lists the chemical shifts observed in ^{1}H NMR spectroscopy, shows that a proton's chemical shift is governed largely by the identity of the surrounding atoms.

> Identify two different types of protons in Table 16-1 whose absorption peaks appear *downfield* and two that appear *upfield* from those produced by the protons in $ClCH_3$.

16.4 YOUR TURN

To begin to understand why each type of proton in Table 16-1 has the characteristic chemical shift it does, recall from Section 16.3 that the surrounding electrons *shield* a proton by inducing a local magnetic field (B_{loc}) that partially cancels the B_{ext}. The signal for such a shielded proton appears at a lower frequency than it would if the proton were bare, thus reducing the chemical shift of the shielded proton.

Two major factors influence the extent to which a hydrogen nucleus is shielded, and thus affect the proton's chemical shift: *inductive effects* and *magnetic anisotropy*. The more straightforward of the two is inductive effects, which was discussed extensively in Section 6.6. Magnetic anisotropy, on the other hand, is a phenomenon we have not previously encountered.

TABLE 16-1 Chemical Shifts in ^{1}H NMR Spectroscopy

Type of Proton	Chemical Shift (ppm)	Type of Proton	Chemical Shift (ppm)	Type of Proton[a]	Chemical Shift (ppm)
1. TMS: H_3C—Si(—CH_3)(—CH_3)—CH_2, H_3C—Si, CH$_3$, H	0	8. R—C≡C—H	2.4	15. (benzene ring)—H	7.3
2. R—CH_2, H	0.9	9. $\underset{Br}{}\overset{H_2}{C}$—H	2.7	16. $\underset{R}{}\overset{O}{C}$—H	9–10
3. $\underset{R}{\overset{R}{}}HC$—H	1.3	10. $\underset{Cl}{}\overset{H_2}{C}$—H	3.1	17. R—N$\underset{H}{}$H	1.5–4
4. $\underset{R}{\overset{R}{}}\underset{R}{C}$—H	1.4	11. R—O—$\overset{H_2}{C}$—H	3.3	18. R—O$\underset{H}{}$	2–5
5. $\underset{R}{\overset{R}{}}C$=CH, H_2C—H	1.7	12. $\underset{F}{}\overset{H_2}{C}$—H	4.1	19. Ar—O$\underset{H}{}$	4–7
6. $\underset{R}{}\overset{O}{C}$—$\overset{}{\underset{H_2}{C}}$—H	2.1	13. $\underset{R}{\overset{R}{}}C$=CH, H	4.7	20. $\underset{R}{}\overset{O}{C}$—O$\underset{H}{}$	10–12
7. (benzene ring)—$\overset{H_2}{C}$—H	2.3	14. $\underset{R}{\overset{R}{}}C$=$\underset{H}{\overset{R}{}}C$	5.3		

[a] Protons in boxes shaded in green undergo hydrogen bonding.

16.7a Inductive Effects

The role of inductive effects on chemical shift can be seen by comparing entries 2 and 12 in Table 16-1 (RCH_2—H and FCH_2—H, respectively). In entry 2, no significantly electronegative atoms are present and the proton has a very low chemical shift of 0.9 ppm. In entry 12, the C atom to which the H atom is attached is itself bonded to a highly electronegative F atom. The presence of that F atom causes a substantial downfield shift of the protons' absorption peak, to 4.1 ppm. A similar phenomenon occurs with other highly electronegative atoms. In general, then,

A proton's absorption peak is shifted downfield by nearby electronegative atoms.

This downfield shift occurs because electronegative atoms are inductively *electron withdrawing*, so they remove electron density from the nearby proton. Consequently, the shielding by those electrons is diminished, and B_{loc} is less effective at canceling B_{ext} (review Equation 16-2). The proton is said to be **deshielded**, resulting in a higher signal frequency, and thus a larger chemical shift.

The extent of this inductive deshielding depends, in part, on the electronegativity of the nearby atom. Notice in Table 16-1, for example, that the chemical shifts of the protons in FCH_3 appear downfield from those in $ClCH_3$. The F atom is more electronegative than Cl, so it more effectively deshields the CH_3 protons.

16.5 YOUR TURN

Look up the chemical shifts for the protons in FCH_3, $ClCH_3$, and $BrCH_3$ in Table 16-1 and compare them to the electronegativities of fluorine, chlorine, and bromine in Figure 1-16.

---

problem **16.13** In which compound will the proton chemical shift be greater, CH_3Br or CH_3I? Explain.

16.7b Magnetic Anisotropy

The second factor that affects B_{eff} is called **magnetic anisotropy**. This effect is responsible for the large chemical shift in an aromatic compound like benzene, whose protons have a chemical shift of 7.3 ppm. Inductive effects cannot explain such a large downfield shift because benzene (C_6H_6) contains no highly electronegative atoms.

Benzene's six π electrons occupy a π system that looks like two donuts—one above the plane of the ring and one below (Fig. 16-10). The B_{ext} forces those π electrons to move in a circular path parallel to the plane of the ring—a so-called **ring current** (Fig. 16-10a). The movement of those charged particles creates an additional magnetic field (Fig. 16-10b). Notice that the H atoms are located where those additional magnetic field lines (represented by the blue arrows) are in the same direction as B_{ext}, so B_{eff} is increased and the hydrogen nuclei are *deshielded*.

16.6 YOUR TURN

In Figure 16-10, identify a *local* magnetic field line oriented in the same direction as the B_{ext}.

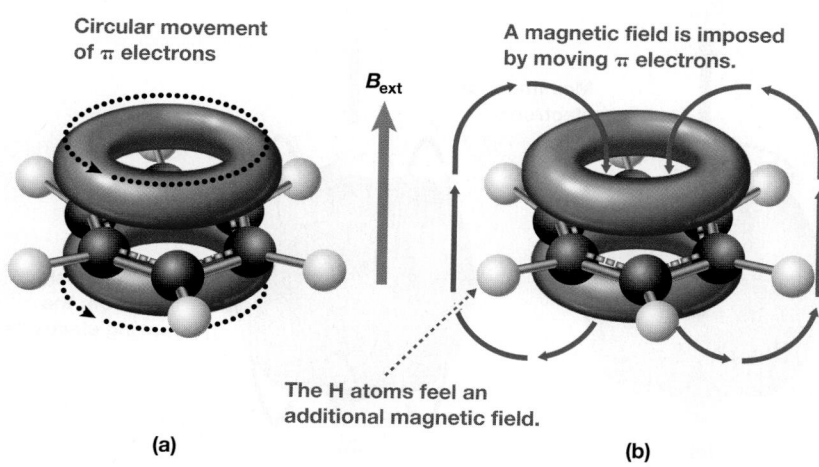

Circular movement of π electrons

A magnetic field is imposed by moving π electrons.

B_{ext}

The H atoms feel an additional magnetic field.

(a)   (b)

FIGURE 16-10 Deshielding of hydrogens in benzene (a) The B_{ext} forces the π electrons of benzene to move in a circular trajectory above and below the plane of the ring, creating ring current (black dotted arrows). (b) Ring current gives rise to a local magnetic field (blue arrows) that is in the same direction as the B_{ext} at the location of the hydrogen atoms. The hydrogen nuclei are therefore deshielded so much that their signal appears well downfield of TMS.

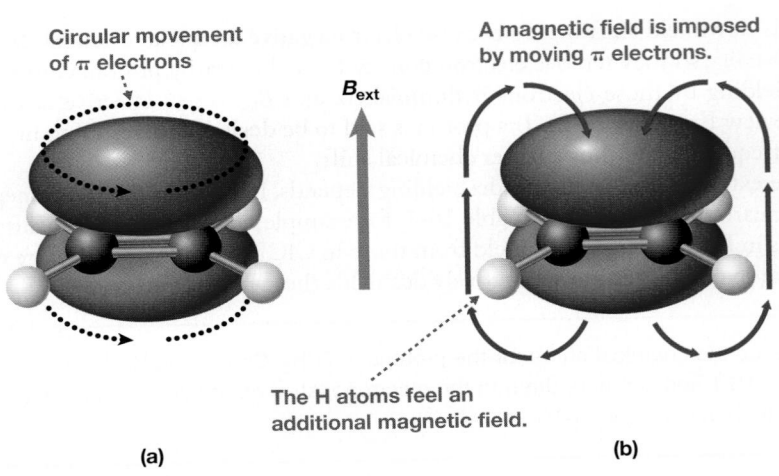

FIGURE 16-11 Deshielding of vinyl hydrogens (a) The B_{ext} forces the π electrons of a double bond to move in a circular trajectory above and below the plane of the double bond, creating ring current (black dotted arrows). (b) Ring current gives rise to a local magnetic field (blue arrows) that is in the same direction as the B_{ext} at the location of the hydrogen atoms. The hydrogen nuclei are therefore deshielded.

Circular movement of π electrons

A magnetic field is imposed by moving π electrons.

B_{ext}

The H atoms feel an additional magnetic field.

(a)   (b)

This phenomenon applies to other aromatic rings, too, not just the benzene ring. That is,

> Protons on the outside of an aromatic ring are substantially deshielded as a result of the local magnetic field induced by the ring current.

Just like the H atoms in benzene, vinyl H atoms (C=C—H) have relatively high chemical shifts at around 5 ppm, despite not having any highly electronegative atoms. Although the π electron density is not donut shaped, the π electrons still undergo a coherent circular motion on either side of the molecular plane, causing an increase in B_{eff} where the vinyl H atom is located (Fig. 16-11). Consequently,

> Magnetic anisotropy deshields hydrogen nuclei associated with simple alkenes and aldehydes.

We might also expect alkyne hydrogens (RC≡CH) to have relatively high chemical shifts, but they do not. Their chemical shifts tend to be relatively low, around 2.4 ppm, indicating that they are much less deshielded than alkene hydrogens. They are less deshielded because the electron density of a triple bond has *cylindrical symmetry* about the bonding axis (Fig. 16-12). As a result, the B_{ext} forces electrons in a triple bond to move in a circular path around the bonding axis. As shown in Figure 16-12b, an alkyne proton lies where B_{loc} opposes B_{ext}, which results in a shielding effect, not a deshielding effect.

FIGURE 16-12 Cylindrical symmetry of a triple bond The two π bonding MOs of ethyne (HC≡CH) are shown. The electron density of a triple bond has cylindrical symmetry about the bonding axis. When placed in a magnetic field, the motion of those electrons (black dotted arrow) generates magnetic field lines (blue arrows) that shield the hydrogen nuclei.

B_{ext} B_{ext}

Movement of electrons

B_{loc}

Magnetic field lines from the moving electrons

(a)   (b)

In Figure 16-12, draw an arrow on top of each hydrogen atom, indicating the direction of the local magnetic field there. (*Hint*: Each arrow should be either with or against B_{ext}.)

16.8 Trends in Chemical Shift

Table 16-1 provides chemical shifts of representative protons; it is not an exhaustive table of chemical shifts. How do we deal with protons in molecules that don't closely match the ones in the table?

We can take advantage of two useful trends. The first can be stated as follows:

The effects that cause deshielding are essentially additive.

We can see the additivity of inductive effects by examining the chemical shifts of the protons in the following molecules:

Increasing number of electronegative atoms

Increasing chemical shift

$\delta =$ 0.9 ppm   3.1 ppm   5.3 ppm

Notice that the chemical shift increases by roughly the same amount as each Cl atom is added to the C bearing the H.

Will the chemical shift of the proton be greater in CH_2Cl_2 or $CHCl_3$?

problem **16.14** Rank the indicated protons in order of increasing chemical shift.

The second useful trend is as follows:

Deshielding effects fall off very rapidly with distance.

When the atom or group of atoms that causes deshielding is more than two carbon atoms away, the effect on chemical shift is generally very small.

In ethylbenzene, for example, the protons that are directly attached to the aromatic ring are deshielded the most—their chemical shift is around 7.2 ppm. The proton on C1 of the ethyl group is farther away from the aromatic ring, so it is deshielded less— its chemical shift is 2.6 ppm. And the proton on C2 is the farthest away from the ring, so it is deshielded the least. In fact, notice that the chemical shift of the proton on C2 is not much greater than 0.9 ppm, the chemical shift of a CH₃ group that is part of an alkane.

- **Closest to the aromatic ring**
- **Most deshielded** δ = 7.2 ppm

- **Farther from the aromatic ring**
- **Less deshielded** δ = 2.6 ppm

- **Farthest from the aromatic ring**
- **Least deshielded** δ = 1.2 ppm

Ethylbenzene

SOLVED problem **16.15** Rank protons **D–H** in the molecule at the right from largest chemical shift to smallest.

Think What groups present are responsible for deshielding nearby protons? Do those groups deshield protons to the same extent? How severely do those deshielding effects fall off with distance?

Solve There are two groups present that cause significant deshielding: the C=O group and the Cl atom. From Table 16-1, we can see that the C=O group has a much greater effect on deshielding, because the chemical shift of an aldehyde proton (H—C=O) is around 9–10 ppm, whereas that of a proton on a C that is attached to a Cl atom is around 3 ppm. These deshielding effects fall off very quickly with distance, so proton **D** will have the highest chemical shift, followed by **G**. Proton **E** is one additional carbon atom away from the C=O group, and protons **F** and **H** are one additional bond away from the Cl atom. Because of the greater deshielding effects by the C=O group, proton **E** will have a higher chemical shift than protons **F** or **H**. Finally, protons **F** and **H** will be deshielded similarly by the Cl atom, but because protons **F** are closer to the C=O group, they will be deshielded slightly more. The chemical shifts of these protons, then, decrease in the order: **D > G > E > F > H**.

---

problem **16.16** Rank protons **I–M** in the molecule at the right from largest chemical shift to smallest.

The integral trace is horizontal over the baseline.

The integral trace rises as it traverses a peak.

FIGURE 16-13 An integral trace
A proton signal is represented by the orange peak. The blue "stairstep" is the integral trace, which represents the cumulative area under the peak (represented by the shading) going from left to right.

16.9 Integration of Signals

One of the most important features of NMR spectrometers is their ability to compute the *area under an absorption peak*, called the signal's **integration**. In modern instruments, the integration is done digitally, and the results are often shown as an **integral trace** superimposed on the spectrum, as shown in blue in Figure 16-13.

We can think of the integral trace as a graphic representation of the *cumulative area* under a peak upon going from left to right in the spectrum. In the regions of the spectrum where absorption is zero—that is, over the baseline—the integral trace appears as a horizontal line. As a peak is traversed, the cumulative area rises, and so does the height of the integral trace. Thus, the integral trace for a single peak resembles a stair step.

The integration represents six protons.

The integration represents four protons.

8 7 6 5 4 3 2 1 0
Chemical shift (ppm)

The importance of signal integration stems from the following rule:

> The area under an absorption peak is proportional to the number of protons that generate that peak.

Consider, for example, the ^{1}H NMR spectrum of 4-methyltoluene (*p*-xylene) shown in Figure 16-14. The integral trace is drawn in blue.

The signal at 2.3 ppm represents the six CH_3 protons, and the signal at 7.1 ppm represents the four aromatic protons. The ratio of these numbers of protons is 6:4, or 1.5:1, which is essentially the same as the ratio of the stair-step heights.

16.9 YOUR TURN

Verify that the integrations of the two signals in Figure 16-14 have a ratio of 1.5:1. To do so, use a ruler to measure the height of each stairstep (e.g., in millimeters). Is the one on the right roughly 1.5 times the height of the one on the left?

problem **16.17** Which signal in this proton NMR spectrum represents the greatest number of protons? Which signal represents the least? Roughly how many times more protons contribute to the former signal than to the latter?

8 7 6 5 4 3 2 1 0
Chemical shift (ppm)

Often, it is convenient to convert the output from integration to the smallest set of whole-number ratios because the number of hydrogen atoms each signal represents *must* be a whole number. The actual number of hydrogen atoms each signal represents is either the same as that in the ratio or a multiple of it. For example, suppose that an NMR spectrum contains three signals and the stairsteps measure 10 mm, 10 mm, and 15 mm, respectively. We can divide each height by the smallest number, yielding 1:1:1.5. The smallest set of whole number ratios is double that, or 2:2:3. The actual number of hydrogen atoms contributing to each signal could be 2, 2, and 3; or 4, 4, and 6; or 6, 6, and 9; and so on.

In spectral problems involving NMR, the smallest set of whole number ratios is sometimes given to you. If the integration is from a proton NMR spectrum, it is common to place an "H" after the numbers. A specific signal might be said to have an integration of 1 H, 2 H, 3 H, and so on.

---

problem **16.18** If the molecule giving the spectrum in Problem 16.17 contains a total of six protons, then how many protons does each signal represent?

16.10 Splitting of the Signal by Spin–Spin Coupling: The *N* + 1 Rule

1,1,2-Trichloroethane (Cl_2CHCH_2Cl) has only two chemically distinct H atoms: the lone H on C1 and the two H atoms on C2. Thus, the ^1H NMR spectrum of 1,1,2-trichloroethane, shown in Figure 16-15, has two signals—one corresponding to a chemical shift of roughly 4.0 ppm and the other corresponding to a chemical shift of roughly 5.8 ppm. Notice in the magnifications, however, that each signal is *split* into more than one peak.

Why are these signals split into multiple peaks, and how can we predict these splitting patterns? We answer these questions here in Section 16.10.

16.10a Spin–Spin Coupling and the *N* + 1 Rule

This **signal splitting** occurs because every proton behaves like a tiny bar magnet that can be in either the α state or the β state. The magnetic field that any particular proton experiences (i.e., B_{eff}) can therefore be altered by the magnetic fields of nearby protons, and these alterations can affect the proton's chemical shift. In these situations, protons are said to interact with each other through **spin–spin coupling**, and the protons involved are said to be **coupled** to each other. In general,

> Protons that are coupled are generally separated from each other by three or fewer single bonds.

(In some cases, protons that are separated by more than three bonds can exhibit coupling. This situation is discussed in Section 16.11.)

FIGURE 16-15 Proton NMR spectrum of 1,1,2-trichloroethane The signal at δ ≈ 4.0 ppm is split into a doublet and the one at δ ≈ 5.8 ppm is split into a triplet. Magnifications (insets) show these splitting patterns more clearly.

Relatively simple splitting patterns like the ones in Figure 16-15 can be predicted by the $N + 1$ rule:

> A proton that is coupled to N protons that are distinct from itself, but equivalent to each other, will give rise to a signal that is split into $N + 1$ peaks.

Various types of splitting patterns that arise from the $N + 1$ rule are summarized in Table 16-2. When there are zero, one, two, three, or four coupled protons, the signal appears as a **singlet** (s), **doublet** (d), **triplet** (t), **quartet** (q), or **quintet** (qn), respectively. With a significantly greater number of coupled protons, the splitting pattern is described as a **multiplet** (m).

Notice how the splitting patterns in Figure 16-15 are consistent with the $N + 1$ rule. The H that is bonded to C1 is coupled to the two H atoms bonded to C2, so $N + 1 = 3$. Therefore, the signal that arises from the proton on C1 ($\delta \approx 5.8$ ppm) is a triplet. The H atoms that are bonded to C2 are coupled to the lone H bonded to C1, so $N + 1 = 2$ for the corresponding signal ($\delta \approx 4.0$), giving rise to a doublet.

TABLE 16-2 Common Splitting Patterns and Relative Peak Heights

Number of Coupled Protons, N	Number of Peaks in Splitting Pattern, $N + 1$	Description of Splitting Pattern	Relative Peak Heights
0	1	Singlet (s)	1
1	2	Doublet (d)	1:1
2	3	Triplet (t)	1:2:1
3	4	Quartet (q)	1:3:3:1
4	5	Quintet (qn)	1:4:6:4:1
Several	Several	Multiplet (m)	N/A

16.10 YOUR TURN

How many protons are coupled to the one that gives rise to each of the signals at the right? Explain.

(a)   (b)   (c)

Keep in mind that, according to the $N + 1$ rule, the protons that are responsible for splitting a signal must be distinct from the protons that give rise to the signal. In other words,

> Chemically equivalent protons do not show the effects of being coupled together.

This is why the six H atoms of benzene give rise to a singlet (Fig. 16-5). If the adjacent protons split each other's signal, then benzene's ^{1}H NMR spectrum would be quite a bit more complex.

SOLVED problem 16.19 An unknown compound produces a ^{1}H NMR spectrum that has just two signals, each of which is a singlet. Which of the compounds, **1–4**, could the unknown be?

Think How many chemically distinct protons are indicated by the spectrum? Which of those protons are coupled to other protons that are chemically distinct from themselves?

Solve The unknown has two chemically distinct protons, because there are only two signals in the spectrum. This rules out **4**, because all six of its protons are chemically equivalent. Each of the protons in the unknown is not coupled to other distinct protons, either, because both signals are singlets. This rules out **1** and **2**, whose two chemically distinct protons are coupled to each other. In **1**, each signal will be split into a triplet by the $N + 1$ rule, because each CH_2 proton is coupled to two protons distinct from itself—those of the adjacent CH_2 group. In **2**, the CH_2 signal will be split by the CH_3 protons into a quartet ($N = 3$), whereas the CH_3 signal will be split by the CH_2 protons into a triplet ($N = 2$). In **3**, on the other hand, both signals will remain singlets. The CH_3 protons are separated from any other protons by at least four bonds, so they remain uncoupled. Additionally, each pair of CH_2 protons is on adjacent C atoms (and is therefore separated by only three bonds), but their signals remain singlets because all four of those protons are chemically equivalent. Therefore, the spectrum is consistent with **3**.

problem **16.20** Determine the number of signals that would be generated in the ^{1}H NMR spectrum of each of the following compounds, and predict the splitting pattern of each signal.

(a)　　　　(b)　　　　(c)　　　　(d)

The origin of the $N + 1$ rule can be understood by considering the various magnetic fields (B_{loc}) generated by a set of protons responsible for splitting a particular signal. Figure 16-16, for example, shows how a doublet arises when a signal is split by one coupled proton. The peak at the greater chemical shift (on the left) is produced when the coupled proton is in the α state, which increases B_{eff} for the proton giving rise to the signal. The peak at the smaller chemical shift is produced when the coupled proton is in the β state, which causes a decrease in B_{eff}.

Figure 16-17 shows how a triplet arises when a signal is split by two coupled protons. The peak on the left is produced when both coupled protons are in the α state, which causes an increase in B_{eff}. The peak on the right is produced when both coupled protons are in the β state, which causes a decrease in B_{eff}. And the peak in the middle is produced when one coupled proton is in the α state and one is in the β state, resulting in no change to B_{eff}.

Finally, Figure 16-18 shows how a quartet arises when a signal is split by three coupled protons, which can produce four different magnetic fields: one in which all three coupled protons are α, a second in which there is one excess α proton, a third in which there is one excess β proton, and a fourth in which all three protons are β. These scenarios correspond to decreasing magnitudes of B_{eff} and, therefore, decreasing chemical shift.

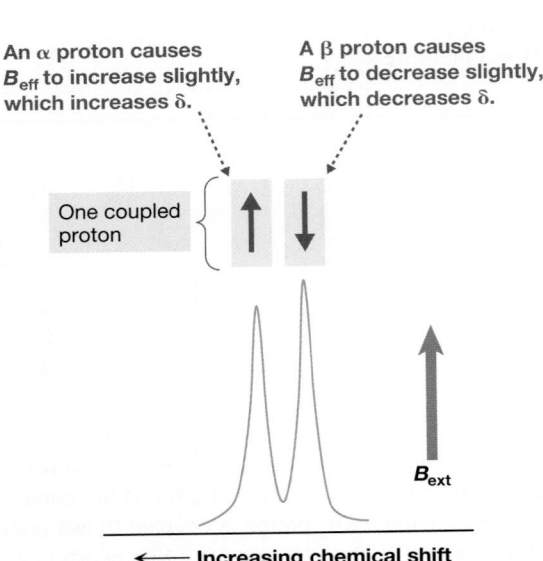

An α proton causes B_{eff} to increase slightly, which increases δ.

A β proton causes B_{eff} to decrease slightly, which decreases δ.

One coupled proton

B_{ext}

← Increasing chemical shift

FIGURE 16-16 **The origin of a doublet** A doublet arises when a signal is split by one coupled proton. The peak at the greater chemical shift (on the left) is due to the coupled proton in the α state, which causes B_{eff} to increase. The peak at lower chemical shift (on the right) is due to the coupled proton in the β state, which decreases B_{eff}.

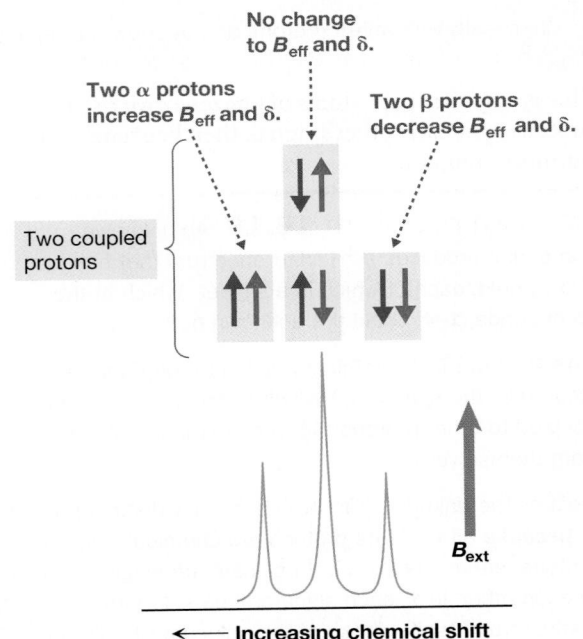

No change to B_{eff} and δ.

Two α protons increase B_{eff} and δ.

Two β protons decrease B_{eff} and δ.

Two coupled protons

B_{ext}

← Increasing chemical shift

FIGURE 16-17 **The origin of a triplet** A triplet arises when a signal is split by two coupled protons. The peak on the far left is due to both coupled protons in the α state, the one on the right is due to both coupled protons in the β state, and the one in the middle is due to one coupled proton in each state.

Notice in Table 16-2 that the peaks in each splitting pattern appear in characteristic height ratios. The peak heights of a doublet are in a 1:1 ratio. Those of a triplet are in a 1:2:1 ratio, and those of a quartet are in a 1:3:3:1 ratio. The reason for these patterns has to do with the fact that each spin combination of a set of coupled protons is nearly equally likely. Therefore, the more spin combinations that contribute to an individual peak in a splitting pattern, the larger the peak. One spin combination contributes to each peak of a doublet (Fig. 16-16), which is why the peaks are roughly equal in height. For a triplet (Fig. 16-17), one spin combination contributes to each of the outer peaks, and two contribute to the middle peak, which is why the middle peak has roughly twice the height. And for a quartet (Fig. 16-18), one spin combination contributes to each outer peak, and three contribute to the inner peaks, giving rise to the 1:3:3:1 height ratio.

These ratios describing the relative peak heights in splitting patterns form a pattern known as **Pascal's triangle** (Fig. 16-19). The first and last numbers in each set of numbers is 1. Each of the remaining numbers is the sum of the two closest numbers in the previous line. For example, consider the relative peak heights for a quintet, 1:4:6:4:1. As shown in Figure 16-19, 4 is the sum of 1 and 3, the two closest numbers in the previous line, and 6 is the sum of the two 3s that appear in the line above it.

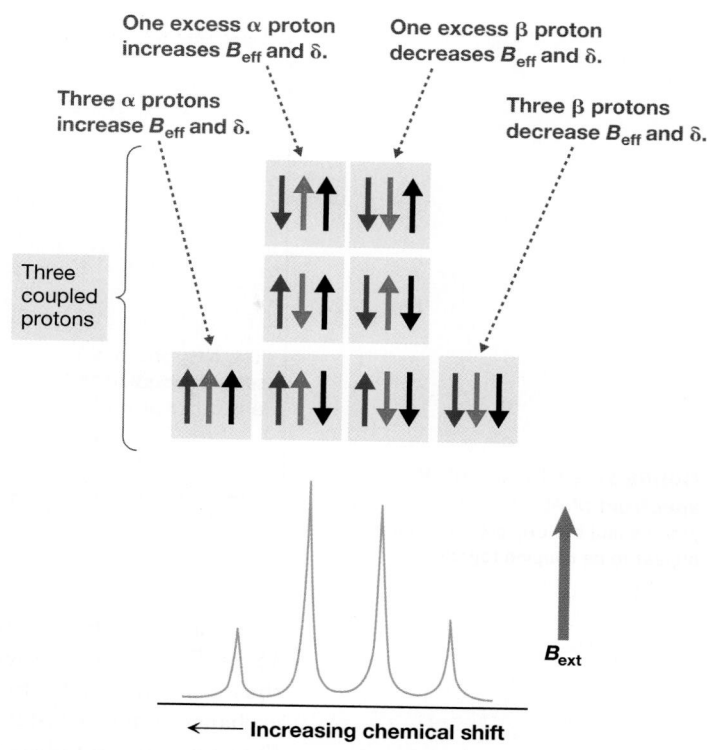

FIGURE 16-18 The origin of a quartet A quartet arises when a signal is split by three coupled protons. From left to right, the four peaks are due to the three coupled protons in the following states: all three α, two α and one β, one α and two β, and all three β.

SOLVED problem **16.21** In what ratios would the peaks of a sextet (a signal with six peaks) appear?

Think What should the first and last numbers in the set of ratios be? How is each remaining number derived from the line above it in Pascal's triangle?

Solve The first and last numbers should be 1. In Pascal's triangle, the ratios for a sextet would appear just below those for a quintet. So, each remaining number in the sextet's ratios is computed by adding each adjacent pair of numbers in the quintet's ratios. That is, $1 + 4 = 5$, $4 + 6 = 10$, $6 + 4 = 10$, and $4 + 1 = 5$. Therefore, the ratios are 1:5:10:10:5:1.

Pascal's triangle

Singlet			1		
Doublet			1 1		
Triplet		1	2	1	
Quartet		1	3 3	1	
Quintet	1	4	6	4	1

4 is the sum of 1 and 3   6 is the sum of 3 and 3

FIGURE 16-19 Pascal's triangle The numbers in each row correspond to the height ratios in various splitting patterns.

---

problem **16.22** In what ratios would the peaks of a septet (a signal with seven peaks) appear?

16.10b Decoupling of OH and NH Protons

The $N + 1$ rule seems to be violated in the ^1H NMR spectrum of ethanamine, which is shown in Figure 16-20. The amine (NH_2) protons ($\delta \approx 3.7$ ppm) and the CH_2 protons ($\delta \approx 2.7$ ppm) don't appear to be coupled together, even though they are separated by only three bonds. The NH_2 protons give rise to a broad singlet, so they appear to be uncoupled to any protons. Moreover, the CH_2 protons appear to be coupled only to the three CH_3 protons, making the signal a quartet. This phenomenon is not limited to ethanamine, but rather occurs commonly with protons bonded to oxygen or nitrogen.

O—H and N—H protons generally give rise to broad singlets, and thus do not appear to couple to other protons.

FIGURE 16-20 Proton NMR spectrum of ethanamine The NH$_2$ protons and the CH$_2$ protons do not appear to be coupled together.

The NH$_2$ protons are not coupled to any other protons.

The reason for this is rapid *proton exchange*, which we first encountered in Section 15.5a. Protons on nitrogen and oxygen undergo extensive hydrogen bonding, and thus can hop rapidly from one nitrogen or oxygen atom to another. This proton exchange is much faster than the time it takes to acquire an NMR spectrum, so the signal that we observe reflects the *average* magnetic environment from all the molecules on which it spends time, resulting in an unsplit singlet. Moreover, protons bonded to oxygen or nitrogen have less well-defined chemical environments than protons bonded to carbon, so they tend to absorb a larger range of frequencies, which also broadens their signals.

We can take advantage of this proton exchange to identify O—H or N—H signals. If a sample is treated with D$_2$O, the sample's O—H and N—H protons are rapidly replaced by D (i.e., by ^{2}H). Because deuterium's nucleus differs from hydrogen's by one neutron, the two types of nuclei absorb at substantially different frequencies, so deuterium signals do not appear in ^{1}H NMR spectra. Thus, the intensities of the O—H and N—H signals decrease dramatically, or disappear entirely.

problem 16.23 What similarities would you expect between proton NMR spectra of compounds **A** and **B**? What differences would you expect? How would each spectrum be affected by the addition of D$_2$O?

A

B

16.11 Coupling Constants and Spectral Resolution

Figure 16-21 shows the ^{1}H NMR spectrum of hexan-3-one taken with a 90-MHz (Fig. 16-21a) and a 300-MHz (Fig. 16-21b) spectrometer. Based on the structure of hexan-3-one, the two sets of CH$_3$ hydrogens are chemically distinct, and should give rise to two different signals, each a triplet. However, the spectrum taken with the 90-MHz spectrometer does not clearly show this. Instead, the two signals overlap, and are therefore not **resolved** from each other. In cases such as this, interpreting the spectrum can be a more formidable task. Taken with the 300-MHz spectrometer, on the other hand, the spectrum clearly shows two distinct triplets.

Overlapping
CH₃ signals

Chemical shift (ppm)

(a)

Signals are
resolved

Chemical shift (ppm)

(b)

FIGURE 16-21 Proton NMR spectra of hexan-3-one (a) Spectrum taken with a 90-MHz spectrometer. (b) Spectrum taken with a 300-MHz spectrometer. The stronger magnet resolves the signals much better.

The 90-MHz and 300-MHz ^{1}H NMR spectra of hexan-3-one demonstrate that

The **resolution** in an NMR spectrum increases as the magnet of the spectrometer becomes more powerful.

This is true primarily because the magnetic fields that cause signal splitting—that is, those produced by the various spin combinations of nearby nuclei—are independent of the B_{ext}. Therefore, regardless of the strength of B_{ext}, the frequency *difference* between peaks of a split signal, called the **coupling constant** (J), remains unchanged. By contrast, recall from Figure 16-3 that for an unsplit signal, increasing B_{ext} causes an increase in ΔE_{spin}, and hence an increase in the signal frequency.

With these two concepts in mind, we can now see how the magnet strength affects spectral resolution. At the top of Figure 16-22a, two unresolved signals are plotted, with frequency on the *x* axis. As the strength of the magnet increases (Fig. 16-22, top), the centers of two signals move apart, but the width of each triplet remains essentially the same. Thus, the signals are resolved from each other.

Recall that to convert signal frequencies to chemical shifts (which are independent of the instrument's magnet strength), we must divide those frequencies by the instrument's *operating frequency* (Equation 16-3). This is shown at the bottom of Figure 16-22. When we do this, the signal width (in units of ppm) becomes narrower with the stronger magnet.

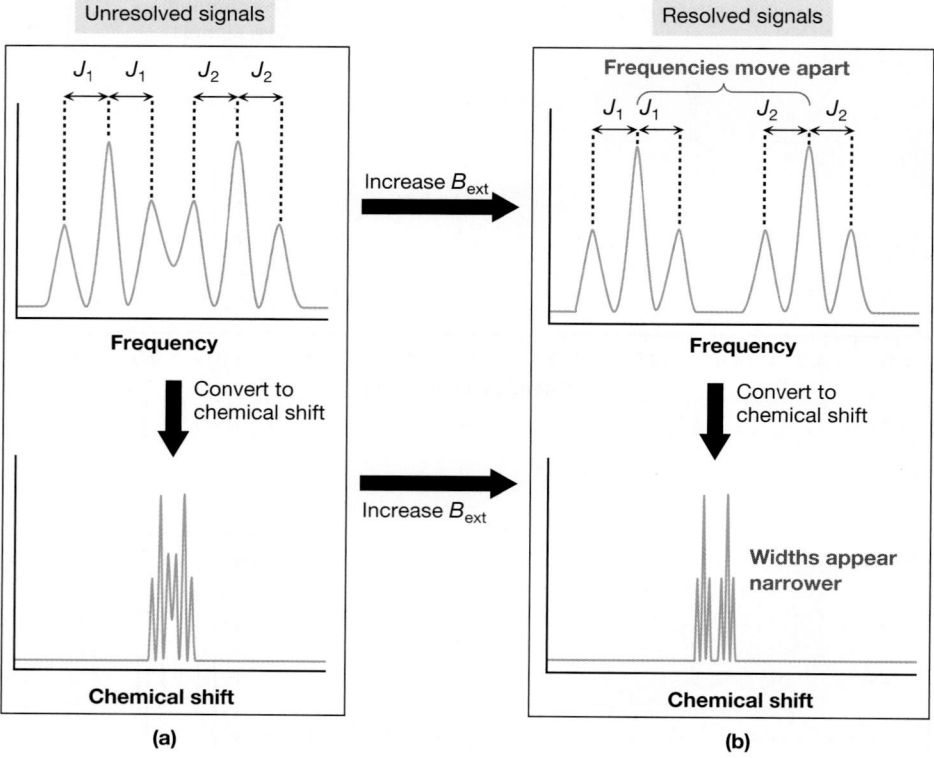

FIGURE 16-22 Dependence of spectral resolution on the spectrometer's magnet strength (a) A representation of two triplets overlapping in an NMR spectrum. The figure at the top has frequency on the x axis, whereas the figure at the bottom has chemical shift on the x axis. J_1 and J_2 are the coupling constants for the respective signals. (b) NMR spectrum of the same compound using a spectrometer with a higher magnetic field strength. The frequency separation between the two signals is increased, but the separation between peaks within the same signal (i.e., the coupling constants J_1 and J_2) remains the same. When we convert to chemical shift, therefore, the signals appear narrower.

Coupling constants can be used to interpret a ^{1}H NMR spectrum if you remember the following:

Signals of protons that are coupled together exhibit the same coupling constant.

That is, the energetic effects of one magnetic dipole on a second are the same as the energetic effects of the second magnetic dipole on the first.

For example, in the spectrum of 1-bromo-4-ethylbenzene (Fig. 16-23), there are four distinguishable signals. Each of the doublets (signals A and B) has a coupling

FIGURE 16-23 Proton NMR spectrum of 1-bromo-4-ethylbenzene Protons A and B are coupled together and their signals have the same coupling constant (i.e., J = 8.5 Hz). Protons C and D are coupled together, too, and their signals have the same coupling constant (i.e., J = 7.6 Hz).

constant of 8.5 Hz, whereas both the quartet (signal C) and the triplet (signal D) have a coupling constant of 7.6 Hz. This suggests that the protons that give rise to signal A are coupled to the protons that give rise to signal B. Likewise, the protons that give rise to signal C are coupled to those that give rise to signal D. Indeed, looking at the structure, we see that protons A and B are coupled together, and protons C and D are coupled together.

16.11 YOUR TURN

In Figure 16.23, do the coupling constants suggest that protons A and C are coupled? Does your answer agree with the molecule's structure?

16.12 YOUR TURN

In the following ^{1}H NMR spectrum, identify each pair of signals that represent coupled protons.

The *magnitude* of the coupling constant can provide structural information about the protons coupled together. As shown in Table 16-3, coupling constants generally range up to about 18 Hz.

There are two examples listed in Table 16-3 in which protons separated by more than three bonds exhibit weak coupling. These situations are said to be **long-range coupling**, and they tend to occur when the protons are connected by a rigid carbon framework.

TABLE 16-3 Commonly Encountered Coupling Constants

Relationship between Protons	Coupling Constant, J (Hz)	Relationship between Protons	Coupling Constant, J (Hz)	Relationship between Protons	Coupling Constant, J (Hz)
(H–C–C–H)	6–9	(H–C=C–H geminal alkene)	13–18	(ortho benzene H)	6–9
(H–C–C–C–H)	~0	(H–C=C cis/gem)	1–3	(meta benzene H)	1–3
(H–C=C–H cis)	7–12	(aldehyde O=C–C–H)	1–3	(para benzene H)	0–1

problem **16.24** Using the ^1H NMR spectrum provided, along with the relative frequencies of absorption noted in Hz, determine which signal(s) represent(s) protons coupled to the protons with δ = 4.1 ppm.

1654.91   1647.71
1661.99   1640.63

930.79   923.10
938.35   915.53

503.17
510.25   495.97

455.57
463.13   447.88

Chemical shift (ppm)

16.12 Complex Signal Splitting

The signal splitting we have discussed so far has been due to the coupling of a given proton to only one chemically distinct type of proton. For these cases, the $N + 1$ rule rigorously holds. It is common, however, for proton signals to be coupled to two or more chemically distinct types of protons. One of the simplest examples is bromoethene, whose ^1H NMR spectrum is shown in Figure 16-24. The $N + 1$ rule does not seem to hold for molecules of this type.

YOUR TURN 16.13

According to the $N + 1$ rule, what should the splitting pattern be for the signal of the H_C proton? Which signals, if any, in Figure 16-24 are consistent with that splitting pattern?

We observe **complex splitting** patterns in these kinds of situations because the strength of spin–spin coupling is different for each type of chemically distinct proton. This results in different coupling constants. When a given proton is coupled to two protons that are distinct from each other, therefore, two independent splitting patterns appear simultaneously.

FIGURE 16-24 Proton NMR spectrum of bromoethene The $N + 1$ rule would predict a simpler spectrum.

In the case of bromoethene, there are three chemically distinct types of protons, and each is coupled to the other two. It helps to view the coupling to each chemically distinct proton one at a time and to construct what is called a **splitting diagram** (Fig. 16-25). We begin with the signal for H_B ($\delta = 5.97$ ppm). H_B is coupled to H_C by a coupling constant of 7.1 Hz. Therefore, if we temporarily ignore the coupling between H_B and H_A, then we would expect the unsplit signal (Fig. 16-25a, top) to be split into a doublet, in which the peaks are separated by 7.1 Hz (Fig. 16-25a, middle). Now, if we take into account the coupling between H_B and H_A, for which the coupling constant is 1.8 Hz, then each of the peaks in the doublet brought about by the first coupling is split again into a doublet, this time with the peaks separated by 1.8 Hz. The result is a **doublet of doublets** (Fig. 16-25a, bottom), consistent with our observations in the spectrum.

The same splitting pattern, a doublet of doublets, is obtained if we reverse the order in which we perform the splitting. As shown in Figure 16-25b, H_B's signal is first split by H_A, and the resulting peaks are then split by H_C. It is generally easier, however, to construct these splitting diagrams by working in order from largest coupling constant to smallest.

For the same reasons, the signals of H_A and H_C are each a doublet of doublets. H_B and H_C separately split H_A's signal into a doublet. Likewise, H_A and H_B separately split the signal from H_C into a doublet. These splitting patterns can be seen in the insets for the two signals in Figure 16-24 at $\delta = 5.85$ ppm and 6.45 ppm.

problem **16.25** Construct a splitting diagram for H_A in bromoethene. Do the same for H_C. (The coupling constant between H_A and H_C is 14.9 Hz. The other relevant coupling constants are provided in the text.)

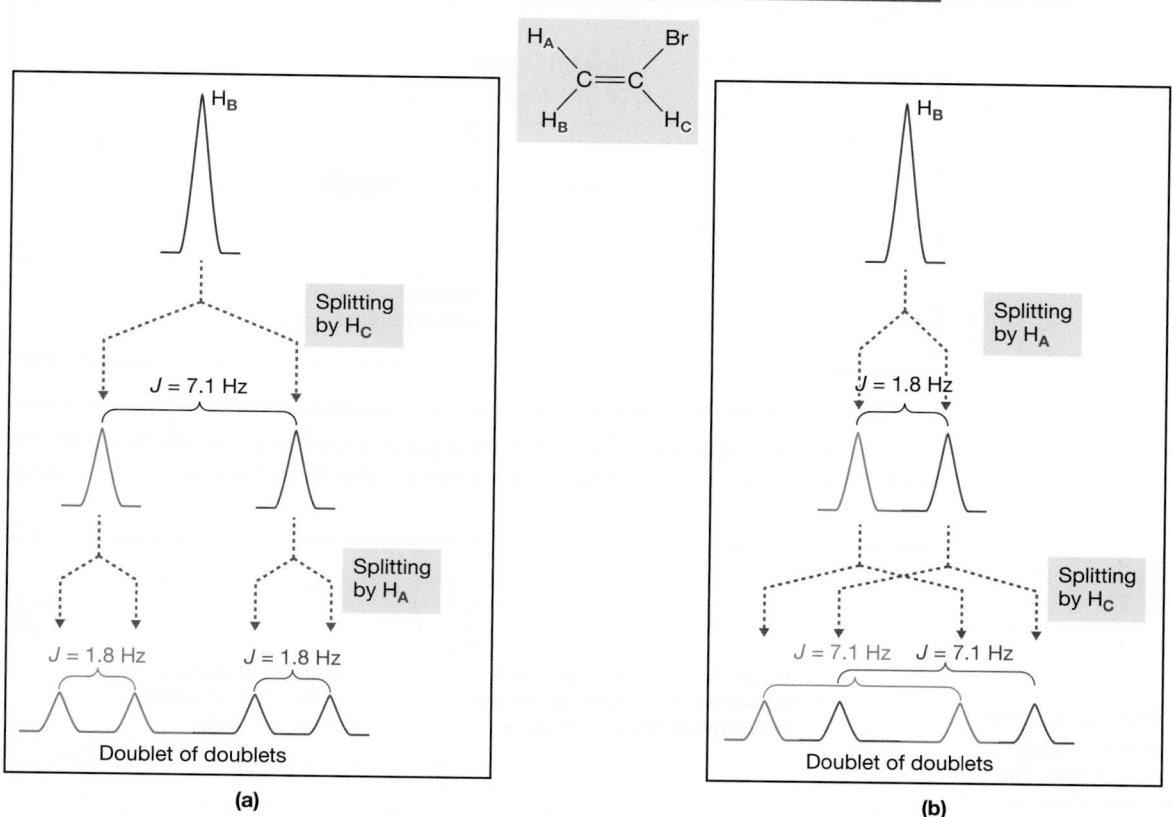

FIGURE 16-25 **Splitting diagram for proton B in bromoethene** (a) The signal from proton B is split first by proton C and then by proton A. (b) The signal from proton B is split first by proton A and then by proton C. The peak locations in the resulting doublet of doublets are independent of the order in which we think about the splitting taking place.

Splits signal of the C2 protons into a triplet

Splits signal of the C2 protons into a doublet

FIGURE 16-26 Origin of a triplet of doublets The signal from the protons on C2 (in red) is split by the protons on C3 into a triplet, and each of the three peaks is separately split by the proton on C1 into a doublet. The result is a triplet of doublets, as shown on the right.

A triplet of doublets

2.45   2.40

Another example of complex splitting appears in the ^{1}H NMR spectrum of butanal, $CH_3CH_2CH_2CH{=}O$. The signal for the protons on C2 are shown in Figure 16-26. The aldehyde proton on C1 splits the signal for the protons on C2 into a doublet. Separately, the protons on C3 split the C2 protons into a triplet. The result is six peaks in a 1:1:2:2:1:1 height ratio, or a **triplet of doublets**.

Sometimes, when two chemically distinct types of protons split the signal of the same proton, the coupling constants for the two different interactions are very nearly equal. When this is the case, the resulting splitting pattern is essentially the same as we would obtain by treating the two chemically distinct types of hydrogens as equivalent and using the $N + 1$ rule. For example, the protons highlighted on C2 in Figure 16-27 are flanked by five protons of two chemically distinct types. Theoretically, then, the signal of the protons on C2 are split twice—once by the CH_2 protons on C1 and once by the CH_3 protons on C3. In theory, this should give a **quartet of triplets** (or a **triplet of quartets**), 12 peaks in all, for the protons on C2. In actuality, however, the protons on C2 in propylbenzene give rise to a signal that appears to be split into a sextet (Fig. 16-27), consistent with the $N + 1$ rule when $N = 5$. Because of the nearly identical coupling constants, some of the peaks that would normally result from such complex splitting overlap instead, thus merging into a single peak (see Problem 16.26).

YOUR TURN 16.14

How would you describe the splitting pattern for the CH_2 proton signal in propanal? Sketch the signal, which should be similar to the one in Figure 16-26.

Splits CH_2 signal into a quartet

Splits CH_2 signal into a doublet

problem 16.26 Construct a splitting diagram similar to the one in Figure 16-25 for a quartet of triplets in which both the quartet and the triplet have the same coupling constant.

FIGURE 16-27 Splitting by distinct protons that give rise to nearly identical coupling constants The protons on C2 (in red) have about the same coupling constant with the protons on C1 as they do with the protons on C3. As a result, what is theoretically a quartet of triplets (or a triplet of quartets) actually appears as a sextet.

These protons split the signal from the C2 protons with nearly the same coupling constants.

Quartet of triplets appears as a sextet (from N+1 rule)

1.75   1.50

The figure at the right is a magnetic resonance imaging (MRI) scan of a human brain. As you may know, MRI is a valuable, noninvasive tool used in medicine to analyze tissues, and is particularly useful for diagnosing cancer. You might not know, however, that MRI is based on the same fundamental concepts that we have applied toward NMR spectroscopy.

In the chamber of an MRI scanner (see photo at the beginning of Chapter 16), a patient is exposed to a very strong magnetic field and subjected to pulses of RF radiation. As we learned here in Chapter 16, a strong magnetic field separates the nuclei of hydrogen atoms into different energy levels, which allows them to absorb RF radiation. In the body, the hydrogen atoms from water are principally responsible for absorption of that RF absorption because water is so abundant in our bodies. This RF absorption causes spin flips, producing a nonequilibrium distribution of nuclear spin. When the RF pulse is turned off, those nuclei produce an RF signal as they relax back into their equilibrium distribution, and that signal is detected. Therefore, an MRI scan can be thought of as a map of water density in the body.

Additional information obtained from the signal is the time it takes for hydrogen nuclei to relax back into an equilibrium distribution of spin states—that is, the relaxation time. This relaxation time is affected by the specific environment in which hydrogen nuclei are found, which depends, in turn, on the type of tissue in which the water is located. As a result, different types of tissues can be mapped differently in an MRI scan.

16.13 ^{13}C NMR Spectroscopy

We have focused so far on ^{1}H NMR spectroscopy, but the nucleus of the ^{13}C isotope (six protons, seven neutrons) has a spin of ½, too, just like ^{1}H. As a result, ^{13}C nuclei can absorb radiation in the RF region when placed in a strong magnetic field, and can thus produce a ^{13}C NMR spectrum. Just as ^{1}H NMR spectroscopy provides valuable information about the environments of a molecule's hydrogen atoms, **^{13}C NMR spectroscopy** provides valuable information about the environments of a molecule's carbon atoms. Section 16.13, therefore, is dedicated to explaining the theory behind ^{13}C NMR spectroscopy and to interpreting ^{13}C NMR spectra.

16.13a The ^{13}C NMR Signal

Because the ^{1}H and ^{13}C nuclei each have a spin of ½, the two types of spectroscopy have a variety of characteristics in common. In ^{13}C NMR spectroscopy, however, the frequency of the RF radiation that is used to irradiate a sample is significantly lower than in ^{1}H NMR spectroscopy, because the *gyromagnetic ratio* of a ^{13}C nucleus is about one-fourth that of a proton. This ensures that proton signals do not appear on carbon spectra, and vice versa.

One of the challenges of ^{13}C NMR spectroscopy is the low natural abundance of the ^{13}C isotope, which is only 1.1%; the remaining 98.9% of carbon is ^{12}C, which has no nuclear spin and therefore cannot absorb radiation. This results in poor signal strength, yielding noisy spectra, as shown in the ^{13}C NMR spectrum of 1-chloropropane (Fig. 16-28a). NMR instruments can combat this problem using **signal averaging**, in which multiple spectra are acquired and averaged together. Noise, because it is random, is reduced by signal averaging, but the signals themselves remain constant. Figure 16-28b shows the result of averaging several ^{13}C NMR spectra of 1-chloropropane.

(a)

Noise

Chemical shift (ppm)

TMS

Signal averaging

(b)

CDCl₃ solvent

TMS

Chemical shift (ppm)

FIGURE 16-28 ¹³C NMR spectrum of 1-chloropropane (a) Without signal averaging. (b) With signal averaging.

Just as a ¹H NMR spectrum can tell us the number of chemically distinct protons in a molecule, a ¹³C NMR spectrum can tell us the number of chemically distinct carbons.

Every chemically distinct carbon atom produces one ¹³C NMR signal.

In 1-chloropropane, there are three chemically distinct carbon atoms, which give rise to three ¹³C NMR signals. (The peak at 0 ppm is due to TMS, which is added as a reference, and the peak at 77 ppm is due to the carbon signal from the solvent, CDCl₃.)

YOUR TURN 16.15

How many chemically distinct carbon atoms are there in the compound with the following ¹³C NMR spectrum? (The signal at 77 ppm is from the solvent, CDCl₃.)

Chemical shift (ppm)

To determine the number of chemically distinct carbon atoms in a molecule, we can apply the chemical distinction test we learned for the hydrogen atom, but substitute an imaginary atom "X" for each carbon atom instead of hydrogen. This is illustrated in Solved Problem 16.27.

SOLVED problem 16.27 How many ^{13}C NMR signals would you expect for this molecule?

Think What molecules are obtained by separately replacing each C atom with X? What are the relationships among the resulting molecules?

Solve Replacing each of the carbon atoms with X, we get compounds **A–F** as follows:

Constitutional isomers, so these
C atoms are chemically distinct.

A   **B**   **C**   **D**   **E**   **F**

Identical molecules, so these
C atoms are chemically equivalent.

Compounds **A–D** are *constitutional isomers*, so the corresponding four C atoms are chemically distinct. Compounds **E** and **F** are identical to **D**, so the C atoms substituted in **E** and **F** are chemically equivalent to the one substituted in **D**. In all, there are four chemically distinct C atoms in the molecule, so we expect four ^{13}C NMR signals.

---

problem 16.28 How many ^{13}C NMR signals would you expect for the molecule at the right? Explain.

16.13b Signal Splitting

Because ^{13}C nuclei have spin, they undergo coupling with other nearby nuclei that have spin, such as protons or other ^{13}C nuclei. However, under the conditions that the spectra are acquired, we do not observe signal splitting. In other words,

> In most of the ^{13}C NMR spectra you will encounter, all of the signals will be singlets.

In the ^{13}C NMR spectrum of 1-chloropropane in Figure 16-28, for example, all three signals are singlets.

We do not observe coupling between two ^{13}C nuclei because the natural abundance of ^{13}C is so low. With ^{13}C comprising approximately 1 out of every 100 carbon atoms, the chance of having two adjacent ^{13}C atoms in a molecule is 1 in 10,000 (i.e., $\frac{1}{100} \times \frac{1}{100}$). This is insufficient to produce observable signal splitting.

We do not observe the splitting of a carbon signal by attached protons because **broadband decoupling** is usually used to essentially "turn off" the coupling between ^{13}C and ^{1}H nuclei. In this technique, a specific range of RF radiation is continuously sent through the sample, forcing all protons to flip back and forth rapidly between the α and β spin states. Therefore, the average magnetic field each proton produces is zero.

With broadband decoupling producing only singlets, we lose valuable information about the number of protons attached to each carbon. This type of information is obtained in other ways, such as with DEPT spectroscopy (see Section 16.14).

16.13c Chemical Shifts

As in ^{1}H NMR spectroscopy, ^{13}C absorptions are reported as chemical shifts, with TMS as the reference compound (its carbon atom is assigned a chemical shift of 0 ppm).

Chemical shifts for a variety of carbon atoms are listed in Table 16-4. Notice that their range is different from that for protons (see Table 16-1). That is, the chemical shifts in ^{1}H NMR generally range from 0 to 12 ppm, whereas those for carbon atoms generally range from 0 to 220 ppm.

Notice, too, that the order of the functional groups in Table 16-4 is roughly the same as the order for ^{1}H NMR chemical shifts in Table 16-1. Saturated alkane carbons have among the lowest chemical shifts (0–50 ppm); carbons of alcohols, alkyl halides, and amines have chemical shifts that are somewhat farther downfield (10–90 ppm), followed by the carbon atoms of alkenes and aromatics (100–170 ppm) and carbonyl carbons (150–220 ppm).

The trends in chemical shift are the same for both carbon and proton signals because shielding and deshielding phenomena affect both types of nuclei in roughly the same way. Sigma (σ) bonding electrons shield the carbon nucleus, and the greater the electron density contributed by those electrons, the more the carbon nucleus is shielded. Nearby

TABLE 16-4 ^{13}C NMR Chemical Shifts

Type of Carbon	Chemical Shift (ppm)	Type of Carbon	Chemical Shift (ppm)	Type of Carbon	Chemical Shift (ppm)
1. $H_3C-\underset{\underset{CH_3}{\overset{CH_3}{|}}}{Si}-CH_3$	0	6. $\underset{Br}{\overset{H_2}{\underset{}{C}}}\!\!-R$	25—35	11. >C=C<	105—150
2. $R-CH_3$	10—25	7. $\underset{Cl}{\overset{H_2}{\underset{}{C}}}\!\!-R$	35—55	12. (benzene ring)	128
3. $\underset{R}{\overset{R}{\diagdown}}CH_2$	20—45	8. $>C-NH_2$	35—40	13. $R-\overset{O}{\overset{\|}{C}}-O-H$, $R-\overset{O}{\overset{\|}{C}}-O-R$, $R-\overset{O}{\overset{\|}{C}}-N<$	160—185
4. $\underset{R}{\overset{R}{\diagdown}}\underset{|}{\overset{}{C}}H$ R	25—45	9. $>C-OH$	55—70	14. $R-\overset{O}{\overset{\|}{C}}-H$, $R-\overset{O}{\overset{\|}{C}}-R$	190—220
5. $\underset{R}{\overset{R}{\diagdown}}\underset{\underset{R}{|}}{\overset{}{C}}-R$	30—35	10. $R-C\equiv C-R$	65—85		

FIGURE 16-29 ^{13}C NMR spectrum of 2-methylbutanal Although one carbon atom contributes to each signal, the heights of the signals differ markedly.

electron-withdrawing substituents (such as halogen, oxygen, or nitrogen atoms) remove electron density from the carbon atom, thereby deshielding it. The result is a downfield shift. In addition, ring current from π electrons adds to the magnetic field experienced by the carbon atom, resulting, once again, in a downfield shift.

Identify two pairs of carbon atoms in Table 16-4 whose chemical shifts are in the same order as the chemical shifts of the analogous protons in Table 16-1.

16.16 **YOUR TURN**

---

problem **16.29** Predict which C atom in the molecule at the right would have the highest chemical shift. Explain.

16.13d Integration of ^{13}C NMR Signals

The ^{13}C NMR spectrum of 2-methylbutanal (Fig. 16-29) shows five signals, each of which is different in height. If the number of carbon atoms contributing to each signal were proportional to the area, then all five peaks would have essentially the same height, because each signal represents exactly one carbon atom in the molecule. However,

In typical ^{13}C NMR spectra, the area under a peak is *not* proportional to the number of carbon atoms contributing to that peak.

There is poor correlation because of the broadband decoupling used to produce only singlets. One of the outcomes of this technique is a magnification of each carbon signal via what is called the nuclear Overhauser effect. Unfortunately for integration, that magnification differs for different types of carbon atoms.

16.14 DEPT ^{13}C NMR Spectroscopy

In Section 16.13b, we saw that proton decoupling is used in ^{13}C NMR spectroscopy to make all signals appear as singlets. This makes it easier for us to interpret spectra, but we lose valuable information about the number of protons attached to each carbon. Fortunately, a separate experiment can be used to obtain this information:

Normal broadband-decoupled ^{13}C NMR spectrum

(a)

DEPT-90

Only the CH signals appear.

(b)

DEPT-135

Only the signals from CH or CH_3 appear as positive signals.

Only the signals from CH_2 appear as negative signals.

(c)

FIGURE 16-30 DEPT ^{13}C NMR spectroscopy (a) The normal broadband-decoupled ^{13}C NMR spectrum of ethylbenzene. (b) The DEPT-90 spectrum of ethylbenzene. Only CH carbon signals appear. (c) The DEPT-135 spectrum of ethylbenzene. CH and CH_3 carbons appear as positive signals, whereas CH_2 carbons appear as negative signals.

FIGURE 16-31 The normal broadband ^{13}C NMR spectrum of ethylbenzene The results from a DEPT experiment have been provided (see Your Turn 16.17).

distortionless enhancement by polarization transfer, or *DEPT, spectroscopy.* In **DEPT spectroscopy**, three separate spectra are acquired, as shown for ethylbenzene in Figure 16-30. One spectrum is the normal broadband-decoupled spectrum (Fig. 16-30a), in which all carbon signals appear as singlets. The other two spectra are acquired after the sample has been subjected to specific pulses sequences of RF radiation, so that the signals given off by the ^{13}C nuclei depend on the number of attached protons. In one of those spectra, called a DEPT-90 (Fig. 16-30b), the only signals are those of carbon atoms bonded to a single proton (CH). Signals of carbons with zero, two, or three protons do not appear. In the other spectrum, called a DEPT-135 (Fig. 16-30c), carbon atoms with one and three protons (CH and CH_3) produce normal signals, whereas carbons with two protons (CH_2) produce negative signals—that is, *below* the baseline. As in DEPT-90, signals from carbon atoms with no protons are absent.

With DEPT spectra, all four types of carbon atoms are readily identified. The signals of CH carbons are easily identified in the DEPT-90 spectrum. CH_2 carbons appear in the DEPT-135 spectrum as negative signals. CH_3 signals are the positive signals that appear in the DEPT-135 spectrum, but they are absent altogether in the DEPT-90. Carbons without protons give rise to signals that are present in the normal broadband-decoupled spectrum, but do not appear in either of the DEPT spectra.

Frequently, information from DEPT spectroscopy is simply summarized in the normal broadband-decoupled ^{13}C NMR spectrum, as shown in Figure 16-31 for the C and CH carbons of ethylbenzene. Providing all three spectra each time can become quite cumbersome.

16.17 YOUR TURN

In Figure 16-31, determine which peak appears as a negative signal in the DEPT-135 spectrum, due to CH_2. Determine which peak appears as a positive signal in the DEPT-135 spectrum, due to CH_3, but does not appear in the DEPT-90 spectrum.

16.15 Structure Elucidation Using NMR Spectroscopy

Here in Section 16.15, we present two problems in which a structure must be deduced from spectral information. We include our interpretation of the data provided to illustrate how the various pieces of information can be used together to arrive at a reasonable structure.

As you go through these examples, keep in mind the following information that ^{1}H and ^{13}C NMR spectra provide:

^{1}H NMR spectra provide information about:

- The total number of chemically distinct hydrogen atoms, from the total number of signals.
- The relative number of each kind of hydrogen, from the integration of those signals.
- The environment for each proton (i.e., extent of deshielding), from the proton's chemical shift.
- The number of neighboring chemically distinct hydrogen atoms, from the signal's splitting pattern.

^{13}C NMR spectra provide information about:

- The total number of chemically distinct carbon atoms, from the total number of signals.
- The relative environment for each carbon nucleus (i.e., the extent of deshielding), from the carbon atom's chemical shift.
- The number of hydrogen atoms on each carbon, from the DEPT spectra.

16.15a Unknown 1

Unknown 1, whose molecular formula is $C_{10}H_{12}O_2$, gives the ^{1}H NMR spectrum shown in Figure 16-32. A completely saturated molecule with 10 carbon and 2 oxygen atoms would have 22 hydrogen atoms, giving Unknown 1 an index of hydrogen deficiency (IHD) of 5. The signals between 7 and 9 ppm suggest the presence of an aromatic ring like a substituted benzene ring, which would account for an IHD of 4. To increase the IHD to 5, there must be another ring or a double bond.

The signals at 1.1, 1.7, and 4.3 ppm have integrations (i.e., stair-step heights) that are roughly in a 3:2:2 ratio, suggesting, perhaps, a CH_3 group and two CH_2 groups, respectively. The splitting pattern of those signals is consistent with these groups being adjacent to each other—that is, a $CH_2CH_2CH_3$ group. The CH_3 signal would be split into a triplet by the middle CH_2 group. Similarly, the signal of the CH_2 group on the left would be split into a triplet by the middle CH_2 group. And the signal of the middle CH_2 group would be split into a multiplet; if its coupling constants with the

FIGURE 16-32 Proton NMR spectrum of Unknown 1

CH$_2$ hydrogens and the CH$_3$ hydrogens are roughly the same, then we would expect a sextet as shown, given the five adjacent hydrogens (the $N + 1$ rule).

If the three signals that are farthest upfield represent three, two, and two hydrogen atoms, for a total of seven, then the aromatic hydrogens would total five, making it a monosubstituted benzene ring (i.e., C$_6$H$_5$—Sub). Assuming we have a C$_6$H$_5$ group and a CH$_2$CH$_2$CH$_3$ group, the atoms that have not been accounted for include one carbon atom and two oxygen atoms. It is unlikely that these would make a ring to give the additional IHD that is necessary, but they could contain a C=O group.

Finally, notice that one set of CH$_2$ protons has a rather high chemical shift—around 4.3 ppm. Those protons are deshielded significantly relative to a normal CH$_2$ group (Table 16-1, entry 3). A bond to a highly electronegative oxygen atom could account for it. That would give us a propoxy group, OCH$_2$CH$_2$CH$_3$.

All that is left is to assemble the pieces. The propoxy group cannot be bonded to the phenyl ring because this would complete the molecule and we would not have enough atoms to give us the specified molecular formula. Instead, each of those groups could be bonded to either side of the C=O group, giving us propyl benzoate:

Unknown 1

Write **A, B, C,** or **D** above the signals in Figure 16-32 to match up with each chemically distinct H atom in propyl benzoate.

16.18 **YOUR TURN**

16.15b Unknown 2

Unknown 2 (C$_5$H$_{12}$O) has the ^{13}C NMR (with DEPT) and IR spectra shown in Figure 16-33.

A compound with five carbon atoms and one oxygen atom is saturated when it has 12 hydrogen atoms. Unknown 2, therefore, has an IHD of 0—that is, no double bonds, no triple bonds, and no rings can appear in the structure.

There are two striking features in the spectra that we should notice right away. First, the ^{13}C NMR spectrum contains only three signals, even though our compound has five total carbon atoms. (The signals at 77 and 0 ppm are from the CDCl$_3$ solvent and the TMS standard, respectively.) This indicates that one or more of the signals represents chemically equivalent carbon atoms. The second conspicuous feature is in the IR spectrum, which unambiguously shows that our compound contains an OH group (see Your Turn 16.19). Since the IHD is 0 and there is only one oxygen atom, Unknown 2 must be an alcohol.

Identify the IR band in Figure 16-33 that indicates an OH group.

16.19 **YOUR TURN**

Knowing that the oxygen atom is part of an alcohol, we can conclude from the ^{13}C NMR spectrum that Unknown 2 is a primary alcohol, because the CH$_2$ signal is the only one that is significantly shifted downfield by the electronegativity of the oxygen atom. Furthermore, we now know that there is only one CH$_2$ group in the molecule—otherwise, the compound would have to contain more than one alcohol group, requiring more than one oxygen atom. The compound must have the general form HO—CH$_2$—(C$_4$H$_{10}$). Based on the DEPT information, moreover, the C$_4$H$_{10}$

FIGURE 16-33 ^{13}C NMR and IR spectra of Unknown 2, $C_5H_{12}O$

group must consist only of quaternary carbon atoms and CH_3 groups. To have the right number of carbons, there are only three possibilities: one CH_3 group and three carbon atoms; two CH_3 groups and two carbon atoms; or three CH_3 groups and one carbon atom. Of these choices, the only one that gives us an IHD of 0 and allows all atoms to have an octet is the last one. That is, the C_4H_{10} group is a *tert*-butyl group, $C(CH_3)_3$.

The structure of the unknown compound must therefore be:

HO—CH₂—C(CH₃)—CH₃ with H₃C and CH₃

Unknown 2

---

problem **16.30** An unknown compound has the formula $C_9H_{10}O_2$. Four signals appear in its ^{1}H NMR spectrum: (1) singlet, δ = 2.3 ppm, 6H; (2) doublet, δ = 7.0 ppm, 2H; (3) triplet, δ = 7.2 ppm, 1H; and (4) very broad singlet, δ = 12.9 ppm, 1H. Six signals appear in its ^{13}C NMR spectrum at δ = 19.4, 127.2, 128.5, 133.6, 135.3, and 170.9 ppm. Propose a structure for this compound.

---

problem **16.31** An unknown compound has the formula $C_8H_{10}O$. There are four distinct signals in its ^{1}H NMR spectrum: (1) doublet, δ = 1.4 ppm, 3H; (2) singlet, δ = 2.4 ppm, 1H; (3) quartet, δ = 4.8 ppm, 1H; and (4) overlapping signals, δ = 7.2–7.4 ppm, 5H. When the sample is treated with D_2O, the signal at 2.4 ppm disappears. Propose a structure for this compound.

16.16 Mass Spectrometry: An Overview

Mass spectrometry provides insight into the mass of a molecule and the fragments that compose it. As with UV–vis, IR, and NMR spectroscopy, this information about a molecule is obtained by interpreting a spectrum, in this case a *mass spectrum*. Unlike spectroscopy, however, mass spectrometry does *not* involve electromagnetic radiation.

Before we examine a mass spectrum, you need to understand how one is generated. One way to do so is shown in Figure 16-34. A small sample (typically on the order of ≤1 μL) is injected into the spectrometer, where it is immediately vaporized. This *vaporization* can be represented by Equation 16-5.

$$M(l) \rightarrow M(g) \tag{16-5}$$

Once vaporized, these gaseous molecules drift through a beam of fast-moving electrons. When an electron from that beam impacts a molecule of $M(g)$, an electron from $M(g)$ is knocked off, producing, a gaseous species that has one fewer electron, and therefore has a positive charge. The process of producing this **molecular ion** $[M^+(g)]$ is called **electron impact ionization** and can be represented by Equation 16-6.

$$M(g) \xrightarrow{\text{Electron beam}} M^+(g) + e^- \tag{16-6}$$

A charged species like $M^+(g)$ can be guided into a **detector** through a curved tube using a magnetic field to bend the ion's path. The extent that a given ion's path is bent depends on the strength of the magnetic field that is applied and the **mass-to-charge ratio** of the particle, represented as *m/z*. If the magnetic field strength is held fixed, then only ions of a specific value of *m/z* can reach the detector. Ions with *m/z* different from that value will collide with the wall of the tube and will be destroyed before reaching the detector—those that are too light will be deflected too much, whereas those that are too massive will not be deflected enough. If multiple ions with different *m/z* values are present, then the magnetic field strength can be adjusted to allow the various ions to be detected.

16.20 YOUR TURN

In Figure 16-34, draw in a dashed line that would represent the path of an ion that is too light to reach the detector and label it "too light." Do the same for an ion that is too massive to reach the detector and label it "too massive."

FIGURE 16-34 Schematic of a mass spectrometer An injected sample (1) is converted to vapor (2) and passed through a beam of highly energetic electrons (3). Collisions with the electrons generate positively charged ions of the entire molecule and of its fragments. These ions are accelerated (4) through a magnet that separates them (5) on the basis of their mass and charge. As the ions impact the detector, a signal is registered. Varying the magnetic field allows the spectrometer to measure the relative abundance of charged particles with different mass-to-charge ratios.

The detector itself is designed to keep track of the number of charged species that collide with it; this number is translated into what is called the **relative abundance** of the ion. The greater the number of charged species detected, the greater the ion's relative abundance. If multiple ions are produced with different values of m/z, then the magnetic field strength can be scanned to determine the relative abundance at each m/z.

The process represented in Figure 16-34 is not the only one that can be used to generate a mass spectrum. A variety of other types of mass spectrometers exist as well, each using its own particular method for generating a mass spectrum. All types of mass spectrometers, however, must have three basic components, providing the means to (1) produce gaseous ions from an uncharged sample, (2) separate ions by their m/z values, and (3) detect the relative abundance of ions having a particular value of m/z.

16.17 Features of a Mass Spectrum; Fragmentation

The **mass spectrum** of a compound is often represented as a bar graph with m/z on the x axis (where m is in atomic mass units, u) and the relative abundance of ions at each m/z on the y axis. When a mass spectrometer uses electron impact to ionize the compound (Equation 16-6), then the charge (z) on each ion that is produced is typically $+1$. In that case, then, m/z is simply m. That is,

> The x axis of a mass spectrum can usually be interpreted as the mass of each ion detected.

In turn, the mass of each ion is just the mass of the atoms that make it up, given that an electron's mass is negligible compared to the mass of a nucleus.

The mass spectrum of hexane is shown in Figure 16-35. The mass of hexane (C_6H_{14}) is 86 u (or 86 g/mol), so the peak representing the molecular ion, M^+, is found at $m/z = 86$. As a result, the mass spectrum of an unknown compound can provide the compound's molecular mass.

Notice in Figure 16-35, however, that ions with other masses are detected as well. The one with the greatest intensity ($m/z = 57$) is called the **base peak** and is assigned a relative abundance of 100%. The relative abundances of all other peaks are assigned values *relative* to that of the base peak. The base peak is *not* the same as the M^+ peak in the case of hexane, but don't be surprised if they are the same in some other compounds.

There are multiple mass peaks in hexane's mass spectrum, despite the fact that a pure sample of hexane is injected, because **fragmentation** occurs after M^+ is formed. Typically, an electron from the electron beam is moving sufficiently fast that it not only

FIGURE 16-35 Mass spectrum of hexane The M^+ peak and the base peak are identified. The values of m/z for selected peaks are provided above the peaks.

FIGURE 16-36 Fragmentation pathways of hexane's molecular ion involving the breaking of carbon–carbon bonds Each fragmentation produces a cation and an uncharged species, ensuring that the total charge after fragmentation is the same as it was before fragmentation—that is, +1.

knocks an electron off of the uncharged molecule when it collides, but also *breaks bonds!* In a molecule such as hexane, there are several different bonds that can be broken; hence, there are several different fragmentation pathways. Three of the fragmentation pathways that involve the breaking of carbon–carbon bonds are shown in Figure 16-36.

When a molecular ion undergoes fragmentation, there must be conservation of charge. As indicated in Figure 16-36, a molecular ion whose charge is +1 invariably fragments into one species bearing a +1 charge and another that is uncharged. *Only the charged fragments can be detected in mass spectrometry!*

16.21 YOUR TURN

Hexane's molecular ion can also undergo the following two fragmentation pathways (not shown in Fig. 16-36):

Both pathways involve the breaking of a C—C bond. In the brackets provided, draw the appropriate structures for the two fragment ions.

Each of the fragment ions in Figure 16-36 appears in hexane's mass spectrum and each is indicated in Figure 16-35. Thus, if we understand fragmentation pathways, then the fragment peaks that appear can help us to determine molecular structure.

Fragmentation pathways can be quite complex. We can begin to appreciate this complexity by noting that many more ion peaks appear in the spectrum of hexane than the ones we examined in Figure 16-36. The complexity of fragmentation processes is due, in part, to the fact that the pathways involve very high-energy species in the gas phase. Additionally, the molecular ion contains an odd number of electrons, making it a *radical cation*. As discussed in Chapter 25, the chemistry of radicals is somewhat different from that of *closed-shell species* in which all electrons are paired. Therefore, we leave a more in-depth discussion of fragmentation in mass spectrometry until Interchapter 2, which comes after we have discussed radicals and radical reactions more fully in Chapter 25.

---

problem **16.32** Using the mass spectrum of ethylbenzene provided, identify the M$^+$ peak and the base peak. What is the formula of the ion responsible for the base peak?

16.18 Isotope Effects: M + 1 and M + 2 Peaks

Although the M$^+$ peak of hexane appears at $m/z = 86$, there is a small peak at $m/z = 87$ in Figure 16-35, too. This peak is called the **M + 1 peak**.

YOUR TURN 16.22

Label the M + 1 peak in Figure 16-35. Also, identify and label the M + 1 peak in the mass spectrum of ethylbenzene in Problem 16.32.

The M + 1 peak represents a molecular ion containing a heavy isotope. Hexane contains only carbon and hydrogen atoms, so in this case it must be due to the appearance of either an atom of ^{2}H (deuterium) or ^{13}C. According to Table 16-5, however, the natural abundance of ^{2}H is negligible, whereas that of ^{13}C is about 1.1%, so hexane's M + 1 peak must be due primarily to an ion that contains five ^{12}C atoms, one ^{13}C atom, and 14 ^{1}H atoms.

YOUR TURN 16.23

Compute the molecular mass of $^{12}C_5{}^{13}C^1H_{14}$. How does this compare to m/z for the M + 1 peak in the mass spectrum of hexane?

The fact that an M + 1 peak appears in a spectrum is not particularly useful by itself, but the *intensity* of that peak relative to the intensity of the M$^+$ peak can be very useful. In the mass spectrum of hexane, for example, the relative intensity of the M + 1 peak is 1.0% and that of the M$^+$ peak is 15.5%. The ratio of these two peaks is $(1.0)/(15.5) = 0.065$, or 6.5%. Consider, now, the mass spectrum of dodecane, $C_{12}H_{26}$,

Element	Most Abundant Isotope	Abundance	Heavy Isotopes	Abundance[a]	Element	Most Abundant Isotope	Abundance	Heavy Isotopes	Abundance[a]
Carbon	^{12}C	98.90%	^{13}C	1.10%	Chlorine	^{35}Cl	75.77%	^{37}Cl	24.23%
Hydrogen	^{1}H	99.985%	^{2}H (D)	0.015%	Bromine	^{79}Br	50.69%	^{81}Br	49.31%
Nitrogen	^{14}N	99.634%	^{15}N	0.366%	Sulfur	^{32}S	95.02%	^{33}S	0.75%
Oxygen	^{16}O	99.76%	^{17}O	0.038%				^{34}S	4.21%
			^{18}O	0.20%	Silicon	^{28}Si	92.23%	^{29}Si	4.67%
								^{30}Si	3.10%

[a]Abundances shown in red are the heavy isotopes most likely observable in a mass spectrum.

which is shown in Figure 16-37. The relative intensity of the M + 1 peak is 0.8%, and that of the M^+ peak is 5.9%. The ratio of these intensities is (0.8)/(5.9) = 0.136, or 13.6%.

In other words, the likelihood that a molecular ion of dodecane contains a ^{13}C nucleus is twice that of a molecular ion of hexane. This is consistent with the fact that dodecane (12 C atoms) contains twice as many carbon atoms as hexane (six C atoms).

Quantitatively, the ratio of the M + 1 peak's intensity to the M^+ peak's intensity can be used to estimate the number of carbon atoms in the molecule, according to Equation 16-7. The 1.1% appears in the equation because that is the probability that a given carbon atom is ^{13}C (Table 16-5).

Number of C atoms ≈ [(intensity of M + 1)/(intensity of M^+)] × (100%/1.1%)

(16-7)

Part of the reason that Equation 16-7 can only be used as an estimate is that atoms other than carbon, such as hydrogen and nitrogen, have isotopes that contribute a small amount to the M + 1 peak.

Use Equation 16-7 to verify that hexane has six carbons and dodecane has 12.

16.24 **YOUR TURN**

FIGURE 16-37 Mass spectrum of dodecane, $C_{12}H_{26}$ The M^+ and the M + 1 peaks are labeled.

problem **16.33** Estimate the number of carbon atoms present in the compound that produced the following mass spectrum.

Based on the relative abundances in Table 16-5, carbon generally contributes significantly primarily to an M + 1 peak. Other elements, however, have relatively abundant isotopes that are 2 u heavier than their most common isotopes. These include sulfur, chlorine, and bromine. The appearance of an **M + 2 peak**, therefore, may indicate the presence of one of these elements, but which one?

With good certainty, we can identify which of these elements is present by the intensity of the M + 2 peak relative to that of the M^+ peak. For example, notice in the mass spectrum of bromobenzene (C_6H_5Br; Fig. 16-38) that the M^+ peak and the M + 2 peak have roughly equal intensities. The M^+ peak represents the molecular ion $[C_6H_5Br]^+$ that contains the ^{79}Br isotope, whereas the M + 2 peak represents the molecular ion $[C_6H_5Br]^+$ that contains the ^{81}Br isotope. The similar intensities of these peaks are consistent with the fact that the ^{79}Br and ^{81}Br isotopes have nearly equal abundances in nature (i.e., 50.69% and 49.31%, respectively).

YOUR TURN 16.25

Verify that the values of m/z for the M^+ and M + 2 peaks in Figure 16-38 are consistent with the presence of the ^{79}Br and ^{81}Br isotopes by calculating the molecular masses of both species.

Fragment ions that contain a bromine atom also appear as pairs of mass peaks of roughly equal magnitude, which differ by 2 in their values of m/z. This is quite evident in the mass spectrum of 1-bromo-4-(1-methylethyl)benzene, shown in Figure 16-39.

A similar phenomenon is apparent in the mass spectrum of chlorobenzene, shown in Figure 16-40. The M^+ peak appears at $m/z = 112$ and an M + 2 peak appears at $m/z = 114$. Because the natural abundances of the ^{35}Cl and ^{37}Cl isotopes have a 3:1 ratio (Table 16-5), these peaks also appear in a 3:1 ratio.

FIGURE 16-38 Mass spectrum of bromobenzene The M^+ and M + 2 peaks correspond to molecular ions with the ^{79}Br and ^{81}Br isotopes, respectively.

FIGURE 16-39 Mass spectrum of 1-bromo-4-(1-methylethyl) benzene Two fragments containing Br are indicated, as evidenced by pairs of peaks of roughly equal intensity, differing by two units of *m/z*.

FIGURE 16-40 Mass spectrum of chlorobenzene The M$^+$ and the M + 2 peaks are in a 3:1 ratio, consistent with the natural abundances of ^{35}Cl and ^{37}Cl, respectively.

problem **16.34** At what value of *m/z* do you expect the M$^+$ peak of C$_6$H$_5$SH to appear? At what value of *m/z* do you expect the M + 2 peak to appear? If the M$^+$ peak is also the base peak, then what intensity would you expect the M + 2 peak to have?

16.19 Determining a Molecular Formula from the Mass Spectrum

The mass spectrum of an unknown compound, Unknown 3, is presented in Figure 16-41. What is the molecular formula of the compound? The relative intensities of the peaks at *m/z* = 129 u and 130 u are 30.0% and 2.4%, respectively.

The M$^+$ and the M + 1 peaks are at *m/z* = 129 u and 130 u, respectively, so the mass of the unknown molecule is 129 u. Right away, we know that the compound must contain 10 or fewer carbons, because the mass of 11 carbons would be >129 u.

FIGURE 16-41 Mass spectrum of Unknown 3

We can more accurately determine the number of carbon atoms by comparing the intensities of the M + 1 and the M^+ peaks, which are about 2.4% and 30%, respectively. Substituting these values into Equation 16-7, we obtain a value of 7.3, so we estimate seven carbons. This accounts for a mass of 84, leaving $(129 - 84) = 45$ u yet to be accounted for.

There must be at least one more nonhydrogen atom in the unknown. Otherwise, the molecular formula would have to be C_7H_{45}, which would be impossible (and nonsensical). There is no significant M + 2 peak, so the compound must not contain bromine, chlorine, or sulfur. It can, however, contain up to two oxygen atoms (at 16 u each) or up to two nitrogen atoms (at 14 u each), while still remaining below the total mass of 129 u.

The **nitrogen rule** can help us determine whether a compound contains nitrogen and, if it does, whether it contains an odd or even number of nitrogen atoms:

The Nitrogen Rule

- A compound containing an odd number of nitrogen atoms typically has an odd molecular mass.

- A compound containing an even number of nitrogen atoms, or no nitrogen atoms at all, typically has an even molecular mass.

For example, NH_3 contains a single nitrogen atom and its mass is 17 u, whereas ethane (H_3CCH_3) and hydrazine (H_2NNH_2) contain zero and two nitrogen atoms, respectively, and have masses of 30 and 32 u.

YOUR TURN 16.26

Compute the molar mass of N,N-dimethylacetamide, CH_3—CO—$N(CH_3)_2$. Is it consistent with the nitrogen rule?

Applying the nitrogen rule to our unknown, we can conclude that it contains a single nitrogen atom. Thus, Unknown 3 contains seven carbon atoms and one nitrogen atom, leaving $(129 - 84 - 14) = 31$ u yet to be accounted for. Because 31 hydrogen atoms are not possible with only eight heavy atoms, there must be one more heavy atom. It cannot be carbon or nitrogen, but it could be oxygen or fluorine. Formulas that are consistent with this mass spectrum could therefore be $C_7H_{15}NO$ or $C_7H_{12}NF$.

---

problem **16.35** Suppose that the IR spectrum of the compound giving rise to the mass spectrum in Figure 16-41 contains a strong absorption at 1670 cm^{-1} and no absorptions above 3000 cm^{-1}. Draw three structures consistent with these data.

---

problem **16.36** An unknown compound has the mass spectrum shown here. Its ^{13}C NMR spectrum has four signals and its IR spectrum has an intense peak near 1720 cm^{-1}. Propose a structure for this compound.

Mass Spectrometry, CSI, and ER

Mass spectrometry is not just an analytical tool in chemistry; it can help investigators solve crimes and can help doctors identify and treat cancer!

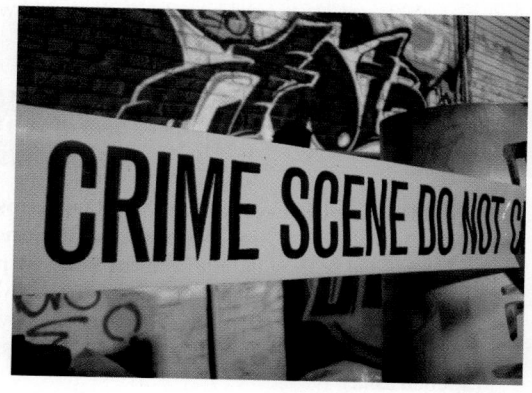

Investigating a crime involving explosives is often challenging when the explosive has been nearly entirely consumed. But a mass spectrum of the residue that is produced can help investigators trace the explosive back to its origin. When investigators suspect arson, they can use mass spectrometry to analyze the partially charred wood for the presence of trace amounts of an accelerant such as gasoline, kerosene, or mineral spirits. And scientists are looking into the possibility of using mass spectrometry as evidence that can place a criminal at the scene of a crime. A recent study showed that human scent can be analyzed by mass spectrometry to produce a unique bar code of an individual's "primary odor" compounds, making it possible to identify an individual by the odor that they leave behind.

In medicine, mass spectrometry can help diagnose brain tumors. Detecting tumors is difficult, in part, because there are >125 different kinds, and pathologists don't always agree on their diagnoses. To help identify a tumor, a spectrum produced by compounds removed from a tissue sample can be compared to a library of spectra generated from various types of tumors. Perhaps more interestingly, researchers are finding that a patient's breath can be analyzed by mass spectrometry to diagnose the specific type and stage of lung cancer with remarkable accuracy. And other breath tests are being developed for breast and colon cancers.

16.20 Wrapping Up and Looking Ahead

Here in Chapter 16 we discussed both NMR spectroscopy and mass spectrometry—two extremely valuable tools for structure elucidation. NMR spectroscopy probes aspects of specific nuclei, ultimately allowing us to map out the connectivity within a molecule. In particular, ^{1}H NMR spectroscopy provides information about a molecule's hydrogen atoms, and ^{13}C NMR spectroscopy provides information about its carbon atoms. From a ^{1}H NMR spectrum, we can determine the total number of chemically distinct types of hydrogen atoms, and for each one, we can determine the type of environment in which it resides, its abundance relative to other hydrogen atoms in the molecule, and the number of nearby hydrogen atoms. Similarly, from a ^{13}C NMR spectrum, we can determine the total number of chemically distinct carbon atoms, as well as the type of environment in which each resides. Adding the information from DEPT spectra, we can also determine the number of hydrogen atoms attached to each carbon.

Mass spectrometry can be used to determine the mass of a molecule and its fragments. Examining peaks that arise from specific isotopes—the M + 1 and M + 2 peaks—we can determine the total number of carbon atoms in a molecule, as well as whether a molecule contains certain heteroatoms, such as bromine, chlorine, or sulfur. Ultimately, a mass spectrum helps us establish a compound's molecular formula.

NMR spectroscopy and mass spectrometry complement each other, and they complement IR and UV–vis spectroscopy as well, which were introduced in Chapter 15. Depending upon the specific information that is sought, the nature of the compound, and the availability of instruments, it may be more advantageous to acquire one type of spectrum rather than another. Moreover, the information obtained from one spectrum can be combined with that from other types of spectra, often allowing us to determine unambiguously a molecule's structure.

Throughout the rest of this book, our focus shifts back to reactions and reaction mechanisms, for which our knowledge of spectroscopy is particularly useful. If we know a reaction's mechanism, for example, then spectroscopy can be used to confirm the products we might predict. Alternatively, knowing the structures of reactants and products can help us establish a reaction's mechanism. Therefore, be prepared to apply your knowledge of spectroscopy in the chapters to come.

Chapter Summary and Key Terms

- In **nuclear magnetic resonance (NMR) spectroscopy**, a sample is placed in a strong **external magnetic field (B_{ext})** and irradiated with radiation from the **radio frequency (RF)** portion of the electromagnetic spectrum. An NMR spectrum plots the intensity of each signal emitted by the sample against *chemical shift*, a quantity related to relative frequency. **(Section 16.1)**

- Absorption of an RF photon causes a nucleus to undergo a **spin flip**. In **^{1}H NMR spectroscopy**, the nuclei that undergo spin flips are those of hydrogen atoms (protons, ^{1}H); in **^{13}C NMR spectroscopy**, carbon-13 atoms undergo spin flips. **(Section 16.2)**

- Signal frequency depends on the identity of the nucleus and increases with increasing B_{ext}. **(Section 16.2)**

- Atoms that are **chemically distinct** have nuclei that reside in different **chemical environments**. Chemically distinct nuclei differ in the extent to which they are **shielded** from B_{ext} and thus generate signals at different frequencies. **(Sections 16.3 and 16.13a)**

- If nuclei in different chemical environments interchange rapidly, they tend to give rise to the same averaged signal. **(Section 16.5)**

- **Chemical shift** is a measure of a signal's frequency relative to the signal frequency generated by **tetramethylsilane (TMS)**. Whereas a signal's frequency is proportional to B_{ext}, chemical shift is independent of B_{ext}. **(Section 16.6)**

- Nearby electron-withdrawing groups **deshield** nuclei inductively, thereby increasing chemical shift. Nearby π electrons from double bonds deshield nuclei via **magnetic anisotropy**. **(Section 16.7)**

- Deshielding from inductive effects and from magnetic anisotropy is additive, and falls off rapidly with distance. A group that is separated from a nucleus by more than two bonds has little effect on the chemical shift of that nucleus. **(Section 16.8)**

- The **integration** of an NMR signal is the area under the peaks in the spectrum and is graphically represented as an **integral trace**. In a ^{1}H NMR spectrum, integration is proportional to the number of protons giving rise to that signal. **(Section 16.9)**

- According to the **$N + 1$ rule**, N protons will split the signal of an adjacent, chemically distinct proton into $N + 1$ peaks. The relative intensities of those peaks are described by **Pascal's triangle**. **(Section 16.10a)**

- Protons on oxygen or nitrogen a tend to exhibit none of the effects of coupling, appearing instead as broad singlets. **(Section 16.10b)**

- The **coupling constant (J)** of an NMR signal is the difference in frequency between adjacent peaks belonging to the same split signal. Coupling constants are independent of B_{ext}, so **resolution** improves with increasing B_{ext}. **(Section 16.11)**

- Signals of protons that are coupled together have the same coupling constant. **(Section 16.11)**

- A proton signal can exhibit **complex splitting** if the proton is coupled to two or more protons that are not equivalent to each other. These splitting patterns can be derived from a **splitting diagram**. **(Section 16.12)**

- Carbon signals are generated by ^{13}C nuclei. **Signal averaging** compensates for the low natural abundance of the ^{13}C nucleus. **(Section 16.13a)**

- In **broadband-decoupled** ^{13}C NMR spectra, all signals appear as singlets, and their integrations do *not* correlate precisely with the number of nuclei. **(Sections 16.13b and 16.13d)**

- **DEPT ^{13}C NMR spectroscopy** gives the number of hydrogen atoms on each carbon. **(Section 16.14)**

- A **mass spectrum** plots the relative abundance of gaseous ions against the **mass-to-charge ratio, m/z**. **(Section 16.17)**

- A molecule's mass can be ascertained from the m/z of the **molecular ion, M^+**. **Fragmentation** gives rise to peaks with smaller m/z values. **(Sections 16.16 and 16.17)**

- The presence of ^{13}C gives rise to an **$M + 1$ peak**. The total number of carbon atoms in a molecule can be computed from the intensity of the $M + 1$ peak relative to that of the M^+ peak. **(Section 16.18)**

- The presence of bromine, chlorine, or sulfur can be identified by the relative intensity of an **$M + 2$ peak**. **(Section 16.18)**

Problems

16.37 For a particular NMR instrument, the operating frequency is 300 MHz for ^{1}H NMR spectroscopy and 75 MHz for ^{13}C NMR spectroscopy. Calculate the gyromagnetic ratio for a ^{13}C nucleus.

16.38 Using the NMR instrument described in Problem 16.37, suppose that a ^{13}C nucleus from a sample generates a signal whose frequency is 11,250 Hz higher than that from the carbons in TMS. What is the chemical shift of that carbon atom from the sample?

16.39 Equation 16-4, shown again below, relates the operating frequency of an NMR spectrophotometer (ν_{op}) to the gyromagnetic ratio of the nucleus of interest (γ) and the magnitude of the external magnetic field (B_{ext}).

$$\nu_{op} = \gamma B_{ext}/2\pi$$

Show how this equation is derived from Equation 16-1 and the condition necessary for photon absorption: $E_{photon} = \Delta E_{spin}$.

16.40 A compound, whose formula is $C_{11}H_{14}$, produces a ^1H NMR spectrum with four signals. The steps made by the integral trace measure 37, 9, 26, and 52 mm. How many protons give rise to each signal?

16.41 A compound generates two ^1H NMR signals. The frequency of the first is 450 Hz higher than that of the signal from TMS, whereas the frequency of the second is 755 Hz higher than that of TMS. Which signal corresponds to the protons that are shielded to a greater extent? Explain.

16.42 A ^1H NMR spectrum exhibits the following set of absorption peaks:

Can those peaks be generated by one chemically distinct type of proton? Why or why not?

16.43 For each of the following molecules, determine how many signals should appear in its ^1H NMR spectrum.

(a) (b) (c) (d) (e)

(f) (g) (h) (i)

16.44 For each of the molecules in Problem 16.43, determine how many signals should appear in its ^{13}C NMR spectrum.

16.45 Explain why TMS has a *lower* chemical shift than its carbon analog, dimethylpropane, $(CH_3)_4C$.

16.46 Determine the splitting pattern for each type of H highlighted in the following molecules. (Ignore long-range coupling.)

(a) (b) (c)

(d) (e) (f)

16.47 ^{13}C NMR spectra of three isomers of C_8H_{18} are provided. Match each of those spectra to one of the following compounds: octane; 2,5-dimethylhexane; and 4-methylheptane. (Remember that the signal at ~77 ppm is due to the solvent.)

16.48 A compound containing only carbon, nitrogen, oxygen, and hydrogen is determined to contain four carbon atoms. If the M^+ peak in its mass spectrum appears at $m/z = 87$, then how many nitrogen atoms does it contain?

16.49 Is the compound giving rise to the following mass spectrum more likely to contain sulfur, bromine, or chlorine? Explain.

16.50 A student wants to determine if the following S_N2 reaction occurs.

(a) Why would this experiment require isotopically labeled compounds?
(b) Suggest how she can answer this question using mass spectrometry.

16.51 A compound has a formula of $C_{19}H_{16}$. Five signals appear in its ^{13}C NMR spectrum—one at 57 ppm and the remaining four between 126 and 144 ppm. In its ^1H NMR spectrum, an unresolved multiplet (15 H) appears between 6.9 and 7.44 ppm and a singlet (1 H) appears at 5.5 ppm. What is the structure of this compound?

16.52 The ^{1}H NMR spectrum of an unknown compound is shown below. In the compound's mass spectrum, the M$^+$ peak appears at m/z = 92 u. An M + 2 peak, whose intensity is roughly one-third that of the M$^+$ peak, also appears. Draw the structure of this compound.

Chemical shift (ppm)

16.53 (a) Draw the structures of the species that correspond to each of the peaks **A–E** in the following mass spectrum of heptane.

    (b) Identify the M$^+$ peak, the M + 1 peak, and the base peak.

16.54 For each of the following mass spectra, determine the formula of a compound that can produce it.

A

B

16.55 Determine the formula of a compound that can give rise to the following mass spectrum.

16.56 For each of the following compounds, sketch both a ^{1}H NMR spectrum and a ^{13}C NMR spectrum. In the ^{1}H NMR spectrum, pay attention to the splitting patterns and include the integral trace. (Ignore long-range coupling.)

(a) (b) (c)

16.57 Estimate the ^{1}H NMR chemical shift for the protons on C2 in the following compound.

16.58 Do you think ^{1}H NMR or ^{13}C NMR spectroscopy would be more suitable for distinguishing among the following three compounds? Explain.

A B C

16.59 How would the ^{1}H NMR spectra of compounds **D** and **E** differ?

D E

16.60 The chemical shifts of two carbon atoms are given for the following molecules:

125 ppm ····· 138 ppm   153 ppm ····· 130 ppm

It appears that the addition of the OH group *increases* the chemical shift of the C atom to which it is attached, but that it *decreases* the chemical shift of the C atom on the opposite side. How do you account for these observations? (*Hint*: Draw all of the pertinent resonance structures.)

16.61 A compound whose formula is $C_9H_{18}O$ produces a ^{1}H NMR spectrum that contains only one signal—namely, a singlet that appears at 1.25 ppm. What is the structure of the compound?

16.62 The formula of a compound is $C_9H_{10}O_3$. Use the IR, ^{1}H NMR, and ^{13}C NMR spectra provided to determine its structure.

16.63 The formula of a compound is C_9H_8O and its IR and ^1H NMR spectra are provided. If its ^{13}C NMR spectrum has seven signals, then what is the compound's structure?

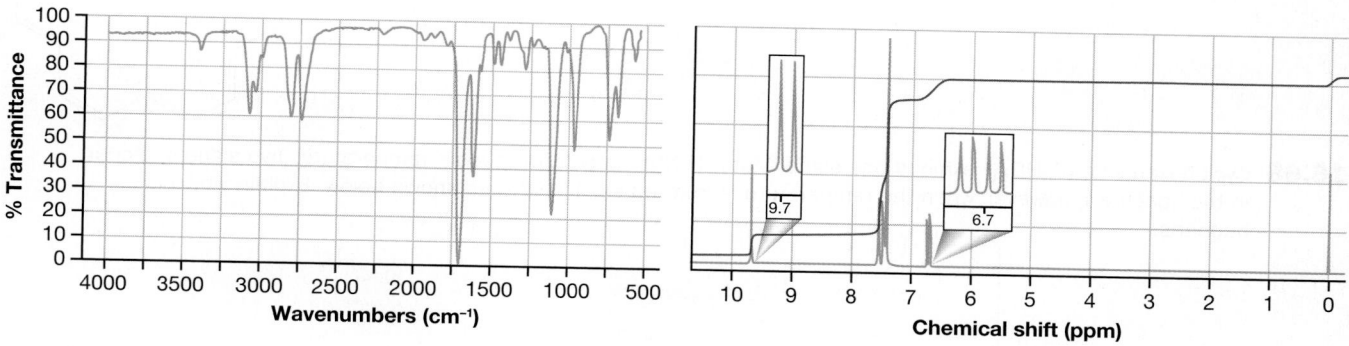

16.64 A compound with the formula $C_4H_{10}O_2$ has the following IR, ^1H NMR, and ^{13}C NMR spectra. Determine the structure of the compound.

16.65 The chemical shifts of the α protons on cyclohexanone and cyclobutanone are as at the right:
(a) Which α carbon has a greater effective electronegativity?
(b) Can you explain why, using arguments of s character and p character?

Cyclohexanone   Cyclobutanone

16.66 At room temperature, N,N-dimethylformamide, HCON(CH$_3$)$_2$, has three ^{1}H NMR signals, appearing at 2.9, 3.0, and 8.0 ppm. As the temperature is increased, the two signals at 2.9 and 3.0 ppm merge into one signal. Explain. (*Hint*: Consider the resonance structures of the compound.)

16.67 Two signals appear in the ^{1}H NMR spectrum of the compound at the right. One has twice the integration of the other. The signal with greater area corresponds to a chemical shift of 9.3 ppm. The signal with less area corresponds to a chemical shift of −2.9 ppm. Explain. (*Hint*: Consider the magnetic field lines from ring current.)

16.68 Cyclohexane-d_{11} (C$_6$HD$_{11}$) exhibits one signal in its ^{1}H NMR spectrum at room temperature. Two signals appear in the spectrum, however, when the temperature is lowered significantly, as shown below. Explain why this happens.

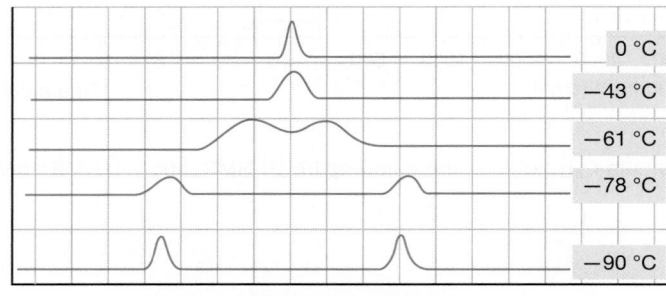

0 °C
−43 °C
−61 °C
−78 °C
−90 °C

⟵ **Chemical shift**

16.69 At room temperature, the ^{1}H NMR spectrum of all-cis 1,2,3,4,5,6-cyclohexanehexacarboxylic acid exhibits two signals for the H atoms directly bonded to the ring. Explain why. (*Hint*: Draw its chair conformation explicitly.)

16.70 Determine the structure of the compound C$_6$H$_{10}$O whose ^{1}H NMR spectrum is shown below.

Chemical shift (ppm)

16.71 Determine the structure of the compound $C_7H_{14}O_2$ whose ^1H NMR spectrum is shown below.

Chemical shift (ppm)

16.72 Determine the structure of the compound C_4H_9NO whose ^1H NMR spectrum is shown below.

Chemical shift (ppm)

16.73 Determine the structure of the compound $C_{10}H_{14}$ whose ^{13}C NMR spectrum is shown below.

Chemical shift (ppm)

16.74 A student runs the following reaction and obtains the ^1H NMR spectrum shown. Determine the product of the reaction and draw the complete, detailed mechanism of the reaction.

Chemical shift (ppm)

16.75 An alcohol was treated with HBr, yielding a mixture of 1-bromopent-2-ene and 3-bromopent-1-ene. The ^{1}H and ^{13}C NMR spectra of the starting alcohol are given here. Determine the structure of the starting alcohol and draw a complete, detailed mechanism to account for each of the products.

16.76 An alcohol was treated with concentrated sulfuric acid and heated. The product that was obtained had the following IR, ^{1}H NMR, and ^{13}C NMR spectra.

(a) Determine the structure of the alcohol.

(b) Draw a complete, detailed mechanism for the reaction that took place.

If your instructor assigns problems in smartwork, log in at **smartwork.wwnorton.com**.

CHAPTER 17

Nucleophilic Addition to Polar π Bonds 1

Addition of Strong Nucleophiles

I n Chapters 11 and 12, we examined reactions involving nonpolar π bonds, such as the C=C bond of an alkene (Fig. 17-1a). Being electron rich, a nonpolar π bond tends to undergo the addition of an electrophile, which is electron poor. Here in Chapter 17, we examine addition reactions involving *polar* π bonds—that is, π bonds joining atoms of significantly different electronegativity—represented as C=Y in Figure 17-1b. Unlike in an alkene, a polar π bond has an atom that is relatively electron poor (i.e., electrophilic), so it tends to undergo the addition of a nucleophile, which is electron rich. These **nucleophilic addition reactions** are some of the most important reactions in organic chemistry, and they are integral in a variety of biological reactions, too.

A nonpolar π bond
(a)

A polar π bond
(b)

FIGURE 17-1 Nonpolar and polar π bonds (a) A nonpolar π bond is relatively electron rich, so it tends to undergo the addition of an electrophile.
(b) A polar π bond has an electron-poor atom that tends to undergo the addition of a nucleophile.

This species of millipede (*Apheloria corrugate*) defends itself by secreting a mixture of hydrogen cyanide and benzaldehyde, produced from a stored cyanohydrin. Cyanohydrins can be made from nucleophilic addition reactions, the topic of Chapter 17.

Upon completing Chapter 17 you should be able to:

- Draw the general mechanism for the addition of a strong nucleophile to a polar π bond.

- Account for the difference between aldehydes and ketones in their relative susceptibility to nucleophilic addition.

- Specify which solvents are compatible with hydride reagents, such as $NaBH_4$ and $LiAlH_4$, and which are compatible with organometallic reagents, such as RMgBr and RLi.

- Determine whether the existence of a proposed Grignard or alkyllithium reagent is feasible, based on functional group compatibility.

- Predict the major organic product of a *Wittig reaction* and draw its complete, detailed mechanism.

- Show how to synthesize a *Wittig reagent* from an alkyl halide precursor.

- Provide the mechanism for the *sulfonium ylide epoxidation* of a ketone or aldehyde and predict the major organic product.

- Draw the general mechanism for both *direct addition (1,2-addition)* and *conjugate addition (1,4-addition)* to an *α,β-unsaturated polar π bond*.

- Determine whether nucleophilic addition to an α,β-unsaturated carbonyl favors the direct addition product or the conjugate addition product, and explain the role of *reversibility* in such an outcome.

A polar π bond appears in a variety of functional groups, as shown in Figure 17-2. Far and away the most common polar π bond that participates in nucleophilic addition reactions is the one in the carbonyl (C=O) group. Carbonyl groups are present in ketones (R_2C=O), aldehydes (RCH=O), carboxylic acids (RCO_2H), esters (RCO_2R), amides ($RCONR_2$), acid halides (RCO_2X), and acid anhydrides (RCO_2COR). Of these, we focus primarily on ketones and aldehydes here in Chapter 17 and in Chapter 18. The remaining carbonyl-containing functional groups behave somewhat differently; we study them in Chapters 20 and 21. Although less common, other polar π bonds encountered in organic molecules include those in the imine (C=N) and nitrile (C≡N) functional groups, which we also study here in Chapter 17.

Recall from Section 9.3a that negatively charged species tend to be strong nucleophiles and uncharged species tend to be weak. Both types of nucleophiles can add to polar π bonds, and the mechanisms of these reactions are quite similar. As we will see, however, the reaction that begins with a weak nucleophile often requires either acid catalysis or base catalysis. Here in Chapter 17, therefore, we focus primarily on the addition of strong nucleophiles, and in Chapter 18, we describe how the reaction differs when the nucleophile being added is weak.

Chapters 17 and 18

Ketone   **Aldehyde**

Imine   **Nitrile**

Chapters 20 and 21

Carboxylic acid   **Ester**   **Amide**   **Acid anhydride**   **Acid halide**

FIGURE 17-2 Some functional groups containing a polar π bond The functional groups in the first row contain the polar C=O bond. The functional groups in the second row contain polar C=N and C≡N bonds. Ketones, aldehydes, imines, and nitriles are discussed here in Chapter 17 and in Chapter 18; the remaining functional groups are discussed in Chapters 20 and 21.

17.1 An Overview of the General Mechanism: Addition of Strong Nucleophiles

As we saw in Chapter 7, a nucleophile tends to form a bond with the atom at the positive end of a **polar π bond**. Whereas the nucleophile is relatively electron rich, the partially positive (δ^+) atom of the π bond is relatively electron poor. An example with a generic strong nucleophile (Nu^-) is shown in Equation 17-1.

To avoid exceeding the octet on the atom attacked by the nucleophile, the π bond is broken and the pair of electrons from the π bond becomes a lone pair on the more electronegative atom of the π bond.

As shown in Equation 17-1, the immediate product of nucleophilic addition (Step 1) is often a strong base because it possesses a relatively unstable negative charge. Sometimes, species already present in the reaction mixture, such as the solvent, are acidic enough to protonate that product (Step 2). Otherwise, we can carry out this protonation in a subsequent *acid workup* (Section 10.7a).

17.1 YOUR TURN

The following reaction is similar to the one in Equation 17-1.

Label each reacting species as either "electron rich" or "electron poor," draw in the appropriate curved arrows, and under each reaction arrow name the type of elementary step involved.

Answers to Your Turns are in the back of the book.

SOLVED problem **17.1** Draw the complete, detailed mechanism and predict the product for the reaction at the right.

Think What nucleophile is generated when KCN dissolves in water? Which atom will it attack? Which bond is most easily broken in the process? What is the role of the acid?

Solve KCN is ionic, consisting of K^+ and NC^- ions. In water, therefore, KCN dissolves as K^+ and NC^-, with NC^- being a strong nucleophile. NC^- will subsequently attack the electron-poor carbonyl carbon, breaking the π bond of the double bond. The resulting O^- is protonated by the acid.

$$\text{(phenyl methyl ketone)} \xrightarrow[\text{CH}_3\text{CH}_2\text{OH}]{\text{NaSCH}_3} \text{?}$$

Stereochemistry is something that must be accounted for in nucleophilic addition when the carbon atom that is attacked becomes a stereocenter in the product. An example is shown in Equation 17-2.

The nucleophile can attack from either side of the carbon's plane.

Racemic mixture of enantiomers

An imine

(17-2)

Because the C atom of the polar π bond is planar, a nucleophile can attack from either side. Thus, as we learned in Chapter 9, the result is a mixture of *R* and *S* stereochemical configurations at the bond formation site. A *racemic mixture* of enantiomers is formed if the reactants are achiral and the reaction takes place in an achiral environment. Otherwise, an unequal mixture of the *R* and *S* configurations is formed.

YOUR TURN 17.2

The reaction in Solved Problem 17.1 results in a new tetrahedral stereocenter. Identify that stereocenter and draw both stereoisomers that would be produced.

Throughout the rest of Chapter 17, we examine specific reactions of ketones, aldehydes, imines, and nitriles (Fig. 17-3a) that proceed by the general nucleophilic addition mechanism shown in Equation 17-1 (or a variation of it). Carboxylic acids, esters, amides, acid halides, and acid anhydrides contain a polar π bond, too (Fig. 17-3b), so they are also susceptible to attack by a nucleophile. In each of these species, however, the C atom of the polar π bond possesses a *leaving group* (highlighted in red). The presence of a leaving group facilitates another type of reaction with nucleophiles, called *nucleophilic addition–elimination reactions*, which are discussed in Chapters 20 and 21.

Each of these species contains a polar π bond but no leaving group on C, so they undergo the general nucleophilic addition reaction in Equation 17-1.

Each of these species contains a leaving group on the C atom of the polar π bond and is discussed in Chapters 20 and 21.

(a)   (b)

FIGURE 17-3 Polar π bonds and leaving groups (a) The polar π bonds in these functional groups do not have a leaving group on C, so they tend to undergo reactions similar to the one in Equation 17-1. (b) In each of these functional groups, a leaving group (red) is attached to the C atom of the polar π bond. Reactions involving these functional groups are discussed in Chapters 20 and 21.

17.2 Substituent Effects: Relative Reactivity of Ketones and Aldehydes in Nucleophilic Addition

Ketones and aldehydes exhibit very similar chemical behavior because they are rather similar structurally, but the nucleophilic addition reactions they undergo differ noticeably both thermodynamically and kinetically:

> Nucleophilic attack at a carbonyl carbon tends to be more energetically favorable and faster for an aldehyde than for a comparable ketone.

These points are exemplified in Table 17-1, which presents both equilibrium and rate data for the *hydration* of three different carbonyl-containing compounds. The key step in the mechanism of a hydration reaction is a nucleophilic addition.

TABLE 17-1 Extent and Rate of Carbonyl Hydration in Ketones and Aldehydes

Reaction	Relative Rate Constants[a]	Percent Hydration at Equilibrium[b]
formaldehyde + H₂O ⇌ hydrate	5×10^5	>99.9%
acetaldehyde + H₂O ⇌ hydrate	2×10^3	57%
acetone + H₂O ⇌ hydrate	1	<1%

[a] Reactions take place in water.
[b] Reactions take place in methanol.

Notice in Table 17-1 that both the rate of hydration and the extent of hydration at equilibrium are greater for the aldehydes than for the ketones. Both of these results can be explained in part by steric effects associated with the bulky alkyl groups of the ketone, as shown in Figure 17-4. In a ketone, the *steric repulsion* from the

FIGURE 17-4 Steric effects in nucleophilic addition (a) In a ketone, steric repulsion by each bulky alkyl group (left) makes it difficult for a nucleophile to attack. Steric strain in the hydrate product (right) decreases stability. (b) In an aldehyde, there is at most one bulky alkyl group, so steric effects are decreased.

Ketone

Steric strain between the alkyl and OH groups destabilizes the hydrate.

Steric repulsion from two R groups in a ketone makes it difficult for the nucleophile to attack.

(a)

Aldehyde

Less steric strain in the hydrate product

Less steric repulsion in an aldehyde, so the nucleophile can attack

(b)

two alkyl groups on the carbonyl carbon (Fig. 17-4a, left) makes it difficult for the nucleophile to attack. Furthermore, each bulky alkyl group in the hydrate product (Fig. 17-4a, right) crashes into the lone pairs on oxygen, generating *steric strain* and decreasing stability. An aldehyde (Fig. 17-4b), on the other hand, has, at most, one alkyl group bonded to the carbonyl carbon, so both of these steric effects are diminished.

Aldehydes are more reactive than ketones, too, because ketones are stabilized more by *inductive effects*, as illustrated in Figure 17-5. Recall from Section 6.6e that alkyl groups are electron donating, so they decrease the concentration of positive charge at the carbonyl carbon. With less concentration of positive charge at that carbon, the carbonyl group is less susceptible to attack by a nucleophile bearing excess negative charge.

FIGURE 17-5 Inductive effects in nucleophilic addition Electrostatic potential maps of (a) formaldehyde, (b) acetaldehyde, and (c) acetone. Additional alkyl groups bonded to the carbonyl carbon decrease its concentration of positive charge by donating electron density, making it less reactive.

YOUR TURN 17.3

Which of the following imines has the greatest concentration of positive charge at the C=N carbon, and which has the least concentration of positive charge? Which is the most reactive imine? The least reactive?

A   B   C

problem **17.3** Chloral, $Cl_3CCH{=}O$, forms a very stable hydrate called chloral hydrate, a sedative and hypnotic drug. When dissolved in water, essentially 100% of chloral is hydrated. In contrast, the extent of hydration for ethanal (acetaldehyde) is much smaller (see Table 17-1). Explain why.

D   **E**

17.3 Hydride Reducing Agents: LiAlH$_4$ and NaBH$_4$

The **hydride anion** is formally written as H$:^-$, making it a hydrogen atom with an extra electron (it is usually further abbreviated to H$^-$). Unlike such species as Cl$^-$ and HO$^-$, however, H$^-$ does not exist on its own at any appreciable concentration in solution. Solutions of Cl$^-$ and HO$^-$ can be prepared by dissolving NaCl and KOH, respectively, in highly polar solvents, including water, ethanol, and DMSO. H$^-$, on the other hand, is extremely reactive, both as a nucleophile and as a base, because its negative charge is not well stabilized. Consequently, H$^-$ can react with many solvents in which it is soluble. To circumvent this problem, we often choose hydride sources in which the hydrogen is part of a covalent bond to a *less* electronegative atom. In these species, the hydrogen atom bears only a *partial* negative charge.

As we learned in Section 7.1b, two common hydride sources in organic chemistry are **sodium borohydride (NaBH$_4$)** and **lithium aluminum hydride (LiAlH$_4$, or LAH)**. As shown in Figure 17-6, NaBH$_4$ consists of Na$^+$ and the polyatomic anion BH$_4^-$, and LiAlH$_4$ consists of Li$^+$ and AlH$_4^-$.

The most important role of these hydride sources is as reducing agents.

The partial negative charges on H make these sources of H$^{\ominus}$.

Sodium borohydride   **Lithium aluminum hydride**
(a)   (b)

FIGURE 17-6 Common hydride sources (a) Sodium borohydride consists of Na$^+$ and BH$_4^-$ ions, and (b) lithium aluminum hydride consists of Li$^+$ and AlH$_4^-$ ions. In both anions, H bears a partial negative charge, so these are sources of hydride ions, H$^-$.

> Both NaBH$_4$ and LiAlH$_4$ act as *reducing agents* in the presence of a compound containing a polar π bond.

For example, when butan-2-one is treated with a solution of NaBH$_4$ dissolved in water, racemic butan-2-ol is formed (Equation 17-3).

Oxidation state = +2   Oxidation state = 0

$$\text{Butan-2-one} \xrightarrow[\text{H}_2\text{O}]{\text{NaBH}_4} \text{Butan-2-ol (racemic)} \quad 83\%$$ (17-3)

Reduction of the ketone is signified by the increase in the number of hydrogen atoms bonded to the carbonyl carbon, resulting in a decrease in that carbon's oxidation state. Similarly, the treatment of butan-2-one with LiAlH$_4$ followed by aqueous acid reduces butan-2-one to the alcohol (Equation 17-4).

$$\text{Butan-2-one} \xrightarrow[\text{2. H}_3\text{O}^+]{\text{1. LiAlH}_4, \text{ ether, reflux}} \text{Butan-2-ol (racemic)} \quad 80\%$$ (17-4)

Although H$^-$ does not physically exist on its own, Equations 17-3 and 17-4 demonstrate that we can *treat* both NaBH$_4$ and LiAlH$_4$ as simply H$^-$. As shown in Equation 17-5, we can envision H$^-$ acting as a nucleophile attacking the carbonyl carbon in the first step of the mechanism.

Simplified mechanism for Equation 17-3

$$ (17\text{-}5) $$

A proton transfer occurs in the second step, thus converting the alkoxide anion into the final alcohol product. The acid in that step is H_2O for the $NaBH_4$ reduction in Equation 17-3, but it is H_3O^+ for the $LiAlH_4$ reduction in Equation 17-4. A racemic mixture of enantiomers is produced in both reactions because the single stereocenter in the product is formed in the first step of the mechanism, in which the reactants and the reaction conditions are achiral.

To take into account the fact that H^- cannot exist on its own, a rigorously formal mechanism must show the nucleophilic attack as a *hydride transfer*. That is, the H atom's bond to either B or Al is broken at the same time the C—H bond is formed. This is illustrated in Equation 17-6 for the $NaBH_4$ reduction of butan-2-one.

Complete mechanism for Equation 17-3

$$ (17\text{-}6) $$

A similar mechanism occurs when $LiAlH_4$ is used as the reducing agent (see Your Turn 17.4). In the interest of simplicity, however, we will generally work with the mechanism in Equation 17-5.

YOUR TURN 17.4

Draw the complete mechanism for the $LiAlH_4$ reduction of butan-2-one (Equation 17-4). (*Hint:* It is similar to the one in Equation 17-6.)

With $NaBH_4$, the reduction can take place in water (Equation 17-3). This is advantageous because water is a weak acid (i.e., it is a source of protons, H^+). Therefore, both the addition of H^- to the carbonyl group and the subsequent protonation can occur in the same synthetic step. Reductions using $LiAlH_4$, on the other hand, require two separate reactions (Equation 17-4). The first is the treatment of the carbonyl-containing species with $LiAlH_4$ dispersed in ether, and the second is the treatment of the product from the first reaction with aqueous acid—an acid workup. The acid workup is necessary because the ether solvent in the first reaction is not acidic enough to facilitate a proton transfer.

Reduction with $LiAlH_4$ must take place in ether, whereas reduction with $NaBH_4$ can take place in a weakly acidic solvent such as water or an alcohol, because

$LiAlH_4$ is much more reactive than $NaBH_4$.

The hydride from LiAlH$_4$ deprotonates the weakly acidic proton from water and alcohols very quickly (Equation 17-7), producing hydrogen gas. The exothermicity of the reaction can ignite the hydrogen, causing an explosion!

The H$_2$ gas that is produced is explosive!

Although NaBH$_4$ can deprotonate water and alcohols, too, this proton transfer is rather slow, especially when the solution is maintained at a slightly basic pH (Equation 17-8).

LiAlH$_4$ is more reactive than NaBH$_4$ because, as shown in Figure 17-7, the electronegativity (EN) of aluminum (1.61) is substantially lower than that of boron (2.04). The EN of H (2.20) is higher than that of either Al or B, giving H a partial negative charge in both the Al—H and the B—H bonds. However, the electronegativity *difference* (ΔEN) is greater in the case of Al, giving H a higher concentration of negative charge in LiAlH$_4$ than in NaBH$_4$. As we learned in Chapter 7, a higher concentration of charge like this in LiAlH$_4$ makes it less stable and thus more reactive.

FIGURE 17-7 The relative reactivities of NaBH$_4$ and LiAlH$_4$ The Al—H bond is more polar than the B—H bond, so the H atoms in LiAlH$_4$ bear a larger concentration of negative charge, making LiAlH$_4$ a more reactive source of H$^-$.

The greater concentration of charge on H makes LiAlH$_4$ more reactive.

EN = 2.04   EN = 2.20
ΔEN = 0.16

EN = 1.61   EN = 2.20
ΔEN = 0.59

SOLVED problem 17.5 Predict the major organic product of the reaction at the right.

Think Can LiAlH$_4$ be treated as a simpler nucleophile? How does NH$_4$Cl behave in solution, and why must the reaction with LiAlH$_4$ be complete before it is added?

1. LiAlH$_4$
2. NH$_4$Cl
?

Solve LiAlH$_4$ can be treated simply as H$^-$, which attacks the carbonyl carbon, as shown in Step 1 of the following sequence:

1. Nucleophilic addition

2. Proton transfer

+ :NH$_3$

NH$_4$Cl is ionic, so in solution it dissociates into its NH$_4^+$ and Cl$^-$ ions. NH$_4^+$ is weakly acidic, so to avoid destroying LiAlH$_4$, it is not added until the nucleophilic addition is complete. Instead, in Step 2, NH$_4^+$ protonates the strongly basic product from Step 1.

---

problem 17.6 Predict the major organic product in each of these reactions.

NaBH$_4$
CH$_3$CH$_2$OH
?

(a)

1. LiAlH$_4$
2. H$_3$O$^+$
?

(b)

Functional groups with polar π bonds other than the carbonyl group react with hydride agents in much the same way. An imine, for example, is reduced with $LiAlH_4$ to form an amine, as shown in Equation 17-9. As before, the nucleophilic addition of H^- takes place in an aprotic solvent such as ether and is followed by an aqueous acid workup.

$$\text{(17-9)}$$

$LiAlH_4$ reduces nitriles to primary amines (Equation 17-10).

$$\text{(17-10)}$$

A nitrile

A primary amine

The first addition of H^- yields an intermediate that resembles a deprotonated imine. Because that intermediate also contains a polar π bond, and because $LiAlH_4$ is such a powerful hydride agent, a second addition of H^- occurs, generating a species that resembles a dianion on the N atom. Subsequent acid workup protonates the N atom twice.

The -2 charge shown in the second intermediate of Equation 17-10 is an outcome of our simplification of $LiAlH_4$ to just H^-. Actually, that N atom bonds to Al (similar to what was shown in Equation 17-6), which provides significant stabilization.

problem **17.7** Draw a complete, detailed mechanism for the reaction in Equation 17-9.

problem **17.8** Draw a complete, detailed mechanism for the reaction in Equation 17-10.

problem **17.9** Predict the major organic product in each of the following reactions.

(a)

(b)

NADH as a Biological Hydride Reducing Agent

Hydride reductions occur in a variety of biochemical processes, but these processes don't involve the $NaBH_4$ or $LiAlH_4$ hydride reducing agents we have examined here in Chapter 17. Rather, the body uses the reduced form of nicotinamide adenine dinucleotide (NADH). The nicotinamide portion of NADH, shown below on the left, is directly involved in reductions. (R″ represents the remainder of NADH, consisting of two sugar units, two phosphate groups, and an adenine base.)

In the simplified mechanism shown above on the left, NADH donates H^+ along with two electrons—the equivalent of an H^- anion—leaving behind NAD^+. Elsewhere, NAD^+ is subsequently used as an oxidizing agent (such as in glycolysis, the citric acid cycle, fatty acid oxidation, and the catabolic elimination of surplus amino acids) and is returned to the NADH form. Thus, NADH and NAD^+ are continuously recycled.

NADH also plays a key role in what is called the *electron transport chain* in mitochondria, as shown above on the right. When NADH is converted to NAD^+, the H^+ and two electrons that make up the equivalent of H^- are separated from each other. The electrons are passed down the transport chain and ultimately are deposited onto an oxygen atom that becomes reduced to H_2O. In the process, protons are passed to the opposite side of the inner membrane of the mitochondrion, thereby establishing a *proton gradient*—i.e., a difference in proton concentration—that sets up the potential to produce chemical energy in the form of adenosine triphosphate (ATP).

17.4 Sodium Hydride: A Strong Base but a Poor Nucleophile

Sodium hydride (NaH) is another common hydride agent used in organic chemistry. Its behavior, however, is somewhat different from that of $NaBH_4$ and $LiAlH_4$.

Sodium hydride, NaH, is a very strong base, but a poor nucleophile.

For example, treatment of pentan-2-one with NaH yields the acetone enolate anion *quantitatively* (i.e., with essentially 100% yield), as shown in Equation 17-11a. Reduction of the carbonyl (Equation 17-11b) does not occur.

NaH is a poor nucleophile.

Because of its properties as a base, NaH is often used to generate carbon nucleophiles, as demonstrated in Solved Problem 17.10.

SOLVED problem 17.10 Predict the major organic product of the sequence of reactions shown at the right.

Think Will the hydride anion from NaH act as a base or a nucleophile? Which atom will it attack? How will the resulting species behave in the presence of CH_3I?

Solve NaH is a strong base but a poor nucleophile, so it will deprotonate the alpha carbon. As we learned in Section 10.3, the resulting enolate anion is a strong nucleophile and will displace I from CH_3I in an S_N2 reaction, yielding an alpha-alkylated aldehyde.

problem 17.11 Predict the major organic product in each of the following reactions.

(a)　　　　　　　　　　(b)　　　　　　　　　　(c)

NaH behaves differently from $NaBH_4$ or $LiAlH_4$ because NaH is an *ionic* hydride, consisting of Na^+ and H^- ions, not a *covalent* hydride. NaH is ionic because Na and H have widely different electronegativities (0.93 vs. 2.20). The B—H and Al—H bonds in $NaBH_4$ and $LiAlH_4$ have more covalent character, on the other hand, because the electronegativities of B (2.04) and Al (1.61) are closer to that of H (2.20).

Because NaH is ionic, it is essentially insoluble in organic solvents; it remains a solid. It is therefore believed that reactions involving NaH take place at the NaH surface. Under these conditions, H^- *directly* participates in reactions, rather than through the hydride transfers we saw for $NaBH_4$ and $LiAlH_4$ in Equation 17-6. Therefore, the carbonyl oxygen is not involved in directing the H^- to the carbonyl carbon. Instead, H^- is free to deprotonate at the alpha carbon, which, as we learned in Chapter 8, is a very fast reaction, and generally wins in competitions with other types of reactions.

17.5 Reactions of Organometallic Compounds: Alkyllithium Reagents and Grignard Reagents

Recall from Section 7.1b that we can usually treat organometallic compounds like **Grignard reagents (RMgBr)** and **alkyllithium reagents (RLi)** as alkyl anions, $:R^-$ (Fig. 17-8).

Alkyllithium reagents can be treated as :$R^\ominus$.

Grignard reagents can be treated as :$R^\ominus$.

Butyllithium

Simplify

Phenylmagnesium bromide

Simplify

(a)

(b)

FIGURE 17-8 Simplifications of alkyllithium and Grignard reagents (a) An alkyllithium reagent has a polar covalent C—Li bond, which can be simplified to $C:^-$ and Li^+. (b) A Grignard reagent has a polar C—MgBr bond, which can be treated as $C:^-$ and $(MgBr)^+$.

Because of the poorly stabilized charge on C, these types of reagents behave both as *strong bases* and as *strong nucleophiles*. For example, both alkyllithium reagents and Grignard reagents react rapidly with water in a substantially exothermic proton transfer (Equations 17-12 and 17-13) to produce an alkane and HO^-. The products are much more stable because the negative charge is transferred from a C atom to a more electronegative O atom.

$$\text{(17-12)}$$

$$\text{(17-13)}$$

Although these organometallic reagents react as if they are R^-, they are not truly ionic. When dissolved in solution, free R^- does not exist. Instead, think of them as R^- *donors,* in much the same way that we view $LiAlH_4$ and $NaBH_4$ as H^- donors. The reaction that takes place is an R^- transfer from the metal atom to H^+, as shown for butyllithium in Equation 17-14.

Mechanism for Equation 17-12

Proton transfer

$$\text{(17-14)}$$

In the space provided here, add the curved arrows to show an R^- transfer mechanism for the reaction in Equation 17-13, similar to what is shown in Equation 17-14 for an alkyllithium reagent.

17.5 YOUR TURN

$$\xrightarrow{\text{H—OH}}$$

When an alkyllithium reagent or a Grignard reagent reacts with a functional group containing a polar π bond, R^- acts as a nucleophile. This behavior is exemplified with a ketone in Equation 17-15 and with a nitrile in Equation 17-16. When a Grignard reagent adds in as a nucleophile, it is called a **Grignard reaction**.

$$(17\text{-}15)$$

74%

Benzonitrile

70%

$$(17\text{-}16)$$

Notice in both of these cases that a new C—C bond is formed! As we learned in Section 13.2, reactions that do so are distinct from functional group transformations and are, therefore, special.

In both Equations 17-15 and 17-16, the nucleophilic addition takes place in ether, followed by workup with a proton source. In each case, it is vitally important to keep the first reaction free of any compounds with acidic protons, such as water or methanol. Otherwise, the alkyllithium or Grignard reagent will be destroyed via proton transfer reactions like those shown in Equation 17-12 or 17-13.

Notice also that the workup in Equation 17-16 is carried out with methanol. If it is carried out with H_3O^+, instead, then the imine that is produced from nucleophilic addition is hydrolyzed to a ketone or an aldehyde, as shown in the parentheses. This reaction is discussed in greater detail in Chapter 18.

The mechanisms for these reactions are shown in Equations 17-17 and 17-18. Specifically, R^- adds to the electron-poor atom of the polar π bond. Once again, because free R^- does not exist in solution, each mechanism would be more appropriately shown as a transfer of R^-.

Mechanism for Equation 17-15

$$(17\text{-}17)$$

Mechanism for Equation 17-16

$$(17\text{-}18)$$

The R⁻ nucleophiles from RLi and RMgBr add only once to a nitrile carbon. This is in contrast to what we saw with $LiAlH_4$ (Equation 17-10), in which H⁻ adds twice. In other words, these R⁻ nucleophiles are not strong enough to generate the effective −2 formal charge on nitrogen that is a necessary outcome of a second nucleophilic addition.

SOLVED problem 17.12 Predict the major product of the reaction at the right.

Think To what R⁻ nucleophile can the Grignard reagent be simplified? Which atom will it attack? What is the role of H_3O^+?

Solve The C_6H_5MgBr Grignard reagent can be treated simply as $C_6H_5^-$, which will undergo nucleophilic addition at the carbonyl C. Once this is complete, H_3O^+ is added in an acid workup to protonate the strongly basic O⁻ generated in the first step.

---

problem 17.13 Predict the major organic product in each of the following reactions.

(a)

(b)

(c)

Although CO_2 is a nonpolar compound overall, each C=O bond is highly polar and the central carbon atom is quite susceptible to nucleophilic attack. Thus, Grignard reagents can add to CO_2 in what is called a **carboxylation** reaction (Equation 17-19). As with any Grignard reaction, a new C—C bond is formed.

New C—C bond

(17-19)

The immediate product of carboxylation—a carboxylate anion—is subsequently protonated through an acid workup to yield a carboxylic acid. Because CO_2 is a gas at room temperature, carrying out this reaction requires either bubbling CO_2 through an ether solution of the Grignard reagent, or pouring the Grignard reagent over dry ice (which is solid CO_2).

---

problem 17.14 Draw the complete, detailed mechanism for the reaction in Equation 17-19.

---

problem 17.15 Predict the major organic product in each of the following reactions.

(a)

(b)

17.6 Limitations of Alkyllithium and Grignard Reagents

Both alkyllithium and Grignard reagents can be synthesized with a wide variety of R groups. Figure 17-9 illustrates this with Grignard reagents specifically, but the same is typically true of alkyllithium reagents. The R group can be a saturated alkyl group, (including the CH_3, CH_2CH_3, and cyclohexyl groups), but it can also be unsaturated. Common unsaturated Grignard reagents include phenylmagnesium bromide (C_6H_5MgBr), and vinylmagnesium bromide (H_2C=CH—$MgBr$).

Some feasible Grignard reagents

FIGURE 17-9 Feasible Grignard reagents These Grignard reagents are feasible because the strongly nucleophilic and strongly basic carbon, highlighted in red, will not react with other functional groups present in the molecule.

These strong organometallic reagents, however, must not contain functional groups that are susceptible to deprotonation or to nucleophilic attack (Fig. 17-10). For example, the hypothetical Grignard reagent shown on the left in Equation 17-20 is unfeasible. The OH group is weakly acidic and can be deprotonated by a strongly basic carbon atom bonded to Mg. As shown in Equation 17-20, this reaction effectively destroys the Grignard reagent; thus, the alkylmagnesium bromide functional group is *incompatible* with the OH group.

The Grignard reagent was destroyed.

(17-20)

A *hypothetical* Grignard reagent

Other groups that are incompatible with Grignard reagents or alkyllithium reagents are polar π bonds such as C=O, C=N, and C≡N, because the alkanide anion, R^-, will add to the electrophilic carbon. Furthermore, saturated alkyl halides are incompatible because they are prone to elimination or nucleophilic substitution reactions, with the R^- species behaving as a strong base or nucleophile, respectively.

FIGURE 17-10 Unfeasible Grignard reagents These Grignard reagents are unfeasible because the groups highlighted in blue would undergo reaction with the strongly nucleophilic and strongly basic carbon highlighted in red.

problem **17.16** For each of the Grignard reagents in Figure 17-10, which are *unfeasible*, draw the complete mechanism of the reaction that would take place. The first is shown in Equation 17-20.

Some unfeasible Grignard reagents

17.7 Wittig Reagents and the Wittig Reaction: Synthesis of Alkenes

Wittig reagents (pronounced *vittig*), named after the German chemist Georg Wittig (1897–1987), are an important class of compounds used to synthesize alkenes. A Wittig reagent (Fig. 17-11) is characterized by a C—P bond in which the C atom bears a −1 formal charge and the P atom bears a +1 formal charge.

Wittig reagent (An ylide)

FIGURE 17-11 **Wittig reagents** A Wittig reagent, also called a phosphonium ylide, is highly nucleophilic at the C$^-$ site. The structure on the left is the major resonance contributor because the C=P π bond in the structure on the right is weak, requiring orbitals from different shells.

A Wittig reagent is an ylide (pronounced *ilid*), which is characterized by adjacent charged atoms that each have a complete octet of electrons. More specifically, because the +1 charge is on P, a Wittig reagent is also called a **phosphonium ylide**. As indicated, the −1 charge on C makes a Wittig reagent highly nucleophilic at the C atom.

Normally, a structure with adjacent positive and negative charges is the weaker contributor of two resonance structures, the stronger contributor being that in which the atoms are connected by a double bond and each atom has a zero formal charge. In the case of a Wittig reagent, however, the C=P resonance structure is the *weaker* contributor, because the π bond of a C=P bond would require the interaction of orbitals from different valence shells—the second shell on carbon and the third shell on phosphorus. This results in a weak interaction and, consequently, little stabilization relative to the atomic orbital energies, leaving the P$^+$—C$^-$ structure as the more accurate representation.

The primary importance of Wittig reagents is in their reaction with ketones and aldehydes—so-called **Wittig reactions**. An example is shown in Equation 17-21.

Wittig reaction

New C—C bond

Triphenylphosphine oxide

97%

(17-21)

As we can see from this example, a Wittig reaction results in the joining of two carbon-containing groups by a C=C bond—one group from the Wittig reagent and the second from the ketone or aldehyde. Notice, in particular, that the newly formed C=C bond is between the original carbonyl carbon and the original P—C carbon. In other words,

> In a Wittig reaction, the C=O bond of a ketone or aldehyde is converted into a C=C bond.

The mechanism shown in Equation 17-22 accounts for the Wittig reaction in Equation 17-21.

Mechanism for Equation 17-21

A betaine

(17-22)

Strained 4-membered ring

Strong P—O bond = 537 kJ/mol (128 kcal/mol)

An oxaphosphetane

In Step 1, the highly nucleophilic C atom of the Wittig reagent attacks the electrophilic carbonyl C atom, yielding a **betaine** (pronounced *beta-ene*)—that is, a species in which a positive and a negative charge are separated by two uncharged atoms. Step 2 is a coordination step, in which a bond is formed between the negatively charged O atom and the positively charged P atom. This results in an **oxaphosphetane** that contains a four-membered ring, which, due to the strain, falls apart into the alkene and triphenylphosphine oxide in Step 3.

The first two steps of the Wittig reaction are driven by the principle of *electron rich to electron poor*. The last step, however, is driven primarily by the formation of the very strong P^+—O^- bond, whose bond energy is 537 kJ/mol (128 kcal/mol).

YOUR TURN 17.6

In each of the first two steps of the mechanism in Equation 17-22, label the pertinent electron-rich and electron-poor sites.

The mechanism for a Wittig reaction is still under debate. Specifically, there is evidence suggesting that the oxaphosphetane forms in one step rather than two. (If the oxaphosphetane forms in a single step, the step would be called cycloaddition, the topic of Chapter 24.)

If *E/Z* isomerism exists about the C=C bond in the Wittig product, then a mixture of the two isomers is produced, as shown in Equation 17-23.

A Wittig reaction produces a mixture of *Z* and *E* isomers.

(17-23)

(*Z*)-2-Phenylbut-2-ene   (*E*)-2-Phenylbut-2-ene

The reason for this can be seen from the first step of the mechanism, in which either face of the carbonyl-containing compound can be attacked. Often, one isomer is heavily favored over the other. Which isomer is favored, however, depends upon the nature of the two reactants as well as on the specific method used to generate the ylide.

---

problem **17.17** Draw the complete, detailed mechanism for the reaction at the right and predict the major organic product(s). Pay attention to stereochemistry.

17.8 Generating Wittig Reagents

One of the reasons that Wittig reactions are so useful is that Wittig reagents can be generated from common precursors—namely, alkyl halides. As shown in Equation 17-24, the Wittig reagent in Equation 17-21 can be synthesized from 2-bromopropane.

2-Bromopropane

1. $(C_6H_5)_3P$
2. $CH_3(CH_2)_3Li$

(17-24)

56%

The alkyl halide is first treated with triphenyl phosphine, $P(C_6H_5)_3$, and the product of that reaction is treated with a very strong base such as butyllithium.

The mechanism for this reaction is shown in Equation 17-25.

Mechanism for Equation 17-24

Step 1 is an S_N2 reaction in which the nucleophile is Ph_3P and the leaving group is the halide anion. The organic product possesses a positive charge on the P atom. Although Ph_3P is uncharged, it is a good nucleophile because the P atom in the product can accommodate the positive charge rather well, due both to its large size (being in the third row of the periodic table) and its modest electronegativity (2.19).

Verify that triphenylphosphine is a very good nucleophile by looking up its relative nucleophilicity and that of Br^- in Chapter 9.

17.7 YOUR TURN

Step 2 of Equation 17-25 is deprotonation by the alkyllithium species. A very strong base such as an alkyllithium is necessary because deprotonation occurs at a carbon atom. Carbon acids are in general extremely weak because a negatively charged carbon atom is normally very unstable. Here, however, deprotonation is facilitated because the negative charge that develops on C is stabilized by the adjacent positive charge on P, and opposite charges attract.

problem **17.18** Draw the complete, detailed mechanism for the reaction sequence at the right and provide structures for both the product (**B**) and the intermediate (**A**).

1. P(C$_6$H$_5$)$_3$
2. Bu—Li
→ **A** → **B**

problem **17.19** Draw an alkyl halide that could be used to synthesize this Wittig reagent.

The C atom that is deprotonated in the proton transfer step in Equation 17-25 is the one originally bonded to the leaving group in the alkyl halide precursor. Therefore,

> For an alkyl halide to be a suitable precursor for a Wittig reagent, the C atom bonded to the leaving group must possess at least one H.

Moreover, having such an H atom ensures that the halogen leaving group is not on a tertiary carbon—something that, as we learned in Section 9.6b, would preclude the S$_N$2 reaction from happening in the first place (Equation 17-26).

A 3° substrate precludes an S$_N$2 reaction from occurring.

(17-26)

SOLVED problem **17.20** Show how you could synthesize oct-4-ene using butanal as your only source of carbon.

Think What Wittig reagent would you need to carry out this reaction? What alkyl halide could serve as a precursor to that Wittig reagent? How can that alkyl halide be synthesized from butanal?

Solve We can carry out a transform on the product, undoing a Wittig reaction.

Undo a Wittig reaction.

+

Wittig reagent

The Wittig reagent can be synthesized from a four-carbon alkyl halide. The alkyl halide cannot be produced directly from butanal, but can be produced from butan-1-ol, and butan-1-ol can be synthesized by reducing butanal using a hydride agent.

Undo a Wittig reagent formation. → Undo a bromination. → Undo a reduction. →

In the forward direction, the synthesis might appear as follows.

problem **17.21** Show how you could carry out the synthesis at the right, beginning with benzaldehyde, and using any other reagents necessary.

17.9 Sulfonium Ylides: Formation of Epoxides

Just like phosphonium ylides, **sulfonium ylides** (Fig. 17-12) are species with opposite charges on adjacent atoms, each with an octet of electrons. In this case, however, the positive charge is on a sulfur atom.

Phosphonium and sulfonium ylides are also similar in that they each have a resonance structure in which no atoms have a formal charge, as shown on the right in Figure 17-12. However, just as we saw with phosphonium ylides, the S=C resonance structure for a sulfonium ylide contributes very little to the resonance hybrid, because it requires a π bond between atomic orbitals in different shells—the second shell on C and the third shell on S. As a result, the S^+—C^- structure shown on the left is a more accurate representation of the species, reflecting the strong nucleophilic character of the C atom.

Sulfonium ylides are synthesized in much the same way as phosphonium ylides, as shown in Equation 17-27. A dialkyl sulfide is treated with an alkyl halide, followed by treatment with a strong base.

This C atom is highly nucleophilic. **There is very little contribution by this resonance structure.**

FIGURE 17-12 Sulfonium ylides A sulfonium ylide is highly nucleophilic at the C^- site. The structure on the left is the major resonance contributor because the C=S π bond in the structure on the right is weak, requiring orbitals from different shells.

$$\text{(17-27)}$$

The mechanism for this reaction, shown in Equation 17-28, is essentially identical to that for the synthesis of a Wittig reagent, shown previously in Equation 17-25.

Mechanism for Equation 17-27

$$\text{(17-28)}$$

Step 1 is an S_N2 reaction in which the sulfide displaces the halide. In Step 2, the C atom attached to S is deprotonated to produce the ylide.

Because of the negative charge on the C atom of the C—S bond, *sulfonium ylides are strong nucleophiles* and they add readily to carbonyls. As we can see in Equation 17-29, however, their overall reaction with ketones and aldehydes is somewhat different from the Wittig reaction. No C=C double bond is formed; instead, the product is an epoxide.

(17-29)

84%

In particular,

In the reaction between a sulfonium ylide and a ketone or an aldehyde, the epoxide ring that forms involves the C and O atoms of the initial C=O bond, and the C atom of the C⁻—S⁺ bond.

YOUR TURN 17.8

Identify the epoxide ring in Equation 17-29.

The mechanism for this epoxidation reaction is shown in Equation 17-30.

Mechanism for Equation 17-29

(17-30)

Similar to the Wittig reaction, Step 1 involves nucleophilic addition of the sulfonium ylide, resulting in a *betaine* (the positive charge on S and the negative charge on O are separated by two uncharged atoms). Step 2 is an internal S_N2 reaction, where the dialkyl sulfide is the leaving group.

problem **17.22** For each of the following reactions, draw the complete mechanism and predict the major organic product.

(a)

(b)

(a) (b) (c)

17.10 Direct Addition versus Conjugate Addition

When a C=C double bond is conjugated to a polar π bond, as in propenal (Fig. 17-13), two electron-deficient sites are present. One of those sites is the carbonyl C atom, given that the highly electronegative O atom is bonded to it, and the other electrophilic site is the C atom *beta* to (i.e., two carbons away from) the carbonyl. We can see why this is so by examining one of its resonance contributors, which places a formal negative charge on the O atom and a formal positive charge on the beta C atom (Fig. 17-13a). Although this resonance contributor is somewhat weak, due to the charges present, its contribution to the resonance hybrid does indeed place a small partial positive charge on the beta C atom, as indicated in the electrostatic potential map in Figure 17-13b.

(a)

Partial positive charges

(b)

FIGURE 17-13 Reactive sites in an α,β-unsaturated carbonyl (a) Resonance structures of propenal are shown on the left. The weak resonance contributor with separated charges generates a small, partial, positive charge on the β carbon in the resonance hybrid on the right. Thus, a nucleophile can attack at either the carbonyl carbon or the β carbon. (b) An electrostatic potential map of propenal, showing the partial positive charge (blue) on both the carbonyl carbon and the β carbon.

Identify the two electron-deficient sites in the following molecule, and label the α and β carbons.

17.9 **YOUR TURN**

Given these *two electrophilic sites* in an **α,β-unsaturated carbonyl**, nucleophiles in general can attack at two positions: either at the carbonyl carbon itself (Equation 17-31) or at the **beta carbon** (Equation 17-32).

Mechanism for direct addition (1,2-addition) to an α,β-unsaturated carbonyl

Mechanism for conjugate addition (1,4-addition) to an α,β-unsaturated carbonyl

Attack of the nucleophile at the carbonyl carbon is called **1,2-addition**, or **direct addition**, because after protonation in Step 2, species have added to adjacent atoms, which are said to be in a 1,2-positioning. Attack at the beta C atom, on the other hand, is called **1,4-addition**, or **conjugate addition**. After protonation in Step 2, species have added to atoms separated by three bonds, which are said to be in a 1,4-positioning.

Notice in Equation 17-32 that the 1,4-addition product is an enol. As we saw in Section 7.9, this enol form will rapidly *tautomerize* to generate the more stable keto form as the major product.

YOUR TURN 17.10

Draw the mechanism that converts the enol in Equation 17-32 into its keto form. (*Hint*: Review Section 7.9.)

If a carbon nucleophile adds via conjugate addition, thereby forming a carbon–carbon bond, the process is called a **Michael reaction** or **Michael addition**, after the American chemist Arthur Michael (1853–1942). Over the years, however, the term *Michael reaction* has evolved into a generic description of conjugate additions involving any nucleophile.

problem 17.24 Draw the complete 1,2- and 1,4-addition mechanisms for each of the reactions at the right.

(a)   (b)

Which of the two possible nucleophilic addition mechanisms dominates (i.e., direct addition or conjugate addition) depends in large part on whether the nucleophile adds to the carbonyl carbon reversibly or irreversibly (Equation 17-33).

$$ \text{(17-33)} $$

In general,

> ■ Nucleophiles that add *reversibly* to the carbonyl carbon yield the *conjugate addition* product as the major product.
>
> ■ Nucleophiles that add *irreversibly* to the carbonyl carbon yield the *direct addition* product as the major product.

Whether or not the nucleophilic addition in Equation 17-33 is reversible depends, in turn, on the charge stability that each side of the reaction exhibits. More specifically, *nucleophilic addition tends to be irreversible if the negative charge that develops in the adduct is substantially better stabilized than it is in the nucleophile*. In practice, these situations are limited to ones in which the nucleophile has the negative charge located on a carbon or hydrogen atom, and the negative charge is not stabilized by resonance or inductive effects. Therefore,

> ■ Nucleophilic addition to a carbonyl carbon tends to be irreversible when it involves a very strong R^- or H^- nucleophile.
>
> ■ Otherwise, the nucleophilic addition tends to be reversible.

These ideas are summarized in Table 17-2.

For example, $C_6H_{11}MgBr$ (in which $C_6H_{11}^-$ is the nucleophile) adds irreversibly to a carbonyl C atom, so the major product is from direct addition (Equation 17-34).

The major product is from 1,2-addition.

$$ \text{(17-34)} $$

71%

On the other hand, CH_3O^- adds reversibly, so the major nucleophilic addition product is from conjugate addition (Equation 17-35).

The major product is from 1,4-addition.

$$ \text{(17-35)} $$

80%

TABLE 17-2 Reversibility in Nucleophilic Addition

Nucleophiles That Add *Reversibly*	Nucleophiles That Add *Irreversibly*
$HO^{\ominus}$, $RO^{\ominus}$	$R-MgBr$ ($R^{\ominus}$)
$H_2N^{\ominus}$, $R_2N^{\ominus}$	$R-Li$ ($R^{\ominus}$)
$Cl^{\ominus}$, $Br^{\ominus}$, $I^{\ominus}$	
	$LiAlH_4$ ($H^{\ominus}$)
$N{\equiv}C^{\ominus}$	$NaBH_4$ ($H^{\ominus}$)

problem **17.25** Draw the major product for the reaction between cyclohex-2-en-1-one and each of the nucleophiles in **(a)–(e)**, and draw the mechanism that leads to each of those products.

CH$_3$SNa

(a)

(b)

(c)

(d)

LiAlH$_4$

(e)

Why does this regioselectivity involving 1,2- versus 1,4-addition depend upon whether the nucleophile adds reversibly to the carbonyl carbon atom? Ultimately, it is because the reversibility governs whether nucleophilic addition takes place under thermodynamic or kinetic control (review Section 9.12).

- If the nucleophile adds reversibly, then the reaction takes place under thermodynamic control and the major product is the thermodynamic product (i.e., the one that is more stable).
- If the nucleophile adds irreversibly, then the reaction takes place under kinetic control and the major product is the kinetic product (i.e., the one that is produced more rapidly).

To apply these ideas, we need to know which product is the thermodynamic product and which is the kinetic product. Generally speaking,

- The conjugate addition product is the one that is more stable, making it the thermodynamic product.
- The direct addition product is the one that is produced more rapidly, making it the kinetic product.

The conjugate addition product is more stable in part because of charge stability. Notice from Equation 17-32 that the immediate product of conjugate addition is an enolate anion, which, as shown in Figure 17-14a, has a resonance-stabilized negative charge. The direct addition product (Equation 17-31) shown in Figure 17-14b, on the other hand, possesses an isolated negative charge, and is therefore less stable.

FIGURE 17-14 Charge stability and direct versus conjugate addition The negative charge produced upon nucleophilic addition is resonance delocalized in the conjugate addition product (a) but not in the direct addition product (b), so the conjugate addition product is more stable.

Resonance-stabilized negative charge

Isolated negative charge

Conjugate addition product

Direct addition product

(a)

(b)

720 kJ/mol
(172 kcal/mol)
= more stable

Conjugate addition product

(a)

619 kJ/mol
(148 kcal/mol)

Direct addition product

(b)

FIGURE 17-15 Bond energies and direct versus conjugate addition The overall product of conjugate addition (a) is more stable than the overall product of direct addition (b) primarily because the average C=O double bond is much stronger than the average C=C double bond.

A second reason that conjugate addition leads to the thermodynamic product is that the *overall* uncharged product of conjugate addition has greater total bond energy than the one produced from direct addition. The greater total bond energy in the conjugate addition product comes primarily from the type of double bond present. As shown in Figure 17-15, the conjugate addition product (Fig. 17-15a) has a C=O double bond, which, on average, is 101 kJ/mol (24 kcal/mol) stronger than the C=C double bond in the direct addition product (Fig. 17-15b).

If the direct addition product is the kinetic product, then it must be formed faster, which means the path to form it has a smaller activation energy (Fig. 17-16). Why this is so is due to the greater concentration of positive charge on the carbonyl C atom than on the beta C atom. Therefore, in the initial stages of nucleophilic attack, the nucleophile is better stabilized by its attraction to the carbonyl C atom than by its attraction to the beta C atom.

Which product of the following nucleophilic addition reaction is the *thermodynamic product* and which is the *kinetic product*?

FIGURE 17-16 Direct versus conjugate addition of a generic nucleophile to propenal The blue curve represents direct addition (Equation 17-31), whereas the red curve represents conjugate addition (Equation 17-32). The energy barrier is lower for direct addition because, in the transition state, the nucleophile is attracted more strongly to the carbonyl carbon than to the beta carbon. Thus, direct addition gives the kinetic product. The more stable product is formed from conjugate addition, so conjugate addition gives the thermodynamic product.

Smaller energy barrier

Stronger attraction

Better stabilized negative charge

Greater total bond energy

Kinetic product

Thermodynamic product

Free energy ⟶

Reaction coordinate ⟶

Michael Addition in the Fight against Cancer

Several cancer drugs work by taking advantage of the fact that cancer cells grow and divide significantly faster than normal cells. In the box on page 554, for example, we saw how one cancer drug—mechlorethamine—uses S_N2 reactions to tether together two strands of DNA, thereby disrupting DNA function. Eriolangin, isolated from *Eriophyllum lanatum*, is part of another class of cancer drugs that uses Michael additions to combat cancer—reactions that we have studied here in Chapter 17.

Rather than target the DNA itself, eriolangin reacts with DNA polymerase, an enzyme that facilitates the replication of DNA. This is made possible by the two α,β-unsaturated carbonyl groups (highlighted in red below) that can react with the nucleophilic thiol (SH) group of a cysteine amino acid in DNA polymerase. As shown on the right, this reaction is conjugate addition (or Michael addition) instead of direct addition because a thiol group will add to the carbonyl carbon reversibly. The DNA polymerase in the product is incapable of copying additional DNA.

As with other cancer drugs, the way in which eriolangin acts to fight cancer is a double-edged sword. Other α,β-unsaturated carbonyl compounds that enter the body can pose a threat to the normal function of DNA polymerase. One notable example is acrolein ($H_2C=CH-CH=O$), a toxin that is found in french fries and other foods fried in oil. In small quantities, however, acrolein does not pose a significant threat, in part because of the presence of glutathione in the body. Glutathione is a type of tripeptide that has a cysteine amino acid, and therefore has a nucleophilic SH group, so it can bind to these kinds of toxins via conjugate addition and safely remove them from the body.

Eriophyllum lanatum

Eriolangin

α,β-Unsaturated carbonyls

Active DNA polymerase

Inactive DNA polymerase

Conjugate addition

SOLVED problem **17.26** Although thus far we have examined conjugate addition only as it pertains to carbonyl compounds, competition between direct addition and conjugate addition can also occur with other α,β-unsaturated polar π bonds. An example involving an α,β-unsaturated nitrile is shown at the right. Predict whether the reaction will take place via direct addition or conjugate addition, and draw the major product.

Think Which group has the polar π bond? Will the nucleophile add reversibly or irreversibly? How does that affect whether direct addition or conjugate addition takes place?

Solve The C≡N group has the polar π bond. The nucleophile is CH_3O^-, which, as we know from Table 17-2, adds reversibly. As a result, the thermodynamic product, which is formed via conjugate addition, is favored. The mechanism is essentially the same as that in Equation 17-32.

---

problem **17.27** Predict the major product of the reaction at the right.

17.11 Lithium Dialkylcuprates and the Selectivity of Organometallic Reagents

Organocopper reagents, also called organocuprates, constitute an important class of organometallic compounds. Among the most common organocuprates are **lithium dialkylcuprates, R_2CuLi**, sometimes referred to as **Gilman reagents** (after the American chemist Henry Gilman, 1893–1986). As with other organometallic reagents, the carbon atoms of lithium dialkylcuprates are electron rich, because carbon (EN = 2.55) has a higher electronegativity than copper (1.90). As a result, these reagents, too, can be treated as simply R^-.

Lithium dialkylcuprates are much less reactive than either alkyllithium reagents (RLi) or Grignard reagents (RMgBr). Largely this is because the C—Cu bond is significantly less polar than either the R—Mg bond or the R—Li bond, as shown in Figure 17-17. Because the electronegativity of Cu (1.90) is greater than that of Mg (1.31) or Li (0.98), R_2CuLi has the smallest concentration of negative charge on C.

Unlike alkyllithium and Grignard reagents, lithium dialkylcuprates tend *not* to undergo direct addition to a polar π bond.

> When an α,β-unsaturated carbonyl compound is treated with R_2CuLi, R^- adds almost exclusively at the beta carbon via conjugate addition.

FIGURE 17-17 Relative reactivities of organometallic reagents As the electronegativity of the metal atom in the metal–C bond decreases, the electronegativity difference between the metal and C atoms increases, resulting in a larger partial negative charge on C and a greater reactivity.

An example is shown in Equation 17-36.

$$\text{(17-36)}$$

98%

The reason for this regioselectivity is not precisely known because the specific mechanism for the reaction of lithium dialkylcuprates is not fully understood. However, chemists generally believe that the species that initially attacks as a nucleophile is *not* R⁻. Instead, many believe that a key step is the reversible binding of R_2Cu^- to the C=C double bond, followed by the transfer of the alkyl group. As we have seen previously with nucleophiles that add reversibly to the carbonyl carbon, the reversible nature of the initial binding may explain the tendency of organocopper reagents to react via conjugate addition instead of direct addition.

SOLVED problem 17.28 Predict the major organic product of the reaction at the right.

Think To what R⁻ nucleophile can this lithium dialkylcuprate be simplified? Will it add predominantly via direct addition or conjugate addition? What role does H_2O play after the addition of R⁻ is complete?

Solve This lithium dialkylcuprate is a source of H_3C^-, which will add to the α,β-unsaturated carbonyl via conjugate addition. The resulting enolate anion is protonated by H_2O.

---

problem 17.29 Predict the major product of each of the following reactions.

(a)

(b)

(c)

Notice in Figure 17-17 that the metal–C bond is more polar in alkyllithium reagents than in Grignard reagents. With a greater partial negative charge on the C atom in alkyllithium reagents,

Alkyllithium reagents (RLi) tend to be significantly more reactive than Grignard reagents (RMgBr).

An ether solution of an alkyllithium reagent, for example, reacts violently with water when exposed to air. Although Grignard reagents react with water, too, they do so much less violently.

The greater concentration of charge in alkyllithium reagents than in Grignard reagents is also reflected in their *selectivity* regarding 1,2- versus 1,4-addition. Even though both reagents favor 1,2-addition,

Alkyllithium reagents generally have a greater tendency toward 1,2-addition to α,β-unsaturated carbonyls than Grignard reagents do.

As shown in Equations 17-37 and 17-38, for example, the Grignard reagent yields a significant (although minor) amount of conjugate addition product, whereas the alkyllithium reagent does not.

problem 17.30 Predict the major product of each of the following reactions.

(a)   (b)   (c)

17.12 Organic Synthesis: Grignard and Alkyllithium Reactions in Synthesis

One of the most important reactions in organic synthesis is the one between a carbonyl-containing compound—such as a ketone or aldehyde—and either a Grignard or an alkyllithium reagent. As shown in Equation 17-39, there are two main reasons why these reactions are so important.

First, they are carbon–carbon bond-forming reactions, so they can play a key role in constructing the desired carbon skeleton of a target compound. Second, an OH group is generated at one of the carbon atoms joined by the new bond, so we can use the characteristic reactivity of the OH group to carry out further changes at that carbon.

For example, an alcohol can be converted into an ether via a *Williamson ether synthesis* (Section 10.6). One way to do so, shown in Equation 17-40, is to add a strong base such as NaH, and treat the resulting alkoxide anion (RO⁻) with an alkyl halide (R—Br).

Alternatively, as shown in Equation 17-41, the product alcohol can be converted to an alkyl halide using PBr$_3$ (Section 10.1), and can then be treated with an alkoxide salt.

Recall from Sections 17.7–17.9, too, that alkyl halides are precursors to Wittig reagents and sulfonium ylides. Therefore, as shown in Equation 17-42, the alkyl halide generated in Equation 17-41 could instead be used to facilitate a Wittig reaction.

From Eq. 17-41 **Formation of a Wittig reagent** **A Wittig reaction**

$$\text{(17-42)}$$

In Chapter 19, we explain how an alcohol can be converted into a Grignard or an alkyllithium reagent, and also how it can be oxidized to a ketone or an aldehyde. In these ways, the alcohol product of a Grignard or alkyllithium reaction can be used to facilitate another such reaction.

Because of the utility of Grignard and alkyllithium reactions in synthesis, it is important to be able to undo these reactions comfortably in your mind to execute transforms efficiently in a retrosynthetic analysis. These transforms can be carried out on essentially any alcohol by disconnecting the bond between the alcohol carbon and an adjacent carbon, as shown in Equation 17-43.

Disconnect the C—C bond adjacent to the OH group.

C=O bond

Undo a Grignard reaction.

C–metal bond

$$\text{(17-43)}$$

One precursor is a ketone or an aldehyde, whose carbonyl group contains the alcohol C atom from the target. In the other precursor, the second C atom from the disconnected bond is bonded to the metal atom.

Notice, too, in Equation 17-43 that the C—C bond on the left side of the alcohol group could have been disconnected instead. In this case, we would have arrived at the same precursors as in Equation 17-43, because the target alcohol is symmetric. In general, however, disconnecting different C—C bonds will result in different precursors. An example is shown in Equation 17-44, in which the alcohol has three different C—C bonds adjacent to the OH group.

$$+ \quad \text{BrMg} - \text{CH}_3 \qquad \text{(17-44a)}$$

$$+ \quad \text{BrMg} \qquad \text{(17-44b)}$$

$$+ \qquad \text{(17-44c)}$$

Disconnecting the C—C bond indicated by the red, blue, or green wavy line yields the precursors in Equations 17-44a, 17-44b, or 17-44c, respectively.

Having the choice of which C—C bond to disconnect may at first seem daunting. With practice, however, carrying out these kinds of transforms will become quite straightforward, and you will appreciate the options that the Grignard reaction affords.

---

problem **17.31** Show how this molecule can be synthesized using each of the Grignard reagents shown in Equation 17-44.

---

problem **17.32** Show how the following compound can be synthesized from two different alkyllithium reagents.

17.13 Organic Synthesis: Considerations of Direct Addition versus Conjugate Additions

As we saw in Section 17.10, a nucleophile can attack an α,β-unsaturated carbonyl at either the carbonyl carbon, yielding the direct addition product (Equation 17-45a), or the beta carbon, yielding the conjugate addition product (Equation 17-45b).

Direct addition (17-45a)
The C=C bond remains.

Conjugate addition (17-45b)
The C=O bond remains.

Direct addition converts the carbonyl group to an OH group, leaving the C=C double bond unaltered. In conjugate addition, on the other hand, the double bond between the alpha and beta carbons is converted into a single bond, leaving the C=O bond unaltered.

When undoing one of these reactions in a retrosynthetic analysis, be sure to look for these clues. In the molecule on the left in Equation 17-46, for example, notice that the cyano group is beta to the C=O bond. Thus, the target can be generated by conjugate addition, in which NC⁻ attacks the beta carbon.

A nucleophile is at the beta position.
An intact C=O bond
Undo a conjugate addition.
+ :CN (17-46)

In the forward direction, the synthetic step might appear as in Equation 17-47.

(17-47)

In the molecule on the left in Equation 17-48, a C=C double bond is adjacent to an alcohol group, so we can envision it as the result of direct addition, with H^- as the nucleophile.

(17-48)

Thus, the forward reaction might appear as in Equation 17-49.

(17-49)

If a particular retrosynthetic analysis suggests the use of an R^- nucleophile in the forward direction, then we must also consider *regioselectivity*. Suppose, for example, that we carry out the transform in Equation 17-50, for which the precursors on the right are an α,β-unsaturated ketone and an R^- nucleophile.

(17-50)

In the forward direction, there are a variety of forms the R^- nucleophile can take, such as an alkyllithium reagent (RLi), a Grignard reagent (RMgBr), or a lithium dialkylcuprate (R_2CuLi). We want the nucleophile to attack at the beta carbon, however, so we should choose a lithium dialkylcuprate, which leads almost exclusively to the conjugate addition product. Thus, the forward reaction would appear as in Equation 17-51.

(17-51)

problem **17.33** Show
how you would synthesize each
of these molecules from
hex-4-en-3-one.

(a)　　　　　　　　(b)

17.14 Organic Synthesis: Considerations of Regiochemistry in the Formation of Alkenes

In Section 17.7, we learned that alkenes can be synthesized using the Wittig reaction. In Chapters 8–10, we also saw examples of alkene formation via E1 and E2 reactions. The Wittig reaction is generally regarded as the more synthetically useful reaction, however, due to its *regiospecificity*. In a Wittig reaction, a C=C double bond forms precisely at the location of the initial C=O bond in the ketone or aldehyde. E1 and E2 reactions, on the other hand, typically form a mixture of alkene isomers, and occasionally the desired isomer is the minor product.

Suppose, for example, that we want to synthesize 1-cyclopentylpent-1-ene. As shown in Equation 17-52, this can be accomplished straightforwardly using the Wittig reaction.

In a Wittig reaction, the C=C bond forms at the C atom of the initial C=O bond.

$$+ \quad (C_6H_5)_3\overset{\oplus}{P} \quad \longrightarrow \quad \text{(17-52)}$$

1-Cyclopentylpent-1-ene
(E + Z mixture)

However, attempting to synthesize this compound using an elimination reaction is problematic. One possible precursor is 1-bromo-1-cyclopentylpentane, as shown in Equation 17-53, but the major product is an undesired isomer—a trisubstituted alkene.

An elimination reaction can produce a mixture of isomeric alkenes.

Trisubstituted alkene   **Disubstituted alkene**

$$\xrightarrow[\Delta]{CH_3CH_2ONa} \quad + \quad \text{(17-53)}$$

Br
1-Bromo-1-cyclopentylpentane

Major product

According to Zaitsev's rule (Section 9.10), elimination favors the more highly substituted alkene. In this case, the desired product is the less highly substituted alkene—a disubstituted alkene.

A similar problem arises if 2-bromo-1-cyclopentylpentane is used as the precursor. As shown in Equation 17-54, both possible elimination products are disubstituted alkenes, so a significant amount of each will be produced.

Disubstituted alkene   **Disubstituted alkene**

Br

$$\xrightarrow[\Delta]{CH_3CH_2ONa} \quad + \quad \text{(17-54)}$$

2-Bromo-1-cyclopentylpentane

The main lesson from these examples is that, when given the option of using either a Wittig reaction or an elimination reaction to synthesize an alkene, a Wittig reaction is usually the better choice.

17.15 Wrapping Up and Looking Ahead

Chapter 17 began by introducing the fundamental mechanism by which a strong nucleophile adds to a polar π bond, called *direct addition* or *1,2-addition*. In this mechanism, nucleophilic attack takes place at the electron-poor atom of a polar π bond, followed by protonation of that bond's electron-rich atom. Throughout the remainder of Chapter 17, we examined the utility of a variety of nucleophiles that undergo nucleophilic addition. $LiAlH_4$ and $NaBH_4$ serve to reduce polar π bonds; most notably, ketones and aldehydes are reduced to alcohols. Grignard and alkyllithium reagents are used to form a new C—C single bond to the electron-poor carbon atom of the polar π bond, whereas lithium dialkylcuprates are used to form a new C—C bond to the polar π bond's beta carbon. Wittig reagents are used to synthesize alkenes, and sulfonium ylides are used to synthesize epoxides.

If the polar π bond is α,β-unsaturated, a second nucleophilic addition mechanism is also available, called *conjugate addition* or *1,4-addition*. Whether the direct or conjugate addition mechanism dominates depends on the reversibility of the reaction. Nucleophiles that add *irreversibly* to the polar π bond favor the direct addition product. These include H^- from either $LiAlH_4$ or $NaBH_4$, R^- from either RMgBr or RLi, and both Wittig reagents and sulfonium ylides. Other nucleophiles that add *reversibly* favor conjugate addition.

In Chapter 18, we examine nucleophilic addition reactions that begin with relatively weak nucleophiles, often requiring acid or base catalysis. Among those are some of the most useful reactions in organic synthesis, such as the *aldol reaction*. The aldol reaction is an important carbon–carbon bond-formation reaction that allows us to join two separate carbon backbones, each of which can be quite elaborate. Other reactions we discuss in Chapter 18 are more specifically aimed at altering particular functional groups.

Chapter Summary and Key Terms

- The general mechanism for the addition of a strong, negatively charged nucleophile (Nu^-) to a **polar π bond** consists of two steps:

Nucleophilic attack occurs in the first step, followed by protonation. **(Section 17.1)**

- Some carbonyl-containing functional groups, including carboxylic acids (RCO_2H), esters (RCO_2R), amides ($RCONR_2$), acid chlorides (RCOCl), and acid anhydrides (RCO_2COR), have a *leaving group* attached to the carbonyl carbon. These functional groups have reactivities that differ from those of ketones and aldehydes; they are discussed in Chapter 20. **(Section 17.1)**

- Nucleophilic addition to the carbonyl group of an aldehyde tends to be more energetically favorable and faster than addition to the carbonyl group of a ketone. **(Section 17.2)**

- **Lithium aluminum hydride ($LiAlH_4$)** and **sodium borohydride ($NaBH_4$)** are sources of H^- that behave as **reducing agents** in the presence of a polar π bond. Both reagents reduce ketones ($R_2C{=}O$) and aldehydes ($RCH{=}O$) to alcohols (ROH). A nitrile (RCN) can be reduced to a primary amine (RCH_2NH_2) by $LiAlH_4$, but not by $NaBH_4$. **(Section 17.3)**

- $LiAlH_4$ is a very strong base, so it is incompatible with protic solvents like water and alcohols. **(Section 17.3)**

- **Sodium hydride (NaH)** is an ionic hydride, and thus a strong base, but it is not nucleophilic. **(Section 17.4)**

- **Grignard reagents (RMgBr)** and **alkyllithium reagents (RLi)** are strong nucleophiles and strong bases that are incompatible with acidic functional groups or those that are prone to nucleophilic attack. **(Section 17.5)**

- A **Wittig reagent**, also called a **phosphonium ylide**, is characterized by a ^+P—C:$^-$ bond and is strongly nucleophilic at the negatively charged carbon atom. In a **Wittig reaction**, the phosphonium ylide undergoes nucleophilic addition to the C=O bond of a ketone or aldehyde and, after subsequent elimination, results in the formation of an alkene. **(Section 17.7)**

- A **sulfonium ylide** is characterized by a ^+S—C:$^-$ bond and is strongly nucleophilic at the negatively charged carbon atom. A sulfonium ylide will add to the C=O bond of a ketone or aldehyde, ultimately producing an epoxide. **(Section 17.9)**

- A polar π bond that is **α,β-unsaturated** is susceptible to nucleophilic attack at both the electron-poor atom of the polar π bond and the **beta carbon**. Attack at the carbonyl carbon yields the **direct addition (or 1,2-addition)** product, whereas attack at the beta carbon yields the **conjugate addition (or 1,4-addition)** product. **(Section 17.10)**

- Direct addition to an α,β-unsaturated polar π bond dominates if the nucleophile adds *irreversibly* to the polar π bond. Otherwise, conjugate addition is favored. **(Section 17.10)**
- Most nucleophiles add reversibly to a polar π bond. Nucleophiles that add irreversibly include H$^-$ from either NaBH$_4$ or LiAlH$_4$ and R$^-$ from either R—MgBr or R—Li. **(Section 17.10)**
- **Lithium dialkylcuprates (R$_2$CuLi)** are weak R$^-$ nucleophiles that favor conjugate addition over direct addition. **(Section 17.11)**

Reaction Tables

TABLE 17-3 Functional Group Conversions

	Starting Functional Group	Typical Reagents and Reaction Conditions	Functional Group Formed	Key Electron–Rich Species	Key Electron–Poor Species	Comments	Discussed in Section
(1)	Ketone or aldehyde	1. NaBH$_4$ or LiAlH$_4$ 2. H$_3$O$^+$	Alcohol	:H$^-$ Hydride anion		Nucleophilic addition	17.3
(2)	Imine	1. LiAlH$_4$ 2. H$_3$O$^+$	Amine	:H$^-$ Hydride anion		Nucleophilic addition	17.3
(3)	Nitrile R—C≡N	1. LiAlH$_4$ 2. H$_3$O$^+$	1° Amine	:H$^-$ Hydride anion	—C≡N	Sequential nucleophilic additions	17.3
(4)	α,β-Unsaturated ketone or aldehyde	1. NaBH$_4$ or LiAlH$_4$ 2. H$_3$O$^+$	Alcohol	:H$^-$ Hydride anion		Nucleophilic addition	17.10
(5)	Ketone or aldehyde	NaH	Enolate anion	:H$^-$ Hydride anion		Proton transfer	17.4

TABLE 17-3 Functional Group Conversions (continued)

Starting Functional Group	Typical Reagents and Reaction Conditions	Functional Group Formed	Key Electron–Rich Species	Key Electron–Poor Species	Comments	Discussed in Section
(6) Alkyl halide	1. $P(C_6H_5)_3$ 2. R–Li	Wittig reagent	Triphenyl phosphine		S_N2 followed by proton transfer	17.7
(7) Alkyl halide	1. $S(CH_3)_2$ 2. NaH	Sulfonium ylide			S_N2 followed by proton transfer	17.9

TABLE 17-4 Reactions That Alter the Carbon Skeleton

Starting Functional Group	Typical Reagent and Reaction Conditions	Functional Group Formed	Key Electron–Rich Species	Key Electron–Poor Species	Comments	Discussed in Section
(1) Ketone or aldehyde	1. R–Li or R–MgBr 2. H_3O^+	Alcohol	:R^-		Nucleophilic addition	17.5
(2) Nitrile	1. R–MgBr 2. CH_3OH	Imine	:R^-	$R-C\equiv N$	Nucleophilic addition	17.5
(3) Grignard reagent	1. $CO_2(s)$ 2. NH_4Cl	Carboxylic acid	:R^-	$O=C=O$	Nucleophilic addition	17.5
(4) α,β-Unsaturated ketone or aldehyde	1. R_2CuLi, THF 2. NH_4Cl	Ketone or aldehyde	:R^-		Conjugate nucleophilic addition	17.11
(5) Ketone or aldehyde	Wittig reagent	Alkene	$^+PPh_3$		Wittig reaction	17.7
(6) Ketone or aldehyde	Sulfonium ylide	Epoxide	$S(CH_3)_2$		Epoxidation	17.9

Problems

17.34 Which of the following nucleophiles will add *reversibly* to a polar π bond? Which will add *irreversibly*?

(a)

(b)

MgBr

(c)

ONa

(d)

ONa

(e)

CuLi

(f)

Li

(g)

H_2O

(h)

(i)

(j)

SNa

(k)

SH

17.35 For each pair of compounds, which has the polar π bond that will undergo nucleophilic addition more rapidly? Why?

(a)

or

(b)

CH₃   or   CF₃

(c)

or

CI

(d)

Cl_3C—C≡N   or H_3C—C≡N

17.36 Draw the mechanism and predict the major product for each of the following reactions.

(a)

MgBr

1.

2. $H_3O^{\oplus}$?

(b)

1.

Li

2. H_2O ?

(c)

CH_3MgBr

1. $CO_2(s)$

2. $H_3O^{\oplus}$?

(d)

H_3C

C≡N

1. LiAlH₄, ether

2. H_2O ?

CH_3

(e)

N

CH₃

1. LiAlH₄, ether

2. H_2O ?

(f)

CH₃

NaCN

H_2O ?

17.37 Draw the mechanism and predict the major product for each of the following reactions.

(a)

1. KCN, H₂O
2. H₂SO₄

?

(b)

1. KCN, H₂O
2. H₂SO₄

?

(c)

1. CH₃MgBr, ether
2. H₂O

?

(d)

1. ⟨ ⟩₂CuLi
2. H₂O

?

(e)

CH₃CH₂SNa
Ethanol

?

17.38 Predict the major organic product and draw the complete, detailed mechanism for each of the following reactions.

(a)

OH

1. PBr₃
2. (C₆H₅)₃P
3. Bu–Li
4. Acetone

?

(b)

1. (CH₃)₂S⁺—CH₂⁻
2. CH₃SNa, H₂O

?

(c)

1. C₆H₅MgBr
2. H₃O⁺

?

1. NaH
2. CH₃CH₂Br

?

(d)

1. NaBH₄, EtOH
2. PBr₃
3. (CH₃)₂S
4. NaH

?

H₃C—CH=O

?

17.39 Show two different syntheses for the compound at the right, one using (CH₃)₂CuLi as a reagent and the other using (CH₃CH₂)₂CuLi.

17.40 Propose three different syntheses of the alcohol at the right, each using a different Grignard reagent.

17.41 The chemical behavior of deuterium (D or ²H) is essentially identical to that of hydrogen (¹H). Therefore, D₂O behaves the same as H₂O, LiAlD₄ behaves the same as LiAlH₄, and CD₃Li behaves the same as CH₃Li. With this in mind, draw the detailed mechanism of each of the following reactions. Using the mechanism, predict the reaction products.

(a)

1. LiAlH₄
2. D₂O

?

(b)

1. LiAlD₄
2. H₂O

?

(c)

1. LiAlD₄
2. D₂O

?

(d)

1. CH₃Li
2. D₂O

?

(e)

1. CD₃Li
2. H₂O

?

(f)

1. CD₃Li
2. D₂O

?

17.42 Consider the following reaction in which an epoxide is formed.

If $^{13}CH_3I$ were used instead of $^{12}CH_3I$, then what percentage of the epoxide product would contain ^{13}C?

17.43 Determine the structures of compounds **A–J** in the following reaction sequences.

17.44 Determine the structures of compounds **K–Y** in the following reaction sequences.

17.45 Alkyllithium and Grignard reagents are highly reactive with protic solvents (e.g., water and alcohols), so they require aprotic solvents (e.g., ethers). Acetone is an aprotic solvent that can be used in nucleophilic substitution and elimination reactions, but it cannot be used as a solvent for reactions involving alkyllithium and Grignard reagents. Explain why.

17.46 Provide the reagents necessary to perform the following transformation.

17.47 Both NaBH$_4$ and LiAlH$_4$ favor direct addition to an α,β-unsaturated carbonyl over conjugate addition, but their selectivities differ. Which of these hydride reducing agents would be more selective toward direct addition? Why? (*Hint*: Review Section 17.11.)

17.48 The conjugate addition of NaBH$_4$ to an α,β-unsaturated carbonyl results in two sequential nucleophilic additions, but the conjugate addition of a lithium dialkylcuprate (R$_2$CuLi) results in only a single nucleophilic addition. Explain.

17.49 Draw the complete, detailed mechanism for the following reaction.

17.50 Unlike most hydrates, the hydrate of cyclopropanone is stable and can be isolated. Explain why this hydrate is stable.

17.51 When carbon disulfide (CS$_2$) is treated with an alcohol in the presence of base, the product is a xanthate salt. If an alkyl halide is also present, a xanthate ester is formed. Propose a mechanism for each of these reactions.

A xanthate salt

A xanthate ester

17.52 The treatment of a nitrile with cyanide yields an α-aminomalonitrile, as shown in the following reaction.

An α-aminomalonitrile

Provide a detailed mechanism for this reaction.

17.53 Phenylmagnesium bromide reacts with sulfur dioxide to produce a reactive intermediate, which, upon further reaction with CH_3Br, produces methylphenylsulfone.

Propose a mechanism for this reaction, and propose a structure for the reactive intermediate.

17.54 An α,β-unsaturated ketone reacts with a conjugated Wittig reagent to produce a relatively highly strained bicyclic compound.

Draw the complete, detailed mechanism for this reaction. (A key intermediate has been provided.)

17.55 Suggest how you would synthesize each of the following from phenylethanone (acetophenone), using any reagents necessary. (*Hint*: Each synthesis may require more than one synthetic step.)

Phenylethanone (Acetophenone)

(a)   (b)   (c)   (d)

17.56 Suggest how you would carry out the following synthesis using any reagents necessary. (*Hint*: The synthesis may require more than one synthetic step.)

17.57 Suggest how you would carry out the following synthesis using any reagents necessary. (*Hint*: The synthesis may require more than one synthetic step.)

17.58 Show how you would carry out the following synthesis using any reagents necessary. (*Hint*: The synthesis may require more than one synthetic step.)

17.59 Suggest how you would carry out the following synthesis using any reagents necessary. (*Hint*: The synthesis may require more than one synthetic step.)

17.60 The following is an example of the Corey–Chaykovsky aziridination reaction.

H₃C—S—CH₃

1. CH₃I
2. NaH
3. NH (R—C(=)—R')

→ HN—CH₂ / R—C—R'

Draw its complete, detailed mechanism.

17.61 The following is an example of the Corey–Chaykovsky cyclopropanation.

Draw its complete, detailed mechanism.

17.62 The reaction of a phosphonium ylide with a ketone or aldehyde produces an alkene, but the reaction of a sulfonium ylide produces an epoxide. Explain why. (*Hint*: What is the major driving force for the Wittig reaction to produce an alkene? The S—O energy in dimethyl sulfoxide is 362 kJ/mol.)

17.63 When propenal is treated with sodium acetylide, a product is formed whose IR spectrum exhibits a broad absorption between 3200 and 3600 cm⁻¹ and shows no absorption near 1650 cm⁻¹.
(a) Draw the structure of the product.
(b) Argue whether the nucleophile adds *reversibly* or *irreversibly* to the carbonyl group.

HC(CH₂=)—CH(=O) 1. HC≡CNa 2. NH₄Cl → ?

17.64 When 5-bromopentanal is treated with sodium borohydride, a compound is produced whose ¹³C NMR spectrum is shown here. In its IR spectrum, no absorption bands appear near 1700 cm⁻¹ or above 3000 cm⁻¹. Propose a mechanism to account for the formation of this product.

Chemical shift (ppm)

17.65 Lithium dimethylcuprate reacts with the beta alkynyl carbonyl shown in the following reaction.

The IR spectrum of the product is shown here and the ^{1}H NMR spectrum has the following four signals: 1.9 ppm, 3H; 2.1 ppm, 3H; 2.2 ppm, 3H; and 6.1 ppm, 1H. What is the structure of the product?

17.66 When phenyl-5-bromopentanone is treated with triphenylphosphine, followed by base, a compound is produced whose formula is $C_{11}H_{12}$.

Its ^{13}C NMR spectrum has nine signals, six of which appear between 120 and 140 ppm and three of which appear below 50 ppm. For this reaction, draw the complete, detailed mechanism, as well as the major organic product.

17.67 Although a Wittig reagent can be prepared from 5-bromo-1,3-cyclopentadiene in the usual way, that Wittig reagent is unreactive toward ketones or aldehydes.

(a) Draw the complete mechanism showing the formation of the Wittig reagent.
(b) Explain why that Wittig reagent does not undergo nucleophilic addition with ketones or aldehydes.

If your instructor assigns problems in smartwork, log in at **smartwork.wwnorton.com**.

CHAPTER 18

Nucleophilic Addition to Polar π Bonds 2

Weak Nucleophiles and Acid and Base Catalysis

Chapter 17 discussed nucleophilic addition reactions involving strong nucleophiles as reactants—for example, $NaBH_4$ and $LiAlH_4$ as sources of H^-, organometallic reagents as sources of R^-, Wittig reagents, and sulfonium ylides. Here in Chapter 18, we maintain our focus on nucleophilic addition, examining reactions that involve weakly nucleophilic reactants. Such reactions generally require catalysis by a strong acid or a strong base. Under such conditions, the nucleophilic addition mechanism is altered slightly from the one we saw in Chapter 17, providing a lower-energy route to forming products. Of these reactions, we spend the greatest amount of time on the *aldol reaction*, because it is one of the most important and versatile reactions in organic synthesis.

Our skin can resume its original shape after being stretched in part because of elastin, a protein that exhibits crosslinked bonding between its long-chain molecules. This crosslinking is the result of an aldol condensation reaction, which we examine here in Chapter 18.

18.1 Addition of Weak Nucleophiles: Acid and Base Catalysis

Thus far, we have dealt mainly with nucleophilic addition reactions in which the starting materials include a strong nucleophile. In general, those nucleophiles bear a full negative charge or (in the case of hydride reagents and organometallic reagents) can be treated *as if* they bear a full negative charge. As we saw in Chapters 8–10, however, a variety of *uncharged* species can act as nucleophiles, including water, alcohols (ROH), amines (RNH_2), thiols (RSH), and phosphines (RPH_2). These nucleophiles can also add to polar π bonds.

Equation 18-1 shows that ethanol, a weak nucleophile, can add to the C=O group of butan-2-one to produce a **hemiacetal**, a compound in which a carbon atom is bonded to both an OH group and an OR group. This reaction tends to be quite slow, however.

(18-1)

This reaction can be sped up dramatically when a small amount of a strong base or a strong acid is added, as shown in Equation 18-2.

(18-2)

In other words,

> The formation of a hemiacetal from a ketone or aldehyde can be base or acid catalyzed; the reaction is said to undergo **base** or **acid catalysis**.

The reasons for these increased reaction rates can be understood by considering the respective mechanisms. The uncatalyzed reaction from Equation 18-1 is shown in Equation 18-3.

Mechanism for Equation 18-1 (uncatalyzed)

The appearance of these new charges makes Step 1 slow.

(18-3)

The nucleophile adds to the C=O carbon in Step 1 to produce an intermediate with two formal charges, and proton transfers in Steps 2 and 3 produce the uncharged product. The mechanism for the base-catalyzed reaction is shown in Equation 18-4.

Mechanism for Equation 18-2 under basic conditions

(18-4)

The strong base deprotonates the weak ROH nucleophile in Step 1, generating RO⁻, a strong nucleophile. In Step 2, RO⁻ adds to the C=O carbon, and the resulting species is protonated by ROH in Step 3. Notice that the RO⁻ nucleophile is regenerated in Step 3, so it can be recycled in Step 2. Because it is responsible for increasing the reaction rate but is not consumed overall, RO⁻ is a *catalyst*.

In the acid-catalyzed mechanism in Equation 18-5, the carbonyl O atom is protonated in Step 1, and the nucleophile attacks in Step 2. The deprotonation in Step 3 produces the uncharged product. Similar to the base-catalyzed mechanism, notice that the strong acid is regenerated in Step 3 and can be recycled in Step 1.

Mechanism for Equation 18-2 under acidic conditions

(18-5)

The dramatic increase in reaction rate under basic and acidic conditions can be understood by considering the nucleophilic addition step in each mechanism. In the mechanism for the uncatalyzed reaction (Equation 18-3), the nucleophilic addition step produces two new formal charges. By contrast, in the mechanisms for the base- and acid-catalyzed reactions (Equations 18-4 and 18-5), the nucleophilic addition step produces no additional formal charges.

YOUR TURN 18.1

At each stage of the mechanism in Equations 18-3, 18-4, and 18-5, count the total *number* of charges that exist, not the total charge (e.g., one positive charge and one negative charge should be counted as two charges).

Answers to Your Turns are in the back of the book.

These differences in the nucleophilic addition steps can be seen in their respective energy diagrams in Figure 18-1. The reactants in the uncatalyzed reaction (Fig. 18-1a) are uncharged, so they appear lower in energy than those in either the

FIGURE 18-1 Nucleophilic addition and acid/base catalysis Free energy diagrams for the nucleophilic addition steps of (a) the uncatalyzed mechanism in Equation 18-3, (b) the base-catalyzed mechanism in Equation 18-4, and (c) the acid-catalyzed mechanism in Equation 18-5. For the uncatalyzed mechanism, the two additional charges produced result in a larger energy barrier and a slower reaction.

Greater concentration of negative charge

The carbonyl group is *activated*.

Greater concentration of positive charge

Neutral conditions

(a)

Basic conditions

(b)

Acidic conditions

(c)

base- or acid-catalyzed reaction (Figs. 18-1b and 18-1c), which have a single charge. By the same token, the nucleophilic addition product in the uncatalyzed reaction is higher in energy than the product in the base- or acid-catalyzed reaction. As a result, the nucleophilic addition step is substantially more favorable when the reaction is base or acid catalyzed. The Hammond postulate (Section 9.3a), therefore, suggests a smaller energy barrier under acid or base catalysis, and thus a faster reaction.

Another way to understand these increases in reaction rate is to see that the electrostatic attraction between the nucleophile and electrophile is increased under basic and acidic conditions, as shown in Figure 18-2. That is,

FIGURE 18-2 Nucleophile and electrophile strength under various conditions (a) The electrostatic attraction between the nucleophile and electrophile is weakest under neutral conditions. (b) Under basic conditions, the nucleophile acquires a full negative charge. (c) Under acidic conditions, the partial positive charge on the carbonyl C atom increases due to the second resonance contributor.

- Under basic conditions, the nucleophile becomes stronger.
- Under acidic conditions, the electrophile becomes stronger.

Under basic conditions, the electrostatic attraction is increased when the uncharged nucleophile is converted into a negatively charged nucleophile. Under acidic conditions, the protonated carbonyl group acquires a larger partial positive charge on its C atom, stemming from its resonance contributor, and the carbonyl group is said to be **activated**.

Which of the following nitrile groups is activated toward nucleophilic attack?

18.2 YOUR TURN

SOLVED problem 18.1 Recall from Chapter 17 that water can add to a ketone or aldehyde to form a *hydrate*. Draw the mechanism for this reaction under **(a)** neutral and **(b)** basic conditions. Under which conditions do you think this hydration reaction will proceed faster? Why?

$$+ \; H_2O \;\rightleftharpoons$$

Cyclohexanone

Cyclohexanone hydrate

Think What nucleophiles are present under neutral conditions? What nucleophiles are present under basic conditions? What is the total number of charges before and after the nucleophilic addition step in each mechanism?

Solve Under neutral conditions, water acts as the nucleophile.

Under basic conditions, HO⁻ is present in a significant concentration and can thus act as the nucleophile.

Under neutral conditions, the nucleophilic addition step increases the total number of charges by two, whereas the total number of charges remains the same under basic conditions. Thus, nucleophilic addition is faster under basic conditions.

---

problem **18.2** Draw the mechanism for the hydration of cyclohexanone under acidic conditions, and argue whether this reaction should be faster or slower than the corresponding hydration under neutral conditions.

---

problem **18.3** Draw the complete, detailed mechanism for the reaction (at the right) under **(a)** basic and **(b)** acidic conditions.

18.1a Addition of HCN: The Formation of Cyanohydrins

When a ketone or aldehyde is treated with aqueous HCN, the product is a **cyanohydrin**, in which an OH group and a cyano (CN) group are bonded to the same carbon atom. An example is shown in Equation 18-6.

Cyclohexanone → **A cyanohydrin** (70%) → **An α-hydroxy acid**

$$(18\text{-}6)$$

Cyanohydrin formation is important for two reasons. First, it is another valuable *carbon–carbon bond-formation reaction*, giving us an additional means of constructing carbon backbones. Second, as discussed in greater detail in Chapter 21, the C≡N functional group is readily transformed into a carboxylic acid via a *hydrolysis* reaction. The overall product is an alpha-hydroxy acid. Not only are these compounds useful in synthesis, but some alpha-hydroxyacids, such as glycolic acid (Fig. 18-3), are used to treat a variety of skin ailments.

In the overall reaction of HCN with the carbonyl group, the cyano group adds to the carbonyl carbon and a hydrogen adds to the carbonyl oxygen. This reaction occurs in three steps, as shown in Equation 18-7.

An α-hydroxy acid

FIGURE 18-3 Glycolic acid Glycolic acid is an α-hydroxy acid used to treat some skin ailments.

Mechanism for Equation 18-6

$$(18\text{-}7)$$

In Step 1, NC⁻ is generated by an equilibrium proton transfer from the weakly acidic HCN to water. In Step 2, NC⁻ attacks the carbonyl group, generating a negative charge on the carbonyl's O atom, and in Step 3, that O atom is protonated.

The equilibrium proton transfer between HCN and water (Step 1 in Equation 18-7) heavily favors reactants, so only a small amount of NC⁻ is generated. In fact, only about 0.001% of HCN molecules dissociate into NC⁻ and H_3O^+. As a result, HCN reacts quite slowly with the carbonyl carbon under normal conditions, even though NC⁻ is a very good nucleophile.

18.3 YOUR TURN

To verify that the first step of Equation 18-7 heavily favors the reactants, determine the appropriate pK_a value for each acid in the equilibrium. (*Hint:* Consult Table 6-1.)

The reaction can be sped up dramatically by the addition of a small amount of either KCN or a strong base such as NaOH (Equation 18-8).

(18-8)

A cyanohydrin

The reaction rate increases because the addition of either KCN or NaOH increases the amount of NC⁻ present in solution. KCN is an ionic compound that dissociates into K^+ and NC⁻ in water, while the HO⁻ from NaOH deprotonates HCN quantitatively to produce NC⁻. Notice in Equation 18-7 that the NC⁻ generated in either of these ways is not consumed during the course of the reaction, because the NC⁻ that is used in Step 2 is regenerated in Step 3. Hence,

Both KCN and NaOH are *catalysts* for the nucleophilic addition of HCN.

18.4 YOUR TURN

Verify that HO⁻ quantitatively deprotonates HCN
(a) by drawing the products of the following reaction, and
(b) by writing the appropriate pK_a below each acid involved and computing the value of K_{eq}.

$$HCN + HO^{\ominus} \rightleftharpoons$$

problem **18.4** Draw the complete, detailed mechanism and predict the products of each of the following reactions.

(a)

(b)

18.1b Direct versus Conjugate Addition of Weak Nucleophiles and HCN

Like negatively charged nucleophiles, uncharged nucleophiles can attack α,β-unsaturated carbonyls at two locations: the carbonyl carbon to give the direct addition product (i.e., the 1,2-addition product) and the β carbon to give the conjugate addition product (i.e., the 1,4-addition product). In general:

> Uncharged nucleophiles heavily favor conjugate addition over direct addition.

An example with ethanethiol (CH_3CH_2SH) as the nucleophile is shown in Equation 18-9.

Cyclohex-2-enone   95%

(18-9)

Weak nucleophiles favor conjugate addition.

The mechanism for this reaction is shown in Equation 18-10.

Mechanism for Equation 18-9

(18-10)

1. Nucleophilic addition
2. Proton transfer
3. Proton transfer

Enol form   Keto form   (Tautomerization)

Uncharged nucleophiles favor conjugate addition because Step 1 of the mechanism, the nucleophilic addition step, is reversible, so it takes place under thermodynamic control (Section 17.10). Conjugate addition not only leads to a resonance-stabilized enolate anion, but also leads to greater total bond strength in the overall, uncharged product.

Frequently, the addition of uncharged nucleophiles to α,β-unsaturated carbonyls does not require acidic conditions. Unlike direct addition, conjugate addition of uncharged nucleophiles can proceed at a reasonable rate, in large part because of resonance stabilization in the enolate anion produced in the first step.

HCN similarly adds to α,β-unsaturated carbonyls via conjugate addition because the addition of NC^- to the carbonyl group is reversible (Table 17-2, p. 839). An example of one such reaction is shown in Equation 18-11.

Conjugate addition of NC⊖

HCN
35 °C

63%

(18-11)

SOLVED problem **18.5** Predict the product of the reaction at the right and draw its complete, detailed mechanism.

CH_3NH_2

HCl, acetone/H_2O

?

Think Does CH_3NH_2 add to the carbonyl group reversibly or irreversibly? Will this result in direct addition or conjugate addition?

Solve CH_3NH_2, an uncharged nucleophile, adds reversibly to a carbonyl group, so it will favor conjugate addition upon attacking an α,β-unsaturated carbonyl. The HCl below the reaction arrow indicates acidic conditions, so no strongly basic species should appear in the mechanism. Therefore, we do not show the nucleophile attacking in the first step, because it would produce a strongly basic O⁻. Instead, the O atom is protonated, generating an activated carbonyl.

Tautomerization

Enol form

Keto form

---

problem **18.6** Predict the major product and draw the complete, detailed mechanism for each of the reactions at the right.

HCN
KOH, H_2O
?

(a)

?

(b)

THF
?

(c)

18.2 Formation and Hydrolysis Reactions Involving Acetals, Imines, Enamines, and Nitriles

In Section 18.1, we saw that weak nucleophiles can add to polar π bonds. Frequently, the immediate product of these reactions can react further under the conditions required for nucleophilic addition. We examine some of those reactions here in

Section 18.2—namely, the formation of acetals, thioacetals, imines, and enamines from ketones or aldehydes. We also examine the hydrolysis of acetals, imines, enamines, and nitriles. As we will see, the hydrolyses of acetals, imines, and enamines are simply the reverse of their formation reactions.

18.2a The Formation and Hydrolysis of Acetals

In Equation 18-1, we saw that a *hemiacetal* forms when a ketone or aldehyde is treated with an alcohol, and in Equation 18-2, we saw that the reaction is catalyzed under either basic or acidic conditions.

> If an aldehyde or ketone is treated with a *large* excess of an alcohol under *acidic* conditions, then the hemiacetal produced from nucleophilic addition reacts further to form an **acetal**, in which two alkoxy (RO) groups are bonded to the same carbon.

The overall reaction, shown in Equation 18-12, is reversible.

The complete mechanism for this acetal formation reaction is shown in Equation 18-13.

The first three steps are identical to the acid-catalyzed nucleophilic addition in Equation 18-5. That is, the C=O bond is protonated in Step 1, which *activates* it toward nucleophilic addition by the weak ROH nucleophile in Step 2. In Step 3, deprotonation produces the uncharged hemiacetal. The remaining steps essentially make up an S_N1 reaction. In Step 4, protonation of the OH group generates a good H_2O leaving group, which departs in Step 5 to produce a resonance-stabilized carbocation. That carbocation is subsequently attacked by another ROH nucleophile in Step 6, and the proton transfer in Step 7 results in the uncharged acetal.

Identify the hemiacetal and the acetal in the mechanism in Equation 18-13.

18.5 **YOUR TURN**

Because acetal formation takes place under equilibrium conditions, the reaction can be shifted according to Le Châtelier's principle. Thus, while formation of the acetal product is favored by using excess alcohol (a reactant), the reverse reaction is favored by using excess water (a product; see Equation 18-12 again) under acidic conditions. This is exemplified in Equation 18-14, whose mechanism is precisely the reverse of that in Equation 18-13.

The mechanism for this reaction is the reverse of the one in Eq. 18-13.

An acetal

(18-14)

Equation 18-14 is a *hydrolysis* reaction, because the addition of water results in the breaking of the C—OCH_3 bonds.

problem **18.7** Draw the complete, detailed mechanism for the reaction in Equation 18-14.

As discussed in greater detail in Chapter 19, the reversibility of acetal formation enables us to use acetals as **protecting groups** for ketones and aldehydes. Whereas ketones and aldehydes are susceptible to attack by nucleophiles (at the carbonyl carbon) and bases (at the α carbon), acetals are inert to both strong nucleophiles and strong bases.

One popular alcohol used to protect ketones and aldehydes is 1,2-ethanediol (ethylene glycol), as shown in Equation 18-15. 1,2-Ethanediol has two alcohol functional groups per molecule, so only one equivalent is needed. It also forms a five-membered ring, which helps favor acetal formation.

1,2-Ethanediol (Ethylene glycol)

An acetal

(18-15)

problem **18.8** Draw a complete, detailed mechanism for the reaction in Equation 18-15.

Although a hemiacetal can form under either basic or acidic conditions, only acidic conditions promote the formation of an acetal from a hemiacetal (compare Equations 18-2 and 18-12). In other words,

A ketone or aldehyde does not readily form an acetal under basic conditions.

Acetal formation is difficult under basic conditions because the nucleophilic substitution that would be required is unfavorable. As shown in Equation 18-16, the leaving group would have to be HO^-, which, as we learned in Chapter 9, is an unsuitable leaving group for an S_N1 or S_N2 reaction.

Poor leaving group for an S_N1 or S_N2 reaction

A hemiacetal

(18-16)

---

problem **18.9** The sulfur analog of an acetal, called a thioacetal, can be produced when a ketone or aldehyde is treated with a thiol (RSH) under acidic conditions. When propane-1,3-dithiol is used, a 1,3-dithiane is produced:

Propane-1,3-dithiol

A 1,3-dithiane

As we will see in Chapter 19, dithianes have important applications in organic synthesis. Draw the complete, detailed mechanism for this 1,3-dithiane formation reaction.

18.2b The Formation and Hydrolysis of Imines and Enamines

Ammonia (NH_3) and amines (RNH_2) are also uncharged nucleophiles that can add reversibly to carbonyl groups under mildly acidic conditions, as shown in Equations 18-17 and 18-18 for ketones and aldehydes, respectively. The product is an **imine**, which has a characteristic $C{=}N$ double bond.

An imine

(18-17)

(+ Z isomer)

An imine

(18-18)

The complete mechanism for the formation of an imine from a ketone is shown in
Equation 18-19.

Mechanism for Equation 18-17

(18-19)

Resonance-stabilized carbocation

The first five steps of the mechanism for imine formation (Equation 18-19) are
identical to the first five steps of the mechanism for acetal formation (Equation 18-13).
Steps 1–3 make up an acid-catalyzed nucleophilic addition of NH_3. In Step 4, the OH
group is protonated, resulting in a good H_2O leaving group, which departs in Step 5.

The mechanisms for acetal formation and imine formation differ in Step 6. In imine
formation, the amino group is deprotonated, thus completing the second step of an E1
mechanism (Chapter 8). In acetal formation, an analogous deprotonation is unavailable,
because the oxygen atom adjacent to the carbocation is not bonded to any hydrogen atoms.

___

problem **18.10** Draw the complete, detailed mechanism for the formation of
the imine from the aldehyde in Equation 18-18.

___

Just as with acetal formation, the reversibility of imine formation makes it possible
to use Le Châtelier's principle to drive the reaction in either direction. Thus, as shown
in Equation 18-20, an imine can be *hydrolyzed* by treating it with excess water under
acidic conditions.

**An imine is hydrolyzed
under acidic conditions.**

$$\text{NH} \quad \xrightleftharpoons[H^{\oplus}]{H_2O} \quad \text{O} \quad + \quad NH_3 \qquad (18\text{-}20)$$

___

problem **18.11** Draw the complete, detailed mechanism for the hydrolysis reaction in Equation 18-20.

___

SOLVED problem 18.12

Can the imine formation at the right take place under *basic* conditions?

Think Under basic conditions, what types of species should *not* appear in the mechanism? Are all of the steps necessary for imine formation under these conditions reasonable?

Solve No strong acids can appear in the mechanism under basic conditions. Thus, the mechanism would be as follows:

Steps 1–3 are plausible. The appearance of a positive charge on N is reasonable under basic conditions because species like H_4N^+ are only weakly acidic. Step 4, however, is implausible. Recall from Chapter 9 that HO^- is an insufficiently good leaving group for an E2 reaction. Thus, *imines do not form under basic conditions.*

---

problem 18.13

Can the imine hydrolysis at the right take place under *basic* conditions? Explain.

Based on the overall reactions in Equations 18-17 and 18-18, *imine formation requires at least two protons on the nucleophilic N atom.* One of those protons is removed in Step 3 of the mechanism (Equation 18-19), and the other is removed in Step 6. Thus,

> Only NH_3 and primary amines (RNH_2) can form imines upon reaction with ketones or aldehydes.

Although a secondary amine (R_2NH) has only one hydrogen on nitrogen and thus cannot form an imine, it can react with a ketone or aldehyde to produce an **enamine**, in which an amino group is attached to an alkene carbon (Equation 18-21).

(18-21)

An enamine

The mechanism for enamine formation (Equation 18-22) is identical to the mechanism for imine formation (Equation 18-19) through the first five steps.

Mechanism for Equation 18-21

The first five steps are identical to those in Equation 18-19.

(18-22)

6. Elimination of H⊕

The only difference is in the last step. In imine formation (Equation 18-19), the second of two H atoms on N is removed. In enamine formation, no such H exists, so a proton is eliminated from an adjacent C instead (i.e., an α carbon).

problem **18.14** Draw the complete mechanism and the major products that are formed in each of the following reactions.

(a)

(b)

(c)

SOLVED problem **18.15** The reaction at the right does *not* form an imine or an enamine. Explain why.

Think Which of the steps in Equations 18-19 or 18-22 can the reaction mechanism include? Are there any steps that are not possible?

Solve Only Steps 1 and 2 of Equation 18-19 (i.e., proton transfer and nucleophilic addition) can take place. Thus, the acid that is present can protonate the carbonyl O in Step 1, and the amine N can attack the carbonyl C in Step 2. Step 3 would normally require a deprotonation of the N atom, but no such proton exists. Thus, an imine or enamine cannot form, and the only product that can be formed cannot be isolated, because it is too unstable.

1. Proton transfer

2. Nucleophilic addition

No proton on N

problem **18.16** The reaction at the right does *not* form an imine or enamine. Explain why.

18.2c The Hydrolysis of Nitriles

Closely related to the hydrolysis of acetals and imines is the hydrolysis of nitriles (R—C≡N). As shown in Equations 18-23 and 18-24,

The treatment of a nitrile with water under either acidic or basic conditions produces a primary amide (R—CO—NH$_2$).

(18-23)

An amide

(18-24)

An amide

As shown in parentheses in the two equations, further hydrolysis converts the amide to a carboxylic acid. Those reactions are described in detail in Chapters 20 and 21.

The mechanism for the hydrolysis of a nitrile under acidic conditions is shown in Equation 18-25.

Mechanism for Equation 18-23

(18-25)

1. Proton transfer
2. Nucleophilic addition
3. Proton transfer
4. Proton transfer
5. Proton transfer

Although this mechanism consists of five elementary steps, the only step that is *not* a proton transfer is the nucleophilic addition step—Step 2. The four proton transfer steps serve to add protons to N and remove them from O, making it possible to avoid generating strongly basic species in the mechanism—a necessary condition for a reaction taking place under acidic conditions. Furthermore, notice that Steps 4 and 5 are essentially tautomerization (Section 7.9).

The mechanism for the hydrolysis of a nitrile under basic conditions is shown in Equation 18-26.

Mechanism for Equation 18-24

As in hydrolysis under acidic conditions, the only step that is not a proton transfer is the nucleophilic addition step—Step 1 in this case. Similar to the acid-catalyzed mechanism, the proton transfer steps in Equation 18-26 serve to add protons to N and remove one from O. Moreover, Steps 3 and 4 make up tautomerization. Because the reaction takes place under basic conditions, these proton transfer steps take place in a way that avoids the generation of strongly acidic species.

problem 18.17 Draw the complete, detailed mechanism and predict the major organic product of each of the following reactions.

(a)

(b)

18.3 The Wolff–Kishner Reduction

When 1-phenylpropan-1-one is treated with hydrazine (H_2N—NH_2), followed by heating in the presence of a base, propylbenzene is produced, as shown in Equation 18-27.

Wolff–Kishner reduction

The carbonyl group is reduced to a CH_2 group.

$$\xrightarrow[\substack{H_2O/\text{triethylene glycol} \\ \Delta}]{H_2N\text{--}NH_2,\ \text{KOH}}$$

+ $N_2(g)$ + H_2O (18-27)

82%
Propylbenzene

1-Phenylpropan-1-one

This is an example of a Wolff–Kishner reduction.

> In a **Wolff–Kishner reduction**, the carbonyl (C=O) group of a ketone or aldehyde is converted to a methylene (CH₂) group under basic conditions.

The Wolff–Kishner reduction is one of three reactions commonly used to reduce a carbonyl group to CH_2. While the Wolff–Kishner reduction takes place under basic conditions, the other two, which are presented in Chapter 19, take place under neutral and acidic conditions, respectively.

YOUR TURN 18.6

> Verify that the reaction in Equation 18-27 is a reduction by computing the oxidation state of the carbonyl C atom in the reactant and the corresponding C atom in the product.

The partial mechanism for the Wolff–Kishner reduction is shown in Equation 18-28. Because each N atom of hydrazine possesses two H atoms, the first several steps are analogous to the imine-formation mechanism in Equation 18-19. The result is a **hydrazone**, characterized by the C=N—NH₂ group.

Mechanism for Equation 18-27

The first several steps are analogous to the imine formation mechanism in Equation 18-19.

A hydrazone

The anion is stabilized by both resonance and inductive effects

1. Proton transfer

2. Proton transfer

3. E2

The stability of N₂ drives this reaction.

4. Proton transfer

(18-28)

The four steps that occur after the hydrazone has formed serve to remove two protons from the NH₂ group and to add two protons to what was originally the carbonyl C. Moreover, Steps 1 and 2 make up tautomerization.

Notice in Step 1 that HO⁻ deprotonates a N atom. HO⁻ is not normally a strong enough base to do this, but the resulting conjugate base is stabilized by resonance (involving the adjacent double bond) and by inductive effects (involving the adjacent electron-withdrawing N atom).

There are two peculiarities about the E2 step (i.e., Step 3) worth noting. First, the proton and the leaving group are removed from adjacent N atoms, whereas we usually see them on adjacent C atoms. Second, the leaving group is quite poor—a negative charge becomes localized on a C atom. Normally this would prevent an E2 reaction from occurring, but the formation of $N_2(g)$ is a substantial driving force. Not only is $N_2(g)$ very stable, but being a gas, it physically leaves the reaction system once it is formed. This makes the overall reaction *irreversible*, which, according to Le Châtelier's principle, helps form additional products.

---

problem **18.18** Draw the complete, detailed mechanism and predict the product for each of the following reactions.

(a)   (b)

---

problem **18.19** What carbonyl-containing compound could be used as a reactant in the reaction at the right?

18.4 Enolate Nucleophiles: Aldol and Aldol-Type Additions

Nearly all of the acid- and base-catalyzed nucleophilic addition reactions we have encountered thus far are ones in which the nucleophilic atom is a heteroatom such as nitrogen, oxygen, or sulfur. Reactions that involve carbon nucleophiles, however, are among the most important reactions in organic synthesis. These include aldol and aldol-type reactions, which we examine throughout the rest of the chapter.

When ethanal (acetaldehyde, $CH_3CH{=}O$) is treated with sodium hydroxide (Equation 18-29), the product is 3-hydroxybutanal, a compound containing a four-carbon chain.

(18-29)

Because the product contains both an *ald*ehyde and an alco*hol* functional group, it is called an **aldol**, and the reaction that forms it is called an **aldol addition**.

Aldol reactions are particularly important because they form a new C—C bond. Notice, too, that the carbonyl and the alcohol groups in the product are separated by a single carbon atom; the molecule is therefore classified as a **β-hydroxycarbonyl**, which is the general form of any aldol addition product.

Every aldol addition produces a β-hydroxycarbonyl.

problem **18.20** Which of the following compounds could be the product of an aldol reaction? Explain.

(a)   (b)   (c)

The mechanism for the aldol addition reaction is shown in Equation 18-30.

Mechanism for Equation 18-29

This proton transfer significantly favors reactants.

A second molecule of the aldehyde acts as the electrophile.

The base is regenerated.

1. Proton transfer

2. Nucleophilic addition

3. Proton transfer

(18-30)

The enolate anion is nucleophilic at this C.

In Step 1, HO⁻ *reversibly* deprotonates the α carbon to yield an enolate anion. In fact, that proton transfer favors the reactant side significantly; at equilibrium, only about 0.01% of the aldehyde is deprotonated. In Step 2, the newly formed enolate anion, which is nucleophilic at the α carbon (Sections 10.3 and 10.4), attacks a second molecule of the aldehyde that still has its proton (and hence is uncharged overall). The immediate product is an alkoxide anion (RO^-), which is subsequently protonated in Step 3.

Verify that the proton transfer in Step 1 of the mechanism in Equation 18-30 significantly favors the reactant side by determining the pK_a values of acetaldehyde and water. Compute the value of K_{eq} for the reaction, too.

The nucleophilic addition step in Equation 18-30 is *reversible* (recall Table 17-2, p. 839), because the enolate anion is stabilized by resonance and so is substantially more stable than the R^- nucleophiles such as alkyllithium and Grignard reagents (Section 17.10).

SOLVED problem 18.21 Draw the complete, detailed mechanism for the reaction at the right and use the mechanism to predict the products.

Think How is this reaction similar to the one in Equation 18-29 (and Equation 18-30)? How is it different? What is the role of NaOH?

Solve Just as in Equation 18-29, an aldehyde is treated with hydroxide. The only difference is the carbon backbone; in this case, we have a four-carbon aldehyde, whereas it was a two-carbon aldehyde in Equation 18-29. The mechanisms are therefore essentially identical.

A β-hydroxycarbonyl

In the first step, HO⁻ acts as a base to generate an enolate anion. (Note that HO⁻ could also act as a nucleophile, but as we saw in Chapter 8, proton transfer reactions are very fast, so the dominant reaction is deprotonation of the aldehyde.) In the second step, the enolate anion acts as a nucleophile, attacking a second molecule of butanal that still has its proton. This step forms a C—C bond between the two species. Finally, the O⁻ is protonated to yield the β-hydroxycarbonyl.

problem 18.22 Draw the complete, detailed mechanism for the reaction at the right and predict the product.

Aldol reactions can be catalyzed by acid, too. Exploring the mechanism for such a reaction is left as an exercise at the end of this chapter (see Problem 18.52).

SOLVED problem 18.23 Show how to synthesize the molecule at the right using an aldol reaction.

Think Can you identify the portion of the molecule that characterizes it as a β-hydroxycarbonyl? Which carbons of the β-hydroxycarbonyl have been joined in the aldol addition? Upon disconnecting that C—C bond in a retrosynthesis, what carbon backbones are required in the precursors?

Solve The α and β carbons of the C=O group are labeled in the following structure:

An OH group is attached to the β carbon, so the molecule is indeed a β-hydroxycarbonyl. To undo the aldol reaction, disconnect the bond between the α and β carbons, as indicated. In the precursor, the C—OH bond is a C=O group. The two precursors are the same aldehyde, so in the forward direction, we simply treat that aldehyde with NaOH.

problem **18.24** Show how to synthesize the molecule at the right using an aldol reaction.

18.5 Aldol Condensations

As we saw in Section 18.4, a *β–hydroxycarbonyl* is the immediate product of an aldol addition. If the reaction is heated, however, then *dehydration* occurs: Water is eliminated from the β-hydroxycarbonyl, giving the overall product a C=C double bond. An example is shown in Equation 18-31, which takes place under basic conditions.

This overall reaction, in which an aldol addition is followed by dehydration, is called an **aldol condensation**. In general, a *condensation* reaction is one in which two larger molecules bond together with the elimination of a smaller molecule; in this case, that smaller molecule is H_2O.

The newly formed C=C double bond appears between the α and β carbons of the aldehyde in Equation 18-31, so the product is called an **α,β-unsaturated carbonyl**. In general,

> The product of an aldol condensation is an α,β-unsaturated carbonyl.

Under basic conditions, dehydration of a β-hydroxycarbonyl takes place via an **E1cb mechanism**, which stands for *conjugate base unimolecular elimination* (Equation 18-32).

Resonance-stabilized enolate anion

$$+ \text{ H}-\ddot{\text{O}}\text{H} \rightleftharpoons \text{ (2. Heterolysis) } + \ddot{\text{O}}\text{H} \quad \text{(18-32)}$$

Conjugated double bonds

Acidic hydrogen

Because of the enhanced acidity at the α carbon, Step 1 is a fast proton transfer, generating a resonance-stabilized enolate anion. In Step 2, the HO^- leaving group departs to yield the overall product.

This elimination may seem to contradict the general rule we encountered in Section 9.9, which states that an HO^- leaving group is generally unsuitable for elimination reactions. In this case, however, elimination is favored, in part, by the proton transfer in the first step, and also by the stability that arises from the conjugation between the C=C and C=O double bonds in the overall product.

problem **18.25** Dehydration occurs if 3-hydroxybutanal is heated under basic conditions, but no dehydration occurs when pent-4-en-2-ol is subjected to the same conditions. Explain.

Dehydration of an aldol product can also take place under acidic conditions, once again producing an α,β-unsaturated carbonyl (Equation 18-33).

Unlike the E1cb mechanism, which takes place under basic conditions, dehydration under acidic conditions proceeds by an E1 mechanism, as shown in Equation 18-34.

Mechanism for the dehydration of an aldol addition product under acidic conditions

(18-34)

In Step 1, protonation of OH generates a very good water leaving group. The leaving group departs in Step 2, and a final deprotonation yields the overall product.

Notice in Step 3 of Equation 18-34 that the α proton is not the only one that can be detached in a dehydration reaction. As shown in Equation 18-35b, a γ proton can also be removed, yielding a β,γ-unsaturated carbonyl. Essentially none of this product is formed, however, because the α,β-unsaturated carbonyl product (Equation 18-35a) possesses *conjugated double bonds*, and is thus more stable than the β,γ-unsaturated carbonyl.

The π bonds are conjugated.

Favored product

(18-35a)

α,β-Unsaturated carbonyl

The π bonds are *not* conjugated.

(18-35b)

β,γ-Unsaturated carbonyl

18.6 Aldol Reactions Involving Ketones

Ketones with α hydrogens can participate in aldol additions in the same way that aldehydes do, as shown in Equation 18-36 for acetone.

The mechanism for this reaction, which is shown in Equation 18-37, is identical to the one for the reaction involving only aldehydes, shown previously in Equation 18-30.

Mechanism for Equation 18-36

According to Equation 18-36, however,

An aldol reaction between ketones tends to form very little product at equilibrium.

This has to do with the fact that nucleophilic addition to a ketone yields a product with considerable steric strain and is therefore less favorable than addition to an aldehyde. Additionally, the electron-donating effects from a ketone's alkyl groups diminish the concentration of positive charge on the carbonyl carbon. Recall from Section 17.2 that a similar phenomenon occurs with the hydration of ketones and aldehydes.

Because aldol additions involving ketones generally favor reactants, chemists must manipulate the equilibrium using Le Châtelier's principle to achieve a reasonable yield. Although we do not discuss the details, one way to do so is to remove the aldol product as it is formed.

18.7 Crossed Aldol Reactions

The aldol additions we have examined thus far have involved only a single aldehyde or a single ketone, in which case the nucleophilic enolate anion is derived from the same aldehyde or ketone it attacks. These kinds of reactions are called **self-aldol additions**. What happens if the nucleophilic species is generated from a *different* aldehyde or ketone than the one that is attacked? Equation 18-38 shows the enolate anion of propanal attacking an uncharged molecule of ethanal (acetaldehyde), yielding 3-hydroxy-2-methylbutanal. This kind of reaction is called a **crossed aldol addition**.

(18-38)

Enolate anion from propanal   **Ethanal (Acetaldehyde)**   **3-Hydroxy-2-methylbutanal**

To carry out a crossed aldol addition, we might imagine treating propanal with NaOH and ethanal (acetaldehyde), as shown in Equation 18-39.

(18-39)

Propanal   **Ethanal (Acetaldehyde)**   **3-Hydroxy-2-methylbutanal**   **3-Hydroxybutanal**   **3-Hydroxy-2-methylpentanal**   **3-Hydroxypentanal**

The hope is that the base will deprotonate propanal, and the resulting enolate anion will attack ethanal to produce 3-hydroxy-2-methylbutanal. Indeed, the desired product is formed, as shown in the equation, but so are three other aldol products. Not only does this compromise the reaction yield, but the four products are difficult to separate due to their similar physical properties. As a result, this kind of reaction is *not* synthetically useful.

The problem stems from the fact that deprotonation of an α hydrogen by NaOH is somewhat unfavorable (recall from Your Turn 18.8 that an aldehyde [$pK_a \approx 19$] is a weaker acid than water [$pK_a = 15.7$], the conjugate acid of HO⁻). Thus, at any given time throughout the reaction, a significant concentration of HO⁻ exists. When the second aldehyde is added, it can be deprotonated to a small extent by the remaining HO⁻. As a result, the enolate anions from *both* aldehydes are present at the same time, and each enolate anion can attack either of two different uncharged aldehydes. As shown in Equation 18-40, there are four different pairings of an enolate anion with an uncharged aldehyde, and each gives rise to a different aldol product. Notice that two of the products are from *self-aldol reactions* (Equations 18-40a and 18-40d) and two are from crossed aldol reactions (Equations 18-40b and 18-40c).

(18-40a)

Propanal

(18-40b) Crossed aldol

**Ethanal
(Acetaldehyde)**

(18-40c) Crossed aldol

(18-40d) Self aldol

---

problem **18.26** Draw the complete, detailed mechanism for the reaction in Equation 18-40c.

One way around this complication is to use an aldehyde that has no α hydrogens, such as methanal (formaldehyde), benzaldehyde, or dimethylpropanal. (Fig. 18-4).

> If an aldehyde *with* α hydrogens is added slowly to a basic solution of an aldehyde *without* α hydrogens, primarily one aldol product is formed.

An example is shown in Equation 18-41.

No α hydrogens   **Add slowly**

Benzaldehyde   +   NaOH

1.

2. $H_3O^{\oplus}$

Cinnamaldehyde
67%

(18-41)

Benzaldehyde has no α hydrogens, so it cannot be deprotonated by NaOH and, therefore, does *not* form an enolate anion. In contrast to Equation 18-39, therefore, only one enolate anion is present at any given time. Moreover, if acetaldehyde (which has α hydrogens) is added slowly, we avoid a buildup of the uncharged aldehyde, and thus minimize the amount of product from the self-aldol reaction between acetaldehyde and its own enolate anion.

**Methanal
(Formaldehyde)**

Benzaldehyde

**Dimethylpropanal
(Pivaldehyde)**

FIGURE 18-4 Aldehydes with no α hydrogens These aldehydes do not form enolate anions, so they can be used effectively in crossed aldol reactions.

Which of the following ketones or aldehyde would be the best choice to use for a crossed aldol reaction with butanal?

SOLVED problem 18.27 Predict the product of the reaction at the right, in which propanal is added slowly to a basic solution of formaldehyde.

Think Which aldehyde can form an enolate anion and which cannot? Which uncharged aldehyde is present in only small concentrations in the reaction mixture?

Solve Formaldehyde has no α hydrogens, so it cannot form an enolate anion. Thus, the only enolate nucleophile that is present is derived from propanal, as shown at the right. Because propanal is added slowly, there is never a substantial concentration of it in its uncharged form, so the major reaction occurs between the propanal enolate anion and formaldehyde:

Because the reaction is heated, dehydration leads to the condensation product.

problem 18.28 Predict the product of the reaction at the right, in which phenylethanal is added slowly to a basic solution of dimethylpropanal.

SOLVED problem 18.29 Show how to synthesize the molecule at the right using an aldol condensation.

Think Can you identify the portion of the molecule that characterizes it as an α,β-unsaturated carbonyl? From what β-hydroxycarbonyl could it have been generated? Upon applying a transform to that β-hydroxycarbonyl to undo an aldol addition, which C—C bond should be disconnected?

Solve The α and β carbons of the C=O group are labeled in the following structure:

The target is an α,β-unsaturated carbonyl, so it can be the product of dehy-
drating the β-hydroxycarbonyl shown. To
apply a transform that undoes an aldol
addition, we must disconnect the bond
between the α and β carbons, as indicated. Because the resulting aldehyde
precursors are different from each other,
more than one aldol product is possible in the forward direction. To produce only the desired product, we should
slowly add the aldehyde *with* α hydrogens to a basic solution of the aldehyde *without* α hydrogens.

---

problem **18.30** Show how to synthesize the molecule at the right using an aldol reaction.

A second way to carry out a crossed aldol reaction selectively is to use a very strong
base such as lithium diisopropylamide (LDA) to generate the enolate anion, as shown
in Equation 18-42.

(18-42)

70%

Because LDA is a very strong base, it deprotonates the α carbon rapidly, quantitatively, and irreversibly. Therefore, essentially 100% of the cyclohexanone is converted
to its enolate anion prior to the addition of the second carbonyl-containing compound. In this way, only one enolate anion is available to react in the second step of
the synthesis.

In an aldol reaction involving ketones, *regiochemistry* becomes a concern when the
ketone has two chemically distinct α carbons, as is the case with methylbutanone. As
shown in Equations 18-43a and 18-43b, the α C that is deprotonated will lead to one
of two enolate anions, and will result in one of two aldol products.

(18-43a)

(18-43b)

**Two different enolate
anions can be formed.**

Different aldol products

Fortunately, by choosing the base appropriately, we can often control which enolate anion is predominantly formed, and so can control the regiochemistry of these kinds of aldol reactions. For example, Equation 18-44 shows that if NaOH is the base, then the major aldol product has the form that appears in Equation 18-43a. On the other hand, if LDA is the base (Equation 18-45), then the major aldol product has the form in Equation 18-43b.

NaOH favors deprotonation here.

Methylbutanone

4-Hydroxy-3,3-dimethylbutanone
72%

(18-44)

LDA favors deprotonation here

Methylbutanone

1. LDA, ether

2. ⟍⟍⟍ , –78 °C

3. NH₄Cl

5-Hydroxy-2-methylheptan-3-one
78%

(18-45)

Recall from Section 10.3a that when NaOH is used as the base, the favored product is the thermodynamic enolate anion, which is formed by deprotonating the more highly substituted α carbon. When LDA is used as the base at low temperature, the favored product is the kinetic enolate anion, which is formed by deprotonating the less highly substituted α carbon.

YOUR TURN 18.10

In each of the following boxes, write the base that would accomplish the respective aldol reactions.

Hypoglycemia and Aldol Reactions

Your brain (and, for that matter, the rest of your central nervous system) runs almost entirely on glucose, yet it has no place to store the sugar. All of the sugar your brain uses must come from the blood. What happens, then, if your blood sugar levels are low—a condition known as hypoglycemia—perhaps from dieting or from a medical condition? Fortunately, your body can respond to hypoglycemia by constructing glucose from other molecules—namely, dihydroxyacetone monophosphate (DHAP) and (R)-glyceraldehyde-3-phosphate (G3P). This process, called gluconeogenesis, takes place primarily in the liver, and involves an aldol reaction similar to the ones we have examined here in Chapter 18.

The overall aldol reaction, shown in Scheme A, produces fructose-1,6-bisphosphate as the immediate product, which is subsequently converted to glucose. Notice that fructose 1,6-bisphosphate is a β-hydroxycarbonyl, which we recognize as the product of an aldol addition.

Aldolase is the enzyme that catalyzes this aldol reaction. An NH_2 group on the side chain of a lysine amino acid from aldolase reacts with the C=O group of DHAP to produce an enamine (Scheme B), which is nucleophilic at the α carbon, analogous to an enolate anion. The enamine undergoes nucleophilic addition to G3P to form the new C—C bond, and the resulting iminium ion is hydrolyzed to produce fructose 1,6-bisphosphate.

Like any catalyst, aldolase can also catalyze the reverse reaction, which, in this case, is called a *retro aldol reaction*. Rather than building up glucose from smaller molecules, this kind of retro aldol reaction is responsible for breaking down glucose as part of sugar metabolism—a process called glycolysis.

18.8 Intramolecular Aldol Reactions

If a molecule includes two carbonyl-containing functional groups, then an *intramolecular* aldol reaction (i.e., one between different parts of the same molecule) is possible. The result is the formation of a *ring*, as shown with hexanedial in Equation 18-46.

Identify the structural features in the product of Equation 18-46 that characterize it as the product of an aldol condensation.

18.11 **YOUR TURN**

The mechanism for this reaction, shown in Equation 18-47, is identical to aldol condensations that occur between two separate aldehyde molecules.

Mechanism for Equation 18-46

(18-47)

In Section 9.11, we learned that cyclization reactions are generally favored over their corresponding intermolecular reactions when the product is a five- or six-membered ring. Aldol reactions are no different.

> Intramolecular aldol reactions are favored when five- or six-membered rings are formed.

Indeed, the product in Equation 18-46 is a five-membered ring. By contrast, dialdehydes such as pentanedial (Equation 18-48) and octanedial (Equation 18-49) do not favor intramolecular aldol reactions, because their aldol products would possess four- and seven-membered rings, respectively.

Formation of the 4-membered ring is *not* favorable.

NaOH

Pentanedial

(18-48)

Formation of the 7-membered ring is *not* favorable.

NaOH

Octanedial

(18-49)

problem **18.31**

problem 18.31 The intramolecular aldol reaction involving heptanedial is favorable. Draw the mechanism for this reaction, along with the major product that is formed. Explain why the reaction is favorable.

Heptanedial

NaOH

?

Equation 18-50 shows a dicarbonyl compound that can form three different enolate anions, in which case we can envision three different cyclic aldol products.

6-Oxo-heptanal

NaOH

Aldehyde attacked, favored

Ketone attacked, not favored

(18-50a)

7-membered ring
not favored

(18-50b)

5-membered ring

Major product

(18-50c)

The product in Equation 18-50b is heavily favored over the others. The aldol product in Equation 18-50a is not favored because it forms a seven-membered ring, whereas the reactions in Equations 18-50b and 18-50c both form five-membered rings. The major difference between the two reactions that form five-membered rings is in the functional group that is attacked: In Equation 18-50b, it is an aldehyde, whereas in Equation 18-50c, it is a ketone. Recall from Section 18.6 that addition to a ketone is generally less favorable than addition to an aldehyde.

SOLVED problem 18.32 Draw the complete, detailed mechanism for the following reaction and predict the major product. (Assume that Le Châtelier's principle is exploited to drive the reaction toward products.)

NaOH

?

Think What enolate anions can be formed? Which of those can attack a carbonyl group to form a five- or six-membered ring? In those nucleophilic additions, what kind of carbonyl group is attacked—a ketone or aldehyde?

Solve The three possible enolate anions resulting from deprotonation of an alpha proton and their corresponding aldol products are as follows:

Ketone attacked

A

Aldehyde attacked

Major product

B

8-membered ring formed

C

Products **A** and **B** are favored over **C** because of the sizes of the rings that are formed— namely, six-membered rings in **A** and **B** and an eight-membered ring in **C**. Moreover, because nucleophilic addition is favored at an aldehyde over a ketone, **B** is favored over **A**. Thus, **B** is the major product.

---

problem **18.33** Predict the major product of the reaction at the right.

NaOH

?

18.9 Aldol Additions Involving Nitriles and Nitroalkanes

Aldol additions are not limited to just ketones and aldehydes. Other functional groups containing a polar π bond have acidic α hydrogens as well. When deprotonated, their resulting enolate anions can act as nucleophiles, just like the enolate anions of ketones and aldehydes.

In compounds containing nitrile (C≡N) and nitro (NO$_2$) groups, for example, the α hydrogens are somewhat acidic. As we can see in Equations 18-51a and 18-51b, the pK_a of a nitrile's α hydrogen is ~25, and the pK_a of a nitro group's α hydrogen is ~10.

(18-51a)

(18-51b)

The acidity of a proton alpha to a nitrile or nitro group, like that of a ketone or aldehyde hydrogen, is due to both resonance and inductive stabilization in the conjugate base. Equations 18-51a and 18-51b show how resonance serves to delocalize the negative charge that develops over two separate atoms: C and N in Equation 18-51a and C and O in Equation 18-51b. Moreover, the CN and NO$_2$ groups are inductively electron withdrawing, which helps reduce the concentration of negative charge on C.

Equations 18-52 and 18-53 show examples of aldol reactions involving a nitrile and a nitroalkane, respectively.

(18-52)

(18-53)

The mechanisms of these reactions are identical to those of aldol reactions involving only ketones or aldehydes (see Solved Problem 18.34 and Problem 18.35). The aldol reaction involving a nitroalkane can be carried out using a weak base such as an amine, however, because the nitroalkane's α proton (pK_a ≈ 10) is so much more acidic than that of an aldehyde (pK_a ≈ 19).

SOLVED problem **18.34** Draw the complete, detailed mechanism for the reaction in Equation 18-52.

Think Which protons are the most acidic? After deprotonation, which polar π bond undergoes nucleophilic addition? How is the C=C double bond formed in an aldol reaction?

Solve The proton alpha to the CN group is the most acidic, so it is removed by the base in the first step. The resulting enolate anion attacks the aldehyde group, and the

O⁻ generated in that step is subsequently protonated. The C=C bond is then formed by dehydration, which takes place by the E1cb mechanism.

problem **18.35** Draw the complete, detailed mechanism for the reaction in Equation 18-53.

problem **18.36** Draw the complete, detailed mechanism, and predict the major organic product for the following reaction.

18.10 The Robinson Annulation

As we learned in Chapter 4, six-membered rings are the most abundant in natural products due to their relative stability. Synthetic methods for constructing six-membered rings are therefore of great value. Sir Robert Robinson (1886–1975), an English chemist, developed one such reaction, which has since become known as the **Robinson annulation** (*annulation* means "ring formation"). An example is shown in Equation 18-54.

A Robinson annulation forms a new six-membered ring.

(18-54)

A Robinson annulation reaction is essentially the conjugate addition of an enolate anion (also called a *Michael reaction*; Section 17.10) followed by an intramolecular aldol condensation (i.e., aldol addition plus dehydration), as shown in Equation 18-55.

A Michael reaction

KOH, ethanol

A 1,5-dicarbonyl

An intramolecular aldol addition

KOH, ethanol

A β-hydroxycarbonyl

Dehydration via an E1cb mechanism

KOH, ethanol
Δ

An α,β-unsaturated carbonyl

(18-55)

---

problem **18.37** Draw the complete, detailed mechanism for the Robinson annulation in Equation 18-54.

For these back-to-back reactions to take place, the reactants of a Robinson annulation must have the general form shown in Equation 18-56.

α, β Unsaturation

α Hydrogens

+

(18-56)

In particular,

1. One of the reactants of a Robinson annulation must be an α,β-unsaturated carbonyl and the second must simply be a carbonyl-containing compound with an acidic α hydrogen.
2. The α,β-unsaturated carbonyl must have two acidic α hydrogens on the side of the carbonyl opposite the C=C double bond (shown above in blue).

The first of these requirements ensures that an enolate nucleophile can be formed and conjugate addition can take place. The second requirement ensures that the intramolecular aldol condensation can take place: Deprotonation of the first α hydrogen generates the enolate nucleophile that leads to the formation of a six-membered ring, while the second α hydrogen is removed in the dehydration steps.

---

problem **18.38** Based on the above requirements, which of the following pairs of reactants are suitable for a Robinson annulation? For each pair that is suitable, draw the complete, detailed mechanism of the Robinson annulation that could occur.

(a)   (b)   (c)   (d)

18.11 Organic Synthesis: Aldol Reactions in Synthesis

Aldol reactions are invaluable tools in organic synthesis because they form carbon–carbon bonds and because the functional groups they involve—ketones and aldehydes—are quite common. Thus, aldol reactions can be used to produce compounds with a wide variety of structures. In Section 18.8, for example, we saw that aldol reactions can form rings. Perhaps more impressively, aldol reactions can be used to link two compounds with elaborate carbon frameworks, as shown in Equation 18-57, which makes up one of the steps in a synthesis of epothilone B, an anticancer agent.

(18-57)

Because of the utility of aldol reactions in synthesis, it is important to be able to efficiently carry out transforms that undo aldol reactions, as part of a retrosynthetic analysis. When doing so, it is helpful to consider the following guidelines:

- The product of an aldol addition is a β-hydroxycarbonyl, and the product of an aldol condensation is an α,β-unsaturated carbonyl.
- The new C—C bond that is formed in an aldol reaction appears between the α and β carbons in the product, so a transform that undoes an aldol reaction involves disconnecting that bond.
- In constructing the precursors to an aldol reaction, the β carbon from the aldol product receives a C=O bond.

These guidelines are illustrated in the two generic transforms illustrated in Figure 18-5.

Suppose, for example, that our target compound is 5,5-dimethylcyclohex-2-enone, which is an α,β-unsaturated ketone (Equation 18-58).

5,5-Dimethylcyclohex-2-enone   **3,3-Dimethyl-5-oxohexanal**

(18-58)

FIGURE 18-5 Retrosynthesis involving aldol reactions To carry out a transform that undoes (a) an aldol addition or (b) an aldol condensation, disconnect the α and β carbons and add a C=O bond to the β carbon.

We can therefore undo an aldol condensation by disconnecting the bond indicated by the wavy line, and giving the β carbon a C=O bond. Thus, 3,3-dimethyl-5-oxohexanal is a suitable precursor.

In the forward direction, we can simply treat the precursor with NaOH and apply heat (to facilitate dehydration), as shown in Equation 18-59:

$$
\text{(structure)} \xrightarrow[\Delta]{\text{NaOH}} \text{(structure)} \qquad \text{(18-59)}
$$

problem 18.39 Show how to synthesize the following compounds using either an aldol addition or aldol condensation.

(a)

(b)

18.12 Organic Synthesis: Imagining an Alternate Target Molecule in a Retrosynthetic Analysis

When a target molecule has conspicuous structural characteristics, it is often clear as to which reactions should be used in a synthesis. If a portion of the target molecule is a β-hydroxycarbonyl or an α,β-unsaturated carbonyl, for example, then we should consider using an aldol reaction. However, further reactions on an aldol product, such as those in Equation 18-60, would disguise these characteristics.

(18-60a)

(18-60b)

In Equation 18-60a, the OH group of a β-hydroxycarbonyl is converted to Br using PBr$_3$ (Section 10.1). In Equation 18-60b, the C=O group is reduced to an OH group using NaBH$_4$ (Section 17.3).

When given a particular target molecule, therefore, we should be prepared to imagine a new target molecule with alternate functional groups that more closely resembles the immediate product of a reaction we wish to use. If we can subsequently determine which reactions will convert the altered molecule into the real target molecule, then the synthesis can be completed.

Suppose, for example, that we need to synthesize the target in Equation 18-61 and that our starting materials are limited to ketones and aldehydes of four or fewer carbon atoms.

(18-61)

Because the target contains seven carbon atoms, we must use a carbon–carbon bond-formation reaction. An aldol reaction can form such a bond, but the direct product of an aldol reaction (a β-hydroxycarbonyl or an α,β-unsaturated carbonyl) is not the 1,3-diene we are seeking. As shown in Equation 18-62, however, we can imagine, as an alternate target, an aldol product that can be formed from a four-carbon aldehyde and a three-carbon ketone.

(18-62)

Thus, the task is to determine how the aldol product can be converted into the target. Because a C—C single bond must be converted to a C=C double bond, we can envision using an E2 reaction on the appropriate alkyl halide (Equation 18-63).

(18-63)

The alkyl halide can be produced from the corresponding alcohol using PBr_3 and the alcohol can be generated by reducing the corresponding ketone using $LiAlH_4$.

Having completed the retrosynthetic analysis, the synthesis in the forward direction might appear as follows (Equation 18-64):

(18-64)

---

problem **18.40** Show how to synthesize the following compound from any acyclic compound.

Although this example focuses on the use of an aldol reaction, you must be prepared to use the same general strategy with other reactions that alter the carbon skeleton, too.

18.13 Wrapping Up and Looking Ahead

Chapter 18 is the second of two chapters devoted to nucleophilic additions to polar π bonds. In Chapter 17, the focus was primarily on the addition of strong nucleophiles added as starting material. Here in Chapter 18, the emphasis has been on reactions involving weaker nucleophiles as starting material, which typically require acid or base catalysis. Whereas base catalysis speeds up nucleophilic addition by converting a weak nucleophile into a strong nucleophile, acid catalysis functions by making the polar π bond more electrophilic—that is, by *activating* the polar π bond. In both types of catalysis, the catalyst is not consumed; it reacts in one step, but is regenerated in another.

In contrast to most of the reactions in Chapter 17, the nucleophilic addition reactions presented here in Chapter 18 are reversible. Thus, their equilibria can generally be driven either toward products or toward reactants by exploiting Le Châtelier's principle, exemplified by the acid-catalyzed formation and hydrolysis of an acetal. Excess alcohol favors the acetal formation, whereas excess water favors hydrolysis back to the ketone or aldehyde. In Chapter 19, we will see how this reversibility is a key to protecting certain functional groups throughout the course of a synthesis.

The latter part of Chapter 18 focused primarily on aldol and aldol-type reactions. These reactions form carbon–carbon bonds and are quite versatile because they involve very common functional groups—ketones and aldehydes. Aldol reactions can be used to link two separate carbon frameworks and can also be used to form new rings.

In the chapters to come, we will apply much of what we have learned about nucleophilic addition reactions. We will see, for example, that nucleophilic addition reactions have many important applications in organic synthesis, the topic of Chapter 19. Frequently, nucleophilic addition reactions that form carbon–carbon bonds are instrumental in constructing a carbon backbone, and the functional group conversions we learned in this and the previous chapter can be extremely useful when a synthesis calls for altering the reactivity at specific locations within a molecule. In Chapter 20, we begin to study the *nucleophilic addition–elimination* mechanism, which, as the name suggests, involves aspects of nucleophilic addition. The nucleophilic addition–elimination mechanism is the basis for a variety of reactions involving carbonyl-containing compounds such as carboxylic acids, esters, and amides.

THE ORGANIC CHEMISTRY OF BIOMOLECULES

18.14 Ring Opening and Closing of Monosaccharides; Mutarotation

In Section 1.15b, we showed that a monosaccharide can have both open-chain and cyclic forms. In water, the various forms equilibrate, as shown in Equations 18-65 and 18-66 for D-glucose and D-ribose, respectively.

(18-65)

(18-66)

The mechanisms showing how D-glucose cyclizes to its six- and five-membered rings are shown in Equations 18-67a and 18-67b, respectively. Step 1 is the nucleophilic addition of an alcohol to the carbonyl group, whereas Steps 2 and 3 are both proton transfers.

Mechanism showing the cyclization of D-glucose to its six- and five-membered ring forms

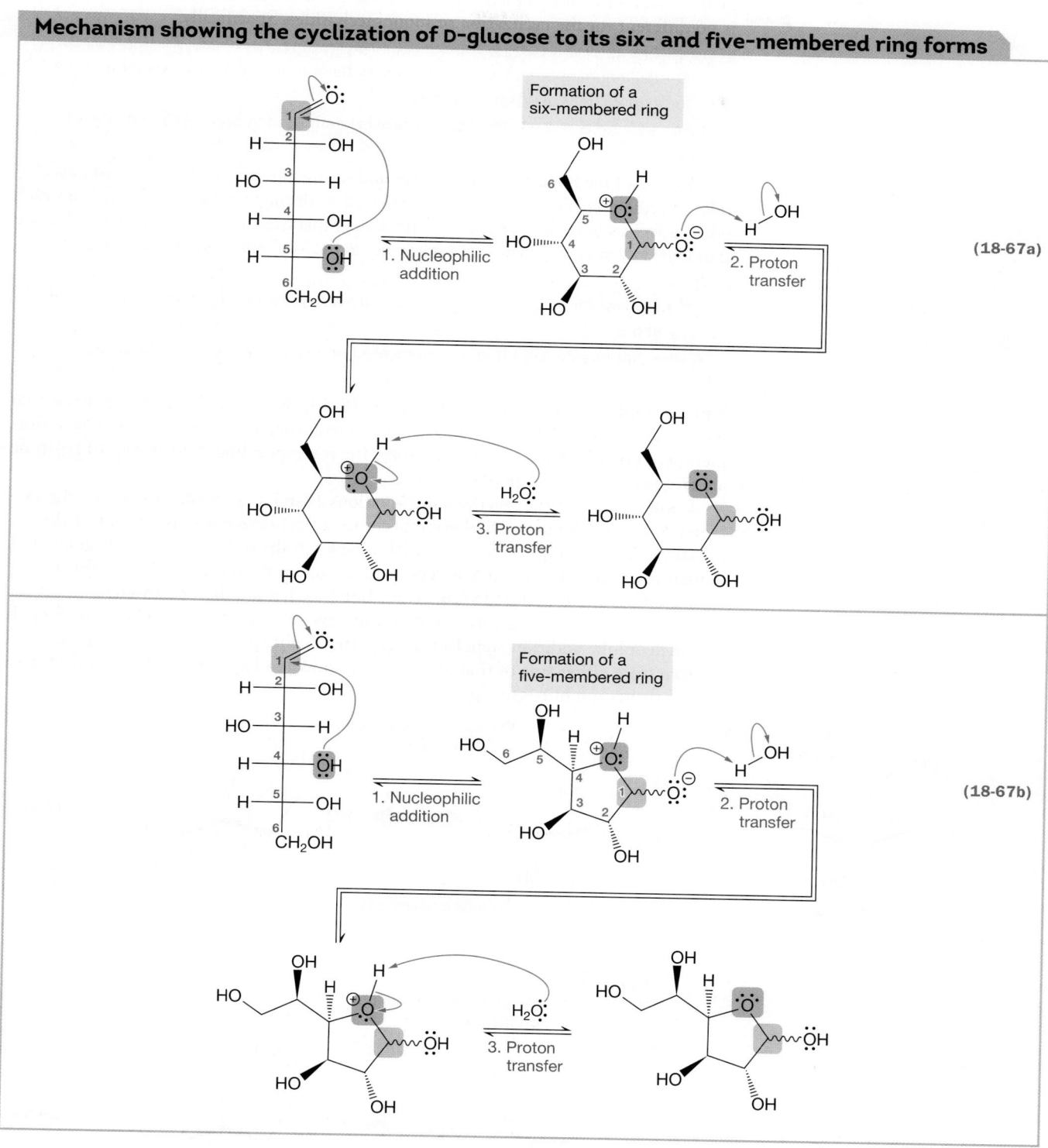

(18-67a)

(18-67b)

These mechanisms are essentially the same as the one shown previously in Equation 18-3. In each case, the carbonyl group of a ketone or aldehyde reacts with an alcohol to produce a hemiacetal. The main difference, however, is that the reaction in Equation 18-3 is *intermolecular*, whereas it is *intramolecular* in Equation 18-67.

Recall from Section 9.11 that cyclization reactions are favored when a five- or six-membered ring is formed, and that the formation of a six-membered ring is usually favored over a five-membered ring. As we can see in Equations 18-65 and 18-66, this is no different for the cyclization of monosaccharides. Specifically,

> ■ At equilibrium, monosaccharides typically favor their five- and six-membered ring forms over their open-chain forms.
> ■ The six-membered ring form is generally favored over the five-membered ring form.

Because of the importance of the five- and six-membered ring forms of monosaccharides, specific nomenclature has been developed to distinguish the two forms from each other, as well as from the open-chain form. Under this system, the monosaccharide's name is modified by inserting "pyran" or "furan" prior to the "ose" suffix, depending upon ring size.

> ■ A monosaccharide that has cyclized to a six-membered ring is designated as a **pyranose**.
> ■ One that has cyclized to a five-membered ring is designated as a **furanose**.

Thus, as indicated previously in Equation 18-65, the cyclic form of D-glucose that has a six-membered ring is called D-glucopyranose, and the one that has a five-membered ring is called D-glucofuranose. The corresponding cyclic forms of D-ribose are D-ribopyranose and D-ribofuranose.

Looking back at the reactions in Equations 18-65 and 18-66, notice that the carbonyl carbon of the monosaccharide becomes a *new stereocenter* as a result of the cyclization. As with any reaction in which a new tetrahedral stereocenter is generated, a mixture of both the *R* and *S* stereochemical configurations is produced. This is why that C—O bond is denoted with a wavy line (∿∿), not a dash or a wedge.

As we can see in Equation 18-67a, the new stereocenter is produced in Step 1, the nucleophilic addition step. In this step, the alcohol group can attack the carbonyl C atom from either side of that atom's plane, shown in Equations 18-68a and 18-68b for the formation of D-glucopyranose.

(18-68a)

(18-68b)

The anomeric OH and CH₂OH
are on opposite sides.

α-D-glucopyranose
m.p. = 146 °C
$[\alpha]_D^{20}$ = +112.2 °

(a)

The anomeric OH and CH₂OH
are on the same side.

β-D-glucopyranose
m.p. = 150 °C
$[\alpha]_D^{20}$ = +18.7 °

(b)

FIGURE 18-6 Anomers of monosaccharides (a) In an α anomer, the anomeric OH is on the side of the ring opposite the CH₂OH group. (b) In a β anomer, the groups are on the same side.

The two stereoisomers differ in configuration at just one of the stereocenters, so they are *diastereomers* of each other. In carbohydrate chemistry, they are more specifically called **anomers** of each other, and the carbon atom that differs in stereochemical configuration—the one that is part of the carbonyl group in the open-chain form—is called the **anomeric carbon**. One anomer is designated the α-anomer and the other the β-anomer (Fig. 18-6).

- In the **α-anomer** of a cyclic monosaccharide, the anomeric OH and the CH₂OH substituents are located on opposite sides of the ring.
- In the **β-anomer**, the two substituents are located on the same side of the ring.

problem **18.41** (a) Draw the Haworth projections of α-D-ribofuranose and β-D-ribofuranose. (b) Draw the mechanisms that show how each of these anomers is produced from the acyclic form of the monosaccharide.

problem **18.42** Name each of the following monosaccharides. (Consult Fig. 5-34 for the name of each sugar in its acyclic form.)

(a)

(b)

Because they are diastereomers, α- and β-anomers have different physical properties. The melting point of α-D-glucopyranose is 146 °C, for example, whereas that of β-D-glucopyranose is 150 °C. Additionally, the two have different optical properties. Whereas the specific rotation of α-D-glucopyranose in water is +112.2°, that of β-D-glucopyranose is +18.7°.

If pure crystalline α-D-glucopyranose is dissolved in water, the specific rotation changes over time, from +112.2° to 52.5°. Likewise, the specific rotation of β-D-glucopyranose changes over time, from +18.7° to 52.5°. This phenomenon is called **mutarotation**; in water, each anomer equilibrates with the open-chain form, so the two anomers equilibrate with each other (Equation 18-69).

α-D-Glucopyranose
36%

β-D-Glucopyranose
64%

(18-69)

Using the specific rotation of an equilibrium mixture of two anomers, we can calculate the relative amounts of the two anomers. For D-glucose, for example, let x be the equilibrium fraction of the α-anomer, and let the remaining amount in the mixture, $1 - x$, be the equilibrium fraction of the β-anomer. The specific rotation of the mixture is the weighted average of their individual specific rotations, according to Equation 18-70.

$$[\alpha]_{\text{D, mixture}}^{20} = x\,(112.2°) + (1-x)(18.7°) = 52.5°$$ (18-70)

Solving for x, we find that the equilibrium amount of the α-anomer is 36%, and that of the β-anomer is 64%.

problem **18.43** The specific rotation of α-D-mannopyranose is +29.3°, and that of β-D-mannopyranose is −16.3°. When either anomer is dissolved in water, the specific rotation slowly changes to +14.5°. Calculate the relative amounts of the two anomers at equilibrium.

Chapter Summary and Key Terms

- Nucleophilic addition of a weak nucleophile to a polar π bond is typically slow. If the nucleophilic atom has a weakly acidic proton, **base catalysis** can increase the rate of the reaction by converting the weak nucleophile into a strong one. Alternatively, **acid catalysis** can increase the reaction rate by making the polar π bond more electrophilic—that is, by **activating** the polar π bond. (Section 18.1)

- A **cyanohydrin** can be formed via the addition of HCN to the carbonyl group of a ketone or aldehyde, and the reaction can be catalyzed by the addition of a strong base or NC⁻. (Section 18.1a)

- When a weak nucleophile adds to an α,β-unsaturated polar π bond, conjugate addition is generally favored over direct addition. (Section 18.1b)

- **Acetals**, **imines**, and **enamines** can be formed reversibly from ketones or aldehydes under acidic conditions. Formation of each of these products is favored (via Le Châtelier's principle) if there is an excess of the respective nucleophiles. (Sections 18.2a and 18.2b)

- Acetals, imines, and enamines can be hydrolyzed to ketones or aldehydes under acidic conditions using an excess of water. Nitriles can be hydrolyzed to amides. (Sections 18-2a–c)

- The C=O group of a ketone or aldehyde can be reduced to a methylene (CH_2) group via the **Wolff–Kishner reduction**, in which the ketone or aldehyde is first treated with hydrazine (H_2NNH_2), followed by heating under basic conditions. (Section 18.3)

- When treated with a strong base, a ketone or aldehyde can react via an **aldol addition**, resulting in a **β-hydroxycarbonyl**. Under these basic conditions, an enolate anion acts as a nucleophile. (Sections 18.4 and 18.6)

- Heating an aldol reaction facilitates an **aldol condensation**, in which the β-hydroxycarbonyl product undergoes dehydration to yield an **α,β-unsaturated carbonyl**. (Section 18.5)

- Aldol reactions involving ketones tend to be unfavorable, and thus require exploiting Le Châtelier's principle to drive the reaction toward products. (Section 18.6)

- A **crossed aldol reaction** forms a carbon–carbon bond between two different carbonyl-containing compounds. These reactions are not synthetically useful if they form a mixture of aldol products. (Section 18.7)

- A crossed aldol reaction can be synthetically useful if one of the carbonyl-containing reactants possesses no α hydrogens. Alternatively, a crossed aldol reaction can be synthetically useful if it involves a ketone that is first deprotonated quantitatively by a very strong base like LDA. **(Section 18.7)**

■ An intramolecular aldol reaction is favored if it forms a five- or six-membered ring. Intramolecular reactions involving the attack of an aldehyde are favored over ones involving the attack of a ketone. **(Section 18.8)**

■ A nitrile or nitroalkane that possesses an acidic α hydrogen can participate in an aldol-type reaction. **(Section 18.9)**

■ The **Robinson annulation** produces a six-membered ring from two separate carbonyl-containing compounds, one of which is an α,β-unsaturated carbonyl. This reaction consists of a conjugate addition to the α,β-unsaturated carbonyl, followed by an intramolecular aldol condensation. **(Section 18.10)**

Reaction Tables

TABLE 18-1 Functional Group Conversions

	Starting Functional Group	Typical Reagents and Reaction Conditions	Functional Group Formed	Key Electron-Rich Species	Key Electron-Poor Species	Comments	Discussed in Section
(1)	Ketone or aldehyde (R or H)(R or H)	H–Nu, Strong base; Nu = OR, RS, NR$_2$	(HO)(Nu) product (R or H)(R or H)	Nu:$^{\ominus}$	C=O, δ+	Base-catalyzed nucleophilic addition	18.1
(2)	Ketone or aldehyde (R or H)(R or H)	H–Nu, Strong acid; Nu = OR, RS, NR$_2$	(HO)(Nu) product (R or H)(R or H)	H–Nu: , δ−	$^{\oplus}$OH	Acid-catalyzed nucleophilic addition	18.1
(3)	α,β-Unsaturated ketone or aldehyde	H–Nu; Nu = OR, RS, NR$_2$	Nu-substituted product	H–Nu: , δ−	C=C, δ+	Conjugate nucleophilic addition	18.1b
(4)	Ketone or aldehyde (R or H)(R or H)	ROH (excess), H$^{\oplus}$	(RO)(OR) Acetal (R or H)(R or H)	R–OH, δ−	$^{\oplus}$OH	Nucleophilic addition, then S$_N$1	18.2a
(5)	Ketone or aldehyde (R or H)(R or H)	Ammonia or 1° amine, NH$_3$ or RNH$_2$ (excess), H$^{\oplus}$	Imine (R or H)(R or H)	R–NH$_2$, δ−	$^{\oplus}$OH	Nucleophilic addition, then E1	18.2b
(6)	Ketone or aldehyde (R or H)(R$_2$)	2° amine, R$_2$NH (excess), H$^{\oplus}$	Enamine (R or H)	R$_2$NH, δ−	$^{\oplus}$OH	Nucleophilic addition, then E1	18.2b
(7)	Acetal (R or H)(R or H)	H$_2$O (excess), H$^{\oplus}$	Ketone or aldehyde (R or H)(R or H)	H$_2$O: , δ−	$^{\oplus}$OR	S$_N$1, then E1	18.2a

(continued)

TABLE 18-1 Functional Group Conversions (continued)

	Starting Functional Group	Typical Reagents and Reaction Conditions	Functional Group Formed	Key Electron-Rich Species	Key Electron-Poor Species	Comments	Discussed in Section
(8)	Imine	H_2O (excess) $\xrightarrow{H^\oplus}$	Ketone or aldehyde	$H_2\ddot{O}^{\delta-}$		Nucleophilic addition, then E1	18.2b
(9)	Nitrile	H_2O (1 eq.) $\xrightarrow{H^\oplus \text{ or } HO^\ominus}$	Amide	$H_2\ddot{O}^{\delta-}$ or $\ddot{\ominus}\ddot{O}H$	$-C\equiv\overset{\oplus}{N}H$ or $\overset{\delta+}{-C}\equiv N$	Nucleophilic addition	18.2c
(10)	Ketone or aldehyde	1. H_2N-NH_2, $H^\oplus$ 2. KOH/H_2O, Δ		$H_2\overset{\delta-}{N}NH_2$	$\overset{\oplus}{O}H$	Wolff-Kishner reduction; nucleophilic addition, then E1, then E2	18.3
(11)	β-Hydroxycarbonyl	$H^\oplus$ or $HO^\ominus$ $\xrightarrow{\Delta}$	α,β-Unsaturated carbonyl			E1 or E1cb	18.5

TABLE 18-2 Reactions That Alter the Carbon Skeleton

	Starting Functional Group	Typical Reagents and Reaction Conditions	Functional Group Formed	Key Electron-Rich Species	Key Electron-Poor Species	Comments	Discussed in Section
(1)	Ketone or aldehyde	H–CN $\xrightarrow{\text{NaOH or KCN}}$	Cyanohydrin	$NC\overset{\ominus}{:}$		Nucleophilic addition	18.1a
(2)	Aldehyde	NaOH	β-Hydroxycarbonyl	Enolate anion		Aldol addition; nucleophilic addition	18.4
(3)	Ketone	NaOH	β-Hydroxycarbonyl	Enolate anion		Aldol addition; nucleophilic addition	18.6
(4)	Aldehyde	1. NaOH 2.	β-Hydroxycarbonyl	Enolate anion		Crossed aldol addition; nucleophilic addition	18.7

(continued)

TABLE 18-2 Reactions That Alter the Carbon Skeleton (continued)

Starting Functional Group	Typical Reagents and Reaction Conditions	Functional Group Formed	Key Electron-Rich Species	Key Electron-Poor Species	Comments	Discussed in Section
(5) Ketone	1. LDA 2.	β-Hydroxycarbonyl	Enolate anion		Crossed aldol reaction; nucleophilic addition	18.7
(6) Nitrile or nitroalkane	NaOH		(or NO₂)		Aldol-type reaction; nucleophilic addition	18.9
(7) Ketone	KOH		Enolate anion		Robinson annulation	18.10

Problems

18.44 Predict the major organic product and draw the complete, detailed mechanism for each of the following reactions.

(a)

$$\text{CH}_3\text{CH}_2\text{OH (excess)}, \ H^\oplus \longrightarrow \ ?$$

(b)

$$\text{HO} \frown \text{OH}, \ H^\oplus \longrightarrow \ ?$$

(c)

$$\text{CH}_3\text{OH (excess)}, \ H^\oplus \longrightarrow \ ?$$

(d)

$$\text{HO} \frown \text{SH}, \ H^\oplus \longrightarrow \ ?$$

(e)

$$\text{HO} \frown \text{OH}, \ H^\oplus \longrightarrow \ ?$$

18.45 Predict the major organic product and draw the complete, detailed mechanism for each of the following reactions.

(a)

$$\text{HO} \frown \text{OH} + \text{(butanone)}, \ H^\oplus \longrightarrow \ ?$$

(b)

$$\text{catechol} + \text{(acetone)}, \ H^\oplus \longrightarrow \ ?$$

(c)

$$\text{HO} \frown \text{OH} + \text{(benzaldehyde)}, \ H^\oplus \longrightarrow \ ?$$

18.46 Predict the major organic product and draw the complete, detailed mechanism for each of the following reactions.

(a)

$\xrightarrow[H^{\oplus}]{\text{NH}_3 \text{ (excess)}}$?

(b)

$\xrightarrow[H^{\oplus}]{\text{CH}_3\text{NH}_2 \text{ (excess)}}$?

(c)

$\xrightarrow[H^{\oplus}]{\text{H}_2\text{O (excess)}}$?

(d)

H₃C—N(H)—CH₃ (excess) $\xrightarrow[H^{\oplus}]{}$?

(e)

$\xrightarrow[H^{\oplus}]{\text{H}_2\text{O (excess)}}$?

18.47 Predict the major organic product and draw the complete, detailed mechanism for each of the following reactions.

(a)

$\xrightarrow[H^{\oplus}]{\text{H}_2\text{O (excess)}}$?

(b)

$\xrightarrow[H^{\oplus}]{\text{H}_2\text{O (excess)}}$?

(c)

$\xrightarrow[H^{\oplus}]{\text{H}_2\text{O}}$?

(d) H₃CH₂CO   OCH₂CH₃

$\xrightarrow[H^{\oplus}]{\text{H}_2\text{O (excess)}}$?

18.48 Predict the major organic product and draw the complete, detailed mechanism for each of the following reactions.

(a)

$\xrightarrow[\text{2. NaOH/H}_2\text{O, }\Delta]{\text{1. H}_2\text{NNH}_2, \text{H}^{\oplus}}$?

(b)

$\xrightarrow[\text{2. NaOH/H}_2\text{O, }\Delta]{\text{1. H}_2\text{NNH}_2, \text{H}^{\oplus}}$?

18.49 Draw three different ketone or aldehyde precursors that could be used to produce hexane via a Wolff–Kishner reduction.

18.50 When acetone is dissolved in either a slightly basic or a slightly acidic solution of oxygen-18 labeled water, $H_2^{18}O$, oxygen-18 labeled acetone, $(CH_3)_2C{=}^{18}O$, is produced. This is a form of an isotopic exchange reaction between acetone and water. Provide a mechanism to account for this reaction in **(a)** basic solution and **(b)** acidic solution. (*Hint:* Is the addition of the nucleophile reversible or irreversible?)

$$\underset{H_3C}{\overset{O}{\underset{}{\|}}}\!\!\!-\!\!\!\underset{CH_3}{} + H_2^{18}O \xrightarrow{\text{Acid or base}} \underset{H_3C}{\overset{^{18}O}{\underset{}{\|}}}\!\!\!-\!\!\!\underset{CH_3}{} + H_2O$$

18.51 Draw the complete, detailed mechanism and predict the major organic product for each of the following reactions.

(a)

NaOH
Δ
?

(b)

NaOH
Δ
?

(c)

NaOH
?

(d)

NaOH
Δ
?

18.52 Draw the mechanism for the aldol addition reaction below, which is catalyzed by acid. (*Hint*: Under acidic conditions, can an enolate anion act as a nucleophile?)

H⊕

Propanal   **3-Hydroxy-2-methylpentanal**

18.53 Draw a complete, detailed mechanism and predict the major product for each of the following reactions.

(a)

1. LDA

2.

?

(b)

1. NaOH

2.

?

(c)

NaOH
Δ
?

(d)

1. LDA

2.

?

18.54 Will compound **A** or compound **B** produce more aldol product at equilibrium? Why?

H₃C   CH₃   FH₂C   CH₂F

A   **B**

18.55 Predict the major product(s) of each of the following reactions.

(a)

1. (CH₃)₂CuLi

2. H₂NNH₂, HO⊖, Δ

?

(b)

1. NaOH

2. NaBH₄, EtOH

?

18.56 Provide the reagent(s) missing from each of the following reactions.

(a)

(b)

(c)

(d)

18.57 Suggest how you would synthesize each of the following, using cyclopentanone as one of the reagents.

(a) (b) (c) (d) (e)

18.58 How would you synthesize the following compound if, as your starting material, you may use any organic reagents that contain exactly two carbon atoms and any inorganic reagents?

18.59 How would you synthesize the following compound using phenylethanal as your only source of carbon atoms?

18.60 How would you synthesize 2-methylhexane from hex-4-en-3-one and any other reagents necessary?

18.61 An enamine, $R_2C=C-NR_2$, behaves as a nucleophile in much the same way that an enolate anion does, with nucleophilic character at the carbon alpha to the one bonded to the amine. This is because an enamine has resonance structures similar to those observed for an enolate anion:

Nucleophilic character

With this in mind, draw the complete mechanism for the following reaction, and provide the structure of the missing product.

18.62 Draw the complete, detailed mechanism for the following reaction and, using the mechanism, predict the major product. TsOH is a relatively acidic compound.

18.63 The following reaction shows that a secondary nitro compound can be hydrolyzed to a ketone upon treatment with aqueous sulfuric acid. Propose a mechanism for this reaction.

18.64 Draw a complete, detailed mechanism for the following reaction.

$$R-C\equiv N \xrightarrow[\text{2. } H_2O, H^{\oplus}]{\text{1. } R'-OH, H^{\oplus}}$$

18.65 The following reaction shows that an aldol-type reaction can be performed with an imine enolate as the nucleophile:

Upon hydrolysis, the product is the normal β-hydroxycarbonyl. Notice that LDA is used as the base instead of HO⁻.

(a) Explain why HO⁻ cannot be used as a base for this reaction.

(b) Provide a detailed mechanism for this reaction, including the structure of the intermediate species not shown.

18.66 The following reaction shows that an aldehyde or a ketone can condense with an α-haloester under basic conditions to give an α,β-epoxy ester, also called a *glycidic ester*. Provide a detailed mechanism for this reaction.

18.67 Treatment of a nitrile with base, followed by acid hydrolysis, yields a β-keto nitrile, as shown here. Provide a detailed mechanism for this reaction.

18.68 An isonitrile is an unusual species that has the form R—N⁺≡C⁻. When an isonitrile is treated with water, an *N*-alkylformamide is formed, as shown here. Propose a mechanism for this reaction.

18.69 The formation of an acetal from a ketone or aldehyde has a very similar mechanism to the formation of an imine (review Equations 18-13 and 18-19). Both require acidic conditions to generate a water leaving group. At very acidic pH, however, the rate of imine formation slows down (see the following graph), whereas the rate of acetal formation does not. Explain this pH dependence of imine formation.

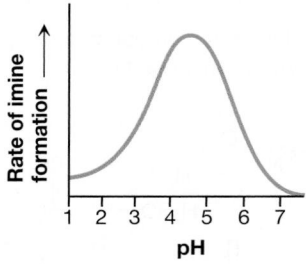

18.70 Hydroxylamine, H₂NOH, has both an alcohol functional group and an amine functional group, so it can feasibly undergo reaction with a ketone or an aldehyde to produce either an acetal or an imine-like compound called an oxime.

(a) Draw each of these mechanisms for the reaction of hydroxylamine with acetone.

(b) Which is the major product? (*Hint*: Which step decides the outcome?)

18.71 When benzaldehyde is treated with a catalytic amount of KCN, the benzoin condensation occurs.

An α-hydroxycarbonyl

Benzaldehyde

KCN

HO⁻/H₂O, EtOH

2-Hydroxy-1,2-diphenylethanone (Benzoin)
88%

Draw a complete, detailed mechanism for this reaction. (*Hint*: The reaction does not take place without the presence of NC⁻.)

18.72 No reaction occurs when benzaldehyde and propenenitrile (acrylonitrile) are combined. In the presence of a catalytic amount of NaCN, however, the following reaction takes place.

Draw a complete, detailed mechanism to account for these results. (*Hint*: See Problem 18.71.)

18.73 The following crossed aldol reaction can be carried out using a weak base such as pyridine.

Diethyl malonate

(a) Draw the complete, detailed mechanism for this reaction.
(b) Explain why a relatively weak base can be used.

18.74 The following reaction is believed to proceed through the intermediate shown.

Draw the complete, detailed mechanism that leads to the formation of that intermediate.

18.75 When a 1,5-diketone is treated with acid, a 2,6-dialkyl pyran is produced. Propose a mechanism for this reaction.

18.76 A carbamate can be prepared by treating an isocyanate with an alcohol, as shown.

An isocyanate **A carbamate**
(Substituted urethane)

This type of reaction is used to synthesize polyurethanes, polymers that have a wide variety of industrial applications, such as surface sealants, high-performance adhesives, and synthetic fibers. Propose a mechanism for this transformation.

18.77 An imino ester is formed when a nitrile is treated with an alcohol in the presence of *dry* HCl (i.e., without H_2O), followed by treatment with a weak base such as sodium bicarbonate. Propose a mechanism for this reaction.

An imino ester

18.78 Pentanedinitrile undergoes a cyclization reaction when treated with ammonia under weakly acidic conditions.

An imidine

The product is an imidine. Propose a mechanism for this reaction.

18.79 Draw the complete, detailed mechanism for the following reaction, which produces an amidine—a nitrogen analog to an ester.

Benzonitrile

1.
2. CH₃OH

An amidine

18.80 After consulting Problem 18.79, suggest how the following amidine can be synthesized from benzonitrile, $C_6H_5C\equiv N$. (*Hint*: More than one synthetic step may be necessary.)

An amidine

18.81 When an α,β-unsaturated carbonyl is treated with H_2O_2 under basic conditions, the C═C double bond is epoxidized, as shown here. Propose a mechanism for this reaction.

18.82 The following transformation is an example of a Darzens reaction. Draw its complete, detailed mechanism.

18.83 The following is an example of a Wittig–Horner reaction.

Draw its complete, detailed mechanism. (*Hint*: A key intermediate is provided.)

18.84 A compound with formula $C_{10}H_{14}O$ has the IR, ^1H NMR, and ^{13}C NMR spectra shown here. The DEPT spectra reveal that the seven carbon signals farthest upfield are produced from CH_2 carbons. $C_{10}H_{14}O$ can be synthesized using an aldol condensation reaction, simply by heating a compound C_5H_8O in the presence of NaOH.

   (a) Provide the structures of C_5H_8O and $C_{10}H_{14}O$.

   (b) Provide the mechanism for the reaction that is described.

18.85 When but-2-enal is treated with 3-phenylpropenal in the presence of a strong base, a compound is formed whose formula is $C_{13}H_{12}O$. In its ^1H NMR spectrum, one signal has a chemical shift around 10 ppm, several overlapping signals have chemical shifts between 7 and 9 ppm, and several other overlapping signals have chemical shifts between 5 and 6 ppm. Integration of those sets of signals gives a 1:5:6 ratio.

   (a) What is the structure of the product?

   (b) Draw a complete, detailed mechanism that accounts for the formation of the product.

18.86 When treated with acid, hexane-2,5-dione forms a compound with the formula C_6H_8O. The ^1H NMR spectrum of the product is shown below. A key intermediate is shown. Identify the structure of the product and propose a mechanism for this reaction.

18.87 When acetonitrile is treated with concentrated sulfuric acid and *tert*-butanol, followed by water, a product is formed whose ^1H NMR spectrum exhibits the following three signals: singlet, 1.3 ppm, 9H; singlet, 2.0 ppm, 3H; and broad singlet, 8.2 ppm, 1H. Its IR spectrum exhibits one broad absorption of medium intensity between 3300 and 3500 cm^{-1}, and a narrow, intense absorption near 1650 cm^{-1}. A key intermediate is shown. Draw the structure of the product, and draw the complete, detailed mechanism for the reaction.

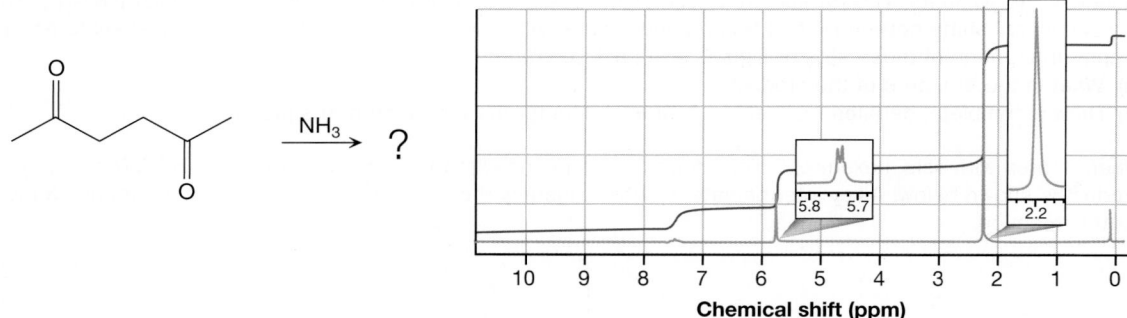

18.88 In mildly acidic conditions, 3-(2-aminophenyl)-propenal reacts to form a compound whose molecular weight is 129 g/mol. The ^1H NMR spectrum for the product is shown below. Its ^{13}C NMR spectrum contains nine peaks, all between 120 and 150 ppm. Propose a mechanism for this reaction and draw the product.

18.89 Reaction of 2,5-hexanedione with ammonia produces a compound whose ^1H NMR spectrum is shown below. Its ^{13}C NMR spectrum exhibits three signals. Draw the complete, detailed mechanism leading to the formation of that product.

If your instructor assigns problems in smartwork, log in at **smartwork.wwnorton.com**.

Organic Synthesis 2

Intermediate Topics in Synthesis Design, and Useful Reduction and Oxidation Reactions

I n Chapter 13, we were introduced to the basics of organic synthesis, focusing mainly on how to construct a molecule with the appropriate carbon skeleton and how to convert one functional group into another. We now continue with organic synthesis, focusing on some higher-level topics that enable us to synthesize more elaborate target molecules.

Much of our focus here is geared toward the formation of new carbon–carbon bonds. Carbon–carbon bonds typically form between oppositely charged carbon atoms, but we are often faced with the challenge of forming these bonds between carbon atoms that are initially of like charge. Therefore, we begin with a strategy to accomplish this. Next, we focus on clues in a target molecule that help us determine the types of carbon–carbon bond-formation reactions that should be considered in a synthesis. We also describe one way to tackle a synthesis in which these kinds of clues do not exist.

Another focus of Chapter 19 is on ways to circumvent *synthetic traps*. (Recall from Section 13.4 that a synthetic trap arises when the conditions necessary for a desired reaction at one site within a molecule will cause an

> Many foods contain synthetic fats derived from naturally occurring oils using a process called catalytic hydrogenation, an important reduction reaction we examine here in Chapter 19.

- Identify reactions that reverse the polarization at a carbon atom, and incorporate these kinds of reactions in organic syntheses.

- Analyze the relative positioning of functional groups in a target molecule to determine which carbon–carbon bond-formation reactions should be considered in a synthesis.

- Devise a synthesis that utilizes reactions in which specific functional groups, such as aldehydes, ketones, and alkenes, are removed entirely from a molecule.

- Identify situations in which a *selective reaction* is called for, and incorporate these kinds of reactions in a synthesis.

- Explain the basic principles behind the use of a *protecting group*, and demonstrate the protection and deprotection of ketones, aldehydes, and alcohols in syntheses.

- Describe the basic mechanism of catalytic hydrogenation, and predict the products of the catalytic hydrogenation of alkenes, alkynes, aldehydes, and ketones.

- Predict the products of oxidation reactions involving Cr^{+6} and Mn^{+7} oxidizing agents, and incorporate these reactions effectively in syntheses.

undesired reaction at another site.) Specifically, we discuss the use of selective reactions that target one functional group over another. We also discuss strategies for protecting functional groups we wish to remain intact.

Finally, we introduce some new reduction and oxidation reactions. In addition to being useful reactions for functional group conversions, they allow us to apply some of the synthesis strategies just mentioned.

19.1 Umpolung in Organic Synthesis: Forming Bonds between Carbon Atoms Initially Bearing Like Charge; Making Organometallic Reagents

In Section 7.1, we learned that a bond can form between two atoms when one atom bears a partial or full negative charge (and thus is electron rich, or nucleophilic) and the other bears a partial or full positive charge (making it electron poor, or electrophilic). This general idea is applicable to a wide variety of bond-formation reactions, including those that form carbon–carbon bonds. Consider, for example, the reaction in Equation 19-1.

Electron-poor C (electrophilic)   **Electron-rich C** (nucleophilic)   **A C—C bond tends to form between an electron-rich and an electron-poor C.**

(19-1)

The bond that is formed involves the carbon atom from NC^- and the carbon atom attached to Cl in the alkyl halide. The NC^- carbon bears a full negative charge, whereas the alkyl halide carbon bears a partial positive charge.

Because carbon–carbon bond-formation reactions usually involve oppositely charged carbon atoms, a problem arises when we want to form a bond between two carbon atoms bearing like charges. For example, Equation 19-2 shows that no bond readily forms between the carbonyl C atom of a ketone and the C atom attached to Br in an alkyl bromide. This is because both C atoms are attached to atoms with relatively high electronegativities (O and Br, respectively), and thus bear partial positive charges.

Both C atoms are relatively electron poor.

No reaction   (19-2)

One way to form a bond between two carbons of like charge is to carry out a separate reaction that first reverses the charge (or *polarity*) at one of the carbons. Thus, one carbon atom would become electron rich while the other would remain electron poor. This general idea of reversing a charge at a particular atom, called **umpolung** (a German term meaning "polarity reversal"), is common practice in organic synthesis. In fact, we have already encountered reactions that use umpolung in the formation of a carbon–carbon bond. One is in the synthesis of a Wittig reagent (Section 17.8), shown in Equation 19-3.

In the alkyl halide reactant, the C atom bonded to the halogen atom bears a partial positive charge and is relatively electron poor. By contrast, that C atom has become electron rich in the Wittig reagent that is produced.

Another class of reactions that accomplishes this charge reversal is the formation of organometallic reagents from their corresponding alkyl halides, examples of which are shown in Equations 19-4 and 19-5. Equation 19-4 shows that an alkyl bromide can be converted into a *Grignard reagent* (RMgBr) simply by treating it with solid magnesium in an ether solvent such as tetrahydrofuran (THF). (Grignard reagents can also be produced from alkyl chlorides and alkyl iodides.) Similarly, Equation 19-5 shows that an *alkyllithium reagent* (RLi) can be synthesized from an alkyl bromide by treating it with solid lithium. (This reaction can also begin with an alkyl chloride.)

Notice that in both cases, the partial positive charge of the alkyl halide carbon becomes a partial negative charge in the organometallic reagent. Thus, umpolung has taken place, which facilitates the formation of a bond with an electron-poor carbon, such as in an epoxide ring-opening reaction (Section 10.7) or a Grignard reaction (Section 17.5).

19.1 YOUR TURN

Write either δ^+ or δ^- next to the C atom bonded to the halogen in the reactant and the C atom bonded to Li in the product.

Br → Li(s)/Ether → Li

Answers to Your Turns are in the back of the book.

A related reaction is the synthesis of *lithium dialkylcuprates* (R$_2$CuLi), reagents that were first introduced in Section 17.11. As shown in Equation 19-6, a lithium dialkylcuprate is synthesized from the corresponding alkyllithium reagent by treating it with copper iodide (CuI). Because the metal-bonded C atom bears a partial negative charge in both the alkyllithium reagent and the lithium dialkylcuprate, umpolung technically does *not* occur in this reaction, but synthesizing a lithium dialkylcuprate from an alkyl halide does reverse the polarity at the C atom. Once again, this facilitates the formation of a carbon–carbon bond to an electron-poor carbon, as we have seen previously in conjugate addition to an α, β-unsaturated carbonyl (Section 17.11).

The polarity at the C atom has been reversed.

A lithium dialkylcuprate

(19-6)

---

problem **19.1** Predict the product of each of the following reactions.

(a) (b) (c) (d)

---

problem **19.2** Show how to synthesize each of the following organometallic compounds from an alkyl halide.

(a) (b) (c) (d)

SOLVED problem 19.3 Using formaldehyde (H$_2$C=O) as your only source of carbon, show how you would synthesize propan-1-ol.

Methanal (Formaldehyde) **Propan-1-ol**

Think Do C—C bonds have to be formed in the synthesis? If so, what should the electron-rich species be? What should the electron-poor species be? Can the synthesis take place in a single reaction? If not, what precursor should we choose?

Solve Since our only source of carbon contains a single C atom and the product contains a chain of three C atoms, C—C bonds must be formed. A C—C bond-formation reaction would be difficult to carry out directly between two molecules of formaldehyde, however, because such a reaction would entail the formation of a bond between two carbon atoms bearing a partial positive charge. We must consider, therefore, a precursor that has a carbon atom bearing a significant negative charge. A Grignard reagent would suffice, especially since the target is an alcohol, the product of a Grignard reaction.

In our retrosynthetic analysis, undo a Grignard reaction as the first transform, disconnecting the

An alcohol is the product of a Grignard reaction.

Undo a Grignard reaction.

New target

C—C bond involving the C that bears the OH group. This shows that the final product can be made from an ethyl Grignard reagent (CH_3CH_2MgBr) and formaldehyde. Formaldehyde is an available starting material, but the ethyl Grignard is not. We must, therefore, devise a synthesis for it.

The ethyl Grignard can be generated from CH_3CH_2Br. Once again, CH_3CH_2Br is not an available starting material, but it can be made from ethanol using PBr_3.

Undo a Grignard formation.

Undo a bromination.

Ethanol is now the possible product of a Grignard reaction between CH_3MgBr and our starting material, $H_2C=O$. CH_3MgBr is generated from CH_3Br, which can be made from CH_3OH using PBr_3. Finally, CH_3OH can be made by reducing our starting material using a hydride reagent such as $NaBH_4$.

Undo a Grignard reaction.

Undo a Grignard formation.

Undo a bromination.

Undo a reduction.

The synthesis is reported as follows:

NaBH₄, EtOH | PBr₃ | Mg(s), Ether | 1. H–C(=O)–H 2. H⁺ | PBr₃

Mg(s), Ether | 1. H–C(=O)–H 2. H⁺

---

problem 19.4 Using acetaldehyde ($CH_3CH=O$) as your only source of carbon, show how to synthesize 3-methylpentan-3-ol.

Ethanal (Acetaldehyde) → ? → 3-Methylpentan-3-ol

---

problem 19.5 Show how you would synthesize 4-phenylbutan-1-ol from bromobenzene, using oxirane, △O, as your only other source of carbon.

Bromobenzene → ? → 4-Phenylbutan-1-ol

19.2 Relative Positioning of Functional Groups in Carbon–Carbon Bond-Formation Reactions

In Chapter 13, we learned that carbon–carbon bond-formation reactions are important in organic synthesis because they allow us to alter the carbon framework of a particular molecule. In many of those reactions, two functional groups in the product are left with very specific relative locations along the carbon skeleton. That relative positioning can be critical to a synthesis because, as we have seen many times before, those functional groups govern the sites at which subsequent reactions can take place.

Consider, for example, the formation of a *cyanohydrin*, shown in Equation 19-7.

These functional groups involve
adjacent carbons, so they
establish a 1,2-positioning.

(19-7)

A cyanohydrin

A C—C bond is formed between the cyanide C and the carbonyl C. Notice, in particular, that the two functional groups in the product—a nitrile group (C≡N) and an alcohol group (C—OH)—are *adjacent* to each other. In other words, the C atoms involved in these functional groups have a **1,2-positioning** relative to each other.

Knowing that this reaction produces a compound with 1,2-positioning of the resulting functional groups is particularly useful when our target molecule is *not* a cyanohydrin but still has functional groups with 1,2-positioning. Suppose, for example, that we want to carry out the following synthesis starting from the ketone given.

1,2-Positioning

The ether and amine functional groups in the target have a 1,2-positioning, so we could consider a cyanohydrin as a synthetic intermediate, as shown in the following retrosynthetic analysis.

A cyanohydrin

What has yet to be solved is the conversion of the cyanohydrin to the target molecule. This entails converting the alcohol to the ether, and converting the nitrile to a primary amine. As shown in the following reaction scheme, the alcohol can be converted to the ether using a Williamson synthesis (Section 10.6), and the nitrile can be converted to the primary amine by reduction with LiAlH$_4$ (Section 17.3).

A cyanohydrin

SOLVED problem 19.6 Show how to synthesize the target at the right, using compounds with three or fewer carbon atoms.

Think Does the target have a 1,2-positioning of the functional groups that could result from the formation of a carbon–carbon bond? If so, from what cyanohydrin could this target be synthesized? How could that cyanohydrin be synthesized from compounds containing three or fewer carbons?

Solve The target does indeed have a 1,2-positioning of the CN and SH functional groups, so we can conceive of obtaining it from a cyanohydrin, as shown in the following retrosynthetic analysis.

To convert the OH group to an SH group, the cyanohydrin could first be treated with PBr₃, followed by NaSH, as shown in the following synthesis.

problem 19.7 Show how to synthesize the following target, using compounds with seven or fewer carbon atoms.

Cyanohydrin formation is not the only carbon–carbon bond-formation reaction that produces functional groups with specific relative positioning along the carbon backbone. Other reactions like this that we have encountered previously are listed in Table 19-1.

19.2 YOUR TURN

For each entry in Table 19-1, number the carbon atoms in the chains of the respective products so that the numbers assigned to the carbon atoms of the functional groups agree with the relative positioning listed. (*Hint*: The "1" and "2" carbons are not necessarily the atoms that are highlighted.)

The reactions in Table 19-1 can be used to strategically synthesize products with 1,2-, 1,3-, 1,4-, and 1,5-positioning of the functional groups that remain. Therefore, if a target has functional groups with one of these relative positionings and the synthesis calls for a carbon–carbon bond-formation reaction, you should consider using an appropriate reaction from Table 19-1. An example is shown in Solved Problem 19.8.

TABLE 19-1 Reactions That Form Carbon–Carbon Bonds

Relative Positioning	Product Formed	Electron-Rich Reactant	Electron-Poor Reactant	Discussed in Section
1,2-	Cyanohydrin	Cyanide anion	Ketone/ aldehyde	18.1a
1,2-	Alkene	Wittig reagent	Ketone/ aldehyde	17.7
1,3-	β-Hydroxy carbonyl	Enolate ion	Ketone/ aldehyde	18.4
1,4-	β-Cyano ketone/ aldehyde	Cyanide anion	α,β-Unsaturated ketone/aldehyde	18.1b
1,5-	1,5-Dicarbonyl	Enolate ion	α,β-Unsaturated ketone/aldehyde	18.10

SOLVED problem 19.8 Show how you can synthesize 2-methylpentane-1,3-diol from compounds containing five or fewer carbons.

Five or fewer carbons $\xrightarrow{\text{?}}$

Think Will a carbon–carbon bond-formation reaction be necessary? What is the relative positioning of the functional groups in the target? Does the appropriate carbon–carbon bond-formation reaction in Table 19-1 leave us with the correct functional groups, or will we need to carry out an additional functional group conversion?

Solve We will have to use a carbon–carbon bond-formation reaction, because the target's carbon skeleton contains six carbons bonded together and we are allowed to start with compounds containing only five. The 1,3-positioning of the two hydroxyl groups in the product suggests that we should use an aldol reaction, which yields a β-hydroxycarbonyl. In a retrosynthetic analysis, therefore, our task becomes to apply a transform that takes our target molecule back to a β-hydroxycarbonyl. We can do this by undoing a hydride reduction on either of the two OH groups. Below, we show this transform applied to the OH group on C3. From there, we disconnect the appropriate C—C bond to take us back to the aldol reactants, pentan-2-one and formaldehyde.

In the forward direction, the synthesis would appear as follows:

problem 19.9 Show how you can synthesize 2-methylpentane-1,3-diol from compounds containing three or fewer carbons.

19.3 Reactions That Remove a Functional Group Entirely from a Molecule: Reductions of C=O to CH₂

When we design a synthesis, we typically begin by looking for structural features in the target that might indicate the need to use certain reactions. For example, in Section 19.2 we saw that if a target calls for a carbon–carbon bond-formation reaction and exhibits a specific relative positioning of functional groups, we should consider using a reaction from Table 19-1. Frequently, however, a synthesis will call for a carbon–carbon bond-formation reaction, but no functional groups in the target provide a clear indication of what that reaction should be.

For cases like this, we can sometimes use reactions that remove a functional group entirely. Thus, a retrosynthetic analysis might require us to consider the existence of a functional group in a precursor at a location where no functional group exists in the target. The presence of that functional group in the intermediate might then facilitate a particular carbon–carbon bond-formation reaction, and the functional group that remains could subsequently be removed.

In this context, we discuss the use of three reactions that convert the C=O group of a ketone or aldehyde into a CH₂ group. In doing so, the ketone or aldehyde is effectively removed entirely. Section 19.3a discusses the Wolff–Kishner reduction, Section 19.3b discusses the Clemmensen reduction, and Section 19.3c discusses the Raney-nickel reduction.

19.3a The Wolff–Kishner Reduction

The **Wolff–Kishner reduction**, first discussed in Section 18.3, converts the carbonyl (C=O) group of a ketone or aldehyde into a methylene (CH₂) group. An example is shown in Equation 19-8.

Thus, a Wolff–Kishner reaction effectively removes a ketone or aldehyde functional group from the molecule.

Prior to its removal, however, a ketone or aldehyde group can facilitate the formation of a new carbon–carbon bond. As shown in Equation 19-9, for example, alkylation can take place at the α carbon via an S_N2 reaction (Section 10.3).

(19-9)

Alternatively, Equation 19-10 shows that conjugate addition to an α,β-unsaturated ketone or aldehyde results in the formation of a new C—C bond at the β carbon (Section 17.11).

(19-10)

With this in mind, suppose that we want to synthesize propylbenzene from starting materials containing eight or fewer carbon atoms:

Given that the target contains nine C atoms, we must use a carbon–carbon bond-formation reaction. That reaction could be an α alkylation if we imagine one of the CH_2 groups being produced from a carbonyl group in a precursor, as shown in the following retrosynthetic analysis.

In the forward direction, the synthesis would appear as follows:

problem 19.10 Show how to synthesize propylbenzene from starting materials containing seven or fewer carbon atoms.

As shown in Solved Problem 19.11, a similar strategy can be applied when the carbon–carbon bond-formation reaction is conjugate addition involving a lithium dialkylcuprate.

SOLVED problem 19.11 Show how to synthesize the following compound beginning with an α,β-unsaturated ketone or aldehyde.

Think What carbon–carbon bond could you disconnect in a transform? If that bond is the result of conjugate addition to an α,β-unsaturated ketone or aldehyde, which CH_2 group could we imagine coming from a C=O group?

Solve The C—C single bond that is not part of the ring could have come from conjugate addition to an α,β-unsaturated ketone, as shown below. This kind of bond formation takes place at the β carbon of an α,β-unsaturated carbonyl, so the topmost CH_2 group on the ring could come from a C=O group.

Therefore, the synthesis in the forward direction might appear as follows:

---

problem 19.12 Show how the compound at the right can be synthesized separately from four different α,β-unsaturated ketones.

19.3b The Clemmensen Reduction

Section 19.3a focused on the use of the Wolff–Kishner reaction, but it is not the only reaction that reduces the C=O group of a ketone or aldehyde to a CH_2 group. Another one is the **Clemmensen reduction**, named after Erik Christian Clemmensen (1876–1941), a Danish chemist. (A third is the Raney-nickel reduction, discussed in Section 19.3c.) An example of a Clemmensen reduction is shown in Equation 19-11.

The Clemmensen reduction uses what is called a **zinc amalgam**, which is an alloy (blend) of zinc with mercury. The ketone or aldehyde is refluxed (i.e., continuously evaporated and recondensed under heat) with the amalgam in an HCl solution. HCl is the source of the protons that form bonds to the carbonyl C. The zinc metal acts as a reducing agent in what is called a **dissolving metal reduction**, whereby the metal dissolves in solution as the reaction ensues. Zinc loses two electrons upon going from elemental Zn to $ZnCl_2$, so it is oxidized. By contrast, the carbonyl C is reduced because it loses the double bond to O and gains two bonds to H.

19.3c The Raney-Nickel Reduction

A third reaction that reduces the carbonyl group of a ketone or aldehyde to a methylene group is the **Raney-nickel reduction**, developed by Murray Raney (1885–1966), an American chemist. An example is shown in Equation 19-12.

$$(19\text{-}12)$$

In a Raney-nickel reduction, the ketone or aldehyde is first converted to a *thioacetal* by treatment with a thiol under acidic conditions (Section 18.2a). Subsequent treatment with **Raney nickel** converts the thioacetal group into a methylene group.

Raney nickel has two roles in this reaction. First, it is the source of hydrogen. Raney nickel is obtained by treating a 50:50 alloy of aluminum and nickel with hot NaOH, which serves to etch the metal. The spongy network that remains is then treated with hydrogen gas, which adsorbs (accumulates via intermolecular interactions) to the metal in part because of the metal's very large surface area. Raney nickel also serves as a *catalyst* for the H_2 reduction, a process that is described in greater detail in Section 19.5. (As a cautionary note, Raney nickel is so reactive that it must be kept under an inert liquid or an inert gas, or it will ignite spontaneously on contact with atmospheric oxygen.)

19.3d Limitations of the Wolff–Kishner, Clemmensen, and Raney-Nickel Reductions

The Wolff–Kishner, Clemmensen, and Raney-nickel reactions reduce the carbonyl group of a ketone or aldehyde to a methylene group, but notice in Equations 19-8, 19-11, and 19-12 that the three reactions take place under different conditions. The Wolff–Kishner reduction takes place under strongly basic conditions, the Clemmensen reduction takes place under strongly acidic conditions, and the Raney-nickel reduction takes place under mild conditions in the presence of a metal catalyst. Thus, if there are functional groups present other than the ketone or aldehyde we want to reduce, then we must take care in deciding upon which of these reduction reactions to use.

- The Wolff–Kishner reduction should be avoided if there is a functional group present that is susceptible to reaction under basic conditions or with good nucleophiles.

- The Clemmensen reduction should be avoided if there is a functional group present that is susceptible to reaction under acidic conditions.

- The Raney-nickel reduction should be avoided if there is a functional group present that is susceptible to reaction with nucleophilic thiols (RSH) or with H_2 in the presence of a metal catalyst.

Although we have not yet discussed reactions involving H_2 and a metal catalyst, we show in Section 19.5 that these conditions can cause alkenes ($R_2C=CR_2$), alkynes ($RC\equiv CR$), and aldehydes ($RCH=O$) to react.

These ideas are explored further in Solved Problem 19.13.

SOLVED problem 19.13 Which reduction reaction(s)—the Wolff–Kishner, Raney-nickel, or Clemmensen—can be used to carry out the following transformation?

Think Are there any functional groups, aside from the carbonyl group, that are susceptible to reaction in the presence of HO^-? HCl? Thiols? H_2 gas and a metal catalyst?

Solve The alkyl halide that is present is susceptible to nucleophilic substitution with HO^- or thiols (both are good nucleophiles), so the Wolff–Kishner and Raney-nickel reductions should be avoided. The aromatic and alkyl halide functional groups are not susceptible to reaction with HCl, so the Clemmensen reduction could be used.

---

problem 19.14 Which reduction reaction(s)—the Wolff–Kishner, Raney-nickel, or Clemmensen—can be used for each of the following transformations?

(a)

(b)

(c)

19.4 Avoiding Synthetic Traps: Selective Reagents and Protecting Groups

As we saw in Section 13.4, a *synthetic trap* arises when a specific transform in a retrosynthesis would lead to an undesired reaction in the *forward* direction. Generally, this is because a functional group we want to leave alone would, in fact, be **labile** under the reaction conditions—that is, the functional group is reactive and readily undergoes a chemical transformation.

Many of the difficulties presented by synthetic traps can be surmounted in one of two ways: (1) by using a *selective reagent* (Section 19.4a) or (2) by using a *protecting group* (Section 19.4b). As discussed in the following sections, both of these strategies exploit differences in reactivity among functional groups.

19.4a Selective Reactions

When two or more outcomes are possible for a given reaction, the actual result often depends upon the specific reagents that are used and the specific reaction conditions. In many cases, we can exploit this specificity in designing a synthesis. When we can, the reaction is called a **selective reaction** and the reagent that causes such selectivity is called a **selective reagent**.

We first encountered an example of a selective reaction in Chapter 9, in the context of the competition between nucleophilic substitution and elimination reactions. Namely, S_N2 and E2 reactions tend to compete with each other, given that strong nucleophiles are often strong bases, but we can choose specific reagents to selectively favor one reaction over another. For example, Cl^- favors the S_N2 reaction (Equation 19-13a), whereas the *tert*-butoxide anion selectively favors the E2 reaction (Equation 19-13b).

We saw another example of selective reagents in Section 10.3a, when we discussed alkylations at the α carbons of ketones and aldehydes. As mentioned in Section 10.3a and as shown in Equation 19-14, lithium diisopropylamide (LDA) is a bulky base that irreversibly deprotonates ketones and aldehydes, so it is selective for alkylation at the *less* highly substituted α carbon. By contrast, $NaOC(CH_3)_3$ deprotonates ketones and aldehydes reversibly, so it is selective for alkylation at the *more* highly substituted α carbon.

Nucleophilic addition to α,β-unsaturated carbonyls presents us with another case of selectivity in reactions. As we saw in Section 17.11, very strong nucleophiles like Grignard reagents (RMgBr) selectively add to the carbonyl carbon in a *direct addition*, or 1,2-addition (Equation 19-15a). Weaker nucleophiles like lithium dialkylcuprates (R_2CuLi), on the other hand, selectively add to the β carbon in a *conjugate addition*, or 1,4-addition (Equation 19-15b).

The selective reactions presented here in Section 19.4a are far from exhaustive; rather, they are intended simply to introduce the concept and to demonstrate how selective reactions can be used in synthesis. As we continue to learn more reactions, we will find several more opportunities for designing syntheses that take advantage of selectivity.

SOLVED problem **19.15** Show how you would carry out the following synthesis using any reagents necessary.

Think What reagents can be used to form a C—C bond to the carbonyl carbon of a ketone? Which ones do so selectively in an α,β-unsaturated ketone? Will such a reaction produce the necessary functional groups that are in the target, or is a functional group transformation necessary?

Solve The ether in the target can be produced from an alcohol using a Williamson ether synthesis. The alcohol can be the product of 1,2-addition of an organometallic reagent to the α,β-unsaturated carbonyl, and Grignard and alkyllithium reagents are selective for 1,2-addition rather than 1,4-addition.

In the forward direction, the synthesis can be written as follows:

---

problem **19.16** Show how you would carry out the synthesis at the right using any reagents necessary.

19.4b Protecting Groups

Selective reactions (Section 19.4a) make it possible to exploit differences in the reactivity of functional groups (or sites on a functional group) when we want to react one group in a molecule but leave another alone. There are times, however, when this is not feasible, when a step in a synthesis may require the use of a reagent that would react with two or more functional groups. In situations like these, we may be able to use a **protecting group** to temporarily make one or more functional groups unreactive under the specific conditions our desired reaction calls for. After making the desired change in the molecule, we must be able to remove the protecting group and restore the original functional group.

Here we focus primarily on the protection of the carbonyl group in ketones and aldehydes, as well as the protection of alcohols. Realize, however, that the use of protecting groups has become a mature subdiscipline of organic synthesis, and several authors have published books in this area. Suffice it to say that many different functional groups can be protected in a variety of ways.

The general idea of using a protecting group is outlined in Figure 19-1. Suppose we have a molecule with a functional group A, and that we want to convert the A group to another functional group B, but there is a labile (i.e., reactive) group that would also be changed. Direct conversion of A to B (i.e., the top reaction in Fig. 19-1) is therefore impossible without affecting the labile group. We can work around this problem by selectively changing the labile group to another functional group that will not react under the conditions that convert A to B.

FIGURE 19-1 The general strategy for using a protecting group The starting material (top left) cannot be converted to the target (top right) directly because the labile group would also react under those conditions. A separate reaction, called a *protection step*, is carried out to convert the labile group into a different group (bottom left) that is unreactive under the conditions that convert A to B. Next, the desired reaction is carried out to convert A to B (bottom right), and a final *deprotection step* converts the protected group back into the original labile group (top right).

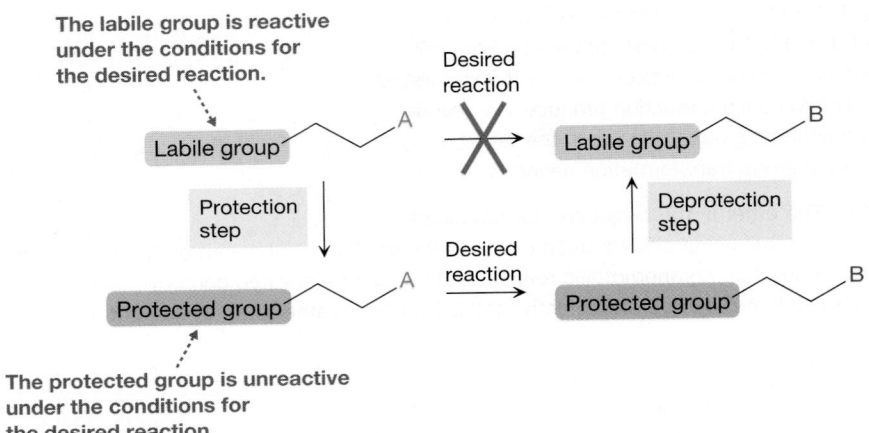

Reversibility is important in reactions that add a protecting group.

> A good protecting group must be unreactive under one set of conditions, but it *must be able to be removed* under another set of mild conditions to restore the original functional group (or perhaps another group that does not interfere with later steps in a synthesis) at the same location.

It would be a poor choice, for example, to protect the carbonyl group of a ketone by converting it into a methylene (CH$_2$) group using the Clemmensen reduction (Equation 19-16). The reactivity at that carbon would be diminished too much; we do not know any reactions that would allow us to selectively convert that CH$_2$ group back into a carbonyl group.

(19-16)

19.4b.1. Protection of ketones and aldehydes

Ketones and aldehydes are most often protected by converting them to **acetals**, as shown in Equation 19-17.

(19-17)

An acetal consists of two C–O–C groups involving the same carbon, each resembling an ether.

> Acetals, like ethers, are resistant to strong bases, strong nucleophiles, oxidizing agents, and reducing agents.

An acetal can be formed by reacting a ketone or aldehyde with excess alcohol under acidic conditions (Section 18.2a). Fortunately, the ketone or aldehyde functional groups can be recovered by hydrolysis under mildly acidic conditions (also discussed in Section 18.2a).

A commonly used alcohol in protecting ketones and aldehydes is ethane-1,2-diol (**ethylene glycol**), which is $HOCH_2CH_2OH$ (Equation 19-18).

(19-18)

Its popularity is due, in part, to the fact that it contains two alcohol groups on the same molecule, so only one molar equivalent is needed to protect a carbonyl (rather than the two molar equivalents that would be required if a mono-alcohol were used). Additionally, the acetal that it produces consists of a five-membered ring, which is relatively stable and helps favor the formation of the acetal.

To see how such a protecting group might be used, let's propose a way to carry out the synthesis in Equation 19-19.

(19-19)

The product has an alcohol, so it may appear at first that we can arrive back at the starting material through a transform that undoes a Grignard reaction, as shown in Equation 19-20.

(19-20)

Notice, however, that the necessary Grignard reagent has a carbonyl group on it. As we learned in Section 17.6, carbonyl groups are incompatible with Grignard reagents. To work around this, we can instead envision using a Grignard reagent in which that carbonyl group is protected, as shown in Equation 19-21.

(19-21)

How, then, do we synthesize the protected Grignard reagent? We know that the cyclic acetal is produced from a C=O group, and we know from Section 19.1 that a Grignard reagent is produced from an analogous alkyl halide. To avoid the presence of a C=O group on the Grignard reagent, the C=O must be protected prior to the addition of Mg, so the retrosynthetic analysis would appear as shown in Equation 19-22.

(19-22)

The synthesis is then reported by reversing the arrows and adding the appropriate reagents (Equation 19-23).

(19-23)

The aldehyde is protected by treating it with ethylene glycol ($HOCH_2CH_2OH$) under acidic conditions, and the Grignard reagent is produced by treating the protected alkyl halide with solid Mg in ether.

---

problem **19.17** Propose a synthesis of hexane-2,5-dione using acetone as your only source of carbon. (*Hint*: Consider how you might use a protecting group.)

19.4b.2. Protection of alcohols

Alcohols are weakly acidic, can react as nucleophiles, are prone to elimination under acidic conditions, and as discussed later here in Chapter 19, are prone to oxidation, too. Depending upon the specific step in a synthesis, we may need to suppress the reactivity of an alcohol by temporarily converting it to a less reactive functional group.

TABLE 19-2 Protecting Groups for Alcohols

	Protection Step	Comments	Deprotection Step	Comments
(1)	R—OH $\xrightarrow[\text{2. R'—X}]{\text{1. NaH}}$ R—O—R'	Williamson ether synthesis (basic conditions)	R—O—R' $\xrightarrow{H_3O^{\oplus}}$ R—OH + HOR'	S_N1 hydrolysis (acidic conditions)
(2)	R—OH $\xrightarrow[\text{2. }Cl—O]{\text{1. NaH}}$ R—O—O	Williamson ether synthesis (basic conditions). The product is a methoxymethyl ether (an acetal).	R—O—O $\xrightarrow{H_3O^{\oplus}}$ R—OH + $\underset{H}{\overset{O}{C}}{}_{H}$ + CH_3OH	S_N1 hydrolysis (acidic conditions)
(3)	**Methylpropene** $H_2C=C(CH_3)CH_3$ R—OH $\xrightarrow{H^{\oplus}}$ R—OC(CH₃)₃	Electrophilic addition of ROH to the alkene (acidic conditions). The product is a *tert*-butyl ether.	R—OC(CH₃)₃ $\xrightarrow{H_3O^{\oplus}}$ R—OH + HOC(CH₃)₃	S_N1 hydrolysis (acidic conditions)
(4)	**Dihydropyran (DHP)** R—OH $\underset{H^{\oplus}}{\rightleftharpoons}$ R—O...O (acetal)	Electrophilic addition of ROH to the alkene (acidic conditions). The product is an acetal.	R—O...O $\underset{}{\overset{H_3O^{\oplus}}{\rightleftharpoons}}$ R—OH + HO...O	S_N1 hydrolysis (acidic conditions)
(5)	***tert*-Butyldimethylsilyl chloride (TBDMS—Cl)** Cl—Si(CH₃)₂—C(CH₃)₃ R—OH → R—O—Si(CH₃)₂—C(CH₃)₃	The product is a silyl ether, characterized by a C—O—Si group. The product is abbreviated R—O—TBDMS.	R—O—Si(CH₃)₂—C(CH₃)₃ $\xrightarrow[\text{THF}]{Bu_4N^{\oplus}F^{\ominus}}$ R—OH	Substitution reaction in which $F^{\ominus}$ is the nucleophile and $RO^{\ominus}$ leaves

Just as we were able to protect ketones and aldehydes, we can protect an alcohol in the form of an *ether* or *acetal*. Table 19-2 shows a variety of ways to carry out these protection steps, as well as their corresponding deprotection steps.

For Entries 1 and 2, the protection step is a Williamson ether synthesis, producing an ether and acetal, respectively. Similarly, the protection steps of Entries 3 and 4 produce an ether and acetal, respectively, but they do so via an acid-catalyzed addition of the alcohol to the alkene. For Entry 5, protection involves a nucleophilic substitution reaction in which the nucleophilic O atom attacks the Si atom that has the Cl leaving group. The result is a silyl ether, characterized by a C—O—Si bond, which, like a regular ether, is unreactive under basic conditions.

The deprotection steps in Entries 1–4 are acid-catalyzed hydrolysis reactions. In Entry 5, the deprotection step is a nucleophilic substitution reaction in which F⁻ attacks the Si atom. Notice that the leaving group is RO⁻, which is usually a terrible leaving group for substitution reactions, but the Si–F bond that is formed is very strong and thus compensates—its bond energy is 553 kJ/mol (132 kcal/mol), making it about 30% stronger than a C—C single bond!

Even though there are multiple options available to protect an alcohol, they are not always interchangeable. You must consider whether the conditions required for the protection and deprotection steps will interfere with functional groups elsewhere in the molecule that contains the alcohol. For example, Entries 1 and 2 in Table 19-2 call for basic conditions in the protection step and acidic aqueous conditions in the deprotection step, so those would be poor choices if other groups in the molecule react under basic or acidic conditions. Entries 3 and 4 would be poor choices if you need to avoid acidic conditions, and Entry 5 should be avoided if the molecule has other sites that are nucleophilic or will react with F⁻.

Identify the acetal functional groups in Entries 2 and 4 in Table 19-2.

problem **19.18** Draw the complete, detailed mechanism for both the protection step and the deprotection step of Entries 1–4 in Table 19-2.

SOLVED problem **19.19** A student attempted to carry out the following reaction, but it did not work.

Suggest why, and propose a synthesis route that would circumvent this problem.

Think What type of reaction does the student intend to carry out? What alternative reaction can take place, and what type of functional group does that reaction involve? Can that functional group be protected?

Solve For the reaction to take place as planned, the Grignard reagent must act as a nucleophile, attacking the carbonyl group. A Grignard reagent is also a very strong base, however, and the alcohol group on the hydroxyaldehyde is mildly acidic. Instead of the intended nucleophilic addition, the following proton transfer reaction occurs, which destroys the Grignard reagent.

To circumvent this problem, we can protect the alcohol group before performing the Grignard reaction, as shown below using dihydropyran (DHP). Once the Grignard reaction is complete, the product can be deprotected. Because the workup step of the Grignard reaction requires acidic aqueous conditions, the acid workup and deprotection occur together.

DHP-protected alchohol

Acid workup and deprotection

problem **19.20** Show how to carry out the following synthesis.

problem 19.21 A student carried out the following sequence of reactions using *tert*-butyldimethylsilyl chloride (TBDMSCI) to protect the alcohol.

Protection step

Protected alcohol

Deprotection step

(a) Could this synthesis be carried out without protecting the alcohol? Why or why not?
(b) Which other protecting groups in Table 19-2 could have been used instead? Explain.

problem 19.22 Show how you would carry out the synthesis at the right.

problem 19.23 Propose how you would carry out the synthesis at the right using a TBDMS protecting group.

Both 1,2-diols and 1,3-diols can be protected using formaldehyde ($H_2C=O$), as shown in Equations 19-24 and 19-25. The protected forms are five- and six-membered ring acetals, respectively. This method of protection is, in fact, the same one we encountered in Section 19.4b.1, where a diol was used to protect a ketone or aldehyde. Deprotection simply requires dilute aqueous acid.

Protection step   Deprotection step

A protected 1,3-diol

(19-24)

Protection step   Deprotection step

A protected 1,2-diol

(19-25)

(a)   **(b)**

Protecting Groups in DNA Synthesis

Oligonucleotides are relatively short strands of DNA, on the order of 10–100 nucleotides (i.e., A, C, T, G) long. They have a number of important applications in molecular biology, including DNA sequencing and synthesizing artificial genes. Consequently, it is important to synthesize oligonucleotides in an efficient and accurate way. As it turns out, protecting groups play a major role in these syntheses, as shown in the following simplified scheme.

Protecting groups are necessary because each nucleotide has multiple possible reactive sites, including the 5′–OH and the 3′–OH groups of each sugar unit, and the NH_2 group of adenine, guanine, and cytosine bases. By using protecting groups appropriately, we can ensure that the correct groups are free to react at any given time, so that the DNA chain grows in a specific way.

The synthesis consists of the five distinct steps shown above. Step 1 connects the 3′ end of the first nucleoside to a silica support via an amide linkage. Notice that the 5′–OH and the NH_2 group of the base (in the case of A, G, and C) are initially protected. The 5′–OH is protected as a p-dimethoxytrityl (DMT) ether, and the NH_2 group of the base is protected as an amide. In Step 2, the 5′–OH is deprotected, making it available to react. In Step 3, that 5′–OH connects to the P atom of a phosphoramidite group [characterized by $(RO)_2PNR_2$] at the 3′ end of a second protected nucleoside, and the oxidation in Step 4 produces a phosphate group. Steps 2–4 are repeated to add more nucleosides, and once the desired chain has been constructed, Step 5 deprotects all the groups and releases the DNA chain from the silica support.

19.5 Catalytic Hydrogenation

In Section 19.3, we learned how a functional group can be removed entirely from a molecule via the reduction of the $C=O$ group of a ketone or aldehyde to a CH_2 group. In Section 19.4, we saw how reactions we have encountered previously can be used to selectively target one functional group over another. In this section, we discuss a new reaction, called **catalytic hydrogenation**, which can accomplish similar tasks. Namely, catalytic hydrogenation can be used to remove an alkene functional group entirely from a molecule. Additionally, it is a selective reaction, involving alkenes, alkynes, and carbonyl-containing groups such as ketones and aldehydes.

There are three main reasons why catalytic hydrogenation is introduced now. First, it is quite useful in synthesis, for reasons mentioned previously. Second, its mechanism is substantially different from essentially every other reaction in this book. Third, catalytic hydrogenation can be used to carry out the reduction of carbonyl ($C=O$) groups, among other types of groups, and it is helpful to see these kinds of reactions introduced shortly after other reactions that reduce ketones and aldehydes—namely, the hydride reductions presented in Section 17.3, and the Wolff–Kishner and Clemmensen, reductions presented in Section 19.3.

19.5a Catalytic Hydrogenation of Alkenes

In the presence of a finely divided, solid metal catalyst such as palladium, platinum, or nickel (all elements in the same column of the periodic table), H_2 gas readily adds to the $C=C$ double bond of an alkene. An example is shown in Equation 19-26.

(19-26)

Notice, in particular, that

> Catalytic hydrogenation reduces the $C=C$ double bond of an alkene to a $C—C$ single bond.

In these kinds of reactions, the metal is called a **heterogeneous catalyst** because it exists in a different phase from the rest of the reaction mixture. The catalyst used during catalytic hydrogenation is *an insoluble solid*, suspended (by stirring) in the solution that contains both hydrogen and the organic substance to be hydrogenated.

The detailed mechanism for catalytic hydrogenation is relatively complex. A simplified picture, however, is shown in Figure 19-2.

Initially, the hydrogen and the alkene adsorb onto the surface of the catalyst, as shown in Figure 19-2a through 19-2c. As that occurs, the H—H single bonds are effectively broken, producing individual H atoms that reside on the metal surface. Similarly, adsorption of the alkene effectively breaks the π bond of the $C=C$ double bond. Eventually, an adsorbed H atom encounters an adsorbed alkene, and the first of two C—H bonds is formed (Fig. 19-2d). Very shortly thereafter, the second C—H bond is formed through a similar process (Fig. 19-2e). As each C—H bond is formed, the corresponding C atom is released from the metal surface. Realizing that the reaction takes place at the metal's surface, you can see the advantage of finely dividing the catalyst—it dramatically increases the surface area of the metal and, hence, the rate of the reaction.

We can take advantage of these alkene reductions in a synthesis that calls for the formation of a new C—C single bond by first incorporating a reaction that forms

H₂ molecule (gas)

H—H   H—H

Alkene dissolved in solution

C=C

H atoms adsorbed onto metal surface

(a)   (b)

Alkene adsorbed onto metal surface

(c)

First C—H bond formed, releasing one C atom

Second C—H bond formed, releasing the second C atom

(d)   (e)

FIGURE 19-2 Simplified mechanism of catalytic hydrogenation (a to b) Hydrogen gas adsorbs to the surface of the metal catalyst, breaking the H—H bonds. (b to c) A molecule of the alkene adsorbs to the surface of the metal catalyst. (c to d) A C—H bond is formed, liberating a C atom from the catalyst. (d to e) The second C atom forms a bond to H, liberating the entire alkane from the catalyst surface. The catalyst surface is now ready for another round of reactions.

a new C=C double bond. Suppose, for example, that we need to carry out the following synthesis, beginning with starting materials having six or fewer carbon atoms.

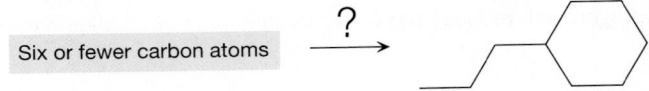

No functional groups appear in the target, so there is no relative positioning of functional groups to provide a clear indication of which carbon–carbon bond-formation reaction should be used. As shown in the following retrosynthetic analysis, however, the target can be produced from an alkene intermediate in which the two rings are joined by a C=C double bond, because catalytic hydrogenation could reduce the C=C double bond to a C—C single bond.

That C=C double bond, in turn, could be produced using a Wittig reaction (Section 17.7), which requires the ketone and Wittig reagent shown. The Wittig reagent could be produced from bromocyclohexane (Section 17.8), which has six carbon atoms. The synthesis in the forward direction would then appear as follows:

problem 19.25 Show how you would carry out the synthesis at the right using a Wittig reaction.

19.5b Catalytic Hydrogenation of Alkynes: Poisoned Catalysts

The C≡C triple bond of an alkyne can undergo catalytic hydrogenation, because it, too, contains carbon–carbon π bonds. Similar to alkenes, the addition of H_2 removes a π bond and causes the total number of bonds between the alkyne carbons to decrease by one. Thus, an alkyne is reduced to an alkene, as shown in Equation 19-27.

One addition of H₂ reduces an alkyne to an alkene.

A second addition of H₂ reduces the alkene to an alkane.

H₂ (excess)
Ni(s), CH₃OH

Hex-1-yne

Hexane
85%

(19-27)

The alkene that is produced is susceptible to catalytic hydrogenation, as indicated in Equation 19-27 and as we saw in Section 19.5a. Therefore,

- If equimolar amounts of H_2 and the alkyne react under normal conditions, we run the risk of producing a mixture of the alkene and alkane products—a generally undesirable result.
- If excess H_2 reacts, on the other hand, an alkyne will be reduced completely to the alkane.
- Reduction of an alkyne can be stopped at the alkene stage by using a *poisoned catalyst*.

A **poisoned catalyst** is simply a metal catalyst that has been specially treated to decrease its catalytic ability, thus making possible a slower and more controlled reaction. One example is **Lindlar's catalyst**—palladium treated with calcium carbonate ($CaCO_3$) and a small amount of quinoline. Barium sulfate ($BaSO_4$) can also be used instead of $CaCO_3$. Representative hydrogenation reactions with these kinds of catalysts are shown in Equations 19-28 and 19-29.

Quinoline

Cis isomer only

(19-28)

But-2-yne-1,4-diol → **(Z)-But-2-en-1,4-diol**

A poisoned catalyst

H₂
Lindlar's catalyst
25 °C

77%

Cis isomer only

(19-29)

1-Phenylprop-1-yne → **(Z)-1-Phenylprop-1-ene**

A poisoned catalyst

H₂
Pd/BaSO₄
25 °C

61%

Equations 19-28 and 19-29 show that catalytic hydrogenation takes place *stereospecifically*. Namely,

> The reduction of an alkyne to an alkene via catalytic hydrogenation favors the cis isomer over the trans.

This stereospecificity can be understood from the simplified picture of the mechanism we saw previously in Figure 19-2. After the first H atom adds to one C of the multiple bond (Fig. 19-2d), the second H atom adds to the other C (Fig. 19-2e) before any significant changes occur in the orientation of the molecule. Thus, the individual H atoms end up on the same side of the newly formed C=C double bond; the H atoms are said to add in a syn fashion.

---

problem **19.26** In Equation 19-29, why are the double bonds in the starting material unaffected by hydrogenation?

SOLVED problem **19.27** Outline a synthesis of (Z)-1-phenylhept-2-ene from bromophenylmethane.

?

Bromophenylmethane
(Benzyl bromide)

(Z)-1-Phenylhept-2-ene

Think Does this synthesis require a carbon–carbon bond-formation reaction? If so, what species possessing an electron-rich carbon could be used? What species containing an electron-poor carbon could be used? Are there any stereochemical issues that must be considered? Can the product be formed in a single reaction, or should we consider making it from a precursor?

Solve The carbon skeleton is different in the products than in the reactants, so a carbon–carbon bond-formation reaction is necessary. Unfortunately, we do not have at our disposal a way to form a carbon–carbon bond between an alkene carbon and an alkyl halide in a single reaction. Therefore, we must perform a *retrosynthetic analysis*, seeking a precursor from which we can readily make the product. As we have seen previously in this section, a cis alkene can be made by hydrogenating an alkyne, suggesting the possibility of an alkyne precursor. That alkyne, in turn, can be the result of an S_N2 reaction that forms the necessary carbon–carbon bond. Overall, our retrosynthetic analysis appears as follows:

The synthesis would then appear as follows:

A strong base deprotonates the terminal alkyne, thus converting it into a strong nucleophile. Addition of bromophenylmethane results in an S_N2 reaction to generate the new C—C bond. Subsequent treatment with H_2 gas in the presence of a poisoned catalyst allows the reduction of the triple bond to stop at the cis double bond.

---

problem **19.28** Show how you would carry out the synthesis at the right.

19.5c Addition of H_2 to Other Functional Groups

Catalytic hydrogenation is not limited to carbon–carbon multiple bonds; the addition of hydrogen to the π bonds of other systems is feasible as well. As Table 19-3 shows, a variety of functional group conversions can be performed using catalytic hydrogenation, though not with equal ease. Pay particular attention to the typical experimental conditions required.

Notice that the reduction of an aldehyde is essentially as easy as the reduction of an alkene or alkyne. Other functional groups are more difficult to reduce and their reactions generally proceed with poorer yields. This is particularly true for the reduction of amides to amines, because the product amine poisons (i.e., deactivates) the metal catalyst. If you try to force the reaction by increasing the temperature and/or pressure of H_2, then the risk of unwanted side reactions increases. Fortunately, $LiAlH_4$ reduces amides to amines in good yield, as discussed in Chapter 20.

19.5d Selectivity in Catalytic Hydrogenation

Catalytic hydrogenation can reduce a variety of different functional groups, including alkenes, alkynes, aldehydes, ketones, esters, and amides. It can also take place selectively,

TABLE 19-3 Reduction of Various Functional Groups via Catalytic Hydrogenation

Reactant	Reaction Conditions	Product
Aldehyde	$\xrightarrow{H_2}$ Pt, 20 °C, 1 atm	1° Alcohol
Ketone	$\xrightarrow{H_2}$ Rh, 50 °C, 3 atm	2° Alcohol
Nitrile	$\xrightarrow{H_2}$ Raney Ni, 80 °C 75 atm	1° Amine
1° Amide	$\xrightarrow{\text{1 equiv. } H_2}$ Pt, 250 °C, 200 atm (<50% yield)	1° Amine

so if two or more of these functional groups appear in a particular molecule, it is often possible to reduce just one of them. Equation 19-30 shows, for example, that treating limonene with one equivalent of H_2 reduces only the terminal C=C double bond.

The less sterically hindered C=C double bond is selectively reduced.

$$\xrightarrow[\text{Pt, 60 °C}]{H_2}$$

Limonene   97%   (19-30)

The terminal alkene is less highly substituted, so its double bond is less sterically hindered. In general,

Catalytic hydrogenation is more favored at a less sterically hindered multiple bond than at a more sterically hindered one.

With more steric bulk surrounding the double bond, it is more difficult for the alkene to adsorb to the surface of the metal catalyst—a critical step in catalytic hydrogenation (review Fig. 19-2c).

problem **19.29** There are two possible syn-addition products in the catalytic hydrogenation of α-pinene, **A** and **B**, but one of them is formed exclusively. Which one? Why?

α-Pinene → (H₂, Pd) → A + B

Of all the functional groups in Table 19-3, only aldehydes are reduced under mild conditions, similar to those under which alkenes and alkynes are reduced. Therefore, alkenes, alkynes, and aldehydes can be selectively reduced over functional groups such as ketones, nitriles, and amides. An example is shown in Equation 19-31.

Catalytic hydrogenation selectively reduces an alkene over a ketone.

Cyclohex-2-enone → (H_2, Pd, Hexane, 25 °C, 2 h) → Cyclohexanone 99%   (19-31)

problem **19.30** Show how you would carry out the following synthesis.

19.6 Oxidations of Alcohols and Aldehydes

In Chapter 17 and here in Chapter 19, we have discussed a variety of reduction reactions, including hydride reductions involving $NaBH_4$ and $LiAlH_4$, as well as catalytic hydrogenation. As shown in Figure 19-3, these reduction reactions increase

Increasing number of C—O bonds
Decreasing number of C—H bonds → **Increasing oxidation state of C**

Oxidation / Reduction

Oxidation / Reduction

← **Decreasing oxidation state of C**

Decreasing number of C—O bonds
Increasing number of C—H bonds

FIGURE 19-3 Oxidations and reductions Oxidation is represented by the conversion of a molecule into one toward the right, and is accompanied by an increase in the number of C—O bonds and/or a decrease in the number of C—H bonds. Reduction is represented by the conversion of a molecule into one toward the left, and is accompanied by a decrease in the number of C—O bonds and/or an increase in the number of C—H bonds.

$$CrO_3 \xrightarrow[H^{\oplus}]{H_2O} H_2CrO_4$$

Chromium trioxide   **Chromic acid**

The oxidation state of Cr is +6.

$$Na_2Cr_2O_7 \xrightarrow[H^{\oplus}]{H_2O} 2\ H_2CrO_4$$

Sodium dichromate   **Chromic acid**

FIGURE 19-4 Chromium oxidizing agents CrO_3, $Na_2Cr_2O_7$, and H_2CrO_4 all contain Cr^{+6}. H_2CrO_4 is produced from either CrO_3 or $Na_2Cr_2O_7$ by treatment with aqueous acid.

the number of C—H bonds or decrease the number of C—O bonds, representing a progression toward the left in Figure 19.3.

To have greater flexibility in designing a synthesis, it is also important to be able to carry out oxidation reactions, which represent a progression toward the right in Figure 19-3. This can be done by carrying out reactions that increase the number of C—O bonds or decrease the number of C—H bonds. We present two types of oxidizing agents that can be used for these kinds of reactions—namely, *chromic acid* and *potassium permanganate*.

19.6a Chromic Acid Oxidations

Chromic acid (H_2CrO_4) is prepared by dissolving either chromium trioxide (CrO_3) or sodium dichromate ($Na_2Cr_2O_7$) in an acidic aqueous solution (Fig. 19-4). A molecule of water adds to the starting material in both reactions. Furthermore, the oxidation state of chromium is +6 in all three forms—CrO_3, $Na_2Cr_2O_7$, and H_2CrO_4.

An example of a chromic acid oxidation is shown in Equation 19-32, in which cyclooctanol is oxidized to cyclooctanone.

Cyclooctanol   **Cyclooctanone** 96%

(19-32)

In general,

> Chromic acid oxidizes a secondary alcohol to a ketone.

YOUR TURN 19.4

Verify that the reaction in Equation 19-32 is an oxidation by calculating the oxidation state of each of the carbon atoms screened in red in the reactant _____ and in the product _____.

The mechanism for this reaction is believed to involve a **chromate ester** as a key intermediate, shown in Equation 19-33. An E2 step on the chromate ester then produces the ketone.

Partial mechanism for Equation 19-32

The first H is removed.

Leaving group

The second H is removed.

Several steps

E2

Chromate ester

$+ HCrO_3^{\ominus}$ (19-33)

The final E2 step might appear a little peculiar. The leaving group in an E2 reaction is usually attached to a carbon atom, but in this case it is attached to O. Nevertheless, the leaving group and a proton on adjacent atoms are eliminated, as usual, leaving behind an additional π bond. In this case, the result is a new C=O double bond rather than a C=C double bond.

Notice in Equation 19-33 that there are two H atoms that must be removed: one from the OH group and one from the adjacent C atom. Therefore,

Oxidation by H_2CrO_4 requires an OH group that is attached to a C atom bonded to at least one H atom:

This is why the tertiary alcohol in Equation 19-34 is *not* oxidized when treated with chromic acid.

No H atom is attached to the alcohol C.

H_2CrO_4 No reaction   (19-34)

A primary alcohol, on the other hand, *is* oxidizable because the alcohol carbon is bonded to at least one hydrogen. An example is shown in Equation 19-35.

A primary alcohol is oxidizable because at least one H is bonded to the alcohol C.

A carboxylic acid is produced from the oxidation of a primary alcohol.

H_2CrO_4
H_2SO_4
H_2O, acetone   (19-35)

Pent-3-yn-1-ol

Pent-3-ynoic acid
36%

The oxidation product, a carboxylic acid, is different from the ketone obtained in the oxidation of a secondary alcohol.

Chromic acid oxidizes a primary alcohol to a carboxylic acid.

The mechanism in Equation 19-33 might suggest that the product should be an aldehyde rather than a carboxylic acid, but water is present in solution, which equilibrates with the aldehyde to produce the corresponding *hydrate* (Section 17.2). As shown in Equation 19-36, an OH is attached to CH in that hydrate, which allows a second oxidation to take place.

H atoms are removed from adjacent OH and CH groups in a chromic acid oxidation.

No OH is present.

H atoms are removed from adjacent OH and CH groups in a chromic acid oxidation.

An aldehyde hydrate

(19-36)

Given the role that water plays in the oxidation of a primary alcohol to a carboxylic acid, the oxidation would have to stop at the aldehyde stage in the absence of water. Chromic acid cannot be used in such an oxidation, however, because water is required to produce chromic acid in the first place (Fig. 19-4). Instead, Equation 19-37 shows that **pyridinium chlorochromate (PCC)** can be used as the oxidizing agent.

Pyridinium chlorochromate (PCC)

Oxidation stops at the aldehyde because no water is present.

91%

(19-37)

Pyridinium chlorochromate oxidizes a primary alcohol to an aldehyde.

PCC contains the same Cr^{+6} species as chromic acid. Unlike chromic acid, however, PCC is soluble in nonaqueous solvents like dichloromethane (CH_2Cl_2), which allows oxidation to take place in the absence of water.

SOLVED problem 19.31 Show how to synthesize this compound using alcohols of seven or fewer carbons as your only carbon source.

Think Which C—C bond can we "disconnect" in a retrosynthetic analysis? By undoing what reaction? Are there any oxidations of alcohols that must be stopped at the aldehyde?

Solve A Grignard reaction forms a carbon–carbon bond, yielding an alcohol. Our target is a ketone that could be the result of oxidizing a Grignard product. Our retrosynthetic analysis may therefore begin with the following transforms.

Undo an oxidation.

Undo a Grignard reaction.

The Grignard reagent can be prepared from an alkyl bromide, which can be produced, in turn, from an alcohol using PBr_3.

Benzaldehyde, on the other hand, can be made from benzyl alcohol via an oxidation.

Undo an oxidation.

There are two oxidations of alcohols required in this synthesis. One is an oxidation to an aldehyde, which requires an aqueous-free solvent and thus PCC as the oxidizing agent.

The other is an oxidation to a ketone, which can take place in water, thus allowing us to use H_2CrO_4 as the oxidizing agent. The overall synthesis can be written as follows:

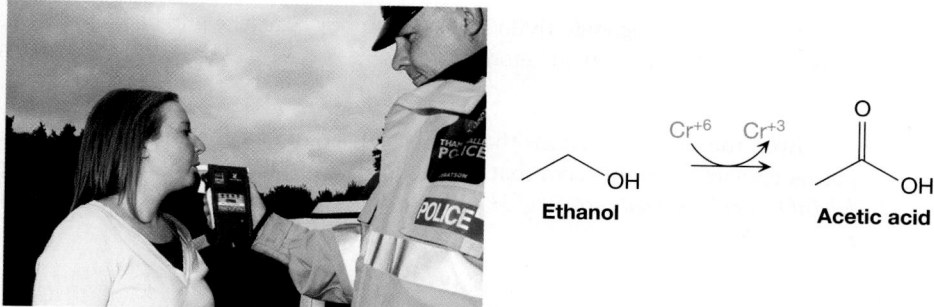

problem **19.32** Show how to synthesize the compound in Solved Problem 19.31 beginning with C_6H_5MgBr and any alcohol.

Chromic Acid Oxidation and the Breathalyzer™ Test

According to the National Highway Traffic Safety Administration, more than 10,000 people died in drunk-driving crashes in the United States in 2010. It is no wonder, then, that there are strict laws against driving under the influence of alcohol, which each state defines as having a blood alcohol content (BAC) > 0.08%. Law enforcement officers can measure the BAC of a suspected drunk driver using a Breathalyzer™ test, the basis of which is the same chromic acid oxidation reaction we have studied here in Chapter 19.

To carry out a Breathalyzer™ test, the suspect blows into a mouthpiece and a fixed volume of breath is collected. The glass neck of a test vial is broken and the breath sample is forced into that vial. The test vial contains a mixture of $K_2Cr_2O_7$ and H_2SO_4 (i.e., H_2CrO_4), so any ethanol from the breath is quickly oxidized to acetic acid. Cr^{+6}, which is bright orange, is simultaneously reduced to Cr^{+3}, which is blue-green. Light is shined through the test vial and through a second vial containing the unreacted mixture, and the amount of transmitted light is detected by photo-cells. A difference in the amount of transmitted light through the two vials results in an electric current, which is used to determine BAC. In more modern instruments, ethanol is oxidized by a fuel cell rather than chromic acid. Oxidation at the anode produces acetic acid, protons, and electrons, and the protons travel to the cathode where atmospheric oxygen is reduced to water. The electric current that is produced is converted to a BAC reading.

Potassium permanganate

Chromic acid

FIGURE 19-5 **KMnO₄ and H₂CrO₄** These oxidizing agents are similar structurally, each with a metal atom in a high, positive oxidation state.

19.6b Permanganate Oxidations of Alcohols and Aldehydes

Potassium permanganate ($KMnO_4$) is another common oxidizing agent; it is an ionic compound consisting of the potassium cation, K^+, and the permanganate anion, MnO_4^-. Structurally, MnO_4^- resembles H_2CrO_4 (Fig. 19-5). In both compounds, a central metal atom is bonded to four O atoms, thus giving the metal atom a high, positive oxidation state—namely, $+7$ for MnO_4^- and $+6$ for H_2CrO_4.

Like chromic acid, potassium permanganate can be used to oxidize alcohols, as shown in Equations 19-38 and 19-39.

(19-38)

(19-39)

Thus,

> Potassium permanganate ($KMnO_4$) oxidizes primary alcohols to carboxylic acids and secondary alcohols to ketones.

Given the similar reactions that can be carried out by chromic acid and potassium permanganate, it might seem that the choice is arbitrary. It's not. All else being equal, *$KMnO_4$ should be used instead of H_2CrO_4 in the oxidation of alcohols.* There are two main reasons why.

> ■ $KMnO_4$ is less expensive than H_2CrO_4.
> ■ H_2CrO_4 and its byproducts are highly toxic and pose serious environmental hazards.

On the other hand, H_2CrO_4 has one major advantage:

> H_2CrO_4 is more selective than $KMnO_4$.

As discussed in Chapters 22 and 24, $KMnO_4$ can oxidize a variety of groups in addition to alcohols and aldehydes, including alkenes, alkynes, and alkylbenzenes. As a result, chromic acid's greater selectivity makes it the oxidizing agent of choice when these other groups are present.

19.7 Wrapping Up and Looking Ahead

We began our discussion of synthesis in Chapter 13, introducing some basic thinking that goes into designing organic syntheses. Here in Chapter 19, we extended that knowledge, discussing some higher-level concepts that can be applied toward designing more complex syntheses.

The first few sections dealt with aspects of incorporating carbon–carbon bond-formation reactions in a synthesis. We discussed ways to facilitate the formation of a bond between two carbon atoms initially with like charge by incorporating a reaction that carries out *umpolung* (polarity reversal) at one of the carbons. One very useful example is the formation of an organometallic compound from an alkyl halide. Then, we saw how the relative positioning of functional groups within a target provides clues as to the type of carbon–carbon bond-formation reaction we should consider incorporating into a synthesis. When such clues are not present, we can sometimes use reactions that remove a functional group entirely from a molecule once a carbon–carbon bond has been formed. We focused on Wolff–Kishner, Clemmensen, and Raney-nickel reductions, which serve to reduce the $C{=}O$ group of a ketone or aldehyde to a CH_2 group.

Next, we discussed selective reactions and protecting groups. A desired reaction may require conditions that would result in an undesired reaction at a particular functional group. In many such cases, we can use a selective reagent to target one functional group (or site) over another. When this is not possible, we might protect a functional group by temporarily converting it into one that is unreactive under the conditions necessary for the desired reaction. After the desired reaction is complete, the original functional group can be restored via a deprotection step. Specifically, we learned how to protect ketones, aldehydes, and alcohols.

Finally, we introduced new reduction and oxidation reactions that are very useful functional group transformations in organic synthesis. Catalytic hydrogenation reduces a variety of functional groups, including alkenes, alkynes, aldehydes, ketones, nitriles, and amides, and they are particularly selective for reducing alkenes, alkynes, and aldehydes. H_2CrO_4 and $KMnO_4$, on the other hand, can oxidize secondary alcohols to ketones and primary alcohols to carboxylic acids, whereas pyridinium chlorochromate (PCC) will oxidize primary alcohols to aldehydes.

As new reactions are discussed in the chapters to come, you will see opportunities to apply and expand upon the concepts you learned here in Chapter 19. In Chapters 20 and 21, for example, you will see that nucleophilic addition–elimination reactions exhibit selectivity with carbonyl-containing functional groups such as amides, esters, carboxylic acids, acid anhydrides, and acid halides. And, in Chapter 25, we will see how free-radical reactions are selective toward specific types of carbon atoms. Moreover, in Chapter 23, we will learn how to use protecting groups to control reactions involving aromatic rings.

Chapter Summary and Key Terms

- The formation of a bond between two carbon atoms is a problem when the atoms are of like charge. To circumvent such a problem, we can first carry out a reaction that reverses the charge (polarity) of one atom, via a process called **umpolung**. (Section 19.1)

- Umpolung takes place at a haloalkane carbon upon treatment with a metal such as Mg or Li, producing a Grignard or alkyllithium reagent, respectively. (Section 19.1)

- Many carbon–carbon bond-formation reactions leave functional groups with a very specific relative positioning. Thus, target molecules that exhibit one such relative positioning provide clues as to which of those reactions should be considered in a synthesis. (Section 19.2)

- If a synthesis requires the formation of a new carbon–carbon bond, but the target does not exhibit a specific relative positioning of functional groups, then the synthesis may require a reaction that removes a functional group entirely from a molecule. (Section 19.3)

  - Reactions that remove a ketone or aldehyde functional group include the **Wolff–Kishner reduction**, the **Clemmensen reduction**, and the **Raney-nickel reduction**. These reactions take place under acidic, basic, and catalytic hydrogenation conditions, respectively, so care must be taken to avoid unwanted side reactions at other functional groups in the molecule. (Section 19.3)

  - An alkene or alkyne functional group can be removed via catalytic hydrogenation. (Section 19.5)

- If two or more reactive functional groups appear in a molecule, we can often carry out a reaction at just one of those groups by using either a *selective reagent* or a *protecting group*. (Section 19.4)

- A **selective reagent** favors reaction with one functional group over another. (Section 19.4a)

- A **protecting group** temporarily converts one functional group into one that is unreactive under the conditions necessary for a step that a synthesis calls for. Ketones, aldehydes, and alcohols can often be protected in the form of an acetal or ether. (Section 19.4b)

- **Catalytic hydrogenation** is a reaction in which H_2 adds to a molecule in the presence of a metal catalyst such as Pt, Pd, or Ni. Catalytic hydrogenation readily reduces an alkene to an alkane. (Section 19.5a)

- Using a **poisoned catalyst**, catalytic hydrogenation reduces an alkyne to a cis alkene. (Section 19.5b)

- Catalytic hydrogenation will readily reduce an aldehyde to an alcohol. Under moderate conditions, a ketone can also be reduced to an alcohol, and under more extreme conditions, other functional groups such as nitriles and amides can be reduced. (Section 19.5c)

- Catalytic hydrogenation is selective toward alkenes, alkynes, and aldehydes. (Section 19.5d)

- **Chromic acid** (H_2CrO_4) will oxidize a primary alcohol to a carboxylic acid or a secondary alcohol to a ketone. (Section 19.6a)

- Oxidation of a primary alcohol by **pyridinium chlorochromate** (**PCC**) stops at the aldehyde because the reaction takes place in a nonaqueous medium. (Section 19.6a)

- Oxidation of an alcohol by **potassium permanganate** ($KMnO_4$) produces the same ketone and carboxylic acid products that would be produced by chromic acid. (Section 19.6b)

Reaction Table

TABLE 19-4 Functional Group Conversions[a]

	Starting Functional Group	Typical Reagents and Reaction Conditions	Functional Group Formed	Comments	Discussed in Section
(1)	R–X Haloalkane	$\xrightarrow[\text{Ether}]{\text{Mg(s)}}$	R–MgX Grignard reagent	Dissolving metal reduction	19.1
(2)	R–X Haloalkane	$\xrightarrow[\text{Ether}]{\text{Li(s)}}$	R–Li Alkyllithium reagent	Dissolving metal reduction	19.1
(3)	R–Li Alkyllithium reagent	$\xrightarrow[\text{Ether}]{\text{CuI}}$	R–Cu(Li)–R Lithium dialkylcuprate	N/A	19.1

[a] X = Cl, Br, I.

(continued)

TABLE 19-4 Functional Group Conversions[a] (continued)

Starting Functional Group	Typical Reagents and Reaction Conditions	Functional Group Formed	Comments	Discussed in Section
(4) Ketone or aldehyde	Zn/Hg, HCl / H₂O, reflux	Alkane	Clemmensen reduction	19.3b
(5) Ketone or aldehyde	1. HSCH₂CH₂SH, H⊕ / 2. Raney Ni (H₂)	Alkane	Raney-nickel reduction	19.3c
(6) Alkene	H₂ / Pt, Pd, or Ni	Alkane	Catalytic hydrogenation	19.5a
(7) Alkyne	H₂ / Lindlar's catalyst	cis-Alkene	Catalytic hydrogenation; poisoned catalyst	19.5b
(8) Ketone or aldehyde	H₂ / Pt, Pd, or Ni	Alcohol	Catalytic hydrogenation	19.5c
(9) Primary alcohol	H₂CrO₄ or 1. KMnO₄, OH⊖ 2. H⊕	Carboxylic acid	Oxidation	19.6a
(10) Secondary alcohol	H₂CrO₄ or KMnO₄, OH⊖	Ketone	Oxidation	19.6a
(11) Primary alcohol	PCC	Aldehyde	Oxidation	19.6a

[a] X = Cl, Br, I.

Problems

19.34 Show how each of the following organometallic compounds can be synthesized from an alkyl halide.

(a)　(b)　(c)　(d)

(e)

(f) MgBr

(g)

(h) Li

(i) MgBr

(j)

ClMg

(k)

—MgI

19.35 Draw the organometallic compound that would be produced by each of the following reactions.

(a)

—Br $\xrightarrow[\text{Ether}]{\text{Li(s)}}$?

(b)

$H_2C{=}C{=}CH$ $\xrightarrow[\text{THF}]{\text{Mg(s)}}$?

Br

(c)

Br $\xrightarrow[\text{THF}]{\text{Mg(s)}}$?

(d)

Cl $\xrightarrow[\text{Ether}]{\text{Mg(s)}}$?

(e)

Br $\xrightarrow{\substack{\text{1. Li(s), THF} \\ \text{2. CuI}}}$?

(f)

Br $\xrightarrow[\text{THF}]{\text{Li(s)}}$?

(g)

Cl $\xrightarrow{\substack{\text{1. Li(s), THF} \\ \text{2. CuI}}}$?

19.36 Predict the product for each of the following reactions. Unless otherwise indicated, you may assume that one molar equivalent of H_2 reacts.

(a)

$\xrightarrow[\text{Pd}]{H_2}$?

Cl

(b)

$\xrightarrow[\text{Pt}]{H_2}$?

OH

(c)

$\xrightarrow[\text{Pt}]{H_2}$?

(d)

H_2N $\xrightarrow[\substack{\text{Lindlar's} \\ \text{catalyst}}]{H_2}$?

(e)

N $\xrightarrow[\text{Pd}]{H_2}$?

(f)

$\xrightarrow[\text{Ni}]{H_2}$?

(g)

H_2N $\xrightarrow[\text{Pd}]{H_2 \text{ (excess)}}$?

19.37 Predict the product of each of the following reactions.

(a)

Br (benzene ring) CH₂—C(=O)—CH₃ $\xrightarrow[\text{HCl}]{\text{Zn/Hg}}$?

(b)

(methylcyclohexenyl ketone) $\xrightarrow[\text{NaOH, H}_2\text{O, }\Delta]{\text{H}_2\text{NNH}_2}$?

(c)

(acetyl dihydroxybenzene with OH groups) $\xrightarrow[\text{2. Raney Ni}]{\text{1. HS}\diagup\diagdown\text{SH , H}^\oplus}$?

19.38 Predict the product of the following reactions.

(a)

HO (chain) $\xrightarrow[\text{H}_2\text{SO}_4, \text{H}_2\text{O}]{\text{Na}_2\text{Cr}_2\text{O}_7}$?

(b)

(cyclopentyl secondary alcohol) OH $\xrightarrow[\text{H}_2\text{SO}_4, \text{H}_2\text{O}]{\text{CrO}_3}$?

(c)

HO (benzene ring) C(OH)(CH₃)₂ $\xrightarrow[\text{H}_2\text{SO}_4, \text{H}_2\text{O}]{\text{Na}_2\text{Cr}_2\text{O}_7}$?

(d)

F (benzene ring with CHO and C(=O)CH₃) $\xrightarrow{\text{H}_2\text{CrO}_4}$?

(e)

(decalin system with CH₂OH and ketone) $\xrightarrow[\text{2. H}^\oplus]{\text{1. KMnO}_4,\ \text{OH}^\ominus}$?

(f)

(decalin system with CH₂OH and ketone) $\xrightarrow{\text{PCC}}$?

19.39 Using acetone, any alcohol with six or fewer carbons, and any inorganic reagents necessary, show how to synthesize each of the following compounds.

(a)

OH (cyclohexyl C(OH)(CH₃)₂)

(b)

(tertiary alcohol chain with OH)

(c)

HO (tertiary alcohol with alkene)

(d)

OH (tertiary alcohol chain)

19.40 Supply the missing intermediates and reagents in the following synthesis.

(benzaldehyde) $\xrightarrow[\text{2. H}_3\text{O}^\oplus]{\text{1. LiAlH}_4}$ **A** $\xrightarrow[\substack{\text{2. Mg(s),}\\ \text{ether}}]{\text{1. PBr}_3}$ **B** $\xrightarrow[\text{2. H}_3\text{O}^\oplus]{\text{1. C}}$ (diphenyl alcohol with OH) $\xrightarrow[\text{2. D}]{\text{1. NaH}}$

19.41 Supply the missing intermediates and reagents in the following synthesis.

(propanol) OH $\xrightarrow{\text{A}}$ **B** $\xrightarrow[\text{Ether}]{\text{Mg(s)}}$ **C**

19.42 Show how to synthesize each of the following compounds, using propanal and any other ketone or aldehyde as your only starting materials containing carbon.

(a)

(b)

(c)

(d)

19.43 Provide the missing intermediates and reagents in the following synthesis.

19.44 Supply the missing reagents and intermediates in the following synthesis.

19.45 Provide the missing intermediates in the following synthesis.

$$\text{styrene} \xrightarrow[\text{2. NaOH, H}_2\text{O}_2]{\text{1. BH}_3\cdot\text{THF}} \mathbf{A} \xrightarrow{\text{PCC}} \mathbf{B} \xrightarrow{(\text{C}_6\text{H}_5)_3\overset{\oplus}{\text{P}}-\overset{\ominus}{\text{CH}}_2} \mathbf{C} \xrightarrow[\text{2. NaBH}_4]{\text{1. Hg(OAc)}_2, \text{H}_2\text{O}} \mathbf{D} \xrightarrow{\text{KMnO}_4} \mathbf{E} \xrightarrow[\text{2. Br}]{\text{1. LDA}} \mathbf{F}$$

19.46 Provide the missing intermediates and reagents in the following synthesis.

$$\xrightarrow[\text{CCl}_4]{\text{Br}_2} \mathbf{A} \xrightarrow[\text{2. H}_2\text{O}]{\text{1. LDA (excess)}} \mathbf{B} \xrightarrow[\text{2. C}]{\text{1. NaH}} \cdots \xrightarrow[\text{Lindlar's catalyst}]{\text{H}_2/\text{Pd}}$$

$$\mathbf{D} \xrightarrow{\text{MCPBA}} \mathbf{E} \xrightarrow[\text{2. H}_3\text{O}^{\oplus}]{\text{1. CH}_3\text{MgBr}} \mathbf{F} \xrightarrow{\text{H}_2\text{CrO}_4} \mathbf{G}$$

19.47 Show how to synthesize each of the following compounds, using the given restrictions on the starting materials.

(a)

Three or fewer carbons $\xrightarrow{?}$

(b)

Six or fewer carbons $\xrightarrow{?}$

(c)

Eight or fewer carbons $\xrightarrow{?}$

(d)

Acyclic compounds $\xrightarrow{?}$

19.48 Show how to carry out each of the following syntheses, using any reagents necessary. (*Hint:* In each case, a ketone or aldehyde functional group is entirely removed.)

(a)

(b)

(c)

(d)

19.49 1,4-Cyclohexanedione monoethylene acetal is commercially available.
  (a) Show how you would use it to synthesize 4-ethylidenecyclohexanone.
  (b) What problems would arise if you tried to synthesize the same target from 1,4-cyclohexanedione?

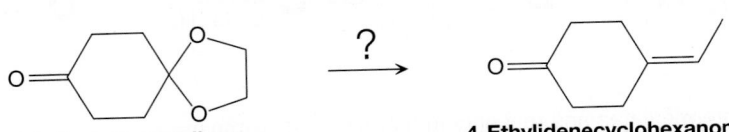

1,4-Cyclohexanedione
monoethylene acetal

4-Ethylidenecyclohexanone

19.50 Here in Chapter 19, we learned that converting a ketone or aldehyde to an acetal is a good way to protect the carbonyl group, because an acetal is composed of ether linkages. In Chapter 17, we saw that a ketone or aldehyde can be converted into an epoxide, which is a cyclic ether. Why would an epoxide be a poor choice as a protecting group?

19.51 A student wants to carry out the following reaction.

$$\text{(structure)} \xrightarrow[\text{2. CH}_3\text{I}]{\text{1. LDA}} \text{(structure)}$$

Explain the problem(s) associated with this synthesis scheme and suggest a way to carry out the transformation efficiently.

19.52 Suppose we want to carry out the following functional group conversions in which A is converted to B. In each case, however, the alcohol group is reactive under the reaction conditions that carry out the transformation, so protecting the alcohol would be necessary. Here in Chapter 19, we learned more than one way to protect alcohols. Which ways do you think will be effective? Which ones will not? Why?

(a) A → B

(b) A → B

19.53 Propose how you would carry out each of the following syntheses.

(a) → ?

(b) → ?

19.54 Suggest how you would synthesize each of the following compounds beginning with pentan-2-one.

(a)   **(b)**   **(c)**   **(d)**

19.55 Show how to synthesize each of the following molecules beginning with phenylmethanol (benzyl alcohol).

(a)

(b)

(c)

(d)

19.56 Show how you would synthesize hexane-3,4-diol using propanal as your only source of carbon atoms.

19.57 Show how you would carry out each of the following syntheses.

(a)

(b)

19.58 Show how you would carry out the following synthesis. (*Hint*: What rearrangement occurs when an enol is formed?)

19.59 Show how you would synthesize the compound at the right using propanal as your only carbon source.

19.60 Show how to carry out each of the following syntheses. (*Hint*: You may need to use selective reagents or protecting groups.)

(a)

(b)

(c)

(d)

(e)

If your instructor assigns problems in smartwork, log in at **smartwork.wwnorton.com**.

Nucleophilic Addition–Elimination Reactions 1

The General Mechanism Involving Strong Nucleophiles

In Chapters 17 and 18, we discussed reactions in which nucleophiles add to functional groups containing polar π bonds, such as ketones, aldehydes, imines, and nitriles. Those nucleophilic addition reactions are driven, in large part, by the flow of electrons from the electron-rich nucleophile to the electron-poor (i.e., electrophilic) atom of the polar π bond.

A variety of other functional groups contain polar π bonds, too, and are therefore susceptible to nucleophilic attack. For example, carboxylic acids have the polar carbonyl group (shown in blue in Fig. 20-1, next page), as do esters, amides, acid anhydrides, and acid halides (collectively known as **carboxylic acid derivatives**).

Although these functional groups are susceptible to nucleophilic attack, they participate in different *overall* reactions from what we saw in Chapters 17 and 18. Whereas the functional groups in Chapters 17 and 18 primarily undergo nucleophilic addition, these functional groups tend to undergo what is called a *nucleophilic addition–elimination* mechanism. This is because they each contain a *leaving group* (highlighted in red in Fig. 20-1) bonded to the electron-deficient atom of the polar π bond.

> The process for making soap involves one of the oldest organic reactions known, called saponification. Here in Chapter 20, we study saponification and other nucleophilic addition–elimination reactions.

A compound can undergo a nucleophilic addition–elimination reaction if it has a leaving group attached to the electron-poor atom of a polar π bond.

Carboxylic acid derivatives are not the only functional groups that can undergo nucleophilic addition–elimination. Several functional groups containing the S=O bond, for example, can undergo these reactions, as we will see in Chapter 21.

By contrast, ketones, aldehydes, imines, and nitriles do not possess a suitable leaving group; the electron-deficient carbon is bonded to H and/or alkyl (R) groups only (Fig. 20-2), so any leaving group would have to depart as H⁻ or R⁻, which are very unstable anions. Consequently, these polar π bonds tend to undergo nucleophilic addition only.

As we will see throughout this chapter and Chapter 21, there are a wide variety of reactions that proceed via a nucleophilic addition–elimination mechanism. Here in Chapter 20, we focus just on nucleophilic addition–elimination reactions that involve strong nucleophiles bearing a negative charge. These include alkoxide anions (RO⁻), hydride anions (H⁻) from various reducing agents, and alkyl anions (R⁻) from various organometallic reagents. In Chapter 21, we shift our focus to reactions involving weak nucleophiles and ones that require acid or base catalysis.

20.1 An Introduction to Nucleophilic Addition–Elimination Reactions: Base-Promoted Transesterification

We begin our discussion of **nucleophilic addition–elimination reactions** with *base-promoted transesterification*, in which an ester is treated with an alkoxide anion (RO⁻). Base-promoted transesterifications are among the simplest nucleophilic addition–elimination reactions, but

FIGURE 20-1 Some functional groups that undergo nucleophilic addition–elimination These functional groups contain a leaving group (highlighted in red) attached to the electron-poor C atom of the polar π bond (highlighted in blue). The functional groups appearing in the box are collectively known as *carboxylic acid derivatives*.

Carboxylic acid derivatives

Polar π bond / Leaving group

Carboxylic acid — Ester — Amide — Acid anhydride — Acid halide

FIGURE 20-2 Functional groups that undergo nucleophilic addition only These functional groups do not have a leaving group attached to the polar π bond (highlighted in blue). The H and R groups that are attached to the electron-poor C atom would have to depart as H⁻ or R⁻, which are highly unstable.

Not suitable leaving groups

Ketone — Aldehyde — Imine — Nitrile

they provide insight into many of the principles that underlie more complex reactions. Thus, base-promoted transesterification acts as a prototype for the other nucleophilic addition–elimination reactions discussed throughout this chapter and Chapter 21.

20.1a The General Nucleophilic Addition–Elimination Mechanism

When an ester such as methyl benzoate is treated with potassium *tert*-butoxide, a different ester, *tert*-butyl benzoate, is produced (Equation 20-1).

Methyl benzoate

tert-Butyl benzoate
85%

(20-1)

This reaction is an example of a **base-promoted transesterification**. It is a *transesterification* because one ester is converted into another, and it is *base promoted* because one of the reactants, potassium *tert*-butoxide, is a strong base and is consumed in the reaction.

The mechanism for this base-promoted transesterification, shown in Equation 20-2, is typical of reactions involving a nucleophile and a carboxylic acid derivative. It is called a *nucleophilic addition–elimination* mechanism, which describes the two steps that take place.

Mechanism for Equation 20-1

Nucleophilic addition–elimination mechanism

A nucleophile attacks the electron-poor C.

The C═O bond is regenerated as the leaving group leaves.

Slow
1. Nucleophilic addition

Fast
2. Nucleophile elimination

(20-2)

Tetrahedral intermediate

Step 1 is *nucleophilic addition* (Section 7.5), in which a nucleophile attacks the electron-poor carbonyl carbon. This forces the pair of electrons from the initial C═O π bond onto the O atom, which generates a negative charge on O. The product of that step is a **tetrahedral intermediate**, which, in Step 2, undergoes *nucleophile elimination* (Section 7.5). A lone pair of electrons on the negatively charged O atom is used to regenerate the C═O double bond, and the leaving group departs as CH_3O^-.

The following is the mechanism for another base-promoted transesterification reaction, but the curved arrows have been omitted.

(a) Complete the mechanism by adding the necessary curved arrows.
(b) Identify the tetrahedral intermediate.
(c) Below each reaction arrow, write the name of the elementary step that takes place.

Answers to Your Turns are in the back of the book.

---

problem **20.1** Draw the complete, detailed mechanism and the major product(s) for each of the following reactions.

NaOCH₂CH₃ / CH₃CH₂OH → ?

(a)

THF → ?

(b)

Overall, nucleophilic addition–elimination results in a *substitution* at the carbonyl carbon that is part of an acyl group (RC=O). Specifically, the –OCH₃ group attached to the carbonyl carbon is replaced by an –OC(CH₃)₃ group. Thus, these kinds of reactions are often referred to as **nucleophilic acyl substitution**. Keep in mind, however, that this mechanism is quite different from the S$_N$2 and S$_N$1 mechanisms we encountered in Chapters 7 and 8. The leaving group in this case is on an *sp²*-hybridized C atom and, as we learned in Chapter 9, these kinds of substrates do not undergo S$_N$2 or S$_N$1 reactions.

20.1b Kinetics of Nucleophilic Addition–Elimination: The Reaction Free-Energy Diagram

Notice from Equation 20-2 that the first step of this nucleophilic addition–elimination mechanism is significantly slower than the second step. In other words,

In a nucleophilic addition–elimination mechanism, the first step (i.e., nucleophilic addition) is generally the rate-determining step.

This indicates that the energy barrier of the first step is much larger than that of the second step, as shown in Figure 20-3.

Part of the reason for the relatively large energy barrier of the first step is the *resonance stabilization* involving the π bond of the carbonyl group and a lone pair of electrons from the leaving group, as shown in Figure 20-3. By contrast, the tetrahedral intermediate is not resonance stabilized (the π bond is absent), so it is significantly higher in energy. According to the Hammond postulate (Section 9.3a), the transition state closely resembles the tetrahedral intermediate, so the transition state's energy, too, is relatively high.

The tetrahedral intermediate is *not* resonance stabilized.

The energy barrier for the first step is much larger than for the second.

$\Delta G^{\circ\ddagger}(1)$

$\Delta G^{\circ\ddagger}(2)$

FIGURE 20-3 Reaction free-energy diagram for nucleophilic addition–elimination The first step, nucleophilic addition, is rate determining (i.e., it is the slow step) because it has a much greater energy barrier than the second step. Largely, this is because resonance involving the lone pair of electrons on the leaving group does not exist in the tetrahedral intermediate.

Free energy ⟶

Resonance stabilization in the reactant

Resonance stabilization in the product

Reaction coordinate ⟶

The reverse is true for the second step. When the leaving group departs to regenerate the C=O bond, resonance is reestablished, which serves to stabilize the overall product. Once again, according to the Hammond postulate, the transition state closely resembles the tetrahedral intermediate, so the energy barrier is relatively small.

YOUR TURN 20.2

Construct a free energy diagram for the reaction in Your Turn 20.1. Include and label the overall reactants, overall products, and the tetrahedral intermediate, and draw a vertical double-headed arrow to represent each step's energy barrier.

20.1c Thermodynamics and Reversibility

Looking back at Equation 20-1, notice that the reactants and products are connected by an equilibrium reaction arrow (⇌). In other words,

Transesterification is a *reversible* reaction.

Thus, if *tert*-butyl benzoate is treated with CH_3O^-, methyl benzoate will be produced via nucleophilic addition–elimination.

YOUR TURN 20.3

Draw the detailed mechanism for the reverse of the reaction in Equation 20-1.

Biodiesel and Transesterification

Biodiesel is an attractive alternative to petroleum-based fuels for a variety of reasons. Its use results in lower net production of CO_2, a greenhouse gas, than petroleum diesel. Derived from plants, it can be produced domestically, so it could significantly decrease our dependence on foreign oil—what many people believe to be a threat to national security. And biodiesel is renewable, which is important because some studies project that the world's oil reserves will begin to run dry within 50 years.

Biodiesel is produced from oils that are made by plants, using essentially the same transesterification process discussed here in Chapter 20. Plant oil is a triglyceride, which is a fatty acid triester of glycerol. Biodiesel, on the other hand, is a fatty acid monoester. To carry out the transesterification that produces the biodiesel, the triglyceride is treated with an alcohol under basic conditions. Methanol is the most popular alcohol used for this process, resulting in a *fatty acid methyl ester*, though other alcohols can be used, depending on cost and the specific properties desired of the biodiesel.

A triglyceride

$\xrightarrow{\text{CH}_3\text{OH}}_{\text{HO}^\ominus}$

Glycerol + **Fatty acid methyl esters**

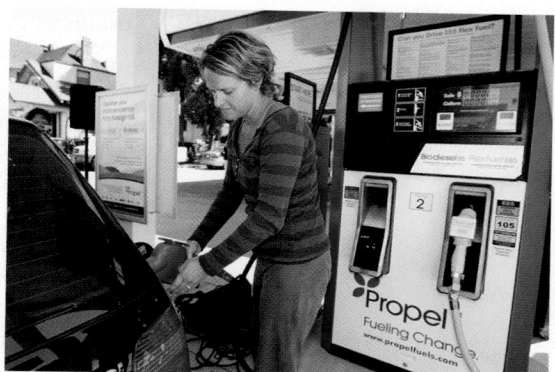

Some issues surrounding biodiesel production still remain. One is that methanol is currently produced primarily from petroleum-based sources. However, ways of making renewable methanol are being developed and optimized. Another is that large amounts of glycerol are produced as a byproduct, but initiatives such as the Glycerol Challenge have been established to promote finding alternative uses for glycerol.

The reversibility of base-promoted transesterification is consistent with the reaction's thermodynamics, depicted previously in the free energy diagram in Figure 20-3. Namely, the overall reactants and products appear at roughly the same energy, so the overall energy barrier that must be traversed in the reverse direction is roughly the same as that in the forward direction. The similar energies of the reactants and products, in turn, stem from the fact that the negative charge that appears on either side of the reaction is stabilized comparably by the alkoxide anions. This is reflected by the fact that the pK_a values of their respective conjugate acids, which are 15.5 and 19, are not dramatically different (Fig. 20-4).

Although base-promoted transesterification reactions are reversible, this is *not* true of all nucleophilic addition–elimination reactions. As we will see, the relative

pK_a of conjugate acid = 15.5

Similar charge stability

pK_a of conjugate acid = 19

FIGURE 20-4 Charge stability and reversibility in nucleophilic addition–elimination These alkoxide anions appear on the reactant and product sides in Equation 20-1. They exhibit similar charge stability, reflected by the similar pK_a values of their respective conjugate acids, which is largely why the transesterification in Equation 20-1 is reversible.

stabilities of the nucleophile and the leaving group greatly affect the reversibility of the reaction.

20.2 Reaction of an Ester with Hydroxide (Saponification) and the Reverse Reaction

Equation 20-3 shows that when ethyl acetate is treated with KOH, followed by acid workup, then nucleophilic acyl substitution *appears* to have taken place.

Saponification

**Ethyl ethanoate
(Ethyl acetate)**

1. KOH, CH₃OH
   60 min, 35 °C
2. HCl

**Ethanoic acid
(Acetic acid)**
87%

(20-3)

The first reaction in this sequence is known as **saponification**, which literally means "soap making" (*sapo* is the Latin root for "soap"). Ancient civilizations made crude soap by boiling animal fat (which is chiefly long-chain esters) together with wood ash, a source of HO⁻ ions. (See Problem 20.53 at the end of the chapter.)

Thus, generally speaking,

> A carboxylic acid is produced when an ester undergoes saponification followed by acid workup.

Although the overall reaction in Equation 20-3 is similar to the base-promoted transesterification in Equation 20-1, there are two key differences. First, saponification is *irreversible*, whereas transesterification is reversible. Second, saponification requires acid workup, whereas transesterification does not.

Both of these differences can be explained by the mechanism shown in Equation 20-4.

Mechanism for Equation 20-3

Steps 1 and 2 are identical to the nucleophilic addition and elimination steps in Equation 20-2. The carboxylic acid that is produced is then *rapidly* and *irreversibly* deprotonated under the basic conditions of the reactions (see Your Turn 20.4), producing a carboxylate anion (RCO_2^-). That proton is then replenished upon addition of a strong acid.

20.4 YOUR TURN

To convince yourself that the proton transfer reaction in Step 3 in Equation 20-4 is irreversible, use the appropriate pK_a values from Table 6-1 to compute the equilibrium constant, K_{eq}, of that step (see Section 6.2b). Which side of that proton transfer step is favored? To what extent?

The *overall* saponification reaction is irreversible because the deprotonation in Step 3 is irreversible. The initial carboxylic acid is produced reversibly and is in equilibrium with the starting ester. As that carboxylic acid is produced, it is continually removed from the equilibrium via Step 3. According to Le Châtelier's principle, the equilibrium is thus continually shifted toward products, which helps drive the reaction to completion.

The irreversible nature of a saponification reaction can be understood better by examining its reaction free energy diagram, shown in Figure 20-5. Notice that the immediate products of nucleophilic addition–elimination—namely, the carboxylic acid and the alkoxide anion—appear at roughly the same energy as the overall reactants, just as we saw in Figure 20-3. By contrast, the products of the proton transfer in Step 3—a carboxylate anion and an alcohol—are much lower

FIGURE 20-5 Free energy diagram of a saponification reaction The direct products of nucleophilic addition–elimination, RCO_2H and RO^-, have roughly the same energy as the overall reactants. The products of the proton transfer in the third step, however, are substantially lower in energy, due largely to the resonance delocalization of the negative charge in the carboxylate anion.

in energy, due in large part to the resonance stabilization in the carboxylate anion. That additional stability gained in Step 3 makes the overall reaction much more favorable energetically.

The irreversible proton transfer between a carboxylic acid and an alkoxide anion becomes a problem when we try to produce an ester from a carboxylic acid. As shown in Equation 20-5, an alkoxide anion simply deprotonates the carboxylic acid and no further reaction takes place.

Consequently,

> An ester cannot be produced by treating a carboxylic acid with an alkoxide anion.

The carboxylate anion does not react under these conditions because it has equivalent resonance structures that heavily stabilize the C=O group's π electrons, and, moreover, it is negatively charged and therefore repels incoming nucleophiles (Fig. 20-6).

FIGURE 20-6 The unreactive nature of carboxylate anions Nucleophiles tend not to add to the carbonyl C of a carboxylate anion. (a) Uncharged nucleophiles like alcohols are not strong enough to overcome the resonance delocalizaton of the charge in the carboxylate anion. (b) Stronger nucleophiles, such as alkoxide anions, are repelled from the carboxylate anion due to the like charges.

ROH is not a strong enough nucleophile to overcome the resonance stabilization in RCO_2^-.

Charge repulsion prevents nucleophilic attack.

(a)   (b)

YOUR TURN 20.5

To illustrate the stabilization of a carboxylate anion, draw the resonance structure and resonance hybrid in the spaces provided.

Resonance hybrid

problem **20.2** Draw the complete, detailed mechanism and predict the overall products for each of the following reactions.

1. NaOH, H_2O
2. $H_3O^{\oplus}$?

(a)

1. NaOH, H_2O
2. $H_3O^{\oplus}$?

(b)

20.3 Acyl Substitution Involving Other Carboxylic Acid Derivatives: The Thermodynamics of Acyl Substitution

The acyl substitution reactions we have examined thus far have been rather limited in scope, involving an ester as the acid derivative, and either RO^- or HO^- as the nucleophile. Numerous other acyl substitutions can be carried out, however, simply by using different combinations of acid derivatives and nucleophiles. Equation 20-6 shows, for example, that CH_3O^- can displace Cl^- from an acid chloride to produce an ester.

Benzoyl chloride

Methyl benzoate
84%

(20-6)

The mechanism for this reaction is shown in Equation 20-7. It consists of the usual nucleophilic addition and elimination steps, and proceeds through a tetrahedral intermediate.

Mechanism for Equation 20-6

Tetrahedral intermediate

1. Nucleophilic addition

2. Nucleophile elimination

(20-7)

An ester such as methyl acetate can react with H_2N^- to produce an amide. The nucleophilic addition–elimination mechanism that describes this reaction is as follows:

20.6 YOUR TURN

(a) Complete the mechanism by adding the curved arrows to show the movement of the electrons.

(b) Under each reaction arrow, write the name of the elementary step taking place.

(c) Label the tetrahedral intermediate.

Not every combination of acid derivative and nucleophile leads to an effective acyl substitution reaction. As shown in Equation 20-8, for example, essentially no acyl substitution takes place when an ester is treated with chloride ion.

Methyl benzoate

$$\xrightarrow{\text{NaCl}} \quad \text{No reaction} \qquad (20\text{-}8)$$

We can understand why by examining the generic, overall acyl substitution reaction in Equation 20-9. Notice that the nucleophile (Nu⁻) bears a negative charge on the reactant side, and the leaving group (L⁻) bears the negative charge in the products.

This side of the reaction is favored if Nu⁻ is substantially more stable than L⁻.

This side of the reaction is favored if L⁻ is substantially more stable than Nu⁻.

$$(20\text{-}9)$$

If L⁻ is more stable than Nu⁻ (reflected by the weaker basicity of L⁻), then the reaction is energetically favorable. This is the case with the reaction in Equation 20-6, because the leaving group (Cl⁻) is more stable than the nucleophile (CH₃O⁻). If Nu⁻ is more stable than L⁻, on the other hand, then the reaction is not energetically favorable, and it does not proceed readily. This is the case with the reaction in Equation 20-8, in which the nucleophile is Cl⁻ and the leaving group is CH₃O⁻.

FIGURE 20-7 The "stability ladder" for carboxylic acid derivatives
A reaction that converts an acid derivative from a higher rung to a lower rung is energetically favorable and generally takes place readily. Going from a lower to a higher rung is unfavorable, and is generally quite difficult to carry out.

Energetically favorable conversions

Less stable acid derivative

Acid Derivative	Leaving Group
Acid chloride	Cl⁻
Acid anhydride	
Ester Carboxylic acid	
Amide	

More stable leaving group (weaker base)

Whether an acyl substitution reaction is energetically favorable can be summarized by the "stability ladder" shown in Figure 20-7. Descending a real ladder is generally easier than ascending one, so

- An acyl substitution that converts an acid derivative from a higher rung on the stability ladder to one on a lower rung of the ladder is energetically favorable.
- An acyl substitution that converts an acid derivative from a lower rung on the stability ladder to one on a higher rung of the ladder is *not* energetically favorable.

For example, the conversion of an acid chloride (RCOCl) to an ester (RCO₂R) represents descending the stability ladder (in this case, two rungs), so it is energetically favorable. This is consistent with what we concluded earlier from the fact that the reaction in Equation 20-6 takes place readily. By the same token, the conversion of an ester to an acid chloride is *not* favorable, consistent with the lack of reaction in Equation 20-8.

20.7 YOUR TURN

Indicate whether each of the following conversions would be energetically favorable or unfavorable.

(a)

(b)

(c)

(d)

SOLVED problem 20.3 Will the following reaction take place? If it will, draw the complete, detailed mechanism and predict the major product(s).

Think What is the nucleophile? What is the leaving group? Is the potential acyl substitution reaction energetically favorable?

Solve The nucleophile is the acetate anion, $CH_3CO_2^-$, and the leaving group is Cl^-, part of the acid chloride. In an acyl substitution reaction between these species, $CH_3CO_2^-$ replaces Cl^- in a nucleophilic addition–elimination mechanism, as follows:

Acid chloride

Acid anhydride

Notice in Figure 20-7 (the stability ladder) that the product acid derivative, an acid anhydride, is below the reactant acid derivative, an acid chloride. As a result, this reaction is energetically favorable and will take place readily.

(a) ?

(b) NaOCH₂CH₃ ?

(c) ?

(d) NaCl ?

(e) ?

(f) ?

The Stability Ladder in Biochemical Systems

Carboxylic acids and carboxylic acid derivatives play major roles in a wide variety of biochemical process-es. Depending upon its particular role, the carboxylic acid derivative might need to be relatively stable. This is the case with proteins, for example, which are long chains of amino acids connected by amide functional groups—so-called peptide linkages (shown below on the left). As we saw in Figure 20-7, amides appear toward the bottom of the stability ladder, so these peptide linkages are rather unreactive—something that helps a protein maintain its integrity, allowing it to carry out its specific function.

An amide serves as a peptide linkage in a protein.

Glutamate

ATP

Acyl phosphate intermediate

:base

In other cases, a particular carbox-ylic acid derivative might be too stable. Consider, for example, the scheme shown on the right, which depicts the biosynthesis of glutamine—one of the 20 naturally occurring amino acids. The carboxylate anion (RCO_2^-) of glutamate is converted into an amide group, catalyzed by the enzyme gluta-mine synthetase. As we saw in Section 20.2, a carboxylate anion is especially stable because its C=O π bond is highly stabilized via resonance and its negative charge repels incom-ing nucleophiles. For this conversion to take place, therefore, the carboxylate group is first phosphory-lated by adenosine triphosphate (ATP) in a nucleophilic substitution reaction (believed to be S_N2). The product of that step is an acyl phosphate intermediate, which has a significantly better leaving group and is considerably less stable. This sets up nucleophilic addition–elimination involving NH_3 in the next two steps, yielding glutamine as the overall product.

Glutamine

20.4 Carboxylic Acids from Amides; the Gabriel Synthesis of Primary Amines

In Section 20.3, we learned that the conversion of an amide ($RCONR_2$) to a carboxylic acid (RCO_2H) is energetically *unfavorable*, consistent with the fact that an amide appears on a lower rung of the stability ladder (Fig. 20-7) than a carboxylic acid. Equation 20-10 shows, however, that we can carry out this transformation by treating an amide with HO^-, followed by acid workup.

(20-10)

This apparent discrepancy can be reconciled by the mechanism for this reaction, which is shown in Equation 20-11.

Mechanism for Equation 20-10

(20-11)

Steps 1 and 2 make up the usual nucleophilic addition–elimination mechanism to produce the initial carboxylic acid. Under the basic conditions of the reaction, this initial carboxylic acid is quickly deprotonated in Step 3, producing a carboxylate anion. Finally, acid workup in Step 4 replenishes the proton on the carboxylate anion so that the carboxylic acid can be isolated.

As indicated in the mechanism, the direct formation of the carboxylic acid from the amide is energetically unfavorable, and the reaction is reversible to that point. This is consistent with the fact that a carboxylic acid appears on a higher rung of the stability ladder than an amide. The proton transfer in Step 3 is irreversible, however, so it is the critical step that drives the reaction toward products: *Without that step, no appreciable amount of product would form.*

FIGURE 20-8 Free energy diagram corresponding to the mechanism in Equation 20-11 The direct products of nucleophilic addition–elimination, RCO₂H and R₂N⁻, are higher in energy than the overall reactants. Thus, the reaction is energetically unfavorable after the first two steps of the mechanism. The third step is energetically very favorable, however, which drives the reaction toward products. This is followed by an acid workup.

This can be better understood from the reaction free energy diagram in Figure 20-8. Notice that the immediate products from nucleophilic addition and elimination are higher in energy than the overall reactants. Thus, direct acyl substitution to form the carboxylic acid is energetically unfavorable. The irreversible proton transfer in the third step, however, is responsible for lowering the energy dramatically, which helps favor products.

YOUR TURN 20.8

The following mechanism, which has the curved arrows omitted, is for another reaction that converts an amide to a carboxylic acid.

(a) Complete the mechanism by drawing the missing curved arrows.
(b) Below each reaction arrow, write the name of the elementary step that takes place and indicate whether the step is reversible or irreversible.
(c) Identify the tetrahedral intermediate.

It is also helpful to consider the parallels between the mechanism in Equation 20-11 and the mechanism for saponification we saw previously in Equation 20-4. In both cases, the acyl substitution product is initially formed after the first two steps under equilibrium (i.e., reversible) conditions. And, in both cases, an irreversible proton transfer in Step 3 drives the reaction toward products, prior to a final acid workup.

(a)   **(b)**

Although the carboxylic acid is usually the desired product of an acyl substitution such as that in Equation 20-10, there are times when isolating the amine is desirable. One such reaction is the **Gabriel synthesis**, shown in Equation 20-12.

> The *Gabriel synthesis* is an effective way to synthesize a primary amine (RNH_2) from an alkyl halide.

The Gabriel synthesis

(20-12)

In the Gabriel synthesis, phthalimide is first treated with a strong base such as potassium hydroxide, followed by an alkyl halide. Subsequent treatment with hydroxide then liberates the primary amine.

In the mechanism shown in Equation 20-13, Step 1 is simply a proton transfer reaction in which HO^- deprotonates phthalimide at the N atom. The resulting anion is strongly nucleophilic at N, so an S_N2 reaction occurs when an alkyl halide is added in Step 2. Steps 3 through 8 make up two successive acyl substitutions, each one similar to the mechanism shown in Equation 20-11.

Partial mechanism for Equation 20-12

(20-13)

The benefit of synthesizing a primary amine via the Gabriel synthesis can be appreciated when we consider the complications that arise when a primary amine is synthesized via the alkylation of ammonia (Equation 20-14). As we saw in Section 10.2, the primary amine product is contaminated with the products from *polyalkylation*. By contrast, the Gabriel synthesis produces only the intended primary amine.

(20-14)

1° amine	2° amine	3° amine	4° ammonium ion
57%	24%	3%	Trace amount

problem 20.6 Predict the product of the sequence of reactions at the right.

1. KOH/EtOH
2. Br (3-bromobenzyl bromide)
3. HO⁻/H₂O, Δ

SOLVED problem 20.7 Suggest how to carry out the synthesis at the right using a Gabriel synthesis.

Think Is the target a primary amine that can be the product of a Gabriel synthesis? What alkyl halide can be used as a precursor in a Gabriel synthesis of the target amine? How can that alkyl halide be produced from the starting material?

Solve The target is a primary amine. To produce it in a Gabriel synthesis, a benzyl halide such as bromophenyl methane is required, as shown in the following retrosynthetic analysis.

That halide can be synthesized from the corresponding primary alcohol, which can be produced, in turn, by reducing the starting aldehyde. To write the synthesis, we reverse the arrows and add the appropriate reagents. First we show how to synthesize the alkyl bromide. Subsequently, that alkyl bromide is incorporated along with phthalimide in a Gabriel synthesis.

problem **20.8** How would you carry out the following transformation using a Gabriel synthesis?

20.5 Haloform Reactions

In all of the acyl substitution reactions we have discussed so far, one reactant is an acid derivative in which the carbonyl carbon is bonded to a leaving group. Equation 20-15 shows, however, that a methyl ketone ($RCOCH_3$) can also undergo acyl substitution when treated with Br_2 in aqueous sodium hydroxide, followed by acid workup.

A haloform reaction

A methyl ketone

One of the products, bromoform ($HCBr_3$), is an example of a **haloform**, which has the general formula HCX_3 (where X is a halogen atom). Thus, the reaction is called a **haloform reaction**. Other haloform reactions can take place when a methyl ketone is treated with Cl_2 or I_2, producing chloroform ($HCCl_3$) or iodoform (HCI_3), respectively. Thus,

In general, a methyl ketone ($RCOCH_3$) can undergo a haloform reaction when treated with a molecular halogen (X_2) under basic conditions, followed by acid workup, producing a carboxylic acid (RCO_2H) and a haloform (HCX_3) byproduct.

20.9 YOUR TURN

The following reactions are the same as the one in Equation 20-15, only the halogen molecule is different. Draw the products in the boxes provided.

1. Cl_2, NaOH/H_2O
2. HCl

(a)

1. I_2, NaOH/H_2O
2. HCl

(b)

It appears in this acyl substitution reaction that HO^- replaces H_3C^-, a very poor leaving group. While this should seem peculiar, all is explained by the mechanism, shown in Equation 20-16.

$$(20\text{-}16)$$

This step is irreversible.

9. Proton transfer

10. Proton transfer

Add H$^{\oplus}$

Insoluble in water

In Steps 1–6 of this mechanism, the three α hydrogens on the methyl group are replaced by bromine atoms via the alpha halogenation mechanism discussed previously in Section 10.4. The resulting tribromomethyl ketone (RCOCBr$_3$) then undergoes attack by HO$^-$ in Step 7 to produce a tetrahedral intermediate, which subsequently eliminates Br$_3$C$^-$ in Step 8 to produce an initial carboxylic acid. In Step 9, that carboxylic acid is deprotonated by Br$_3$C$^-$, yielding a relatively stable carboxylate anion. In the acid workup in Step 10, the proton is replenished on the carboxylate anion to produce the final carboxylic acid product.

As a result, the leaving group in the nucleophilic addition–elimination mechanism is Br$_3$C$^-$, *not* H$_3$C$^-$. This is important, because H$_3$C$^-$ is very unstable, making it unsuitable as a leaving group. Br$_3$C$^-$ is a suitable leaving group, however, because the three Br atoms significantly stabilize the negative charge on C.

All three α halogen atoms are necessary to make the carbanion a suitable leaving group. As shown in Equation 20-17, an ethyl ketone cannot undergo acyl substitution because it gains only two α halogens.

With only two halogen atoms on the α carbon, the leaving group is insufficiently stable for acyl substitution.

Br$_2$, NaOH

$$(20\text{-}17)$$

YOUR TURN 20.10

Which of the following ketones can undergo a haloform reaction?

A   B   C   D   E

problem 20.9 Draw the complete, detailed mechanism and the major product(s) for each of the following reactions.

(a)

(b)

SOLVED problem 20.10 Show how to carry out the following synthesis.

Think What reaction can convert a methyl ketone into a compound with a leaving group on the carbonyl carbon? Can the product of that reaction be converted directly into the amide target or must another precursor be used?

Solve The starting compound is a methyl ketone, which does not have a suitable leaving group. However, a haloform reaction can be used to convert that methyl ketone into a carboxylic acid.

We must now determine how to convert the carboxylic acid into the amide target. Thinking retrosynthetically, the amide can be produced from an ester via an acyl substitution. The ester, in turn, can be produced using diazomethane (Section 10.5).

To complete the synthesis, reverse the direction and supply the appropriate reagents.

problem 20.11 Show how to carry out the following synthesis using any reagents necessary.

A negative iodoform test

(a)

A positive iodoform test

(b)

FIGURE 20-9 The iodoform test
(a) A sample was treated with a
basic solution of I_2, but no precipitate
formed. **(b)** Iodoform, a yellow
precipitate, formed when a basic
solution of I_2 was added, indicating
that the sample contained a methyl
ketone.

Br_3C^-, despite its three Br atoms, is less stable than HO^-, as reflected by its stronger basicity (pK_a of $Br_3CH \approx 20$; pK_a of $H_2O = 15.7$). Consequently, the acyl substitution that takes place (Steps 7 and 8 in Equation 20-16) is somewhat unfavorable energetically and is thus reversible. The subsequent proton transfer step (Step 9 in Equation 20-16) is energetically quite favorable, however, which helps drive the reaction toward products. Moreover, the haloform that is produced in that step is insoluble in water, so it is effectively removed from the reaction mixture. According to Le Châtelier's principle, this drives the acyl substitution reaction even further toward products.

Prior to the advent of spectroscopy, the iodoform (HCI_3) reaction was commonly used as a test for methyl ketones. As shown in Figure 20-9, iodoform is a bright yellow solid at room temperature and is insoluble in water. Therefore, if a bright yellow precipitate appears upon treating an organic compound with I_2 in basic solution, it is likely that the compound is a methyl ketone.

problem **20.12** A compound whose molecular formula is $C_6H_{12}O$ produces a bright yellow solid when it is treated with iodine and a basic solution of water. The IR spectrum of the compound shows a distinct peak at 1708 cm^{-1} and its ^1H NMR spectrum contains only two singlets: one at 1.2 ppm and one at 2.2 ppm. Integration shows that the upfield signal has three times the area of the downfield signal.

(a) Draw the structure for this molecule.
(b) Draw the products of the reaction that is described.

20.6 Hydride Reducing Agents: Sodium Borohydride (NaBH$_4$) and Lithium Aluminum Hydride (LiAlH$_4$)

In all of the nucleophilic addition–elimination reactions we have examined so far, the nucleophile that attacks the polar π bond becomes a potential leaving group in the product. In these reactions, therefore, one of the products is a carboxylic acid or acid derivative.

However, not all nucleophiles that can attack a carbonyl group are capable of behaving as leaving groups. Hydride ions (H^-) from sources such as $NaBH_4$ and $LiAlH_4$, for example, are excellent nucleophiles but very poor leaving groups. These kinds of nucleophiles can still react with acid derivatives in nucleophilic addition–elimination reactions, but some key differences exist.

Consider Equations 20-18 and 20-19, which show that $NaBH_4$ and $LiAlH_4$ can readily reduce an acid chloride.

NaBH$_4$ reduces an
acid chloride to a
primary alcohol.

1. NaBH$_4$, THF
2. CH$_3$OH
(dropwise)

Benzoyl chloride

**Phenylmethanol
(Benzyl alcohol)**
92%

(20-18)

LiAlH$_4$ reduces an
acid chloride to a
primary alcohol.

1. LiAlH$_4$, ether
2. H$_2$SO$_4$,
ethyl acetate

83%

(20-19)

Notice that no carbonyl group is present in the product. Instead, a primary alcohol is produced. Equation 20-20 shows how this occurs for the reduction using $NaBH_4$, and, as shown in Your Turn 20.11, the mechanism for the $LiAlH_4$ reduction is essentially the same.

Mechanism for Equation 20-18

(20-20)

Steps 1 and 2 make up the usual nucleophilic addition–elimination mechanism, producing an aldehyde as an intermediate. Under these reduction conditions, however, the aldehyde reacts rapidly with another equivalent of hydride (Section 17.3) to produce an alkoxide anion. Subsequent acid workup yields the alcohol.

20.11 YOUR TURN

The following scheme outlines the mechanism for the $LiAlH_4$ reduction of the acid chloride in Equation 20-19.

(a) Supply the appropriate curved arrows.
(b) Write the name of the elementary step underneath each reaction arrow.
(c) Identify the tetrahedral intermediate.

problem 20.13 Predict the major product and draw the complete, detailed mechanism for each of the following reactions.

(a)

(b)

Other acid derivatives and carboxylic acids can undergo hydride reduction, too. As with acid chlorides,

> The hydride reduction of an acid derivative or carboxylic acid typically involves the addition of two equivalents of H^-.

These reactions are summarized in Table 20-1. As with acid chlorides, many of these reductions produce a primary alcohol and proceed by the same mechanism shown previously in Equation 20-20.

TABLE 20-1 Hydride Reductions of Acid Derivatives and Carboxylic Acids

Reduction with NaBH$_4$	Reduction with LiAlH$_4$
Acid chloride → Primary alcohol	Acid chloride → Primary alcohol
Acid anhydride → Primary alcohol	Acid anhydride → Primary alcohol
This reduction is very slow. Ester → Primary alcohol (+ HOR′)	Ester → Primary alcohol (+ HOR′)
Carboxylic acid → No acyl substitution	Carboxylic acid → Primary alcohol
Amide → No acyl substitution	Amide → Amine

20.6a The Selectivity of NaBH$_4$

Notice in Table 20-1 that NaBH$_4$ reduces esters very slowly and does not reduce amides at all. This is a reflection of the resonance stabilization involving the carbonyl group and the leaving group, shown previously in Figure 20-3, which is more pronounced in esters and amides than in acid halides or acid anhydrides. (This is discussed further in Chapter 21.)

We can take advantage of the low reactivity of esters and amides to *selectively* reduce a more reactive carbonyl-containing functional group, such as an acid chloride, acid anhydride, aldehyde, or ketone. In the β-ketoester in Equation 20-21, for example, the ketone functional group is readily reduced to the alcohol, but the ester functional group is unaffected.

The ketone is selectively reduced over the ester.

(20-21)

90%

SOLVED problem **20.15** Predict the major product of the following reaction.

Think Which functional groups in the reactant can be reduced by NaBH$_4$? Does the reduction of one of those groups take place more readily than the other(s)?

Solve The acid chloride and ester groups can be reduced by NaBH$_4$ to produce primary alcohols, but the ether group cannot be reduced by NaBH$_4$. Whereas NaBH$_4$ reduces an acid chloride readily, the reduction of an ester is sluggish, so the acid chloride will be reduced selectively as follows:

problem 20.16 Draw the complete, detailed mechanism and predict the major product for each of the following reactions.

(a)

(b)

Notice, too, in Table 20-1 that $NaBH_4$ doesn't reduce a carboxylic acid even though it reduces esters slowly. This might seem peculiar because carboxylic acids are on the same rung of the stability ladder as esters (Fig. 20-7). $NaBH_4$ is somewhat basic, however, and a carboxylic acid is moderately acidic. Therefore, as shown in Equation 20-22, a rapid proton transfer converts the carboxylic acid into its carboxylate anion. Recall from Section 20.2 that carboxylate anions are especially resistant to nucleophilic attack.

A carboxylate anion is very highly resonance stabilized, so no further reaction occurs with $NaBH_4$.

A carboxylic acid   A carboxylate anion   (20-22)

problem 20.17 Show how the carboxylic acid below can be converted into the corresponding alcohol using $NaBH_4$ as the reducing agent. (*Hint*: Can you convert the carboxylic acid into a different acid derivative first?)

20.6b The Greater Reactivity of $LiAlH_4$

Whereas $NaBH_4$ cannot reduce carboxylic acids, $LiAlH_4$ is a much stronger reducing agent (Section 17.3) and will reduce a carboxylic acid to a primary alcohol. An example is shown in Equation 20-23.

$LiAlH_4$ readily reduces a carboxylic acid to a primary alcohol.

1. $LiAlH_4$, ether
   0 °C, 3 h
2. H_2O

95%   (20-23)

The mechanism for the reduction of a carboxylic acid is shown in Equation 20-24. Step 1 is a rapid proton transfer that produces a carboxylate anion, just as we saw in Equation 20-22 when a carboxylic acid is treated with $NaBH_4$. At the same time, the negatively charged O atom of the carboxylate anion forms a relatively strong bond to the Al atom. (This is analogous to the hydride transfer step we saw previously in Section 17.3.) Subsequently, Steps 2–5 make up the back-to-back hydride reductions we saw previously for more reactive acid derivatives (Equation 20-20).

Mechanism for Equation 20-23

(20-24)

The formation of the O—Al bond in Step 1 of Equation 20-24 facilitates the $LiAlH_4$ reduction of a carboxylic acid in two ways. First, as indicated in Equation 20-24, this bond removes the equivalence in the resonance structures normally observed for a carboxylate anion (RCO_2^-). Thus, the C=O is not as highly resonance stabilized as in a regular carboxylate anion and is therefore more susceptible to nucleophilic attack by H^-. Second, a better leaving group is generated; without the formation of the O—Al bond, the leaving group in a regular carboxylate anion would be an isolated oxygen atom bearing a -2 charge—a species that is much too unstable to depart on its own.

Although not shown, the $O—AlH_3^-$ group produced in Step 1 can donate H^- to another species, becoming $O—AlH_2$. Thus, the leaving group would depart as $^-O—AlH_2$, which carries just a single negative charge and is even more stable than the leaving group in Equation 20-24.

$LiAlH_4$ and $NaBH_4$ also differ in the way they react with an amide. Whereas $NaBH_4$ cannot reduce an amide, $LiAlH_4$ can, as shown in Equation 20-25.

LiAlH$_4$ reduces an amide to an amine.

(20-25)

98%

More specifically,

> LiAlH$_4$ reduces an amide (RCONR$_2$) to an amine (RCH$_2$NR$_2$), preserving the overall carbon backbone of the original compound.

It is particularly striking that the product is an amine, because LiAlH$_4$ reduces all of the other acid derivatives to primary alcohols. This difference stems, once again, from the relatively strong O—Al bond that forms, as shown in the mechanism in Equation 20-26.

Mechanism for Equation 20-25

A decent leaving group

Tetrahedral intermediate

1. Nucleophilic addition

2. Nucleophile elimination

(20-26)

From LiAlH$_4$

3. Nucleophilic addition

Iminium ion

In Step 1, H$^-$ adds to the carbonyl group of the amide, and the O—Al bond forms at the same time. The tetrahedral intermediate that is produced contains the same O—AlH$_3^-$ leaving group we saw in the mechanism describing the reduction of carboxylic acids (Equation 20-24). In Step 2, the O—AlH$_3^-$ leaving group departs, yielding an iminium ion, which is subsequently attacked by H$^-$ in Step 3 to yield the final product.

Similar to the lithium aluminum hydride reduction of a carboxylic acid, the O—AlH$_3^-$ group in the product of Step 1 can donate an H$^-$ to become O—AlH$_2$. The leaving group in the subsequent nucleophilic addition–elimination, therefore, would be more stable.

The following mechanism is for the LiAlH$_4$ reduction of another amide, but the curved arrows have been omitted.

(a) Supply the missing curved arrows.
(b) Below each reaction arrow, write the name of the elementary step that is taking place.
(c) Identify the tetrahedral intermediate.

---

problem **20.18** Draw the complete, detailed mechanism and predict the major product for each of the following reactions.

(a)

(b)

SOLVED problem **20.19** How would you carry out the transformation at the right? (*Hint*: Consider using a protecting group.)

Think Which of the groups in the starting compound must be reduced? Which should remain unchanged? Will the desired reduction also reduce the latter functional group? How can a protecting group be used to prevent the undesired reduction?

Solve The synthesis calls for the reduction of the ester to a primary alcohol, while the ketone remains unchanged. This kind of a reaction would proceed via nucleophilic addition–elimination at the ester carbon and will open the ring. However, as shown at the right, reduction using either LiAlH$_4$ or NaBH$_4$ will reduce the ketone as well.

This problem can be circumvented by using a protecting group on the ketone, as follows:

problem **20.20** Show how to carry out this synthesis. (*Hint*: Consider using a protecting group.)

20.7 Specialized Reducing Agents: Diisobutylaluminum Hydride (DIBAH) and Lithium Tri-*tert*-butoxyaluminum Hydride

A source of hydride (H⊖)

Lithium tri-*tert*-butoxyaluminum hydride [LiAlH(O-*t*-Bu)₃]

(a)

A source of hydride (H⊖)

Diisobutylaluminum hydride (DIBAH)

(b)

FIGURE 20-10 Two specialized hydride reducing agents These hydride reducing agents can be used to reduce a carboxylic acid derivative to an aldehyde. (a) Lithium tri-*tert*-butoxyaluminum hydride can reduce an acid chloride to an aldehyde. (b) Diisobutylaluminum hydride can reduce an ester to an aldehyde.

The mechanism in Equation 20-20, which describes the hydride reduction of an acid derivative to a primary alcohol, proceeds through an aldehyde intermediate. That is, once the aldehyde is formed, it quickly reacts with H^- in a second reduction. In some situations, however, it might be advantageous to stop at the aldehyde. Two specialized reducing agents are commonly used for such purposes: **lithium tri-*tert*-butoxyaluminum hydride** and **diisobutylaluminum hydride** (**DIBAH**) (Fig. 20-10). Both of these compounds possess an Al—H bond and thus are hydride anion (H^-) sources, much like LiAlH₄.

Examples of how these specialized reducing agents are used are shown in Equations 20-27 and 20-28.

LiAlH(O-*t*-Bu)₃ reduces an acid chloride to an aldehyde.

Benzoyl chloride

$$\xrightarrow[\substack{(CH_3OCH_2CH_2)_2O \\ -75\ °C,\ 1\ h \\ 20\ °C,\ 1\ h}]{LiAlH(O\text{-}t\text{-}Bu)_3}$$

Benzaldehyde
73%

(20-27)

DIBAH reduces an ester to an aldehyde.

Ethyl 4-methylpent-4-enoate

$$\xrightarrow[\substack{2.\ HCl,\ H_2O/CH_3OH}]{\substack{1.\ DIBAH,\ CH_2Cl_2 \\ -78\ °C,\ 2.5\ h}}$$

4-Methylpent-4-enal
95%

(20-28)

Specifically,

> ▪ Lithium tri-*tert*-butoxyaluminum hydride is commonly used to reduce an acid chloride (RCOCl) to an aldehyde (RCH=O).
>
> ▪ Diisobutylaluminum hydride (DIBAH) is commonly used to reduce an ester (RCO₂R) to an aldehyde (RCH=O).

The mechanism by which lithium tri-*tert*-butoxyaluminum hydride reduces an acid chloride to an aldehyde is shown in Equation 20-29.

Mechanism for Equation 20-27

From LiAlH(O-*t*-Bu)₃

1. Nucleophilic addition

2. Nucleophile elimination

The reaction stops at the aldehyde because a second addition of H⊖ from LiAlH(O-*t*-Bu)₃ is much slower than the first.

Slow

(20-29)

This mechanism is essentially the same nucleophilic addition–elimination mechanism as the one by which LiAlH₄ operates, the only difference being the rate.

> Reduction by lithium tri-*tert*-butoxyaluminum hydride occurs much more slowly than by LiAlH₄, allowing the reaction to take place in a controlled fashion.

Thus, an aldehyde is less reactive than an acid chloride, so the second reduction is slower than the first. Once the first reduction has come to completion, the reaction can be stopped before the second reduction can proceed.

There are two reasons why reduction with lithium tri-*tert*-butoxyaluminum hydride is significantly slower than reduction with LiAlH₄. One is the bulkiness of the *tert*-butoxy groups, which introduces significant steric hindrance in the nucleophilic addition step. The second reason is that the reaction is carried out at very cold temperatures, around −75 °C. At room temperature, lithium tri-*tert*-butoxyaluminum hydride can rapidly reduce an acid chloride twice, all the way down to the alcohol (Equation 20-30).

At room temperature, LiAlH(O-*t*-Bu)₃ reduces an acid chloride twice, producing a primary alcohol.

1. LiAlH(O-*t*-Bu)₃
   25 °C
2. HCl

(20-30)

The reduction of an ester by DIBAH stops at the aldehyde stage *not* because the second reduction is slower than the first, but because the tetrahedral intermediate that is formed (Equation 20-31) is relatively stable at −78 °C. In that tetrahedral intermediate, the O atom from the initial carbonyl group has formed a bond to Al, much as we have seen previously in other mechanisms. This species persists until H_3O^+ is added. Effectively, H_3O^+ "washes away" the Al-containing species, generating a hemiacetal that equilibrates with the aldehyde. Consequently,

> In the DIBAH reduction of an ester, the aldehyde product and the reducing agent are not present in the reaction mixture at the same time, so the aldehyde cannot be further reduced to the alcohol.

This tetrahedral intermediate is stable at −78 °C and remains until H₃O⊕ is added.

(20-31)

Nucleophilic addition

Excess water converts a hemiacetal into the aldehyde.

Hemiacetal

Aldehyde

The relative stability of the tetrahedral intermediate in the DIBAH reduction of an ester can be rationalized in part by charge stability. As shown in Equation 20-32, elimination of the alkoxy leaving group from the tetrahedral intermediate would generate two additional charges: a negative charge on the O atom of the leaving group (RO⁻) and a positive charge on the O atom bonded to Al.

Elimination of RO⊖ is relatively slow in part because two additional charges appear in the products.

Slow

(20-32)

SOLVED problem 20.21 Show how you would convert the carboxylic acid at the right into the corresponding aldehyde.

Think What reactions do we know that can convert an acid derivative into an aldehyde? Can this be accomplished directly from a carboxylic acid?

Solve We just learned that an aldehyde can be produced from an acid chloride using lithium tri-*tert*-butoxyaluminum hydride, or from an ester using DIBAH. Thus, we must convert the initial carboxylic acid into an acid chloride or an ester. We have not yet learned a way to convert a carboxylic acid into an acid chloride, but we have seen that diazomethane can convert the carboxylic acid into a methyl ester.

The synthesis would then be written as follows:

problem 20.22 Show how to carry out the following synthesis.

20.8 Organometallic Reagents

Like hydride (H^-) anions, alkyl anions (R^-) can act as nucleophiles but not as leaving groups. As we have seen in previous chapters, sources of R^- are organometallic compounds such as alkyllithium reagents (RLi), Grignard reagents (RMgBr), and lithium dialkylcuprates (R_2CuLi). Thus, as with hydride reagents, the product of nucleophilic addition–elimination involving one of these organometallic reagents is *not* an acid derivative. We can see this explicitly in Equations 20-33 and 20-34, in which an acid derivative is treated with an alkyllithium and a Grignard reagent, respectively.

Two equivalents of $R^{\ominus}$ have added in.

(20-33)

Methyl hept-6-enoate

2-Methyloct-7-en-2-ol
99%

Two equivalents of $R^{\ominus}$ have added in.

(20-34)

Benzoyl chloride

3-Phenylpentan-3-ol

In general,

Acid chlorides ($RCOCl$), acid anhydrides (RCO_2COR), and esters (RCO_2R) can be treated with an alkyllithium ($R'Li$) or a Grignard ($R'MgBr$) reagent to produce a tertiary alcohol ($R_2'RCOH$).

In each case, two equivalents of the R⁻ nucleophile must add to the acid derivative to produce a tertiary alcohol. This is explained by Equation 20-35, the mechanism of the reaction in Equation 20-34.

Mechanism for Equation 20-34

(20-35)

3° Alcohol

This mechanism is essentially the same as the one in Equation 20-20, which describes the hydride reduction of an acid chloride to a primary alcohol. The first two steps make up the usual nucleophilic addition–elimination mechanism, which produces a ketone in this case. Once that ketone is produced, it reacts with a second equivalent of R⁻ in Step 3 to produce a tertiary alkoxide anion, RO⁻. Acid workup in Step 4 yields the tertiary alcohol as the final product. Notice that Steps 3 and 4 compose the mechanism for a typical alkyllithium or Grignard reaction involving a ketone, which we previously discussed in Section 17.5.

YOUR TURN 20.13

Draw the mechanism for the reaction in Equation 20-33.

Although similar mechanisms can be drawn involving carboxylic acids or amides, these reactions are generally avoided because amides are significantly less reactive than esters, and carboxylic acids are acidic, so they can protonate the R⁻ nucleophile.

problem **20.23** Draw the complete, detailed mechanism and predict the major products for each of the following reactions.

(a)

(b)

As with the hydride reductions in Section 20.7, it can be advantageous to carry out reactions in which only one equivalent of R⁻ adds to an acid derivative, thus producing a ketone that can be isolated. One way to carry out this kind of a reaction involves lithium dialkylcuprates, as shown in Equation 20-36.

A ketone does not react further with R₂CuLi.

Ethanoyl chloride
(Acetyl chloride)

(20-36)

Hept-6-en-2-one
74%

Thus,

> An acid chloride (RCOCl) can generally be treated with a lithium dialkylcuprate (R₂CuLi) to produce a ketone (R₂C=O).

As we saw in Chapter 17, lithium dialkylcuprates are relatively weak R⁻ nucleophiles, so they do not directly add to the carbonyl group of a ketone or aldehyde. They will, however, attack the carbonyl group of an acid chloride, because it is much more reactive than a ketone or aldehyde. The nucleophilic addition–elimination product is a ketone, which does not react further with the lithium dialkylcuprate.

Lithium dialkylcuprates do not react with esters or amides, which are significantly more stable than acid chlorides. Thus, to convert an ester or amide into a ketone, we must first convert it into an acid chloride. We discuss how to do so in Chapter 21.

problem **20.25** Draw the complete, detailed mechanism and the product for each of the following reactions. If no reaction takes place, write "no reaction."

(a)

(b)

(c)

SOLVED problem **20.26** Show how to carry out the synthesis at the right.

Think What precursor can be used to produce the α,β-unsaturated ketone? To make that precursor from the starting compound, which carbonyl-containing functional group must gain an additional carbon–carbon bond? What reagent can be used to selectively react with that functional group?

Solve The α,β-unsaturated ketone can be produced from an aldol condensation, as shown in the following retrosynthetic analysis.

To make the precursor from the starting material, the Cl of the acid chloride must be replaced by a CH_3 group, which can be accomplished using $(CH_3)_2CuLi$. In the forward direction, the synthesis might be written as follows:

problem **20.27** Show how to carry out the synthesis at the right.

20.9 Wrapping Up and Looking Ahead

The main thrust of Chapter 20 is the general nucleophilic addition–elimination mechanism involving a carbonyl (C=O) group, shown previously in Equation 20-2. With a sufficient leaving group bonded to the carbonyl carbon, a strong nucleophile can add to the carbonyl carbon in a first step, temporarily producing a tetrahedral intermediate. In a second step, the tetrahedral intermediate collapses to regenerate the carbonyl group and eliminate the leaving group. The net result is a nucleophilic acyl substitution.

Much of Chapter 20 focused on the conversion of one acid derivative to another, which is observed if the reactant nucleophile can also act as a leaving group on a carbonyl carbon. For these conversions, the "stability ladder" in Figure 20-7 indicates whether the reaction is energetically favorable. An interconversion is energetically favorable, and thus relatively easy to carry out, if the acid derivative that is produced is on a lower rung of the ladder than the starting acid derivative. Conversely, an interconversion is unfavorable, and thus more difficult to carry out, if the product derivative is on a higher rung of the ladder. Reactions in which the initial and final derivatives are on the same rung of the ladder, such as transesterification, tend to take place under equilibrium conditions and are reversible.

We also saw acyl substitutions in which either the carbonyl-containing reactant or the carbonyl-containing product is *not* an acid derivative. For example, the haloform reaction converts a methyl ketone into a carboxylic acid, and is facilitated by alpha halogenation, which first transforms the very poor H_3C^- leaving group into a decent X_3C^- leaving group. Other examples involve H^- and R^- nucleophiles, which cannot act as leaving groups in the product. Acyl substitution involving one of these nucleophiles produces either an aldehyde or a ketone, respectively, and under normal conditions, the initial aldehyde or ketone is attacked a second time to produce a primary alcohol or tertiary alcohol.

Chapter 20 is the first of two chapters that discuss the nucleophilic addition–elimination mechanism. In Chapter 21, we extend our discussion to include reactions involving weak nucleophiles. Many of the principles we learned here in Chapter 20 are applied to those reactions as well, although the mechanisms involving weak nucleophiles are somewhat different. Moreover, the lower reactivity of a weak nucleophile has a significant impact on the feasibility of acyl substitution reactions, as well as on the conditions necessary to carry out some of those reactions.

Chapter Summary and Key Terms

- The general **nucleophilic addition–elimination mechanism** consists of two steps:

First a nucleophile attacks the electron-poor atom of a polar π bond in a nucleophilic addition step, producing a high-energy **tetrahedral intermediate**. Second, the tetrahedral intermediate eliminates a leaving group (L) initially bonded to the electron-poor atom of the polar π bond. (**Sections 20.1 and 20.1a**)

- When the leaving group is bonded to a carbonyl group, nucleophilic addition–elimination is called a **nucleophilic acyl substitution**. (**Section 20.1a**)

- In a **base-promoted transesterification reaction**, an ester reacts with an alkoxide anion (RO⁻) to produce a new ester. (**Section 20.1a**)

- The first step of a nucleophilic addition–elimination reaction is usually the rate-determining step. (**Section 20.1b**)

- A transesterification reaction is reversible because the reactants and products have roughly equal stabilities. (**Section 20.1c**)

- In a **saponification reaction**, an ester reacts with HO⁻ to produce an initial carboxylic acid that is rapidly and irreversibly deprotonated under the basic conditions of the reaction. (**Section 20.2**)

- The stability of a carboxylic acid derivative decreases in the order: amide > ester ≈ carboxylic acid > acid anhydride > acid chloride. (**Section 20.3**)

- The conversion of one carboxylic acid derivative into another can be carried out with relative ease if the acid derivative on the product side is more stable than the one on the reactant side. By contrast, the conversion is difficult if the acid derivative on the product side is higher in energy than the one on the reactant side. (**Section 20.3**)

- The conversion of an amide (RCONR₂) to a carboxylic acid (RCO₂H) via nucleophilic addition–elimination is energetically unfavorable, but it can be carried out with relative ease by treating the amide with HO⁻. The initial carboxylic acid that is formed is rapidly and irreversibly deprotonated under the basic conditions of the reaction. (**Section 20.4a**)

- The **Gabriel synthesis** produces a primary amine from a corresponding alkyl halide. (**Section 20.4**)

- A **haloform reaction** converts a methyl ketone (RCOCH₃) into a carboxylic acid (RCO₂H). The CH₃ group is an unsuitable leaving group, but under the conditions of the reaction, it is first converted into CX₃ (X = Cl, Br, I), which is a suitable leaving group. Subsequent acyl substitution involving HO⁻ as the nucleophile produces an initial carboxylic acid that is rapidly and irreversibly deprotonated under the basic conditions of the reaction. (**Section 20.5**)

- NaBH₄ readily reduces high-energy acid derivatives, such as acid chlorides (RCOCl) and acid anhydrides (RCO₂COR), to primary alcohols. Esters, which are significantly more stable, are reduced to primary alcohols more slowly. All of these reactions proceed through an aldehyde intermediate, which is reduced a second time to the alcohol. (**Sections 20.6**)

- LiAlH₄ is a more powerful reducing agent than NaBH₄, and can thus reduce acid chlorides, acid anhydrides, and esters to primary alcohols, too. (**Section 20.6b**)

- LiAlH₄ reduces carboxylic acids (RCO₂H) to primary alcohols, and reduces amides (RCONR₂) to amines (RCH₂NR₂). These reactions are facilitated by the strong O—Al bond that forms in the mechanism. (**Section 20.6b**)

- **Diisobutylaluminum hydride (DIBAH)** and **lithium tri-*tert*-butoxyaluminum hydride** are two specialized reducing agents that can be used to reduce an acid derivative to an aldehyde. DIBAH reduces an ester to an aldehyde, whereas lithium tri-*tert*-butoxyaluminum hydride reduces an acid chloride to an aldehyde. (**Section 20.7**)

- When an acid derivative such as an acid chloride, acid anhydride, or ester is treated with an alkyllithium reagent (RLi) or a Grignard reagent (RMgBr), R⁻ adds twice to produce a tertiary alcohol. (**Section 20.8**)

- When an acid chloride (RCOCl) is treated with a lithium dialkylcuprate, R⁻ adds once to produce a ketone. (**Section 20.8**)

Reaction Tables

TABLE 20-2 Functional Group Conversions

	Starting Functional Group	Typical Reagents and Reaction Conditions	Functional Group Formed	Key Electron–Rich Species	Key Electron–Poor Species	Comments	Discussed in Section
(1)	Ester	NaOR″ / Ether	Ester	$^{\ominus}OR''$		Nucleophilic addition–elimination (transesterification)	20.1a
(2)	Ester	1. NaOH / 2. H$^{\oplus}$	Carboxylic acid	$^{\ominus}OH$		Nucleophilic addition–elimination (saponification)	20.2
(3)	Acid chloride	LiOR′	Ester	$^{\ominus}OR'$		Nucleophilic addition–elimination	20.3
(4)	Acid anhydride	NaOR′	Ester	$^{\ominus}OR'$		Nucleophilic addition–elimination	20.3
(5)	Ester	LiNR″$_2$	Amide	$^{\ominus}NR''_2$		Nucleophilic addition–elimination	20.3
(6)	Amide	1. NaOH / 2. H$^{\oplus}$	Carboxylic acid	$^{\ominus}OH$		Nucleophilic addition–elimination	20.4
(7)	Phthalimide	1. KOH/EtOH / 2. RBr / 3. KOH/H$_2$O	Primary amine		$R{-}Br$	S$_N$2, then nucleophilic addition–elimination (Gabriel synthesis)	20.4
(8)	Acid chloride	1. NaBH$_4$ or LiAlH$_4$ / 2. H$^{\oplus}$	Primary alcohol	$^{\ominus}H$		Nucleophilic addition–elimination, then addition (reduction)	20.6
(9)	Acid anhydride	1. NaBH$_4$ or LiAlH$_4$ / 2. H$^{\oplus}$	Primary alcohol	$^{\ominus}H$		Nucleophilic addition–elimination, then addition (reduction)	20.6
(10)	Ester	1. NaBH$_4$ or LiAlH$_4$ / 2. H$^{\oplus}$	Primary alcohol	$^{\ominus}H$		Nucleophilic addition–elimination, then addition (reduction); very slow with NaBH$_4$	20.6

(continued)

TABLE 20-2 Functional Group Conversions (continued)

	Starting Functional Group	Typical Reagents and Reaction Conditions	Functional Group Formed	Key Electron–Rich Species	Key Electron–Poor Species	Comments	Discussed in Section
(11)	Carboxylic acid	1. LiAlH$_4$ 2. H$^{\oplus}$	Primary alcohol	$^{\ominus}$H		Nucleophilic addition–elimination, then addition (reduction)	20.6
(12)	Amide	LiAlH$_4$ Ether	Amine	$^{\ominus}$H		Nucleophilic addition–elimination, then addition (reduction)	20.6
(13)	Acid chloride	LiAlH(O-t-Bu)$_3$ −78 °C	Aldehyde	$^{\ominus}$H		Nucleophilic addition–elimination (reduction); note cold T	20.7
(14)	Ester	1. DIBAH −78 °C 2. H$_3$O$^{\oplus}$	Aldehyde	$^{\ominus}$H		Nucleophilic addition–elimination (reduction); note cold T	20.7

TABLE 20-3 Reactions That Alter the Carbon Skeleton[a]

	Starting Functional Group	Typical Reagents and Reaction Conditions	Functional Group Formed	Key Electron–Rich Species	Key Electron–Poor Species	Comments	Discussed in Section
(1)	Methyl ketone	1. X$_2$, NaOH 2. H$^{\oplus}$	Carboxylic acid	$^{\ominus}$OH		Halogenation, then nucleophilic addition–elimination (haloform reaction)	20.5
(2)	Acid chloride	1. R′—Li or R′—MgBr 2. H$^{\oplus}$	Tertiary alcohol	$^{\ominus}$R′		Nucleophilic addition–elimination, then addition (alkyllithium or Grignard reaction)	20.8
(3)	Acid anhydride	1. R′—Li or R′—MgBr 2. H$^{\oplus}$	Tertiary alcohol	$^{\ominus}$R′		Nucleophilic addition–elimination, then addition (alkyllithium or Grignard reaction)	20.8
(4)	Ester	1. R′—Li or R′—MgBr 2. H$^{\oplus}$	Tertiary alcohol	$^{\ominus}$R′		Nucleophilic addition–elimination, then addition (alkyllithium or Grignard reaction)	20.8
(5)	Acid chloride	R′$_2$CuLi	Ketone	$^{\ominus}$R′		Nucleophilic addition–elimination	20.8

[a] X = Cl, Br, I

Problems

20.28 Predict the product for the reaction between *m*-ethylbenzoyl chloride and each of the following compounds. Draw the complete, detailed mechanism for each reaction. If no reaction is expected to occur, write "no reaction."

m-Ethylbenzoyl chloride

(a) NaOH, then H_3O^+ (b) CH_3NHLi (c) CH_3CH_2OK
(d) $C_6H_5CO_2K$ (e) $(CH_3CH_2)_2CuLi$ (f) $LiAlH(O\text{-}t\text{-}Bu)_3$, -78 °C
(g) CH_3Cl (h) $NaBH_4$, EtOH (i) C_6H_5MgBr (excess), then H^+
(j) CH_3OCH_3

20.29 Predict the product for the reaction between acetic anhydride and each of the following compounds. If no reaction is expected to occur, write "no reaction." For those reactions that do occur, draw the complete, detailed mechanism.

(a) NaOH, then H_3O^+ (b) CH_3NHLi (c) CH_3CH_2OK
(d) $C_6H_5CO_2K$ (e) NaBr (f) $CH_3CH_2OCH_2CH_3$
(g) 3-Chloropentane (h) Hexanal

20.30 Predict the product for the reaction between methyl cyclohexylmethanoate and each of the following. If no reaction is expected to occur, write "no reaction." For those reactions that do occur, draw the complete, detailed mechanism.

(a) NaOH, then H_3O^+ (b) $CH_3CH_2CH_2ONa$, $CH_3CH_2CH_2OH$ (c) $LiAlH_4$, then H_3O^+
(d) $CH_3CH_2CH_2Li$ (excess), then H_3O^+ (e) $(CH_3CH_2)_2CuLi$ (f) NaBr
(g) 3-Chloropentane (h) Hexanal (i) CH_3Cl

20.31 Predict the product of the following sequence of reactions.

1. $LiAlH_4$

2.

20.32 Starting with acetyl chloride and any other reagents necessary, how would you synthesize each of the following compounds?

(a)   (b)   (c)   (d)

(e)   (f) (g) (h)

20.33 Using acetyl chloride as your only source of carbon, propose a synthesis for each of the following compounds. You may use any inorganic reagents necessary.

(a) Ethyl acetate (b) Butan-2-ol (c) 3-Methylpentan-3-ol
(d) Butan-2-one (e) Ethanamine (f) Acetic anhydride

20.34 Which of the following esters, **A** or **B**, will undergo saponification faster? Why?

A   **B**

20.35 Show how to synthesize the following molecule from any compounds containing two carbons. Draw the complete, detailed mechanism for the reaction.

20.36 Show how to synthesize 2,4-diphenylbut-2-ene using phenylethanoic acid (phenylacetic acid) as your only source of carbon atoms.

Phenylethanoic acid
(Phenylacetic acid)

2,4-Diphenylbut-2-ene

20.37 *N,N*-Diacylamides can be prepared by treating an acyl chloride with lithium nitride in a 3-to-1 ratio:

An *N,N*-diacylamide

Draw a complete, detailed mechanism for this reaction.

20.38 Barbituric acid can be prepared from malonic ester and urea as follows:

Barbituric acid

Provide a complete, detailed mechanism for this reaction.

20.39 Enamines can react with acyl chlorides via nucleophilic addition–elimination, such as in the synthesis of the following 1,3-diketone.

Provide a complete, detailed mechanism for this transformation.

20.40 Amidines have the general form, R–C(=NH)–NH₂, so they are nitrogen analogs of carboxylic acids. As shown here, they can be hydrolyzed under basic conditions to form amides.

Propose a mechanism showing the conversion of an amidine to an amide.

20.41 **(a)** Propose a mechanism for the substitution reaction at the right.

(b) Explain why the reaction at the right does not lead to a similar substitution.

No reaction

20.42 Propose a mechanism for the following reaction.

20.43 If a lactam (cyclic amide) contains an alkyl group with an amine functional group, treatment with lithium diisopropyl amide (LDA) results in a ring-expanded lactam, as shown here.

Provide a detailed mechanism for this reaction.

20.44 Draw the complete, detailed mechanism for the following reaction.

20.45 Draw a complete, detailed mechanism to account for the incorporation of ¹⁸O twice into the acetate anion in the following reaction.

20.46 Which of the following compounds will form a yellow solid when dissolved in a basic, aqueous solution of I₂?

(a) Butanoic acid   **(b)** Pentan-2-one   **(c)** Pentan-3-one
(d) Cyclohexanone   **(e)** Pentanal

20.47 Show how to synthesize each of the following amines from an alkyl halide via a Gabriel synthesis.

(a)

(b)

(c)

20.48 Supply the missing compounds **A** through **D**.

$$\xrightarrow[\text{2. H}_3\text{O}^{\oplus}]{\text{1. I}_2, \text{ NaOH}} \textbf{A} \xrightarrow[\text{2. H}_3\text{O}^{\oplus}]{\text{1. LiAlH}_4, \text{ THF}} \textbf{B} \xrightarrow{\text{PBr}_3} \textbf{C}$$

1. NaOH
2. **C** ⟵ ········ → **D**
3. NaOH, Δ   Same compound as above

20.49 Supply the missing compounds **A** through **G**.

$$\xrightarrow[\text{2. H}_3\text{O}^{\oplus}]{\text{1. DIBAH}} \textbf{A} \xrightarrow[\text{NaOH, }\Delta]{\textbf{B}}$$

1. **C**
2. NH₄Cl

$$\xrightarrow[\text{EtOH}]{\text{NaBH}_4} \textbf{D} \xrightarrow{\text{PBr}_3} \textbf{E}$$

1. NaOH
2. **E** ⟵ ········
3. NaOH, Δ   Same compound as above

→ **F** $\xrightarrow[\text{NaOH, }\Delta]{\text{CH}_3\text{Br (excess)}}$ **G**

20.50 Supply the missing compounds **A** through **D**.

$$\xrightarrow[\text{2. H}_3\text{O}^{\oplus}]{\text{1. NaOH, }\Delta} \textbf{A} \xrightarrow{\text{CH}_2\text{N}_2}$$

B

$$\xrightarrow[\text{2. H}_3\text{O}^{\oplus}]{\text{1. } \nwarrow \text{MgBr}} \textbf{C} \xrightarrow[\text{2.}]{\text{1. NaH}} \textbf{D}$$

20.51 Supply the missing compounds **A** through **G**.

$$\text{(3-methyl-1-butene)} \xrightarrow[\text{2. NaOH, H}_2\text{O}_2]{\text{1. BH}_3\bullet\text{THF}} \mathbf{A} \xrightarrow{\text{KMnO}_4} \mathbf{B} \xrightarrow{\text{CH}_2\text{N}_2} \mathbf{C} \xrightarrow[\text{2. H}_3\text{O}^\oplus]{\text{1. C}_6\text{H}_5\text{MgBr}} \mathbf{D} \xrightarrow[\Delta]{\text{conc. H}_3\text{PO}_4}$$

$$\mathbf{E} \xrightarrow[\text{2. NaOH, H}_2\text{O}_2]{\text{1. BH}_3\bullet\text{THF}} \mathbf{F} \xrightarrow[\text{2. } \mathbf{C}]{\text{1. NaOH}} \mathbf{G} \text{-------- Same compound as above}$$

20.52 The following is a proposed synthesis of a thioester from an ester.

(a) Draw the mechanism for this reaction.
(b) Would this reaction be energetically favorable? Why or why not?

20.53 Ancient civilizations synthesized soap by heating animal fat with wood ash, a source of HO⁻. Animal fat consists of triesters, known as triglycerides, an example of which is shown here on the left.

$$\text{(triglyceride)} \xrightarrow{\text{KOH (excess)}} \textbf{?}$$

(a) Assuming that the triglyceride reacts completely with HO⁻, what are the products of this reaction?
(b) Draw the complete, detailed mechanism that leads to those products.
(c) Explain how those products can serve as soap. (*Hint*: Review Section 2.10.)

20.54 In a Gabriel synthesis, the yield of the amine can be compromised when hydrolysis is carried out under very basic conditions. The problem, in particular, is with the second hydrolysis (Steps 6–8 in Equation 20-13), not the first (Steps 3–5). Explain why.

20.55 A pain reliever has the formula $C_8H_9NO_2$. Its IR spectrum is as follows:

When it is heated in water under basic conditions, followed by an acid workup, two compounds are formed—acetic acid and a compound whose formula is C_6H_7NO and whose ^1H NMR spectrum is shown below. Draw the structure for this compound and for the pain reliever.

Chemical shift (ppm)

20.56 When methyl 5-oxopentanoate is treated with vinyl Grignard, a compound is produced whose formula is $C_7H_{10}O_2$.

Methyl 5-oxopentanoate

In the IR spectrum of $C_7H_{10}O_2$, an intense absorption appears at 1740 cm^{-1} and a weaker absorption appears at 1650 cm^{-1}. Seven signals appear in the ^{13}C NMR spectrum of $C_7H_{10}O_2$. The DEPT spectrum shows that there is one carbon that is bonded to no hydrogens, two CH carbons, and four CH_2 carbons. Draw the structure of $C_7H_{10}O_2$.

20.57 When dimethyl phthalate is treated with lithium aluminum hydride, a compound is produced whose ^1H NMR spectrum is shown below. Determine the product, and draw the complete, detailed mechanism of the reaction that produces it.

Dimethyl phthalate

Chemical shift (ppm)

If your instructor assigns problems in smartwork, log in at **smartwork.wwnorton.com**.

Problems / **1005**

Nucleophilic Addition–Elimination Reactions 2

Weak Nucleophiles

Chapter 20 discussed reactions that proceed by the nucleophilic addition–elimination mechanism, which consists of the following two steps:

Nucleophilic addition–elimination mechanism

A nucleophile attacks the electron-poor atom.

The leaving group is eliminated, which regenerates the π bond.

1. Nucleophilic addition

2. Nucleophile elimination

Tetrahedral intermediate

In Step 1, a nucleophile attacks the electrophilic atom of a polar π bond to generate an unstable tetrahedral intermediate. In Step 2, a leaving group is expelled from the tetrahedral intermediate, thus regenerating the π bond.

The specific reactions we examined in Chapter 20 were limited to nucleophilic acyl substitutions involving *strong nucleophiles* as overall reactants. In a transesterification reaction, for example, an alkoxide anion (RO⁻) serves as the nucleophile, and in a Grignard reaction involving an ester, an alkyl or aryl anion (R⁻) serves as the nucleophile.

Kevlar® is a strong, lightweight, flexible material commonly used to make body armor. Kevlar® is produced from an aminolysis reaction, a type of nucleophilic addition–elimination reaction we discuss here in Chapter 21.

Here in Chapter 21, we expand our discussion of nucleophilic addition–elimination reactions to include ones that involve *weak nucleophiles* as overall reactants. Fundamentally, nucleophilic addition–elimination reactions involving weak nucleophiles are quite similar to ones involving strong nucleophiles, but there are some key differences. Nucleophilic addition–elimination reactions involving weak nucleophiles tend to proceed much slower, for example, than ones involving strong nucleophiles. As a result, many of these reactions require acid or base catalysis. Mechanisms that describe reactions involving weak nucleophiles, moreover, tend to consist of more steps.

Most of our discussion will concern reactions in which a nucleophile attacks the carbonyl carbon of an acid derivative or a carboxylic acid, leading to acyl substitution. However, we will also examine reactions that involve nucleophilic attack at other polar π bonds, such as S=O. Some reactions we will examine are functional group conversions, and some alter a molecule's carbon skeleton. Given this variety of reactions, it will be particularly important for you to maintain focus on the mechanism, so that you can see clearly how all of these reactions are related.

21.1 The General Nucleophilic Addition–Elimination Mechanism Involving Weak Nucleophiles: Alcoholysis and Hydrolysis of Acid Chlorides

In Section 20.3, we learned that an ester such as methyl benzoate can be produced from benzoyl chloride using methoxide anion (CH_3O^-), a relatively *strong* nucleophile (Equation 21-1). The mechanism for this reaction is simply the two-step nucleophilic addition–elimination mechanism just presented in the Introduction.

Benzoyl chloride

Methyl benzoate
84%

An ester can also be produced from an acid chloride by treating the acid chloride with an alcohol, a relatively *weak* nucleophile. An example is shown in Equation 21-2. Similarly, treating an acid chloride with water (another relatively weak nucleophile) produces a carboxylic acid, as shown in Equation 21-3.

Upon completing Chapter 21 you should be able to:

- Draw the general mechanism for the nucleophilic addition–elimination reaction of a weak nucleophile with a highly reactive polar π bond.

- Rank the relative reactivities of carboxylic acid derivatives in nucleophilic addition–elimination reactions and explain what gives rise to that order.

- Explain why the aminolysis of an acid chloride requires two molar equivalents of an amine unless one molar equivalent of a base such as pyridine or triethylamine is present.

- Show how to convert a carboxylic acid to an acid halide and explain the significance of this reaction in the context of producing other carboxylic acid derivatives and halogenating the α carbon of a carboxylic acid.

- Draw the mechanism for the production of a sulfonate ester from the corresponding sulfonyl chloride and an alcohol, and explain the significance of this reaction in the context of nucleophilic substitution and elimination reactions.

- Determine whether a nucleophilic addition–elimination reaction involving an amide, ester, or carboxylic acid can be acid or base catalyzed, and draw the mechanisms for these reactions.

- Predict the products of a Baeyer–Villiger oxidation and explain the driving force for that reaction in the context of an acid-catalyzed nucleophilic addition–elimination reaction.

- Draw the mechanism for and predict the products of a Claisen condensation reaction.

- Incorporate Claisen and crossed Claisen reactions effectively in a synthesis.

- Determine whether an intramolecular Claisen condensation reaction will take place.

- Design a synthesis of an α-substituted acetone and an α-substituted acetic acid using an acetoacetic ester synthesis or a malonic ester synthesis, respectively.

- Design a synthesis that requires the protection of a carboxylic acid or amine.

An alcoholysis reaction

$$\text{Benzoyl chloride} \quad \xrightarrow{\text{HO} \left(\!\!\left.\right)_3\right.} \quad \text{4-Pentenyl benzoate} \quad + \quad \text{HCl} \quad \textbf{(21-2)}$$

Benzoyl chloride

4-Pentenyl benzoate
71%

A hydrolysis reaction

$$\text{Butanoyl chloride} \quad \xrightarrow{\text{H}_2\text{O}} \quad \text{Butanoic acid} \quad + \quad \text{HCl} \quad \textbf{(21-3)}$$

Butanoyl chloride

Butanoic acid
77%

Equation 21-2 is called **alcoholysis** because the addition of the alcohol results in the breaking of a bond (*lysis* in Greek means "breaking")—in this case, the C—Cl bond of the acid chloride. Equation 21-3 is called **hydrolysis** because the bond in the acid chloride is broken as a result of the addition of water.

The mechanism for the hydrolysis of an acid chloride is shown in Equation 21-4. The mechanism for alcoholysis is essentially the same, and is presented in Your Turn 21-1.

Mechanism for Equation 21-3

This is the slow step and is therefore rate determining.

Tetrahedral intermediate

The nucleophilic addition–elimination product bears a positive charge.

1. Nucleophilic addition

2. Nucleophile elimination

3. Proton transfer

(21-4)

After deprotonation, the organic product becomes uncharged.

Steps 1 and 2 of the mechanism in Equation 21-4 consist of the usual nucleophilic addition and elimination steps. H_2O attacks the carbonyl carbon in Step 1 to produce the tetrahedral intermediate, which subsequently expels the Cl^- leaving group in Step 2 to reform the carbonyl bond. In Step 3, however, a proton transfer takes place, which is necessary to produce an uncharged product; without it, the OH group in the product would remain an OH_2^+ group.

HCl appears as a product in the overall reaction in Equation 21-3, but it does not appear in the mechanism in Equation 21-4 because the reaction involves water, in which HCl, a strong acid, exists almost entirely as hydronium ion (H_3O^+) and chloride ion (Cl^-). This is consistent with the production of Cl^- in Step 2 of the mechanism and H_3O^+ in Step 3.

21.1 YOUR TURN

The mechanism for the alcoholysis reaction in Equation 21-2 is shown here, but the curved arrows have been omitted.

(a) Draw in the appropriate curved arrows.
(b) Write the name of the elementary step below each reaction arrow.
(c) Identify the tetrahedral intermediate.

Answers to Your Turns are in the back of the book.

---

problem **21.1** Draw the complete, detailed mechanism and predict the major organic products for each of the following reactions.

(a)   (b)

Just as we saw for the nucleophilic addition–elimination reactions in Chapter 20, the nucleophilic addition step (Step 1 in Equation 21-4) is usually slow and, therefore, is rate determining. The additional proton transfer step is fast, so it does not significantly affect the rate of the overall reaction. (See Your Turn 21.2.)

The free energy diagram for the hydrolysis reaction in Equations 21-3 and 21-4 is as follows:

Reaction coordinate ⟶

(a) Draw the species that appear in the mechanism (Equation 21-4) in the appropriate boxes provided, and label the tetrahedral intermediate. (*Hint*: Review Fig. 20-3.)

(b) Explain how this diagram is consistent with the nucleophilic addition step being the slow step.

---

problem **21.2** Draw the free energy diagram for the alcoholysis reaction in Equation 21-2. Include the species that appear in its mechanism (see Your Turn 21.1) and label the tetrahedral intermediate.

21.2 Relative Reactivities of Acid Derivatives: Rates of Hydrolysis

Section 21.1 highlighted alcoholysis and hydrolysis reactions involving acid chlorides only, but these reactions are not limited to just acid chlorides. As shown in Equations 21-5 and 21-6, acid anhydrides can readily undergo alcoholysis and hydrolysis, producing esters and carboxylic acids, respectively. The mechanisms for these reactions are essentially the same as the one in Equation 21-4 that describes the hydrolysis of an acid chloride.

Alcoholysis of an acid anhydride

(21-5)

80%

Hydrolysis of an acid anhydride

Phthalic anhydride $\xrightarrow{H_2O}$ **Phthalic acid**

(21-6)

---

problem **21.3** Draw the complete, detailed mechanisms for the reactions in Equations 21-5 and 21-6.

We can also envision similar reactions taking place when other acid derivatives, such as esters or amides, are treated with water. As indicated in Table 21-1, however,

Ester and amide hydrolysis reactions are much too slow under normal conditions to be useful for synthesis.

The relative hydrolysis rates of the species in Table 21-1 are essentially the rates of the rate-determining step—that is, nucleophilic addition of H_2O to the $C=O$

TABLE 21-1 Relative Reactivities of Various Carbonyl–Containing Species

Carbonyl–Containing Species[a]	Relative Rate of Hydrolysis
R–C(=O)–Cl	10^{11} — Hydrolysis takes place readily.
R–C(=O)–O–C(=O)–R	10^7
R–C(=O)–H R–C(=O)–R	–
R–C(=O)–OH R–C(=O)–O–R	1 — Too slow to be useful for synthesis.
R–C(=O)–NH₂ R–C(=O)–NHR R–C(=O)–NR₂	10^{-2}
R–C(=O)–O⊖	–

Susceptibility to nucleophilic attack

[a] The leaving groups are indicated in red.

FIGURE 21-1 Relative reactivities of the acid derivatives Free energy diagrams for just the nucleophilic addition steps in the hydrolysis of various acid derivatives. A slower rate of hydrolysis (Table 21-1) reflects a larger energy barrier for nucleophilic addition.

- Increasing $\Delta G^{\circ\ddagger}$
- Slower reaction

Acid chloride

Acid anhydride

Carboxylic acid/Ester

Amide

Carboxylate anion

Tetrahedral intermediate

Carboxylate anions have equivalent resonance structures, so they are the most stable.

(a)

Contribution by this resonance structure decreases in the order:

$$L = NH_2 > OH \approx OR > OCR > Cl$$

(b)

FIGURE 21-2 Resonance stabilization in the carboxylic acid derivatives The stabilization of the various carboxylic acid derivatives depends on the extent to which the π electrons from the C=O bond participate in resonance with a lone pair on the leaving group. (a) Carboxylate anions (RCO_2^-) are stabilized the most because they have equivalent resonance structures. (b) For uncharged acid derivatives, the contribution by the resonance structure—and therefore the stability of the carboxylic acid derivative—decreases in the order: amides > carboxylic acids ≈ esters > acid anhydrides > acid chlorides.

group. Therefore, as shown in Figure 21-1, the energy barrier for nucleophilic addition increases in the order: acid chlorides < acid anhydrides < carboxylic acids and esters < amides < carboxylate anions. The Hammond postulate (Section 9.3a) suggests, therefore, that the stabilities of the acid derivatives should increase in the same order.

These relative stabilities of the acid derivatives are an outcome of the extent to which the π electrons in the C=O group participate in resonance with a lone pair of electrons on the leaving group. As shown in Figure 21-2a, those π electrons are delocalized the most in a carboxylate anion because it has two equivalent resonance structures. For the uncharged carboxylic acid derivatives, that resonance structure contributes less because it has both a positive and a negative formal charge, as shown in Figure 21-2b. Moreover, the contribution by that resonance structure decreases in the order: amides > carboxylic acids ≈ esters > acid anhydrides > acid chlorides.

The contribution by that resonance structure is greater for an amide than an ester because a positive charge is better accommodated by N than by O. In an acid anhydride, that contribution is even less because the central O atom's lone pair is tied up in resonance with *the other* carbonyl group simultaneously. Finally, the contribution is the least in an acid chloride because the lone pair on Cl belongs to a different shell ($n = 3$) than the valence electrons of the C and O atoms making up the C=O group ($n = 2$). This leads to a relatively poor interaction among the valence orbitals, and thus diminishes any resulting stabilization.

Notice in Table 21-1 that acid chlorides and acid anhydrides are, in fact, more reactive than aldehydes and ketones. With relatively little resonance delocalization in acid chlorides and acid anhydrides, this can be explained by *inductive effects* from the leaving group, as illustrated in Figure 21-3. These *inductive effects increase the concentration of positive charge on the carbonyl carbon, making it more susceptible to nucleophilic attack.* By the same token, recall from Chapter 17 that inductive effects are also partly responsible for ketones being *less* reactive than aldehydes—the additional alkyl group in a ketone is electron donating and therefore serves to decrease the concentration of positive charge on the carbonyl carbon.

SOLVED problem 21.4 Is the hydrolysis of phenyl acetate faster or slower than the hydrolysis of ethyl acetate? Explain.

Phenyl acetate   **Ethyl acetate**

Think How do resonance and inductive effects play a role in the reactivity of carboxylic acid derivatives? Are the carbonyl groups stabilized differently by resonance? By inductive effects? How so?

Solve Phenyl acetate reacts faster because its carbonyl group is less stabilized by resonance than ethyl acetate's carbonyl group. In phenyl acetate, the lone pair of electrons on the O of the leaving group is partly tied up in resonance with the phenyl ring, making those electrons less available for resonance with the carbonyl group.

Resonance with the phenyl ring makes these electrons less available for resonance with the C=O group.

The inductive effects, on the other hand, are not expected to be much different in the two compounds, because both carbonyl groups are bonded to an alkyl group and an O atom.

The electronegativities of Cl and O increase the positive charge on C.

Electron–donating effects of R decrease the positive charge on C.

FIGURE 21-3 Inductive effects in various carbonyl-containing groups The relative reactivities of acid chlorides, acid anhydrides, aldehydes, and ketones are governed largely by inductive effects. The leaving groups in acid chlorides and acid anhydrides are electron withdrawing, so the concentration of positive charge on the carbonyl C is higher than it is in aldehydes. Thus, the C attracts nucleophiles more strongly, enhancing the rate of nucleophilic addition. By contrast, the electron-donating effect of an alkyl group decreases the concentration of positive charge on the carbonyl C, thus slowing the rate of nucleophilic addition.

problem 21.5 Which acid derivative in each of these pairs do you think undergoes hydrolysis more quickly? Explain.

A

or

B

(a)

C

or

D

(b)

E

or

F

(c)

21.3 Aminolysis of Acid Derivatives

Amines are even more nucleophilic than water and alcohols, so they, too, will react with acid chlorides and acid anhydrides. As shown in Equations 21-7 and 21-8, these **aminolysis** reactions produce amides.

Aminolysis

(21-7)

Benzoyl chloride

N-Decylbenzamide

(21-8)

Acetic anhydride

N,N–Diethylethanamide

The mechanisms for Equations 21-7 and 21-8 are similar to the one for the hydrolysis of an acid chloride, shown previously in Equation 21-4. Equation 21-9 shows the mechanism for the aminolysis of an acid chloride. Step 1 is nucleophilic addition of the amine to produce the tetrahedral intermediate, followed by elimination of Cl⁻ in Step 2. Finally, a proton transfer produces the uncharged product in Step 3.

Mechanism for Equation 21-7

(21-9)

An important consideration for aminolysis reactions is the number of equivalents of the nucleophile required.

Aminolysis reactions require two equivalents of the amine.

According to the mechanism, the first equivalent is used in Step 1, in which the amine, acting as a nucleophile, attacks the carbonyl carbon of the acid chloride. The second equivalent is used in Step 3, where it irreversibly deprotonates the elimination product from Step 2.

If the amine is readily available and inexpensive, it may be acceptable to carry out an aminolysis reaction that consumes two equivalents of the amine. If the amine is difficult to obtain or expensive, however, then it is advantageous to use only one equivalent. In these cases, another amine such as pyridine or triethyl-amine can be added to the reaction mixture to act as a base, freeing up the desired amine to act as the nucleophile. For example, the aminolysis reaction in Equation 21-10 requires only one equivalent of the amine when one equivalent of pyridine is also present. Similarly, the aminolysis reaction in Equation 21-11 requires only one equivalent of the desired amine in the presence of one equivalent of triethylamine.

Pyridine and triethylamine are good choices as bases in these reactions because neither compound's N atom is bonded to H. Thus, even though they can undergo nucleophilic addition–elimination with an acid chloride or acid anhydride, a final proton transfer cannot take place to produce a stable uncharged amide. (This is explored further in Problem 21.65 at the end of the chapter.)

problem **21.7** Draw the mechanism for each of the following reactions and predict the major product.

(a)

(b)

21.4 Synthesis of Acid Halides: Getting to the Top of the Stability Ladder

Recall from Section 20.3 that an acid chloride appears at the top of the stability ladder and can thus be converted into any of the other acid derivatives with relative ease. This is true not only for reactions involving strong nucleophiles, but also those involving weaker nucleophiles, such as water, alcohols, and amines. In light of this, it can be extremely useful to produce acid chlorides from compounds that are more readily available.

As shown in Equations 21-12 and 21-13, **thionyl chloride** ($SOCl_2$) and phosphorus trichloride (PCl_3) can be used to convert carboxylic acids into acid chlorides. Of these, thionyl chloride is the more widely used reagent, due largely to the fact that the two byproducts, SO_2 and HCl, are gases that irreversibly bubble out of solution as they are produced.

Gaseous SO_2 and HCl bubble out of solution *irreversibly*.

3,5-Dinitrobenzoic acid

3,5-Dinitrobenzoyl chloride
90%

(21-12)

(21-13)

97%

The mechanism for Equation 21-12 is shown in Equation 21-14. It consists of back-to-back nucleophilic addition–elimination mechanisms. The first addition–

elimination mechanism involves the carboxylic acid and $SOCl_2$, in which the carboxylic acid behaves as the nucleophile, with its OH group attacking the polar $S{=}O$ bond of $SOCl_2$. In the next step, a Cl^- leaving group is eliminated from the intermediate, thereby regenerating the $S{=}O$ bond. Thus, Steps 1 and 2 in this mechanism resemble Steps 1 and 2 in the acid chloride hydrolysis we saw previously in Equation 21-4.

Mechanism for Equation 21-12

(21-14)

The product of Step 2 contains a very good leaving group, HOSOCl, bonded to the carbonyl C. This sets the stage for a second nucleophilic addition–elimination reaction (Steps 3 and 4), in which the Cl^- anion generated in Step 2 replaces HOSOCl. Finally, in Step 5, HOSOCl decomposes to SO_2 and HCl.

The mechanism for the reaction involving PCl_3 is shown in Your Turn 21.3, and is similar to the one involving $SOCl_2$ (Equation 21-14). The difference is that PCl_3 does not possess a polar π bond, so Cl^- is generated from an S_N2 step instead of an addition–elimination mechanism. This is essentially identical to the way in which Cl^- is produced from PCl_3 in the chlorination of an alcohol, which we examined previously in Section 10.1. Once the Cl^- nucleophile is generated, however, it enters a nucleophilic addition–elimination mechanism with the carbonyl group to produce the carboxylic acid chloride.

The mechanism for the reaction in Equation 21-13 is shown here, but the curved arrows have been omitted.

(a) Supply the missing curved arrows and label each elementary step below its respective reaction arrow.
(b) Identify the leaving group in the product of the first step.

SOLVED problem 21.8 Propose how you would carry out the synthesis at the right, in which an amide is converted into an acid anhydride.

Think What reactions do we know that transform an acid derivative lower on the stability ladder to one higher on the ladder? Can this transformation be done in a single step?

Solve We know that SOCl$_2$ will transform a carboxylic acid into an acid chloride, from which any other acid derivative can be formed. In this case, we could treat benzoyl chloride with sodium acetate to arrive at the final anhydride, via an acyl substitution. Benzoyl chloride could be formed from benzoic acid using SOCl$_2$. Benzoic acid, in turn, could be formed from benzamide via hydrolysis.

The overall synthesis would then be written as follows:

---

problem 21.9 Propose a synthesis that would carry out the transformation in Solved Problem 21.8, but does *not* involve benzoyl chloride as a synthetic intermediate.

21.5 The Hell–Volhard–Zelinsky Reaction: Synthesizing α-Bromo Carboxylic Acids

In Section 10.4, we learned that ketones and aldehydes undergo bromination at the α carbon by treating the ketone or aldehyde with molecular bromine (Br$_2$) in the presence of acid or base. In light of this, it might seem that similar conditions could be used to brominate carboxylic acids at the α carbon to produce an α-bromo carboxylic acid. As indicated in Equation 21-15, however, this does *not* occur.

Bromination of the α carbon does *not* take place under these conditions.

(21-15)

An α-bromoacid

We can understand why if we recall from Section 10.4 that α-bromination requires a key intermediate in which the α carbon is nucleophilic. Under basic conditions, that intermediate is an enolate anion, whereas it is an uncharged enol under acidic conditions. With carboxylic acids, however, neither of these intermediates is feasible. As shown in Figure 21-4a, the carboxyl hydrogen is deprotonated under basic conditions instead of the α hydrogen. And, under acidic conditions (Fig. 21-4b), the enol is unfavorable due to the resonance stabilization of the carbonyl group involving the hydroxyl group.

The enolate anion is not formed because the OH group is much more acidic.

The enol is not formed because the C=O group is involved in resonance with the OH group.

(a)

(b)

FIGURE 21-4 The diminished reactivity of carboxylic acids at the α carbon (a) Under basic conditions, the α H is not removed because the carboxyl H is more acidic. (b) Under acidic conditions, resonance disfavors the enol form.

A clever solution to this problem is to carry out the bromination in the presence of a phosphorus trihalide or elemental phosphorus, as shown in Equations 21-16 and 21-17. This kind of bromination is called a **Hell–Volhard–Zelinsky (HVZ) reaction**.

The Hell–Volhard–Zelinksky reaction

$$\text{(21-16)}$$

$$\text{(21-17)}$$

The role of PCl_3 in the reaction in Equation 21-16 is to convert the carboxylic acid to an acid chloride (Section 21.4), as shown in Equation 21-18.

Carboxylic acids are not enolizable. **Acid halides are enolizable.** **Hydrolysis**

$$\text{(21-18)}$$

In the acid chloride form, the enol becomes more favorable, which facilitates the substitution reaction in the subsequent steps to produce the α-bromo acid chloride. Then, treatment with water hydrolyzes the acid chloride to reform the carboxylic acid (Section 21.1).

Likewise, in Equation 21-17, the carboxylic acid is first converted to an acid bromide. Molecular bromine (Br_2) and elemental phosphorus react to form PBr_3, which behaves much like PCl_3.

The excellent leaving-group ability of Br^- makes α-bromoacids useful synthetic intermediates. For example, Equation 21-19 shows that they can be used to synthesize α-amino acids, the building blocks of proteins, via a nucleophilic substitution reaction with aqueous ammonia.

An α-amino acid

$$\text{(21-19)}$$

2-Bromo-3-phenylpropanoic acid   **Phenylalanine**
                                                   60%

(a)

(b)

(c)

21.6 Sulfonyl Chlorides: Synthesis of Mesylates, Tosylates, and Triflates

The three compounds in Figure 21-5 are sulfonyl chlorides, having the general form R—SO$_2$Cl. When one of these sulfonyl chlorides is treated with an alcohol, a **sulfonate ester** is produced via a sulfonation reaction, as shown in Equation 21-20.

(21-20)

Methanesulfonyl chloride (Mesyl chloride, MsCl)

(a)

Trifluoromethanesulfonyl chloride (Triflyl chloride, TfCl)

(b)

p-Toluenesulfonyl chloride (Tosyl chloride, TsCl)

(c)

FIGURE 21-5 Some common sulfonyl chlorides These sulfonyl chlorides can be used to make sulfonate esters, which have excellent leaving groups.

There is significant evidence suggesting that these sulfonation reactions proceed through an S$_N$2 mechanism, but there is also strong evidence supporting the nucleophilic addition–elimination mechanism in Equation 21-21.

Mechanism for Equation 21-20

(21-21)

This mechanism is essentially identical to the one shown previously in Equation 21-4 for the hydrolysis of acid chlorides, but in this case the nucleophilic OH group attacks the S=O group rather than a C=O group. Notice, too, that pyridine is added to the reaction to act as a base in Step 3. As we saw in Section 21.3, pyridine will not compromise the desired nucleophilic addition–elimination reaction because pyridine has no N—H proton that can be removed.

Recall from Chapter 9 that the OMs, OTf, and OTs groups are very good leaving groups for S$_N$2, S$_N$1, E2, and E1 reactions. Therefore,

Sulfonation of an alcohol converts a bad OH leaving group into a very good leaving group.

Moreover, notice in Equation 21-21 that no bonds to the alcohol C are broken or formed. Consequently,

> Sulfonation of an alcohol occurs with retention of configuration at the alcohol carbon.

A specific example is shown in Equation 21-22.

(21-22)

100%

Sulfonation of an alcohol, therefore, complements the PBr$_3$ bromination of an alcohol (Section 10.1), which converts a bad OH leaving group into a good Br leaving group via *inversion* of stereochemistry at the alcohol C.

SOLVED problem 21.11 What is the major product of the following sequence of reactions?

Think What is the product of sulfonation of an alcohol, and what is the stereochemistry that is associated with that reaction? What type of reaction is promoted by a strong, bulky base in the second step of this synthesis? What is the stereochemistry associated with that reaction?

Solve The first reaction is a sulfonation in which tosyl chloride (TsCl) converts the alcohol to a tosylate. Because the tosylate is an excellent leaving group, the presence of the strong, bulky base, (CH$_3$)$_3$CONa, promotes an E2 reaction. Recall that the alkene that is favored in an E2 reaction is produced from the substrate conformation in which the H and the leaving group on adjacent carbons are anti to each other. That conformation is the one shown below, so the major product is the *E* alkene.

---

problem 21.12 Show how to carry out the following synthesis.

21.7 Base and Acid Catalysis in Nucleophilic Addition–Elimination Reactions

Thus far, we have seen that weak nucleophiles, such as water, alcohols, and amines, react readily with acid halides, acid anhydrides, thionyl chloride, and sulfonyl chlorides. Even though the nucleophiles are weak, the polar π bonds (C=O and S=O) in these compounds are reactive enough to compensate. By contrast, we saw in Section 21.2 that esters are significantly less reactive at the carbonyl carbon due to resonance involving the leaving group. Thus, Equation 21-23 shows that esters do not undergo nucleophilic addition–elimination under normal conditions with weak nucleophiles like alcohols.

Transesterification takes place readily, however, when a small amount (i.e., a catalytic amount) of a base (sodium methoxide, CH_3ONa, in Equation 21-24) is added. Moreover, the base is not consumed overall in the reaction, so the reaction is *base catalyzed*.

The mechanism for this reaction, presented in Equation 21-25, shows why the reaction is catalyzed by the methoxide anion base.

Mechanism for Equation 21-24

The reaction rate is increased because the nucleophile that attacks the carbonyl carbon in the nucleophilic addition step (Step 1, the rate-determining step) is CH_3O^-, a strong nucleophile. In the absence of CH_3O^- (e.g., in Equation 21-23), the nucleophile would have to be methanol, a weak nucleophile. Moreover, there is no net consumption of CH_3O^- because it is regenerated in the proton transfer step (Step 3).

Because water and alcohols have similar behaviors, it might also seem that a base could be used to catalyze the interconversion between esters and carboxylic acids. But it can't. Although the presence of hydroxide facilitates the nucleophilic addition–elimination reaction, as shown in Equation 21-26, the carboxylic acid that is initially produced is acidic and thus reacts further with hydroxide. Consequently, the base is consumed in the overall reaction, making it a *base-promoted* reaction instead of a base-catalyzed reaction. In fact, this is an example of a *saponification reaction*, discussed previously in Section 20.2. As shown in Equation 21-27, a similar proton transfer takes place when a carboxylic acid is treated with an alkoxide anion, making it unfeasible to convert a carboxylic acid into an ester under basic conditions.

A rapid, irreversible proton transfer consumes the base.

(21-26)

A rapid, irreversible proton transfer consumes the base.

No further reaction

(21-27)

Transesterification can also be *acid* catalyzed. Equation 21-28 shows, for example, that a catalytic amount of a strong acid, such as sulfuric acid, facilitates the production of hexyl acetate from methyl acetate and hexan-1-ol.

Acid-catalyzed transesterification

(21-28)

Methyl acetate   Hexan-1-ol H_2SO_4 5 h, 50–95 °C   Hexyl acetate 54%   + $HOCH_3$

Catalytic amount of acid

The mechanism for this reaction is shown in Equation 21-29.

(21-29)

Sulfuric acid is a strong acid, so the predominant acid in solution, according to the leveling effect (Section 6.2c), is the protonated alcohol, $CH_3(CH_2)_5OH_2^+$. In Step 1 of the mechanism, that acid protonates the ester's carbonyl group, and in Step 2, the alcohol attacks the carbonyl C atom, yielding a tetrahedral intermediate. Steps 3 and 4 are proton transfers. In Step 3, the O atom from the incoming nucleophile becomes uncharged, and thus is stabilized. In Step 4, the singly bonded O atom of the original ester gains a positive charge, which increases its leaving group ability. In Step 5, the leaving group departs and the C=O bond is reformed, and in Step 6, the carbonyl O is deprotonated, yielding the overall uncharged product.

Why does the presence of a strong acid catalyze this reaction? Notice in Step 2 (the rate-determining step) that the nucleophile is a relatively weak alcohol, but the rate of nucleophilic addition is enhanced because the carbonyl group has become *activated* (Section 18.1) by the proton transfer in Step 1—that is, the carbonyl C has become more electrophilic. Notice, too, that the strong acid used in Step 1 is regenerated in Step 6, so it is not consumed in the overall reaction.

Each step of the acid-catalyzed transesterification reaction in Equation 21-29 is reversible, so the overall reaction is reversible. Therefore, according to Le Châtelier's principle, the product yield can be increased by adding reactants or removing products. Often the product alcohol (CH_3OH in Equation 21-28) is volatile and can be removed by distillation.

Unlike what we saw previously under basic conditions, the conversion of an ester to a carboxylic acid can also be acid catalyzed. Equation 21-30 shows an example of this *ester hydrolysis*.

Acid-catalyzed ester hydrolysis

The ester is hydrolyzed to a carboxylic acid.

$$\xrightarrow[\substack{H_2SO_4, \text{ acetic acid} \\ \text{reflux}}]{H_2O}$$

94%

(21-30)

The *esterification* of a carboxylic acid can be acid catalyzed, too, as shown in Equation 21-31. This is an example of a **Fischer esterification reaction**, named after Emil Fischer (1852–1919), the German chemist and Nobel laureate who first reported these kinds of reactions.

Fischer esterification

$$\xrightarrow[\substack{H_2SO_4, H_2O \\ 2 \text{ h, reflux}}]{CH_3CH_2OH}$$

86%

(21-31)

A carboxylic acid is converted to an ester.

YOUR TURN 21.4

The mechanism for the Fischer esterification reaction in Equation 21-31 is shown here, but the curved arrows have been omitted.

(a) Supply the curved arrows for each step.
(b) Write the name of each elementary step below its reaction arrow.

Acyl substitutions under acidic conditions are not limited to esters and carboxylic acids. Amides, which are even less reactive than esters or carboxylic acids, do not react with weak nucleophiles under normal conditions either. Under acidic conditions, however, amides can undergo acyl substitution via an addition–elimination mechanism. This is exemplified by the amide hydrolysis shown in Equation 21-32.

Amide hydrolysis

N-Ethyl-N-phenylmethanamide

**Methanoic acid
(Formic acid)**

N-Ethylaniline
88%

(21-32)

The mechanism for this reaction, presented in Equation 21-33, is essentially the same as the one that describes acid-catalyzed transesterification, shown previously in Equation 21-29.

Mechanism for Equation 21-32

(21-33)

The carbonyl group is activated in Step 1, which better facilitates nucleophilic addition of the weak nucleophile in Step 2. In Step 3, the charge on the O atom from the nucleophile goes from +1 to 0, and thus the O is stabilized. The reverse is true for the N atom of the leaving group in Step 4. In Step 5, the amine leaving group departs and the C=O bond is simultaneously regenerated. Finally, in Step 6, the newly formed amine, which is weakly basic, irreversibly deprotonates the carbonyl O.

---

problem **21.14** Draw the complete detailed mechanism and the overall products for each of the following reactions.

(a)

(b)

(c)

(d)

(e)

(f)

21.8 Baeyer–Villiger Oxidations

When diphenylmethanone (benzophenone) is treated with a **peroxyacid** (RCO_3H) such as *meta*-chloroperbenzoic acid (MCPBA), phenyl benzoate is produced, as shown in Equation 21-34. This is an example of a **Baeyer–Villiger oxidation**, named after the German chemist Adolf von Baeyer (1835–1917) and the Swiss chemist Victor Villiger (1868–1934).

Baeyer–Villiger oxidation

MCPBA
(a peroxyacid)

CF_3SO_3H, CH_2Cl_2

(21-34)

**Diphenylmethanone
(Benzophenone)**

Phenyl benzoate
100%

The mechanism for this reaction is shown in Equation 21-35 and is very similar to the mechanism for an acid-catalyzed transesterification (Equation 21-29).

(21-35)

After the carbonyl group of the ketone has been activated by protonation in Step 1, the relatively weak peroxyacid nucleophile attacks the carbonyl C in Step 2. A proton is then removed from the peroxyacid's O atom in Step 3, and an aryl group departs from the carbonyl C in Step 4. Simultaneously, the aryl group forms a bond to an O atom from the peroxyacid, breaking the peroxyacid's O—O bond in the process. Finally, in Step 5, the ester's carbonyl group is deprotonated, yielding the overall uncharged ester product.

The leaving group ability of $C_6H_5^-$ is extremely poor, so why does it act as the leaving group in this mechanism? There are two factors that facilitate the breaking of the C—C_6H_5 bond. First, a negative charge never appears on C_6H_5 in the products of Step 4. Instead, it appears on a carboxylate anion (RCO_2^-), and is thus resonance stabilized over two O atoms. Second, there is a significant driving force due to the weak O—O bond of the peroxyacid. In Step 4, where the C_6H_5 group leaves, the weak O—O bond is replaced by a much stronger C—O bond.

The ketone oxidized in Equation 21-34 is *symmetric*—that is, the carbonyl group is bonded to identical groups. If the groups bonded to the carbonyl C are different, such as in an *unsymmetric* ketone or aldehyde, then the major product depends on which group leaves from the carbonyl C. The ability of a group to do so is called its **migratory aptitude**. Migratory aptitudes have been determined empirically for a variety of groups, as follows:

Migratory Aptitude in a Baeyer–Villiger Oxidation

Methyl group < 1° Alkyl group < 2° Alkyl group < 3° Alkyl group < Aryl group < H

The reason for these relative migratory aptitudes is not fully known, but notice that the relative order is similar to the one for cation stability. This information is useful in predicting the major Baeyer–Villiger product, as shown in Solved Problem 21.15.

SOLVED problem 21.15 Predict the major product of the reaction at the right. (CF_3CO_3H is relatively acidic.)

Think What type of reagent is CF_3CO_3H? How does it tend to react with a ketone? Does one side of the ketone favor reaction over the other? If so, which side?

Solve CF_3CO_3H is a peroxyacid and is acidic, so it will react with a ketone in a Baeyer–Villiger oxidation. In such a reaction, an O atom from the peroxyacid is inserted between the carbonyl C and one of the groups initially bonded to the carbonyl C. In this case, the carbonyl C is bonded to a primary and a secondary alkyl group. The secondary alkyl group has a greater migratory aptitude, so its bond will preferentially break, producing the lactone (cyclic ester) shown below.

A secondary alkyl group has a greater *migratory aptitude* than a primary alkyl group.

1° Alkyl group   2° Alkyl group

$$\xrightarrow[\text{CH}_2\text{Cl}_2]{\text{CF}_3\text{CO}_3\text{H}}$$

---

problem **21.16** Predict the major product of each of the following reactions.

(a) $\xrightarrow[\text{CH}_2\text{Cl}_2]{\text{CF}_3\text{CO}_3\text{H}}$?

(b) $\xrightarrow[\text{CH}_2\text{Cl}_2]{\text{CF}_3\text{CO}_3\text{H}}$?

(c) $\xrightarrow[\text{CH}_2\text{Cl}_2]{\text{MCPBA} \atop \text{CF}_3\text{SO}_3\text{H},}$?

---

problem **21.17** In the mechanism in Equation 21-35, the hydroxyl (OH) oxygen of the peroxyacid is shown as the nucleophile. Why is that oxygen more nucleophilic than the adjacent oxygen?

21.9 Claisen Condensations

When ethyl ethanoate (ethyl acetate) is treated with sodium ethoxide, followed by acid workup, ethyl 3-oxobutanoate, a β-**keto ester**, is produced (Equation 21-36).

Claisen condensation   A β-keto ester

$$\xrightarrow[\text{2. H}_3\text{O}^\oplus]{\substack{\text{1. NaOCH}_2\text{CH}_3, \\ \text{CH}_3\text{CH}_2\text{OH} \\ \text{reflux 4 h}}}$$

+ HO⁀   (21-36)

Ethyl ethanoate   **Ethyl 3-oxobutanoate**
(Ethyl acetate)   **(Ethyl acetoacetate)**
                                      68%

This is an example of a **Claisen condensation reaction**, named after Rainer Ludwig Claisen (1851–1930), who developed it. It is a *condensation* reaction because, overall, two ester molecules are fused together and a smaller molecule—in this case, a molecule of ethanol—is eliminated.

The mechanism for the Claisen condensation reaction in Equation 21-36, shown in Equation 21-37, is very similar to the mechanism for base-catalyzed transesterification (review Equation 21-25). Normally an ester is not strongly nucleophilic, but due to the basic conditions of the reaction, $CH_3CH_2O^-$ deprotonates the ester at its α carbon in Step 1 to produce a strongly nucleophilic enolate anion. In Step 2, that enolate anion attacks the carbonyl carbon of a second molecule of the ester, forming a new C—C bond. $CH_3CH_2O^-$ is then eliminated in Step 3, producing an initial β-keto ester, which is quickly deprotonated in Step 4. Acid workup in Step 5 replenishes the β-keto ester.

Mechanism for Equation 21-36

(21-37)

The first two proton transfer steps (i.e., Steps 1 and 4) play important roles in this mechanism. The first (Step 1) is reversible and favors the reactant side, much more than we saw for aldol reactions (Section 18.4). Thus, only a small fraction of the initial ester exists in the form of the enolate anion at any given time, leaving a substantial amount of the uncharged ester available to react with the enolate anion. The second proton transfer (Step 4) is irreversible, so it heavily favors the product side. This is important because the first three steps establish an equilibrium that favors the overall reactants. The deprotonation in Step 4 effectively removes the β-keto ester product from that equilibrium and, according to Le Châtelier's principle, the equilibrium shifts continuously toward products.

The necessity for these proton transfer steps places a significant restriction on an ester that can participate in a Claisen condensation.

A successful Claisen condensation generally requires at least two α protons on the initial ester.

Without the first α proton, the enolate anion could not be produced. Without the second α proton, the equilibrium would favor the overall reactants and, as shown in Equation 21-38, no significant amount of the β-keto ester would be isolated.

Only one α H atom

$$\xrightarrow[\text{CH}_3\text{CH}_2\text{OH}]{\text{CH}_3\text{CH}_2\text{ONa}}$$

No reaction   (21-38)

Ethyl 2-methylpropanoate

---

problem **21.18** Draw the complete, detailed mechanism for the reaction at the right and predict the major product.

$$\xrightarrow{\begin{array}{l}\text{1. CH}_3\text{ONa}\\\text{2. H}^{\oplus}\end{array}}$$

SOLVED problem **21.19** Draw the ester that can be used to synthesize the compound at the right using a Claisen condensation reaction.

Think Which C—C bond in the β-keto ester product would have been formed in a Claisen condensation? Which C atom would have been bonded to the alkyoxy leaving group?

Solve The new C—C bond is the one between the α and β carbons, as shown in the following structure. Moreover, the leaving group would have been attached to the β C atom.

An alkoxy leaving group would have been attached to this C.   This is the C—C bond that would have formed in a Claisen condensation.

Therefore, we can apply a transform that undoes a Claisen condensation by disconnecting that C—C bond and reattaching the leaving group. Doing so yields two ester precursors that are identical.

Reattach an alkoxy leaving group.

Undo a Claisen condensation.

Disconnect this C—C bond.

21.9a The Importance of the Solvent and Base in Claisen Condensations

In the Claisen condensation reaction in Equation 21-36, the base that is added (i.e., the ethoxide anion, $CH_3CH_2O^-$), is the same as the leaving group that departs in Step 3 of the mechanism in Equation 21-37. *Other choices of base could lead to undesired products.* If the base were an alkoxide other than ethoxide, for example, as in Equation 21-39, then transesterification would produce an ester that is different from the starting material. Alternatively, if hydroxide were used as the base, as in Equation 21-40, then *saponification* (Section 20.2) would irreversibly convert the ester into a carboxylate anion.

Transesterification

This alkoxide is a poor choice of base because it produces a different ester via a transesterification reaction.

(21-39)

Saponification

NaOH is a poor choice of base because it converts the starting ester into a carboxylate anion *irreversibly*.

(21-40)

> **21.5 YOUR TURN**
> Draw the complete, detailed mechanism for the reactions in Equations 21-39 and 21-40.

These problems are avoided when ethoxide is the base. Transesterification still occurs, but the transesterification product would be identical to the initial ester.

> **21.6 YOUR TURN**
> Draw the complete, detailed mechanism for the transesterification reaction that takes place among the species in Equation 21-36.

Notice, too, that the solvent in Equation 21-36 (i.e., ethanol, CH_3CH_2OH) is the conjugate acid of the base that is added. *Other choices of solvent can also lead to an unwanted transesterification or saponification reaction.* If, for example, the solvent for the reaction in Equation 21-36 were another alcohol, such as propan-1-ol ($CH_3CH_2CH_2OH$), then the added ethoxide anion would deprotonate propan-1-ol to generate the 1-propoxide anion ($CH_3CH_2CH_2O^-$), thereby causing the transesterification reaction in Equation 21-39 to ensue. Alternatively, if water were used as the solvent, then HO^- would be produced, and the saponification reaction in Equation 21-40 would take place. With ethanol as the solvent, this kind of proton transfer still takes place, but that proton transfer is not a problem because the base that is produced from it is the same as the base that is added as an overall reactant.

YOUR TURN 21.7

Write the proton-transfer equilibrium that would take place if propan-1-ol were used as the solvent for the reaction in Equation 21-36. Do the same if water were used as the solvent and if ethanol were used as the solvent.

In light of the issues that can arise with the wrong choice of base or solvent, Claisen condensation reactions are often carried out according to the following guideline:

For a Claisen condensation reaction, unwanted transesterification and saponification reactions can be avoided if the base that is added is the same as the alkoxide leaving group on the ester, and if the alcohol solvent is the conjugate acid of that base. In other words, the following alkyl groups highlighted in red should be identical.

Same R groups

SOLVED problem 21.21 Identify the appropriate base–solvent pair to accomplish a Claisen condensation with the ester at the right.

Think What is the leaving group on the ester? What choice of base would ensure that a nucleophilic addition–elimination would leave us with the same ester? What solvent could be used so that, if it were deprotonated, the resulting base would be the same as the base that is added?

Solve The leaving group is $CH_3CH_2CH_2O^-$. Therefore, if we were to add $NaOCH_2CH_2CH_3$ as the base, then nucleophilic addition–elimination would leave us with the same ester. $CH_3CH_2CH_2OH$ should therefore be the solvent, because if it were to be deprotonated, it would yield $CH_3CH_2CH_2O^-$, the same as the added base. Overall, the reaction would be as follows:

problem 21.22 Identify the appropriate base–solvent pair to accomplish a Claisen condensation with each of the following esters.

(a) (b)

problem 21.23 The following Claisen condensation proceeds without an unwanted transesterification, even though the choice of base is not the same as the leaving group. Explain why.

1. (CH₃)₃CONa/(CH₃)₃COH
2. H⊕

21.9b Crossed Claisen Condensation Reactions

The Claisen condensation reactions discussed in Section 21.9a involve two molecules of the same ester—that is, they are *self condensations*. A Claisen condensation can also involve two *different* esters—a so-called **crossed Claisen condensation reaction**. These reactions require special considerations, however, because crossed Claisen reactions can potentially lead to a mixture of condensation products (just as we saw with crossed *aldol* reactions in Section 18.7).

The reaction in Equation 21-41, for example, would produce a mixture of four β-keto esters, because two different ester enolates can be generated as nucleophiles and two different, uncharged esters can be attacked by those nucleophiles.

Four different β-keto esters can be produced from this crossed Claisen condensation reaction.

1. CH₃ONa, CH₃OH
2. H⊕

(21-41a)

(21-41b)

(21-41c)

(21-41d)

problem 21.24 Draw a complete, detailed mechanism for the formation of *each* product in Equation 21-41.

Diethyl carbonate **Ethyl formate** **Ethyl benzoate** **Diethyl oxalate**

FIGURE 21-6 Esters with no α hydrogens Without α hydrogens, these esters will not produce enolate anions that would act as nucleophiles in Claisen condensation reactions.

One way to circumvent this problem is to ensure that one of the esters has no α hydrogens. (We previously used this strategy in Section 18.7 with crossed aldol reactions.) Some common examples of these kinds of esters are shown in Figure 21-6.

Because these esters have no α hydrogens, they can be dissolved in solution with an alkoxide base and no Claisen condensation takes place. A second ester that has α hydrogens can then be added slowly to this solution. As the ester with α hydrogens comes in contact with the base, its enolate anion is produced and the crossed Claisen condensation is initiated. An example is shown in Equation 21-42 using ethyl formate as the ester with no α hydrogens.

This ester has no α hydrogens.

This ester has two α hydrogens.

(21-42)

45%

Aldehyde and ketone enolate anions can also undergo nucleophilic addition–elimination with an ester, as shown in Equations 21-43 and 21-44. The reaction in Equation 21-43 is synthetically useful because the ester has no α hydrogens. In Equation 21-44, the ester has α hydrogens, but the reaction is synthetically useful because an ester is significantly less acidic than a ketone. Therefore, when the ester is added to the ketone enolate that was first produced quantitatively by H_2N^-, the α protons largely remain on the ester. (See Your Turn 21.8.)

The ester has no α hydrogens.

+ *t*-BuOK/DMF

1.

50 °C, 0.45 h

2. H_2SO_4, H_2O

80%

(21-43)

The ester's α protons are less acidic than the ketone's.

+ NaNH$_2$/ether

1.

reflux, 2 h

2. $H^{\oplus}$

57%

(21-44)

Once the ketone enolate anion is produced in the reaction in Equation 21-44, the following proton transfer reaction is possible when the ester is added, and could result, therefore, in undesired condensation products.

Estimate the K_{eq} for this reaction by looking up or estimating the appropriate pK_a values, and explain why this proton transfer reaction is not a major concern.

SOLVED problem 21.25 Show how you would synthe-size the compound at the right via a Claisen condensation reaction.

Think Which C—C bond can be "disconnected" in a transform that undoes a Claisen condensation? Does this transform suggest a self-Claisen condensation reaction or a crossed Claisen? Can we use a base that reversibly deprotonates an α hydrogen or should we use one that brings about an irreversible deprotonation?

Solve A transform that undoes a Claisen conden-sation disconnects a car-bonyl's α C atom from the carbonyl group at the β position, as shown at the right; the carbonyl at the β position is part of an ester in the reactants. Given that the two carbonyl-containing precursors are not identical esters, a crossed Claisen conden-sation reaction is in order. One of the precursors is an ester that has no α hydrogens, which helps make a crossed Claisen reaction feasible.

No α hydrogens

Butan-2-one

In the forward direction, butan-2-one would be converted into an enolate nucleophile via deprotonation at the α carbon. Because butan-2-one is unsymmetric, two isomeric enolate anions can be produced. This synthesis calls for the kinetic enolate, which can be produced by using LDA as the base. Overall, the synthesis would be written as follows:

LDA/ether

1.
2. H⊕

---

problem 21.26 Devise another synthesis of the target in Solved Problem 21.25 without using [ester structure].

21.9c Intramolecular Claisen Condensation Reactions: The Dieckmann Condensation

The Claisen condensation, like the aldol reaction, has the potential for an *intramolecular* reaction, ultimately producing a ring. When it does, the reaction is called a **Dieckmann condensation**, named after the German chemist Walter Dieckmann (1869–1925). An example of one is shown in Equation 21-45.

Dieckmann condensation

New C—C bond

1. *t*-BuOK/toluene
2. H⊕

(21-45)

63%

Deprotonation of an
α proton produces the
enolate nucleophile.

Nucleophilic attack by
the enolate anion forms
a new ring.

A Dieckmann condensation should be considered a possibility whenever two ester functional groups appear on the same molecule. As with most cyclization reactions, Dieckmann condensations occur most readily when the reaction leads to the formation of a five- or six-membered ring.

problem **21.27** Draw the complete, detailed mechanism for the reaction in Equation 21-45.

An intramolecular nucleophilic addition–elimination reaction on an ester can also occur if the enolate anion is that of a ketone or aldehyde. For example, methyl 6-oxo-6-phenylhexanoate cyclizes when treated with sodium amide to yield the five-membered ring product shown in Equation 21-46.

Ketones are more
acidic than esters.

NaNH₂
Benzene

(21-46)

Methyl 6-oxo-6-phenylhexanoate

90%

Notice in the reactant in Equation 21-46 that there are H atoms on the α carbons of both the ketone and the ester functional groups. Deprotonation takes place almost exclusively at the α carbon of the ketone, however, generating the ketone enolate anion instead of the ester enolate anion, because *ketones (and aldehydes) are substantially more acidic than esters.*

problem **21.28** Draw the complete, detailed mechanism for the following reaction and predict the major product.

1. NaOEt, EtOH
2. H⊕

?

problem **21.29** Draw the acyclic dicarbonyl compound that can be used to synthesize the cyclic β-keto aldehyde at the right.

Biological Claisen Condensations

When people consume more calories in their diet than they burn, the excess fuel (typically from carbohydrates) is stored as fat. A fat is a triglyceride (Section 2.12a), which is a triester formed between glycerol (a triol) and three fatty acids. Each fatty acid is a carboxylic acid whose carbon chain has an even number of carbons, most commonly 14–20 carbons long. How do carbohydrates become converted to fat, and why do the carbon chains of fatty acids appear in increments of two carbons? The answer has to do with biological Claisen condensation reactions. An example between malonyl acyl carrier protein (ACP) and acetyl synthase is shown in Steps 1 and 2 of the following scheme.

In Step 1, the α C of malonyl ACP undergoes nucleophilic addition to the carbonyl C of acetyl synthase to produce a tetrahedral intermediate. In Step 2, the S–synthase leaving group is eliminated and the C=O bond is regenerated, producing acetoacyl ACP, a β-hydroxycarbonyl. Through several subsequent steps, acetoacyl ACP is converted to butyryl ACP, which is the sulfur analog to the fatty acid butyric acid. Butyryl ACP is converted to butyryl synthase, which can participate in another Claisen condensation with malonyl ACP. Each time this process is repeated, the carbon chain grows by two, which explains the even number of carbons.

This scheme is validated in part by an experiment in which organisms are fed ^{14}C-labeled acetic acid (CH_3CO_2H). A cell uses acetic acid to synthesize acetyl-coenzyme A, from which both malonyl ACP and acetyl synthase are derived, so the ^{14}C label appears in the synthesized fatty acid. When the carbonyl C of acetic acid was labeled, the ^{14}C label appeared at the locations highlighted in blue above. When the label was at the α C, the label appeared at the locations highlighted in red.

**Ethyl 3-oxobutanoate
(Acetoacetic ester)**

**Diethyl malonate
(Malonic ester)**

**FIGURE 21-7 Two synthetically
useful active methylene
compounds** Ethyl 3-oxobutanoate and
diethyl malonate are starting materials
for the acetoacetic ester and malonic
ester syntheses, respectively.

**The ester group can
be removed entirely.**

**The α carbon can be
alkylated relatively easily.**

**FIGURE 21-8 The utility of malonic ester and
acetoacetic ester in synthesis** α-Alkylation,
followed by removal of the ester group, produces an
α-alkylated carbonyl.

21.10 Organic Synthesis: Decarboxylation, the Malonic Ester Synthesis, and the Acetoacetic Ester Synthesis

The main focus of Section 21.9 was producing β-keto esters via Claisen condensation reactions. Ethyl 3-oxobutanoate (acetoacetic ester), shown in Figure 21-7, is a β-keto ester that is particularly useful as a starting material in organic syntheses. Diethyl malonate (malonic ester), a β-diester, is structurally similar to acetoacetic ester and is also quite useful synthetically.

Their usefulness stems from their ability to undergo the following reactions:

> ■ The α carbon of malonic ester and acetoacetic ester can be alkylated relatively easily.
>
> ■ The ester group in each of these compounds can be removed entirely from the molecule.

Thus, malonic ester and acetoacetic ester can undergo the general transformation shown in Figure 21-8.

Alkylation is relatively easy to carry out because the protons on the α CH_2 group are significantly more acidic than normal esters or ketones; malonic ester and acetoacetic ester are therefore called **active methylene compounds**. Each of the two C=O groups flanking the α CH_2 group significantly enhances the acidity via resonance and inductive stabilization in the enolate conjugate base. The α protons are sufficiently acidic that even a moderately strong base such as $CH_3CH_2O^-$ will produce the enolate anion *quantitatively* (Equation 21-47). That negatively charged enolate anion is highly nucleophilic and will react with an alkyl halide in an S_N2 reaction to produce the alkylated product.

**A moderately strong base such
as RO⊖ will remove this α proton
irreversibly and quantitatively.**

**This S_N2 reaction yields
the alkylated α carbon.**

Proton
transfer

S_N2

(21-47)

YOUR TURN 21.9

Verify the preceding statement that an *active methylene compound* can be quantitatively deprotonated by using the appropriate pK_a values to estimate the K_{eq} for the following reaction.

pK_a = [ ]

pK_a = [ ]

K_{eq} = [ ]

To remove the ester group entirely from a β-keto ester or a β-diester, we can simply heat it under aqueous, acidic conditions, according to Equation 21-48.

The ester group is removed entirely.

(21-48)

The partial mechanism for this reaction is shown in Equation 21-49.

Partial mechanism for Equation 21-48

Hydrolysis takes place in several steps to produce the carboxylic acid.

A β-keto acid or β-diacid undergoes decarboxylation when heated.

β-Keto ester or β-diester

β-Keto acid or β-diacid

(21-49)

Bond rotation

Gaseous CO_2 bubbles out of solution.

Enol form

+ $CO_2(g)$

Tautomerization

Keto form

The acidic, aqueous conditions hydrolyze esters to carboxylic acids (review Section 21.7)—more specifically, a β-keto ester is converted to a **β-keto acid** and a β-diester is converted to a **β-diacid**. Heating the β-keto acid or β-diacid drives a **decarboxylation** in which gaseous CO_2 is lost and an enol is formed. The enol rapidly tautomerizes to the keto form.

Beginning with acetoacetic ester, we can put all of these steps together in what is called an **acetoacetic ester synthesis**, as shown in Equation 21-50. Similarly, a **malonic ester synthesis** is shown in Equation 21-51.

Acetoacetic ester synthesis

An α-alkylated acetone

Acetoacetic ester

1. NaOEt, EtOH
2.

Br

OEt

1. NaOH, H_2O
2. H_2SO_4, H_2O, Δ

(21-50)

55%

21.10 Organic Synthesis: Decarboxylation, the Malonic Ester Synthesis, and the Acetoacetic Ester Synthesis / **1041**

Malonic ester synthesis

An α-alkylated acetic acid

(21-51)

The acetoacetic ester and malonic ester syntheses have characteristic products.

- An acetoacetic ester synthesis produces an α-alkylated acetone.
- A malonic ester synthesis produces an α-alkylated acetic acid.

If your target is one of these derivatives, then consider using acetoacetic ester or malonic ester as your starting material.

SOLVED problem 21.30 Show how you would synthesize hexan-2-one beginning with acetoacetic ester.

Hexan-2-one

Think What would the precursor look like before decarboxylation? How could that precursor be produced from acetoacetic ester?

Solve Hexan-2-one is an α-alkylated acetone, so we can first perform a transform that undoes a decarboxylation, given that decarboxylation is the last step in an acetoacetic ester synthesis.

Undo a decarboxylation.

Continuing the retrosynthesis, let's undo a hydrolysis and an α-alkylation to bring us back to the starting acetoacetic ester.

Undo a hydrolysis.

Undo an α-alkylation.

In the forward direction, the synthesis would be written as follows:

1. NaOEt
2.

1. NaOH
2. H⊕

Δ

---

problem **21.31** Outline a synthesis of
pent-4-enoic acid from either acetoacetic ester or malonic
ester.

Pent-4-enoic acid

Both the acetoacetic ester synthesis and the malonic ester synthesis also lend
themselves to *dialkylation* of the active methylene group, given that there are two
acidic hydrogens. Examples are shown in Equations 21-52 and 21-53. Notice that the
two alkyl groups can be the same or different.

**A stronger base is used
for the second alkylation.**

**Hydrolysis and
decarboxylation**

(21-52)

**Two acidic
α hydrogens**   **The first H has been
replaced by an alkyl group.**   **The second H has been
replaced by an alkyl group.**   A dialkylated acetone

**A stronger base is used
for the second alkylation.**

**Hydrolysis and
decarboxylation**

(21-53)

**Two acidic
α hydrogens**   **The first H has been
replaced by an alkyl group.**   **The second H has been
replaced by an alkyl group.**   A dialkylated acetic acid

In both cases, alkylation of the methylene C is carried out twice prior to hydrolysis and decarboxylation. The base in the second alkylation is the *tert*-butoxide anion,
$(CH_3)_3CO^-$, however, because the second deprotonation is more difficult than the
first, so it requires a stronger base. The second deprotonation is more difficult because
the alkyl group added in the first alkylation is electron donating, and thus destabilizes
the resulting enolate anion.

In the figure at the right, draw an arrow representing the inductive electron donation by the alkyl group. Do the alkyl group's effects increase or decrease the concentration of negative charge?	**21.10 YOUR TURN**

problem **21.32** Show how you would synthesize 2-ethylhexanoic acid from malonic ester.

21.11 Organic Synthesis: Protecting Carboxylic Acids and Amines

Chapter 19 introduced the concept of the *protecting group*. The general approach was as follows: In the anticipation of an undesired reaction involving a particular functional group, (1) protect that functional group by temporarily converting it into another one that is unreactive under the conditions for the desired reaction, (2) perform the desired reaction, and (3) deprotect the functional group by converting it back to its original form. In Chapter 19, we specifically examined how to protect ketones, aldehydes, and alcohols. Table 21-2 provides examples of how to protect and deprotect carboxylic acids and amines.

The acidic nature of a carboxylic acid can interfere with a synthesis, so carboxylic acids are often protected by converting them to esters (Entries 1–3 in Table 21-2). Unlike carboxylic acids, esters have no O—H protons and are therefore much weaker acids.

Amines are susceptible to oxidation and to deprotonation by very strong bases, such as Grignard and alkyllithium reagents. They are also moderate nucleophiles. To suppress these reactivities, amines can be protected by converting them to amides (Entry 4 in Table 21-2). Amides are less susceptible to oxidation and are less nucleophilic because of the resonance stabilization involving the carbonyl group and the lone pair on nitrogen (Fig. 21-2). Furthermore, secondary amines result in tertiary amides, which have no N—H bonds and are therefore *not* susceptible at all to deprotonation of the nitrogen atom.

You must be aware of the conditions necessary to protect and deprotect to ensure that no other undesired reactions take place. Notice, for example, that acidic conditions are required for protection via a Fischer esterification and deprotection via acid-catalyzed hydrolysis, which could result in undesired reactions with alcohols and alkenes. Protection via diazomethane could result in undesired reactions with alkenes

TABLE 21-2 Protecting Groups for Carboxylic Acids and Amines

(Section 12.2), and deprotection via catalytic hydrogenation could result in undesired reactions with alkenes, alkynes, and aldehydes (Section 19.5).

SOLVED problem 21.33 Show how to synthesize non-6-ynoic acid from 5-bromopentanoic acid.

5-Bromopentanoic acid

Non-6-ynoic acid

Think Without considering undesired side reactions, how would you normally carry out this synthesis? Under the conditions for that synthesis, are there any functional groups that would undergo an undesired side reaction? If so, how can that functional group be protected?

Solve It might appear at first that the synthesis could be carried out by using $NaC{\equiv}CCH_2CH_3$ in an S_N2 reaction.

The desired S_N2 reaction should take place here.

An undesired deprotonation will take place here.

The desired product

However, this alkynide anion is a relatively strong base and will rapidly deprotonate the carboxylic acid. The resulting carboxylate anion is nucleophilic, and could then attack the alkyl halide in an intramolecular S_N2 reaction, yielding an undesired cyclic product.

To circumvent this problem, we can first convert the carboxylic acid into an ester, as follows:

No acidic proton

Remains free to act as a nucleophile

Ester hydrolysis regenerates the carboxylic acid.

Therefore, no acidic protons remain when the alkynide anion is introduced, leaving the alkynide anion free to displace Br⁻ in an S_N2 reaction. Once the substitution reaction has come to completion, the ester can be hydrolyzed back into a carboxylic acid.

problem 21.34 Show how to carry out the following synthesis.

21.12 Wrapping Up and Looking Ahead

Chapter 21 was the second of two chapters discussing nucleophilic addition–elimination reactions. Whereas Chapter 20 focused on reactions involving *strong* nucleophiles, the focus here in Chapter 21 was on reactions involving *weak* nucleophiles as overall reactants. Highly reactive acid derivatives such as acid chlorides and acid anhydrides react readily with weak nucleophiles, including water, alcohols, and amines, to produce carboxylic acids, esters, and amides, respectively. The mechanism that describes these reactions is nearly identical to the one that describes reactions involving strong nucleophiles, the only difference being that reactions involving weak nucleophiles require an additional proton transfer step to produce an uncharged product.

Carboxylic acids, esters, and amides are less reactive than acid chlorides or acid anhydrides, and thus do not react as readily with weak nucleophiles under normal conditions. These reactions are frequently base or acid catalyzed. Under basic conditions, the rate of nucleophilic addition–elimination is enhanced by deprotonating the nucleophile, thereby converting it into a strong nucleophile. Under acidic conditions, the reaction rate is enhanced by protonating the oxygen atom of a carbonyl group, which *activates* the carbonyl carbon toward nucleophilic attack—that is, makes it more electrophilic.

Similar to what we saw in Chapter 20, many nucleophilic addition–elimination reactions discussed here in Chapter 21 serve to convert one carboxylic acid derivative into another. However, the same fundamental mechanism describes other important reactions as well. For example, treating an alcohol with a sulfonyl chloride (RSO_2—Cl) produces a sulfonate ester (RSO_2—OR′), which possesses an excellent leaving group (RSO_2—O$^-$) for nucleophilic substitution and elimination reactions. The mechanisms for many of these reactions are essentially identical to the ones that describe the alcoholysis of acid chlorides.

With the completion of Chapter 21, we have essentially completed our set of chapters on nucleophilic reactions. Chapters 9 and 10 discussed nucleophilic substitution and elimination reactions; Chapters 17 and 18 discussed nucleophilic addition reactions; and Chapters 20 and 21 discussed nucleophilic addition–elimination reactions. For each of these types of reactions, a key step in the mechanism shows the formation of a bond involving a lone pair of electrons from a nucleophile.

In Chapters 22 and 23, we shift our focus back to electrophilic reactions. Recall that Chapters 11 and 12 introduced electrophilic addition reactions, which are characterized by the formation of new σ bonds to atoms that are initially connected by a nonpolar π bond. Chapters 22 and 23 also deal with reactions involving nonpolar π bonds, but these reactions lead to overall substitution rather than addition.

THE ORGANIC CHEMISTRY OF BIOMOLECULES

21.13 Determining a Protein's Primary Structure via Amino Acid Sequencing: Edman Degradation

Recall from Section 1.15a that proteins are large molecules constructed from α-amino acids, and each of the 20 possible amino acids is distinguished by the identity of its side group, R. As shown in Equation 21-54, each pair of amino acids in a protein is joined via an amide functional group, also called a **peptide linkage** or a **peptide bond**.

Amino acids in a protein are connected by an amide functional group, also called a peptide linkage or peptide bond.

Complete hydrolysis of the amides liberates the individual amino acids.

The N-terminus

The C-terminus

(21-54)

The fact that amino acids are joined by amide groups is critical to the stability of proteins. Recall from Figure 21-1 that amides are among the least reactive acid derivatives, so they require relatively extreme conditions to be hydrolyzed. As shown in Equation 21-54, for example, the *complete* hydrolysis of a protein requires strongly acidic conditions and elevated temperatures.

A complete hydrolysis of a protein can be used to determine the relative amounts of each type of amino acid in a protein. To do so, the product mixture is injected into an *amino acid analyzer*. However, complete hydrolysis does not reveal the **primary structure** of the protein—namely, the specific sequence of amino acids from the **N-terminus** (the end with the free amino group) to the **C-terminus** (the end with the carboxyl group). This sequence is important, because it is what gives a protein its key function; frequently, proteins that differ by the identity of just one amino acid at a key site, or **residue**, have very different biological properties.

One way to determine the amino acid sequence of a protein is with an **Edman degradation**. The basis of an Edman degradation is to remove and detect one amino acid at a time, from the N-terminus. The removal of each N-terminal amino acid requires the three steps shown in Figure 21-9.

FIGURE 21-9 Partial mechanism for the Edman degradation (Step 1) The amino group of the N-terminal amino acid undergoes nucleophilic addition to phenyl isothiocyanate to yield the corresponding PTC derivative. (Step 2) The PTC derivative cyclizes when treated with HCl to yield the corresponding thiazolinone. (Step 3) Treatment of the thiazolinone with HCl produces the corresponding PTH derivative, which can be detected using chromatography or spectroscopy. Steps 1–3 can be repeated with each of the remaining N-terminal amino acids.

Step 1

Nucleophilic addition to the polar C=N bond

Phenyl isothiocyanate

PTC derivative

Step 2

Nucleophilic addition–elimination to the polar C=O bond

PTC derivative

Thiazolinone

Step 3

Nucleophilic addition–elimination to the polar C=O bond

Nucleophilic addition–elimination to the polar C=O bond

Thiazolinone

PTH derivative

In Step 1, the protein is treated with phenyl isothiocyanate. The free amine of the N-terminal amino acid adds to the polar C═N bond in a nucleophilic addition reaction, producing a phenylthiocarbamoyl (PTC) derivative that contains the side group (in this case, R_1). In Step 2, under mildly acidic conditions, the N-terminal amino acid cleaves in the form of a thiazolinone, via a nucleophilic addition–elimination mechanism. The thiazolinone is then extracted into an organic solvent, and when treated with HCl, it undergoes another nucleophilic addition–elimination mechanism, rearranging to a more stable phenylthiohydantoin (PTH) derivative. The PTH derivative is then identified using either chromatographic or mass spectrometric techniques. The entire process is repeated with the remainder of the protein.

Theoretically, the entire sequence of a protein can be determined in this manner. However, sequencing proteins via Edman degradation becomes impractical when the protein exceeds roughly 30 amino acids, because side products accumulate and interfere with the results. In these cases, the protein can be *partially* hydrolyzed using dilute acid, in which case hydrolysis takes place at essentially random locations in the protein. The smaller polypeptides that are produced then can be sequenced using an Edman degradation, and the sequence of the original protein subsequently can be pieced together from the results.

Suppose, for example, that the partial hydrolysis of a 15-amino-acid polypeptide yielded three smaller polypeptides, **A–C**, whose sequences were determined to be the ones shown on the left of Figure 21-10. As shown on the top right of Figure 21-10, certain portions of the peptides have the same sequence—specifically, the Ser-His sequence in peptides **A** and **C** (red), and the Gln-His-Leu sequence in peptides **B** and **C** (blue). Assuming those sequences come from the same portions of the original protein, we can piece together the overall sequence, as shown at the bottom right of Figure 21-10.

FIGURE 21-10 Determining the sequence of a protein from partial hydrolysis (Left) Peptides A–C are obtained from partial hydrolysis of a protein, and their sequences are determined by Edman degradation. (Top right) Peptides A–C have portions of their sequences in common. (Bottom right) The sequence of the original protein is determined.

---

problem **21.35** Suppose that an 18-amino-acid polypeptide is partially hydrolyzed into four smaller peptides, **A–D**, which are analyzed by Edman degradation to have the following sequences.

> **A** Asp-Asp-Ser
>
> **B** Ile-Val-Met-Pro-Val
>
> **C** Asp-Ser-Met-Trp-Pro-Cys-Pro-Asn
>
> **D** Pro-Asn-Gln-Asp-Cys-Phe-Ile-Val-Met-Pro-Val

Determine the sequence of the initial protein.

21.14 Synthesis of Peptides

Historically, people with type 1 diabetes have required routine injections of a protein called insulin that was isolated from pigs. In 1978, however, human insulin was synthesized, and soon after became less expensive and more readily available than pig insulin.

To synthesize a protein, amino acids must be joined together by amide bonds, produced from the amine group of one amino acid and the carboxylic acid group of another. Under normal conditions, these reactions are slow (see Section 21.2), so to facilitate them, **dicyclohexylcarbodiimide** (**DCC**) is typically used, as shown in Equation 21-55.

The mechanism for this reaction is shown in Equation 21-56.

In Steps 1 and 2, DCC undergoes a nucleophilic addition reaction with the carboxylic acid group to produce an *O*-acyl isourea derivative, which contains a good leaving group on the carbonyl carbon. Subsequently, the *O*-acyl isourea undergoes a nucleophilic addition–elimination reaction with the amine group, producing the amide and *N,N'*-dicyclohexyl urea.

Thus, DCC can be used to couple together amino acids, but each amino acid has both an amine and a carboxylic acid group, so multiple different couplings can take place. For example, if a mixture of glycine and alanine is treated with DCC, four different couplings will take place to produce a mixture of four dipeptides, as shown in Equation 21-57.

A mixture of dipeptides is produced.

(21-57)

Benzyl chloroformate = Z—Cl

To carry out one of these couplings selectively, the groups that we don't want to react must be *protected*. Suppose, for example, that we want to couple together alanine (Ala) and glycine (Gly) to produce Ala-Gly (Equation 21-58).

(21-58)

The amine group of Ala and the carboxylic acid group of Gly must be protected. The amine is commonly protected with benzyl chloroformate (Z—Cl) in a reaction that is essentially an aminolysis (Section 21.3) to produce a carbamate, which has reactivity similar to an amide. The carboxylic acid group is protected as an ester by treatment with benzyl alcohol under acidic conditions; this is a Fischer esterification (Section 21.7). Once the necessary groups are protected, DCC is used to couple the

Benzyl alcohol = Bn—OH

The Merrifield Synthesis of a Peptide

FIGURE 21-11 The Merrifield synthesis of a peptide The solid-phase synthesis protocol developed by R. B. Merrifield. The individual steps are described in the text.

amino acids together. Conveniently, both protecting groups can be removed in a single step using catalytic hydrogenation.

These steps can be repeated to attach additional amino acids. However, because the process takes place entirely in solution, the products must be purified after each step. This is tedious and it leads to significant loss of product. R. B. Merrifield (1921–2006) solved these problems by designing a *solid-phase* synthesis, for which he was awarded the 1984 Nobel Prize in Chemistry.

The basis of the **Merrifield synthesis**, outlined in Figure 21-11, begins with polymer beads that terminate with benzyl chloride groups. The C-terminal amino acid, whose amine is protected with a *tert*-butoxycarbonyl (BOC) group, is then added to the polymer beads (Step 1). An S_N2 reaction ensues, joining the amino acid to the polymer bead via an ester linkage. In Step 2, the amine group of the newly joined amino acid is deprotected under acidic conditions, and in Step 3, a second BOC-protected amino acid is added in the presence of DCC. A new peptide bond is produced, according to the mechanism in Equation 21-56, resulting in a BOC-protected dipeptide attached to the polymer bead. Steps 2 and 3 are then repeated for each amino acid that is to be added to the peptide, working from the C-terminus to the N-terminus. Once the protected N-terminal amino acid has joined, hydrogen bromide is added, which simultaneously releases the peptide from the bead and deprotects the N-terminal amino acid.

Solid-phase synthesis is so simple in its execution that the process has been automated. It is now possible to use a "peptide synthesizer" to make peptides of >100 amino acid residues. The addition of each residue takes about 4 hours, which is about 250 times slower than the natural process that makes a protein in a cell. The Merrifield synthesis is quite effective, however, at making peptides when it is not possible to clone the protein using recombinant DNA techniques.

Chapter Summary and Key Terms

■ The general mechanism for nucleophilic addition–elimination involving a weak nucleophile (H–Nu) is similar to the one involving a strong nucleophile, but includes a proton transfer step to produce an uncharged product. This mechanism describes the **hydrolysis**, **alcoholysis**, and **aminolysis** of both an acid chloride and an acid anhydride. **(Sections 21.1 and 21.3)**

■ The reactivity of an acid derivative decreases as the resonance stabilization of the carbonyl group increases and as the electron-withdrawing effects of the leaving group decrease **(Section 21.2)**. Overall, the relative reactivities of acid derivatives decrease in the order:

Acid chloride > acid anhydride > ester
$\approx$ carboxylic acid > amide

■ **Thionyl chloride** (SOCl₂) converts a carboxylic acid to an acid chloride (the acid derivative at the top of the stability

ladder), from which any other acid derivative can be produced. **(Section 21.4)**

■ The **Hell–Volhard–Zelinsky (HVZ) reaction** produces an α-bromo carboxylic acid from a carboxylic acid. The key is to convert the carboxylic acid into an acid bromide, which is enolizable, and is thus nucleophilic at the α carbon. **(Section 21.5)**

■ Treating an alcohol with a **sulfonyl chloride** produces a **sulfonate ester**. Thus, a poor HO⁻ leaving group is converted into an excellent alkylsulfonate leaving group for nucleophilic substitution and elimination reactions. Examples include mesylate (OMs), tosylate (OTs), and triflate (OTf) leaving groups. (Section 21.6)

- Carboxylic acids, esters, and amides are relatively stable and thus do not react readily with weak nucleophiles under normal conditions. However, nucleophilic acyl substitution can be base catalyzed or acid catalyzed. Base catalysis involves deprotonating the nucleophile to convert it into a strong nucleophile. Acid catalysis involves protonating the carbonyl oxygen to activate the carbonyl carbon. (Section 21.7)

- The **Fischer esterification reaction** produces an ester from a carboxylic acid under highly acidic conditions, and is described by the acid-catalyzed acyl substitution mechanism. (Section 21.7)

- The **Baeyer–Villiger oxidation** produces carboxylic acids from aldehydes, and esters from ketones. In these reactions, a hydrogen or alkyl group departs from the carbonyl carbon, facilitated by the breaking of the weak O—O bond from a **peroxyacid** (RCO_3H). (Section 21.8)

- In a **Claisen condensation reaction**, an ester with at least two α protons is treated with a strong base to produce a **β-keto ester**. In these reactions, the base deprotonates the α carbon of one ester to produce an ester enolate anion, which acts as a nucleophile and attacks the carbonyl carbon of a second ester. (Section 21.9)

- If the base in a Claisen condensation is nucleophilic, it should be identical to the alkoxide leaving group on the ester. If the solvent can be deprotonated to become strongly nucleophilic, it should be the conjugate acid of the base that is used. (Section 21.9a)

- In a **crossed Claisen condensation reaction**, the enolate anion that acts as the nucleophile derives from an ester that is different from the ester it attacks. Crossed Claisen condensations are synthetically useful if only one ester enolate anion is present, and only one uncharged ester can be attacked. (Section 21.9b)

- A **Dieckmann condensation** is an intramolecular Claisen condensation, and is favored when a five- or six-membered ring can be formed. (Section 21.9c)

- A **β-keto acid** or **β-diacid** can undergo **decarboxylation** upon heating under acidic conditions. The result is loss of CO_2, leaving behind a ketone or carboxylic acid. Decarboxylation is used in the **acetoacetic ester synthesis** to produce an alkyl-substituted ketone, as well as the **malonic ester synthesis** to produce an alkyl-substituted carboxylic acid. (Section 21.10)

- Carboxylic acids are frequently protected as a methyl or benzyl ester. Deprotection takes place upon treatment of either of these esters with hydroxide in a saponification reaction. Alternatively, deprotection of a benzyl ester can take place by catalytic hydrogenation. (Section 21.11)

- Amines are frequently protected as amides. Deprotection typically requires acid-catalyzed hydrolysis. (Section 21.11)

Reaction Tables

TABLE 21-3 Functional Group Conversions

	Starting Functional Group	Typical Reagents and Reaction Conditions	Functional Group Formed	Key Electron-Rich Species	Key Electron-Poor Species	Comments	Discussed in Section
(1)	Acid chloride	H_2O	Carboxylic acid	$H_2\ddot{O}:$	Acid chloride	Nucleophilic addition–elimination (hydrolysis)	21.1
(2)	Acid chloride	R'OH	Ester	$R'\ddot{O}H$	Acid chloride	Nucleophilic addition–elimination (alcoholysis)	21.1
(3)	Acid anhydride	H_2O	Carboxylic acid	$H_2\ddot{O}:$	Acid anhydride	Nucleophilic addition–elimination (hydrolysis)	21.2
(4)	Acid anhydride	R'OH	Ester	$R'\ddot{O}H$	Acid anhydride	Nucleophilic addition–elimination (alcoholysis)	21.2

(continued)

TABLE 21-3 Functional Group Conversions (continued)

	Starting Functional Group	Typical Reagents and Reaction Conditions	Functional Group Formed	Key Electron–Rich Species	Key Electron–Poor Species	Comments	Discussed in Section
(5)	Acid chloride	R'_2NH / Et_3N or pyridine	Amide	$R'_2\overset{\delta-}{\ddot{N}}H$	Acid chloride ($\delta+$)	Nucleophilic addition–elimination (aminolysis)	21.3
(6)	Acid anhydride	R'_2NH / Et_3N or pyridine	Amide	$R'_2\overset{\delta-}{\ddot{N}}H$	Acid anhydride ($\delta+$)	Nucleophilic addition–elimination (aminolysis)	21.3
(7)	Carboxylic acid	$SOCl_2$	Acid chloride	$:\!\ddot{\underset{..}{Cl}}\!:^{\ominus}$	(intermediate)	Back-to-back nucleophilic addition–elimination	21.4
(8)	Carboxylic acid	Br_2, P	α-Bromo acid	(enol)	$\overset{\delta+}{Br}\!-\!\overset{\delta-}{Br}$	Hell–Volhard–Zelinsky reaction	21.5
(9)	Alcohol	$Cl\!-\!\overset{O}{\underset{O}{S}}\!-\!R'$	Sulfonate ester	$R\overset{\delta-}{\ddot{O}}H$	$Cl\!-\!\overset{O}{\underset{O}{\underset{\delta+}{S}}}\!-\!R'$	Nucleophilic addition–elimination	21.6
(10)	Ester	$R''OH$ / $R''O^{\ominus}$	Ester	$:\!\ddot{O}R''^{\ominus}$	Ester ($\delta+$)	Base-catalyzed nucleophilic addition–elimination (transesterification)	21.7
(11)	Ester	$R''OH$ / $H^{\oplus}$	Ester	$R''\overset{\delta-}{\ddot{O}}H$	Ester ($\overset{\oplus}{O}H$)	Acid-catalyzed nucleophilic addition–elimination (transesterification)	21.7
(12)	Ester	H_2O / $H^{\oplus}$	Carboxylic acid	$H_2\overset{\delta-}{\ddot{O}}:$	Ester ($\overset{\oplus}{O}H$)	Acid-catalyzed nucleophilic addition–elimination (hydrolysis)	21.7
(13)	Carboxylic acid	$R'OH$ / $H^{\oplus}$	Ester	$R'\overset{\delta-}{\ddot{O}}H$	Carboxylic acid ($\overset{\oplus}{O}H$)	Acid-catalyzed nucleophilic addition–elimination (Fischer esterification)	21.7
(14)	Aldehyde	$R'\!-\!\overset{O}{\underset{}{C}}\!-\!O\!-\!OH$	Carboxylic acid	$R'\!-\!\overset{O}{\underset{}{C}}\!-\!O\!-\!\overset{\delta-}{\ddot{O}}H$	Aldehyde ($\overset{\oplus}{O}H$)	Acid-catalyzed nucleophilic addition–elimination (Baeyer–Villiger oxidation)	21.8

TABLE 21-4 Reactions That Alter the Carbon Skeleton[a]

Starting Functional Group	Typical Reagent and Reaction Conditions	Functional Group Formed	Key Electron–Rich Species	Key Electron–Poor Species	Comments	Discussed in Section
(1) Ketone		Ester			Acid-catalyzed nucleophilic addition–elimination (Baeyer–Villiger oxidation)	21.8
(2) Ester	1. NaOR' 2. H⊕	β-Keto ester			Base-promoted nucleophilic addition–elimination (Claisen condensation)	21.9
(3) Acetoacetic ester	1. NaOEt 2. R—X 3. H₃O⊕, Δ	Alkyl-substituted acetone		R—Br	Acetoacetic ester synthesis	21.10
(4) Malonic ester	1. NaOEt 2. R—X 3. H₃O⊕, Δ	Alkyl-substituted acetic acid		R—Br	Malonic ester synthesis	21.10

[a] X = Cl, Br, I

Problems

21.36 Show how each of the following compounds can be synthesized from an acid chloride and either water, an alcohol, or an amine. For each reaction, provide the complete, detailed mechanism.

(a)

(b)

(c)

(d)

21.37 Draw the complete, detailed mechanism and the products for each of the following reactions.

(a)

?

(b)

(2 equiv.)

?

(c)

(d)

21.38 A *thioester* is the sulfur analog of an ester. Do you think a thioester would undergo hydrolysis faster or slower than an ester under normal conditions? Explain your reasoning.

A thioester

21.39 Predict the product of the following sequence of reactions.

1. NaBH₄, EtOH

2.

?

21.40 Predict the product of the reaction between *m*-ethylbenzoyl chloride and each of the following compounds. Draw the complete, detailed mechanism for each reaction. If no reaction is expected to occur, write "no reaction."

(a) H₂O   **(b)** CH₃NH₂   **(c)** (S)-Butan-2-ol   **(d)** Diethyl ether

***m*-Ethylbenzoyl chloride**

21.41 Predict the product of the reaction between acetic anhydride and each of the following compounds. If no reaction is expected to occur, write "no reaction." Draw the complete, detailed mechanism for each reaction that does occur.

(a) H₂O   **(b)** CH₃NH₂, pyridine
(c) (S)-Butan-2-ol   **(d)** Diethyl ether

21.42 Predict the product of the reaction between methyl cyclohexylmethanoate and each of the following. If no reaction is expected to occur, write "no reaction." For those reactions that do occur, draw the complete, detailed mechanism. (*Hint*: Pay attention to the reaction conditions.)

(a) H₂O, H⊕   **(b)** H₂O, OH⊖, then H₃O⊕   **(c)** CH₃CH₂CH₂OH, CH₃CH₂CH₂ONa

(d) Propan-2-ol, H⊕   **(e)** Propan-2-amine (excess), H⊕   **(f)** Propan-1-ol

21.43 A crossed Claisen reaction between methylpropanoate and methyl acetate is not synthetically useful because it produces a mixture of products. The reaction between methyl 2-methylpropanoate and methyl acetate, however, is synthetically useful. Explain why.

1. CH₃ONa, CH₃OH

2. H⊕

Mixture of products

Methyl propanoate

1. CH₃ONa, CH₃OH

2. H⊕

Methyl 2-methylpropanoate

78%

21.44 Starting with acetic acid and using any other reagents necessary, show how you would synthesize each of the following compounds.

(a) **(b)** **(c)**

(d) **(e)** **(f)** **(g)**

21.45 Using acetic acid as your only source of carbon, propose a synthesis of each of the following compounds. You may use any inorganic reagent necessary.

(a) Ethyl acetate **(b)** Butan-2-ol **(c)** 3-Methylpentan-3-ol **(d)** Butan-2-one
(e) Ethanamine **(f)** N-Ethylacetamide **(g)** N,N-Diethylacetamide

21.46 Aspirin (acetylsalicylic acid) is made by treating salicylic acid with acetic anhydride:

Salicylic acid → [Acetic anhydride] → **?** Acetylsalicylic acid

Draw the complete, detailed mechanism for this reaction and draw the product.

21.47 In each of the following reactants, a carboxylic acid is separated from an alcohol group by the same number of carbons. When heated in the presence of acid, only the first compound forms a lactone. Explain why.

[structure with OH and CO₂H] H⊕ → [bicyclic lactone]

[structure with OH and CO₂H] H⊕ → No lactone

21.48 The hydrolysis of an ester can be sped up by both acidic and basic conditions. Aminolysis of an ester can be sped up by acidic conditions, but not by basic conditions. Explain why.

21.49 One method for synthesizing lactones (cyclic esters) involves treating a hydroxyacid with 2-pyridinethiol, followed by heating under reflux. The mechanism proceeds through a 2-pyridinethiol ester, as indicated.

[reaction scheme: HO-...-CO₂H + HS-pyridine, H⊕ → A 2-pyridinethiol ester → Δ, Xylene → cyclic lactone]

A 2-pyridinethiol ester

(a) Provide a complete, detailed mechanism for this reaction.
(b) If benzenethiol (C₆H₅SH) is used instead of 2-pyridinethiol, the conversion is much less effective. This suggests that the N atom is instrumental in the mechanism. Explain the role of the N atom.

21.50 An acid anhydride can be formed under equilibrium conditions by reacting an ester with a carboxylic acid, as shown below. Reasonable yield is achieved if the equilibrium can be shifted using Le Châtelier's principle.

(a) Provide a complete, detailed mechanism for this reaction.
(b) In general, the above equilibrium favors the reactants. However, if the ester that is used is the one at the right, then the equilibrium favors the products. Explain why.

21.51 When a methyl ester is hydrolyzed under acidic conditions in oxygen-18 labeled water, the ^{18}O isotope ends up in the carboxylic acid. When a *tert*-butyl ester is hydrolyzed under the same conditions, the labeled oxygen ends up in the alcohol product.

(a) Propose mechanisms to account for these observations.
(b) Explain why each ester undergoes the respective mechanism.

21.52 Treating a δ-lactone with ammonia yields a hydroxyamide. If a β-lactone is treated with ammonia, however, then a β-amino acid is formed.

Provide the detailed mechanism for each of these reactions and explain these observations.

21.53 Propose a mechanism for the following reaction.

21.54 Draw the complete, detailed mechanism for each of the following reactions and predict the major products.

(a)

(b)

1. LDA
2. Ethyl acetate
3. $H_3O^{\oplus}$

?

21.55 Draw the complete, detailed mechanism for each of the following reactions and predict the major product.

(a) Ethyl 3-methylbutanoate $\xrightarrow[\text{EtOH}]{\text{NaOEt}}$?

(b) Ethyl propanoate + Ethyl benzoate $\xrightarrow[\text{EtOH}]{\text{NaOEt}}$?

(c) Ethyl butanoate + Diethyl carbonate $\xrightarrow[\text{EtOH}]{\text{NaOEt}}$?

21.56 Draw structures for compounds **A** through **H**.

(a) Ethyl acetoacetate $\xrightarrow[\text{2. 1-Chloro-4-methylpentane}]{\text{1. NaOEt, EtOH}}$ **A** $\xrightarrow[\text{2. H}_3\text{O}^{\oplus}, \Delta]{\text{1. NaOH, H}_2\text{O}}$ **B**

Same as above

(b) **A** $\xrightarrow[\text{2. 1-Chloropropane}]{\text{1. NaOC(CH}_3)_3, \text{(CH}_3)_3\text{COH}}$ **C** $\xrightarrow[\text{2. H}_3\text{O}^{\oplus}, \Delta]{\text{1. NaOH, H}_2\text{O}}$ **D**

(c) Diethyl malonate $\xrightarrow[\text{2. Bromophenylmethane}]{\text{1. NaOEt, EtOH}}$ **E** $\xrightarrow[\text{2. H}_3\text{O}^{\oplus}, \Delta]{\text{1. NaOH, H}_2\text{O}}$ **F**

Same as above

(d) **E** $\xrightarrow[\text{2. 1-Iodopentane}]{\text{1. NaOC(CH}_3)_3, \text{(CH}_3)_3\text{COH}}$ **G** $\xrightarrow[\text{2. H}_3\text{O}^{\oplus}, \Delta]{\text{1. NaOH, H}_2\text{O}}$ **H**

21.57 Draw the complete, detailed mechanism for the following reaction.

21.58 Draw the structures of **A**, **B**, and **C** in the following sequence of reactions.

21.59 Draw the major product of each of the following sequences of reactions performed on methyl 3-cyclohexylpropanoate.

(a)

Same as above

(b) **A** $\xrightarrow[\text{2. H}_3\text{O}^{\oplus}]{\text{1. PBr}_3, \text{Br}_2}$ **E** $\xrightarrow{\text{NaCN}}$ **F** $\xrightarrow[\Delta]{\text{H}_3\text{O}^{\oplus}}$ **G**

(c)

$$\xrightarrow[\text{2. H}_3\text{O}^{\oplus}]{\text{1. DIBAH, }-78°\text{C}} \mathbf{H} \xrightarrow[\text{2. H}^{\oplus}]{\text{1. }} \mathbf{I} \xrightarrow{\text{H}_2\text{CrO}_4} \mathbf{J}$$

(d)

$$\xrightarrow[\text{2. H}^{\oplus}]{\text{1. LiAlH}_4} \mathbf{K} \xrightarrow[\text{Pyridine}]{\text{TsCl}} \mathbf{L} \xrightarrow[\text{DMSO}]{\text{CH}_3\text{CH}_2\text{SH}} \mathbf{M}$$

21.60 Suggest how you should carry out each of the following syntheses.

(a)

(b)

21.61 Devise a synthesis of 2-methylpentane-1,3-diol. You may use any inorganic reagents, but your only carbon source must be alcohols containing three or fewer carbons.

21.62 Show how you would synthesize 3-oxo-2-methylpentanal. You may use any inorganic reagents, but your only carbon source must be alcohols containing three or fewer carbons.

21.63 Show how you would synthesize pentanoic acid from 1,3-propanedioic acid, using any reagents necessary.

21.64 Show how you would carry out the synthesis at the right, using any reagents necessary.

21.65 In an aminolysis of an acid chloride, pyridine is often used to minimize the amount of the amine that is required, as shown at the right. Pyridine is a moderately strong nucleophile, however, so it can react with an acid chloride via a nucleophilic addition–elimination mechanism. Nevertheless, this reaction involving pyridine does not interfere with the production of the desired amide.

Pyridine

(a) Draw the mechanism and the product for the nucleophilic addition–elimination reaction of pyridine with an acid chloride.
(b) Explain why this reaction involving pyridine does not interfere with the production of the desired amide product. (*Hint*: What reaction would ensue between the above amine and the product of nucleophilic addition–elimination involving pyridine?)

21.66 Explain why the following Claisen condensation does *not* work.

$$\xrightarrow[\text{2. H}^{\oplus}]{\text{1. C}_6\text{H}_5\text{ONa, C}_6\text{H}_5\text{OH}}$$

21.67 An example of the McFayden–Stevens reaction is shown below, in which an acyl chloride is converted to an aldehyde.

Benzoyl chloride   **1-Benzoyl-2-benzenesulfonylhydrazide**   **Benzaldehyde**

First, benzoyl chloride is reacted with hydrazine, H_2NNH_2, the product of which is reacted with benzenesulfonyl chloride. The result is a 1-benzoyl-2-benzenesulfonylhydrazide, which, when heated under basic conditions, decomposes into the aldehyde. Provide the detailed mechanism showing the conversion of benzoyl chloride into 1-benzoyl-2-benzenesulfonyl-hydrazide.

21.68 Show how to synthesize each of the following compounds from 3-methylpentanoic acid.

(a)   **(b)**   **(c)**

21.69 An imino chloride can be prepared from an amide according to the following reaction.

An imino chloride

Propose a mechanism for this reaction.

21.70 Draw a complete, detailed mechanism for the following reaction.

$NaOCH_3$

21.71 The Gabriel synthesis of primary amines discussed in Chapter 20 involves hydrolysis under basic conditions to release the amine. As shown in the following reaction, that hydrolysis can take place under acidic conditions, too.

1. NaOH
2. R—Br
3. $H_3O^{\oplus}$, Δ

+ H_2N—R

Draw complete, detailed mechanisms for this set of reactions.

21.72 The Gabriel–malonic ester synthesis is used to make α-amino acids:

Draw complete, detailed mechanisms for this set of reactions and draw the structure of the intermediate **A**.

21.73 Diazomethane can be used to bring about a *ring expansion* of a cyclic ketone:

(a) Propose a mechanism for this reaction.
(b) Suggest why this reaction is capable of converting a more stable six-membered ring to a less stable seven-membered ring.

21.74 A **sulfonamide** is produced when a sulfonyl chloride is treated with an amine:

A sulfonamide

This reaction is a key step in the synthesis of **sulfa drugs**, which constitute an important class of antibiotics. Draw a complete, detailed mechanism for this reaction.

21.75 The following is an example of the *Favorskii reaction*, which involves an R⁻ leaving group in a nucleophilic addition–elimination reaction.

(a) Draw the complete, detailed mechanism for this reaction and explain why R⁻ can act as a leaving group.
(b) Suggest how you can synthesize an ester from cyclopropanone using only this reaction.

21.76 During the hydrolysis of proteins, some amino acids, such as tryptophan, do not survive the reaction conditions. Other amino acids, such as asparagine and glutamine, are modified. Referring to Table 1-7, which shows the structures of the 20 common amino acids, write the structures of the two amino acids that are formed when asparagine and glutamine decompose in hot, concentrated HCl.

21.77 When phosgene is treated with excess methanol, a product is formed whose ¹H NMR spectrum shows one peak—a singlet at 3.8 ppm. Provide a complete, detailed mechanism for this reaction.

Phosgene

21.78 A student carries out the following sequence of reactions. The IR and ^{13}C NMR spectra are shown for the product **C**. Draw structures for **A–C**. (Remember that the ^{13}C NMR signal at 77 ppm is due to the $CDCl_3$ solvent.)

21.79 A student treated methyl 2-methylpropanoate with sodium methoxide dissolved in methanol. After the solution was refluxed for 2 hours, the mixture was analyzed by NMR, IR, and mass spectrometry. Those spectra are shown below.

(a) Draw the mechanism of the reaction the student was expecting and draw the expected product.
(b) What do the spectra suggest occured?
(c) Explain these results.

21.80 A liquid, which is insoluble in water, reacts in acidic water to form an insoluble solid product. That product is soluble in water under basic conditions. The IR and ^{1}H NMR spectra of the reactant are shown below. Provide a structure for both the reactant and the product.

21.81 A compound is treated with excess acetyl chloride. The product is a compound whose formula is $C_6H_{10}O_4$. The IR and ^{1}H NMR spectra of the product are shown below. Draw structures for both the product and the reactant.

If your instructor assigns problems in smartwork, log in at **smartwork.wwnorton.com**.

CHAPTER 22

Electrophilic Aromatic Substitution 1

Substitution on Benzene; Useful Accompanying Reactions

The Lewis structure of benzene (also called its Kekulé structure) contains three conjugated double bonds arranged in a ring. We might expect, therefore, that benzene would undergo electrophilic addition like the various alkenes and alkynes in Chapters 11 and 12. But it doesn't. No reaction occurs, for example, when benzene is treated with HCl (Equation 22-1) or with Cl_2 (Equation 22-2) under normal conditions.

A polyurethane

The wheels on these inline skates are made of polyurethane, which has uses ranging from rigid foams, to varnish, to electronic components. Polyurethane is typically made from aniline, $C_6H_5NH_2$, which is produced, in turn, from benzene using reactions that we will study here in Chapter 22.

Upon completing Chapter 22 you should be able to:

- Explain why aromatic species resist electrophilic addition reactions.
- Draw the general mechanism for *electrophilic aromatic substitution*. Describe the characteristics and the importance of the *arenium ion intermediate*.
- Explain why electrophilic aromatic substitution tends to take place under *kinetic control*.
- Draw the detailed mechanism for the *halogenation* of benzene. Provide examples of *Lewis acid catalysts* and describe their role in the mechanism.
- Draw the detailed mechanism for a *Friedel–Crafts alkylation* and a *Friedel–Crafts acylation* reaction. Specify the limitations of each kind of reaction.
- Draw the detailed mechanism for a *nitration* reaction, and describe the role of sulfuric acid.
- Draw the detailed mechanism for a *sulfonation* reaction. Describe the role of SO_3 in *fuming sulfuric acid*, and explain why sulfonation is *reversible*.
- Show how to synthesize a primary alkylarene using a Friedel–Crafts acylation reaction.
- Show how to synthesize aromatic carboxylic acids by using $KMnO_4$ oxidations of carbon side chains on aromatic rings.
- Show how to synthesize aromatic amines via nitration.
- Devise syntheses that incorporate substitution reactions on aryldiazonium ions. Identify which of these reactions are classified as *Sandmeyer reactions*.

$$\text{(benzene)} \xrightarrow{\text{HCl}} \text{No reaction} \qquad \text{(22-1)}$$

$$\text{(benzene)} \xrightarrow{Cl_2} \text{No reaction} \qquad \text{(22-2)}$$

This lack of reactivity is attributed to the stability exhibited by benzene's aromatic π system (see Chapter 14). Therefore,

> Electrophilic addition reactions that *permanently* destroy benzene's aromaticity—such as those in Equations 22-1 and 22-2—are heavily disfavored.

When iron or iron(III) chloride, $FeCl_3$, is added to the reaction mixture, however, benzene does react with Cl_2, as shown in Equation 22-3.

Cl has replaced H.

$$\text{Benzene} \xrightarrow[\text{Fe or FeCl}_3]{Cl_2} \text{Chlorobenzene} + \text{HCl} \qquad \text{(22-3)}$$

The Cl in the product appears where there was initially a H, so this is a *substitution* reaction, *not* an addition. Because an aromatic species is involved, the reaction is called **electrophilic aromatic substitution**.

YOUR TURN 22.1

Draw the electrophilic addition products for the reactions in Equations 22-1 and 22-2 and show that they are, indeed, nonaromatic.

Answers to Your Turns are in the back of the book.

Here in Chapter 22, we study the general mechanism for the reaction in Equation 22-3 and other electrophilic aromatic substitution reactions involving benzene. Although the reactions can include a variety of different reagents, they follow the same general mechanism.

Aromaticity is not limited to benzene. Other aromatic compounds we examined in Chapter 14 can also undergo electrophilic aromatic substitution. The mechanisms for the reactions involving these other aromatic compounds are no different from the ones involving benzene, but they require consideration of reaction rates and regiochemistry. These aspects add a significant level of complexity, so we discuss them in Chapter 23. Here in Chapter 22, we focus just on variations in the electrophilic aromatic substitution mechanism.

22.1 The General Mechanism of Electrophilic Aromatic Substitutions

Just like alkenes and alkynes, the π system of benzene is relatively electron rich. In contrast, an electrophile is relatively electron poor, and may generically be represented as "E^+." (The positive charge emphasizes the electron-poor nature of the electrophile.) When benzene encounters an electrophile, therefore, the two species can form a new bond, using a pair of π electrons from benzene. This is an electrophilic addition step, shown in Step 1 of Equation 22-4, the general mechanism for electrophilic aromatic substitution. It is the same first step we have seen previously in the addition of a Brønsted acid across an alkene (Chapter 11), in which the electrophile is a proton (H^+).

The general mechanism for electrophilic aromatic substitution

Step 1 is SLOW and thus is the rate-determining step.
• Aromaticity is destroyed.
• A carbocation is produced.

Step 2 is FAST.
• Aromaticity is gained.
• A carbocation is neutralized.

1. Electrophilic addition

2. Electrophile elimination

Arenium ion intermediate
Resonance stabilized

(22-4)

The product of Step 1 is called an **arenium ion** or a **Wheland intermediate**. It is a carbocation intermediate consisting of five sp^2-hybridized C atoms and one sp^3-hybridized C atom. This intermediate participates in an electrophile elimination step in Step 2 of the mechanism. A proton (H^+) is eliminated, and is assisted by the formation of a bond to a base that is present. Overall, then, E^+ replaces H^+, converting benzene into a substituted benzene.

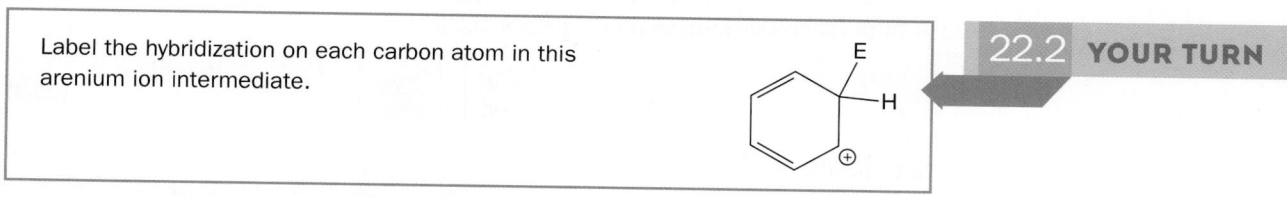

Label the hybridization on each carbon atom in this arenium ion intermediate.

22.2 **YOUR TURN**

The arenium ion intermediate is not aromatic, and it possesses a positively charged carbon that lacks an octet. As a result, this intermediate is much less stable than the starting compound, benzene, as illustrated in the reaction energy diagram in Figure 22-1. Consequently, the first step is highly unfavorable. By contrast, elimination of H^+ from the sp^3-hybridized C atom in Step 2 is highly favorable, because aromaticity is regained and the carbocation is neutralized.

Given the loss of aromaticity and the gain of a carbocation in Step 1, you might wonder why electrophilic aromatic substitution takes place at all. The arenium ion intermediate is stabilized by the resonance delocalization of the positive charge over three different C atoms (Equation 22-4), so it is significantly more stable than it otherwise would be.

YOUR TURN 22.3

Draw all of the resonance structures and the resonance hybrid of the arenium ion intermediate from Equation 22-4.

Hybrid

According to Figure 22-1, the energy barrier for Step 1 is much larger than it is for Step 2, making Step 1 the much slower step. That is,

In electrophilic aromatic substitution, the electrophilic addition step (Step 1 in Equation 22-4) is the rate-determining step.

Because benzene and the electrophile both appear as reactants in the rate-determining step, the rate of the overall reaction depends upon the concentration of both species, as indicated in Equation 22-5.

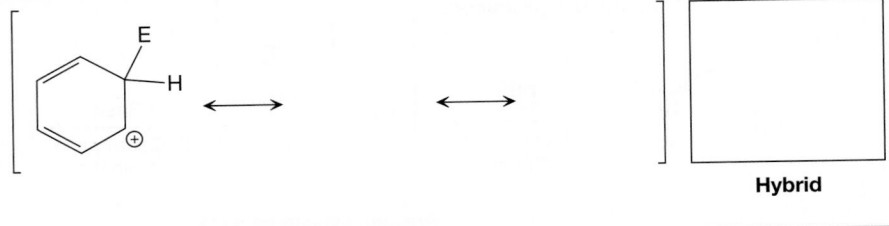

$$\text{Rate} = k \left[\text{C}_6\text{H}_6 \right] \left[E^{\oplus} \right] \tag{22-5}$$

Specifically,

The general electrophilic aromatic substitution reaction is first order with respect to the concentration of benzene and first order with respect to the concentration of the electrophile.

In situations where electrophilic aromatic substitution is involved in a competition, it is important to know whether that competition takes place under *kinetic control* or *thermodynamic control* (review Section 9.12). As indicated in Figure 22-1, the overall products appear significantly lower in energy than the reactants, so the overall reaction tends to be *irreversible*. Consequently,

> Electrophilic aromatic substitution reactions generally take place under *kinetic control*.

These reactions tend to be irreversible due to the conditions necessary for the reaction to take place. Because Step 1 in Equation 22-4 temporarily destroys the benzene ring's aromaticity, *a very strong electrophile is required*. Consequently, electrophiles (E^+) that are suitable for aromatic substitution tend to be significantly less stable than the H^+ that is removed, making the reaction highly favorable. Moreover, removing the H^+ in Step 2 is facilitated by the formation of a bond to a base, which further favors products.

So far, we have dealt only with the two-step electrophilic aromatic substitution mechanism in a general sense: An electrophile first adds to the benzene ring and a proton is then eliminated. In the sections that follow, we focus on specific types of substitution reactions. Each of the reactions we encounter involves the same two-step electrophilic aromatic substitution mechanism, but the identity of "E^+" differs from reaction to reaction.

The high reactivity required of each of these electrophiles generally makes it difficult or impossible to add them directly as reactants. Instead,

> Electrophiles in electrophilic aromatic substitution reactions typically must be generated in situ from more stable precursors that can be added as starting materials.

Therefore, as you encounter the mechanisms for the various kinds of electrophilic aromatic substitution reactions, pay particular attention to what the actual electrophile is and how it is generated.

22.2 Halogenation

Recall from the Introduction that benzene does not react with molecular chlorine, Cl_2 (Equation 22-2) under normal conditions. In the presence of Fe or $FeCl_3$, however, **chlorination** occurs to yield chlorobenzene (Equation 22-3); this is a type of **aromatic halogenation** reaction. Similarly, molecular bromine, Br_2, does not react with benzene under normal conditions (Equation 22-6), but **bromination** produces bromobenzene (Equation 22-7) in the presence of Fe or $FeBr_3$.

Bromobenzene
97%

The mechanism for the *bromination* of benzene is shown in Equation 22-8.

(22-8)

FeBr$_3$ acts as a Lewis acid in Step 1, thus complexing Br$_2$ in a coordination step. FeBr$_3$ can be added directly, or it can be produced in situ by combining Fe and Br$_2$. In Step 2, heterolysis of the Br—Br bond takes place *slowly*, producing Br$^+$ and FeBr$_4^-$. Finally, Steps 3 and 4 formally make up the electrophilic aromatic substitution (i.e., the mechanism presented previously in Equation 22-4), with Br$^+$ as the electrophile. Notice that FeBr$_3$ is regenerated in Step 4, so it is not consumed overall—it is a catalyst.

YOUR TURN 22.4

The mechanism of the chlorination of benzene is as follows, but the curved arrows have been omitted.

Supply the missing curved arrows, and below each reaction arrow, write the name of the reaction that is taking place. Which two steps make up the electrophilic aromatic substitution mechanism presented earlier in Equation 22-4?

SOLVED problem 22.2 Show how to carry out the synthesis at the right using an aromatic halogenation reaction.

Think What reaction can be used to form the necessary C—C bond? Does the product of that reaction contain the appropriate functional groups? What precursors are necessary to carry out the C—C bond formation reaction? How must halogenation be incorporated to produce the appropriate precursor?

Solve The new C—C bond can be formed using a Grignard reaction, the product of which is an alcohol that can be oxidized to the ketone target. To carry out the Grignard reaction, a phenyl Grignard reagent must be used, which can be produced from bromobenzene. Bromobenzene can be produced by brominating benzene in an electrophilic aromatic substitution reaction.

In the forward direction, the synthesis would be written as follows:

problem 22.3 Show how to carry out the synthesis at the right using an aromatic halogenation reaction.

We can also envision the fluorination and iodination of benzene through a similar mechanism. However, just as we saw with the addition of molecular halogens across double bonds (Chapter 12), fluorine is too reactive to be of much synthetic use and iodination is too slow. Iodination, though, can be carried out in the presence of an oxidizing agent like Cu^{2+} or hydrogen peroxide (Equation 22-9). Oxidation of I_2 generates I^+, which can then enter into the electrophilic aromatic substitution mechanism.

(22-9)

22.3 Friedel–Crafts Alkylation

In 1877, Charles Friedel (1832–1899) and James Crafts (1839–1917) reported that *alkylation* can take place when benzene is treated with an alkyl chloride in the presence of aluminum trichloride ($AlCl_3$), a strong Lewis acid catalyst (Equations 22-10 and 22-11). In these **Friedel–Crafts alkylation** reactions, the electrophile is a carbocation, R^+. In Equation 22-10, the electrophile is a *tert*-butyl cation, $(CH_3)_3C^+$; in Equation 22-11, it is the cyclohexyl cation, $C_6H_{11}^+$.

Friedel–Crafts alkylation

Benzene $\xrightarrow[\substack{AlCl_3 \\ 2\ h,\ 0\text{--}5\ °C}]{Cl}$ ***tert*-Butylbenzene**
72%

(22-10)

Benzene $\xrightarrow[\substack{AlCl_3 \\ 3\ h,\ 30\ °C}]{Cl}$ **Cyclohexylbenzene**
93%

(22-11)

The production of these carbocation electrophiles is illustrated in Equation 22-12, the mechanism of the alkylation in Equation 22-10.

Mechanism for Equation 22-10

$AlCl_3$ is a strong Lewis acid catalyst.

1. Coordination

2. Heterolysis

An alkyl cation is a highly reactive electrophile.

3. Electrophilic addition

4. Electrophile elimination

$\left(\quad + \quad HCl \quad + \quad AlCl_3 \quad \right)$

(22-12)

The AlCl$_3$ catalyst functions in much the same way as the FeBr$_3$ catalyst does in aromatic bromination, which we saw previously in Equation 22-8. In Step 1, the alkyl halide coordinates to the Al atom of AlCl$_3$. In Step 2, the Cl—C bond breaks in a heterolysis step, which takes place *slowly*, yielding the carbocation electrophile. The final two steps, once again, are the electrophilic aromatic substitution steps originally shown in Equation 22-4.

22.5 YOUR TURN

The mechanism of the reaction in Equation 22-11 is as follows, but the curved arrows have been omitted.

Supply the missing curved arrows, and below each reaction arrow, write the name of the elementary step taking place. Identify the two steps that make up the general electrophilic aromatic substitution mechanism shown previously in Equation 22-4.

Friedel–Crafts alkylation reactions can also take place under other conditions that generate carbocation intermediates. Examples are shown in Equations 22-13 and 22-14. In Equation 22-13, the acidic conditions convert the OH group into an excellent leaving group, which departs as H$_2$O. The carbocation is produced upon loss of the water leaving group. In Equation 22-14, on the other hand, protonation of the double bond yields a carbocation intermediate directly. These are the same steps that constitute the beginnings of S$_N$1/E1 and electrophilic addition reactions, respectively.

(22-13)

(22-14)

68%

problem **22.5** Propose a mechanism for the reaction in Equation 22-13.

SOLVED problem **22.7** Show how to synthesize the following compound in a single reaction, using benzene as one reagent, and (a) an alkyl halide, (b) an alkene, and (c) an alcohol as the other reagent.

Think What is the electrophile that must substitute for H⁺? How can that electrophile be produced from an alkyl halide, an alkene, or an alcohol?

Solve The electrophile that must substitute for H^+ is a carbocation, as follows:

This carbocation electrophile must be produced.

The carbocation can be produced from an alkyl halide using $AlCl_3$, a strong Lewis acid catalyst, or it can be produced from an alkene or an alcohol under acidic conditions.

problem **22.8** Show how you can synthesize the alkylbenzene at the right in a single reaction from benzene and (a) an alkyl halide, (b) an alkene, and (c) an alcohol.

22.4 Limitations of Friedel–Crafts Alkylations

Many different alkyl halides can be used to carry out a Friedel–Crafts reaction on benzene, making it possible to synthesize a variety of alkylbenzenes. Some alkylbenzenes, however, cannot be produced *directly* from these kinds of reactions. For example, propylbenzene cannot be synthesized in good yield via a Friedel–Crafts alkylation (Equation 22-15a) because the reaction using 1-chloropropane yields predominantly isopropylbenzene instead (Equation 22-15b).

Propylbenzene is a minor product because, as with any reaction in which a carbocation is produced,

> Friedel–Crafts alkylations are susceptible to carbocation rearrangements.

The carbocation that is initially formed in Equation 22-15—the 1-propyl cation—is a primary carbocation, so it rapidly rearranges to the 2-propyl cation, a secondary carbocation. As a result, it is the 2-propyl cation that is the actual electrophile that enters the electrophilic aromatic substitution mechanism (Equation 22-16).

Partial mechanism for Equation 22-15

In some cases, no Friedel–Crafts reaction takes place at all, as shown in Equations 22-17 and 22-18. In general,

A Friedel–Crafts alkylation reaction does not occur readily unless the halogen atom of the alkyl halide is bonded to an sp^3-hybridized carbon.

The sp^2 hybridization increases the strength of the C—Cl bond and decreases the stability of the carbocation that would be produced.

No reaction (22-17)

No reaction (22-18)

In Equations 22-17 and 22-18, which involve an aryl and a vinyl halide, respectively, the halogen atom is bonded to an sp^2-hybridized C. With the increased s character of an sp^2-hybridized C relative to an sp^3-hybridized C, the strength of the C—Cl bond is greater and so is the C atom's effective electronegativity (Section 3.11). Thus, the carbocation that is produced is excessively unstable. Recall from Section 9.6a that aryl and vinylic halides are resistant to nucleophilic substitution and elimination reactions for very similar reasons.

YOUR TURN 22.6

Verify the preceding statement by looking in Chapter 3 to find the C—H bond strength of an sp^3-hybridized carbon and comparing it to the bond strength of an sp^2-hybridized carbon.

sp^3 _____ sp^2 _____

SOLVED problem 22.9 Show how you can synthesize 2-phenylbut-2-ene from benzene.

Think Can the target be produced directly from a simple Friedel–Crafts alkylation? Is there another reaction that can be used to form the appropriate C—C bond?

Solve For the target to be produced directly from a Friedel–Crafts alkylation, the precursors would be those in Equation 22-18. For reasons discussed previously, this kind of a reaction does not occur readily. Another route must be sought.

Undo a Friedel–Crafts alkylation.

Another C—C bond formation reaction we have used before is the Grignard reaction. The product of a Grignard reaction is an alcohol, which can be dehydrated to form the final product. Here we can envision a Grignard reaction between phenylmagnesium bromide and butan-2-one. The Grignard reagent can be produced from bromobenzene, which can be synthesized, in turn, using a bromination reaction.

In the forward direction, the synthesis would be written as follows (note that the dehydration step would produce a mixture of both E and Z isomers):

problem **22.10** Butylbenzene cannot be synthesized in good yield directly from benzene using a Friedel–Crafts alkylation. Why not? Propose an alternate synthesis of butylbenzene that does not use a Friedel–Crafts reaction.

22.5 Friedel-Crafts Acylation

Friedel–Crafts acylation of benzene is carried out in much the same way as Friedel–Crafts alkylation. The difference is that the aromatic species is treated with an *acid chloride*, also called *acyl chloride* (an acyl group is R—C=O), instead of an alkyl chloride. The product, as shown in Equation 22-19, is an *aromatic ketone*, in which a C atom from benzene forms a bond to the C atom of the acyl group.

Friedel–Crafts acylation

An acid chloride, also called an acyl chloride

An aromatic ketone

(22-19)

1-Phenylpropan-1-one
78%

The mechanism of Friedel–Crafts acylation, presented in Equation 22-20, is essentially identical to the one for Friedel–Crafts alkylation, shown previously in Equation 22-12.

Mechanism for Equation 22-19

AlCl₃ is a strong Lewis acid catalyst.

An acylium ion is a relatively highly reactive electrophile.

1. Coordination
2. Heterolysis
3. Electrophilic addition
4. Electrophile elimination

$$\left(\quad + \;\; HCl \;\; + \;\; AlCl_3 \quad \right)$$

(22-20)

The strong Lewis acid catalyst, AlCl₃, is responsible for generating the cationic electrophile, called an **acylium ion**. In Step 1, the Cl atom of the acyl chloride coordinates to the electron-deficient Al atom. In Step 2, the C—Cl bond undergoes heterolysis. Step 3 is electrophilic addition of the acylium ion to the benzene ring, producing the arenium ion intermediate. Finally, in Step 4, removing the proton from the arenium ion yields the aromatic ketone product.

YOUR TURN 22.7

The mechanism for the Friedel–Crafts acylation involving acetyl chloride is as follows, but most of the curved arrows have been omitted, as have the overall products.

Supply the missing curved arrows and draw the overall products. Also, write the name of each elementary step below the appropriate reaction arrow and label the acylium ion.

problem 22.11 As we learned in Chapters 20 and 21, carboxylic acid chlorides behave quite similarly to carboxylic acid anhydrides. Not surprisingly, then, aromatic acylation can be carried out using an acid anhydride in the presence of AlCl₃, analogous to the Friedel–Crafts acylation reaction in Equation 22-19. The following example shows this using acetic anhydride to produce phenylethanone. Propose a mechanism for this reaction.

**Phenylethanone
(Acetophenone)**

problem 22.12 The compound at the right can be synthesized from benzene and a carboxylic acid anhydride in a single Friedel–Crafts reaction. What acid anhydride must be used? (*Hint:* The two phenyl rings come from two different compounds.)

An acylium ion

FIGURE 22-2 Resonance stabilization of the acylium ion The lone pair of electrons on O participates in resonance to delocalize the positive charge over O and C.

It might seem peculiar that Friedel–Crafts acylation can take place at all, given that the Cl atom in the acid chloride reactant is bonded to an sp^2-hybridized C. (Recall from Section 22.4 that this, in fact, prevents a Friedel–Crafts *alkylation* from taking place.) The acylium ion that is produced in Friedel–Crafts acylation, however, is *resonance stabilized* by a lone pair of electrons on the adjacent oxygen (Fig. 22-2). With this additional stability, the cationic electrophile can be produced and enter into the electrophilic aromatic substitution mechanism.

Draw the resonance hybrid of a generic acylium ion.	**22.8 YOUR TURN**

Unlike Friedel–Crafts alkylations,

> Friedel–Crafts acylation reactions are *not* susceptible to carbocation rearrangements.

The acylium ion is a relatively stable carbocation because its positive charge is delocalized by resonance and because it has a resonance structure in which all atoms have their octet. A carbocation rearrangement would generally lead to an isolated positive charge that is less stable.

Friedel–Crafts acylation reactions can be used with a variety of acyl halides to produce different aromatic ketones. However,

> Friedel–Crafts acylation as a means of *formylation*, whereby H is replaced by a formyl group, HC=O, requires in situ formation of the very unstable Cl—CH=O.

For an example, see Problem 22.37 at the end of the chapter.

SOLVED problem 22.13 Show how to synthesize phenylbutanone from benzene and any alcohol, using a Friedel-Crafts acylation.

Think The product is an aromatic ketone, suggesting a Friedel–Crafts acylation. What precursors are necessary to produce the target from a Friedel–Crafts acylation reaction? How can those precursors be produced from the starting material?

Solve In a Friedel–Crafts acylation, an aromatic carbon forms a bond to an acyl carbon. We begin our retrosynthetic analysis by disconnecting that bond to arrive at the appropriate precursors—an aromatic ring and an acyl chloride. The acyl chloride can be produced by chlorinating the corresponding carboxylic acid, which can be produced, in turn, by oxidizing a primary alcohol.

1-Phenylbutan-1-one

In the forward direction, the synthesis would be written as follows:

problem 22.14 Show how to synthesize benzophenone from benzene and any alcohol.

Benzophenone

22.6 Nitration

When benzene is treated with concentrated nitric acid (HNO_3), nitrobenzene is produced (Equation 22-21). Benzene is said to have undergone **nitration**.

Nitration

(22-21)

Nitrobenzene
83%

The mechanism of this reaction is presented in Equation 22-22.

Mechanism for Equation 22-21

(22-22)

Steps 1 and 2 of the mechanism are responsible for generating the powerful NO_2^+ electrophile, called the **nitronium ion**. First, HNO_3 is protonated to create a good H_2O leaving group. Next, H_2O departs via heterolysis. During the heterolysis step, an additional N=O double bond is formed, using a lone pair of electrons from an O atom. This aids the elimination of the water leaving group by preventing the N atom from losing its octet. Once the NO_2^+ electrophile is created, it enters into the two-step electrophilic aromatic substitution mechanism (i.e., Steps 3 and 4).

Although this nitration reaction is effective,

With concentrated nitric acid alone, the nitration of benzene tends to proceed relatively slowly.

The rate is slow because the amount of NO_2^+ present in concentrated HNO_3 is small—about 4%—and, as we saw in Equation 22-5, the rate of electrophilic aromatic substitution is directly proportional to the concentration of the electrophile.

To increase the rate of nitration, the reaction can be run at high temperatures. Alternatively,

The rate of nitration increases by adding concentrated sulfuric acid.

Sulfuric acid (H_2SO_4) is much more acidic than HNO_3, so H_2SO_4 generates a substantially higher concentration of NO_2^+, the active electrophile.

Draw the complete mechanism for the nitration of benzene in which concentrated sulfuric acid has been added.

22.7 Sulfonation

Treating benzene with concentrated sulfuric acid yields benzene sulfonic acid (Equation 22-23), whereby a **sulfonyl group**, SO$_3$H, has been incorporated onto the ring. This kind of substitution reaction is called **sulfonation**.

Sulfonation

Benzenesulfonic acid
66%

$$(22\text{-}23)$$

Under these reaction conditions, sulfur trioxide (SO$_3$) is believed to be the electrophile involved in the electrophilic addition step of electrophilic aromatic substitution. This is illustrated in the mechanism presented in Equation 22-24.

Mechanism for Equation 22-23

$$(22\text{-}24)$$

Concentrated H$_2$SO$_4$ contains a small amount of SO$_3$ at equilibrium. It is produced in much the same way as NO$_2^+$ is generated in concentrated HNO$_3$ (Equation 22-22). Protonation of an OH group in Step 1 generates a very good H$_2$O leaving group, which subsequently leaves in Step 2 to produce HSO$_3^+$. Deprotonation in Step 3 then yields SO$_3$, which enters into the two-step electrophilic aromatic substitution mechanism (i.e., Steps 4 and 5). Finally, in Step 6, a proton transfer produces the overall uncharged product.

Evidence for the involvement of SO$_3$ comes from the fact that sulfonation is much more rapid with **fuming sulfuric acid** than it is with sulfuric acid alone. Fuming sulfuric acid is simply concentrated sulfuric acid infused with SO$_3$.

Sulfa drugs, discovered in 1932, were among the first clinically useful antibiotics—a discovery that ultimately sparked a revolution in medicine. The general structure of these antibiotics, shown below on the left, exhibits sulfonamide ($-SO_2NHR'$) and amine groups that are para to each other on the phenyl ring. To synthesize the sulfonamide portion, a sulfonic acid group ($-SO_3H$) is first attached to the aromatic ring via sulfonation, similar to the one we have seen here in Chapter 22. Then, the sulfonic acid is converted to a sulfonyl chloride ($-SO_2Cl$), which is condensed with an amine (H_2NR') to produce the final sulfonamide.

Sulfa drugs resemble *para*-aminobenzoic acid (PABA, $H_2N-C_6H_4-CO_2H$), and it is this resemblance that is believed to be responsible for their antimicrobial properties. Bacteria, like humans, need folic acid to synthesize nucleic acids and proteins. Bacteria synthesize their folic acid from PABA in an enzymatic reaction. Sulfa drugs compete with PABA for the active site of that enzyme, and therefore inhibit the synthesis of folic acid. We humans, on the other hand, do not synthesize folic acid. Rather, it is a vitamin that is an essential part of our diets, so we can tolerate sulfa drugs.

RHN— A sulfa drug

$R = C_{12}H_{25}$

A sodium dodecylbenzenesulfonate

Several detergents, too, are synthesized using aromatic sulfonation reactions. As we learned in Section 2.10, detergents have the general structure $RSO_3^- Na^+$, where R is a relatively large, nonpolar group. Some of the most common detergents have a benzene ring separating the SO_3^- portion from an alkyl portion, as shown above on the right. The alkyl portion can be linear or branched, but typically consists of 12 carbons in total, and is thus called a sodium dodecylbenzenesulfonate.

Although electrophilic aromatic substitution reactions are generally irreversible, sulfonations are not.

> The sulfonation of an aromatic ring is *reversible*.

Notice that water is a product in the sulfonation reaction in Equation 22-23. Therefore, the reverse of a sulfonation reaction, called **desulfonation** (Equation 22-25), can be carried out by treating an aryl sulfonic acid with large amounts of water under acidic conditions.

Desulfonation

The sulfonyl group has been removed.

$+ H_2O \rightleftharpoons$ dilute H_2SO_4 $+ H_2SO_4$ **(22-25)**

Because a sulfonyl group can be added and removed from an aromatic ring in this way, sulfonation can be used as a *protecting group* at a specific site on the benzene ring. This strategy is discussed more fully in Chapter 23.

The mechanism for the desulfonation in Equation 22-25 is shown below, but the curved arrows have been omitted.

Draw the missing curved arrows and write the name of each elementary step below the corresponding reaction arrow.

22.8 Organic Synthesis: Considerations of Carbocation Rearrangements and the Synthesis of Primary Alkylbenzenes

Recall from Section 22.4 that one of the limitations of Friedel–Crafts alkylations is the potential for carbocation rearrangements. As we saw in Equation 22-15, for example, propylbenzene cannot be synthesized directly in good yield from a Friedel–Crafts alkylation because rearrangement of the resulting primary carbocation intermediate yields isopropylbenzene as the major product instead.

To circumvent this problem, we can use Friedel–Crafts *acylation* instead, which forms a C—C bond to the aromatic ring without competing carbocation rearrangements (Section 22.5). Because the immediate product of an acylation reaction is an aromatic ketone, the carbonyl group must be reduced to produce an alkylbenzene. An example using the Clemmensen reduction (Section 19.3b) is shown in Equation 22-26.

Friedel–Crafts acylation forms the new C—C bond without risking a carbocation rearrangement.

The ketone has been reduced to a methylene group to produce the primary alkylbenzene.

(22-26)

SOLVED problem 22.15 Show how you would synthesize the compound at the right, using benzene as the only aromatic starting compound.

Think The product is an alkylbenzene. Can a Friedel–Crafts alkylation be used to form the C—C bond to benzene? If not, why not? How can a Friedel–Crafts acylation reaction be incorporated into this synthesis?

Solve A Friedel–Crafts alkylation reaction would require a primary carbocation intermediate, as shown on the next page. This is unfeasible, however, because a primary carbocation intermediate, if produced, would rearrange to a more stable secondary carbocation.

A primary carbocation, if produced, would undergo rearrangement.

Instead, we could incorporate a Friedel–Crafts acylation reaction to produce the new C—C bond. The aromatic ketone product could then be reduced using a Clemmensen reduction.

The final synthesis would be written as follows:

---

problem **22.16** Show how to carry out the synthesis in Solved Problem 22.15 without using the Clemmensen reduction. (*Hint*: Are there other methods to reduce a ketone?)

---

problem **22.17** Show how you would synthesize the compound at the right, using benzene as the only aromatic starting compound.

22.9 Organic Synthesis: Common Reactions Used in Conjunction with Electrophilic Aromatic Substitution Reactions

Electrophilic aromatic substitution reactions allow us to incorporate a variety of substituents onto an aromatic ring, but several substituents cannot be put in place directly via an electrophilic aromatic substitution reaction. Here we consider methods to incorporate these kinds of substituents by first using an electrophilic aromatic substitution reaction to incorporate a *different* substituent that is subsequently transformed into the desired substituent.

22.9a Oxidation of Carbon Side Chains: Synthesis of Benzoic Acids

Using the reactions we have learned thus far, how would you incorporate a carboxylic acid group into a benzene ring? Equation 22-27 shows that benzene can be brominated, and the resulting bromobenzene can undergo dissolving metal reduction to produce a Grignard reagent. Bubbling carbon dioxide through the mixture leads to a Grignard reaction that produces the carboxylate, which is protonated via an acid workup.

Bromination of the benzene ring   **Dissolving metal reduction**   **Grignard reaction**

$$\text{(22-27)}$$

An alternative way to incorporate a carboxylic acid group into the ring begins with any of a variety of monosubstituted benzenes in which a C atom from the substituent is directly bonded to the ring, as shown in Equations 22-28a through 22-28e. Oxidation of these kinds of substituted benzenes using $KMnO_4$ heated in a basic solution produces benzoic acid after acid workup. Several of these precursors could be produced using a Friedel–Crafts reaction.

$$\text{(22-28)}$$

Toluene	An alkylbenzene	An alkenylbenzene	An alkynylbenzene	An acylbenzene
(a)	(b)	(c)	(d)	(e)

1. $KMnO_4$, $HO^{\ominus}$, Δ
2. $H^{\oplus}$

Benzoic acid

These oxidation reactions are believed to proceed through *free-radical intermediates* (species with an unpaired valence electron), which are discussed in detail in Chapter 25. The C atom bonded directly to the phenyl ring—the **benzylic carbon**—is particularly reactive under these conditions. This is why the product of each of the reactions in Equation 22-28 is the same, essentially independent of the groups that are bonded to the benzylic carbon.

This permanganate oxidation fails, however, when the benzylic C is *quaternary* (i.e., bonded to four alkyl groups). 2-Methyl-2-phenylpropane, for example, is *not* oxidized to benzoic acid (Equation 22-29).

This benzylic carbon is quaternary.

$KMnO_4$ $HO^{\ominus}$, Δ No reaction $$\text{(22-29)}$$

2-Methyl-2-phenylpropane

Thus, when a substituent on a phenyl ring is attached by C, $KMnO_4$ oxidation requires the benzylic C to be bonded to at least one H atom or be part of a π bond.

In a synthesis, the benzoic acid product need not be part of the target molecule. Instead, as Solved Problem 22.18 demonstrates, the carboxylic acid functional group can be transformed into other functional groups, such as esters.

SOLVED problem 22.18 Propose a synthesis of ethyl benzoate from benzene and any other compounds containing two or fewer carbon atoms.

Think How can an ester be synthesized from a precursor with fewer carbons? How can the appropriate precursor(s) be synthesized using compounds with two or fewer carbons?

Solve In Chapter 21 we learned that an ester can be synthesized from a carboxylic acid and an alcohol. Here, the necessary alcohol is ethanol, which contains two C atoms. The carboxylic acid that is required is benzoic acid, which can be generated from the $KMnO_4$ oxidation of any of a variety of substituted benzenes, as shown in Equation 22-28. In the following retrosynthetic analysis, we have chosen phenylethanone.

In the forward direction, the synthesis would be written as follows:

---

problem 22.19 Show how you could synthesize benzoic anhydride using benzene as the only aromatic starting compound.

22.9b Reduction of Nitrobenzenes to Aromatic Amines

An amino group (—NH_2) is typically not incorporated into a benzene ring directly via electrophilic aromatic substitution. Instead, the benzene ring is often nitrated first (Equation 22-21), followed by reduction of the nitro group to an amine, either by catalytic hydrogenation (Equation 22-30a) or by treatment with a metal (e.g., Fe or Sn) under acidic conditions (Equation 22-30b). The acidic conditions resemble those of the Clemmensen reduction we first encountered in Chapter 19, in which the carbonyl group of a ketone or aldehyde is reduced to a methylene (CH_2) group by treatment with a Zn(Hg) amalgam in HCl.

The reduction of a nitro substituent to an amino group should be considered if the target is an aromatic amine. Moreover, the amine could be used as a synthetic intermediate toward the synthesis of a different target. In Section 22.9c, we will see that aromatic amines are precursors to arenediazonium ions ($Ar-N_2^+$), which can be transformed into a variety of functionalized aromatic species. And in Chapter 23, we will see that amine groups can be important in influencing the incorporation of additional groups onto the aromatic ring.

SOLVED problem 22.20 Show how to synthesize *N*-phenylbenzamide, using benzene as the only aromatic starting compound.

Think What acid derivative and amine precursors will produce the amide target? Does each of those precursors have the same carbon backbone as the benzene ring, or will you need to produce a new carbon–carbon bond? What reactions add a carbon substituent to an aromatic ring?

***N*-Phenylbenzamide**

Solve An *N*-substituted amide can be produced from an amine and an acid chloride—in this case, aniline and benzoyl chloride. Aniline can be produced by the nitration of benzene, followed by reduction of the nitrobenzene product. Benzoyl chloride can be produced by chlorinating benzoic acid, which can be produced, in turn, by acylation of benzene, followed by $KMnO_4$ oxidation.

The synthesis in the forward direction would then be written as follows:

---

problem **22.21** Show how to synthesize each of these compounds, using benzene as the only aromatic starting compound.

(a)   (b)

22.9c The Benzenediazonium Ion and the Sandmeyer Reactions

Bromobenzene is produced when aniline is treated with sodium nitrite ($NaNO_2$) under acidic conditions, followed by copper(I) bromide (CuBr) (Equation 22-31).

1. $NaNO_2$, HBr
   <5 °C

2. CuBr
   2 h, reflux

Aniline   **Bromobenzene**
                        50%

(22-31)

Overall, Br has replaced NH_2, but the mechanism is not a simple electrophilic aromatic substitution. As shown in Equation 22-32, the first of the two reactions is a multistep mechanism, which ultimately produces a **benzenediazonium ion**.

(22-32)

In Step 1, the nitrite anion is protonated, which produces nitrous acid, HONO. Protonation in Step 2 converts a poor hydroxide leaving group into an excellent H_2O leaving group, which departs in Step 3 to produce a nitrosonium ion, NO^+. The aromatic amine attacks the nitrosonium ion in Step 4, in which a N—N bond is formed. Steps 5–8 are proton transfers that serve to remove protons from the initial amino N and to produce an excellent H_2O leaving group on the adjacent N, analogous to how protons are shuttled in a keto–enol tautomerization. Finally, in Step 9, the H_2O leaving group departs, producing the benzenediazonium ion.

The benzenediazonium ion possesses an N_2^+ group, which can behave as an excellent $N_2(g)$ leaving group. Thus, the ion is highly reactive, and must be kept below $\sim 10\ °C$ to prevent decomposition or explosion.

At the same time, the high reactivity of the benzenediazonium ion makes it possible to carry out a subsequent substitution reaction simply by choosing the appropriate nucleophile. Examples are shown in Equations 22-33a through 22-33g. Some of these reactions require the presence of a Cu^+ catalyst; collectively, they are known as the **Sandmeyer reactions**, after the Swiss chemist Traugott Sandmeyer (1854–1922).

Sodium Nitrite and Foods: Preventing Botulism but Causing Cancer?

In Section 22.9c, we have seen how sodium nitrite ($NaNO_2$) is used to convert an aromatic amine to its corresponding diazonium ion. A very similar reaction is responsible for making most cured meats, such as ham and bacon, carcinogenic.

Sodium nitrite is added to a variety of cured meats to inhibit the growth of *Clostridium botulinum*, the bacterium whose toxin is responsible for botulism poisoning. It works by interfering with the metabolic processes of the bacteria. Just as we saw in Equation 22-32, however, sodium nitrite will also react with amines present in the meat, which are supplied by proteins. In Equation 22-32, a primary amine (RNH_2) is shown to proceed through an *N*-nitrosamine intermediate on its way to the diazonium ion. A secondary amine (R_2NH), on the other hand, stops at the *N*-nitrosamine, the actual compound believed to be carcinogenic. Since the 1950s, in fact, hundreds of these *N*-nitrosamines have been studied, and about 90% of them have been found to cause cancer in a variety of animals.

Notice in Equation 22-32 that the reaction requires an acidic environment, so it might seem that adding sodium nitrite to cured meats shouldn't promote the formation of *N*-nitrosamines. The stomach, however, provides a highly acidic environment. Furthermore, the production of *N*-nitrosamines is facilitated by high temperatures, such as those typically used to fry bacon.

(22-33)

Even though these reactions involve nucleophiles and an excellent leaving group, they do not universally proceed by ionic nucleophilic substitution mechanisms. Some of these reactions do indeed resemble S_N1 mechanisms, but others are believed to involve *radical intermediates*, which contain unpaired electrons (Chapter 25). Because of the complex nature of these reactions, their mechanisms are not discussed here.

problem **22.22** How would you synthesize deuterobenzene from aniline?

Some of the products in Equation 22-33 can be produced using electrophilic aromatic substitution reactions we have seen previously here in Chapter 22. For example, bromobenzene and chlorobenzene can be produced using halogenation reactions (Section 22.2). However, the benzenediazonium ion intermediate provides much more flexibility in synthesis, allowing us to synthesize compounds that cannot be synthesized directly via electrophilic aromatic substitution. Recall, for example, that difficulties arise with the iodination and fluorination of benzene via electrophilic aromatic substitution. Beginning with the arenediazonium ion, however, iodobenzene (Equation 22-33d) and fluorobenzene (Equation 22-33e) can be synthesized rather straightforwardly. Similarly, cyanobenzene and phenol cannot be produced directly from benzene via electrophilic aromatic substitution, but can be produced by incorporating the benzenediazonium ion as a synthetic intermediate. This is exemplified by Solved Problem 22.23.

SOLVED problem **22.23** Show how to synthesize cyanobenzene from benzene.

Think Can a cyano group replace H on a benzene ring directly? If not, what precursor is required? Can that precursor be synthesized from benzene using electrophilic aromatic substitution?

Cyanobenzene

Solve We have not encountered a reaction in which a cyano group directly replaces an H^+ on benzene. However, cyanobenzene can be made from a benzenediazonium ion precursor, C_6H_5—N_2^+. The precursor to the benzenediazonium ion is an aromatic amine, which can be made, in turn, by reducing nitrobenzene. Finally, nitrobenzene can be made directly from benzene via nitration.

In the forward direction, the synthesis would be written as follows:

problem **22.24** Show how to synthesize phenol from benzene.

22.10 Wrapping Up and Looking Ahead

Chapter 22 has introduced a variety of reactions in which a proton (H$^+$) on an aromatic ring is replaced by another group. The mechanisms for these reactions have two steps in common: addition of an electrophile (E$^+$) to the aromatic ring, producing an arenium ion intermediate, followed by elimination of H$^+$ to regenerate the aromatic ring. As a result, these reactions make up a class of reactions called *electrophilic aromatic substitution*.

Because electrophilic aromatic substitution reactions temporarily destroy the aromaticity of the ring, they require very strong electrophiles. These electrophiles are generally too reactive to be isolated, so they cannot be added directly to the reaction mixture. Rather, they must be produced in situ from appropriate precursors. Thus, the mechanisms for the various electrophilic aromatic substitution reactions differ not only in the identity of the electrophile, but also in the mechanism that describes how the electrophile is produced.

Some electrophilic aromatic substitution reactions have significant limitations that must be considered in a synthesis. Friedel–Crafts alkylations, for example, are susceptible to carbocation rearrangements, preventing the direct formation of a primary alkylbenzene other than methylbenzene and ethylbenzene. A primary alkylbenzene can be produced, however, by first carrying out a Friedel–Crafts acylation, followed by a reduction of the carbonyl group in the resulting ketone to a methylene group.

Some additional reactions were also introduced, which can be used to modify functional groups on an aromatic ring after they have been incorporated via electrophilic aromatic substitution. Namely, we studied the oxidation of carbon side chains on a benzene ring, the reduction of nitrobenzenes to aromatic amines, and the production and reaction of benzenediazonium ions. Using these reactions, we can incorporate functional groups into an aromatic ring that are difficult or impossible to incorporate directly using electrophilic aromatic substitution. For example, a carboxylic acid group can be added to a benzene ring by first alkylating or acylating the ring, and then oxidizing the new carbon side chain. An aromatic amine is produced by first nitrating the aromatic ring and then reducing the nitro group to an amino group. The aromatic amine that is produced can be converted to a benzenediazonium ion, which is a precursor to a wide variety of substitution products.

The scope of Chapter 22 was limited primarily to reactions involving benzene. In Chapter 23, we focus on aromatic substitution reactions involving substituted benzenes and other aromatic compounds, in which multiple isomeric products are possible. We will find that the resulting product mixture is usually dominated by one or two of those products. Predicting which products those are involves careful consideration of the rates of the reactions that lead to each product.

Chapter Summary and Key Terms

- The general mechanism for an **electrophilic aromatic substitution reaction** consists of two steps. First, an electrophile (E$^+$) adds to an aromatic ring to produce an **arenium ion intermediate**, also called a **Wheland intermediate**. Second, a proton (H$^+$) is eliminated, resulting in a substituted aromatic ring. **(Section 22.1)**

- The reaction temporarily destroys the aromaticity of the ring, so the electrophiles must be powerful (in which case the electrophiles must be generated in situ). Moreover, electrophilic aromatic substitution reactions are generally irreversible. **(Section 22.1)**

- The first step of electrophilic aromatic substitution—addition of the electrophile—is slow, so it is the rate-determining step of the reaction. Thus, the reaction is first order with respect to both the aromatic ring and the electrophile. **(Section 22.1)**

- **Aromatic halogenation** typically requires a molecular halogen, Br_2 or Cl_2, in the presence of a strong **Lewis acid catalyst**, such as $FeBr_3$ or $FeCl_3$. These catalysts aid in the production of the electrophile, Br^+ or Cl^+. **(Section 22.2)**

- A **Friedel–Crafts alkylation** involves the formation of a carbocation (R^+) as the electrophile, which can be produced from an alkyl halide in the presence of a strong Lewis acid catalyst, such as $AlCl_3$. **(Section 22.3)**

- Friedel–Crafts alkylation reactions are susceptible to carbocation rearrangements, so they cannot be used to produce primary alkylbenzenes other than methylbenzene and ethylbenzene. **(Section 22.4)**

- In a Friedel–Crafts alkylation, the halogen atom of the alkyl halide must be bonded to an sp^3-hybridized carbon atom. **(Section 22.4)**

- A **Friedel–Crafts acylation reaction** requires an acid halide or acid anhydride in the presence of a strong Lewis acid catalyst. The catalyst is responsible for the production of the electrophile, an **acylium ion**, $R—C{\equiv}O^+$. These reactions are not susceptible to carbocation rearrangements. **(Section 22.5)**

- **Nitration** of an aromatic ring requires concentrated nitric acid (HNO_3), which produces the **nitronium ion** (NO_2^+) as the electrophile. The addition of concentrated sulfuric acid (H_2SO_4) increases the concentration of the nitronium ion, and therefore increases the rate of the electrophilic aromatic substitution reaction. **(Section 22.6)**

- **Sulfonation** of an aromatic ring requires concentrated sulfuric acid to produce sulfur trioxide (SO_3) as the electrophile. The rate of sulfonation is greater with **fuming sulfuric acid**, which is enriched with SO_3. **(Section 22.7)**

- Sulfonation is *reversible*. A sulfonic acid group on an aromatic ring can be replaced by H^+ upon treatment with H_3O^+. **(Section 22.7)**

- Although most primary alkylbenzenes cannot be produced directly from a Friedel–Crafts alkylation, they can be produced using a Friedel–Crafts acylation, followed by a reduction of the ketone's carbonyl group to a methylene (CH_2) group. **(Section 22.8)**

- A variety of carbon side chains on an aromatic ring can be oxidized to a carboxylic acid by heating the aromatic compound in the presence of basic $KMnO_4$, followed by acid workup. **(Section 22.9a)**

- An aromatic amine can be produced by reducing a nitro-substituted aromatic ring, using either catalytic hydrogenation [e.g., $H_2(g)$, Pd] or dissolving metal reduction (e.g., Fe, HCl). **(Section 22.9b)**

- A **benzenediazonium ion** ($C_6H_5—N_2^+$) is produced when an amino-substituted benzene is treated with sodium nitrite ($NaNO_2$) under acidic conditions. The benzenediazonium ion can undergo a variety of substitution reactions readily, due to the excellent $N_2(g)$ leaving group. Reactions involving a Cu^{+1} catalyst are called **Sandmeyer reactions**. **(Section 22.9c)**

Reaction Tables

TABLE 22-1 Functional Group Conversions

	Starting Functional Group	Typical Reagent and Reaction Conditions	Functional Group Formed	Key Electron–Rich Species	Key Electron–Poor Species	Comments	Discussed in Section
(1)	Arene	Br_2 / $FeBr_3$	Aryl bromide (Br)		$\overset{\oplus}{.\ddot{Br}.}$	Electrophilic aromatic substitution	22.2
(2)	Aryl amine (NH_2)	1. $NaNO_2$, H_2SO_4 2. CuBr	Aryl bromide (Br)	CuBr	$N_2^{\oplus}$	Proceeds through a diazonium ion	22.9c

(continued)

TABLE 22-1 Functional Group Conversions (continued)

	Starting Functional Group	Typical Reagent and Reaction Conditions	Functional Group Formed	Key Electron–Rich Species	Key Electron–Poor Species	Comments	Discussed in Section
(3)	Arene	Cl_2 / $FeCl_3$	Aryl chloride	(arene)	$:\overset{\cdot\cdot}{Cl}{}^{\oplus}$	Electrophilic aromatic substitution	22.2
(4)	Aryl amine (NH_2)	1. $NaNO_2$, H_2SO_4 2. CuCl	Aryl chloride (Cl)	CuCl	aryl $N_2^{\oplus}$	Proceeds through a diazonium ion	22.9c
(5)	Aryl amine (NH_2)	1. $NaNO_2$, H_2SO_4 2. KI	Aryl iodide (I)	KI	aryl $N_2^{\oplus}$	Proceeds through a diazonium ion	22.9c
(6)	Aryl amine (NH_2)	1. $NaNO_2$, H_2SO_4 2. HBF_4, then Δ	Aryl fluoride (F)	HBF_4	aryl $N_2^{\oplus}$	Proceeds through a diazonium ion	22.9c
(7)	Aryl amine (NH_2)	1. $NaNO_2$, H_2SO_4 2. H_2O, $Cu^{\oplus}$	Aryl alcohol (OH)	H_2O	aryl $N_2^{\oplus}$	Proceeds through a diazonium ion	22.9c
(8)	Aryl amine (NH_2)	1. $NaNO_2$, H_2SO_4 2. H_3PO_2	Arene (H)	N/A	aryl $N_2^{\oplus}$	Proceeds through a diazonium ion	22.9c
(9)	Nitroarene (NO_2)	HCl / Fe	Aryl amine (NH_2)	N/A	N/A	Reduction	22.9b
(10)	Arene	HNO_3 / H_2SO_4	Nitroarene (NO_2)	(arene)	$^{\oplus}NO_2$	Electrophilic aromatic substitution; nitration	22.6
(11)	Arene	SO_3 / H_2SO_4	Arene sulfonic acid (SO_3H)	(arene)	SO_3	Electrophilic aromatic substitution; sulfonation	22.7
(12)	Arene sulfonic acid (SO_3H)	$H_3O^{\oplus}$	Arene (H)	arene sulfonic acid (SO_3H)	$H-\overset{\oplus}{O}H_2$	Electrophilic aromatic substitution	22.7

TABLE 22-2 Reactions That Alter the Carbon Skeleton

	Starting Functional Group	Typical Reagent and Reaction Conditions	Functional Group Formed	Key Electron–Rich Species	Key Electron–Poor Species	Comments	Discussed in Section
(1)	Arene	$R-Cl$ / $AlCl_3$	Alkylarene (with R)	(arene)	$R^{\oplus}$	Electrophilic aromatic substitution; Friedel–Crafts alkylation	22.3
(2)	Arene	$\overset{O}{\underset{Cl}{\parallel}}{C}{-}R$ / $AlCl_3$	Acylarene	(arene)	Acylium ion	Electrophilic aromatic substitution; Friedel–Crafts acylation	22.5
(3)	Alkylarene (with R)	1. $KMnO_4$, $OH^{\ominus}$, Δ 2. $H^{\oplus}$	Aromatic carboxylic acid (CO_2H)	N/A	N/A	Oxidation	22.9a
(4)	Acylarene	1. $KMnO_4$, $OH^{\ominus}$, Δ 2. $H^{\oplus}$	Aromatic carboxylic acid (CO_2H)	N/A	N/A	Oxidation	22.9a
(5)	Aryl amine (NH_2)	1. $NaNO_2$, H_2SO_4 2. CuCN	Aryl nitrile (CN)	CuCN	(diazonium, $N_2^{\oplus}$)	Proceeds through a diazonium ion	22.9c

Problems

22.25 Identify compounds A–G in the following synthesis scheme.

22.26 Identify compounds **A–G** in the following synthesis scheme.

22.27 Identify compounds **A–I** in the following synthesis scheme.

22.28 Propose a mechanism for the isotopic exchange reaction at the right.

22.29 In the acid-catalyzed aromatic alkylation involving 1-methylcyclohexene and benzene, two isomeric products are possible, but only one is formed:

Draw the complete mechanism that leads to each product, and explain why only one isomer is formed.

22.30 When benzene is treated with sulfur dichloride (SCl$_2$) in the presence of a Friedel–Crafts catalyst like AlCl$_3$, diphenyl sulfide is produced:

Diphenyl sulfide

Propose a mechanism for this reaction.

22.31 The following reaction is a halosulfonation, which is a useful variation of the sulfonation reaction.

Benzenesulfonyl chloride
77%

Draw the complete mechanism for this reaction.

22.32 Propose a mechanism for the following reaction. Note that the reaction does *not* take place without the presence of AlCl$_3$.

22.33 Propose a mechanism for the following reaction.

22.34 Two isomeric products can be produced from the reaction at the right.
(a) Draw the complete, detailed mechanism showing the formation of each of those products.
(b) Which of those products will be formed in greater abundance?

22.35 Halogenation using the mixed halogen ICl can feasibly lead to two different products:

Draw the complete, detailed mechanism for the formation of each product. Which do you think is the major product? Why?

22.36 Draw the complete, detailed mechanism for the reaction at the right. Will the product be optically active? Explain.

22.37 Previously we mentioned that formylation of benzene (i.e., replacement of H by HC=O) cannot be carried out through a standard Friedel–Crafts acylation because methanoyl chloride (formyl chloride) cannot be added directly. The **Gattermann–Koch synthesis** circumvents this problem by making methanoyl chloride in situ, using a gaseous mixture of carbon monoxide and hydrochloric acid at high pressures.

Formyl chloride is unstable, so it is produced only temporarily.

Formyl chloride

With this in mind, draw the detailed mechanism for the electrophilic aromatic substitution that takes place in the following reaction.

$$\text{benzene} + CO(g) + HCl(g) \xrightarrow{AlCl_3} \textbf{?}$$

22.38 The reaction at the right is an example of a *deiodination* reaction. Without $AlCl_3$ present, no reaction occurs. Draw a complete, detailed mechanism for this reaction.

22.39 An example of a deiodination reaction is provided in Problem 22.38. Halogens other than iodine may be replaced on a benzene ring, but the reactions rates are slower; rates of substitution are observed to be: Ar—F < Ar—Cl < Ar—Br < Ar—I. Explain.

$$\text{3-iodobenzaldehyde} \xrightarrow[AlCl_3]{HI} \text{benzaldehyde} + I_2$$

22.40 No reaction occurs when benzene is treated with Br_2 in CCl_4. When anthracene is treated with Br_2 in CCl_4, however, then addition of Br_2 occurs, as shown here. Explain why.

$$\text{Anthracene} \xrightarrow[CCl_4]{Br_2} \text{9,10-dibromo-9,10-dihydroanthracene}$$

Anthracene

22.41 Predict the product of the following reaction.

$$\xrightarrow{H^{\oplus}} \textbf{?}$$

22.42 Draw the complete, detailed mechanism for the following reaction.

$$\xrightarrow{H^{\oplus}}$$

22.43 Benzene can be *hydroxylated* by treating it with hydrogen peroxide and a strong acid such as trifluoromethanesulfonic acid (TfOH).

$$\text{benzene} \xrightarrow[TfOH]{H_2O_2} \text{phenol}$$

Propose a mechanism for this reaction.

22.44 Show how to carry out the following synthesis using benzene and any alcohol as your only source of carbon.

$$\text{benzene} \xrightarrow{\textbf{?}} \text{product}$$

22.45 *N*-Phenyl benzamide (benzanilide) is believed to have anti-atherosclerotic properties—that is, it counters the effects of atherosclerosis (artery hardening), a cardiovascular disease.

**N-Phenylbenzamide
(Benzanilide)**

Show how to synthesize benzanilide using benzene and any alkyl halide as your only source of carbon.

22.46 Dibenzyl ether is used as a flavor and fragrance agent.

Dibenzyl ether

Show how to synthesize dibenzyl ether using benzene and any ketone or aldehyde as your only source of carbon.

22.47 Stilbene is used in the manufacture of dyes and also has estrogenic activity.

Stilbene

Show how to synthesize stilbene using benzene and any carboxylic acid as your only sources of carbon.

22.48 Which of the following isomers of trimethylbenzene will produce exclusively one monobrominated product when treated with Br$_2$ and FeBr$_3$? Explain.

A　　　　　　　**B**　　　　　　　**C**

22.49 An isomer of tetramethylbenzene undergoes nitration to yield a single product. Based on this information, which isomer of tetramethylbenzene could the starting material have been?

22.50 Compounds **A**, **B**, and **C** are isomers of xylene (dimethylbenzene). When each of these isomers undergoes a single nitration, compound **A** produces just one product, **B** produces a mixture of two products, and **C** produces a mixture of three products. Identify which compound is the ortho isomer, which is the meta isomer, and which is the para isomer.

22.51 The reactant in the reaction at the right is aromatic and will undergo electrophilic aromatic substitution in much the same way that benzene does. This reaction leads to a single product only. Explain why. Predict the product of the reaction and draw the complete, detailed mechanism for its formation.

22.52 In each of the following reactions, the aromatic ring has just one chemically distinct, aromatic H, so a single electrophilic aromatic substitution will lead to just a single product. With this in mind, predict the product of each of these reactions.

(a)

(b)

(c)

(d)

(e)

22.53 In each of the following reactions, the aromatic ring has just one chemically distinct, aromatic H, so a single electrophilic aromatic substitution will lead to just a single product. With this in mind, supply the missing reagents needed to carry out each transformation.

(a)

(b)

(c)

(d)

(e)

(f)

(g)

22.54 An aromatic imine is formed when 3,5-dihydroxyphenol is treated with ethanenitrile (CH_3CN) in HCl in the presence of a $ZnCl_2$ catalyst.

Propose a mechanism for this reaction.

22.55 Propose a mechanism for the Pictet–Spengler reaction, an example of which is shown here. Note that a key intermediate is provided.

22.56 (a) Predict the product of the following set of reactions.
(b) Draw the complete, detailed mechanism for the formation of the synthetic intermediate that is not shown.

22.57 When diphenyl ether is reacted under the same conditions as in Problem 22.30, a compound is produced whose ^{13}C NMR spectrum shows six signals. Draw that product.

22.58 When the following acid chloride is treated with $AlCl_3$, a product is formed whose ^1H NMR and IR spectra are shown here. Draw the product and propose a complete, detailed mechanism for the first of the two reactions.

22.59 The product of the following set of reactions exhibits eight signals in its ^{13}C NMR spectrum.

Draw a complete, detailed mechanism for the formation of that product.

22.60 When benzene is treated with dichloromethane in the presence of aluminum trichloride, a product is formed whose IR, ^{13}C NMR, and ^{1}H NMR spectra are shown below.

Draw the product and propose a mechanism for this reaction.

If your instructor assigns problems in smartwork, log in at **smartwork.wwnorton.com**.

Electrophilic Aromatic Substitution 2

Substitution on Mono- and Disubstituted Benzene and Other Aromatic Rings

In Chapter 22, we examined a variety of *electrophilic aromatic substitution* reactions, focusing primarily on substitution involving benzene (C_6H_6), which has no substituents. Electrophilic aromatic substitution on a **monosubstituted benzene** (C_6H_5—Sub), however, involves additional issues.

One of those issues is *regiochemistry*. In benzene itself, all six hydrogen atoms are equivalent, so substitution of any one of them leads to precisely the same product. A monosubstituted benzene, however, has three chemically distinct hydrogens—the ortho, meta, and para hydrogens. Substitution, therefore, can lead to three different possible products—namely, the ortho-, meta-, and para-disubstituted benzenes (Equation 23-1), depending upon which hydrogen is replaced. *(Notice the convention that is used to denote a generic disubstituted benzene without designating the specific isomer.)*

A generic monosubstituted benzene

| Ortho | Meta | Para |

(23-1)

A generic disubstituted benzene

Aspirin, or acetylsalicylic acid, is one of the most widely produced and consumed medicines. A key step in its industrial synthesis is a regioselective electrophilic aromatic substitution reaction, one of the main concepts of Chapter 23.

As we will see here in Chapter 23, the substituent already in place on the benzene ring prior to substitution dictates the regiochemistry of the reaction. Some substituents are designated as ortho/para directors because they lead to product mixtures consisting primarily of the ortho- and para-disubstituted products. Other substituents are designated as meta directors because they promote the formation of the meta-disubstituted product.

A second issue with monosubstituted benzenes is the effect the substituent has on the *rate* of electrophilic aromatic substitution. Some substituents slow the reaction and therefore are called *deactivating groups*, whereas other substituents, called *activating groups*, cause the reaction to speed up.

One of the main goals of Chapter 23 is to solidify your understanding of the factors that control both regiochemistry and the kinetics of electrophilic aromatic substitution. Ultimately, you should be able to apply this knowledge toward designing efficient syntheses of aromatic rings with multiple substituents.

23.1 Regiochemistry of Electrophilic Aromatic Substitution: Identifying Ortho/Para and Meta Directors

The nitration of phenol, a monosubstituted benzene, can lead to the ortho-, meta-, and para-disubstituted products. As shown in Equation 23-2, however, the ortho- and para-disubstituted products dominate the product mixture. The hydroxyl (—OH) group, therefore, is an **ortho/para director**.

The ortho and para products dominate the product mixture, so OH is designated as an ortho/para director.

$$\text{Phenol} \xrightarrow[\text{CH}_3\text{CO}_2\text{H, 60 °C}]{\text{HNO}_3} \text{Ortho} + \text{Meta} + \text{Para} \quad (23\text{-}2)$$

	Ortho	Meta	Para
Product ratio:	50%	~0%	50%

When nitrobenzene undergoes nitration, on the other hand, *meta*-dinitrobenzene is the major product (Equation 23-3). Thus, the NO$_2$ group initially attached to the ring is a **meta director**.

The meta product dominates the product mixture,
so NO₂ is designated as a meta director.

$$\text{Nitrobenzene} \xrightarrow[\substack{\text{30 min, 0 °C;} \\ \text{24 h, 25 °C}}]{\substack{\text{HNO}_3 \\ \text{H}_2\text{SO}_4}} \text{Ortho} + \text{Meta} + \text{Para} \qquad \textbf{(23-3)}$$

Product ratio: Ortho 7%   Meta 91%   Para 2%

The results from these and other nitration reactions are summarized in Table 23-1. The OH group and other ortho/para-directing substituents appear on the left side of the table; the NO₂ group and other meta-directing substituents appear on the right. To help illustrate which type of director each substituent is, the sum of the relative percentages of the ortho and para products ("O+P") is also listed in the table.

YOUR TURN 23.1

The iodo (—I) and formyl (—CHO) groups are not listed in Table 23.1. The relative amounts of ortho and meta nitration products are as follows for each substituent:

Substituent	O	M	P	O+P	Type of Director
—I	45	1			
—CHO	19	72			

(a) Supply the missing information pertaining to the relative amounts of the para isomers that are produced, as well as the sum of the amounts of ortho and para products.

(b) Based on that information, determine whether each substituent is an ortho/para director or a meta director.

Answers to Your Turns are in the back of the book.

TABLE 23-1 Product Distribution of the Nitration of Various Monosubstituted Benzenes

Ortho/Para-Directing Substituents					Meta-Directing Substituents				
Substituent	O	M	P	O+P	Substituent	O	M	P	O+P
—OH	50	0	50	100	—NO₂	7	91	2	9
—NHCOCH₃	19	2	79	98	—N(CH₃)₃⁺	2	87	11	13
—CH₃	63	3	34	97	—CO₂H	22	76	2	24
—F	13	1	86	99	—CN	17	81	2	19
—Cl	35	1	64	99	—CO₂Et	28	66	6	34
—Br	43	1	56	99	—COCH₃	26	72	2	28

In the next few sections, we will examine why these substituents have the effects on regiochemistry that they do. We will look at three prototypical substituents in particular: the OH, CH_3, and NO_2 groups. What we learn can be extended to other similar groups.

---

problem 23.1 Based on the information in Table 23.1, predict the major product(s) for each of the following reactions. Draw the complete, detailed mechanism that leads to the formation of each major product.

(a)   (b)   (c)   (d)

23.2 Nitration of Phenol: Why Is a Hydroxyl Group an Ortho/Para Director?

According to Table 23-1, the OH group of phenol (HO—C_6H_5) directs essentially 100% of incoming electrophiles to the ortho and para positions, making it an ortho/para director. To begin to understand why, recall from Section 22.1 that electrophilic aromatic substitution reactions usually run under *kinetic control*. Therefore,

> The relative amounts of ortho-, meta-, and para-disubstituted products that are produced in an electrophilic aromatic substitution reaction are proportional to the rates at which they are produced.

For example, phenol leads to a distribution of products that is 50% ortho, ~0% meta, and 50% para, so the rate of formation of the ortho product is about the same as that of the para product, but the formation of the meta product is orders of magnitude slower.

Recall also that, of the two steps that formally make up electrophilic aromatic substitution, the first step is *rate determining* (Section 22.1). That step is substantially endothermic, producing the unstable arenium ion intermediate (Fig. 22-1). Therefore, the rates of formation of the ortho and para arenium ions are about the same for the reaction in Equation 23-2, and the formation of the meta arenium ion is much slower. Figure 23-1 shows how these ideas translate into energy diagrams for the production of the ortho and meta intermediates. Namely, the energy barrier to forming the ortho arenium ion is smaller than the barrier to forming the meta one.

Because the ortho and meta intermediates are produced by the same elementary step, we can invoke the *Hammond postulate* (Section 9.3a) to say that the relative sizes of the energy barriers correspond to the ortho arenium ion being more stable than the meta one. Thus, we arrive at an important concept to help us predict the outcome of an electrophilic aromatic substitution:

> In general, the major product of electrophilic aromatic substitution is the one that is produced from the most stable arenium ion intermediate.

FIGURE 23-1 Energy diagrams for the rate-determining step in the nitration of phenol Nitration at the ortho position is represented by the red curve, and nitration at the meta position is represented by the blue curve. Because the ortho product is produced in greater abundance than the meta product (Table 23-1), we can say that the $\Delta G^{\circ\ddagger}$ leading to the ortho intermediate is smaller. The *Hammond postulate* thus allows us to say that the ortho arenium ion is more stable than the meta arenium ion.

problem **23.2** Consider these arenium ion intermediates that are formed in the nitration of fluorobenzene. Using Table 23-1, determine which one is formed the fastest. Which is formed the slowest? Which is the most stable? Which is the least stable?

A   B   C

The relative stabilities of the isomeric arenium ion intermediates are governed in large part by differences in charge stability. We can see how this applies to the ortho and meta intermediates in the nitration of phenol, whose resonance structures are shown in Equations 23-4 and 23-5.

All atoms have octets, making this resonance structure especially stable.

Ortho intermediate

(23-4)

The lone pairs on O cannot participate in resonance to stabilize the positive charge.

Meta intermediate

(23-5)

In particular, the positive charge that develops is better stabilized in the ortho intermediate than it is in the meta intermediate for two reasons. First, the ortho intermediate has four resonance structures, serving to distribute the positive charge over four different atoms, whereas the meta intermediate has only three. As indicated above, the additional resonance structure of the ortho intermediate arises from the fact that a lone pair of electrons from the OH oxygen can participate in resonance to help delocalize the positive charge. This is not true in the meta intermediate; the positive charge does not appear on the ring atom attached to the O in any resonance structure, so the positive charge cannot be delocalized onto O. Second, the additional resonance structure that the ortho arenium ion has (highlighted in green) is substantially more stable than the other three. It is the only resonance structure in which *all atoms have their octet*. Therefore, the additional stability provided by that single resonance structure is extremely important.

YOUR TURN 23.2

Identify all of the atoms in Equations 23-4 and 23-5 that do *not* have an octet.

Similar reasoning allows us to rationalize why the nitration of phenol at the para position is also much faster than at the meta position, thereby making the OH group an ortho/para director (see Solved Problem 23.3). The para arenium ion intermediate has four resonance structures, including one in which the positive charge is on the O atom. As with the ortho arenium ion, all atoms in that additional resonance structure have a complete octet of valence electrons.

SOLVED problem 23.3 (a) Draw a diagram similar to Figure 23-1 that illustrates the formation of the para and meta intermediates during the nitration of phenol. (b) Which of those two arenium ions is more stable and why?

Think Which of the isomeric products—meta or para—is formed faster? What does that say about the stability of the respective intermediates? What role is played by resonance delocalization of the charge?

Solve According to Table 23-1, the para arenium ion is formed faster than the meta, so the para intermediate must be more stable. This gives rise to the energy diagram on the right.

 Whereas the meta intermediate has only three resonance structures, the para intermediate has four, similar to the ortho intermediate. Furthermore, the one highlighted in green is the most important of the four, because all of its atoms have complete octets.

All atoms have octets

problem 23.4 Draw a plot similar to Figure 23-1 that shows the formation of the ortho and meta intermediates during the nitration of fluorobenzene. Use resonance arguments to explain the relative stabilities of the two isomeric intermediates. (*Hint:* See Problem 23.2.)

> Substituents that are bonded to the benzene ring by an atom possessing a lone pair of electrons tend to be ortho/para directors.

In addition to the OH group, this is also true for the $NHCOCH_3$, F, Cl, and Br groups, because the lone pair of electrons is vital in providing the fourth—and most important—resonance structure in the arenium ion intermediate upon electrophilic attack at either the ortho or para positions.

problem 23.5 Draw the mechanism that leads to the ortho-, meta-, and para-disubstituted products for each of the following reactions, and identify the major products.

(a) $\xrightarrow[\text{FeCl}_3]{\text{Cl}_2}$?

(b) $\xrightarrow[\text{AlCl}_3]{}$?

Iodized Salt and Electrophilic Aromatic Substitution

When table salt (NaCl) is sold as "iodized," it contains a small amount of potassium iodide (KI). Have you ever wondered why? What is the importance of iodide in our diet?

Your body needs trace amounts of iodide for proper growth and development. A deficiency in iodide can result in goiter—a condition that is characterized by an enlarged thyroid gland—and other symptoms related to hypothyroidism (a deficiency in the production of thyroid hormone), including fatigue, increased sensitivity to cold, and unexplained weight gain. A deficiency of iodide in a pregnant or nursing woman puts the child at risk for brain damage.

Thyroglobulin

Thyroxine

Iodide is important in our diets because it is necessary for the biosynthesis of thyroxine, a thyroid hormone, which is shown above on the right. The biosynthesis of thyroxine begins with the iodination of phenolic rings of tyrosine residues in the protein thyroglobulin. The iodination reactions proceed by an electrophilic aromatic substitution mechanism, in which the electrophile behaves like I^+. The electrophile is produced, in turn, from the oxidation of I^-, which is catalyzed by the enzyme iodoperoxidase. Notice that all of the substitutions occur ortho to the OH group in each phenolic ring, just as we would expect from the ortho/para-directing capability of the OH group.

23.3 Nitration of Toluene: Why Is CH$_3$ an Ortho/Para Director?

Table 23-1 indicates that the nitration of toluene also yields a mixture that is predominantly the ortho- and para-disubstituted products. The CH$_3$ group is thus an ortho/para director. Unlike the OH group in phenol, however, the CH$_3$ group in toluene is bonded to the ring by an atom (an uncharged C) that does *not* possess a lone pair of electrons. Therefore, the reason cannot be due to the presence of an *additional* resonance structure in the arenium ion intermediate. What, then, can it be?

Because the dominant overall products are the ortho and para isomers, the ortho and para arenium ion intermediates must be more stable than the meta arenium ion.

By drawing all of the resonance structures, as is done for the ortho and meta intermediates in Equations 23-6 and 23-7, we can see why.

This resonance structure is especially stable because the CH₃ group is electron donating and stabilizes the adjacent positive charge.

Ortho intermediate

(23-6)

None of the resonance structures has the CH₃ group adjacent to the positive charge, so there is not as much stabilization.

Meta intermediate

(23-7)

There are three resonance structures in each of the isomeric intermediates. If the NO_2^+ electrophile adds ortho to the methyl group (Equation 23-6), then two of the resonance structures have the positive charge on an unsubstituted C atom of the ring, and the third (highlighted in green) has the positive charge on a C atom attached to a CH_3 group. The resonance structure with the positive charge on the C atom attached to CH_3 is the most stable of the three, owing to the electron-donating ability of the CH_3 group, which stabilizes nearby positive charges. If the NO_2^+ electrophile adds meta to the methyl group (Equation 23-7), instead, then all three of the intermediate's resonance structures have the formal positive charge on an unsubstituted C atom. Without contribution from a resonance structure with the positive charge adjacent to the CH_3 group, the meta arenium ion is not as stable as the ortho arenium ion.

As with the nitration of phenol, the arguments that hold for the ortho intermediate also hold for the para intermediate (see Your Turn 23.3). Thus, in the nitration of toluene, the para intermediate is also more stable than the meta intermediate.

23.3 YOUR TURN

The following resonance structures are for the para arenium ion intermediate produced during the nitration of toluene.

Para intermediate

Identify the one that is responsible for making this intermediate more stable than the meta arenium ion intermediate.

All alkyl groups stabilize carbocations in the same way, so, given that CH$_3$ is an ortho/para director,

All alkyl groups are ortho/para directors.

problem **23.6** Draw all possible products that can be produced in the nitration of ethylbenzene. Which are the major products? Explain.

23.4 Nitration of Nitrobenzene: Why Is NO$_2$ a Meta Director?

According to Table 23-1, the major product in the nitration of nitrobenzene is *meta*-dinitrobenzene. As a result, the NO$_2$ group initially attached to the phenyl ring is a meta director. But why?

To answer this question, we must again examine the stabilities of the various arenium ion intermediates. This is done for the ortho and meta intermediates in Equations 23-8 and 23-9, respectively.

Ortho intermediate

The attached NO$_2$ is electron withdrawing and significantly destabilizes the positive charge on C.

(23-8)

Meta intermediate

In no resonance structure is the positive charge on C destabilized by an attached NO$_2$.

(23-9)

Although each arenium ion has three resonance structures, the ortho intermediate is significantly less stable than the meta intermediate, so the formation of the ortho-disubstituted product is *slower* than the formation of the meta-disubstituted product. This decrease in stability arises because the ortho intermediate has a resonance structure (highlighted in red in Equation 23-8) in which the positive charge on the ring is adjacent to the NO$_2$ group that was initially present. An NO$_2$ group is highly electron withdrawing, in large part due to the positive charge that appears on N, which destabilizes the adjacent positive charge on C. That resonance structure, as a result, is significantly higher in energy than the other two, so it does not contribute as much to the resonance hybrid. By contrast, none of the three resonance structures of the meta intermediate has the positive charge of the ring adjacent to the NO$_2$ group.

Nitration of nitrobenzene at the para C atom proceeds through an arenium ion intermediate that has the following three resonance structures.

Para intermediate

Identify the *least* stable resonance structure.

Knowing why NO_2 is a meta director allows us to make the following generalization:

Meta-directing groups are generally electron withdrawing and are attached to the ring by an atom that has no lone pair of electrons.

The absence of a lone pair on the atom at the point of attachment prevents stabilization of the arenium ion via resonance. The electron-withdrawing nature of the group, moreover, makes one resonance structure of the intermediate particularly unstable when the incoming electrophile attaches at either the ortho or para position.

SOLVED problem **23.7** The trifluoromethyl group (CF_3) is not listed in Table 23-1. Predict whether CF_3 is an ortho/para director or a meta director.

Think Which of the arenium ions exhibit resonance structures in which the positive charge on the ring is directly adjacent to the CF_3 substituent? What effect does the CF_3 group have on an adjacent positive charge—is it stabilizing or destabilizing?

Solve The ortho and meta arenium ions, with all of their resonance structures, are as follows:

Ortho intermediate

The positive charge is destabilized.

Meta intermediate

The ortho isomer has a resonance structure (highlighted in red) in which the positive charge is adjacent to the CF_3 group. Because CF_3 is electron withdrawing, it destabilizes the adjacent positive charge. Overall, then, the ortho intermediate is less stable than the meta intermediate. Analysis of the para intermediate in the same way shows that the para intermediate is also less stable than the meta intermediate. Hence the CF_3 group is a meta director.

---

problem **23.8** Verify the statement in Solved Problem 23.7 that the para intermediate is less stable than the meta intermediate. Begin by drawing all resonance structures of the para intermediate, and note the relative stability of each structure.

23.5 The Activation and Deactivation of Benzene toward Electrophilic Aromatic Substitution

The discussion to this point in Chapter 23 has focused on how a substituent affects the rates of formation of the ortho, meta, and para products, *relative to one another*, for a single electrophilic aromatic substitution reaction. We have yet to discuss how substituents affect the *overall* rate of an electrophilic aromatic substitution reaction—that is, the rate of disappearance of the aromatic reactant relative to unsubstituted benzene. Let's begin by examining Table 23-2, which contains the relative overall rates of nitration for a variety of monosubstituted benzenes.

TABLE 23-2 Relative Rates of Nitration of Monosubstituted Benzenes

Substituent	Relative Rate	Type of Group
—NH$_2$	N/A[a]	Strongly activating
—OH	1000	Strongly activating
—CH$_3$	25	Weakly activating
—H (benzene)	1 (reference)	–
—I	0.18	Weakly deactivating
—Cl	0.033	Weakly deactivating
—CO$_2$Et	0.0037	Moderately deactivating
—NO$_2$	6×10^{-8}	Strongly deactivating
—$\overset{\oplus}{N}(CH_3)_3$	1.2×10^{-8}	Strongly deactivating

[a]Aromatic amines are susceptible to protonation and oxidation under nitration conditions. The NH$_2$ group is determined to be a strongly activating group using other electrophilic aromatic substitution reactions.

Formed the slowest

The NO₂ group destabilizes the arenium ion intermediate and raises the energy barrier.

Formed the fastest

Resonance involving the OH group stabilizes the arenium ion intermediate and lowers the energy barrier.

FIGURE 23-2 Activation and deactivation of an aromatic ring Free energy diagrams for the addition of NO_2^+ to benzene (black curve), phenol (green curve), and nitrobenzene (red curve). The OH group in phenol stabilizes the arenium ion intermediate by involving the lone pair of electrons on the O atom in resonance. Hence, the OH group increases the rate of formation of the arenium ion intermediate, making it an *activating group*. A nitro group, on the other hand, destabilizes the arenium ion intermediate. Hence, the nitro group slows the rate of formation of the arenium ion, making it a *deactivating group*.

According to Table 23-2, phenol ($HO–C_6H_5$) undergoes nitration about 1000 times faster than benzene itself. Consequently, the OH group is classified as an **activating group**—we say that the group *activates the benzene ring* toward electrophilic aromatic substitution. The nitro (NO_2) group is a **deactivating group**, on the other hand, because nitrobenzene ($O_2N—C_6H_5$) undergoes nitration over 10 million times *slower* than benzene itself. Relative to H, the NO_2 group *deactivates the benzene ring* toward electrophilic aromatic substitution.

Why is the OH group so strongly activating? When the electrophile adds to the ortho or para position on phenol's ring, a lone pair of electrons from the O atom of the OH group participates in resonance in the arenium ion intermediate. This lowers the energy barrier leading to the formation of the arenium ion, as shown in Figure 23-2 (green curve), which increases the rate of formation of the arenium ion.

Conversely, the NO_2 group of nitrobenzene destabilizes the arenium ion intermediate that is formed. This increases the energy barrier to the formation of the arenium ion (Fig. 23-2, red curve), which slows the overall reaction rate. As a result, the NO_2 group is a *deactivating group*.

---

problem 23.9 Use Table 23-2 to estimate the relative rate of nitration of bromobenzene under the same conditions as the other entries. Is it activating or deactivating?

SOLVED problem 23.10 Why does toluene undergo electrophilic aromatic substitution faster than benzene, thereby making CH_3 an *activating group*?

Think Are toluene's arenium ion intermediates more stable or less stable than those of benzene? How does that affect the energy barriers for the formation of toluene's arenium ions and, hence, the rates?

The arenium ion is stabilized by the electron donation of CH_3.

Solve The CH_3 group is electron donating, so it stabilizes the arenium ion intermediate. This lowers the activation energy for formation of the intermediates, which increases the rate at which they form. Hence, a CH_3 group increases the overall substitution rates.

Table 23-3 organizes substituents according to their ortho/para- or meta-directing ability as well as their activating/deactivating ability. The characteristics of the various groups suggest the following trends:

- Activating groups are generally ortho/para directors.
- Deactivating groups are generally meta directors.

These trends break down this way because both the regiochemistry and the extent of activation of the aromatic ring derive from essentially the same phenomena. Substituents that are electron donating, either via resonance or inductive effects, stabilize the arenium ion intermediate and increase the overall reaction rate, so they are activating. These groups also stabilize the ortho and para intermediates better than they do the meta intermediate, so they are ortho/para directors. By the same token, substituents that destabilize the arenium ion intermediate by withdrawing electron density are deactivating groups. These groups destabilize the ortho and para intermediates more than they do the meta, making them meta directors.

TABLE 23-3 Comparison of the Activating/Deactivating and the Ortho/Meta/Para-Directing Nature of Substituents

Substituent	Activating/ Deactivating Nature	Ortho/Para- or Meta-Directing	Substituent	Activating/ Deactivating Nature	Ortho/Para- or Meta-Directing
—O⁻	Strongly activating	Ortho/para	—C(=O)—OR (or OH)	Moderately deactivating	Meta
—NH₂ —NR₂	Strongly activating	Ortho/para	—C(=O)—R (or H)	Moderately deactivating	Meta
—OH —OR	Strongly activating	Ortho/para	—C≡N	Strongly deactivating	Meta
—N(H)—C(=O)—R (or H)	Moderately activating	Ortho/para	—S(=O)(=O)—OH	Strongly deactivating	Meta
—O—C(=O)—R (or H)	Moderately activating	Ortho/para	—N⁺(R or H)(R or H)(R or H)	Strongly deactivating	Meta
—R	Weakly activating	Ortho/para	—N⁺(=O)—O⁻	Strongly deactivating	Meta
—H (Benzene)	–	–			
—Cl —Br —I	Weakly deactivating	Ortho/para			

The lone exceptions in Table 23-3 are the halogen atoms.

Halogen substituents are weakly deactivating, but they are ortho/para directors.

Halogens are ortho/para directors because, like the OH group, halogens possess a lone pair of electrons. Consequently, a halogen atom substituent stabilizes the arenium ion intermediate via resonance, but only when an electrophile attacks at either the ortho or para position. Conversely, halogens are deactivators because they are inductively electron-withdrawing groups.

We can gain more insight into the activating/deactivating character of a substituent by examining the effect the substituent has on the electron density of the aromatic ring prior to electrophilic attack. As examples, the electrostatic potential maps of benzene, aniline, and nitrobenzene are shown in Figure 23-3.

The aromatic ring of aniline, which has an NH₂ activating group, has *greater* electron density (i.e., is more red) than the aromatic ring of benzene. The NH₂ group donates electron density to the ring via resonance, making it more electron rich than

FIGURE 23-3 Effect of activating and deactivating groups on electron density in the benzene ring Electrostatic potential maps of (a) benzene, (b) aniline, and (c) nitrobenzene. The electron-donating ability of NH₂ via resonance increases the electron density of the aromatic ring (more red) and, hence, activates the ring toward electrophilic aromatic substitution. Conversely, the electron-withdrawing ability of the NO₂ group decreases electron density in the ring (more blue/green) and, hence, deactivates the ring toward electrophilic aromatic substitution.

benzene itself and giving it greater affinity for an incoming electrophile (E^+). This serves to enhance the rate of electrophilic attack, thus increasing (activating) the overall rate of substitution compared to benzene.

Conversely, the aromatic ring of nitrobenzene, which has an NO_2 deactivating group, is less electron rich (i.e., is more blue/green) than benzene itself. The NO_2 group withdraws electron density from the ring, so nitrobenzene's aromatic ring attracts an incoming electrophile less strongly than benzene, causing nitrobenzene to react more slowly.

YOUR TURN 23.6

The electrostatic potential maps of benzene and a monosubstituted benzene are shown at the right. Is the aromatic ring of the monosubstituted compound activated or deactivated relative to benzene?

problem **23.12** Does the electrostatic potential map of the monosubstituted benzene in Your Turn 23.6 represent toluene or trifluoromethylbenzene? Explain.

23.6 The Impacts of Substituent Effects on the Outcomes of Electrophilic Aromatic Substitution Reactions

In Chapter 22, we learned how to carry out electrophilic aromatic substitution reactions on benzene, which has no substituents. As we saw in Section 23.5, the presence of a substituent on the ring can activate or deactivate the ring toward a subsequent electrophilic aromatic substitution, and these substituent properties can have significant impacts on the outcome of a desired electrophilic aromatic substitution reaction.

Compare the nitrations of benzene (Equation 23-10) and nitrobenzene (Equation 23-11).

Conc. HNO_3
15 °C

Nitrobenzene
83%

(23-10)

The ring is deactivated, so a stronger acid is necessary to carry out a second nitration.

HNO_3
H_2SO_4
25 °C, 1 h

Nitrobenzene

m-Dinitrobenzene
58%

(23-11)

The first nitration requires only concentrated HNO_3, but H_2SO_4 is added to carry out the second reaction. H_2SO_4 is added because the presence of the NO_2 group in nitrobenzene deactivates the ring substantially, making the second nitration slower and more difficult than the first. H_2SO_4 is a stronger acid than HNO_3 and increases the equilibrium concentration of the NO_2^+ electrophile (Section 22.6).

2,4,6-Trinitrotoluene (TNT)

2,4,6-Trinitrotoluene (TNT) is one of the most widely known explosives. Originally synthesized in the 1860s, TNT has had a long and rich history in military and industrial applications. It is a very useful explosive, in part because it is not very shock sensitive, which makes it relatively safe to handle. Moreover, it is a solid that melts well below the temperature at which it will spontaneously detonate, and, as a liquid, it can be cast into molds and mixed with other explosives.

In the industrial synthesis of TNT, toluene undergoes three successive nitration reactions, as we learned in Chapter 22. The first and second nitrations can be carried out with a nitric acid/sulfuric acid mixture. The third nitration is significantly more difficult, however, because the two NO_2 groups strongly deactivate the ring (Section 23.5), so fuming sulfuric acid is used to compensate.

TNT is explosive because its combustion reaction produces hot, gaseous products (e.g., N_2 and CO_2) *very* rapidly. The fast expansion of these gases is what produces the destructive shock wave, as can be seen over the water in the above photo. The exothermicity of the reaction stems from the fact that the CO_2 and N_2 products are highly stable. The fast reaction rate is facilitated by the NO_2 groups. In the combustion of a hydrocarbon—which is also highly exothermic—the reaction rate is limited to a large extent by the availability of gaseous O_2, which must come from the surroundings. TNT, on the other hand, does not rely as much on O_2 from the surroundings because the molecule itself has a rather high density of oxygen.

problem **23.13** Explain why a single sulfonation of benzene can be carried out with just concentrated H_2SO_4, but fuming sulfuric acid is needed to produce the disulfonated product. (*Hint*: Review Section 22.7.)

Many electrophilic aromatic substitution reactions can still be carried out despite the presence of a strongly deactivating group on the ring, but this is not true for Friedel–Crafts reactions.

Friedel–Crafts reactions do not readily take place on strongly deactivated aromatic rings.

So, for example, no alkylation takes place in Equation 23-12 because the $C{=}O$ group is strongly deactivating.

The $C{=}O$ group strongly deactivates the ring, which prevents Friedel–Crafts reactions.

$$\xrightarrow[\text{AlCl}_3]{\text{Cl}} \quad \text{No reaction} \qquad (23\text{-}12)$$

The opposite problem exists when we want to carry out a Friedel–Crafts alkylation on benzene itself (Equation 23-13).

The alkyl group is activating, so it promotes subsequent alkylations.

$$\xrightarrow[\text{AlCl}_3]{\text{Cl}-\text{R}} \qquad (23\text{-}13)$$

The immediate product is a monoalkyl-substituted benzene, which, due to the presence of the alkyl group, is activated toward a subsequent electrophilic aromatic substitution. Therefore, the second alkylation is faster than the first.

Friedel–Crafts alkylations are subject to polyalkylation.

In light of this problem, it is often better to add an alkyl group by first acylating the ring and then reducing the $C{=}O$ group to a CH_2 group, as shown in Equation 23-14, and as we saw previously in Section 22.8.

No subsequent acylation takes place because the $C{=}O$ is strongly deactivating.

$$\xrightarrow[\text{AlCl}_3]{} \qquad \xrightarrow[\text{Zn(Hg)}]{\text{HCl}} \qquad (23\text{-}14)$$

We are not concerned about multiple acylations because the ring in the product of the first acylation is deactivated.

23.7 The Impact of Reaction Conditions on Substituent Effects

A substituent's effect on the regiochemistry and reaction rate of electrophilic aromatic substitution is not necessarily absolute. For some substituents, the reaction conditions play a significant role. For example, when phenol is treated with bromine in water, it undergoes multiple brominations, even in the absence of a Lewis acid catalyst (Equation 23-15). When acetic acid is added to the reaction mixture, however, only a single bromination occurs (Equation 23-16).

Three brominations occur, even without a strong Lewis acid catalyst.

(23-15)

97%

Only a single bromination occurs under mildly acidic conditions.

(23-16)

The nature of the substituent is affected by the pH of the solution. Phenol is a weak acid ($pK_a = 10.0$), so in the absence of acetic acid, phenol is in equilibrium with a small amount of its conjugate base, the phenoxide anion (C_6H_5—O^-), as shown in Equation 23-17. Even though very little of the phenoxide anion is present, the O^- substituent is a *very* powerful activating group—far better than the OH group—so it dramatically increases the rate of electrophilic aromatic substitution. As a result, the phenoxide anion is the dominant reactant in the reaction.

The O$^{\ominus}$ substituent is a *very* powerful activator.

(23-17)

The presence of acetic acid decreases the pH of the solution. The resulting increased concentration of H^+ substantially decreases the concentration of the phenoxide anion (Le Châtelier's principle), effectively cutting off the route by which the phenoxide anion participates in electrophilic aromatic substitution. The only route available is the one in which the electrophile attacks uncharged phenol itself—a much slower reaction that can be easily stopped after just a single bromination.

The pH can also dramatically affect the nitration of *N,N*-dimethylaniline. If *N,N*-dimethylaniline is treated with HNO_3 in acetic acid (Equation 23-18), then the ortho and para nitro products are predominantly formed. This is as expected, since the lone pair on the substituent's N atom makes it ortho/para-directing and activating. However, if nitration is carried out in sulfuric acid (i.e., at a much lower pH), then the major product is the meta isomer (Equation 23-19).

The amine group is an ortho/para director under *mildly* acidic conditions.

(23-18)

The amine group is a meta director under *strongly* acidic conditions.

(23-19)

Sulfuric acid is much stronger than acetic acid, so it *quantitatively* protonates the amino group, generating an anilinium ion in which a +1 formal charge exists on N (Equation 23-20). In this form, the substituent is highly electron withdrawing, much like an NO_2 group, so it is a meta director and a deactivator.

The positive charge on the N atom makes the substituent a deactivator and a meta director, much like the NO_2 group.

(23-20)

A similar phenomenon can occur when a strong Lewis acid is present. For example, even though aniline possesses a highly activating NH_2 group, it is incompatible with Friedel–Crafts reactions, as shown in Equation 23-21. The amino group is a relatively strong Lewis base, so it readily coordinates to the $AlCl_3$ Lewis acid. In that complexed form, the N atom possesses a formal positive charge, making it a highly *deactivating* group that precludes Friedel–Crafts reactions altogether (Section 23.6).

The strong Lewis base coordinates to the Lewis acid.

The positive charge on the N atom makes the substituent a deactivating group, which is incompatible with Friedel–Crafts reactions.

$$\text{(23-21)}$$

No reaction

23.8 Electrophilic Aromatic Substitution on Disubstituted Benzenes

When electrophilic aromatic substitution takes place on a disubstituted benzene, effects from both of the substituents on the ring must be considered—both their activating/deactivating abilities and their ortho/meta/para-directing capabilities. Fortunately, these effects tend to be essentially additive.

The regiochemistry of these reactions is straightforward if the two substituents are "in agreement." For example, the nitration of *p*-nitrotoluene produces almost exclusively 2,4-dinitrotoluene (Equation 23-22).

Both substituents "agree" with the regiochemistry.

$$\text{(23-22)}$$

p-Nitrotoluene

2,4-Dinitrotoluene
98%

The CH_3 group is an ortho/para director, so it favors substitution at the 2 and 6 positions on the ring (indicated by the blue arrows). Furthermore, the NO_2 group is a meta director, so it favors substitution at the 2 and 6 positions on the ring (indicated by red arrows), too. With agreement like this, we observe substitution primarily at the C atom adjacent to the CH_3 group.

The situation is not quite so straightforward without this agreement among the substituents. In those situations, we can apply the following rule:

When two substituents attached to an aromatic ring "disagree" about where to direct the incoming electrophile, regiochemistry is usually dictated by the substituent that is the stronger activating group—that is, the group that appears nearer the beginning of Table 23-3.

For example, when *m*-nitrotoluene undergoes nitration (Equation 23-23), the CH_3 group directs substitution to the 2, 4, and 6 positions (indicated by the blue arrows), whereas the NO_2 group directs substitution to the 5 position (indicated by the red arrow).

The CH₃ and NO₂ groups are *not* in agreement about where to direct the incoming electrophile.

CH₃ is the more activating group, so it directs the incoming electrophile more strongly than the NO₂ group.

m-Nitrotoluene → 3,4-Dinitrotoluene + 2,5-Dinitrotoluene   (23-23)

Experimentally, we observe that substitution occurs mainly at the 4 and 6 positions on the ring, suggesting that the CH_3 group has greater influence on the regiochemistry than the NO_2 group. This is consistent with the fact that CH_3 is a more strongly activating substituent than NO_2.

We can understand why the more activating group dictates the regiochemistry by examining the resonance contributors of the various arenium ion intermediates that are possible. When the electrophile adds meta to the methyl group (Equation 23-24), then all resonance contributors of the resulting arenium ion have about the same stability—in each case, the positive charge appears on an unsubstituted carbon.

In each structure, the positive charge is little affected by the substituents on the ring.

(23-24)

When the electrophile adds para to the CH_3 group (Equation 23-25), the resulting arenium ion has one resonance structure that is especially stable (highlighted in green) due to the electron-donating effect of the CH_3 group to the positively charged C atom. There is also one resonance structure that is particularly unstable (highlighted in red) due to the electron-withdrawing effect of the NO_2 group.

Destabilized, so this structure contributes little.

Stabilized, so this structure contributes a lot.

(23-25)

It is tempting to presume that the destabilization in the first resonance structure counters the effects from the stabilization in the second resonance structure. However, resonance theory (Chapters 1 and 6) tells us otherwise, because the greatest contribution to the resonance hybrid is from the most stable resonance structure. In other words, contribution by the destabilized resonance structures is minimized. Thus, the arenium ion that is produced is more stable when the electrophile adds para to the methyl group (Equation 23-25) than it is when the electrophile adds meta (Equation 23-24).

The following resonance structures are for the arenium ion produced when NO_2^+ adds ortho to the CH_3 group in *m*-nitrotoluene.

Identify the most stable and least stable resonance structures. Overall, is this arenium ion more stable or less stable than the one in Equation 23-24?

Although the methyl group in *m*-nitrotoluene directs the incoming electrophile to the 2, 4, and 6 positions, substitution is favored at just the 4 and 6 positions in Equation 23-23. This is due to *steric hindrance*. The 2 position is flanked by both the CH_3 group and the NO_2 group, making that site the least accessible to the incoming electrophile. The 4 and 6 positions are more accessible because each is flanked by only one substituent—either the CH_3 group or the NO_2 group.

The situation is not as clean if the two substituents on benzene disagree with the placement of the incoming electrophile *and* have roughly the same activating/deactivating ability—that is, if they are in roughly the same location in Table 23-3. This is the case when *p*-ethyltoluene undergoes electrophilic aromatic substitution, because the two alkyl groups have about the same activating capabilities. As indicated in Equation 23-26, a mixture of isomers is produced because the two alkyl groups are nearly equally good at directing the incoming electrophile to the position on the ring ortho to themselves.

The CH_3 and CH_2CH_3 groups are similar in their activating abilities, but direct the incoming electrophile to different carbons.

p-Ethyltoluene $\xrightarrow[\substack{H_2SO_4, \\ \text{acetic anhydride} \\ 0\,°C,\ 1.5\ h}]{HNO_3}$ 4-Ethyl-2-nitrotoluene + 4-Ethyl-3-nitrotoluene (23-26)

Relative percentages: 56% 44%

Substituent effects on the *overall* rate of electrophilic aromatic substitution are also additive. We can see this explicitly in Figure 23-4, which shows the rates of chlorination for several methylated benzenes. In general, the rate of chlorination increases

FIGURE 23-4 Additivity of substituent effects on reaction rate With each additional alkyl group, the ring becomes increasingly activated toward electrophilic aromatic substitution, which causes the reaction rate to increase.

Increasing reaction rate

Increasingly activated ring

Relative rate of chlorination: 1   300   400,000   800,000   4,000,000   720,000,000

23.8 Electrophilic Aromatic Substitution on Disubstituted Benzenes / **1125**

with each additional methyl group, given that each methyl group is *activating*. (As explored further in Problem 23.42, the reaction rates are also sensitive to the particular locations of the substituents relative to one another about the ring.)

SOLVED problem 23.15 Rank the following aromatic rings from the slowest to the fastest rate of electrophilic aromatic substitution.

A   B   C   D   E

Think Are the substituents on the ring activating or deactivating? How many are there of each? Are all the activating and deactivating substituents of equal strength?

Solve According to Table 23-3, the NO_2 group is strongly deactivating, the $CH_3C{=}O$ group is moderately deactivating, the CH_3 group is weakly activating, and the OCH_3 group is strongly activating. **A** will react the slowest because it is the only one with two deactivating groups. **D** and **E** will both react faster because each has one deactivating group and one activating group. **E** will react faster than **D**, however, because the deactivating groups are the same but the activating group in **E** is stronger. Finally, **B** and **C** will react faster still because they each have two activating groups, but **C** will react faster than **B** because both substituents in **C** are strongly activating. Therefore, the order of increasing reaction rate is: A < D < E < B < C.

---

problem 23.16 Rank the following aromatic rings from the slowest to the fastest rate of electrophilic aromatic substitution.

A   B   C   D   E   F

FIGURE 23-5 Some aromatic rings other than benzene These compounds have aromatic rings and, therefore, can undergo electrophilic aromatic substitution.

Heterocyclic aromatics

Naphthalene   **Furan**   **Pyrrole**   **Pyridine**

23.9 Electrophilic Aromatic Substitution Involving Aromatic Rings Other Than Benzene

Benzene is but one of many aromatic compounds. Naphthalene is an aromatic hydrocarbon consisting of two fused rings, whereas furan, pyrrole, and pyridine are *heterocyclic* aromatic compounds (Fig. 23-5).

Because of the aromatic nature of these compounds, their π systems, similar to benzene's, are resistant to electrophilic addition. Instead, treating these compounds with electrophiles tends to lead to substitution to preserve their aromaticity.

These compounds undergo many of the same electrophilic aromatic substitution reactions as benzene. Unlike benzene, however, each of these compounds has at least two chemically distinct C atoms, so more than one isomeric product can be produced even without a substituent initially attached to the ring.

For example, as shown in Equation 23-27, naphthalene has two chemically distinct H atoms—one designated α and the other β—and substitution of each yields a different product.

The general mechanisms for these reactions, presented in Equations 23-28a and 23-28b, are essentially identical to the one involving benzene. In Step 1, the electron-rich aromatic species undergoes electrophilic addition with the electrophile (E^+), producing an *arenium ion intermediate*, in which the positive charge is delocalized via multiple resonance structures. Step 2 is the elimination of H^+ to give the overall product.

Mechanism for Equation 23-27

Similarly, pyrrole has two chemically distinct H atoms and can undergo electrophilic aromatic substitution to produce two isomeric products, as shown in Equation 23-29. And pyridine has three chemically distinct H atoms, so it can undergo electrophilic aromatic substitution to produce three isomeric products, as shown in Equation 23-30.

Just as with benzene and naphthalene, the mechanism to produce each of these isomeric products involves two steps. This is shown for pyrrole in Equation 23-31 and for pyridine in Equation 23-32.

Mechanism for Equation 23-29

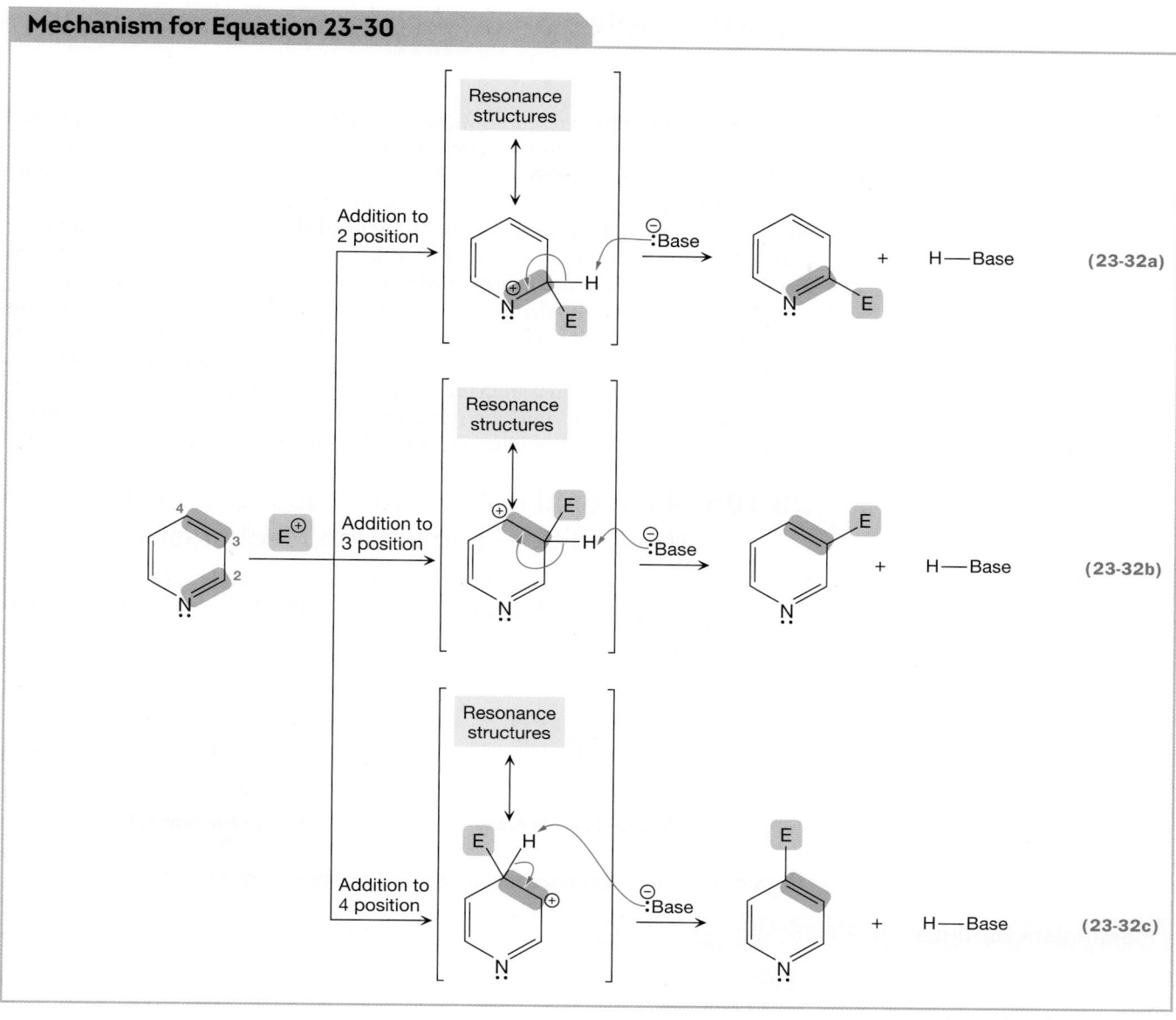

Draw all substitution products of the following general electrophilic aromatic substitution, and draw the mechanism that leads to each one.

Furan

Although electrophilic aromatic substitution reactions involving naphthalene and heterocyclic aromatics can lead to isomeric products, those products are not all favored equally. Rather, just as we saw with electrophilic aromatic substitution reactions involving substituted benzene rings, certain of the isomeric products tend to be strongly favored. Moreover, a reaction involving naphthalene or a heterocyclic aromatic compound can proceed at a rate that is substantially different from the analogous reaction involving benzene; that is, these other aromatic rings can be *activated* or *deactivated* relative to benzene. Some of these aspects of regiochemistry and ring activation/deactivation are explored further in the problems at the end of the chapter.

23.10 Nucleophilic Aromatic Substitution Mechanisms

The aromatic substitution reactions we have examined in Chapters 22 and 23 proceed by mechanisms involving an aromatic ring attacking an electrophile—so-called *electrophilic aromatic substitution* reactions. Here in Section 23.10, we examine two other types of aromatic substitution reactions, in which the aromatic ring is attacked by a nucleophile instead; these are examples of **nucleophilic aromatic substitution** reactions. One of these reactions proceeds by a *nucleophilic addition–elimination* mechanism and the other by a mechanism in which a *benzyne intermediate* is produced.

These reactions are fundamentally different from the ones classified as electrophilic aromatic substitution. However, nucleophilic aromatic substitution reactions are presented here because they are frequently used in syntheses in conjunction with, or as an alternative to, electrophilic aromatic substitution reactions. Having had a thorough discussion of electrophilic aromatic substitution, we are now poised to have a better appreciation for what nucleophilic aromatic substitution has to offer.

23.10a Nucleophilic Aromatic Substitution via the Addition–Elimination Mechanism

When 1-chloro-2-nitrobenzene is treated with ethanamine, $CH_3CH_2NH_2$, *N*-ethyl-2-nitroaniline is produced (Equation 23-33). Overall, Cl is replaced by an amine group ($NHCH_2CH_3$).

$$\text{(23-33)}$$

1-Chloro-2-nitrobenzene → **N-Ethyl-2-nitroaniline** 99%

(H_2N, H_2O, 24 h, 105 °C)

The mechanism for this aromatic substitution reaction is shown in Equation 23-34.

Mechanism for Equation 23-33

The C atom bonded to Cl is relatively electron poor.

1. Nucleophilic addition

Other resonance structures

This intermediate is especially stabilized by the NO_2 group attached to a C atom that shares the −1 formal charge.

2. Nucleophile elimination

3. Proton transfer

(23-34)

In Step 1, the amine acts as a nucleophile, undergoing nucleophilic addition to the C atom bonded to Cl. This is the slow step and is therefore rate determining. In Step 2, Cl departs as a leaving group. Step 3, a proton transfer, is necessary to produce the overall uncharged product.

Step 1 is assisted partly by the resonance delocalization of the negative charge in the intermediate (see Your Turn 23.9), called a **Meisenheimer complex**, and is also assisted by the electron-poor nature of the aromatic ring. Not only does the leaving group impart a partial positive charge on the carbon atom to which it is attached, thereby attracting the electron-rich nucleophile, but the Meisenheimer complex is stabilized by the powerfully electron-withdrawing NO_2 group. This contrasts with electrophilic aromatic substitution, which is favored by electron-rich aromatic rings.

23.9 **YOUR TURN**

Draw two additional resonance structures of the Meisenheimer complex in Equation 23-34.

For maximum stabilization of the Meisenheimer complex, the leaving group must be ortho or para to a strongly electron-withdrawing group. Thus,

Nucleophilic aromatic substitution involving a substituted benzene generally requires at least one strongly electron-withdrawing group ortho or para to the leaving group.

This ensures that the negative charge that develops is adjacent to the electron-withdrawing group in at least one of the resonance structures. Notice that this is, indeed, the case with the resonance structure shown in Equation 23-34. (See Your Turn 23.10.)

23.10 **YOUR TURN**

The nucleophilic addition of $CH_3CH_2NH_2$ to 1-chloro-4-nitrobenzene produces the following Meisenheimer complex.

Draw the additional resonance structures of the intermediate and show that in one of them, the negative charge is adjacent to the electron-withdrawing NO_2 group.

The importance of the number and locations of the strongly electron-withdrawing groups relative to the leaving group is demonstrated by the reaction in Equation 23-35, which is a substitution similar to the one in Equation 23-33. The reaction requires less heat because the leaving group is ortho to one NO_2 group and para to the other.

The additional NO$_2$ group further stabilizes the negatively charged intermediate.

Less extreme conditions are required.

(23-35)

1-Chloro-2,4-dinitrobenzene

N-Methyl-2,4-dinitroaniline
96%

---

problem **23.17** Draw the detailed mechanism for this reaction, and predict the major product. Will this reaction be faster or slower than the one in Equation 23-35?

Nucleophilic aromatic substitution can be carried out with a variety of leaving groups—even F$^-$ (Equation 23-36). The major restriction is that the leaving group needs to be more stable (i.e., less basic) than the incoming nucleophile, because the more stable of the two will depart in the elimination step.

(23-36)

---

problem **23.18** Draw the mechanism and the major product of each of the following reactions.

(a)

(b)

(c)

23.10b Nucleophilic Aromatic Substitution via a Benzyne Intermediate

Even without a strongly electron-withdrawing group attached to the ring, a halobenzene can still undergo nucleophilic substitution. For example, chlorobenzene can react with sodium hydroxide to produce phenol (Equation 23-37), or it can react with potassium amide to produce aniline (Equation 23-38). These reactions typically require extreme conditions, however, such as the presence of a very strong base, high temperatures, or both.

$$(23\text{-}37)$$

Chlorobenzene → Phenol 94%
NaOH (15%), 300 °C, 20 h

$$(23\text{-}38)$$

Chlorobenzene → Aniline 60%
KNH$_2$, Liq. NH$_3$, −33 °C

The requirement for these extreme conditions is explained by Equation 23-39, the mechanism for the reaction in Equation 23-37.

Mechanism for Equation 23-37

In Step 1, the base removes a proton from the aromatic ring to produce a carbanion, and in Step 2, the leaving group departs. These two steps make up the same E1cb mechanism we saw in Section 18.5. The product of Step 2 is called a **benzyne intermediate**, because it has a C≡C triple bond in its Lewis structure. In Step 3, the benzyne intermediate undergoes nucleophilic addition to produce a deprotonated form of phenol. In Step 4, a proton from H$_2$O adds to produce the overall uncharged product.

Both Steps 1 and 2 are relatively slow. Step 1 is slow because a very unstable carbanion is produced, which is why a strong base is required. Step 2 is slow because the benzyne product is highly unstable due to the large amount of angle strain caused by the triple bond. The ideal bond angle for each triply bonded C atom is 180°, but the constraints of the ring require it to be ~120°. This has a major effect on the strength of the π bond that is in the plane of the ring, as seen in Figure 23-6. Adjacent p orbitals are parallel in a normal π bond, but they are ~60° apart in this case. As a result,

FIGURE 23-6 The instability of the benzyne intermediate The p orbitals in the plane of the ring intersect at ~60°. Thus, their overlap is less than ideal, making the resulting π bond rather weak and benzyne highly reactive.

the overlap of those adjacent p orbitals is significantly decreased, making the π bond abnormally weak and benzyne highly reactive.

YOUR TURN 23.11

The mechanism for the reaction in Equation 23-38 is as follows, but the curved arrows have been omitted.

Supply the missing curved arrows, and write the name of the elementary step below each reaction arrow.

Because benzyne is highly reactive, it has not been isolated and purified, though some experiments indirectly support its existence. One of the most compelling is a carbon-13 labeling experiment, shown in Equation 23-40. When the Cl atom in chlorobenzene is bonded to a ^{13}C atom, reaction with KNH$_2$ produces a roughly equal mixture of two isomers of aniline, where the NH$_2$ group is bonded to the ^{13}C atom in one isomer and to a ^{12}C atom in the other.

Nucleophilic addition occurs with roughly equal likelihood at these two carbons.

A roughly equal mixture of these two isomers is produced.

Benzyne intermediate

(23-40)

These results are consistent with benzyne's symmetry about the triple bond. Thus, nucleophilic addition to benzyne occurs with essentially equal likelihood at both alkyne C atoms. Addition to the alkyne ^{13}C atom produces one isomer, and addition to the alkyne ^{12}C atom produces the other.

problem **23.19** Draw the mechanism for the formation of benzyne from bromobenzene and sodium hydroxide.

Suppose, instead, that the following deuterium-labeled bromobenzene were used:

What percentage of the product would you expect to contain the deuterium atom? Explain.

23.11 Organic Synthesis: Considerations of Regiochemistry; Attaching Groups in the Correct Order

As we learned in Section 23.1, a substituent that is already on the aromatic ring greatly influences the site of attack in a subsequent electrophilic aromatic substitution. In a synthesis that requires successive substitutions, then, we must choose wisely the order in which the reactions are carried out. For example, how would you synthesize *m*-chloronitrobenzene from benzene?

m-Chloronitrobenzene

Benzene can be nitrated by treating it with concentrated HNO_3 in H_2SO_4, and it can be chlorinated with Cl_2 in $FeCl_3$. The synthesis therefore must involve both a nitration and a chlorination, but in which order?

If chlorination takes place first, then we encounter a problem with regiochemistry, as shown in Equation 23-41. The Cl substituent in chlorobenzene is an ortho/para director (Table 23-1), so a subsequent nitration would yield *o*-chloronitrobenzene and *p*-chloronitrobenzene as the major products. Our target, however, is *m*-chloronitrobenzene.

(23-41)

If the nitration of benzene is carried out first (Equation 23-42), then the correct isomer is, indeed, produced as the major product, because the nitro group in nitrobenzene is a meta director.

(23-42)

SOLVED problem 23.20 How would you synthesize *p*-isopropylnitrobenzene from benzene?

Think Is the NO_2 group an ortho/para director or a meta director? Is the isopropyl group an ortho/para director or a meta director? Which group should be on the ring prior to the second substitution?

Solution To obtain the target, benzene must be both alkylated and nitrated. Because an alkyl group is ortho/para directing and the NO_2 group is meta directing (Table 23-1), the order in which these substitutions are carried out is important. Given that we want to synthesize the para isomer, an ortho/para-directing substituent should be on the ring prior to the second substitution reaction. In other words, the second reaction must be a nitration of the alkylbenzene. That leaves alkylation as the first reaction. The overall synthesis would be written as follows:

---

problem **23.21** Show how to synthesize the compound at the right from benzene.

The order in which substitutions are carried out can be an issue, too, because some substituents on the ring make certain electrophilic aromatic substitution reactions unfeasible. Friedel–Crafts reactions, for example, do not occur with highly deactivated rings (Section 23.6). Therefore, we must avoid placing a strongly deactivating group on the ring prior to carrying out a Friedel–Crafts reaction.

How, then, would you synthesize *m*-nitroacetophenone from benzene?

This synthesis requires two substitutions—a nitration and a Friedel–Crafts acylation. Because both the NO_2 group and the acetyl group are meta directors (Table 23-1), regiochemistry will not depend on the order in which these substitutions are carried out. We do need to avoid attempting a Friedel–Crafts reaction on nitrobenzene, however, because the benzene ring would be highly deactivated (Equation 23-43).

Benzene   ***m*-Nitroacetophenone**

Highly deactivated ring NO_2

$$\xrightarrow[H_2SO_4]{HNO_3}$$ No reaction   (23-43)

Instead, nitration should be the final step, leaving the Friedel–Crafts acylation as the first step. Equation 23-44 shows the best way to carry out this synthesis.

$$\xrightarrow[AlCl_3]{}$$ $$\xrightarrow[H_2SO_4]{HNO_3}$$ (23-44)

NO_2

23.12 Organic Synthesis: Interconverting Ortho/Para and Meta Directors

In Section 23.11, we learned that when devising a synthesis that requires carrying out successive electrophilic aromatic substitution reactions, the order in which ortho/para- and meta-directing groups are added to the ring can impact whether the synthesis is successful. With this in mind, we revisit some reactions we first encountered in Chapter 22, which can be used to interconvert ortho/para- and meta-directing substituents. Thus, the regiochemistry of a substitution that takes place later in a synthesis is not automatically predetermined by the ortho/meta/para-directing ability of a substituent added earlier.

Recall from Section 22.9a, for example, that treating an alkylbenzene with a hot, basic solution of $KMnO_4$ yields benzoic acid upon acid workup (see Equation 23-45). The substituent in the reactant is an alkyl group, which is an ortho/para director, but the substituent in the product is a CO_2H group, which is a meta director.

(23-45)

How can we use this reaction in the synthesis of p-nitrobenzoic acid from benzene?

These substituents are para to each other but are both meta directors.

The CO_2H substituent and the NO_2 substituent are both meta directors (Table 23-1), but the desired product is the para isomer. We can circumvent this problem by making sure that the second electrophilic aromatic substitution takes place with an ortho/para director already on the ring, such as an alkyl group. After the para-disubstituted benzene is produced, the alkyl group can then be oxidized to the carboxyl group. The retrosynthetic analysis is shown in Equation 23-46:

(23-46)

The final synthesis would then be written as in Equation 23-47.

$$(23\text{-}47)$$

YOUR TURN 23.12

How would you alter the synthesis in Equation 23-47 to produce *m*-nitrobenzoic acid instead?

---

problem **23.22** Show how to synthesize benzene-1,4-dicarboxylic acid from benzene, using ethanol as your only other source of carbon.

Benzene   Benzene-1,4-dicarboxylic acid

We can achieve even greater flexibility in synthesis with the ability to convert a meta director into an ortho/para director. We have already encountered two types of reductions that accomplish this. One, which was discussed in Section 22.9b, is the reduction of a nitro (NO_2) group to an amine (NH_2). The second is the reduction of a carbonyl group on a ketone or aldehyde to a methylene (CH_2) group. Recall from Section 19.3 that this can be done in either acidic conditions (Clemmensen reduction), basic conditions (Wolff–Kishner reduction), or neutral conditions (Raney-nickel reduction).

The utility of these reactions is demonstrated in the synthesis of *m*-bromoethylbenzene from benzene. The substituents in the target are meta to each other, so the second of two electrophilic aromatic substitution reactions must be carried out with a meta director already on the ring. Both the Br and the CH_2CH_3 substituents are ortho/para directors (Table 23-1), however, so a meta director must be converted into an ortho/para director after the second substitution reaction is carried out.

Thinking retrosynthetically, the meta director could be an acetyl group that is reduced to the ethyl group, as in Equation 23-48.

These substituents are meta to each other but are both ortho/para directors.

Benzene   *m*-Bromoethylbenzene

$$(23\text{-}48)$$

A meta director

In the forward direction, the synthesis would be written as in Equation 23-49:

(23-49)

---

problem **23.23** Show how to synthesize *m*-chloroaniline from benzene.

23.13 Organic Synthesis: Considerations of Protecting Groups

In the realm of electrophilic aromatic substitution, there are two ways in which to incorporate a protecting group. One is to temporarily suppress the reactivity of a substituent attached to the ring, by reversibly converting it to a different functional group. The second is to block a particular site on the ring where substitution is not desired.

To illustrate the first scenario, how would you carry out the transformation at the right, where aniline is acylated at the para position?

A simple Friedel–Crafts acylation would seem to be required, given that the NH_2 group on aniline is an ortho/para director. As we learned in Section 23.7, however, the reaction conditions that are required would convert the activating amino substituent into a strongly deactivating group (Equation 23-50), thereby making a direct Friedel–Crafts acylation unfeasible.

(23-50)

The problem here is that the NH_2 group is a good Lewis base; its lone pair of electrons is available to form a bond to $AlCl_3$, a strong Lewis acid. To circumvent this problem, the amino group can be converted *temporarily* into a different functional group (e.g., an amide group) in which the lone pair of electrons is less available. As we learned in Chapters 20 and 21, amides are relatively unreactive, due in large part to the resonance of the lone pair on the N atom with the adjacent carbonyl group.

Treating aniline with an acid chloride or an acid anhydride in the *absence* of a strong Lewis acid catalyst converts it to the amide (Equation 23-51). Even in the protected form, the substituent on the ring is an ortho/para director and a moderate activator (Table 23-3), due to the presence of the lone pair of electrons on N. (It would be undesirable for a protecting group to alter the regiochemistry of the reaction or to strongly deactivate the ring.) In the protected form, the acylation can then be carried out. Afterward, the amine is deprotected by hydrolysis.

(23-51)

problem **23.24** Show how to carry out the synthesis at the right.

In some instances, the reaction we want to carry out leads to possible substitution at two or more sites on the ring. This demonstrates the second use of protecting groups in electrophilic aromatic substitution—namely, we can guide the regiochemistry by protecting one of those sites. For example, how would you convert phenol into o-bromophenol, if the para product is undesired?

The hydroxy group is an ortho/para director, so simple bromination will yield a substantial amount of both the ortho and para isomers, as shown in Equation 23-52.

(23-52)

One method that can be used to protect the para position is to place a sulfonyl group there temporarily (Equation 23-53).

The sulfonyl group "blocks" the para position.

(23-53)

As we learned in Section 22.7, sulfonation is *reversible*. Furthermore, it will take place almost exclusively at the para position, so subsequent bromination will take place ortho to the OH group (an ortho/para director). The OH group dictates regiochemistry because it is the more activating group on the ring. It can't direct Br to the para position because the SO_3H group is already there, blocking the para position. After bromination, the SO_3H group is removed with aqueous acid.

An NH_2 group can also be used as a blocking group, as shown in the synthesis of *m*-chlorotoluene from *p*-methylaniline (Equation 23-54):

p-Methylaniline

Prevents overhalogenation

(23-54)

m-Chlorotoluene

In the first step, the NH_2 group (a strongly activating group) is converted to the amide (a moderately activating group) to prevent overhalogenation. In the second step, halogenation takes place ortho to the amide group (an ortho/para director), given that the amide is more activating than the CH_3 group. Next, the amine is deprotected by hydrolysis. Diazotization followed by treatment with H_3PO_2 converts the NH_2 group into an H (Section 22.9c). Therefore, even though the original NH_2 group is not in the final product, it was instrumental in directing the regiochemistry of chlorination.

23.14 Wrapping Up and Looking Ahead

Chapter 23 continued the discussion of electrophilic aromatic substitution reactions that began in Chapter 22. Whereas Chapter 22 focused primarily on the specific mechanisms of these reactions involving benzene, the main areas of focus here in Chapter 23 were regiochemistry, kinetics, and synthesis involving substituted benzenes and other aromatic compounds.

With regard to regiochemistry, we saw that if a certain group is already attached to an aromatic ring, and it favors subsequent electrophilic substitution at positions that are ortho and para to itself, the group is designated as an ortho/para director. Other groups favor subsequent electrophilic substitution at positions meta to where they are attached, making those groups meta directors. With regard to kinetics, certain groups attached to an aromatic ring tend to increase the rate of subsequent electrophilic aromatic substitution reactions, so those substituents are said to activate the aromatic ring. Other groups deactivate the ring.

An activating group speeds up electrophilic aromatic substitution because it stabilizes the arenium ion intermediate that is produced. This stabilization is usually more pronounced when an incoming electrophile adds either ortho or para to an activating group, which is why activating groups tend to be ortho/para directors. Conversely, a deactivating group *destabilizes* the arenium ion intermediate, and usually does so more effectively when the incoming electrophile adds ortho or para to itself. Thus, deactivating groups tend to be meta directors.

Using similar analyses, we also examined electrophilic aromatic substitution reactions involving disubstituted benzenes. The activating/deactivating nature of the substituents is essentially additive in disubstituted benzenes, and regiochemistry is governed by the more activating substituent present.

Finally, we introduced two types of aromatic substitution reactions that do not proceed by the electrophilic mechanism. Rather, they are examples of *nucleophilic* aromatic substitution reactions. One proceeds by a nucleophilic addition–elimination mechanism, and the other proceeds through a benzyne intermediate.

Many of the reactions in the chapters to come involve the addition of species to nonpolar π bonds, reminiscent of reaction steps we have seen throughout Chapters 11, 12, 22, and 23. In Chapter 24, for example, we study the Diels–Alder reaction, in which a diene adds to an alkene or alkyne. In Chapter 25, we examine examples of the addition of free radicals to alkenes and alkynes. Therefore, there are plenty of opportunities to apply the knowledge we have gained from the last few chapters.

Chapter Summary and Key Terms

- In electrophilic aromatic substitution reactions, certain substituents attached to a phenyl ring favor reaction at the ortho and para positions, and are characterized as **ortho/para directors**. Other substituents favor reaction at the meta position and are called **meta directors**. (Section 23.1)

- A substituent attached to a phenyl ring by an atom possessing a lone pair of electrons tends to be an ortho/para director. When the incoming electrophile adds to a carbon at the ortho or para position of the ring, the arenium ion intermediate that is produced is stabilized by these substituents via resonance. (Section 23.2)

- Alkyl groups are ortho/para directors, because they stabilize ortho and para arenium ion intermediates via electron-donating effects. (Section 23.3)

- Meta directors such as the nitro group are electron withdrawing and are attached to the aromatic ring by an atom that has no lone pair of electrons. They destabilize the arenium ion intermediate when the incoming electrophile adds to the ortho or para position. (Section 23.4)

- In general, ortho/para directors increase the rate of electrophilic aromatic substitution, and are called **activating groups**. Conversely, meta directors tend to slow the reaction rate, and are called **deactivating groups**. Halogen substituents are exceptions—they are ortho/para directors, but deactivating. (Section 23.5)

- Activating groups speed up an electrophilic aromatic substitution reaction because they *stabilize* the arenium ion intermediate that is produced. Deactivating groups *destabilize* the arenium ion intermediate. (Section 23.5)

- A deactivating group on an aromatic ring requires more extreme conditions to carry out an electrophilic aromatic substitution reaction such as nitration and sulfonation, and precludes Friedel–Crafts reactions altogether. The electron-donating property of an alkyl group makes polyalkylation a potential problem in Friedel–Crafts alkylations. (Section 23.6)

- The conditions under which an electrophilic aromatic substitution reaction takes place can alter a substituent's ortho/meta/para-directing capabilities, as well as its activating/deactivating capabilities. For example, the aromatic ring of phenol is much more activated under neutral conditions than it is under acidic conditions. More strikingly, an amine group is an ortho/para director under mildly acidic conditions, but it is a meta director under strongly acidic conditions. (Section 23.7)

- When multiple substituents are attached to a phenyl ring, their activating/deactivating qualities are additive, and the regiochemistry tends to be governed by the most activating of those substituents. (Section 23.8)

- Aromatic rings other than benzene can also undergo electrophilic aromatic substitution. These include naphthalene, furan, pyrrole, and pyridine. Just as in substituted benzenes, the regiochemistry and relative reaction rates are governed by the stabilities of the arenium ion intermediates that are produced. (Section 23.9)

- Aromatic rings with halogen substituents can undergo **nucleophilic aromatic substitution**, whereby the halogen atom is replaced by a nucleophile. (Section 23.10)
  - Rings that are strongly deactivated toward reaction with an electrophile (and thus electron poor) react via a nucleophilic addition–elimination mechanism under relatively mild conditions. (Section 23.10a)
  - Rings that are not very electron poor typically require strong bases, high temperatures, or both, and proceed through a **benzyne intermediate**. (Section 23.10b)

- When incorporating successive electrophilic aromatic substitution reactions in a synthesis, it is important to consider the order in which those reactions are carried out. The incorrect order could lead to undesired regiochemistry, or could preclude some reactions entirely. (Section 23.11)

- Reactions that transform an ortho/para director into a meta director, and vice versa, can be instrumental in synthesis. An example of the former is the $KMnO_4$ oxidation of an alkyl side chain to a carboxylic acid. An example of the latter is the reduction of a nitro group to an amine. (Section 23.12)

- The reactivity of an aromatic amine can be temporarily decreased by converting the amine group into an amide group. Additionally, the para position of a phenyl ring can be temporarily blocked by sulfonating that position. (Section 23.13)

Reaction Table

TABLE 23-4 Functional Group Conversions[a,b]

	Starting Functional Group	Typical Reagent and Reaction Conditions	Functional Group Formed	Key Electron–Rich Species	Key Electron–Poor Species	Comments	Discussed in Section
(1)	Deactivated aromatic halide	H_2NR	Substituted aromatic amine	$H_2\ddot{N}R$		Nucleophilic addition–elimination mechanism; EWG ortho or para to X	23.10a
(2)	Deactivated aromatic halide	NaOH	Substituted phenol	$^-\!\ddot{:}\!OH$		Nucleophilic addition–elimination mechanism; EWG ortho or para to X	23.10a

[a] X = F, Cl, Br, I
[b] EWG = electron-withdrawing group

(continued)

TABLE 23-4 Functional Group Conversions[a,b] (continued)

	Starting Functional Group	Typical Reagent and Reaction Conditions	Functional Group Formed	Key Electron– Rich Species	Key Electron– Poor Species	Comments	Discussed in Section
(3)	Deactivated aromatic halide	NaOR	Substituted aromatic ether			Nucleophilic addition– elimination mechanism; EWG ortho or para to X	23.10a
(4)	Aromatic halide	NaOH Δ	Phenol			Proceeds through benzyne intermediate	23.10b
(5)	Aromatic halide	KNH₂	Aniline			Proceeds through benzyne intermediate	23.10b

[a] X = F, Cl, Br, I
[b] EWG = electron-withdrawing group

Problems

23.26 Predict the major product(s) of each of the following reactions. Draw the complete, detailed mechanism that leads to the formation of each of those products.

(a)

(b)

(c)

(d)

(e)

23.27 For each pair of aromatic compounds, determine which will undergo electrophilic aromatic substitution faster.

(a)

(b)

(c)

(d)

(e)

(f)

(g)

(h)

23.28 Draw the structures of compounds **A–F** in the following synthesis scheme.

23.29 Draw the structures of compounds **A–H** in the following synthesis scheme.

23.30 Draw the structures of compounds **A–H** in the following synthesis scheme.

23.31 An amide group (—NHCOR) is an ortho/para director and is moderately activating.

(a) Explain why it is an ortho/para director.
(b) Explain why it is an activating group.
(c) Explain why it is less activating than NH_2.

23.32 The OH group on phenol is an activating group, but the ring in phenylmethanol is deactivated. Explain.

23.33 The electrostatic potential maps of benzene and pyridine are shown at the right. Is the electrostatic potential map of pyridine consistent with the ring being activated or deactivated relative to benzene? Explain.

23.34 The electrostatic potential maps of benzene and pyrrole are shown at the right. Is the electrostatic potential map of pyrrole consistent with the ring being activated or deactivated relative to benzene? Explain.

23.35 In Section 23.3, we learned that all alkyl groups are ortho/para directors. However, the relative amounts of ortho and para products depend on the specific identity of the alkyl group, as shown below for the nitration of various alkylbenzenes:

	CH₃	CH₂CH₃	CH(CH₃)₂	C(CH₃)₃
% Ortho	63	45	30	16
% Meta	3	6	8	11
% Para	34	49	62	73

What trend do you observe? What factor accounts for that trend?

23.36 Predict the most likely site of attack in electrophilic aromatic substitution.

(a) OCH₃ ... OCH₃

(b) OH ... Cl

(c) (acetyl, Br)

(d) (acetyl, Br)

(e) NO₂ (with CH₃)

(f) (tert-butyl)

(g) NH₂ (tetrahydronaphthalene)

23.37 A Br substituent is an ortho/para director, so the halogenation of bromobenzene predominantly yields the ortho and para products, as shown in the following bromination and chlorination reactions:

Br $\xrightarrow{\text{Br}_2, \text{FeBr}_3}$ (ortho Br, 13%) + (para Br, 85%)

Br $\xrightarrow{\text{Cl}_2, \text{FeCl}_3}$ (ortho Cl, 42%) + (para Cl, 53%)

Explain why bromination yields more of the para product than chlorination.

23.38 Which of the following reactions do you think will produce the para product in the greater amount? Explain. (*Hint:* See Problem 23.37.)

A   **B**

23.39 We learned that halogen atoms are one of the few substituents that are ortho/para directing but deactivating. The nitroso group (—N=O) is ortho/para directing and deactivating, too.

(a) Explain why the nitroso group is ortho/para directing.
(b) Explain why it is deactivating.

23.40 Predict the most likely sites of electrophilic aromatic substitution in each of the following molecules.

(a)

(b)

(c)

(d)

23.41 Predict the most likely site of attack of electrophilic aromatic substitution in the compound at the right. (*Hint:* How do you determine whether the substituent is an ortho/para-directing group or a meta-directing group?)

23.42 Alkyl groups are activating, so both of the following trisubstituted benzenes undergo chlorination much faster than benzene.

Relative rate of chlorination: 1   680,000   800,000

The 1,2,3-trisubstituted one, however, undergoes chlorination faster than the 1,2,4-trisubstituted one. Explain.

23.43 For each pair of isomers, determine the one that will undergo electrophilic aromatic substitution faster. Explain. (*Hint:* See Problem 23.42.)

(a) *o*-Dimethylbenzene or *m*-dimethylbenzene
(b) *m*-Dimethylbenzene or *p*-dimethylbenzene
(c) 1,2,3,4-Tetramethylbenzene or 1,2,3,5-tetramethylbenzene

23.44 Show how you would carry out the following synthesis.

23.45 Electrophilic aromatic substitution on N,N-dimethylaniline is faster than on the unsubstituted aniline.

In other words, the aromatic ring in N,N-dimethylaniline is more activated than the ring is in aniline itself. If the ring is methylated at the 2 and 6 positions, however, then the N,N-dimethyl-substituted compound reacts more slowly in electrophilic aromatic substitution than the unsubstituted compound.

Explain both of these results. (*Hint*: It does not have to do with the number of H atoms on the ring.)

23.46 Draw a complete, detailed mechanism for the following reaction. A key intermediate is provided.

23.47 Show how you would synthesize each of the following compounds from benzene.

23.48 The phenyl group, C_6H_5, is known to be an ortho/para-directing group.
  (a) With that in mind, predict the product of the reaction at the right.
  (b) Justify why it is an ortho/para director by examining the ortho, meta, and para arenium ion intermediates that would be formed during the course of the reaction.

23.49 Draw a complete, detailed mechanism for the following reaction.

23.50 Many synthetic dyes are produced from a process called diazo coupling. In the following example, benzenediazonium chloride reacts with *N,N*-dimethylaniline to produce a dye called butter yellow.

Benzenediazonium chloride   **N,N-Dimethylaniline**   **Butter yellow (An azo dye)**

H$_2$O
Mildly acidic

Draw the complete mechanism for this reaction. Would you expect a significant amount of the meta product? Why or why not?

23.51 Show how you would synthesize the compound at the right, using propan-1-ol and benzene as your only sources of carbon.

23.52 Show how you would synthesize each of the following compounds, using butan-1-ol and benzene as your only sources of carbon.

(a)   **(b)**

23.53 Aspirin, or acetylsalicylic acid, is one of the most widely produced medications. Show how to synthesize aspirin from benzene.

Acetylsalicylic acid (Aspirin)

23.54 Here in Chapter 23, we learned that aniline becomes highly deactivated in the presence of a strong Lewis acid, due to coordination of the N atom to the Lewis acid. Thus, Friedel–Crafts reactions involving aniline are unfeasible.

Lewis acid catalyst

No reaction

As shown in the following reaction, though, this does not appear to be a problem with the N atom in pyrrole. Explain why.

Lewis acid catalyst

23.55 According to Table 23-1, the nitration of phenol results in a product mixture that is 50% ortho and 50% para. What would the major products be if one of the ortho positions of phenol were labeled with deuterium (D)? What would you expect the relative amounts of each of those products to be?

23.56 The bromination of benzene requires a Lewis acid catalyst such as $FeBr_3$, but the bromination of naphthalene does not. Explain why.

23.57 In Section 23.9, we learned that an electrophilic aromatic substitution on naphthalene can lead to two different isomeric products: an α-substituted product and a β-substituted product. When the reaction is reversible, such as in a sulfonation reaction, the major product is the one that is most stable. With this in mind, predict the major product of the following reaction.

23.58 When naphthalene undergoes an irreversible electrophilic aromatic substitution, such as a Friedel–Crafts acylation, the major product is the kinetic product, which proceeds through the most stable arenium ion intermediate. With this in mind, predict the major product of the reaction at the right. (*Hint*: Draw out all resonance structures for each arenium ion intermediate. Does each one have the same number of resonance structures? How many resonance structures of each intermediate preserve the aromaticity?)

23.59 Electrophilic aromatic substitution on a monosubstituted naphthalene tends to take place on the same ring as the substituent when the substituent is an activator like CH_3, and tends to take place on the unsubstituted ring when the substituent is a deactivator like NO_2. Explain.

23.60 When 2-methylnaphthalene undergoes an irreversible electrophilic aromatic substitution, the electrophile predominantly attacks the 1 position instead of the 3 position.

2-Methylnaphthalene

Major product

This suggests that the arenium ion that is formed from addition of the electrophile to the 1 position is more stable than the arenium ion formed from addition of the electrophile to the 3 position. Explain why this is so. (*Hint*: Simple resonance theory can explain why.)

23.61 Draw the complete, detailed mechanism for the reaction at the right and predict the major product. (*Hint*: See Problems 23.58 through 23.60.)

23.62 Draw the complete, detailed mechanism and predict the major product for each of the following reactions. (*Hint*: See Problems 23.58 through 23.60.)

(a)

(b)

(c)

(d)

23.63 In Section 23.9, we learned that an electrophilic aromatic substitution on pyrrole can take place at either the 2 position or the 3 position. In a nitration reaction, substitution takes place primarily at the 2 position, as shown at the right.
Explain why. (*Hint*: Draw all resonance structures for the 2- and the 3-substituted arenium ion intermediates.)

Pyrrole

2-Nitropyrrole
55%

23.64 In Section 23.9, we learned that an electrophilic aromatic substitution on pyridine can take place at positions 2, 3, or 4. In a nitration reaction, substitution takes place primarily at the 3 position.

Pyridine

3-Nitropyridine
20%

Explain why. (*Hint*: Draw all resonance structures for the 2-, 3-, and 4-substituted arenium ion intermediates.)

23.65 In an electrophilic aromatic substitution reaction, pyrrole tends to react significantly faster than benzene, whereas pyridine tends to react much slower. In other words, pyrrole has an activated ring, whereas pyridine has a deactivated ring.

Increasing rate of electrophilic aromatic substitution

Pyridine   **Benzene**   **Pyrrole**

Explain why. (*Hint*: Draw all resonance structures for each compound's most stable arenium ion intermediate. Consult Problems 23.63 and 23.64 to see which ones those are for pyrrole and pyridine, respectively.)

23.66 **(a)** For which aromatic compound do you expect nitration to take place faster: furan or thiophene?
(b) For each of these compounds, at which C atom do you expect electrophilic aromatic substitution to predominantly take place? Explain your reasoning. (*Hint*: See Problem 23.63.)

Furan   **Thiophene**

23.67 A thiophene ring is sufficiently activated that bromination may take place without the presence of a Lewis acid catalyst. With this in mind, draw the complete mechanism for the reaction at the right and predict the major product.

23.68 Draw the complete, detailed mechanism and predict the major product for each of the following reactions.

(a) $\xrightarrow{Br_2}$?

(b) $\xrightarrow[FeBr_3]{Br_2}$?

(c) $\xrightarrow[\substack{Acetic \\ anhydride}]{HNO_3}$?

(d) $\xrightarrow[Acetic\ acid,\ \Delta]{HNO_3}$?

23.69 Predict the site on each molecule that is most likely to undergo electrophilic aromatic substitution.

(a)   **(b)**

23.70 In nucleophilic aromatic substitution reactions that proceed by the nucleophilic addition–elimination mechanism, the reaction rate increases as the electronegativity of the halogen leaving group increases: Ar—I < Ar—Br < Ar—Cl < Ar—F. Which step does this suggest is the rate-determining step of the mechanism—the addition step or the elimination step? Explain.

23.71 Draw the detailed mechanism and predict the major product for each of the following reactions. (*Hint*: See Problem 23.70.)

(a) $\xrightarrow{CH_3NH_2}$?

(b) $\xrightarrow{NaOH}$?

23.72 The following reaction yields a compound whose molecular formula is $C_7H_6O_4$.

The ^1H NMR and ^{13}C NMR spectra of $C_7H_6O_4$ are shown. Draw the product and the complete, detailed mechanism for this reaction.

23.73 Compound **A**, whose formula is C_9H_{10}, dimerizes in the presence of acid to produce the compound at the right. The ^1H NMR and ^{13}C NMR spectra of **A** are shown below. Determine the structure of **A** and draw the complete mechanism for the reaction. (in the ^{13}C-NMR spectrum, there are two signals >130 ppm.)

23.74 Heating compound **B** under acidic conditions produces compound **C**, whose formula is $C_{15}H_{12}$.

The ^{13}C NMR spectrum of **C** exhibits one signal near 20 ppm and 14 signals between 120 and 140 ppm. The ^1H NMR spectrum exhibits one signal at 2.7 ppm and several overlapping signals between 7 and 9 ppm. The integrations of the two sets of signals are in a 1:3 ratio. Determine the structure of **C** and draw the complete mechanism for the reaction.

If your instructor assigns problems in smartwork, log in at **smartwork.wwnorton.com**.

Problems / **1153**

The Diels–Alder Reaction and Other Pericyclic Reactions

Ethene reacts with buta-1,3-diene, a conjugated diene, to form cyclohexene, as illustrated in Equation 24-1. This is the simplest example of what is called the **Diels–Alder reaction**. This reaction is so important in organic chemistry that Otto Diels (1876–1954) and Kurt Alder (1902–1958), the two German chemists after whom the reaction is named, shared the 1950 Nobel Prize in Chemistry.

The Diels–Alder reaction

Buta-1,3-diene **Ethene** 200 °C, 17 h High pressure **Cyclohexene** 18%

New C—C bonds

(24-1)

The Diels–Alder reaction is especially useful because it forms two new carbon–carbon bonds. (As we stressed in Chapter 13, these are the most important types of reactions in organic synthesis.) Notice, too, that the product of the Diels–Alder reaction is a six-membered ring, which is widely abundant in natural products. The reaction is also quite robust;

Tetrodotoxin

The organs of this puffer fish contain tetrodotoxin, a neurotoxin with no known antidote. In 1972, Y. Kishi and coworkers devised a total synthesis of tetrodotoxin that incorporated a Diels–Alder reaction, one of the main topics of Chapter 24.

it can be carried out with a variety of other functional groups present, and many Diels–Alder reactions can be carried out under mild conditions. Finally, the reaction is often *regioselective* and *stereospecific*, which can provide chemists substantial control over the products that are formed.

The mechanism of the Diels–Alder reaction cannot be described by the elementary steps we learned in Chapters 6 and 7. Therefore, we introduce a new elementary step and we explore it with the same systematic approach we took with the introduction of the previous elementary steps. We begin with the curved arrow notation and examples. We then investigate the reaction's driving force, followed by its stereochemistry and regiochemistry.

Once we have thoroughly examined the Diels–Alder reaction, we turn our attention to syn hydroxylation and oxidative cleavage reactions—important reactions that involve elementary steps closely related to those of the Diels–Alder mechanism. Much of your knowledge gained from studying the Diels–Alder reaction is applicable to these reactions as well.

24.1 Curved Arrow Notation and Examples

The mechanism for the simplest possible Diels–Alder reaction (shown earlier in Equation 24-1) is provided in Equation 24-2. This serves as the prototype for all other Diels–Alder reactions.

Mechanism for Equation 24-1, the Diels–Alder reaction

The **dienophile is** relatively electron poor.

Six electrons are delocalized around the ring in the transition state.

The **diene is** relatively electron rich.

(24-2)

The entire reaction takes place in a single step, so

The Diels–Alder reaction is **concerted**, whereby all of the bonds are formed and broken simultaneously.

For this to occur, six electrons must flow in a cyclic fashion. Thus, the reaction proceeds through a *cyclic transition state*, which makes it a **pericyclic reaction**. More specifically, because this cyclic movement of electrons joins together two separate species to form a ring, it is also called a **cycloaddition reaction**.

A Diels–Alder reaction always involves a *diene* and a *dienophile*:

- A **diene** contributes four π electrons from a pair of *conjugated* π bonds.
- A **dienophile** (meaning "diene loving") contributes two π electrons from a single π bond.

Consequently, a Diels–Alder reaction can be described as a **[4+2] cycloaddition**, where the numbers indicate the number of π electrons contributed by each species involved.

Recall from the elementary steps in Chapters 6 and 7 that electrons flow from an electron-rich site to an electron-poor one. In the case of the Diels–Alder reaction, then, which species—the diene or the dienophile—is electron rich and which is electron poor? According to the discussion of nonpolar π bonds in Chapters 11 and 12, both species appear to be electron rich because each has at least one alkene group in which four electrons (two σ and two π electrons) are concentrated in a bonding region. The diene is more electron rich, however, because it has four electrons in a single π system, whereas the dienophile has just two. Therefore,

In a Diels–Alder reaction, the diene is relatively electron rich and the dienophile is relatively electron poor.

Viewed in this way, a curved arrow is drawn from the diene to the dienophile to initiate the flow of electrons. The other two curved arrows (e.g., in Equation 24-2) are needed to maintain octets of the atoms involved.

The Diels–Alder reaction can take place with an alkyne as the dienophile instead, as shown in Equation 24-3.

Buta-1,3-diene **Ethyne (Acetylene)** **Cyclohexa-1,4-diene**

(24-3)

Two electrons are contributed from one π bond of the triple bond, leaving the other π bond intact in the product.

YOUR TURN 24.1

The reaction in Equation 24-3 is repeated here:

Supply the curved arrows, label the diene and dienophile, and identify which reactant is relatively electron rich and which is relatively electron poor. (Remember that the π bonds of a triple bond are part of different π systems.)

Answers to Your Turns are in the back of the book.

What is consistent among all Diels–Alder reactions is the cyclic flow of six π electrons. The number of electrons is important, because those electrons are delocalized over a complete ring in the transition state. Recall from Chapter 14 that six electrons is a *Hückel number* of electrons (i.e., an odd number of pairs), and the species is *aromatic* when the electrons are delocalized over an entire ring. Therefore,

The Diels–Alder reaction proceeds through an **aromatic transition state**.

This substantially lowers the reaction's energy barrier, and is in large part what enables the reaction to take place without the need for severe conditions. We say that the reaction is "allowed" under normal conditions.

For comparison, let's examine two other possible cycloaddition reactions. In Equation 24-4, two molecules of ethene combine in a **[2+2] cycloaddition**, and in Equation 24-5, a molecule of hexa-1,3,5-triene and a molecule of ethene combine in a **[6+2] cycloaddition**. Neither of these reactions occurs readily under normal conditions, because they would have to proceed through an **antiaromatic transition state**. In both cases, an even number of pairs of electrons is delocalized over the entire ring—two pairs for the [2+2] cycloaddition and four pairs for the [6+2] cycloaddition. Because these reactions do not occur readily, we say that they are "forbidden."

Six electrons
delocalized

"Allowed"

This is an *antiaromatic* transition state, characterized by four electrons delocalized over the entire ring.

(24-4)

Ethene   Ethene

This is an *antiaromatic* transition state, characterized by eight electrons delocalized over the entire ring.

(24-5)

Hexa-1,3,5-triene   Ethene

24.2 **YOUR TURN**

The cycloaddition reaction involving two molecules of cyclopentene is as follows:

Is the transition state *aromatic* or *antiaromatic*?

Biological Diels–Alder Reactions

The Diels–Alder reaction was first documented in 1928, but nature has been using the reaction since long before then. One example is in the biosynthesis of lovastatin, a natural cholesterol-lowering drug produced by the fungus *Aspergillus terreus* that is marketed under the trade name Mevacor® . The Diels–Alder step is shown in Scheme A below, which produces a bicyclic compound that goes on to form lovastatin. Cholesterol is synthesized in our bodies through a key step in which 3-hydroxy-3-methylglutaryl coenzyme A (HMG-CoA) is converted to mevalonate. Lovastatin binds to the active site of HMG-CoA reductase, which inhibits its enzymatic activity.

Scheme A

ACP = acyl carrier protein

Diels–Alder

Lovastatin

Scheme B

Two thymine nitrogenous bases

UV light

A thymine dimer

UV light

Thymine dimer

Kink

Cycloaddition reactions also play a role in causing skin cancer. In DNA, the thymine nitrogenous base has a C=C double bond, so two adjacent thymine nucleotides can feasibly undergo a [2+2] cycloaddition to produce a *thymine dimer* (Scheme B, left). As we saw in Section 24.1, this kind of [2+2] cycloaddition is "forbidden" under normal conditions. However, when one of the molecules absorbs a photon of UV light (such as from the sun), the reaction becomes "allowed." (This is discussed in detail in Section 24.10.) The resulting thymine dimer causes a kink in the DNA strand (Scheme B, right), which interferes with the normal function of DNA.

Fortunately, our bodies can remove these thymine dimers through a process called nucleotide excision repair. Once the thymine dimer is recognized, a short segment of damaged single-stranded DNA is removed. DNA polymerase uses the remaining undamaged single-stranded DNA as a template to produce the complementary sequence that is replaced using an enzyme called DNA ligase.

SOLVED problem 24.1 Is this cycloaddition reaction allowed or forbidden?

Think In the transition state, are there electrons delocalized over an entire ring? If so, how many electrons? Is that a Hückel number or an anti-Hückel number? How does that correspond to whether the reaction is allowed or forbidden?

Solve The following transition state is somewhat of an average of the reactants and products.

Six electrons are delocalized over an entire ring in the transition state (two electrons are represented by each of the three curved arrows). Since six is a Hückel number of electrons, the transition state is aromatic, making the reaction allowed.

---

problem 24.2 Draw the transition state for the cycloaddition reaction at the right. Does this reaction take place readily? Why or why not?

24.2 Conformation of the Diene

The concerted nature of the Diels–Alder reaction means that both of the new carbon–carbon σ bonds form simultaneously. Thus, both ends of the diene must be relatively close to the dienophile in the transition state. The appropriate distances are achieved only if the diene attains a geometry in which the two ends of the π system are pointing in the same direction—the so-called *s*-**cis conformation** (Fig. 24.1a). (The *s* indicates that the cis designation describes the orientation of groups about a single bond, rather than a double bond or the plane of a ring.) In the *s*-**trans conformation** (Fig. 24.1b), the ends of the two reacting species are farther apart, which raises the energy of the transition state and makes the Diels–Alder reaction unfeasible. Thus,

> For a Diels–Alder reaction to take place, the diene must be in the *s*-cis conformation.

This requirement makes it somewhat difficult for buta-1,3-diene to react in a Diels–Alder reaction because rotation can occur about the central single bond (Equation 24-6) and the *s*-cis conformation is about 10 kJ/mol (2.4 kcal/mol) higher in energy than the *s*-trans conformation. At equilibrium, therefore, 98% of the molecules are in the *s*-trans conformation (Equation 24-6a), and the remaining 2% are in the *s*-cis conformation (Equation 24-6b).

With the diene in the *s*-cis conformation, the ends of the diene are at the appropriate distance from the dienophile to produce two new C—C bonds.

$$HC{=}CH_2 \cdots CH_2$$
$$HC{-}CH_2 \cdots CH_2$$

s-Cis conformation

(a)

With the diene in the *s*-trans conformation, the ends of the diene are too far from the dienophile.

$$H_2C \quad CH_2$$
$$CH \quad CH_2$$
$$HC \quad CH_2$$
$$CH_2$$

s-Trans conformation

(b)

FIGURE 24-1 The diene conformation in Diels–Alder reactions A diene can exist in (a) the *s*-cis conformation and (b) the *s*-trans conformation. Diels–Alder reactions are feasible only when the diene is *s*-cis.

s-Trans

98%

(24-6a)

No reaction

**Rotation about
C—C single bond**

2%

s-Cis

(24-6b)

This means that only about 2% of buta-1,3-diene molecules can participate in the Diels–Alder reaction at any given time, which slows the reaction substantially, and is part of the reason why the Diels–Alder reaction involving buta-1,3-diene and ethene requires such high temperatures (Equation 24-1).

YOUR TURN 24.3

Which of the following conformations is s-cis and which is s-trans? Which can undergo a Diels–Alder reaction?

A   **B**

The requirement for the *s*-cis conformation precludes the Diels–Alder reactions in Equations 24-7 and 24-8 altogether.

**The diene is locked in the
s-trans conformation.**

No reaction

(24-7)

No reaction

(24-8)

The diene in Equation 24-7 is locked in the *s*-trans conformation by the fused ring system. The diene in Equation 24-8 is *not* locked in the *s*-trans conformation, but it experiences severe steric strain in the *s*-cis conformation, as illustrated in Equation 24-9.

**Steric
strain**

***s*-Trans**   ***s*-Cis**

(24-9)

SOLVED problem 24.3 Which molecule, **A** or **B**, will react faster with ethene in a Diels–Alder reaction? Explain.

Think Are the dienes in an s-cis conformation? If not, can they attain an s-cis conformation? Which is more stable in the s-cis conformation?

Solve Neither diene is in the s-cis conformation as written, but the single bond in each molecule that connects the two double bonds can rotate.

No substantial strain exists in the s-cis conformation of molecule **A**. By contrast, molecule **B** experiences severe steric strain from the overlapping *tert*-butyl groups when in the s-cis conformation. Thus, the equilibrium heavily favors the s-trans conformation for **B**, which really hinders the Diels–Alder reaction. Consequently, diene **A** will undergo a Diels–Alder reaction much faster than diene **B**.

---

problem 24.4 Which of these molecules will react faster with ethene in a Diels–Alder reaction? Explain.

C   **D**

---

problem 24.5 Although benzene's Lewis structure exhibits a pair of conjugated double bonds locked in an s-cis conformation, benzene does not react with ethene. Draw the product of this hypothetical reaction and explain why it does not occur.

By the same token, Diels–Alder reactions take place more readily when the diene is locked in the s-cis conformation. This is the case for cyclopentadiene in Equation 24-10 and cyclohexa-1,3-diene in Equation 24-11.

The diene is locked in the s-cis conformation.

Bridgehead carbons

$$\text{Cyclopentadiene} + \begin{array}{c} \text{CH}_2 \\ \| \\ \text{CH}_2 \end{array} \longrightarrow \text{Bicyclo[2.2.1]hept-2-ene} \quad (24\text{-}10)$$

Cyclopentadiene   **Bicyclo[2.2.1]hept-2-ene**

The diene is locked in
the s-cis conformation.

Bridgehead carbons

Cyclohexa-1,3-diene   Bicyclo[2.2.2]octadiene   (24-11)

Both of these reactions produce a **bicyclic compound**, in which two **bridgehead carbons** are part of multiple rings. Notice that the bridgehead carbons are the C atoms from the reactant diene that form new bonds to the dienophile.

problem **24.6** Draw the mechanism and the product for the Diels–Alder reaction at the right. Do you think the reaction will take place as readily as the one in Equation 24-10? Why or why not?

24.3 Substituent Effects on the Driving Force

The prototypical Diels–Alder reaction (Equation 24-1) between ethene and buta-1,3-diene is quite sluggish, requiring temperatures around 200 °C to proceed at a reasonable rate. Even then, the yield is only 18%. Fortunately, Diels–Alder reactions can be helped by the presence of certain substituents. Recall from Section 24.1 that a Diels–Alder reaction is driven by the flow of electrons from the relatively electron-rich diene to the relatively electron-poor dienophile. This driving force can be increased, therefore, when a substituent makes the diene more electron rich or the dienophile more electron poor.

FIGURE 24-2 **Electron-donating groups on the diene** Electron-donating groups (highlighted in red) make the diene more electron rich, which facilitates Diels–Alder reactions.

- Electron-donating substituents bonded directly to the diene (Fig. 24-2) facilitate Diels–Alder reactions. These are the same substituents that were classified as "activating" groups in electrophilic aromatic substitution (Table 23-3, p. 1116).
- Electron-withdrawing substituents bonded directly to the dienophile (Fig. 24-3) facilitate Diels–Alder reactions. These are the same substituents that were classified as "deactivating" groups in electrophilic aromatic substitution.

In the reaction in Equation 24-12, for example, two electron-donating CH_3 groups are attached directly to the diene. The reaction takes place at a slightly lower temperature than the one in Equation 24-1, and the yield is much higher.

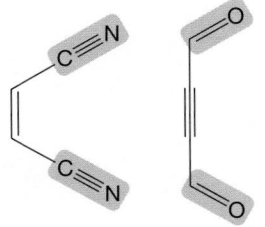

FIGURE 24-3 **Electron-withdrawing groups on the dienophile** Electron-withdrawing groups (highlighted in blue) make the dienophile more electron poor, which facilitates Diels–Alder reactions.

Electron-donating groups on the diene facilitate a Diels–Alder reaction.

CH_3 + CH_2 ‖ CH_2 →[185 °C, 17 h] (24-12)

CH_3

CH_3

cis-3,6-Dimethylcyclohexene
60%

A similar outcome is observed with the reaction in Equation 24-13, in which a highly electron-withdrawing NO_2 group is attached to the dienophile.

An electron-withdrawing group on the dienophile facilitates a Diels–Alder reaction.

$$\xrightarrow{\text{107 °C, 17 h}}$$

(24-13)

4-Nitrocyclohexene
84%

SOLVED problem 24.7 Does diene **A** or diene **B** react faster in a Diels–Alder reaction? Explain.

Think Is the diene relatively electron rich or electron poor in a Diels–Alder reaction? Will electron-donating groups or electron-withdrawing groups on the diene speed up the reaction? What is the relative electron-donating/electron-withdrawing capability of each substituent?

Solve The diene is relatively electron rich in a Diels–Alder reaction, so the more electron donating a substituent is, the faster the reaction. In both dienes **A** and **B**, the substituent is electron donating, because each is attached to the diene by an atom with a lone pair of electrons. According to Table 23-3, the alkoxy group in **B** is more strongly activating than the amido group in **A** (due to the presence of the carbonyl group). Therefore, diene **B** is more electron rich than diene **A**, and **B** will react faster.

---

problem 24.8 Does dienophile **C** or dienophile **D** react faster in a Diels–Alder reaction? Explain.

24.4 Stereochemistry of Diels–Alder Reactions

In the prototypical Diels–Alder reaction between buta-1,3-diene and ethene, four carbon atoms rehybridize from sp^2 to sp^3—two from the diene and two from the dienophile (see Your Turn 24.4). These atoms, therefore, have the potential to become tetrahedral stereocenters in the product.

24.4 **YOUR TURN**

In the prototypical Diels–Alder reaction at the right, identify the C atoms in the products that have been transformed from sp^2 to sp^3 hybridization and are therefore *potential* stereocenters.

Two tetrahedral stereocenters are produced, for example, in each of the reactions in Equation 24-14 and 24-15.

A cis configuration in the dienophile   **A cis configuration in the ring**

(24-14)

Racemic mixture of enantiomers

(24-15)

A trans configuration in the dienophile   **A trans configuration in the ring**

Only the cis configuration is produced in Equation 24-14, however, whereas only the trans configuration (a racemic mixture of enantiomers) is produced in Equation 24-15. These outcomes are due to the configurations of the substituents in the dienophile: They are cis to each other about the initial C=C double bond in Equation 24-14, but they are trans to each other in Equation 24-15. In other words,

> The stereochemical configuration in the dienophile is conserved throughout the course of a Diels–Alder reaction.

problem **24.9** Predict the product of the Diels–Alder reaction at the right, paying particular attention to stereochemistry of the C atoms from the dienophile.

problem **24.10** Draw the dienophile that would be required to generate each of the molecules at the right from buta-1,3-diene in a Diels–Alder reaction.

(a)   (b)

Similarly, the stereochemistry in the diene leads to specific stereochemical configurations in the products, as shown in Equations 24-16 through 24-18.

Both double bonds are trans.

(24-16)

A cis configuration in the ring

Both double
bonds are cis.

$$\text{(diene)} + \underset{CH_2}{\overset{CH_2}{\|}} \longrightarrow \qquad (24\text{-}17)$$

A cis configuration in the ring

Trans double
bond   CH₃

Racemic mixture of enantiomers

$$\text{(diene)} + \underset{CH_2}{\overset{CH_2}{\|}} \longrightarrow \qquad + \qquad (24\text{-}18)$$

CH₃

Cis double bond

A trans configuration in the ring

Namely,

> - If the double bonds in the diene are either both trans (Equation 24-16) or both cis (Equation 24-17), then the substituents at the ends of the diene become cis to each other with respect to the ring in the product.
> - Conversely, if one of the diene's double bonds is cis and the other is trans (Equation 24-18), then the two substituents become trans to each other in the product.

We obtain the stereochemical outcomes in Equations 24-14 through 24-18 when the diene and dienophile approach each other as shown in Equation 24-19. The Diels–Alder reaction is concerted—the two σ bonds formed at each end of the diene and dienophile are formed *simultaneously*—so the substituents about the dienophile (A, B, C, and D in Equation 24-19) cannot rotate relative to each other to scramble the stereochemistry. This explains our observations in Equations 24-14 and 24-15.

The new C—C bonds form simultaneously, so the relative orientation of the substituents A, B, C, and D cannot change.

$$(24\text{-}19)$$

The concerted nature of the Diels–Alder reaction also explains the stereochemistry observed in Equations 24-16 through 24-18, as shown in Equation 24-20 for the generic diene with substituents W, X, Y, and Z attached to the terminal carbon atoms. Those terminal carbons must rotate slightly as the new bonds are formed, allowing the six-membered ring to relax into its equilibrium geometry. More specifically, the carbon atoms rotate in *opposite directions* during the course of the reaction (if viewed from one end of the diene, one carbon rotates clockwise and the other counterclockwise). If substituents W and Z are both CH₃ groups and X and Y are both H atoms, for example, then both CH₃ groups will be cis to each other in the product, just as we saw previously in Equation 24-16.

(24-20)

YOUR TURN 24.5

Match the generic groups (W, X, Y, and Z) on the diene in Equation 24-20 to the specific groups on the diene in Equation 24-18. Does the stereochemistry of the product in Equation 24-20 agree with that in Equation 24-18?

SOLVED problem 24.11
Predict the product of the Diels–Alder reaction at the right, paying particular attention to stereochemistry.

Think Do the carbons at the ends of the diene become tetrahedral stereocenters in the products? What are the configurations about each double bond in the diene? Do the alkene carbons of the dienophile become tetrahedral stereocenters? What is the configuration about the C=C double bond in the dienophile?

Solve If we draw the product without worrying about stereochemistry, notice that each end carbon in the diene becomes a tetrahedral stereocenter, and so does each alkene carbon in the dienophile.

Each D—C=C—C configuration is cis along the diene skeleton, and each H—C=C—C configuration is trans. Therefore, the two D atoms and the two H atoms will both be cis to each other in the product. The alkene in the dienophile is in the cis configuration, so the carbonyl groups will be cis to each other in the product, too. Overall, then, there are four results to consider (**A–D**), which, due to the plane of symmetry along the horizontal axis of the molecule, reduce to two distinct stereoisomers (**A** and **B**).

A

B

C
Redundant to B

D
Redundant to A

problem 24.12 Predict the Diels–Alder product for each of the following reactions in which the dienophile is isotopically labeled. Pay particular attention to stereochemistry.

(a)

(b)

problem 24.13 If the compound at the right is the product of a Diels–Alder reaction with ethene, then what is the structure of the diene that would have been required?

So far, the stereochemical aspects we have considered for the Diels–Alder reaction deal with the diene and dienophile independently. Another aspect of stereochemistry to consider pertains to the configurations of the carbon atoms from the diene *relative* to those from the dienophile. This is shown in Equation 24-21, in which a mixture of *diastereomers* is produced.

The endo product is heavily favored.

Exo product
3%

Endo product
97%

(24-21)

Equations 24-22 and 24-23 show that these two products result from the two different approaches possible for the dienophile—one with the substituents pointed toward the diene and the other with it pointed away. When the dienophile's substituents point toward the diene (Equation 24-22), then it is called an **endo approach**, and the resulting product is called the **endo product**. Conversely, when the dienophile's substituents point away from the diene (Equation 24-23), then it is called an **exo approach**, and the resulting product is called the **exo product**. In essence, if the dienophile's substituents (highlighted in blue) reside opposite the diene's substituents (highlighted in red) on the six-membered ring that is produced, then you have the endo product, and if the dienophile's substituents reside on the same side of the ring as the diene's substituents, then you have the exo product.

The dienophile's substituents point toward the diene.

Endo approach

Endo product

The dienophile's substituents reside on the side opposite those from the diene.

(24-22)

The dienophile's substituents point away from the diene.

Exo approach

Exo product

The dienophile's substituents reside on the same side as those from the diene.

(24-23)

Although both endo and exo approaches are feasible, the product mixture is usually dominated by one product, according to the **endo rule**:

> In a Diels–Alder reaction, the endo approach is generally favored over the exo approach, so the major product is usually endo.

Understanding the preference for the endo approach requires molecular orbitals, so it is discussed in Section 24.10.

problem 24.14 Draw the major product of the Diels–Alder reaction at the right, paying particular attention to stereochemistry.

24.5 Regiochemistry of Diels–Alder Reactions

If both the diene and the dienophile of a Diels–Alder reaction are *unsymmetric*, then two constitutional isomers can be produced. An example is shown in Equation 24-24.

Both the diene and dienophile are unsymmetric.

A mixture of two constitutional isomers is produced.

Hydroquinone

74%

(24-24)

The isomers are produced from different relative orientations of the two reactants when they approach each other, as shown in Equations 24-25 and 24-26. The major product is produced via the approach in Equation 24-25.

One isomer is produced from this approach.

Major product

(24-25)

(24-26)

The other isomer is produced from this approach.

Minor product

This regiochemistry is governed by the following rule:

> When the diene and the dienophile of a Diels–Alder reaction are both unsymmetric, the major isomeric product is the one produced by the approach that exhibits the more favorable electrostatic attraction among atoms undergoing bond formation.

We can determine which approach gives rise to a more favorable electrostatic attraction by examining the resonance hybrids of the two reactants, as shown in Figure 24-4. The resonance structures and the resonance hybrids are shown in Figure 24-4a. With the approach shown at the top of Figure 24-4b, there is a favorable electrostatic interaction between the atoms undergoing bond formation. The approach shown at the bottom of Figure 24-4b, however, does *not* exhibit such favorable interactions, so the approach at the top of the figure leads to the major product.

(a)   (b)

FIGURE 24-4 Predicting the regiochemistry in Diels–Alder reactions (a) The resonance structures and the resonance hybrid of the diene are shown at the top, with regions of excess negative charge highlighted in red. The resonance structures and the resonance hybrid of the dienophile are shown at the bottom, with regions of excess positive charge highlighted in blue. (b) Interactions between the diene and dienophile. With the approach shown at the top, there is a favorable electrostatic interaction between atoms forming a bond. With the approach shown at the bottom, there are no such favorable electrostatic interactions, so the approach at the top leads to the major product.

The following resonance structures and resonance hybrid are of a diene that is an isomer of the one in Figure 24-4.

A and B represent the two ways this diene can approach the dienophile from Figure 24-4.

A

B

Which of these interactions will lead to the major Diels–Alder product?

SOLVED problem 24.15 Draw the mechanism and predict the major product for the Diels–Alder reaction at the right.

Think Where are the regions of excess negative charge and excess positive charge in these molecules? Which approach leads to favorable electrostatic interactions among the atoms undergoing bond formation?

Solve The resonance hybrids of these two species are shown in Your Turn 24.6. Interaction **B** in Your Turn 24.6 has a favorable electrostatic interaction among the C atoms undergoing bond formation, whereas interaction **A** does not. Therefore, the major product is from interaction **B**, yielding the following major product.

Favorable electrostatic interaction

Draw the mechanism that leads to the minor product of the Diels–Alder reaction in Solved Problem 24.15.

---

problem 24.16 Without taking stereochemistry into account, draw the mechanism and major product of each of these reactions.

(a)

(b)

24.6 The Reversibility of Diels–Alder Reactions; the Retro Diels–Alder Reaction

Under normal conditions, the prototypical Diels–Alder reaction in Equation 24-1 (shown again in Equation 24-27) has a substantially negative standard free-energy change ($\Delta G^\circ_{rxn} = -113$ kJ/mol $= -27$ kcal/mol), giving it a very large equilibrium constant ($K_{eq} = 3.5 \times 10^{19}$). As a result, this reaction is *irreversible* (review Section 9.12).

Loss of 3 π bonds **Gain of 1 π bond and 2 σ bonds**

$$\Delta G^\circ_{rxn} = -113 \text{ kJ/mol}$$
$$= -27 \text{ kcal/mol}$$
$$K_{eq} = 3.5 \times 10^{19}$$

(24-27)

As with any reaction, ΔG°_{rxn} can be broken down into a ΔH°_{rxn} term and a $T\Delta S^\circ_{rxn}$ term (i.e., $\Delta G^\circ_{rxn} = \Delta H^\circ_{rxn} - T\Delta S^\circ_{rxn}$). For the Diels–Alder reaction in Equation 24-27, $\Delta H^\circ_{rxn} = -168$ kJ/mol (-40 kcal/mol) and $T\Delta S^\circ_{rxn} = -55$ kJ/mol (-13 kcal/mol) at 25 °C. Thus, the large, negative ΔH°_{rxn} term dominates to make ΔG°_{rxn} substantially negative.

The large negative ΔH° for this reaction stems primarily from how the bond energies in the product differ from those in the reactants. Notice in Equation 24-27 that there are three C=C double bonds and one C—C single bond in the reactants, whereas there are five C—C single bonds and one C=C double bond in the product. In essence, then, three π bonds are converted into two σ bonds and one π bond, and, as we learned in Chapter 3, σ bonds are stronger than their corresponding π bonds.

24.8 **YOUR TURN**

To estimate ΔH°_{rxn} in Equation 24-27 quantitatively, complete and solve the following equation, which accounts for the changes in the numbers of C—C and C=C bonds. Use the average C—C and C=C bond energies in Table 1-3.

Three C=C bonds   **Five C—C bonds**

One C—C bond   **One C=C bond**

$\Delta H^\circ_{rxn} = [3(\underline{\quad}\text{kJ/mol}) + 1(\underline{\quad}\text{kJ/mol})] - [5(\underline{\quad}\text{kJ/mol}) + 1(\underline{\quad}\text{kJ/mol})] = -\underline{\quad}\text{kJ/mol}$

How does this estimate of ΔH°_{rxn} compare to the experimentally measured value of -168 kJ/mol (-40 kcal/mol)?

The entropy term in a Diels–Alder reaction is negative because the reaction decreases the number of independent molecules (i.e., the reaction leads to less "disorder"). This means that increasing the temperature makes ΔG°_{rxn} less negative (because $\Delta G^\circ_{rxn} = \Delta H^\circ_{rxn} - T\Delta S^\circ_{rxn}$). Even at 200 °C, however, the prototypical reaction in Equation 24-27 is not easily reversible due to its large, negative ΔH°_{rxn}. The Diels–Alder reaction in Equation 24-28, on the other hand, has a significantly smaller value of ΔH°_{rxn} (-77 kJ/mol $= -18$ kcal/mol), so it is easily reversible at moderately high temperatures.

This Diels–Alder reaction is *reversible*.

$$\Delta G^\circ_{rxn} = -17 \text{ kJ/mol}$$
$$= -4 \text{ kcal/mol}$$
$$K_{eq} = 120$$

(24-28)

Cyclopentadiene   **Dicyclopentadiene**

Supply the curved arrows missing from Equation 24-28.

For the following Diels–Alder reaction, $\Delta H°_{rxn} = -121$ kJ/mol (-29 kcal/mol).

Do you think this reaction would be easily reversible at 150 °C?

problem **24.17** What temperature would be required to achieve $K_{eq} = 120$ for the prototypical Diels–Alder reaction in Equation 24-27? (The values of $\Delta H°$ and $\Delta S°$ can be obtained from the text.)

Chemists take advantage of the reversibility of the Diels–Alder reaction in Equation 24-28 to prepare a fresh sample of cyclopentadiene. At room temperature, K_{eq} heavily favors the dicyclopentadiene product, and over about 1 week's time, cyclopentadiene *dimerizes* entirely into dicyclopentadiene. When a sample of dicyclopentadiene is heated to 150 °C or higher, the reaction is driven in the reverse direction, which allows the low-boiling cyclopentadiene (b.p. = 42 °C) to be removed and collected via distillation, whereas the higher-boiling dicyclopentadiene (b.p. = 170 °C) remains behind. Le Châtelier's principle continues to drive the reaction toward the side that has cyclopentadiene—that is, in the reverse direction.

Because this process "undoes" the Diels–Alder reaction in Equation 24-28, it is called a **retro Diels–Alder reaction** or a **[4+2] cycloelimination**. The mechanism for this reaction is simply the reverse of a Diels–Alder mechanism, so the entire reaction consists of a single (concerted) step in which six electrons move in a cyclic fashion, as shown in Equation 24-29. Notice that the three curved arrows are all oriented in the same direction about the six-membered ring, and one curved arrow originates from the double bond.

Dicyclopentadiene   **Cyclopentadiene**
                                    70%

When cyclohexene is heated to very high temperatures, the reverse of the reaction in Equation 24-27 takes place, producing ethene and buta-1,3-diene. Draw the curved arrows for this reaction.

problem **24.18** When the compound at the right is heated, a single product with the formula C_6H_8 is produced. Draw the product and the curved arrows necessary to produce it.

24.7 Syn Dihydroxylation of Alkenes and Alkynes Using OsO$_4$ or KMnO$_4$

What distinguishes the Diels–Alder reaction from other reactions we have studied previously is the cyclic movement of six electrons in a concerted fashion. Another reaction that incorporates a similar concerted elementary step is the **syn dihydroxylation** of alkenes and alkynes (Equation 24-30), in which two OH groups add across the C=C double bond in a syn fashion.

Two OH groups add across the C=C bond in a syn fashion.

(24-30)

(E)-But-2-enoic acid

(2S,3R)- and (2R,3S)-1,2-Dihydroxybutanoic acid

53%

Care must be taken with this reaction because OsO$_4$ is toxic and expensive.

A partial mechanism for this reaction is shown in Equation 24-31.

Partial mechanism for Equation 24-30

Both C—O bonds form on the same face of the initial alkene.

1. [4+2] Cycloaddition

Several steps

Hydrolysis

OsO$_4$ is regenerated.

$\left(+ \text{ H}_2\text{OsO}_4 \xrightarrow[\text{Oxidation}]{\text{H}_2\text{O}_2} \text{OsO}_4 \right)$

(+ Enantiomer)

Osmate ester

(+ Enantiomer)

(24-31)

Step 1 is the addition of OsO$_4$ across the C=C double bond in a syn fashion, producing an **osmate ester**. Then, over the course of several steps, the osmate ester is hydrolyzed: The O—Os bonds are replaced by O—H bonds, thus producing a cis-1,2-diol and a reduced form of the osmium. Hydrogen peroxide, H$_2$O$_2$, oxidizes the reduced form back to OsO$_4$, making OsO$_4$ a *catalyst*, because it is not consumed overall.

Step 1 is key in this mechanism, because it attaches the two O atoms in a syn fashion to the initial alkene C atoms. This step involves the concerted movement of six electrons in a cyclic fashion, so all bond formation and all bond breaking occur simultaneously; it is, therefore, a type of *pericyclic reaction*. In particular, the step involves the flow of four electrons from OsO$_4$ and two from the alkene, resulting in a new ring, so it is a [4+2] cycloaddition, much like the Diels–Alder reaction. Moreover, the concerted movement of the electrons requires the new bonds to the alkene to form on the same side (also like the Diels–Alder reaction).

Unlike in a Diels–Alder reaction, the electron movement in this [4+2] cycloaddition involves just five atoms—three atoms from one of the reacting species (in this case, OsO$_4$) and two atoms from the other (in this case, the alkene). As a result, two of the six electrons moving cyclically end up as a lone pair on the middle of the three atoms contributed from one reacting species (in this case, the Os atom from OsO$_4$).

An older method of producing cis-1,2-diols involves treating an alkene with a cold, basic solution of potassium permanganate (KMnO$_4$), as shown in Equation 24-32. It is not very synthetically useful, however, because the yields are typically low and the diol product can be further oxidized by KMnO$_4$ (see Section 24.8).

OH groups add across the C=C double bond in a syn fashion.

Cyclopentene — KMnO₄, HO⁻, Cold → *cis*-Cyclopentane-1,2-diol 31%

(24-32)

YOUR TURN 24.12

The mechanism for the reaction in Equation 24-32 is believed to be similar to the one for OsO_4. The key step is the [4+2] cycloaddition of MnO_4^- across the C=C double bond, making a **manganate ester**.

A manganate ester → Hydrolysis →

Supply the curved arrows that are necessary to depict the formation of the manganate ester from the starting materials.

The C≡C triple bond of an alkyne reacts with $KMnO_4$, too, generating a 1,2-diketone (not a 1,2-diol), as shown in Equation 24-33.

A 1,2-diketone

The product of two dihydroxylations has two carbonyl hydrates.

Diphenylethyne — $KMnO_4$, HO⁻, cold → [OH OH / OH OH] — 2 H_2O → Diphenylethanedione 88%

(24-33)

We can envision the alkyne as having undergone two syn dihydroxylations to yield a 1,1,2,2-tetraol intermediate. This intermediate has two adjacent hydrates (characterized by two OH groups attached to the same carbon atom). Recall from Chapter 17 that hydrates like this equilibrate with a C=O group through the loss of H_2O, resulting in the dicarbonyl product.

problem **24.19** Which of the following diols can be produced from an alkene using either OsO_4 or $KMnO_4$? For each one that can, draw the alkene that can be used to produce it.

(a)　(b)　(c)　(d)　(e)

24.8 Oxidative Cleavage of Alkenes and Alkynes

Here we introduce a variety of **oxidative cleavage** reactions, whereby the carbon–carbon bond of an alkene or alkyne is broken and the initial alkene or alkyne carbons are oxidized in the process. The general process is illustrated in Equations 24-34 and 24-35, but specific reactions can generate ketones, aldehydes, carboxylic acids, or carbon dioxide, depending on the reagents used and the exact conditions of the reaction.

Oxidative cleavage

The C=C bond of the alkene is broken.

The initial alkene C atoms become carbonyl C atoms.

(24-34)

The C≡C bond of the alkyne is broken.

The initial alkyne C atoms become carbonyl C atoms.

(24-35)

We will examine oxidative cleavage reactions involving three different reagents: potassium permanganate ($KMnO_4$; Section 24.8a), the periodate anion (IO_4^-; Section 24.8b), and ozone (O_3; Section 24.8c). The mechanisms for these reactions involve at least one pericyclic step similar to the ones we have encountered thus far. Therefore, whereas the structural changes that take place in oxidative cleavage reactions might seem to differ significantly from what we have seen in the Diels–Alder and syn dihydroxylation reactions, all of these reactions are related through their mechanisms.

24.8a Oxidative Cleavage Involving $KMnO_4$

In Section 24.7, we saw that the C=C double bond of an alkene undergoes syn dihydroxylation when treated with cold, basic $KMnO_4$. When the $KMnO_4$ solution is concentrated and the reaction mixture is heated, however, then *oxidative cleavage* of the C=C bond takes place, as shown in Equation 24-36.

The alkene is cleaved.

1. $KMnO_4$, conc.
   $HO^{\ominus}$, H_2O, 50 °C
2. HCl

(24-36)

1-Methylcyclohexene

6-Oxoheptanoic acid
61%

The mechanism for this reaction is shown in Equation 24-37.

Mechanism for Equation 24-36

A manganate ester

1. [4+2] Cycloaddition   2. Cycloelimination

(24-37)

Aldehydes are oxidized to carboxylic acids in the presence of $KMnO_4$.

+ MnO_2 Multiple steps

Just like the dihydroxylation reaction (Your Turn 24.12), MnO_4^- undergoes a [4 + 2] cycloaddition to the double bond in Step 1, yielding a manganate ester. The added heat, however, provides enough energy to break the C—C bond in Step 2 via another cyclic rearrangement of six electrons, thus eliminating MnO_2. Step 2 is also a pericyclic reaction that involves the cyclic movement of six electrons, but it results in the breaking apart of one species to produce two separate species, so it is a cycloelimination.

At this stage, the initial alkene C atoms have already become carbonyl C atoms in a ketone and an aldehyde. Under these oxidation conditions, however, the aldehyde group reacts further to produce a carboxylic acid (Section 19.6b).

When one of the alkene carbons is part of a CH_2 group (i.e., when the alkene is terminal), then cleavage of the C=C double bond initially produces formaldehyde, $H_2C=O$, as shown in Equation 24-38. Under the oxidation conditions of the reaction, formaldehyde is further oxidized by $KMnO_4$ to CO_2.

Phenylethene

$KMnO_4$, conc. $HO^\ominus$, H_2O, 70 °C

1. $KMnO_4$
2. HCl

Benzoic acid
50%

Formaldehyde is oxidized to carbon dioxide.

$KMnO_4$

$O=C=O$
Carbon dioxide

(24-38)

Alkynes are also susceptible to oxidative cleavage with $KMnO_4$. In this case, though, oxidative cleavage produces only carboxylic acids and (in the case of a terminal alkyne) CO_2, as shown in Equation 24-39.

Hex-1-yne →
1. KMnO₄, conc. HO⁻, H₂O, hot
2. HCl
→ **Pentanoic acid** 61% + $O=C=O$ **Carbon dioxide** (24-39)

Ketones cannot be formed under these conditions because, after cleavage of the C≡C bond, each alkyne C atom can be bonded to, at most, one alkyl group.

SOLVED problem 24.20 A hydrocarbon having the formula C_8H_{14} is treated with hot, basic, concentrated $KMnO_4$. Bubbling is observed and the compounds at the right are recovered. What is a possible structure for the original hydrocarbon?

Think Is the number of C atoms in the product compounds the same as in the reactant? What could give rise to the bubbling? What functional group is the precursor to a carboxylic acid under oxidative cleavage conditions?

Solve There are a total of seven C atoms in the two compounds that were recovered, whereas there are eight C atoms in the original hydrocarbon. The other C atom is lost in the form of CO_2, which bubbles out of solution.

Thinking backward, the two carboxylic acid functional groups must have come from the initial formation of aldehydes, and CO_2 must have come from formaldehyde ($H_2C=O$).

Thinking backward again, each pair of C=O bonds must have come from the cleavage of a C=C double bond. Therefore, a possible structure for the original hydrocarbon is as shown below.

Undo oxidative cleavage.

---

problem 24.21 What is another possible structure for the original hydrocarbon in Solved Problem 24.20?

24.8b Oxidative Cleavage Involving IO_4^-

An alkene can also be cleaved oxidatively by treatment with OsO_4, followed by the **periodate anion**, IO_4^-. The periodate anion is most commonly introduced as $NaIO_4$ (Equation 24-40), but can also be introduced in the form of **periodic acid**, HIO_4 (pronounced *per-eye-OH-dik*), which produces an equilibrium amount of IO_4^-.

The alkene is cleaved.

1-Methylcyclohexene
1. OsO_4
   acetone, H_2O
   10 min, 25 °C
2. $NaIO_4$
   12 h, 25 °C
3. $Na_2S_2O_3$, H_2O

The aldehyde is *not* oxidized to the carboxylic acid.

6-Oxoheptanal (24-40)

Unlike oxidative cleavage involving $KMnO_4$, however,

Aldehydes that are produced upon oxidative cleavage involving IO_4^- are *not* oxidized further to either a carboxylic acid or carbon dioxide.

A partial mechanism for this reaction is shown in Equation 24-41.

Partial mechanism for Equation 24-40

Syn dihydroxylation

OsO_4, H_2O_2

A cis-1,2-diol

$IO_4^\ominus$

A periodate ester

Cycloelimination

+ HIO_3

(24-41)

The first several steps make up the syn dihydroxylation mechanism from Equation 24-31 (Section 24.7), thus producing a cis-1,2-diol. Then, the addition of IO_4^- produces a cyclic **periodate ester**. The final step in the mechanism is key, because it cleaves the C—C bond and produces the two carbonyl groups. This step involves the cyclic rearrangement of six electrons over five atoms, resulting in an additional lone pair of electrons on the iodine atom. The resulting decomposition of one species into two separate species makes this step a cycloelimination, much like Step 2 in Equation 24-37.

Notice in Equation 24-41 that the mechanism proceeds through a 1,2-diol. Therefore,

Periodate oxidative cleavage can be carried out directly on a 1,2-diol.

Examples are shown in Equations 24-42 and 24-43.

HIO_4

Ether, THF

98%

(24-42)

$NaIO_4$, H_2O

CH_2Cl_2

Silica gel

97%

(24-43)

A variety of 1,2-diols can undergo this reaction, the major requirement being the ability of the 1,2-diol to produce the cyclic periodate ester in Equation 24-41. (See Problem 24.22.)

problem 24.22

(a) Explain why **A** will undergo periodate oxidative cleavage but **B** will not.

(b) Draw the product of the reaction between **A** and HIO_4, including stereo-chemistry.

24.8c Oxidative Cleavage Involving Ozone: Ozonolysis

Oxidative cleavage with **ozone**, O_3, a process called **ozonolysis**, is shown in Equation 24-44 for 1-methylcyclohexene. Treatment with ozone is followed by dimethyl-sulfide, CH_3SCH_3, though zinc metal in acetic acid could be used instead of CH_3SCH_3. The double bond is cleaved to yield precisely the same product as oxidative cleavage with IO_4^-, shown previously in Equation 24-40.

The alkene is cleaved.

The aldehyde is *not* oxidized to the carboxylic acid.

1. O_3
   CH_2Cl_2, –78 °C
2. CH_3SCH_3, 23 h
   (or Zn, HOAc)

(24-44)

1-Methylcyclohexene

6-Oxoheptanal
30%

The mechanism for this reaction is shown in Equation 24-45.

Mechanism for Equation 24-44

A carbon–carbon bond is broken.

1. [4+2] Cycloaddition

A molozonide

2. Cycloelimination

3. [4+2] Cycloaddition

(24-45)

H_3C S CH_3

+ $(CH_3)_2SO$

An ozonide

In Step 1, ozone undergoes a [4+2] cycloaddition to the alkene to produce a **molozonide**, much like OsO_4 adds to an alkene to produce an osmate ester (Equation 24-31) or MnO_4^- adds to an alkene to produce a manganate ester (Equation 24-37).

Ethene, KMnO₄, and Fruit Ripening

One of the marvels of today's supermarkets is their ability to offer wide selections of fruits from all over the world, regardless of the season. How is this possible? One of the keys is the reaction that potassium permanganate ($KMnO_4$) undergoes with alkenes, just as we have seen here in Chapter 24.

Ethene ($H_2C=CH_2$) is the particular alkene of interest. It is a gas, and although it is very simple in structure, it is a ripening hormone and has a profound effect on the development of fruit. When ethene is detected by an ethene receptor in the fruit, certain biochemical pathways are turned on, which change the fruit's color, odor, and texture. The fruit industry takes advantage of this by picking fruit when it is still green, shipping it to a warehouse, and then treating it with ethene a few days before its delivery to nearby supermarkets.

Because of its effect on ripening, ethene can pose a serious problem during the shipping process. Some fruits, such as tomatoes, bananas, and apples, are climactic fruit, meaning that when they detect ethene, they quickly make more. (If you place a green banana in a paper bag with a ripe apple, the banana will ripen in about a day!) One way to deal with this is to expose the fruit to 1-methylcyclopropene, which competes with ethene for the receptor. Treating the fruit with 1-methylcyclopropene for several hours can block the receptor for more than 10 days. Another way to deal with ethene is to ship the fruit in containers in which $KMnO_4$ has been added. The $KMnO_4$ will react with ethene and effectively remove it from the surroundings.

Step 1, more specifically, is called a **1,3-dipolar cycloaddition** because ozone is a 1,3-dipolar compound, characterized by the resonance delocalization of charge over atoms that have a 1,3-positioning. Step 2 is a cycloelimination, which breaks the carbon–carbon single bond. Step 3 is another [4+2] cycloaddition, producing an **ozonide**.

Ozonides are unstable and can be explosive, so they are generally not isolated. Rather, when the ozonide is treated with dimethyl sulfide, the ozonide decomposes into the final product. Dimethyl sulfide is a reducing agent, moreover, so it prevents aldehydes in the product from being oxidized to carboxylic acids.

If oxidizing the aldehyde product to a carboxylic acid is desirable, then hydrogen peroxide—an oxidizing agent—can be added to the ozonolysis reaction, as indicated in Equation 24-46.

The alkene is cleaved.

H_2O_2 oxidized the initial aldehyde to the carboxylic acid.

1. O_3
   CH_2Cl_2, −78 °C
2. H_2O_2

(24-46)

1-Methylcyclohexene

6-Oxoheptanoic acid

problem **24.23** Draw the molozonide, the ozonide, and the products that are formed in each of these reactions.

1. O_3
2. H_2O_2
?

(a)

1. O_3
2. Zn, HOAc
?

(b)

24.9 Organic Synthesis: The Diels–Alder Reaction in Synthesis

As we have emphasized throughout Chapter 24, the Diels–Alder reaction is a valuable asset to organic synthesis, in large part because it makes it possible to synthesize a six-membered carbon ring from acyclic precursors. The prototypical Diels–Alder reaction—between buta-1,3-diene and ethene—is repeated in Equation 24-47; the fundamental motif in the product is a cyclohexene ring.

The *alkene* functionality that is left in the product is an important feature of the Diels–Alder reaction. This allows further reaction to take place specifically at that site. For example, the C=C double bond can be hydrogenated (Section 19.5) to produce a cyclohexane ring. Or, it can undergo electrophilic addition to produce a new functional group that can be used further (Chapters 11 and 12); an example is provided in Equation 24-48, in which the Diels–Alder product is converted into a Grignard reagent.

Another option is to alter the size of the ring, as shown in Equation 24-49.

Oxidative cleavage (Section 24.8) opens the six-membered ring at the C=C double bond. If this is followed by an intramolecular aldol condensation (Section 18.8), then a five-membered ring is formed.

The Diels–Alder reaction is not the first reaction we have encountered that can produce a six-membered carbon ring. In Section 18.10 we used the Robinson annulation reaction, which is shown again in Equation 24-50. The immediate product is an unsaturated cyclohexenone, but the carbonyl group can be selectively reduced to produce the cyclohexene ring.

Although a cyclohexene ring can be produced by the Robinson annulation reaction, the Diels–Alder reaction generally offers numerous advantages. One advantage is the relatively mild conditions required for a Diels–Alder reaction; by contrast, the Robinson annulation requires significantly basic conditions that may interfere with other functional groups present. Furthermore, reduction of a carbonyl group to a methylene (CH$_2$) group in the second step in Equation 24-50 can require harsh conditions, such as the presence of strong acid or strong base (see again Section 19.3).

Other advantages include the *stereospecificity* and the *regioselectivity* of the Diels–Alder reaction. In 1952, for example, R. B. Woodward and coworkers pioneered the use of the Diels–Alder reaction as a key step in the total synthesis of cholesterol, as indicated in Equation 24-51.

(24-51)

+ Enantiomer

The Diels–Alder reaction was used to generate the six-membered ring that later becomes the "D-ring" in the target. It was also used to establish the initial relative stereochemistry of the H and CH_3 groups indicated in red: Immediately after the Diels–Alder reaction they are cis to each other, and through a later step in the synthesis they become trans to each other.

Notice, too, the regiochemistry of the Diels–Alder reaction in Equation 24-51. There are two portions that can act as the dienophile in the starting material—namely, the two C=C double bonds on either side of the six-membered ring. Nevertheless, buta-1,3-diene selectively reacts with only one of them (the one to which the CH_3 group is attached), leaving the other one (to which the OCH_3 group is attached) available for further reaction.

problem **24.24** In Equation 24-51, why does buta-1,3-diene react selectively with the C=C double bond containing the CH_3 group instead of the one containing the OCH_3 group?

Equation 24-51 provides a glimpse of the utility of the Diels–Alder reaction in total synthesis. If you are interested in learning more about its use in total synthesis, see the review by Nicolaou and coworkers (*Angew. Chem. Int. Ed.* **2002**, *41*, 1668–1698).

24.10 A Molecular Orbital Picture of the Diels–Alder Reaction

Lewis structures and resonance theory can describe the Diels–Alder and other pericyclic reactions quite well, but these theories alone cannot account for some aspects of the reaction, such as the *endo rule* for stereochemistry. Moreover, the presence of UV light changes the outcome of the Diels–Alder reaction and this, too, cannot be explained by simple resonance theory.

These results can be accounted for, however, by *frontier molecular orbital* (*FMO*) theory. As we saw in Interchapter 1, the focus of FMO theory is on the symmetries of the HOMO and LUMO of the reactants. A reaction is *allowed* if the HOMO and LUMO have the appropriate symmetries to interact and is *forbidden* if they do not.

Figure 24-5 shows the π MO energy diagrams for both buta-1,3-diene and ethene, and highlights one frontier orbital interaction between the HOMO of buta-1,3-diene and the LUMO of ethene.

FIGURE 24-5 Frontier orbitals involved in a Diels–Alder reaction (a) MO energy diagram of buta-1,3-diene. (b) MO energy diagram of ethene. The green screen illustrates the HOMO–LUMO interaction of interest.

(a)

(b)

As the diene and dienophile approach each other, Figure 24-6 shows that the ends of the HOMO and LUMO π systems begin to overlap. Each region of overlap results in constructive interference. Thus,

> The frontier orbitals involved in a Diels–Alder reaction have the appropriate symmetries to interact, so the reaction is "allowed."

This is in agreement with our analysis in Section 24.1, where we mentioned that the Diels–Alder reaction proceeds through an aromatic transition state.

24.13 YOUR TURN

Figure 24-6 highlights one of two possible frontier orbital interactions, involving the HOMO of buta-1,3-diene and the LUMO of ethene. The drawing at the right highlights another, involving the HOMO of ethene and the LUMO of buta-1,3-diene.

Determine each type of interference resulting from the indicated regions of overlap. Are they both the same or are they different? Do the orbitals have the appropriate symmetries to interact?

LUMO of diene

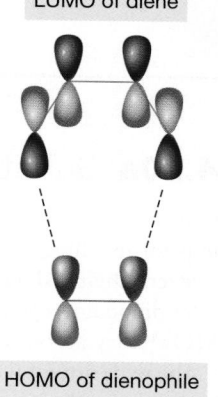

HOMO of dienophile

LUMO of dienophile

HOMO of diene

FIGURE 24-6 Frontier orbital interaction in a Diels–Alder reaction The HOMO of buta-1,3-diene (red) and the LUMO of ethene (blue) have the appropriate symmetries to interact. Specifically, the regions of overlap at the ends of the π systems result in the same kind of interference—in this case, constructive interference.

The story is quite different for the [2+2] cycloaddition reaction between two molecules of ethene. Figure 24-7 shows the MOs of π symmetry for each reactant, and highlights one possible frontier orbital interaction.

The overlap that takes place with these orbitals is illustrated in Figure 24-8. Unlike the Diels–Alder reaction (a [4+2] cycloaddition; Fig. 24-6), symmetry prevents the orbitals of the [2+2] cycloaddition from having a net interaction. Overlap on one end leads to constructive interference, but overlap on the other end leads to destructive interference.

> Because the frontier orbitals in a [2+2] cycloaddition do not have the appropriate symmetries to interact, the reaction is "forbidden."

This outcome is consistent with the fact that a [2+2] cycloaddition proceeds through an antiaromatic transition state, as we saw in Section 24.1.

FIGURE 24-8 Frontier orbital interaction in a [2+2] cycloaddition reaction The net interaction for the HOMO–LUMO interaction is zero. Whereas overlap on the right leads to constructive interference, overlap on the left leads to destructive interference. The two effects cancel each other.

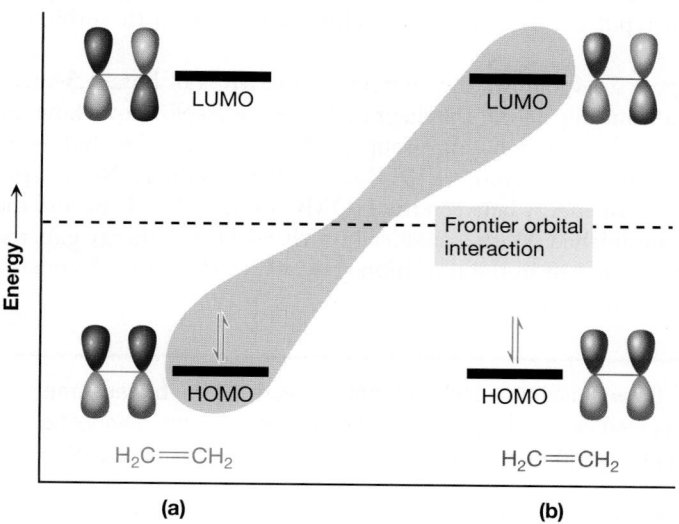

(a) (b)

FIGURE 24-7 Frontier orbitals involved in a [2+2] cycloaddition (a) MO energy diagram of one molecule of ethene. (b) MO energy diagram of the second molecule of ethene. The green screen indicates a HOMO–LUMO interaction.

The second possible HOMO–LUMO interaction in Figure 24-7 involves the HOMO of the blue molecule and the LUMO of the red one. The overlap of those orbitals is shown at the right.

Determine whether the overlap will lead to constructive or destructive interference. Do the orbitals have the appropriate symmetries to interact?

24.10a Substituent Effects on the Driving Force

Recall that the prototypical Diels–Alder reaction involving buta-1,3-diene and eth-ene is rather sluggish (see again Equation 24-1). As we learned earlier in the chapter, however, Diels–Alder reactions are facilitated by electron-donating groups attached to the diene and electron-withdrawing groups attached to the dienophile. How does FMO theory account for these results?

The experimental difficulties encountered with the prototypical Diels–Alder reac-tion can be understood by examining Figure 24-5 once again. Even though symmetry allows the frontier orbitals to interact, there is a substantial difference in energy be-tween the HOMO of buta-1,3-diene and the LUMO of ethene. This energy differ-ence limits the extent of interaction that can take place, and thus limits the resulting stabilization of the transition state.

Electron-donating and electron-withdrawing groups can change this picture, however, because

- The energy of an orbital is usually *raised* by a nearby electron-donating group.
- The energy of an orbital is usually *lowered* by a nearby electron-withdrawing group.

These effects can be understood in terms of charge repulsion between electrons. Increased electron density from a nearby electron-donating group results in increased electron–electron repulsion. Conversely, a nearby electron-withdrawing group re-duces electron–electron repulsion by removing electron density from the orbital of interest.

The diagram in Figure 24-9a shows the energies of the MOs of buta-1,3-diene and ethene (reproduced from Fig. 24-5). The diagram in Figure 24-9b shows how the picture changes when an electron-donating group (EDG) is attached to buta-1,3-diene, and an electron-withdrawing group (EWG) is attached to ethene. Notice spe-cifically that the difference in energy between the HOMO of buta-1,3-diene and the LUMO of ethene has diminished. With a smaller HOMO–LUMO energy gap, the orbitals interact to a greater extent in the transition state, which decreases the energy barrier between reactants and products.

problem **24.25** Draw a diagram similar to that in Figure 24-9b, but assume instead that an EWG is placed on the diene and an EDG is placed on the dienophile. How does the HOMO(diene)–LUMO(dienophile) energy gap compare to the one in Figure 24-9a?

FIGURE 24-9 Substituent effects on a frontier orbital interaction (a) MO energy diagrams of unsubstituted buta-1,3-diene and ethene. (b) MO energy diagrams of buta-1,3-diene with an attached electron-donating group (EDG) and ethene with an attached electron-withdrawing group (EWG). The EDG raises the MO energies of the diene and the EWG lowers the MO energies of the dienophile, which decreases the HOMO–LUMO energy gap.

24.10b The Endo Rule for Stereochemistry

When both endo and exo products can be made in a Diels–Alder reaction, the endo products tend to be favored (Section 24.4). Simple resonance theory cannot account for these results, but FMO theory can.

The Diels–Alder reaction between cyclopentadiene and propenal can produce both endo and exo products (Equation 24-52).

$$\text{(24-52)}$$

Endo product   **Exo product**

The π MOs of the two molecules are shown in Figure 24-10. Notice that the HOMO and LUMO of cyclopentadiene (Fig. 24-10a) are essentially the same as those of buta-1,3-diene. (The presence of the extra CH_2 group modifies the energies only slightly, so we will ignore its effect.)

The HOMO and LUMO of propenal (Fig. 24-10b), on the other hand, are somewhat different from those of ethene, because the C=O bond in propenal is *conjugated* to the C=C bond. There are, therefore, four π orbitals in propenal—two bonding and two antibonding—similar to buta-1,3-diene. Propenal's MOs are lower in energy due to the presence of the electron-withdrawing C=O group.

Figure 24-11 explicitly shows the interaction between the HOMO of cyclopentadiene and the LUMO of propenal, in both the endo approach (Fig. 24-11a) and the exo approach (Fig. 24-11b). Even though propenal has contributions from four *p* AOs in each of its π MOs, only the two *p* orbitals that make up the C=C double bond (i.e., those on C2 and C3) are directly involved in the reaction. (C2 and C3 are the ones that gain a σ bond during the course of the reaction.) Both the endo and exo approaches show the overlap of these two *p* orbitals on the dienophile with the *p* orbitals on C1 and C4 of the diene—the so-called **primary orbital overlap** for the reaction.

In the endo approach, the C=O group of the dienophile is oriented toward the π system of the diene, allowing the *p* orbital on the carbonyl carbon to interact with

FIGURE 24-10 Frontier orbitals in the Diels–Alder reaction involving cyclopentadiene and propenal (a) π MOs of cyclopentadiene. (b) π MOs of propenal. The important HOMO–LUMO interaction between them is indicated by the green screen.

FIGURE 24-11 Primary and secondary orbital overlap in a Diels–Alder reaction Interaction between the HOMO of cyclopentadiene with the LUMO of propenal when propenal approaches in (a) an endo fashion and (b) an exo fashion. Both approaches exhibit primary overlap of the orbitals used to form the two new σ bonds in the reaction, but only the endo approach exhibits secondary orbital overlap (dashed line), using an orbital on the dienophile that is not directly involved in bond formation or bond breaking. This lowers the energy for the transition state, thereby favoring the endo approach over the exo approach.

an unused *p* orbital on the diene (indicated by the dashed line in Fig. 24-11a). This **secondary orbital overlap** is not possible in the exo approach. Even though secondary overlap does not specifically lead to bond breaking or bond making in the reaction, it does provide stabilization.

In a Diels–Alder reaction, the transition state produced via the endo approach tends to be lower in energy than the one produced via the exo approach. Thus, the endo approach tends to be favored, making the endo product the major product.

YOUR TURN 24.15

Figure 24-11b is reproduced here. Identify the two *p* orbitals that, in Figure 24-11a, are involved in secondary orbital overlap. Are those *p* orbitals closer together in Figure 24-11a or Figure 24-11b?

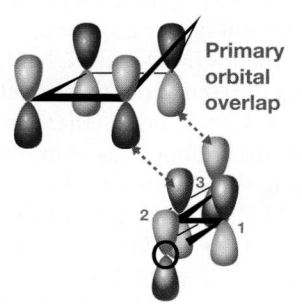

24.10c Thermal versus Photochemical Reactions

In all of the cycloaddition reactions we have considered thus far, the reactants have been in their ground state (i.e., lowest energy) electron configuration, where all electrons reside in the lowest-energy orbitals and are paired. When all reactants

(a)

(b)

FIGURE 24-12 Impact of UV photon absorption on the identities of the HOMO and LUMO (a) Ground state electron configuration of ethene. (b) Excited electronic state of ethene, denoted by *. A UV photon with sufficient energy ($h\nu$) is capable of promoting an electron from ethene's HOMO to its LUMO to generate the excited state. The excited state's HOMO is the ground state's LUMO.

are in their ground state, the reaction is said to be **thermal**. Thus, the Diels–Alder reaction—a [4+2] cycloaddition reaction—is **thermally allowed**, due to the favorable frontier orbital interactions. By contrast, the [2+2] cycloaddition is **thermally forbidden**, because the frontier orbital interactions do not lead to net stabilization (see again Figs. 24-6 and 24-8).

It turns out, though, that [2+2] cycloaddition reactions, such as the one between two molecules of ethene, take place readily when illuminated with UV light (Equation 24-53).

$$H_2C{=}CH_2 \xrightarrow[\text{UV light}]{} \begin{array}{c} H_2C{-}CH_2 \\ | \qquad | \\ H_2C{-}CH_2 \end{array} \qquad (24\text{-}53)$$

When ethene absorbs a photon with sufficient energy, an electron is promoted from the HOMO to the LUMO (Section 15.3a), thereby generating an **excited electronic state** (Fig. 24-12).

The new HOMO in that excited state is the π^* MO. Therefore, when an excited molecule of ethene encounters another ethene molecule still in the ground electronic state, the frontier orbital interaction involves the π^* orbital from each species (Fig. 24-13).

These HOMO and LUMO orbitals do, indeed, have the appropriate symmetries to interact, as illustrated in Figure 24-14.

FIGURE 24-13 Frontier orbitals in a photochemical [2+2] cycloaddition reaction The π MOs of ethene in an excited state are shown in red, and the π MOs of ethene in its ground state are shown in blue. The resulting frontier orbital interaction is between the π^* MO in each species.

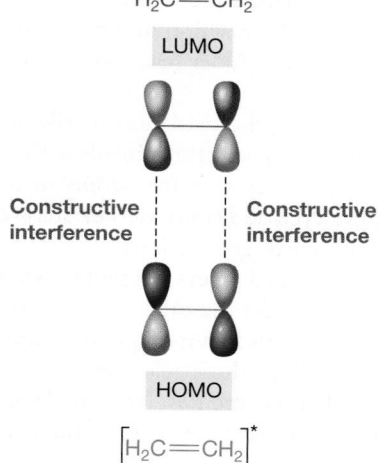

FIGURE 24-14 Frontier orbital interaction in a photochemical [2+2] cycloaddition reaction Both ends of the π systems exhibit the same type of interference—in this case, constructive interference. Therefore, the orbitals have the appropriate symmetries to interact, thus making this reaction *photochemically allowed*.

Consequently, the [2+2] cycloaddition reaction is *thermally forbidden*, but **photochemically allowed**.

Unlike the [2+2] cycloaddition reaction, the Diels–Alder reaction is impeded by UV light, as indicated in Equation 24-54.

(24-54)

When a molecule of buta-1,3-diene absorbs a photon with sufficient energy, an excited state of buta-1,3-diene is generated, in which an electron from what was originally the HOMO now occupies what was originally the LUMO. When this excited state species encounters a molecule of ethene in the ground state, the HOMO and LUMO no longer have the appropriate symmetries to interact, making the reaction forbidden. Consequently, the Diels–Alder reaction is *thermally allowed*, but **photochemically forbidden**.

---

problem **24.26** Redraw Figure 24-13 for buta-1,3-diene in its excited state and ethene in its ground state. Do the same for buta-1,3-diene in its ground state and ethene in its excited state. Are these results consistent with a Diels–Alder reaction being photochemically forbidden?

24.11 Wrapping Up and Looking Ahead

The primary focus of Chapter 24 was the Diels–Alder reaction, a type of cycloaddition reaction. In this reaction, a diene and a dienophile are joined via the formation of two new σ bonds, producing a six-membered ring of carbon atoms. Like all cycloaddition reactions, the Diels–Alder reaction is pericyclic, meaning that it is concerted and involves the cyclic movement of electrons. Thus, it is an elementary step that differs from the ones we examined in Chapters 6 and 7.

We began Chapter 24 by developing the Diels–Alder reaction in a systematic way. We first established how, in the curved arrow notation for the step, electrons flow from a relatively electron-rich diene to a relatively electron-poor dienophile. Next, we learned why the Diels–Alder reaction exhibits its specific stereochemistry and regiochemistry. Finally, we examined the driving force for the reaction in terms of free energy, enthalpy, and entropy, and we learned how a Diels–Alder reaction can be either reversible or irreversible.

We also looked at [4+2] cycloadditions involving the oxidation of alkenes and alkynes. Like a Diels–Alder reaction, each of these cycloadditions involves the cyclic movement of six electrons in a single step, but involves only five atoms instead of six. In Sections 24.7 and 24.8, we saw that these cycloadditions are key steps of the mechanisms for syn dihydroxylation and oxidative cleavage reactions.

Although the reactions we examined here introduced a new elementary step, that step involves the movement of *pairs* of electrons, just like the other elementary steps examined previously. In Chapter 25, we examine reactions that incorporate elementary steps requiring the movements of *single* electrons. Thus, those reactions involve species in which not all electrons are paired—so-called *free radicals*. Nevertheless, many aspects of the reactions involving free radicals are similar to aspects of the reactions we have studied thus far.

Chapter Summary and Key Terms

- The **Diels–Alder reaction** joins a conjugated **diene** and a **dienophile** (either an alkene or an alkyne) via the formation of two new σ bonds. The product is a six-membered ring of carbon atoms. **(Introduction and Section 24.1)**

- The Diels–Alder reaction is concerted; all bonds that are formed and broken do so simultaneously. It requires the cyclic movement of electrons, so it is classified as a **pericyclic reaction**. **(Section 24.1)**

- The Diels–Alder reaction is thermally "allowed," because it is a **[4+2] cycloaddition**, proceeding through an **aromatic transition state**. Reactions that proceed through an **antiaromatic transition state**, such as the **[2+2] cycloaddition** and the **[6+2] cycloaddition**, are thermally "forbidden." **(Section 24.1)**

- For a Diels–Alder reaction to take place, the diene must be able to attain the **s-cis conformation**. Diels–Alder reactions cannot take place with the diene in the **s-trans conformation**. **(Section 24.2)**

- Diels–Alder reactions are facilitated by electron-donating groups attached to the diene and electron-withdrawing groups attached to the dienophile. **(Section 24.3)**

- Diels–Alder reactions are stereospecific. **(Section 24.4)**
  - Substituents that are cis to each other about the C=C double bond of the dienophile end up cis to each other in the new ring that is produced. Otherwise, the substituents end up trans to each other in the ring.
  - Substituents attached to the diene end up cis to each other in the new ring that is produced if the double bonds of the diene are both cis or both trans. Otherwise, the substituents end up trans to each other in the ring.

- Diels–Alder reactions tend to favor an **endo product** over an **exo product**. **(Section 24.4)**

- When the diene and dienophile are both unsymmetric, two isomeric products can be produced. The major product is the one produced by the approach that exhibits the most favorable electrostatic attraction among atoms undergoing bond formation. **(Section 24.5)**

- Enthalpy tends to heavily favor the products of a Diels–Alder reaction, whereas entropy favors reactants. Therefore, under normal conditions, Diels–Alder reactions tend to be irreversible and proceed under kinetic control. At high temperatures, though, some can undergo the reverse reaction—a **retro Diels–Alder reaction** or **[4+2] cycloelimination** **(Section 24.6)**

- An alkene can undergo **syn dihydroxylation** with OsO_4 or $KMnO_4$ (cold), resulting in a cis-1,2-diol. An alkyne that undergoes such a reaction results in a 1,2-dicarbonyl compound. **(Section 24.7)**

- **Oxidative cleavage** of an alkene or alkyne completely breaks the C=C double bond or C≡C triple bond, with the initial alkene or alkyne carbons becoming part of C=O groups in the final products. **(Section 24.8)**
  - Aldehydes that are initially produced from oxidative cleavage are further oxidized to carboxylic acids when cleavage is initiated with a hot, concentrated solution of $KMnO_4$ **(Section 24.8a)** or with **ozonolysis** followed by workup with H_2O_2 **(Section 24.8c)**.
  - Aldehydes that are initially produced from oxidative cleavage are *not* oxidized further when cleavage involves the treatment of a cis-1,2-diol with $NaIO_4$ or HIO_4 **(Section 24.8b)** or is initiated with ozonolysis followed by workup with CH_3SCH_3 or Zn/HOAc **(Section 24.8c)**.

Reaction Tables

TABLE 24-1 Functional Group Conversions

	Starting Functional Group	Typical Reagent and Reaction Conditions	Functional Group Formed	Key Electron–Rich Species	Key Electron–Poor Species	Comments	Discussed in Section
(1)	Alkene	OsO_4 / H_2O_2	Syn 1,2-diol			An alternate method uses a cold, basic solution of $KMnO_4$.	24.7
(2)	Alkyne	OsO_4 / H_2O_2	1,2-Dione			An alternate method uses a cold, basic solution of $KMnO_4$.	24.7

TABLE 24-2 Reactions That Alter the Carbon Skeleton

	Starting Functional Group	Typical Reagent and Reaction Conditions	Functional Group Formed	Key Electron–Rich Species	Key Electron–Poor Species	Comments	Discussed in Section(s)
(1)	Conjugated diene	Dienophile	Substituted cyclohexene			Diels–Alder reaction. Facilitated by an electron-donating group on the diene and an electron-withdrawing group on the dienophile	24.1–24.6
(2)	Substituted cyclohexene	Δ	Conjugated diene + Alkene	N/A	N/A	Retro Diels–Alder reaction	24.6
(3)	Alkene	$KMnO_4$ $HO^{\ominus}$, Δ	Ketone and/or carboxylic acid			Oxidative cleavage; formic acid product oxidizes to CO_2.	24.8
(4)	Alkyne	$KMnO_4$ $HO^{\ominus}$, Δ	Carboxylic acid			Oxidative cleavage; formic acid product oxidizes to CO_2.	24.8
(5)	Alkene	1. O_3 2. CH_3SCH_3 or Zn, HOAc	Ketone and/or aldehyde			Ozonolysis	24.8
(6)	Alkene	1. O_3 2. H_2O_2	Ketone and/or carboxylic acid			Ozonolysis; oxidative cleavage; formic acid product oxidizes to CO_2.	24.8
(7)		$NaIO_4$ or HIO_4	Ketone and/or aldehyde	N/A	N/A	Oxidative cleavage	24.8

Problems

24.27 The following are several isomers of $C_{10}H_{14}$ with two fused six-membered rings.

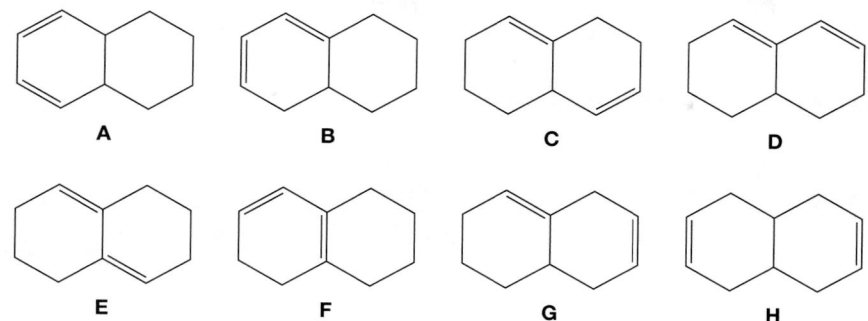

A   **B**   **C**   **D**

E   **F**   **G**   **H**

(a) Identify which will react with ethene in a Diels–Alder reaction and which will not.

(b) Draw one more isomer with two fused six-membered rings that *will* react with ethene in a Diels–Alder reaction.

24.28 Rank the following compounds in order from slowest to fastest rate of reaction in a Diels–Alder reaction with ethene.

I   **J**   **K**

24.29 Rank the following in order from slowest to fastest rate of reaction in a Diels–Alder reaction with buta-1,3-diene.

L   **M**   **N**   **O**   **P**

24.30 Consider the following dienes in a Diels–Alder reaction with ethene. Which will react the fastest? Which will react the slowest? Explain.

Q   **R**   **S**

24.31 Which of the following will react faster as a diene in a Diels–Alder reaction with ethene? Explain.

T   **U**

24.32 Which of the following will react faster as a dienophile in a Diels–Alder reaction with buta-1,3-diene? Explain.

V   **W**

24.33 Which of the following will react faster as a dienophile in a Diels–Alder reaction with buta-1,3-diene? Explain.

X   Y

24.34 Draw the product of each of the following reactions.

(a)
1. KMnO₄
NaOH, Δ
2. H⁺
?

(b)
1. KMnO₄
NaOH, Δ
2. H⁺
?

(c)
1. O₃, –78 °C
2. Zn, HOAc
?

(d)
1. O₃, –78 °C
2. H₂O₂
?

(e)
1. OsO₄, H₂O₂
2. HIO₄
?

(f)
KMnO₄
NaOH, Δ
?

24.35 Draw the structure of the reactant that produces the molecules shown for each of the following reactions.

(a)
C₉H₁₆
1. KMnO₄
NaOH, Δ
2. H⁺

(b)
C₆H₁₀
KMnO₄
NaOH, Δ
+ CO₂

(c)
C₁₀H₁₈
1. O₃, –78 °C
2. (CH₃)₂S
+

(d)
C₇H₁₂
1. O₃, –78 °C
2. H₂O₂
+ CO₂

24.36 Draw the major product of the following reaction.

HIO₄
?

24.37 Rubber degrades when it is exposed to ozone for an extended period of time—a phenomenon called ozone cracking. Rubber is a natural polymer (a long-chain molecule with a regular repeating structural unit), a portion of which is as follows:

Rubber

Explain why ozone cracking occurs.

24.38 Anthracene readily undergoes a Diels–Alder reaction with tetracyanoethene, even though anthracene is aromatic.

(a) Draw two possible products that can form from this reaction.

(b) Explain why anthracene can readily undergo a Diels–Alder reaction, whereas benzene does not.

24.39 T. R. Hoye and coworkers reported the synthetic utility of the alkyne analog of a Diels–Alder reaction. An example is shown below.

A [4+2] cycloaddition is believed to produce a benzyne intermediate that is quickly "trapped." Show the benzyne intermediate that would be produced in the above reaction.

24.40 The compound at the right can be produced from two *different* Diels–Alder reactions.

(a) Draw the reactants that can be used for each Diels–Alder reaction.

(b) Which would be the better route? Why?

24.41 The compound at the right can be produced from two different Diels–Alder reactions.

(a) Draw the reactants that would be required for each reaction.

(b) Which set of reactants would be the better choice? Why?

24.42 Draw a mechanism to account for the following reaction that scrambles the isotopic labeling.

24.43 The following is an example of a hetero Diels–Alder reaction, because non-carbon atoms (in this case, N atoms) are involved in bond formation and bond breaking.

Draw the curved arrows necessary to account for this transformation.

24.44 The following are examples of hetero Diels–Alder reactions (see Problem 24.43).

(a)

(b)

Draw the mechanism and predict the product for each reaction, paying particular attention to stereochemistry.

24.45 Two constitutional isomers can be produced from the following Diels–Alder reaction.
(a) Draw the mechanism for the formation of each isomer.
(b) Determine which isomer is the major product. (*Hint:* Is a phenyl ring electron rich or electron poor?)

24.46 A reaction takes place when *cis*-1,3,9-decatriene is heated. The product of that reaction, when treated with excess Br_2 in CCl_4, yields a compound whose formula is $C_{10}H_{16}Br_2$. What is the product of the first reaction?

24.47 Strong support for the mechanism of the nucleophilic aromatic substitution reaction that proceeds through a benzyne intermediate comes from the reaction at the right, in which bromobenzene is treated with KNH_2 in the presence of cyclopentadiene. A product that is isolated has the formula $C_{11}H_{10}$. Draw the structure of that product and explain how it validates the production of a benzyne intermediate.

24.48 Draw the structures of compounds **A–I** in the following synthesis scheme.

24.49 Draw the structures of compounds **A–I** in the following synthesis scheme.

24.50 Draw the structures of compounds **A–F** in the following synthesis scheme.

24.51 Draw the structures of compounds **A–H** in the following synthesis scheme.

24.52 Show how you would synthesize each of the following compounds using ethene and buta-1,3-diene as your only sources of carbon.

(a)

(b)

(c)

(d)

24.53 The compound at the right cannot be synthesized *directly* from a Diels–Alder reaction.
(a) Why not? (*Hint*: Examine the reactants that would be required.)
(b) What change(s) can be made to the synthesis to get around this problem?

24.54 Show how you would synthesize the compound at the right beginning with benzene.

24.55 For each of the following reactions,
(a) draw the curved arrows necessary for a concerted mechanism,
(b) draw the transition state,
(c) determine whether each transition state is *aromatic* or *antiaromatic*, and
(d) based on your answer to part (c), determine whether the reaction is allowed or forbidden under normal (thermal) conditions.

(i)

(ii)

(iii)

24.56 Draw the product of the following reaction, assuming that it takes place via an [8+2] cycloaddition.

24.57 Draw the product of the following reaction, assuming that it takes place via a [6+4] cycloaddition.

24.58 For each of the following reactions,
  (a) draw the HOMO of one reactant and the LUMO of the other, assuming that both reactants are in their ground state and
  (b) illustrate the HOMO–LUMO interaction and determine if such an interaction leads to a thermally allowed or forbidden reaction.

(i)

$H_2C{=}CH_2$ + ⟶

(ii)

+ ⟶

(iii)

+ ⟶

24.59 Repeat Problem 24.58, assuming that *one* of the reactants is in its lowest excited state. Based on your answers, which of the reactions are photochemically allowed?

24.60 Show how to carry out the transformation at the right using any reagents necessary.

24.61 Gibberellic acid is a plant hormone that controls the development of plants.

Gibberellic acid

At various points in their synthesis of gibberellic acid, E. J. Corey and coworkers incorporated the following Diels–Alder reactions.

(a)

(b)

Draw the mechanism and the product of each of these reactions. (*Hint*: In the first reaction, which alkene is more electron poor?)

24.62 Myrocin C is an antitumor antibiotic.

Myrocin C

In their synthesis of myrocin C, S. J. Danishefsky and coworkers incorporated the following Diels–Alder reactions.

(a)

(b)

Draw the mechanism and the product of each of these reactions.

24.63 When biphenylene is warmed to 40 °C, it decomposes into a compound whose ¹H NMR spectrum exhibits one signal—a singlet at 7.3 ppm.
(a) Draw the mechanism for this reaction.
(b) Draw the transition state for this reaction, showing that it proceeds through an antiaromatic transition state.
(c) Explain why this reaction proceeds at mild temperatures, even though it proceeds through an antiaromatic transition state.

Biphenylene

24.64 When the compound at the right is heated to 700 °C, a product with the formula C_4H_6 is collected. The ¹H NMR and ¹³C NMR spectra of C_4H_6 are shown below. Draw the structure of C_4H_6, and draw the mechanism of the reaction that accounts for its formation.

24.65 When the compound at the right is heated, ethene gas is evolved and a product with the formula $C_{14}H_8O_2$ is formed. The ¹H NMR and ¹³C NMR spectra of $C_{14}H_8O_2$ are shown below. (There are two signals >150 ppm in the ¹³C NMR spectrum. Recall that the ¹³C NMR signal at 77 ppm is from the $CDCl_3$ solvent.)
(a) Draw the structure of $C_{14}H_8O_2$.
(b) Draw the mechanism that accounts for its formation.
(c) What is the main driving force that favors the products of this reaction?

24.66 A compound with the formula C_6H_{10} is known to react with one molar equivalent of Br_2 in carbon tetrachloride. When C_6H_{10} was treated with a hot, basic solution of potassium permanganate, followed by acid workup, a product was formed whose IR spectrum exhibits a broad absorption of medium intensity from 2500 to 3300 cm^{-1} and a sharp, intense absorption near 1700 cm^{-1}. The ^1H and ^{13}C NMR spectra of the product are shown. (*Note:* The multiplet at 39 ppm is from the solvent, DMSO-d_6.) Draw the structure of C_6H_{10}.

24.67 Heating the compound below at 225–235 °C for 8 h produced a new compound with the formula $C_{23}H_{16}N_2O$.

$$H_5C_6 \diagdown \quad \diagup N\diagdown_N \qquad OCH_2CN$$

$$H_5C_6 \diagup \qquad N \xrightarrow{\Delta} \boxed{C_{23}H_{16}N_2O}$$

The mechanism for this reaction is believed to consist of a [4+2] cycloaddition followed by a [4+2] cycloelimination. The ^1H NMR spectrum of the product contained the following signals: δ 8.33–8.25 and 7.58–6.97 (m, 14 H), 5.46 (s, 2H).

(a) Draw the mechanism for this reaction.

(b) Draw the product.

If your instructor assigns problems in **smartwork**, log in at **smartwork.wwnorton.com**.

CHAPTER 25

Reactions Involving Free Radicals

Throughout this book, our focus has been on reaction mechanisms whose individual steps are driven by the flow of electrons from a site that is electron rich to one that is electron poor. Alkanes and alkyl (i.e., R) groups lack functional groups—they consist of only carbon and hydrogen atoms connected by relatively nonpolar single bonds—so they do not have sites that are especially electron rich or electron poor.

However, alkanes are not inert under all conditions. In fact, alkanes react *violently* via combustion. Methane, for example, is used as natural gas to heat our homes, propane is used in gas grills and camp stoves, and a mixture of relatively small hydrocarbons makes up the gasoline used in car engines. These reactions proceed by mechanisms that differ from what we have encountered thus far.

Although we will not explore combustion in depth, we will examine other, related reactions that are important to synthetic organic chemistry. These include the halogenation of alkanes, alkene additions, and dissolving metal reductions. All of these reactions involve very highly reactive intermediates called *free radicals*.

Hydrocortisone

Hydrocortisone is a steroid hormone that is used to provide relief from itching, such as that resulting from exposure to poison ivy. The biosynthesis of hydrocortisone involves a free radical reaction, the type of reaction we will examine here in Chapter 25.

Upon completing Chapter 25 you should be able to:

- Distinguish a *free radical* from a *closed-shell species*.
- Explain why free radicals are typically very unstable and highly reactive.
- Use single-barbed curved arrows to describe the flow of electrons in elementary steps involving free radicals.
- Determine the relative stabilities of free radicals, given appropriate bond dissociation energies.
- Predict the relative stabilities of alkyl radicals, given only their Lewis structures.
- Draw all resonance structures of a given free radical.
- Describe the geometry and hybridization of a carbon atom that has an unpaired electron.
- Outline the major characteristics of a free radical *chain reaction*.
- Identify *initiation*, *propagation*, and *termination* steps of a radical chain-reaction mechanism.
- Recognize common *radical initiators* and explain their relevance in a free radical chain reaction.
- Draw the mechanism for the *radical halogenation* of an alkane and predict the major product.
- Explain the difference in selectivity between free radical bromination and free radical chlorination.
- Explain the advantage of using *N*-bromosuccinimide in the free radical bromination of an allylic carbon.
- Draw the complete mechanism for the radical addition of HBr to an alkene, and predict the major product of such a reaction, including regiochemistry.
- Draw the complete mechanism for a *dissolving metal reduction* and predict the major product, including stereochemistry.
- Incorporate free radical reactions in synthesis.

A **free radical** is a species that possesses at least one unpaired electron.

Examples include the bromine atom and the methyl radical (Fig. 25-1):

FIGURE 25-1 Examples of free radicals Each of these species is a free radical because it has one unpaired electron.

Free radicals such as Br• and H_3C• behave somewhat differently from other species in which all of the electrons are paired—so-called **closed-shell species**. Here in Chapter 25, we present a picture of how free radicals behave and develop the picture in a systematic way. We first discuss free radical structure and stability. Then, we take a close look at the most common elementary steps that free radicals undergo. Finally, we delve into the mechanisms of the useful free radical reactions mentioned above.

25.1 Homolysis: Curved Arrow Notation and Radical Initiators

Because free radicals contain an unpaired electron, it is impossible for all of their atoms to have a complete octet. Therefore,

Free radicals are generally very unstable and highly reactive, so they usually cannot be isolated for any substantial length of time.

Once a radical is produced, it will typically react very quickly with other species present, including the solvent.

By and large, radicals are produced from uncharged, closed-shell precursors via *homolytic bond dissociation*, or *homolysis*.

Homolytic bond dissociation, or **homolysis**, is the breaking of a covalent bond, whereby the electrons making up that bond are distributed equally to the atoms that are disconnected.

A bromine radical (Br•), for example, can be produced from molecular bromine, Br_2, by homolysis of the Br—Br bond (Equation 25-1). Homolysis of a C—H bond in methane (Equation 25-2), on the other hand, produces a methyl radical (H_3C•) and a hydrogen radical (H•).

Homolysis of the
Br—Br bond

Two radicals
are produced.

Br—Br $\xrightarrow{\Delta}$ Br• + •Br (25-1)

Homolysis of the
C—H bond

Two radicals
are produced.

(25-2)

The curved arrow notation used to describe homolysis differs from that used in previous chapters to describe elementary steps. Specifically, *single-barbed arrows* are used in homolysis.

A **single-barbed arrow** ( ) represents the movement of a *single electron*.

The curved arrow notation in Equations 25-1 and 25-2 explicitly shows that one electron from the covalent bond moves to one atom and the second electron moves to the other atom. In contrast, we have used *double-barbed arrows* (⤳) in previous elementary steps to show the movement of *pairs* of electrons in closed-shell species.

problem **25.1** Draw the appropriate curved arrows and draw the products for the homolysis of each of the following bonds.

(a) The C—C bond in ethane
(b) A C—H bond in ethane
(c) The C—Br bond in 2-bromopropane
(d) A C—H bond in benzene

The diagram in Figure 25-2 shows that it takes 435 kJ/mol (104 kcal/mol) of energy for a C—H bond in CH_4 to undergo homolysis and thereby produce H• and $H_3C•$. This corresponds to a vanishingly small equilibrium constant of $<10^{-76}$, and should give you some idea of just how unstable these free radicals are.

Many other single bonds have similar homolytic bond dissociation energies, as shown in Tables 25-1 through 25-3. Table 25-1 lists bond dissociation energies of bonds involving hydrogen, Table 25-2 lists bond dissociation energies of C—C bonds, and Table 25-3 lists bond dissociation energies of bonds involving carbon and a heteroatom, or two heteroatoms.

FIGURE 25-2 Reaction energy diagram for the homolysis of a C—H bond in CH_4 The homolytic bond dissociation energy of 435 kJ/mol (104 kcal/mol) is the difference in energy between the energy minimum and the energy of the separated free radicals at the right of the figure.

TABLE 25-1 Homolytic Bond Dissociation Energies Involving Bonds to Hydrogen[a]

Bond	Homolytic Bond Dissociation Energy		Bond	Homolytic Bond Dissociation Energy	
	kJ/mol	kcal/mol		kJ/mol	kcal/mol
H—H	436	104	H—CH$_3$	435	104
H—F	569	136	H—CH$_2$CH$_3$	410	98
H—Cl	431	103	H—CH(CH$_3$)$_2$	395	94
H—Br	368	88	H—C(CH$_3$)$_3$	381	91
H—I	297	71	H—CH=CH$_2$	464	111
H—OH	498	119	H—C≡CH	556	133
H—OCH$_3$	427	102	H—CH$_2$—CH=CH$_2$	356	85
H—OC$_6$H$_5$	356	85	H—CH$_2$—C$_6$H$_5$	356	85
H—SH	377	90	H—CH$_2$Br	406	97
			H—CH$_2$OH	385	92

[a] Adapted from Lowry, T.H.; Richardson, K.S. *Mechanism and Theory in Organic Chemistry*, 3rd ed.; Benjamin Cummings Publishing Co.: San Francisco, 1997.

TABLE 25-2 Homolytic Bond Dissociation Energies Involving C—C Bonds[a]

Bond	Homolytic Bond Dissociation Energy		Bond	Homolytic Bond Dissociation Energy	
	kJ/mol	kcal/mol		kJ/mol	kcal/mol
H$_3$C—CH$_3$	368	88	H$_3$C—C(=O)—CH$_3$	343	82
H$_3$C—CH$_2$CH$_3$	356	85	H$_3$C—CH$_2$—C(=O)—CH$_3$	331	79
CH$_3$—H$_2$C—CH=CH$_2$	372	89	H$_3$C—C≡N	510	122
CH$_3$—H$_2$C—C$_6$H$_5$	377	90	H$_2$C=CH—C(=O)—CH$_3$	372	89

[a] Adapted from Lowry, T.H.; Richardson, K.S. *Mechanism and Theory in Organic Chemistry*, 3rd ed.; Benjamin Cummings Publishing Co.: San Francisco, 1997.

(continued)

TABLE 25-2 Homolytic Bond Dissociation Energies Involving C—C Bonds[a] (continued)

Bond	Homolytic Bond Dissociation Energy		Bond	Homolytic Bond Dissociation Energy	
	kJ/mol	kcal/mol		kJ/mol	kcal/mol
$H_2C{=}HC{-}CH{=}CH_2$	418	100	$H_3C-C({=}O){-}C({=}O){-}CH_3$	347	83
$HC{\equiv}C{-}C{\equiv}CH$	628	150	$H_3C{-}CH{=}CH_2$	385	92
$H_5C_6{-}C_6H_5$	418	100	$H_3C{-}C_6H_5$	389	93
$H_2C{=}HC{-}C_6H_5$	414	99	$H_3C{-}CH_2{-}C_6H_5$	301	72
			$H_3C{-}CH_2{-}CH{=}CH_2$	301	72

[a] Adapted from Lowry, T.H.; Richardson, K.S. *Mechanism and Theory in Organic Chemistry*, 3rd ed.; Benjamin Cummings Publishing Co.: San Francisco, 1997.

TABLE 25-3 Homolytic Bond Dissociation Energies Involving Carbon-Heteroatom Bonds and Heteroatom-Heteroatom Bonds[a]

Bond	Homolytic Bond Dissociation Energy		Bond	Homolytic Bond Dissociation Energy	
	kJ/mol	kcal/mol		kJ/mol	kcal/mol
$F{-}CH_3$	451	108	$HO{-}CH_3$	383	92
$Cl{-}CH_3$	355	85	$HO{-}C_6H_5$	464	111
$Cl{-}CH_2CH_3$	343	82	$H_2N{-}CH_3$	331	79
$Cl{-}CH(CH_3)_2$	339	81	$HO{-}OH$	213	51
$Cl{-}C(CH_3)_3$	334	80	$CH_3O{-}OCH_3$	151	36
$Cl{-}CH{=}CH_2$	376	90	$H_3C{-}C({=}O){-}O{-}O{-}C({=}O){-}CH_3$	123	29
$Cl{-}C{\equiv}CH$	464	111	$F{-}F$	159	38
$Cl{-}CH_2{-}CH{=}CH_2$	293	70	$Cl{-}Cl$	243	58
$Cl{-}CH_2{-}C_6H_5$	284	68	$Br{-}Br$	192	46
$Br{-}CH_3$	297	71	$I{-}I$	151	36
$I{-}CH_3$	234	56			

[a] Adapted from Lowry, T.H.; Richardson, K.S. *Mechanism and Theory in Organic Chemistry*, 3rd ed.; Benjamin Cummings Publishing Co.: San Francisco, 1997.

Because of the large amount of energy required for homolysis, the concentration of free radicals in solution is essentially zero under normal conditions for most species.

> To produce any reasonable concentration of free radicals from a closed-shell species, energy is supplied in the form of heat or UV light.

Therefore, $\xrightarrow{\Delta}$ or $\xrightarrow{hv}$ often indicates that radicals are being produced, as illustrated previously in Equations 25-1 and 25-2.

For a molecule that has only one type of covalent bond, such as Br_2 or CH_4, only one set of homolysis products is possible. For molecules that have more than one type of covalent bond, however, more than one set of homolysis products is possible, but one homolysis reaction generally dominates.

> The major products of homolysis derive from breaking the weakest bond in the molecule—that is, the bond with the smallest bond dissociation energy.

For example, bromomethane (CH_3Br) has three C—H bonds and one C—Br bond. As shown in Equation 25-3, the C—Br bond (297 kJ/mol, 71 kcal/mol) is weaker than the C—H bond (406 kJ/mol, 97 kcal/mol), so homolysis leads almost exclusively to Br• and H_3C•.

SOLVED problem 25.2 Predict the major homolysis products of ethane.

$$CH_3CH_3 \xrightarrow{hv} \text{?}$$

Think What are the distinct types of bonds in the molecule? Which is the weakest? What are the products upon homolysis of that bond?

Solve There are two distinct types of bond. One is H—CH_2CH_3, and the other is H_3C—CH_3. The bond dissociation energy of H—CH_2CH_3 is 410 kJ/mol (98 kcal/mol) (Table 25-1), and that of H_3C—CH_3 is 368 kJ/mol (88 kcal/mol) (Table 25-2). Therefore, H_3C—CH_3 is weaker, and its homolysis leads to the major products.

$$H_3C\text{—}CH_3 \xrightarrow{hv} H_3C\bullet + \bullet CH_3$$

---

problem 25.3 Predict the major homolysis products of methanol.

$$CH_3OH \xrightarrow{hv} \text{?}$$

Free radicals can conceivably be produced from the homolysis of a covalent bond in virtually any molecule, but in practice, some precursors are particularly well suited to producing free radicals. These **radical initiators** include molecular halogens (Cl_2, Br_2, and I_2), peroxides (RO—OR), **N-bromosuccinimide (NBS)** (Equation 25-4), and **2,2′-azobisisobutyronitrile (AIBN)** (Equation 25-5). They are often added to a reaction mixture to initiate a reaction that proceeds by a free radical mechanism. (See Section 25.4.)

N-Bromosuccinimide
(NBS)

(25-4)

N$_2$ is a very stable gas that escapes from the reaction mixture, making radical initiation effectively irreversible.

2,2′-Azobisisobutyronitrile
(AIBN)

(25-5)

Radical initiators generally have a rather weak bond, making the bond relatively easy to break.

The bond dissociation energy of the N—Br bond in NBS, for example, is 194 kJ/mol (46 kcal/mol), which is roughly half the strength of a typical C—H bond. AIBN is advantageous, too, because homolysis of the two C—N single bonds produces gaseous N$_2$, which permanently leaves the reaction mixture and makes homolysis effectively irreversible.

25.2 **YOUR TURN**

Verify that halogens and peroxides have weak bonds by comparing the bond energies of the following bonds to that of a typical C—H bond.

Cl—Cl _____   Br—Br _____   I—I _____

HO—OH _____   H$_3$C—H _____

25.2 Structure and Stability of Alkyl Radicals

Although free radicals tend to be very unstable, we can assign different *relative* stabilities to various radicals, based on the energy required to produce them via homolysis (Tables 25-1 through 25-3). For example, Equation 25-6 shows the C—Cl homolysis reactions of a variety of chloroalkanes, yielding the Cl• atom and an alkyl radical (R•) as the products.

Methyl radical

355 kJ/mol
(85 kcal/mol)

(25-6a)

1° radical

343 kJ/mol
(82 kcal/mol)

(25-6b)

$$H_3C-\overset{\overset{\displaystyle H}{|}}{\underset{\underset{\displaystyle CH_3}{|}}{C}}-Cl \xrightarrow{\;h\nu\;} H_3C-\overset{\displaystyle \bullet}{\underset{\underset{\displaystyle CH_3}{|}}{C}}H \;+\; \bullet Cl \qquad \begin{array}{l}\text{339 kJ/mol}\\ \text{(81 kcal/mol)}\end{array} \qquad \text{(25-6c)}$$

2° radical

$$H_3C-\overset{\overset{\displaystyle CH_3}{|}}{\underset{\underset{\displaystyle CH_3}{|}}{C}}-Cl \xrightarrow{\;h\nu\;} H_3C-\overset{\overset{\displaystyle CH_3}{|}}{\underset{\underset{\displaystyle CH_3}{|}}{C}}\bullet \;+\; \bullet Cl \qquad \begin{array}{l}\text{334 kJ/mol}\\ \text{(80 kcal/mol)}\end{array} \qquad \text{(25-6d)}$$

3° radical

The free energy diagrams for these homolysis reactions are shown in Figure 25-3 (we have assumed that the various alkyl chlorides have essentially the same stability).

A Cl• radical is produced in all four reactions, so any difference in product energy must come from the different stabilities of the alkyl radicals that are produced. For example, it takes 12 kJ/mol (3 kcal/mol) of additional energy to produce a methyl radical than it does to produce an ethyl radical (355 kJ/mol vs. 343 kJ/mol; 85 kcal/mol vs. 82 kcal/mol), so •CH$_3$ is 12 kJ/mol (3 kcal/mol) higher in energy than •CH$_2$CH$_3$ in Figure 25-3. In other words, •CH$_3$ is 12 kJ/mol (3 kcal/mol) less stable than •CH$_2$CH$_3$. In turn, •CH$_2$CH$_3$ is 4 kJ/mol (1 kcal/mol) less stable than •CH(CH$_3$)$_2$, which is 5 kJ/mol (1 kcal/mol) less stable than •C(CH$_3$)$_3$.

YOUR TURN 25.3

The following figure represents homolysis for the H—Cl and H—Br bonds, but it is incomplete.

Write the homolytic bond dissociation energies in the boxes provided under the curves. Also, write the Br• and Cl• products in the appropriate boxes at the top of the figure. Which radical is more stable, Br• or Cl•?

The relative stabilities shown in Figure 25-3 establish the following trend:

Alkyl radical stability increases in the order: Methyl < 1° < 2° < 3°.

The stability of alkyl radicals follows the same order as the stability of *carbocations* (R$^+$) (Section 6.6e). This suggests that the carbon atom that has the unpaired electron can be viewed as *electron poor*, in much the same way as a carbon atom bearing a positive charge in a carbocation. With this in mind, each additional alkyl group stabilizes a free radical, due to its electron-donating capability,[1] as shown in Figure 25-4.

---

[1]This electron donation occurs inductively and also via hyperconjugation, involving the half-filled *p* AO on C•.

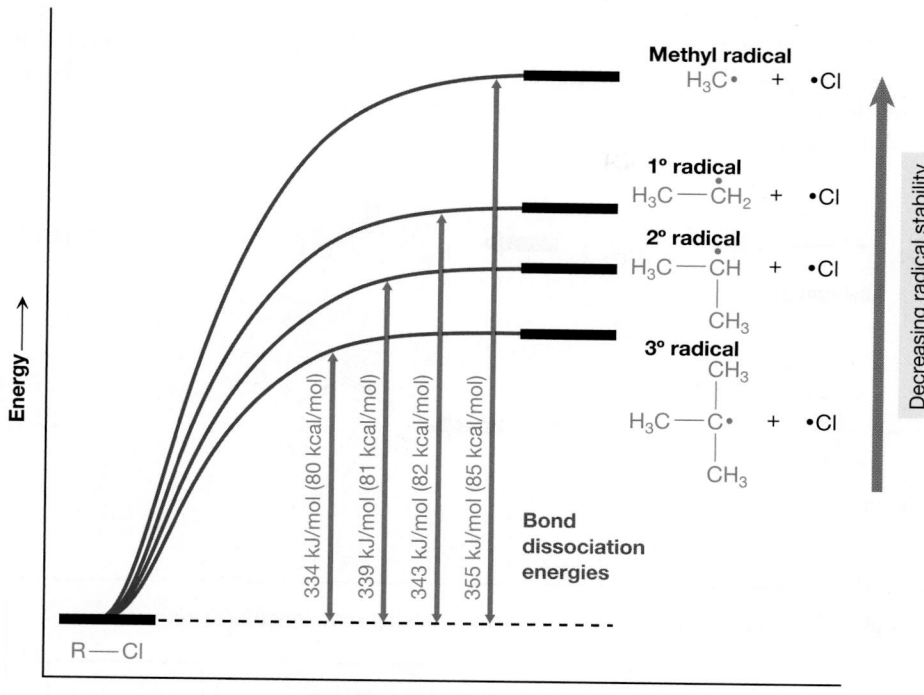

FIGURE 25-3 Relative stabilities of alkyl radicals Reaction energy diagrams for C—Cl homolysis in a variety of chloroalkanes. Alkyl radical stability increases in the order $H_3C\bullet < H_3CCH_2\bullet < (CH_3)_2CH\bullet < (CH_3)_3C\bullet$.

---

problem **25.4** Homolysis of a C—H bond can occur at three distinct locations on 1,3,5-trimethylcyclohexane to yield three different free radicals. Draw each of the radicals and rank them in order from least stable to most stable.

Hybridization also impacts the stability of an alkyl radical. Notice in Table 25-3 that the C—Cl bond dissociation energies for Cl—CH$_2$CH$_3$ and Cl—CH=CH$_2$ are 343 kJ/mol and 376 kJ/mol (i.e., 82 kcal/mol and 90 kcal/mol), respectively. According to these data, the ethyl radical ($\bullet$CH$_2$CH$_3$) is more stable than the vinyl radical ($\bullet$CH=CH$_2$). Thus,

> Alkyl radical stability decreases as the hybridization of the carbon atom possessing the unpaired electron goes from sp^2 to sp—that is, as the s character of the carbon atom increases.

The stability of the radical decreases because as a carbon atom's s character increases, so does its effective electronegativity (Section 3.11). Just as its ability to accommodate a positive charge decreases (Section 6.6c), so, too, does its ability to accommodate an unpaired electron.

FIGURE 25-4 Relative stabilities of alkyl radicals A radical carbon (in blue) is stabilized by the electron-donating effect of attached alkyl groups (in red), so alkyl radical stability increases in the order: methyl radical < 1° radical < 2° radical < 3° radical.

The similarities between alkyl radicals and carbocations are further demonstrated with the allyl (CH_2=CH—CH_2•) and benzyl (C_6H_5—CH_2•) radicals. Recall that the allyl cation and the benzyl cation are more stable than ordinary primary carbocations, due to resonance delocalization of the positive charge, as shown in Equations 25-7 and 25-8.

(25-7)

Allyl cation

(25-8)

Benzyl cation

YOUR TURN 25.4

Complete Equation 25-8 by drawing the three missing resonance structures.

Benzyl cation

The allyl radical and the benzyl radical are more stable than ordinary primary alkyl radicals, too. This is evidenced by the fact that the C—Cl bonds in CH_2=CH—CH_2—Cl and C_6H_5—CH_2—Cl are weaker than that in chloroethane by 50 kJ/mol and 59 kJ/mol (i.e., by 12 kcal/mol and 14 kcal/mol), respectively.

problem **25.5** Draw a plot similar to that in Figure 25-3 and Your Turn 25.3, illustrating the relative stabilities of the ethyl radical (CH_3—CH_2•) and the allyl radical.

Equations 25-9 and 25-10 show that the additional stabilization of the allyl and benzyl radicals is also due to resonance. Even though the curved arrow notation is different from that used to indicate resonance in the cations, the unpaired electron in each radical is shared among the same carbon atoms as the positive charge is in the analogous carbocations.

The unpaired electron is shared
over these two C atoms.

(25-9)

(25-10)

The unpaired electron is
shared over four C atoms.

Equation 25-10 is missing the curved arrow notation that shows how the fourth resonance structure is converted to the fifth. Supply that curved arrow notation in the structure on the left that follows.

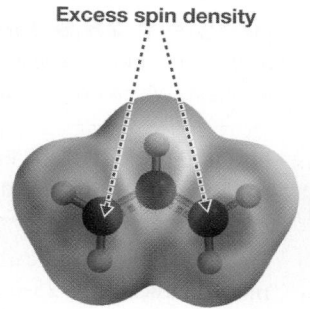

The *resonance hybrids* of the allyl and benzyl radicals (Equations 25-9 and 25-10) illustrate the delocalization of the unpaired electron about the entire species. Another way to depict this is through **electron spin density plots** like those in Figure 25-5. In these plots, *electron spin density* (i.e., unpaired electron character) is represented by the blue area—the deeper the blue color, the greater the spin density. Notice how well the electron spin density plots match the resonance hybrids.

problem **25.6** Which carbon atom in the benzyl radical possesses the greatest spin density? Why does that carbon bear the most spin density? (*Hint*: Examine the resonance structures of the benzyl radical.)

SOLVED problem **25.7** Which is the weakest C—C single bond in the molecule at the right?

Think What are the distinct types of C—C bond? What are the products upon homolysis of each of those bonds? Can the stabilities of the product radicals be differentiated based on resonance and/or alkyl substitution?

Solve There are four distinct C—C single bonds, labeled **A–D** below on the left. The corresponding homolysis products are shown on the right.

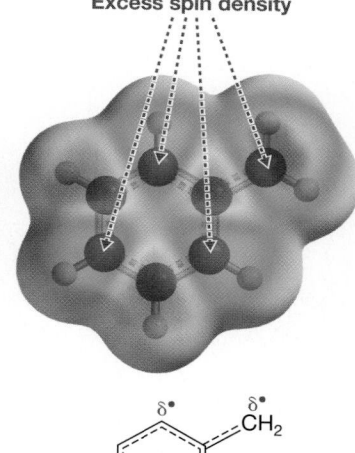

FIGURE 25-5 Electron spin density plots The electron spin density plots of the allyl radical (top left) and the benzyl radical (top right). The deeper the blue, the greater the electron spin density. Spin density in the allyl radical is delocalized over the two terminal carbon atoms. Spin density in the benzyl radical is delocalized over four carbon atoms.

The alkyl radicals produced upon homolysis of bonds **B** and **C** have the greatest alkyl substitution on the carbon atom with the unpaired electron—one is a primary radical, and the other is a secondary. However, the primary alkyl radical produced upon homolysis of bond **B** is stabilized substantially by resonance. Thus, homolysis of bond **B** gives the most stable products, making bond **B** the weakest of the C—C bonds.

problem **25.8** Which is the weakest C—H bond in the molecule at the right?

problem **25.9** For each of the molecules at the right, indicate which C—H bond is the weakest. Which do you think is the weaker of those two C—H bonds? Explain.

problem **25.10** Predict the most likely homolysis product when a *mixture* of cyclohexene, HBr, and hydrogen peroxide (H_2O_2) is heated.

Alkyl radicals and carbocations are structurally similar, too. The methyl cation and the methyl radical (Fig. 25-6), for instance, are both *entirely planar*. Because each sp^2-hybridized orbital is used to make a bond to hydrogen, the unpaired electron must reside in the carbon atom's unhybridized p orbital. By comparison, that p orbital is empty in H_3C^+.

$H_3C\cdot$ and $H_3C^\oplus$ are both planar, so each one's C atom is sp^2 hybridized, possessing a single unhybridized p orbital.

FIGURE 25-6 Structural similarities between the methyl radical and the methyl cation The three-dimensional structure of $H_3C\cdot$ is shown on the left, and that of H_3C^+ is shown on the right. Both structures are entirely planar, indicating that the C atom is sp^2 hybridized. The single, unpaired electron in $H_3C\cdot$ resides, therefore, in an unhybridized p orbital. The corresponding p orbital in H_3C^+ is empty.

The unhybridized p orbital in $H_3C\cdot$ contains the unpaired electron (red arrow).

In $H_3C^\oplus$, the unhybridized p orbital is empty.

This planarity is common to other carbon radicals, too.

> A carbon atom that has an unpaired electron tends to be planar and sp^2 hybridized.

Most carbon radicals are slightly pyramidal, however, but often behave as if they were planar, due to rapid inversion of the pyramid.

25.3 Common Elementary Steps That Free Radicals Undergo

Just as there are a small number of common elementary steps involving closed-shell species (Chapters 6 and 7), there are relatively few elementary steps that are common for free radicals, too. We examine those steps briefly here in Section 25.3, in particular looking at each step's curved arrow notation and aspects of its driving force. We will then be better equipped to tackle the mechanisms of the synthetically useful reactions beginning in Section 25.4, which are nothing more than specific sequences of these elementary steps.

25.3a Radical Coupling

Perhaps the simplest of steps that a free radical can undergo is *radical coupling*, an example of which is shown in Equation 25-11.

Radical coupling

A covalent bond is formed.

$$H_3C\bullet \quad \longrightarrow \quad H-C-H \qquad (25\text{-}11)$$

Radical coupling involves two free radicals. In this case, the unpaired electron on C joins the unpaired electron on H to form a new C—H bond. This is indicated by the two single-barbed arrows pointing to the bond-forming region. (Just as with double-barbed arrows, the atoms are assumed to follow their own electrons.) In general,

> In a **radical coupling** step, an unpaired electron from one atom joins an unpaired electron from a second atom, forming a new covalent bond that connects the two atoms.

Supply the missing curved arrows for the following radical coupling step involving two $H_3C\bullet$ radicals.

$$H_3C\bullet \quad + \quad \bullet CH_3 \quad \longrightarrow \quad H_3C-CH_3$$

25.6 YOUR TURN

Radical coupling is essentially the reverse of homolysis: Whereas homolysis produces two new free radicals by *breaking* a covalent bond, radical coupling removes two free radicals by *forming* a new covalent bond. Thus, in contrast to homolysis being very unfavorable,

> Radical coupling is usually very favorable and, therefore, *irreversible*.

FIGURE 25-7 Reaction energy diagram for radical coupling The completely separated $H_3C\bullet$ and $H\bullet$ radicals appear on the left, and the fully formed C—H bond appears on the right. This energy diagram is essentially the mirror image of the one for homolysis, shown previously in Figure 25-2. Unlike other reaction steps we have seen, radical coupling typically proceeds with no energy barrier between reactants and products.

problem 25.11 Two diphenylamino radicals undergo radical coupling *reversibly* to produce tetraphenylhydrazine.

Explain why this reaction is an exception to the general rule just given.

Because radical coupling is the reverse of homolysis, the free energy diagram for a radical coupling step is precisely the reverse of one for homolysis. For example, the free energy diagram for the radical coupling of a methyl radical and a hydrogen radical, shown in Figure 25-7, is the reverse of CH_4 homolysis, shown previously in Figure 25-2.

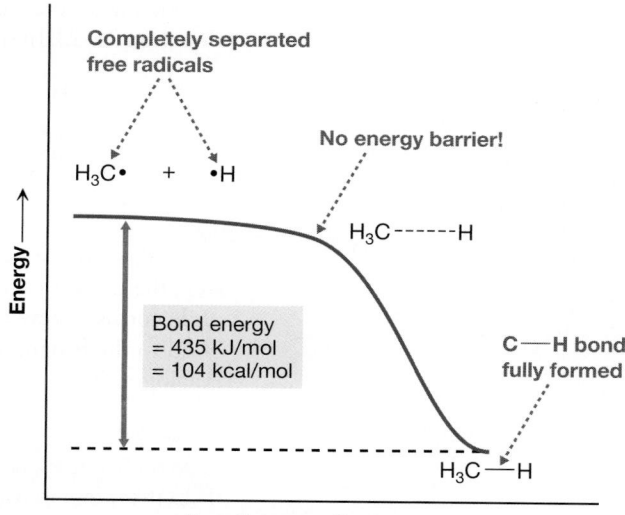

Notice that at every distance the two radicals are separated, shortening that distance (i.e., moving to the right in the figure) results in a lower energy. In other words,

> Radical coupling steps generally have no energy barrier!

In this regard, free radical coupling is distinct from every other elementary step we have seen. This is further evidence of just how reactive and unstable free radicals are.

25.3b Bimolecular Homolytic Substitution (S_H2)

As we saw in Section 25.3a, a radical coupling step involves the reaction of one free radical with another. It is also possible, however, for a free radical to react with a *closed-shell species*. One example is the **bimolecular homolytic substitution (S_H2)** step shown in Equation 25-12, which involves Br• and a molecule of methane, CH_4.

The curved arrows show explicitly that the C—H bond is broken, and that one of the two electrons from that bond ends up on the C atom. The other electron joins the unpaired electron on Br• to form a new H—Br bond. Thus, a H atom is transferred from the C atom to the Br atom. For this reason, the S_H2 step in Equation 25-12 is more specifically called a **hydrogen atom abstraction** (Br has *abstracted* a H atom).

YOUR TURN 25.7

Supply the missing curved arrows for the following hydrogen atom abstraction step.

$$\underset{H_3C}{\overset{H_2}{C}}{\diagdown}_H + \ •Cl \longrightarrow \underset{H_3C}{\overset{•}{C}H_2} + \ H—Cl$$

An example of an S_H2 step that is *not* a hydrogen atom abstraction is shown in Equation 25-13. In this case, a chlorine radical, Cl•, is abstracted from a molecule of Cl_2.

$$\underset{H}{\overset{H}{\diagup}}\overset{•}{C}{\diagdown}_H + \quad Cl—Cl \longrightarrow H—\underset{H}{\overset{H}{C}}—Cl \ + \ •Cl \qquad (25\text{-}13)$$

Although the curved arrow notation for an S_H2 step differs from that of an S_N2 step, the dynamics of the two steps are essentially the same. In an S_N2 step, a nucleophile forms a new bond to an atom attached to a leaving group, thus breaking the bond to the leaving group. The leaving group typically leaves in the form of an anion. Analogously,

> In an S_H2 step, a free radical forms a new bond to an atom, causing another bond to that atom to break. In the process, a new free radical is displaced from that atom.

problem **25.12** Draw the three possible S_H2 steps, including curved arrows, which can occur between Cl• and ethane. (*Hint*: Which atom[s] can be attacked by Cl•? Which bond[s] can be broken?)

$$C_2H_6 \quad + \quad \cdot Cl \quad \longrightarrow \quad ?$$

25.3c Radical Addition to an Alkene or Alkyne

Recall from Chapter 11 that the C=C double bond of an alkene is relatively *electron rich*. When an alkene encounters an *electron-poor* free radical, therefore, a **radical addition** step can take place, such as the one shown in Equation 25-14. Likewise, a free radical can add to the C≡C triple bond of an alkyne, as shown in Equation 25-15.

Radical addition

(25-14)

(25-15)

One electron from the multiple bond of an alkene or alkyne joins the unpaired electron from the free radical, and a second electron from the multiple bond ends up as an unpaired electron on carbon. Thus,

In a radical addition step, one carbon atom from an alkene or alkyne forms a new bond to a free radical, and the second alkene or alkyne carbon gains an unpaired electron.

Supply the curved arrows necessary for the following radical addition step.

25.8 YOUR TURN

25.4 Radical Halogenation of Alkanes: Synthesis of Alkyl Halides

When an alkane such as cyclohexane is treated with a molecular halogen such as Cl_2, no reaction occurs (Equation 25-16). However, if the same mixture is illuminated with UV light, then *halogenation* takes place, producing chlorocyclohexane and HCl (Equation 25-17).

$$\text{Cyclohexane} + Cl_2 \longrightarrow \boxed{\text{No reaction}} \qquad (25\text{-}16)$$

$$\text{Cyclohexane} + Cl_2 \xrightarrow[hv]{40\text{-}75\,°C,\ \text{UV light}} \text{Chlorocyclohexane (93\%)} + HCl \qquad (25\text{-}17)$$

As we will see in Section 25.4a, the mechanism involves free radical intermediates, so the reaction in Equation 25-17 is more specifically called a **radical halogenation**.

Radical halogenation can take place with a variety of alkanes, as well as other molecular halogens. In general,

> **Radical halogenation** is a substitution reaction in which a hydrogen atom from an alkane is replaced by a halogen atom from a molecular halogen, X_2, producing a hydrogen halide as a by-product.

Here, we discuss some important aspects of radical halogenations, beginning with a close look at the mechanism. Once we understand the mechanism, we then examine issues pertaining to the choice of halogen, including kinetics, thermodynamics, and selectivity.

25.4a The Mechanism of Radical Halogenation: An Introduction to Chain-Reaction Mechanisms

The mechanism for radical halogenation involves free radicals, but no radicals appear in the overall reaction in Equation 25-17. Recall from Section 25.1, however, that Cl_2 is a radical initiator and UV light can induce homolysis to produce free radicals. This is called the **initiation step** of the free radical mechanism and, in this case, produces two Cl• radicals (Equation 25-18).

> **Initiation step for Equation 25-17**
>
> UV light brings about the homolysis of the Cl—Cl bond.
>
> $$Cl{-}Cl \xrightarrow{hv} Cl\bullet \ + \ Cl\bullet \qquad (25\text{-}18)$$

YOUR TURN 25.9

Cyclohexane can be halogenated with Br_2 instead of Cl_2:

$$\text{Cyclohexane} + Br_2 \xrightarrow{hv} \text{Bromocyclohexane} + HBr$$

Draw the initiation step, including the necessary curved arrows.

The next two steps of the mechanism, shown in Equations 25-19a and 25-19b, are responsible for producing the overall products.

Propagation steps for Equation 25-17

(25-19a)

The initial chlorine radical is regenerated.

(25-19b)

Overall reaction

(25-19c)

Equation 25-19a is an S_H2 step in which Cl• (formed by the homolysis in Equation 25-18) abstracts a H atom from cyclohexane (an overall reactant), producing a molecule of HCl (an overall product) and a cyclohexyl radical. Then, in Equation 25-19b, the cyclohexyl radical abstracts a Cl• from Cl_2 (an overall reactant) in another S_H2 step, producing a molecule of chlorocyclohexane (an overall product) and leaving behind another Cl•.

The Cl• that is produced in Equation 25-19b is available to react with another molecule of cyclohexane, so Equations 25-19a and 25-19b can be repeated many times. Each time these two elementary steps take place, one molecule of cyclohexane and one molecule of Cl_2 are converted to one molecule of chlorocyclohexane and one molecule of HCl. We can see this more clearly in Equation 25-19c, which is the sum of Equations 25-19a and 25-19b (the redundant free radical intermediates have been canceled, as indicated by the red dashed lines above), and is also the same conversion that is represented by the balanced *overall* reaction shown previously in Equation 25-17. In other words, Equations 25-19a and 25-19b are responsible for *propagating* the overall reaction, and are therefore called **propagation steps**; together they make up a cyclic process called a **propagation cycle**.

25.10 **YOUR TURN**

Draw the propagation steps for the reaction in Your Turn 25.9. (*Hint:* They are very similar to the ones in Equation 25-19.)

Each free radical that is introduced into a propagation cycle is generally capable of converting many thousands of reactant molecules into product. There are two ways for a propagation cycle to cease, however. One is for the overall reactants to be completely consumed, as is the case for any chemical reaction. The second is for the radicals involved in the propagation steps to be destroyed.

Any reaction that is responsible for destroying free radicals that are involved in a propagation cycle is called a **termination step**.

Equations 25-20a and 25-20b are two possible termination steps for the radical chlorination of cyclohexane. Both of these are *radical coupling* steps that convert free radicals from the propagation cycle into closed-shell molecules.

Termination steps for Equation 25-17

$$Cl\bullet \quad \bullet Cl \xrightarrow{\text{Radical coupling}} Cl-Cl \qquad \text{(25-20a)}$$

(25-20b)

YOUR TURN 25.11

Draw two termination steps for the reaction in Your Turn 25.9. (*Hint:* They are similar to the ones in Equation 25-20.)

problem **25.13** Draw a termination step for the reaction in Equation 25-17 that is different from the ones shown in Equation 25-20.

SOLVED problem **25.14** Methane can undergo radical halogenation:

$$CH_4 \ + \ Br_2 \xrightarrow{h\nu} H_3C-Br \ + \ H-Br$$

Draw the mechanism for this reaction, including the initiation step, the propagation steps, and two different termination steps.

Think For the initiation step, what is the weakest bond in the reactants that can undergo homolysis? For the propagation steps, what S_H2 step can the initial radical undergo to produce the HBr product? What S_H2 step can the resulting radical undergo to produce the CH_3Br product and regenerate the initial radical? For the termination steps, what possible radical coupling steps can take place?

Solve The weakest bond present is the Br—Br bond, which undergoes homolysis in an initiation step to produce two Br• radicals.

$$Br \overset{\frown}{\underset{\smile}{-}} Br \xrightarrow{h\nu} Br\bullet \ + \ Br\bullet$$

In the first of two propagation steps, a Br• abstracts a hydrogen from CH_4 to produce one overall product, HBr, and a methyl radical, $H_3C\bullet$. In the second propagation step, the $H_3C\bullet$ abstracts a Br atom from Br_2 to produce the second overall product, CH_3Br, and regenerate the initial Br•.

$$Br\bullet \quad H-CH_3 \longrightarrow Br-H \ + \ \bullet CH_3$$

$$H_3C\bullet \quad Br-Br \longrightarrow H_3C-Br \ + \ Br\bullet$$

Termination steps can be a radical coupling between any two radicals that appear, such as the following two:

$$H_3C\bullet \quad \bullet Br \longrightarrow H_3C-Br$$

$$H_3C\bullet \quad \bullet CH_3 \longrightarrow H_3C-CH_3$$

problem **25.15** Ethane can undergo radical halogenation. Draw the mechanism for this reaction, including the initiation step, the propagation steps, and two different termination steps.

$$CH_3CH_3 + Cl_2 \xrightarrow[h\nu]{} CH_3CH_2Cl + HCl$$

Figure 25-8 is a schematic representation of the mechanism for the radical chlorination reaction in Equation 25-17. Study this diagram and match each step to either Equation 25-18, 25-19a, 25-19b, 25-20a, or 25-20b.

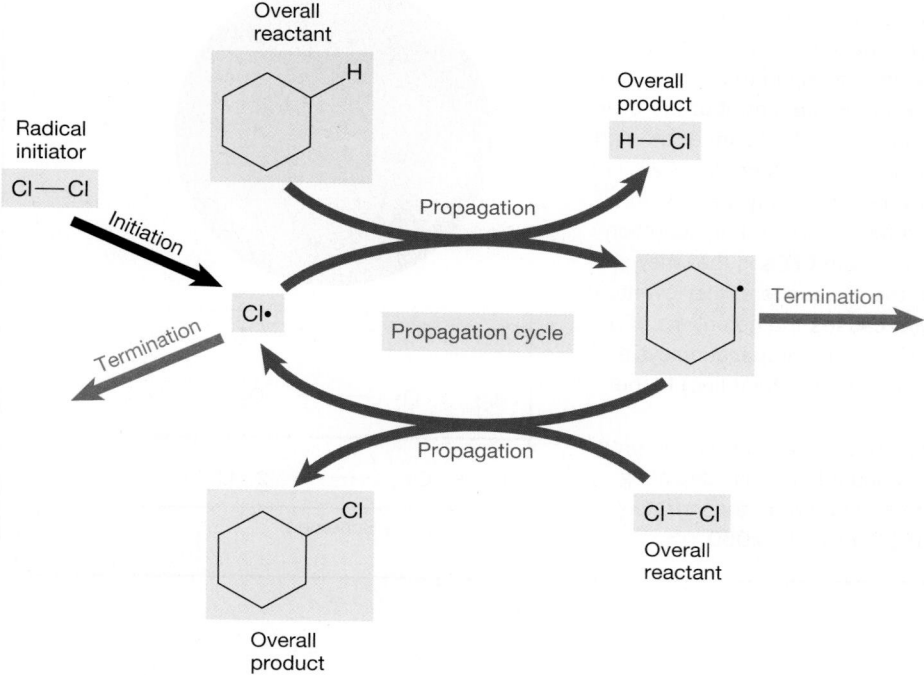

FIGURE 25-8 Schematic representation of the radical chlorination mechanism The initiation step (black arrow) represents the introduction of free radicals into the propagation cycle (blue arrows). Each time a full propagation cycle is completed, one molecule of cyclohexane (top left) and one molecule of Cl_2 (bottom right) are converted to a molecule of HCl (top right) and a molecule of chlorocyclohexane (bottom left). Termination steps (red arrows) represent the destruction of the radicals that participate in the propagation cycle.

Construct a schematic representation of the free radical bromination of methane in Solved Problem 25.14. Use Figure 25-8 as a guide.

25.12 YOUR TURN

Free radical halogenation is not the only type of reaction whose mechanism consists of initiation, propagation, and termination steps. Rather, it belongs to a broader class of reactions that are called *chain reactions.*

Chain-reaction mechanisms generally consist of *initiation*, *propagation*, and *termination* steps.

As we saw for free radical halogenation, the free radical intermediates that participate in any chain-reaction mechanism are produced by the initiation steps and are destroyed by the termination steps. Furthermore, the propagation cycle is responsible for converting the overall reactants into the overall products. That is,

The sum of the propagation steps of a chain-reaction mechanism yields the balanced overall reaction.

In Section 25.5, we will see how these ideas are applied to the radical addition of HBr to alkenes, which also proceeds by a chain-reaction mechanism. And we will apply these ideas toward a number of other chain-reaction mechanisms in the problems at the end of the chapter.

The Earth is constantly bombarded by harmful UV radiation from the sun, but a very small amount (<10 ppm) of gaseous ozone (O_3), located roughly 10–20 miles above sea level (in the region of the atmosphere called the stratosphere), protects us from much of that radiation. Ozone is believed to convert UV photons into heat via the free radical mechanism depicted in Scheme A.

Since the 1970s, however, the total amount of stratospheric ozone has decreased an average of about 5% per decade. Moreover, each year from mid-August to late November (i.e., winter/spring in the southern hemisphere), there is a dramatic decrease in the amount of ozone over Antarctica. This "ozone hole" is depicted in blue in the figure at the right.

Chlorofluorocarbons (CFCs) receive much of the blame. In 1928, certain CFCs, such as CCl_2F_2 and CCl_3F, were patented as types of Freon™, and were the dominant coolants used in refrigerators and air conditioners throughout the 20th century. The problem with CFCs is that they are *very* stable, so when they are released in the atmosphere, they eventually migrate to the stratosphere, where UV light causes the homolysis of a C—Cl bond to produce Cl•. Cl• can then enter the propagation cycle shown in Scheme B, the sum of which (shown on the third line) is the catalytic breakdown of O_3 to O_2.

The Montreal Protocol was established in 1987 to combat the problem by phasing out the production of CFCs and other ozone-depleting substances. If the Montreal Protocol can be adhered to, then stratospheric ozone is projected to reach pre-1980 levels by 2050.

Scheme A

$$O_3 + h\nu \longrightarrow O_2 + O\bullet$$

$$O_2 + O\bullet \longrightarrow O_3 + heat$$

$$Cl\bullet + O_3 \longrightarrow ClO\bullet + O_2$$

$$ClO\bullet + O\bullet \longrightarrow O_2 + Cl\bullet$$

$$O_3 + O\bullet \longrightarrow 2\ O_2$$

Scheme B

25.4b Kinetics and Thermodynamics of Radical Halogenation

Although the mechanism of halogenation is the same regardless of the identity of the molecular halogen used, some halogenations are more feasible to carry out than others. One important aspect to consider is the relative reaction rates, which exhibit a periodic table trend, as shown in Equations 25-21a through 25-21d.

Explosive

Decreasing reaction rate

$$CH_4 + F_2 \xrightarrow{h\nu} CH_3F + HF \qquad \text{(25-21a)}$$

$$CH_4 + Cl_2 \xrightarrow{h\nu} CH_3Cl + HCl \qquad \text{(25-21b)}$$

$$CH_4 + Br_2 \xrightarrow{h\nu} CH_3Br + HBr \qquad \text{(25-21c)}$$

$$CH_4 + I_2 \xrightarrow{h\nu} CH_3I + HI \qquad \text{(25-21d)}$$

Extremely slow

Namely,

The rate of free radical halogenation decreases in the order: $F_2 > Cl_2 > Br_2 > I_2$.

Fluorination is explosive, even with dilute concentrations. Chlorination is slower, but is still potentially explosive. Bromination is slower still and therefore quite controllable. Iodination is so slow that the reaction has to be heated for it to proceed at a reasonable rate.

To begin to understand this trend, recall from the discussion of kinetics in Chapter 8 that a reaction's rate depends on the size of the energy barrier between the reactants and products. To construct the energy diagrams for each of these halogenation reactions, we focus on the two propagation steps in each reaction, because those are the ones that are directly responsible for converting the overall reactants into overall products (review Equation 25-19). The value of ΔH°_{rxn} for each of these propagation steps and for each net reaction is listed in Table 25-4. (These values are calculated from the bond dissociation energies provided earlier in Tables 25-1, 25-2, and 25-3.)

TABLE 25-4 Reaction Enthalpies for the Propagation Steps of $CH_4 + X_2 \longrightarrow CH_3X + HX$

Reaction	ΔH°_{rxn}							
	$X\bullet = F\bullet$		$X\bullet = Cl\bullet$		$X\bullet = Br\bullet$		$X\bullet = I\bullet$	
	kJ/mol	kcal/mol	kJ/mol	kcal/mol	kJ/mol	kcal/mol	kJ/mol	kcal/mol
$X\bullet + CH_4 \longrightarrow HX + CH_3\bullet$	-134	-32	$+4$	$+1$	$+69$	$+16$	$+138$	$+33$
$CH_3\bullet + X_2 \longrightarrow CH_3X + X\bullet$	-292	-70	-106	-25	-101	-24	-83	-20
$CH_4 + X_2 \longrightarrow CH_3X + HX$	-426	-102	-102	-24	-32	-8	$+55$	$+13$

From these values, we can construct the reaction energy diagrams in Figure 25-9. The size of the energy barrier for these reactions decreases in the order: $I_2 > Br_2 > Cl_2 > F_2$. This is primarily because the first step of the mechanism becomes increasingly exothermic in the same order, which, according to the Hammond postulate (Section 9.3a), should result in a smaller barrier. This is consistent with fluorination being the fastest of the halogenation reactions and iodination being the slowest.

FIGURE 25-9 Energy diagrams for the halogenation of CH_4 The purple curve represents fluorination, the red curve represents chlorination, the blue curve represents bromination, and the green curve represents iodination. The overall energy barriers between reactants and products are indicated by the respective double-headed arrows. Fluorination has the smallest energy barrier, followed by chlorination, bromination, and iodination.

Another major factor contributing to the differences in halogenation rates is the *overall* ΔH°_{rxn}. The fluorination reaction is very exothermic *overall*—by $\sim$426 kJ/mol (102 kcal/mol). Having this much heat generated raises the temperature of the reaction mixture, which further increases the reaction rate and contributes to making fluorination explosive. Iodination, on the other hand, is endothermic overall, which is why it must be heated to proceed.

25.4c Selectivity of Chlorination and Bromination

The chlorination and bromination of propane are shown in Equations 25-22 and 25-23.

(25-22)

(25-23)

In both cases, the 2° halide is favored over the 1° halide. Why does halogenation exhibit this selectivity?

We must consider the first of the two propagation steps—the step in which the H atom is abstracted by the halogen radical. In Equation 25-24, for example, Cl• abstracts a H at the 2° C, and in Equation 25-25, Cl• abstracts a H at the 1° C.

(25-24)

(25-25)

The only difference between the two reactions is in the alkyl radical that is produced—a 2° alkyl radical in Equation 25-24 and a 1° radical in Equation 25-25. Because a 2° alkyl radical is more stable (Section 25.2), there is a greater driving force for the hydrogen abstraction in Equation 25-24. Generally speaking,

> The rate of halogenation increases with increasing stability of the alkyl radical that is produced upon hydrogen abstraction.

Thus, not only does radical halogenation tend to be faster at 2° carbons than 1° carbons, but it also tends to be faster at 3° carbons than 2° carbons. Halogenation is particularly favored when hydrogen abstraction produces a resonance-stabilized alkyl radical.

YOUR TURN 25.13

Which of the following radical abstraction steps will proceed faster?

(a)

(b)

Looking back at Equations 25-22 and 25-23, notice that the *extent* to which the 2° alkyl halide product is favored is much greater for bromination than it is for chlorination. In other words,

> Free radical bromination of an alkane is much more selective than free radical chlorination.

Bromination is so much more selective than chlorination primarily because the H—Br bond is substantially *weaker* than the H—Cl bond. This has a significant impact on the hydrogen abstraction step of the mechanism. As shown in Figure 25-10, the hydrogen abstraction by Cl• is significantly exothermic, whereas the hydrogen abstraction by Br• is significantly endothermic. According to the Hammond postulate (Section 9.3a), therefore, when Br• abstracts a hydrogen atom, the transition states leading to the primary and secondary products should differ significantly in energy (Fig. 25-10b), and the respective rates should differ significantly, too. When Cl• abstracts those H atoms, the transition states are similar in energy (Fig. 25-10a), as are the reaction rates.

25.14 YOUR TURN

Verify that the H–Br bond is substantially weaker than the H—Cl bond by looking up their bond dissociation energies in Table 25-1.

H—Br _____   H—Cl _____

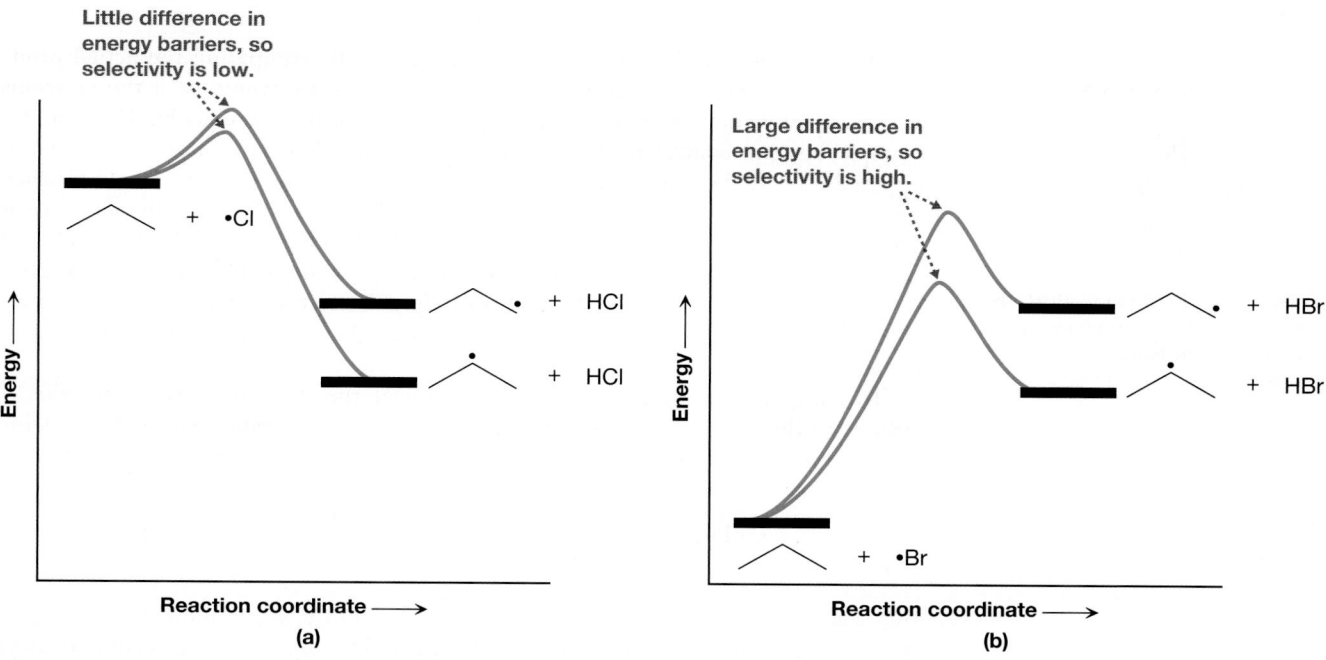

FIGURE 25-10 Selectivity in radical halogenation reactions (a) Hydrogen abstraction by Cl• at a 2° and 1° C atom. (b) Hydrogen abstraction by Br• at a 2° and 1° C. The energy barriers are similar in size in (a), so there is little selectivity in chlorination. The energy barriers are significantly different in size in (b), so bromination occurs with high selectivity.

SOLVED problem 25.16 Predict the major product of the reaction at the right and draw the complete, detailed mechanism. Include all initiation and propagation steps.

Think What is the initial radical that is formed? By what process? How will that radical interact with the uncharged molecules present? Is regiochemistry a concern?

Solve Br₂ will undergo homolysis (shown at the right), as suggested by the presence of light, so the initial radical is Br•. With the alkane present, radical halogenation will occur via two propagation steps. The Br• abstracts a H atom to yield an alkyl radical, R•. That alkyl radical then abstracts a Br atom from Br₂ to yield R—Br and Br•.

Regiochemistry is a concern because there are multiple, distinct H atoms that can be abstracted by Br•. Recall, however, that Br• is highly selective toward hydrogen atoms attached to 3° carbons, as shown.

problem 25.17 Predict the major product of the reaction at the right and draw the complete, detailed mechanism. Include all initiation and propagation steps. Draw two plausible termination steps as well.

Two 2° H atoms

Propane

Six 1° H atoms

FIGURE 25-11 Equivalent protons in propane The six 1° H atoms are equivalent and the two 2° H atoms are equivalent.

We can quantify the selectivity of halogenation by comparing the actual product distribution of each reaction to what the distributions would be if the reactions were completely nonselective (i.e., if all positions were favored equally). Propane, for example, has two equivalent 2° H atoms and six equivalent 1° H atoms, as shown in Figure 25-11. If there were equal likelihood for each of the eight H atoms to react, then there would be a 25% chance (two of eight) that the reaction would take place at one of the 2° H atoms to produce 2-chloropropane, and there would be a 75% chance (six of eight) that the reaction would take place at one of the 1° H atoms to produce 1-chloropropane. In other words, a completely nonselective reaction would produce 25% 2-chloropropane and 75% 1-chloropropane, giving a ratio of 1:3, not the observed ratio of 60:40 (i.e., 3:2).

To compute the selectivity of chlorination at the 2° carbon relative to the 1° carbon, we divide the projected nonselective ratio (i.e., 1:3) into the one that is actually observed (i.e., 60:40).

$$\text{Chlorination selectivity } (2°:1°) = \frac{60:40}{1:3} = 60:13.\bar{3} = 4.5:1$$

This selectivity of 4.5:1 means that if a molecule had the same number of 1° and 2° hydrogens, then the product mixture would have 4.5 times more of the secondary alkyl chloride than the 1°.

Repeating this for the bromination in Equation 25-23, we obtain:

$$\text{Bromination selectivity } (2°:1°) = \frac{99:1}{1:3} = 99 : \tfrac{1}{3} = 297:1$$

Based on the two ratios for halogenation at a 2° versus a 1° carbon, bromination is about 297/4.5 = 66 times more selective than chlorination.

Product distributions from other halogenation reactions have also enabled us to determine the relative selectivities of halogenating 2° versus 3° carbons. These results can be combined with the ones from above to give the following relative selectivities for 3°, 2°, and 1° carbons.

Notice how selective bromination is toward a 3° carbon!

25.4d Radical Bromination Using *N*-Bromosuccinimide

If an alkene undergoes radical bromination, then substitution will most likely take place at the allylic position, due to the resonance stabilization in the allylic radical that would be produced upon hydrogen abstraction. Thus, radical bromination of cyclohexene would produce 3-bromocyclohexene, as shown in Equation 25-26a.

(25-26a)

(25-26b)

A problem arises, however, if we carry out a bromination by simply treating the alkene with molecular bromine and UV light. As we learned in Chapter 12, Br_2 will also add to the C=C double bond in an *electrophilic addition* reaction (Equation 25-26b).

To avoid this problem, chemists carry out free radical brominations using *N*-bromosuccinimide (NBS) (Fig. 25-12). As shown in Equation 25-27, NBS reacts with cyclohexene to give only the allylic bromide.

(25-27)

NBS = Br—N

FIGURE 25-12 *N*–Bromosuccinimide and bromination NBS is often used to carry out radical bromination at allylic positions without producing unwanted electrophilic addition products.

N-Bromosuccinimide does not alter the mechanism of radical bromination, but it controls the rates of the competing reactions in Equation 25-26 by controlling the concentration of Br_2 (and, thus, the concentration of Br•).

In the presence of a compound with alkane C—H bonds, *N*-bromosuccinimide (NBS) is the source of a low, steady concentration of Br_2.

The way in which Br_2 is produced from NBS is shown in the mechanism in Equation 25-28.

Mechanism for the production of Br₂ from NBS

(25-28)

Step 1 is the homolysis of the N—Br bond to produce Br•. This step requires only modest heating because the N—Br bond is particularly weak, but could also be accomplished by irradiation with light. In Step 2, the newly formed Br• abstracts a H atom from the alkane, producing HBr, a strong acid. In Step 3, therefore, a carbonyl O on NBS is protonated, which also produces Br⁻, a strong nucleophile. Finally, Step 4 is an S_N2 step, in which the Br⁻ nucleophile attacks a Br atom from a second molecule of NBS, yielding Br₂.

YOUR TURN 25.15

Step 1 in Equation 25-28, the homolysis of NBS, produces a resonance-stabilized free radical. In the space provided here, draw the additional resonance structures involving the unpaired electron, and include the appropriate curved arrows.

Given that NBS maintains a low concentration of Br₂, why is radical bromination favored over electrophilic addition? Recall from Section 12.3 that the electrophilic addition of Br₂ to an alkene proceeds through a bromonium ion intermediate (Equation 25-29):

(25-29)

The second step is *highly* unlikely because the bromonium ion and Br⁻ intermediates are both present in *very* low concentrations, making it *extremely* unlikely that they will find each other. The concentrations of both species are so low because at any given time, only a fraction of the Br₂ present will have reacted to produce them, and the concentration of Br₂ is already low.

The story is somewhat different for radical bromination using NBS, whose initiation and propagation steps are shown in Equation 25-30.

Br₂ is maintained in a small concentration from NBS.

Very small concentration

Very small concentration

$$Br-Br \longrightarrow Br\bullet \;+\; Br\bullet \quad\quad\quad\quad \tag{25-30}$$

None of these steps requires the reaction between two species produced from Br₂, so there are no steps that require the reaction between two species whose concentrations are excessively low.

---

problem 25.18 Predict the major product of the reaction at the right.

$$\xrightarrow[\Delta]{\text{NBS}} \;?$$

Free Radicals in the Body: Lipid Peroxidation and Vitamin E

In the body, free radicals play important roles in cell signaling and protecting against invading bacteria. Because they are highly reactive, however, these free radicals can lead to unwanted side reactions that damage or kill cells. It is no wonder that free radicals are implicated in the aging process and are linked to diseases such as diabetes and cancer.

The hydroxyl radical (HO•) is particularly damaging via a process called lipid peroxidation. In Section 2.12b, we saw that the lipid bilayer of a cell membrane is made up of phospholipids, each containing two nonpolar hydrocarbon tails. As shown below on the left, cis C=C double bonds are common features of the hydrocarbon tails, and, as we learned here in Chapter 25, the allylic H atoms are susceptible to hydrogen abstraction.

A phospholipid

Lipid

Lipid radical

α-Tocopherol (Vitamin E)

Lipid peroxyl radical

Lipid peroxide

As shown in the scheme above on the right, hydrogen abstraction produces an allylic radical, which subsequently undergoes radical addition to a molecule of O₂ to produce a lipid peroxyl radical. The lipid peroxyl radical then abstracts a H atom from another phospholipid, resulting in a lipid peroxide and another lipid radical that can go on to continue the propagation cycle.

Lipid peroxides and the compounds they produce in subsequent reactions are very toxic. Fortunately, vitamin E (one form of which is α-tocopherol, shown above) is a natural antioxidant that is one of the body's most important defenses against lipid peroxidation. A lipid radical can readily abstract the phenolic H from vitamin E, leaving behind a relatively unreactive phenoxy radical. In so doing, the lipid radical is converted to a closed-shell species and the propagation cycle is terminated.

25.5 Radical Addition of HBr: Anti-Markovnikov Addition

Recall from Section 11.3 that a hydrogen halide adds across the double bond of an alkene in a Markovnikov fashion under normal conditions. Thus, HBr will add to propene to produce 2-bromopropane, as shown in Equation 25-31.

Markovnikov addition of HBr

Propene → (HBr, No peroxides) → 2-Bromopropane   (25-31)

Overall, H^+ adds to the less alkyl-substituted C atom, whereas Br^- adds to the more alkyl-substituted C atom. The reason, as discussed in Chapter 11, is that the first step of the mechanism is protonation of the C=C double bond, which proceeds to give the *more stable carbocation intermediate*.

If a small amount of peroxides (RO—OR) is added, however, then 1-bromopropane is produced as the major product instead, as indicated in Equation 25-32.

Anti-Markovnikov addition of HBr

Propene → (HBr, Peroxides) → 1-Bromopropane   (25-32)

In this reaction, HBr still adds across the C=C double bond, but with a regiochemistry opposite to that in Equation 25-31—that is, the H atom adds to the more substituted C, and the Br atom adds to the less substituted one. This is an example of an **anti-Markovnikov addition**.

Peroxides are *radical initiators* because of the weak O—O bond. Hence, the reaction in Equation 25-32 proceeds by a radical mechanism, which is outlined in Equation 25-33.

Mechanism for Equation 25-32

RO—OR → (1. Homolysis) → 2 RO•   (25-33a)

RO• + H—Br → (2. S_H2) → RO—H + Br•   (25-33b)

Br• + (alkene) → (3. Radical addition) → Br—(radical)   (25-33c)

Br—(radical) + H—Br → (4. S_H2) → Br—(product) + Br•   (25-33d)

Propagation steps

Step 1 is homolysis of the peroxide to produce two alkoxy radicals (RO•). In Step 2, an RO• abstracts a H atom from HBr to produce a Br•. That Br• adds to the alkene in Step 3, and the resulting alkyl radical abstracts a H atom from HBr, yielding the overall product.

This is a *chain reaction*, because Steps 3 and 4 make up a propagation cycle—that is, the Br• that reacts in Step 3 is regenerated as a product in Step 4. Summing Steps 3 and 4 yields the net reaction.

25.16 YOUR TURN

The propagation steps from Equation 25-33 are as follows:

Br• ⟶ 3. Radical addition ⟶ Br

Br + H—Br ⟶ 4. S_H2 ⟶ Br H + Br•

Cross out the redundant species, and sum the two steps to arrive at the net reaction.

The initial Br• that is used in the propagation steps in Equation 25-33 is produced by the sequence of Steps 1 and 2. Hence, Steps 1 and 2 can be viewed as a set of initiation steps. As usual, a termination step is any step that removes a free radical from the propagation cycle (see Your Turn 25.17).

Draw two plausible termination steps for the reaction in Equation 25-32.

25.17 YOUR TURN

Just as we saw with radical halogenation, the mechanism for the radical addition of HBr can be represented schematically, as shown in Figure 25-13.

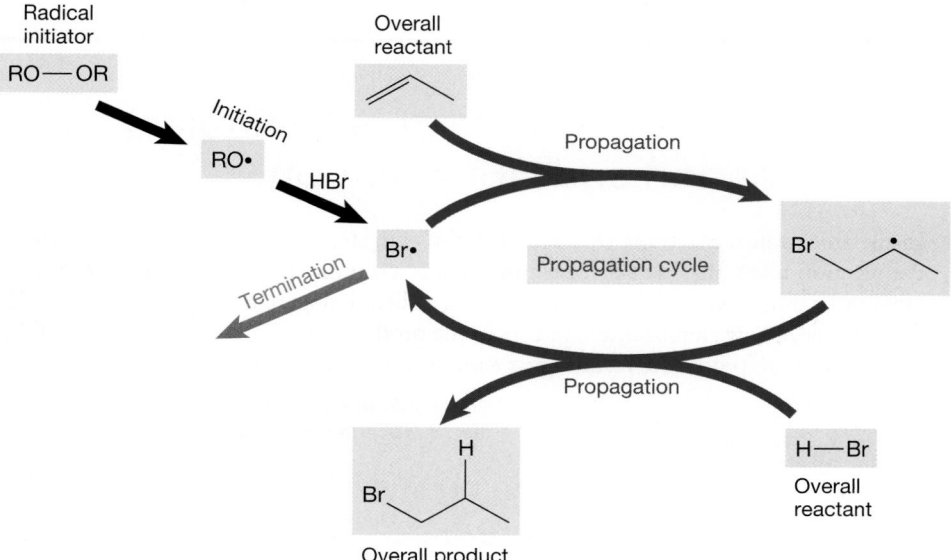

FIGURE 25-13 Schematic representation of the mechanism for the radical addition of HBr The initiation steps (black arrows) represent the introduction of free radicals into the propagation cycle (blue arrows). Each time a full propagation cycle is completed, one molecule of propene (top left) and one molecule of HBr (bottom right) are converted to a molecule of 1-bromopropane (bottom left). Termination steps (red arrows) represent the destruction of the radicals that participate in the propagation cycle.

Two steps (black arrows) are required to introduce Br• into the propagation cycle (blue arrows). With each completion of the propagation cycle, one molecule of propene and one molecule of HBr are converted to one molecule of 1-bromopropane.

Termination steps (red arrows) include any steps that destroy the radicals that participate in the propagation cycle.

The reason that the mechanism in Equation 25-33 proceeds in an anti-Markovnikov fashion has everything to do with Step 3, which is the first of the two propagation steps. Step 3 is a radical addition to the C=C double bond. As shown in Equations 25-34a and 25-34b, Br· can add to the terminal C or to the central C.

This alkyl radical is more stable.

(25-34a)

A 2° alkyl radical

3. Radical addition

(25-34b)

A 1° alkyl radical

When Br· adds to a terminal C (Equation 25-34a), a 2° alkyl radical is produced, and when it adds to the central C (Equation 25-34b), a 1° alkyl radical is produced. The greater stability of the 2° radical provides more driving force.

> Radical addition to an alkene generally takes place so as to produce the more stable alkyl radical intermediate.

problem **25.19** Predict the major product of each of the following reactions.

(a)

(b)

25.6 Stereochemistry of Free Radical Halogenation and HBr Addition

In the free radical reactions we have seen thus far, stereochemistry can be an issue if the reaction takes place at a carbon atom that is a tetrahedral stereocenter in the reactants or becomes one in the products. In the radical halogenation in Equation 25-35, for example, a new stereocenter appears in the products and both configurations of the stereocenter are produced, resulting in a mixture of enantiomers.

Both configurations of the stereocenter are produced.

(25-35)

Racemic

The reaction results in a mixture of stereochemical configurations because the stereocenter is produced directly from the S$_H$2 step shown in Equation 25-36 (review Equation 25-19).

This radical intermediate is essentially planar and achiral.

The new C—Br bond can form on either side of the plane.

Racemic   (25-36)

Equation 25-37 shows an example in which halogenation takes place at a C atom that is initially a stereocenter.

The stereocenter becomes a planar radical intermediate after H is abstracted.

Both configurations of the stereocenter are produced.

(25-37)

Once again, the reaction produces a mixture of configurations at the C atom that undergoes halogenation. Even though the C atom is initially a stereocenter, it becomes a planar free radical intermediate after the H atom is abstracted.

Similarly, if the free radical addition of HBr to an alkene results in a new tetrahedral stereocenter, then both configurations of the stereocenter will be produced. An example is shown in Equation 25-38.

Br· can add to either side of the planar C.

Both configurations of the stereocenter are produced.

Racemic   (25-38)

In this case, the stereocenter is produced when Br· adds to the alkene C (review Equation 25-33). The alkene C is planar, so Br· can add to either side of the plane.

---

problem **25.20** Draw the major product of each of the following reactions, paying attention to stereochemistry.

(a)   (b)

25.7 Dissolving Metal Reductions: Hydrogenation of Alkenes and Alkynes

To this point in Chapter 25, we have examined only free radical reactions that proceed by chain-reaction mechanisms. Here in Section 25.7, we will study two types of hydrogenation reactions that are categorized as **dissolving metal reductions**, and neither reaction involves a chain-reaction mechanism.

Hydrogenation via a dissolving metal reduction takes place under conditions that differ from the catalytic hydrogenation discussed in Section 19.5. A solid metal catalyst such as Pt(s), Pd(s), or Ni(s) is used in catalytic hydrogenation. Thus, the catalyst is in a different phase from the reactants dissolved in solution, making catalytic hydrogenation with a solid metal catalyst an example of **heterogeneous catalysis**.

In a *dissolving metal reduction*, on the other hand, a metal such as Na(s) or Li(s) is *dissolved* in solution along with the reactants. When this takes place in a solvent such as liquid ammonia, $NH_3(\ell)$, for example, the metal donates an electron, and the ammonia solvent stabilizes the electron via extensive solvation. The result, called a **solvated electron**, is depicted in Equation 25-39.

$$\text{Na} \cdot \xrightarrow{\text{NH}_3(\ell)} \text{Na}^{\oplus} + e^{\ominus} \text{(solvated)} \tag{25-39}$$

In light of this,

> The active radical species in a dissolving metal reduction involving $NH_3(\ell)$/Na(s) is simply an electron that does not formally belong to any atoms.

In the discussion that follows, we will see examples of how this solvated electron is involved in the reduction of a multiple bond. Section 25.7a discusses the reduction of alkynes to trans alkenes, and Section 25.7b discusses the Birch reduction of benzene rings.

25.7a Anti-Hydrogenation: Synthesis of Trans Alkenes

When oct-4-yne is treated with sodium metal dissolved in liquid ammonia (–78 °C), the triple bond is reduced to a double bond (Equation 25-40).

Conditions for a solvated electron

The reaction produces only the trans alkene.

Na(s), NH$_3$(ℓ)
–78 °C, 2–3 h

Oct-4-yne

(*E*)-Oct-4-ene
>80%

(25-40)

The reaction is highly stereoselective.

> A dissolving metal reduction of an alkyne produces the *E* alkene almost exclusively.

This is an **anti-hydrogenation** reaction in which two H atoms, overall, have added to the triple bond in a trans fashion. It complements catalytic hydrogenation (Section 19.5), which exclusively forms the cis product.

YOUR TURN 25.18

Draw the product of the following reaction, in which oct-4-yne undergoes catalytic hydrogenation. (Review Section 19.5.)

$$\xrightarrow[\text{Pd, BaSO}_4]{\text{H}_2(g)} \ ?$$

The mechanism for the dissolving metal reduction is shown in Equation 25-41.

Mechanism for Equation 25-40

Solvated electron

The trans radical is more stable than the cis radical.

1. Radical addition

2. Proton transfer

(25-41)

Solvated electron

3. Radical coupling

4. Proton transfer

The mechanism begins with a solvated electron, which was produced upon dissolving solid sodium in liquid ammonia. In Step 1, a solvated electron adds to an alkyne C, thus converting the triple bond into a double bond. One of the initial alkyne C atoms gains an unpaired electron and the other gains a negative charge. The carbanion is a very strong base and deprotonates NH_3 in Step 2 (notice the double-barbed curved arrows). This produces an uncharged vinyl radical, which, in Step 3, undergoes radical coupling with a second solvated electron to produce a carbanion. Finally, in Step 4, the carbanion deprotonates a second molecule of NH_3, yielding the overall product.

The stereochemistry of this reaction is established in the first step—the addition of a solvated electron to the C≡C triple bond. Although both the cis and trans forms of the vinyl radical can be produced, the two configurations are in rapid equilibrium (Fig. 25-14). The trans form has less steric strain, though, so it is more stable, and hence leads to the major product.

The equilibrium between the cis and trans forms is analogous to nitrogen inversion (Section 5.4d), in which the two pyramidal configurations of an uncharged nitrogen atom equilibrate. Unlike nitrogen inversion, however, which proceeds through a trigonal planar transition state, the equilibrium involving the vinyl radical proceeds through a linear transition state.

Dissolving metal reductions are much more sluggish with alkenes than with alkynes, in part because of charge stability. In the reduction of an alkyne, the C atom that gains the negative charge has more *s* character—and thus a higher effective electronegativity (Section 3.11)—than the C atom that gains the negative charge in the reduction of an alkene. As a result,

Steric strain makes this configuration less stable.

FIGURE 25-14 Stereoselectivity of anti-hydrogenation This radical anion, which is produced in Step 1 of Equation 25-41, exists in rapid equilibrium between the cis and trans forms. The trans form is more stable, which is why the reaction produces the trans alkene almost exclusively.

A dissolving metal reduction will selectively reduce an alkyne over an alkene.

An example is shown in Equation 25-42. This kind of selectivity can be very useful in synthesis.

The alkyne is selectively reduced over the alkene.

$$\xrightarrow[\text{NH}_3(\ell),\ -78\ °\text{C}]{\text{Na}(s)}$$ (25-42)

---

problem 25.21 Draw the alkyne from which the compound at the right can be produced via a dissolving metal reduction.

$$? \xrightarrow[\text{NH}_3(\ell),\ -78\ °\text{C}]{\text{Li}(s)}$$

problem 25.22 A dissolving metal reduction can be used to reduce a ketone to an alcohol, as shown in the reduction at the right. Draw a complete, detailed mechanism for this reaction.

$$\xrightarrow[\text{THF, dilute ROH}]{\text{Na}(s)}$$

25.7b The Birch Reduction

Benzene can be reduced to cyclohexane using catalytic hydrogenation under high temperature and pressure (Equation 25-43). These extreme conditions are necessary because aromaticity makes benzene's π system quite stable.

$$\xrightarrow[\substack{135\ °\text{C, 160 atm} \\ 10\ \text{min}}]{\text{H}_2(g),\ \text{Ni}(s)}$$ (25-43)

Benzene **Cyclohexane**
100%

Under these conditions, it is impractical to stop the reaction at an intermediate stage of reduction—that is, at a diene or an alkene—because the second and third reductions, which involve alkenes that are no longer aromatic, are faster than the first.

A **Birch reduction** (Equation 25-44), named after the Australian chemist Arthur Birch (1915–1995), on the other hand, reduces benzene to cyclohexa-1,4-diene. Not only does the reaction stop after just a single hydrogenation, but reduction takes place *regioselectively*, yielding the 1,4-diene.

Birch reduction   **Selectively produces the 1,4-diene**

$$\xrightarrow[\substack{t\text{-BuOH, NH}_3,\ \text{THF} \\ 25\ °\text{C, 5 h}}]{\text{Li}(s)}$$ (25-44)

Benzene   **Cyclohexa-1,4-diene**
84%

The conditions for a Birch reduction are very similar to those for the dissolving metal reductions discussed in Section 25.7a. The difference is that a small amount of an alcohol [typically *tert*-butyl alcohol, $(\text{CH}_3)_3\text{COH}$] is used as the proton source. As a result, the mechanism for the Birch reduction, shown in Equation 25-45, is similar to that for the dissolving metal reduction of an alkyne, shown previously in Equation 25-41.

Mechanism for the Birch reduction in Equation 25-44

(25-45)

Step 1 is radical addition of a solvated electron to a C=C double bond of the benzene ring. The resulting radical anion intermediate is strongly basic, so it deprotonates the alcohol in Step 2 to produce an uncharged alkyl radical. Radical coupling takes place in Step 3, this time producing a closed-shell carbanion species in which the negative charge is delocalized over three C atoms. The carbanion then deprotonates a second molecule of the alcohol to produce the overall product.

Notice in the final step of the mechanism that protonation can occur at any of the three C atoms sharing the negative charge, but it takes place almost exclusively at the one indicated in resonance structure **Y**. Why this happens is not entirely clear, but it's probably because resonance structure **Y** is more stable than **X** or **Z**. In **Y**, the negative charge is adjacent to two sp^2-hybridized C atoms, each of which has a relatively high effective electronegativity (Section 3.11). This allows some electron density to be drawn away from the C atom bearing the negative charge. In **X** and **Z**, on the other hand, the negative charge is adjacent to only one sp^2-hybridized C, so there is less inductive stabilization.

---

problem **25.23** Benzoic acid undergoes the Birch reduction to yield a carboxyl-substituted cyclohexa-1,4-diene. There are two possible products, however, depending upon the location of the carboxyl group relative to the C=C double bonds. (a) Draw the mechanism that leads to the formation of *each* of these products, and predict which one is the major product. (b) Will the Birch reduction of benzoic acid occur faster or slower than the Birch reduction of benzene itself? (*Hint*: Is –CO_2H an electron-donating or an electron-withdrawing group?)

---

problem **25.24** Why does a Birch reduction require ROH as the proton source, whereas a dissolving metal reduction of an alkyne uses NH_3 as the proton source? (*Hint*: Which acid is stronger?)

25.8 Organic Synthesis: Radical Reactions in Synthesis

The free radical reactions we have examined in Chapter 25 are valuable synthetically for two main reasons. First, they allow us to carry out transformations that are impractical with reactions whose mechanisms involve only closed-shell species and Lewis acid–Lewis base interactions. Second, they allow us to carry out transformations we have seen in previous chapters, but with different regiochemistry or stereochemistry.

Free radical halogenation, for example, is one of the few practical reactions that can *functionalize* (i.e., add a functional group to) the 3-position of cyclohexene, as indicated in Equation 25-46, and it is the only reaction we have discussed in this book that does so.

(25-46)

This transformation can be achieved, in particular, with bromination using NBS, similar to what we previously saw in Equation 25-27.

(25-47)

Allylic bromination such as this has been used in the synthesis of natural products, such as (+)-koninginin D. A key step in the synthesis is shown in Equation 25-48.[2]

(25-48)

(+)-Koninginin D

Under the relatively mild conditions that are used, other functional groups such as ketones and esters are unaffected.

Radical bromination with NBS is also a popular method to functionalize the benzylic position of aromatic compounds in synthesis. Equation 25-49[3] shows how it has been used in a key step in the synthesis of flavones.

1. NBS (excess), ROOR
2. Et$_3$N, reflux

(25-49)

A flavone

[2]Liu, G.; Wang, Z. *Chem. Commun.* **1999**, *12*, 1129–1130.
[3]Silva, A. M. S.; Silva, A. M. G.; Tomé, A. C.; Cavaleiro, J. A. S. *Eur. J. Org. Chem.* **1999**, *1*, 135–139.

problem **25.25** (a) Draw the product that is immediately formed in Equation 25-49 after reaction with excess NBS. (b) What kind of reaction takes place in the second reaction in Equation 25-49?

Another synthetically useful aspect of radical reactions is *regiochemistry* that can differ from that in other reactions we have encountered previously. An example is the hydrohalogenation of an alkene. Recall from Chapter 11 that HBr adds across a C=C double bond in a Markovnikov fashion, proceeding through the more stable carbocation intermediate. The addition of HBr in the presence of a radical initiator (such as peroxides), however, takes place in an *anti*-Markovnikov fashion, proceeding through the more stable *radical* intermediate. The importance of this difference is demonstrated in Solved Problem 25.26.

SOLVED problem 25.26 Show how you would carry out the transformation at the right using at least one radical chain reaction.

Think What functional group can be used to halogenate a carbon that is two carbons away from a phenyl ring? How can that functional group be introduced on a carbon that initially is not functionalized?

Solve As shown in the following retrosynthetic analysis, the alkyl bromide product can be made from phenylethene via the anti-Markovnikov radical addition of HBr.

The C=C double bond can be made from 1-bromo-1-phenylethane in an E2 reaction, and the alkyl bromide can be made, in turn, from a radical halogenation of phenylethane. Reversing the reactions and adding the appropriate reagents, the synthesis can be written as follows:

Finally, free radical reactions can offer *stereochemistry* that differs from other reactions we have seen previously. An important example of this is the dissolving metal reduction, which can reduce an alkyne to a trans alkene (Equation 25-50).

A trans alkene

This contrasts with catalytic hydrogenation (Section 19.5), which reduces an alkyne to a cis alkene (Equation 25-51).

$$R-C\equiv C-R' \quad \xrightarrow[\text{Pd/BaSO}_4]{\text{H}_2} \quad \underset{\underset{H}{\quad}}{\overset{R}{}}C=C\underset{\underset{H}{\quad}}{\overset{R'}{}} \qquad (25\text{-}51)$$

A cis alkene

An application of the dissolving metal reduction of an alkyne is shown in Equation 25-52.[4] This is a step in the total synthesis of sphingosine, which undergoes phosphorylation in vivo to produce a potent signaling lipid. (Notice in Equation 25-52 that the reaction conditions also remove a protecting group.)

Reduction of the alkyne to an alkene

Stereoselectively produces the trans alkene.

$$\xrightarrow[\text{NH}_3(\ell),\ \text{THF}]{\text{Li(s)}}$$

Sphingosine

(25-52)

Protecting group

25.9 Wrapping Up and Looking Ahead

Here in Chapter 25, you learned about the chemistry of free radicals. Because a free radical has at least one unpaired electron, one of its atoms lacks an octet, making the radical highly unstable and very reactive. Moreover, the atom that lacks an octet is relatively electron poor. Thus, there are many similarities between carbon radicals and carbocations. For example, a carbon atom bearing an unpaired electron tends to have a planar geometry, analogous to the positively charged carbon atom of a carbocation. Furthermore, radicals can be stabilized by resonance delocalization of the unpaired electron, as well as by alkyl groups attached to the radical carbon. Likewise, carbocations can be stabilized by resonance delocalization of the positive charge, as well as by alkyl groups attached to the positively charged carbon.

Like reactions involving closed-shell species, reactions involving free radicals consist of relatively few different types of elementary steps. The ones we examined are homolysis, radical coupling, bimolecular homolytic substitution (S_H2), and radical addition. To describe these steps, we introduced the single-barbed arrow, which corresponds to the movement of a single electron.

Free radical reactions can be classified according to whether they undergo chain reaction mechanisms, which are characterized by the participation of free radicals in a propagation cycle. Dissolving metal reductions involve free radicals, but are not chain reactions, whereas the radical halogenation of alkanes and the radical hydrobromination of alkenes do proceed by chain-reaction mechanisms.

The mechanisms of free radical reactions are different from mechanisms involving only closed-shell species, so free radical reactions can offer flexibility in organic synthesis. Free radical halogenation, for example, can be used to functionalize alkane

[4]Boutin, R. H.; Rapoport, H. *J. Org. Chem.* **1986**, *51*, 5320–5327.

carbons. Furthermore, free radical reactions can be used to carry out transformations with stereochemistry or regiochemistry that differs from reactions involving only closed-shell species.

Chapter 25 merely provided an introduction to free radical chemistry. We will see more examples of reactions involving free radicals in Chapter 26, specifically when it comes to synthesizing polymers. However, the field of free radical chemistry is vast—more than can be covered in this book. If you are interested in learning more about the chemistry of free radicals, read *Advanced Free Radical Reactions for Organic Chemistry* by Hideo Togo (Elsevier Science Ltd.: Stockholm, 2004).

Chapter Summary and Key Terms

- A species that possesses at least one unpaired electron is called a **free radical**. In a **closed-shell species**, all electrons are paired. **(Introduction)**

- Free radicals lack an octet, so they are usually very unstable and highly reactive. They are generally produced by **homolysis** of a covalent bond in a closed-shell precursor. **(Section 25.1)**

- Homolysis generally breaks the weakest bond in the compound. **(Section 25.1)**

- A **single-barbed arrow** (⌒) describes the movement of a single electron in an elementary step. **(Section 25.1)**

- A **radical initiator** is a precursor from which a free radical is produced and generally possesses a relatively weak covalent bond. Common initiators are molecular halogens, *N*-**bromosuccinimide (NBS), 2,2′-azobisisobutyronitrile (AIBN)**, and peroxides (RO—OR). **(Section 25.1)**

- Radical stability can be determined from homolytic bond dissociation energies. In general, the greater the energy required to produce the radical via homolysis, the more unstable the radical is. **(Section 25.2)**

- The stability of a radical increases with:
  - Decreasing *s* character on the atom with the unpaired electron.
  - Increasing resonance delocalization of the unpaired electron.
  - Additional alkyl groups attached to the atom with the unpaired electron. **(Section 25.2)**

- A carbon atom that has an unpaired electron tends to be sp^2 hybridized and has a planar geometry. **(Section 25.2)**

- In a **radical coupling** step, an unpaired electron from a radical joins an unpaired electron from a second radical to produce a new covalent bond. **(Section 25.3a)**
  - Radical coupling is typically irreversible, and proceeds to products with no energy barrier.

- In a **bimolecular homolytic substitution (S_H2)** step, a free radical forms a bond to an atom of a closed-shell species and displaces another free radical from that atom. **(Section 25.3b)**

- In a **radical addition** step, an alkene or alkyne carbon forms a bond to a free radical, and the other alkene or alkyne carbon gains an unpaired electron. **(Section 25.3c)**

- A free radical **chain reaction** involves free radicals in a cyclic sequence of steps, called **propagation steps**, which are responsible for converting overall reactants into overall products. **(Section 25.4a)**
  - Throughout the sequence of propagation steps, also called the **propagation cycle**, free radicals are not consumed overall.
  - The overall reaction is obtained by summing just the propagation steps.
  - An **initiation step** is responsible for producing the free radical that enters the propagation cycle.
  - A **termination step** removes a free radical from the propagation cycle and is responsible for slowing the overall reaction.

- **Radical halogenation** is a chain reaction that replaces an alkane hydrogen with a halogen atom. It takes place when free radicals are generated in the presence of a molecular halogen and a compound containing an alkane hydrogen. **(Section 25.4)**
  - The identity of the molecular halogen plays a major role in the kinetics and thermodynamics of radical halogenation. Both the rate and exothermicity of halogenation decrease in the order: $F_2 > Cl_2 > Br_2 > I_2$.
  - Bromination is highly regioselective, whereas chlorination is only slightly selective.
  - *N*-Bromosuccinimide is commonly used to brominate an allylic carbon to avoid the addition of Br_2 to the alkene.

- HBr adds to an alkene via **anti-Markovnikov addition** when free radicals are present. **(Section 25.5)**

- Free radical halogenation and the radical addition of HBr to an alkene produce a mixture of stereochemical configurations of the carbon atom that gains the halogen. **(Section 25.6)**

- In a **dissolving metal reduction**, a **solvated electron** behaves as the free radical species. **(Section 25.7)**
  - The dissolving metal reduction of an alkyne produces a trans alkene. **(Section 25.7a)**
  - The dissolving metal reduction of a benzene ring, called a **Birch reduction**, produces cyclohexa-1,4-diene. **(Section 25.7b)**

Reaction Table

TABLE 25-5 Functional Group Conversions

	Starting Functional Group	Typical Reagent and Reaction Conditions	Functional Group Formed	Key Electron–Rich Species	Key Electron–Poor Species	Comments	Discussed in Section
(1)	R—H \n Alkane	$\xrightarrow{X_2}$ \n $h\nu$	R—X \n Alkyl halide	X—X	R•	X = Cl, Br	25.4
(2)	Alkene	NBS \n Δ or $h\nu$	Allylic bromide	Br—Br		Favors allylic substitution over bromination of C=C	25.4d
(3)	Phenyl alkane	NBS \n Δ or $h\nu$	Benzylic bromide	Br—Br		Functionalizes a relatively unreactive carbon	25.4d
(4)	Alkene	HBr \n Peroxides	Alkyl bromide		Br•	Anti-Markovnikov regiochemistry	25.5
(5)	R—C≡C—R \n Alkyne	Na(s) or Li(s) \n NH₃(ℓ), −78 °C	Alkene			Forms trans alkene only	25.7a
(6)	Benzene	Li(s), ROH \n NH₃(ℓ)	Cyclohexa-1,4-diene			Birch reduction forms cyclohexa-1,4-diene only	25.7b

Problems

25.27 This is an example of an S_H2 step. Supply the curved arrow notation for this step.

25.28 The reaction at the right is an example of a McLafferty rearrangement, which takes place in the gas phase in a single step. It is an important fragmentation process in mass spectrometry. Provide the curved arrow notation for this step.

25.29 The radical cation product in Problem 25.28 has another resonance structure. Draw that resonance structure, using the proper curved arrow notation.

25.30 In each of the following compounds, which H would most likely be abstracted by a bromine radical, Br•?

(a)   (b)   (c)   (d)

25.31 Upon treatment with $Cl_2/h\nu$, a compound, C_9H_{12}, yields only a single monochloride. What is the structure of the compound?

25.32 The energy barrier for rotation of a CH_2 group in an allyl radical ($H_2C=CH-CH_2$•) is about 66 kJ/mol (16 kcal/mol). The energy barrier for rotation of a CH_3 group in propane is ~14 kJ/mol (3.3 kcal/mol). Explain this difference.

25.33 A hydrogen radical, H•, is known to abstract a hydrogen atom from an alkane in the same manner that halogen atoms do. Like halogen atoms, H• abstracts a hydrogen atom from a secondary carbon over a hydrogen atom from a primary carbon in a 5:1 ratio.
(a) Draw the two isomeric propyl radicals that are formed upon hydrogen abstraction from propane by H•.
(b) Draw the curved arrow notation for the formation of each of those isomeric radicals.
(c) Compute the percentage of each propyl radical that is formed, taking into account the different numbers of each type of hydrogen on propane.

25.34 At which carbon in the molecule at the right will radical substitution predominantly take place? Explain.

25.35 As we saw in this chapter, dissolving metal reductions reduce alkynes to trans alkenes. Why, then, do these reactions fail with a terminal alkyne? (*Hint*: What species appear in the mechanism?)

25.36 Predict the major product of each of the following reactions and provide the complete, detailed mechanism.

(a)   (b)   (c)

$\xrightarrow[h\nu]{Br_2}$? $\xrightarrow[h\nu]{NBS}$? $\xrightarrow[CCl_4]{Br_2}$?

(d)   (e)

$\xrightarrow[ROOR]{HBr}$? $\xrightarrow{HBr}$?

25.37 Draw the complete mechanisms leading to the two major products and determine whether the product mixture will be optically active for each of the following reactions.

(a)   (b)

$\xrightarrow[h\nu]{Br_2}$? $\xrightarrow[h\nu]{Br_2}$?

25.38 Similar to alkanes, hydrogen gas can undergo radical halogenation according to the reaction at the right, where X = F, Cl, Br, or I. Propose a chain-reaction mechanism for this reaction, including an initiation step, propagation steps, and two plausible termination steps.

$$H-H + X-X \xrightarrow{h\nu} 2 \ H-X$$

25.39 In the halogenation of H_2, described in Problem 25.38, which halogen will react fastest? Defend your answer using the appropriate bond energies from Tables 25-1, 25-2, and 25-3.

25.40 Propose a mechanism for the reaction at the right, which proceeds by a radical chain reaction. Include an initiation step, propagation steps, and two plausible termination steps.

$$H_2C{=}CH_2 + CCl_4 \xrightarrow{h\nu}$$

25.41 When an ether is exposed to air for prolonged periods of time, it undergoes a radical chain reaction with oxygen gas to form *hydroperoxides* that are explosive. The overall reaction between diethyl ether and O_2 is shown at the right. Provide a complete, detailed mechanism for this reaction, including an initiation step, propagation steps, and two plausible termination steps.

A hydroperoxide of diethyl ether

25.42 Notice in Problem 25.41 that the hydroperoxide forms at the C atom that is α to the O atom and not at the one that is β to it. Explain why. (*Hint*: Write out the respective radicals that are formed upon hydrogen abstraction.)

25.43 The following aromatic substitution reaction has been proposed to proceed via a radical chain reaction.

The mechanism at the right has been proposed, where Ar is the aromatic ring.
(a) Identify each step as either initiation, propagation, or termination.
(b) Sum the propagation steps to verify that the net equation matches that given in the problem statement.
(c) Propose a possible termination step.

$$ArI + e^{\ominus} \longrightarrow ArI^{\ominus}$$
$$ArI^{\ominus} \longrightarrow Ar{\bullet} + I^{\ominus}$$
$$Ar{\bullet} + H_2N^{\ominus} \longrightarrow \overset{\bullet}{Ar}NH_2^{\ominus}$$
$$\overset{\bullet}{Ar}NH_2^{\ominus} + ArI \longrightarrow ArNH_2 + ArI^{\ominus}$$

25.44 At the right are proposed steps of a radical chain-reaction mechanism, but are in no particular order.
(a) Draw in the appropriate curved arrows for each step.
(b) Label each step as either initiation, propagation, or termination.
(c) Using this mechanism, write the balanced net reaction.

25.45 At the right are proposed initiation and propagation steps of a radical chain-reaction mechanism, but are in no particular order.
(a) Draw in the appropriate curved arrows for each step.
(b) Label each step as either initiation or propagation.
(c) Write the balanced net reaction.
(d) Provide two plausible termination steps, including curved arrows.

25.46 The following are proposed initiation and propagation steps of a radical chain-reaction mechanism, but are in no particular order.

(a) Draw in the appropriate curved arrows for each step.
(b) Label each step as either initiation or propagation.
(c) Write the balanced net reaction.
(d) Provide two plausible termination steps, including curved arrows.

25.47 The selectivity of chlorination at 1°, 2°, and 3° carbons is about 1:4:6, a ratio that is relatively independent of the type of most solvents. If the reaction takes place in benzene, however, it is believed that Cl_2 forms a weak complex to the aromatic ring, thereby stabilizing the chlorine molecule. Does this lead to a greater selectivity or a lesser selectivity in chlorination? Explain.

25.48 **(a)** Propose a chain-reaction mechanism for the elimination of HI to form an alkene, showing reasonable initiation, propagation, and termination steps.
(b) Based on this mechanism, calculate ΔH°_{rxn} for each propagation step and for the net reaction.

25.49 Propose a chain-reaction mechanism for the decomposition of dimethyl ether to form methane and formaldehyde.

25.50 The following reaction proceeds by a chain-reaction mechanism.

Propose an initiation step, propagation steps, and two plausible termination steps.

25.51 Both bicyclo[3.3.1]nonane and bicyclo[2.2.1]heptane are saturated hydrocarbons composed of only 2° and 3° carbons.

Bicyclo[3.3.1]nonane   100%   **Bicyclo[2.2.1]heptane**   0%

Recall that radical substitution generally takes place preferentially at a 3° carbon. Indeed, when bicyclo[3.3.1]nonane is treated with the bromotrichloromethane under irradiation, substitution takes place 100% at the 3° carbon. Under the same conditions, however, no substitution is observed at the 3° carbon in bicyclo[2.2.1]heptane. Explain. (*Hint*: Build molecular models of each of these compounds.)

25.52 Propose a mechanism for the chain reaction in Problem 25.51, including initiation and propagation steps. Provide two plausible termination steps.

25.53 Acyl peroxides can be used to initiate alkyl radicals.

Propose a mechanism for this reaction.

25.54 Radical halogenation generally yields a mixture of stereoisomers. When a Br atom is adjacent to the site of hydrogen abstraction, however, then retention of stereochemistry is observed, as shown at the right. Provide a detailed mechanism for this reaction and propose a key intermediate that leads to the reaction's stereochemistry. With that key intermediate, explain why retention of stereochemistry is observed.

25.55 This addition of HBr to (*Z*)-2-bromobut-2-ene takes place regioselectively, with the Br preferentially adding to the alkene C that does *not* already possess a Br atom.
   (a) Provide a detailed mechanism for this reaction, including initiation and propagation steps.
   (b) Explain why this regiochemistry is observed.

25.56 The conversion of *cis*-1,2-diphenylethene to *trans*-1,2-diphenylethene is catalyzed when I_2 is added and the reaction mixture is irradiated with UV light. Provide a detailed mechanism for this reaction.

25.57 The radical addition of HBr can be used to cyclize a carbon chain:

Provide a detailed mechanism for this reaction. Include an initiation step, propagation steps, and two plausible termination steps.

25.58 Provide a detailed mechanism to account for the following reaction.

Include an initiation step, propagation steps, and two plausible termination steps. (*Hint*: Consult Problem 25.57.)

25.59 Suggest how you would synthesize each of the following compounds beginning with propylbenzene. You may use any other reagents necessary.

(a)

(b)

(c)

(d)

(e)

25.60 Supply the structures of compounds **A–I** in the following synthesis scheme.

25.61 Supply the structures of compounds **J–U** in the following synthesis scheme.

25.62 How would you synthesize each of the following compounds beginning with 2-methylpropane? You may use any other reagents necessary.

(a)

(b)

(c)

(d)

(e)

(via)

(f)

25.63 Show how you would synthesize each of the following from 1-cyclopentylprop-1-yne. You may use any other reagents necessary.

(a)

(b)

(c)

(d)

(e)

(f)

25.64 A student wanted to determine whether alkyl radicals rearrange, so she ran the reaction below.
   (a) Why is it plausible to think that radical rearrangement might take place in this reaction? To help answer this question, draw the mechanism of the radical reaction that would potentially take place.
   (b) The student acquired a ^{1}H NMR spectrum of the product, which is shown below. What should she conclude regarding alkyl rearrangements? Explain.

Chemical shift (ppm)

If your instructor assigns problems in smartwork, log in at **smartwork.wwnorton.com**.

Fragmentation Pathways in Mass Spectrometry

Chapter 16 provided an introduction to mass spectrometry. There, we learned that once a sample is injected into a mass spectrometer, the acquisition of a mass spectrum typically consists of four main stages, as shown in Figure IC2-1.

First, the sample is heated to produce gaseous molecules, $M(g)$. Second, the vaporized sample passes through an electron beam, where ionization takes place. When a high-energy electron collides with $M(g)$, an electron is ejected from the molecule, producing a molecular ion, $M^+(g)$. Third, ions are guided to the detector. Finally, the electronic signal generated by the collisions of the ions with the detector is converted into the ion's relative abundance, and the spectrum is produced.

In the ionization stage (depending on the type and parameters of the instrument), collisions between the electrons and the uncharged molecules can carry enough energy to break covalent bonds as well, resulting in fragmentation of the molecular ion. The fragment ions thus produced are also detected, which is why numerous mass peaks often appear in a mass spectrum, even for pure samples.

In Chapter 16, we interpreted mass spectra primarily to determine the molar mass of the compound that produces the spectrum. Doing so largely involves identifying the M^+ peak, as well as various $M + 1$ and $M + 2$ isotope peaks. Now we will learn how to interpret additional structural information from the specific fragment peaks in a mass spectrum.

Interpreting fragment peaks requires an understanding of the chemical processes responsible for producing the corresponding fragment ions. Notably, these **fragmentation pathways** consist of elementary steps that involve free radicals. The ionization process causes the loss of a single electron from a closed-shell, uncharged molecule, so the molecular ion must have an unpaired electron, making it a free radical. More specifically, it also carries a positive charge, so the molecular ion is a **radical cation**, depicted as $M^{+\cdot}$.

Because our focus in this interchapter is on gathering structural information about a compound, our discussion is divided into sections according to the functional groups involved. We begin by discussing the fragmentation of alkanes, which contain no functional groups at all. We then discuss functional groups containing double bonds, including alkenes and aromatic rings. Following that, we discuss functional groups with heteroatoms consisting of only single bonds, including alkyl halides, amines, ethers, and alcohols. Finally, we discuss functional groups containing carbonyl groups, including ketones, aldehydes, and carboxylic acids.

FIGURE IC2-1 How an electron ionization mass spectrum is produced (1) A sample, $M(\ell)$ is vaporized to $M(g)$. (2) $M(g)$ is ionized by an electron beam, producing $M^+(g)$. With enough energy from the electron beam, $M^+(g)$ will fragment into smaller ions. (3) The various ions are guided to the detector. (4) The signal generated by the detector is converted into a mass spectrum.

IC2.1 Alkanes

Let's begin by examining the mass spectrum of hexane (Fig. IC2-2), which we discussed previously in Section 16.17.

FIGURE IC2-2 Mass spectrum of hexane The M⁺ peak and the base peak are indicated. The values of *m/z* for selected peaks are indicated above the peaks.

The M⁺ peak is identified at 86 u, which corresponds to hexane's molecular mass. Because the molecular ion is produced by the loss of an electron from one of the many σ bonds, the molecular ion is ambiguously represented by placing square brackets around the parent molecule, along with a dot and a positive charge to indicate a radical cation (Equation IC2-1).

$$\text{(IC2-1)}$$

Several key fragment peaks can also be identified—at 71, 57, 43, 29, and 15 u—which, as we saw in Section 16.17, correspond to the breaking of the various C—C bonds in hexane. For example, the base peak (i.e., the most intense mass peak) at 57 u corresponds to the breaking of the C2—C3 bond to produce both a butyl cation and an ethyl radical (Equation IC2-2).

The uncharged radical is not detected.   **The fragment ion is detected.**

$$\text{(IC2-2)}$$

Ethyl radical   **Butyl cation** *m/z* = 57

The peak at 15 u corresponds to the breaking of the C1—C2 bond to produce a methyl cation and a pentyl radical (Equation IC2-3).

The fragment ion is detected.   **The uncharged radical is not detected.**

$$\text{(IC2-3)}$$

Methyl cation *m/z* = 15   **Pentyl radical**

Because the intensity of the peak at 57 u is substantially more intense than the one at 15 u, we know that the fragmentation pathway in Equation IC2-2 is much more likely than the one in Equation IC2-3. Why should this be so?

The answer has to do with the stability of the fragmentation products. In each case, both a cation and an alkyl radical are produced. Cation stability favors Equation IC2-2, because the positively charged C is more highly alkyl substituted (Section 6.6e). Radical stability favors Equation IC2-3, however, because the C atom bearing the unpaired electron is more highly alkyl substituted. Yet the cation stability wins out, as cations are more electron-deficient.

> In general, the more stable a fragment ion is, the more likely is the fragmentation pathway that produces that fragment ion.

This general rule can be very useful, particularly when distinguishing isomers by mass spectrometry. Consider 2-methylpentane, whose mass spectrum is shown in Figure IC2-3. The $M^{+\bullet}$ peak appears at 86 u, the same as in the mass spectrum of hexane. The base peak for 2-methylpentane is at 43 u, however, whereas the one for hexane is at 57 u. The fragmentation that accounts for the base peak at 43 u is shown in Equation IC2-4. It is favored because the cation that is produced is secondary.

The secondary carbocation is more stable than a primary or methyl carbocation.

Isopropyl cation
m/z = 43

Propyl radical

(IC2-4)

problem **IC2.1** The mass peak in Figure IC2-3 at 71 u is more intense than the one at 29 u. Explain why, accounting for the fragment ions that correspond to those peaks.

Base peak = 43

$M^{+\bullet} = 86$

FIGURE IC2-3 **Mass spectrum of 2-methylpentane** The values of *m/z* for the molecular ion and base peaks are indicated above the peaks.

Another difference between the spectra in Figures IC2-2 (hexane) and IC2-3 (2-methylpentane) is the relative sizes of the $M^{+\bullet}$ peaks. The one for 2-methylpentane is much smaller than the one for hexane. Thus, fragmentation is more likely for the molecular ion of 2-methylpentane because, with additional branching, fragment ions that are more stable can be produced. With even more branching, as in 2,2-dimethylbutane (Fig. IC2-4), the $M^{+\bullet}$ peak essentially disappears entirely.

> If a fragment ion is sufficiently stable, then the $M^{+\bullet}$ peak can essentially disappear entirely from a compound's mass spectrum.

FIGURE IC2-4 Mass spectrum of 2,2-dimethylbutane The M⁺ peak is almost entirely absent from the spectrum, because the significant branching of the alkyl chain means that fragmentation gives rise to relatively highly stable fragment ions.

The M⁺ peak almost entirely disappears.

---

problem IC2.2 Mass spectra of butylcyclopentane and *tert*-butylcyclopentane were acquired. Spectrum **A** exhibited significant mass peaks at 126, 97, 83, 69, 55, and 41 u. Spectrum **B** exhibited significant peaks at 111, 69, 57, and 41 u. Match each spectrum with its compound.

IC2.2 Alkenes and Aromatic Compounds

The π electrons of alkenes and aromatic compounds are relatively high in energy, so one of them is usually ejected in the ionization process to produce the molecular ion. An example is shown for hex-2-ene in Equation IC2-5.

Loss of an electron from the π bond

Hex-2-ene → *m/z* = 84 (IC2-5)

Because of the relatively well-defined nature of an alkene's molecular ion, fragmentation pathways tend to be more clear-cut. In fact,

> An alkene's molecular ion tends to expel an alkyl radical from an allylic carbon, producing an allylic cation.

Equation IC2-6 shows that the molecular ion of hex-2-ene can expel an ethyl radical to produce a resonance-stabilized 2-butenyl cation.

The allylic cation that is produced is resonance stabilized.

Allylic C from the original alkene

m/z = 84 → **2-Butenyl cation** *m/z* = 55 + (IC2-6)

The driving force for this process is the stability that is gained by the resonance delocalization of the positive charge in the resulting allylic cation. This fragmentation

FIGURE IC2-5 Mass spectrum of hex-2-ene The $M^{\ddagger}$ peak appears at 84 u. Loss of an ethyl radical produces a resonance-stabilized allylic cation that gives rise to the base peak at 55 u.

pathway is the most likely one for the molecular ion of hex-2-ene because the base peak in the mass spectrum of hex-2-ene (Fig. IC2-5) is found at 55 u, which corresponds to the 2-butenyl cation in Equation IC2-6.

Similar fragmentations are observed for aromatic compounds.

> The molecular ion of an alkylbenzene tends to expel an alkyl radical from a benzylic carbon, producing a benzylic cation.

The mass spectrum of butylbenzene is shown in Figure IC2-6. The molecular ion, produced by the loss of a π electron, gives rise to the peak at 134 u. Loss of a propyl radical from the benzylic carbon (Equation IC2-7) produces the benzylic cation, whose mass peak appears at 91 u.

Benzylic cation
$m/z = 91$

(IC2-7)

The benzylic cation in Equation IC2-7 is believed to rearrange to the more stable **tropylium ion**, as shown in Equation IC2-8.

$m/z = 91$ → Rearrangement → **The tropylium ion** $m/z = 91$ (IC2-8)

FIGURE IC2-6 Mass spectrum of butylbenzene The $M^{\ddagger}$ peak appears at 134 u. Loss of a propyl radical produces a resonance-stabilized benzylic cation that gives rise to the base peak at 91 u.

The additional stability is attributed to the greater charge delocalization in the tropylium ion due to its seven equivalent resonance structures, all while maintaining aromaticity in the ring.

---

problem **IC2.3** Draw all resonance structures and the resonance hybrid of the tropylium ion.

---

problem **IC2.4** A peak appears at 69 u in the mass spectrum of hex-3-ene. Show how the fragmentation of hex-3-ene's molecular ion produces the ion that gives rise to this peak.

---

problem **IC2.5** Mass spectra were acquired for 1,4-diethylbenzene and 1-methyl-4-propylbenzene. The base peak of spectrum **A** is at $m/z = 105$ and that of spectrum **B** is at $m/z = 119$. Match each spectrum to its compound.

IC2.3 Alkyl Halides, Amines, Ethers, and Alcohols

Many functional groups contain *heteroatoms* (i.e., atoms other than carbon or hydrogen), which can play major roles in fragmentation pathways, and, therefore, help govern the fragmentation patterns observed in a mass spectrum. In mass spectrometry, one of the most important features of a heteroatom is its lone pair of electrons. As nonbonding electrons, lone pairs are typically the least tightly bound electrons in the molecule. Therefore,

> In the ionization of a molecule containing a heteroatom, a lone-pair electron is generally the most likely electron removed to produce the molecular ion, $M^{+\cdot}$.

Loss of a nonbonding electron from the Cl atom in 2-chloro-2-methylbutane produces the molecular ion, with $m/z = 106$, as shown in Equation IC2-9.

The peak corresponding to $M^{+\cdot}$ does not appear in the spectrum in Figure IC2-7, however, because fragmentation takes place very readily. One fragmentation path-

FIGURE IC2-7 Mass spectrum of 2-chloro-2-methylbutane The $M^{+\cdot}$ peak, which would appear at 106 u, is absent. The masses of other significant fragment ions are labeled above their corresponding peaks.

way, shown in Equation IC2-9, is heterolysis of the C—Cl bond, producing an alkyl cation fragment with $m/z = 71$.

Another common fragmentation pathway is elimination of an alkyl group from the carbon atom bonded to the heteroatom—a process known as **α-cleavage**. There are two α-cleavage pathways for the molecular ion of 2-chloro-2-methylbutane, as shown in Equations IC2-10 and IC2-11.

These account for the mass peaks at 91 and 77 u. Notice that in the resonance structure shown for each fragment ion, *all atoms have an octet*, which provides substantial driving force for the process.

These two fragmentation pathways—heterolysis and α-cleavage—are also common to other functional groups with heteroatoms, such as amines, ethers, and alcohols. Depending on the identity of the functional group, as well as the specific structure of the compound, one of these fragmentation pathways can be highly favored over the other.

α-Cleavage is very common for amines, typically leading to the fragment ion that corresponds to the base peak. This is the case for *N,N*,2-trimethylpropan-2-amine, whose mass spectrum is shown in Figure IC2-8. The value of m/z for the molecular

FIGURE IC2-8 Mass spectrum of *N,N*,2-trimethylpropan-2-amine The M⁺ peak appears at 101 u and is relatively small. The base peak at 86 u corresponds to a fragment ion produced by α-cleavage of the molecular ion.

ion is 101 u. Loss of a CH_3 group via α-cleavage (Equation IC2-12) produces the fragment ion giving rise to the base peak at $m/z = 86$.

FIGURE IC2-9 Mass spectrum of diisopropyl ether The M⁺ peak appears at 102 u. The peak at 87 u corresponds to a fragment ion produced by α-cleavage of the molecular ion.

Ethers and alcohols exhibit α-cleavage as well, but typically not as prominently. In the mass spectrum of diisopropyl ether (Fig. IC2-9), for example, fragmentation via α-cleavage gives rise to the mass peak at 87 u, which is significantly smaller than the base peak. Likewise, in the mass spectrum of pentan-1-ol (Fig. IC2-10), α-cleavage gives rise to the relatively small peak at 31 u.

FIGURE IC2-10 Mass spectrum of pentan-1-ol The M⁺ peak would appear at 88 u, but is absent. The peak at 87 u corresponds to a fragment ion produced by loss of a hydrogen atom. The peak at 70 u corresponds to loss of water from the molecular ion. The peak at 31 u corresponds to loss of an alkyl radical from M⁺ via α-cleavage.

---

problem **IC2.6** Draw the fragmentation pathway of pentan-1-ol's molecular ion that accounts for the mass peak at 31 u.

---

problem **IC2.7** The mass spectrum of an amine ($C_6H_{15}N$) exhibits a base peak at $m/z = 30$. Which of the following amines could give rise to that mass spectrum?

A   **B**   **C**

There are two other fragmentation pathways characteristic of an alcohol.

An alcohol's molecular ion will typically undergo
- Loss of a hydrogen atom to give rise to an $(M - H)^+$ peak that is 1 u lighter than the molecular ion.
- Loss of a water molecule to give rise to an $(M - H_2O)^+$ peak that is 18 u lighter than the molecular ion.

For example, the molecular ion of pentan-1-ol is 88 u, but there is a small $(M - H)^+$ peak at 87 u in Figure IC2-10, which is lighter than the molecular ion by 1 u. This is accounted for by the α-cleavage mechanism in Equation IC2-13, in which H• is expelled. There is a prominent peak at 70 u, too, which is lighter than the molecular ion by 18 u, and corresponds to the $(M - H_2O)^+$ ion. The curved arrow notation for this step is shown in Equation IC2-14.

$$m/z = 88 \quad \xrightarrow{\alpha\text{-Cleavage}} \quad m/z = 87 \quad + \quad \bullet H \quad \text{(IC2-13)}$$

$$m/z = 88 \quad \longrightarrow \quad m/z = 70 \quad + \quad H_2\ddot{O}\colon \quad \text{(IC2-14)}$$

IC2.4 Carbonyl-Containing Compounds

Similar to the functional groups discussed in Section IC2.3, carbonyl-containing groups, such as ketones, aldehydes, and carboxylic acids, have an oxygen atom with lone pairs of electrons. As a result, these compounds undergo ionization and fragmentation processes similar to the ones that alkyl halides, amines, and ethers undergo. Namely,

- The molecular ion of a carbonyl-containing compound is typically produced by loss of a nonbonding electron from the carbonyl oxygen.
- A common fragmentation pathway of this molecular ion is α-cleavage.

Equation IC2-15, for example, shows that hexan-2-one undergoes ionization to produce the molecular ion with $m/z = 100$ (Fig. IC2-11).

Hexan-2-one → Loss of nonbonding e⁻ → $m/z = 100$ → α-Cleavage → An acylium ion + (IC2-15) $m/z = 43$

Subsequent α-cleavage can expel a butyl radical from the carbonyl carbon to produce a relatively stable *acylium ion*, giving rise to the base peak at 43 u. (This is the same

FIGURE IC2-11 **Mass spectrum of hexan-2-one** The M⁺ peak appears at 100 u. The peaks at 43 and 85 u correspond to fragment ions produced upon α-cleavage of the molecular ion. The mass peak at 58 u corresponds to a fragment ion produced by a McLafferty rearrangement.

type of ion that appears as an intermediate in a Friedel–Crafts acylation reaction; Section 22-5.) Alternatively, α-cleavage of a methyl radical produces the ion corresponding to the peak at 85 u.

---

problem **IC2.8** Show the fragmentation pathway that gives rise to the peak at 85 u in the mass spectrum of hexan-2-one (Fig. IC2-11).

---

problem **IC2.9** The mass spectrum of a ketone ($C_7H_{14}O$) exhibits a base peak at $m/z = 57$. Is the ketone most likely heptan-2-one, heptan-3-one, or heptan-4-one? Explain.

In addition to α-cleavage, there is another fragmentation pathway characteristic of carbonyl-containing compounds.

> A carbonyl-containing compound can undergo a **McLafferty rearrangement** if an alkyl group attached to the carbonyl carbon possesses a γ-carbon with at least one hydrogen atom.

This fragmentation pathway is named after Professor Fred W. McLafferty (b. 1923). As shown in Equation IC2-16, a H atom from the γ-carbon shifts to the carbonyl O atom, and the bond joining the α and β carbons is broken.

$$m/z = 100 \qquad \text{An enol radical cation} \quad m/z = 58$$

(IC2-16)

The result is the ejection of an uncharged alkene molecule. In the case of hexan-2-one's molecular ion, the McLafferty rearrangement produces an enol radical cation of acetone, giving rise to the mass peak at 58 u.

The McLafferty rearrangement is not limited to just ketones; it is characteristic of several functional groups containing the carbonyl group, including aldehydes, carboxylic acids, esters, and amides. For example, the molecular ion of pentanoic acid undergoes a McLafferty rearrangement, giving rise to the mass peak at 60 u in Figure IC2-12.

FIGURE IC2-12 **Mass spectrum of pentanoic acid** The mass peak at 60 u corresponds to a fragment ion produced by a McLafferty rearrangement.

---

problem **IC2.10** The base peak in the mass spectrum of hexanamide appears at 59 u. Draw the ion that corresponds to this mass peak, and show how it is produced from the molecular ion.

CHAPTER 26

Polymers

I n Chapter 25, we studied free radicals, with particular emphasis on the mechanism of chain reactions. Free radical chain reactions allow alkanes, which are unreactive under normal laboratory conditions, to undergo halogenation. We also saw that the free radical chain reaction of HBr with alkenes increases a chemist's control of regioselectivity, because the reaction provides the product of the anti-Markovnikov addition of HBr.

Here in Chapter 26, we apply the free radical chain reaction to the synthesis of *polymers*. **Polymers** are large molecules formed from smaller molecules. Trademarked products such as Plexiglas, Kevlar, and Teflon are all polymers with special properties that make them desirable for specific uses, as shown in Figure 26-1.

The uses of polymers are determined principally by their physical properties, many of which depend upon the intermolecular interactions present (Chapter 2) and the range of motion allowed by rotation about their single bonds (Chapter 4). We will see a variety of reactions from different chapters (particularly Chapters 21 and 25) that allow chemists to synthesize polymers and subsequently alter polymer molecules to introduce desirable properties. We begin Chapter 26 with an examination of polystyrene, a model for free radical polymerization and a polymer you are familiar with, and then study the relationship between the structure of polymer molecules and their properties.

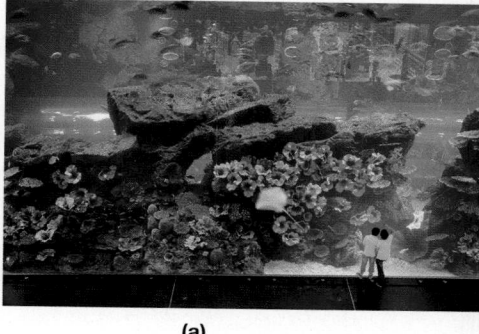

(a)

(b)   (c)

FIGURE 26-1 Examples of commercially available polymers (a) The Dubai Mall aquarium walls are made of Plexiglas. (b) Body armor made from Kevlar fiber. (c) A nonstick skillet with a Teflon lining. Teflon is the only known substance that geckos cannot climb (see p. 76).

Microneedles of poly(*N*-vinyl pyrrolidone) (PVP) are used to encapsulate vaccines and offer an alternative method for the administration of vaccines and medications. PVP is water soluble and biocompatible; the polymer dissolves and releases the vaccine after the needles pierce the skin. Applications for the administration of influenza, measles, and other vaccines are under development. Here in Chapter 26, we describe how polymers such as PVP are made and how their properties and uses are related.

(a) **(b)**

FIGURE 26-2 Uses of styrofoam (a) Styrofoam cups for hot beverages and (b) styrofoam "peanuts" for preventing damage to fragile objects during shipping.

26.1 Polystyrene: A Synthetic Polymer

Styrofoam (Fig. 26-2) is a foam made from the synthetic polymer polystyrene. It is used in "packing peanuts" and in insulated, disposable coffee cups. Polystyrene has been commercially available since the 1930s.

Polystyrene was discovered in the 19th century when several chemists found that styrene, a liquid at room temperature, can produce polystyrene, which was originally obtained as a jelly-like substance (Equation 26-1).

A monomer

A polymer

The molecule continues in both directions.

(26-1)

Styrene

Polystyrene

Only a partial structure of polystyrene is shown; the dotted lines indicate that the structure continues in both directions. A single molecule of polystyrene is so large, in fact, that it is difficult to depict on a single page.

26.1a The Structure of Styrene and Polystyrene

Equation 26-1 shows how polystyrene, the polymer, consists of many units of styrene, the **monomer**, held together by covalent bonds. That is, styrene is the single part (*mono* + *mer*) that produces a large molecule with many parts (*poly* + *mer*). Because of the particular way that the monomers are linked, polystyrene forms a long, continuous chain of carbons, called the **polymer chain**, **main chain**, or **backbone**. In Figure 26-3, the polymer backbone is screened in red and the phenyl rings, shown in blue, are attached to the polymer backbone.

The phenyl rings can be described as substituents (the phenyl ring is "substituted" for a hydrogen, the default atom attached to a carbon), or **side groups**, also called side chains or pendant groups. Pendant means "hanging," so it appropriately describes the phenyl groups that are hanging off the polymer chain.

The *polymer chain*, or *polymer backbone*

The phenyl groups are *substituents* attached to the polymer chain.

Polystyrene

26.1 YOUR TURN

Identify the side chains in the following partial structure of polypropylene.

What adjective, similar to phenyl, can you use to describe the side groups?

Answers to Your Turns are in the back of the book.

You can see the relationship between styrene and polystyrene more clearly by mentally cutting the main chain of the polymer into two-carbon units, as shown on the left in Figure 26-4. This gives you the *repeating unit* of the polymer, as shown on the right of the figure.

The *repeating unit* of polystyrene

Same as:

The *degree of polymerization*

Polystyrene

FIGURE 26-4 The repeating unit of polystyrene Mentally cutting the backbone of polystyrene, as indicated on the left, produces the repeating unit shown on the right. The value of *n* is called the *degree of polymerization*.

■ The **repeating unit** is the group of atoms that occurs over and over (repeatedly) in a polymer.

■ The number of repeating units determines the **degree of polymerization (DP)**.

Using the notation on the right of Figure 26-4, DP is the value of the subscript n.

The monomer and the repeating unit are similar but not identical. One skill you need to develop is the ability to identify the relationship between monomers and the repeating unit of the corresponding polymers. In the case of polystyrene (Fig. 26-5), the double bond between two C atoms in the monomer (i.e., the *vinyl group*) has disappeared in the repeating unit, and an additional single bond to each C atom has appeared. As we explain later, this is because the π electrons of the $C{=\!=}C$ double bond are used to join the monomers together.

Many polymers have been made from vinyl monomers like styrene. The relationship between a vinyl monomer and the corresponding polymer (Fig. 26-5), therefore, applies to a large number of polymers.

FIGURE 26-5 Monomers versus repeating units The double bond that makes up the vinyl group in (a) styrene, the monomer, does not appear in the repeating unit of (b) polystyrene, the polymer. Each C atom along the polymer backbone gains a single bond.

YOUR TURN 26.2

Given the following partial structure of polypropylene, draw its repeating unit and provide the appropriate value of *n* based on the atoms shown.

$$\cdots-\underset{H}{\overset{H}{C}}-\underset{CH_3}{\overset{H}{C}}-\underset{H}{\overset{H}{C}}-\underset{CH_3}{\overset{H}{C}}-\underset{H}{\overset{H}{C}}-\underset{CH_3}{\overset{H}{C}}-\cdots$$

problem 26.1 Given the repeating units at the right for *n* = 3, draw the structure of each polymer, similar to the representation in Your Turn 26.2.

$$\left(\underset{F}{\overset{F}{\underset{|}{C}}}-\underset{F}{\overset{F}{\underset{|}{C}}}\right)_{n}$$ (a)

$$\left(\underset{H}{\overset{H}{\underset{|}{C}}}-\underset{Cl}{\overset{Cl}{\underset{|}{C}}}\right)_{n}$$ (b)

SOLVED problem 26.2 Poly(vinyl chloride) (PVC) is used for plumbing materials (e.g., pipe) and has the following partial structure:

$$\cdots-\underset{H}{\overset{H}{C}}-\underset{Cl}{\overset{H}{C}}-\underset{H}{\overset{H}{C}}-\underset{Cl}{\overset{H}{C}}-\underset{H}{\overset{H}{C}}-\underset{Cl}{\overset{H}{C}}-\cdots$$

Based on the partial structure of the polymer, what is the structure of vinyl chloride?

Think What is the repeating unit of PVC? How is the repeating unit of a vinyl polymer related to the monomer?

Solve The C atoms in the structure make up the main chain of the molecule. The Cl atoms are substituents, or side groups, analogous to the phenyl rings in polystyrene. The repeating unit is as follows:

$$\cdots-\underset{H}{\overset{H}{C}}-\underset{Cl}{\overset{H}{C}}\sim\underset{H}{\overset{H}{C}}-\underset{Cl}{\overset{H}{C}}\sim\underset{H}{\overset{H}{C}}-\underset{Cl}{\overset{H}{C}}-\cdots$$ Same as: $$\left(\underset{H}{\overset{H}{\underset{|}{C}}}-\underset{Cl}{\overset{H}{\underset{|}{C}}}\right)_{n}$$

To determine the monomer from the repeating unit, remove the single bonds on the outside of the repeating unit and change the C—C single bond to a C═C double bond, as shown at the right.

$$\underset{H}{\overset{H}{\diagdown}}C{=}C\underset{Cl}{\overset{H}{\diagup}}$$

problem **26.3** Poly(methyl methacrylate), or Plexiglas, has the following partial structure:

Plexiglas can be made from a vinyl monomer. What is the structure of the monomer?

FIGURE 26-6 The benzoyloxyl radical Resonance delocalization of the unpaired electron in the benzoyloxyl radical facilitates the homolysis of benzoyl peroxide.

26.1b Free Radical Polymerization of Styrene

Manufacturers create polystyrene from styrene by means of **free radical polymerization**, which is a type of free radical chain reaction. An example is shown in Equation 26-2.

The mechanistic steps of this reaction are identical to the initiation, propagation, and termination steps you studied in Section 25.4a. The *initiation* step is needed to create free radicals. Monomers such as styrene are usually incapable of producing free radicals readily, so we use a radical initiator (see Section 25.1). Benzoyl peroxide, drawn over the reaction arrow in Equation 26-2, is a common initiator. Because oxygen–oxygen single bonds are particularly weak (see Table 25-3), heating benzoyl peroxide results in the homolysis of the bond between the two O atoms (Equation 26-3):

Benzoyl peroxide   **The benzoyloxyl radical**

The homolysis of benzoyl peroxide is further promoted by the additional resonance stabilization in the resulting radical, as shown in Figure 26-6.

Draw in the curved arrows that show the movement of electrons in Equation 26-3. Do the same for the conversion of the top resonance structure in Figure 26-6 into the bottom structure.

26.3 YOUR TURN

After initiation, *propagation* occurs via radical addition (Section 25.4a). A radical, such as the benzoyloxyl radical produced directly from the initiator, adds to a styrene molecule, as shown in Equation 26-4.

Free radical from initiator   **Monomer**   **Reactive polymer chain (*n* = 1)**

What occurs next explains how the polymer chain grows—namely, the radical on the product side of Equation 26-4 adds to another molecule of styrene, increasing the length of the carbon chain, as shown in Equation 26-5.

Reactive polymer chain (n = 1)   Monomer   Reactive polymer chain (n = 2)

(26-5)

This radical can continue to propagate, adding to a third styrene molecule (Equation 26-6).

Reactive polymer chain (n = 2)   Monomer   Reactive polymer chain (n = 3)

(26-6)

YOUR TURN 26.4

Add the curved arrows to, and draw the product of the next step in the polymerization reaction:

?

The synthesis of polystyrene occurs by **chain-growth polymerization** because individual sites (in this case, the initiating radicals) start the formation of a large chain. This reaction is also described as **addition polymerization**. Growth of polymer chains is depicted schematically in Figure 26-7, where the series of diagrams shows how propagation creates chains over time. The diagram begins with unreacted starting material in (a), proceeds to initiation in (b), and then shows how propagation and initiation continue in (c) through (e).

Propagation could continue, in theory, until there are no remaining molecules of monomer. In reality, though, the growth of the chain usually ends via *termination*.

Free radical polymerization can terminate by *combination* or *disproportionation*.

FIGURE 26-7 Growth of polymer chains during the polymerization of styrene (a) A solution of styrene (red dots) and benzoyl peroxide (blue squares) prior to heating. (b) During heating, the benzoyl peroxide molecules undergo homolysis (initiation) and propagation. The gray lines represent bonds formed between an initiated radical and a styrene molecule. (c) Propagation continues for each growing chain. Initiation continues, but at a slower rate than propagation. (d) Propagation continues and each chain increases in length. Benzoyl peroxide molecules that remain also initiate chain growth. (e) Later in the reaction, initiation is minimal. Propagation continues and polymer chains grow longer.

Combination, shown in Equation 26-7, is another term for radical coupling (see Section 25.3a).

Combination

$$(26\text{-}7)$$

Disproportionation, shown in Equation 26-8, describes any reaction in which two identical species react to produce two different products.

Disproportionation

The radical in black abstracts a H atom from the radical in red.

A π bond was formed from electrons that were on adjacent atoms.

(26-8)

In the case of polystyrene, one of the radicals can abstract a hydrogen atom from another radical (see Section 25.3b). Because the hydrogen abstraction occurs at a carbon adjacent to a free radical, a double bond forms in the species that gives up the hydrogen atom. In this regard, this reaction is analogous to the second step of an E1 reaction (see Section 8.2).

Both combination and disproportionation produce closed-shell species from free radicals, so these steps are typically quite rapid. Which step dominates can depend on the nature of the monomer and the reaction conditions.

Figure 26-8 shows how a particular polystyrene molecule might appear after a combination step. Two growing polymer chains—one with n repeating units and the other with m repeating units—join to produce a longer chain with $n + m$ total repeating units. Figure 26-9 shows the two polystyrene molecules that might be produced after disproportionation.

Two growing radicals combine.

Combination

End groups

This carbon–carbon bond forms in the interior of the chain.

End groups

FIGURE 26-8 Termination via combination (Top) Two growing polystyrene chains meet and the unpaired electrons join to produce a new carbon–carbon bond. One growing chain has n repeating units, and the other has m. (Bottom) The terminated polystyrene molecule has a total of $n + m$ repeating units, and has two benzoate end groups (blue).

(a)

(b)

FIGURE 26-9 Termination via disproportionation A disproportionation involving two growing polystyrene chains might produce these two polystyrene molecules. Both have a benzoate end group on one end. On the other end, one molecule (a) is capped by H and the other (b) is capped by a vinyl group.

Repeating units in decane

End groups in decane

(a)

Repeating units in decane

$$CH_3(CH_2)_8CH_3$$

End groups in decane

(b)

FIGURE 26-10 End groups in decane Decane represented as (a) a Lewis structure and (b) a condensed formula. Decane can be viewed as having a chain of eight repeating CH_2 units, capped by two CH_3 end groups.

Notice in Figure 26-8 that when the polymer is terminated by combination, each end is capped by a benzoate group—called an **end group**—because each growing polymer chain was initiated by a benzoyloxyl radical. This idea is similar to viewing a molecule of decane (Fig. 26-10) as having a chain made of repeating CH_2 groups, capped on each end by a CH_3 end group.

When polystyrene is terminated by disproportionation (Fig. 26-9), on the other hand, each polymer molecule should have only one benzoate end group. The other end is capped by either a hydrogen atom or by a vinyl group.

Unlike in a molecule of decane, the end groups of a polymer are variable and contribute little to the polymer's overall structure (two units out of >1900 in a typical molecule in commercial polystyrene). Therefore,

> Only the repeating unit of a polymer is used to evaluate the polymer's structure.

Consequently, most representations of polymers omit the end groups.

26.1c Other Structural Aspects of Polystyrene

The goal of most organic syntheses is to make a single product in high yield. In the synthesis of a polymer like polystyrene, however, this is difficult or impossible due to the variations that exist in the chain-growth mechanism. As we learned in Section 26.1b, for example, the lengths of the chains that are produced cannot be precisely controlled. Instead, a polymer is a mixture (a distribution) of chains of different lengths and can be described with an *average* molar mass or, alternatively, an average degree of polymerization. Commercial grade polystyrene, for example, has an average molar mass of >200,000 g/mol (i.e., an average degree of polymerization of >1920).

A polymer sample can be characterized further using **gel permeation chromatography** (**GPC**), which works by size exclusion. Polymer samples in solution pass

FIGURE 26-11 Gel permeation
chromatogram of a polystyrene
sample with average molar mass
≈ 24,000 g/mol (DP ≈ 230)
The magnitude of the signal on the y
axis is proportional to the number of
molecules that elute at a particular time.
The signal can be used to determine the
proportion of polymer molecules with
a particular molar mass. The elution
time on the x axis can be correlated
with molar mass. Standard samples of
polymers are run through the column
to calibrate the column. These data are
used to determine the molar mass of
molecules that elute at a particular time.

through a column with a porous gel. Smaller polymer molecules spend more time
interacting with the gel, so they elute more slowly and are detected at later times.
Consider Figure 26-11, for example, which shows the results of GPC of a sample
of polystyrene with an average molar mass of 24,000 g/mol, or an average degree of
polymerization of 230. Molecular size varies widely, from ~2000 g/mol (DP ≈ 19)
to ~520,000 g/mol (DP ≈ 5000) and the sample consists of two separate distributions, one centered at ~8000 g/mol (DP ≈ 77) and the other at ~130,000 g/mol
(DP ≈ 1250).

Another way in which polymer molecules can vary has to do with the way each
monomer links together in the propagation steps of the chain-growth mechanism. To
differentiate between the two carbons in the vinyl group, the less-substituted carbon
in the double bond is called the *tail* and the more-substituted carbon is called the
head. As a result, there are two possible ways the radical can add to the vinyl group of
styrene: **head-to-tail** (Equation 26-9) and **head-to-head** (Equation 26-10).

Head-to-tail addition

(26-9)

Head-to-head addition

(26-10)

The head-to-tail addition (i.e., Equation 26-9) is favored in almost all free radical
polymerizations, for the two reasons shown in Figure 26-12. The first reason is kinetic.

Head-to-tail addition

Less steric repulsion

This radical is stabilized by resonance.

(a)

Head-to-head addition

Steric repulsion between the two phenyl groups

This radical cannot be stabilized by resonance.

(b)

FIGURE 26-12 Head-to-tail and head-to-head addition in the polymerization of styrene (a) Head-to-tail addition is favored both kinetically (less steric repulsion) and thermodynamically (the product is resonance stabilized) over (b) head-to-head addition.

Steric repulsions between the two phenyl groups (one at the end of the propagating radical and one in the monomer) make it less likely that the two species will success-fully collide head to head. Second, the radical formed in the head-to-tail addition is usually favored thermodynamically. In the formation of polystyrene, for example, the radical produced from head-to-tail addition is resonance stabilized, while the radical produced from head-to-head addition is not.

Both thermodynamics and kinetics usually favor the head-to-tail addition of vinyl monomers in a free radical polymerization.

26.5 **YOUR TURN**

Use curved-arrow notation to show the movement of electrons in the head-to-tail addition of a growing chain of poly(ethyl acrylate) to a molecule of ethyl acrylate. Draw resonance structures to explain why head-to-tail addition occurs instead of head-to-head addition.

Propagating chain of poly(ethyl acrylate)   **Ethyl acrylate**   **Product of head-to-tail addition**

problem **26.4** Crotonic acid polymer-izes much more slowly than acrylic acid. How do you explain the difference in reactivity?

Crotonic acid   **Acrylic acid**

26.2 General Aspects of Polymers

The terms and concepts that were applied to the free radical polymerization of styrene in Section 26.1 apply equally well to numerous other polymers and polymerizations. For example, all polymers are created from monomers and contain repeating units, but different polymers fall into different classifications. Here in Section 26.2, we learn how to classify polymers and how to name simple ones.

26.2a Polymers, Monomers, and Repeating Units

The repeating unit of a polymer may contain all or only a portion of the atoms in the monomer, depending on how the monomers combine to form the polymer. For example, when Wallace Carothers (1896–1937) synthesized nylon-6,6 in 1935, he combined adipic acid (hexane-1,6-dioic acid) and hexane-1,6-diamine, as shown in Equation 26-11.

Hexane-1,6-dioic acid (Adipic acid) **Hexane-1,6-diamine** (26-11)

Nylon-6,6

Based on the structures of the starting materials and the product, all of the carbon and nitrogen atoms in the monomers are incorporated into the repeating unit, but some of the oxygen and hydrogen atoms are missing. That is because H_2O is produced as a by-product.

The production of a second product is not unusual in polymerizations. The product is most frequently water, but there are other possibilities, too (see Your Turn 26.6).

YOUR TURN 26.6

What is the other product of the following reaction, which produces poly(ethylene terephthalate)? That is, what is a plausible formula for a product that accounts for the atoms in the reactants that do not appear in the repeating unit?

Dimethyl terephthalate   **Poly(ethylene terephthalate)**

---

problem **26.5** Nylon-6,6 (see Equation 26-11) can also be synthesized from adipoyl chloride and hexane-1,6-diamine:

What are the two products of this reaction?

26.2b Classes of Polymers

Polymers can be classified according to the following criteria:

- Type of chain
- Functional group(s) in the chain
- Monomer(s) used to make the polymer
- Number of different monomers used

When polymers are classified by type of chain, we distinguish between polymers with only carbons in the main chain (**carbon-chain polymers**) and those with heteroatoms in the chain (**heterochain polymers**). Polystyrene (Equation 26-1, p. 1256) is a carbon-chain polymer, whereas nylon (Equation 26-11) is a heterochain polymer.

A more specific way to classify polymers is by functional group. **Poly(ethylene terephthalate)** (**PET**), for example, which is used to make bottles for carbonated beverages, is a **polyester** because the main chain contains multiple ester groups. This is shown in Figure 26-13, both for a portion of the PET molecule (Fig. 26-13a) and its repeating unit (Fig. 26-13b). Notice that the repeating unit for PET contains two ester groups; the one on the left is a complete ester group, and the one on the right is completed by a carbon atom from an adjacent repeating unit.

The presence of multiple ester groups makes PET a polyester.

$$\cdots-CH_2CH_2-O-\overset{\overset{\displaystyle O}{\|}}{C}-\underset{}{\bigcirc}-\overset{\overset{\displaystyle O}{\|}}{C}-O-CH_2CH_2-O-\overset{\overset{\displaystyle O}{\|}}{C}-\underset{}{\bigcirc}-\overset{\overset{\displaystyle O}{\|}}{C}-O-CH_2CH_2-O-\overset{\overset{\displaystyle O}{\|}}{C}-\underset{}{\bigcirc}-\overset{\overset{\displaystyle O}{\|}}{C}-O-\cdots$$

(a)

A carbon atom from another repeating unit completes this ester group.

$$\left(-CH_2CH_2-O-\overset{\overset{\displaystyle O}{\|}}{C}-\underset{}{\bigcirc}-\overset{\overset{\displaystyle O}{\|}}{C}-O-\right)_n$$

(b)

FIGURE 26-13 Poly(ethylene terephthalate) or PET (a) A portion of PET and (b) its repeating unit. PET is a polyester because of the repeating ester groups in the main chain, screened in red.

26.7 YOUR TURN

> Based on its functional groups, to what class of polymers does nylon-6,6 (Equation 26-11) belong? How many times does the functional group appear in the repeating unit?

Carbon-chain polymers are usually classified according to the monomer used to make them. For example, the polymers in Figure 26-14 are classified as **polyacrylates** or **acrylics** because the monomers are derivatives of acrylic acid.

(a) Monomers

Acrylic acid · Methyl methacrylate · Acrylonitrile

(b) Polymers

Poly(acrylic acid) · Poly(methyl methacrylate) · Poly(acrylonitrile)

FIGURE 26-14 Polyacrylates (a) These acrylic acid derivatives can polymerize to form (b) the corresponding polyacrylate (or acrylic) polymers.

Alkenes are commonly referred to as **olefins**, so polyethylene, polypropylene, and other polymers made from alkenes or dienes are referred to as **polyolefins**. Some examples are shown in Figure 26-15.

FIGURE 26-15 Polyolefins
(a) These alkene (or olefin) monomers can polymerize to form (b) the corresponding polyolefins.

Polymers such as polystyrene and polyethylene are said to be **homopolymers** because each is synthesized from one monomer (i.e., styrene or ethylene, respectively). Other homopolymers are shown in Table 26-1 for comparison.

TABLE 26-1 A Comparison of Homopolymers and Their Descriptions

Monomer(s)	Polymer	Carbon–Chain or Heterochain Polymer	Description Based on Monomer	Description Based on Functional Group in Polymer
Methyl vinyl ether (Methoxyethene)	Poly(methyl vinyl ether)	Carbon–chain	A poly(vinyl ether)	–
Isobutylene (2-Methylpropene)	Polyisobutylene	Carbon–chain	A polyolefin	–
1,4-Dichlorobenzene + Na_2S Sodium sulfide	Poly(phenylene sulfide)	Heterochain	–	A polysulfide

The prefix *poly* is not solely used to describe polymers; it can also be used to describe collections of similar molecules with several of a particular functional group. Figure 26-16, for example, shows a collection of polycarboxylic acids, which are not polymers. In this case, the prefix refers to pendant groups, and not the monomer used to make a polymer or to the functional group in a polymer chain.

FIGURE 26-16 A series of benzene polycarboxylic acids Each molecule has the same core or backbone (benzene) but has multiple carboxylic acid groups attached. Since the number varies from one molecule to another, the term *poly* is used to indicate that each member of the series has more than one carboxylic acid group. None of these molecules is a polymer because none is formed by combining monomers to create a repeating unit.

SOLVED problem 26.6 Tannins are biomolecules found in plants and are components of tea and wine; they contribute significantly to the taste of each drink.

Although the tannins found in beverages are relatively small compared to most polymers, a tannin can be referred to as a polyphenol or as a polyester.

Evaluate the structure of tannin and explain how each term applies. Which term applies to the polymeric nature of tannin?

Think What functional groups appear in the tannin? Which is the functional group that joins monomers together?

Solve The phenol groups (shown in blue) are on the exterior of the molecules and the ester groups (shown in red) are in the interior, connecting portions of the molecule together.

The term *polyester* refers to the polymeric nature of the tannin and describes the functional group in the polymer chain. The tannin can be viewed as a polymer of gallic acid:

Gallic acid

The carboxylic acid on one gallic acid molecule reacts with an –OH of another molecule to form an ester.

The term *polyphenol* refers to the many phenol groups present in the molecule. It does not refer to the functional group that joins monomers together. It also does not refer to the monomer that created the polymer. Although the monomer has the phenol functional group, it also contains a carboxylic acid, and *both* functional groups are necessary to form the polymer chain. Consequently, the tannin is a poly(hydroxybenzoic acid).

---

problem **26.7** Polyserine, the polymer made from the amino acid serine, can be described by several terms that begin with the prefix *poly*. Provide two different descriptions and indicate which one refers to the polymeric nature of polyserine.

Polyserine

If two different monomers are polymerized together, then the resulting polymer is called a **copolymer**. A copolymer of styrene and butadiene, called styrene–butadiene rubber (SBR), was used during World War II as a suitable replacement for natural rubber, which was in limited supply. The formation of SBR is shown in Equation 26-12.

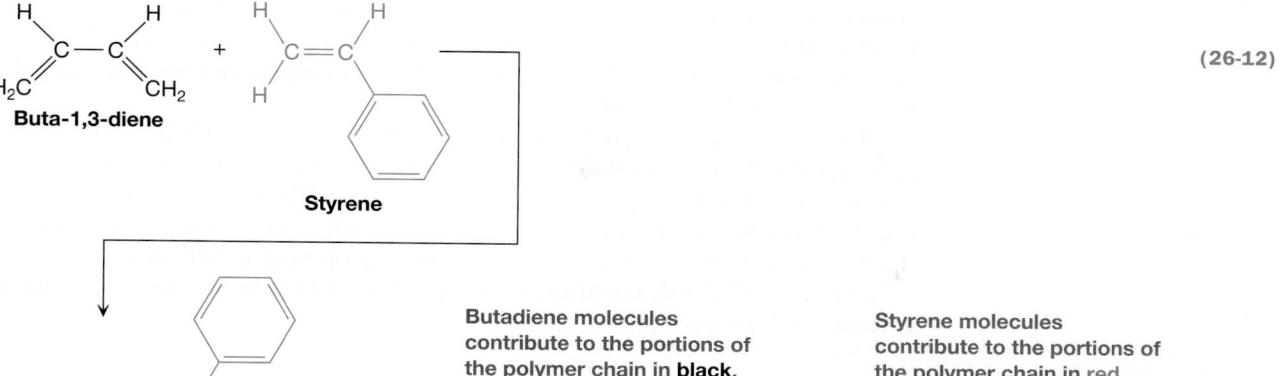

Buta-1,3-diene

Styrene

(26-12)

Butadiene molecules contribute to the portions of the polymer chain in **black**.

Styrene molecules contribute to the portions of the polymer chain in red.

Styrene–butadiene rubber (SBR)
A copolymer of buta-1,3-diene and styrene

Copolymers are possible when each monomer is capable of forming a homopolymer. For example, buta-1,3-diene is capable of forming a homopolymer (polybutadiene), as is styrene (polystyrene). The reaction of the two different monomers creates a copolymer. Nylon-6,6 (see Section 26.2a), on the other hand, is not a copolymer. Although two reactants are used to synthesize nylon-6,6, neither adipic acid nor hexane-1,6-diamine is capable of forming a polymer in the absence of the other reactant.

26.2c Linear Polymers, Branched Polymers, Network Polymers, and Crosslinking

Polystyrene, nylon-6,6, and poly(ethylene terephthalate) are **linear polymers**. They are analogous to straight chain or linear alkanes, because there are no branches in the main chain of the polymer (Fig. 26-17a). **Branched polymers** are analogous to alkyl-substituted alkanes, because there are carbon chains that are not part of the main

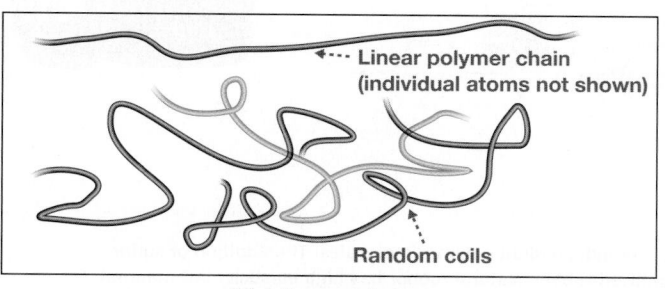

- - - - Linear polymer chain (individual atoms not shown)

Random coils

(a) A linear polymer

A branch coming off the main polymer chain

(b) A branched polymer

A crosslink between two polymer chains

(c) A crosslinked polymer

FIGURE 26-17 Branching and crosslinking in polymers (a) In a linear polymer, there is one continuous chain that contains the pendant groups. Linear polymer chains usually exist as random coils. (b) In a branched polymer, multiple smaller chains split off from the main chain. (c) In a crosslinked polymer, multiple main chains are tethered via covalent bonds.

26.2 General Aspects of Polymers / **1271**

chain of the polymer (Fig. 26-17b). Low-density polyethylene (LDPE) is a branched polymer. If branches connect two polymer chains (Fig. 26-17c), then the short connections between longer polymer chains are called **crosslinks** and the resulting polymer is said to be **crosslinked**.

Rubber provides one particularly notable application of crosslinking. In the 19th century, natural rubber had limited uses because it became brittle at low temperatures and shapeless at high temperatures. Charles Goodyear (1800–1860) discovered, however, that processing rubber at high temperatures with sulfur gave a material that had the desired elasticity of rubber but did not lose its properties at extreme temperatures. This process, called **vulcanization**, introduced crosslinks between the rubber chains, as shown in Figure 26-18.

FIGURE 26-18 Vulcanization of rubber (a) Natural rubber consists of independent polymer molecules. The addition of sulfur (S_8) results in crosslinking among the main chains to produce (b) vulcanized rubber. Natural rubber has high elasticity but minimal durability. Tires made of vulcanized rubber, on the other hand, are less elastic but much more durable.

It is also possible to introduce crosslinks as part of the polymerization process. One of the best known examples is the reaction of phenol with formaldehyde to produce the hard material known as **Bakelite** (Equation 26-13). If you compare the number of crosslinks per monomer unit in vulcanized rubber and in Bakelite, you will see that Bakelite has a much higher degree of crosslinking. It is so high, in fact, that Bakelite is referred to as a **network polymer** because there is no recognizable linearity in the polymer. As a result, a sample of Bakelite can be described as a single, *very* large molecule.

$$\text{Phenol} + \text{Formaldehyde} \xrightarrow{\Delta} \text{Bakelite} + H_2O \quad (26\text{-}13)$$

26.2d Polymer Nomenclature

Just as with other forms of chemical nomenclature, an IUPAC name can differ from its trivial or common name.

> The IUPAC name of a polymer is based on its repeating unit, but its trivial name is based on the monomer from which the polymer is made.

For example, if the repeating unit of a polymer is $-CH_2-$, then the IUPAC name is polymethylene, but the trivial (and more common) name is polyethylene, because it is made by polymerizing ethylene (Equation 26-14).

Ethylene → **Polymethylene** or **Polyethylene** (26-14)

The IUPAC name is based on the simplest repeating unit.

The common (trivial) name is based on the name of the monomer.

The IUPAC recognizes the trivial names for approximately 20 common polymers. We use these trivial names in the text because you are more likely to encounter them in everyday life and because they help establish the relationship between the polymer and the monomer from which it is made.

In general,

> - The names for homopolymers follow a polymonomer format (Fig. 26-19, p. 1275).
> - If the name of the monomer consists of two or more words, then the name of the monomer appears in parentheses in the name of the polymer.

Thus, *polystyrene*, but *poly(methyl methacrylate)*.

Stereochemistry, Polypropylene, and the Nobel Prize

Karl Ziegler (1898–1973) and Giulio Natta (1903–1979) shared the Nobel Prize in Chemistry in 1963 for developing systems that facilitated the synthesis of linear polymers made from alkenes (also called polyolefins) and led to the control of stereochemistry in polypropylene. Known as Ziegler–Natta catalysts, these systems initiate the polymerization of alkenes and provide an active site for the addition of monomer to the growing polymer chain.

The system used by Ziegler in 1953 polymerized ethylene (ethene) at low pressures (1 bar) and moderate temperatures—a vast improvement over uncatalyzed systems, which required high pressures (>1000 bar) and temperatures (>150 °C). In 1954, Natta used the system to polymerize propylene (propene) and produced isotactic polypropylene:

Note that carbons with –H and –CH$_3$ attached are stereocenters. The product is **stereoregular** because the H atoms and CH$_3$ groups occur in a regular, repeating pattern. The arrangement is **isotactic** because the methyl side groups are on the same side of the chain. This stereoregularity leads to a highly crystalline polymer with uses in products such as Rubbermaid containers for food storage and the lids for boxes of Tic Tac breath mints.

Previously, only **atactic** polypropylene had been available, in which there is no regularity in the positions of the methyl groups. Atactic polypropylene is not crystalline, has little strength as a plastic, and is limited to uses in applications such as adhesives for roofing (see photo on the right).

Besides IUPAC and trivial names, there are also trademarked names for polymers, such as Teflon and Plexiglas, but these names provide no information about the structure of the repeating unit. Some polymers are referred to by their acronyms:

> A polymer's acronym is usually derived from the prefix *poly* and the initial letters of the different portions of the monomer.

Teflon, for example, is also known as PTFE (*poly*tetra*f*luoro*e*thylene).

Some polymers use the name of the repeating unit as the basis for the name of the polymer. For example, the most common polyester is poly(ethylene terephthalate), or

Monomer	Monomer Name	Polymer Structure	Polymer Names
F₂C=CF₂ (Tetrafluoroethylene structure)	Tetrafluoroethylene	$\left(\!-CF_2-CF_2-\!\right)_n$	Polytetrafluoroethylene PTFE Teflon
Methyl methacrylate structure	Methyl methacrylate	PMMA polymer structure	Poly(methyl methacrylate) PMMA Plexiglas

The prefix *poly* is added to the name of the monomer to create the name of the polymer.

Because the monomer name is composed of two separate words, the name is placed in parentheses before the prefix *poly* is added.

The initial letters of *poly* and of the other structural features of the monomer are used to form the acronym PMMA.

FIGURE 26-19 Nomenclature of homopolymers In these examples, the name of a polymer is related to the name of the monomer.

PET (Section 26.2b). In PET, the repeating unit is named as an ester of terephthalic acid and the ethylene group (Equation 26-15).

Terephthalic acid + **Ethylene glycol** $\xrightarrow[\Delta]{\text{Acid catalyst}}$ **Poly(ethylene terephthalate)**
PET
An ester of terephthalic acid and ethylene glycol

(26-15)

problem 26.8 The derivative of acrylic acid at the right undergoes free radical polymerization to make a polymer that is water soluble. Draw the condensed formula for the polymer and provide its name and acronym.

2-Hydroxyethyl acrylate

26.3 Other Polymerization Reactions

So far, we have focused primarily on free radical polymerization of vinyl monomers. In theory, however, any reaction that forms a bond between two molecules can be used to make a polymer, involving other mechanisms and other types of monomers. Despite these variations, we will see that each polymerization reaction can be classified as either *chain-growth polymerization* or *step-growth polymerization*.

26.3a Chain-Growth Polymerization

Most (but not all) chain-growth polymerizations occur via free radical chain reactions, so the steps in chain-growth polymerization are generally the same as those in free radical polymerization—namely, *initiation*, *propagation*, and *termination*. Chain-growth polymerizations can also be initiated by anions or cations, resulting in **anionic polymerization** or **cationic polymerization**, respectively. For example, styrene can undergo anionic polymerization. The initiator, butyllithium, reacts with styrene to form an anion on the growing chain instead of a free radical, as shown in Equation 26-16.

The anion from butyllithium initiates the reaction by donating a pair of electrons to the vinyl group in styrene.

A new, more stable anion is formed.

$$CH_3CH_2CH_2CH_2^{\ominus} + \underset{\text{styrene}}{C=C} \xrightarrow{\text{Initiation}} CH_3CH_2CH_2CH_2-\overset{H}{\underset{H}{C}}-\overset{H}{C:^{\ominus}} \quad (26\text{-}16)$$

YOUR TURN 26.8

Why is the anion on the product side of the reaction in Equation 26-16 more stable than the anion on the reactant side?

In the propagation steps, monomers still combine with the growing chain to form a reactive species similar to the reactant—that is, the reactive species are anions, too, not free radicals, as shown in Equation 26-17.

$$CH_3CH_2CH_2CH_2-\overset{H}{\underset{H}{C}}-C:^{\ominus} + C=C \xrightarrow{\text{Propagation}} CH_3CH_2CH_2CH_2-\overset{H}{\underset{H}{C}}-\overset{H}{C}-\overset{H}{\underset{H}{C}}-C:^{\ominus} \quad (26\text{-}17)$$

Termination occurs in anionic polymerization via a proton transfer step. That is, an acid is added to the reaction mixture, which neutralizes the negative charge and ends the propagation, as shown in Equation 26-18.

$$CH_3CH_2CH_2CH_2-\overset{H}{\underset{H}{C}}-\overset{H}{C}-\overset{H}{\underset{H}{C}}-C:^{\ominus} + H-B \xrightarrow{\text{Termination}} CH_3CH_2CH_2CH_2-\overset{H}{\underset{H}{C}}-\overset{H}{C}-\overset{H}{\underset{H}{C}}-C-H \quad :B^{\ominus} \quad (26\text{-}18)$$

YOUR TURN 26.9

Why are combination and disproportionation not possible mechanisms for termination in the anionic polymerization of styrene?

Cationic polymerization is similar to anionic polymerization except that the initiator and propagating species are positively charged. For examples of cationic polymerization, see the synthesis of PVK in the box on page 579 and Problem 26.37 at the end of this chapter.

Equation 26-19 shows an example of chain-growth polymerization in which the monomer has no vinyl groups. In this case, ethylene oxide (oxirane) undergoes an anionic polymerization that can also be described as a **ring-opening polymerization**.

$$H_2C-CH_2 \xrightarrow{\text{CaO}} \left(\overset{H}{\underset{H}{C}}-\overset{H}{\underset{H}{C}}-O\right)_n \quad (26\text{-}19)$$

Equation 26-20 shows that the oxide anion in calcium oxide initiates polymerization, acting as a nucleophile.

The alkoxide anion forms.

(26-20)

The oxide anion acts as a nucleophile and initiates the reaction.

The alkoxide anion can now act as a nucleophile and be the site of chain growth, as shown in Equation 26-21.

(26-21)

Propagation

Propagation steps repeat

The alkoxide anion reacts with another ethylene oxide molecule.

Termination occurs when an acid is added to the reaction mixture, because the alkoxide anion reacts with the acid in a proton transfer step, similar to the step that occurs in Equation 26-18.

> **26.10 YOUR TURN**
>
> The polymer chain that is produced in Equation 26-21 will undergo two steps to complete its termination. Write the mechanism for these two steps.

26.3b Step-Growth Polymerization

Many heterochain polymers, including nylon-6,6 and PET, are made via step-growth polymerization.

- **Step-growth polymerization** occurs when molecules throughout a mixture join with any other molecule that has an appropriate functional group available to react, whether it is a monomer unit or another polymer molecule.
- This is in contrast to chain-growth polymerization, in which chains only grow through reactions with monomers.

Consider the synthesis of nylon-6,6 shown in Equation 26-22.

The first steps in the growth of the polymer chain

Hexane-1,6-dioic acid
Adipic acid

Hexane-1,6-diamine

(26-22)

The reaction mixture contains two different compounds, adipic acid and hexane-1,6-diamine, each of which is bifunctional. A carboxylic acid group from adipic acid reacts with an amine group from hexane-1,6-diamine to form a C—N bond and a molecule of water. The reaction continues and the chain length increases. One possible way for this to occur is shown in Equation 26-23.

$$HO-\overset{\overset{O}{\|}}{C}CH_2CH_2CH_2CH_2\overset{\overset{O}{\|}}{C}\underset{\underset{H}{|}}{N}CH_2CH_2CH_2CH_2CH_2CH_2NH_2 \quad + \quad HO-\overset{\overset{O}{\|}}{C}CH_2CH_2CH_2CH_2\overset{\overset{O}{\|}}{C}\underset{\underset{H}{|}}{N}CH_2CH_2CH_2CH_2CH_2CH_2NH_2$$

$$\downarrow \Delta \qquad\qquad\qquad\qquad (26\text{-}23)$$

$$HO-\overset{\overset{O}{\|}}{C}CH_2CH_2CH_2CH_2\overset{\overset{O}{\|}}{C}\underset{\underset{H}{|}}{N}CH_2CH_2CH_2CH_2CH_2CH_2\underset{\underset{H}{|}}{N}-\overset{\overset{O}{\|}}{C}CH_2CH_2CH_2CH_2\overset{\overset{O}{\|}}{C}\underset{\underset{H}{|}}{N}CH_2CH_2CH_2CH_2CH_2CH_2NH_2 \quad + \quad H_2O$$

Figure 26-20 shows schematically what occurs over time. At each step, growth of the polymer chain occurs when a carboxylic acid group of adipic acid (AA, represented by the red dots) reacts with an amine group of hexane-1,6-diamine (HD, represented

FIGURE 26-20 The growth of polymer chains during the synthesis of nylon-6,6 (a) A mixture of adipic acid (AA, shown as red dots) and hexane-1,6-diamine (HD, shown as blue squares). (b) When the reaction begins, chains begin to grow when a single AA reacts with a single HD. (c) The AA–HD units can grow at either end, reacting with either another AA or another HD. (d) The polymer chains grow stepwise, adding an AA or HD to the end of the chain. At the same time, single units of AA and HD continue to react to start new chains. (e) While the chains continue to grow through the addition of a single AA or HD, existing chains also react with each other.

by the blue squares). Early in the reaction (Fig. 26-20a through 26-20d), the size of the chain may increase by the length of a single AA molecule or a single HD molecule. As the reaction progresses, however, long chains may join (Fig. 26-20e), greatly increasing the size of the chain in a single step. This step is similar to combination in chain-growth polymerization (Equation 26-7), but the chains in step-growth polymerization are capable of further reactions.

The distinctions between step-growth polymerization and chain-growth polymerization are important. Chain-growth polymerizations are fast, irreversible reactions, having products that are thermodynamically very favored. Step-growth polymerizations are slower and, because the reactions are typically reversible, removal of water or other products can be necessary for the polymer to reach a high molecular weight. In the synthesis of nylon-6,6, for example, the reaction is reversible and so slow that heat is required to drive the reaction to completion. This method for synthesizing an amide is different from those you studied in Chapter 21.

Historically, this type of reaction (e.g., Equation 26-22) was referred to as **condensation polymerization**, because a small molecule (in this case, water) is formed as a product.

Many nucleophilic addition–elimination reactions can be used to make step-growth polymers. For example, the Fischer esterification (Section 21.7) of terephthalic acid and ethylene glycol may be used to make polyesters, such as PET, as shown in Equation 26-24.

(26-24)

Terephthalic acid   **Ethylene glycol**

or

+ H_2O

Poly(ethylene terephthalate) or PET
An ester of terephthalic acid and ethylene glycol

Alternatively, transesterification can be used to make PET from the dimethyl ester of terephthalic acid, as shown in Equation 26-25.

+ CH_3OH (26-25)

---

problem **26.9** In a third reaction for the synthesis of PET, the reactants were a diacid chloride and ethylene glycol. Provide the structures of the reactants and products for this reaction.

In the two examples just given, water (Equation 26-24) or methanol (Equation 26-25) was a product of the reaction and each must be removed from the product mixture to isolate the polymer. Some step-growth polymerizations, however, do *not* produce a second product, as shown for the polyurethanes in Equation 26-26.

The isocyanate functional group

$$O=C=N \quad \cdots \quad N=C=O \; + \; HO-CH_2CH_2-OH \quad\quad (26\text{-}26)$$

Toluene diisocyanate   **Ethane-1,2-diol**
                                          (Ethylene glycol)

The urethane functional group

A polyurethane

$$\cdots-CH_2CH_2-O-C(=O)-N(H)- \cdots -N(H)-C(=O)-O-CH_2CH_2-O-\cdots$$

or

$$\left(C(=O)-N(H)- \cdots -N(H)-C(=O)-O-CH_2CH_2-O \right)_n$$

FIGURE 26-21 Poly(vinyl alcohol) (PVA) (a) A portion of the PVA polymer and its repeating unit. (b) Vinyl alcohol is the monomer that would be derived from the repeating unit. However, as described in the text, PVA is not made from vinyl alcohol, an enol that exists primarily in its keto form.

Poly(vinyl alcohol) (PVA)

(a)

Vinyl alcohol

(b)

problem 26.10 The steps in the reaction in Equation 26-26 are (1) addition of a nucleophile to a polar double bond and (2) two proton transfer steps. Write the mechanism for this reaction.

26.4 Chemical Reactions after Polymerization

We have seen how to synthesize polymers, but they may undergo subsequent reactions as well. These reactions impart special properties to the polymers that render them more suitable for their purpose. They usually include modifications of the side groups in the polymer or crosslinking polymer chains to create network polymers.

26.4a Modification of Side Groups

In many cases, the monomer for a particular polymer is unavailable or it forms a polymer with inferior properties. Changes to the polymer after it has been made, particularly in its side chains, can address these problems.

26.4a.1 Transesterification of poly(vinyl acetate) to poly(vinyl alcohol)

Elmer's glue, contact lens solution, and other products contain **poly(vinyl alcohol)**, or **PVA** (Fig. 26-21a). Because of the large number of hydroxyl groups off the main chain, PVA is soluble in water and is used in a large number of water-based products.

You cannot make PVA directly from vinyl alcohol (Fig. 26-21b), however, because vinyl alcohol does not exist in significant quantities. Instead, vinyl alcohol is an enol that readily tautomerizes to acetaldehyde, its keto form (Equation 26-27):

$$(26\text{-}27)$$

Enol form — Vinyl alcohol · Tautomerization · Keto form — Ethanal (Acetaldehyde)

26.11 YOUR TURN

Explain why vinyl alcohol is disfavored thermodynamically and isomerizes to a different product.

To prepare PVA, chemical plants begin with the synthesis of poly(vinyl acetate) from vinyl acetate, as shown in Equation 26-28.

$$(26\text{-}28)$$

Vinyl acetate · Benzoyl peroxide, Δ · Poly(vinyl acetate) · or

Poly(vinyl acetate) can then undergo transesterification to remove the acetyl groups and produce PVA (Equation 26-29).

$$(26\text{-}29)$$

Poly(vinyl acetate) · NaOCH₃, CH₃OH, Δ · Poly(vinyl alcohol)

The properties of PVA depend upon the degree of deacetylation. A minimum of 80% of the acetyl groups are removed, but the amount can be as high as 99.3%.

26.12 YOUR TURN

Provide the mechanism for the reaction that occurs in Equation 26-29. What is the second product of the reaction?

26.4a.2 De-*tert*-butylation of poly(ether ether ketone)

Polymers that can withstand high temperatures without degrading are highly desirable, so chemists have tried to introduce thermal stability into polymers. Most linear polymers, however, oxidize at high temperatures and effectively burn up.

Oxidation, like other chemical reactions, requires collisions between oxygen and the reacting molecules. One way to prevent the oxidation of polymers is to create rigid chains that do not rotate readily. These **rigid-rod polymers** are highly crystalline

The continuous conjugated system imparts rigidity to the polymer chain.

Poly(ether ether ketone) or **PEEK**
Thermally stable >500 °C

(a)

Flexible portion of polymer chain   Conjugated portion of polymer chain

PET
Decomposes at ~300 °C

(b)

FIGURE 26-22 Polymer rigidity and thermal stability
(a) The repeating unit of PEEK. (b) The repeating unit of PET. The entire repeating unit of PEEK is conjugated (indicated by the blue screen), making the polymer rigid and thermally stable at >500 °C. A portion of PET's repeating unit is flexible (indicated by the red screen), so it decomposes at ~300 °C.

and maintain their shape until they are subjected to extremely high temperatures. An example is **poly(ether ether ketone)**, or **PEEK** (Fig. 26-22a), which is stable to temperatures >500 °C. A more flexible polymer, such as PET (Fig. 26-22b), decomposes at ~300 °C.

PEEK is rigid because the highly conjugated main chain does not undergo rotation readily. The chains are also long and rigid, so they can organize and pack easily into crystals, as shown in Figure 26-23. These crystals are stable, and a large amount of energy (available only at high temperatures) is required to separate the chains and make them available for chemical reactions (e.g., oxidation).

These polymer molecules belong to planes that lie beneath the plane of molecules in black.

The polymer molecules shown in black are in the same plane.

FIGURE 26-23 A crystal of a poly(ether ether ketone) or PEEK, a highly crystalline, rigid-rod polymer The rigid molecules orient themselves to maximize interactions between the polar C=O (carbonyl) groups, as shown in the molecules in black. These rigid planes then stack in such a way that the carbonyl group in one plane is facing the opposite direction of the carbonyl group underneath. This maximizes the attractions between the partial charges on the atoms of carbonyl groups in different molecules.

problem **26.11** All of the bond angles in PEEK are approximately 120°. Explain why the C–O–C bonds between the phenyl rings have a bond angle closer to 120° than to 109.5°.

The advantages of PEEK are also its disadvantages. When chemists began to synthesize PEEK, they found that the polymer chains would crystallize out of solution before reaching a substantial length. That is, the formation of the crystalline polymer is thermodynamically favored over the polymer in solution, where it could continue to react. To address this problem, chemists incorporated a monomer with a *tert*-butyl group attached to the aryl ring, as shown in Figure 26-24. Because the *tert*-butyl groups prevent the polymer from crystallizing as it forms, the reaction is able to proceed to a higher degree of polymerization.

This *tert*-butyl group helps prevent crystallization.

FIGURE 26-24 PEEK with a *tert*-butyl substituent PEEK is synthesized with a *tert*-butyl substituent attached to the ring to help prevent crystallization. This allows the polymer to be produced with a higher degree of polymerization. Subsequently, the *tert*-butyl group can be removed.

To isolate PEEK, the *tert*-butyl groups must be removed. The reverse reaction of a Friedel–Crafts alkylation allows this to occur. When the reaction occurs in toluene, the solvent acts as a substrate for Friedel–Crafts alkylation. The overall reaction is then a trans *tert*-butylation, in which the *tert*-butyl group moves from the polymer to the solvent, as shown in Equation 26-30.

The *tert*-butyl group originally on the polymer...

PEEK with a *tert*-butyl substituent

PEEK

...is transferred to a solvent molecule.

(26-30)

problem **26.12** Propose a mechanism for the reaction that occurs in Equation 26-30. Do you think the reaction will be as successful if the *tert*-butyl group were a linear butyl group instead? Why or why not?

26.4b Crosslinking

In Section 26.4a, we discussed chemical reactions that change the structure and properties of linear polymers, but the polymer chains remained linear. Many polymers, however, undergo reactions that connect linear polymer chains to create networks.

Recall from Section 26.4a.1 that PVA is extremely water soluble. Its hydroxyl groups can be crosslinked, however, to decrease the solubility of PVA in water. This is what happens in the formation of "slime" from PVA and borax (Equation 26-31).

Linear molecules of PVA dissolve in water.

A crosslink that connects molecules of PVA

$$\xrightarrow[\text{H}_2\text{O}]{\text{Borax} \atop \text{Na}_2\text{B}_4\text{O}_7}$$

(26-31)

The crosslinking decreases the solubility of the PVA and also traps water molecules in the polymer, creating a gel. That is, instead of the PVA being a solute and the water a solvent, the water molecules now act as a solute in the crosslinked polymer. The result is a gel because liquid water is dispersed throughout the solid, crosslinked PVA.

PVA can also be crosslinked with formaldehyde through the formation of an acetal, as shown in Equation 26-32.

$$\xrightarrow{\text{Acid catalyst}}$$

A crosslink

(26-32)

problem **26.13** Write the mechanism for the crosslinking reaction between two PVA molecules and formaldehyde. Assume some acid is present to catalyze the reaction. (Only the reactive portions of the PVA molecules are shown in detail.)

FIGURE 26-25 A crystal lattice of solid benzene Solid benzene is depicted with (a) stick models and (b) space-filling models.

26.5 Properties of Polymers

The final properties of a polymer are important, so polymers are frequently "designed," as are drugs and other molecules, to perform a specific function. Here in Section 26.5, we consider several properties of polymers in terms of the structural properties discussed previously in this book, such as intermolecular forces (Chapter 2) and conformational isomerism (Chapter 4). We begin with thermal transitions in polymers.

26.5a Thermal Transitions in Polymers

The molecules in liquids can readily flow past one another, whereas the molecules in solids are in fixed positions relative to each other. **Crystalline solids** are ones in which the molecules are arranged in a regular pattern that repeats throughout the solid. This pattern is referred to as the **crystal lattice**, a term which may be familiar to you from general chemistry. An example of the crystal lattice of solid benzene is shown in Figure 26-25, with two different representations of the benzene molecule.

Because of their high order,

> Crystalline solids have a well-defined **melting point** (T_m).

In polymers, T_m is the temperature at which the polymer chains have sufficient kinetic energy to leave the crystal lattice and flow as a liquid.

In contrast to crystalline solids, there is no long-range order to the arrangement of the species in an **amorphous solid**. The most common example of an amorphous solid is glass, which is composed primarily of silicon dioxide (SiO_2). Figure 26-26 shows a representation of the arrangement of atoms in SiO_2.

Amorphous solids do not have a well-defined melting point. Instead,

> Amorphous solids have a **glass transition temperature** (T_g), at which the atoms in the solid have sufficient kinetic energy that they can move past each other and flow.

In polymers, T_g is the temperature at which the molecules begin to display significant molecular motion and exhibit a loss in mechanical strength.

The energy required to change a crystalline solid into a liquid is greater than the energy required for a glass transition. Therefore,

> T_g of an amorphous solid is lower than T_m of a crystalline solid of the same substance.

FIGURE 26-26 The structure of silicon dioxide, an amorphous solid There are four oxygen atoms attached to each silicon atom; each of those oxygen atoms is attached to two different silicon atoms. Note the lack of a regular or repeating arrangement of atoms.

FIGURE 26-27 A comparison
of a polymer in crystalline and
amorphous forms (a) The molecules
in an organized crystalline solid require
significant kinetic energy to overcome
intermolecular forces and to increase
the entropy of the system as it enters
the liquid phase. (b) The molecules in an
amorphous solid are already at higher
entropy and the intermolecular forces are
not maximized, as in a crystalline solid.
Therefore, less kinetic energy is required
for the molecules to act as a liquid and
flow past each other.

Crystalline solid	Amorphous solid
• Stronger intermolecular forces	• Weaker intermolecular forces
• Lower entropy	• Higher entropy

(a)   (b)

Figure 26-27 schematically depicts the difference between two such solids and briefly explains the difference in the amount of kinetic energy needed by each sample for the molecules to be able to flow past each other.

The T_m and the T_g of polymers are important data to know because polymers are generally heated before they are formed into a particular shape. In addition, these temperatures determine the range of temperatures over which the polymer can function without losing its shape or function.

For example, at room temperature, the rubber in a rubber band or a racquetball is an **elastomer** because it can experience large deformations yet return to its original shape. A racquetball is deformed by the impact of the racquet or a wall, yet it resumes its shape and, in the process, forces itself away from the racquet or the wall (Fig. 26-28a). The rubber can do this because its T_g is well below room temperature. At room temperature, therefore, the molecules are not particularly constrained and can be deformed readily.

If you cool a racquetball in liquid nitrogen (77 K) to a temperature below its T_g (Fig. 26-28b) then the rubber is in a glassy, amorphous state and the molecules cannot flow past each other. If you throw a frozen racquetball against the wall, it shatters like glass because the molecules no longer have the mobility they had at room temperature (Fig. 26-28c). If you warm the broken pieces to room temperature, however, then the rubber passes through its glass transition state and regains its elasticity.

Polymers range from being completely amorphous to being highly crystalline. They are rarely entirely crystalline for several reasons. For example, polymers are mixtures of long chains that vary in length, thus making the creation of a regular, repeating crystal lattice difficult. Moreover, the length of the chains makes it difficult for a single molecule to arrange itself entirely in an orderly fashion.

Polymers can have both crystalline regions and amorphous regions. The crystalline regions are more organized, reflect the order found in crystalline solids, and have a T_m.

FIGURE 26-28 A comparison of
a racquetball's performance as
an elastomer at high and low
temperatures (a) At room temperature,
the racquetball has its normal elastic
quality. (b) A ball immersed in liquid
nitrogen is cooled below its T_g. (c) The
polymer molecules in the ball no longer
bend and deform as easily; the ball
shatters when thrown against a wall.

(a)   (b)   (c)

Sample 1

Lower T_g

The zigzag portions of the polymer chain (red screens) represent regions of high crystallinity.

Sample 2

Higher T_g

FIGURE 26-29 **The effect of crystallinity on T_g** Sample 2 has a greater degree of crystallinity than Sample 1. The noncrystalline regions in Sample 2 have a higher T_g than the noncrystalline regions in Sample 1, because the crystalline regions in Sample 2 constrain the noncrystalline regions and increase the amount of kinetic energy needed for those segments to flow.

The amorphous regions lack regular order and undergo a glass transition. The T_g of a polymer can vary and depends on how the polymer was processed and the conditions under which it was cooled. The amount of crystallinity affects the T_g, too, because the crystalline regions "anchor" the amorphous regions and impede the movement of the chains in the amorphous regions. As a result, the more crystalline a polymer, the higher its T_g. Figure 26-29 illustrates this point.

In polymers, both melting and undergoing a glass transition are endothermic processes, whereas crystallization is an exothermic process. That is, when polymer molecules arrange themselves in an organized crystal lattice, they release energy as they go from a higher energy state to a lower energy state. This lower energy is achieved by maximizing intermolecular forces and minimizing steric repulsions.

We can see these points illustrated more clearly with differential scanning calorimetry (DSC), a technique that measures the heat that flows into or out of a sample as the temperature of the sample is changed. The curve shown in Figure 26-30 is the DSC result for a sample of PEEK. Around 140 °C, the polymer absorbs heat and goes through its glass transition. This gives the polymer chains enough mobility to become more ordered and, therefore, crystallize at 169 °C. When they do, heat is given off because they go from a higher energy state to a lower, more stable crystalline form. At about 339 °C, the crystalline form melts and the polymer molecules are able to flow.

(a) Glass transition, an endothermic process

~140 °C

(c) Melting, an endothermic process

339 °C

(b) Crystallization, an exothermic process

169 °C

Heat flow →

Temperature →

FIGURE 26-30 **A differential scanning calorimetry (DSC) scan of a polymer (PEEK)** **(a)** Above this temperature, the polymer molecules have sufficient mobility to organize into a crystalline solid. **(b)** As the polymer molecules organize into a crystalline solid, which is more ordered and at lower energy, they give off energy in the form of heat. **(c)** At the melting point, the polymer molecules absorb sufficient energy that they are not constrained in a crystal lattice, but change into a liquid and flow past each other.

26.5b Factors Affecting Thermal Transitions in Polymers

Synthetic polymers can be molded into useful objects. **Thermoplastic** polymers can be heated above their T_g or T_m and then forced into molds or pushed through slots to form films or threads (see Fig. 26-31). This is the origin of the term "plastic" to describe polymers.

Recall from Chapter 2 that the stronger the intermolecular interactions in a substance, the higher the temperature required for a phase transition, either from solid to liquid or liquid to gas. This is also true for polymers:

> The stronger the intermolecular interactions between polymer molecules, the higher the T_g or T_m.

In general, strong intermolecular interactions lead to a more stable crystal lattice, one with an arrangement of molecules that maximizes attractions between the molecules and minimizes steric repulsions and other unfavorable arrangements. For the molecules in a solid to move past each other and flow, the solid must absorb sufficient

(a)

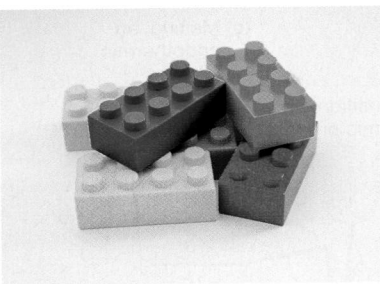

(b)

FIGURE 26-31 Injection molding of thermoplastic polymers (a) Upper and lower sides of the molds used to produce Legos. (b) Legos that are produced from the injection molding process.

heat to overcome the attractions between molecules and provide the molecules with enough kinetic energy to move away from the crystal lattice. Seven major factors affect the amount of heat required for this change.

1. An increase in the chain length of linear molecules raises T_g and T_m.

As discussed in Section 2.6d, more energy (or a higher temperature) is needed to melt the solid as the length of the hydrocarbon chain increases. That is, the dispersion forces experienced by each molecule increase, and these stronger intermolecular interactions stabilize the solid, so a higher temperature is required for melting. Similarly, as the molecular mass or degree of polymerization increases for a linear polymer, the T_m also increases.

2. Structural features or functional groups that increase intermolecular interactions increase the T_g and T_m.

Polar molecules generally have higher melting points than nonpolar molecules with similar molar masses. Similarly, poly(vinyl chloride), a polar polymer, has a significantly higher T_g than polypropylene, a nonpolar polymer, even though both polymers have similar molar masses (Fig. 26-32).

Nonpolar bond　　　　　**Polar bond**

$$\left(\!\!\begin{array}{c} H \;\; H \\ | \;\;\;\; | \\ C\!-\!C \\ | \;\;\;\; | \\ H \;\; CH_3 \end{array}\!\!\right)_n \qquad \left(\!\!\begin{array}{c} H \;\; H \\ | \;\;\;\; | \\ C\!-\!C \\ | \;\;\;\; | \\ H \;\; Cl \end{array}\!\!\right)_n$$

Polypropylene　　　　　**Poly(vinyl chloride)**
$T_g = -18\ °C$　　　　　$T_g = 87\ °C$

FIGURE 26-32 Polarity and thermal transitions Because the C—Cl bond in poly(vinyl chloride) is polar, poly(vinyl chloride) has a higher T_g than polypropylene.

3. Periodic or random branching of the polymer chain decreases the T_g and T_m.

Branched alkanes have lower melting points and boiling points than straight-chain alkanes, and polymers follow the same trend. In addition to reducing surface area for induced-dipole–induced-dipole attractions, branching interferes with the molecules' ability to organize into a regular crystal lattice; there is no longer a simple, regular pattern for organizing the molecules into the solid. Moreover, in the solid form, branches push the neighboring molecules away from the main chain, decreasing the dispersion forces between them. Branched polyethylene has a melting point of 112 °C, for example, while linear polyethylene has a melting point of 135 °C.

4. The regular occurrence of side groups along a polymer chain increases T_g and T_m.

If the polymer has side groups that occur regularly, then the T_m will increase because these groups increase the barrier to rotation about single bonds. For example, polypropylene has a higher melting point than polyethylene (Fig. 26-33).

If the side groups are long, however, as in poly(vinyl *n*-butyl ether) and poly(vinyl *tert*-butyl ether) (Fig. 26-34), then they push the polymer molecules apart, thus decreasing the intermolecular attractions and decreasing the T_m or T_g. Notice that poly(vinyl *n*-butyl ether) has a longer side chain, and thus a lower T_g than poly(*tert*-

The bulkiness of the CH₃ groups makes rotation about this bond more difficult.

Polyethylene
$T_m = 135\ °C$
(a)

Same as

Polypropylene (isotactic)
$T_m = 160–185\ °C$
(b)

FIGURE 26-33 Regularity of side groups and thermal transitions (a) Polyethylene. (b) Polyproplyene (isotactic). In each case, the repeating unit is shown on the left and a portion of the polymer chain is shown as a Newman projection on the right. The bulkiness of the CH₃ group in polypropylene hinders rotation about the single bonds in the polymer's main chain, and the cumulative effect along the polypropylene chain results in a significantly higher melting point.

butyl ether). Furthermore, the bulky *tert*-butyl groups greatly hinder rotation around the single bonds in the polymer chain. The *n*-butyl groups do not have as dramatic an effect on bond rotation.

• The longer side group keeps the main chains farther apart.
• Lower T_g

Poly(vinyl *n*-butyl ether)
$T_g = –52\ °C$
(a)

Pol(vinyl *tert*-butyl ether)
$T_g = 88\ °C$
(b)

FIGURE 26-34 Side group length and thermal transitions Because the butyl group is longer than the *tert*-butyl group, the chains of (a) poly(vinyl *n*-butyl ether) are kept farther apart than those of (b) poly(vinyl *tert*-butyl ether). This diminishes the intermolecular attractions and lowers the glass transition temperature.

5. Rigid chains have higher T_g and T_m than flexible chains.

If rotation about single bonds is not possible, then more energy is required for the molecule to move away from the crystal structure. We have already seen this effect in Section 26.4a.2, where we compared the thermal stabilities of PEEK and PET. PEEK, with its more rigid chain, has the higher T_m (Fig. 26-35).

FIGURE 26-35 Chain rigidity and thermal transitions The chain of (a) PEEK is more rigid than that of (b) PET (indicated by the blue screens), so PEEK has the higher melting point.

• The continuous conjugated system imparts rigidity to the polymer chain.
• Higher T_m

Flexible portion of polymer chain

Conjugated portion of polymer chain

PEEK
$T_m = 365\ °C$
(a)

PET
$T_m = 270\ °C$
(b)

6. Factors that allow molecules to pack into an orderly crystal structure usually increase the T_m.

When polymer molecules have regularity in their structures, they organize more readily into crystal structures. A measure of regularity is tacticity, which describes the configurations of the stereocenters in the polymer chains (see the box on p. 1274). For example, isotactic polypropylene (PP) (Fig. 26-36a) has a relatively well-defined

Isotactic polypropylene
Melting point = 160–185 °C
(a)

Atactic polypropylene
No true melting point
(b)

FIGURE 26-36 A comparison of isotactic and atactic polypropylene (PP) (a) In isotactic polypropylene, the methyl groups are all on the same side of the chain. Isotactic PP has a melting point of 160–185 °C. (b) In atactic polypropylene, the positions of the methyl groups are random, so atactic PP is not crystalline and has no true melting point.

Poly(acrylic acid)
$T_g = 106$ °C
(a)

Moderately strong hydrogen bonding

Poly(sodium acrylate)
Decomposes at $T > 260$ °C
(b)

Very strong ion–ion interaction

FIGURE 26-37 Electrostatic attraction and thermal transitions (a) Hydrogen bonds between molecules of poly(acrylic acid) give it a relatively high T_g. (b) The much stronger electrostatic forces between the ions in poly(sodium acrylate), however, prevent the polymer from softening before it reaches a temperature at which it decomposes.

melting point, whereas atactic polypropylene (Fig. 26-36b) does not; it is tacky at room temperature and is used in adhesives.

In isotactic PP, the regular distribution of the methyl groups on one side of the chain allows the molecules to crystallize readily. The random arrangement of the methyl groups in atactic PP prevent the molecules from assembling into an orderly arrangement.

7. Ion–ion interactions among polymers dramatically increase T_g and T_m.

The attraction between cations and anions is extremely strong. In fact, some organic compounds with ionic groups decompose before they melt, because the energy required to overcome the electrostatic forces is greater than the energy required to break the covalent bonds. Likewise, polymers with ionic side groups frequently decompose before they melt. For example, as shown in Figure 26-37, poly(acrylic acid) has a T_g of 106 °C, whereas poly(sodium acrylate) does not soften when heated and decomposes at temperatures >260 °C.

problem **26.14** For each pair of polymers, predict which one has the higher T_g. Justify your choice.

(a)

(b)

problem **26.15** For each pair of polymers, predict which one has the higher T_g. Justify your choice.

(a)

(b)

(c)

26.5c Solubility

The solubility of polymers generally follows the same patterns that we discussed in Chapter 2. When the intermolecular interactions involving the solvent molecules are similar to those involving the polymer molecules, then the polymer will dissolve. For example, polystyrene dissolves in toluene ($C_6H_5CH_3$), an aromatic solvent that readily solvates the aromatic side groups in polystyrene. Poly(vinyl alcohol) dissolves in water because its hydroxyl groups can hydrogen bond with water. Polymers with ionic side groups, such as poly(sodium acrylate), dissolve even more readily in water due to ion–dipole interactions.

Any factors that limit the interactions between the solvent molecules and the polymer will decrease solubility. Network polymers (Section 26.2c) are rarely soluble, for example, because the crosslinks prevent sufficient interactions between the solvent and polymer. Linear poly(vinyl alcohol) (Section 26.4b) is soluble in water and is used in water-based glues, but it can be used as a sponge when it is crosslinked. The crosslinked PVA retains its ability to hydrogen bond with water, but is no longer soluble in water.

Highly crystalline polymers also have lower solubilities, because solvent–polymer interactions may not be strong enough to overcome polymer–polymer interactions. Again, we saw this in the discussion of PEEK in Section 26.4a.2.

---

problem **26.16** Among the polymers shown in Problems 26.14 and 26.15, there are three that are water soluble and are used in pharmaceutical applications. Which polymers are they? Justify your choices.

26.6 Uses of Polymers: The Relationship between Structure and Function in Materials for Food Storage

Historically, food products were shipped in containers made of glass, metal, or ceramic. In an effort to improve shelf life, lower costs, and expand the range of foods that can be packaged, industry has turned to polymers for food packaging. One important property of materials for food storage is a lack of chemical reactivity—that is, you don't want to store food in a material that will react with it and affect its taste. The materials used for food storage need to be insoluble in water, too, and should show limited permeability toward water and air. You don't want the material to dissolve on the surface of the food, and you want to prevent spoilage. Consequently, polyolefins such as polyethylene and polypropylene, which are essentially alkanes, are logical choices for food storage.

26.6a Polyethylene

Polyethylene, the simplest polyolefin, comes in three types: low-density polyethylene (LDPE), linear low-density polyethylene (LLDPE), and high-density polyethylene (HDPE). LDPE is a flexible polymer that is used for packaging foods. As shown in Figure 26-38, it has a low T_g ($-120\ °C$) and fairly low T_m ($\sim112\ °C$). As noted in Section 26.5b, it has substantial branching, which contributes to its low T_g. The branching results from hydrogen atom abstraction (see Section 25.3b) during the free radical polymerization, as shown schematically in Figure 26-39.

Branching prevents crystallinity and close packing of the polymer.

Low-density polyethylene (LDPE)

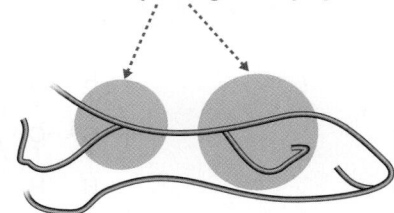

Density (g/cm³)	T_g (°C)	T_m (°C)
0.92	−120	112

FIGURE 26-38 Structure, properties, and use of LDPE The branching of the polymer prevents crystallization, which results in a relatively low density, T_g, and T_m. It also makes the plastic flexible, making it ideal for use as plastic wrap.

An unpaired electron at the end of a growing chain

An unpaired electron in the middle of a growing chain

An unpaired electron at the end of a growing chain

(a)   (b)   (c)

FIGURE 26-39 Formation of branches in low-density polyethylene Hydrogen atom abstraction during free radical polymerization transfers the reactive radical from the end of one chain (a) to somewhere in the middle of another (b). Further polymerization at this radical (c) creates a branch in the chain, giving rise to low-density polyethylene.

YOUR TURN 26.13

The following two equations are specific examples of the elementary steps shown schematically in Figure 26-39. Use curved arrows to show the movement of electrons in both steps, and provide the product of the second step.

(a)

(b)

In contrast to LDPE, HDPE is synthesized using a catalytic system (see box on p. 1274) that prevents branching, so only linear polymer is produced. Due to its increased crystallinity, HDPE has a higher T_m (135 °C) and higher tensile strength than LDPE. This makes HDPE suitable for applications in which LDPE would fail, such as milk jugs and other containers for liquids (Fig. 26-40).

Linear low-density polyethylene (LLDPE) combines properties of both LDPE and HDPE (Fig. 26-41). LLDPE is made by copolymerizing ethylene and a small percentage of 1-alkenes such as but-1-ene. The copolymerization takes place with a catalyst that produces a linear polymer. The linearity of the polymer increases intermolecular forces, leading to a higher T_m ($T_m = 125$ °C) and greater strength than that of LDPE. LLDPE is used extensively in the manufacture of garbage bags, which need to be flexible and also need to have high tensile strength to prevent ripping.

Highly linear polymer chains result in maximum crystallinity.

High-density polyethylene (HDPE)

Density (g/cm³)	T_g (°C)	T_m (°C)
0.94	~–120	135

FIGURE 26-40 Structure, properties, and use of HDPE The polymer is relatively highly crystalline, so HDPE is more rigid than LDPE. The high crystallinity of the plastic also gives it a higher density and T_m than LDPE. This makes HDPE ideal as material for containers such as milk jugs.

Regular branches from 1-alkenes that are copolymerized with ethylene

Linear low-density polyethylene (LLDPE)

Density (g/cm³)	T_g (°C)	T_m (°C)
0.92	~–120	125

FIGURE 26-41 Structure, properties, and use of LLDPE The incorporation of 1-alkenes pushes the polymer chains apart; this leads to lower density than in HDPE. However, the absence of random branches of different lengths leads to higher crystallinity, which is reflected in a melting point that is higher than that of LDPE. The regular branching of the polymer allows for more crystallinity in LLDPE and, consequently, a stronger plastic, making it ideal for use in manufacturing garbage bags.

The phenyl groups hinder rotation of the polymer chain, which raises both T_g and T_m and creates a more rigid material.

Polystyrene

Density (g/cm³)	T_g (°C)	T_m (°C)
1.05	100	N/A

FIGURE 26-42 Structure, properties, and uses of polystyrene The large pendant groups limit the flexibility of the polymer and make it rigid. Polystyrene is used to make "jewel" cases for compact discs and is also used to make styrofoam.

26.6b Polystyrene

Polyethylene (PE) is extremely flexible, whereas polystyrene (PS) is not, due largely to the phenyl rings that increase the barrier to rotation around the main chain. Consequently, the T_g of PS (100 °C) is much higher than that of PE (−120 °C). Commercial polystyrene is completely amorphous due to the random arrangement of the phenyl rings on either side of the main chain (atactic).

The mass of the phenyl groups makes PS denser (1.05 g/cm³) than HDPE (0.94 g/cm³). The rigidity and low cost of PS make it a choice for a variety of applications, such as the "jewel" cases for compact discs (Fig. 26-42). As we discussed earlier in the chapter, PS is also used in foams (styrofoam) for insulation and for insulated food packaging such as coffee cups, "to-go" boxes, and egg cartons.

problem 26.17 Why would HDPE be an inappropriate material for a CD case? Why is PS an inappropriate material for a milk jug?

26.6c Poly(Ethylene Terephthalate)

PET is the material used in bottles for carbonated beverages. Faced with the shipping costs of relatively heavy glass bottles and loss from breakage, manufacturers of carbonated beverages have found PET to be an effective replacement. However, an important requirement for a pop bottle is impermeability—that is, the carbon dioxide in carbonated beverages must be contained in the bottle at moderate pressures and not seep through the plastic bottle over time.

Plastics manufacturers have maximized the impermeability and strength of PET by maximizing its crystallinity (Fig. 26-43). When crystallinity is imposed on PET mechanically, the density of the plastic increases from 1.33 to 1.39 g/cm^3. This compares to 0.94 g/cm^3 for HDPE and 1.05 g/cm^3 for PS. The high T_m of PET (260–265 °C) reflects its strength. PET has four times the tensile strength of HDPE and more than twice the tensile strength of PS.

FIGURE 26-43 Molding of poly(ethylene terephthalate) bottles (a) Process for the manufacture of PET bottles. In the second stage, pressure is used to expand the PET, which stretches the polymer chains and facilitates crystallization. (b) Depiction of the crystallization that occurs when the PET is stretched.

26.7 Degradation and Depolymerization

Up to now, we have discussed ways to form bonds among a large number of small molecules to create a polymer. Under different circumstances, however, the ability to break a polymer into smaller units is also useful or, under certain conditions, un-avoidable. A reaction that breaks a large molecule into smaller molecules is generally referred to as a **degradation reaction**. A reaction that causes polymer molecules to revert to monomer molecules is called a **depolymerization**.

The retro Diels–Alder reaction of dicyclopentadiene (see Section 24.6) is analogous to depolymerization, because the Diels–Alder adduct is a dimer that reverts to its constituent monomers. Recall that the retro Diels–Alder reaction is entropy driven and occurs at high temperatures. Consequently, dicyclopentadiene, which forms readily at room temperature, reverts to cyclopentadiene when heated to 200 °C (Equation 26-33).

A dimer, a molecule of two parts, made from two dicyclopentadiene molecules

Upon heating, the dimer reverts to the monomers from which it was made.

$$(26\text{-}33)$$

In 1860, Greville Williams (1829–1910) carried out a depolymerization of natural rubber by heating it in the absence of oxygen. The volatile product that he isolated turned out to be isoprene (Equation 26-34).

or

$$(26\text{-}34)$$

Natural rubber   **Isoprene**

Although polyisoprene (i.e., natural rubber) has been synthesized from isoprene, the rubber plant does not use isoprene to biosynthesize rubber. Nevertheless, the isolation of isoprene from rubber is significant because many other natural products are built from the same building blocks that the rubber plant uses. Essential oils contain terpenes and terpenoids, for example, most of which can be broken down into isoprene units (see Sections 2.12c and 11.12).

In general, reactions in which the products have a higher entropy than the reactants are likely to occur at high temperatures. Elimination, for example, is favored at high temperatures because of the increase in entropy as the reactants form products (see Section 9.8). An analogous situation occurs in PVC, in which HCl can be eliminated to form double bonds, as shown in Equation 26-35. Currently, chemists believe that the double bonds form via an E1 mechanism (see Chapters 8 and 9 to review this reaction).

$$(26\text{-}35)$$

Once double bonds begin to form, subsequent eliminations of HCl from the polymer chain occur readily, and the molecule "unzips" to form a polyene. PVC, therefore, poses a safety hazard when fires occur, because the HCl that is formed can be breathed in, leading to lung damage. This danger is offset by the addition of metal salts that react with HCl.

---

problem **26.18** Provide the mechanism for the reaction that forms the first double bond in the product in Equation 26-35.

---

problem **26.19** In the series of reactions shown in Equation 26-35, the second and subsequent double bonds form more quickly than the first double bond. Explain why this occurs.

Polymer degradation is a major issue in polymer recycling. The heat and physical processing required to reform plastic also promote degradation reactions that reduce the desirable properties of plastics.

THE ORGANIC CHEMISTRY OF BIOMOLECULES

26.8 Biological Macromolecules

Recall from Section 1.15 that proteins, carbohydrates (saccharides), and nucleic acids are constructed from relatively small molecular building blocks. Proteins are constructed from amino acids, carbohydrates from monosaccharides, and nucleic acids from nucleotides. Thus, proteins, carbohydrates, and nucleic acids can be thought of as **biopolymers** made up of amino acid, monosaccharide, and nucleotide monomers. More specifically, a protein is a **polyamino acid** or **polypeptide** (due to the peptide bond that links together amino acids), a large carbohydrate is a **polysaccharide**, and a nucleic acid is a **polynucleotide**.

In Section 9.14, we saw how bonds between monosaccharides—glycosidic linkages—can be formed and broken, and in Sections 21.13 and 21.14 we discussed the formation and breaking of bonds that connect amino acids, so-called peptide linkages. Here in Section 26.8, we turn our attention to the macroscopic properties of these biopolymers. Just as we saw with synthetic polymers throughout Chapter 26, the properties of biopolymers are governed by the nature of their monomers. We discuss examples of how this is so with polypeptides in Section 26.8a and with polysaccharides in Section 26.8b.

26.8a Polypeptides: Primary, Secondary, Tertiary, and Quaternary Structures

The specific function of a protein is governed by the four levels of its structure, which are designated as *primary*, *secondary*, *tertiary*, and *quaternary*. We will examine each of these levels of structure in order.

26.8a.1 Primary structure

The **primary structure** of a protein is its specific sequence of amino acids.

> A protein is characterized by its primary structure; different proteins have different primary structures.

Some amino acids can form *crosslinks*, whereby one portion of the protein's backbone is covalently bonded to another portion of the backbone (see Section 26.2c). This occurs when the thiol (RSH) groups of two cysteine amino acids are oxidized, creating a **disulfide bond** (also called a **disulfide bridge**), as shown in Equation 26-36.

$$(26\text{-}36)$$

Vasopressin, a hormone protein that is responsible for the body's retention of water and is involved in the "fight or flight" response of mammals, has a disulfide bond. The sequence of amino acids that characterizes vasopressin is Cys-Tyr-Phe-Gln-Asn-Cys-Pro-Arg-Gly-NH$_2$. When oxidized, the two cysteine amino acids form a disulfide bond, giving rise to the structure in Figure 26-44.

FIGURE 26-44 A disulfide bond in vasopressin (a) Vasopressin represented as a line structure (top) and in its abbreviated form (bottom). The backbone of the protein is highlighted in blue in the line structure. In both representations, the disulfide bond is highlighted in red. (b) Vasopressin in a ball-and-stick representation. In the disulfide bond, the sulfur atoms (yellow) are connected by a single bond.

problem **26.20** Determine how many different primary sequences can be produced if any of the naturally occurring amino acids can appear at each location and the sequence is three amino acids long. Do the same for a protein that is 300 amino acids long.

problem **26.21** Glutathione is an antioxidant that helps prevent cell damage from free radicals. Under oxidizing conditions, glutathione forms a dimer via a disulfide bond. Draw the structure of that dimer.

Glutathione

26.8a.2 Secondary structures

The secondary structure of a protein is the three-dimensional structural *pattern* of specific portions of the protein's backbone. The two most common secondary structures are the α-helix and the β-pleated sheet, shown in Figures 26-45a and 26-45b, respectively.

An **α-helix** has the protein's backbone wound tightly into a spiral. There are two factors that favor the formation of an α-helix. The first is the large number of hydrogen bonds that form, each involving the NH bond of an amino group from one amino acid and the O atom of a carbonyl group from an amino acid that is four residues further down the chain. The second factor is sterics. The side groups of the amino acids in an α-helix project outward, maximizing the distances between them.

Whereas each α-helix involves a single portion of a protein's backbone, a **β-pleated sheet** comprises several portions of the backbone that align next to each other in the same region of space. These portions of the backbone can be many amino acids apart in the primary structure. As with the α-helix, a β-pleated sheet is stabilized by hy-

FIGURE 26-45 Secondary structures within a protein (a) An α-helix. Hydrogen bonds are established between the NH bonds and O atoms from different amino acids. Side groups (represented by the green spheres) project outward from the center of the helix. (b) A β-pleated sheet. Different portions of the protein's backbone align next to each other in the same region of space. Hydrogen bonds are established between NH bonds and O atoms from different amino acids.

(a) α-Helix

Side chains
(R groups)

Peptide
bonds

Hydrogen
bonds

(b) β-Pleated sheet

drogen bonds, each involving an NH amino bond and a carbonyl O atom from different amino acids. The β-pleated sheet is also favored by the planarity that each amide functional group is able to achieve. This planarity results from the contribution of a resonance structure that gives the C—N bond of an amide significant double-bond character, as shown in Figure 26-46a. Adjacent amide planes are slightly rotated relative to each other, however, giving the chain kinks that establish the pleats in the sheet (Figure 26-46b).

Although there is more extensive hydrogen bonding in a β-pleated sheet than in an α-helix, the nature of the amino acid side groups determines whether or not one of these secondary structures is favored.

- An α-helix tends to be disrupted when adjacent or nearby amino acids in a protein's sequence (within about three or four residues) have side groups that are either bulky or carry the same charge (remember, like charges repel).
- A β-pleated sheet tends to be disrupted if it requires bulky or like-charged side groups from different segments of the primary structure to occupy similar locations in space.

The double-bond character favors planarity of an amide group.

(a)

A kink between each amide group establishes the "pleats."

Amide plane   Amide plane

Side chain

(b)

FIGURE 26-46 Planarity of a peptide bond (a) Resonance structures of an amide group, showing the double bond character of the C—N bond. (b) In a β-pleated sheet, the planarity of each amide group is maintained, and the kink between each amide bond is what gives rise to the "pleats".

---

problem **26.22** Suggest whether alanine or isoleucine should be found more commonly in an α-helix? Explain why. (Consult Table 1-7 on p. 42 for their structures.)

26.8a.3 Tertiary Structures

Even after a protein has been synthesized with a definite primary structure and segments of the protein have adopted specific secondary structures, the protein may still lack proper biological activity. That's because the function of a protein depends critically on its **tertiary structure**, which is the location of all of its atoms in space. Achieving the proper tertiary structure requires the protein to *fold* in a particular way.

A **ribbon structure** is typically used to depict the tertiary structure of a protein, as shown in Figure 26-47a for phosphotriose isomerase, an enzyme used in glycolysis to catalyze the isomerization of three-carbon sugars with an attached phosphate group (see the box on p. 376).

In a ribbon structure, a trace of the backbone is shown explicitly, and the side groups are omitted. Regions of secondary structure are represented by helices (for α-helices) and by arrows (for β-sheets), where the arrows point in the direction from the N-terminal amino acid to the C-terminal amino acid. Secondary structures are further highlighted using a consistent color scheme. In Figure 26-47a, for example, the α-helices are indicated using magenta and the β-sheets are indicated using gold.

The advantage of using a ribbon structure can be seen by comparing Figure 26-47a to Figure 26-47b, which is a cross-section of phosphotriose isomerase in which all atoms are shown explicitly as a space-filling model.

FIGURE 26-47 Tertiary structure of phosphotriose isomerase (a) Ribbon structure of phosphotriose isomerase. Magenta indicates an α-helix and gold indicates a β-sheet. (b) Cross-section of a space-filling model of phosphotriose isomerase. Red indicates nonpolar amino acids and green indicates polar amino acids.

α-Helix

β-Sheet

(a)

Polar amino acids (green) interact favorably with the aqueous environment.

Nonpolar amino acids (red) are hydrophobic and tend to be buried inside the protein.

(b)

Hemoglobin

β-chain

α-chain   Fe³⁺   Heme

(a)

Aspartate transcarbamoylase

Catalytic site   3-fold axis

C_1 C_2 C_3

R_1' R_2'

R_1 R_2

C_4 C_5 C_6

Regulatory site

(b)

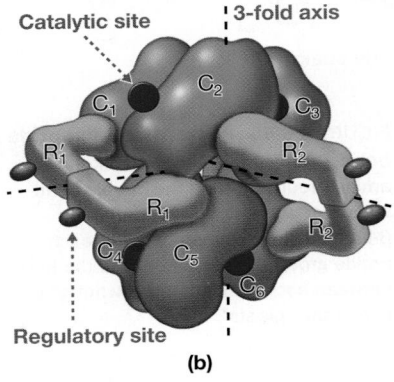

FIGURE 26-48 Quaternary structures The quaternary structure of (a) hemoglobin and (b) aspartate transcarbamoylase (ATC). Four distinct peptide chains have associated in hemoglobin, whereas ATC contains six peptide chains with a catalytic role (C_1 to C_6).

α-1,4′-Glycosidic linkage

FIGURE 26-49 Partial structure of amylose Amylose consists of a large number of D-glucose units that are connected by α-1,4′-glycosidic linkages. One glycosidic linkage is highlighted in red, and the numbering systems for the rings are shown in blue.

The space-filling model makes it difficult to distinguish the secondary structures from the rest of the enzyme.

The tertiary structure of a protein tends to be the one that gives the lowest possible free energy, G. Factors such as intermolecular forces and steric crowding among amino acids contribute to a protein's overall free energy, similar to how they contribute to the stability of a particular secondary structure. Another important factor is the **hydrophobic effect**. Proteins typically reside in an aqueous environment. Amino acids such as serine and aspartic acid have polar side groups (see Section 1.15a), so they favor such an environment—they are *hydrophilic*. Others, such as valine and phenylalanine, have nonpolar side groups, so they do not favor an aqueous environment—they are *hydrophobic*. To maximize the favorable interactions between the amino acids and the aqueous environment, and to minimize the unfavorable ones, a protein tends to fold in such a way as to have its polar amino acids on the exterior and its nonpolar amino acids on the interior. This is shown in Figure 26-47b.

problem 26.23 For some analyses, proteins are first denatured (i.e., unfolded) by treatment with a detergent called sodium dodecyl sulfate (SDS). Explain why detergents like SDS have this effect.

26.8a.4 Quaternary Structures

Many proteins in living cells consist of multiple, distinct peptide chains, not just one chain. Quaternary structure describes the way in which these protein *subunits* combine to form an even larger structure. Subunits are individual polypeptide chains, each with a specific primary, secondary, and tertiary structure.

The methods used to depict quaternary structure are typically less detailed than those used for tertiary structure. Sometimes only the general shape of a polypeptide chain is drawn, with few of the finer structural details. Examples are shown in Figure 26-48 for hemoglobin and aspartate transcarbamoylase.

problem 26.24 Would you expect the surface of each subunit of hemoglobin to have a relatively large or small number of nonpolar amino acids? Explain your reasoning.

26.8b Polysaccharides

Some polysaccharides, such as *starch* and *glycogen*, are used to store energy, and are called **storage polysaccharides**. Others, such as cellulose, are responsible for maintaining the structure of organisms, and are called **structural polysaccharides**.

Starch, a main component of foods such as rice, potatoes, corn, and wheat, is actually a mixture of two different polysaccharides—namely, *amylose* (~20%) and *amylopectin* (~80%).

Amylose consists of unbranched chains of D-glucose, which, as shown in Figure 26-49, are connected by α-1,4′-glycosidic linkages (Section 9.14). That is, the glycosidic linkage involves C1 of one glucose unit and C4 of the next, and the substituent on C1 is axial. These glycosidic linkages are hydrolyzed readily by α-glucosidase, an enzyme that is found in every known mammal. Hydrolysis of a glycosidic linkage releases a molecule of glucose, which is subsequently broken down further for energy.

The α configuration of the anomeric carbon also has implications for amylose's physical properties. Because of the α configuration, amylose favors a helical form, as shown in Figure 26-50a, which is reminiscent of the α-helix adopted by proteins. Based on the cross section in Figure 26-50b, there are numerous hydroxyl groups on the exterior of the helix, where they are available for extensive hydrogen bonding with water. Amylose, therefore, is highly soluble in water.

Amylopectin is very similar to amylose, having a main chain that consists of glucose units joined by α-1,4'-glycosidic linkages. The essential difference between the two is that amylopectin's chain is relatively highly branched. Branching takes place via α-1,6'-glycosidic linkages (Fig. 26-51), which occur roughly every 25–30 glucose units. As a result of this branching, there are more terminal glycosidic linkages, so α-glucosidase is capable of releasing glucose units much faster. This is beneficial when energy is needed moderately quickly.

(a)

Extensive hydrogen bonding with water

(b)

FIGURE 26-50 **The helical structure of amylose** (a) Representation of amylose as a helix. Most H atoms have been omitted. (b) Cross section of amylose, showing that the exterior of the helix has numerous OH groups available for hydrogen bonding with water.

α-1,6'-Glycosidic linkage

Branching occurs as a result of glucose units attached at both C4 and C6.

FIGURE 26-51 **Partial structure of amylopectin** Glucose units in the main chain are joined by α-1,4'-glycosidic linkages, whereas glucose units at the branch points are joined by α-1,6'-glycosidic linkages.

In animals, excess glucose is stored in the form of *glycogen*, which is structurally very similar to amylopectin. The glucose units in glycogen are connected primarily by α-1,4'-glycosidic linkages to establish a main chain, and branching occurs via α-1,6'-glycosidic linkages. In glycogen, branching occurs even more frequently than in amylopectin—roughly every 10 glucose units—allowing hydrolysis to take place even more quickly. This is important for organisms that need quick bursts of energy.

Cellulose, a structural polysaccharide, is the main structural component of plants' cell walls, making up about one-third of all plant matter. The specific cellulose content of a plant, however, depends on the specific nature of the plant. Wood, for example, is about 50% cellulose, whereas cotton is about 90% cellulose.

Cellulose consists of long, unbranched chains of D-glucose. How, then, is cellulose different from amylose? The glucose units in amylose are connected by α-1,4'-glycosidic linkages, whereas the glucose units in cellulose are connected by β-1,4'-glycosidic linkages—that is, the substituent attached to C1 is equatorial in each unit, as shown in Figure 26-52.

Although the difference between an α linkage and a β linkage may seem slight on the small scale, the differences are dramatic on the large scale. As shown in Figure 26-52, the β linkage makes extensive intramolecular hydrogen bonding possible, which makes cellulose more rigid than amylose. Moreover, the free OH groups are situated for hydrogen bonding with other cellulose chains to create a stable polymer network. This is what makes cellulose such a useful structural polysaccharide.

This glucose unit is inverted relative to the adjacent ones.

β-1,4′-Glycosidic linkage

Internal hydrogen bonds add rigidity to the molecule.

Another difference between cellulose and amylose is that cellulose is insoluble in water. Whereas the helical shape of amylose allows its OH groups to remain free for hydrogen bonding with water, the OH groups in cellulose are tied up in hydrogen bonding with adjacent glucose units of the same chain, or with OH groups from a separate chain. Without substantial hydrogen bonding with water, there are not enough favorable interactions to make cellulose soluble.

Yet another difference between cellulose and amylose is that mammals cannot digest cellulose directly. Mammals lack β-glucosidase, the enzyme required for hydrolyzing the β-1,4′-glycosidic linkage of cellulose. In the digestive tracts of some animals, however, such as horses and cows, there exist certain bacteria that do have the enzyme, and therefore are capable of breaking down cellulose. In this way, grazing animals can obtain glucose from grass and hay.

---

problem **26.25** The exoskeletons of insects and arthropods incorporate chitin, which is structurally almost identical to, but stronger than, cellulose. The difference between the two polymers is that in chitin, the OH group at each 2-position has an acetamide ($-NHCOCH_3$) group instead. Explain why this makes chitin stronger.

26.9 Wrapping Up and Looking Ahead

In Chapter 26, we discussed several aspects of polymers, including their fundamental structure, various representations, nomenclature, and physical properties, and we learned how polymers can be synthesized using a variety of reactions. Polymers are large molecules constructed from relatively small monomers, and they generally have an easily recognizable repeating unit. We saw that a polymer can be linear, with any of a wide variety of pendant groups attached to the main chain. The polymer's main chain can also branch, and separate chains can be crosslinked. Network polymers, on the other hand, don't have any long-range linearity; rather, an entire sample can be viewed as a single molecule.

Polymers can be produced via chain-growth or step-growth polymerization. In chain-growth polymerization, the polymer grows at a specific reactive site one monomer at a time, and the mechanism consists of three types of steps: initiation, propagation, and termination. Most chain-growth polymerization mechanisms involve free radicals, but they can also be anionic or cationic, depending on the nature of the monomer and the initiator. In step-growth polymerization, on the other hand, the polymer can grow at more than one type of reactive site, so a single monomer can join a growing polymer chain, or two polymer chains can join. Step-growth polymerization doesn't require separate initiation or propagation steps, and often involves condensation reactions, whereupon the joining of two molecules results in the expulsion of a smaller molecule—usually water.

We saw that the macroscopic properties of polymers are established by relatively few variables in the polymer's structure. These variables include the average size of the polymer, the extent of branching in a main chain, the extent of crosslinking, the functional groups that make up the main chain, and the size and polarity of the pendant groups. In general, the glass transition and melting temperatures of a polymer increase with an increase in the polymer's average size, rigidity along the main chain, and the ease in which polymer molecules (or portions of them) can arrange in an orderly fashion. A polymer's solubility in water typically increases with a greater proportion of hydrophilic functional groups either along its main chain or as part of its pendant groups, whereas solubility in an organic solvent generally increases with a greater proportion of hydrophobic groups.

This was the final chapter of the textbook, and we have covered a great deal of material over these 26 chapters. The first several chapters dealt with aspects of molecular structure and stability, including bonding, Lewis structures, MO theory, intermolecular interactions, and isomerism. We then introduced a handful of elementary steps involving closed-shell species, which make up the majority of organic reaction mechanisms. We saw how each elementary step is described using curved arrow notation, and we saw, further, how charge stability and bond energies contribute to the driving force for an elementary step. With those elementary steps under our belts, we learned how mechanisms can be constructed in reasonable ways, and we applied that understanding toward the variety of reaction types thereafter, including nucleophilic substitution and elimination, electrophilic addition, nucleophilic addition, nucleophilic addition–elimination, and electrophilic aromatic substitution. Finally, we introduced pericyclic and free radical elementary steps and studied reactions in which they are involved.

In addition to our focus on reaction mechanisms and the products they make, we learned some specific strategies to help us incorporate these reactions in synthesis. Furthermore, we learned the fundamentals of UV–vis, IR, and NMR spectroscopy, as well as mass spectrometry. These techniques are often used to establish various aspects of molecular structure, perhaps to characterize a newly discovered compound or to verify that a particular target was synthesized.

Even though this marks the end of the textbook, this may not mark the end of your study of organic chemistry. You might go on to take a course in biochemistry or medicinal chemistry, in which the mechanistic aspects of organic chemistry are routinely applied. You might continue your academic work in synthetic organic or natural product chemistry. Or you might pursue a career in which you work with polymers or other organic materials. Even if you do not apply organic chemistry directly in your future academic or career endeavors, what you have learned in organic chemistry will allow you to better understand the world around you, and that, alone, can give you tremendous appreciation for the efforts you have put forth.

Best of luck in whatever you decide!

Chapter Summary and Key Terms

- **Polymers** are large molecules made from smaller molecules called **monomers**; they contain a **main chain** and **side groups**. Polymers are usually depicted by showing their **repeating units**, because the chains are so large that a full structure is impractical to draw. (**Sections 26.1a and 26.2a**)

- The number of monomers making up a polymer's chain is the **degree of polymerization**. (**Section 26.1a**)

- Numerous polymers are made by **free radical polymerization**, which is a chain reaction with *initiation*, *propagation*, and *termination* steps. Termination can occur by **combination** or **disproportionation**. Most monomers add in a

head-to-tail fashion during propagation. (**Sections 26.1b and 26.1c**)

- Polymers can be classified according to the type of chain (**carbon-chain** or **heterochain polymers**), the type of functional group in the chain, the type of monomer used to make the polymer, and by the number of different kinds of monomers used to make the polymer (**homopolymers** vs. **copolymers**). Polymers can also be classified by the types of chain (**linear** vs. **branched polymers**) and the degree of crosslinking (**crosslinked** and **network** polymers). (**Sections 26.2b and 26.2c**)

- Names for polymers can be derived from the repeating unit or from the monomer used to create the polymer. **(Section 26.2d)**
- Free radical polymerization is an example of **chain-growth polymerization**, one general class of polymerization reactions. The other class of reactions is **step-growth polymerization**. **(Sections 26.3a and 26.3b)**
  - In chain-growth polymerization, reaction occurs at the end of the growing chain, and each reaction lengthens the polymer chain by a single repeating unit.
  - In step-growth polymerization, reaction occurs between the ends of molecules, either monomers or polymer chains, and the increase in chain size depends on the size of the two molecules that react.
- Polymers can undergo chemical reactions after polymerization is terminated. Most involve modifications to the side groups, an important one being **crosslinking**, in which polymer chains are joined together through the reaction of two side groups or a functional group within a chain with another reactant. **(Sections 26.2c, 26.4a, and 26.4b)**
- The **melting point** (T_m) and **glass transition temperature** (T_g) are thermal properties that quantify the crystalline and amorphous character of a polymer. **(Section 26.5a)**

- Thermal transitions and the solubility properties of a polymer depend on the length of the polymer molecules, the extent of branching, the rigidity of the polymer, the symmetry of the polymer, and the intermolecular forces. **(Section 26.5b and 26.5c)**
- The structure of a polymer determines how it can function—that is, its properties dictate its uses.
  - Polymers with low T_g, such as LDPE, are flexible and can be used as packaging films. **(Section 26.6a)**
  - Polymers with higher T_g are more rigid and are used for objects that need to retain their shapes. Polystyrene, for example, can be used for coffee cups and "to-go" boxes. **(Section 26.6b)**
  - Introducing crystallinity increases rigidity and density, often with retention of flexibility, so HDPE is suitable for milk jugs and PET is suitable for carbonated beverage bottles. **(Section 26.6c)**
- At high temperatures, depolymerization (the reverse of polymerization) and degradation (the loss of atoms from the polymer) are thermodynamically favorable because of the higher entropy of the products. **(Section 26.7)**

Problems

26.26 Each of the monomers at the right can undergo free radical polymerization.
  (i) Using benzoyl peroxide as the initiator, show the mechanism for the first two propagation steps.
  (ii) Draw the condensed formula for the polymer, showing the repeating unit.
  (iii) Provide a name for the polymer.

(a) H, Cl / C=C / H, Cl

Vinylidene chloride

(b) H, H / C=C / H, C=O, H₂N

Acrylamide

(c) H, H / C=C / H, OCH₃

Methyl vinyl ether

26.27 Using each of the following polymers,

(i) Draw the condensed formula for the polymer.
(ii) Propose a structure for the monomer from which the polymer was synthesized.
(iii) Propose names for the monomer and the polymer.

26.28 Which polymer in Problem 26.27 will be most soluble in ethanol? Justify your choice.

26.29 Which polymer in Problem 26.27 will be most soluble in hexane? Justify your choice.

26.30 Identify the repeating unit in each of the following polymers and write the condensed formula for the polymer.

(a)

Poly(lactic acid) or polylactide

(b)

Poly(propylene oxide)

(c)

Polyglycine

26.31 Identify the class of polymer, based on functional groups, for the polymers in Problem 26.30.

26.32 A polymer closely related to PET is PBT, which is made from terephthalic acid and butane-1,4-diol.
    (a) Propose a structure for PBT and write the condensed formula for the structure.
    (b) What does PBT stand for? That is, what do you expect the trivial name of PBT to be?
    (c) Would you expect PBT to have a higher or lower melting point than PET? Justify your answer.
    (d) PBT, rather than PET, is used in molding because it crystallizes faster. Explain why PBT crystallizes more quickly than PET.

26.33 Teflon can be made from tetrafluorethylene by means of a free radical polymerization. In the reaction, termination occurs only by combination and not by disproportionation. Explain this observation.

26.34 When polyethylene is made by free radical polymerization, the resulting polymer has a significant amount of branching, resulting in low-density polyethylene (see Section 26.6a). When styrene undergoes free radical polymerization, on the other hand, very little branching occurs. Explain this phenomenon.

26.35 Polystyrene can *autoinitiate* free radical polymerization and form polystyrene in the absence of an initiator. At temperatures >100 °C, radicals form due to homolysis of the π bond in the vinyl group. In the study of this reaction, researchers have found 1,2-diphenylcyclobutane in samples of heated styrene. The following mechanism for the formation of 1,2-diphenylcyclobutane has been proposed:

Step 1: Homolysis of the vinyl π bond in a styrene molecule
Step 2: Homolysis of the vinyl π bond in a second styrene molecule
Step 3: Tail-to-tail addition of the radicals from Steps 1 and 2
Step 4: Radical coupling of the diradical formed in Step 3

    (a) Use curved-arrow notation to show the steps in this mechanism.
    (b) Are the steps in this mechanism consistent with those for free radical polymerization?
    (c) Why is the use of an initiator such as benzoyl peroxide more efficient in the synthesis of polystyrene than autoinitiated polymerization?

26.36 Polyoxetane is a polyether similar to poly(ethylene oxide) and is made from oxetane via a chain-growth polymerization. (See Section 26.3a.)
    (a) Propose a synthesis of polyoxetane and show the mechanism for the first propagation steps.
    (b) Provide a condensed formula for polyoxetane.
    (c) Do you expect polyoxetane to be more or less soluble in water than poly(ethylene oxide)? Justify your answer.
    (d) Do you expect polyoxetane to have a higher or lower melting point than poly(ethylene oxide)? Justify your answer.
    (e) Explain why tetrahydrofuran reacts much more slowly than oxetane if you try to use it to synthesize a polyether.

Oxetane

Tetrahydrofuran (THF)

26.37 Polyisobutylene is made by cationic polymerization. A source of H⁺ is a common initiator for the reaction.

Isobutylene

(a) Propose a mechanism for the initiation step.
(b) Propose a mechanism for the first step of propagation.
(c) Draw the structure of polyisobutylene.
(d) Why is isobutylene a better candidate for cationic polymerization than ethylene (ethene) or propylene (propene)?
(e) Do you think isobutylene would be a good candidate for anionic polymerization? Why or why not?

26.38 Cyanoacrylates like methyl 2-cyanoacrylate are used to make adhesives such as Super Glue. They readily undergo polymerization with an initiator such as hydroxide anion or an alkoxide anion.

(a) Provide a mechanism for the initiation step.
(b) Propose a mechanism for the first step of propagation.
(c) Provide the structure of poly(methyl 2-cyanoacrylate)
(d) Why is methyl 2-cyanoacrylate a better candidate for anionic polymerization than methyl acrylate?
(e) Do you think methyl 2-cyanoacrylate would be a good candidate for cationic polymerization? Why or why not?

Methyl 2-cyanoacrylate

26.39 (a) Which monomer is most likely to undergo anionic polymerization? Justify your choice.
(b) Which one is most likely to undergo cationic polymerization? Justify your choice.

Ethyl vinyl ether   **But-1-ene**   **Acrolein**

26.40 Propylene does not undergo free radical polymerization readily because there are two competing steps after initiation: propagation and hydrogen atom extraction.

(a) Using a generic radical R• as a reactant with propylene, draw the mechanism and products for the two competing steps.
(b) Which step produces the more stable product?
(c) How do your results explain propylene's poor reactivity in free radical polymerization?
(d) Which monomer(s) in Problem 26.39 will also be poor candidates for free radical polymerization? Justify your choice.

26.41 Poly(vinyl alcohol), PVA, is produced from poly(vinyl acetate) by converting the acetate groups into hydroxyl groups (Section 26.4a.1). As the number of hydroxyl groups in PVA increases, the solubility of the polymer in water decreases. How do you explain this behavior?

26.42 Dry-erase boards, or whiteboards, in classrooms consist of a network polymer similar to Bakelite that is made from formaldehyde and melamine.

(a) Propose a structure for the polymer formed.
(b) Why is this type of polymer suitable for use in whiteboards?

Melamine

26.43 Consider the following series of poly(phenylene oxide) polymers and their glass transition temperatures.

A
$T_g = 82\ °C$

B
$T_g = 211\ °C$

C
$T_g = 169\ °C$

D
$T_g = 99\ °C$

Explain how the differences in structure can account for the differences in T_g.

26.44 Consider polycarbonate polymers **E–G**.

E

F

G

(a) Which one will have the highest T_g? **(b)** Which will have the lowest T_g?

26.45 Two of the compounds **X**, **Y**, and **Z** will react with phosgene to form a polycarbonate; the third one will not.

X **Y** **Z** **Phosgene**

Which compound will *not* form a polymer, and what will the product of its reaction with phosgene be?

26.46 One method for the purification of polystyrene samples is to dissolve the polymer in benzene and then add methanol slowly to precipitate solid polystyrene. Explain why polystyrene precipitates under these conditions.

26.47 Polycarbophil is a dietary fiber supplement. It is the calcium salt of a copolymer of acrylic acid and divinyl glycol, a crosslinking agent. In the stomach, the calcium ions exchange for hydrogen ions and, in the higher pH of the intestine, the polymer absorbs 70 times its mass in water.
(a) Propose a structure for polycarbophil.
(b) Why is crosslinking of the polymer necessary for this application?
(c) Why is polycarbophil effective at providing bulk in the intestine? That is, what occurs on a molecular level to polycarbophil?

Divinyl glycol

26.48 In the discussion of poly(ether ether ketone)s, or PEEKs, in Section 26.4a.2, we observed that the addition of a *tert*-butyl group to the aromatic ring increases the solubility of PEEKs in organic solvents. However, when researchers added a second *tert*-butyl group, as shown here, they found that the solubility of PEEKs decreased. Explain this observation.

A PEEK with two *tert*-butyl groups per repeating unit

26.49 Urea has the following structure:

Urea

Polyureas are used to make truck bed liners. Examine the synthesis of polyurethanes in Section 26.3b, then **(a)** propose a synthesis of a polyurea and **(b)** provide the condensed formula for the polymer.

26.50 Research in the "unzipping" of poly(vinyl chloride) indicates that certain defects in the polymer contribute to the rate of unzipping. These defects are arrangements of atoms that are different from the arrangements in the repeating unit shown in the condensed formula. Researchers have found that defects involving allylic chlorides and tertiary chlorides promote the reaction. Review the mechanism for the unzipping and explain this observation.

26.51 PEEK (see Section 26.4a.2) can be synthesized as follows:

PEEK

Propose a mechanism for this reaction.

26.52 Explain how IR spectroscopy could be used to monitor the conversion of poly(vinyl acetate) to poly(vinyl alcohol). That is, how can a chemist use IR to tell if poly(vinyl acetate) has been converted to 99+% poly(vinyl alcohol)?

26.53 Methyl acrylate and methyl methacrylate react with radical initiators (R•) as follows:

The difference in their reactivities in free radical polymerizations is dramatic. For example, methyl acrylate is less reactive with radicals than methyl methacrylate, but free radicals formed from methyl acrylate are more reactive than those formed from methyl methacrylate. In other words, **C** is more reactive than **A**, and **B** is more reactive than **D**. Explain these observations. (*Hint:* Draw an energy diagram for the two reactions.)

If your instructor assigns problems in smartwork, log in at **smartwork.wwnorton.com**.

Values of K_a and pK_a for Various Acids

Acid	Conjugate Base	K_a	pK_a	Acid	Conjugate Base	K_a	pK_a
Trifluoromethanesulfonic acid		1×10^{13}	-13	Benzoic acid		6.3×10^{-5}	4.2
Sulfuric acid		1×10^{9}	-9	Ethanoic acid (Acetic acid)		1.8×10^{-5}	4.75
HCl Hydrochloric acid	Cl$^{\ominus}$	1×10^{7}	-7	Bicarbonate anion		2×10^{-6}	6.3
Methanesulfonic acid		1×10^{2}	-2	Thiophenol		2.5×10^{-7}	6.6
$H_3O^{\oplus}$ Hydronium ion	H_2O	55	-1.7	4-Nitrophenol		6.3×10^{-8}	7.2
Trifluoroethanoic acid (Trifluoroacetic acid)		1	0	2,4-Pentanedione		1.2×10^{-9}	8.9
Trichloroethanoic acid (Trichloroacetic acid)		0.17	0.77	N≡C—H Hydrocyanic acid	N≡C$^{\ominus}$	6.3×10^{-10}	9.2
Chloroethanoic acid (Chloroacetic acid)		1.3×10^{-3}	2.87	$H_4N^{\oplus}$ Ammonium ion	NH_3	4×10^{-10}	9.4
HF Hydrofluoric acid	F$^{\ominus}$	6.3×10^{-4}	3.2	Phenol		1×10^{-10}	10
Methanoic acid (Formic acid)		1.8×10^{-4}	3.75				

(continued)

Acid	Conjugate Base	K_a	pK_a	Acid	Conjugate Base	K_a	pK_a
O_2N-CH_3 Nitromethane	$O_2N-\overset{\ominus}{CH_2}$	6.3×10^{-11}	10.2	Methylpropan-2-ol (*tert*-Butyl alcohol) OH	$O^{\ominus}$	1×10^{-19}	19
Ethanethiol SH	$S^{\ominus}$	2.5×10^{-11}	10.6	Propanone (Acetone)		1×10^{-20}	20
$H_3C-\overset{\oplus}{NH_3}$ Methylammonium ion	H_3C-NH_2	2.3×10^{-11}	10.63	Ethyl ethanoate (Ethyl acetate) EtO	EtO $CH_2^{\ominus}$	1×10^{-25}	25
Ethyl 3-oxobutanoate (Acetoacetic ester) OEt	OEt	1×10^{-11}	11	$N{\equiv}C-CH_3$ Ethanenitrile (Acetonitrile)	$N{\equiv}C-\overset{\ominus}{CH_2}$	1×10^{-25}	25
2,2,2-Trifluoroethanol OH	$O^{\ominus}$	4×10^{-13}	12.4	$HC{\equiv}CH$ Ethyne (Acetylene)	$HC{\equiv}C^{\ominus}$	1×10^{-25}	25
2-Chloroethanol Cl OH	Cl $O^{\ominus}$	1.3×10^{-13}	12.9	Aniline (Phenylamine) NH_2	$\overset{\ominus}{NH}$	1×10^{-27}	27
Diethyl propanedioate (Diethyl malonate) EtO OEt	EtO OEt	3.2×10^{-14}	13.5	H_2 Hydrogen gas	$H^{\ominus}$	1×10^{-35}	35
Pyrrole NH	$N^{\ominus}$	1×10^{-15}	15	Dimethyl sulfoxide H_3C CH_3	H_3C $CH_2^{\ominus}$	1×10^{-35}	35
CH_3OH Methanol	$CH_3O^{\ominus}$	3.2×10^{-16}	15.5	NH_3 Ammonia	$H_2N^{\ominus}$	1×10^{-36}	36
H_2O Water	$HO^{\ominus}$	2×10^{-16}	15.7	*N*-Methylmethanamine (Dimethylamine) H_3C CH_3	H_3C CH_3 $N^{\ominus}$	1×10^{-38}	38
Ethanol OH	$O^{\ominus}$	1×10^{-16}	16	Toluene (Methylbenzene) CH_3	$CH_2^{\ominus}$	1×10^{-40}	40
Cyclopentadiene CH_2	$\overset{\ominus}{CH}$	1×10^{-16}	16	Benzene CH	$C^{\ominus}$	1×10^{-43}	43
Propan-2-ol (Isopropyl alcohol) OH	$O^{\ominus}$	3.2×10^{-17}	16.5				
Ethanamide (Acetamide) NH_2	$\overset{\ominus}{NH}$	1×10^{-17}	17				

(*continued*)

Acid	Conjugate Base	K_a	pK_a	Acid	Conjugate Base	K_a	pK_a
$H_2C{=}CH_2$ Ethene (Ethylene)	$H_2C{=}\overset{\ominus}{C}H$	1×10^{-44}	44	CH_4 Methane	$H_3C^{\ominus}$	1×10^{-48}	48
Ethoxyethane (Diethyl ether)		$\sim 1 \times 10^{-45}$	~45	CH_3CH_3 Ethane	$CH_3\overset{\ominus}{C}H_2$	1×10^{-50}	50

Characteristic Reactivities of Particular Functional Groups

Alkanes

An alkane H is replaced by a halogen atom, X, when treated with $X_2/h\nu$. (25.4)

Alkenes

HBr, in the presence of radicals, will add across the double bond in an anti-Markovnikov fashion to produce an alkyl halide. (25.5)

OsO_4 or $KMnO_4$ adds to produce a *cis*-1,2-diol. (24.7)

A strong Brønsted acid, such as HCl, HBr, HI, or $H_3O^{\oplus}$, will add across the double bond to produce an alkyl halide or alcohol. (11.2)

Br_2 or Cl_2 will add to the double bond to produce a vicinal dihalide. (12.3)

Under conditions for catalytic hydrogenation, H_2 will add to the C=C bond to reduce an alkene to an alkane. (19.5)

Oxidized to a benzoic acid using $KMnO_4$ when R = C_6H_5. (22.9a)

An allylic hydrogen is replaced by a halogen atom, X, when treated with $X_2/h\nu$. (25.4)

BH_3 will add across the double bond to produce an alkylborane that can be oxidized to an alcohol. (12.6)

$KMnO_4$ or O_3 will cleave the C=C bond to produce separate carbonyl-containing groups. (24.8)

RCO_3H will add an oxygen atom to the C=C bond to produce an epoxide. (12.5)

Oxymercuration with $Hg(OAc)_2$ in water followed by reduction with $NaBH_4$ can produce an alcohol. (12.4)

A carbene will add to the C=C bond to produce a cyclopropane ring. (12.2)

Alkynes

Treatment with Li(s) or Na(s) in liquid NH_3 will reduce the alkyne to a *trans*-alkene. (25.7a)

OsO_4 or $KMnO_4$ adds to produce an α-diketone. (24.7)

A strong Brønsted acid, such as HCl, HBr, HI, or $H_3O^{\oplus}$, will add across the double bond to produce a vinyl halide or ketone. (11.7 and 11.8)

Br_2 or Cl_2 will add to the triple bond to produce a vinyl dihalide or a 1,1,2,2-tetrahalide. (12.3a)

Under conditions for catalytic hydrogenation, H_2 will add in a *syn* fashion to the C≡C to convert the alkyne into a *cis*-alkene. A second addition produces an alkane. (19.5b)

A strong base can deprotonate a terminal alkyne, generating a highly nucleophilic C. (9.3b)

Oxidized to a benzoic acid using $KMnO_4$ when R = C_6H_5. (22.9a)

$KMnO_4$ or O_3 will cleave the C≡C bond to produce separate carboxylic acid groups. (24.8)

BH_3 will add across the double bond to produce a vinylborane that can be oxidized to an aldehyde. (12.7)

Oxymercuration with $Hg(OAc)_2$ in water followed by reduction with $NaBH_4$ can produce a ketone. (12.4)

A carbene will add to produce a cyclopropene ring. (12.2)

Alkyl halides

Elimination of a halogen and a proton on an adjacent carbon results in an alkene. (7.4)

A metal such as Mg(s) or Li(s) produces an organometallic compound, making the C atom strongly nucleophilic and basic. (19.1)

A nucleophile will replace the halogen in a nucleophilic substitution reaction. (7.2)

X (Cl, Br, I)

Ethers

Elimination of H and ROH under acidic conditions produces an alkene. (8.6)

Under acidic conditions, a nucleophile will replace ROH. (9.5b)

OR

Alcohols

Elimination of H and H_2O under acidic conditions produces an alkene. (8.6)

The carbon–carbon bond of a *cis*-1,2-diol is cleaved by HIO_4, producing separate ketone and/or aldehyde groups. (24.8b)

H_2CrO_4 or $KMnO_4$ will oxidize a primary alcohol to a carboxylic acid, and a secondary alcohol to a ketone. (19.6a and 19.6b)

PCC will oxidize a primary alcohol to an aldehyde. (19.6a)

Deprotonation by a strong base converts ROH to $RO^{\ominus}$, a strong nucleophile. (9.3)

Under acidic conditions, a nucleophile will replace H_2O. (9.5b)

OH is converted to OTs, OTf, or OMs, using the sulfonyl chlorides TsCl, TfCl, or MsCl. (21.6)

OH is converted to Br or Cl with PBr_3 or PCl_3. (10.1)

Aromatic rings

If the substituent is a halogen leaving group, it can be replaced by a nucleophile if the aromatic ring is highly deactivated, or if the nucleophile is also a strong base. (23.10)

If the subsitutent is a sulfonyl group, it is replaced by a proton under acidic aqueous conditions. (22.7)

This aromatic proton is replaced by a strong electrophile if the substituent is an ortho/para director. (23.2 and 23.3)

A Birch reduction (a type of dissolving metal reduction) will reduce the aromatic ring to a 1,4-cyclohexadiene. (25.7b)

Sub

A benzylic hydrogen is replaced by a halogen atom, X, when treated with $X_2/h\nu$. (25.4)

This aromatic proton is replaced by a strong electrophile if the substituent is a meta director. (23.4)

A benzylic C is oxidized to a benzoic acid using $KMnO_4$. (22.9a)

This aromatic proton is replaced by a strong electrophile if the substituent is an ortho/para director. (23.2 and 23.3)

Ketones

The electron-rich O can be protonated by a strong acid, which activates the C=O carbon toward attack by a weak nucleophile. (18.1)

After nucleophilic addition at C, this O can be converted to an H_2O leaving group under acidic conditions, which can produce an acetal or imine via substitution or elimination, respectively. (18.2)

Under conditions for catalytic hydrogenation, H_2 adds to C=O under moderate conditions to produce an alcohol. (19.5)

Oxidized to a benzoic acid using $KMnO_4$ when R = C_6H_5. (22.9a)

A strong base can deprotonate an α carbon, generating a highly nucleophilic C. (18.4)

Oxidation with a peroxyacid can produce an ester. (21.8)

A methyl ketone is converted to a carboxylate anion by treatment with Cl_2, Br_2, or I_2 in basic, aqueous conditions. (20.5)

Nucleophiles add to the electron-poor carbon, converting C=O into C—O. Addition of an $H^{\ominus}$ or $R^{\ominus}$ nucleophile reduces a ketone to an alcohol. Addition of an $R^{\ominus}$ nucleophile results in the formation of a new C—C bond. (17.1, 17.3, 17.5)

Aldehydes

The electron-rich O can be protonated by a strong acid, which activates the C=O carbon toward attack by a weak nucleophile. (18.1)

After nucleophilic addition at C, this O can be converted to an H_2O leaving group under acidic conditions, which can produce an acetal or imine via substitution or elimination, respectively. (18.2)

Under conditions for catalytic hydrogenation, H_2 adds to C=O under moderate conditions to produce an alcohol. (19.5)

A strong base can deprotonate an α carbon, generating a highly nucleophilic C. (18.4)

In the presence of water, the hydrate is oxidized to a carboxylic acid by H_2CrO_4 or $KMnO_4$. (19.6)

A methyl ketone is converted to a carboxylate anion by treatment with Cl_2, Br_2, or I_2 in basic, aqueous conditions. (20.5)

Oxidation with a peroxyacid can produce carboxylic acid. (21.8)

Nucleophiles add to the electron-poor carbon, converting C=O into C—O. Addition of an $H^{\ominus}$ or $R^{\ominus}$ nucleophile reduces an aldehyde to an alcohol. Addition of an $R^{\ominus}$ nucleophile results in the formation of a new C—C bond. (17.1, 17.3, 17.5)

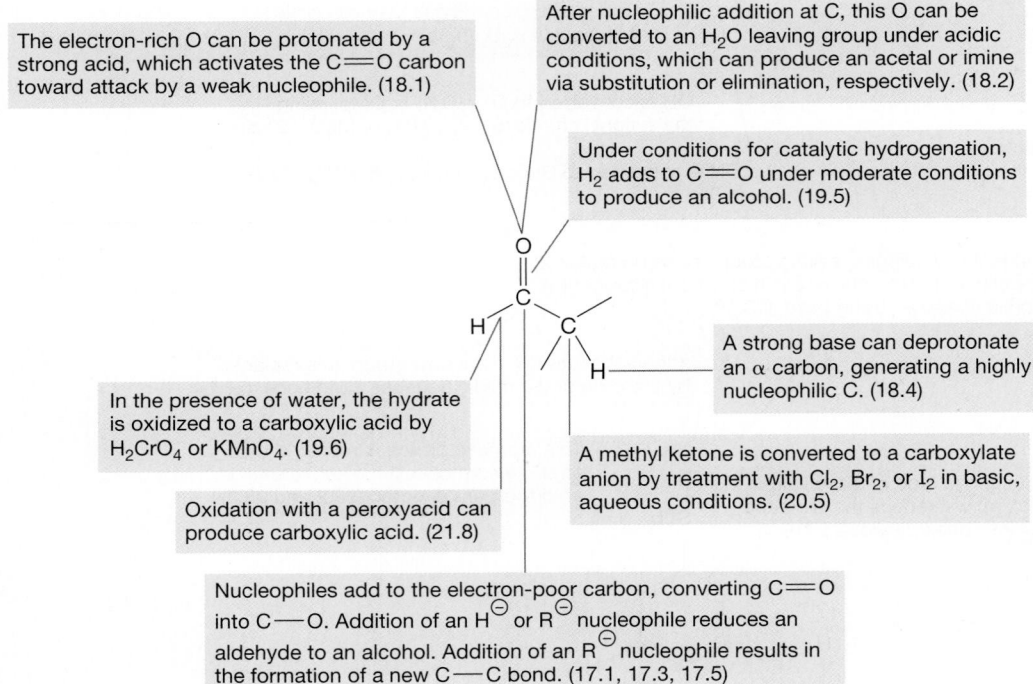

α,β-Unsaturated carbonyls

Direct addition is favored by nucleophiles that add irreversibly to the C=O group. (17.10)

Conjugate addition is favored by nucleophiles that add reversibly to the C=O group. (17.10)

Amines

Elimination of the H and the N-containing group can take place in the presence of a base when N is quaternary. (10.9)

Aromatic Ar—NH_2 produces an arenediazonium ion when treated with $NaNO_2/H^{\oplus}$. (22.9c)

N is moderately nucleophilic. (9.3)

N is weakly basic. (6.2)

Acid halides

Attack by $R^{\ominus}$ from an alkyllithium or Grignard reagent produces a 3° alcohol. (20.8)

Attack by $R^{\ominus}$ from lithium dialkylcuprate produces a ketone. (20.8)

Attack by $H^{\ominus}$ from $NaBH_4$ or $LiAlH_4$ produces a 1° alcohol. (20.6)

Attack by $H^{\ominus}$ from $LiAl(O\text{-}t\text{-Bu})_3$ produces an aldehyde. (20.7)

X (X = Cl, Br)

Alpha halogenation takes place with the enol form. (21.5)

Acyl substitution by a strong or weak nucleophile produces an acid anhydride, ester, carboxylic acid, or amide. (21.1)

Acid anhydrides

Attack by $R^{\ominus}$ from an alkyllithium or Grignard reagent produces a 3° alcohol. (20.8)

Attack by $H^{\ominus}$ from $NaBH_4$ or $LiAlH_4$ produces a 1° alcohol. (20.6)

Acyl substitution by a strong or weak nucleophile produces an ester, carboxylic acid, or amide. (21.1)

Esters

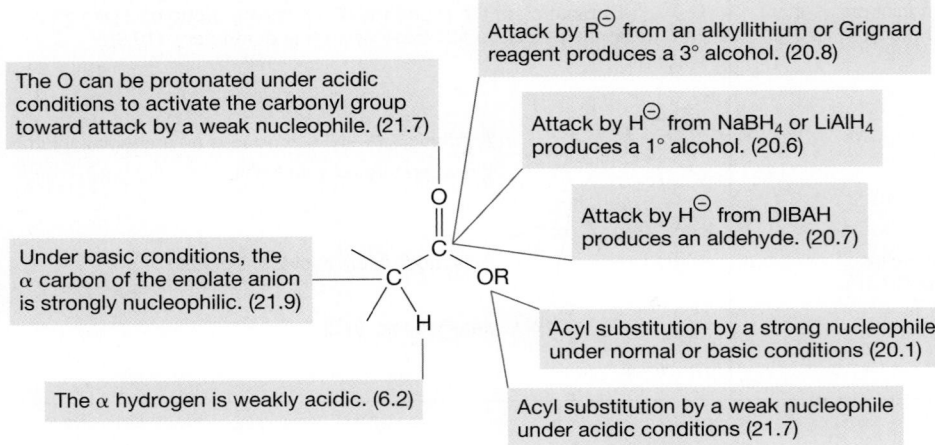

The O can be protonated under acidic conditions to activate the carbonyl group toward attack by a weak nucleophile. (21.7)

Attack by R⊖ from an alkyllithium or Grignard reagent produces a 3° alcohol. (20.8)

Attack by H⊖ from NaBH₄ or LiAlH₄ produces a 1° alcohol. (20.6)

Attack by H⊖ from DIBAH produces an aldehyde. (20.7)

Under basic conditions, the α carbon of the enolate anion is strongly nucleophilic. (21.9)

Acyl substitution by a strong nucleophile under normal or basic conditions (20.1)

The α hydrogen is weakly acidic. (6.2)

Acyl substitution by a weak nucleophile under acidic conditions (21.7)

Carboxylic acids

The O can be protonated under acidic conditions to activate the carbonyl group toward attack by a weak nucleophile. (21.7)

Attack by H⊖ from LiAlH₄ produces a 1° alcohol. (20.6)

Attack by H⊖ from DIBAH produces an aldehyde. (20.7)

This H is acidic and can be deprotonated to produce a carboxylate anion, a moderately strong nucleophile. (9.3)

SOCl₂ converts the OH to a Cl. (21.4)

Acyl substitution by a weak nucleophile under acidic conditions (21.7)

Amides

The O can be protonated under acidic conditions to activate thecarbonyl group toward attack by a weak nucleophile. (21.7)

Attack by H⊖ from LiAlH₄ produces an amine. (20.6)

Acyl substitution by a weak nucleophile under acidic conditions (21.7)

This H is mildly acidic and can be deprotonated by a strong base to produce a nucleophilic N. (6.2, 20.4)

Reactions That Alter the Carbon Skeleton

	Starting Functional Group	Typical Reagent and Reaction Conditions	Functional Group Formed	Key Electron–Rich Species	Key Electron–Poor Species	Comments	Discussed in Section (s)
(1)	Alkyne	1. NaH 2. R'—X (X = Cl, Br, I)	Alkyne	Alkynide anion	Alkyl halide	S_N2	7.2, 8.4, 8.5, 9.3b, 9.9
(2)	Alkyl halide	NaCN	Nitrile	Cyanide anion	Alkyl halide	S_N2	7.2, 8.4, 8.5, 9.9
(3)	Ketone or aldehyde	1. Base 2. R''—X (X = Cl, Br, I)	α-Alkylated ketone or aldehyde	Enolate anion	Alkyl halide	S_N2	10.3
(4)	Epoxide	1. R''—Li or R''—MgBr 2. H_3O^+	Alcohol (2-alkyl substituted)	:R''	Epoxide	S_N2	10.7a
(5)	Epoxide	NaCN H_2O	Nitrile (α-hydroxy)	NC:	Epoxide	S_N2	10.7a
(6)	Alkyne (terminal)	1. NaH 2. epoxide 3. H_3O^+	Alcohol (3-alkyn-1-ol)	R''C≡C:	Epoxide	S_N2	10.7a
(7)	Oxetane	1. R—Li or R—MgBr 2. H_3O^+	Alcohol	:R	Oxetane	S_N2	10.7a
(8)	Alkene	CH_2N_2 Δ or $h\nu$	Cyclopropyl ring	Alkene	:CH_2	Syn addition; retention of cis/trans configuration; not very useful in synthesis	12.2
(9)	Alkene	$CHCl_3$ $(CH_3)_3CONa$	Dichlorocyclopropane	Alkene	:CCl_2	Syn addition; retention of cis/trans configuration	12.2

(continued)

	Starting Functional Group	Typical Reagent and Reaction Conditions	Functional Group Formed	Key Electron–Rich Species	Key Electron–Poor Species	Comments	Discussed in Section (s)
(10)	Ketone or aldehyde	1. R'—Li or R'—MgBr 2. $H_3O^{\oplus}$	Alcohol	:R'$^{\ominus}$	δ^+	Nucleophilic addition	17.5
(11)	Nitrile	1. R'—MgBr 2. CH_3OH	Imine	:R'$^{\ominus}$	δ^+ —C≡N	Nucleophilic addition	17.5
(12)	Grignard reagent	1. $CO_2(s)$ 2. NH_4Cl	Carboxylic acid	:R$^{\ominus}$	$O=C=O$ δ^+	Nucleophilic addition	17.5
(13)	α,β-Unsaturated ketone or aldehyde	1. R'$_2$CuLi, THF 2. NH_4Cl	Ketone or aldehyde	:R'$^{\ominus}$	δ^+	Conjugate nucleophilic addition	17.11
(14)	Ketone or aldehyde	$^{\oplus}P(C_6H_5)_3$ Wittig reagent	Alkene	:R$^{\ominus}$	δ^+	Wittig reaction; cannot be used to produce tetraalkyl-substituted alkenes	17.7
(15)	Ketone or aldehyde	Sulfonium ylide H_3C—$\overset{\oplus}{S}$—CH_3	Epoxide	:R$^{\ominus}$	δ^+	Epoxidation	17.9
(16)	Ketone or aldehyde	H—CN NaOH or KCN	Cyanohydrin	NC:$^{\ominus}$	δ^+	Nucleophilic addition	18.1a
(17)	Aldehyde	NaOH	β-Hydroxycarbonyl	Enolate anion	δ^+	Nucleophilic addition	18.4
(18)	Ketone	NaOH	β-Hydroxycarbonyl	Enolate anion	δ^+	Nucleophilic addition	18.6
(19)	Aldehyde	1. NaOH 2.	β-Hydroxycarbonyl	Enolate anion	δ^+	Nucleophilic addition	18.7

(continued)

	Starting Functional Group	Typical Reagent and Reaction Conditions	Functional Group Formed	Key Electron–Rich Species	Key Electron–Poor Species	Comments	Discussed in Section (s)
(20)	Ketone	1. LDA 2.	β-Hydroxycarbonyl	Enolate anion		Nucleophilic addition	18.7
(21)	Nitrile or nitroalkane (CN or NO₂)	NaOH	(CN or NO₂)	(or NO₂)		Nucleophilic addition	18.9
(22)	Ketone	KOH		Enolate anion		Robinson annulation	18.10
(23)	Methyl ketone	1. X₂, NaOH 2. H⊕ (X = Cl, Br, I)	Carboxylic acid	:ÖH	CX₃	Nucleophilic addition–elimination (haloform reaction)	20.5
(24)	Acid chloride	1. R′—Li or R′—MgBr 2. H⊕	Tertiary alcohol	:R′	Cl	Nucleophilic addition–elimination (Grignard reaction)	20.8
(25)	Acid anhydride	1. R′—Li or R′—MgBr 2. H⊕	Tertiary alcohol	:R′		Nucleophilic addition–elimination (Grignard reaction)	20.8
(26)	Ester	1. R′—Li or R′—MgBr 2. H⊕	Tertiary alcohol	:R′	OR′	Nucleophilic addition–elimination (Grignard reaction)	20.8
(27)	Acid chloride	R′₂CuLi		:R′	Cl	Nucleophilic addition–elimination (Grignard reaction)	20.8
(28)	Ketone		Ester		⊕ÖH	Acid-catalyzed nucleophilic addition–elimination (Baeyer–Villiger oxidation)	21.8
(29)	Ester	1. NaOR″	β-ketoester		OR	Base-promoted nucleophilic addition–elimination (Claisen condensation)	21.9

	Starting Functional Group	Typical Reagent and Reaction Conditions	Functional Group Formed	Key Electron–Rich Species	Key Electron–Poor Species	Comments	Discussed in Section (s)
(30)	Acetoacetic ester	1. NaOEt 2. R—Br 3. $H_3O^{\oplus}$, Δ	Alkyl-substituted acetone		R—Br	Acetoacetic ester synthesis	21.10
(31)	Malonic ester	1. NaOEt 2. R—Br 3. $H_3O^{\oplus}$, Δ	Alkyl-substituted acetic acid		R—Br	Malonic ester synthesis	21.10
(32)	Arene	R—Cl / $AlCl_3$	Alkylarene		$R^{\oplus}$	Electrophilic aromatic substitution	22.3
(33)	Arene	Cl—C(=O)R / $AlCl_3$	Acylarene		Acylium ion	Electrophilic aromatic substitution	22.5
(34)	Akylarene	1. $KMnO_4$, Δ 2. $H^{\oplus}$	Aromatic carboxylic acid	N/A	N/A	Oxidation	22.9a
(35)	Acylarene	1. $KMnO_4$, Δ 2. $H^{\oplus}$	Aromatic carboxylic acid	N/A	N/A	Oxidation	22.9a
(36)	Aryl amine	1. $NaNO_2$, H_2SO_4 2. CuCN	Aryl nitrile	CuCN		Proceeds through a diazonium ion	22.9c
(37)	Conjugated diene	Dienophile	Substituted cyclohexene			Diels–Alder reaction. Facilitated by an electron-donating group on the diene and an electron-withdrawing group on the dienophile	24.1–24.6
(38)	Substituted cyclohexene	Δ		N/A	N/A	Retro Diels–Alder reaction	24.6
(39)	Alkene	$KMnO_4$ / $HO^{\ominus}$ Δ	Ketone and/or carboxylic acid			Oxidative cleavage	24.8

(continued)

	Starting Functional Group	Typical Reagent and Reaction Conditions	Functional Group Formed	Key Electron–Rich Species	Key Electron–Poor Species	Comments	Discussed in Section (s)
(40)	R—C≡C—R' **Alkyne**	1. $KMnO_4$, $HO^{\ominus}$, Δ 2. $H^{\oplus}$	Carboxylic acid	R—C≡C—R'	$O=Mn=O$	Oxidative cleavage	24.8
(41)	**Alkene**	1. O_3 2. CH_3SCH_3 or Zn, HOAc	Ketone and/or aldehyde			Ozonolysis	24.8
(42)	**Alkene**	1. O_3 2. H_2O_2	Ketone and/or carboxylic acid			Ozonolysis; oxidative cleavage	24.8
(43)	**1,2-Diol**	$NaIO_4$ or HIO_4	Ketone and/or aldehyde	N/A	N/A	Oxidative cleavage	24.8

Synthesizing Particular Functional Groups via Functional Group Transformations

TABLE D-1 Reactions That Produce or Alter Alkenes and Alkynes

	Starting Functional Group	Typical Reagent and Reaction Conditions	Functional Group Formed	Key Electron–Rich Species	Key Electron–Poor Species	Comments	Discussed in Section(s)
(1)	Vinyl alkane	NBS, $h\nu$	Allylic bromide	Br—Br		Favors allylic substitution over bromination of C=C	25.4d
(2)	Conjugated diene	HX (1 equiv), cold, X = Cl, Br, I	Allylic halide			Kinetic control; 1,2-addition	11.9, 11.10
(3)	Conjugated diene	HX (1 equiv), warm, X = Cl, Br, I	Allylic halide			Thermodynamic control; 1,4-addition	11.9, 11.10
(4)	Bezene	Li(s), ROH, $NH_3(\ell)$	Cyclohexa-1,4-diene		H—OR	Birch reduction forms cyclohexa-1,4-diene only	25.7b
(5)	R—C≡C—R Alkyne	H_2, Poisoned catalyst	cis-Alkene	N/A	N/A	Catalytic hydrogenation	19.5b
(6)	R—C≡C—R Alkyne	Na(s) or Li(s), $NH_3(\ell)$, −78 °C	trans-Alkene		H—NH₂	Forms trans alkene only	25.7a
(7)	—C≡C— Alkyne	HX (1 equiv), X = Cl, Br, I	Vinyl halide			Markovnikov addition, predominantly trans; not useful for synthesis	11.7
(8)	Alkyl halide	$(CH_3)_3CONa$, Δ	Alkene	$(CH_3)_3CO^{\ominus}$		E2	7.4, 8.4, 8.5, 9.3d, 9.9

(continued)

TABLE D-1 Reactions That Produce or Alter Alkenes and Alkynes (continued)

	Starting Functional Group	Typical Reagent and Reaction Conditions	Functional Group Formed	Key Electron–Rich Species	Key Electron–Poor Species	Comments	Discussed in Section(s)
(9)	Alcohol	H_3PO_4 or H_2SO_4, Δ	Alkene	$H_2\ddot{O}$ δ^-	(carbocation)	E1	7.3, 7.6, 8.2, 8.4, 8.5, 9.5a, 9.9
(10)	Amine	CH_3Br (excess), NaOH, Δ	Alkene	$\ominus\ddot{O}H$	$\oplus N(CH_3)_3$	E2	10.9
(11)	Ketone or aldehyde	R_2NH (excess), $H^\oplus$, 2° amine	Enamine	$R_2\ddot{N}H$ δ^-	$\oplus OH$	Nucleophilic addition, then E1	18.2b
(12)	β-Hydroxycarbonyl	$H^\oplus$ or $OH^\ominus$, Δ	α,β-Unsaturated carbonyl	—	—	E1 or E1cb	18.5
(13)	Vinyl halide	1. $NaNH_2$ 2. $H^\oplus$	Terminal alkyne R—C≡CH	$\ominus\ddot{N}H_2$	(vinyl halide)	E2	10.8
(14)	Vinyl halide	NaOH, Δ	Internal alkyne R—C≡C—CH_3	$\ominus\ddot{O}H$	(vinyl halide)	E2	10.8

TABLE D-2 Reactions That Produce Alkyl and Aryl Halides

	Starting Functional Group	Typical Reagent and Reaction Conditions	Functional Group Formed	Key Electron–Rich Species	Key Electron–Poor Species	Comments	Discussed in Section(s)
(1)	Alkane R—H	X_2, $h\nu$	Alkyl halide R—X	X—X	R•	X = Cl, Br	25.4
(2)	Phenyl alkane	NBS, $h\nu$	Benzylic bromide	Br—Br	(benzylic radical)		25.4d
(3)	Alkene	HX, CCl_4, X = Cl, Br, I	Alkyl halide	$\ddot{X}^\ominus$	(carbocation)	Markovnikov addition	11.1
(4)	Alkene	X_2 (X = Cl, Br), CCl_4	Vicinal dihalide	$\ddot{X}^\ominus$	(halonium ion)	Anti-addition	12.3a
(5)	Alkene	NBS, $h\nu$	Allylic bromide	Br—Br	(allylic radical)	Favors allylic substitution over bromination of C=C	25.4d

(continued)

	Starting Functional Group	Typical Reagent and Reaction Conditions	Functional Group Formed	Key Electron–Rich Species	Key Electron–Poor Species	Comments	Discussed in Section(s)
(6)	Arene	Cl_2 / $FeBr_3$	Aryl chloride		$:\overset{\cdot\cdot}{\underset{\cdot\cdot}{Cl}}:^{\oplus}$	Electrophilic aromatic substitution	22.2
(7)	Arene	Br_2 / $FeBr_3$	Aryl bromide		$:\overset{\cdot\cdot}{\underset{\cdot\cdot}{Br}}:^{\oplus}$	Electrophilic aromatic substitution	22.2
(8)	Alkene	HBr / Peroxides	Alkyl bromide		Br•	Anti-Markovnikov regiochemistry	25.5
(9)	Conjugated diene	HX (1 equiv), cold / X = Cl, Br, I	Allylic halide	$:\overset{\cdot\cdot}{\underset{\cdot\cdot}{X}}:^{\ominus}$		Kinetic control; 1,2-adduct	11.9, 11.10
(10)	Conjugated diene	HX (1 equiv), warm / X = Cl, Br, I	Allylic halide	$:\overset{\cdot\cdot}{\underset{\cdot\cdot}{X}}:^{\ominus}$		Thermodynamic control; 1,4-adduct	11.9, 11.10
(11)	Alkyne	HX (1 equiv) / X = Cl, Br, I	Vinyl halide	$:\overset{\cdot\cdot}{\underset{\cdot\cdot}{X}}:^{\ominus}$		Markovnikov addition, predominantly trans; not useful for synthesis	11.7
(12)	Alkyne	HX (2 equiv) / X = Cl, Br, I	Geminal dihalide	$:\overset{\cdot\cdot}{\underset{\cdot\cdot}{X}}:^{\ominus}$		Markovnikov addition twice	11.7
(13)	Alkyne	X_2 (X = Cl, Br) / CCl_4	1,1,2,2-Tetrahalide	$:\overset{\cdot\cdot}{\underset{\cdot\cdot}{X}}:^{\ominus}$		2 equivalents of halogen	12.3a
(14)	L = Cl, Br, I, OTs, OMs, or OTf	NaX	Alkyl halide	$:\overset{\cdot\cdot}{\underset{\cdot\cdot}{X}}:^{\ominus}$	$R-\overset{H_2}{\underset{\delta^+}{C}}-L$ or $R^{\oplus}$	S_N1 or S_N2	7.2, 7.3, 8.1, 8.4, 8.5, 9.9
(15)	Alcohol	HX	Alkyl halide	$:\overset{\cdot\cdot}{\underset{\cdot\cdot}{X}}:^{\ominus}$	$R-\overset{\oplus}{O}H_2$ or $R^{\oplus}$	S_N1 or S_N2	7.2, 7.3, 8.1, 8.4, 8.5, 9.5a, 9.9
(16)	(H or) R 1° or 2° alcohol	PBr_3	(H or) R 1° or 2° alkyl halide	$:\overset{\cdot\cdot}{Br}:^{\ominus}$	$R'-\overset{\oplus}{O}PBr_2$ (H or) R	Back-to-back S_N2 reactions; overall inversion of stereochemistry	10.1

(continued)

TABLE D-2 Reactions That Produce Alkyl and Aryl Halides (continued)

	Starting Functional Group	Typical Reagent and Reaction Conditions	Functional Group Formed	Key Electron–Rich Species	Key Electron–Poor Species	Comments	Discussed in Section(s)
(17)	Aryl amine (–NH₂)	1. NaNO₂, H₂SO₄ 2. CuBr	Aryl bromide (–Br)	CuBr	diazonium (–N₂⁺)	Proceeds through a diazonium ion	22.9c
(18)	Aryl amine (–NH₂)	1. NaNO₂, H₂SO₄ 2. CuCl	Aryl chloride (–Cl)	CuCl	diazonium (–N₂⁺)	Proceeds through a diazonium ion	22.9c
(19)	Aryl amine (–NH₂)	1. NaNO₂, H₂SO₄ 2. KI	Aryl iodide (–I)	KI	diazonium (–N₂⁺)	Proceeds through a diazonium ion	22.9c
(20)	Aryl amine (–NH₂)	1. NaNO₂, H₂SO₄ 2. HBF₄, then Δ	Aryl fluoride (–F)	HBF₄	diazonium (–N₂⁺)	Proceeds through a diazonium ion	22.9c
(21)	Ketone or aldehyde	X_2 (X = Cl, Br, I) NaOH	α-Halogenated ketone or aldehyde	Enolate anion	$\overset{\delta+}{X}-\overset{\delta-}{X}$	Multiple S_N2 reactions	10.4
(22)	Ketone or aldehyde	X_2 (X = Cl, Br, I) H⁺	α-Halogenated ketone or aldehyde	Enol	$\overset{\delta+}{X}-\overset{\delta-}{X}$	Single S_N2 reaction	10.4
(23)	Carboxylic acid	Br₂, P	R–CHBr–COOH	(enol with OH)	$\overset{\delta+}{Br}-\overset{\delta-}{Br}$	Hell–Volhard–Zelinsky reaction	21.5

TABLE D-3 Reactions That Produce Alcohols

	Starting Functional Group	Typical Reagent and Reaction Conditions	Functional Group Formed	Key Electron–Rich Species	Key Electron–Poor Species	Comments	Discussed in Section(s)
(1)	Alkene	H₂O H⁺	Alcohol	$H_2\overset{..}{O}:$	$C=C\oplus$	Markovnikov addition, acid catalysis	11.6
(2)	Alkene	X_2 (X = Cl, Br) H₂O	Halohydrin	$H_2\overset{..}{O}:$	$X^{\oplus}$ (bridged)	OH bonds to more highly substituted carbon	12.3b

(continued)

Appendix D   Synthesizing Particular Functional Groups via Functional Group Transformations / **A-17**

	Starting Functional Group	Typical Reagent and Reaction Conditions	Functional Group Formed	Key Electron–Rich Species	Key Electron–Poor Species	Comments	Discussed in Section(s)
(3)	Alkene	1. Hg(OAc)₂, H₂O 2. NaBH₄	Alcohol	H₂O:		Markovnikov addition of water; no carbocation rearrangements	12.4
(4)	Alkene	1. B₂H₆ or BH₃•THF 2. OH⊖, H₂O₂	Alcohol		BH₃	Anti-Markovnikov addition of water	12.6
(5)	Alkene	OsO₄ H₂O₂	Syn 1,2-diol			An alternate method uses a cold, basic solution of KMnO₄.	24.7
(6)	Primary alkyl halide	NaOH	1° alcohol	HÖ:⊖		S_N2 reaction	7.2, 8.4, 8.5, 9.9
(7)	Aromatic halide X = F, Cl, Br, I	NaOH Δ	Aromatic alcohol	HÖ:⊖		Proceeds through benzyne intermediate	23.10b
(8)	Tertiary alkyl halide	H₂O H⊕	3° alcohol	H₂Ö		S_N1 reaction	7.3, 8.1, 8.4, 8.5, 9.5a, 9.9
(9)	Ether	H₂O H⊕	Alcohol	H₂Ö	R⊕	S_N1 reaction	7.3, 8.1, 8.4, 8.5, 9.5a, 9.9
(10)	Epoxide	:Nu⊖ Neutral or basic	Alcohol (2-substituted)	:Nu⊖		S_N2	10.7a
(11)	Epoxide	H—Nu Acidic	Alcohol (2-substituted)	:Nu⊖		S_N2	10.7b
(12)	Aryl amine	1. NaNO₂, H₂SO₄ 2. H₂O, Cu⁺	Aryl alcohol	H₂Ö:		Proceeds through a diazonium ion	22.9c
(13)	Ketone or aldehyde	1. NaBH₄, or LiAlH₄ 2. H₃O⊕	Alcohol	:H⊖ Hydride anion		Nucleophilic addition	17.3

(continued)

TABLE D-3 Reactions That Produce Alcohols (continued)

	Starting Functional Group	Typical Reagent and Reaction Conditions	Functional Group Formed	Key Electron–Rich Species	Key Electron–Poor Species	Comments	Discussed in Section(s)
(14)	α,β-Unsaturated ketone or aldehyde	1. NaBH₄, or LiAlH₄ 2. H₃O⁺	Alcohol	:H⁻ Hydride anion		Nucleophilic addition	17.10
(15)	Ketone or aldehyde	H₂ Pt, Pd, or Ni	Alcohol	N/A	N/A	Catalytic hydrogenation	19.5c
(16)	Carboxylic acid	1. LiAlH₄ 2. H⁺	Primary alcohol	:H⁻ Hydride anion		Nucleophilic addition–elimination (reduction)	20.6
(17)	Ester	1. NaBH₄, or LiAlH₄ 2. H⁺	Primary alcohol	:H⁻ Hydride anion		Nucleophilic addition–elimination (reduction) Very slow with NaBH₄	20.6
(18)	Acid chloride	1. NaBH₄, or LiAlH₄ 2. H⁺	Primary alcohol	:H⁻ Hydride anion		Nucleophilic addition–elimination (reduction)	20.6
(19)	Acid anhydride	1. NaBH₄, or LiAlH₄ 2. H⁺	Primary alcohol	:H⁻ Hydride anion		Nucleophilic addition–elimination (reduction)	20.6

TABLE D-4 Reactions That Produce Ethers, Epoxides, and Acetals

	Starting Functional Group	Typical Reagent and Reaction Conditions	Functional Group Formed	Key Electron–Rich Species	Key Electron–Poor Species	Comments	Discussed in Section(s)
(1)	Alkene		Epoxide	C=C		Conservation of cis/trans configurations	12.5
(2)	Alkyl halide	NaOR′	Ether (symmetric or unsymmetric)	R′Ö:⁻		S_N2 (X = Cl, Br, I)	10.6
(3)	Halohydrin	NaOH	Epoxide			Intramolecular S_N2	10.7

(continued)

TABLE D-4 Reactions That Produce Ethers, Epoxides, and Acetals (continued)

	Starting Functional Group	Typical Reagent and Reaction Conditions	Functional Group Formed	Key Electron–Rich Species	Key Electron–Poor Species	Comments	Discussed in Section(s)
(4)	R—OH Alcohol	$\xrightarrow{\text{H}^{\oplus}}$ Δ	R—O—R Ether (symmetric)	R—$\overset{..}{\overset{..}{O}}$H	R—$\overset{\oplus}{O}$H$_2$	S_N1 or S_N2 (dehydration)	7.3, 8.1, 8.4, 8.5, 9.5a, 9.9, 10.6
(5)	Ketone or aldehyde	R'OH (excess) $\xrightarrow{\text{H}^{\oplus}}$	R'O OR' R—C—R (or H) (or H) Acetal	R'—$\overset{..}{\overset{\delta^-}{O}}$H	$\overset{\oplus}{O}$H C	Nucleophilic addition, then S_N1	18.2a

TABLE D-5 Reactions That Produce Amines

	Starting Functional Group	Typical Reagent and Reaction Conditions	Functional Group Formed	Key Electron–Rich Species	Key Electron–Poor Species	Comments	Discussed in Section(s)
(1)	Nitroarene (Ar—NO$_2$)	$\xrightarrow{\text{HCl}}$ Fe	Aryl amine (Ar—NH$_2$)	N/A	N/A	Reduction	22.9b
(2)	Deactivated aromatic halide (X = F, Cl, Br, I; with EWG)	$\xrightarrow{\text{RNH}_2}$	Aromatic amine (NHR, EWG)	R$\overset{\delta^-}{N}$H$_2$	X = F, Cl, Br, I; $\overset{\delta^+}{}$, EWG	Nucleophilic addition–elimination mechanism; EWG ortho or para to X	23.10a
(3)	Aromatic halide (X = F, Cl, Br, I)	$\xrightarrow{\text{LiNHR}}$	Aromatic amine (NHR)	R$\overset{..}{N}$H$^{\ominus}$	benzyne	Proceeds through benzyne intermediate	23.10b
(4)	R—X 1° Alkyl halide	1 equiv NH$_3$ $\xrightarrow{}$	R—NH$_2$ + other amines 1° amine	:NH$_3$ $\overset{\delta^-}{}$	$\overset{\delta^+}{R}$—X	S_N2 reaction; not synthetically useful	10.2
(5)	Ketone or aldehyde	R$_2$NH (excess) $\xrightarrow{\text{H}^{\oplus}}$ 2° Amine	Enamine (NH$_2$, CHR$_2$)	R$_2\overset{\delta^-}{N}$H	$\overset{\oplus}{O}$H C	Nucleophilic addition, then E1	18.2b
(6)	Imine	1. LiAlH$_4$ 2. H$_3$O$^{\oplus}$	Amine	:H$^{\ominus}$ Hydride anion	N C $\overset{\delta^+}{}$	Nucleophilic addition	17.3
(7)	Amide (R—C(=O)—NR'$_2$)	$\xrightarrow{\text{LiAlH}_4}$ Ether	Amine (R—CH$_2$—NR'$_2$)	:H$^{\ominus}$ Hydride anion	$\overset{\oplus}{N}$ R—CH	Nucleophilic addition–elimination (reduction)	20.6

(continued)

TABLE D-5 Reactions That Produce Amines (continued)

	Starting Functional Group	Typical Reagent and Reaction Conditions	Functional Group Formed	Key Electron–Rich Species	Key Electron–Poor Species	Comments	Discussed in Section(s)
(8)	Phthalimide structure	1. KOH/EtOH 2. RBr 3. KOH/H_2O	H_2N—R Primary amine	Phthalimide anion $\ddot{N}$:$^{\ominus}$	$\overset{\delta^+}{R}$—Br	S_N2 then nucleophilic addition–elimination (Gabriel synthesis)	20.4
(9)	R—C≡N Nitrile	1. $LiAlH_4$ 2. $H_3O^{\oplus}$	H H C with R and NH_2 1° Amine	:H$^{\ominus}$ Hydride anion	$\overset{\delta^+}{C}$≡N	Sequential nucleophilic additions	17.3

TABLE D-6 Reactions That Produce or Alter Ketones and Aldehydes

	Starting Functional Group	Typical Reagents and Reaction Conditions	Functional Group Formed	Key Electron–Rich Species	Key Electron–Poor Species	Comments	Discussed in Section(s)
(1)	—C≡C— Alkyne	H_2O TfOH or Tf_2NH	Ketone	$H_2\ddot{O}$:	C=C$^{\oplus}$	Markovnikov addition of H_2O, keto–enol tautomerization; acid catalyst significantly stronger than H_2SO_4	11.8
(2)	—C≡C— Alkyne	$Hg(OAc)_2$, H_2O	Ketone	$H_2\ddot{O}$:	OAc $\overset{\oplus}{Hg}$ C=C	Markovnikov addition of water	12.4
(3)	—C≡C— Alkyne	1. $(C_5H_{11})_2BH$ 2. $OH^{\ominus}$, H_2O_2	Ketone or aldehyde	—C≡C—	$(C_5H_{11})_2BH$	Anti-Markovnikov addition of water	12.7
(4)	—C≡C— Alkyne	OsO_4 H_2O_2	1,2-Dione	—C≡C—	O=Os=O	An alternate method uses a cold, basic solution of $KMnO_4$.	24.7
(5)	OH CH R R Secondary alcohol	H_2CrO_4 or $KMnO_4$	Ketone	N/A	N/A	Oxidation	19.6
(6)	OH CH_2 R Primary alcohol	PCC	Aldehyde	N/A	N/A	Oxidation	19.6

Starting Functional Group	Typical Reagents and Reaction Conditions	Functional Group Formed	Key Electron–Rich Species	Key Electron–Poor Species	Comments	Discussed in Section(s)
(7) Acetal	H_2O (excess) / $H^\oplus$	Ketone or aldehyde	$H_2\ddot{O}$: δ^-	$^{\oplus}OR$	S_N1, then E1	18.2a
(8) Ketone or aldehyde	X_2 (X = Cl, Br, I) / NaOH	α-Halogenated ketone or aldehyde	Enolate anion	$X{-}X$ δ^+ δ^-	Multiple S_N2 reaction	10.4
(9) Ketone or aldehyde	X_2 (X = Cl, Br, I) / $H^\oplus$	α-Halogenated ketone or aldehyde	Enol	$X{-}X$ δ^+ δ^-	Single S_N2 reaction	10.4
(10) α, β-Unsaturated ketone or aldehyde	H—Nu / Nu = OR, RS, NR_2	β-Substituted ketone or aldehyde	H—Nu: δ^-	δ^+	Conjugate nucleophilic addition	18.1b
(11) Imine	H_2O (excess) / $H^\oplus$	Ketone or aldehyde	$H_2\ddot{O}$: δ^-	$^{\oplus}HN$	Nucleophilic addition, then E1	18.2b
(12) Acid chloride	$LiAlH(O\text{-}t\text{-}Bu)_3$ / −78 °C	Aldehyde	:$H^\ominus$ Hydride anion	δ^+	Nucleophilic addition–elimination (reduction)	20.7
(13) Ester	1. DIBAH, −78 °C / 2. $H_3O^\oplus$	Aldehyde	:$H^\ominus$ Hydride anion	δ^+	Nucleophilic addition–elimination (reduction)	20.7

TABLE D-7 Reactions That Produce or Alter Carboxylic Acids

Starting Functional Group	Typical Reagents and Reaction Conditions	Functional Group Formed	Key Electron–Rich Species	Key Electron–Poor Species	Comments	Discussed in Section(s)
(1) Primary alcohol	H_2CrO_4 or 1. $KMnO_4$, $OH^\ominus$ 2. $H^\oplus$	Carboxylic acid	N/A	N/A	Oxidation	19.6

TABLE D-7 Reactions That Produce or Alter Carboxylic Acids (continued)

	Starting Functional Group	Typical Reagents and Reaction Conditions	Functional Group Formed	Key Electron–Rich Species	Key Electron–Poor Species	Comments	Discussed in Section(s)
(2)	Aldehyde	(peroxyacid) →	Carboxylic acid			Acid-catalyzed nucleophilic addition–elimination (Baeyer–Villiger oxidation)	21.8
(3)	Carboxylic acid	Br_2, P →	α-Halo acid		$Br\!-\!Br$	Hell–Volhard–Zelinsky reaction	21.5
(4)	Ester	1. NaOH 2. $H^{\oplus}$ →	Carboxylic acid	$^{\ominus}:\!OH$		Nucleophilic addition–elimination (saponification)	20.2
(5)	Ester	H_2O / $H^{\oplus}$ →	Carboxylic acid	$H_2\overset{..}{O}{:}$		Acid-catalyzed nucleophilic addition–elimination (hydrolysis)	21.7
(6)	Amide	1. NaOH →	Carboxylic acid	$^{\ominus}:\!OH$		Nucleophilic addition–elimination	20.4
(7)	Acid chloride	H_2O →	Carboxylic acid	$H_2\overset{..}{O}{:}$		Nucleophilic addition–elimination (hydrolysis)	21.1
(8)	Acid anhydride	H_2O →	Carboxylic acid	$H_2\overset{..}{O}{:}$		Nucleophilic addition–elimination (hydrolysis)	21.2

TABLE D-8 Reactions That Produce or Alter Esters

	Starting Functional Group	Typical Reagents and Reaction Conditions	Functional Group Formed	Key Electron–Rich Species	Key Electron–Poor Species	Comments	Discussed in Section(s)
(1)	Carboxylic acid	R'OH / $H^{\oplus}$ →	Ester	$R'\overset{..}{O}H$		Acid-catalyzed nucleophilic addition–elimination (Fischer esterification)	21.7
(2)	Carboxylic acid	CH_2N_2 →	Ester	Carboxylate anion	$H_3C\!-\!N_2^{\oplus}$	S_N2	10.5
(3)	Ester	NaOR'' / R''OH →	Ester	$^{\ominus}:\!\overset{..}{O}R''$		Nucleophilic addition–elimination (transesterification)	20.1a, 21.7

(continued)

Appendix D   Synthesizing Particular Functional Groups via Functional Group Transformations / **A-23**

TABLE D-8 Reactions That Produce or Alter Esters (continued)

	Starting Functional Group	Typical Reagents and Reaction Conditions	Functional Group Formed	Key Electron–Rich Species	Key Electron–Poor Species	Comments	Discussed in Section(s)
(4)	Ester (R–C(=O)–OR')	R"OH, H⁺	Ester (R–C(=O)–OR")	$\overset{\delta-}{ROH}$	R–C(=O⁺H)–OR'	Acid-catalyzed nucleophilic addition–elimination (transesterification)	21.7
(5)	Acid chloride (R–C(=O)–Cl)	LiOR'	Ester (R–C(=O)–OR')	:ÖR"⁻	R–C($\delta+$)(=O)–Cl	Nucleophilic addition–elimination	20.3
(6)	Acid chloride (R–C(=O)–Cl)	R"OH	Ester (R–C(=O)–OR")	$\overset{\delta-}{R"ÖH}$	R–C($\delta+$)(=O)–Cl	Nucleophilic addition–elimination (alcoholysis)	21.1
(7)	Acid anhydride (R–C(=O)–O–C(=O)–R)	NaOR'	Ester (R–C(=O)–OR')	:ÖR"⁻	R–C($\delta+$)(=O)–O–C(=O)–R	Nucleophilic addition–elimination	20.3
(8)	Acid anhydride (R–C(=O)–O–C(=O)–R)	R'OH	Ester (R–C(=O)–OR')	$\overset{\delta-}{R'ÖH}$	R–C($\delta+$)(=O)–O–C(=O)–R	Nucleophilic addition–elimination (alcoholysis)	21.2

TABLE D-9 Reactions That Produce Amides

	Starting Functional Group	Typical Reagents and Reaction Conditions	Functional Group Formed	Key Electron–Rich Species	Key Electron–Poor Species	Comments	Discussed in Section(s)
(1)	Ester (R–C(=O)–OR')	LiNR"₂	Amide (R–C(=O)–NR"₂)	:NR"₂⁻	R–C($\delta+$)(=O)–OR'	Nucleophilic addition–elimination	20.3
(2)	Nitrile (R–C≡N)	H_2O (1 equiv.), H⁺ or HO⁻	Amide (R(or H)–C(=O)–NH₂)	$H_2\overset{\delta-}{Ö}$: or :ÖH⁻	$-\overset{\delta+}{C}≡N$ or $-C≡\overset{\oplus}{NH}$	Nucleophilic addition	18.2c
(3)	Acid chloride (R–C(=O)–Cl)	R'₂NH, Et₃N or pyridine	Amide (R–C(=O)–NR'₂)	$\overset{\delta-}{R'_2\ddot{N}H}$	R–C($\delta+$)(=O)–Cl	Nucleophilic addition–elimination (aminolysis)	21.3
(4)	Acid anhydride (R–C(=O)–O–C(=O)–R)	R'₂NH, Et₃N or pyridine	Amide (R–C(=O)–NR'₂)	$\overset{\delta-}{R'_2\ddot{N}H}$	R–C($\delta+$)(=O)–O–C(=O)–R	Nucleophilic addition–elimination (aminolysis)	21.3

TABLE D-10 Reactions That Alter Aromatic Compounds

	Starting Functional Group	Typical Reagents and Reaction Conditions	Functional Group Formed	Key Electron–Rich Species	Key Electron–Poor Species	Comments	Discussed in Section(s)
(1)	Arene	Br_2 / $FeBr_3$	Br / Aryl bromide		$:\overset{..}{Br}:^{\oplus}$	Electrophilic aromatic substitution	22.2
(2)	NH_2 / Aryl amine	1. $NaNO_2$, H_2SO_4 2. CuBr	Br / Aryl bromide	CuBr	$N_2^{\oplus}$	Proceeds through a diazonium ion	22.9c
(3)	Arene	Cl_2 / $FeCl_3$	Cl / Aryl chloride		$:\overset{..}{Cl}:^{\oplus}$	Electrophilic aromatic substitution	22.2
(4)	NH_2 / Aryl amine	1. $NaNO_2$, H_2SO_4 2. CuCl	Cl / Aryl chloride	CuCl	$N_2^{\oplus}$	Proceeds through a diazonium ion	22.9c
(5)	NH_2 / Aryl amine	1. $NaNO_2$, H_2SO_4 2. KI	I / Aryl iodide	KI	$N_2^{\oplus}$	Proceeds through a diazonium ion	22.9c
(6)	NH_2 / Aryl amine	1. $NaNO_2$, H_2SO_4 2. HBF_4, then Δ	F / Aryl fluoride	HBF_4	$N_2^{\oplus}$	Proceeds through a diazonium ion	22.9c
(7)	NH_2 / Aryl amine	1. $NaNO_2$, H_2SO_4 2. H_2O, $Cu^{\oplus}$	OH / Aryl alcohol	H_2O	$N_2^{\oplus}$	Proceeds through a diazonium ion	22.9c
(8)	SO_3H / Arene sulfonic acid	$H_3O^{\oplus}$	H / Arene	SO_3H	$H-\overset{\oplus}{O}H_2$	Electrophilic aromatic substitution	22.7
(9)	NH_2 / Aryl amine	1. $NaNO_2$, H_2SO_4 2. H_3PO_2	H / Arene	N/A	$N_2^{\oplus}$	Proceeds through a diazonium ion	22.9c
(10)	NO_2 / Nitroarene	HCl / Fe	NH_2 / Aryl amine	N/A	N/A	Reduction	22.9b

(continued)

TABLE D-10 Reactions That Alter Aromatic Compounds (continued)

	Starting Functional Group	Typical Reagents and Reaction Conditions	Functional Group Formed	Key Electron–Rich Species	Key Electron–Poor Species	Comments	Discussed in Section(s)
(11)	Arene	HNO_3 / H_2SO_4	Nitroarene (NO_2)		$^{\oplus}NO_2$	Electrophilic aromatic substitution	22.6
(12)	Arene	SO_3 / H_2SO_4	Arene sulfonic acid (SO_3H)		SO_3	Electrophilic aromatic substitution	22.7
(13)	Deactivated aromatic halide (X = F, Cl, Br, I; EWG)	:Nu$^{\ominus}$	Nu / EWG	:Nu$^{\ominus}$	X = F, Cl, Br, I; EWG	Nucleophilic addition–elimination mechanism; EWG ortho or para to X	23.10a
(14)	Aromatic halide (X = F, Cl, Br, I)	:Base$^{\ominus}$	Base	:Base$^{\ominus}$		Proceeds through benzyne intermediate	23.10b
(15)	Phenyl alkane	NBS / $h\nu$	Benzylic bromide (Br)	Br—Br	$\dot{C}H_2$		25.4d

TABLE D-11 Reactions That Produce Alkanes

	Starting Functional Group	Typical Reagents and Reaction Conditions	Functional Group Formed	Key Electron–Rich Species	Key Electron–Poor Species	Comments	Discussed in Section(s)
(1)	Alkene	H_2 / Pt, Pd, or Ni	Alkane	N/A	N/A	Catalytic hydrogenation	19.5a
(2)	Ketone or aldehyde (R or H, R or H)	1. H_2N—NH_2, $H^{\oplus}$ 2. KOH/H_2O, Δ	Alkane	$H_2N\overset{\delta-}{\ddot{N}}H_2$	$^{\oplus}OH$	Nucleophilic addition, then E1, then E2	18.3
(3)	Ketone or aldehyde (R, R or H)	Zn/Hg, HCl / H_2O, reflux	Alkane	N/A	N/A	Clemmensen reduction	19.3b
(4)	Ketone or aldehyde (R, R or H)	1. $HSCH_2CH_2SH$, $H^{\oplus}$ 2. Raney nickel (H_2)	Alkane	N/A	N/A	Raney-nickel reduction	19.3c

TABLE D-12 Reactions That Produce Other Species

	Starting Functional Group	Typical Reagents and Reaction Conditions	Functional Group Formed	Key Electron–Rich Species	Key Electron–Poor Species	Comments	Discussed in Section(s)
(1)	Amine	Excess CH$_3$Br	Quaternary ammonium salt			Multiple S$_N$2 reactions	10.2
(2)	Alkyl halide	1. P(C$_6$H$_5$)$_3$ 2. R—Li	Wittig reagent	Triphenyl phosphine		S$_N$2	17.7
(3)	Alkyl halide	1. S(CH$_3$)$_2$ 2. NaH	Sulfonium ylide			S$_N$2	17.9
(4)	Haloalkane	Mg(s) / Ether	Grignard reagent	N/A	N/A	Dissolving metal reduction	19.1a
(5)	Haloalkane	Li(s) / Ether	Alkyllithium reagent	N/A	N/A	Dissolving metal reduction	19.1a
(6)	Alkyllithium reagent	CuI / Ether	Lithium dialkyl cuprate	N/A	N/A	N/A	19.1
(7)	Carboxylic acid	SOCl$_2$	Acid chloride			Back-to-back nucleophilic addition–elimination	21.4
(8)	Alcohol		Sulfonate ester			Nucleophilic addition–elimination	21.6
(9)	Ketone or aldehyde	NH$_3$ or R'NH$_2$ (excess) H$^\oplus$	Imine			Nucleophilic addition, then E1	18.2b

Glossary

Absorbance (A) a measure of light absorbed by a sample for spectroscopy, ranging from 0 (no absorption) to infinity (complete absorption). $A = \log(I_{detected}/I_{source}) = 2 - \log(\%T)$

Absorption bands (also known as *peaks*) features of a spectrum where absorbance is high.

Acetal a functional group in which two alkoxy (—OR) groups are attached to the same carbon atom.

Acetoacetic ester synthesis a synthesis scheme that converts acetoacetic ester into an alkyl- or dialkyl-substituted acetone.

Acetylide anion $RC\equiv C^-$, a strong nucleophile.

Achiral term used to describe a molecule that does not have an enantiomer; the molecule's mirror image is superimposable on itself.

Achiral environment environment that is superimposable on its mirror image.

Acid workup a synthetic step in which acid is added to reprotonate a product or remove leftover basic components (e.g., extra Grignard or alkyllithium reagents) after the main reaction is completed.

Acid-catalyzed alkoxylation reaction a reaction in which an alcohol adds across a $C\equiv C$ or $C\equiv C$ bond in the presence of acid.

Acid-catalyzed hydration reaction the addition of water to a $C\equiv C$ or $C\equiv C$ bond that occurs more readily in the presence of strong acid.

Acidity constant (K_a) an experimentally obtained constant that reflects the strength of an acid (i.e., an acid's propensity to donate a proton).

Acrylics See *polyacrylates*.

Activated carbonyl group a $C\equiv O$ group made more susceptible to nucleophilic attack, generally by protonation of the carbonyl O atom.

Activating group a substituent that increases the reactivity of a molecule in a particular reaction; often refers to substituents that make an aromatic ring more susceptible to electrophilic aromatic substitution.

Active methylene compound a CH_2 group with two adjacent $C\equiv O$ groups; can be converted into a highly nucleophilic enolate anion relatively easily.

Acylium ion a cation of the form $R\text{—}C\equiv O^+$.

1,2-Addition (also known as *direct addition*) a type of reaction in which two substituents add to two adjacent centers.

1,4-Addition (also known as *conjugate addition*) a type of reaction in which two substituents add to two different centers that have a 1,4 relative positioning.

Addition polymerization See *chain-growth polymerization*.

Adduct the addition product of two molecules.

Alcoholysis a reaction in which an alcohol reactant is responsible for breaking bonds; used to convert a carboxylic acid derivative into an ester.

Aldohexose a six-carbon sugar with an aldehyde group.

Aldol a compound characterized by a β-hydroxy carbonyl group.

Aldol addition a reaction in which an enolate anion adds to the carbonyl group of a ketone or an aldehyde to produce a β-hydroxy carbonyl.

Aldol condensation a reaction consisting of an *aldol addition* followed by the elimination of water, resulting in the production of an α,β-unsaturated carbonyl.

Aldopentose a five-carbon sugar with an aldehyde group.

Aldose a sugar whose carbonyl group is part of an aldehyde.

Alkanes acyclic compounds consisting of only C—C and C—H single bonds.

Alkenes functional groups having the bonding arrangement, $C\equiv C$; compounds containing an alkene functional group.

Alkoxy group the group —OR.

Alkoxymercuration–reduction the sequence of reactions in which an Hg^{+2} species adds to an alkene or alkyne, followed by reduction (typically with $NaBH_4$), resulting in the *Markovnikov addition* of an alcohol.

Alkyl group a portion of a molecule that consists only of C—C and C—H single bonds.

Alkyllithium reagents reagents with a C—Li bond wherein the C atom is strongly nucleophilic and strongly basic.

1,2-Alkyl shift a carbocation rearrangement wherein an alkyl group migrates to an adjacent atom.

Alkyl substituent a substituent consisting only of C—C and C—H single bonds.

Alkyl substitution the number of alkyl groups bonded to an isomer's alkene carbon atoms.

Alkylation reaction in which a hydrogen atom is replaced by an alkyl group.

Alkylborane ($R\text{—}BH_2$) a functional group or compound in which an alkyl group has replaced hydrogen on borane, BH_3.

Alkynes functional groups having the bonding arrangement, $C\equiv C$; compounds containing an alkyne functional group.

All-anti conformation (also known as *zigzag conformation*) the lowest energy conformation of an alkyl chain; conformation wherein an alkyl chain has the *anti conformation* at each C—C bond.

Allyl the group $\text{—}CH_2\text{—}CH\equiv CH_2$.

α-Amino acids the relatively few types of small organic molecules from which proteins are constructed; an α-amino acid contains an amine, a carboxylic acid, and a side chain that are all attached to the a carbon of the amino acid.

α-Anomer a diastereomeric form of a cyclic sugar in which the OH group attached to C1 is on the same side of the ring as the CH_2OH group.

α,β-Unsaturated carbonyl a carbonyl-containing molecule that has a double or triple bond connecting the α and β carbons.

α Carbon a carbon atom that is attached to a carbonyl ($C\equiv O$) group.

α-Cleavage a fragmentation pathway in mass spectrometry in which a C—C bond adjacent to the radical is cleaved.

α-Elimination type of reaction in which a proton and a leaving group are eliminated from the same atom; mechanism by which dichlorocarbene is produced from trichloromethane.

α-Helix a type of secondary structure in a protein characterized by a tight winding of the protein's backbone around an axis.

α Spin state a proton state when it has a spin of $+\frac{1}{2}$ a.u.

Aminolysis a reaction in which an amine reactant is responsible for breaking bonds; used to convert a carboxylic acid derivative into an amide.

Amorphous solid a solid that does not have a well-defined crystal structure.

Analyte a sample that one wishes to analyze.

Angle strain the increase in energy that results from the deviation of a bond angle from its ideal valence shell electron pair repulsion (VSEPR) angle.

Anion a negatively charged ion.

Anionic polymerization a polymerization reaction in which the initiator is anionic and the propagation steps involve anionic species.

Annulenes a class of compounds that are monocyclic, fully conjugated, and contain only carbon and hydrogen.

Anode positively charged pole of an electrochemical cell, such as the one used in electrophoresis.

Anomeric carbon the carbon atom of a cyclic sugar that is part of an acetal or a hemiacetal and is the carbonyl carbon of the sugar in its acyclic form. Depending on the stereochemical configuration of the anomeric carbon, the cyclic sugar can exist as one of two diastereomers called *anomers*.

Anomers diastereomers of a cyclic sugar that differ in the stereochemical configuration at the *anomeric carbon*. A cyclic sugar can exist as the α or β anomer.

Antiaromatic transition state a transition state that is unusually unstable due to the cyclic delocalization of a Hückel number of electrons.

Anti conformation staggered conformation wherein bulky groups are 180° apart in a Newman projection.

Anti-Hückel numbers the set of numbers: 4, 8, 12, 16, and so on, which can be described as either even numbers of pairs or numbers that are consistent with $4n$, where n is a positive integer.

Anti-Markovnikov addition a reaction in which the incoming halide in H—X, a Brønsted acid, adds to an alkene or alkyne at the less-substituted carbon (with more protons) and the proton adds to the other. See *Markovnikov's rule* and *Markovnikov addition*.

Anti-Markovnikov regiochemistry the regiochemistry obtained from *anti-Markovnikov addition*.

Anti-Zaitsev product the less-substituted alkene product of an elimination reaction, produced when the reaction does not follow Zaitsev's rule (also known as the *Hofmann product*).

Antiaromatic term used to describe compounds that have cyclic π systems that are unusually unstable; the cyclic π system contains an *anti-Hückel number* of electrons.

Antibonding MO a molecular orbital (MO) that is higher in energy than its contributing atomic orbitals.

Anticoplanar (also known as *antiperiplanar*) the conformation occurring when the H and leaving group on adjacent atoms are anti to each other and all four atoms reside in the same plane.

Antiperiplanar See *anticoplanar*.

Aprotic not possessing a hydrogen-bond donor.

Aprotic solvent a solvent that does not possess a hydrogen-bond donor.

Arenium ion (also known as a *Wheland intermediate*) a cationic intermediate in an electrophilic aromatic substitution produced by the addition of an electrophile to an aromatic ring.

Aromatic term used to describe compounds that have cyclic π systems that are unusually stable; the cyclic π system contains a *Hückel number* of electrons.

Aromatic halogenation an electrophilic aromatic substitution reaction that replaces an aromatic proton with a halogen.

Aromaticity the property of a molecule containing an unusually stable π system, and thus is consistent with Hückel's rules.

Arrow pushing See *curved arrow notation*.

Aryl a substituent or functional group made from an aromatic ring. See *phenyl* or *benzyl*.

Asymmetric stretch a vibrational mode involving two bonds within a molecule whereby one bond lengthens while the other shortens.

Asymmetric syntheses syntheses that favor the formation of a particular enantiomer or diastereomer.

Atactic term used to describe a polymer that has no regularity in the stereochemical configurations along its main chain.

Atomic number (Z) the number of protons in the nucleus.

Atomic orbitals regions of space where the probability of finding a given electron is high.

Attacking species any species that can act as a nucleophile or base to displace bonding electrons and/or a leaving group from an atom.

Aufbau principle the principle that states that each successive electron must fill the lowest energy orbital available.

Axial term used to describe a bond that is perpendicular to the plane that is roughly defined by the ring to which it is attached (i.e., the *equator* of the molecule).

2,2′-Azobisisobutyronitrile (AIBN) the molecule $(CH_3)_2C(CN)N\!\!=\!\!NC(CN)(CH_3)_2$, a common radical initiator.

Backbone the carbon chain or ring of a molecule to which substituents are attached. In polymers, also known as the *polymer chain* or *main chain*.

Backside attack occurs when a nucleophile approaches the substrate from the side opposite the leaving group, leading to an inversion of stereochemical configuration at the attacked carbon; occurs in all S_N2 reactions.

Baeyer–Villiger oxidation an oxidation reaction in which a peroxyacid (RCO_3H) converts a ketone or an aldehyde into an ester or a carboxylic acid.

Base peak the peak produced by the ion with the greatest intensity on a mass spectrum; it is assigned a relative abundance of 100% and used as a reference to assign the relative abundances of other peaks.

Base-promoted transesterification a reaction that converts one ester into another under basic conditions.

Beer–Lambert law law describing the direct relationship of absorbance (A) to sample concentration (C), molar absorptivity (ε), and light path length (l) in spectroscopy.

Bending one of two basic, independent types of vibrational motion. In bending, an angle—either a bond angle or a dihedral angle—becomes larger and smaller throughout one vibrational cycle.

Benzene C_6H_6, a nonpolar, aromatic ring; six-carbon annulene.

Benzenediazonium ion the reactive species $C_6H_5N_2{}^+$, used to produce a variety of substituted benzenes.

Benzyl a —CH_2—C_6H_5 group.

Benzylic carbon   a carbon atom attached directly to a phenyl ring.

Benzyne intermediate   a six-membered carbon ring whose Lewis structure exhibits two C=C bonds and one C≡C bond; an intermediate in a type of nucleophilic aromatic substitution reaction.

β-Carbon   a carbon atom located two bonds away from an atom or group of interest—usually a C=O group.

β-Diacid   a species having two carboxylic acid functional groups separated by a single carbon atom: HO_2C—C—CO_2H.

β-Hydroxylcarbonyl   a species in which a hydroxyl (—OH) group is bonded to the β carbon of a carbonyl-containing functional group.

β-Keto ester   a species in which the β carbon of an ester is part of a ketone.

β-Pleated sheet   a type of secondary structure in a protein in which several portions of the backbone align next to each other in the same region of space.

β spin state   a proton state when it has a spin of $-\frac{1}{2}$ a.u.

Betaine   a species in which a positive and a negative charge are separated by two uncharged atoms.

B_{ext}   abbreviation for *external magnetic field*.

Bicyclic compound   a compound consisting of two rings.

Bimolecular elimination (E2) step   elementary step wherein a proton and a leaving group are eliminated from adjacent atoms, and a double bond or triple bond is formed.

Bimolecular homolytic substitution (S_H2)   an elementary step in which a radical forms a bond to a closed-shell species and displaces another radical.

Bimolecular nucleophilic substitution (S_N2) step   elementary step wherein a molecular species undergoes substitution in which a leaving group is replaced by a nucleophile.

Biomolecules   the particular organic molecules found almost exclusively in or produced by living organisms.

Biopolymers   polymers (i.e., molecules consisting of monomers) that are produced biosynthetically; examples include polypeptides (proteins), polysaccharides (carbohydrates), and polynucleotides (nucleic acids) such as DNA and RNA.

Birch reduction   a dissolving metal reduction that reduces a benzene ring to a cyclohexa-1,4-diene.

Bond dipole   a separation of partial positive and negative charges along a covalent bond.

Bond energy   See *bond strength*.

Bonding pairs   pairs of electrons that make up covalent bonds.

Bond length   the internuclear distance at which energy is a minimum.

Bond strength   (also known as *bond energy*) the energy that would be required to increase the distance between two bonded atoms from bond length to infinity.

Bonding MO   a molecular orbital (MO) that is lower in energy than its contributing atomic orbitals.

Borate ester   a class of compounds or functional groups, $B(OR)_3$.

Branched polymer   a polymer in which hydrogens or substituents on the main chain of the polymer are replaced by another chain of the polymer.

Bridgehead carbons   carbon atoms that are simultaneously part of more than one ring.

Broadband decoupling   a technique used to "turn off" the coupling between ^{13}C and ^1H nuclei so that carbon signals are not split by attached protons in nuclear magnetic resonance spectroscopy; radiofrequency radiation forces protons to rapidly switch between α and β spin states, producing a zero average magnetic field produced by each proton.

Bromination   a reaction that adds bromine atoms to a molecule; can be accomplished by substituting a bromine atom for a hydrogen atom or by adding molecular bromine (Br_2) to a C=C or C≡C bond.

Bromohydrin   a type of compound or functional group in which a bromine and hydroxyl group (—OH) are attached to two adjacent carbons.

Bromonium ion intermediate   a positively charged intermediate containing a three-membered ring composed of two carbon atoms and a bromine atom.

Brønsted–Lowry acid   a species that donates a proton in a proton transfer reaction.

Brønsted–Lowry acid–base reaction   See *proton transfer reaction*.

Brønsted–Lowry base   a species that accepts a proton in a proton transfer reaction.

^{13}C NMR spectroscopy   a spectroscopy method in which radiofrequency electromagnetic radiation is used to cause spin flips in the nuclei of carbon-13 atoms.

Cahn–Ingold–Prelog convention   a set of tie-breaking rules applied to substituents to determine their relative priorities in establishing stereochemical configurations.

Carbanion   an *anion* in which a negative formal charge appears on carbon.

Carbene   an uncharged species containing a carbon atom that possesses two bonds and a lone pair of electrons.

Carbocation rearrangement   an elementary step in which the connectivity within a carbocation changes. Examples include a *1,2-hydride shift* and a *1,2, methyl* shift.

Carbocations   species that contain a positively charged carbon atom (C^+); key reactive intermediates in a variety of chemical reactions.

Carbohydrates   compound with the formula $C_xH_{2y}O_y$. (see *saccharides*).

Carbon chain polymers   polymers in which the main chain consists only of carbon atoms.

Carbon skeleton   the bonding arrangement (connectivity) of the carbon atoms.

Carboxylation   a reaction in which a substrate gains a carboxylic acid group.

Carboxylic acid derivatives   functional groups that can be produced relatively easily from carboxylic acids; generally characterized by the presence of a leaving group attached to a carbonyl (C=O) carbon.

Catalytic hydrogenation   a reaction in which hydrogen atoms are added to a molecule in the presence of a catalyst.

Cathode   negatively charged pole of an electrochemical cell, such as the one used in electrophoresis.

Cation   a positively charged ion.

Cationic polymerization   a polymerization reaction in which the initiator is cationic and the propagation steps involve cationic species.

Chain-growth polymerization   (also known as *addition polymerization*) a polymerization reaction in which the polymer grows one monomer at a time and the reaction takes place at specific sites on the growing polymer.

Chain reaction   a reaction whose mechanism consists of initiation, propagation, and termination steps and whose net reaction is the sum of its propagation steps.

Chair conformation a cyclohexane conformation in which all bond angles of the ring are about 111°, and all C — C bonds are staggered.

Chair flip process of single-bond rotations that converts one chair conformation of cyclohexane (or a substituted cyclohexane) into the other.

Chemical distinction test protocol used to determine the number of chemically distinct atoms in a molecule.

Chemical reaction the transformation of one substance (reactant) into another substance (product), typically through changes in chemical bonds.

Chemical shift (δ) a measure of the extent to which a nuclear magnetic resonance (NMR) signal's frequency differs from that of a reference compound (usually tetramethylsilane [TMS]).

Chemically distinct protons See *heterotopic protons*.

Chemical environments electron distributions surrounding a particular location in space, governing the chemical behavior of an atom.

Chemically equivalent protons See *homotopic protons*.

Chiral term used to describe a molecule that has an *enantiomer*.

Chiral environment environment that is nonsuperimposable upon its mirror image.

Chlorination a reaction that adds chlorine atoms to a molecule; can be accomplished by substituting a chlorine atom for a hydrogen atom or by adding molecular chlorine (Cl_2) to a C=C or C≡C bond.

Chlorohydrin a type of compound or functional group in which a chlorine and hydroxyl group (— OH) are attached to two adjacent carbons.

Chloronium ion intermediate a positively charged species containing a three-membered ring composed of two carbon atoms and a chlorine atom.

Chromate ester a functional group characterized by a chromium atom that has both Cr=O and Cr—OR bonds; an intermediate in the chromium oxidation of alcohols and aldehydes.

Chromic acid the compound H_2CrO_4, which is used to oxidize secondary alcohols to ketones and primary alcohols to carboxylic acids.

Chromophore the species responsible for absorbing light or other electromagnetic radiation.

Cis term that describes the configuration wherein two atoms are on the same side of a double bond or plane of a ring.

Claisen condensation reaction a reaction in which an enolate anion undergoes a nucleophilic addition–elimination with an ester, typically producing a β-keto ester.

Clemmensen reduction a reaction that uses a zinc amalgam under acidic conditions to reduce the C=O group of a ketone or an aldehyde to a methylene (CH_2) group; named after Erik Christian Clemmensen.

Closed-shell species a molecule or an ion in which all electrons are paired.

Combination See *radical coupling*.

Common names (also known as *trivial names*) nonsystematic names given to molecules based on their properties or origin.

Complementary nucleic acids two DNA or RNA strands are complementary when their facing nitrogenous bases pair, forming strong hydrogen bonds and a characteristic double helix. In DNA, adenine pairs with thymine, and guanine with cytosine; in RNA, adenine pairs with uracil, and guanine pairs with cytosine.

Complex splitting in NMR spectroscopy, splitting that arises when a nucleus is coupled to more than one distinct type of

other nuclei; the resulting splitting patterns do not follow the $N + 1$ rule.

Condensation polymerization a type of *step-growth polymerization* in which a small molecule is eliminated when two monomers or growing polymer chains bond together.

Condensed formulas a way of representing molecules using a line of text with hydrogens written immediately to the right of the atom to which they are bonded.

Configurational isomers isomers that have the same connectivity, but differ in a way other than by rotations about single bonds. See the two types: *enantiomers* and *diastereomers*.

Conformational analysis a plot of a molecule's energy as a function of one or more of its dihedral angles.

Conformational isomers (also known as *conformers*) isomers that have the same connectivity (see *stereoisomers*) and differ by rotations about single bonds.

Conformers See *conformational isomers*.

Conjugate acid the species that a base becomes after acquiring a proton.

Conjugate addition See *1,4-addition*.

Conjugate base the species that an acid becomes after losing a proton.

Conjugated an adjective used in reference to *p* atomic orbitals or π molecular orbitals, and meaning that they are adjacent and overlap in a side-by-side fashion.

Connectivity (also bonding scheme) information that conveys which atoms are bonded together and by what types of bonds (single, double, or triple).

Conservation of number of orbitals the generalized concept that when *n* orbitals are mixed, *n* different orbitals must be produced. See *molecular orbital theory*.

Constitutional isomers (also known as *structural isomers*) isomers that share the same molecular formula but differ in their connectivity.

Constructive interference interference that occurs when waves or orbitals overlap with the same phase; constructive interference produces a new wave or orbital that has been *built up* by the addition of the amplitudes from the contributing waves.

Contact surface area the area over which two molecules interact.

Convergent synthesis a method of synthesizing molecules, wherein portions of the target are synthesized separately and then assembled at a later stage, usually resulting in a higher yield.

Coordination step a step in a mechanism in which a single bond is formed using two electrons from the same atom and no bonds are broken.

Core electrons inner, lower-energy shell electrons that are not used for bonding.

Coupled term used to describe two or more atoms that exhibit *spin-spin coupling*.

Coupling constant (J) the frequency difference between peaks of a split NMR signal, independent of the external magnetic field (B_{ext}).

Covalent bond one of two types of fundamental bonds in chemistry; it is characterized by the sharing of valence electrons between two or more atoms.

Crossed aldol addition an *aldol addition* reaction in which the carbonyl-containing reactants are different.

Crossed Claisen condensation reaction a *Claisen condensation reaction* in which the carbonyl-containing reactants are different.

Crosslink a portion of a polymer that is responsible for joining two separate main chains.

Crosslinked term that describes two separate polymer main chains that are joined together.

Crystal lattice the regular array into which ions or molecules in a solid are arranged.

Crystalline solid a solid that has a well-defined crystal structure (i.e., a structure in which the locations of the atoms form a regular pattern).

C-terminus the end of a polypeptide (protein) chain that is part of a carboxyl group.

Curved arrow the symbol that illustrates the electron movement necessary to change one resonance structure into another, or to form or break bonds in an elementary step of a mechanism.

Curved arrow notation (also known as *arrow pushing*) notation used by organic chemists to describe the movement of electrons necessary to change one resonance structure into another or to form or break bonds in an elementary step of a mechanism.

Cyanohydrin a functional group characterized by a hydroxyl (—OH) and a cyano (—CN) group attached to the same carbon.

Cycloaddition reaction a concerted reaction in which two separate species come together to produce a new ring.

[2+2] Cycloaddition reaction a concerted reaction in which two separate species come together to produce a new ring, with two electrons supplied by each species.

[6+2] Cycloaddition reaction a concerted reaction in which two separate species come together to produce a new ring; six electrons are supplied by one species, and two electrons are supplied by the other.

Cycloalkanes cyclic molecules consisting of only carbon and hydrogen atoms.

[4+2] Cycloelimination a concerted reaction in which a cyclic molecule is cleaved into two separate species; one species receives four electrons, and the other receives two.

Dash–wedge notation a system of drawing to represent three-dimensional molecules using dashes (bonds point away from the viewer) and wedges (bonds point toward the viewer).

Deactivating group a substituent that decreases the reactivity of a molecule in a particular reaction; often refers to substituents that make an aromatic ring less susceptible to electrophilic aromatic substitution.

Debye conventional unit of the dipole moment.

Decarboxylation a reaction that removes a carboxyl (—CO_2H) group from a molecule.

Degenerate orbitals orbitals that have identical energies.

Degradation reaction a reaction that breaks larger molecules into smaller molecules

Degree of polymerization (DP) the number of *repeating units* in a polymer.

Dehydration the removal of water from a substance; often describes a reaction in which water is a product.

Delocalization the phenomenon of electrons or charges being less confined to a particular atom or bonding region. In a species for which resonance occurs, an electron may be delocalized over multiple atoms. Delocalization is generally stabilizing.

Delocalization energy See *resonance energy*.

Delocalized term used to describe an electron that is not confined to a particular atom or bonding region.

Deoxyribonucleic acid (DNA) a double-helical pair of intertwined nucleic acids that stores genetic information.

Depolymerization a reaction that causes a polymer to lose its monomers; the reverse of a *polymerization reaction.*

DEPT (distortionless enhancement by polarization transfer) spectroscopy technique used in ^{13}C NMR spectroscopy to determine the type of carbon (CH_3, CH_2, CH, or C) responsible for producing a particular signal.

Deshielded term used to describe protons whose shielding (see *shielded*) becomes diminished by nearby electronegative atoms or magnetic anisotropy. Deshielding results in a higher signal frequency, and thus a larger *chemical shift.*

Destructive interference interference that occurs when waves or orbitals overlap with opposite phase; destructive interference between two orbitals produces a new orbital that is diminished in size.

Desulfonation a reaction that removes a sulfonyl (—SO_3H) group from a molecule.

Detector a device used to measure the presence or absence of a specific property.

Detergents a class of substances with long-chain molecules that are very hydrophilic on one end and very hydrophobic on the other. They form micelles to emulsify dirt, grease, and oils, and unlike soap, detergents are effective cleansers in *hard water.*

Dextrorotatory (from Latin, meaning "rotating to the right") term used to describe chiral compounds that rotate plane-polarized light clockwise (in the + direction).

Dialkylborane (R_2BH) a functional group or compound with two alkyl groups attached to boron.

Diastereomers *configurational isomers* that are *not* mirror images of each other.

Diastereotopic term used to describe atoms such that Step 1 of the *chemical distinction test* yields *diastereomers.*

1,3-Diaxial interactions interactions between atoms or groups separately attached to a cyclohexane ring at relative positions numbered 1 and 3, and occupying axial positions.

Diazomethane the compound CH_2N_2; used as a reagent to produce a methyl ester from a carboxylic acid or to produce a cyclopropane ring from an alkene.

Diborane (B_2H_6) explosive, gaseous dimer of borane (BH_3) that consists of three-center, two-electron bonds.

Dichlorocarbene (also called dichloromethylene) the compound Cl_2C.

Dicyclohexylcarbodiimide the molecule $(C_6H_{11})N$═C═$N(C_6H_{11})$, which is used as a reagent to couple a carboxylic acid and an amine to produce an amide.

Dieckmann condensation an intramolecular *Claisen condensation reaction.*

Diels–Adler reaction a reaction between a conjugated diene and a dienophile that produces a cyclohexene ring.

Diglyceride a fatty acid diester of glycerol.

Dihedral angle (θ) the angle between a bond on the front atom of a Newman projection and a bond on the rear atom of the Newman projection; each angle of rotation defines a particular dihedral angle.

Diisobutylsluminum hydride (DIBAH) the compound $(i$-$Bu)_2Al$—H, which reduces esters to aldehydes.

1,3-Dipolar cycloaddition a concerted reaction in which two separate species come together to produce a new ring and in which one species is characterized by the resonance delocalization of charge over atoms that have a 1,3-positioning.

Dipole arrow a labeling arrow depicting the direction from lower (the origin) to higher (the point) electronegativity.

Dipole moment (μ)   a measure of the magnitude of a molecule's dipole.

Dipole–dipole interaction   the intermolecular interaction that arises because the positive end of one molecule's *net dipole* is attracted to the negative end of another's.

Direct addition   See *1,2 addition*.

Disiamylborane $(C_5H_{11})_2BH$, a bulky dialkylborane used in the hydroboration of an alkyne to help ensure that the alkyne undergoes a single addition.

Disproportionation   a reaction in which two identical species react to form two different products.

Dissolving metal reduction   a reduction reaction during which a metal dissolves in solution. Commonly used to reduce alkynes to *trans*-alkenes.

Disubstituted   having two substituents.

Disubstituted benzenes   benzene rings that have two substituents attached.

Disulfide bond   a link characterized by two sulfur atoms singly bonded together (i.e., —S—S—); often found in proteins linking two different cysteine residues.

Disulfide bridge   See *disulfide bond*.

Diterpene   a *terpene* that contains two pairs of isoprene units (i.e., four isoprene units), for a total of 20 carbons.

Doublet (d)   the splitting pattern of an NMR signal into two peaks of essentially equal height.

Doublet of doublets   the splitting pattern of a signal when each of the peaks in a doublet brought about by a first coupling is split again into a doublet by a second coupling.

Downfield   appearing in an NMR spectrum in a region of higher chemical shift.

Driving force   a thermodynamic concept (not an actual force) describing the extent to which a reaction favors products over reactants under any given conditions; it tends to increase with increasing stability of products relative to reactants; charge stability and total bond energy are important contributors.

Duet   a set of two electrons in the $n = 1$ shell of an atom.

Eclipsed conformation   rotational conformation in which the bonds to the front atom in a Newman projection cover, or "eclipse," the bonds to the rear atom.

Edman degradation   a reaction that removes the N-terminal amino acid from a polypeptide for analysis.

Effective electronegativity   property describing the tendency of an atom to hold electrons more tightly as an outcome of the atom's hybridization.

Effective magnetic field (B_{eff})   the magnetic field that is "felt" by the nucleus; the sum of B_{ext} and B_{loc}.

Elastomer   a polymer that has the ability to retain its original shape after large deformations; vulcanized rubber is a common example.

Electron configuration   the way in which electrons are arranged in atomic or molecular orbitals.

Electron density   the probability of finding an electron at a given location in space.

Electron donating   term used to describe a substituent that adds electron density to an atom to which it is bonded; electron-donating ability is considered relative to hydrogen.

Electron geometry   the orientation of the electron groups about a particular atom.

Electron impact ionization   the process of producing a molecular ion by colliding an electron beam with a gaseous molecule; used in mass spectrometry.

Electron spin density plot   a plot of the electron cloud about a molecule, color coded to show regions where there is high probability of finding an unpaired electron.

Electron transition   the change in the orbital occupied by an electron.

Electron withdrawing   term used to describe a substituent that removes electron density from an atom to which it is bonded; electron-withdrawing ability is considered relative to hydrogen.

Electronegativity   the ability of an atomic nucleus to attract electrons in a covalent bond.

Electrophile   a strongly electron-deficient species.

Electrophile elimination step   a common elementary step in which an electrophile is eliminated from a carbocation, generating a new π bond; the reverse of an electrophilic addition step.

Electrophilic addition reaction   a reaction whose mechanism involves an electrophilic addition step as a key step.

Electrophilic addition step   an elementary step in which electrons from a nonpolar π bond (as part of a double or triple bond) are used to form a new σ bond to an electrophile.

Electrophilic aromatic substitution   a reaction that consists of the addition of an electrophile to an aromatic ring followed by the elimination of another electrophile, resulting in a net substitution.

Electrophoresis   a common way of separating a mixture of amino acids or proteins (often at a given pH) by using an electric field; it is often done in a gel-like medium (gel electrophoresis).

Electrostatic forces   the forces by which opposite charges attract one another and like charges repel one another.

Electrostatic potential map   the depiction of a molecule's electron cloud in colors that indicate its *relative* charge; it is one way to illustrate the distribution of charge along a covalent bond.

Elementary steps   distinct events that usually involve the breaking and/or formation of specific bonds and constitute the *mechanism* of a chemical reaction.

Empirical rate laws   experimentally determined rate laws.

Emulsify   to disperse a substance in a solvent in which it is normally insoluble. See *detergents*.

Enamine   a functional group characterized by a carbon atom that is part of both an alkene and an amine.

Enantiomeric excess (ee)   the fraction of a mixture of enantiomers that is not racemic; it is the fraction of the mixture that contributes to the rotation of plane-polarized light.

Enantiomers   configurational isomers that are mirror images of each other.

Enantioselective synthesis   synthesis carried out in a way that would favor one enantiomer over another.

Enantiotopic   term used to describe hydrogen atoms such that Step 1 of the *chemical distinction test* yields *enantiomers*.

Endergonic   term describing a reaction that gains free energy upon producing products—i.e., $\Delta G_{rxn} > 0$— and is generally nonspontaneous.

End group   the collection of atoms that characterize the end of the main chain of a polymer.

Endo approach   in a Diels–Alder reaction, the orientation of the dienophile in which the substituents on the dienophile point toward the diene, resulting in the *endo product*.

Endo product the diastereomeric product of a Diels–Alder reaction in which the substituents on the diene and the dienophile end up on opposite sides of the new cyclohexene ring.

Endo rule the generality describing the tendency of a Diels–Alder reaction to favor the *endo product* over the *exo product*.

Endothermic term that describes a reaction or process that absorbs heat; $\Delta H_{rxn} > 0$.

Enol functional group or compound in which an alk*ene* functional group (C=C) and an alcoh*ol* functional group (ROH) are connected to the same C atom; the enol form of a molecule is the tautomer of its *keto form*.

Enolate anion a deprotonated, negatively charged *enol*.

Entropy a thermodynamic quantity that increases with the number of equivalent ways the energy in a system can be arranged. Many people like to think of entropy as a measure of disorder; a system with greater entropy (i.e., one that is more disordered) tends to be more likely to occur than one with less entropy.

Envelope conformation the lowest energy conformation of cyclopentane where four of its five carbon atoms lie essentially in one plane, with the fifth carbon outside that plane—resembling an envelope.

Enzymes proteins that act as catalysts for biological reactions—that is, enzymes facilitate biological reactions, but are not consumed while doing so.

Epimers compounds that differ in stereochemical configuration at only one tetrahedral stereocenter.

Epoxidation reactions reactions wherein an epoxide is produced, often involving an alkene and peroxy acid like *meta*-chloroperbenzoic acid (MCPBA).

Epoxides three-membered ring ethers.

Equatorial term used to describe bonds in a cyclic molecule that lie almost in the plane that is roughly defined by the ring (i.e., the *equator* of the molecule) and point outward from the center of the ring.

Equilibrium constant (K_{eq}) an experimentally obtained constant that describes a reaction's tendency to form products.

Essential fatty acids naturally occurring fatty acids that cannot be synthesized in the human body by any known chemical pathways; instead, they must be consumed.

Ethoxy group the group —OCH_2CH_3; also written as —OEt.

Ethyl group the group —CH_2CH_3; also written as — Et.

Ethylene glycol the compound $HOCH_2CH_2OH$, often used to protect a ketone or aldehyde.

Excited electron state an electron configuration that is not the lowest-energy configuration.

Exergonic term describing a reaction that releases free energy upon producing products—i.e., $\Delta G_{rxn} < 0$—and is generally spontaneous.

Exo approach in a Diels–Alder reaction, the orientation of the dienophile in which the substituents on the dienophile point away from the diene, resulting in the *exo product*.

Exo product the diastereomeric product of a Diels–Alder reaction in which the substituents on the diene and the dienophile end up on the same side of the new cyclohexene ring.

Exothermic term that describes a reaction or process that releases heat; $\Delta H_{rxn} < 0$.

Epoxidation reactions reactions that produce *epoxides*.

Expanded octet a group of more than eight valence electrons belonging to a single atom; only atoms in the third row of the periodic table and below can have an expanded octet.

External magnetic field (B_{ext}) a magnetic field applied to a sample by a magnet, such as a superconductor magnet in NMR spectroscopy.

Fats a subclass of lipids whose most common biological function is to store energy. They contain three adjacent ester groups, each of which can be produced from a *fatty acid* and an alcohol group from *glycerol*. Thus, a fat or oil is often described as a *triacyl glycerol* or a *triglyceride*. Animal fats, such as lard, are generally solids at room temperature. Fats from plants, however, are generally liquids at room temperature, and are thus more properly called *oils*.

Fingerprint region the region of a sample's infrared (IR) spectrum below ~1400 cm^{-1}. A fingerprint region is often difficult to analyze but is unique to each compound.

Fischer projections relatively convenient, two-dimensional representations of configurations about tetrahedral stereocenters in a given molecule; for each stereocenter represented in a Fischer projection, horizontal bonds point toward the viewer and vertical bonds point away from the viewer.

Fischer esterification reaction an acid-catalyzed reaction between an alcohol and a carboxylic acid that produces an ester.

Flagpole interaction the steric strain resulting from atoms or groups attached to C1 and C4 of cyclohexane in its boat conformation.

Formal charge one of two methods used with Lewis structures to determine the charges of atoms involved in covalent bonds: each lone pair is assigned to the atom on which it appears in the Lewis structure, and in a given covalent bond, half the electrons are assigned to each atom involved in the bond.

Fragmentation the process of breaking molecules into smaller pieces, such as with an electron beam for mass spectrometry.

Fragmented pathways in mass spectrometry, the reactions that account for the breaking apart of the molecular ion (M^+) into smaller pieces.

Free energy of activation ($\Delta G^{o\ddagger}$) the energy barrier that must be surmounted for reactants to form products; the difference in standard free energy between the reactants and the transition state.

Free radical a species that contains at least one unpaired electron.

Free radical polymerization a polymerization reaction in which the initiator produces a *radical* and the propagation steps involve radical species.

Free rotation rotation about a bond that can occur relatively rapidly under a given set of conditions; free rotation can occur about single bonds at room temperature but not about double bonds. Rotation about a double bond can occur only if the π bond breaks.

Frequency (ν) the number of complete oscillations that occur in a given period of time. It is given in units of hertz (Hz), cycles/second, or s^{-1}, all of which are equivalent.

Friedel–Crafts acylation an electrophilic aromatic substitution reaction in which a proton on an aromatic ring is replaced by an acylium ion (R—C≡O$^+$); named after Charles Friedel and James Crafts.

Friedel–Crafts alkylation an electrophilic aromatic substitution reaction in which a proton on an aromatic ring is replaced by a carbocation (R$^+$); named after Charles Friedel and James Crafts.

Frontier molecular orbitals the HOMO and LUMO orbitals of reacting species in an elementary step.

Frontier molecule orbital (FMO) theory a theory that invokes HOMO–LUMO interactions in the reacting species to establish whether an elementary step is allowed or forbidden.

Frontside attack occurs in a nucleophilic substitution reaction when a nucleophile approaches the substrate from the same side as the leaving group, leading to retention of stereochemical configuration.

Frost method a method to derive the relative energies of an annulene's π MOs; developed by namesake Arthur A. Frost.

Functional group conversions or **transformations** reactions that change one functional group into another.

Functional groups common bonding arrangements of relatively few atoms. Functional groups dictate the behavior of entire molecules—that is, molecules with the same functional groups tend to behave similarly.

Furanose cyclic form of a monosaccharide characterized by a five-membered ring.

Fused rings rings that have bonds in common.

Gabriel synthesis a synthesis scheme that involves the reaction of phthalimide and an alkyl halide (RX) to produce a primary amine (RNH_2).

Gauche conformation staggered conformation wherein bulky groups are 60° apart in a Newman projection.

Gel permeation chromatography (GPC) a technique used to analyze the distribution of molecular size within a sample; smaller molecules are detected later in the analysis because they interact more strongly with the gel.

Geminal dihalide description of a molecule in which two halide substituents are attached to the same carbon atom.

Gilman reagents See *lithium dialkylcuprates*.

Glass transition temperature (T_g) the temperature at which the species that make up an amorphous solid begin to flow past each other.

Glycolysis a metabolic pathway that breaks down simple carbohydrates for their energy.

Glycosides a class of compounds characterized by sugars with glycosidic linkages to another group.

Glycosidic linkage the high-energy C—O—C bond that connects a sugar unit to another group, often another sugar.

Grignard reaction reaction in which a Grignard reagent adds in as a nucleophile to the electrophilic atom of a group containing a polar π bond.

Grignard reagent an alkylmagnesium halide (R—MgX, where X = Cl, Br, or I) nucleophile that favors direct addition to a polar π bond over conjugate addition. They are often used to make new carbon–carbon bonds.

Ground state the most stable (i.e., the lowest energy) electron configuration.

Gyromagnetic ratio (γ) a value that relates the energy separation between the spin states of a nucleus and an applied magnetic field.

^{1}H NMR spectroscopy See *proton NMR spectroscopy*.

Haloform a compound having the general form HCX_3, where X = F, Cl, Br, or I.

Haloform reaction the reaction of a methyl ketone or an aldehyde with a molecular halogen (e.g., Cl_2, Br_2, or I_2) under basic conditions, producing a *haloform* and a carboxylate anion that can be protonated to a form a carboxylic acid.

Halogenation the process of adding halogens to a molecule.

Halohydrins a class of molecules in which a halogen atom and an alcohol group are on adjacent carbon atoms.

Hammond postulate postulate proposed in 1955 by the American chemist George S. Hammond: "If two states...occur consecutively during a reaction process and have nearly the same energy content, their interconversion will involve only a small reorganization of the molecular structures." In other words, if, in a reaction free-energy diagram, two species lie near each other along the reaction coordinate (the *x* axis) and are similar in energy, then they will have very similar structures.

Hard water water that contains a significant concentration of Mg^{2+} or Ca^{2+} ions.

Haworth projection a two-dimensional representation wherein a ring is depicted as being planar, occupying the plane perpendicular to the page, and substituents are drawn perpendicular to that plane; they illustrate cis and trans relationships well, but they do not accurately convey three-dimensional relationships or steric strain.

Head-to-head in a polymerization reaction, the orientation in which the heads of two monomers are facing each other.

Head-to-tail in a polymerization reaction, the orientation in which the head of one monomer faces the tail of another monomer.

Heat of hydrogenation the amount of heat released when $H_2(g)$ adds to double or triple bonds. It is used to gain a sense of the stability of π systems.

Heat of combustion the energy given off in the form of heat (ΔH) during a combustion reaction.

Heisenberg uncertainty principle principle that states that the uncertainty in a measurement of an electron's position is inversely proportional to the uncertainty in a measurement of its momentum. In other words, the more precisely we know the electron's position, the less precisely we know where it is going, and vice versa.

Hell–Volhard–Zelinksky (HVZ) reaction a reaction in which the α carbon of a carboxylic acid is brominated to produce an α-bromo acid.

Hemiacetal a functional group having one hydroxy (—OH) and one alkoxy (—OR) group attached to the same carbon atom.

Henderson-Hasselbalch equation the equation that relates the pH of a solution to the ratio of the concentration of an acid's conjugate base to that of the acid. $pH = pK_a + \log([A^-]/[HA])$.

Heteroaromatic compounds aromatic hydrocarbons with atoms other than carbon incorporated into the ring.

Heteroatoms noncarbon and nonhydrogen atoms.

Heterochain polymers polymers in which the main chain consists of carbon atoms and noncarbon atoms.

Heterocyclic aromatic compounds a class of aromatic compounds that have heteroatoms incorporated into their aromatic ring structures.

Heterogeneous catalysis a reaction in which the catalyst is in a different phase from the rest of the reaction mixture.

Heterogenous catalyst a catalyst that is in a different phase from the rest of the reaction mixture.

Heterolysis See *heterolytic bond dissociation step*.

Heterolytic bond dissociation step an elementary step in which only a single bond is broken and both electrons from that bond end up on one of the atoms initially involved in the bond.

Heterotopic protons (also known as *chemically distinct protons*) protons surrounded by different electron distributions—that is, ones that reside in different chemical environments and so absorb at different frequencies in NMR spectroscopy.

Hexose a six-carbon sugar; nomenclature referring to the class of six-carbon sugars.

Highest occupied molecular orbital (HOMO) the highest-energy MO that contains an electron.

Hofmann elimination reaction reaction creating a double bond and leading to an anti-Zaitsev elimination product: the less stable, less alkyl substituted alkene.

Hofmann product product resulting from the Hoffman elimination reaction; the less substituted product (also known as an *anti-Zaitsev product*).

Homolysis (also known as *homolytic bond dissociation*) an elementary step in which only a single bond is broken and the two electrons from that bond are split evenly between the atoms that were initially bonded together.

Homolytic bond dissociation See *homolysis*.

Homopolymers polymers made from a single monomer.

Homotopic protons (also known as *chemically equivalent protons*) protons surrounded by identical electron distributions—that is, ones that reside in identical chemical environments and so absorb at the same frequency in NMR spectroscopy.

Hückel numbers the set of numbers: 2, 6, 10, 14, and so on, which can be described as either odd numbers of pairs or numbers that are consistent with $4n + 2$, where n is a nonnegative integer.

Hückel's rules empirical rules for defining aromaticity, developed by German physicist Erich Hückel.

Hund's rule principle stating that all orbitals at the same energy must contain a single electron before a second electron can be paired in the same orbital.

Hybridized atomic orbital a cross between two or more pure atomic orbitals from the valence shell of a single atom.

Hydrazone a functional group characterized by the bonding arrangement $C=N-NH_2$.

Hydride anion (H:⁻) a hydrogen atom with an extra electron.

Hydride reagent reagents that contain or have the tendency to donate hydride. Common hydride reducing agents include lithium aluminum hydride ($LiAlH_4$) and sodium borohydride ($NaBH_4$). Sodium hydride (NaH) is a strong base but a poor nucleophile.

1,2-Hydride shift a carbocation rearrangement wherein a hydride ion (H^-) migrates to an adjacent atom.

Hydroboration the net addition of BH_3 across a double or triple bond; hydrogen adds to one atom of the multiple bond and BH_2 adds to the other.

Hydroboration–oxidation a sequence of reactions (hydroboration followed by oxidation) that results in the anti-Markovnikov addition of water to an alkene or alkyne.

Hydrocarbon tail the hydrophobic, nonpolar end of a species that is distinguished from its hydrophilic end; often used to describe the nonpolar end of a fatty acid carboxylate.

Hydrogen atom abstraction an elementary step in which a hydrogen atom is removed from a molecule; a type of S_H2 step in which a radical forms a bond to a hydrogen atom in a molecule, and the initial bond to hydrogen is broken, generating a new radical.

Hydrogen bond (H bond) intermolecular interaction that involves a *hydrogen-bond donor* (NH, FH, or OH) and a *hydrogen-bond acceptor* (typically N, O, or F).

Hydrogen-bond acceptor any atom with a large concentration of negative charge and a lone pair of electrons, such as N, O, or F.

Hydrogen-bond donor a hydrogen atom covalently bonded to either F, O, or N.

Hydrolysis a type of reaction in which water is a reactant that facilitates the breaking of bonds; water is generally a nucleophile in these reactions.

Hydroperoxide ion the HOO^- anion. It is a key nucleophile in the oxidation of a trialkylborane to produce an alcohol.

Hydrophilic "water loving"; tending to dissolve in water.

Hydrophobic "water fearing"; not tending to dissolve in water.

Hydrophobic effect an effect characterized by the tendency of hydrophobic side chains to reside in the interior of a folded protein to minimize interactions of those side chains with an aqueous environment.

Hyperconjugation the delocalization of electrons in a σ-bonding molecular orbital into an adjacent orbital of π symmetry.

Ideal bond angle the theoretical bond angle predicted by valence shell electron pair repulsion (VSEPR) theory.

Imine the functional group characterized by the bonding arrangement $-C=N-R$ (or H).

In-plane bending vibration vibrational motion of a molecule where bond angles change but the atoms involved in the vibration all remain in the same plane.

Index of hydrogen deficiency (IHD) or **degree of unsaturation** *half* the number of hydrogen atoms missing from that molecule compared to an analogous, completely saturated molecule.

Induced dipole temporary dipole that arises within a molecule as a result of the electron distribution being altered by a nearby charge.

Induced dipole–dipole interactions (also known as *London dispersion forces*) the dominant intermolecular interactions between nonpolar molecules resulting from the temporary distortions of their electron distributions.

Induction the distortion of electron density along covalent bonds, brought about by the replacement of a hydrogen atom with another substituent.

Inductive effect the effect that *induction* has on stability or another property of a species.

Inorganic compound a compound that is not organic; that is, a compound that is *not* composed primarily of carbon and hydrogen.

Instantaneous dipole temporary dipole that arises when there are more electrons on one side of a molecule than there are on the other at some instant.

Integral trace the line overlaid on an NMR spectrum that corresponds to the cumulative area under the peaks in the spectrum. The height of a stairstep in an integral trace is proportional to the number of nuclei that give rise to the corresponding peak.

Integration the operation of calculating the area under a curve, such as a peak in a spectrum; used to generate an integral trace in an NMR spectrum.

Intermediate transitory species that appears in a reaction mechanism but does not appear in the overall reaction.

Intermolecular term used to describe the involvement of two separate species.

Intermolecular forces See *intermolecular interactions*.

Intermolecular interactions (also known as *intermolecular forces*) interactions that arise between molecules due to their charge distributions.

International Union of Pure and Applied Chemistry (IUPAC) international chemistry authority that standardizes chemical nomenclature, constants, and processes.

Intramolecular term used to describe the involvement of separate portions of a single species.

Intramolecular proton transfer a reaction that results in the loss of a proton from one part of a species and the gain of a proton at another part of the same species.

Inverse centimeters (cm^{-1}) (also known as *wavenumbers*) a unit of frequency that is calculated by taking the reciprocal of the wavelength—that is, by dividing 1 by the wavelength in cm. Physically, it corresponds to the number of waves that fit in 1 cm.

Ion a species that bears a net charge.

Ion pair a pair of interacting, oppositely charged ions.

Ion–dipole interaction attraction between a positive or negative ion and a net molecular dipole; the intermolecular interaction whereby free ions can interact with the molecules of a polar solvent like water.

Ion–ion interaction the attraction between two oppositely charged ions.

Ionic bonding one of two types of fundamental bonds in chemistry; in an ionic bond, the more electronegative atom acquires electrons given up by the less electronegative atom, forming oppositely charged ions. The electrostatic attraction between the positively charged cations and the negatively charged anions constitutes the ionic bond.

Ionic head group the charged (hydrophilic) end of a species that is distinct from its uncharged, nonpolar (hydrophobic) end; often used to describe the carboxylate portion of a fatty acid carboxylate.

Ionizable term used to describe a species that is able to gain or lose a proton to become charged.

Irreversible term describing a reaction that does not readily take place in the reverse direction.

Irreversible reaction arrows (⟶) reaction arrows pointing in one direction, signifying that the reaction is irreversible.

Isoalkyl groups an alkyl group terminating in —CH(CH$_3$)$_2$.

Isoelectric focusing a type of electrophoresis experiment whereby amino acids or proteins are separated by charge using a pH gradient.

Isoelectric pH (also known as *isolelectric point*) the pH at which a substance (usually an amino acid or protein) has an average charge of zero.

Isoelectric point (pI) See *isoelectric pH*.

Isomer a molecular species that has the same formula as another molecular species but is different in some way; see *constitutional isomers, stereoisomers, conformational isomers, configurational isomers, diastereomers,* and *enantiomers*.

Isomerism the relationships among molecules that have the same formula but are not identical.

Isoprene unit a five-carbon unit into which a terpene's carbon backbone can be divided; every carbon atom in a terpene can be assigned to separate isoprene units.

Isotactic a term used to describe a polymer in which all tetrahedral stereocenters along the main chain have the same configuration.

Kekulé structures a specific resonance structure of a species, usually referring to benzene; after namesake August Kekulé.

Keto form constitutional isomer of an enol, containing a carbonyl group (C═O) as either a ketone or an aldehyde.

Keto–enol tautomerization the transformation of one keto or enol tautomer (form) into the other; usually the process is in equilibrium.

Ketohexose a six-carbon sugar with a ketone group.

Ketopentose a five-carbon sugar with a ketone group.

Ketose a sugar whose carbonyl group is part of a ketone.

Kinetic control a reaction condition in which the major product is the one produced the fastest.

Kinetic enolate anion the enolate ion that is produced most rapidly by deprotonation of the corresponding ketone.

Labile a term used to describe a species that is reactive under a given set of conditions.

λ$_{max}$ the wavelength of light at which *absorbance* of a particular peak is a maximum.

Law of conservation of energy law that states that energy cannot be created or destroyed.

Le Châtelier's principle principle that states that if a reaction at equilibrium experiences a change in reaction conditions (e.g., concentrations, temperature, pressure, or volume), then the equilibrium will shift to counteract that change.

Leaving group the functional group displaced from the substrate in a reaction.

Leaving group ability the relative rate at which a leaving group will separate from its substrate in a reaction.

Leveling effect phenomenon dictating the maximum strength of an acid or base that can exist in solution; the strongest acid is the protonated solvent and the strongest base is the deprotonated solvent.

Levorotatory (from Latin, meaning "rotating to the left") term used to describe chiral compounds that rotate plane-polarized light counterclockwise (ie., in the − [negative] direction).

Lewis acid an electron-pair acceptor.

Lewis base an electron-pair donor.

Lewis dot structures representations of molecules that convey connectivity and depict all valence electrons.

Lindlar's catalyst a catalyst prepared by treating palladium with calcium carbonate and a small amount of quinoline; it is a type of *poisoned catalyst* that is used to reduce an alkyne to a *cis*-alkene.

Linear alkanes (also known as *straight-chain alkanes*) alkanes (containing only C—C and C—H single bonds with no functional groups) that have no branching along the chain.

Linear combination an expression of sums and differences, often used to describe the various ways in which contributing waves or orbitals can mix.

Linear combinations of atomic orbitals (LCAOs) sums and differences of various atomic orbitals, which generate molecular orbitals.

Linear polymers polymers that consist of a single main chain—i.e., they have no branching.

Linear synthesis a synthesis composed of sequential synthetic steps.

Line structures standard chemical representation method wherein carbons are implied at each intersection of two or more lines, hydrogens bonded to carbon are not drawn but any bonded to another atom are drawn, and heteroatoms are drawn.

Lipid bilayer the double-layer structure into which phospholipids organize in an aqueous environment; it is the basis of a cell membrane.

Lipids biomolecules that are relatively insoluble in water. Thus, most lipids are highly nonpolar, consisting primarily of carbon and hydrogen, with very little oxygen or nitrogen content.

Consequently, they tend to be soluble in relatively nonpolar organic solvents, such as ether.

Lithium aluminum hydride (LiAlH$_4$) a very strong hydride reducing agent that favors direct addition over conjugate addition.

Lithium dialkylcuprates (also known as *Gilman reagents*) R$_2$CuLi, which are sources of weak R$^-$ nucleophiles that tend to favor conjugate addition over direct addition.

Lithium-*tert*-butoxyaluminum hydride the compound LiAlH(O—*t*-Bu)$_3$, commonly used to reduce a carboxylic acid chloride to an aldehyde.

l$_{max}$ the wavelength at which a peak in a spectrum has its maximum absorption.

Local magnetic field (B_{loc}) magnetic field that adds to or subtracts from an external magnetic field (B_{ext}); when subjected to B_{ext}, the electrons that surround an atom can undergo coherent motion to contribute to B_{loc}.

Local minimum a location along a curve in a plot where the value on the y axis (often energy) rises upon an increase or decrease of the value on the x axis.

Localized term used to describe an electron or charge that is confined to a particular location within a molecular species.

Localized electron an electron that is confined to a particular location within a molecular species.

London dispersion forces See *induced dipole– induced dipole interaction*.

Lone pairs pairs of nonbonding electrons confined to a particular atom.

Long-range coupling in NMR spectroscopy, the situation in which nuclei separated by more than three bonds exhibit weak coupling.

Lowest unoccupied molecular orbital (LUMO) the lowest-energy MO that is empty.

M + 1 peak a peak on a mass spectrum representing a molecular ion containing a heavy isotope, the mass of which is 1 u heavier than the most abundant isotope. The relative intensity of this peak can be used to estimate the number of carbon atoms in a sample molecule.

M + 2 peak a peak on a mass spectrum representing a molecular ion containing a heavy isotope, the mass of which is 2 u heavier than the most abundant isotope. The relative intensity of this peak can be used to determine the presence of atoms such as Br, Cl, and S.

m/z See *mass-to-charge ratio*.

Magnetic anisotropy in NMR spectroscopy, the phenomenon that contributes to the B_{loc} of nearby nuclei, resulting from ring current in double bonds and aromatic rings; causes substantial deshielding of hydrogen and carbon nuclei associated with simple alkenes, aromatic rings, and aldehydes.

Main chain the longest continuous chain of carbon atoms.

Malonic ester synthesis a synthesis scheme that converts malonic ester (diethyl malonate) into an alkyl- or dialkyl-substituted carboxylic acid.

Markovnikov addition a reaction in which the incoming halide in H—X adds to an alkene or alkyne at the more-substituted carbon and the proton adds to the adjacent alkene or alkyne carbon with more protons.

Markovnikov's rule empirical rule describing the *regiochemistry* of addition of a Brønsted acid across a C=C bond: The addition of a hydrogen halide to an alkene favors the product in which the proton adds to the alkene carbon that is initially bonded to the greater number of hydrogen atoms.

Mass spectrometry an experimental process that provides insight into the mass of a molecule and the fragments that compose it; this information about a molecule is obtained by interpreting a *mass spectrum*.

Mass spectrum a plot of the relative abundances of gaseous ions produced from an analyte against the mass-to-charge ratio, *m/z*.

Mass-to-charge ratio (*m/z*) the ratio of a particle's mass to its charge; it is the property that a mass spectrometer distinguishes.

McLafferty rearrangement in mass spectrometry, a type of fragmentation pathway exhibited by molecular ions of compounds that have a carbonyl-containing functional group.

Measured angle of rotation (α) the angle by which plane-polarized light is rotated upon passing through a sample of a chiral compound.

Mechanism the precise sequence of elementary steps that results in the conversion of the original reactants to the final products in a chemical reaction.

Melting point (T_m) the temperature at which the species that form a crystal lattice begin to flow past each other and behave as a liquid.

Meisenheimer complex a reactive intermediate in nucleophilic aromatic substitution reactions that proceed by the nucleophilic addition–elimination mechanism.

Mercurinium ion intermediate a positively charged, three-membered ring intermediate in which the ring consists of two carbon atoms and a mercury atom.

Merrifield synthesis solid-phase synthesis of a peptide or protein; named after R. B. Merrifield.

Meso term used to describe a molecule that contains at least two tetrahedral stereocenters but has a plane of symmetry that makes it achiral overall.

***meta*-Chloroperbenzoic acid (MCPBA)** a peracid reagent used in epoxidation reactions.

Meta-director the substituent attached to a benzene ring that directs an incoming electrophile to the meta carbon of the ring.

Methoxy group the group, —OCH$_3$; also written as —OMe.

Methyl cation (H$_3$C$^+$) carbocation that has only hydrogen attached to the positively charged carbon.

Methyl group the group —CH$_3$; also written as —Me.

1,2-Methyl shift a carbocation rearrangement wherein a methyl group migrates to an adjacent atom.

Methylene substituent the substituent =CH$_2$.

Micelles spherical aggregates formed among surfactant species (such as soaps or detergents) in solution to maximize the intermolecular interactions involving the surfactant and solvent.

Michael reaction a reaction in which a carbon nucleophile adds via conjugate addition to form a carbon–carbon bond; general description of conjugate additions involving any nucleophile; after namesake chemist Arthur Michael. See *conjugate addition*.

Migratory aptitude the tendency of a bond between an H or alkyl group and a carbonyl carbon to be broken in a Baeyer–Villiger oxidation.

Molar absorptivity (ε) an experimentally derived quantity that is characteristic of a given species' probability of absorbing a photon at a given wavelength of radiation.

Molecular geometry the arrangement of atoms or groups of bonding electrons about a particular atom.

Molecular ion [M$^+$(g)] an ion formed when an electron is ejected from a molecule, usually via *electron impact ionization* for *mass spectrometry*.

Molecular modeling kit   kit with which you can construct real models of molecules and manipulate them in your hands.

Molecular orbital theory   the theory whose central concept is that all electrons in a molecule can be thought of as occupying orbitals called molecular orbitals (MOs), or *linear combinations of atomic orbitals* (LCAOs), which are delocalized over the entire molecule.

Molecular orbitals (MOs)   orbitals resulting from *linear combinations of atomic orbitals* (LCAOs), delocalized over the entire molecule; each MO accommodates up to two electrons.

Molecularity   the number of separate reacting species in an elementary step.

Molozonide   the product of the cycloaddition of one molar equivalent of ozone (O_3) to an alkene or alkyne in a [4+2] cycloaddition, characterized by a five-membered ring consisting of three adjacent O atoms and two adjacent C atoms. In an ozonolysis reaction, it is reacted further to produce ketones, aldehydes, and carboxylic acids from an initial alkene.

Monomer   one of the small molecules that make up polymers.

Monosaccharides   (also known as *simple sugars*) carbohydrates whose number of oxygen atoms is the same as their number of carbon atoms, giving them the general formula $C_xH_{2x}O_x$; polysaccharides are constructed from monosaccharides.

Monosubstituted   term used to describe a molecule in which a hydrogen atom has been replaced by one substituent.

Monosubstituted benzene   a benzene ring that is bonded to one substituent, having the form C_6H_5–Sub.

Monosubstituted cyclohexane   the result when one of the hydrogen atoms in cyclohexane is replaced by another substituent.

Monoterpene   a terpene that contains one pair of isoprene units (i.e., two isoprene units), for a total of 10 carbons.

Monounsaturated   term that describes a fatty acid or other molecule that has just one C=C double bond.

Multiple-center molecular orbitals   molecular orbitals (MOs) that span multiple atoms and involve the simultaneous interaction of multiple *atomic orbitals (AOs)*.

Multiplet (m)   the NMR splitting pattern of a signal when there is a large number (more than five) of coupled protons.

Multistep mechanisms   a mechanism composed of two or more *elementary steps.*

Mutarotation   a change in the rotation of plane-polarized light due to the equilibration between epimers (stereoisomers that differ by the configuration at one stereocenter); observable when one anomer of a particular sugar equilibrates with its other anomer when dissolved in water.

N + 1 Rule   a rule stating that if a proton is coupled to N protons that are distinct from itself but equivalent to each other, then the signal produced by that proton will be split into $N + 1$ separate peaks.

n → π* Transition   term describing an electron from a nonbonding orbital being promoted to a π antibonding (π*) orbital.

Natural product   a class of compounds synthesized in biological systems having a biological function or role; often terpenes or terpenoids.

N-Bromosuccinimide (NBS)   a radical initiator with a particularly weak Br—N bond, which is the source of a small, steady concentration of Br_2. Commonly, NBS is used to brominate allylic and benzylic carbons.

Net molecular dipole   (also known as *permanent dipole*) a separation of partial positive charge and partial negative charge that sum to zero; the dipole is said to point from the center of positive charge to the center of negative charge.

Net reaction   See *overall reaction.*

Newman projections   a two-dimensional representation of a molecule viewed down the bond of interest; one convenient way to illustrate rotational conformations.

Nitration   a type of electrophilic aromatic substitution reaction in which a proton is replaced by a nitro (—NO_2) group.

Nitrogen inversion   the process (an umbrella flip) by which the configurations about nitrogen interconvert in the species NRR′R″.

Nitrogen rule   principle that states that a compound tends to have an odd molecular mass if it contains an odd number of nitrogen atoms, and an even molecular mass if it contains an even number of nitrogen atoms or none at all.

Nitronium ion   the ion $^+NO_2$, which is the strong electrophile that replaces a proton in an aromatic nitration reaction.

Noble gas   an element in group 8A, characterized by completely filled valence shells that provide substantial stability.

Nodal plane   a plane in which the electron density is zero—that is, a plane in which an electron has zero probability of being found.

Node   a location where electron density is zero.

Nomenclature   the naming of atoms and molecules.

Nonaromatic   catchall term used to describe compounds that are neither aromatic nor antiaromatic and, thus, do not have unusually stable or unusually unstable π systems.

Nonbonding MO   a molecular orbital (MO) that is similar in energy to its contributing atomic orbitals.

Nonpolar   term to describe a molecule having no net dipole moment.

Nonsuperimposable   term used to describe two molecules for which there is no orientation such that *all* atoms of both molecules can be superimposed (i.e., lined up perfectly).

Normal modes of vibration   independent types of vibrational motion: stretching and bending

N-terminus   the end of a polypeptide (protein) chain that is part of an amine group.

Nuclear magnetic resonance (NMR) spectroscopy   experimental process used for structure determination, in which radiofrequency radiation is used to cause spin flips in the nuclei of atoms that have an odd number of protons, an odd number of neutrons, or both.

Nuclear spin   property of some atomic nuclei, which generates a small magnetic field, an outcome expected of a charge in motion.

Nucleic acid   a large molecular chain that is primarily associated with the storage and transfer of genetic information; it is constructed from relatively small molecular units called nucleotides.

Nucleophile   "nucleus-loving" species (usually with a lone pair of electrons) that seeks a substrate with a full or partial positive charge to stabilize its full or partial negative charge

Nucleophile elimination step   elementary step wherein a new π bond is produced and the eliminated leaving group, in turn, has the characteristics of a nucleophile; the reverse of a *nucleophilic addition step.*

Nucleophilic acyl substitution   a nucleophilic addition–elimination mechanism in which a nucleophile replaces the leaving group attached to a C=O carbon.

Nucleophilic addition–elimination reaction   a reaction whose mechanism contains subsequent nucleophilic addition and nucleophile elimination steps, resulting in the net substitution of a nucleophilc for a leaving group.

Nucleophilic addition reaction a type of reaction in which a key step in the mechanism is a nucleophilic addition step.

Nucleophilic addition step an elementary step in which a nucleophile forms a bond to the electron-deficient atom of a polar π bond.

Nucleophilic aromatic substitution a reaction in which a nucleophile replaces a leaving group attached to an aromatic ring; the reaction could have a nucleophilic addition–elimination mechanism or could proceed through a benzyne intermediate.

Nucleophilicity (also known as nucleophile strength) the tendency of a nucleophile to promote an S_N2 reaction; typically determined by the rate of an S_N2 reaction between the nucleophile and a methyl halide.

Nucleotides relatively small molecular units from which nucleic acids are constructed. All nucleotides have three distinct components: an inorganic phosphate (PO_4) group, a cyclic monosaccharide (or sugar), and a nitrogenous base.

Octet a group of eight valence electrons; a complete valence shell of atoms in the second row of the periodic table.

Oils triglycerides that are liquid at room temperature; typically produced from plants.

Olefins See *alkenes.*

Operating frequency the frequency at which a bare proton absorbs electromagnetic radiation when subjected to an NMR spectrometer's magnetic field.

Optically active term describing a sample that rotates plane-polarized light. Optically active species are chiral.

Optically inactive term describing an achiral species that does not rotate plane-polarized light. Achiral species are optically inactive, as are racemic mixtures of enantiomers.

Orders the exponents to which reactant (or product) concentrations are raised in the rate law of a reaction.

Organic chemistry the branch of chemistry involving *organic compounds* (compounds containing carbon atoms).

Organic compound a compound composed primarily of carbon and hydrogen.

Organometallic term used to describe compounds that contain a metal atom bonded directly to a carbon atom.

Orbital description of the particular state of an electron in an atom or molecule; an electron that belongs to a particular orbital has a high probability of being located in a region of space whose size and shape are characteristic of that orbital.

Ortho/para director a substituent attached to a benzene ring that directs an incoming electrophile to the ortho and meta carbons of the ring.

Osmate ester a functional group having the bonding arrangement RO—Os=O. A cyclic osmate ester is an intermediate in the oxidation of an alkene using osmium tetroxide (OsO_4).

Out-of-plane bending vibration normal mode of vibration in which the set of atoms undergoing vibrational motion occupy different planes throughout an oscillation.

Overall product species that appears as a product in the overall (i.e., net) reaction; it is generally a product that can be isolated upon the completion of the reaction.

Overall reactant species that appears as a reactant in the overall (i.e., net) reaction; it is generally a starting material that is added to a reaction.

Overall reaction (also known as *net reaction*) the reaction that includes only the reactants that are added as starting materials and the products that are isolated; the sum of the elementary steps of the mechanism.

Oxaphosphetane intermediate in a Wittig reaction characterized by a four-membered ring consisting of two carbon atoms, one phosphorus atom, and one oxygen atom.

Oxetanes four-membered ring ethers.

Oxidation state one of two methods used to assign charge to atoms involved in covalent bonds. Lone pairs are assigned to the atom on which they appear in the Lewis structure; in a given covalent bond, all electrons are assigned to the more electronegative atom; if the two atoms are identical, the electrons are split evenly.

Oxidative cleavage an oxidation reaction that results in the severing of C=C or C≡C bonds into separate carbonyl-containing functional groups.

Oxymercuration–demercuration See *oxymercuration–reduction.*

Oxymercuration–reduction (also known as *oxymercuration–demercuration*) the sequence of reactions in which a Hg^{+2} species adds to an alkene or alkyne followed by reduction (typically with $NaBH_4$), resulting in the *Markovnikov addition* of water.

Ozone the compound O_3, which is a key reagent in ozonolysis reactions.

Ozonide a key intermediate in an ozonolysis reaction, characterized by a five-membered ring consisting of three O atoms and two C atoms, and in which only two of the O atoms are adjacent. An ozonide is reacted further to produce ketones, aldehydes, and carboxylic acids from an initial alkene.

Ozonolysis a reaction in which an alkene is reacted with ozone to cleave the C=C bond, producing two separate carbonyl-containing functional groups.

p character the resemblance of an orbital's characteristics (e.g., shape and energy) to those of a p orbital; the fraction of a hybridized atomic orbital that comes from a p orbital.

Pascal's triangle pattern formed by the ratios describing the relative peak heights in splitting patterns in accordance with the $N + 1$ rule.

Pauli's exclusion principle the principle that states that no more than two electrons (i.e., zero, one, or two electrons) can occupy a single orbital; two electrons in the same orbital must have opposite spins.

Peaks See *absorption bands.*

Pentose a five-carbon sugar.

Peptide bond (also known as *peptide linkage*) the amide functional group that serves to connect two amino acids together in a peptide or protein.

Peptide linkage See *peptide bond.*

Percent dissociation the extent to which an acid or base dissociates into its ions in a solution (usually water); related to the $[A^-]/[HA]$ fraction in the *Henderson-Hasselbalch* equation.

Pericyclic reaction a concerted reaction in which all of the electrons involved in bond breaking and forming are delocalized in a ring in the transition state.

Periodate ester a functional group characterized by the bonding arrangement RO—I=O. A cyclic periodate ester is a key intermediate in the reaction between a 1,2-diol and the periodate ion (IO_4^-) to produce ketones and aldehydes.

Permanent dipole See *net molecular dipole.*

Peroxyacid A molecule of the form RCO_3H.

pH gradient a nonconstant distribution of pH throughout a particular space.

Phase (of an electron) an immeasurable quantity that plays an important role in how orbitals interact with one another via *constructive interference* or *destructive interference*; the relative phase of an orbital in a region of space is indicated by whether the orbital is shaded or unshaded there.

Phenyl the substituent —C_6H_5, often abbreviated Ph.

Phenylmethyl the substituent —$CH_2C_6H_5$, also called *benzyl*, abbreviated Bn.

Phospholipid a subclass of lipids; it can efficiently store energy and is integral in the formation of cell membranes. Whereas a fat or an oil has three fatty acids, making it a *triglyceride*, a phospholipid has only two fatty acids, making it a *diglyceride*.

Phosphonium ylide See *Wittig reagent*.

Photochemically allowed a term used to describe a reaction that proceeds at a reasonable rate when irradiated with ultraviolet light; one or more of the reactants have an excited electron configuration.

Photons the particles that make up light, each of which carries a specific quantity of energy that can be associated with its frequency and wavelength.

$\pi \rightarrow \pi^*$ Transition occurs when an electron from a π molecular orbital (MO) is promoted to a π^* MO; the HOMO–LUMO transition in an alkene or alkyne.

π Stacking the interaction (believed to be stabilizing) that takes place between separate π systems when the face of one π system is in close proximity to the face of the other.

π Symmetry the symmetry of a molecular orbital that is generated by atomic-orbital overlap on *either side of* the bonding axis; the bonding axis itself contains a nodal plane.

Planck's constant (h) the constant of proportionality that appears in the dependence of a photon's energy on its frequency: 6.626×10^{-34} J·s.

Plane of symmetry plane that bisects a molecule in such a way that one half of the molecule is the mirror image of the other half.

Plane-polarized light light whose photons have their electric fields oscillating in the same plane.

Poisoned catalyst a specially treated catalyst that prevents a reaction from occurring at the rate at which it would occur with an untreated catalyst. A poisoned catalyst is often used to reduce an alkyne to a *cis*-alkene.

Polar covalent bond a covalent bond between atoms of different electronegativity, such that the electrons that make up the bond are shared unequally.

Polar π bond a π bond between atoms of different electronegativity, such that the electrons that make up the bond are shared unequally.

Polarizability the ease with which a species' electron cloud is distorted; tends to increase with the total number of electrons a species has.

Polarizable term that describes the ability of a species' electron cloud to be distorted.

Polarizer device that generates plane-polarized light by allowing through only photons whose electric field is oscillating in a specified plane, effectively filtering out light whose electric field oscillates in any other plane.

Polyacrylates (also known as *acrylics*) polymers derived from monomers that are derivatives of acrylic acid.

Polyamino acid See *proteins*.

Polyatomic ions ions that contain more than one atom. Polyatomic ions typically consist only of nonmetals, and their atoms are held together by covalent bonds.

Polycyclic aromatic hydrocarbons (PAHs) a class of molecules, characterized by fused, unsubstituted, aromatic rings and generally having a single π system, that are known atmospheric pollutants.

Poly(ether ether ketone) (PEET) a polymer in which two ether groups and a ketone group make up the main chain in the repeating unit.

Poly(ethylene terephthalate)(PET) a polyester with the repeating unit (—O_2C—C_6H_4—CO_2—)$_n$; commonly used in two-liter bottles for carbonated beverages.

Polyhalogenation the process of replacing multiple hydrogens on a single molecule with halogen atoms.

Polymer chain (also known as *main chain* or *backbone*) the chain of atoms bonded together in a polymer, to which substituents are attached.

Polymerization reaction a reaction that produces a polymer from monomers.

Polymers large molecules that are made from relatively small monomers and that have a recognizable repeating unit.

Polynucleotide See *nucleic acid*.

Polyolefins carbon–chain polymers made from the polymerization of alkene (i.e., olefin) monomers.

Polypeptide See *proteins*.

Polysaccharide a type of large carbohydrate made from many saccharides, including starch and cellulose.

Polysaccharides (types of *carbohydrates*) very large molecules constructed from just a few types of smaller molecules, called *monosaccharides* or *simple sugars*.

Polyunsaturated term that describes a molecule that has more than one C=C double bond; often used to describe fatty acids.

Poly(vinyl alcohol) a polymer with the repeating unit (—[OH]CH—CH_2—)$_n$.

1,2-positioning positioning in which substituents are connected to adjacent atoms.

Potassium permanganate the compound $KMnO_4$, used principally as an oxidizing agent. $KMnO_4$ can oxidize primary alcohols and aldehydes to carboxylic acids and secondary alcohols to ketones, and can cleave alkenes, alkynes, and C—C bonds to a benzylic carbon.

Precursors the preceding species from which a synthetic target can be synthesized.

Primary (1°) carbocation carbocation that has one alkyl group directly attached to C^+.

Primary orbital overlap an overlap among orbitals that are directly involved in bond formation in a Diels–Alder reaction.

Primary structure the sequence of amino acids in a protein; it determines a protein's identity.

Principal quantum number (n) the number that defines a shell in an atom, given that n can assume any integer value from 1 to infinity.

Principle of microscopic reversibility the notion that a reaction has the same free energy diagrams in the forward direction and the reverse direction.

Propagation cycle in a chain reaction, the collection of propagation steps that sum to the net reaction.

Propagation steps in a chain reaction, the elementary steps that are principally responsible for converting the overall reactants into the overall products.

Propyl group the group —$CH_2CH_2CH_3$.

Protecting groups modified functional groups that are made unreactive under the conditions for a desired chemical reaction. The original functional groups are restored by removing the protecting groups.

Proteins versatile biomolecules that play diverse roles in biological systems; proteins are constructed from amino acids and can be hundreds or thousands of amino acids long.

Protic possessing a hydrogen-bond donor.

Protic solvent a solvent that possesses a hydrogen-bond donor.

Proton nuclear magnetic resonance spectroscopy (also known as *^1H NMR spectroscopy*) spectroscopy method in which radio-frequency electromagnetic radiation is used to cause spin flips in the nuclei of hydrogen atoms.

Proton transfer reaction (also known as *Brønsted–Lowry acid–base reaction*) reaction in which a proton is transferred from a *Brønsted–Lowry acid* to a *Brønsted–Lowry base* in a single elementary step—that is, one bond is broken and another is formed simultaneously.

Pseudorotation partial rotations about the single bonds that compose a ring structure; envelope conformations of cyclopentane rapidly interconvert via pseudorotation.

Pyranose a cyclic form of a monosaccharide characterized by a six-membered ring.

Pyridinium chlorochromate (PCC) a specialized oxidizing agent that can be used in organic solvents and, therefore, can oxidize a primary alcohol to an aldehyde.

Quantum mechanics branch of physics whose central idea is that very small particles, like electrons, have traits that are characteristic of both particles and waves.

Quartet (q) the splitting pattern of an NMR signal into four peaks with a 1:3:3:1 height ratio; occurs when $N = 3$ in the $N+1$ rule.

Quartet of triplets the NMR splitting pattern in which each peak of a quartet is split into a triplet, resulting in a total of 12 peaks.

Quaternary ammonium salt An ionic compound in which the cation is of the form R_4N^+.

Quintet (qn) the splitting pattern of an NMR signal into five peaks; occurs when $N = 4$ in the $N+1$ rule.

R group See *side chain*.

Racemic mixture mixture that contains equal amounts of the (+) and (−) enantiomers of a chiral molecule.

Radical addition an elementary step in which a radical adds to an atom of a nonpolar π bond.

Radical cation a positively charged ion that has an unpaired electron.

Radical coupling (also known as *combination*) an elementary step in which unpaired electrons from two radicals join to make a new covalent bond.

Radical halogenation a reaction that results in the addition of a halogen atom to a molecule and proceeds by a mechanism that involves radicals.

Radical initiators compounds that have especially weak covalent bonds and can be used to produce radicals via heating or illumination with ultraviolet light.

Radio frequency (RF) the region of the electromagnetic spectrum having wavelengths between ~10^{-3} m and 10^5 m. RF radiation is used in NMR spectroscopy to cause spin flips among nuclei.

Raney nickel a specialized catalyst used in the Raney-nickel reduction of ketones and aldehydes to methylenes. It is produced by treating a 50:50 alloy of aluminum and nickel with hot sodium hydroxide, followed by treatment with hydrogen gas.

Rate constant the constant of proportionality that relates the concentrations of reactants or products (raised to certain exponents) to the rate of a given reaction.

Rate-determining step the elementary step that establishes the rate of an *overall reaction*; usually the slowest step of a mechanism.

Reaction coordinate variable that corresponds to geometric changes, on a molecular level, of the species involved in a reaction as reactants are transformed into products; as the reaction coordinate increases, the geometries of the species involved in the reaction increasingly resemble those of the products.

Reaction free-energy diagram a plot of Gibbs free energy as a function of the *reaction coordinate*.

Rearrangement a reaction in which the reactant and product species are isomers of each other.

Reducing agents reagents used to donate electrons, thereby causing the oxidation state of an atom in another species to decrease.

Regiochemistry describes the outcome (i.e., the product distribution) of a reaction that can take place at more than one site within a molecule.

Regioselectivity the tendency of a reaction to take place at one site within a molecule over another. See *regiochemistry*.

Relative abundance the amount of a substance that is present relative to that of a standard; in mass spectrometry, the quantity that is plotted on the *y* axis.

Repeating unit the bonding scenario that occurs over and over within a polymer.

Resolution the extent to which items (such as peaks in a spectrum or enantiomers) are separated from each other. See *resolved*.

Resolved term used to describe items that are separated; enantiomers that are separated are said to be resolved, and peaks in a spectrum that are separated are said to be resolved.

Resonance contributor See *resonance structure*.

Resonance effects the impact that resonance has on a property of a molecule or ion, such as stability or acid/base strength.

Resonance energy (also known as *delocalization energy*) the extent by which a species is stabilized due to the delocalization of electrons.

Resonance hybrid a weighted average of all resonance structures that represent a species.

Resonance structure (also known as *resonance contributor*) one of two or more valid Lewis structures that represent a species. Resonance structures are imaginary and are related by the movement of valence electrons, not atoms; the one, true species is represented by the resonance hybrid.

Resonance theory the theory that reconciles the differences between a species' multiple Lewis structures and its observed characteristics. *Resonance* exists when two or more valid Lewis structures can be drawn for a given molecular species.

Retro Diels–Alder reaction a [4+2] cycloelimination reaction that is the reverse of a Diels–Alder reaction and produces a diene and a dienophile.

Retrosynthetic analysis the strategy for designing a synthesis in which the chemist begins by focusing on the target and asks questions pertaining to feasible precursors that can be used to produce the target. Each precursor becomes the new target until the chemist arrives at precursors that can be used as starting materials.

Retrosynthetic arrow (⟹) an open arrow used to indicate a transform in retrosynthetic analysis and is drawn from the target to the precursor.

Reversible descriptor for reactions that proceed in the reverse direction at a rate that is comparable to or faster than that in the forward direction.

Reversible reaction arrows (⇌) reaction arrows pointing in both directions, signifying that the reaction is reversible. Distinct from an *irreversible reaction arrow*.

Ribbon structure a trace of a three-dimensional protein's backbone that highlights secondary structures such as α-helices and β-pleated sheets.

Ribonucleic acid (RNA) a pair of intertwined nucleic acids that participate in protein synthesis.

Rigid-rod polymers polymers having rigid chains that do not rotate readily.

Ring current motion of π electrons along a circular path parallel to the plane of the molecular ring to which they belong, giving rise to *magnetic anisotropy*; the motion is driven by forces from an external magnetic field.

Ring-opening polymerization a polymerization reaction in which the propagation steps involve the breaking of a bond in the monomer.

Ring strain the increase in energy due to geometric constraints of the bonds and substituents involved in a ring; it is partly responsible for the instability of rings other than five- and six-membered rings; it is quantified using *heats of combustion*.

Robinson annulation a reaction that is made up of a Michael addition followed by aldol condensation and that results in the formation of a new cyclohexenone ring.

Root the name of the alkane containing the same number of carbon atoms as the main chain for nomenclature.

Rotational conformations distinct angles of rotation about single bonds.

Rotational energy barrier the amount of energy needed to convert one conformational isomer into another via rotation about a specified bond.

Rotaxanes a class of compounds in which a dumbbell-shaped molecule is threaded through a large cyclic molecule called a *macrocycle*.

Royal purple a clothing dye obtained by ancient Phoenicians from a type of aquatic snail called *Murex brandaris*.

s Character the resemblance of a hybridized orbital's characteristics (e.g., shape and energy) to those of an *s* orbital; the fraction of a hybridized atomic orbital that comes from an *s* orbital.

Saccharides (also known as *carbohydrates*) molecules that are characterized by the molecular formula $C_xH_{2y}O_y$.

Sandmeyer reactions reactions in which a nucleophile replaces the N_2 leaving group in an arenediazonium ion and in which a Cu^{+1} catalyst is involved.

Saponification a reaction between hydroxide (HO^-) and an ester, resulting in the formation of a carboxylate anion. This is the same type of reaction used by ancient civilizations to make soap from animal fat.

Saturated term used to describe a molecule that has the maximum number of hydrogen atoms possible, consistent with the number and type of each nonhydrogen atom in the molecule, the octet rule, and the duet rule. A saturated molecule has no double bonds, triple bonds, or rings.

S-cis conformation a rotational conformation in which substituents are on the same side of a single bond.

Secondary (2°) carbocation carbocation that has two alkyl groups attached directly to C^+.

Secondary orbital overlap in a Diels–Alder reaction, overlap among orbitals that are *not* directly involved in bond formation.

Selective reaction a reaction in which multiple products can be produced, but one product is favored over the others.

Selective reagent a reagent that is used to favor the production of one product over another.

Self-aldol additions *aldol addition* reactions in which the carbonyl-containing reactants are the same.

Sesquiterpene a terpene that contains one and one-half pairs of isoprene units (i.e., three isoprene units), for a total of 15 carbons.

Shell the principal quantum number (*n*) associated with an atomic orbital; also referred to as the energy level of the orbital.

Shielded term used in NMR spectroscopy to describe nuclei in a molecule that have a local magnetic field that opposes an external magnetic field; shielded nuclei have lower chemical shifts.

Side chain (often referred to as *R group*) a group connected to α carbon(s) in an amino acid or protein.

σ Bond a bond wherein a pair of electrons occupies a molecular orbital (MO) of σ symmetry—that is, an MO in which the overlap of atomic orbitals takes place along a bonding axis. A single bond in a Lewis structure is generally a σ bond.

σ Symmetry the symmetry of a molecular orbital (MO) that is generated by the interaction of atomic orbitals along bonding axes. A σ *bond* occurs when two electrons occupy a σ MO.

Signal averaging technique used by modern instruments in ^{13}C NMR to acquire and average together multiple spectra, ultimately reducing noise.

Signal splitting the splitting of an NMR signal into multiple peaks due to the spin–spin coupling between the nucleus giving rise to the signal and nuclei of nearby atoms.

Signals in NMR spectroscopy, re-emitted frequencies that correspond to frequencies that have been absorbed by a sample.

Simple sugars See *monosaccharides*.

Single-barbed arrows (⤴) curved arrows used to show the movement of a single electron.

Singlet the splitting pattern of an NMR signal when there are no coupled protons; appears as just a single peak.

Soap scum the precipitate of long-chain fatty acid carboxylates with cations from *hard water*.

Sodium borohydride ($NaBH_4$) a moderately strong hydride reducing agent that favors direct addition over conjugate addition.

Sodium hydride (NaH) an ionic hydride reagent that is a strong base but a poor nucleophile.

Solvated electron an electron that is stabilized by solvent molecules and, therefore, does not formally belong to any one atom. Solvated electrons are used in dissolving metal reductions.

Solvated (ions) ions that are involved in multiple ion–dipole interactions with their solvent.

Solvation the phenomenon wherein an individual ion participates in multiple ion–dipole interactions with its solvent. When this occurs, the ions are said to be solvated.

Solvent-mediated proton transfer transfer of a proton from one site to another assisted by a sufficiently acidic or basic solvent.

Solvolysis a reaction in which the solvent participates in breaking a bond (lysis) as an attacking species; the reaction can be classified as nucleophilic substitution or elimination.

sp Hybridization   the mixing of the single s orbital and one of the three p orbitals—p_x, p_y, or p_z—from the valence shell to form what are called sp-hybridized atomic orbitals.

sp-Hybridized atomic orbitals   the hybridized atomic orbitals that result from sp hybridization.

sp^2 Hybridization   the mixing of the single s orbital and two of the three p orbitals—p_x, p_y, or p_z—from the valence shell to form sp^2-hybridized orbitals.

sp^2-Hybridized orbital   the hybridized atomic orbital that results from sp^2 hybridization.

sp^3 Hybridation   the mixing of the single s orbital and all three p orbitals—p_x, p_y, and p_z—from the valence shell to form sp^3-hybridized orbitals.

sp^3-Hybridized orbital   the hybridized atomic orbital that results from sp^3 hybridization.

Species   a particular collection of protons, neutrons, and electrons.

Specific rotation ($[\alpha]_\lambda^T$)   constant term that is unique for a given chiral compound and describes the compound's propensity for rotating plane-polarized light; it is the angle of rotation of plane-polarized light of a given wavelength in nanometers, λ, if it passes through a sample whose concentration is 1 g/mL, whose length is 1 dm, and whose temperature is $T°$C.

Spectator ion   ion that is relatively inert in solution and tends not to react.

Spectroscopy   the study of the interaction of electromagnetic radiation with matter.

Spin flip   a change in spin from one spin state to another (e.g., from the α state to the β state, or vice versa).

Spin–spin coupling   the interaction between nearby nuclei that have spin; the cause of signal splitting in NMR spectroscopy.

Splitting diagram   diagram that illustrates the complex splitting of an NMR signal in a stepwise fashion.

Staggered conformations   rotational conformations in which each C—H bond on the front carbon atom in a Newman projection bisects a pair of C—H bonds on the rear carbon.

Standard enthalpy difference ($\Delta H°_{rxn}$)   the enthalpy difference between reactants and products under standard conditions; associated with a reaction's tendency to absorb or release heat.

Standard entropy difference ($\Delta S°_{rxn}$)   the entropy difference between reactants and products under standard conditions; related to the gain or loss of the number of different states available to a system.

Standard Gibbs free energy difference ($\Delta G°_{rxn}$)   the Gibbs free energy difference between reactants and products under standard conditions; associated with a reaction's tendency to favor products at equilibrium (i.e., to be spontaneous).

Starting materials   reactants that are available to use in a synthesis.

Step-growth polymerization   a polymerization reaction in which molecules throughout a mixture join with any other molecules that have appropriate functional groups available to react, whether they are monomer units or other polymer molecules.

Stereocenter   an atom with the property that interchanging any two of its attached groups produces a different stereoisomer.

Stereochemical configurations   classifications of the three-dimensional arrangements of substituents attached to a tetrahedral stereocenter or a double bond. The configuration of a tetrahedral stereocenter is designated as R or S, and that of a double bond is designated as E or Z.

Stereochemistry   the study of the arrangement of atoms in space.

Stereoisomers   isomers that have the same connectivity.

Stereospecific   term used to describe reactions that produce one stereochemical configuration or arrangement of atoms (stereochemistry) exclusively.

Stereoregular   a term used to describe a polymer in which the tetrahedral stereocenters that are part of the main chain establish a recognizable pattern.

Steric hindrance   the spatial restriction of movement (in rotation or of an incoming reacting species) from bulky substituents on a molecule.

Steric strain   an increase in energy that results from electron repulsion between atoms or groups of atoms that are not directly bonded together but occupy the same space.

Storage polysaccharides   polysaccharides, such as starch, whose main function is to store energy.

Straight-chain alkanes   See *linear alkanes*.

S-trans conformation   a rotational conformation in which substituents are on opposite sides of a single bond.

Stretching   one of two basic, independent types of vibrational motion. In stretching, the distance between two atoms in a chemical bond grows longer and shorter throughout one period of vibration.

Strong absorption   absorption whose percent transmittance is near zero (near the bottom of the spectrum if percent transmittance is plotted on the y axis).

Structural isomers   See *constitutional isomers*.

Structural polysaccharides   polysaccharides, such as cellulose, whose main function is to provide structural integrity.

Substituent   an atom or group of atoms attached to a molecule in place of hydrogen.

Substrate   a molecular species that undergoes a particular reaction with an attacking species; in nucleophilic substitution or elimination, the species that contains the leaving group.

Sulfonate ester   a functional group having the bonding arrangement R—SO₃R′. Sulfonate esters typically possess very good leaving groups for nucleophilic substitution reactions.

Sulfonation   a reaction in which the group —SO₃H or the group —SO₃R is added to a molecule; one type of electrophilic aromatic substitution reaction.

Sulfonium ylide   species characterized by the $^+$S—C$^-$ bond.

Sulfonyl group   a group with the bonding arrangement —SO₂—.

Symmetric stretch   type of vibration in which two or more bonds stretch simultaneously and compress simultaneously.

Synthesis   a specific sequence of chemical reactions that converts starting materials into a desired compound, called the target of the synthesis (or the *synthetic target*).

Synthetic step   an overall (or net) reaction that is carried out in a synthesis.

Synthetic target   the desired end product compound of a *synthesis*.

Synthetic trap   a difficulty preventing the execution of a *synthesis* in the forward direction as planned.

Target   See *synthetic target*.

Termination steps   elementary steps in a chain reaction that result in a decrease in the number of reactive intermediates involved in the propagation cycle.

Termolecular   term used to describe an elementary step that involves three reactant species simultaneously.

Terpene naturally produced hydrocarbon whose carbon backbone can be divided into separate and unique five-carbon units called *isoprene units*.

Terpenoids natural products that are produced from chemical modifications to terpenes, in which the carbon backbone is altered and/or atoms other than just carbon or hydrogen are introduced.

Tertiary (3°) carbocation carbocation that has three alkyl groups attached directly to C^+.

Tertiary structure a term used to describe the three-dimensional location of the amino acid residues that make up a particular protein.

Tesla (T) the SI unit for the strength of a magnetic field.

Tetramethylsilane (TMS) $(CH_3)_4Si$, usually added to samples to provide a reference signal in NMR spectroscopy.

Tetrasubstituted term used to describe an atom, bond, or molecule that has four attached substituents.

Theoretical rate law the equation that relates the rate of a reaction to the concentrations of the species involved in the reaction, which is derived from a proposed mechanism.

Thermal a term used to describe a molecule that is in its ground-state electron configuration.

Thermal energy the average energy available through molecular collisions, which increases as temperature increases.

Thermally allowed a term used to describe a reaction that can take place readily when the reactants are in their ground-state electron configurations.

Thermally forbidden a term used to describe a reaction that cannot take place readily when the reactants are in their ground-state electron configurations.

Thermodynamic control the execution of a chemical reaction under equilibrium conditions, such that the major product is the one that is most stable.

Thermodynamic enolate anion the more stable of the enolate ions that can be produced by deprotonation of the corresponding ketone.

Thermoplastic a term used to describe a polymer that, when heated above its glass-transition or melting temperature, can be forced into molds or pushed through slots to form films or threads.

Thionyl chloride the compound $SOCl_2$, used to convert alcohols to alkyl chlorides and carboxylic acids to carboxylic acid chlorides.

Three-center, two-electron bonds bonds formed by two electrons and shared by three atoms; diborane is an example.

Torsional stain an increase in energy (i.e., decrease in stability) that appears in an eclipsed conformation.

Trans term that describes the configuration wherein two atoms are on opposite sides of a double bond or plane of a ring.

Transform in a retrosynthetic analysis, the proposed undoing of a single reaction or set of reactions to arrive at a potential precursor.

Transition state a high-energy species that corresponds to an energy maximum along the reaction coordinate on a reaction free-energy diagram.

Transition state theory theory that relates *free energy of activation* of a reaction to its rate constant.

Transmittance (%T) in spectroscopy, the portion of light sent through an analyte that reaches a detector: $\%T = (I_{detected}/I_{source}) \times 100$.

Triacylglycerol a description of a fat or oil that contains three adjacent ester groups, each of which can be produced from a *fatty acid* and an alcohol group from *glycerol*.

Trialkylborane (R_3B) a compound with three alkyl groups attached to boron.

Trialkylborate ester (RO_3B) a compound with three alkoxy groups attached to boron.

Triglyceride a fat or oil containing three adjacent ester groups, each of which can be produced from a fatty acid and an alcohol group from glycerol. See *triacylglycerol*.

Triplet (t) the NMR splitting pattern of a signal into three peaks in a 1:2:1 height ratio.

Triplet of doublets the NMR splitting pattern in which each peak of a triplet is split into a doublet, resulting in a total of six peaks.

Triplet of quartets the NMR splitting pattern in which each peak of a triplet is split into a quartet, resulting in 12 peaks.

Triterpene a terpene that contains three pairs of isoprene units (i.e., six isoprene units), for a total of 30 carbons.

Trivial names See *common names*.

Tropylium ion the aromatic cation $C_7H_7^+$, which has a seven-membered ring. The benzylic cation $(C_6H_5CH_2^+)$ rearranges into $C_7H_7^+$ as part of a fragmentation pathway in mass spectrometry.

Ultraviolet–visible (UV-vis) spectroscopy spectroscopy that uses light from the ultraviolet (UV) and visible regions of the electromagnetic spectrum; typically causes electron transitions between orbitals.

Umpolung a German word that means "polarity reversal"; describes reactions that convert an electron-poor carbon atom into an electron-rich one.

Unimolecular having a molecularity of 1; describing an elementary step in which there is one reactant species.

Unimolecular elimination (E1) reaction two-step reaction consisting of heterolysis followed by electrophile elimination, resulting in the net loss of a proton and a leaving group, and the production of a new π bond.

Unimolecular nucleophilic substitution (S_N1) reaction two-step reaction consisting of heterolysis followed by coordination, in which a leaving group is replaced by a nucleophile.

Unpaired electron an electron occupying an orbital without the presence of a second electron in that orbital; unpaired electrons are found on species called free radicals.

Unpolarized term to describe light whose photons travel in the same direction but have their electric fields oscillating in different planes.

Unsaturated term used to describe a molecule that has fewer than the maximum number of hydrogen atoms possible; such molecules have double bonds, triple bonds, and/or rings.

Upfield appearing in an NMR spectrum in a region of lower chemical shift.

UV–vis spectrum a plot of absorbance against wavelength of ultraviolet (UV) and visible light, showing *absorption bands* or *peaks*.

Valence bond (VB) theory theory that describes orbitals in molecules as being produced by mixing just two atomic orbitals (AOs) at a time, one from each of two adjacent atoms; often one or both of the AOs involved in each mixing is a hybridized AO.

Valence electrons   electrons available to participate in bonding interactions, occupying the highest (outermost) energy shell.

Valence shell electron pair repulsion (VSEPR) theory   theory used to predict the geometry about an atom, based on the idea that the lowest-energy geometry about an atom is achieved by minimizing the electrostatic repulsions among the electron groups to which valence electrons can be assigned.

Vector   a geometric entity that has both magnitude and direction.

Vinyl   the substituent, $—CH=CH_2$.

Vitalism   the outdated belief that organic compounds could *not* be made in the laboratory; instead, only living systems could summon up a mysterious "vital force" needed to synthesize them.

Vulcanization   the reaction of natural rubber with sulfur that results in the crosslinking of the polymer chains.

Walden inversion   the phenomenon that occurs upon nucleophilic attack in an S_N2 reaction that causes inversion of stereochemistry.

Wavelength (λ)   the distance between two successive maxima or minima in a wave's oscillating amplitude.

Wavenumbers (ν)   See *inverse centimeters*.

Waxes   secretions from plants and animals that are solid at room temperature but melt at relatively low temperatures; often long-chain hydrocarbons or esters.

Weak absorption   absorption whose percent transmittance is near 100% (near the top of the spectrum if percent transmittance is plotted on the *y* axis).

Wheland intermediate   See *arenium ion*.

Williamson ether synthesis   a reaction used to synthesize either symmetrical or unsymmetrical ethers via an S_N2 reaction between an alkoxide anion (RO^-) and an alkyl halide ($R—X$) in basic conditions.

Wittig reactions   reactions using Wittig reagents with ketones and/or aldehydes to convert $C=O$ bonds into $C=C$ double bonds.

Wittig reagents   (also known as *phosphonium ylides*) a class of compounds having a $^+P—C^-$ bond, used to synthesize alkenes; characteristically, it is highly nucleophilic at the carbon attached to phosphorus.

Wolff–Kishner reduction   a reaction that reduces the carbonyl group of a ketone or an aldehyde to a methylene (CH_2) group through treatment with hydrazine followed by heating under basic conditions.

Zaitsev product   the most highly substituted alkene product that can be produced in an elimination reaction.

Zaitsev's rule   empirical rule for elimination reaction regioselectivity, producing the *Zaitsev product*: The major elimination product is the one produced by deprotonating the carbon atom initially attached to the fewest hydrogen atoms. In other words, elimination usually takes place so as to produce the most highly alkyl-substituted alkene.

Zigzag conformation   See *all-anti conformation*.

Zinc amalgam   an alloy of zinc and mercury, which is a key reactant in the Clemmensen reduction.

Zwitterion   a species that has both a positive and a negative formal charge, but a net charge of zero.

Answers to Your Turns

Chapter 1

YOUR TURN 1.1
The 2s and 2p are in the second shell. Electrons (3)–(10) in Figure 1-7 are in the second shell.

YOUR TURN 1.2
The valence electrons are second shell, $2s^2$ and $2p^2$, and the core electrons are in the first shell, $1s^2$.

YOUR TURN 1.3
Bond energy = 450 kJ/mol − 0 kJ/mol = 450 kJ/mol

YOUR TURN 1.4
(a) 586 kJ/mol (140 kcal/mol). (b) The Si—F bond.
(c) 138 kJ/mol (33 kcal/mol). (d) The O—O bond.
(e) 1072 kJ/mol (256 kcal/mol). (f) The C≡O triple bond.

YOUR TURN 1.5

YOUR TURN 1.6

YOUR TURN 1.7
For NaCl, which is ionic, the difference is large: 3.16 − 0.93 = 2.23. For CH_4, which is covalent, the difference is 2.55 − 2.20 = 0.35.

YOUR TURN 1.8
Valence electrons (total # electrons, # protons, charge): 3 (5, 6, +1), 4 (6, 6, 0), 5 (7, 6, −1)

YOUR TURN 1.9
Sum of formal charges: $0 + 0 + 0 + (-1) = -1$. Sum of oxidation states: $(+2) + (+1) + (-2) + (-2) = -1$

YOUR TURN 1.10

YOUR TURN 1.11

YOUR TURN 1.12

YOUR TURN 1.13

YOUR TURN 1.14

YOUR TURN 1.15

General ketone functional group   Cyclohexanone   Hexan-3-one

YOUR TURN 1.16

Serine
(a)

Aspartic acid
(b)

Lysine
(c)

Phenylalanine
(d)

YOUR TURN 1.17

Alcohol (R—OH) groups are circled. The C=O group is part of an aldehyde.

Glucose, $C_6H_{12}O_6$

YOUR TURN 1.18

Uracil
(U)

Guanine
(G)

Adenine
(A)

Alkene (red) C=C
Amide (blue) RC(O)N

Cytosine
(C)

Thymine
(T)

Chapter 2

YOUR TURN 2.1

CH_3^+ has three single bonds and no lone pairs, thus three groups. CH_3^- has three single bonds and one lone pair, thus four groups.

C = 3 groups   C = 4 groups

YOUR TURN 2.2

First structure   Second structure

YOUR TURN 2.3

YOUR TURN 2.4

YOUR TURN 2.5

YOUR TURN 2.6

YOUR TURN 2.7

The thin, red arrows indicate the bond dipoles. The thick, red arrows indicate the vector sum of each pair of bond dipoles. The thick, red arrows are equal in magnitude and point in opposite directions, so the molecule has no net dipole moment and, therefore, is nonpolar.

YOUR TURN 2.8

Methanoic acid has a carboxylic acid; ethanol has an alcohol; ethanal has an aldehyde; dimethyl ether has an ether; propene has an alkene; propane and ethane have no functional groups.

YOUR TURN 2.9

Higher boiling point, larger dipole moment

YOUR TURN 2.10

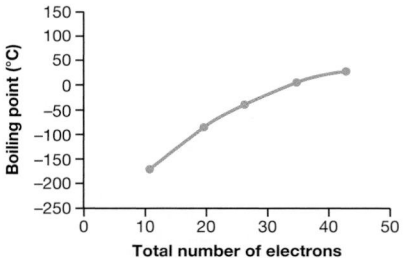

YOUR TURN 2.11

Cholesterol has one potential H-bond acceptor (the O atom) and one potential donor (the OH hydrogen). Octanoic acid has two potential H-bond acceptors (each O atom) and one potential donor (the OH hydrogen).

YOUR TURN 2.12

Boiling point increases as the number of electrons increases.

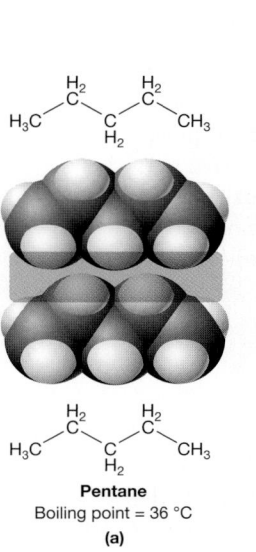

YOUR TURN 2.13

As shown below, pentane has the greater contact surface area.

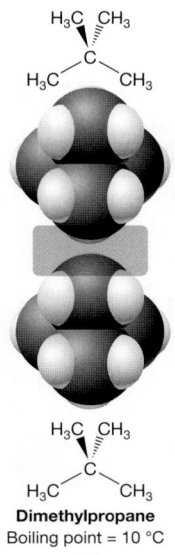

Pentane
Boiling point = 36 °C
(a)

Dimethylpropane
Boiling point = 10 °C
(b)

YOUR TURN 2.14

B is more soluble in H_2O because RNH_2 is capable of hydrogen bonding.

YOUR TURN 2.15

Anion attracted to the dipole's partial positive charge

Cation attracted to the dipole's partial negative charge

(a) (b)

YOUR TURN 2.16

A molecule must possess an H—O, H—N, or H—F bond to be capable of hydrogen-bonding interactions. The partial positive end—namely, the H—is the hydrogen-bond donor region. The only solvents in Table 2-7 that have a hydrogen-bond donor are water, ethanol, and ethanoic acid.

Structure	Name
	Water
	Ethanol
	Ethanoic acid (acetic acid)

YOUR TURN 2.17

The CO_2^- (red/orange region) is hydrophilic and the hydrocarbon tail (blue/green region) is hydrophobic.

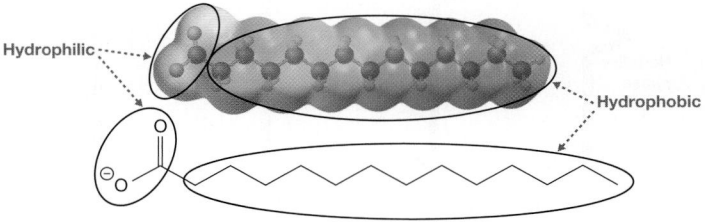

Hydrophilic

Hydrophobic

YOUR TURN 2.18

An ester functional group

Fat or oil
(triacyl glycerol)
or triglyceride)

Ester functional groups

Chapter 3

YOUR TURN 3.1

High electron density

Low electron density

Encompasses 90% of the dots

YOUR TURN 3.2

Node

YOUR TURN 3.3

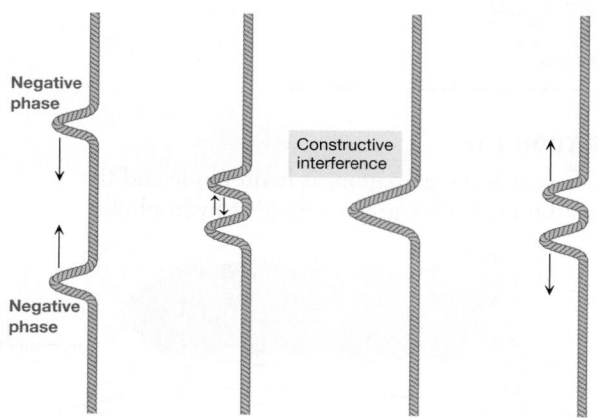

Negative phase

Negative phase

Constructive interference

YOUR TURN 3.4

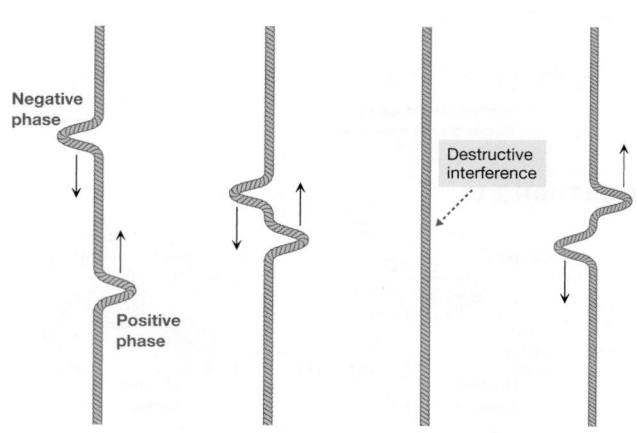

Negative phase

Positive phase

Destructive interference

YOUR TURN 3.5

The σ* MO is the HOMO.

YOUR TURN 3.6

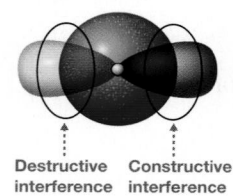

Destructive interference   Constructive interference

YOUR TURN 3.7

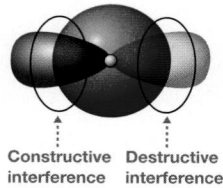

Constructive interference   Destructive interference

YOUR TURN 3.8

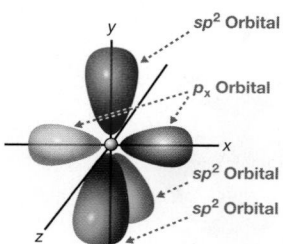

sp^2 Orbital

p_x Orbital

sp^2 Orbital

sp^2 Orbital

YOUR TURN 3.9

Nodal plane

π* MO   π MO

YOUR TURN 3.10

The HOMO is the π MO, and the LUMO is the π* MO.

YOUR TURN 3.11

The HOMO is one of the π MOs, and the LUMO is one of the π* MOs.

YOUR TURN 3.12

If the molecule on the left is flipped vertically 180°, it appears to be identical to the one on the right, so the two molecules are identical.

YOUR TURN 3.13

The first molecule is planar. The second molecule is not; one CH_2 plane is perpendicular to the other.

(a)   (b)

YOUR TURN 3.14

From left to right, the values are: $sp^3 = 25\%$, $sp^2 = 33.33\%$, and $sp = 50\%$.

Chapter 4

YOUR TURN 4.1

YOUR TURN 4.2

If the molecule in Figure 4-2 is rotated 90° clockwise (as viewed from the top), the CBr_3 group is in the front and the CH_3 group is in the back, yielding the first molecule below. If the molecule is rotated 90° counterclockwise (as viewed from the top), the CBr_3 group is in the back and the CH_3 group is in the front, yielding the second molecule below. Both are acceptable answers.

YOUR TURN 4.3

YOUR TURN 4.4

In this example, all eclipsed conformations are indistinguishable (−120°, 0°, and +120°) and all three staggered conformations are indistinguishable (±180°, −60°, and 60°).

YOUR TURN 4.5

Your model should resemble the structures in Figure 4-5. The front and back C—H bonds are closer to each other in an eclipsed conformation, and, thus, are *higher* in energy.

YOUR TURN 4.6

13 kJ/mol (3.1 kcal/mol)

YOUR TURN 4.7

Eclipsed conformations (−120°, 0°, and +120°), staggered gauche conformations (−60° and +60°), staggered anti conformations (±180°).

YOUR TURN 4.8

Eclipsed conformation energy difference, $40 - 27 = 13$ kJ/mol (3.1 kcal/mol), staggered conformation energy difference, 12 kJ/mol (2.9 kcal/mol).

YOUR TURN 4.9

YOUR TURN 4.10

$27 - 0 = 27$ kJ/mol (6.5 kcal/mol)

YOUR TURN 4.11

8 CH_2 groups. 620.3 kJ/mol. 5.2 kJ/mol. 41.6 kJ/mol. This value is less than cyclopropane and cyclobutane, but more than cyclopentane, cyclohexane, and cycloheptane.

YOUR TURN 4.12

YOUR TURN 4.13

This C—C bond is eclipsed.

YOUR TURN 4.14

This C—C bond rotates upward.

Pseudorotation

This C—C bond rotates downward.

YOUR TURN 4.15

All of the C—C bonds undergo rotation as axial and equatorial groups change their positions. The "head" of the chair C′ rotates to become the foot of the chair and the foot of the chair C″ rotates to become the head of the chair.

All six C—C bonds rotate.

Ring flip

YOUR TURN 4.16

Angle strain arises from three sp^3-hybridized C atoms having bond angles of ~120°. Torsional strain arises from eclipsed C—C bonds.

Angle strain

~120°
~120°
~120°

Torsional strain

Eclipsed conformation

Eclipsed conformation

YOUR TURN 4.17

(a) Start with two C—C bonds.
(b) Add in V.
(c) Add ∧.
(d) Draw in axial bonds at head (up) and foot (down).

(e) Fill in remaining axial bonds, alternating up and down.
(f) **(g)** Drawn in equatorial bonds. Wherever an axial bond points up, an equatorial bond points down, and vice versa.

YOUR TURN 4.18

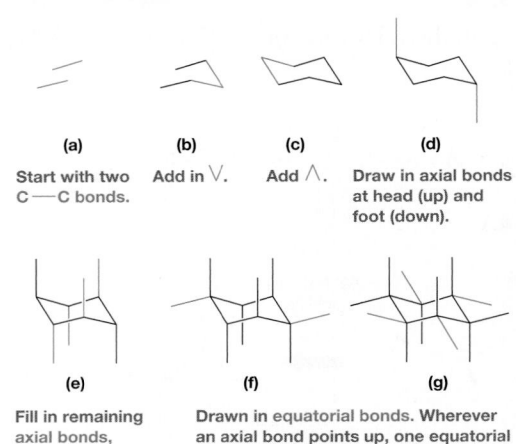

(a) Start with two C—C bonds.
(b) Add in V.
(c) Add ∧.
(d) Draw in axial bonds at head (up) and foot (down).

(e) Fill in remaining axial bonds, alternating up and down.
(f) **(g)** Drawn in equatorial bonds. Wherever an axial bond points up, one equatorial bond points down, and vice versa.

YOUR TURN 4.19

The CH_3 group is circled for clarity. Axial H atoms are labeled. All other H atoms bonded directly to the ring are equatorial.

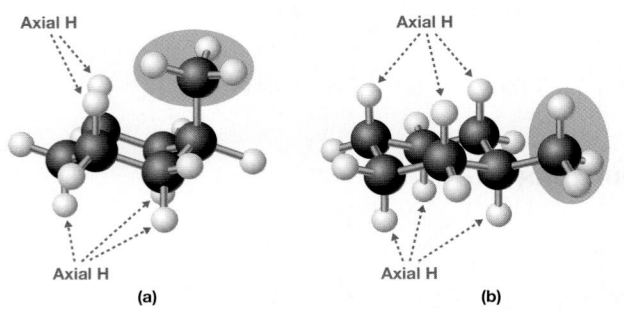

Axial H

Axial H

Axial H

Axial H

(a)

(b)

YOUR TURN 4.20

These H atoms are involved in 1,3-diaxial interactions with the CH₃ group.

This C atom is gauche to the CH₃ group.

YOUR TURN 4.21

(a)

(b)

View from top

View from top

H_3C—————⫼CH_3

H_3C—————⫼CH_3

(c)

(d)

YOUR TURN 4.22

To translate the conformation of *trans*-1,3-dimethylcyclohexane on the left into the "ring-flipped" conformation on the right, without flipping the chair, carry out the following rotations of the molecule.

Rotate 180°   Rotate 60°

Figure on left of 4-33b

Same as figure on right in Figure 4-33b

YOUR TURN 4.23

The longest continuous chain of carbons has six carbons. The first substituent is a CH₃ group located on C2. The second substituent is a CH₃ located on C4. This is the same molecule as the previous two molecules. Recall that there is free rotation about single C—C bonds. These three molecules are *conformational* isomers.

YOUR TURN 4.24

H_3C—CH_3 H_3C—CH_3 H_3C—CH_2 H_3C—CH_2
                                          OH OH

(a)   (b)   (c)   (d)

YOUR TURN 4.25

G = alkene, alcohol. **H** = alkene, alcohol. **I** = aldehyde. **K**, **L** = alkene, alcohol. **M** = ketone.

Chapter 5

YOUR TURN 5.1

The two molecules are not superimposable. They are enantiomers.

Enantiomers

Rotate 109.5°

Molecule on left in Figure 5-2a

Molecule on right in Figure 5-2a

YOUR TURN 5.2

The mirror image is superimposable; therefore, the mirror image is the same as the original molecule. Note: H_A and H_B are labels assigned to the mirror image to aid in your visualization of the transformations performed.

From the mirror

109.5° rotation

YOUR TURN 5.3

Use model kits for verification.

YOUR TURN 5.4

Chiral. The red and black molecules in Solved Problem 5.6 are mirror images of each other. From the solved problem, we have already verified that the two molecules are not the same and are, therefore, enantiomers.

YOUR TURN 5.5

Use model kits for verification.

YOUR TURN 5.6

Construct molecular models and carry out the rotations/chair flips shown in Figure 5-7.

YOUR TURN 5.7

Use model kits for verification.

YOUR TURN 5.8

The plane of symmetry is shown below. It reflects C1 and C2, C6 and C3, C5 and C4, and the two C—F bonds that both are pointed down.

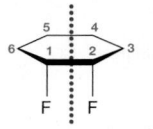

cis-1,2-Difluorocyclohexane

YOUR TURN 5.9

The F atoms' mirror images would be located at the "O"s that are shown. No F atoms are located at those positions.

trans-1,2-Difluorocyclohexane

YOUR TURN 5.10

Butan-2-ol 3-Chloro-1,1-dimethylcyclohexane

YOUR TURN 5.11

YOUR TURN 5.12

The first molecule does not line up perfectly with the second, regardless of rotations about single bonds or the orientations of the molecules.

YOUR TURN 5.13

YOUR TURN 5.14

YOUR TURN 5.15

The configurations differ at the circled C atoms.

D-Allose D-Glucose

YOUR TURN 5.16

YOUR TURN 5.17

Chapter 6

YOUR TURN 6.1

Bond formation   Bond breaking

YOUR TURN 6.2

The curved arrows are shown below. The second K_{eq} is boxed because it is a larger number.

YOUR TURN 6.3

HCl ($pK_a = -7$) is a stronger acid than H_3O^+ ($pK_a = -1.7$). The difference in pK_a values is $-1.7 - (-7) = 5.3$, which corresponds to a difference in acid strength of $>10^5$. Thus, HCl is $>100{,}000$ times stronger an acid than H_3O^+.

YOUR TURN 6.4

H_2O ($pK_a = 15.7$) is a stronger acid than $(CH_3)_2NH$ ($pK_a = 38$). The difference in pK_a values is $38 - 15.7 = 22.3$, which corresponds to a difference in acid strength of $10^{22.3}$. Thus H_2O is 2.0×10^{22} times stronger an acid than $(CH_3)_2NH$.

YOUR TURN 6.5

$(CH_3)_2NH$ ($pK_a = 38$) is a stronger acid than diethyl ether $(CH_3CH_2)_2O$ (pK_a ~45). The reactant side of the reaction is favored because the stronger acid is on the product side. This is not the same side that is favored in Equation 6-13. Therefore, diethyl ether is a suitable solvent for $(CH_3)_2N^-$ because the equilibrium lies to left, indicating that diethyl ether is relatively inert in the presence of $(CH_3)_2N^-$.

YOUR TURN 6.6

The acid is nearly 100% dissociated at ~2 pH units above the pK_a, or pH = 7. It is nearly 100% associated ~2 pH units below the pK_a, or pH = 3.

YOUR TURN 6.7

YOUR TURN 6.8

The distance between Cl and H **decreases** and the distance between the O and H **increases**.

YOUR TURN 6.9

The free energy quantities are indicated in the diagram below. The free energy of activation is larger in Figure 6-2b compared to the reaction in Fig. 6-2a.

YOUR TURN 6.10

The pK_a of HCl is -7, and that of H_2S is approximately the same as CH_3CH_2SH, which is 10.6. HCl is a stronger acid.

YOUR TURN 6.11

The pK_a of H_3O^+ is -1.7, and that of H_4N^+ is 9.4. H_3O^+ is a stronger acid.

YOUR TURN 6.12

The pK_a of H_3C—CH_3 is ~50, that of H_2C=CH_2 is ~44, and that of HC≡CH is ~25.

YOUR TURN 6.13

The pK_a of ethanoic acid (acetic acid) is 4.75. That of ethanol is 16.

YOUR TURN 6.14

Resonance hybrid

YOUR TURN 6.15

Resonance hybrid

YOUR TURN 6.16

YOUR TURN 6.17

YOUR TURN 6.18

Protonated amines and alcohols are stronger acids compared to the uncharged species. R—NH_3^+ (9.4); R—NH_2 (38 estimated from 2° amine R_2NH); R—OH_2^+ (≈ -2, similar to H_3O^+); R—OH (16).

YOUR TURN 6.19

R—OH (16); R—NH_2 (38 estimated from 2° amine R_2NH); R—CH_3 (50), RC≡CH (25).

YOUR TURN 6.20

Chapter 7

YOUR TURN 7.1

Electron rich to electron poor

- Electron rich
- Lewis base
- Electron poor
- Lewis acid

YOUR TURN 7.2

This is unacceptable because H cannot form two covalent bonds.

YOUR TURN 7.3

Negatively charged nucleophiles

Uncharged nucleophiles

YOUR TURN 7.4

YOUR TURN 7.5

YOUR TURN 7.6

YOUR TURN 7.7

YOUR TURN 7.8

YOUR TURN 7.9

The curved arrow originates from the electron-rich double bond and points to the electron-poor H^+.

YOUR TURN 7.10

To show the C—H bond breaking, a curved arrow originates from the center of the C—H bond. To show the pair of electrons ending up in the C=C double bond, the curved arrow points to the center of the C—C bond.

YOUR TURN 7.11

YOUR TURN 7.12

The single curved arrow shows a C—H bond breaking from the tertiary carbon and simultaneously forming to the secondary carbon.

YOUR TURN 7.13

HCl $pK_a = -7$; HF $pK_a = 3.2$. The difference in pK_a values $= 3.2 - (-7) = 10.2$, so $K_{eq} = 10^{10.2} = 1.6 \times 10^{10}$.

YOUR TURN 7.14

HCl $pK_a = -7$; NH_3 $pK_a = 36$. The difference in pK_a values $= 36 - (-7) = 43$, so $K_{eq} = 10^{43}$.

YOUR TURN 7.15

ΔBond energy (C=O)−(C=C) = 720 − 619 kJ/mol = 101 kJ/mol; ΔBond energy (C—C) − (C—O) = 339 − 351 kJ/mol = −12 kJ/mol; ΔBond energy (C—H) − (O—H) = 418 − 460 kJ/mol = −42 kJ/mol. From this it appears that, because the difference is greatest between the C=O and C=C bond energies, the C=O bond (being the stronger of the two) is the one that has the most influence on the outcome of the reaction.

Interchapter 1

YOUR TURN IC1.1

H is not hybridized and is s; O and Cl are both sp^3 hybridized. End-on overlap occurs in both H—O⁻ and H—Cl bonding and, therefore, results in a sigma bond. When the two AOs overlap, two MOs are produced: sigma bonding (σ) and sigma antibonding (σ^*). The three sets of lone pairs on Cl and O are in sp^3 hybrid, nonbonding orbitals. The HOMO from HO⁻ is, therefore, a nonbonding orbital, and the LUMO from HCl is a σ^* MO, in agreement with Figure IC1-3.

The nonbonding orbital of HO⁻ comes from a hybridized AO. The σ^* MO of HCl is the result of destructive interference, leaving a node in the internuclear region:

YOUR TURN IC1.2

Constructive interference occurs when two orbitals overlap with the same phases. Destructive interference occurs when two orbitals overlap with opposite phases.

YOUR TURN IC1.3

C and Cl are both sp^3 hybridized and each H atom contributes a $1s$ orbital. The 11 AOs produce 11 new MOs. End-on overlap occurs in the C—Cl and C—H bonding and each overlap produces two MOs: σ bonding and σ^* antibonding which accounts for eight MOs. The remaining three orbitals are from the noninteracting sp^3 AOs from Cl, producing nonbonding orbitals. The LUMO is, therefore, a σ^* MO, in agreement with Figure IC1-4.

The C—Cl σ^* can be thought of as resulting from the destructive interference between two sp^3 AOs, producing a node in the internuclear region:

YOUR TURN IC1.4

Constructive interference occurs when two orbitals overlap with the same phases. Destructive interference occurs when two orbitals overlap with opposite phases.

(a) HOMO LUMO   (b) LUMO

YOUR TURN IC1.5

The central C is sp^2 hybridized and has an unhybridized pure p orbital. The CH_3 carbons are all sp^3 hybridized and the H atoms each contribute a $1s$ orbital. End-on overlap occurs in the C—H and C—C bonding and, therefore, results in 12 σ bonds. When the 24 AOs overlap, 24 MOs are produced: 12 σ bonding and 12 σ* antibonding.

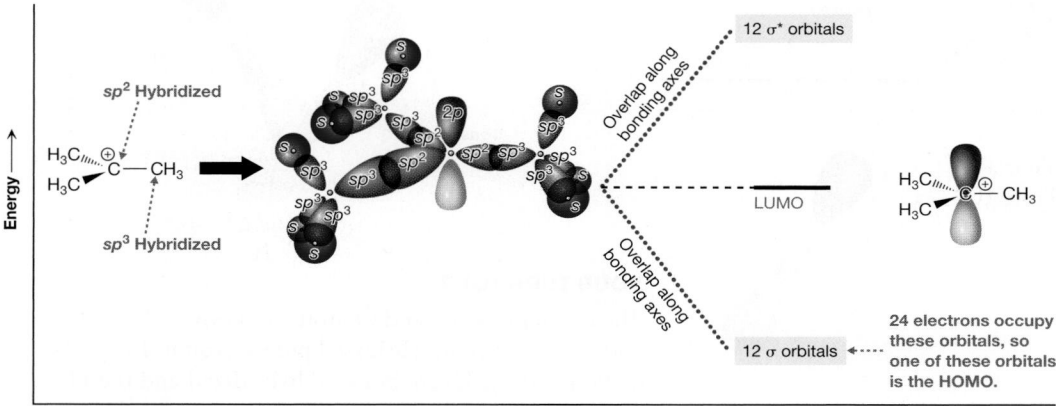

Yes, these figures are in agreement with Figure IC1-5, which shows the LUMO of $(H_3C)_3C^+$ as

YOUR TURN IC1.6

Constructive interference occurs when two orbitals overlap with the same phases. Destructive interference occurs when two orbitals overlap with opposite phases. Both IC 1-5a and 5b exhibit constructive interference.

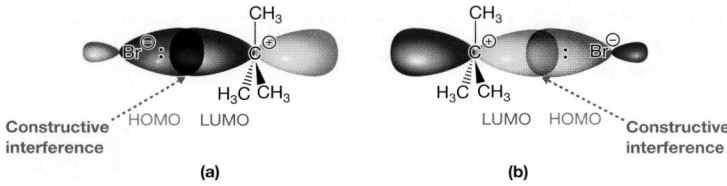

(a)   (b)

The two C atoms and the Br atom are all *sp*³ hybridized, and the H atoms each contribute a 1*s* orbital. End-on overlap occurs in the C—H, C—C, and C—Br bonding and, therefore, results in 7 σ bonds. When the 14 AOs overlap, 14 MOs are produced: 7 σ bonding and 7 σ* antibonding. This leaves three nonbonding orbitals.

(a)

The LUMO of the molecule is a σ* MO, in agreement with Figure IC1-7a:

Although the HOMO of the entire molecule is a nonbonding orbital, the highest-occupied orbital involving the CH₃ group is a σ MO, in agreement with Figure IC1-7a:

Constructive interference occurs when two orbitals overlap with the same phases. Destructive interference occurs when two orbitals overlap with opposite phases. Predominantly constructive interference takes place in both IC 1-7a and 7b.

The central C atom and O atom are both *sp*² hybridized and each has an unhybridized pure *p* orbital. The CH₃ carbon and the Cl are both *sp*³ hybridized and the H atoms each contribute a 1*s* orbital. End-on overlap occurs in the C—H, C—C, C—Cl, and C—O bonding and, therefore, results in 6 σ bonds. Side-on overlap occurs between the two pure *p* orbitals of C and O. This results in π bonding and π* antibonding. When the 14 AOs overlap, 14 MOs are produced: six σ bonding, six σ* antibonding, one π bonding, and one π* antibonding.

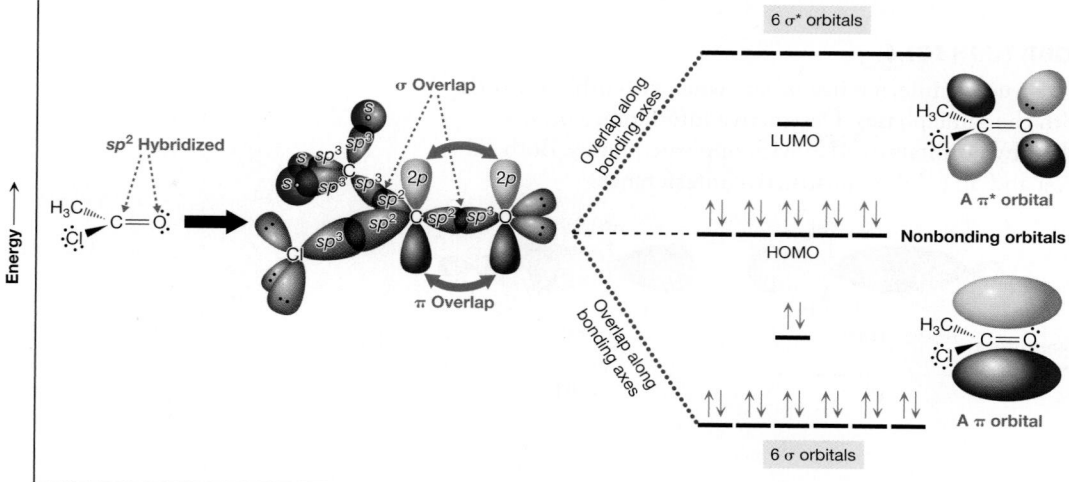

The LUMO is the π* MO, in agreement with the LUMO shown in Figure IC1-8a.

YOUR TURN IC1.10

Constructive interference occurs when two orbitals overlap with the same phases. Destructive interference occurs when two orbitals overlap with opposite phases. Constructive interference predominantly takes place in both IC1-8a and 8b.

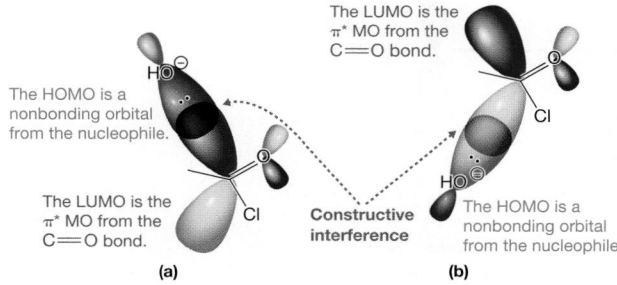

(a)   (b)

YOUR TURN IC1.11

The two central C atoms are both sp^2 hybridized and each has an unhybridized pure p orbital. The CH_3 carbons are sp^3 hybridized and the H atoms each contribute a $1s$ orbital. End-on overlap occurs in the C—H and C—O bonding and, therefore, results in 11 σ bonds. Side-on overlap occurs between the two pure p orbitals of the two sp^2 hybridized C atoms. This results in π bonding and π* antibonding. When the 24 AOs overlap, 24 MOs are produced: 11 σ bonding, 11 σ* antibonding, one π bonding, and one π* antibonding. When the 24 total valence electrons fill the MOs, the π MO is the HOMO and the π* MO is the LUMO.

Yes, these figures are in agreement with Figure IC1-9, which shows the following as the HOMO of $CH_3CH{=}CHCH_3$.

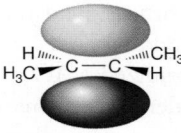

YOUR TURN IC1.12

Constructive interference occurs when two orbitals overlap with the same phases. Destructive interference occurs when two orbitals overlap with opposite phases. Constructive interference predominantly takes place in Figure IC1-9.

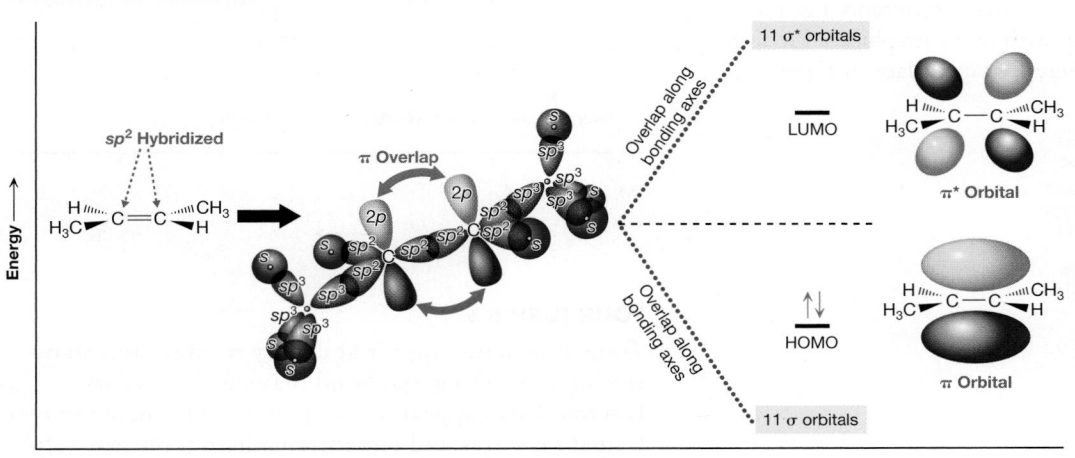

YOUR TURN IC1.13

The carbocation C is sp^2 hybridized and has an unhybridized pure p orbital. All of the other carbons are sp^3 hybridized and the H atoms each contribute a $1s$ orbital. End-on overlap occurs in the C—H and C—C bonding and, therefore, results in 15 σ bonds. The unhybridized p AO remains as a nonbonding orbital. The 30 valence electrons fill the 15 σ bonding MOs, so the HOMO is a σ bonding MO and the p AO is the LUMO.

Yes, these figures are in agreement with Figure IC1-10, which shows the HOMO and LUMO of $(CH_3)_2CHC^+HCH_3$ as

HOMO

LUMO

YOUR TURN IC1.14

Constructive interference occurs when two orbitals overlap with the same phases. Destructive interference occurs when two orbitals overlap with opposite phases. Predominantly constructive interference takes place in Figure IC1-10.

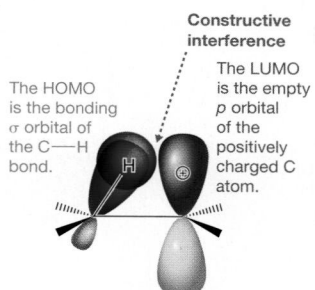

Constructive interference

The HOMO is the bonding σ orbital of the C—H bond.

The LUMO is the empty p orbital of the positively charged C atom.

Chapter 8

YOUR TURN 8.1

Equation 8-2: (a) heterolysis, (b) coordination

YOUR TURN 8.2

The allylic C^+ appears only in the reaction mechanism, not in the overall reaction, and is, therefore, an intermediate.

(a)

Overall reactant   Intermediate   Overall product

Intermediate   Overall reactant   Overall product

(b)

YOUR TURN 8.3

Transition states appear at energy maxima and show the bond breaking and bond making that occurs. Intermediates appear at energy minima. There are two transition states and one intermediate for a two-step mechanism.

YOUR TURN 8.4

For a mechanism that contains n number of steps, there must be n number of transition states and $n-1$ number of intermediates. Transition states occur at energy maxima, and there are three transition states for this mechanism. Intermediates occur at energy minima, and there are two intermediates for this mechanism. Therefore, there are three elementary steps in this mechanism.

YOUR TURN 8.5

Equation 8-5: (a) heterolysis, (b) electrophile elimination (could also be designated as proton transfer)

YOUR TURN 8.6

The carbocation C^+ appears only in the reaction mechanism, not in the overall reaction, and is, therefore, an intermediate. $B{:}^-$ and the C—L compound are overall reactants. B—H, C=C, and $:L^-$ are overall products.

YOUR TURN 8.7

Transition states appear at energy maxima and show the bond breaking and bond making in progress. Intermediates appear at energy minima. There are two transition states and one intermediate for a two-step mechanism.

YOUR TURN 8.8

(a) The red curve (100 °C) has a value of ~2% at an energy barrier of 15 kJ/mol, and (b) a value of <0.1% at an energy barrier of 25 kJ/mol.

YOUR TURN 8.9

(a) The blue curve (0 °C) at 20 kJ/mol on the x axis has a value of ~0.1% on the y axis. (b) The red curve (100 °C) has a value of ~0.4% on the y axis.

YOUR TURN 8.10

The starting material and both products each have one tetrahedral stereocenter (the C bonded to the halogen) and are, therefore, chiral. The carbocation intermediate has a plane of symmetry and is, therefore, achiral.

YOUR TURN 8.11

The first step yields a strong acid, R_2OH^+. This strong acid is incompatible in the basic, HO^- conditions.

YOUR TURN 8.12

The first step yields a strong base, RO^-. This strong base is incompatible in the acidic, H_3O^+ conditions.

YOUR TURN 8.13

The second step is an intramolecular proton transfer mechanism. This is an unreasonable step because there are typically solvent molecules that reside between the acidic and basic sites at any given time. This makes *direct* transfer of protons within a molecule difficult and not likely to occur.

Chapter 9

YOUR TURN 9.1

NH_3 acts as a base (mechanism **A**) when it bonds to the H and a nucleophile (mechanism **B**) when it bonds to the electron-poor C.

YOUR TURN 9.2

YOUR TURN 9.3

YOUR TURN 9.4

YOUR TURN 9.5

YOUR TURN 9.6

sp^3 C—H = 410 kJ/mol; sp^2 C—H = 431 kJ/mol; and sp C—H = 523 kJ/mol.

YOUR TURN 9.7

YOUR TURN 9.8

Br^- and N_3^- are reversed in ethanol compared to DMF. In DMF, N_3^- is a stronger nucleophile; but in ethanol, Br^- is the stronger nucleophile. This suggests that N_3^- is more strongly solvated by ethanol than Br^- is. Similarly, $CH_3CO_2^-$ is a stronger nucleophile in DMF than either Br^- or Cl^-; but in ethanol the opposite is true, suggesting that ethanol solvates $CH_3CO_2^-$ stronger than it solvates Br^- or Cl^-.

YOUR TURN 9.9

The top reaction (products are: $PhCH(OH)CH_3 + TsO^-$) results from substitution. The bottom reaction (products are: $PhCH=CH_2 + TsO^- + H_2O$) results from elimination and has greater entropy (three products compared to two).

YOUR TURN 9.10

TABLE 9-13 Summary Table for the Reaction in Equation 9-40

Factor	S_N1	S_N2	E1	E2
Strength				✓
Concentration		–		✓
Leaving group	✓		✓	
Solvent		✓		✓
Total	1	1	1	3

Chapter 10

YOUR TURN 10.1

YOUR TURN 10.2

The nucleophilicities for NH_3 and Cl^- are 320,000 and 23,000, respectively. So NH_3 is, in fact, a stronger nucleophile in water than Cl^- is, making NH_3 a moderate nucleophile, too.

YOUR TURN 10.3

YOUR TURN 10.4

Alkyl groups are electron donating and, therefore, stabilize a nearby positive charge. Having four versus three electron-donating groups stabilizes the N^+ more in R_4N^+ compared to R_3NH^+ (see the solution to Your Turn 10.3.). This is an example of the inductive effect.

YOUR TURN 10.5

YOUR TURN 10.6

1. Proton transfer

2. S_N2

Enolate anion

YOUR TURN 10.7

Step 1 is proton transfer; Step 2 is S_N2; Step 3 is proton transfer, Step 4 is S_N2.

YOUR TURN 10.8

B is faster due to the electron-withdrawing Cl. Cl stabilizes the negative charge on the enolate anion that is produced.

YOUR TURN 10.9

1. Proton transfer

2. Proton transfer

Electron rich C=C bond

Enol form

3. S_N2

4. Proton transfer

Electron-withdrawing effects from the Br atom destabilize the positive charge.

Halogenation stops here.

YOUR TURN 10.10

A is faster due to the electron-withdrawing Br. Br destabilizes the positive charge that develops when the carbonyl O is protonated.

YOUR TURN 10.11

1. Proton transfer

2. S_N2 + $N_2(g)$

YOUR TURN 10.12

1. Proton transfer

2. S_N2

Tetrahydropyran (THP)

YOUR TURN 10.13

1. Proton transfer

2. S_N2

YOUR TURN 10.14

1. Proton transfer

2. S_N2

3. Proton transfer

YOUR TURN 10.15

YOUR TURN 10.16

The pK_a of NH_3 is 36, and the pK_a of $R-C\equiv C-H$ is about the same as that of $HC\equiv CH$, which is 25. A lower pK_a indicates a stronger acid; thus $R-C\equiv C-H$ is a stronger acid compared to NH_3 by a factor of 10^9. Therefore, the side opposite the terminal alkyne—the product side—is favored by a factor of 10^9. This makes the reverse reaction difficult, so the forward reaction is irreversible.

YOUR TURN 10.17

The pK_a of H_2O is 15.7, and that of $R-C\equiv C-H$ is about 25. $R-C\equiv C-H$ is a weaker acid than H_2O by a factor of about 10^9. Therefore, the side opposite H_2O—the reactant side—is favored by a factor of $\sim 10^9$, which makes the reverse reaction easier than the forward reaction. Thus, the forward reaction is reversible.

YOUR TURN 10.18

As shown below, there is less steric strain in conformation **B**.

Chapter 11

YOUR TURN 11.1

HBr adds across the $C=C$ of (E)-but-2-ene to produce 2-bromobutane.

YOUR TURN 11.2

YOUR TURN 11.3

$\Delta H_{rxn}^{\circ} = (619\ kJ/mol + 431\ kJ/mol) - (339\ kJ/mol + 418\ kJ/mol + 331\ kJ/mol) = -38\ kJ/mol$, exothermic, which favors products.

YOUR TURN 11.4

The primary carbocation is less stable (higher energy) and goes in the box with the red line:

The secondary carbocation is more stable (lower energy) and goes in the box with the blue line:

YOUR TURN 11.5

YOUR TURN 11.6

YOUR TURN 11.7

YOUR TURN 11.8

YOUR TURN 11.9

A less stable, vinylic carbocation would have to be generated, as shown below.

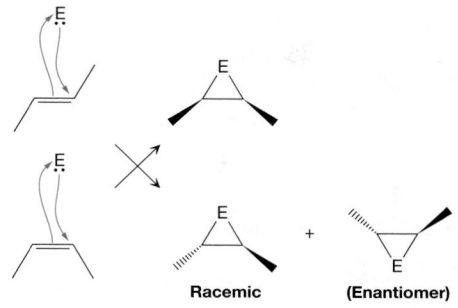

YOUR TURN 11.10

The pK_a values are: H_2SO_4 __−9__ TfOH __−13__

YOUR TURN 11.11

The first is from 1,2-addition and the second is from 1,4-addition.

YOUR TURN 11.12

Equation 11-30: The second product is the thermodynamic product (major product under thermodynamic control). Equation 11-31: The first product is the kinetic product (major product under kinetic control).

YOUR TURN 11.13

First product: thermodynamic, more stable alkene; second product: kinetic, 1,2-addition.

Chapter 12

YOUR TURN 12.1

YOUR TURN 12.2

The carbene C is assigned one electron from each of the two bonds, and two electrons from the lone pair. The carbene C has four valence electrons in all, the same as in an isolated C atom, so the formal charge is 0.

YOUR TURN 12.3

YOUR TURN 12.4

YOUR TURN 12.5

(reaction scheme)

Chloronium ion intermediate

1. Electrophilic addition

2. S_N2

3. Proton transfer

$+ H_3O^{\oplus}$

YOUR TURN 12.6

(reaction scheme)

1. Electrophilic addition

2. 1,2-Hydride shift

3. Coordination

4. Proton transfer

YOUR TURN 12.7

Equation 12-26

Equation 12-27

YOUR TURN 12.8

O—O = 138 kJ/mol; C—C = 339 kJ/mol

YOUR TURN 12.9

More steric repulsion

Minor

Major

YOUR TURN 12.10

Each C—B bond is replaced by a C—OH bond, and the stereochemistry is retained.

H_2O_2, NaOH

H_2O

Chapter 13

YOUR TURN 13.1

+ NaCl

2-Phenyl-2-tosylpropane

2-Chloro-2-phenylpropane

(a)

1. Heterolysis

2. Coordination

(b) S_N1

YOUR TURN 13.2

H_2O is the solvent and the reaction condition is 70 °C. Therefore, both H_2O and 70 °C are written below the arrow.

YOUHR TURN 13.3

First, butanoic acid is treated with sodium hydroxide. Once that reaction has come to completion, bromoethane is added, which produces ethyl butanoate.

YOUR TURN 13.4

1. $Hg(OAc)_2$, H_2O
2. $NaBH_4$

(Z)-Hex-3-ene → Hexan-3-ol

YOUR TURN 13.5

Review the reaction tables in Chapters 9–12 and note any difficulties you have.

YOUR TURN 13.6

Epoxide → Functional group conversion → **Alcohol and ether** (OH, OCH3)

Functional group conversion

Alkyl halide (Ether) (Br, OCH3) → Alters carbon skeleton → **Addition of C≡CH (Ether)** (CH, OCH3)

Functional group conversion

Aldehyde (Ether) (O, OCH3) → Alters carbon skeleton → **Addition of CH2CH3 (Ether)** (O, OCH3)

YOUR TURN 13.7

Contains 16 carbon atoms

Contains 7 carbon atoms

Contains 9 carbon atoms

H_2C—C≡C—CH_2 ⟹ H_2C—Br + H—C≡C—CH_2

Undo C–C bond formation reaction, entry 1 of Table 13-1

YOUR TURN 13.8

≡—O— ⟹ ≡—ONa + CH_3—Br

≡—ONa + CH_3—Br → ≡—O—

1. NaH
2. CH_3CH_2Br

YOUR TURN 13.9

Ethanol, CH_3CH_2OH, would be an appropriate solvent. That way, if the nucleophile deprotonates the solvent, the product species would be identical to the reactant species, as shown below.

$CH_3CH_2\ddot{O}$—H $^{\ominus}$:$\ddot{O}CH_2CH_3$ → $CH_3CH_2\ddot{O}$:$^{\ominus}$ + H—$\ddot{O}CH_2CH_3$

YOUR TURN 13.10

If each step's yield is 80%, the yield of a six-step synthesis would be $(0.80)^6 = 0.26$, or 26%.

Chapter 14

YOUR TURN 14.1

$H_2\overset{\oplus}{C}$—CH
$\quad\quad\quad$ ‖
$\quad\quad\quad CH_2$

YOUR TURN 14.2

The nodes are indicated by blue, vertical, dotted lines and planes in the MOs below.

YOUR TURN 14.3

In this case, all of the antibonding interactions among atomic orbitals are indicated by the presence of a nodal plane (dotted blue line). The bonding interactions are indicated by green dashed ovals below.

YOUR TURN 14.4

The resonance energy of 152 kJ/mol is 0.448, or 44.8%, as strong as the average C—C single-bond energy of 339 kJ/mol (from Chapter 1).

YOUR TURN 14.5

These numbers are 132 pm for a C—C bond and 154 pm for a C—C bond, compared to 132 pm and 160 pm, respectively, for cyclobutadiene. Overall, the numbers are in very good agreement, suggesting that the π electrons are NOT significantly resonance delocalized.

YOUR TURN 14.6

(b)

YOUR TURN 14.7

Number of nodes: $\pi_1 = 0$; $\pi_2 = 1$; $\pi_3 = 1$; $\pi_4 = 2$; $\pi_5 = 2$; $\pi_6 = 3$ (highest energy, greatest number of nodal planes)

Two of the nodal planes are the same as the ones in Figure 14-20d. The third is the one shown in Figure 12-20e that is parallel to the plane of the paper.

YOUR TURN 14.8

YOUR TURN 14.9

The H atoms are included in the figure below. Notice that they do not crash into each other in [18]annulene, as they do in the other molecules in Figure 14-24.

[18]Annulene

YOUR TURN 14.10

The dotted lines indicate a closed loop of orbitals. This is viewed as a single aromatic π system.

YOUR TURN 14.11

The resonance structures do not affect the locations of the p AOs.

Chapter 15

YOUR TURN 15.1

$\%T = 20\%$ represents more light being absorbed than $\%T = 40\%$. A = 0.750 represents more light being absorbed than A = 0.500.

YOUR TURN 15.2

If molar absorptivity increases by a factor of 3, absorbance also increases by a factor of 3.

YOUR TURN 15.3

The wavelength of light is represented by λ_{photon}, which is inversely proportional to E_{photon}. So E_{photon} increases as λ_{photon} decreases. Therefore, a 375-nm photon has more energy than a 530-nm photon.

YOUR TURN 15.4

The additional transitions are shown in blue to the right of the figure below.

YOUR TURN 15.5

Of all the arrows drawn in Your Turn 15.4 to represent the various transitions, the one representing the HOMO–LUMO transition is the shortest, meaning that the HOMO–LUMO transition corresponds to the smallest-energy absorbed photon.

YOUR TURN 15.6

The electron is circled below, and the transition is indicated by the curved blue arrow. The lowest two π MOs are bonding because they are below the energy of the isolated p orbitals (the horizontal dashed line), whereas the highest two π MOs are antibonding, because they are above that line.

YOUR TURN 15.7

Refer to Table 15-1. One C═C π bond (no conjugation) ~180 nm, two conjugated C═C π bonds ~225 nm, three conjugated C═C π bonds ~275 nm, four conju-

gated C═C π bonds ~290. As the number of conjugated π bonds increases, the λ_{max} value increases. The HOMO–LUMO energy gap, therefore, decreases.

YOUR TURN 15.8

The peak at 1320 cm^{-1} is labeled below. However, there are several other peaks that could have just as equally been selected. The frequency value of the absorbed photon is the value of the peak at the x axis and the units are wavenumbers, cm^{-1}. These are the same frequencies for the mode of vibration responsible for photon absorption.

YOUR TURN 15.9

The C═O peak generally appears at 1720 cm^{-1} as a strong sharp peak. The benzaldehyde C═O peak specifically occurs around 1700 cm^{-1}.

YOUR TURN 15.10

The fingerprint region encompasses the region below ~1400 cm^{-1}. In Figure 15-12, the two regions that are in this range are the single-bond stretches and the bending modes.

YOUR TURN 15.11

The general region for a C═C stretch is 1620–1680 cm^{-1}. Unlike the IR spectrum for hept-1-ene, hept-3-ene exhibits no peak in this region. This is due to the larger extent of symmetry about the C═C bond in hept-3-ene.

YOUR TURN 15.12

The H—O stretch bands are intense, broad, and centered around 3300 cm^{-1} for RO—H (alcohols) and 3000 cm^{-1} ROO—H (carboxylic acid). In a carboxylic acid, the OH stretch is shifted to lower frequency and is much broader than in an alcohol; therefore, it overlaps the alkane C—H stretches (2800–3000 cm^{-1}).

YOUR TURN 15.13

The small bump around 3300 cm^{-1} is due to a water impurity in the hept-3-ene sample when the IR spectrum was taken. The OH stretch would appear much more intense if the OH were part of the molecule itself, instead of an impurity.

YOUR TURN 15.14

Triple-bond stretching modes appear between 2000 and 2500 cm^{-1}. In the first spectrum, that triple-bond absorption at ~2250 cm^{-1} is strong, whereas in the second spectrum, it has weak intensity at ~2050 cm^{-1}.

YOUR TURN 15.15

From Table 15-2, alkane, sp^3 C—H stretches occur between 3000 and 2800 cm^{-1} and vary in intensity. Alkene, sp^2 C—H stretches occur between 3100 and 3000 cm^{-1} and are generally weak. (a) Alkane C—H stretches are intense and occur at 3000–2800 cm^{-1}. (b) Alkane C—H stretches are moderately intense and occur at 3000–2800 cm^{-1}. (c) Alkene C—H stretches are weak and occur around 3100 cm^{-1}; no alkane C—H stretch is present.

YOUR TURN 15.16

The C—H stretch near 2700 cm^{-1} is indicative of an aldehyde C—H.

YOUR TURN 15.17

The N—H stretching modes are the moderately intense, moderately broad peaks near 3300 cm^{-1}. There are two in the first spectrum (primary amide), one in the second (secondary amide), and none in the third.

YOUR TURN 15.18

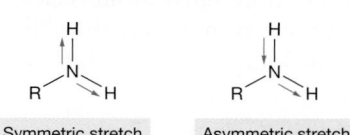

Symmetric stretch   Asymmetric stretch

YOUR TURN 15.19, 15.20, 15.21, 15.22

YOUR TURN 15.23, 15.24, 15.25, 15.26

Two possibilities for Unknown 2

YOUR TURN 15.27

A saturated molecule with 6 C's and 1 N is shown below. It has 15 H atoms.

YOUR TURN 15.28, 15.29, 15.30

Possible structure for Unknown 3

Chapter 16

YOUR TURN 16.1

A stronger magnetic field appears to the right of the states already shown in the figure. As shown below, this creates a larger energy difference between the α and β spin states.

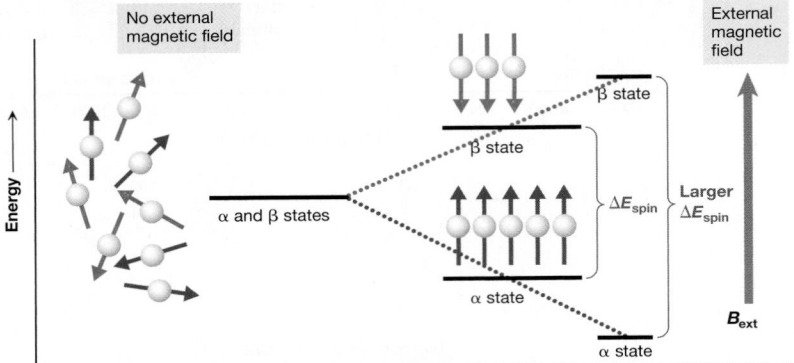

YOUR TURN 16.2

The nucleus at the left (Proton 1) is shielded to a greater extent. This is reflected by the fact that its spin states are closer together in energy, meaning that the nucleus experiences less of the external magnetic field.

YOUR TURN 16.3

One plane of symmetry that benzene has is indicated by the dashed line below. As shown, with respect to that plane of symmetry, H2 and H6 are mirror images, as are H3 and H5. As indicated below on the right, after benzene is rotated 60°, the molecule appears unchanged, but all six H atoms end up occupying a location that a different H atom occupied before rotation.

YOUR TURN 16.4

When comparing signals in Table 16-1 to the CH_3Cl proton signal, the chemical shifts with smaller values are upfield relative to CH_3Cl and the chemical shifts with larger values are downfield relative to CH_3Cl. Entries 1–9, therefore, are upfield relative to CH_3Cl, whereas entries 11–16, 19, and 20 are downfield.

YOUR TURN 16.5

FCH_3 (δ = 4.1 ppm, EN = 3.98), $ClCH_3$ (δ = 3.1 ppm, EN = 3.16), $BrCH_3$ (δ = 2.7 ppm, EN = 2.96). Increased electronegativity leads to increased deshielding of a nearby CH proton. The electronegative atom is electron withdrawing and removes electron density from the nearby proton. Increased deshielding causes the signal to shift downfield.

YOUR TURN 16.6

The local magnetic field lines are the blue lines in Figure 16-10. The ones that are in the same direction as the external magnetic field are circled below. Notice that these local magnetic field lines appear where benzene's H atoms are located. This leads to additional deshielding of the protons on benzene and the signal appears downfield.

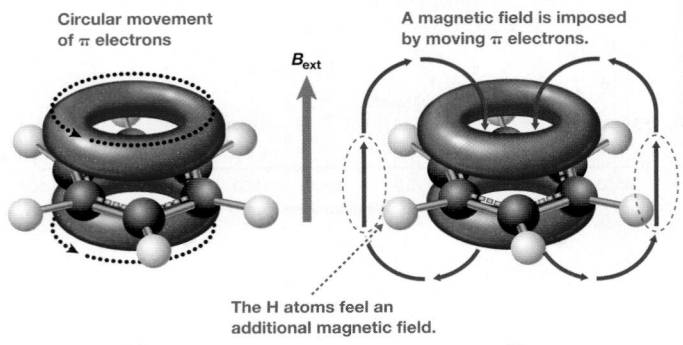

Circular movement of π electrons

A magnetic field is imposed by moving π electrons.

B_{ext}

The H atoms feel an additional magnetic field.

(a)   (b)

YOUR TURN 16.7

On each alkyne proton, draw an arrow pointing directly downward, which opposes B_{ext}. This results in a shielding, not deshielding, effect.

YOUR TURN 16.8

CH_2Cl_2 has two electronegative Cl atoms, which deshield the CH proton. This results in a downfield chemical shift (5.3 ppm) compared to CH_3Cl (Table 16-1; 3.1 ppm). $CHCl_3$ has three electronegative Cl atoms, which deshield the proton even more, resulting in a chemical shield further downfield (7.3 ppm).

YOUR TURN 16.9

The signal on the right is ~39 mm (~1.5 in.) and the one on the left is ~28 mm (~1.1 in.), which gives the ~1.5:1 height ratio. The molecule has a total of 10 protons, and thus a ratio of 6:4 correlates to a 1.5:1 ratio.

YOUR TURN 16.10

When a proton is coupled to N protons that are distinct from itself, the resulting splitting pattern has $N + 1$ peaks.

(a) Doublet = 2 = $N + 1$, $N = 1$, (b) quintet = 5 = $N + 1$, $N = 4$, (c) singlet = 1 = $N + 1$, $N = 0$.

YOUR TURN 16.11

The coupling constants suggest that protons **A** and **C** are *not* coupled because the coupling constants, 8.5 Hz and 7.6 Hz, are different.

YOUR TURN 16.12

The quartet at 4.15 ppm and triplet at 1.35 ppm both have a coupling constant equal to 7.1 Hz and, therefore, are coupled. The quartet at 2.33 ppm and triplet at 1.15 ppm both have a coupling constant equal to 7.5 Hz and, therefore, are coupled.

YOUR TURN 16.13

There are a total of two H atoms coupled to the H_c proton, so according the $N + 1$ rule, the H_c signal should be a triplet. No signals in the spectrum appear to consist of three signals in a 1:2:1 ratio. So the $N + 1$ rule does not hold for this case.

YOUR TURN 16.14

The CH would split the CH_2 into a doublet with a coupling constant value ~2 Hz. The CH_3 would split the CH_2 into a quartet with a coupling constant value ~6 Hz. Therefore, the signal is a quartet of doublets.

YOUR TURN 16.15

There are six signals in the ^{13}C NMR (not including the solvent or TMS) and, thus, six chemically distinct carbon atoms in the compound.

YOUR TURN 16.16

Several examples are possible. For example CH_3R has a proton chemical shift of 0.9 ppm and a carbon chemical shift of 10–25 ppm. CH_2R_2 has a proton chemical shift of 1.3 ppm and a carbon chemical shift of 20–45 ppm. Thus CH_3R and CH_2R_2 have the same order for H and C chemical shifts (different scales). Other examples include $BrCH_3$ and $ClCH_3$, alkenes, and alkynes.

YOUR TURN 16.17

The signal at 30 ppm corresponds to the CH_2 and the signal at 18 ppm corresponds to CH_3.

YOUR TURN 16.18

Protons **A** correspond to the signals near 7.5–8 ppm; protons **B** correspond to the signal at ~4.3 ppm; protons **C** correspond to the signal at ~1.8 ppm; and protons **D** correspond to the signal at ~1.0 ppm.

YOUR TURN 16.19

In general, the H—O signal occurs between 3500 and 3000 cm^{-1} and gives rise to a broad and strong signal. In this example, the H—O signal occurs around 3400 cm^{-1}.

YOUR TURN 16.20

YOUR TURN 16.21

$CH_3CH_2^+ = m/z\ 29$; $CH_3^+ = m/z\ 15$

YOUR TURN 16.22

The M + 1 peak in Figure 16-35 is the small peak at m/z 87. The M + 1 peak for ethylbenzene is at m/z 107, according to the following calculation. Molar mass $C_8H_{10} = (8 \times 12) + (10 \times 1) = M = 106$. M + 1 = 107

YOUR TURN 16.23

The mass of ^{12}C is 12 amu and of ^{13}C is 13 amu. There are five ^{12}C (= 60 amu) and one ^{13}C (= 13 amu), and 14 amu for the 14 H atoms present. This adds up to 87. This is the M + 1 peak in the mass spectrum of hexane (C_6H_{14}).

YOUR TURN 16.24

Hexane: M = 86 (15.5%) and M + 1 = 87 (~1%)
$[(1.0)/(15.5)] \times (100\%)/(1.1\%) = 5.87$, which rounds to 6.
Dodecane: M = 160 (5.9%) and M + 1 = 0.8%
$[(0.8)/(5.9)] \times (100\%)/(1.1\%) = 12.33$, which rounds to 12.

YOUR TURN 16.25

The mass of ^{79}Br is 79 amu and the mass of ^{81}Br is 81 amu. The mass of the C_6H_5 portion is 77 amu, which is added to either 79 amu or 81 amu, depending on the isotope of Br. The mass of $C_6H_5^{79}Br$ is 156 amu and does correspond to the M peak. The mass of $C_6H_5^{81}Br$ is 158 amu and does correspond to the M + 2 peak.

YOUR TURN 16.26

$CH_3-CO-N(CH_3)_2 = C_4H_9NO = 12(4) + 9(1) + 14 + 16 = 87$ amu. There is only one nitrogen atom and the mass is an odd number. This is consistent with the nitrogen rule.

Chapter 17

YOUR TURN 17.1

YOUR TURN 17.2

Nucleophile can attack from either side of the carbon's plane.   **Attack from behind the plane**   **Attack from in front of the plane**

YOUR TURN 17.3

The most reactive imine is **B**, followed by **A**, then **C**.

A   B   C

Highest concentration of positive charge   **Lowest concentration of positive charge**

YOUR TURN 17.4

1. Nucleophilic addition   2. Proton transfer

YOUR TURN 17.5

YOUR TURN 17.6

YOUR TURN 17.7

In Table 9-10, the nucleophilicity of Ph$_3$P is listed as 10,000,000, and that of Br$^-$ is listed as 620,000.

YOUR TURN 17.8

YOUR TURN 17.9

YOUR TURN 17.10

The enol can convert to the keto form in either acidic or basic conditions. Both are shown below. A proton must move from the O to the alkene C in two separate proton transfer steps. A single step is not reasonable because it would constitute an intramolecular proton transfer.

YOUR TURN 17.11

Conjugate addition
Thermodynamic product

Direct addition
Kinetic product

Chapter 18

YOUR TURN 18.1

Equation 18-3: 0, 2, 1, and 0 charges; Equation 18-4: 1, 1, 1, and 1 charge; Equation 18-5: 1, 1, 1, and 1 charge.

YOUR TURN 18.2

The positively charged species is activated toward nucleophilic attack because nucleophiles form bonds to electron-poor species. It has a resonance structure that puts a positive charge on the C atom, making that C atom more susceptible to nucleophilic attack.

YOUR TURN 18.3

HCN has a $pK_a = 9.2$ and H_3O^+ has a $pK_a = -1.7$. The stronger acid (lower pK_a) is on the product side, so the reactants are favored.

YOUR TURN 18.4

HCN has a $pK_a = 9.2$ and H_2O has a $pK_a = 15.7$. The stronger acid (lower pK_a) is on the reactant side, so the products are favored. $K_{eq} = 10^{(15-9.2)} = 3.2 \times 10^6$. This large K_{eq} means that the acid will be deprotonated quantitatively.

YOUR TURN 18.5

Hemiacetal (after Step 3) and acetal (after Step 7) are shown below.

YOUR TURN 18.6

In 1-phenylpropan-1-one, the ketone C has an oxidation state of $+2$, and in propylbenzene, the C has an oxidation state of -2.

YOUR TURN 18.7

YOUR TURN 18.8

$pK_a (CH_3CH{=}O) = 19$; $pK_a (H_2O) = 15.7$;
$K_{eq} = 10^{\Delta pK_a} = 10^{(15.7-19)} = 5 \times 10^{-4}$.
$K_{eq} \ll 1$, so the reactants are favored heavily.

YOUR TURN 18.9

A, because it is the only choice that contains no α protons.

YOUR TURN 18.10

NaOH is a proper choice of base in the first reaction, and LDA is a proper choice of base in the second reaction.

YOUR TURN 18.11

The product of an aldol condensation is an α,β-unsaturated carbonyl.

Chapter 19

YOUR TURN 19.1

YOUR TURN 19.2

1,2 positioning   1,2 positioning   1,3 positioning   1,4 positioning

1,5 positioning

YOUR TURN 19.3

Entry 2   Entry 4

YOUR TURN 19.4

The secondary alcohol C is assigned four valence electrons and has an oxidation state of 0. The ketone C is assigned two valence electrons and has an oxidation state of $+2$. The oxidation state of C increased by two, verifying that C was oxidized.

Chapter 20

YOUR TURN 20.1

YOUR TURN 20.2

YOUR TURN 20.3

YOUR TURN 20.4

CH_3CO_2H has a $pK_a = 4.75$ and CH_3CH_2OH has a $pK_a = 16$. The stronger acid (lower pK_a) is on the reactant side, so the products are favored. $K_{eq} = 10^{\Delta pK_a} = 10^{(16-4.75)} = 1.8 \times 10^{11}$. This shows that the reaction very heavily favors the products, making the reaction irreversible.

YOUR TURN 20.5

YOUR TURN 20.6

YOUR TURN 20.7

(a) Acid anhydride to acid chloride = up = unfavorable
(b) Amide to ester = up = unfavorable
(c) Acid chloride to ester = down = favorable
(d) Acid anhydride to carboxylic acid = down = favorable

YOUR TURN 20.8

YOUR TURN 20.12

YOUR TURN 20.9

(a)   (b)

YOUR TURN 20.10

Only methyl ketones or aldehydes can undergo a haloform reaction, which includes compounds **A**, **B**, and **D**.

YOUR TURN 20.11

YOUR TURN 20.13

Methyl hept-6-enoate

2-Methyloct-7-en-2-ol
99%

Two equivalents of $CH_3^{\ominus}$ have added in.

Chapter 21

YOUR TURN 21.1

1. Nucleophilic addition

2. Nucleophile elimination

Tetrahedral intermediate

3. Proton transfer

1. S$_N$2

Leaving group

2. Nucleophilic addition

3. Nucleophile elimination

+ HOPCl$_2$

YOUR TURN 21.2

The species are added below. The tetrahedral intermediate is the product of nucleophilic addition, and has a relatively high energy. The first step has the largest activation energy, $\Delta G^{\circ\ddagger}(1)$; which is consistent with the first step (nucleophilic addition) being the slow step.

$\Delta G^{\circ\ddagger}(1)$

$\Delta G^{\circ\ddagger}(2)$

$\Delta G^{\circ\ddagger}(3)$

Free energy

Reaction coordinate

YOUR TURN 21.4

YOUR TURN 21.5

YOUR TURN 21.6

YOUR TURN 21.7

In each case, $CH_3CH_2O^-$ is the base and picks up a proton from each of the uncharged acids. The products are different from the reactants unless ethanol is the acid.

Propan-1-ol

Water

Ethanol (Notice that the products are identical to the reactants.)

YOUR TURN 21.8

The pK_a for an ester is 25 and the pK_a for a ketone is 20, so $K_{eq} = 10^{(20-25)} = 10^{-5}$. The p$K_a$ of the product is lower and, therefore, the reactants are favored, so this reaction will not interfere with the intended Claisen condensation reaction.

YOUR TURN 21.9

The exact activated methylene compound is not found in Table 6-1 or Appendix A, but acetoacetic ester is found in Appendix A, and has a pK_a of 11. The pK_a of ethanol is 16, so $K_{eq} = 10^{(16-11)} = 10^5$, which favors products significantly. This signifies an irreversible reaction.

YOUR TURN 21.10

The alkyl group is electron donating, as shown below, so it increases the concentration of negative charge on the nearby carbon, which destabilizes that anion.

Chapter 22

YOUR TURN 22.1

The products of HCl and Cl_2 addition are shown below. In both cases, aromaticity has been lost because the π system is no longer fully conjugated around the ring and no longer contains a Hückel number of electrons.

YOUR TURN 22.2

YOUR TURN 22.3

YOUR TURN 22.4

Steps 1 and 2 involve the formation of the Cl^+ electrophile. Steps 3 and 4 formally make up the electrophilic aromatic substitution mechanism with Cl^+ as the electrophile.

Formation of the Cl^+ electrophile

YOUR TURN 22.5

Steps 1 and 2 involve the formation of the cyclohexyl C^+ electrophile. Steps 3 and 4 formally make up the electrophilic aromatic substitution mechanism with the cyclohexyl C^+ as the electrophile.

YOUR TURN 22.6

The C—H bond energy for an sp^3-hybridized C is 410 kJ/mol, and that for an sp^2-hybridized C is 431 kJ/mol.

YOUR TURN 22.7

YOUR TURN 22.8

A resonance hybrid is an average of all resonance structures. The carbon–oxygen bond is between a double and triple bond. The positive charge is delocalized over both C and O.

YOUR TURN 22.9

YOUR TURN 22.10

Desulfonation is the reverse of Steps 4–6 in Equation 22–24.

Chapter 23

YOUR TURN 23.1

Substituent	O	M	P	O+P	Type of Director
–I	45	1	54	99	Ortho/para
–CHO	19	72	9	28	Meta

YOUR TURN 23.2

The positively charged C in each resonance structure does not possess an octet. In each case, the C^+ has only three bonds and no lone pairs, and thus has a share of only six electrons.

YOUR TURN 23.3

The second resonance structure is especially stable due to the electron-donating CH_3 group attached directly to the C^+. This is because the CH_3 group decreases the concentration of positive charge there.

YOUR TURN 23.4

When the positive charge on the ring is adjacent to the positive charge on the N, the arenium ion is destabilized (second structure). The second resonance structure, as a result, is higher in energy than the other two and is the least stable, contributing the least to the resonance hybrid.

YOUR TURN 23.5

The inductive effect leads to a larger concentration of positive charge on the intermediate because Cl is an electron-withdrawing group and pulls electron density toward itself.

YOUR TURN 23.6

Deactivated, because the ring is displaying less negative charge (i.e., less red color).

YOUR TURN 23.7

NO_2 groups are electron withdrawing, and the structure where the C^+ is adjacent to the positive charge on the N (the second resonance structure) is the least stable resonance structure. CH_3 groups are electron donating and the resonance structure that has C^+ adjacent to the CH_3 (the third resonance structure) is the most stable. Overall, this intermediate is more stable, due to the stabilization provided by the third structure and the minimization of the contribution by the second structure.

YOUR TURN 23.8

The usual electrophilic aromatic substitution mechanism consists of two steps: electrophilic addition followed by elimination of H^+. Substitution can occur at both the 2 and 3 positions on the ring.

YOUR TURN 23.9

YOUR TURN 23.10

In the second resonance structure, the C^- is attached to NO_2.

YOUR TURN 23.11

YOUR TURN 23.12

In an attempt to produce the *meta* product, the meta-directing NO_2 group should be on the ring before alkylation. However, the alkylation will not work because the nitrobenzene ring is deactivated.

Chapter 24

YOUR TURN 24.1

The diene is relatively electron rich. The dienophile is relatively electron poor.

YOUR TURN 24.2

Antiaromatic, because there are four electrons involved in the cyclic transition state, and 4 is an anti-Hückel number (it is $4n$ with $n = 1$, and there are an even number of pairs).

YOUR TURN 24.3

Conformation **B** is *s*-cis, illustrated by the fact that the double bonds are on the same side of the single bond connecting them. Therefore, **B** can undergo a Diels–Alder reaction. Conformation **A** is *s*-trans.

YOUR TURN 24.4

The two terminal C atoms of the diene and both C atoms of the dienophile are transformed from sp^2 to sp^3 hybridization.

YOUR TURN 24.5

The H atoms in Equation 24-18 match up with W and Y in Equation 24-20, whereas the CH_3 groups match up with X and Z. Because X and Z end up on opposite sides of the ring in Equation 24-20, so, too, do the CH_3 groups in Equation 24-18, resulting in the trans product.

YOUR TURN 24.6

B has a favorable interaction where opposite partial charges δ^+ and δ^- are located on atoms undergoing bond formation, whereas **A** does not. Therefore, **B** will lead to the major Diels–Alder product.

YOUR TURN 24.7

YOUR TURN 24.8

$[3(619 \text{ KJ/mol}) + 1(339 \text{ KJ/mol})] - [5(339 \text{ kj/mol}) + 1(619 \text{ KJ/mol})] = -118 \text{ kJ/mol} = -118 \text{ kJ/mol}$. This value is 50 kJ/mol more positive than the experimentally measured value of -168 kJ/mol.

YOUR TURN 24.9

Note: Arrows are drawn clockwise, but counterclockwise arrows would be equally correct.

Diene   Dienophile

YOUR TURN 24.10

This Diels–Alder reaction takes place at the same temperature as the one in Equation 24-28, but ΔH°_{rxn} is much more negative (-121 kJ/mol vs. -17 kJ/mol). With ΔH°_{rxn} much more negative, ΔG°_{rxn} is much more negative. A significantly negative ΔG°_{rxn} is associated with an irreversible reaction, so whereas the reaction in Equation 24-28 is reversible, this one is irreversible.

YOUR TURN 24.11

YOUR TURN 24.12

YOUR TURN 24.13

In both regions of overlap, the orbitals have the same phase, which leads to constructive interference. Because both regions exhibit the same kind of interference, the orbitals have the appropriate symmetries to interact.

YOUR TURN 24.14

On the left, the orbitals have opposite phases, resulting in destructive interference. On the right, the orbitals have the same phase, resulting in constructive interference. Therefore, there is no net overlap and the orbitals do not have the appropriate symmetries to interact.

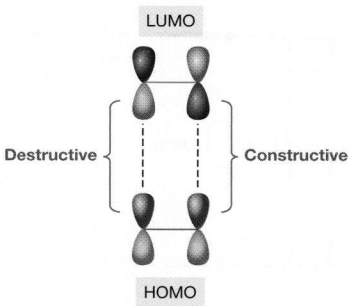

YOUR TURN 24.15

The p orbitals on C2 and C3 of the dienophile are involved in primary orbital overlap in both Figures 24-11a and 24-11b. In Figure 24-11a, the p orbital on C1 of the dienophile is involved in secondary orbital overlap with the leftmost orbital of the diene. The same two orbitals in the figure below (circled) are farther away than they are in Figure 24-11a.

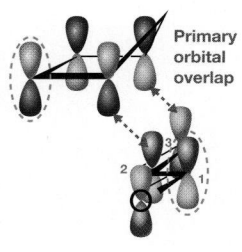

Chapter 25

YOUR TURN 25.1

Homolysis of the Cl—Cl bond → Two radicals are produced.

Cl—Cl ⟶ Cl• + •Cl

YOUR TURN 25.2

Cl—Cl = 243 kJ/mol, Br—Br: 192 kJ/mol, I—I = 151 kJ/mol, HO—OH = 213 kJ/mol, H_3C—H = 435 kJ/mol. The larger the bond energy, the stronger the bond. This shows that the C—H bond is the strongest bond in this list.

YOUR TURN 25.3

The H—Br and H—Cl bond energies are 368 and 431 kJ/mol, respectively. The H—Cl bond is stronger and requires more energy to cleave homolytically. The product chlorine radical (•Cl) is therefore higher in energy and less stable than the bromine radical (•Br).

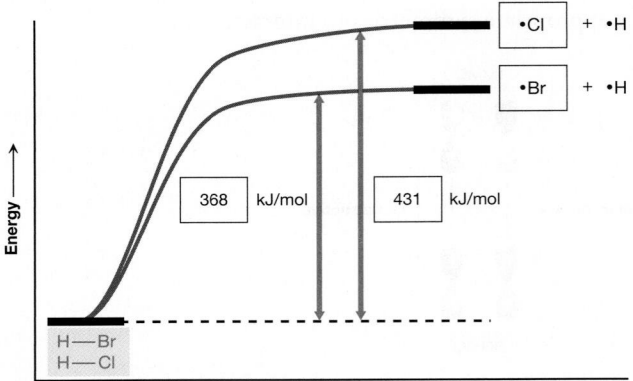

YOUR TURN 25.4

Benzyl cation

YOUR TURN 25.5

YOUR TURN 25.6

H_3C• + •CH_3 ⟶ H_3C—CH_3

YOUR TURN 25.7

H_3C—CH_2—H + •Cl ⟶ H_3C—•CH_2 + H—Cl

YOUR TURN 25.8

+ •Cl ⟶

YOUR TURN 25.9

The Br—Br is the weakest bond present and therefore will be broken in the initiation step.

Initiation   Br—Br $\xrightarrow{h\nu}$ Br• + Br•

YOUR TURN 25.10

Br• + H ⟶ S_H2 H—Br +

+ Br—Br ⟶ S_H2 + Br•

YOUR TURN 25.11

Br• + •Br $\xrightarrow{\text{Radical coupling}}$ Br—Br

+ •Br $\xrightarrow{\text{Radical coupling}}$ Br

YOUR TURN 25.12

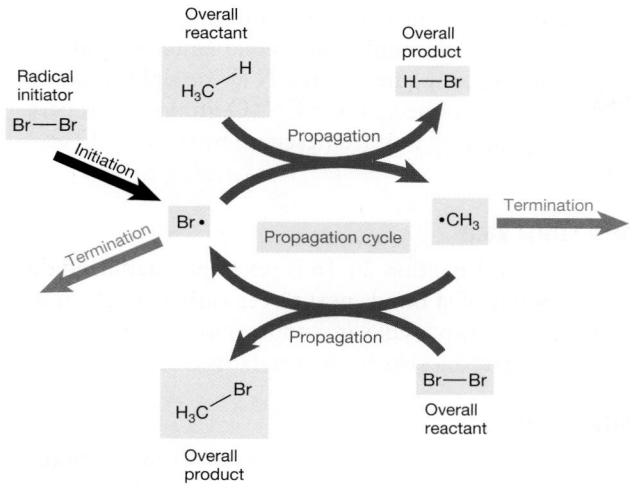

YOUR TURN 25.13

The allylic radical produced in (**b**) is more stable than the propyl radical produced in (**a**), due to resonance delocalization of the unpaired electron. The rate of free radical halogenation increases with increasing stability of the radical that is formed, so the radical abstraction in (**b**) will proceed faster.

YOUR TURN 25.14

H—Br = 368 kJ/mol and H—Cl = 431 kJ/mol. With the lower bond energy, the HBr bond is weaker than the HCl bond.

YOUR TURN 25.15

YOUR TURN 25.16

YOUR TURN 25.17

YOUR TURN 25.18

Catalytic hydrogenation produces the cis alkene.

Chapter 26

YOUR TURN 26.1

Methyl groups are the side chains
attached to the main polymer chain.

YOUR TURN 26.2

The repeating unit is shown below. For the structure given, $n = 3$.

YOUR TURN 26.3

YOUR TURN 26.4

YOUR TURN 26.5

The head-to-tail addition is shown below, requiring three single-barbed curved arrows.

Propagating chain of poly(ethyl acrylate)   **Ethyl acrylate**   **Product of head-to-tail addition**

The product radical has the following two resonance structures:

Product of head-to-tail addition

YOUR TURN 26.6

The atoms that appear on the reactant side are: 12 C; 16 H; 6 O. The atoms that appear on the product side are: 10 C; 8 H; 4 O. Therefore, on the product side, there appear to be missing: 2 C; 8 H; 2 O. This comes from two molecules of CH_3OH.

YOUR TURN 26.7

Nylon is a polyamide. The amide group appears twice in the repeating unit. One complete amide appears in the center of the repeating unit. The second amide is split between two repeating units—the N at the right of one repeating unit is connected to a C=O group that is part of the next repeating unit (represented by the C=O at the far left).

YOUR TURN 26.8

The product in Equation 26-16 is resonance stabilized due to the presence of π bonds next to the carbanion electron pair. Resonance is possible around the entire phenyl ring. No such resonance exists in the reactants.

YOUR TURN 26.9

In a radical mechanism, combination and disproportionation each require two radicals reacting. The equivalent species in an anionic mechanism are anions. Neither of these steps is likely because the like charges of the two anions would repel each other.

YOUR TURN 26.10

Each growing polymer will be protonated twice.

YOUR TURN 26.11

The total bond strength in an enol such as vinyl alcohol is less than in the analogous keto form. This is largely due to the strong C=O bond.

YOUR TURN 26.12

An ester group is part of the repeating unit of the polymer, and during the course of the reaction, the bond between O and C=O must break. This can occur when CH_3O^- acts as a nucleophile and RO^- acts as a leaving group from the C=O bond. Because the leaving group is attached to a C=O, this reaction takes place in a nucleophilic addition–elimination mechanism. The second product will be methyl acetate.

YOUR TURN 26.13

Because radicals are present, single-barbed arrows are used to show electron movement. The first step is a hydrogen atom abstraction, whereby one C—H bond is formed and a second one is broken simultaneously. The second step is a radical addition, whereby a C—C single bond is formed and a C=C π bond is broken. Both steps require three single-barbed curved arrows.

Art Credits

Chapter 16

p. 757: Shutterstock RF; **p. 787:** Shutterstock RF; **p. 805:** iStock RF.

Chapter 17

p. 815: Photo by Ted Kinsman; **p. 842:** Alamy.

Chapter 18

p. 861: Shutterstock RF.

Chapter 19

p. 919: John Kelly Green Photography; **p. 951:** TRL Ltd./Science Source.

Chapter 20

p. 962: iStockphoto RF; **p. 967:** (left) © Dave Reede/All Canada Photos/Corbis; (right) © Paul Chinn/San Francisco Chronicle/Corbis; **p. 982:** Joel Karty; **p. 1005:** Problem 20.55b: © Sigma-Aldrich Co. LLC. Used with permission.

Chapter 21

p. 1006: Shutterstock RF; **p. 1063:** Problem 21.78: © Sigma-Aldrich Co. LLC. Used with permission. Problem 21.79: © Sigma-Aldrich Co. LLC. Used with permission. **p. 1064:** Problem 21.80: © Sigma-Aldrich Co. LLC. Used with permission. Problem 21.81: © Sigma-Aldrich Co. LLC. Used with permission.

Chapter 22

p. 1065: Alamy; **p. 1083:** (left) © Andrey Kiselev/Fotolia RF; (right) Alamy; **p. 1091:** (left) © Tetra Images/Alamy; (right) © the food passionates/Corbis; **p. 1102:** Problem 22.58: © Sigma-Aldrich Co. LLC. Used with permission.

Chapter 23

p. 1104: Problem 22.60b: © Sigma-Aldrich Co. LLC. Used with permission. **p. 1104:** Last Resort/Getty Images; **p. 1110:** GIPhotoStock/Getty Images.

Chapter 24

p. 1154: iStockphoto; **p. 1180:** Shutterstock RF; **p. 1197:** Problem 24.64b: © Sigma-Aldrich Co. LLC. Used with permission. Problem 24.65b: © Sigma-Aldrich Co. LLC. Used with permission. **p. 1198:** Problem 24.66: © Sigma-Aldrich Co. LLC. Used with permission.

Chapter 25

p. 1199: (left) Shutterstock RF; (right) Leonard Lessin/Getty Images; **p. 1218:** Nasa; **p. 1244:** Problem 25.64b: © Sigma-Aldrich Co. LLC. Used with permission.

Chapter 26

p. 1255: (opener) Jeong-Woo Lee, Georgia Institute of Technology; (top) Alamy; (left) Shutterstock RF; (right) Shutterstock; **p. 1256:** (left) iStockphoto; (right) iStockphoto RF; **p. 1272:** Shutterstock RF; **p. 1274:** Shutterstock; **p. 1284:** (left) Joel Karty; (right) Alamy; **p. 1285:** Shutterstock; **p. 1286:** (left) iStockphoto; (center and right) D. Dockstader/M. Noble/S. Pruett; **p. 1288:** (top) Wikimedia Commons; (bottom) Shutterstock; **p. 1291:** iStockphoto; **p. 1293:** (top left and right) Alamy; (bottom left) iStockphoto; (bottom right) Photodisc/Getty Images; **p. 1296:** TORSTEN BLACKWOOD/AFP/Getty Images.

Index

Note: Material in figures or tables is indicated by italic page numbers. Footnotes are indicated by n after the page number.

A

absorbance (A), 717–18
absorption bands or peaks, *717*
acetaldehyde (ethanal)
    aldol reactions, 886–87
    dipole–dipole interactions, 90–91
    electrostatic potential map, *820*
    equilibrium with vinyl alcohol, 1281
    index of hydrogen deficiency (IHD), 207
    physical properties, *88*, 90–91
    structure, *415, 923*
acetals
    acetal groups in carbohydrates, 512, 514
    formation, 870–72, 873, 935
    hydrolysis, 871, 935
    as protecting groups for alcohols, 937
    as protecting groups for aldehydes or ketones, 871, 935–36, 939
    stability of five-membered rings, 871, 935
    structure, 512, 935
    Williamson ether synthesis, 937
    See also ethers
acetic acid. *See* ethanoic acid
acetic anhydride (ethanoic anhydride), *414, 1014,* 1079
acetoacetic ester (ethyl 2-oxobutanoate), 1040–44
acetoacyl acyl carrier protein (ACP), 1039
acetone (propanone), 78, 304, *415, 820,* 885
acetylcholine, 40
acetylcholinesterase, 40
acetyl coenzyme A, 115
acetylene. *See* ethyne
acetylide ion, 476, 657
acetylsalicylic acid (aspirin), *1104*
achiral environments, 252–53, 263–64
achiral molecules, definition, 230
acid anhydrides
    from acid chlorides, 973
    alcoholysis reactions, 1010
    aminolysis reactions, 1014–16
    hydride reduction, 985
    hydrolysis reactions, 1010–11
    nomenclature, *399,* 400, *414*
    polar π bond, *816,* 818
    synthesis, 1016–18
acid–base chemistry
    acidity constant (K_a), 299–300, *301*
    acid strength and pK_a, 300, 302, 330–32
    acid strengths of charged and uncharged acids, 312–14, 330, 331–32

Brønsted–Lowry base, definition, 296
    conjugate acids, 296, 297
    conjugate bases, 296, 297
    functional groups and acidity, 310–12
    Henderson–Hasselbalch equation, 304
    mechanically generated acid, 561
    percent dissociation of acids, 306
    pH, 306–8
    ranking acid and base strengths, 329–33
    reactivity of charged species, 312–14
    relative effects of resonance structures, 333–35
    relative importance of charge-stability factors, 329–33
    See also Brønsted–Lowry acids; Lewis acids; Lewis bases; pK_a; proton transfer reactions
acid-catalyzed amide hydrolysis, 1027–28
acid-catalyzed ester hydrolysis, 1025–26
acid-catalyzed hydration of alkenes, 583–85
acid-catalyzed hydration of alkynes, 589–91
acid-catalyzed nucleophilic addition to polar π bonds, overview, 862, 863–65
acid chlorides
    acid anhydride formation by acyl substitution, 973
    acyl chlorides in Friedel–Crafts acylation, 1077–78, 1080
    alcoholysis reactions, 1007–9
    aminolysis reactions, 1014–16
    ester formation by acyl substitution, 971, 973
    hydrolysis reactions, 1008–9
    nomenclature, *399,* 401, *410, 414*
    reaction with lithium dialkylcuprates, 993, 995–96
    reduction by hydride reagents, 982–84
    reduction by lithium tri-*tert*-butoxyaluminum hydride, 990–92
acid derivatives, nomenclature, 413–14
    See also acid anhydrides; acid chlorides; amides; carboxylic acids; esters; nitriles
acid halides
    leaving groups, *818, 963,* 985
    polar π bonds, 818
    structure, *816, 818, 963*
    synthesis, 1016–18
    See also acid chlorides; carboxylic acid derivatives
acidity constant (K_a), 299–300, *301*
acid rain, *295*
acid workup
    carboxylic acid formation from amides, 975–76
    in Claisen condensations, 1030–31
    E2 reactions, 556–57
    haloform reactions, 979–80
    hydride reduction of acid chlorides, 983
    after nucleophilic addition to polar π bonds, 817, 822, 824, 828–29

after organometallic treatment of acid chlorides, 994
    oxetanes, 553
    after saponification reactions, 968
    after S$_N$2 reactions, 548–49
acrylic acid (propenoic acid), *406, 407, 412, 1265,* 1267
acrylics, 1267
acrylonitrile, *1267*
activation energy ($\Delta G^{\circ\ddagger}$), 432–33, 510
    See also Gibbs free energy (G°); transition states
active methylene compounds, 1040, 1043
acyl carrier protein (ACP), 1039
acyl chlorides in Friedel–Crafts acylation, 1077–78, 1080
    See also acid chlorides
acylium ions, 1078–79
acyl substitution, thermodynamics of, 971–74
    See also nucleophilic addition–elimination reactions
addition polymerization, 1260
adenine, *46,* 706, 707, 708
adenosine diphosphate (ADP), 445
adenosine triphosphate (ATP), 445, 825, 974
adipic acid (hexane-1,6-dioic acid), 1266, 1271, 1277–78
adipoyl chloride, 1266
alcohols
    acid-catalyzed dehydration, 486, 545
    from acid-catalyzed hydration of alkenes, 583–84, 585
    chemical shift in ^{13}C NMR, 790
    chromic acid oxidations, 948–51, 952
    conversion of hydroxide leaving group to water, 485, 525
    conversion to alkyl halides, 485, 525–29, 846
    converting bad leaving groups to good leaving groups, 486
    decoupling of NMR spectra, 779–80
    dehydration to form ethers, 545–47
    from epoxides under acidic conditions, 551–53
    from epoxides under basic conditions, 548–51
    hemiacetal formation by reaction with ketones or aldehydes, 862–63, 870–72, 903–4
    hydroxyl group as poor leaving group, 483, 525
    mass spectrometry, 1252–53
    nomenclature, *399,* 400–401, 416–17
    numbering system, 400–401
    O—H stretch absorption bands in IR spectroscopy, *729,* 734–35
    permanganate oxidation of, 952–53
    primary (1°) alcohols, 416

ppm scale, 768
shielding, 762
signal averaging, 787–88
signal splitting and ^{13}C NMR, 789
signal splitting by spin–spin coupling, 776–80
spin flips, 760, 761, 787
structure elucidation, 793–96
symmetry effects, 766, 772
tert-butyl carbocation $(CH_3)_3C^+$, 427
tetramethylsilane (TMS) standard, *762, 763, 767*, 768–69, 788, 790
time scale of NMR spectroscopy, 766–68
typical NMR experimental setup, 758, *759*
See also chemical shifts; proton nuclear magnetic resonance spectroscopy; spin–spin coupling
nuclear Overhauser effect, 791
nucleic acids
  base pairs, 706–7
  bases, 46, 706–8
  DNA double helix, 44, 706, 707
  hydrogen bonding, 707
  nucleotides, composition, 45, 46, 706
  overview, 44–46
  π stacking in DNA, 707
  as polynucleotides, 1296
  ribonucleic acid (RNA), 44, 46
  size hierarchy, *45*
nucleophile elimination steps
  curved arrow notation, *363,* 364
  frontier molecular orbital theory, 395
  HOMO–LUMO interactions, 395
  importance in organic chemistry, 363
  microscopic reversibility, 396
  nucleophilic addition–elimination reactions, 964
  overview, 363–65
nucleophiles
  attraction to atoms with positive charge, 357
  attributes of nucleophiles, 357
  carbon nucleophiles, 476–77, 549
  common negatively charged nucleophiles, 357
  common uncharged nucleophiles, 357, 862
  and competition between S_N2 and S_N1 reactions, 474–75
  intrinsic strength of nucleophiles, 475–76
  Lewis bases, 359, 364
  nucleophile strength in acid or basic conditions, 865
  nucleophile strength in S_N2 and S_N1 reactions, 471–76
  in nucleophilic addition steps, 362–64
  relative nucleophilicity of common species, 496–98, 816
  in S_N1 reaction mechanisms, 421–22, 430, 434–37, 455
  and S_N1 reaction rates, 470, 474–75
  in S_N2 reaction mechanisms, 357–59, *421,* 437–39, 447
  and S_N2 reaction rates, 470, 475
  weakening by protic solvents, 495, 496–98
nucleophilic acyl substitution, 965
  See also nucleophilic addition–elimination reactions
nucleophilic addition–elimination reactions, 962–1005, 1006–64
  acetoacetic ester synthesis, 1040–44
  acid-catalyzed amide hydrolysis, 1027–28

acid-catalyzed ester hydrolysis, 1025–26
acid-catalyzed transesterification, 1024–25
alcoholysis, 1007–10
aminolysis reactions, *1006,* 1014–16
Baeyer–Villiger oxidations, 1028–30
base-catalyzed transesterification, 1023–24
base-promoted transesterification, 963–68
carboxylic acids from amides, 975–77
charge stability of reactants and products, 967
decarboxylation, 1040–44
Dieckmann condensation, 1037–39
Fischer esterification reaction, 543, 1026, 1044, 1050
free energy diagrams, nucleophilic addition for acid derivatives, 1012
functional group transformations, *963, 998–99, 1053–54*
Gabriel synthesis of primary amines, 977–79, 1061
haloform reactions, 979–82
Hell–Volhard–Zelinsky (HVZ) reaction, 1019–21
hydride reductions of carboxylic acids and derivatives, 982–90
hydrolysis reactions, 1008–9, 1010–13
kinetics, 965–66, 1009–10
leaving groups, 818, 962–66, *973*
malonic ester synthesis, 1040–44
nucleophile elimination step, 964
nucleophilic aromatic substitution, 1130–34
organometallic reagents, 993–96
rate-determining nucleophilic addition step, 965, 1009
reaction free-energy diagram, *966*
reaction mechanism with strong nucleophile, general, 964–65, 971, 1006
reaction mechanism with weak nucleophile, general, 1008–9
reactions that alter the carbon skeleton, *999, 1055*
saponification, *962,* 968–70, 976, 1024, 1033–34
specialized reducing agent reductions of acid chlorides and esters, 990–93
stability ladder of carboxylic acid derivatives, *972,* 973, 974, 975, 986, 1016–18
strong nucleophiles and, 962–1005
sulfonation reactions of alcohols, 1021–23
tetrahedral intermediates, 964–66, 971, 980, *983,* 988, 991–92
thermodynamics and reversibility, 966–68, 971–74
weak nucleophiles and, 1006–64
See also Claisen condensations
nucleophilic addition steps
  curved arrow notation, 363–64
  frontier molecular orbital theory, 395
  HOMO–LUMO interactions, 395
  importance in organic chemistry, 363
  Lewis acids and bases, 364, *395*
  nucleophiles in, 362–64
  overview, 362–65
nucleophilic addition to polar π bonds, 815–60, 861–918
  acetal formation, 870–72, 873
  acid-catalyzed addition overview, 862, 863–65
  acid workup afterward, 817, 822, 824, 828–29

base-catalyzed addition, overview, 862–63, 864–66
carbonyl hydration in ketones and aldehydes, 819–20, 865–66
charge stability and total bond energy as driving forces, 840–41
conjugate addition by weak nucleophiles, 868–69
cyanohydrin formation by HCN addition, 866–67
direct *vs.* conjugate addition to α,β-unsaturated carbonyls, 837–43, 848–50, 868–69, 932–33
enamine formation, 874–75
epoxides from sulfonium ylide addition, 835–37
functional group transformations, 852–53, *907–8*
general reaction mechanism for strong nucleophiles, 816–18
hemiacetal formation from ketones or aldehydes, 862–63, 870, 903–4
hydride transfers, 822, 826
imine formation, 828, 872–74
inductive effects in, 820
Michael reactions, 838, 842, 896–97
nucleophile and electrophile strength in acids or bases, 865
racemic mixtures in products, 818, 821–22
reactions that alter the carbon skeleton, *853, 908–9*
regioselectivity involving 1,2- *vs.* 1,4-addition, 838–41, 849, 932–33
relative reactivity of ketones and aldehydes, 819–21
reversibility in nucleophilic addition, 838–41
steric repulsion in ketones, 819–20
thermodynamic and kinetic control of addition reactions, 840–41, 843
weak nucleophile addition with acid and base catalysis, 862–69
Wittig reaction, 831–33
Wolff–Kishner reduction, 877–79, 927–29, 930–31
See also aldol reactions; alkyllithium reagents; Grignard reactions
nucleophilic aromatic substitution
  benzyne intermediates, 1132–34
  leaving groups, 1131–32, 1133
  Meisenheimer complex, 1131
  nucleophilic addition–elimination mechanism, 1130–32
nucleophilicity
  alkyllithium reagents, 476, 549, 553
  ammonia, 497, 529, 530
  decrease in strong, bulky bases, 480
  Grignard reagents, 549, 553
  organometallic compounds, 475–76, 827–29
  relative nucleophile strength of common species, 496–98, 816
  S_N2 reaction rate and, 472
nucleotides, 45, 46, 264, 706
nylon-6,6, 1266, 1267, 1271, 1277–79

O

octa-3,5-dien-1-yne, 703
octanedial, *892*
octanoic acid (caprylic acid), *93*

I-23